SENSING AND CONTROLLING
MOTION
VESTIBULAR AND SENSORIMOTOR FUNCTION

Dr. C. Wahl-Loew
College of Veterinary Medicine
Department of Anatomy
Cornell University
Ithaca, NY 14853
USA

ANNALS OF THE NEW YORK ACADEMY OF SCIENCES
Volume 656

SENSING AND CONTROLLING MOTION

VESTIBULAR AND SENSORIMOTOR FUNCTION

Edited by Bernard Cohen, David L. Tomko, and Fred Guedry

The New York Academy of Sciences
New York, New York
1992

The micrograph of the hair cell shown on the cover of this volume was provided by R. A. Jacobs and A. J. Hudspeth. Reprinted with permission from the *Journal of Neuroscience*.

Library of Congress Cataloging-in-Publication Data

Sensing and controlling motion : vestibular and sensorimotor function
/ edited by Bernard Cohen, David L. Tomko, and Fred Guedry.
    p. cm. — (Annals of the New York Academy of Sciences, ISSN
0077-8923 ; v. 656)
    Includes bibliographical references and index.
    ISBN 0-89766-733-6 (cloth : alk. paper).—ISBN 0-89766-734-4
(paper : alk. paper)
    1. Vestibular apparatus—Congresses. 2. Sensorimotor integration–
–Congresses. I. Cohen, Bernard, 1929– . II. Tomko, David L.
III. Guedry, Fred E. IV. Series.
    [DNLM: 1. Cerebellar Nuclei—physiology—congresses. 2. Movement–
–congresses. 3. Reflex. Vestibulo-Ocular—physiology—congresses.
4. Sensation—Congresses. 5. Space Perception—physiology–
–congresses. 6. Vestibular Nuclei—physiology—congresses.
7. Vestibule—physiology—congresses. W1 AN626YL v. 656 / WV 255
5474]
Q11.N5   vol.656
[QP47]
500 s—dc20
[591.1'8]
                                                                    92-13003
                                                                         CIP

SP
Printed in the United States of America
ISBN 0-89766-733-6 (cloth)
ISBN 0-89766-734-4 (paper)
ISSN 0077-8923

ANNALS OF THE NEW YORK ACADEMY OF SCIENCES

Volume 656
May 22, 1992

SENSING AND CONTROLLING MOTION

VESTIBULAR AND SENSORIMOTOR FUNCTION[a]

Editors and Conference Organizers
BERNARD COHEN, DAVID L. TOMKO, and FRED GUEDRY

CONTENTS

[a]This volume is the result of a conference entitled Sensing and Controlling Motion: Vestibular and Sensorimotor Function held from July 7 to July 11, 1991 in Palo Alto, California and sponsored by the New York Academy of Sciences.

Oral Presentations

Financial assistance was received from:

Contributors
- NATIONAL AERONAUTICS AND SPACE ADMINISTRATION
- NATIONAL INSTITUTE ON DEAFNESS AND OTHER COMMUNICATION DISORDERS/NIH
- OFFICE OF NAVAL RESEARCH GRANT N00014-91-J-1310

Introduction

BERNARD COHEN,[a] DAVID L. TOMKO,[b]
AND FRED GUEDRY[c]

[a]*Department of Neurology*
Mount Sinai School of Medicine
One Gustave Levy Place
New York, New York 10029-6574

[b]*Vestibular Research Facility*
Life Sciences Division
National Aeronautics and Space Administration
Ames Research Center
Moffett Field, California 94035-1000

[c]*Naval Aerospace Medical Research Laboratory*
Naval Air Station
Pensacola, Florida 32508-5700

There have been dramatic changes over the last 40 years in the way in which we consider the vestibular system. Studies in the 50s and 60s emphasized oculomotor effects of stimulation of single semicircular canals, electrophysiology in anesthetized or encephale isolè cats, studies of caloric and positional nystagmus, and perceptual effects of rotation and linear acceleration. This work laid the foundation for studies in the 70s and 80s of peripheral vestibular physiology, of the organization of the central vestibular system, and of the structural basis of vestibuloocular and vestibulospinal pathways. A key contribution was the discovery of the importance of vision and the vestibulocerebellum in adapting vestibuloocular reflexes. More recently, the field has moved to considerations of molecular mechanisms of hair cell transduction, transmitters utilized in the central vestibular system and cerebellum, multisensory function in alert behaving animals, and studies of perception of complex motion in human subjects. Mathematical models and corresponding model-based studies can now provide a solid theoretical framework for further experimentation.

During this time we have gained considerable understanding about the ways in which motion is sensed, and how compensatory gaze movements are generated and controlled. Indeed, there is probably no motor system in the body about which we know more than the oculomotor system. To this end, a number of conferences have been held over the last 10 years in which the vestibuloocular reflex (VOR) has been utilized as a tool to study the generation of slow and fast eye movements. Until now, however, there has not been an opportunity to consider the vestibular system in depth as a sensory motor entity that senses and controls motion, both in a terrestrial environment and in altered states of gravity, such as the microgravity of spaceflight.

Several events set the stage for this conference. First, the National Institute on Deafness and Other Communication Disorders expressed a specific interest in supporting vestibular research. Second, the Vestibular Research Facility (VRF) at the NASA-Ames Research Center was designated as a national resource, and was made available for use to the general community. At this point, it became clear that there was significant interest in a survey of the field of vestibular research, especially if it could provide vestibular researchers with an opportunity to visit the VRF at the NASA-Ames Research Center.

This conference was held at Palo Alto, California from July 7 to 11, 1991 and was sponsored by the New York Academy of Sciences. Over a four-day period, topics were considered that are of importance for summarizing our understanding of the vestibular system and suggesting fruitful areas for future vestibular research. These included vestibular transduction, visual motion processing responsible for the linear VOR, spatial orientation and three-dimensional organization of the vestibular system, perception of motion and position in spaceflight, mechanisms responsible for stabilization of gaze, neural processing in the vestibular system and cerebellum, and vestibular and cerebellar transmitters.

The conference was attended by anatomists, physiologists, pharmacologists, psychologists, clinicians, and bioengineers working on the vestibular, oculomotor, visual, and body postural systems. An important goal of the conference was to provide a forum for young investigators. In addition, clinical studies were combined with basic research to provide a balanced view and to encourage cross-fertilization of ideas. How well these goals were met will be the judgment of the reader.

We would like to acknowledge the support of Frank Sulzman, Janis Stoklosa, and Arnauld Nicogossian of NASA Life Sciences, and Daniel Sklare and James Snow of the NIDCD, who encouraged us to plan the conference. An advisory committee that included Jay Goldberg, Stephen Highstein, Fred Miles, Theodore Raphan, Igor Vodyonoy, Victor Wilson, Laurence Young, and David Zee made helpful suggestions about topics and speakers. We are grateful for the financial support of the National Institute on Deafness and Other Communication Disorders, the National Aeronautics and Space Administration, the Office of Naval Research, and the New York Academy of Sciences that allowed us to hold the conference. Geraldine Busacco and Renée Wilkerson of the New York Academy of Sciences were of particular help in organizing and running the conference, and Sheila Treitler and Bill Boland provided expert editorial help in assembling the book that contains the proceedings.

As will be evident from perusal of this book, much has been learned, but much remains to be learned. We hope that this volume will summarize some of the former and contribute significantly to the latter.

The Hair Cell's Mechanoelectrical Transducer Channel[a]

R. FETTIPLACE,[b] A. C. CRAWFORD,[c] AND M. G. EVANS[c]

[b]Department of Neurophysiology
University of Wisconsin Medical School
273 Medical Sciences Building
Madison, Wisconsin 53706

[c]Physiological Laboratory
Cambridge CB2 3EG, United Kingdom

INTRODUCTION

The hair cell of the inner ear is an exquisitely sensitive mechanoreceptor capable of converting imposed forces of a few piconewtons into electrical signals. The basis of the mechanical sensitivity has been substantially elucidated over the last 10 years for both vestibular and auditory hair cells.[1-5] Mechanoreception resides in the bundle of modified microvilli that protrude from the cell's apical surface, each component stereocilium containing no more than a few transduction channels to sense deformation of the bundle. When opened by hair-bundle displacement, these channels allow influx of small cations such as Na^+, K^+, and especially Ca^{2+} to which they are most permeable.[3,6] This article summarizes some attributes of the transduction channel, including its size, kinetics, and displacement sensitivity, exemplifying these features with results obtained on hair cells isolated from the turtle's cochlea.[5,7] It also describes the regulatory role of intracellular Ca^{2+} which has emerged from recent studies of transducer adaptation.[8,4,5,9]

METHODS

Experiments were performed on hair cells that had been isolated from the basilar papilla of the turtle *Pseudemys scripta elegans* employing previously documented methods.[10,5] Cell bodies were immobilized by plating hair cells onto a coverslip precoated with 2 mg/ml concanavalin A, which allowed the stereociliary bundle to be manipulated using a rigid glass stylus (tip diameter 1 μm) attached to a piezoelectric bimorph. Provided the glass probe was clean, it would adhere strongly to the bundle, which could be stepped from one position to another in less than 100 μseconds. Cells oriented such that the hair bundle's plane of bilateral symmetry was parallel to the floor of the recording chamber, so that the bundle could be deflected towards or away from the kinocilium, were the only ones usable (see insert to FIGURE 2). The stimuli were calibrated by projecting the Nomarski image of the probe and stereociliary bundle onto a photodiode array.[5] These calibrations demonstrated that the hair bundle could be rotated and held in a new position relative to the stationary apical pole of the cell body, and that there was no sign of the bundle slipping during a prolonged (100-msecond) stimulus. Transduction currents were recorded under

[a]This work was supported by grants from the Medical Research Council and Royal Society.

1

whole-cell voltage clamp employing patch electrodes[11] filled with a solution containing (in mM) KCl (or CsCl), 125; $MgCl_2$, 3; HEPES, 5; EGTA, 5; Na_2ATP, 2.5, adjusted to pH 7.2. In experiments to buffer rapidly the intracellular Ca^{2+}, 10 mM BAPTA[12] was substituted for EGTA and the KCl was reduced to 116 mM. The external control solution contained (in mM) NaCl, 130; $MgCl_2$, 2.2; $CaCl_2$, 2.8; HEPES, 5, glucose, 4, pH 7.6. In some experiments, the solution around the stereociliary bundle was altered by ejecting a test solution of reduced Ca^{2+} content using a U-tube superfusion method.[13,10] All measurements were made at room temperature, about 23°C.

RESULTS

General Properties of Hair Cell Transduction

The mechanically sensitive organelle of all hair cells is the hair bundle, which in turtle cochlear receptors comprises 50 to 100 closely packed stereocilia that increase in height from one side of the bundle to the other (FIGURE 1). In addition, a single abneurally placed kinocilium abuts the tallest row of stereocilia which project approximately 6 μm above the cell's apical facet. The hair bundles *in vivo* are lodged in pockets in an overlying tectorial membrane, some vestiges of which can be seen still attached to the longest stereocilia. In scanning electron micrographs like that shown in FIGURE 1, filamentous links are visible passing from the tip of each stereocilium to the shank of its next tallest neighbor.[14] It has been suggested that such "tip links" might be the mechanical attachments to the transduction channels[15,16] which would therefore be located near the distal ends of the stereocilia.[17,18] The tip links all run roughly parallel to the bundle's plane of symmetry, so that rotation of the bundle towards the kinocilium, resulting in a vertical shear between adjacent rows of stereocilia, would stretch the tip links and exert force on the channels (for review see Reference 19). This would explain why deflections of the hair bundle along its axis of symmetry (towards or away from the kinocilium) are transduced whereas orthogonal displacements of the bundle are ineffective.[20]

In order to record transducer currents while stimulating the hair bundle in a controlled manner, experiments were performed on isolated hair cells. A family of transducer currents, generated in response to a range of step displacements to the hair bundle, is shown in FIGURE 2 for a cell immersed in control saline. At the holding potential of −74 mV, steps towards the kinocilium (denoted as positive in the stimulus monitor) evoked an increase in inward current whose initial amplitude was graded with the size of displacement; steps away from the kinocilium produced a smaller decrease in inward current and an overshoot at the step's termination. There are several features of note in these records: (1) the peak amplitude of the current saturated for both positive and negative stimuli over a range of tip deflexions of about 1 μm equivalent to 10° of bundle rotation. The maximum current obtained in this experiment was 210 pA, and in other cells currents up to 400 pA were observed which, assuming a reversal potential of about 0 mV,[5,3] translates into a maximum transducer conductance of 5.5 nS; (2) the responses were asymmetric in that only about one-tenth of the current was activated at the resting position of the bundle where the current-displacement relationship had its steepest slope; (3) the onset and offset of these currents occurred within 1 msecond and, from fitting the onsets at faster sweeps, it was shown that the current developed with a time constant of 0.1 to 0.4 msecond;[5] (4) the currents for positive steps relaxed back from their peak value thus adapting to the maintained stimulus. For small displacements, the current

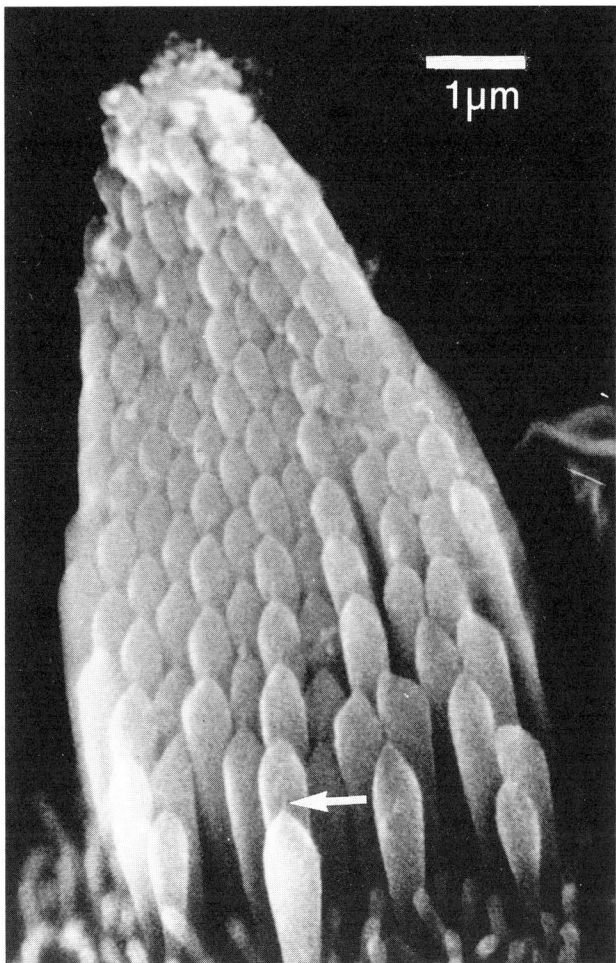

FIGURE 1. Scanning electron micrograph of a hair bundle from the central region of the turtle's basilar papilla. Papillae in which the tectorial membrane had previously been removed were fixed in 2.5% glutaraldehyde in buffer and prepared for micrography using an osmium impregnation technique.[16] Note the increasing height of the stereociliary ranks and the fine filaments that pass from the tip of each stereocilium to the side of the next tallest one in front (arrowed). Some strands of tectorial membrane are still attached to the longest stereocilia. Micrograph kindly supplied by C. M. Hackney and D. N. Furness.

relaxed with a principal time constant of roughly 4 mseconds, but the adaptation became progressively slower with larger stimuli (FIGURE 2). Most of these features of the transducer current have also been inferred from experiments on bullfrog saccular hair cells.[2,4]

Adaptation does not result from inactivation of opened transduction channels but rather is due to a resetting of the bundle's operating range. This can be

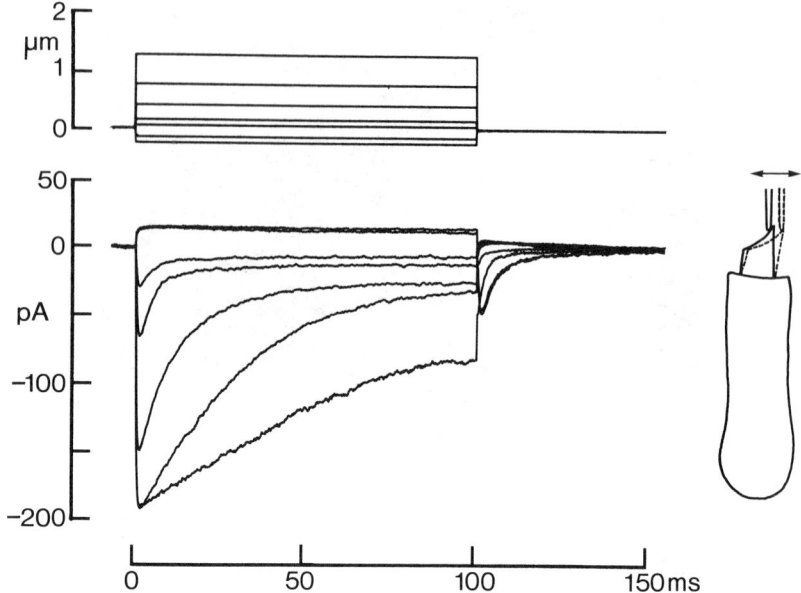

FIGURE 2. Averaged mechanoelectrical transducer currents measured in an isolated hair cell under whole-cell voltage clamp. Top traces give timing of displacement steps delivered to tip of the hair bundle; positive steps are towards the tallest rank of stereocilia which evoke inward currents shown as downward-going traces. The method of stimulation is illustrated in the inset, the cell body being fixed to the floor of the recording chamber. Holding potential, −74 mV.

demonstrated by determining the instantaneous current-displacement relationship at the beginning and end of an adapting step.[8,4,5] The maximum current evokable in the adapted state is identical to that of the control, but the entire current-displacement relationship is shifted in the direction of positive displacements. One interpretation of these results is that adaptation acts to reduce the mechanical input to the transducer channels. On this hypothesis, it might be argued that such effects are an artifact of slipping of the hair bundle relative to the stimulating probe. This explanation seems unlikely however since there was no evidence of such slippage in the photometric calibrations. Furthermore adaptation could be abolished reversibly by depolarizing the hair cell to near the Ca^{2+} equilibrium potential.[5]

Role of Calcium in Adaptation

There is ample evidence that adaptation is sensitive to the amount of Ca^{2+} in the bathing medium.[4,7–9] When the Ca^{2+} concentration around the turtle's hair bundle was reduced from its control value of 2.8 mM to 1 mM, adaptation became slower, and if the concentration was lowered further to 50 μM, adaptation was abolished entirely. The effects of lowering the external Ca^{2+} are visible in the transducer currents in FIGURE 3, where the responses in 50-μM Ca^{2+} show no hint of adaptation but instead display an additional slow component of growth at the onset and a slower offset. Quite apart from loss of adaptation, there were other changes caused by lowering the Ca^{2+} concentration: the maximum current approximately doubled in amplitude, and the proportion of the current activated at rest changed from 10% to over 50%. The latter effect is quantified in FIGURE 4, where it may be seen that the current-displacement relationship in 50-μM Ca^{2+} has been shifted in the direction of negative displacements and its slope has been substantially reduced relative to that of the control. Thus lowering extracellular Ca^{2+} produces an effect that is the opposite of adaptation, increasing the apparent mechanical stimulus to the transducer channels. However, the reduction in the slope of the current-displacement relationship in low Ca^{2+} is not consistent with a simple biasing of the mechanical input, where the current-displacement relationship should have been translated along the displacement axis without a change in shape.

There are two lines of evidence that argue that calcium's action to regulate adaptation and the position of the current-displacement relationship is produced through changes in its intracellular concentration. These derive from experiments that suppressed Ca^{2+} entry or buffered it intracellularly. Effects identical to those of reducing the concentration of Ca^{2+} extracellularly could be achieved by depolarizing the hair cell to +66 mV which is near the Ca^{2+} equilibrium potential. To make this measurement, the recording electrode was filled with CsCl rather than KCl so as to prevent activation of the large Ca^{2+}-activated K^+ current.[10] Changing the holding potential to +66 mV caused a loss of adaptation and a negative shift and a reduction in slope of the current-displacement relationship.[5,4] These effects might be explained if Ca^{2+} ions normally enter the stereocilia via the transducer channels[3] and then induce adaptation at an intracellular site. Depolarization would reduce the driving force on the entry of Ca^{2+} thus preventing its intracellular accumulation.

Secondly, the changes in intracellular Ca^{2+} that occur during a bundle displacement could be minimized by including a high concentration of the Ca^{2+} chelator

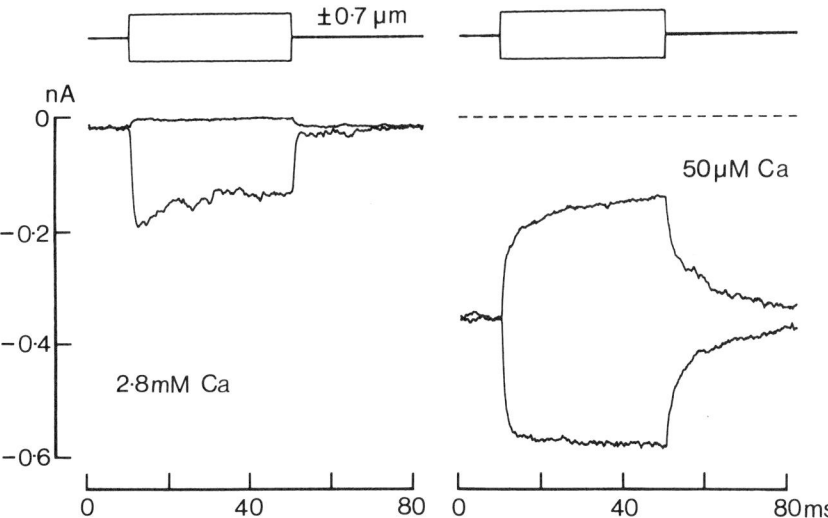

FIGURE 3. Averaged transducer currents for ±0.7 μm steps in a control saline containing 2.8-mM Ca^{2+} (left) and one in which Ca^{2+} was reduced to 50 μM (right). Apart from the changes in the form of the transducer currents, there was also an increase in holding current in the low Ca^{2+}. All effects were reversible. Holding potential, −85 mV. (Reproduced from Reference 7.)

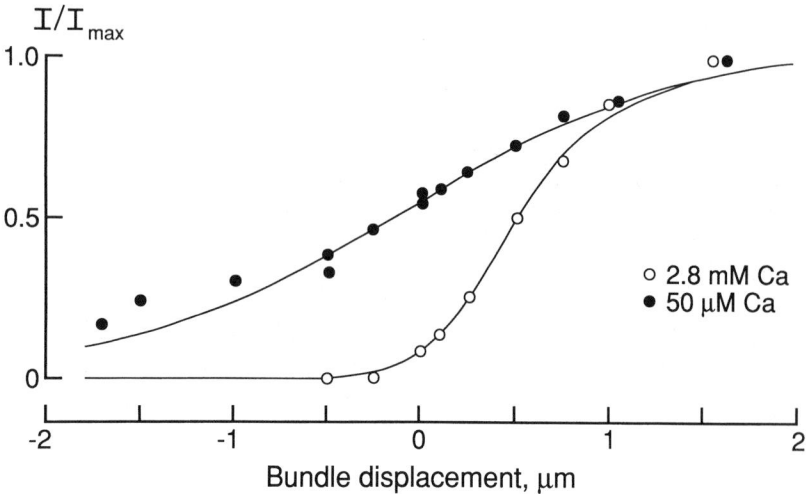

FIGURE 4. Current-displacement relationships in control saline with 2.8-mM Ca^{2+} (open symbols) and in a saline containing reduced, 50-μM, Ca^{2+} (filled symbols). Transducer currents, I, were peak values measured from the current level produced by the largest negative deflexion, and have been normalized to I_{max}, their maximum values: 115 pA (control), 261 pA (50-μM Ca^{2+}). The smooth curves were calculated using a theoretical model described in the Discussion in which Ca^{2+} stabilizes the transducer channel in its closed state. (Reproduced with modification from Reference 7.)

BAPTA in the filling solution for the patch electrodes. Lack of adaptation in transducer currents recorded with a BAPTA-filled electrode is illustrated in FIGURE 5. Since the electrode-filling solution rapidly equilibrated with the cytoplasmic contents, it was not possible first to obtain a control measurement, though BAPTA's ability to buffer rapidly the intracellular Ca^{2+} was confirmed by the simultaneous abolition of the Ca^{2+}-activated K^+ current. As in other experiments where adaptation was removed, the currents acquired a secondary slow component in their onset and markedly slowed offsets when the hair bundle was returned to its resting position. In this and other experiments it was observed that the offset time course became slower as the size and the duration of the preceding bundle displacement were increased, and for prolonged or large stimuli, highly-asymmetric responses could be obtained (FIGURE 5). Such behavior may indicate that the open state of the channel is being progressively stabilized throughout the stimulus.

Transducer Channels

In the course of experiments to reduce the extracellular Ca^{2+} concentration, it was unexpectedly found that if the hair bundle was exposed to 1 μM Ca^{2+} for more than about 20 seconds, the transducer current vanished irreversibly. Occasionally when the Ca^{2+} in the bathing solution was restored to its control value (2.8 mM), a very small transducer current consisting of only one or a few channels remained. This fortuitous observation allowed some of the single channel properties to be estimated. Examples of inward single-channel currents are shown in FIGURE 6, which it should be stressed were obtained from *whole-cell* recordings. The measurements were made at a holding potential of −85 mV, within a few millivolts of the K⁺ equilibrium potential (thus eliminating any contamination from K⁺ channels), where they had an amplitude of about −9 pA. In six experiments, the channel amplitude was determined as −9.3 ± 1.3 pA [mean ± standard deviation (SD)], yielding a single-channel conductance of approximately 100 pS. This large size distinguishes them from other channels that have been found in hair cells, such as voltage-sensitive Na⁺ or Ca²⁺ channels which would also generate inward currents at this holding potential. For comparison, the size of the voltage-sensitive Ca²⁺ channel in hair cells has been estimated as −1.2 pA at at −65 mV.[21]

A more compelling argument that these represent the mechanoelectrical transducer channels is that they were activated by small positive (but not negative) deflections of the hair bundle. Channel activity in FIGURE 6 is clearly higher during the step; this was confirmed from an ensemble average on 55 sweeps from which it was estimated that the 0.15-μm displacement of the hair bundle increased the probability of the channel being open from 0.15 to 0.5. In another cell the probability of the channel being open was shown to be graded with displacement amplitudes up to about 1 μm. The channels also displayed appropriately fast kinetics to be consistent with the onsets of the macroscopic transducer current. For the cell illustrated in FIGURE 6, the distribution of open times could be well fitted by a single exponential; the mean open time in the resting state was 1.1 msecond, this value increasing with bundle displacement along with a concomitant reduction in the mean closed time.

There have been two prior estimates of the amplitude of transducer channels in hair cells. One was based on observations of unitary events in chick vestibular cells,[3] many of which had an amplitude in the region of 100 pS, though in some recordings channels of about half this size were detected. The other estimate was derived from

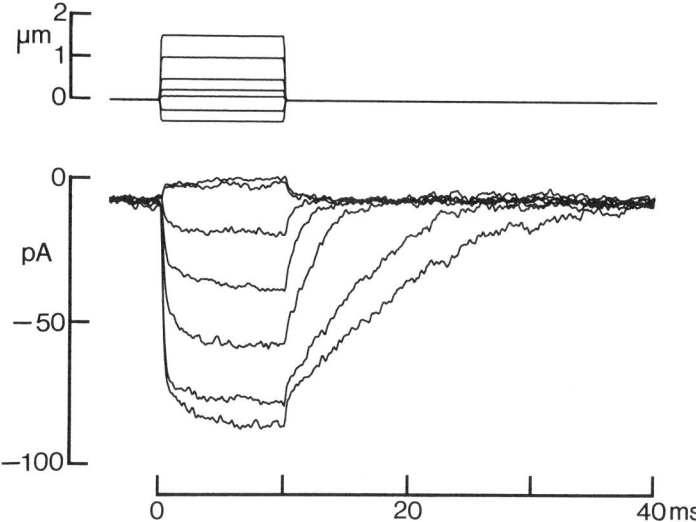

FIGURE 5. Averaged transducer currents recorded with an electrode containing 10 mM BAPTA as the Ca²⁺ buffer. Note the absence of adaptation during the stimulus and the slow tails at the offset. Holding potential, −74 mV.

the variance of current noise and gave a value of 17 pS when corrected to room temperature.[22] This latter value is not necessarily inconsistent with the present observation since such fluctuation measurements often underestimate the true channel size due to the limited bandwidth of the recording system. If the channel amplitude is assumed to be 100 pS and the maximum transducer conductance in normal saline is 5.5 nS (see above), then the total number of channels per hair cell is fifty-five, comparable to the number of stereocilia in the bundle (50 to 100). Given the possibility that even in the best recordings, some channels were lost during hair cell isolation, there are still likely to be no more than a few channels per stereocilium, a conclusion that has also been reached from measurements of the channel's gating compliance.[23]

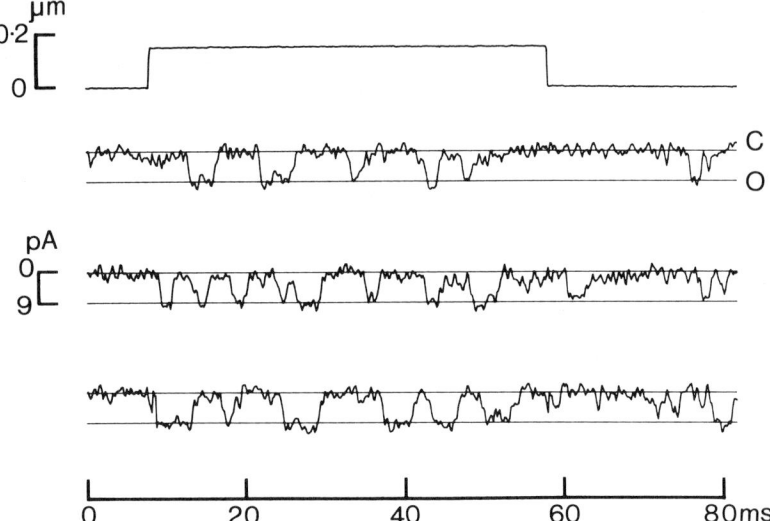

FIGURE 6. Responses of a single channel to 0.15-μm deflections of the hair bundle towards the kinocilium. Recording made in the whole-cell condition in a hair cell that had previously been exposed to a saline containing 1-μM Ca^{2+} and was then returned to normal saline. Pairs of lines of 9-pA separation are superimposed on the individual traces, **C** and **O** indicating closed and open levels. From amplitude histograms, channel size was -8.3 ± 2.9 pA. Holding potential, -85 mV, data filtered at 1.2 kHz. (Reproduced with modification from Reference 7.)

DISCUSSION

The main issue to be discussed is the mechanism by which intracellular Ca^{2+} can influence the time course of the transducer current and the position of the current-displacement relationship. Two types of explanation are available, one involving an action on the gating springs thought to deliver force to the transducer channels[24,4,9] and the other postulating a direct interaction with the channel itself.[5] At present the most likely structures to be identified as the gating springs are the tip links connecting adjacent stereocilia,[15] but since these are extracellular elements, it is not obvious how their mechanical properties might be susceptible to *intracellular* Ca^{2+}. However, if the anchorage point of the transducer channel were a cytoplasmic motor that could drag the channel up the stereocilium,[24] then intracellular Ca^{2+} might control the motor's activity so as to set the bias point on the mechanical input. While this hypothesis accommodates the shifts of the current-displacement relationship, it does not explain the reduced slope in low Ca^{2+}. Nevertheless, a force generator within the stereocilia may still be needed to account for the movements of the hair bundle produced by changes in membrane potential.[25,4]

An alternative approach is to propose a direct interaction of Ca^{2+} with the transducer channel. The form of the transducer's current-displacement relationship can be described by a scheme where the channel has two closed states and one open state.[2,22,5] Both equilibrium constants (K_1 and K_2) are assumed to be regulated by displacement of the bundle, x, according to a Boltzmann relation $[K_i (x) = K_{i0} \exp[-zx]$, where K_{i0}, z are constants$]$, and, provided the two reactions have different displacement sensitivities, this can account for the asymmetric saturating relationship between current and bundle deflection. To explain the role of intracellular Ca^{2+}, it will be assumed that Ca^{2+} binds to the second closed state thus stabilizing the channel in its closed configuration. Under these assumptions it is possible to fit the changes in both position and slope of the current-displacement relationship that occur during adaptation or reduced extracellular Ca^{2+}.[7] As an example, the smooth curves drawn through the experimental points in FIGURE 4 for 2.8 mM and 50 μM Ca^{2+} have been calculated on such a scheme.

It may be seen qualitatively that with this channel mechanism, the action of Ca^{2+} is to counter the effects of bundle displacement. Thus as intracellular Ca^{2+} is lowered consequent to the reduction in its extracellular concentration, or during depolarization of the hair cell, the concentration of the unbound closed state (C) increases which raises the probability of the channel being open. Conversely, as intracellular Ca^{2+} grows during a displacement step, it binds to the closed state and pulls the reaction to the left thus causing adaptation. It is worth noting that the time course of adaptation is likely to be complicated since the internal Ca^{2+} will change as the channels close during adaptation. However, near saturation where the probability of the channels being in the open state (p_o) is close to unity, the rate of adaptation will be approximately proportional to the probability of the closed state C (p_c, which will be low). Increasing displacements will produce a small relative increase in p_o but a large fractional reduction in p_c, and this may partially explain why adaptation becomes much slower when the current is almost saturated (FIGURE 2).

An unusual feature of the current kinetics is the secondary slow component that is unmasked on removing adaptation (FIGURES 3 and 5). This behavior cannot be explained by the three-state model and may require an additional open state with a long lifetime;[5] such a scheme could generate the variable slow component in the current offset, since the larger or longer the displacement, the more channels would exist in this second open state. Furthermore, the simplest explanation for the offsets being slow in low Ca^{2+} is that the transition to the second open state also involves Ca^{2+} binding. For this to be the case, the Ca^{2+} may need to have a greater affinity for the open state than for the closed state since the slow kinetics are apparent under circumstances where the intracellular Ca^{2+} is too low to cause adaptation. On this interpretation, the fast component of the current will mainly reflect residence in the first open state and the slower component will reflect mixed residence in both open states. Postulating two open states is not necessarily contradicted by the observation of only one time constant in the distribution of single-channel open times, since in high Ca^{2+} the occupancy of the first state may be too brief to be measurable under whole-cell conditions. A similar problem was encountered in studying the gating kinetics of Ca^{2+}-activated K^+ channels.[26] In summary the scheme for the hair cell's transducer channel that accounts for many of the properties reported here is that

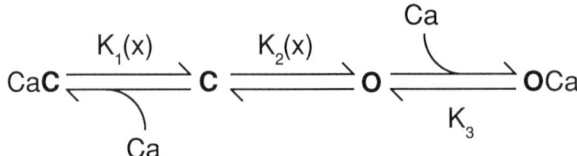

FIGURE 7. Proposed scheme for the mechanoelectrical transducer channel in hair cells. C and O denote closed (nonconducting) and open (conducting) states of the channel. The equilibrium constants K_1 and K_2 for the first two transitions are assumed to be dependent on bundle displacement (x) and in addition both the closed and open states can bind calcium (Ca).

shown in FIGURE 7, which is similar to one proposed previously on different grounds.[27] Confirmation of this kinetic model for the transducer channel may require detailed characterization of the effects of Ca^{2+} on single-channel behavior.

ACKNOWLEDGMENTS

We wish to thank Carole Hackney and Dave Furness for providing the electron micrograph in FIGURE 1.

REFERENCES

1. HUDSPETH, A. J. & D. P. COREY. 1977. Sensitivity, polarity and conductance change in the response of vertebrate hair cells to controlled mechanical stimuli. Proc. Natl. Acad. Sci. USA **74:** 2407–2411.
2. COREY, D. P. & A. J. HUDSPETH. 1983. Kinetics of the receptor current in bullfrog saccular hair cells. J. Neurosci. **3:** 962–976.
3. OHMORI, H. 1985. Mechano-electrical transduction currents in isolated vestibular hair cells of the chick. J. Physiol. **359:** 189–217.
4. ASSAD, J. A., N. HACOHEN & D. P. COREY. 1989. Voltage dependence of adaptation and active bundle movements in bullfrog saccular hair cells. Proc. Natl. Acad. Sci. USA **86:** 2918–2922.
5. CRAWFORD, A. C., M. G. EVANS & R. FETTIPLACE. 1989. Activation and adaptation of transducer currents in turtle hair cells. J. Physiol. **419:** 405–434.
6. COREY, D. P. & A. J. HUDSPETH. 1979. Ionic basis of the receptor potential in a vertebrate hair cell. Nature **281:** 675–677.
7. CRAWFORD, A. C., M. G. EVANS & R. FETTIPLACE. 1991. The actions of calcium on the mechanoelectrical transducer current of turtle hair cells. J. Physiol. **434:** 369–398.
8. EATOCK, R. A., D. P. COREY & A. J. HUDSPETH. 1987. Adaptation of mechanoelectrical transduction in hair cells of the bullfrog's sacculus. J. Neurosci. **7:** 2821–2836.
9. HACOHEN, N., J. A. ASSAD, W. J. SMITH & D. P. COREY. 1989. Regulation of tension on hair-cell transduction channels: displacement and calcium dependence. J. Neurosci. **9:** 3988–3997.
10. ART, J. J. & R. FETTIPLACE. 1987. Variation of membrane properties in hair cells isolated from the turtle cochlea. J. Physiol. **385:** 207–242.
11. HAMILL, O. P., A. MARTY, E. NEHER, B. SAKMANN & F. J. SIGWORTH. 1981. Improved patch-clamp techniques for high-resolution current recording from cells and cell-free membrane patches. Pflugers Arch. **391:** 85–100.
12. TSIEN, R. Y. 1980. New calcium indicators and buffers with a high selectivity against magnesium and protons: design, synthesis and properties of prototype structures. Biochemistry **19:** 2396–2404.

13. KRISHTAL, O. A. & V. I. PIDOPLICHKO. 1980. A receptor for protons in the nerve cell membrane. Neuroscience **5:** 2325–2327.
14. HACKNEY, C. M., R. FETTIPLACE & D. N. FURNESS. 1989. The ultrastructure of hair cells from the basilar papilla of the red-eared turtle, *Chrysemys scripta elegans.* Br. J. Audiol. **23:** 146.
15. PICKLES, J. O., S. D. COMIS & M. P. OSBORNE. 1984. Cross-links between stereocilia in the guinea-pig organ of Corti, and their possible relation to sensory transduction. Hearing Res. **15:** 103–112.
16. FURNESS, D. N. & C. M. HACKNEY. 1985. Cross-links between stereocilia in the guinea-pig cochlea. Hearing Res. **18:** 177–188.
17. HUDSPETH, A. J. 1982. Extracellular current flow and the site of transduction in vertebrate hair cells. J. Neurosci. **2:** 1–10.
18. HACKNEY, D. M. & D. N. FURNESS. 1991. Antibodies to amiloride-sensitive sodium channels reveal the possible location of the mechanoelectrical transducer channels in guinea pig cochlear hair cells. J. Physiol. **248:** 124 P.
19. HOWARD, J., W. M. ROBERTS & A. J. HUDSPETH. 1988. Mechanoelectrical transduction by hair cells. Annu. Rev. Biophys. Biophys. Chem. **17:** 99–124.
20. SHOTWELL, S. L., R. JACOBS & A. J. HUDSPETH. 1981. Directional sensitivity of individual vertebrate hair cells to controlled deflection of their hair bundles. Ann. N.Y. Acad. Sci. **374:** 1–10.
21. ROBERTS, W. M., R. A. JACOBS & A. J. HUDSPETH. 1990. Colocalization of ion channels involved in frequency selectivity and synaptic transmission at presynaptic active zones of hair cells. J. Neurosci. **10:** 3664–3684.
22. HOLTON, T. & A. J. HUDSPETH. 1986. The transduction channel of hair cells from the bullfrog characterized by noise analysis. J. Physiol. **375:** 195–227.
23. HOWARD, J. & A. J. HUDSPETH. 1988. Compliance of the hair bundle associated with gating of mechanoelectrical transduction channels in bullfrog's saccular hair cell. Neuron **1:** 189–199.
24. HOWARD, J. & A. J. HUDSPETH. 1987. Mechanical relaxation of the hair bundle mediates adaptation in mechanoelectrical transduction by the bullfrog's saccular hair cell. Proc. Natl. Acad Sci. USA **84:** 3064–3068.
25. CRAWFORD, A. C. & R. FETTIPLACE. 1985. The mechanical properties of ciliary bundles of turtle cochlear hair cells. J. Physiol. **364:** 359–379.
26. MOCZYDLOWSKI, E. & R. LATORRE. 1983. Gating kinetics of Ca^{2+}-activated K^+ channels from rat muscle incorporated into planar lipid bilayers. J. Gen. Physiol. **82:** 511–542.
27. JARAMILLO, F., J. HOWARD & A. J. HUDSPETH. 1990. Calcium ions promote rapid mechanical evoked movements of hair bundles. *In* The Mechanics and Biophysics of Hearing. P. Dallos, C. D. Geisler, J. W. Matthews, M. A. Ruggero & C. R. Steele, Eds.: 26–31. Springer Verlag. Berlin, Germany.

Morphological and Electrophysiological Properties of Hair Cells in the Bullfrog Utriculus[a]

R. A. BAIRD

Department of Neurootology
and
R. S. Dow Neurological Sciences Institute
Good Samaritan Hospital and Medical Center
1120 Northwest 20th Avenue
Portland, Oregon 97209

INTRODUCTION

With few exceptions,[1,2] previous studies of mechanoelectric transduction in vestibular otolith hair cells have been confined to the amphibian sacculus (for reviews, see References 3 and 4), a sensor of substrate-borne vibration.[5,6] Hair cells in the amphibian sacculus are specifically adapted to sense small-amplitude, high-frequency linear accelerations. The receptor current of saccular hair cells rapidly adapts to maintained hair bundle displacements, sharply limiting their sensitivity to static deflections and low-frequency linear accelerations.[7] An electrical resonance phenomenon,[8–11] determined by the interplay of basolateral calcium and calcium-activated potassium conductances,[12,13] further sharpens the responses of these hair cells to high frequencies.

Little is known about the comparative transduction mechanisms of hair cells in other otolith end organs, which generally are adapted to sense lower frequencies than hair cells in the bullfrog sacculus. Hair cells in the bullfrog utriculus, a sensor of static gravity and low-frequency linear acceleration,[14–17] have a variety of hair bundle morphologies distinct from those in the bullfrog sacculus.[14,18] Moreover, morphophysiological studies in the bullfrog reveal that the response dynamics of utricular afferents are correlated with both the macular location and hair bundle morphology of their innervated hair cells (FIGURE 8),[6,14] implying that utricular hair cells exhibit regional variations in their transduction mechanisms. Similar studies in the mammalian utriculus[19] and in the semicircular canals, both in mammals[20] and lower vertebrates,[21] have also noted that the discharge regularity and response dynamics of vestibular afferents are more closely related to the epithelial location than to the number or kind (type I vs. type II) of hair cells that they contact.

To learn more about regional variations in hair cell transduction mechanisms and the particular functions of hair cell types in the vertebrate utriculus, we studied the responses of hair cells in the isolated utriculus to intracellular current and hair bundle displacement. The primary aim of these studies was to see if hair cells in different macular zones differed in their membrane properties or their mechanical responses.[22] We were particularly interested in determining how utricular hair cells

[a]Funding for this work was provided by National Institute on Deafness and Communicative Disorders grant DC-00355, by National Aeronautics and Space Administration grant NCC 2-651, and by grants from the Oregon Lions Sight and Hearing Foundation.

in different macular zones adapted to maintained displacements of their hair bundles and assessing, as in recent studies of isolated vestibular hair cells,[23-26] the contribution, if any, of an electrical resonance to the responses of utricular hair cells. A second aim was to assess the contribution of specific hair cell transduction mechanisms to the response properties of utricular afferents.

Our results reveal that the utricular macula is highly organized, with utricular hair cells in different macular zones varying dramatically in their response properties. In particular, hair cells with varying hair bundle morphology differ in their membrane properties, suggesting that they differ in their complement of membrane conductances. Moreover, hair cells in different macular zones differ in their response dynamics, particularly in the rate and extent of their adaptation to maintained hair bundle displacement.

METHODS

Wholemount utricular maculae were removed from the bullfrog inner ear and maintained in cold, artificial perilymph consisting of (in mM) Na^+, 120; K^+, 2; Ca^{2+}, 4; Cl^-, 128; D-glucose, 3; and HEPES, 5 (pH = 7.25). Following proteolytic digestion in 30 μg/ml subtilopeptidase BPN (Sigma), the otolith membrane was removed by gentle mechanical agitation. Excised utricular maculae were then trimmed of excess nervous and connective tissue to improve the visibility of utricular hair bundles and mounted flat, hair cells uppermost, in a small chamber on the fixed stage of a limb-focusing microscope (Zeiss, model 16).

Using Nomarski optics, utricular maculae were viewed from above with a ×40 water-immersion objective and contrast-enhancement video camera (Hamamatsu, C2400). With these optics, we were able to resolve the component kinocilia and stereocilia of the hair bundles of utricular hair cells (FIGURE 1, bottom). The macular location of selected hair cells was measured from their position relative to the striolar border and the line dividing hair cells of opposing polarities. Hair bundle morphology was characterized by the size of the hair bundle, the relative lengths of the kinocilium and longest stereocilia, and the absence or presence of a bulbed kinocilium.

Hair cells were impaled, under visual guidance, with microelectrodes pulled from thin-walled aluminosilicate glass tubing, and bent near their tips to allow near-vertical penetrations of hair cells. The membrane properties of hair cells were determined from their voltage responses to brief (100-msecond) intracellular current steps. Hair cells were mechanically stimulated by displacing the distal tips of their longest stereocilia toward the kinocilium with a glass stimulating probe. Response dynamics were studied by sinusoidally displacing hair bundles in the frequency range 0.5–150 Hz and analyzing the gains and phases of the resulting sinusoidal responses with Fourier techniques. Low-frequency response dynamics were further assessed with step displacements of varying amplitude and duration. The motion of stimulating probes was monitored visually and calibrated against a stage micrometer at high magnification (×10,000).

RESULTS

Hair Cell Morphology

The utricular macula in the bullfrog is divided into medial and lateral parts by the striola, a narrow ribbon-shaped zone, that runs for the length of the sensory

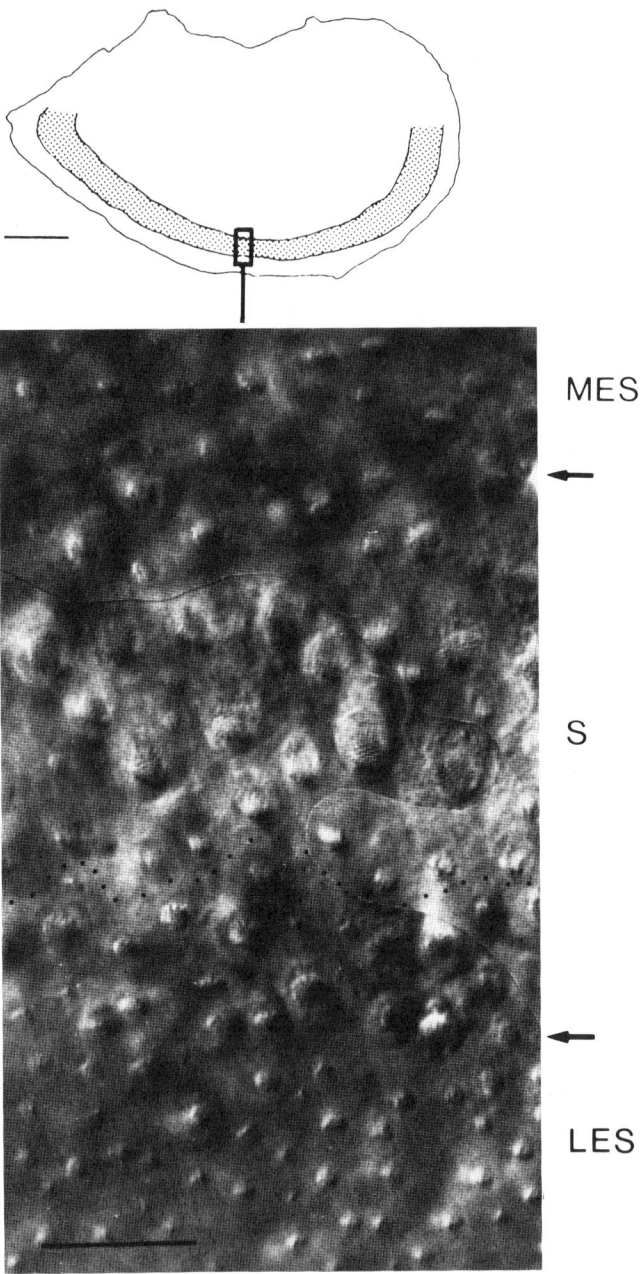

FIGURE 1. Surface reconstruction of wholemount utricular macula (top) and photomicrograph of utricular striola and surrounding extrastriolar regions (bottom). The striola, a ribbon-shaped zone, separates the extrastriola into medial and lateral parts. Arrows indicate the borders of the striolar region. Dotted line indicates reversal of hair cell polarization. MES, medial extrastriola; S, striola; LES, lateral extrastriola. Bars, 250 μm (top); 25 μm (bottom).

epithelium near its lateral border (FIGURE 1, top). The striola is distinguished from the medial and lateral extrastriola by the larger size and wider spacing of its hair cells (FIGURE 1, bottom) and the taller height of its apical surface (FIGURE 2). A line dividing hair cells of opposing polarities, located near the lateral border of the striola, separates the striola into medial and lateral parts. On average, the striola consists of 5–6 medial rows and 2–3 lateral rows of hair cells.

Vestibular hair cells in the bullfrog have previously been defined as type II according to cell body and synapse morphology.[27] Utricular hair cells, following the scheme of Lewis and Li,[18] were further classified into four types by hair bundle morphology (FIGURE 2). Type B hair cells have uniformly short stereocilia and kinocilia 2–5× as long as their longest stereocilia. These hair cells, the most common hair cell type in the utriculus, are found throughout the medial and lateral extrastriola and, more rarely, in the striolar region (FIGURE 3). The remaining three hair cell types are confined to the striolar region. Type C hair cells, the second most common utricular hair cell, resemble an enlarged version of the predominant type B hair cell (FIGURE 2). These cells are found throughout the outer rows of the medial and lateral striola and, more rarely, in other parts of the striolar region (FIGURE 3). Moving inward, these hair cells are gradually replaced by type F hair cells, hair cells with visibly larger hair bundles and kinocilia equal to or slightly longer in length than their longest stereocilia (FIGURE 2). Type F hair cells, unlike type C hair cells, are restricted to the middle rows of the medial and lateral striola (FIGURE 3). Type E hair cells, slightly smaller than type F hair cells, have prominent kinociliary bulbs and resemble hair cells from the bullfrog sacculus (FIGURE 2). These hair cells are restricted to the innermost striolar rows, lying astride both sides of the line dividing hair cells of opposing polarities (FIGURE 3).

Hair Cell Physiology

The mean resting potentials of utricular hair cells did not vary significantly in different macular zones or between hair cells with differing hair bundle morphology. Input resistance, on the other hand, varied from 200–2500 MΩ, type C hair cells having consistently lower input resistances than other hair cell types. Input resistance was not correlated with the size or membrane potential of hair cells, suggesting that this variation was not due to injury due to microelectrode impalement.

The current-voltage (I-V) relations of utricular hair cells were relatively linear near resting potential but differed at larger depolarizing and hyperpolarizing voltages (FIGURE 4, left). These differences were correlated with differences in hair bundle morphology. Type B and type C hair cells had near-linear I-V relations. Type E and type F hair cells, on the other hand, were strongly outwardly rectifying for depolarizing voltages. With the possible exception of type F hair cells, the I-V relations of utricular hair cells had no detectable "N shape" for depolarizing voltages, suggesting that, as in semicircular canal hair cells,[28,29] Ca^{2+}-activated potassium currents provide only a small contribution to the membrane current of these cells. Utricular hair cells, with the exception of type E hair cells, had linear I-V relations for hyperpolarizing voltages. In type E hair cells, an inward (anomalous) rectification was observed for voltages more negative than −90 mV (FIGURE 4, left).

There were also marked differences in the voltage responses of utricular hair cells to depolarizing current (FIGURE 4, right). In type B and type C hair cells, depolarizing currents produced passive exponential changes in membrane potential. Type F hair cells, on the other hand, displayed active potential changes lasting <25 mseconds following the onset of depolarizing current. For type F hair cells, I-V

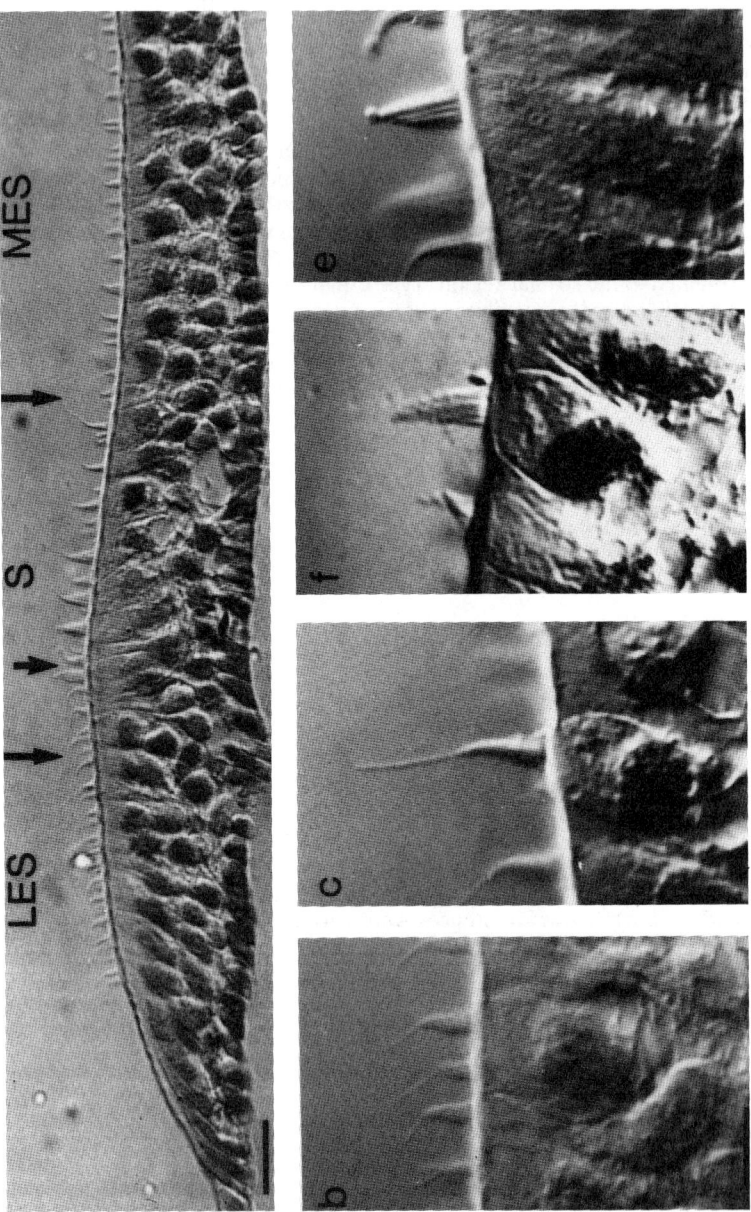

FIGURE 2. Photomicrographs of toluidine blue stained cross section of the utricular macula (top) and of individual hair cells in the extrastriolar and striolar regions (bottom). Outer arrows denote the medial (left) and lateral (right) borders of the striolar region; middle arrow marks the line dividing hair cells of opposing polarities. MES, medial extrastriola; S, striola; LES, lateral extrastriola. Bars, 25 μm (top); 10 μm (bottom).

curves remained relatively linear for latencies > 25 mseconds, but became increasing nonlinear at shorter latencies, displaying increasing amounts of outward rectification for larger depolarizing currents. At these shorter latencies, there was a depolarizing peak whose amplitude and duration was graded with current size. Termination of this depolarizing peak could result from a transient underlying conductance, the

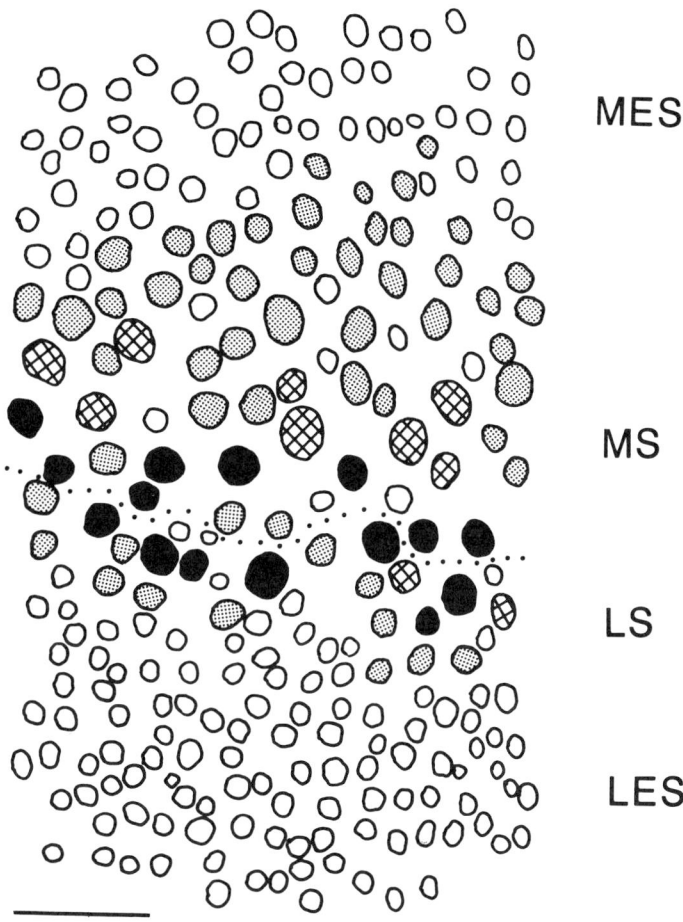

FIGURE 3. Macular distribution of utricular hair cell types. Dotted line indicates line of reversal of hair cell polarization. MES, medial extrastriola; S, striola; LES, lateral extrastriola. White, type B; stippled, type C; black, type E; cross-hatched, type F. Bar, 25 μm.

delayed activation of a different current, or both. The linearity of the I-V curve after 25 mseconds and the lack of a hyperpolarizing undershoot at the termination of depolarizing current steps suggest that the termination of the peak is the result of a transient conductance change. Hyperpolarizing current steps in type F hair cells produced a rebound depolarization at the termination of the current step. Ampli-

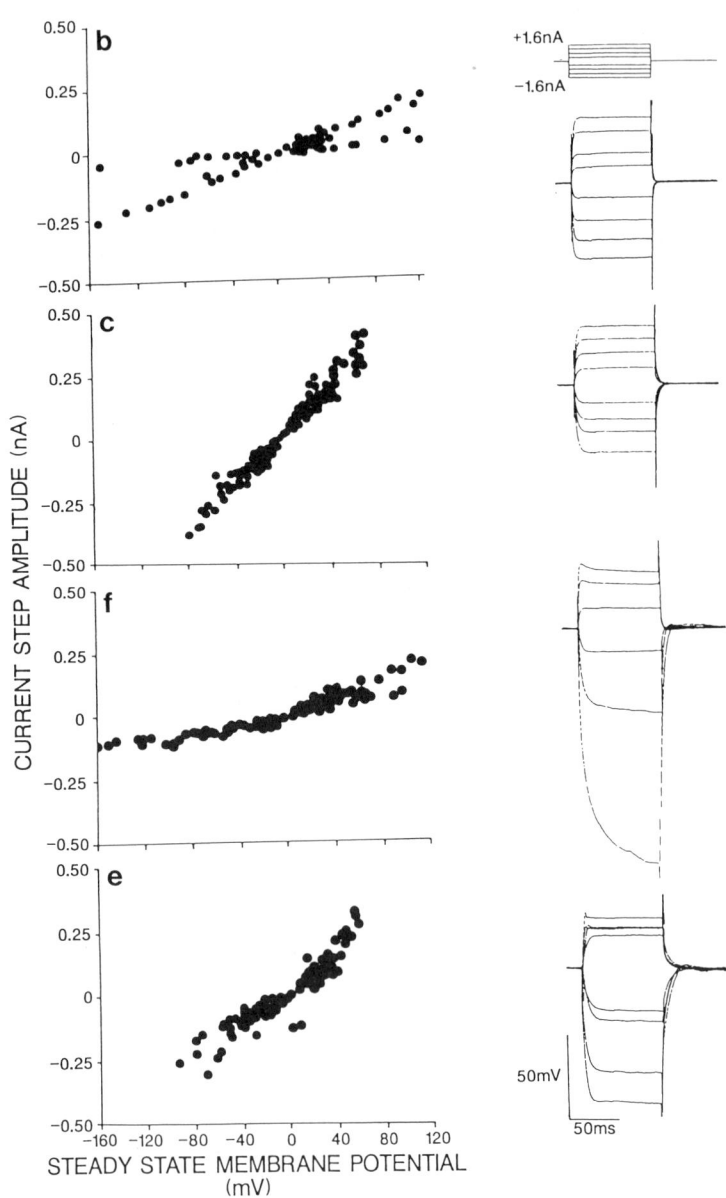

FIGURE 4. Current-voltage (I-V) plots (left), measured 100 mseconds after the onset of an intracellularly injected current step, and representative voltage responses (right) of utricular hair cell types.

tude of the rebound depolarization was dependent on the amplitude of the hyperpolarizing steps that evoked it. When of sufficient amplitude, the rebound depolarization appeared to trigger a local response resembling that triggered by direct depolarization.

Type E hair cells, unlike other utricular hair cells, were electrically resonant, producing 1–3 cycles of oscillatory potential changes at the onset of depolarizing current steps. These oscillatory changes resembled those seen in hair cells in the bullfrog sacculus but were smaller and more highly damped (FIGURES 4 and 5). Unlike saccular hair cells, there was no hyperpolarizing undershoot at the termination of depolarizing current steps. Hyperpolarizing current steps in type E hair cells produced a small sag in membrane potential from an initial maximum and oscillatory potential changes after current termination. Unlike hair cells in the bullfrog sacculus, these oscillations were of lower frequency than those seen at the onset of depolarizing current steps, suggesting the involvement of a transient rectifier potassium current in their production. The amplitude and frequency of oscillatory changes increased with both depolarizing and hyperpolarizing current (FIGURE 5).

The sensitivities of utricular hair cells to natural stimulation were examined with sinusoidal hair bundle displacement. The sinusoidal gains of type B and type C hair cells were significantly lower than those of type F and type E hair cells (FIGURE 6, top). This difference was due, in large part, to the way in which hair cells were mechanically stimulated in this study. To normalize responses from different hair cell types, theoretical models were used to predict the contribution of geometric factors associated with hair bundles to natural sensitivity. Three factors were considered in this analysis. The first factor is a measure of the mechanical gain produced by the lever arm formed by the different anchoring points of the stereocilia and otolith membrane on the kinocilium. The second factor, largely dependent upon the length and spacing of adjacent stereocilia, determines the rotary deflection of the hair bundle per unit linear stereociliary deflection. The third factor, the number of stereocilia in the hair bundle, is a measure of the number of transduction channels. The normalized sensitivities of utricular hair cells was similar, suggesting that differences in hair bundle morphology compensate for intrinsic differences in sensitivity between utricular hair cells. In general, short hair bundles with many stereocilia (type E and type F) are more sensitive than long hair bundles with fewer stereocilia (type B and type C), since small linear deflections of these hair bundles produce large angular rotations and more stereocilia mean more transducer channels.

The response dynamics of utricular hair cells were correlated with their hair bundle morphology (FIGURE 6, top and bottom). Type B hair cells, in both the striolar and extrastriolar regions, had low-pass filter characteristics, with gains that were nearly constant and phases hovering near zero for frequencies <0.5 Hz. For frequencies >0.5 Hz, gains fell -20 dB/decade. Phases showed increasing phase lags up to an asymptotic value of $-90°$. Type C hair cells, on the other hand, had very high (>50 Hz) bandwidths. These cells also displayed varying degrees of low-frequency adaptation, with reductions in sinusoidal gain of -20 dB/decade and phase leads of 25–40° for frequencies <5 Hz. Type F hair cells had little or no low-frequency adaptation and were most responsive to midband (10–20 Hz) frequencies. Type E hair cells, unlike other utricular hair cells, had tuned filter characteristics, displaying highly damped resonances around 1–5 Hz at their resting membrane potentials.

The low-frequency response dynamics of hair cells were further examined with step displacements of varying amplitude and duration (FIGURE 7). Hair cells differed markedly in the rate and extent of their adaptation to step displacement. These

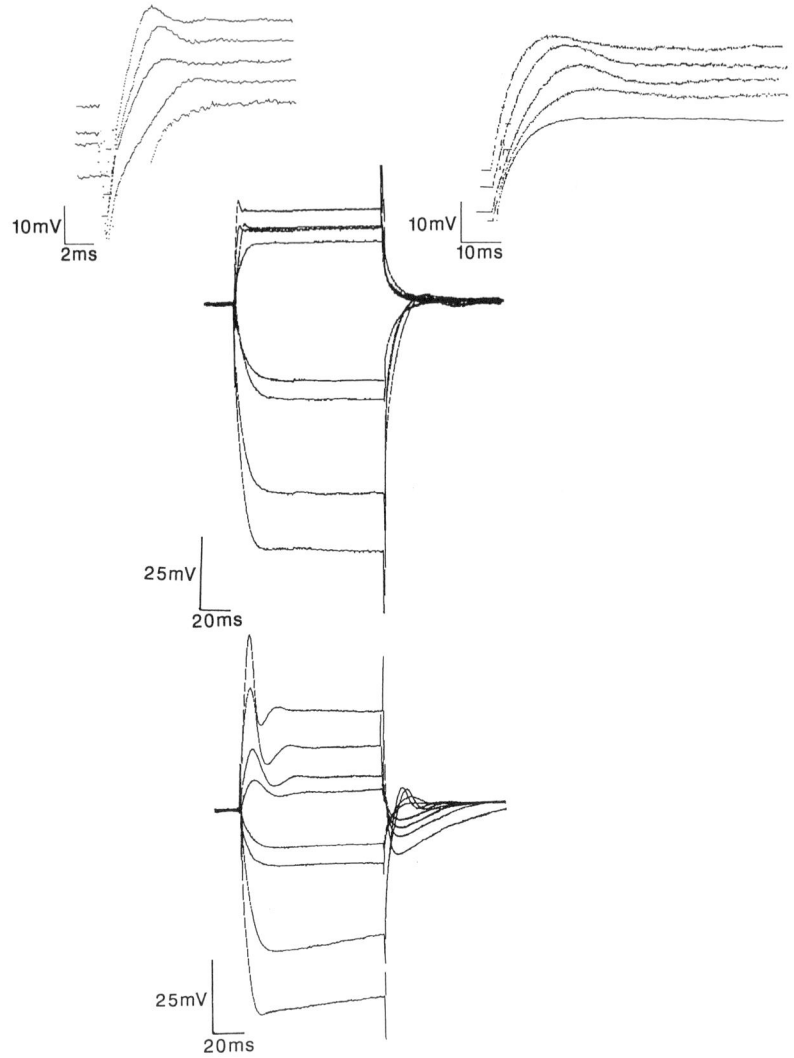

FIGURE 5. Voltage responses of electrically resonant utricular (Type E) (top) and saccular (bottom) hair cells to depolarizing and hyperpolarizing current steps. Insets of oscillatory potential changes in the utricular type E hair cell at the onset of depolarizing and the termination of hyperpolarizing current are shown at the top left and top right, respectively.

differences were correlated with macular location and hair bundle morphology. Type B hair cells (FIGURE 7, far left) had nonadapting responses to mechanical steps, displaying slow potential changes at the beginning and termination of mechanical steps. Type C hair cells, on the other hand, adapted rapidly (100–200 mseconds) and completely to resting levels (FIGURE 7, top middle). Moving inward within the striola, the step responses of these hair cells (FIGURE 7, middle) became increasingly

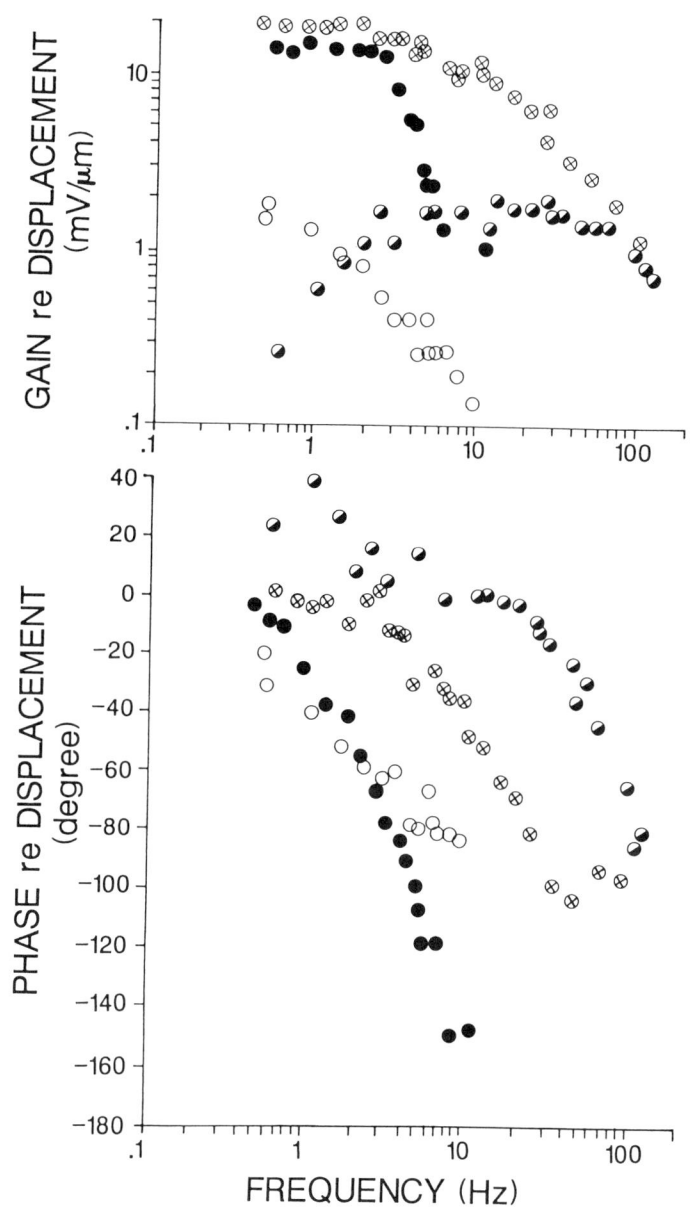

FIGURE 6. Bode plots of sinusoidal gain (top) and phase (bottom) of utricular hair cells to sinusoidal displacement of their hair bundles. Open circles, type B; half-filled circles, type C; filled circles, type E; circles with crosses, type F.

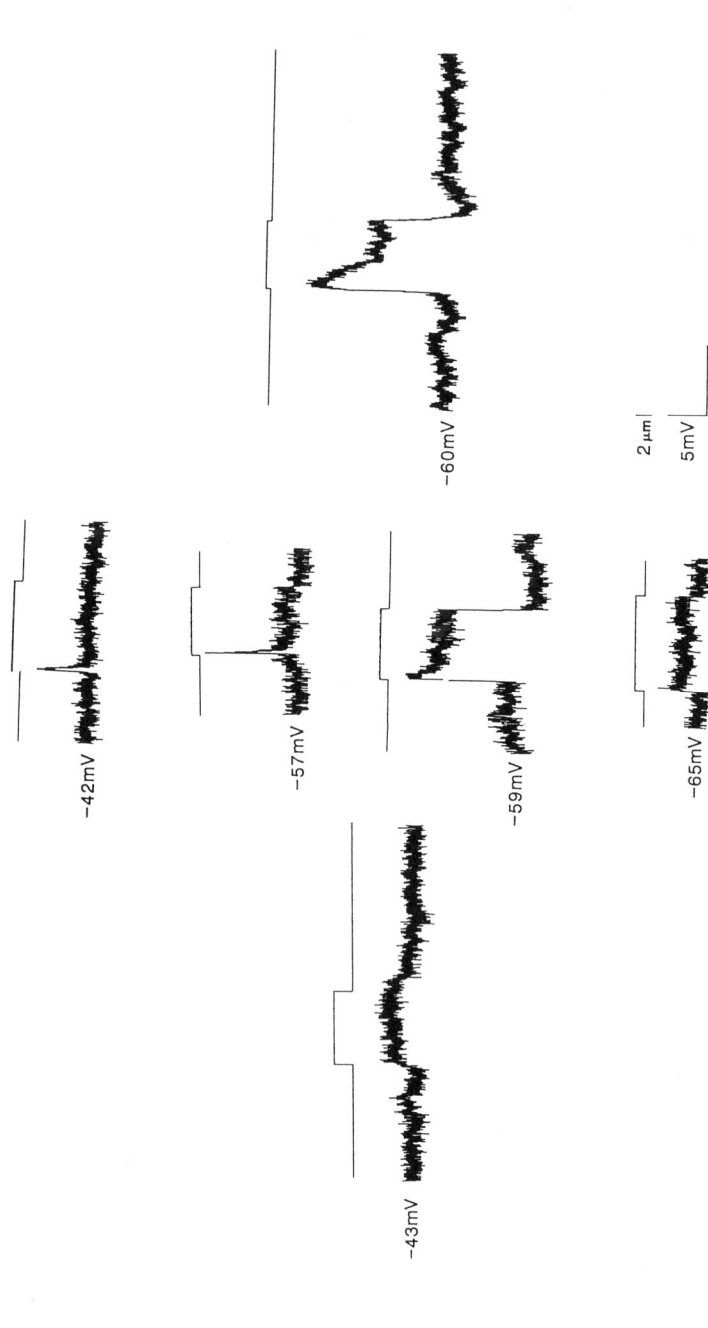

FIGURE 7. Responses of utricular hair cells to excitatory step displacements of their hair bundles. Left, type B; top middle, type C; bottom middle, type F; right, type E.

slower and less complete. Type E hair cells, like type F cells, adapted slowly and incompletely to step displacements (FIGURE 7, bottom middle and far right).

DISCUSSION

Hair cells in different regions of the bullfrog utriculus differ markedly in their responses to intracellular current and mechanical displacement. More specifically, the active membrane properties of utricular hair cells in different macular zones vary significantly, suggesting that these cells differ in their complement of membrane conductances. In particular, type F hair cells appear to uniquely possess a Ca^{2+}-activated potassium conductance. Type E hair cells uniquely possess an inward rectifier potassium conductance similar to that previously observed in vestibular[26,29] and cochlear[30] hair cells. They also exhibit an electrical resonance that strongly resembles that seen in pigeon semicircular canal hair cells.[23] This resonance, unlike that seen in auditory[9,10] and vibratory[8,11] hair cells, may be determined by Ca^{2+}-insensitive potassium conductances. Using whole-cell patch-clamp techniques, we are now confirming these conclusions in isolated utricular hair cells and determining whether further differences exist in the size and gating kinetics of particular ionic currents in specific hair cell types.

Utricular hair cells also differ in their response dynamics to mechanical displacement. In particular, hair cells differ in the rate and extent of their low-frequency adaptation. Type B hair cells, for example, are nonadapting or very slowly adapting. With the exception of these cells, utricular hair cells adapt to long-duration mechanical displacement. The rate and extent of this adaptation are correlated with macular location. Because the low-frequency response dynamics of hair cells to mechanical displacement differ from those to intracellular current, it seems likely that differences in low-frequency response dynamics are due not to voltage-dependent membrane conductances but rather to the nature of the adaptation process associated with mechanoelectric transduction in different hair cell types. We suspect that regional differences in the calcium sensitivity of the adaptation process exist, providing a mechanism for regulating the rate or extent of adaptation in different hair cell types.[31]

Differences in response properties are correlated with differences in hair bundle morphology rather than macular location per se. Type B hair cells, for example, have similar response properties, whether they are located in the striolar or extrastriolar region. Moreover, hair cells in similar macular locations, but with different hair bundle morphology, differ in their responses to intracellular current and mechanical displacement. This suggests that the nature of the adaptation process and the expression of voltage-gated membrane channels are intimately associated with the development of the sensory hair bundle.

Utricular afferents in the bullfrog have been previously classified as gravity or vibratory sensitive.[6,14] Gravity afferents have been further classified into three classes according to their responses to head tilt. Tonic gravity afferents respond to head position, phasic gravity afferents respond to head velocity, and phasic-tonic afferents respond to both head position and velocity.

Morphophysiological studies have shown that the response dynamics of utricular afferents are correlated with the macular location and hair bundle morphology of their innervated hair cells.[6,14] Tonic gravity afferents, for example, innervate type B hair cells in the medial and lateral extrastriolar regions (FIGURE 8, left). The other types of utricular afferents innervate hair cells in the striolar region. Phasic and phasic-tonic gravity afferents innervate type C or type C and type F hair cells, respectively (FIGURE 8, top right). Vibratory afferents innervate type E hair cells in the innermost striolar rows (FIGURE 8, bottom right).

Our results suggest that it is unnecessary to postulate nonhomogeneous motions of the otolithic membrane or differential coupling of hair cells to this membrane to

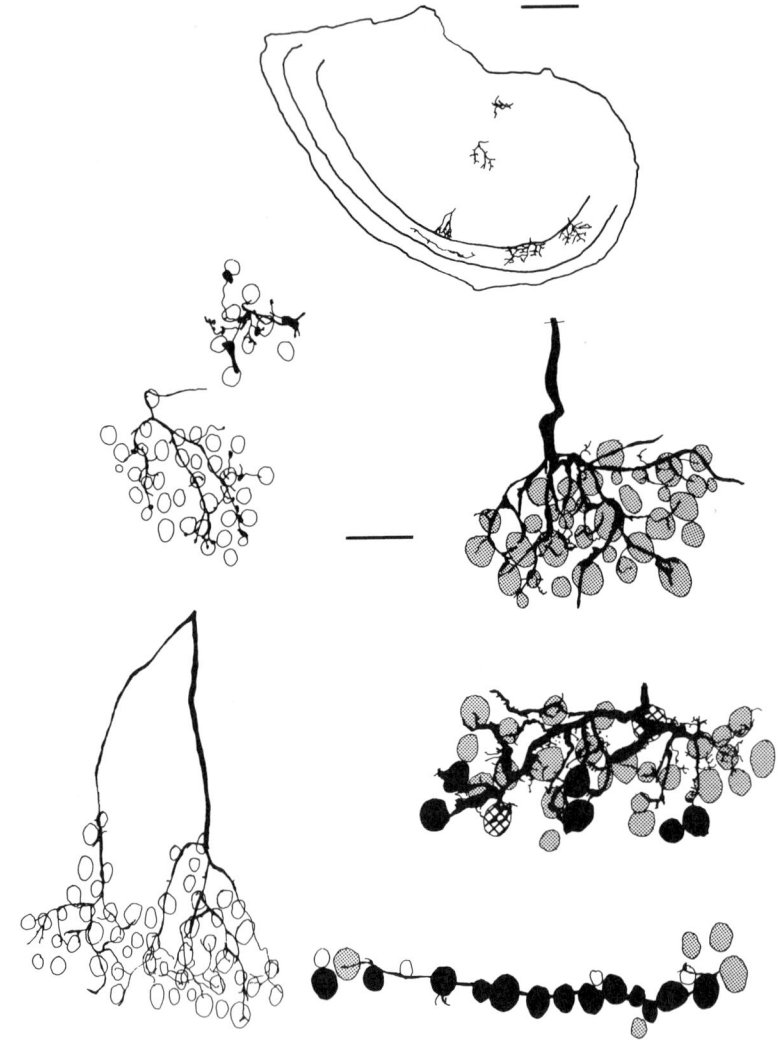

FIGURE 8. Macular location (top) and peripheral innervation patterns (bottom) of typical vestibular nerve afferents in the bullfrog utriculus. Bar, 200 μm (top); 20 μm (bottom).

account for differences in the response dynamics of utricular afferents. Rather, these differences may be derived largely, if not completely, by differences in the transduction mechanisms of their innervated hair cells. Under this scheme, vibratory afferents owe their vibration sensitivity to the electrical resonance of type E hair cells.

Phasic-tonic afferents presumably owe their head velocity sensitivity to rapidly adapting type C hair cells and their head position sensitivity to slowly adapting type F hair cells.

In conclusion, the utricular macula of the bullfrog is highly organized. Hair cells in the extrastriolar regions of this end organ are nonadapting and most sensitive to static gravity and very low frequency linear accelerations. Striolar hair cells, on the other hand, are specialized to encode higher frequency information. These hair cells adapt to maintained hair bundle displacement, reducing their sensitivity to low-frequency linear accelerations. Moreover, they have higher frequency sensitivities than extrastriolar hair cells, and, in the case of type E hair cells, are electrically tuned to further enhance their high-frequency sensitivity. With the exception of vibratory afferents, utricular afferents in higher vertebrates[32] have peripheral innervation patterns and physiological response properties similar to those seen in amphibians. This regional dichotomy may therefore be generally applicable to other vertebrate species, including mammals.

REFERENCES

1. OHMORI, H. 1985. Mechanoelectric transduction currents in isolated vestibular hair cells of the chick. J. Physiol. London **350:** 561–581.
2. OHMORI, H. 1987. Gating properties of the mechanoelectrical transducer channel in the dissociated vestibular hair cell of the chick. J. Physiol. London **359:** 189–217.
3. HOWARD, J., W. M. ROBERTS & A. J. HUDSPETH. 1988. Mechanoelectric transduction by hair cells. Annu. Rev. Biophys. Biophys. Chem. **17:** 99–124.
4. ROBERTS, W. M., J. HOWARD & A. J. HUDSPETH. 1988. Hair cells: transduction, tuning, and transmission in the inner ear. Annu. Rev. Cell Biol. **4:** 63–92.
5. KOYAMA, H., E. R. LEWIS, E. L. LEVERENZ & R. A. BAIRD. 1982. Acute seismic sensitivity in the bullfrog ear. Brain Res. **250:** 168–172.
6. LEWIS, E. R., R. A. BAIRD, E. L. LEVERENZ & H. KOYAMA. 1982. Inner ear: dye injection reveals peripheral origins of specific sensitivities. Science **215:** 1641–1643.
7. EATOCK, R. A., D. P. COREY & A. J. HUDSPETH. 1988. Adaptation of mechanoelectric transduction in hair cells of the bullfrog's sacculus. J. Neurosci. **7:** 2821–2836.
8. ASHMORE, J. F. 1983. Frequency tuning in a frog vestibular organ. Nature **304:** 536–538.
9. CRAWFORD, A. C. & R. FETTIPLACE. 1981. An electrical tuning mechanism in turtle cochlear hair cells. J. Physiol. London **312:** 377–412.
10. FETTIPLACE, R. 1987. Electrical tuning of hair cells in the inner ear. Trends Neurosci. **10:** 421–425.
11. LEWIS, R. S. & A. J. HUDSPETH. 1983. Voltage- and ion-dependent conductances in solitary vertebrate hair cells. Nature **304:** 538–541.
12. ART, J. J., A. C. CRAWFORD & R. FETTIPLACE. 1986. Electrical resonance and membrane currents in turtle cochlear hair cells. Hear. Res. **22:** 31–36.
13. HUDSPETH, A. J. & R. S. LEWIS. 1988. A model for electrical resonance and frequency tuning in saccular hair cells of the bullfrog, *Rana catesbeiana*. J. Physiol. London **400:** 275–297.
14. BAIRD, R. A. & E. R. LEWIS. 1986. Correspondences between afferent innervation patterns and response dynamics in the bullfrog utricle and lagena. Brain Res. **369:** 48–64.
15. BLANKS, R. H. I. & W. PRECHT. 1976. Functional characterization of primary vestibular afferents in the frog. Exp. Brain Res. **25:** 369–390.
16. MACADAR, O., G. E. WOLFE, D. P. O'LEARY & J. P. SEGUNDO. 1975. Response of the elasmobranch utricle to maintained spatial orientation, transitions, and jitter. Exp. Brain Res. **22:** 1–12.
17. FERNANDEZ, C. & J. M. GOLDBERG. 1976. Physiology of peripheral neurons innervating otolith organs of the squirrel monkey. I. Response to static tilts and to long-duration centrifugal force. J. Neurophysiol. **39:** 970–984.

18. LEWIS, E. R. & C. W. LI. 1975. Hair cell types and distributions in the otolithic and auditory organs of the bullfrog. Brain Res. **83:** 35–50.
19. GOLDBERG, J. M., G. DESMADRYL, R. A. BAIRD & C. FERNANDEZ. 1990. The vestibular nerve in the chinchilla. V. Relation between afferent response properties and peripheral innervation patterns in the utricular macula. J. Neurophysiol. **63:** 781–790.
20. BAIRD, R. A., G. DESMADRYL, C. FERNANDEZ & J. M. GOLDBERG. 1988. The vestibular nerve of the chinchilla. II. Relation between afferent response properties and peripheral innervation patterns in the semicircular canals. J. Neurophysiol. **60:** 182–203.
21. HONRUBIA, V., L. F. HOFFMAN, S. T. SITKO & I. R. SCHWARTZ. 1989. Anatomic and physiological correlates in bullfrog vestibular nerve. J. Neurophysiol. **61:** 688–701.
22. ART, J. J. & R. FETTIPLACE. 1987. Variation of membrane properties in hair cells isolated from the turtle cochlea. J. Physiol. London **385:** 207–242.
23. CORREIA, M. J., B. N. CHRISTENSEN, L. E. MOORE & D. G. LANG. 1989. Studies of solitary semicircular canal hair cells in the adult pigeon. I. Frequency- and time-domain analysis of active and passive membrane properties. J. Neurophysiol. **62:** 924–934.
24. HOUSLEY, G. D., C. H. NORRIS & P. S. GUTH. 1989. Electrophysiological properties and morphology of hair cells isolated from the semicircular canal of the frog. Hear. Res. **38:** 259–276.
25. RENNIE, K. J. & J. F. ASHMORE. 1991. Ionic currents in isolated vestibular hair cells from the guinea-pig crista ampullaris. Hear. Res. **51:** 279–291.
26. SUGIHARA, I. & T. FURUKAWA. 1989. Morphological and functional aspects of two different types of hair cells in the goldfish sacculus. J. Neurophysiol. **62:** 1330–1342.
27. WERSALL, J. & D. BAGGER-SJOBACK. 1974. Morphology of the vestibular sense organs. *In* Vestibular System: Basic Mechanisms. Handbook of Sensory Physiology. H. H. Kornhuber, Ed. Springer-Verlag. New York, N.Y.
28. LANG, D. G. & M. J. CORREIA. 1989. Studies of solitary semicircular canal hair cells in the adult pigeon. II. Voltage-dependent ionic conductances. J. Neurophysiol. **62:** 935–945.
29. OHMORI, H. 1984. Studies of ionic currents in the isolated vestibular hair cells of the chick. J. Physiol. London **350:** 561–581.
30. FUCHS, P. A. & M. G. EVANS. 1990. Potassium currents in hair cells isolated from the cochlea of the chick. J. Physiol. London **429:** 529–551.
31. HACOHEN, N., J. A. ASSAD, W. J. SMITH & D. P. COREY. 1989. Regulation of tension on hair-cell transduction channels: displacement and calcium dependence. J. Neurosci. **9:** 3988–3997.
32. FERNANDEZ, C., J. M. GOLDBERG & R. A. BAIRD. 1990. The vestibular nerve of the chinchilla. III. Peripheral innervation patterns in the utricular macula. J. Neurophysiol. **63:** 767–780.

Sensory Coding in the Saccule[a]

Patch Clamp Study of Ionic Conductances in Isolated Cells

ANTOINETTE STEINACKER[b] AND LISETTE PEREZ[c]

[b]Institute of Neurobiology
University of Puerto Rico
Medical Sciences Campus
201 Boulevard del Valle
San Juan, Puerto Rico 00901

[c]Department of Otolaryngology
University of Puerto Rico
Medical Sciences Campus
Rio Piedras, Puerto Rico 00936

INTRODUCTION

Sensory coding in all acousticolateralis hair cells occurs through three distinct processes: sensory transduction at the hair cell bundle, the activation of ion specific conductances of the basolateral surface, and the synaptic release process at the hair cell base. The work described here concerns only the properties of the ionic conductances of the basolateral surface. These conductances serve as intermediaries between the transduction at the apical pole and the synaptic release process at the basal pole. For this reason, knowledge of the properties of the ionic conductances of the basolateral membrane is critical to understanding the process of sensory coding in acousticolateralis systems. For any individual hair cell, the ionic conductances present and their magnitude and kinetics of activation, deactivation, and inactivation will vary in accordance with the sensory feature that cell is designed to encode.

Emphasis to date on sensory coding in the basolateral membrane of the hair cell has focused on a membrane resonance that has been shown to be at or near the characteristic frequency of the hair cell.[1] This resonance is proposed to result from an interaction of passive membrane properties and two ionic currents, an outward calcium activated potassium current (IKCa) and an inward calcium current (ICa) whose magnitude and kinetics have been proposed to determine the characteristic frequency of the cell.[2,3] We have found that the majority of the saccular hair cells have a second potassium current active with IKCA and that this current endows the hair cell with additional coding capabilities, i.e., the ability of some cells to produce spiking potentials in response to a stimulus. Two other currents, an A current (IA) and an inward rectifier, also have a differential distribution in hair cells. Specific combinations of these currents produce spiking, rather than resonance, in the hair cell. Perhaps other aspects of the stimulus, in addition to stimulus frequency, are processed at the hair cell level.

[a]This work was supported by the National Institutes of Health 1F35GM14163 and the Deafness Research Foundation. Part of this work was carried out at the Marine Biological Laboratory, Woods Hole, Mass.

27

METHODS

The skull was opened in a spinalized, anesthetized (Finquil, Ayerst, Mass.) toadfish, *Opanus tau,* the cranial nerves cut, and the brain removed. The saccules were transferred to a zero calcium Ringer and perfused first with a solution in zero Ca^{2+} salt water teleost Ringer containing (in mM) NaCl, 165; KCl, 5; $MgCl_2$, 1; Hepes, 10; pH 7.2 with either collagenase Sigma type A at 0.25 mg/ml (Sigma, St. Louis, Mo.) or protease Sigma type XIV 0.03 mg/ml (Sigma, St. Louis, Mo.) for 10 minutes and then with a solution of papain at 0.5 mg/ml (Fluka, Switzerland) for 20 minutes. Pipettes of Kimble hematocrit or Corning 7052 glass were Sylgard coated, fire polished, and filled with (in mM) KCl, 165; $CaCl_2$, 0.1; $MgCl_2$, 1.5; Hepes K^+ salt, 5.0; EGTA, 10; Mg^{2+} or K^+ ATP 2.5 buffered to pH 7.2 with KOH. Bath solution was the above zero Ca^{2+} Ringer with 4.0 mM $CaCl_2$ added. Whole cell recording mode was carried out in the conventional manner using an Axopatch 1 A amplifier (modified by Axon Instruments, Burlingame, Calif., to speed up the current clamp recording).

In voltage clamp mode, cell leakage current and capacity were compensated manually and then digitally using a P/4 routine.[4] The current response to a −20-mV pulse from a −60-mV holding potential was used to record the remaining transient to calculate residual series resistance and capacity from the time constant of decay and area of the transient. The membrane input resistance was calculated from the steady-state current during the pulse. Exponential fits to the tail currents were done using a least-squares algorithm. All voltage and current clamp protocols were run on line for 6 to 8 trials, averaged, and stored. Quality factor (*Q*) of the resonance was calculated as

$$Q = [(\pi f_0 \tau_0)^2 + \frac{1}{4}]^{1/2}$$

where f_0, the frequency of the oscillations measured as cycles/second, and τ_0, the time constant of decay of the oscillations, are measured following current onset.[2]

Data acquisition and digitizing were done using the Axolab interface (Axon Instruments, Burlingame, Calif.). Data were digitized on line at 10–50 kHz, depending on the protocol. pCLAMP, a software package from Axon Instruments, was used to run the experiments on an Everex 386 computer. Acetylcholine (ACh) application was via a puffer pipette (10^{-4} M) placed near the cell and triggered by a pulse from the computer timed to precede voltage test pulses to the cell. Subtraction of control records from ACh records show the ACh effect.

HAIR CELLS DO NOT HAVE A UNIFORM COMPOSITION OF POTASSIUM CONDUCTANCES

Sensory transduction in hair cells has been modeled on the interaction of passive membrane properties with an inward calcium current (ICa) and an outward IKCa.[2,3] In this model, after depolarization by a transducer current, ICa is activated and the cell is depolarized. The IKCa, being both calcium and voltage sensitive, responds with a rapid opening of its ion-specific channels, resulting in a hyperpolarization. IKCa is then rapidly deactivated by the hyperpolarization. This process is repeated for every cycle of the stimulus. The rate with which these two currents turn on and off, coupled with the number of ion-specific channels available and the passive properties of the membrane, determines the frequency of resonance.

In the toadfish saccule, these criteria are met for only a small percentage of the cells. In approximately 10% of the cells, only one outward current is present. FIGURE 1 illustrates this cell type in which, after tetraethyl ammonium (TEA), only the inward ICa remains (FIGURE 1B). It could be rapidly blocked by external TEA (0.5–25 mM), known to block the IKCa in hair cells.[2,3] It was also blocked by CdCl$_2$ (0.5 mM), BaCl$_2$ (5 mM), and charbydotoxin and by internal Cs$^+$ substituted for K$^+$ in the pipette. It was only partially blocked by zero Ca^{2+} solutions. This current resembles the IKCa in the hair cell.[2,5,6]

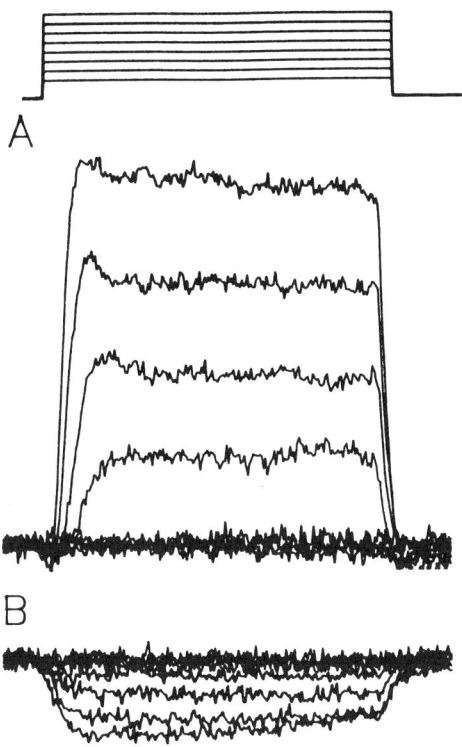

FIGURE 1. A small percentage of the saccular hair cells show only one outward current (A) in response to voltage commands. This current is rapidly blocked by 25 mM TEA, leaving the underlying inward calcium current (B). Voltage commands above are 5-mV incremental steps from −50-mV holding potential to −5 mV.

In the majority of cells, after 25 mM TEA, a second outward current was found whose rate of activation and tail current decay differed from that of IKCa (FIGURE 2). This current was not blocked by any of the above agents except internal Cs$^+$ and charbydotoxin. Both currents shift along the voltage axis of the current-voltage (I/V) plot as expected for a K$^+$-selective channel when 20 or 50 mM KCl is substituted for normal extracellular K$^+$. This voltage-gated potassium current has the properties expected of a delayed rectifier and is referred to here as IK.[7,8] This current has been found in the pigeon semicircular canal hair cells,[9] in chick cochlea,[10] and in inner hair cells of guinea pig.[11]

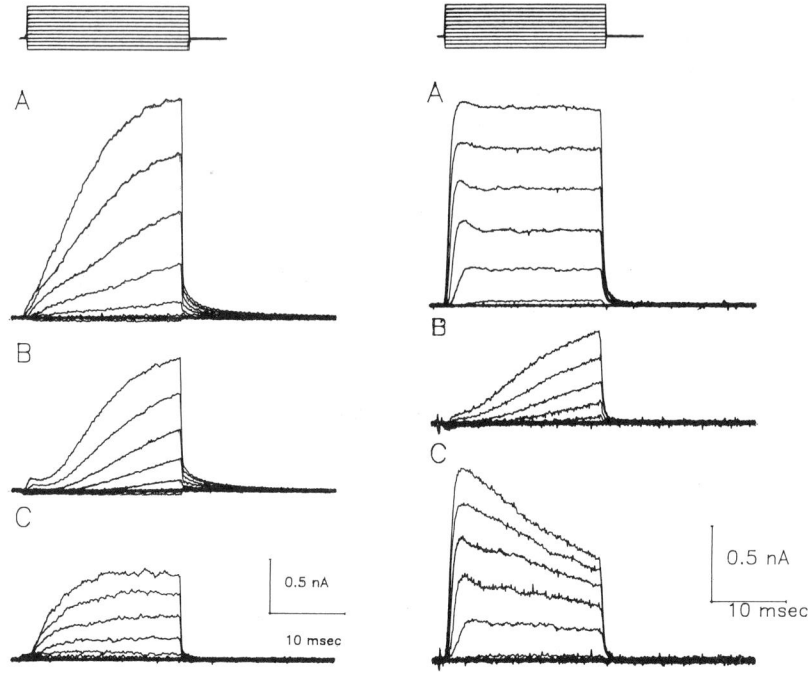

FIGURE 2. Total outward current may have slow (left panel) or fast (right panel) activation and deactivation kinetics. In most hair cells, the total outward current is composed of two major components. **A:** Total outward current. **B:** TEA-insensitive current (IK). **C:** IKCA blocked by TEA, obtained by subtracting B from A. Voltage commands from the −60-mV holding potential are 10-mV incremental steps from −90 to +20 mV. The current in B right shows a small IA.

To understand the role an ionic conductane may play in a cell, it is necessary to characterize its conductance and the kinetics of activation, deactivation, and inactivation. The rate of activation and deactivation of IK is much slower than that of the IKCa. Not only is IK slower to activate than IKCa, but it often continues to grow after the latter current has peaked. This is illustrated in FIGURE 3 by the use of cadmium (0.2 mM) to block IKCa and the underlying ICa. This enables the measurement of the activation rate without the underlying ICa present during TEA block. FIGURE 3C shows the two conductances measured as steady-state current. In FIGURE 3D, the voltage sensitivity of the activation rate for the two conductances is plotted. It is apparent that two conductances are present with quite different voltage sensitivities of the activation rates. Activation is measured from the time course of the current recorded in whole cell mode following the onset of a voltage command pulse [measured by the fit of a modified Hodgkin-Huxley equation to the rising phase of the current (FIGURE 4)] and is primarily a reflection of the opening of the individual ion-specific channels that make up the macroscopic current. Deactivation is measured by the time course of the tail current following the end of the voltage command pulse and reflects primarily the closing of the individual channels. It is most precisely measured by changing the potassium concentration of the extracellular fluid which shifts the equilibrium or reversal potential for potassium currents to more positive values. The voltage is jumped from the holding potential to $+20$ mV and returned in 5-mV steps to -115 mV. The resulting inward tail currents can be fit with two exponentials whose decay time is related to the open time or closing rates of the channels and whose amplitude reflects the conductance of the current. Since each jump is to the same voltage, the same number of channels are open for each voltage step. The amplitude of the tail current then equals the number of channels available times the probability of channel opening times the single-channel conductance. These conductances are seen in FIGURE 5C. In FIGURE 5D, the decay time of the currents is plotted against voltage and the difference between deactivation, or closing rates, of the two conductances is seen. The absolute values and relationship between activation rates and deactivation rates of these two currents differ widely between cells but are related within the individual cell.[8] Inactivation is the process of

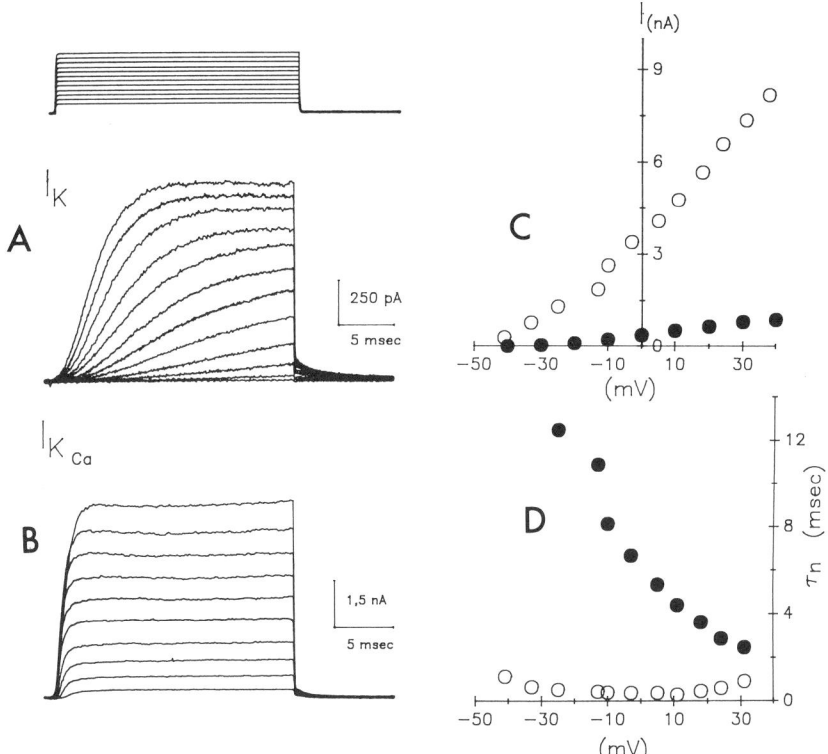

FIGURE 3. The true activation rate of IK can be determined after 0.2-mM cadmium block of IKCa and ICa. Records in A show outward current (IK) remaining after cadmium block. Records in B show IKCa obtained by subtracting current in A from total outward current (not shown). Data corrected for series resistance error (3.2 mohm) and currents selected for series resistance corrected voltage commands up to $+30$ mV steps from a holding potential of -60 mV. C illustrates the steady-state current/voltage relationship for the two currents (IK, filled circles; IKCa, open circles). The voltage sensitivity of the activation rate for the two currents is plotted in D. IK activation rate (filled circles) is much more steeply voltage dependent than that of IKCa (open circles). See FIGURE 4 for activation rate fitting.

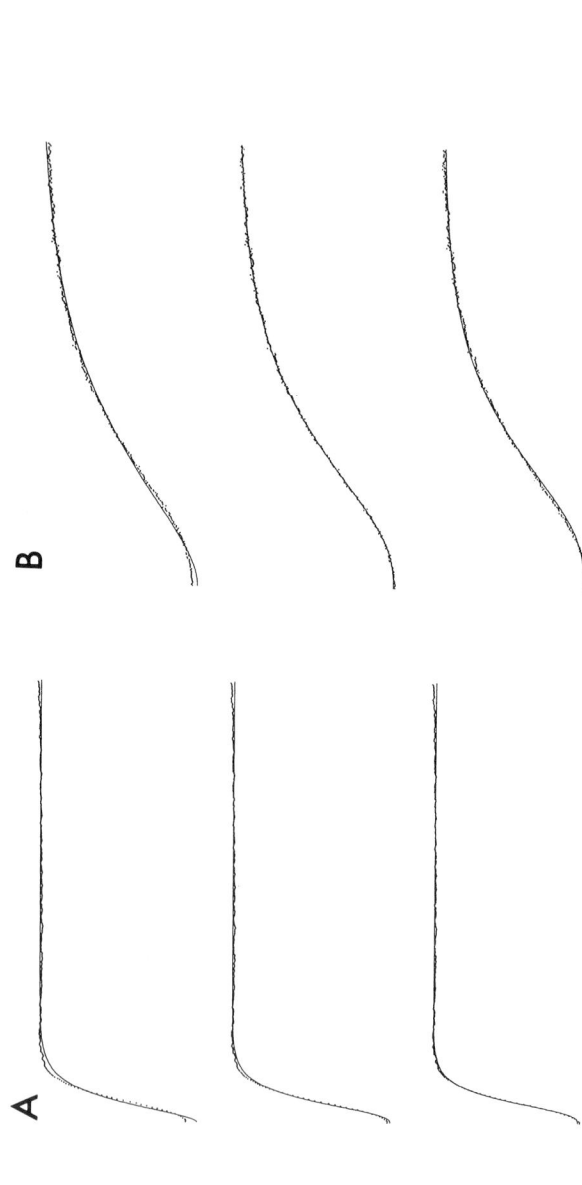

FIGURE 4. Activation of the 2 currents was fit with a modified Hodgkin and Huxley type equation. Activation rate of IKCa was best fit with a power of 4 (A) and that of IK with a power of 3 (B). Smooth lines of the fit are drawn through the dotted trace of the actual data which were taken from 3 successive voltage steps of the data in FIGURE 3.

channel closure into a state from which they cannot readily open following a maintained stimulus. When the voltage command pulse was lengthened to more than 100 mseconds, inactivation was apparent and had 2 phases. These three rates are important because they determine how the currents of the basolateral surface respond to the transduction current.

In cell-attached recordings, single-channel openings of two potassium channel species have been found that match the conductance and kinetics of the two currents

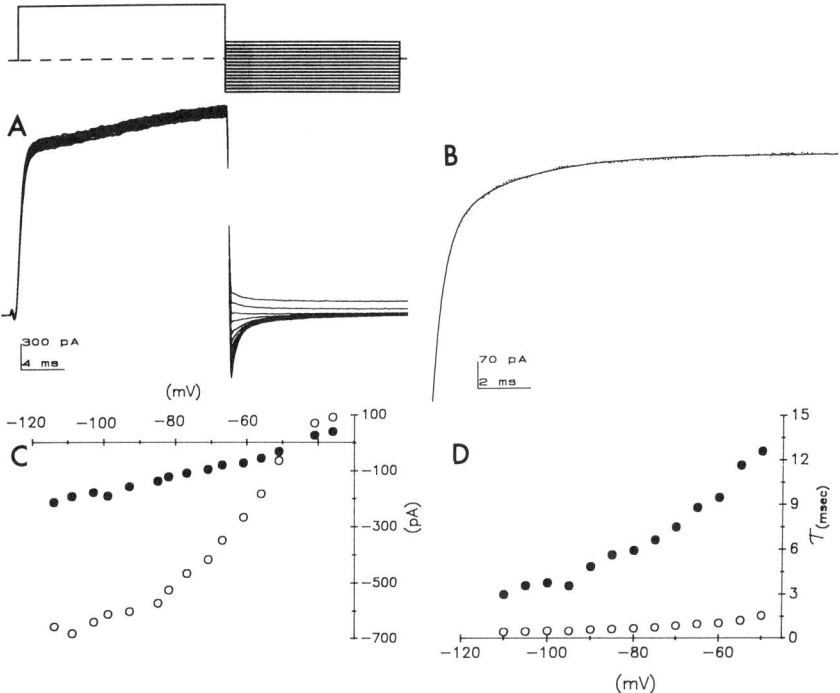

FIGURE 5. Instantaneous tail current analysis using two exponential least-squares fit to describe the conductance and deactivation rates of the two currents. Twenty-millimolar external potassium was used to produce inward tail current records to increase accuracy of tail current measurements. Voltage protocol is a step to +20 mV from a −60-mV holding potential and subsequent repolarizations in 5-mV steps from 115 mV to −35 mV (A). In B, a single tail current trace is shown fit by 2 exponentials. Lower plots are (C) the tail current amplitude of the two exponentials corresponding to the low-conductance IK (filled circles) and high-conductance IK_{Ca} (open circles) and (D) voltage dependence of deactivation rate of the two currents determined from the time constant of decay of the fast component (IK_{Ca}) and the slow component (IK).

above, found by whole-cell recording (FIGURE 6) These two potassium channels are active over the same voltage range but with considerably different kinetics. The mean open time of the high-conductance channel is markedly shorter than that of the low-conductance channel (see arrows, FIGURE 6). The high-conductance potassium channel showed the sensitivity to externally applied calcium in cell-detached inside-out patches that would be expected of a calcium-activated potassium channel and is thus identified as IKCa. The low-conductance channel was insensitive to the same calcium levels, and its low conductance and long open times identify it as IK.

In whole-cell recording, a fast transient potassium current, the A current (IA), was found in only a subset of the hair cells, even when -120-mV prepulses were used to prevent inactivation. This current has been reported in other hair cells where its level and holding potential dependence varies considerably.[5,11-13] An IA, when present, was seen following steps from -120 and -90 mV and sometimes from

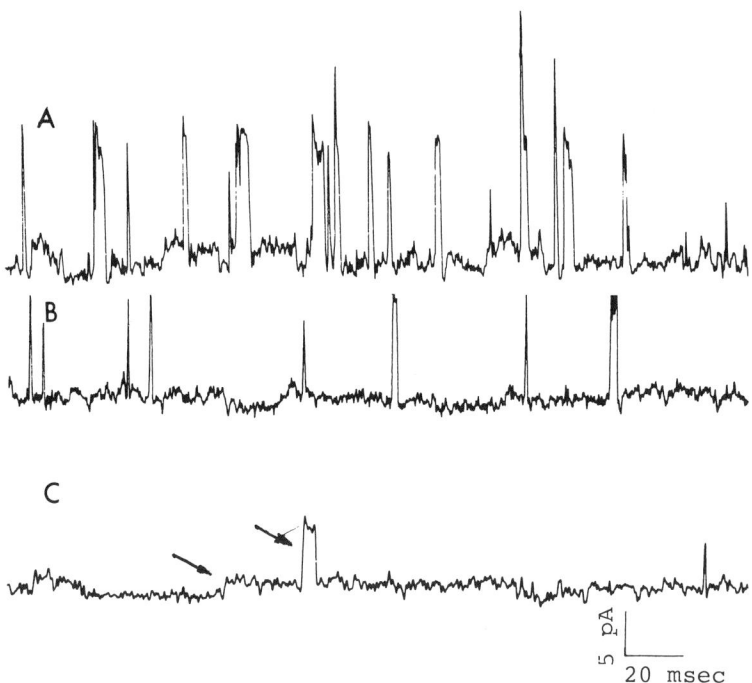

FIGURE 6. Single-channel records using cell-attached recording mode shows two channel types whose conductance and open times correspond to the two currents found using whole-cell analysis. In C, at zero holding potential, the characteristics of the two channels can most easily be seen in the long open times of the low-amplitude openings and the short open time of the higher-amplitude opening. Upper two traces are at -10 (B) and -20 (A) mV pipette potentials. Outward current upward. Recordings made with a Dagan 8900 patch clamp amplifier and Bessel filtered at 2 KHz.

the -60-mV prepulse. An *inward rectifier* was also found only in some cells. At the potassium concentrations used here inside and outside the pipette, this current is found in measurable amounts only as an inward current negative to -90 mV, that is, negative to the potassium equilibrium potential.

INWARD CURRENTS

The Ca^{2+} current (ICa) in these cells is often seen as an inward transient before the much larger outward K^+ current begins. The total ICa current is revealed when

Cs^+ is substituted for K^+ in the pipette or in 10% of the cells when all outward current is blocked with TEA. Little exploration of the ICa has been done in toadfish saccular hair cells because the major emphasis has been on the diversity of the potassium conductances. ICa in toadfish hair cells appears to be similar to that reported in other hair cells.[2,5,14] It displays rapid activation and deactivation kinetics and shows no inactivation. These three characteristics are of importance in sensory coding since they will provide a continuous underlying depolarization for the duration of the hair cell receptor potential. This depolarization can then be shaped by the different combinations of potassium currents into the desired output signal.

RESONANCE OF THE MEMBRANE POTENTIAL

Membrane potential resonance was first described in the squid axon.[15-17] The figures in Mauro et al. from squid axon are of higher Q resonance[17] than many of those published recently for hair cells. Resonance in hair cells was first described in intracellular recordings from the auditory hair cell of the turtle where a membrane potential oscillation was found in response to sound or current injection.[1,18,19] The frequency and Q of the resonance match closely the characteristic frequency and Q of that cell in response to sound stimuli. Since the tuning curves recorded from many auditory nerve fibers are sharper than could be accounted for by the physical properties of the basilar membrane, a second source for the frequency selectivity has been sought. The resonance of the hair cell membrane potential was thought to be that "second filter." (However, more refined measurements of basilar membrane movement have shown that it can be as sharply tuned as auditory afferent fibers innervating it.)[20,21]

Membrane potential ringing or resonance has been found in patch clamp recordings from frog sacculus,[5,6] turtle basilar papilla,[2] chick cochlea,[10] alligator cochlea,[22] goldfish saccule,[23] and toadfish saccule.[8] Published records of hair cell resonance differ widely in the quality of resonance. Some cells show high-quality symmetrical resonance in response to small current pulses, while others use huge current pulses and evoke only one or two cycles with no response at the current offset. One of the reasons for this may be that many laboratories try to evoke the resonance of the membrane potential at the zero-current holding potential, considered the resting potential of the cell. However, the zero-current potential is largely set by the choice of potassium concentrations inside and outside the cell. This may not be the true in vivo resting potential (see Reference 23).

Resonance has been modeled as an interaction between passive membrane properties (capacitance and leak resistance) and the magnitude and kinetics of IKCa and ICa.[2,3] However, the majority of the cells in the toadfish saccule have another current, IK, that is active over the same voltage range as IKCa. In order to assess the influence of IK on resonance, current clamp mode was used to characterize the ringing response. The outward current composition was then determined in voltage clamp mode using TEA block. An example of resonance in a toadfish saccular hair cell having only IKCa as the outward current is seen in FIGURE 7. Four such cells were recorded from the total population of cells in which ionic current composition and resonance were tested ($n = 40$). The ringing in these cells is symmetrical, that is, there is ringing at both command onset and termination. There is also a much more marked resonance at the baseline holding current before and after the command pulse. The majority of the cells with both IKCa and IK also showed a high Q resonance but only at current command onset and not the termination of the

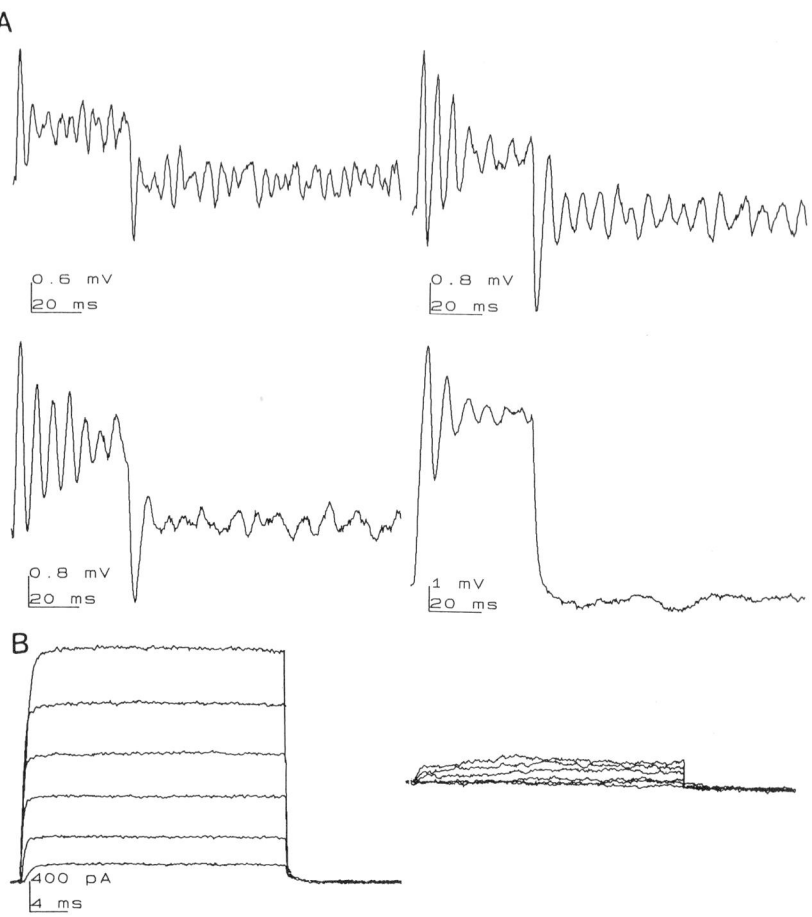

FIGURE 7. **A:** Resonance in a hair cell in which all outward current is IKCa. Holding currents in A (from left to right) −44 mV, −48 mV, −50 mV, −55 mV. Command step is 4 pA. Q at −48 mV = 4.4; frequency = 174 Hz. Note quality of resonance at baseline holding potential in first three panels. This is characteristic of cells lacking IK. **B:** Current response of the same cell to voltage commands from −40 to +60 mV consists of IKCa only. After application of 25 mM TEA, the current is rapidly blocked (lower panel). The cell was lost shortly after TEA application, but from the rapid block of the current and the lack of a slow component in the tail of either the total outward current or the remaining current, the current can be identified as IKCa.[8] Measured using a −20-mV pulse from −60-mV holding potential; input resistance = 1.4 gohm, series resistance = 8.1 mohm, capacitance = 12 pF.

command where a deep hyperpolarization was seen (FIGURES 8 and 9). Only once, and then only at a single holding potential, did a cell with IKCa and IK (n = 28) show symmetrical resonance. All the other recordings of cells with both IKCa and IK showed a prolonged voltage-dependent hyperpolarization at command pulse termination. Even the cell in FIGURE 8 that has a very small component of IKCA shows a high-Q, high-frequency ringing. Resonant frequencies, measured at the holding

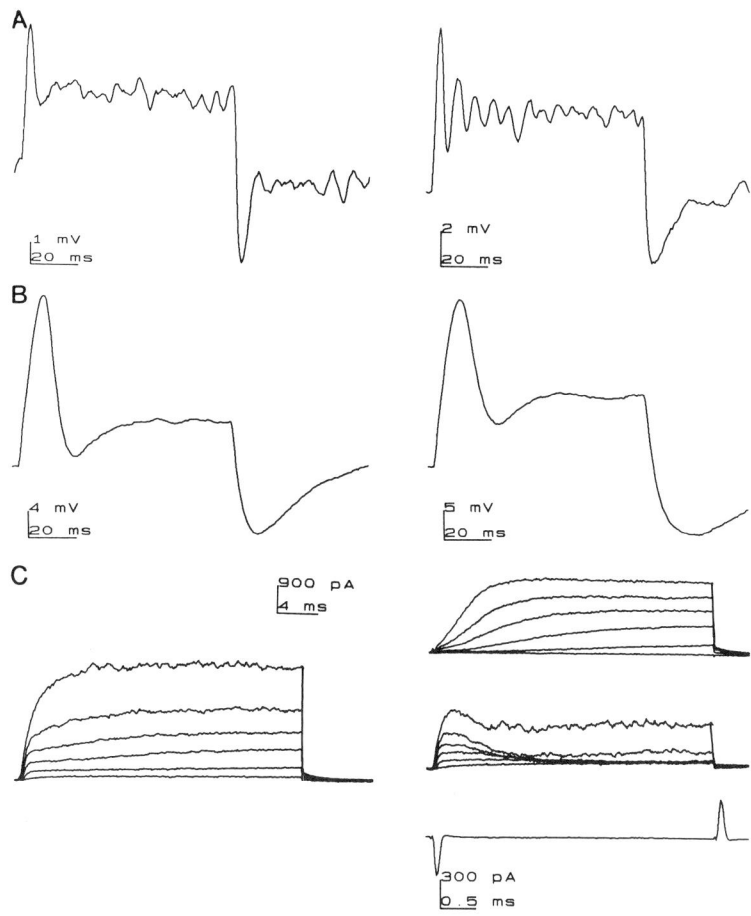

FIGURE 8. Resonant responses from a hair cell having IKCa and a large component of IK. **A:** Current clamp records of resonance evoked in response to 4-pA command pulses at two different holding potentials. Note difference in the current calibrations indicative of the difference in input resistance as conductances decrease at more negative holding potentials and with TEA. Holding currents (from left to right) −47 mV and −48 mV. Quality factor at −48 mV is 3.1; input resistance at −48 mV = 1.2 gohm; frequency = 136 Hz. **B:** Spiking appears after block of IKCa with 25 mM TEA. Holding potentials −48 mV and −52 mV. **C:** Voltage clamp recordings of left panel, total outward current in response to 20-mV steps from a holding potential of −60 mV to +60 mV; upper right panel, current response to same voltage commands in the presence of 25 mM TEA separates IK and IKCa (lower right) obtained by subtracting IK from total outward current (left panel). Trace at bottom is current response to a −20-mV command pulse from −60-mV holding potential run at intervals throughout all experiments in order to measure residual series resistance, capacitance and input resistance. Input resistance = 1.8 gohm; capacitance = 12 pF; series resistance = 13 mohm.

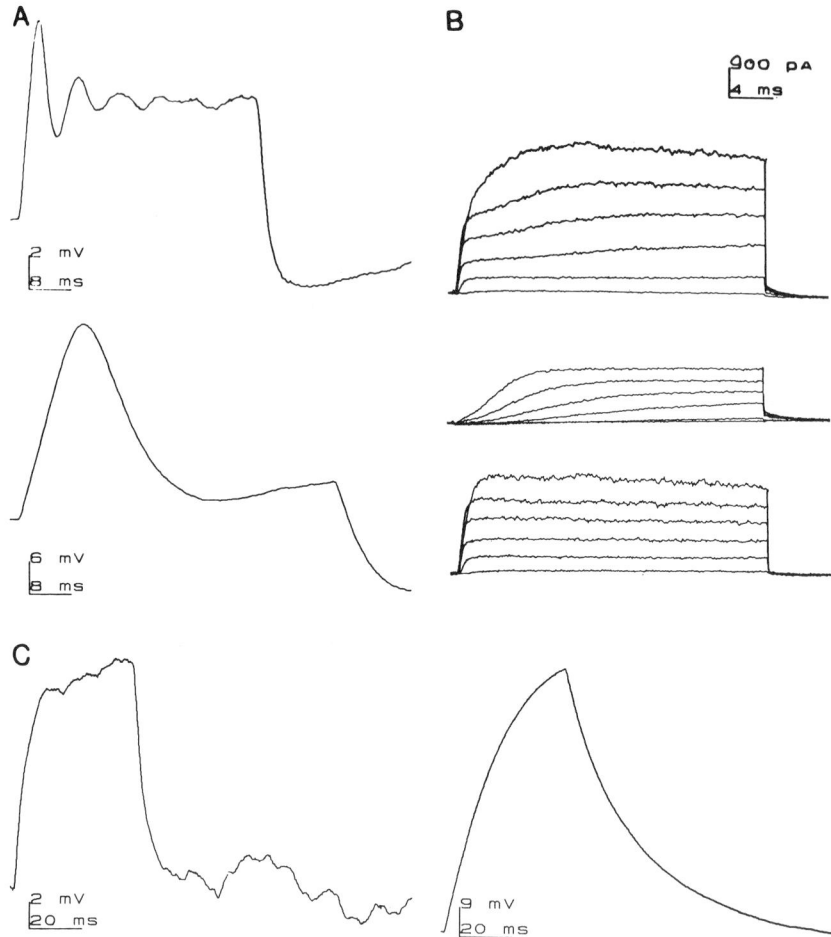

FIGURE 9. Hair cell with both IKCa and IK shows resonance at −48-mV holding potential; frequency = 130; Q = 2.6 (A, upper panel) that is converted to a spike (lower panel) after TEA treatment. Command pulse = 4 pA. In B, the current composition is shown as total outward current (upper panel), TEA-resistant IK (middle panel), and IKCa (lower panel) in response to 20-mV steps from a −60-mV holding potential to +60 mV. Measured using a −20-mV step from −60 mV: cell capacitance = 14 pF; series resistance = 20 mohm; input resistance = 0.9 gohm. After treatment with 0.2-mM cadmium (C), the spike is blocked and only the passive properties of the membrane shape the response to the current command. Holding potential in C: left trace, −48 mV; and right trace, −55 mV.

potential that gave the highest Q, range from 107 to 175 Hz (n = 32). When IKCa is blocked by TEA, a spike is produced in response to the same current command. This spike is of higher amplitude and longer duration than the ringing cycle, reflecting both the increased input impedance and the slower activation and deactivation kinetics of IK, compared to the rapid rates of IKCa that terminate the ringing cycles.

When IKCa and ICa are blocked by cadmium, the spike is also blocked and only the passive properties of the membrane remain (FIGURE 9C).

A third class of cells ($n = 8$) with IKCa and IK produced a spike at all holding potentials tested (FIGURES 10–12). Nothing resembling a ringing response was found in these cells at any holding potential. The spiking cells were characterized by 3 subsets of ionic current composition. In one subset, IK was of large magnitude and relatively rapidly activating while IKCA was of low magnitude (FIGURE 10). In

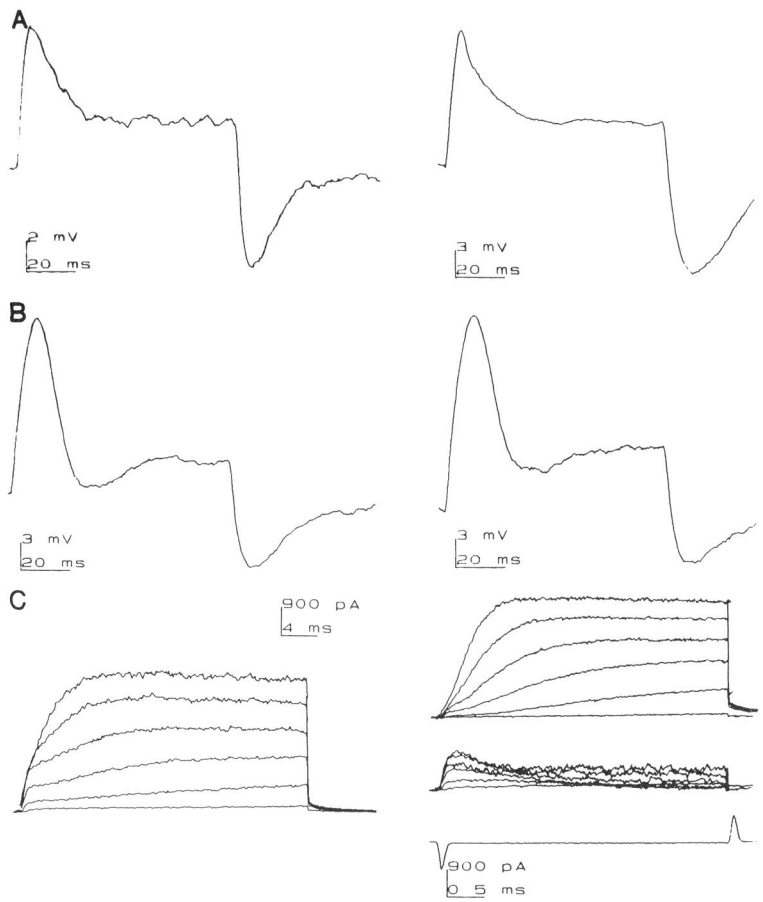

FIGURE 10. One of the three types of spiking cells, based on outward current characteristics. In the cell shown here, spiking is associated with a low level of IKCa and high levels of IK. **A:** Before TEA application, spiking response to a 4-pA command from a holding potential of −45 mV and −55 mV. **B:** Response to command of 4 pA from holding potentials of −45 and −52 mV after 25 mM TEA. **C:** Right panel, total outward current; left upper panel, TEA-resistant current; left lower panel, current blocked by TEA and obtained by subtraction. Voltage command pulses are 20-mV steps to +60 mV from −60 mV holding potential. Lower panel is response to a −20-mV step to −80 mV before TEA. Series resistance = 6.9 mohm; capacitance = 10.7; input resistance = 1.2 gohm.

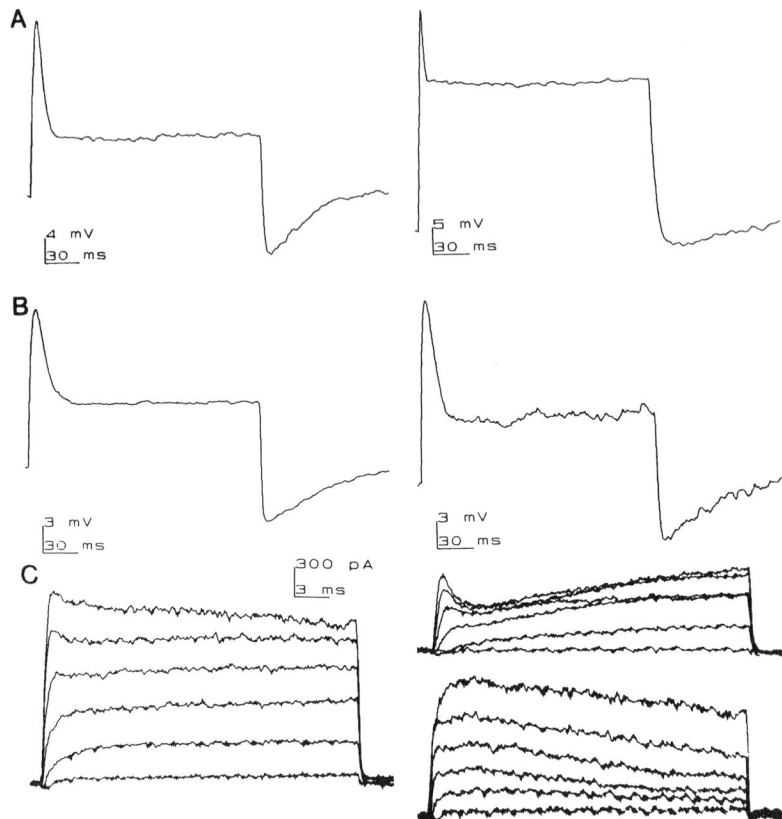

FIGURE 11. Spiking in a cell with fast activation kinetics and an A current, IKCa, and IK. **A:** Cell responded with spike at all voltages tested between −42 and −75 mV, shown here. Command pulse = 10 pA. **B:** In 25 mM TEA, at the same voltages, spike is of longer duration. **C:** Upper panel, total outward current; middle panel, TEA-resistant current (IK); lower panel, current blocked by TEA (IKCa). Command pulses +20-mV steps to +60 mV from −60-mV holding potential. Capacitance = 7.7 pF; input resistance = 1.3 gohm and series resistance = 19 mohm measured from a −20-mV pulse from −60-mV holding potential before TEA application.

another set, an IA was present, in addition to IK and a rapidly activating IKCa (FIGURE 11). In a third set of cells, there were low levels of IKCa and both IKCa and IK showed slow activation rates (FIGURE 12). (Four of the eight spiking cells had an IA. None of the resonating cells had an IA.) TTX was without effect on the spike duration or amplitude, indicating that a sodium current played no part in spike genesis.

ACETYLCHOLINE MODULATION OF HAIR CELL IONIC CONDUCTANCES

The following data on efferent transmitters and hair cell ionic conductances are from work in the preliminary stage, but it is possible to see the complexity of the efferent action at the level of ionic currents. The effect of ACh on resonance was tested satisfactorily in one cell. It dampened the resonance of the membrane potential and also decreased the IKCa (FIGURE 13). The increase in input resistance after ACh can be calculated by comparing the plateau voltage in the current clamp before and after ACh. This increase in input resistance may convert a resonating response to a spiking response or simply dampen the resonance. This decrease in IKCa after ACh agrees with that seen in earlier experiments using voltage clamp recordings. One population of cells responded to ACh with a decrease in outward current (FIGURE 14) and another with an increase in outward current (FIGURE 15).

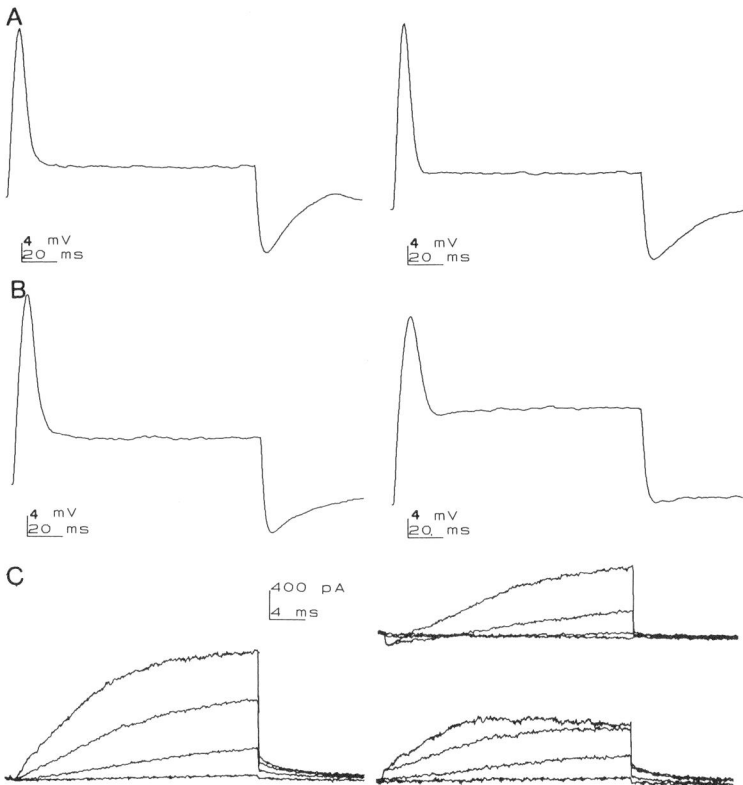

FIGURE 12. Hair cell that shows only spiking in response to current commands has slow kinetics of activation of both IKCa and IK. Command pulses of 10 pA. **A:** Before TEA, only spiking was seen at all holding potential levels, here at −49- and −64-mV holding current. **B:** After TEA, spiking was elicited between −39-mV (at left) and −62-mV (at right) holding potential. **C:** Left panel, total outward current in response to 20-mV steps from −60 to +20 mV; upper right panel, IK after TEA application; lower right panel, IKCa obtained by subtraction. Slow tail current is absent in IK and appears in IKCA record because the cell was in 30-mM external K$^+$ prior to TEA (between record sets A and B) in order to measure inward tail currents for another purpose. Residual K$^+$ in the TEA solution sets the K$^+$ equilibrium potential near the holding potential. After subtraction, the slow tail current, which was present in total outward current, ends up in IKCa record. Input resistance measured before TEA using a −20-mV step to −80 mV is = 2 gohm; series resistance = 6.8 mohm; C = 11 pF.

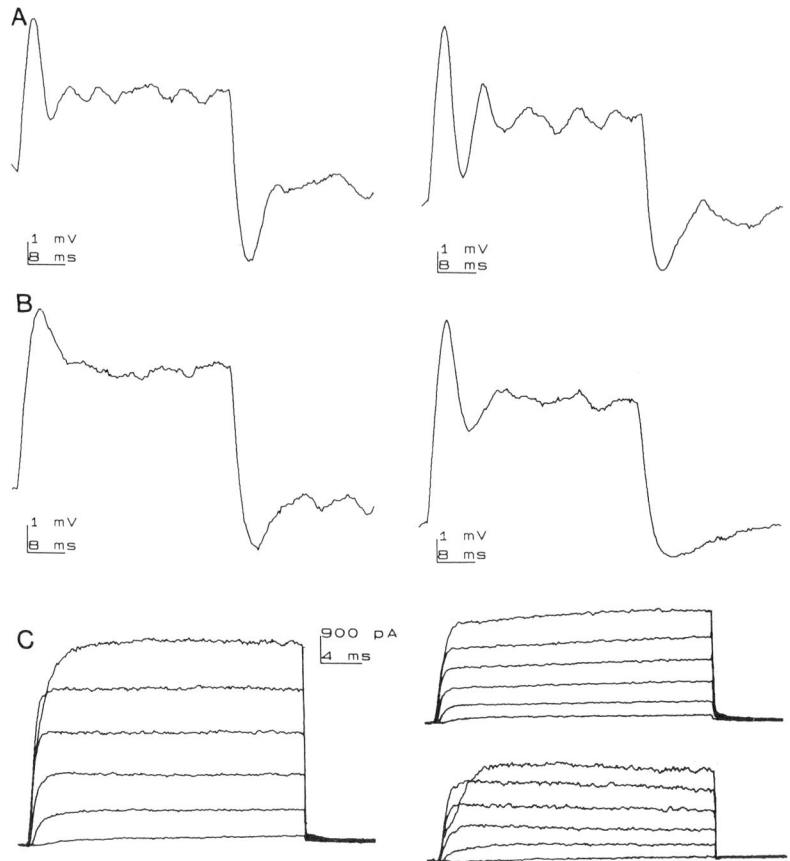

FIGURE 13. ACh blocks IKCa and converts high-Q resonating cell into low-Q, lower frequency cell. **A:** Resonance in response to 4-pA command pulse before ACh, **B:** After ACh. Resonance holding potentials for both sets, −46 and −50 mV. Frequency = 122 Hz; Q = 3.5 for response of control at −50 mV. **C:** Left panel, total outward current; left upper panel, current post ACh; bottom right panel, current blocked by ACh. The lack of a slow tail current decay in these currents indicates that it is IKCa that ACh blocks in a cell that has only IKCa as the outward current. Voltage commands are +20-mV steps from −60-mV holding potential to +60 mV. Capacitance = 6.8 pF; series resistance = 20 mohm; input resistance = 1.3 gohm measured by −20-mV pulse from −60 mV.

However, it appears in some cases (but not all) that what appears as a decrease in an outward current is actually an increase in an inward current with the I/V characteristics of a calcium current. An apparent slower activation of post-ACh current may be the arithmetic subtraction of an increased inward calcium current from the rising phase of the larger outward current (FIGURE 15). It is difficult in this record to identify the outward current that increased after ACh, but the tail current decay rate suggests that it is IK. However, IK sometimes "runs up" during the experiment, increasing slowly over long experiments (40–50 minutes) and this cell was not held long enough to reverse the effect by washing, as done for FIGURE 14. Thus, the action of ACh on a potassium current here is inconclusive. Data suggesting an ACh-evoked increase in ICa without a concomitant increase in IKCa are not without precedent. In both Betz cells of the cortex and hippocampal neurons, muscarinic stimulation increases ICa while decreasing IKCa.[24-26] The diversity in response to ACh is consistent with recordings of the complex responses of toadfish saccular afferents to efferent stimulation (A. Steinacker, unpublished data).

DISCUSSION

Several significant points were derived from the data above. High Q resonance of the membrane potential using low-level current commands can be obtained from cells having both IKCa and IK. The frequency range of the resonance in these cells (107 to 175) is rather limited. For a gravistatic response, one would expect much lower frequencies, and for an auditory response, a larger range and higher resonant frequency would be expected. Toadfish auditory response extends to 700 HZ and their calls up to 250 Hz.[27] A resonating cell with IKCa and IK could be rapidly converted to a spiking cell by blocking IKCa with TEA. The slow activation kinetics of IK were reflected in the long duration of the spike in TEA as compared to the higher frequency ringing produced by the rapid kinetics of IKCa (FIGURES 8 and 9). Thus, it appears that the kinetics of IKCa play the major role in setting the resonant frequency, as predicted by the models, A certain proportion of the cells produce only a calcium spike ($n = 8$). Unlike the data from goldfish,[23] there was no regional localization of these cells. It is not difficult to see why these cells show spiking

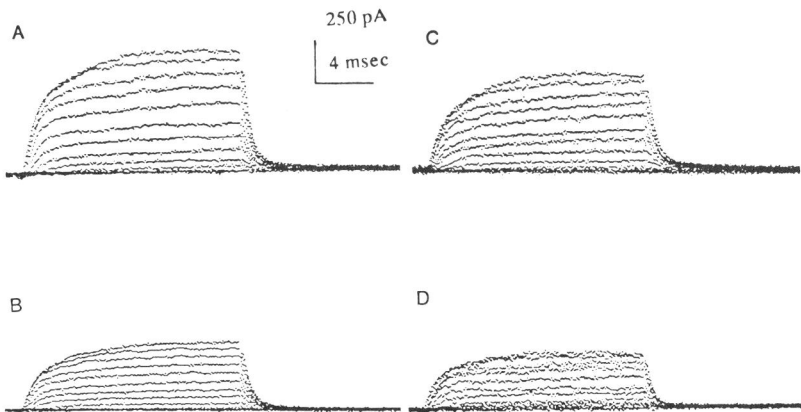

FIGURE 14. ACh blocks an outward current. **A:** Control total outward current. **B:** Total outward current after ACh. **C:** After ACh is washed out of the bath, the current is restored to near control levels. **D:** Current blocked by ACh found by subtracting post-ACh current from pre-ACh current. The identity of this current cannot be defined, except that it is a potassium current, since in these experiments, series resistance compensation was not recorded and use of tail current decay for identification is thus unreliable. However, the current in D, derived by subtraction as the current blocked, has the same tail current decay as the other 3 currents, which would suggest that there is only one current component and thus that current would have to be IKCa. Command voltages, 5-mV steps from a −60-mV holding potential.

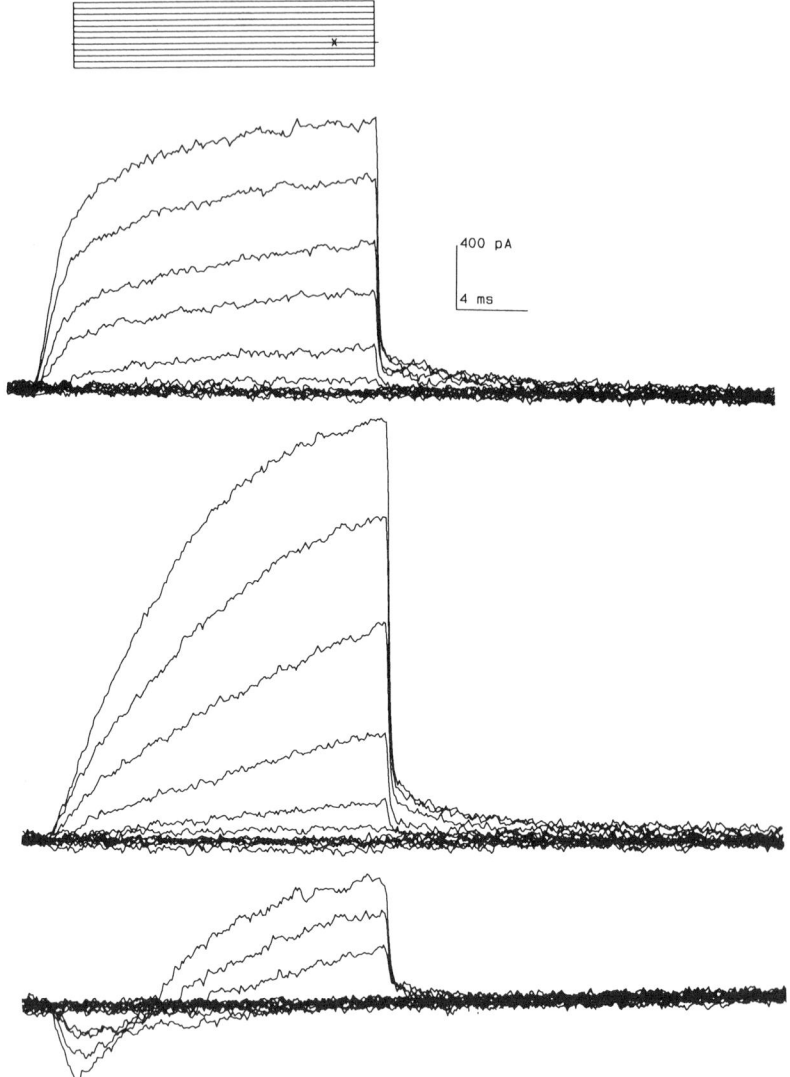

FIGURE 15. ACh increases an inward current. **Top panel:** Control total outward current. **Middle panel:** Current post-ACh. **Lower panel:** Current induced by ACh, obtained by subtraction of post-ACh current from pre-ACh current. Ten-millivolt steps from −90 mV to +10 mV from a −60-mV holding potential. Outward current identity is uncertain (see text).

behavior. They all have a higher input resistance at the holding potentials at which they spike. (Compare the steady-state voltage levels in resonating cells and spiking cells. Input resistance can be determined by dividing the current command into the steady-state voltage level during the command.) Also, the rapid activation and slow deactivation of an IA, in the presence of a low level of IKCa, favors spiking since the slower deactivation rate of an A current will keep the membrane hyperpolarized and prevent ringing. In fact, this can be seen as the difference in spike duration with and without IKCa in FIGURE 11A and B. Though the mode of response of the hair cell to command pulses varies, these responses (resonance, spiking, afterhyperpolarization) are the result of the combination of passive membrane properties and the active conductances in the cell. These combinations of hair cell membrane properties must be configured to transmit through the cell certain aspects of the signal. The scientific literature offers many such examples in other cell types in which the different configurations of ionic conductances exist to transmit certain messages across the cell (for an excellent recent review of this subject, see Reference 28). The hair cell is governed by these same rules, and to understand sensory coding at this level, it is necessary to understand the ionic conductances of the hair cell.

Data dealing with efferent transmitter action on the hair cell directly are scant. Using intracellular recordings from the auditory hair cell of the turtle, a hyperpolarization suggestive of a modulation of a K^+ conductance was recorded intracellularly in response to efferent stimulation.[29-31] An ACh decrease in a current presumed to be IKCa has been reported from frog semicircular canals.[32] In cell-attached patch recordings from isolated hair cells from this laboratory, an effect of ACh and oxotremorine (a muscarinic agonist) on a K^+ channel was demonstrated.[33] A muscarinic-induced increase in intracellular Ca^{2+} has recently been demonstrated in chick hair cells using Fura-2.[34] The data above on ACh action on toadfish saccular hair cell conductances give an indication of the complexity to be expected when the actions of ACh, enkephalin, and chorionic gonadatrophin releasing peptide (CGRP) are integrated in modulation of hair cell ionic currents.

The basic properties of the potassium currents that will determine their functional role can be summed up as follows. IKCa is able to turn on and off rapidly and will remain on for the duration of the stimulus due to its lack of inactivation. Functionally, this produces a fast termination of an excitory response and a rapid return to prestimulus conditions. IK will be slow to turn on and may not be activated during short stimuli. However, once turned on, it is relatively slow to turn off. During a maintained stimulus, it will slowly decrease due to inactivation. Functionally, this means a slow repolarization and slow return to the prestimulus condition. The IA is fast to turn on and then slowly inactivates for the duration of the stimulus. It is thus suited for production of spikes and response to transients. Although it is largely inactivated at normal resting potentials, this inactivation may be removed when the hair cell is stimulated in the off direction. The inward rectifier will only be activated when the membrane potential is negative to the potassium equilibrium potential. Activation of this current depolarizes the membrane potential back to the resting potential.

What does this diversity in the composition and kinetics of the ionic currents among different hair cells in the same end organ tell us. Certainly, if these cells were coding only, and precisely, characteristic frequency, a division into spiking and nonspiking cells would not be necessary. Spiking cells are not an artifact of cell isolation or patch clamp electrode dialysis of the cells' interior since they were reported in intracellular work done before the use of isolated cells and patch clamp methods.[1,35] Therefore, one must ask what the difference in the message transmitted to the afferent fiber would be in a spiking and resonating cell and between a cell with IKCa or multiple potassium conductances in the outward current. The delay between sensory stimulus and the spike or first ringing response should be nearly the same since it is governed by the activation rate of the calcium current (and the passive membrane properties). The action potentials are of longer duration and higher amplitude than the single cycles of the resonance. However, they are not repetitive in the span of time used here. Thus, the long-duration, large-amplitude depolarization of the hair cell would produce a large episodic transmitter release from the cell at the beginning of the stimulus. This would be followed by a reduced level of release for the duration of the stimulus. A long silent period may follow signal termination due to deep hyperpolarization seen at the termination of stimuli in cells that have IK. This could conceivably be the set of properties needed for localization of a sound or the acceleration needed to initiate a vestibularly guided eye movement. In cells with only IKCa as the outward current, in the absence of a stimulus, there may be a small continuous release of transmitter corresponding to the resonance of the membrane potential at resting potential. Following a stimulus, the kinetics of IKCa should accentuate the characteristic frequency with a smoothly graded synaptic release following the frequency of the stimulus. This may produce the phase locking seen in some afferent fibers. However, phase locking in a single primary afferent of the goldfish saccule occurs over a wide frequency range and that is hard to attribute to a resonant frequency tightly locked to potassium current kinetics matching the characteristic frequency.[36] In the frog saccule, where resonance of the membrane potential is found in the hair cells,[5,6] no sign of resonant behavior was found in the firing of the afferent fibers.[37] Additionally, some of the fastest ionic current kinetics are found in the hair cells of the semicircular canals[9,13] which, as a system, have one of the slowest response frequencies found in acousticolateralis systems. This is difficult to reconcile with the concept that the kinetics of the ionic conductances are coding only characteristic frequency. There may be a specialization among the cells for specific parameters of the sensory signal. Conceivably, different cell types may feed into specific classes of afferent fiber. The resolution of this question awaits data recorded from hair cell and afferent fiber simultaneously.

ACKNOWLEDGMENTS

The authors wish to thank Drs. Harvey Fishman and D. C. Zuazaga for their critical reading of the manuscript.

REFERENCES

1. CRAWFORD, A. C. & R. FETTIPLACE. 1981. An electrical tuning mechanism in turtle cochlear hair cells. J. Physiol. **312:** 377–412.
2. ART, J. J. & R. FETTIPLACE. 1987. Variation of membrane properties in hair cells isolated from the turtle cochlea. J. Physiol. London **385:** 207–242.
3. HUDSPETH, A. J. & R. S. LEWIS. 1988. A model for electrical resonance and frequency tuning in saccular hair cells of the bull-frog, *Rana catesbeina*. J. Physiol. London **400:** 275–297.
4. ARMSTRONG, C. M. & F. BEZANILLA. 1974. Charge movement associated with the opening and closing of the activation gates of the Na channels. J. Gen. Physiol. **63:** 533–552.
5. HUDSPETH, A. J. & R. S. LEWIS. 1988. Kinetic analysis of voltage- and ion-dependent conductances in saccular hair cells of the bull-frog, *Rana catesbeina*. J. Physiol. London **400:** 237–274.
6. LEWIS, R. S. & A. J. HUDSPETH. 1983. Voltage and ion-dependent conductances in solitary vertebrate hair cells. Nature **304:** 538–541.
7. STEINACKER, A. & A. ROMERO. 1989. Characterization of outward currents in toadfish saccule hair cell. Soc. Neurosci. Abstr. **55:** 546a.
8. STEINACKER, A. & A. ROMERO. Characterization of the voltage gated potassium current in toadfish saccular hair cell. Brain Res. (In press).

9. LANG, D. L. & M. J. CORREIA. 1989. Studies of solitary semicircular canal hair cells in the adult pigeon. II. Voltage-dependent ionic conductances. J. Neurophysiol. **62:** 935–945.
10. FUCHS, P. A. & M. G. EVANS. 1990. Potassium currents in hair cells isolated from the cochlea on the chick. J. Physiol. **429:** 529–551.
11. KROS, C. J. & A. C. CRAWFORD. 1990. Potassium currents in inner hair cells isolated from the guinea pig cochlea. J. Physiol. **421:** 263–291.
12. OHMORI, H. 1984. Studies of ionic currents in the isolated vestibular hair cell of the chick. J. Physiol. **350:** 561–581.
13. HOUSLEY, G. D., C. H. NORRIS & P. S. GUTH. 1989. Electrophysiological properties and morphology of hair cells isolated from the semicircular canal of the frog. Hear. Res. **38:** 259–276.
14. FUCHS, P. A., M. G. EVANS & B. W. MURROW. 1990. Calcium currents in hair cells isolated from the cochlea of the chick. J. Physiol. **429:** 553–568.
15. COLE, K. S. & R. F. BAKER. 1941. Longitudinal impedance of squid giant axon. J. Gen. Physiol. **24:** 771–788.
16. HODGKIN, A. L. & A. F. HUXLEY. 1952. A quantitative description of membrane current and its application to conduction and excitation in nerve. J. Physiol. **117:** 500–544.
17. MAURO, A., F. CONTI, F. DODGE & R. SCHOR. 1970. Subthreshold behavior and phenomenological impedance of the squid axon. J. Gen. Physiol. **55:** 497–523.
18. FETTIPLACE, R. & A. C. CRAWFORD. 1978. The coding of sound pressure and frequency in cochlear hair cells of the terrapin. Proc. R. Soc. London Ser. B **203:** 209–218.
19. CRAWFORD, A. C. & R. FETTIPLACE. 1981. Non-linearities in the responses of turtle hair cells. J. Physiol. **315:** 317–338.
20. KHANNA, S. M. & D. G. B. LEONARD. 1982. Basilar membrane tuning in the cat cochlea. Science **215:** 305–306.
21. SELLICK, P. M., R. PATUZZI & B. M. JOHNSTONE. 1982. Measurements of basilar membrane motion in the guinea pig using the Mossbauer technique. J. Acoust. Soc. Am. **72:** 131–141.
22. FUCHS, P. A. & M. G. EVANS. 1988. Voltage oscillations and ionic conductances in hair cells isolated from the alligator cochlea. J. Comp. Physiol. **164:** 151–163.
23. SUGIHARA, I. & T. FURUKAWA. 1989. Morphological and functional aspects of two different types of hair cells in the goldfish sacculus. J. Neurophysiol. **62:** 1330.
24. SCHWINDT, P. C., W. J. SPAIN, R. C. FOEHRING, M. C. CHUBB & W. E. CRILL. 1988. Slow conductances in neurons from cat sensorimotor cortex in vitro and their role in slow excitability changes. J. Neurophysiol. **59:** 450–467.
25. SCHWINDT, P. C., W. J. SPAIN, R. C. FOEHRING, M. C. CHUBB & W. E. CRILL. 1988. Slow conductances in neurons from cat sensorimotor cortex in vitro and their role in slow excitability changes. J. Neurophysiol. **59:** 450–467.
26. MADISON, D. V., B. LANCASTER & R. A. NICOLL. 1987. Voltage clamp analysis of cholinergic action in the hippocampus. J. Neurosci. **7:** 783–741.
27. FINE, M. L. 1981. Mismatch between sound production and hearing in the oyster toadfish. *In* Hearing and Sound Communication in Fishes. M. N. Tavolga, A. N. Popper & R. R. Fay, Eds. Springer Verlag, New York, Heidelberg & Berlin.
28. BAXTER, D. A. & J. H. BYRNE. Ionic conductance mechanisms contributing to the electrophysiological properties of neurons. Curr. Opinion Neurobiol. **1.** (In press.)
29. ART, J. J., A. C. CRAWFORD, R. FETTIPLACE & P. A. FUCHS. 1982. Efferent regulation of hair cells in the turtle cochlea. Proc. R. Soc. London Ser. B **216:** 377–384.
30. ART, J. J., A. C. CRAWFORD, R. FETTIPLACE & P. A. FUCHS. 1985. Efferent modulation of hair cell tuning in the cochlea of the turtle. J. Physiol. **360:** 397–422.
31. ART, J. J., R. FETTIPLACE & P. A. FUCHS. 1984. Synaptic hyperpolarization and inhibition of cochlear hair cells in the turtle *Pseudemys Scripta elegans.* J. Physiol. **356:** 525–550.
32. HOUSLEY, G. D., C. H. NORRIS & P. S. GUTH. 1990. Cholinergically-induced changes in outward currents in hair cells isolated from the semicircular canal of the frog. Hear. Res. **43:** 121–134.
33. STEINACKER, A. & L. ROJAS. 1988. Acetylcholine modulated K^+ channel in the toadfish saccular hair cell. Hear. Res. **35:** 265–270.

34. SHIGEMOTO, T. & H. OHMORI. 1990. Muscarinic agonists and ATP increase the intracellular Ca^{2+} concentration in chick cochlear hair cells. J. Physiol. **420:** 127–148.
35. HUDSPETH, A. J. & D. P. COREY. 1977. Sensitivity, polarity, and conductance change in the response of vertebrate hair cells to controlled mechanical stimuli Proc. Natl. Acad. Sci. USA **74:** 2407–2411.
36. FURUKAWA, T. & Y. ISHII. 1967. Neurophysiological studies on hearing in goldfish J. Neurophysiol. **30:** 1377–1403.
37. LEWIS, E. R. 1988. Tuning in the bullfrog ear. Biophys. J. **53:** 441–447.

Filtering Properties of Hair Cells[a]

MANNING J. CORREIA

Departments of Otolaryngology and Physiology & Biophysics
University of Texas Medical Branch
Galveston, Texas 77550

The stereocilia, where mechanoelectric transduction (MET) apparently occurs (reviewed elsewhere),[1,2] probably can be considered as ideal cables. Therefore, the membrane of the cell body of the hair cell contributes most to filtering the MET voltage. The passive filtering properties of the hair cell are determined by the cell's plasma membrane. The lipid bilayer provides resistance to ionic flow and is a thin insulator separating the intracellular and extracellular fluids and therefore acts as an electrical resistor and an electrical capacitor.[3] Additional properties of the hair cell membrane such as ionic channels contribute to filtering. Ionic, voltage- and time-dependent conductances associated with these channels determine the resting membrane potential (RMP) and the potential change resulting from the MET current. Within the context of signal processing, it is interesting to ask three questions about the filtering properties of hair cells. First, are the filtering properties of the hair cells in a specific end organ (cochlea, semicircular canals, and otolith organs) specialized to permit the receptor to perform its function more efficiently? Second, are there regional variations of filtering properties of hair cells in the same receptor organ? Finally, do hair cells (e.g., vestibular type I and type II) with different morphology and different afferent and efferent innervation have different filtering properties?

FACTORS THAT CAN INFLUENCE FILTERING BY THE MEMBRANE OF A HAIR CELL

The frequency response of small membrane potential changes in the hair cell about a given membrane potential to MET currents can be described by the hair cell's impedance (Z) function. Alternatively, described as the admittance function

$$Y_T = Z^{-1} = (R_e + 1/Y')^{-1} \tag{1}$$

where R_e is the series resistance and

$$Y' = j2\pi f C_{in} + G_{in} + \Sigma_i [(G_i)(1 + j2\pi f \tau_i)^{-1}] \tag{2}$$

where $j = (-1)^{1/2}$, $\pi = 3.14 \ldots$, and f = frequency in Hertz. As has been pointed out,[4,5] Y' is a generalized linearized formulation of the Hodgkin-Huxley[6] quantitative description of membrane current. From the above equation, it can be seen that the active and passive properties of the membrane contribute to its filtering capabilities. The passive membrane properties include the input capacitance (C_{in}) and input resistance ($R_{in} = G_{in}^{-1}$) of the membrane. The active membrane properties include

[a]This work is supported in part by grants from the Office of Naval Research (N00014-87-K-0358) and the National Institutes of Health (DC01273; Claude Pepper investigator award to M. J. Correia).

conductances (G_i) with relaxation time constants (τ_i). Positive G_is are conductances that represent ionic processes with a driving force that produces a net outward current over a restricted range around a particular membrane potential; negative G_is are conductances that represent ionic processes with a driving force that produces a net inward current. Thus, in effect, the filtering properties of a hair cell at a particular membrane potential and at a particular frequency depend upon the input resistance and capacitance of the hair cell, the net sum of the direction and magnitude of the ionic currents, activation/inactivation kinetics of the currents, and how many channels are not steady-state inactivated at that membrane potential.

Input capacitance can vary by an order of magnitude in hair cells. Short oscillatory hair cells, dissociated from the rostral saccule of the goldfish,[7] have a mean input capacitance, C_{in} = 2.9 pF; type II hair cells dissociated from the semicircular canals of the pigeon[5] have an input capacitance, C_{in}, around 13 pF; and C_{in} in guinea pig outer hair cells is between 24 and 32 pF.[8] The input capacitance of other hair cells generally falls within this range. Hair cells also vary considerably in size. Short oscillatory hair cells in the goldfish saccule[7] are < 15 μm long but guinea pig outer hair cells are 55–80 μm long. But, as with other neural tissue, the specific capacitance of many hair cells[9,10] appears to be around 1 μF/cm². The RMP and the input resistance of hair cells around the RMP are difficult to accurately measure since they appear to depend on method of measurement, possibly the contents of patch pipettes and whether the cells are *in situ* or isolated. But when measurements are made using the same three conditions for different types of hair cells, differences are seen. For example, on average, pigeon semicircular canal type I hair cells have an RMP near −70 mV,[11] which is significantly different from pigeon semicircular canal type II hair cells which, have a mean RMP near −57 mV[5] (but cf. Reference 12). Goldfish rostral saccule oscillatory type hair cells have a mean RMP of −78 mV, but spike type caudal cells have a mean RMP of −101 mV.[7] Chick cochlea hair cells[13] near the base of the cochlea have a mean RMP of around −61 mV, while those near the apex have a mean RMP of around −78 mV. It seems that the RMP must provide one of the mechanisms by which the filtering properties of the hair cell are determined since this voltage level is the set point on the current-voltage (I/V) curve for determining the percent of steady-state activation/inactivation of ionic channels and how much depolarization or hyperpolarization is necessary to activate different ionic currents with different kinetics and directions. The input resistance at the RMP (zero-current potential) also determines the filtering properties of the hair cell in several ways. First, the $R_{in}C_{in}$ product determines the cell membrane time constant τ_{in} which partially determines the low-pass cutoff frequency [$f_c = (2\pi\tau_{in})^{-1}$] of signals, particularly at voltage levels where ionic currents are not activated. Second, R_{in} determines the input conductance at the RMP and therefore the attenuation of current at the RMP and the slope conductance at small levels of membrane depolarizations. The slope conductance for small levels of depolarization determines the sensitivity of the cell. Using whole-cell patch clamp techniques on enzymatically dissociated cells, average input resistance values around the RMP have been observed to be in the range from 1–10 GΩ in chick vestibular hair cells (ca. 4 GΩ),[14] frog type II semicircular canal hair cells (ca. 2 GΩ),[9] pigeon type II semicircular canal hair cells (ca. 0.8–2 GΩ),[5] and guinea pig cochlea inner hair cells (ca. 0.4–2 GΩ).[15] Average input resistance values around the RMP have been observed to be in the range from 0.01–1.0 GΩ in guinea pig cochlea outer hair cells (0.025–0.040 GΩ[8] cf. 0.1 GΩ),[16] and pigeon semicircular canal type I hair cells (ca. 0.4 GΩ).[11] Passive membrane time constants ($\tau_{in} = C_{in}R_{in}$) for the chick, frog, and pigeon vestibular hair cells are variable (varying by an order of magnitude) but are on the average about 15 mseconds (f_c near 11 Hz). For chick and pigeon, these measurements were made at

20° below body temperature. Assuming a Q_{10} of 2, the average f_c may be closer to 44 Hz for these species.[5,13] However, as the cell depolarizes, the slope conductance can change radically and the membrane time constant can shorten. For example, Kros and Crawford suggest that activation of the potassium conductance in guinea pig inner hair cells changes the corner frequencies from 8–40 Hz to 480–940 Hz as the cell depolarizes from the RMP to −55 mV.[15]

During depolarization, the active membrane properties of hair cells filter the membrane potential response. Frog saccule hair cells, pigeon semicircular canal type II hair cells, goldfish saccule hair cells, chick basal cochlea tall hair cells, turtle basilar papilla hair cells, and frog amphibian papilla hair cells show membrane potential oscillations (resonance or ringing),[5,7,17–30] other hair cells (alligator apical cochlea tall hair cells, chick apical cochlea tall hair cells, goldfish caudal saccule long hair cells) demonstrate spike-plateau responses,[7,13,31,32] and yet other hair cells (guinea pig cochlea outer hair cells) show elongation or contraction of the cell body.[33,34] Hair cells that show resonance demonstrate both outward and inward currents. The outward currents are an A-type potassium current, $I_{K(A)}$,[7,9,12,13,20,21,25,28,29,35,38] a calcium-activated potassium current, $I_{K(Ca)}$,[7,9,12–15,19–21,23,25,28,29,35–37,39] and a delayed rectifier type potassium current $I_{K(D)}$.[35,36] The inward currents are an L-type calcium current, $I_{Ca(L)}$,[7,9,14,19,20,21,25,28,35,36,40] a T-type calcium current, $I_{Ca(T)}$;[12] an inward H-type hyperpolarization-activated current, I_H,[7] and an anomolous rectifier potassium current $I_{K(An)}$.[7,14,19] Evidence[19,21,29] is accumulating that $I_{K(Ca)}$ magnitude and activation kinetics determine the hair cell resonant frequency whereas the quality of resonance is influenced by the magnitude of I_{Ca}. Hair cells that demonstrate the spike plateau response show weak $I_{K(D)}$,[7] $I_{(Na)}$,[7,23,31] $I_{K(An)}$,[7] and $I_{Ca(L)}$.[7,13,31] Guinea pig cochlear outer hair cells that show elongation and contraction have three classes of I_K[8,43] and two classes of I_{Cl}[43] currents. Guinea pig outer hair cells also demonstrate an $I_{Ca(L)}$[41] and an ATP-induced current through a large cation channel.[42]

Thus in different receptors in the same animal (e.g., cf. chick vestibular hair cells[14] with chick cochlear hair cells[13,38]), in the same receptors in different animals (e.g., cf. goldfish saccule[7,21] with frog saccule[20,25,29]), and in different hair cell types in the same receptor (e.g., cf. oscillatory and spike-type hair cells in the goldfish saccule[7,21] and type I and type II hair cells in the pigeon[11] and guinea pig[12] semicircular canals), there is a differential expression of active and passive membrane properties. The membrane potential responses resulting from filtering by these properties range from spiking to resonance (frequency selectivity).

DIFFERENT FILTERING PROPERTIES OF HAIR CELLS OF DIFFERENT RECEPTORS

Since the bandwidths of the frequencies of sound stimulation, seismic stimulation, and angular head motion form a continuum from high to low, it would seem that hair cells in the receptors sensing these stimuli should have specialized membrane properties to enable them to efficiently tune to the frequencies in these different ranges. To varying degrees, tuning has been demonstrated in auditory,[7,13,17–19,21,23,27,30] seismic,[22,25,29] and angular head motion detectors.[5,24] In auditory hair cells, for example, it has been shown that different species use electrical tuning to varying degrees to amplify a "best" or resonant frequency depending upon the hair cell's place along the length of the neuroepithelium from the apex to the base of the cochlea.[13,19] While resonance has been demonstrated for hair cells in the cochlea, saccule, and semicircular canals, the quality of the resonance seems to depend both on species and on receptor type. For example, the turtle cochlea hair cells possess a high quality of resonance factor (Q_e) of 9–35. Yet, Q_e for the alligator and chick cochlea is typically an order of magnitude lower. While turtle cochlea hair cells have high Q_e values, hair cells from the saccule of the bullfrog (seismic detector), demonstrate Q_e values in the range from 2–13, hair cells from the pigeon semicircular canal (angular head motion detector)[5] show a low Q_e, ranging from 0.6–3.0, and frog semicircular canal hair cells[9] demonstrate little or no resonance at membrane potentials near rest. The filtering effects of different values of Q_e on the hair cell's membrane potential can be deduced from its impedance function (Equation 2). For example, as has been pointed out elsewhere,[5] a Q_e of 0.6 produces a resonant peak in the impedance function so that when current of equal magnitude but different frequency (ranging from 1–500 Hz.) is injected (or flows as in the case of the MET current) into the hair cell, the membrane potential at the frequency where the resonant peak occurs is roughly twice what it is at the frequency that corresponds to the low-frequency flat portion of the impedance function. Parenthetically, the turtle[17,19] not only demonstrates the largest value of Q_e but it is the only animal whose range of resonant frequencies in the high-frequency range approach those determined by recording from the primary afferents.

Thus, it does appear that hair cells of receptors sensing acoustic, seismic, and head rotation have different expressions of ionic channels that facilitate filtering of MET currents in particular frequency ranges and augmenting responses at certain critical frequencies or frequency ranges. It will be interesting to see if utricular hair cells, particularly those innervated by afferents reflecting increased tonic discharge properties during static head tilt, will display further differentiation of ionic channels to accomplish appropriate filtering of MET currents in response to the static stimulus.

DIFFERENT FILTERING PROPERTIES OF HAIR CELLS IN DIFFERENT REGIONS OF THE SAME RECEPTOR

At present, the strongest support for different filtering properties of hair cells in different regions of the same receptor comes from studies of resonance in cochlea hair cells of the chick[13] and turtle.[19] In the turtle, for example, resonant frequency of ringing in cells isolated from different regions of the papilla increased as the cells were taken from regions moving along the papilla from apex to base.[19] In the chick, tall hair cells were taken from the base and apex of the cochlea. Topographical organization of resonant frequency like that of the turtle was observed.[13] However, the cells in the different regions showed subtle morphological differences. In the turtle, the ciliary length[19] and in the chick, cell body length[13] varied along the length of the papillae. The length of the cell bodies of the tall hair cells from the base and the apex were statistically different.[13] It is not understood why hair cells of different lengths express different currents but they do in several hair cell systems.[7,12,38] Therefore, the variable of difference in topography may be confounded with difference in morphology (see next section). In the toadfish saccule,[36] it appears that while there were different types of potassium currents with diverse kinetics in various morphological types of hair cells in different regions of saccule, hair cells of the same shape and with the same filtering characteristics do not seem to cluster in a given region. It will be interesting to learn if hair cells from specialized regions of the vestibular neuroepithelium, such as the striola of the utricle, have the same filtering properties as hair cells with the same morphology but from the extrastriolar regions.

DIFFERENT FILTERING PROPERTIES OF DIFFERENT HAIR CELL TYPES IN THE SAME RECEPTOR ORGAN

There is strong evidence that hair cells of different shapes and different innervation express different ionic currents (e.g., tall[37] and short[38] chick cochlea hair cells; type I and type II pigeon[11,35] and guinea pig[12] semicircular canal hair cells; short and tall hair cells in the goldfish[7,21] saccule) and behave differently to current injection (e.g., outer[33] versus inner[15] hair cells in the guinea pig cochlea, oscillatory vs. spike type cells in the goldfish[7,21] saccule).

Results from patch clamp studies[5,11,35] of solitary type I and type II hair cells in the pigeon semicircular canals illustrate the diversity of ionic currents and the difference in active and passive membrane properties. FIGURE 1 summarizes an anatomical and physiological comparison of type I and type II hair cells from the pigeon semicircular canal crista.

Scanning electron photomicrographs of solitary mechanically dissociated pigeon semicircular canal type I (left panel) and type II (center panel) hair cells illustrate features we used to quantatively separate the two types. The ratio of the minimum width of the neck to the maximum width of the apical surface (incorporating the majority of the cuticular plate) divides the hair cells into two groups.[5] In pigeon, type II hair cells have neck to plate ratios (NPRs) that are 1.7 times those of type I hair cells and hair cells with NPRs of 0.7 or above are classified as type II.[5] The NPR difference is maintained *in situ* as illustrated in the transmission electron photomicrograph of a section through the anterior crista (right panel). Based on the NPR criteria, type I and type II hair cells were studied[11] using whole-cell, tight-seal, ruptured-patch techniques. The membrane potential responses of type I and type II hair cells to current pulses are shown in the middle panel. The 100-msecond current pulses, incremented by 10 pA over the range from −40 to 80 pA, were chosen to be in the same range as MET currents measured in chick vestibular hair cells.[44] A comparison of the raw traces and the V(I) plots in the right column of the middle panel of FIGURE 1 shows that while the type I and type II responses to current pulses are qualitatively similar, type II hair cells have a greater steady-state slope impedance particularly during hyperpolarizing current pulses. The V(I) plot summarizes the average membrane potential response of several type I and type II hair cells. Steady-state slope conductances were calculated from the average values in the V(I) plot using a linear regression over the range 0 to −40 pA and over the range 0 to 80 pA. The steady-state slope conductance was about 30% larger for type I hair cells than type II hair cells over the entire range of positive currents and about 200% larger over the entire range of negative currents. This difference was statistically significant. This difference in conductance at depolarized values (e.g., −30 mV) is caused, in part, by an outward current in type I hair cells that does not appear to be present in type II hair cells. The bottom panel shows traces of responses to 400-msecond voltage pulses at several voltage levels from a holding potential of −30 mV. The prominent $I_{K(A)}$ and $I_{K(D)}$ currents seen in type II hair cells[5,35] at other more depolarized potentials are greatly reduced due to steady-state inactivation and lack of activation at this membrane potential. The outward current in the type I hair cell, on the other hand, like the M-type potassium current in frog sympathetic ganglion,[45] is present at a membrane potential of −30 mV and it is larger at more depolarized potentials. In addition to higher conductance at depolarized and hyperpolarized membrane potentials, a sample of pigeon type I hair cells was found to have significantly more hyperpolarized mean zero-current potentials (13 mV more hyperpolarized) than type II hair cells.[11] Thus, if, *in situ,* type II hair cells have more

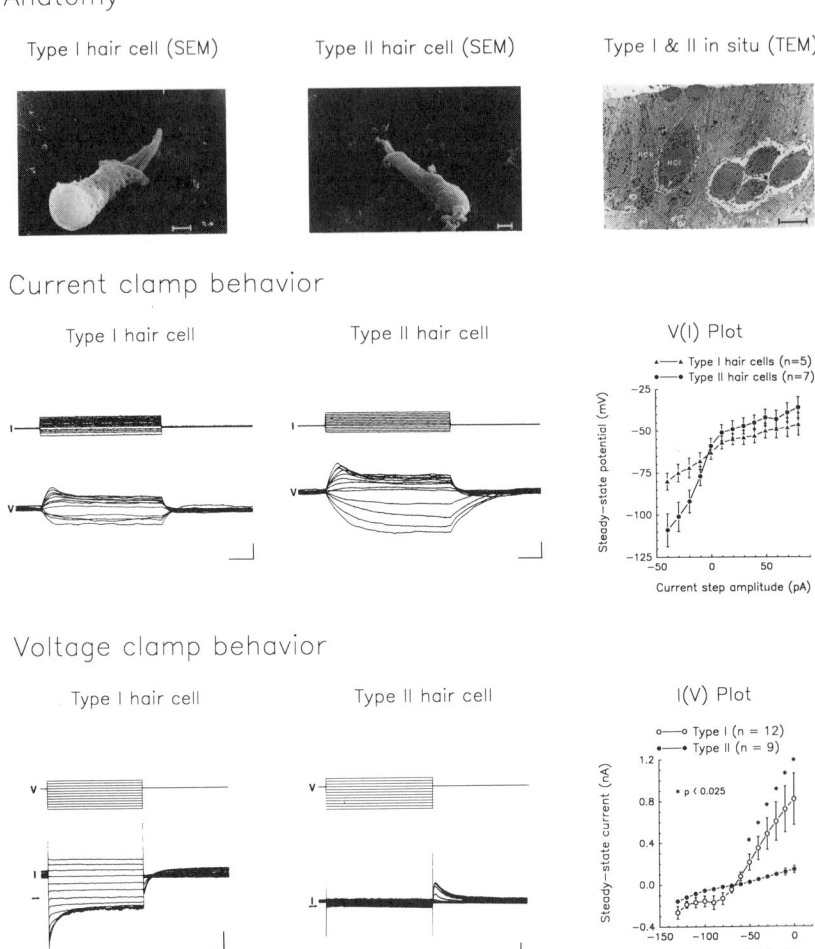

FIGURE 1. A collage of photomicrographs, traces, and graphs comparing the anatomy, current clamp, and voltage clamp behavior of type I and type II hair cells from the pigeon semicircular canal crista. Bars below the photomicrographs in the top panel represent 10 μm. Bars below the traces in the middle and bottom panels (redrawn from Correia and Lang, FIGURES 1 and 3)[11] represent 20 mseconds and 20 mV and 50 mseconds and 0.5 nA, respectively. The arrows in the traces obtained under voltage clamp (bottom panel) point at the zero current potential level.

depolarized RMPs and much less conductance around the RMP than type I hair cells, then as discussed in a previous section, they should be more sensitive to small MET currents and have a greater capability to increase their bandwidth (decrease the membrane time constant) during small depolarizations. The filter characteristics of type I hair cells, on the other hand, may be modifiable by cholinergic agonists acting on channels that produce its dominant outward current, if the current can be shown to be $I_{K(M)}$.

Sugihara and Furukawa studied hair cell membrane voltage responses to current injection in different morphological types of hair cells in the goldfish saccule.[7] They injected half wave rectified sine wave currents, while recording in current clamp mode, into short oscillatory hair cells and long spike type hair cells from the caudal part of the saccule. They varied frequency from 25 Hz to 1 KHz in the case of the oscillatory cells and from 10 Hz to 300 Hz in the case of the spike type cells. They observed that *at* the resonant frequency of the oscillatory hair cell, the membrane potential showed a large response with almost symmetrical hyperpolarizations and depolarizations about a steady-state membrane potential. This somewhat *symmetrical* response to a rectified sinusoidal current (asymmetrical stimulus) was not observed at other stimulation frequencies. At higher frequencies of current injection, the membrane response to individual cycles of sinusoids decreased and the membrane potential followed the envelope of the burst of sinusoidal waveforms. The membrane potential of the spike type long hair cells was a summation of the depolarizations of each of the cycles of the half wave rectified current until a threshold was reached. Then a spike followed by a plateau was seen. The plateau was maintained as long as the burst of sinusoidal current was maintained. Superimposed on the plateau were sinusoidal membrane potential oscillations corresponding in frequency to the injected current sinusoids.

Thus, it appears that different types of hair cells, whether they are differentiated on the basis of their morphology, innervation, or location on the receptor neuroepithelium, might act as filters with different characteristics to either increase the efficiency of the mechanical response of the end organ (e.g., the role of outer hair cell deformation in the mechanical response of the basilar membrane) or increase the efficiency with which MET currents are transformed into neurotransmitter release at the hair cell–primary afferent interface.

SUMMARY

Three questions were asked about filtering properties of hair cells. First, it was asked if hair cells in different receptors showed different filtering properties to match the receptor response to the particular characteristics of the stimulus? It seems that the answer is yes, since for example, the resonant frequencies of pigeon semicircular hair cells at membrane potentials around the RMP are in a range $(12–280 \text{ Hz})^5$ that could match angular head frequencies. Since critical frequency tuning of one frequency probably is not necessary for the semicircular canals it is quite reasonable that the largest quality factor (Q_e) of resonance in pigeon semicircular canal hair cells for membrane potentials around the RMP is 4 times lower than the largest Q_e in bullfrog saccular hair cells[28,29] and 10 times lower than Q_e in the turtle basilar papilla.[19,27] Second, it was asked if hair cells of the same morphological type but in different regions of the neuroepithelium have different filtering properties? It seems that this question needs more careful study since most available data on hair cells that show different filtering properties depending on their location in the neuroepithelium[13,19] show subtle morphological differences and therefore the hair cells could be classified as different types. Finally, it was asked if different hair cell types (based on morphology and innervation) have different filtering properties. The currents in pigeon[11] and guinea pig[12] type I and type II semicircular canal hair cells, guinea pig inner[15] and outer[8,33] cochlear hair cells, goldfish[7,21] short and tall saccule hair cells and chick[37,38] short and tall cochlea hair cells suggest that the answer to this question is yes. The challenge is to continue to precisely specify the filtering properties of different types of hair cells in different places on the neuroepithelium of different receptors.

ACKNOWLEDGMENTS

Daniel G. Lang Ph.D. contributed significantly to the studies of pigeon type I and type II hair cells discussed in this paper.

REFERENCES

1. OHMORI, H. 1989. Mechano-electrical transduction of the hair cell. Jpn J. Physiol. **39:** 643–657.
2. HUDSPETH, A. J. 1983. Mechanoelectrical transduction by hair cells in the acousticolateralis sensory system. Annu. Rev. Neurosci. **6:** 187–215.
3. HILLE, B. 1984. Ionic Channels of Excitable Membranes. Sinauer Associates. Sunderland, Mass.
4. MAURO, A., F. CONTI, F. DODGE & R. SCHOR. 1970. Subthreshold behavior and phenomological impedance of the squid giant axon. J. Gen. Physiol. **55:** 497–523.
5. CORREIA, M. J., B. N. CHRISTENSEN, L. E. MOORE & D. G. LANG. 1989. Studies of solitary semicircular canal hair cells in the adult pigeon. I. Frequency- and time-domain analysis of active and passive membrane properties. J. Neurophysiol. **62**(4): 924–934.
6. HODGKIN, A. L. & A. F. HUXLEY. 1952. A quantitative description of membrane current and its application to conduction and excitation in nerve. J. Physiol. London **117:** 500–544.
7. SUGIHARA, I. & T. FURUKAWA. 1989. Morphological and functional aspects of two different types of hair cells in the goldfish sacculus. J. Neurophysiol. **62**(6): 1330–1343.
8. ASHMORE, J. F. & R. W. MEECH. 1986. Ionic basis of membrane potential in outer hair cells of guinea pig cochlea. Nature **322**(6077): 368–371.
9. HOUSLEY, G. D., C. H. NORRIS & P. S. GUTH. 1989. Electrophysiological properties and morphology of hair cells isolated from the semicircular canal of the frog. Hear. Res. **38:** 259–276.
10. OHMORI, H. 1985. Mechano-electrical transduction currents in isolated vestibular hair cells of the chick. J. Physiol. London **359:** 189–217.
11. CORREIA, M. J. & D. G. LANG. 1990. An electrophysiological comparison of solitary type I and type II vestibular hair cells. Neurosci. Lett. **116:** 106–111.
12. RENNIE, K. J. & J. F. ASHMORE. 1991. Ionic currents in isolated vestibular hair cells from the guinea-pig crista ampullaris. Hear. Res. **51:** 279–292.
13. FUCHS, P. A., T. NAGAI & M. G. EVANS. Electrical tuning in hair cells isolated from the chick cochlea. J. Neurosci. **8:** 2460–2467.
14. OHMORI, H. 1984. Studies of ionic currents in the isolated vestibular hair cell of the chick. J. Physiol. London **350:** 561–581.
15. KROS, C. J. & A. C. CRAWFORD. 1990. Potassium currents in inner hair cells isolated from the guinea-pig cochlea. J. Physiol. London **421:** 263–291.
16. SANTOS-SACCHI, J. & J. P. DILGER. 1988. Whole cell currents and mechanical responses of isolated outer hair cells. Hear. Res. **35:** 143–150.
17. CRAWFORD, A. C. & R. FETTIPLACE. 1981. An electrical tuning mechanism in turtle cochlear hair cells. J. Physiol. London **312:** 377–412.
18. ASHMORE, J. F. & S. PITCHFORD. 1985. Evidence for electrical resonant tuning in hair cells of the frog amphibian papilla. J. Physiol. **364:** 39P.
19. ART, J. J. & R. FETTIPLACE. 1987. Variation of membrane properties in hair cells isolated from the turtle cochlea. J. Physiol. London **385:** 207–242.
20. LEWIS, R. S. & A. J. HUDSPETH. 1983. Voltage- and ion-dependent conductances in solitary vertebrate hair cells. Nature **304**(5926): 538–541.
21. FURUKAWA, T. & I. SUGIHARA. 1990. Multiplicity of ionic currents underlying the oscillatory-type activity of isolated goldfish hair cells. Neurosci. Res. (Suppl. 12): S27–S38.
22. ASHMORE, J. F. 1983. Frequency tuning in a frog vestibular organ. Nature **304**(5926): 536–538.

23. FUCHS, P. A. & M. G. EVANS. 1988. Voltage oscillations and ionic conductances in hair cells isolated from the alligator cochlea. J. Comp. Physiol. A **164**: 151–163.
24. ANGELAKI, D. E. & M. J. CORREIA. 1991. Models of membrane resonance in pigeon semicircular canal type II hair cells. Biol. Cybern. **65**: 1–10.
25. LEWIS, R. S. & A. J. HUDSPETH. 1983. Frequency tuning and ionic conductances in hair cells of the bullfrog's sacculus. Proceedings of the 6th International Symposium on Hearing, Bad Nauheim, Germany: 17–24.
26. ASHMORE, J. F. & D. ATTWELL. 1985. Models for electrical tuning in hair cells. Proc. R. Soc. London Ser. B **226**: 325–344.
27. ART, J. J., A. C. CRAWFORD & R. FETTIPLACE. 1986. Electrical resonance and membrane currents in turtle cochlear hair cells. Hear. Res. **22**: 31–36.
28. HUDSPETH, A. J. & R. S. LEWIS. 1988. Kinetic analysis of voltage- and ion-dependent conductances in saccular hair cells of the bull-frog, *Rana Catesbeiana*. J. Physiol. London **400**: 237–274.
29. HUDSPETH, A. J. & R. S. LEWIS. 1988. A model for electrical resonance and frequency tuning in saccular hair cells of the bull-frog, *Rana Catesbeiana*. J. Physiol. London **400**: 275–297.
30. PITCHFORD, S. & J. F. ASHMORE. 1987. An electrical resonance in hair cells of the amphibian papilla of the frog. Hear. Res. **27**: 75–83.
31. EVANS, M. G. & P. A. FUCHS. 1987. Tetrodotoxin-sensitive, voltage-dependent sodium currents in hair cells from the alligator cochlea. Biophys. J. **52**: 649–652.
32. HUDSPETH, A. J. & D. P. COREY. 1977. Sensitivity, polarity and conductance change in the response of vertebrate hair cells to controlled mechanical stimuli. Proc. Natl. Acad. Sci. USA **74**(6): 2407–2411.
33. BROWNELL, W. E., C. R. BADER, D. BERTRAND & Y. DE RIBAUPIERRE. 1985. Evoked mechanical responses in isolated cochlear outer hair cells. Science **227**: 194–196.
34. ASHMORE, J. F. 1987. A fast motile response in guinea-pig outer hair cells: the cellular basis of the cochlear amplifier. J. Physiol. London **388**: 323–347.
35. LANG, D. G. & M. J. CORREIA. 1989. Studies of solitary semicircular canal hair cells in the adult pigeon. II. Voltage-dependent ionic conductances. J. Neurophysiol. **62**: 935–945.
36. STEINACKER, A. & A. ROMERO. 1990. Toadfish saccular hair cells do not all contain identical species of ionic currents. Abstr. Assoc. Res. Otolaryngol. **13**: 249.
37. FUCHS, P. A. & M. G. EVANS. 1990. Potassium currents in hair cells isolated from the cochlea of the chick. J. Physiol. London **429**: 529–551.
38. MURROW, B. W. & P. A. FUCHS. 1990. Preferential expression of transient potassium current (I_A) by 'short' hair cells of the chick's cochlea. Proc. R. Soc. London **242**: 189–195.
39. FUCHS, P. A. & B. H. A. SOKOLOWSKI. 1990. The acquisition during development of Ca-activated potassium currents by cochlear hair cells of the chick. Proc. R. Soc. London Ser. B **241**: 122–126.
40. FUCHS, P. A., M. G. EVANS & B. W. MURROW. 1990. Calcium currents in hair cells isolated from the cochlea of the chick. J. Physiol. London **429**: 553–568.
41. NAKAGAWA, T., S. KAKEHATA, N. AKAIKE, S. KOMUNE, T. TAKASAKA & T. UEMURA. 1991. Calcium channel in isolated outer hair cells of guinea pig cochlea. Neurosci. Lett. **125**: 81–84.
42. NAKAGAWA, T., N. AKAIKE, T. KIMITSUKI, S. KOMUNE & T. ARIMA. 1990. ATP-induced current in isolated outer hair cells of guinea pig cochlea. J. Neurophysiol. **63**: 1068–1074.
43. GITTER, A. H., H. P. ZENNER & E. FROEMTER. 1986. Membrane potential and ion channels in isolated outer hair cells of guinea pig cochlea. Otorhinolaryngol. **48**: 68–75.
44. OHMORI, H. 1987. Gating properties of the mechano-electric transducer channel in the dissociated vestibular hair cell of the chick. J. Physiol. London **387**: 589–609.
45. ADAMS, P. R., D. A. BROWN & A. CONSTANTI. 1982. M-currents and other potassium currents in bullfrog sympathetic neurones. J. Physiol. London **330**: 537–572.

Ionic Currents of Mammalian Vestibular Hair Cells[a]

RUTH ANNE EATOCK AND MICHAEL J. HUTZLER

Department of Physiology
University of Rochester
601 Elmwood Avenue
Rochester, New York 14642-8642

INTRODUCTION

The sensory hair cells of the vestibular organs of some vertebrates comprise two distinct morphological classes, called type I and type II.[1] The most salient difference between the two classes of hair cell is in the form of their synaptic contacts with afferent nerve fibers. The afferents form bouton terminals on type II cells, and calyceal (cuplike) endings on type I cells. A single afferent fiber frequently contacts both type I and type II cells. Type II hair cells occur in all hair cell organs in all vertebrate groups, whereas type I cells occur only in the vestibular organs of mammals, birds, and some reptiles (reviewed in Lewis *et al.*).[2] It is natural to suppose that type I cells serve a distinct role in vestibular sensation, but functional differences between type I and type II cells have proven difficult to establish.

Different categories of spontaneous and evoked firing patterns have been described for the afferent neurons.[3–6] One approach to the question of the special role, if any, of the type I hair cell/afferent synapse has been to morphologically identify the synaptic terminals of physiologically characterized primary vestibular afferents.[7,8] These studies have for the most part failed to establish a link between afferent firing patterns and the type I/type II dichotomy in the end organs. The exception was a correlation between type I endings and low gain of the afferent response to head-movement stimuli, but this was found only in semicircular canal organs,[7] not otolith organs.[8] Differences in the pattern of spontaneous activity or in the rate of adaptation to maintained vestibular stimuli were found to correlate more with region of innervation within the end organ than with the type I/type II classification.

A complementary approach to evaluating functional differences between type I and type II hair cells is to investigate their electrophysiological properties in isolation. In this report, we compare voltage-dependent currents and voltage responses to injected current in the two cell types. These experiments suggest that type I and type II cells do differ in some aspects of posttransduction stimulus processing. Some of these data have been presented in abstract form.[9]

METHODS

Preparation

Vestibular inner-ear organs from young Long-Evans (hooded) rats (8–30 days old) were used. Each rat was deeply anesthetized with sodium pentobarbitol (65

[a]This study was supported by the Office of Naval Research.

58

TABLE 1. External Solutions

Solution	Na	K	Mg	Ca	Ba	Cl	PO$_4$	SO$_4$	HEPES	D-glu	pH	mmol/kg
Standard external	150	5.8	0.8	3.3	0.0	159	0.8	0.8	5.0	5.6	7.3	307
Ba^{2+} external	150	5.8	0.8	0.0	3.3	160	0.0	0.0	5.0	5.6	7.3	307
Low-Ca^{2+} external	150	5.8	0.8	0.1	0.0	154	0.8	0.0	5.0	5.6	7.3	299

mg/kg intraperitoneally), then decapitated. The temporal bones were removed and placed in a dish of chilled, oxygenated Hank's balanced salt solution (HBSS) buffered with N-2-hydroxyethylpiperazine-N′-2-ethanesulfonic acid (HEPES) and supplemented with 2 mM CaCl$_2$ and 10 mM NaCl. This solution was also the standard external medium during recording (TABLE 1). Osmolalities were measured with a vapor-pressure osmometer (Model 5500, Wescor, Logan, Utah).

The otolith and semicircular canal organs were excised and then treated with protease XXVII (Sigma; 50 μg/ml in 4 ml of the standard external medium) for 10 minutes. The otolithic membranes were then removed and the organs were placed for 40 minutes in low-Ca^{2+} external solution (TABLE 1) containing crude papain (Sigma; 500 μg/ml) and 2.5-mM L-cysteine. The papain treatment was followed by a 5-minute exposure to HBSS containing bovine serum albumen (Sigma; 500 μg/ml). The organs were then transferred to low-Ca^{2+} HBSS for mechanical dissociation onto glass coverslips. Enzyme treatments were performed at 37°C, and surgery and experiments at 22–25°C.

Recording

The tight-seal whole-cell recording technique was employed.[10] Electrodes were pulled from borosilicate glass and had impedances of 5–10 MΩ. Pipette solutions used to record K$^+$ currents are shown in TABLE 2. In most experiments we used "KCl I" or "KCl II." To record Ca^{2+} currents, we used a "Cs CP/CPK" internal solution containing an ATP-regenerating system (creatine phosphate and creatine phosphoki-nase).[11] The solution contained, in mM, 113 Cs$^+$, 0.12 Li$^+$, 5 Mg^{2+}, 1 Ca^{2+}, 42 Cl$^-$, 20 creatine phosphate, 10 ethylene glycol-bis(β-aminoethyl ether)N,N,N′,N′-tetraace-tic acid (EGTA), 10 HEPES, 5 adenosine 5′-triphosphate (ATP), 0.04 guanosine 5′-triphosphate (GTP), 84 D-glucose, and 50 units/ml of creatine phosphokinase (pH 7.3, 271 mmol/kg). In one experiment (FIGURE 6), a "K CP/CPK" internal solution was used, which resembled the Cs CP/CPK solution except that it contained K$^+$ instead of Cs$^+$. Membrane voltages are corrected for liquid junction potentials. In

TABLE 2. Internal Solutions

Solution	K	Li	Mg	Ca	Cl	EGTA	HEPES	ATP	GTP	pH	mmol/kg
KCl I	139	0.12	2	0.1	110.2	11	10	2	0.04	7.3	260
KCl II	142	0.12	2	0.1	120.2	8	10	2	0.04	7.3	259
KCl III	140		2	1.0	132.0	5	5	2		7.3	262

experiments in which the K channel blocker tetraethylammonium (TEA) or 4-ami-nopyridine (4-AP) was included in the external medium, osmolarity was held constant by reducing NaCl. Charybdotoxin was obtained from IBF-Biotechnics, Inc., Columbia, Md.

The patch clamp amplifier (L/M EPC7, Adams and List Associates, Ltd., Great Neck, N.Y.) was used in either the voltage clamp or current clamp mode. In voltage clamp mode, membrane voltage steps were generated with a 12-bit digital/analog converter, and membrane currents were sampled with a 12-bit analog/digital converter. Both converters were controlled by a 286 or 386 computer using a Scientific Data Acquisition System (Scientific Associates, Inc., Rochester, N.Y.). Membrane currents were low-pass filtered with a cutoff frequency of either 3 kHz or 10 kHz using the patch clamp amplifier's intrinsic filters (3-pole Bessel); records originally filtered at 10 kHz were digitally filtered at 3 kHz off-line. A holding potential of -60 mV was used. In some cases, linear currents were subtracted from total current using an on-line P, +P/4 pulse procedure.[12] Typically, 50–75% series resistance compensation was employed.

Current clamp mode was used to investigate electrical resonance. Current steps were delivered from zero holding current, and membrane voltage was sampled using the system described above. Series resistance was not compensated in current clamp mode; the resulting voltage errors were generally less than 10 mV. The frequency of the electrical resonance (f_e) was estimated from the discrete Fourier transform of the voltage response to a current step. f_e and Equation 1, which describes the response of a resonant electrical (LRC) circuit to a current step,[13] were then used to fit the waveform of the voltage response. $V(t)$ is membrane voltage at time t, I is the applied current step, t_0 and t_1 are the times at onset and offset of the current step, τ is the time constant of decay of

$$V(t) = IR\left[1 + \pi f \tau e^{-t/\tau} \sin\left(2\pi f (t - t_0) - \tan^{-1}\frac{1}{\pi f \tau} + \phi\right) \right.$$
$$\left. - \left(1 + \pi f \tau e^{-t/\tau} \sin\left(2\pi f (t - t_1) - \tan^{-1}\frac{1}{\pi f \tau} + \phi\right)\right)\right] + IR_e \quad (1)$$

the oscillations, ϕ is a parameter that accounts for the delay in the onset of the resonance, and R_e is a constant needed to match the steady-state membrane potential. See Figure 9B for an example of a fit. From the value for τ provided by the fit, the quality of the electrical resonance (Q_e) was estimated according to Equation 2.[13]

$$Q_e = \sqrt{(\pi f \tau)^2 + 0.25} \quad (2)$$

RESULTS

Identification of Hair Cell Type

Type I cells (FIGURE 1A and B) were distinguished by their flask shape: the round nuclear region narrows apically into a "neck" and widens again at the base of the hair bundle.[14] Those cells that were classified as type II can be broadly divided into round cells (FIGURE 1C) and cells with more elongate somata, which we will call "cylindrical" (FIGURE 1D). Although cylindrical hair cells may become rounder as they deteriorate, the round cells from which we recorded maintained their form for

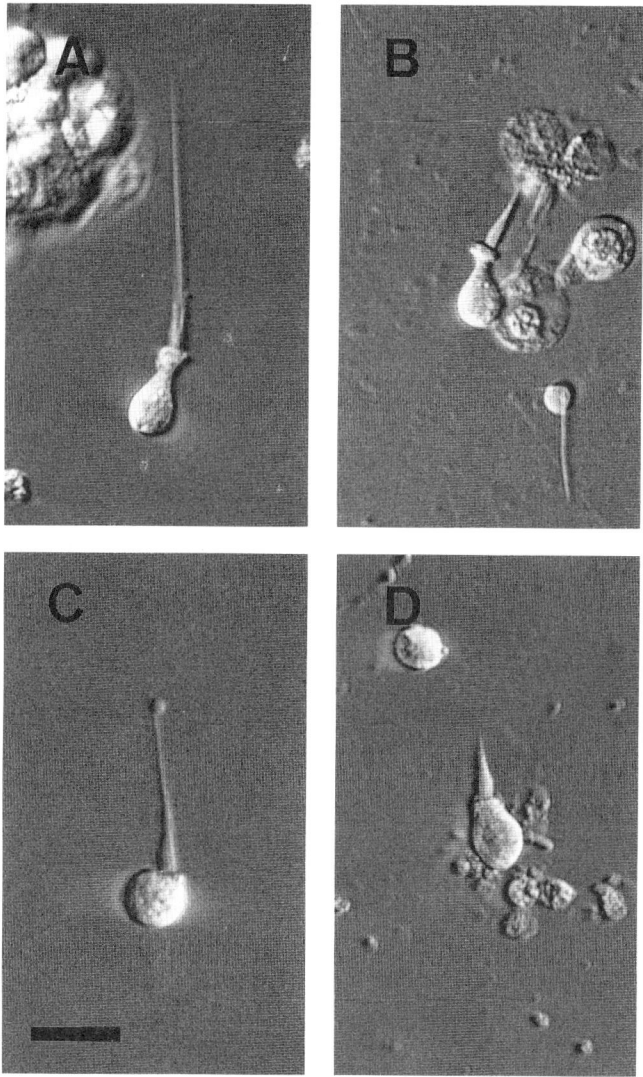

FIGURE 1. Dissociated type I (A and B) and type II (C and D) hair cells. The cells in A and C are from semicircular canal organs; those in B and D are from the utricle. Scale bar: 20 μm.

hours and did not display the usual indicators of deterioration, such as large intracellular vacuoles or grainy appearance, depolarized zero-current potentials, or low input resistances. The mean membrane capacitances of type I cells, round type II cells, and cylindrical type II cells were, in pF, 4.6 ± 0.59 [standard error of the mean (SEM); $n = 3$], 4.0 ± 0.26 ($n = 13$), and 4.5 ± 0.63 ($n = 6$), respectively. (These values are for hair cells with hair bundles. The mean capacitance for 4 type I cells that lacked hair bundles was 2.7 ± 0.14 pF.)

Utricular maculae and semicircular canal cristae were dissociated in the same experimental chamber. Those cells that retained hair bundles were identified as originating from cristae or maculae on the basis of hair bundle height (FIGURE 1). TABLE 3 shows the numbers of hair cells used in this study, classified by cell type and end organ.

Whole-Cell Currents

FIGURE 2 shows examples of whole-cell currents recorded from a type I cell (FIGURE 2A) and a type II cell (FIGURE 2B) while the membrane potential was stepped from a holding potential of −60 mV to hyperpolarizing and depolarizing potentials over a wide range. The corresponding steady-state current-voltage (I-V) relations are shown in FIGURE 2C. As suggested by these examples, depolarizing voltage steps activated outward current in both type I and type II hair cells. Some type I cells (but not that shown in FIGURE 2A) had channels that were open at −60 mV. Voltage steps to potentials below −80 mV evoked inwardly rectifying current in many hair cells of both types. Finally, depolarizing voltage steps elicited Ca^{2+} currents in both type I and type II cells. In Figure 2 (A and B), these appear as small

TABLE 3. Categories of Hair Cells

Hair Cell Type	Semicircular Canal Cristae	Utricular Maculae	Unidentified Origin	Total
Type I	2	1	11	14
Type II	7	11	15	33
Unidentified	0	1	1	2
Totals	9	13	27	49

inward currents that precede the large outward currents. These currents will be discussed in turn.

Outwardly Rectifying Current

Instantaneous I-V relations[15] and the use of K channel blockers showed that most of the outward current was carried by K^+. In 5 type II cells (3 from utricles, 1 from a semicircular canal organ, and 1 of unidentified origin), instantaneous I-V relations were obtained by extrapolating the peak current at a particular voltage following a pulse to a very positive potential (FIGURE 3). The reversal potential derived from these instantaneous I-V relations was −71 ± 4.6 mV (mean ± SEM; $n = 5$), much closer to the equilibrium potential for K^+ (−80 mV) than to that for Cl^- (−7 to −9 mV) (internal solutions KCl I and II, TABLE 2; standard external solution, TABLE 1).

In both type I and type II cells, the outward current was almost completely blocked (~90% at +90 mV) by bath application of the K channel blocker 4-aminopyridine (10 mM) (FIGURE 4). In type II cells, the outward current was 80–90% blocked by bath application of either 4-AP (10 mM), TEA (10 mM), barium (3.3 mM), or charybdotoxin (100 nM). Thus, based on its reversal potential and pharmacology, the outward current was largely through K channels. The residual current

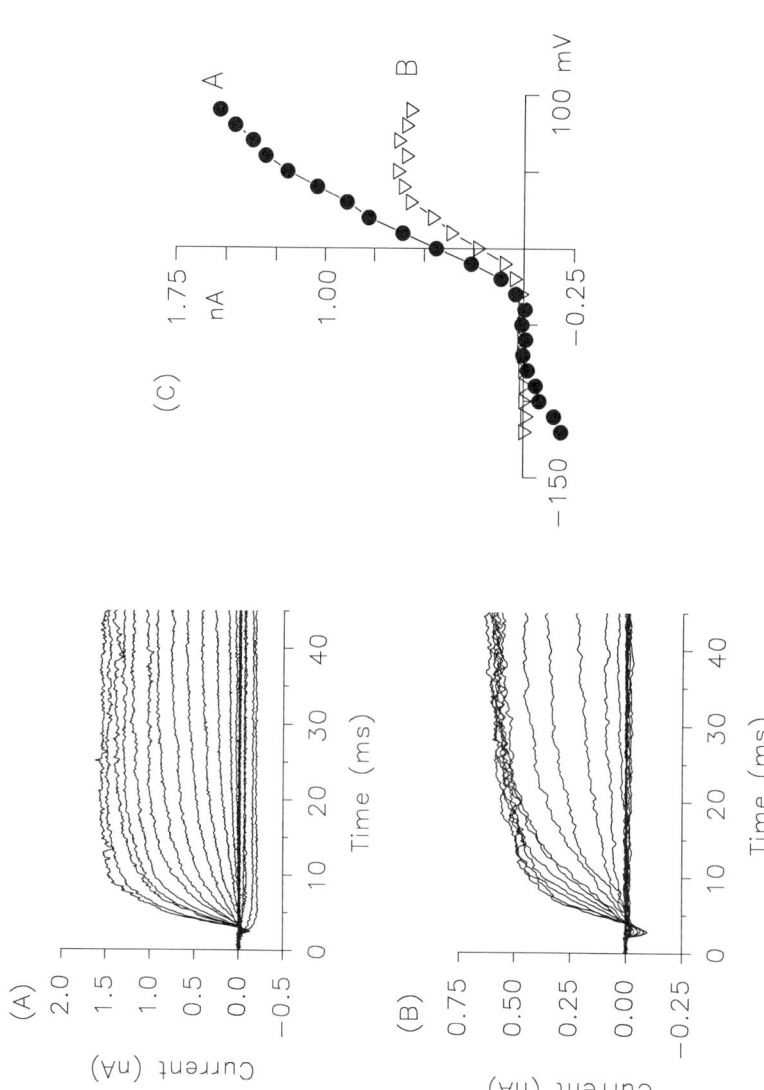

FIGURE 2. Voltage-dependent whole-cell currents in rat vestibular cells. Data are from A, a type I cell from a semicircular canal organ; and B, a round type II cell from the utricle. The holding potential was −60 mV. In A and B currents evoked by a series of voltage steps (from −120 to +90 mV in 10-mV increments) are superimposed. The corresponding steady-state I-V relations are shown in C; currents were measured near the end of the 45-msecond steps. Internal solution: KCl II. Standard external solution.

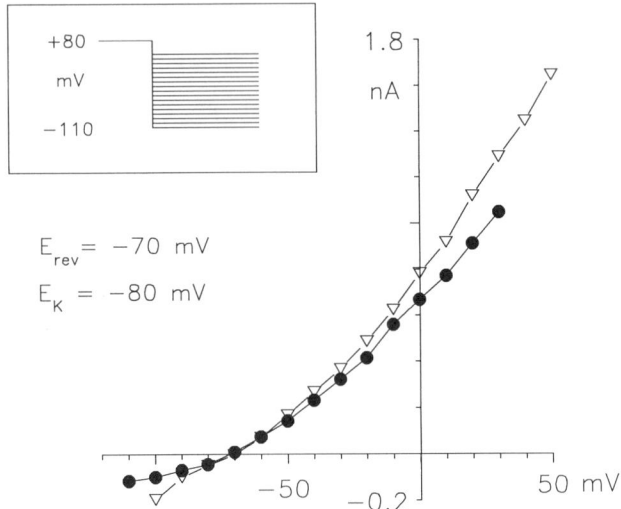

FIGURE 3. Reversal potential of the outward current. After a 20-msecond voltage step from the holding potential (-60 mV) to $+80$ mV, voltage was stepped to different levels between -110 and $+50$ mV (inset). The peak tail current at each potential was extrapolated from an exponential fit to the tail current and plotted as a function of voltage. Data shown are from two type II cells from the utricle. Internal solution: KCl I. Standard external solution.

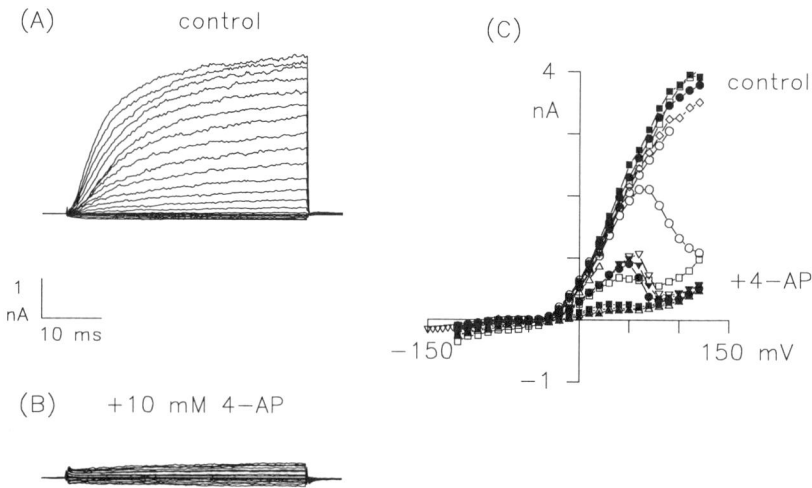

FIGURE 4. Block of the outward current in a type I hair cell by the K-channel blocker, 4-AP. **A:** Superimposed currents in response to voltage steps from a holding potential of -60 mV. Voltage was stepped between -120 and $+90$ mV, in increments of 10 mV. Internal solution: KCl I. Standard external solution. **B:** Currents from the same cell as in A after perfusion with external medium containing 10 mM 4-AP. Most of the outwardly rectifying current was blocked. **C:** Steady-state I-V relations for the data in A and B and at different times between the records of A and B. The block of the outward current increased with time after the beginning of 4-AP perfusion. Note the N-shaped I-V relations at intermediate stages of block.

that was not blocked by K channel blockers may reflect nonchannel pathways for K current or "leak."

An estimate of the time course of activation of the outward current was obtained by measuring the time to reach 63.21% $[100^*(1 - 1/e)]$ of the steady-state value after stepping the membrane potential from -60 mV to $+10$ mV. For type I cells, the mean time was 11 ± 3.7 mseconds [standard deviation (SD); $n = 9$] and for type II cells it was 12 ± 5.4 mseconds ($n = 21$). These values are not significantly different, nor was there a significant difference between the values for round and cylindrical type II cells.

To compare the amplitude of the outward current across cells, we measured the steady-state current evoked by a step from -60 to $+10$ mV. The mean values were 1104 ± 196.8 (SEM) pA for type I cells ($n = 9$), 382 ± 46.1 pA for round type II cells ($n = 10$), and 851 ± 111.5 pA for cylindrical type II cells ($n = 7$). The difference between the means for type I and round type II cells is significant at the 0.005 level, and the difference between the means for round type II cells and cylindrical type II cells is significant at the 0.05 level. However, the latter difference disappears when current densities (currents normalized by cell capacitance) are compared. The mean current densities for round and cylindrical type II cells with hair bundles were 103 ± 24.0 pA ($n = 7$) and 192 ± 48.3 pA ($n = 7$), respectively; these are not significantly different ($p = 0.11$). This suggests that the difference in mean outward currents between round and cylindrical type II cells reflects the smaller size of the round type II cells. The current densities of type I and II cells cannot be compared because hair bundles were present on most of the type II cells and absent on all but 3 of the type I cells for which we have capacitance measurements. The hair bundle adds significant capacitance but presumably does not have any voltage-dependent channels, only mechanoelectric transduction channels.[16]

While we have not systematically analyzed the components of the K current in either cell type, the following observations can be made:

1. Some cells had N-shaped I-V relations, suggesting the presence of Ca^{2+}-dependent K current[17] (FIGURE 4C, at intermediate levels of block by 4-AP). Such currents grow with increasing depolarization until the increased driving force on K^+ is more than offset by the decreased driving force on Ca^{2+} entry. However, in some cells the I-V relations were not N shaped, but simply saturated at high positive potentials.

2. For the holding potential of -60 mV that was used in these experiments, the outward current did not appreciably inactivate over 50 mseconds (FIGURE 1A and B and FIGURE 5A).

3. Two type I hair cells had appreciably larger currents at $+10$ mV (> 1.75 nA) than the other cells (all < 1.65 nA). This reflected the presence in the two cells (and in a third, for which we only recorded current responses to negative voltage steps) of a component that in some ways resembled the M current of amphibian sympathetic neurons.[18] The current was outward for depolarizations from a holding potential of -60 mV and inward for hyperpolarizations (FIGURE 5A). FIGURE 5B shows I-V relations for the data in FIGURE 5A; the early current, 500 μseconds after the step onset, is compared with the steady-state current, 40 mseconds after the step onset. The steady-state I-V relation diverged from the early I-V relation above -50 mV and below -90 mV. The divergence above -50 mV may reflect the voltage- and time-dependent relaxation of the M-like current to a new, higher activation level, or the recruitment of another voltage-dependent outward current, or both. The

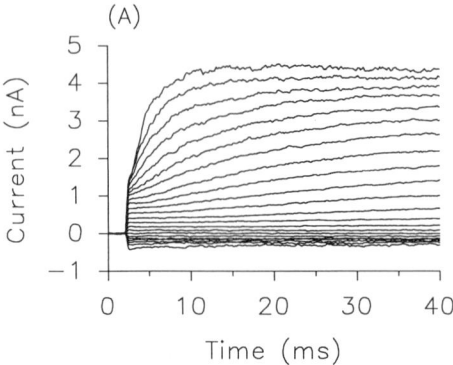

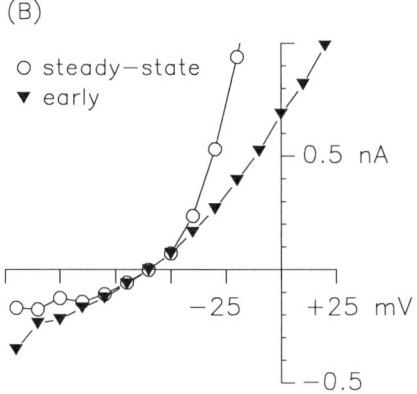

FIGURE 5. M-like current in a type I hair cell. **A:** Whole-cell currents for a series of hyperpolarizing and depolarizing voltage steps (between −120 and +90 mV) from a holding potential of −60 mV. **B:** Current-voltage relations from the data in A. Currents were measured 500 μseconds (triangles) and 40 mseconds after the step onset (circles). The early and late currents diverged above −50 mV and below −90 mV. Internal solution: KCl II. Standard external solution.

divergence below −90 mV results from time-dependent deactivation of the M-like current in this voltage range. This time- and voltage-dependent deactivation during hyperpolarizing steps is shown more clearly in an experiment on a third type I cell (FIGURE 6A) in which membrane potential was stepped to very negative voltages (outward current data were not obtained from this cell). The deactivation time course was well fit by a single exponential with a time constant that depended on voltage (FIGURE 6B), having a maximum value of 7.3 mseconds at −104 mV and falling to 2.9 mseconds at −164 mV.

4. In type I cells without M-like current, the voltage dependence of the outwardly rectifying current was shifted negatively relative to that in type II cells (FIGURE 7). Outward current was detectable above about −50 to −40 mV in type I cells ($n = 10$), but only above about −30 mV in type II cells ($n = 26$).

Inwardly Rectifying Current

Voltage steps to potentials more negative than −70 to −80 mV evoked inwardly rectifying currents in 50% of type I cells and 61.5% of type II cells (FIGURE 2). Whether this current was present in type II cells did not depend on cell shape. These currents were readily distinguished from the M-like current by their time-dependent activation (rather than deactivation) during a hyperpolarizing voltage step. All cells with no demonstrable inwardly rectifying current (5 type I and 10 type II cells) were from animals ≤21 days old, raising the possibility that they would have acquired the inwardly rectifying current with time. However, in those cells with inwardly rectifying current, the current's steady-state amplitude at −120 mV and the age of the animal were only weakly correlated ($r = 0.3$).

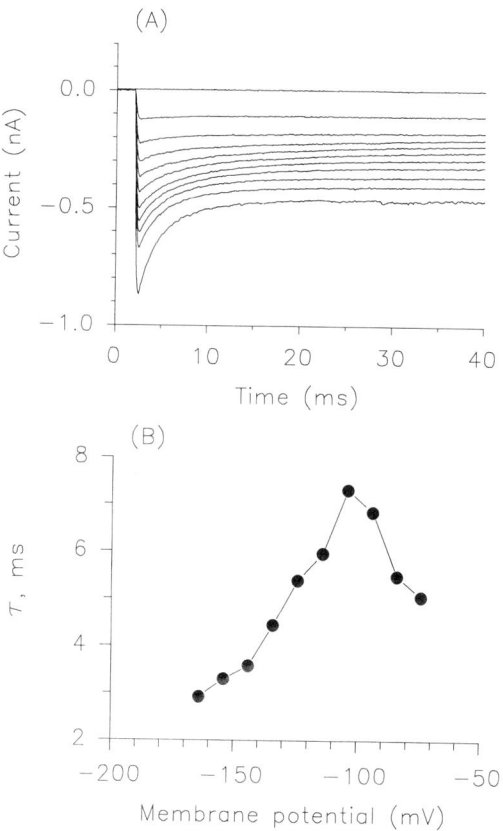

FIGURE 6. Time- and voltage-dependent deactivation of the M-like current evoked by hyperpolarizing voltage steps. Type I hair cell. **A:** Currents in response to voltage steps from a holding potential of −64 mV to potentials between −164 and −74 mV. The deactivation following the onset of the negative step was fit by a single exponential. **B:** The time constant of deactivation (τ) as a function of voltage during the step, for the data in A. Internal solution: K CP/CPK. Standard external solution.

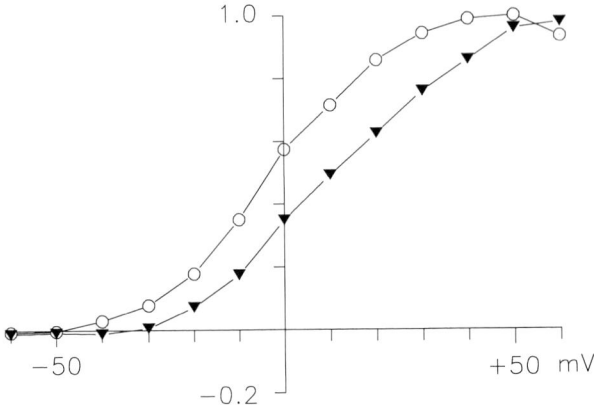

FIGURE 7. The voltage dependence of the outwardly rectifying current in type I and type II cells. Shown are steady-state I-V relations from a type I cell that lacked the M-like current (circles) and a type II cell (triangles) from the same animal. The currents have been normalized by dividing by the maximum current. Internal solution: KCl I. Standard external solution.

Calcium Current

In some cells, a fast small inward current was visible for depolarizing stimuli, before the outward current was activated (FIGURE 2B). Based on results from other hair cells, it seemed likely that the early inward current was through voltage-dependent Ca^{2+} channels. Replacement of internal K^+ with Cs^+ and of external Ca^{2+} with Ba^{2+} revealed small inward currents that activated with depolarization above -55 mV (FIGURE 8). These currents peaked at -10 mV and declined to zero by $+40$ or $+50$ mV, and did not inactivate within 40 mseconds. With 10-mM Ca^{2+}, rather than 3.3-mM Ba^{2+}, in the external medium, the I-V relation was shifted about 10 mV positive to the Ba^{2+} I-V relation but no other differences were observed.

Current Clamp Results

In current clamp mode, no voltage oscillations were apparent at the zero-current potentials, which had mean values of -46 ± 3.4 mV (SEM; $n = 7$) and -44 ± 3.0 mV ($n = 15$) for type I and type II cells respectively. The effects of current steps on membrane potential were examined in 3 type I cells that lacked M-like current and in 10 type II cells. Depolarizing current steps generally elicited a single initial peak that declined to a steady-state value (FIGURE 9A), presumably as a consequence of activation of the outwardly rectifying current. Sometimes small secondary and tertiary oscillations were observed.

The initial peak in the voltage response indicates that the cell membrane is electrically resonant, while the absence of large repeated oscillations shows that the resonance is heavily damped. A measure of the sharpness or quality of tuning of a resonant electrical circuit is provided by its quality factor (Q_e) which is related to the time constant of decay of the oscillations (Equation 2, Methods). The sharper the tuning, the longer the time course of decay and the higher the Q_e value. To estimate Q_e, we fit the voltage response using the equation for a resonant electrical circuit

(Equation 1) (FIGURE 9B). Both Q_e and the electrical resonant frequency were voltage dependent. Peak Q_e values were generally observed for current steps of $+100$ pA (mean $\pm$ SEM = 104 $\pm$ 3.6 pA; resolution 25–50 pA), corresponding to a mean steady-state voltage of -17 ± 2.3 (SEM) mV. Peak Q_e values had similar mean values and ranges for type I and type II cells. For all cells, Q_e values ranged from 0.5 to 1.5, with a mean of 1.0 $\pm$ 0.27 (SD; $n = 13$). By convention,[13] a hair cell's characteristic electrical resonant frequency (CF_e) is defined as that recorded in response to the current step that evokes the highest quality tuning (the peak Q_e value). For type II cells, CF_e values had a small range, 28 Hz to 43 Hz (mean $\pm$ SD = 36 $\pm$ 4.9 Hz). For the 3 type I cells tested, CF_e's were 8.5, 41.2, and 130.0 Hz.

The mean slope conductances determined from the steady-state voltage response to small current steps (between -50 and $+50$ pA) were 1.1 $\pm$ 0.18 (SEM) nS for the

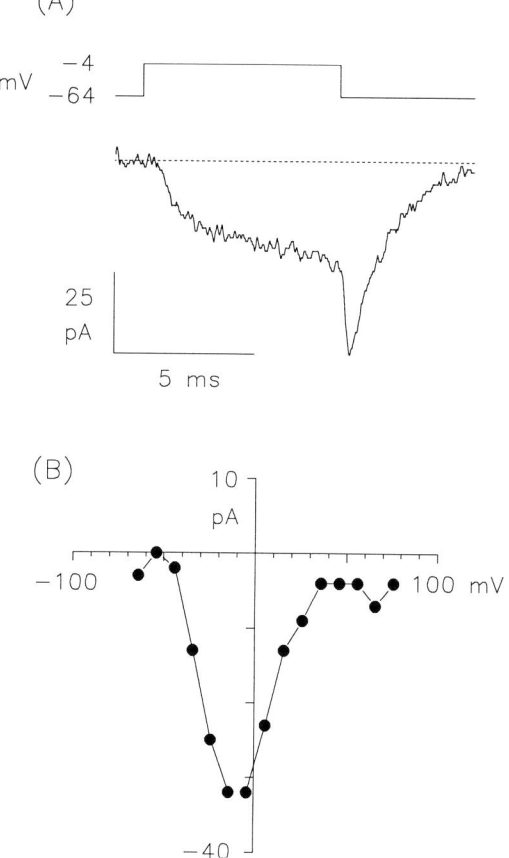

FIGURE 8. Ba^{2+} currents in a type II cell. **A:** Whole-cell current evoked by a voltage step to -4 mV from a holding potential of -64 mV. Dashed line: zero-current level. **B:** I-V relation for the data in A, measured at the end of the 7-msecond step. Internal solution: Cs CP/CPK. Ba^{2+} external solution.

3 type I cells and 1.2 ± 0.21 nS for the 10 type II cells. For 9 type II cells and 1 type I cell, slope conductance increased dramatically for larger positive current steps (mean ± SEM = 21.1 ± 2.33 nS), reflecting activation of the outwardly rectifying current. For the remaining 3 cells (2 type I cells and 1 type II cell), the voltage-current relation for currents > 50 pA was too complex to extract a single meaningful slope conductance.

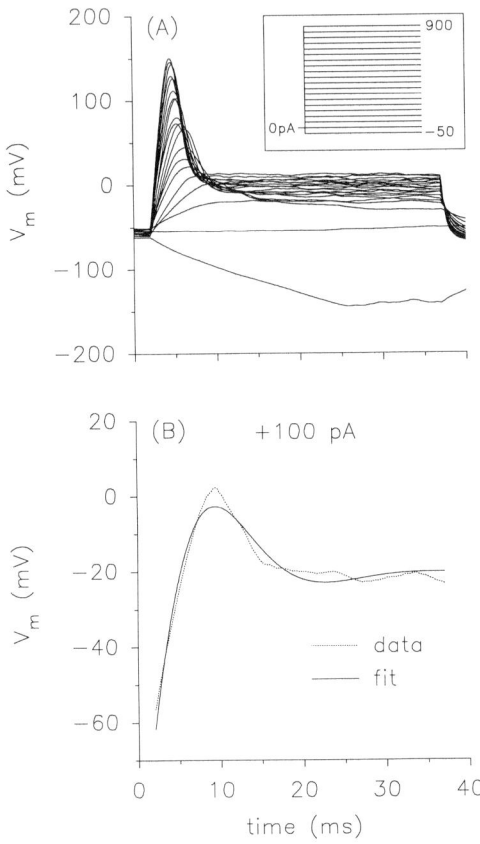

FIGURE 9. Voltage responses evoked by current steps in current clamp mode. Type II cell. **A:** Superimposed voltage responses to current steps from −50 to +900 pA in increments of 50 pA (inset). Zero holding current. **B:** Dashed line: the voltage response to a +100-pA step (from the data set in A). Bold line: fit to the data using Equation 1 (Methods). From the fit, the quality (Q_e) and frequency of the electrical resonance were estimated as 0.94 and 38 Hz, respectively. Internal solution: KCl I. Standard external solution.

DISCUSSION

Our classification of type II cells into "cylindrical" and "round" cells is reminiscent of the "tall" and "short" categories of type II cells from the semicircular canal organs of adult guinea pigs.[19] However, comparison of the photographs of the guinea pig cells with our own suggests that the short guinea pig cells more closely resemble the cylindrical rat cells. We do not believe it to be likely that the round cells in our study were deteriorating cylindrical cells, on the basis of other morphological and physiological indicators. Because the cells are from young animals, an alternative possibility is that the round cells are immature.

The heterogeneity of some features of the voltage-dependent currents in the rat vestibular hair cells was not obviously related to cell type. These variable features included the activation time course of the outwardly rectifying current, the shape of the I-V relation at high positive potentials, and whether inwardly rectifying current was present. Moreover, these features did not relate systematically to the type of organ (semicircular canal or otolith), at least for type II cells. (Most of the type I cells were not identified as to organ of origin.)

However, some systematic differences between type I and type II cells have emerged. M-like currents were found only in type I cells (3/14). A similar current has been described in type I hair cells from semicircular canal organs of pigeons[20] and guinea pigs.[19] In the type I cell from which currents were recorded at very negative potentials (FIGURE 6), the deactivation rate during hyperpolarizing steps was well fit by a single exponential and had a voltage dependence consistent with simple, first-order kinetics.[21] In these respects, the M-like current resembled the M current of bullfrog sypathetic neurons and the M-like "I_{Kx}" current of salamander rods.[22] However, deactivation of the M-like current in the type I hair cell was more rapid, by 1–2 orders of magnitude, and had a voltage dependence that was shifted negatively relative to amphibian M current and I_{Kx}.

Because the M-like current was evident at rest, like a large leak conductance, it would help to set resting potential and reduce the membrane time constant. The latter effect would both speed the voltage response to transduction current and diminish the steady-state amplitude of the response. Whether the time- and voltage-dependent deactivation of the M-like current at very negative potentials is functionally significant is in doubt, given that these potentials are below the equilibrium potential for K^+.

Type I cells without M-like current had a voltage-dependent outwardly rectifying current that activated above −50 to −40 mV, 10–20 mV more negative than the voltage at which outward current was first detectable in type II cells. One possible source of this difference is Ca^{2+} currents that activate at different voltages, and therefore activate Ca^{2+}-dependent K currents at different voltages. Alternatively, outward K currents may have different intrinsic voltage dependence in the two cell types. Whether this difference is functionally significant depends on whether the outwardly rectifying currents are activated under physiological conditions, which in turn depends on the normal resting potential and the maximum amplitude of receptor potentials. Our only estimate of resting potential is the zero-current potential in current clamp mode. The mean zero-current potentials for type I and type II cells were −46 and −44 mV, respectively. Receptor potentials have not been measured in these cells, but receptor potentials in excess of 10 mV have been measured in hair cells of the frog saccule,[23] the turtle cochlea,[24] and the mammalian cochlea.[25] Therefore, it is reasonable to suggest that moderate to large vestibular stimuli will activate the outward currents in rat vestibular hair cells. If so, the outward currents would tend to repolarize the cell (FIGURE 9), which might constitute a significant form of sensory adaptation in this system.

The nature of the K currents that contribute to the outwardly rectifying current in these hair cells is not known. In some cells, "N"-shaped I-V relations suggested a Ca^{2+}-dependent component. Small Ca^{2+}-dependent K currents were observed in pigeon type II cells[26] and in both type I and II cells from guinea pig semicircular canal organs.[19] It is also possible that some of the outward current was delayed rectifier current. The outward current of pigeon type II cells was shown to include a slowly inactivating delayed rectifier K current and a rapidly inactivating A current, in addition to the small Ca^{2+}-dependent current. The tall type II cell from guinea pig cristae also had considerable A current when held at relatively negative potentials. In both pigeon and guinea pig cells, the A current was ~80% inactivated at −60 mV, the holding potential that we have used in this study. Therefore, it is possible that the rat's vestibular cells have A channels that were largely inactivated in our experiments.

The Ca^{2+}-channel currents that we recorded activated above −55 mV (in 3.3 mM Ba^{2+}) and peaked at about −10 mV. This voltage dependence appears optimal for modulation of synaptic transmission about a resting potential of approximately −45 mV. In other hair cells, activation voltages of −60 mV,[27,19] −50 mV,[26,28,29] and −30 mV[30] have been reported. In some of these studies, the apparent voltage dependence may have been shifted positively by the use of high external concentrations of divalent ions (20 mM[26] and 100 mM[30]).[31]

The heavily damped voltage resonances that we observed upon injecting depolarizing currents in current clamp mode resemble those reported in pigeon vestibular hair cells,[20,32] in type II cells from the guinea pig,[19] and in hair cells from the lagena of the lizard.[33] In hair cells from frog crista[34] and guinea pig cochlea,[35] no electrically resonant behavior occurs at physiologically relevant current levels. The type I cells from which we collected resonance data lacked obvious M-like current. This may explain why their resonances were sharper than those of type I cells from guinea pig semicircular canal organs;[19] the high resting conductance of cells with M-like current could interfere with electrical resonance. Qualitatively, results in vestibular cells are consistent with models that implicate Ca^{2+}-dependent K current in high-quality electrical resonance.[36,37] It is argued that the interplay between the Ca^{2+} current and the Ca^{2+}-dependent K current, in particular the delayed response of the Ca^{2+}-dependent current, is necessary to achieve high Q_e values (>5). While most vestibular cells appear to have some Ca^{2+}-dependent K current, it does not dominate the outward current to the degree that it does in sharply resonant hair cells.[13,28,38–42] The presence of K currents that are not directly linked to Ca^{2+}-current activation, such as A currents and delayed rectifier currents, may prevent high-quality resonance. One might speculate that evolution has acted to enhance electrical tuning in some hair cells by emphasizing the Ca^{2+}-dependent K current, or to suppress electrical tuning in other cells by emphasizing other K currents.

The functional significance of the heavily damped resonances in the rat cells is not clear, because it is not known how much stimulus energy is normally present at the resonant frequencies. CF_e values varied between 28 and 43 Hz in type II cells and between 8 and 130 Hz in type I cells. The electrical resonant frequencies would be higher at mammalian body temperatures; extrapolation from thermal effects on tuning in auditory nerve fibers (reviewed in Manley)[43] suggests that the frequencies might be two- to threefold higher *in vivo*. Because information about the responses of mammalian vestibular afferents to head movements is restricted to lower frequencies (<10 Hz; reviewed in Goldberg and Fernandez),[44] we cannot say whether the electrical resonance observed *in vitro* might contribute to tuning to natural stimuli. It may be more profitable to consider the resonant behavior in the time domain (FIGURE 9) in which it is seen to repolarize the membrane following a perturbation.

ACKNOWLEDGMENTS

We thank Drs. T. Begenisich and J. H. R. Maunsell for helpful comments on the manuscript.

REFERENCES

1. WERSÄLL, J. 1956. Studies on the structure and innervation of the sensory epithelium of the crista ampullares in the guinea pig. Acta Otolaryngol. **126**(Suppl.): 1–85.
2. LEWIS, E. R., E. L. LEVERENZ & W. S. BIALEK. 1985. The Vertebrate Inner Ear. CRC Press, Inc. Boca Raton, Fla.
3. GOLDBERG, J. M. & C. FERNANDEZ. 1971. Physiology of peripheral neurons innervating semicircular canals of the squirrel monkey. III. Variations among units in their discharge properties. J. Neurophysiol. **34**: 676–684.
4. FERNANDEZ, C. & J. M. GOLDBERG. 1976. Physiology of peripheral neurons innervating otolith organs of the squirrel monkey. I. Response to static tilts and to long-duration centrifugal force. J. Neurophysiol. **39**: 970–984.
5. FERNANDEZ, C. & J. M. GOLDBERG. 1976. Physiology of peripheral neurons innervating otolith organs of the squirrel monkey. II. Directional selectivity and force-response relations. J. Neurophysiol. **39**: 985–995.
6. FERNANDEZ, C. & J. M. GOLDBERG. 1976. Physiology of peripheral neurons innervating otolith organs of the squirrel monkey. III. Response dynamics. J. Neurophysiol. **39**: 996–1008.
7. BAIRD, R. A., G. DESMADRYL, C. FERNANDEZ & J. M. GOLDBERG. 1988. The vestibular nerve of the chinchilla. II. Relation between afferent response properties and peripheral innervation patterns in the semicircular canals. J. Neurophysiol. **60**: 182–203.
8. GOLDBERG, J. M., G. DESMADRYL, R. A. BAIRD & C. FERNANDEZ. 1990. The vestibular nerve of the chinchilla. V. Relation between afferent discharge properties and peripheral innervation patterns in the utricular macula. J. Neurophysiol. **63**: 791–804.
9. EATOCK, R. A. 1990. Whole-cell currents in mammalian vestibular hair cells. Soc. Neurosci. Abstr. **16**: 1079.
10. HAMILL, O. P., A. MARTY, E. NEHER, B. SAKMANN & F. J. SIGWORTH. 1981. Improved patch-clamp techniques for high-resolution current recording from cells and cell-free membrane patches. Pflugers Arch. **391**: 85–100.
11. FORSCHER, P. & G. S. OXFORD. 1985. Modulation of calcium channels by norepinephrine in internally dialyzed avian sensory neurons. J. Gen. Physiol. **85**: 743–763.
12. ARMSTRONG, C. M. & F. BEZANILLA. 1974. Charge movement associated with the opening and closing of the activation gates of the Na channels. J. Gen. Physiol. **63**: 533–552.
13. CRAWFORD, A. C. & R. FETTIPLACE. 1981. An electrical tuning mechanism in turtle cochlear hair cells. J. Physiol. **312**: 377–412.
14. WERSÄLL, J. & D. BAGGER-SJÖBÄCK. 1974. Morphology of the vestibular sense organ. In Handbook of Sensory Physiology. Vestibular System. Basic Mechanisms. H. H. Kornhuber, Ed.: 123–170. Springer-Verlag. New York, N.Y.
15. HODGKIN, A. L. & A. F. HUXLEY. 1952. The components of membrane conductance in the giant axon of *Loligo*. J. Physiol. **116**: 473–496.
16. COREY, D. P. & A. J. HUDSPETH. 1983. Analysis of the microphonic potential of the bullfrog's sacculus. J. Neurosci. **3**: 942–961.
17. MEECH, R. W. & N. B. STANDEN. 1975. Potassium activation in *Helix aspersa* under voltage clamp: a component mediated by calcium influx. J. Physiol. **249**: 211–239.
18. ADAMS, P. R., D. A. BROWN & A. CONSTANTI. 1982. M-currents and other potassium currents in bullfrog sympathetic neurons. J. Physiol. **330**: 537–572.
19. RENNIE, K. J. & J. F. ASHMORE. 1991. Ionic currents in isolated vestibular hair cells from the guinea-pig crista ampullaris. Hear. Res. **51**: 279–291.
20. CORREIA, M. J. & D. G. LANG. 1990. An electrophysiological comparison of solitary type I and type II vestibular hair cells. Neurosci. Lett. **116**: 106–111.
21. HODGKIN, A. L. & A. F. HUXLEY. 1952. A quantitative description of membrane current and its application to conduction and excitation in nerve. J. Physiol. **117**: 500–544.
22. BEECH, D. J. & S. BARNES. 1989. Characterization of a voltage-gated K^+ channel that accelerates the rod response to dim light. Neuron **3**: 573–581.
23. HUDSPETH, A. J. & D. P. COREY. 1977. Sensitivity, polarity, and conductance change in the response of vertebrate hair cells to controlled mechanical stimuli. Proc. Natl. Acad. Sci. USA **74**: 2407–2411.

24.	FETTIPLACE, R. & A. C. CRAWFORD. 1978. The coding of sound pressure and frequency in cochlear hair cells of the terrapin. Proc. R. Soc. London Ser. B **203:** 209–218.

25.	RUSSELL, I. J. & P. M. SELLICK. 1983. Low-frequency characteristics of intracellularly recorded receptor potentials in guinea-pig cochlear hair cells. J. Physiol. **338:** 179–206.

26.	LANG, D. G. & M. J. CORREIA. 1989. Studies of solitary semicircular canal hair cells in the adult pigeon. II. Voltage-dependent ionic conductances. J. Neurophysiol. **62:** 935–945.

27.	HUDSPETH, A. J. & R. S. LEWIS. 1988. Kinetic analysis of voltage- and ion-dependent conductances in saccular hair cells of the bull-frog, *Rana catesbeiana.* J. Physiol. **400:** 237–274.

28.	ART, J. J. & R. FETTIPLACE. 1987. Variation of membrane properties in hair cells isolated from the turtle cochlea. J. Physiol. **385:** 207–242.

29.	FUCHS, P. A., M. G. EVANS & B. W. MURROW. 1990. Calcium currents in hair cells isolated from the cochlea of the chick. J. Physiol. **429:** 553–568.

30.	OHMORI, H. 1984. Studies of ionic currents in the isolated vestibular hair cell of the chick. J. Physiol. **350:** 561–581.

31.	HILLE, B. 1984. Ionic Channels of Excitable Membranes: 316–326. Sinauer Associates, Inc. Sunderland, Mass.

32.	CORREIA, M. J., B. N. CHRISTENSEN, L. E. MOORE & D. G. LANG. 1989. Studies of solitary semicircular canal hair cells in the adult pigeon. I. Frequency- and time-domain analysis of active and passive membrane properties. J. Neurophysiol. **62:** 924–945.

33.	EATOCK, R. A. & M. SAEKI. 1991. Electrical resonance in vestibular and auditory hair cells of the alligator lizard. Biophys. J. **59:** 592a.

34.	HOUSLEY, G. D., C. H. NORRIS & P. S. GUTH. 1989. Electrophysiological properties and morphology of hair cells isolated from the semicircular canal of the frog. Hear. Res. **38:** 259–276.

35.	KROS, C. J. & A. C. CRAWFORD. 1990. Potassium currents in inner hair cells isolated from the guinea-pig cochlea. J. Physiol. **421:** 263–291.

36.	ASHMORE, J. F. & D. ATTWELL. 1985. Models for electrical tuning in hair cells. Proc. R. Soc. London Ser. B **226:** 325–344.

37.	HUDSPETH, A. J. & R. S. LEWIS. 1988. A model for electrical resonance and frequency tuning in saccular hair cells of the bull-frog, *Rana catesbeiana.* J. Physiol. **400:** 275–297.

38.	ASHMORE, J. F. 1983. Frequency tuning in a frog vestibular organ. Nature **304:** 536–538.

39.	LEWIS, R. S. & A. J. HUDSPETH. 1983. Frequency tuning and ionic conductances in hair cells of the bullfrog's sacculus. *In* Hearing—Physiological Bases and Psychophysics. R. Klinke & R. Hartmann, Eds.: 17–24. Springer-Verlag. Berlin, Germany.

40.	PITCHFORD, S. & J. F. ASHMORE. 1987. An electrical resonance in hair cells of the amphibian papilla of the frog *Rana temporaria.* Hear. Res. **27:** 75–83.

41.	FUCHS, P. A., T. NAGAI & M. G. EVANS. 1988. Electrical tuning in hair cells isolated from the chick cochlea. J. Neurosci. **8:** 2460–2467.

42.	FUCHS, P. A. & M. G. EVANS. 1988. Voltage oscillations and ionic conductances in hair cells isolated from the alligator cochlea. J. Comp. Physiol. A **164:** 151–163.

43.	MANLEY, G. A. 1990. Peripheral Hearing Mechanisms in Reptiles and Birds. Springer-Verlag. Berlin & Heidelberg.

44.	GOLDBERG, J. M. & C. FERNANDEZ. 1984. The vestibular system. *In* Handbook of Physiology: the Nervous System. I. Darian-Smith, Ed. **3**(Part 2): 977–1022. American Physiological Society. Bethesda, Md.

Computer-Assisted Three-Dimensional Reconstruction and Simulations of Vestibular Macular Neural Connectivities

MURIEL D. ROSS,[a] THOMAS CHIMENTO,[a]
DAVID DOSHAY,[b] AND REI CHENG[b]

[a]National Aeronautics and Space Administration
Ames Research Center
Moffett Field, California 94035

[b]Sterling Software
Palo Alto, California 94303

INTRODUCTION

For nearly three decades, NASA has supported research to learn the influence of gravity on the organization and functioning of the nervous system, and several investigations have been carried out on sensorimotor and vestibular neural adaptation in the microgravity of space. Our research team studies the fundamental properties of vestibular macular organization in earth's normal gravity and the mechanisms whereby the neural circuits adapt to altered gravitational environments. We believe that mammalian gravity receptors are microcosms of the brain, and that knowledge of their basic, functional organization will help us to understand information processing in more complex neural tissues.

The research began with the task of learning whether exposure to the microgravity of space causes morphological changes in vestibular maculas in response to diminution of gravitational bias. In addition to seeking an answer to the fundamental question of the role gravity plays in shaping a gravity-sensing organ, the work was considered a first step in determining whether physiological changes at the periphery might help explain the etiology of space-adaptation syndrome. Interest turned to computer-assisted methods to uncover macular three-dimensional organization when electron microscopic study of several long series of sections of rat maculas failed to demonstrate an independent innervation for the type II hair cell, as had been commonly accepted. Instead, the studies indicated that type I and type II hair cells are integrated into the same neural circuitry and that type II hair cells distribute their output to terminals of more than one nerve fiber.[1] It was subsequently concluded that mammalian macular end organs, like more complex parts of the central nervous system, are morphologically organized for weighted, parallel distributed processing of information.[2–4]

These results and interpretations are contrary to textbook descriptions of macular organization and functioning that, since the advent of ultrastructural studies of maculas,[5,6] have stressed the independence of type I and type II hair cell outputs and their central rather than peripheral integration. Our findings refocused attention on those few previous studies that indicated that the vestibular end organ neural architecture might be more complex than was commonly appreciated. In their ultrastructural investigation, Ades and Engström indicated that type II hair cells synapse with calyces and calyceal processes.[7] This finding is in agreement with a much earlier report by Lorente de Nó in which macular calyceal and nerve branch collaterals ("zarte Kollateralfasern," p. 192) were first illustrated from Golgi preparations of mouse maculas.[8] Lorente de Nó described three major macular regions (a, b, and c) where innervation patterns differed, and he noted the existence of a transverse plexus of nerve fibers previously described by Poljak[9] as of extravestibular origin. Lorente de Nó further suggested that a relationship exists between otoconial load and the kinds of hair cells and nerve fibers in the neuroepithelium below.

Our computer-assisted research has now shown that macular architecture is even more complicated and variable than these earlier investigators realized. The ultimate goals are to learn the three-dimensional architecture of an entire macula, and to increase our knowledge of the relationship between this architecture and neural coding through computer simulations. This will permit detailed comparisons across species, provide a data base for studies of adaptation of the entire organ to novel environments or experimental conditions, and allow predictions of functioning that can then be tested experimentally.

This report will summarize previous results of the reconstruction work and will describe new findings of synaptic connections across the striola. It will also include preliminary results of computer simulations of nerve fiber collateral functioning that is a prelude to a three-dimensional (3-D) simulation of a functioning macular neural network.

MATERIAL AND METHODS

Three-dimensional reconstruction of an entire macula requires computer-automated methods of photography, digitization, and imaging that are under development in our laboratory, but not yet realized. In the meantime, a more labor-intensive approach (described more fully in References 3 and 4) provides information about nerve fiber terminal/receptive fields and small parts of the neural network in maculas of the Sprague-Dawley rat, our model systems for mammalian species. The method involves digitization of objects traced from montages of serial section electron micrographs, assembly of the contours on a personal computer, using software provided by Kinnamon,[10] and reassembly as shaded solid or transparent objects on a Silicon Graphics IRIS high-performance workstation using software developed in our laboratory.[11] In the transparent mode, synapses or other internal features can be shown in their original locations.

Our reconstructions provide the morphological basis for mathematical analysis and computer simulations, to help elucidate aspects of macular dynamics that are presently obscure. A symbolic, 2-D model,[12] based on mathematical interpretations of the activity of only three geometrically different units, did not include connectivities of collaterals, and a 3-D simulation is now under development. To simulate collateral functioning, we use dimensions obtained from photographic negatives and electron micrographs, mathematical approaches to compartmental modeling applied by Segev and Rall to dendritic spines,[13] and software (NEURON) designed by Hines for computer simulations of neural activity.[14]

NEURAL NETWORK CONNECTIVITIES

Terminal/Receptive Fields

Our results indicate that vestibular nerve fibers have terminal fields of one of three general types, M, M/U, and U.[1] The fields are grouped according to the myelination of the parent nerve fiber, but there are related differences in the patterns of distribution of their terminals (see below). Receptive fields are rounded, oblong, or highly elongated, depending upon the geometry of the terminal field. Nevertheless, no two terminal/receptive fields are identical in details and a continuous spectrum of field geometries exists.

M-type nerve fibers dimple the macula so that the basal lamina of the myelin is continuous with that of the neuroepithelium at the base of the calyx. M-type nerve fibers are invariably unbranched, their calyces typically contain several type I cells and lack processes, and their receptive fields are rounded. The myelin of M/U-type nerve fibers terminates at the base of the macula where its basal lamina is continuous with that of the neuroepithelium. The unmyelinated nerve fiber continues intramacularly for a short distance (typically, 1.0–2.0 μm) before ending as a single calyx or, more commonly, by branching to terminate in two calyces. Collaterallike processes are infrequent. The receptive fields of M/U-type terminals may be rounded but usually are oblong. In U-type nerve fibers, myelin terminates well below the basal lamina and processes of macular supporting cells enclose the fibers as they approach the neuroepithelium. The intramacular course of the unmyelinated segment may be long (100 μm or more). U-type nerve fibers typically branch to terminate in two or three calyces and additionally may emit several collaterals from both the nerve branches and their calyces. Many of the collaterals are vesiculated and efferent in type, and similar vesicles are present in the calyceal cytoplasm. However, U-type nerve fibers may terminate in a single calyx, with or without collaterallike processes, so that their receptive fields are rounded, oblong, or highly elongated. There are related differences in the ratios of type I to type II hair cells in the various kinds of terminal/receptive fields, with the most hair cells and highest ratios present in receptive fields of M- and M/U-type nerve fibers.

Topologically, while M-type nerve fibers are confined to the striola and the M/U type to striolar and parastriolar portions, the U type is found everywhere. All three kinds of nerve fiber terminal patterns are integrated in the neural network at sites along the striola. Different combinations of nerve terminal/receptive fields, with varying ratios of type I to type II hair cells, exist elsewhere. These combinations function together in an intact macular network to describe the three-dimensional linear acceleratory space in which the animal lives and functions.

Because otoconial loading also differs from site to site, it is clear that parallel processing of information begins at the otoconial layer and proceeds in parallel across the macula through subarrays of terminal/receptive fields. This arrangement favors segregation of information processing along different pathways for functional purposes as is known to occur in the visual system,[15] and as Lisberger has proposed for vestibuloocular circuits.[16]

Each receptive field of a subarray consists of hair cells with stereociliary tufts differing in directional polarizations. Although the tufts of hair cells of a receptive field have the same general configurations, those of type II cells are shorter, their stereocilia are more slender, and stereocilia of the tufts differ in number, height, and thickness from site to site. The geometry of the tufts is apparently related to their functioning. Fourier analysis of the hexagonal organization of the tufts, taking into account the number of stereocilia and their graded heights, indicates that the stereociliary tufts are organizationally comparable to optimally engineered (fixed) phased array antennas.[17] The results of mathematical analysis of certain tuft morphologies match those obtained experimentally in frog saccular auditory hair cells studied *in vitro*.[18] These tufts show maximal excitatory responses in the direction of the kinocilium, ~20% inhibition in the reverse direction, and no response when the stimulus is broadside.[19] Tufts having other morphologies, when analyzed mathematically, show broader or narrower bandwidths of directional tuning and different asymmetries in excitatory and inhibitory responses.

Further asymmetries in the way type I and type II hair cells process information can be inferred from their differing subcellular organizations and synaptic connectivities. Type I cells have a paucity of synapses with the calyx that contains them, and form junctions only at those sites along the cell membrane where electron-opaque intercellular substance is lacking. One of our studies of 64 type I cells showed an average of 2.3 synapses/type I cell, with the highest number being 5 in that series. By contrast, in 67 type II cells there was an average of 7.2 synapses/cell, with a few cells having 19–24 synapses. Synapses in type II hair cells are frequently squeezed into a narrow compartment between the nucleus and cell membrane. Multiple synapses, in which intervening vesicles are shared by two or more electron-opaque bodies, are more common in type II cells. Most are double or triple, and one quintuple synapse was noted.

Type II hair cells differ further in the number of terminals ending on them and in proportions of terminal types. For example, a study of 88 type II hair cells in one of our series showed that 44% had synaptic junctions only with more highly vesiculated, efferent-type terminals, another 48% synapsed only with calyces and afferent-type terminals, and another 8% synapsed with both afferent- and efferent-type terminals. Although not all terminals are accounted for on some of the cells, these differences are striking. The meaning of apparent variability in type II hair cell connectivities is presently obscure but should be clarified as simulations based on accurate anatomical data are produced.

Type II cells are often inserted into feedforward-feedback calyceal loops, through efferent- and afferent-type calyceal collaterallike processes, and are subject to modulation by efferent boutons. Calyces protect type I cells from external modulation, but efferent-type boutons synapse on all other vestibular neural elements: calyces, collaterallike processes, intramacular nerve fiber branches, and even other boutons. The boutons are nonspecific with respect to termination. For example, a single bouton may (1) synapse asymmetrically with a calyx and also with a nearby type II hair cell opposite a subsynaptic cistern (c synapse),[20] or (2) synapse with a type II hair cell (c synapse) and a nearby nerve fiber branch or collateral (asymmetric synapse). The properties of the subsynaptic, or postsynaptic, site determine bouton action: asymmetric synapses, as those on calyces and nerve fiber branches, should be excitatory while those ending opposite or near subsynaptic cisterns of type II hair cells (c synapses) likely produce cell hyperpolarization[21,22] (see discussion of c synapses below).

The Question of Synapses across the Striola

An important issue arising from the discovery that maculas are functionally organized as neural networks is whether there are two separate networks within each macula, because of the presence of a striola where directional polarization of the hair cells reverses. Reconstructions of hair cells at the striola were obtained from two new series with the specific objective of resolving this question. The two reconstructions could be subdivided into smaller parts to demonstrate that type II cells at the striola typically come into apposition with nearby calyces that enclose type I cells having opposite directional polarizations. Calyceal processes also stretch across the striola. Not all such intimate transstriolar appositions, whether brief or extensive, exhibit synapses, but some do. Two such synapses between type II cells and calyces are illustrated in FIGURE 1 (see arrows), in which the type II cells are rendered in transparent mode. The electron micrograph (FIGURE 2) illustrates one of the synapses (indicated by right arrow in FIGURE 1).

An unresolved question is whether the extracellular compartment between any of the type II hair cells and calyces is ever of the proper dimensions for electrical interactions to take place through potassium accumulation. This would favor hair cell/terminal synchronization should the potassium ion concentration be raised within the compartment for any reason (see Reference 23). The same question can be asked in regard to type II hair cells that lie in close apposition and even interdigitate,[1] and unmyelinated segments of U-type nerves that intertwine and display ruffled borders (see also Reference 24). The advantage of such an electrical relationship is that it permits rapid cell-to-cell interaction in either direction whereas a synapse produces a unidirectional voltage change that must then be integrated spatiotemporally with other inputs to produce a response.

Endoplasmic Reticulum and Its Potential Role in Macular Information Processing

An early hypothesis was that the hair cell endoplasmic reticulum (ER) and its subsurface and subsynaptic cisterns function like sarcoplasmic reticulum in muscle, to distribute voltage changes rapidly to the cell membrane.[25] This hypothesis lacks experimental verification but remains tenable based on our studies of ER distribution in type I and type II vestibular hair cells (unpublished). Recently, however, one specific role of ER in macular information processing has begun to become clear through work on c synapses.

C synapses[20] (and L synapses[26]) consist of large vesiculated boutons that terminate opposite or near flattened subsynaptic cisterns in many different kinds of neurons. The cisterns are assumedly disklike, often are studded with ribosomes on their cytoplasmic side, are bulbous peripherally, and appear to be attached to the neuronal membrane by discontinuous, spotlike, electron-opaque material. Where they have been studied experimentally, terminals ending as c synapses have been shown to be cholinergic, to form typical asymmetric synapses, and to function by releasing calcium from subsynaptic stores.[20,26,27] The result may be either membrane hyperpolarization,[21] by action on K^+ channels, or neurotransmitter release.[28]

Support for the concept that vesiculated boutons hyperpolarize the type II hair cell comes from recent findings of Shigemoto and Ohmori[22] and Ohmori,[29] who demonstrate that extracellularly applied calcium will enter a stimulated hair cell in an *in vitro* preparation of chick vestibular hair cells. The calcium rapidly distributes along the cell membrane in patches (assumedly in ER cisterns), and causes an extremely long hyperpolarization of the hair cell. The role of terminals of vesiculated collaterallike processes in this functioning is presently unknown, but they, too, end opposite cisterns.

Comparable cisterns are present in thinned regions of the calyx, where they are inserted between inner and outer calyceal membranes or between type I cells in close apposition. These cisterns, like those in hair cells, could be involved in intra- or intercell communication.

The Inner Epithelial Plexus

While this report emphasizes peripheral distributions and connectivities of the vestibular nerve, it would be remiss not to include a few words about the inner epithelial plexus.[8,9] Very fine solitary nerve fibers that appear to branch from other

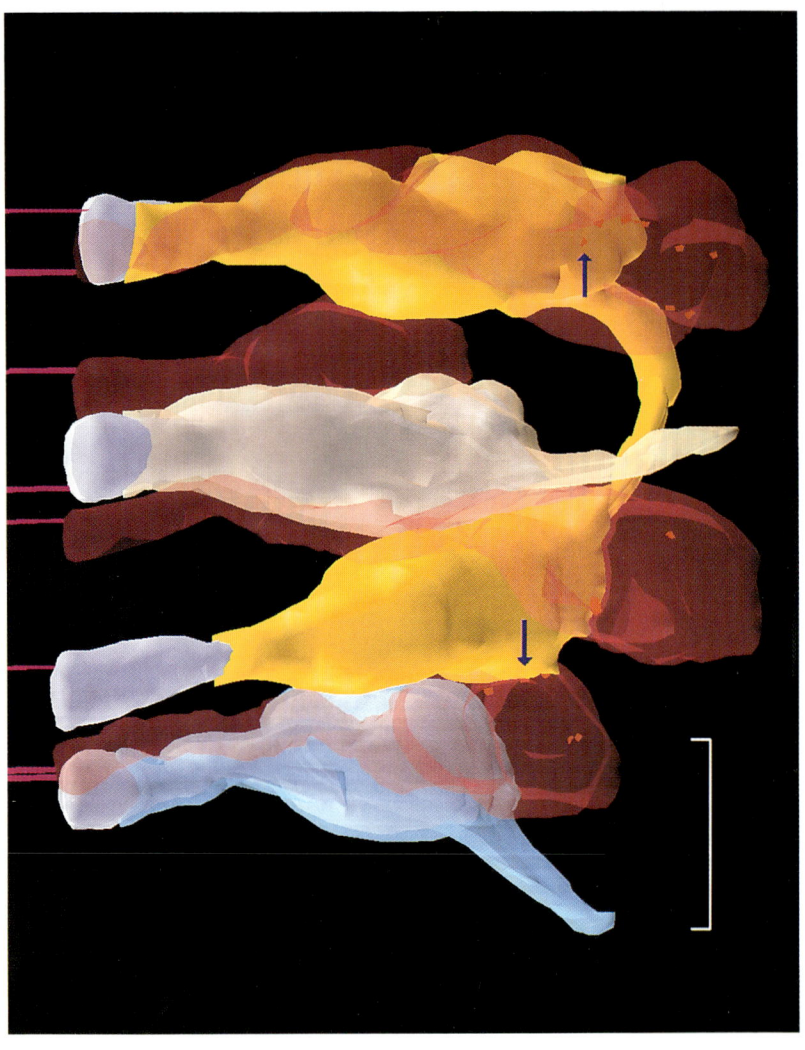

FIGURE 1

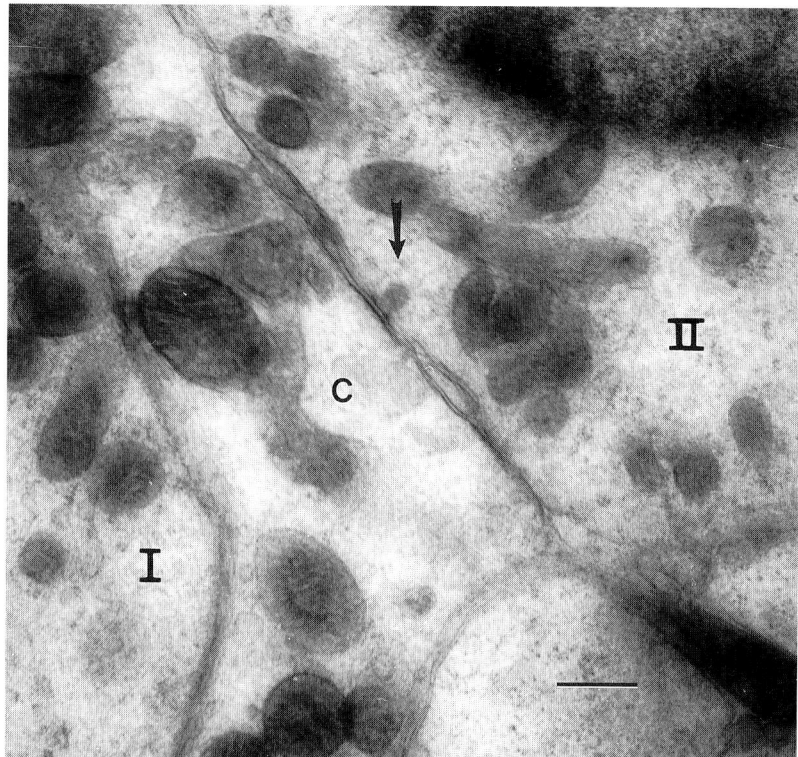

FIGURE 2. This transmission electron micrograph shows the synapse (arrow) between the type II hair cell (II) and the calyx shown in yellow at far right in FIGURE 1. C, calyx; I, type I cell. The bar equals 0.5 μm.

fine fibers below the macula enter the neuroepithelium alongside vestibular nerve fibers.[1] The small diameter fibers, which branch freely within the neuroepithelium, likely correspond to those described as contributing to an inner epithelial plexus in the macula.[9] Further contributions come from vestibular nerve branches, which pass through the web as Lorente de Nó thought,[8] and from efferent-type boutons. Some of the fine fibers end abruptly, shortly after entering the macula; others wander in the

---

FIGURE 1. This computer-assisted reconstruction illustrates the finding that some type II hair cells synapse with calyces on the opposite side of the striola. Calyces (in light blue, at left, in yellow, left center and at the right, and pale yellow, right center) are shown in the transparent mode. The outlines of the type I cells (all shown as solid objects in light lavender) are visible within the calyces. Type II hair cells are all displayed in the transparent mode, in red, so that their synapses (orange-red dots) are visible. Note the opposing orientations of the kinocilia (in purple) of the type II cells and of the type I cells in the calyces with which the type II hair cells synapse across the striola (arrows, right and left). However, other appositions between type II hair cells and calyces containing type I cells with opposite directional polarization lack synapses (type II cell and calyx at left; two type II cells and calyx in light yellow at center right.). The bar equals 10 μm.

neuroepithelium. None form boutons or other specialized terminals. Based on their morphologies, these fibers correspond to general afferents, possibly of facial nerve and geniculate ganglion origin. Many early investigators, including Ramón y Cajal,[30] described a connection from the intermediate nerve of the facial to the vestibular, and Cajal illustrated similar fibers wandering in Scarpa's ganglion (Reference 30, Figure 315, p. 756). While nothing is known of their function, these fibers provide a pathway to nucleus parasolitarius and present us with the tantalizing possibility that they may participate in the etiology of motion sickness and space adaptation syndrome.

MODELS AND SIMULATIONS

The Circuit Model

The level of complexity of innervation displayed by even a relatively simple end organ like the macula may seem bewildering. However, synthesis of the morphologi-

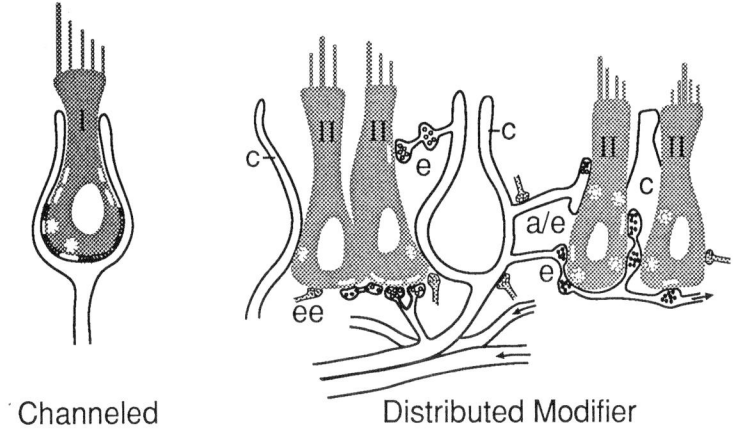

Channeled   Distributed Modifier

FIGURE 3. The macular neural network can be described as consisting of two major circuits: channeled (left) and distributed modifier (right). I, type I cell; II, type II cell; a/e, collaterallike process with mixed afferent and efferent properties; c, calyx; e, efferent-type process; ee, efferent boutons, possibly of extrinsic origin. Starlike symbols indicate ribbon or spherule synapses; bars indicate subsynaptic and subsurface cisterns. The arrows indicate direction of flow of output along the nerve branches toward the impulse initiation zone. Typically, three or more unmyelinated nerve branches come into close apposition during their course through the neuroepithelium, as drawn here.

cal findings permits a simple model of macular organization to emerge (FIGURE 3). There are two major circuits in this model: (1) highly channeled and (2) distributed modifying. Type I hair cell output is highly channeled to the calyx. All remaining neural elements (except for those afferents presumably of intermediate nerve origin) can be assigned to the distributed modifying circuitry since they alter the final output of the nerve fibers. It is possible that timing between the two circuits is essential to produce the coded message.

The above interpretations greatly simplify concepts of macular organization and indicate that the macular neural network is organized similarly to the olfactory system and retina. While the terminology applied here to the two major circuits is new and was selected as more modernly descriptive of the morphology observed, it has a basis in the earlier vertical and horizontal circuits of Shepard[31] and the direct and local circuits of Rakic[32,33] and of Schmitt.[34] In agreement with Shepard, we believe that it is the distributed modifying circuit that is the more adaptive to altered environments and functions in memory and learning.

Computer Simulations

The circuit model described above will be incorporated into our developing 3-D simulation of a functioning macula. Understanding the contribution made by intrinsic collaterals to modulation of nerve fiber output is vital to this effort and is relevant to a broader objective, which is to understand the functional role(s) of distributed modifying circuits in adaptation, memory and learning. The intrinsic collaterals are feedforward (presynaptic to the type II cell) as well as feedback (postsynaptic to the type II cell) with some reciprocal connections. A role of feedforward-feedback circuits might be to influence type II cell output and the timing of type I and type II hair cell inputs to the calyx and thus determine which group of calyces are stimulated sufficiently to send a signal to the next station. Here, potential delays in type II cell processing due to shorter and thinner stereocilia become important. Any delay in response of the type II hair cell will be distributed temporally according to the spatial arrangement and weighting of its synapses. Additionally, the type II hair cell response is subject to continuous modification by collateral activity which could conceivably alter the timing and/or efficacy of type II hair cell input to the calyx. A basic question is whether efferent-type collaterals function to turn down, or off, certain type II hair cells to focus on the dynamics of the type I cells and other type II cells of the group.

Currently, there are no experimental observations on which to base simulations of collateral activity. This is due to the exceedingly small size of the neural elements, the difficulty of identifying them for electrophysiological study in an intact system, and the level of technology currently available. However, previous research has established that much can be learned about the possible functioning of similar but more minute structures, dendritic spines, through compartmental analysis and modeling.[13,35,36] Thus, our knowledge of the dimensions and intrinsic morphology of the collaterals can be used to help interpret their role(s) in macular information processing and to learn more about the relationship between geometry and function. As technology improves, experimental results can be incorporated into the 3-D model.

To test for possible collateral function, we turn to mathematical modeling and computer simulations. The model presently consists of two fundamental components, a nerve branch and collateral process, with the collateral process consisting of a stem and a head. (Processes from calyces require more complicated mathematical treatment[37] for which numerical routines have not yet been implemented.) Simulations are presently biased toward collaterals with short stems, because these are easier to trace completely over several sections and measure accurately.

In TABLE 1, parameters c (capacitance), r_m (membrane resistance), and r_i (cytoplasmic resistance) are calculated from our measurements of collaterallike processes using the following equations:

$$c = C_m A \qquad (1)$$

where $C_m = 1 \ \mu F/cm^2$ and A is head or stem membrane area calculated from our

TABLE 1. Afferents[a]

Head					Stem		
Number of Synapses	Percent of Synapse Area	r_m (MΩ)	c_m (pF)	r_i (MΩ)	r_m (MΩ)	c_m (pF)	r_i (MΩ)
1	0.6%	6.8×10^3	0.73	1.1	1.2×10^5	0.04	1.8
1	6.5%	5.1×10^4	0.10	4.5	7.7×10^5	0.01	11.0
5	1.2%	1.3×10^4	0.39	3.3	7.6×10^4	0.07	2.2
dbl	6.7%	7.1×10^4	0.07	1.0	1.8×10^5	0.03	27.0
1	1.2%	4.5×10^4	0.11	0.63	6.3×10^5	0.008	2.8
trpl	2.3%	1.7×10^4	0.30	0.48	1.3×10^6	0.004	0.35

[a]Dbl and trpl indicate double and triple synapses, in which the two and three central rods share vesicles lying between them. Numbers indicate the total number of discrete synapses along the head of the process. For other parameters, see text.

data.

$$r_m = \frac{R_m}{A} = \frac{1}{g_1} \qquad (2)$$

where $R_m = 5000 \ \Omega cm^2$.

$$r_i = R_i \int_{y_i}^{y_2} \frac{1}{S(y)} \, dy \qquad (3)$$

where $R_i = 70 \ \Omega cm$ and $S(y)$ is the cross-sectional area of the collateral head or stem, calculated from our data, which extends from y_1 to y_2, the starting and ending points of the measured compartment.

Our calculations of membrane resistance, membrane capacitance, and cytoplasmic resistance for actual heads and stems of collaterallike processes taken from our data set are illustrated in TABLE 1. Calculations of surface and cross-sectional areas utilized equations for cylinders, frustrums, and spheres, depending on the geometry of the processes, to calculate electrical properties. Later, these calculations will be carried out by the reconstruction software.

The synaptic area is computed as a disk with a diameter equal to the greatest length of the electron-dense postsynaptic thickening measured in the photographic negatives of the electron micrographs. The percent of synaptic area on each head, given in the table, demonstrates the interesting fact that doubling, tripling, or quintupling the ribbon junction does not necessarily result in a corresponding increment in synaptic area, although it does provide for more vesicles. As the table shows, stem membrane resistance is commonly one or two orders of magnitude greater than resistance in the head. Additionally, a constriction of the stem near the head is common, increasing resistance at that site. It should be noted further that stem cytoplasmic resistance is probably much higher than indicated, because the calculations are based on the premise that the cytoplasm is homogeneous throughout. In fact, many collateral stems have elongated mitochondria within them, nearly filling the cytoplasmic space and assumedly increasing resistance.

Computer simulations of the collaterals utilize a model[14] that requires the frustrumlike or ellipsoidal shapes of biological heads and stems (FIGURE 4) to be treated as a series of stacked cylinders (FIGURE 5), each of which corresponds to an isopotential compartment. For our simulations, the nerve fiber branch is kept standard while stem and head dimensions are taken from negatives of the electron micrographs. The collaterallike process is placed in the middle of a uniform, passive nerve branch (FIGURE 4a and b). The nerve branch is 2.5 μm in diameter, 1118 μm long (2 λ, the electrotonic length constant), and is subdivided into 200 compartments. Stems and heads are subdivided into sections for measurement, with a change in taper rate or an abrupt change in diameter determining borders of the sections. Each section is subdivided into 30 compartments for simulation purposes. The number of compartments used was determined empirically. For example, when

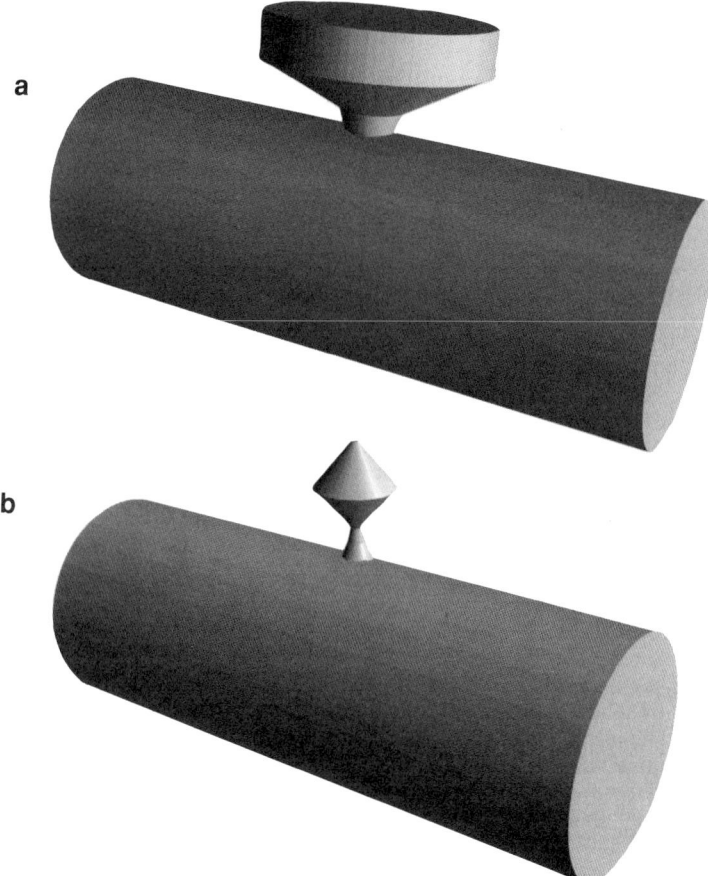

FIGURE 4. **a** and **b** are 3-D renderings of the geometries of two different kinds of processes studied through compartmental modeling. For initial simulations, all processes are considered to emerge from cylindrical nerve fiber branches of fixed dimensions (see text). In **a**, the head of the process traced in a series of electron micrographs is essentially funnel shaped while its stem is cylindrical. The head is described geometrically as consisting of two segments, a cylinder and frustrum, while the stem consists of one cylindrical segment. In **b**, the head of the process resembles the shape of a fingertip and is configured as two segments, both of which are frustrums. The stem of the process has a constriction near the head but flares at its base so that it, too, is considered as a frustrum. Each segment was modeled with 30 cylindrical compartments (see FIGURE 5).

a   b

FIGURE 5. This figure illustrates how a frustrum (left) is treated as a series of cylinders (right) for compartmental modeling. For clarity, only 5 of the 30 compartments modeled are illustrated.

compartment size is greater than 0.01λ for that section, results differ between simulations according to comparative size of the compartments. At 0.01λ or less, no difference is observed. Measurements of simulated voltage are taken at (1) the synaptic site at the distal end of the head; (2) the base of the process; (3) 33.5 μm along the nerve branch; and (4) 145 μm along the nerve branch.

Here we highlight only four simulations of passive current flow in afferent-type collaterals. The gating characteristics of the synapse are identical for all cases. In two examples (processes 11 and 21), with geometries similar to that shown in FIGURE 4b, electrical outputs are virtually identical (FIGURE 6). Process 21 had a slightly longer head but both processes had a stem constriction near the head. Both constrictions were of about equal dimensions [0.12 μm (p 11) and 0.10 μm (p 21)]. In the third example (process 13), with a geometry comparable to that of FIGURE 4a, all voltages are higher (FIGURE 7). The geometry of the head of the collateral is different, and the diameter of the stem is uniform and larger (0.2 μm). The fourth example (process 4) has approximately the same shape as process 13. The surface area and volume of the head are slightly larger than in 13, but the diameter of the stem is approximately doubled (0.42 μm). In this example, the voltages simulated are higher than in all the other cases illustrated (FIGURE 7).

The results of simulations of the four examples shown indicate that a change in diameter, such as a constriction, can dramatically alter voltage at the base of a collaterallike process and along the nerve branch while moderate changes in size or shape of the head are relatively unimportant. The diameter of the stem of a spine or of a collaterallike process is, therefore, a critical factor in information processing as has been discussed previously.[35,38] It would seem possible that the diameter of the stem could be increased or decreased, particularly at constriction sites, or that stem resistance could be altered by inserting or removing a mitochondrion, to change the voltage in a neural circuit undergoing adaptation.

SUMMARY AND CONCLUSIONS

The macular neuroepithelium is morphologically organized as a weighted neural network for parallel distributed processing of information. The network is continuous across the striola, where some type II hair cells synapse with calyces containing type I cells with tufts of opposite directional polarities. Whether other hair cell to calyx appositions that lack synapses interact because of intercellular potassium accumulation remains an open question. A functionally important inference of macular organization is that just as arrays of hair cells communicate an entire piece of information to a nerve fiber, so do macular subarrays of nerve fibers (not single units) carry the whole coded message to the brain stem. Moreover, the size of the network subarray can expand or become more limited depending upon the strength and/or duration of the input. It is the functioning of the network and its subarrays that must be understood if we are to learn how maculas carry out their work and adapt to new environments.

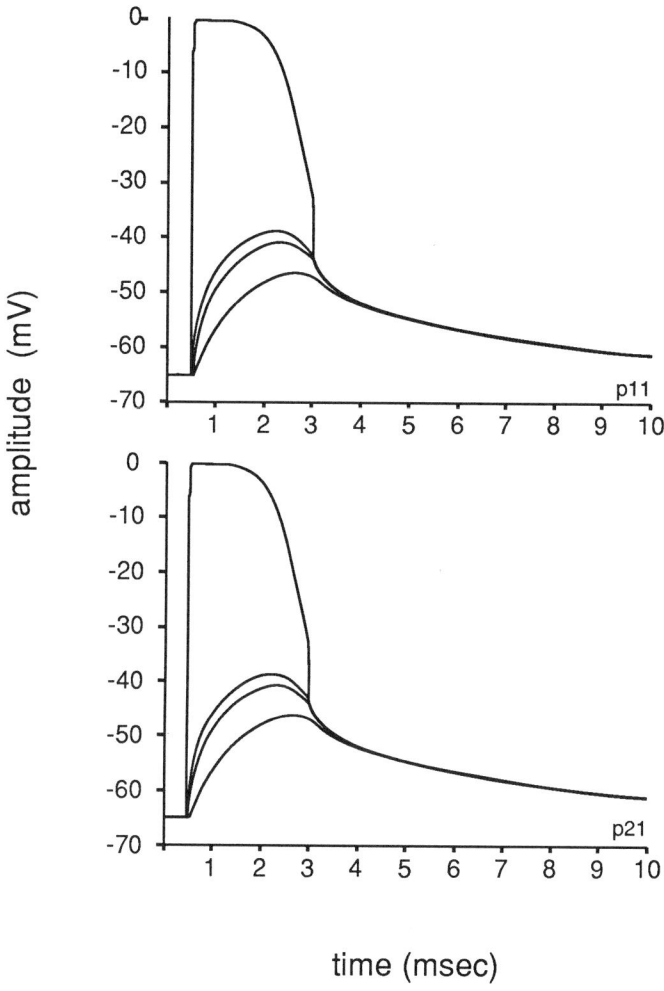

FIGURE 6. Two collaterallike processes (p 11, upper graph and p 21, lower graph) with frustrumlike segments. The two heads differed in length but the processes had nearly identical stem diameters (0.12 μm, p 11 and 0.10 μm, p 21). The processes, considered as having passive electrical properties and as emerging from a passively conducting nerve fiber branch, were treated as a series of stacked cylinders for modeling purposes (see FIGURE 5 and text). The graphs illustrate simulated voltage versus time for (1) the most distal end of the process (upper trace); (2) the base of the process (second trace); (3) a point 33.5 μm along the nerve fiber (third trace); and (4) a point 145 μm along the nerve fiber. The two simulations are similar, showing the importance of stem diameter in determining the voltage delivered to the base of the process and the timing of its arrival.

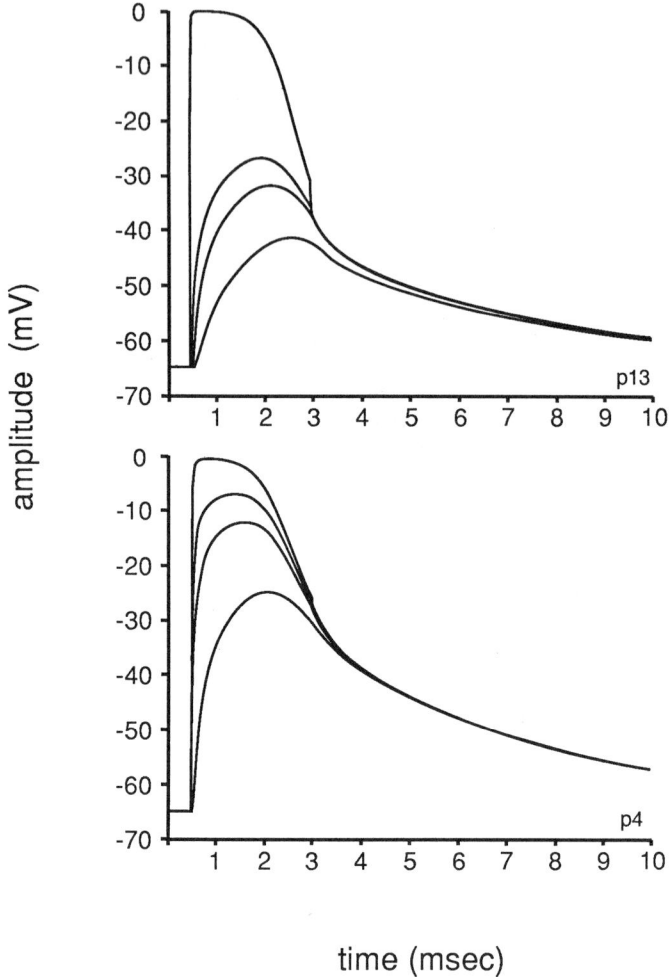

FIGURE 7. The two collaterallike processes whose simulated voltage versus time is graphed here had a different geometry from those presented in FIGURE 6. Their heads were more funnel shaped, with cylindrical and frustrum segments. The stems were cylindrical but differed in diameter (0.2 μm for p 13, and 0.42 μm for p 4). Both graphs show greater voltage delivered to the base with shorter intervals to peak for these two processes compared to those shown in FIGURE 6. The shortest time to peak and the highest voltage at the base occurred in the process with the largest stem diameter (p 4).

Simulations of functioning maculas, or subparts, based on precise morphology and known physiology are useful tools to gain insights into macular information processing. The current simulations of afferent collateral electrical activity are a prelude to development of a 3-D model. The simulations demonstrate a relationship between geometry and function, with the diameter of the stem apparently being a major determinant of electrical activity transmitted to the base in the case of collaterals with short stems. Thus, while changes in synaptic number and/or size may be an important adaptive mechanism in an altered g environment, changes in diameter of the stem is another means of altering outflow. Research on the effects of microgravity should be extremely useful in examining the validity of this and other concepts of neural adaptation, since maculas are biological linear accelerometers ideally suited to the task.

Maculas are also extremely interesting to study in detail because of the richness of connectivities and submicroscopic organization they present. Many of their features are common with more complex parts of the brain. It seems possible that knowledge of the three-dimensional geometric relationships operative in a functioning macula will contribute much to the understanding of the dynamics underlying more complex behavior. Computerized approaches greatly facilitate this task and provide an objective method of analysis. It is likely that, in the end, simple rules will be found to govern optimal neural architectural organization, even at higher cognitive levels. The architecture only appears complex because we do not yet grasp its meaning. Once the rules are learned, we shall doubtless say, as did Wheeler on another occasion in the realm of physics, "Oh, how beautiful. How could it have been otherwise?"[39]

ACKNOWLEDGMENTS

Dr. Thomas Chimento is a National Research Council Research Associate at NASA-Ames Research Center. We thank Dr. Michael Hines for his gracious help with our implementation of Neuron.

REFERENCES

1. Ross, M. D., C. M. Rogers & K. M. Donovan. 1986. Innervation patterns in rat saccular macula. Acta Otolaryngol. Stockh. **102:** 75–86.
2. Ross, M. D. 1988. Morphological evidence for parallel processing of information in rat macula. Acta Otolaryngol. Stockh. **106:** 213–218.
3. Ross, M. D., L. Cutler, G. Meyer, T. Lam & P. Vaziri. 1990. 3-D components of a biological neural network visualized in computer generated imagery. I. Receptive field organization. Acta Otolaryngol. Stockh. **109:** 83–92.
4. Ross, M. D., G. Meyer, T. Lam, L. Cutler & P. Vaziri. 1990. 3-D components of a biological neural network visualized in computer generated imagery. II. Macular neural network organization. Acta Otolaryngol. Stockh. **109:** 235–244.
5. Wersäll, J. 1956. Studies on the structure and innervation of the sensory epithelium of the cristae ampullaris in the guinea pig. A light and electron microscopic investigation. Acta Otolaryngol. Stockh. Suppl. 126: 85.
6. Smith, C. 1956. Microscopic structure of the utricle. Ann. Otol. **65:** 450–469.
7. Ades, H. W. & H. Engström. 1965. Form and innervation of the vestibular epithelia. *In* The Role of the Vestibular Organs in the Exploration of Space. A. Graybiel, Ed.: 23–41. NASA. Washington, D.C.
8. Lorente de Nó, R. 1931. Ausgewahlte Kapitel aus der vergleichenden Physiologie des Labyrinthes. Ergebn. Physiol. **32:** 73–242.
9. Polyak, S. L. (As quoted in Reference 8.)
10. Kinnamon, J. C., S. J. Young & T. A. Sherman. 1986. 3DEd, Recon. and Show. Laboratory for High Voltage Electron Microscopy, University of Colorado. Boulder, Colo.
11. Ross, M. D., R. Cheng & T. Lam. 1990. ROSS 2.3 version of CARMA, Computer-Aided Reconstruction of Macular Accelerometers.

12. Ross, M. D., J. Dayhoff & D. H. Mugler. 1990. Toward modeling a dynamic neural network. Math. Comput. Modelling **13:** 97–106.

13. Segev, I. & W. Rall. 1988. Computational study of an excitable dendritic spine. J. Neurophysiol. **60:** 499–523.

14. Hines, M. 1989. A program for simulation of nerve equations with branching geometries. Int. J. Bio-Med. Comput. **24:** 55–68.

15. Livingston, M. & D. Hubel. 1988. Segregation of form, color, movement, and depth: anatomy, physiology and perception. Science **240:** 470–490.

16. Lisberger, S. 1986. Properties of pathways subserving long-term adaptive plasticity of the vestibulo-ocular reflex in monkeys. *In* The Biology of Change in Otolaryngology. R. J. Ruben, T. R. Van Der Water & E. W. Rubel, Eds.: 171. Elsevier. San Francisco, Calif.

17. Mugler, D. & M. D. Ross. 1990. Vestibular receptor cells and signal detection: bioaccelerometers and the hexagonal sampling of 2-D signals. Math. Compat. Modelling **123:** 85–92.

18. Shotwell, S. L., R. Jacobs & A. J. Hudspeth. 1981. Directional sensitivity of individual vertebrate hair cells to controlled deflection of their hair bundles. Ann. N.Y. Acad. Sci. **374:** 1–10.

19. Mugler, D. H. & M. D. Ross. The directional sensitivity of vestibular hair cells based on planar phased array design. Submitted.

20. Conradi, S. 1969. Ultrastructure and distribution of neuronal and glial elements on the motoneuron surface in the lumbosacral spinal cord of the adult cat. Acta Physiol. Scand. **332:** 85–111.

21. Fujimoto, S. D., K. Yamamoto, K. Kuba, K. Morita & E. Kato. 1980. Calcium localization in the sympathetic ganglion of the bullfrog and effects of caffeine. Brain Res. **202:** 21–32.

22. Shigemoto, T. & H. J. Ohmori. 1990. Muscarinic agonists and ATP increase the intracellular Ca^{++} concentration in chick cochlear hair cells. J. Physiol. London **420:** 127–148.

23. McBain, C. J., S. F. Traynelis & R. Dingledine. 1990. Regional variations of extracellular space in the hippocampus. Science **249:** 674–677.

24. Scheibel, M. E. & A. B. Scheibel. 1970. Organization of spinal motoneuron dendrites in bundles. Exp. Neurol. **28:** 106–112.

25. Rosenbluth, J. 1965. Subsurface cisterns and their relationship to the neuronal plasma membrane. J. Cell Biol. **13:** 405–421.

26. Bodian, D. 1966. Synaptic types on spinal motoneurons: an electron microscope study. Bull. Johns Hopkins Hosp. **119:** 16–45.

27. Connaughton, M., J. V. Priestley, M. V. Sofroniew, F. Eckenstein & A. C. Cuello. 1986. Inputs to motoneurones in the hypoglossal nucleus of the rat: light and immunocytochemistry for choline acetyltransferase, substance P and enkephalins using monoclonal antibodies. Neuroscience **17:** 205–224.

28. Iverson, L. L. 1979. Neurotransmitter interactions in the substantia nigra: a model for local circuit chemical interactions. *In* The Neurosciences: Fourth Study Program. F. O. Schmitt & F. G. Worde, Eds. 1085–1092. MIT Press. Cambridge, Mass.

29. Ohmori, H. 1990. Mechano-electrical transduction in hair cell. Presented at the Sixteenth Barany Society Meeting, Tokyo, Japan.

30. Ramón y Cajal, S. 1952. Histologie du Systeme Nerveux de l'Homme et des Vertebres, Institutio Ramon y Cajal. Madrid, Spain. (Translated from Spanish by L. Azoulay.)

31. Shepard, G. M. 1970. The olfactory bulb as a simple cortical system: experimental analysis and functional implications. *In* The Neurosciences: Second Study Program. F. O. Schmitt, Ed.: 539–552. Rockefeller Press. New York, N.Y.

32. Rakic, P. 1975. Local circuit neurons. Neurosci. Res. Program Bull. **13:** 291–446.

33. Rakic, P. 1979. Genetic and epigenetic determinants of local neuronal circuitry in the mammalian nervous system. *In* The Neurosciences: Fourth Study Program. F. O. Schmitt & F. Worden, Eds.: 109–127. MIT Press. Cambridge, Mass.

34. Schmitt, F. O. 1979. The role of structural, electrical and chemical circuitry in brain function. *In* The Neurosciences: Fourth Study Program. F. O. Schmitt & F. G. Worden, Eds.: 5–20. MIT Press. Cambridge, Mass.

35. KOCH, C. & T. POGGIO. 1983. A theoretical analysis of electrical properties of spines. Proc. R. Soc. London Ser. B Biol. Sci. **218:** 455–477.
36. GAMBLE, E. & C. KOCH. 1987. The dynamics of free calcium in dendritic spines in response to synaptic input. Science. **236:** 1311–1315.
37. JACK, J. J. B., D. NOBLE & R. W. TSIEN. 1983. Electric Current Flow in Excitable Cells. Clarendon Press. Oxford, England.
38. RALL, W. 1981. Functional aspects of neuronal geometry. *In* Neurones without Impulses. A. Roberts & B. M. Bush, Eds.: 223–254. Cambridge University Press. Cambridge, England.
39. WHEELER, J. A. (As quoted in Ferris, T. 1989. Coming of Age in the Milky Way: 346. Doubleday. New York, N.Y.)

Structure and Function of Vestibular Nerve Fibers in the Chinchilla and Squirrel Monkey[a]

JAY M. GOLDBERG,[b] ANNA LYSAKOWSKI,[b]
AND CÉSAR FERNÁNDEZ[c]

[b]Department of Pharmacological and Physiological Sciences
[c]Department of Surgery (Otolaryngology–Head & Neck Surgery)
University of Chicago
Chicago, Illinois 60637

INTRODUCTION

The afferent innervation patterns in the vestibular organs of mammals were investigated in silver-stained material by early anatomists.[1,2] Recently, newer axonal labeling techniques have been used to reinvestigate the problem and to relate the physiology of afferents with the morphology of their peripheral terminations. A major purpose of this paper is to summarize the more recent work, which was done in the cristae[3,4] and utricular macula of chinchilla.[5,6] Two other topics will also be considered. First, the axonal labeling studies raised questions concerning the synaptic organization of the end organs and these are being addressed in a quantitative ultrastructural study of the chinchilla cristae.[7] Second, axonal labeling methods have now been applied to the squirrel monkey.[8,9]

AFFERENT INNERVATION PATTERNS IN THE CHINCHILLA

The afferent innervation has been surveyed by the extracellular injections of horseradish peroxidase (HRP) into the vestibular nerve, which labels several fibers as far as their peripheral terminals.[3,5] In describing results for the cristae, it has been useful to divide the neuroepithelium into central, intermediate, and peripheral zones of equal areas (FIGURE 1A, inset). Similarly, the utricular macula can be divided into a striola, a juxtastriola, and an extrastriola (FIGURE 1B, inset). Three classes of afferents can be recognized in both the cristae and the macula (FIGURE 1A and B). *Calyx units* are thick fibers seen in central (striolar) zones. The axon is usually unbranched, giving rise to a single calyx ending surrounding almost the entire basolateral surfaces of 1–3 neighboring, flask-shaped type I hair cells. *Bouton units,* found in peripheral (extrastriolar) zones, have thin axons, whose fine collateral branches provide bouton endings to several cylindrically shaped type II hair cells. *Dimorphic units* innervate all parts of the neuroepithelium and supply both kinds of hair cells. The medium-sized axons of dimorphs give rise to one or more relatively thick collateral branches terminating as calyx endings. In addition, bouton endings

[a]The authors' research is supported by National Institute on Deafness and Other Communication Disorders Grant DC 00070, NASA Grant NAG 2-148, and Office of Naval Research Contract 00014-88-k-0381. A. Lysakowski was a NASA Research Associate.

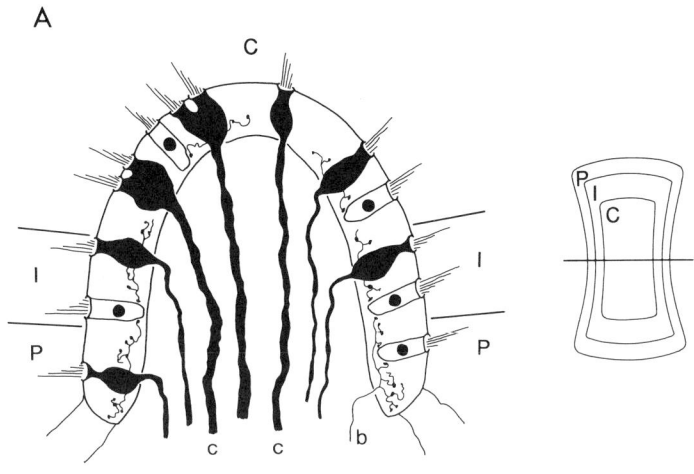

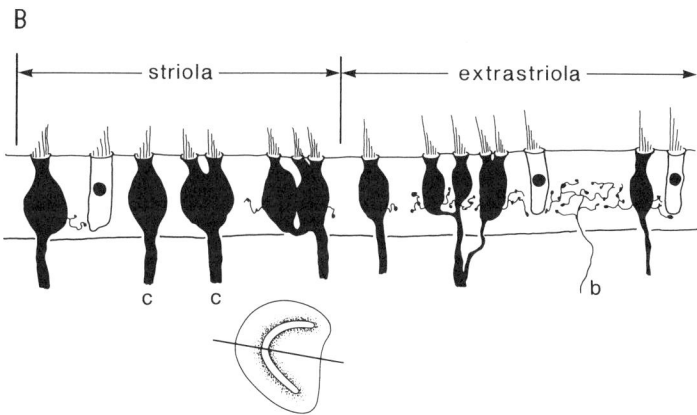

FIGURE 1. Afferent innervation patterns in the crista ampullaris (A) and utricular macular (B). The crista is divided into central (C), intermediate (I), and peripheral zones (P) (see inset to A). The macula is divided into three zones (see inset to B): a narrow, ribbon-shaped striola, a juxtastriola (indicated by stippling in the inset) immediately surrounding the striola, and a broad extrastriola. The section planes of the main figures are indicated in the insets. Fibers include bouton (b), calyx (c), and dimorphic (unlabeled) fibers.

are present on thin collaterals that emerge from the parent axon, the thick branches, or the calyx endings. In much of the recent ultrastructural literature,[10] dimorphic units have been considered as rare. They are, in fact, the most numerous kind of afferent, making up 70% of the fibers going to the cristae and more than 80% of those supplying the utricular macula. To understand the contribution of dimorphic fibers, it has to be remembered that type I and type II hair cells are present throughout the neuroepithelium.[11] Given the restricted distribution of calyx and bouton units, dimorphs must provide the main source of afferents to type II hair cells in central (striolar) zones, to type I hair cells in peripheral (extrastriolar) zones, and to both kinds of hair cells in intermediate (juxtastriolar) zones.

The morphological observations can be used to estimate the number of calyx, dimorphic, and bouton units in each zone of the neuroepithelium. A so-called

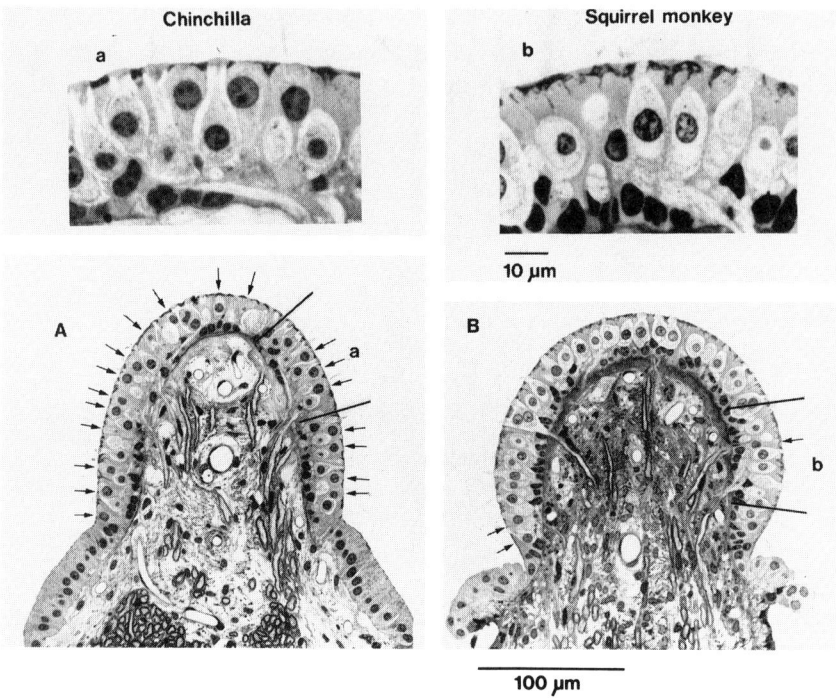

FIGURE 2. Transverse sections through the middle of the horizontal cristae in a chinchilla (A) and a squirrel monkey (B). In the neuroepithelium, supporting cells are located just above the basement membrane and are characterized by irregularly shaped nuclei. The insets (a and b), which are magnifications of the parts of the main figures delimited by lines, show that the two kinds of hair cells are easily distinguished because type I hair cells are surrounded by calyx endings, whereas type II hair cells are not. All type II hair cells are indicated in the main figures by arrows. Of the 35 hair cells with nuclei in each section, 19 type II hair cells are found in the chinchilla section and only three such hair cells in the squirrel monkey section. Plastic-embedded 2-μm sections; Richardson's stain (1% azure II, 2% methylene blue).

afferent reconstruction has been done for both the cristae[3] and the utricular macula.[5] Conclusions are summarized here for the chinchilla cristae. Counts of the numbers of type I and type II hair cells in the various zones are made from sections where the two kinds of hair cells are easily distinguished (FIGURE 2a). Results are summarized in TABLE 1 (upper). Confirming previous counts obtained from surface preparations of the guinea pig cristae,[11] it is found that there are approximately equal numbers of type I and type II hair cells in each of the three zones. A sample of HRP-labeled

TABLE 1. Type I and Type II Hair-Cell Counts, Chinchilla and Squirrel Monkey Cristae[a]

Zone	Type I	Type II	Both	Ratio
Chinchilla				
Central	590 ± 40	500 ± 90	1090 ± 90	1.2 ± 0.3
Intermediate	1310 ± 30	980 ± 100	2290 ± 120	1.3 ± 0.1
Peripheral	1220 ± 220	1270 ± 220	2490 ± 440	1.0 ± 0.1
All zones	3120 ± 170	2750 ± 200	5880 ± 350	1.1 ± 0.1
Squirrel monkey				
Central	1350 ± 90	260 ± 10	1610 ± 80	5.2 ± 0.6
Intermediate	2010 ± 160	460 ± 170	2470 ± 200	4.3 ± 0.9
Peripheral	1310 ± 160	880 ± 100	2200 ± 230	1.5 ± 0.2
All zones	4660 ± 270	1610 ± 130	6270 ± 310	2.9 ± 0.3

[a]All values, mean ± standard error (SE). Based on counts in three cristae (chinchilla) and seven cristae (squirrel monkey).

afferents is used to estimate the average number of type I hair cells innervated by individual calyx and dimorphic units and the relative proportions of the three unit classes.[3,5] Combining these estimates for each zone with the hair-cell counts of TABLE 1 (upper) gives the numbers in TABLE 2 (upper). The total number of afferents innervating a crista is predicted to be 2500, close to the actual count of 2200 fibers.[12] Three additional points can be made: (1) Of the entire population, 70% are dimorphic units, 20% are bouton units, and 10% are calyx units. (2) A considerably

TABLE 2. Reconstruction of Afferent Innervation, Chinchilla and Squirrel Monkey Cristae[a]

	Zones of Crista			All Zones
	Peripheral	Intermediate	Central	(Percent Total)
Chinchilla				
Number of afferent fibers				
Calyx	0	50	170	220 (9%)
Dimorphic	710	900	180	1,790 (71%)
Bouton	420	100	0	520 (21%)
All (% total)	1,130 (45%)	1,050 (42%)	350 (14%)	2,530 (100%)
Number of bouton end-				
ings	31,000	22,100	2,500	55,600
Bouton endings per type				
II hair cell	24.4	22.6	4.9	20.2
Squirrel monkey				
Number of afferent fibers				
Calyx	0	270	510	780 (27%)
Dimorphic	700	800	240	1,740 (60%)
Bouton	310	60	0	370 (13%)
All (% Total)	1,010 (35%)	1,130 (39%)	750 (26%)	2,890 (100%)
Number of bouton end-				
ings	36,600	12,600	1,400	50,500
Bouton endings per type				
II hair cell	41.5	27.4	5.3	31.4

[a]Percentages are of total reconstructed afferent sample.

larger number of afferents innervate the peripheral zone, as compared to the central zone. The proportions are in line with the hair-cell counts for the two zones. (3) There should be a substantial difference in the number of afferent boutons innervating type II hair cells in the central and peripheral zones.

AFFERENT RESPONSE PROPERTIES

Afferents, including those innervating semicircular canals[4,13] and otolith organs,[6,14] differ in their physiological properties. Some units have a regular spacing of action potentials. In other units, the spacing is irregular. Fibers first classified as regularly or irregularly discharging differ in several other respects as well (TABLE 3).

Which of these other properties are causally related to discharge regularity, rather than being merely correlated with it? To address this issue, we need to consider the physiology of the afferent terminal where spike discharge is encoded. Terminals differ in their postspike afterhyperpolarizations (AHPs)[15] and, consequently, in their postspike recovery of excitability.[16] AHPs are a major determinant of both discharge regularity and the encoder's sensitivity to synaptic and other depolarizing inputs; the faster the AHP, the more irregular is the discharge and the more sensitive is the encoder.[17] The enhanced encoder sensitivity of irregular afferents contributes to their relatively high sensitivities to head rotations and linear forces, to electrical activation of efferent pathways, and to externally applied galvanic currents (TABLE 3).

Regular and irregular afferents also differ in the dynamics of their response to natural stimulation. Regular afferents may be termed tonic receptors as their discharge parallels the expected macromechanical displacement of the cupula or otolithic membrane. Over a frequency range encompassing normally occurring head movements (0.5–5 Hz), the firing of regular semicircular-canal fibers is in phase with angular head velocity[4,13] and the firing of regular otolith afferents is in phase with applied linear force.[6,14] Irregular afferents are more phasic. In the same frequency range, they show phase leads and gain enhancements, indicating that their discharge reflects the velocity of cupular or otolithic displacement, as well as the displacement

TABLE 3. Comparison of Discharge Characteristics, Regularly and Irregularly Discharging Mammalian Vestibular Nerve Fibers

Regularly Discharging	Irregularly Discharging
Medium-sized and thin axons, ending as dimorphic and bouton units in intermediate (juxtastriolar) and peripheral (extrastriolar) zones of the neuroepithelium.	Thick and medium-sized axons, ending as calyx and dimorphic units in central (striolar) and intermediate (juxtastriolar) zones of the neuroepithelium.
Tonic response dynamics, resembling the expected displacement of the cupula or otolithic membrane.	Phasic-tonic response dynamics, including a sensitivity to the velocity of cupular or otolithic displacement.
Low sensitivity to head rotations or linear forces.	High sensitivity to head rotations or linear forces. (Calyx units in the cristae have relatively low sensitivities.)
Small excitatory responses to efferent activation, usually including only slow response components.	Large excitatory responses to efferent activation, usually including both fast and slow response components.
Small responses to externally applied galvanic currents.	Large responses to externally applied galvanic currents.

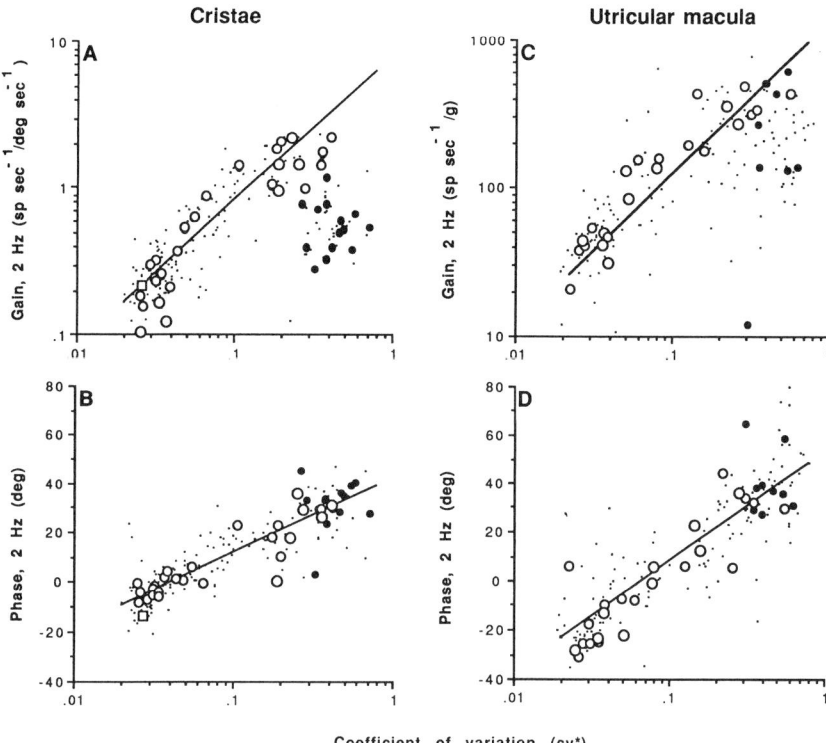

FIGURE 3. Gains and phases of responses to 2-Hz sinusoidal stimulation are plotted against the cv*, a normalized coefficient of variation. Data are from vestibular nerve fibers in the chinchilla. **A** and **B**: Semicircular canal units, 40°/second head rotations. **C** and **D**: Utricular units, 0.05g centrifugal forces. Gains and phases are taken with respect to angular head velocity (A and B) or linear force (C and D); phase leads are positive. Small symbols, extracellularly recorded units. Large symbols, intraaxonally labeled units, including calyx (filled circles), dimorphic (open circles), and a bouton unit (square). The lines are power-law regressions between gain and cv* for noncalyx units (A and C) and semilogarithmic relations between phase and cv* for all units (B and D). Modified from References 4 and 6.

itself. The relation between discharge regularity and response dynamics is illustrated in FIGURE 3B and D. The former is measured by a normalized coefficient of variations (cv*); the latter, by the phase lead occurring during 2 Hz sinusoidal stimulation. For both canal (FIGURE 3B) and otolith units (FIGURE 3D), the phase lead increases the higher is the cv* and, hence, the more irregular is the discharge.

How do the phasic response dynamics of irregular afferents arise? There is no evidence for a regional difference in the response dynamics of cupular motion.[18] Nor is there a difference in response dynamics when the spike encoders in afferent terminals are directly activated by galvanic currents.[19] From this last finding, it may be concluded that discharge regularity and response dynamics are causally unrelated. Several distinct stages in the transduction process are interposed between macromechanics and spike encoding. Any of a number of these could be involved in shaping the afferents' response dynamics.

RELATION BETWEEN AFFERENT MORPHOLOGY AND PHYSIOLOGY

Do afferents with distinctive discharge properties differ in their branching patterns and in the kinds of hair cells they innervate? The question has recently been studied by intraaxonal labeling of physiologically characterized afferents in the cristae[4] and utricular macula[6] of the chinchilla. Results are illustrated in FIGURE 3. Calyx units, which supply central (striolar) zones, are the most irregularly discharging and most phasic afferents found in either kind of end organ. The discharge properties of dimorphic units depend on the epithelial region they innervate (FIGURE 4). Central (striolar) dimorphs are irregularly discharging and phasic. Peripheral (extrastriolar) dimorphs are regularly discharging and tonic. Dimorphs innervating the two zones differ in their physiology, even in cases where afferent innervation patterns are similar.[4,6] Only one physiologically characterized bouton unit was labeled, possibly because such units have thin axons and are difficult to impale. The labeled bouton unit supplied the peripheral zone of a crista and was regularly discharging and tonic. To obtain additional information on bouton units, extracellular recordings were done in the squirrel monkey and these units were identified by their slow conduction velocities (see below).

The results, particularly those relating to dimorphic units, emphasize the importance of regional differences in determining afferent physiology. Presumably, there are regional variations in transduction mechanisms that are shared by type I and type II hair cells. Is there anything distinctive about the type I hair cell? Attention naturally turns to calyx afferents, since they get most of their input from such hair cells. In the mammalian cristae, the one feature that distinguishes calyx units from irregularly discharging dimorphs is their relatively low gains to head rotations (FIGURE 3A). As both types of units contact similar numbers of type I hair cells, it can be supposed that the higher gains of dimorphs are a result of the additional inputs they receive from type II hair cells. The difference in gains can be used to estimate that, on average, a simple calyx ending gets three times the synaptic input of an afferent bouton.[4] A similar conclusion is possible in the utricular macula even though linear-force gains can be similar for irregular dimorphs and for calyx units (FIGURE 3C) (see Reference 6 for details). A 3:1 difference in the effective synaptic weights of calyx and bouton endings can be compared to a 100:1 difference in the appositional contacts that each of the endings makes with its hair cell. Viewed in this light, the calyx ending has a surprisingly low synaptic gain.

REGIONAL VARIATIONS IN SYNAPTIC ULTRASTRUCTURE

Our extracellular HRP studies[3,5] confirm the results of earlier anatomists[1,2] that there are differences in the innervation of central (striolar) and peripheral (extrastriolar) zones. Furthermore, units supplying the various zones differ in their physiological properties.[4,6] This has led us to investigate the possibility that there are regional differences in synaptic organization. The work was prompted by four considerations. (1) An afferent reconstruction of the chinchilla crista (TABLE 2, upper) predicts that there are large differences in the number of afferent boutons innervating type II hair cells in the various zones. Such a difference had not been described previously. (2) Morphophysiological results lead to the suggestion that the synaptic input to a simple calyx ending is about three times larger than that to a bouton ending. It is possible that this might be related to the number of ribbon synapses made onto the two kinds of endings. Unfortunately, there is a serious discrepancy in the literature. Favre and Sans[20] and Gulley and Bagger-Sjöbäck[21] have found that type I hair cells make few synapses with calyx endings, whereas Hamilton has observed as many as 10 synapses per hair cell in serial-sectioned material.[22] (3) In interpreting our morphophysiological data, we have assumed that afferent synaptic transmission only occurs between type I hair cells and calyx endings and between type II hair cells and bouton endings, arrangements that can be evaluated by light microscopy. Two synaptic configura

Discharge regularity

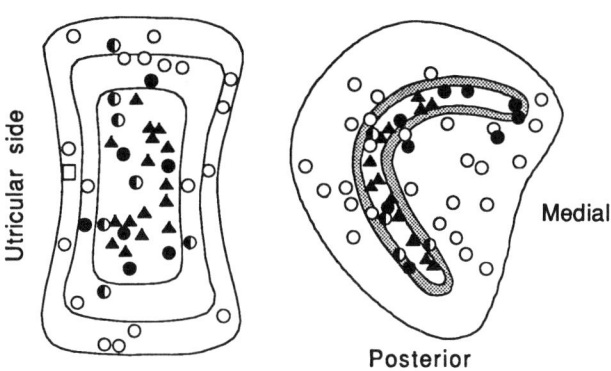

Response dynamics

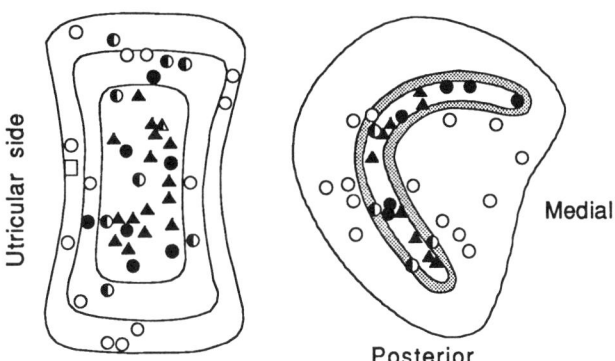

FIGURE 4. Locations of intraaxonally labeled units are indicated on standard surface maps of the cristae (left) and the utricular macula (right), each divided into three zones as in FIGURE 1. Calyx units (triangles), dimorphic units (circles), and a bouton unit (square). **Top:** Units are sorted by their normalized coefficients of variation (cv*) as regular (cv* < 0.10, unfilled symbols), intermediate (cv* between 0.10–0.20, half-filled symbols), or irregular (cv* > 0.20, filled symbols). **Bottom:** Units are sorted by the phase leads ($\phi_{2\,Hz}$) of their responses to 2-Hz sinusoidal head rotations or linear forces as tonic ($\phi_{2\,Hz}$ < 0°, unfilled symbols), intermediate ($\phi_{2\,Hz}$ between 10 and 20°, half-filled symbols), or phasic ($\phi_{2\,Hz}$ > 20°, filled symbols). Based on data from References 4 and 6.

tions, type II synapses onto the outer face of calyx endings[23-25] and reciprocal synapses,[25,26] can only be appreciated at an ultrastructural level. (4) Responses to electrical activation of efferent fibers differ in regularly and irregularly discharging afferents (TABLE 3).[27] We now know that irregular units are concentrated in the central zone and regular units in the peripheral zone.[4,6] It is therefore of interest to compare the efferent innervation in the two zones.

An ultrastructural study is being done in the chinchilla to investigate these questions, and preliminary data have been presented.[7] Conclusions are based on blocks taken from four cristae, two of them sectioned transverse to the long axis of the organ and the other two sectioned longitudinally. Findings are illustrated by micrographs of type II hair cells in the peripheral (FIGURE 5A, cells 48 and 49) and central zones (FIGURE 5B, cell 30). Two type I hair cells are also shown (FIGURE 5C, cell 3 in the peripheral zone, and FIGURE 5D, cell 21 in the central zone). Quantitative data are summarized in TABLE 4. As will now be described, there are striking regional variations in synaptic organization.

More afferent boutons innervate individual type II hair cells in the peripheral zone as compared to the central zone. So, for example, the two peripheral type II hair cells (FIGURE 5A) each had more than 30 such boutons, while the central type II hair cell had only 13 (FIGURE 5B). On average, the number of afferent boutons supplying type II hair cells is 2–3 times greater in the peripheral zone (TABLE 4). A comparison of TABLE 2 (upper) and TABLE 4 indicates that the afferent reconstruction accurately predicts the number of afferent boutons per type II hair cell in the peripheral and intermediate zones, but gives too small a number in the central zone. A likely explanation for the discrepancy is that the sample of HRP-labeled afferent fibers on which the reconstruction is based underestimated the proportion of dimorphic units in the central zone or the number of bouton endings provided by such units. Despite the fact that they are contacted by relatively few afferent endings, type II hair cells in the central zone have as many ribbon synapses as do those in the peripheral zone. One reason for this is that many central boutons make two or more synapses (FIGURE 5B). In contrast, a single synapse is usually seen at the contact between a peripheral afferent bouton and its type II hair cell (FIGURE 5A).

Many type II hair cells in the central zone have large areas of apposition with the outer faces of neighboring calyx endings and ribbon synapses are usually seen between the two structures (FIGURE 5D). The potential importance of this kind of synaptic input is indicated by the observation that central calyx endings receive 10–20% of their afferent synapses from type II hair cells (TABLE 4). Most calyx units

---

FIGURE 5. Electron micrographs taken from a superior-canal crista in a chinchilla. **A**: Two type II hair cells (48 and 49) in the peripheral zone that were serially reconstructed. The boutons contacting each of them were numbered consecutively based on their first appearance in the set of sections. Here and in the other panels, long arrows point to synaptic ribbons seen in the section; short arrows mark the locations of ribbons appearing in nearby sections. Most peripheral afferent boutons make only a single ribbon synapse with their type II hair cells. The large bouton, labeled 27 and 19, synapses on both hair cells. **B**: A serially reconstructed type II hair cell (30) in the central zone innervated by 13 afferent boutons. Three of the boutons, including two shown here (3 and 11), get two ribbon synapses. **C** : A serially reconstructed type I hair cell (3) in the peripheral zone has eight synaptic ribbons. Most of the ribbons appear to be associated with calyceal invaginations; the asterisk (*) marks a case where both structures are seen in the same section. An efferent bouton contacts the calyx ending at its base. There is no appositional contact between the calyx and a neighboring type II hair cell (4). An apposition between cell 4 and the calyx of cell 5 is interrupted by interdigitating processes from a supporting cell (arrowheads). **D**: A type I hair cell in the central zone (21) has 14 synaptic ribbons. The calyx ending is also contacted by a type II hair cell (22), which contains six ribbons forming synapses with the outer face of the ending (arrows); the two synapses seen in this section of cell 22 (long arrows) are shown at higher power in the inset. Modified from Reference 7.

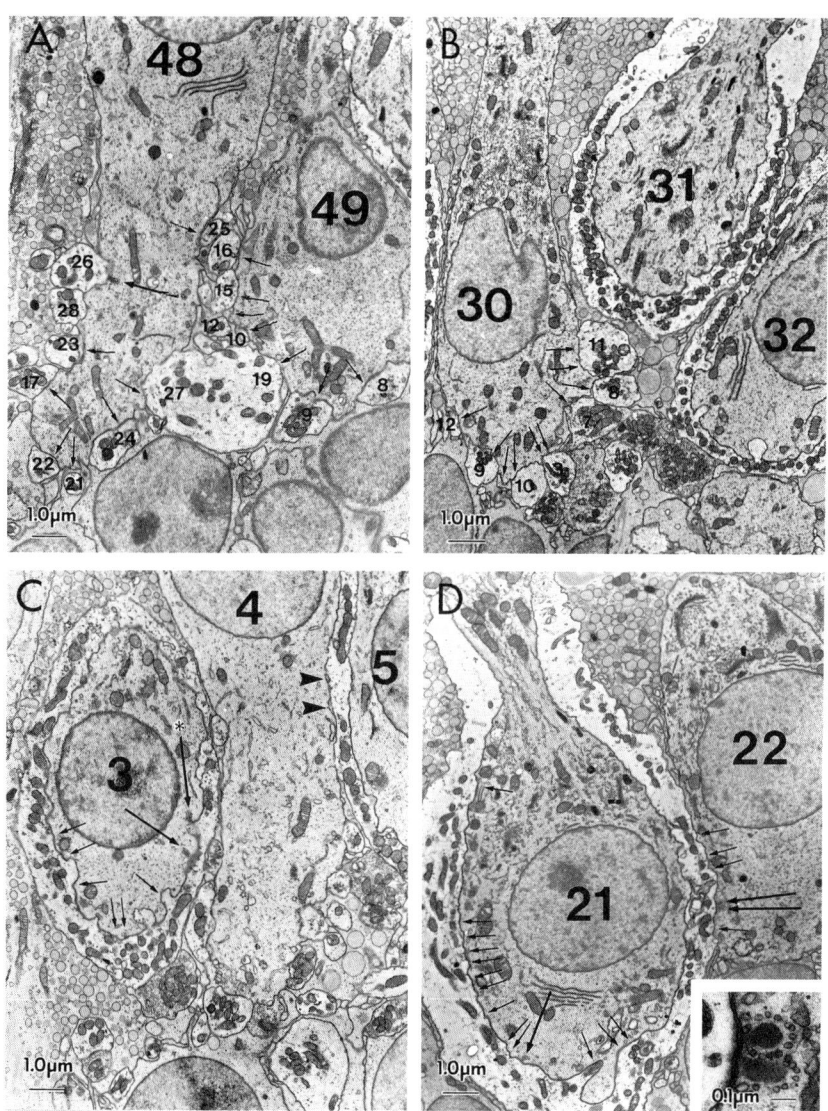

are located in the central zone. Many of them could receive inputs from both type I and type II hair cells and, in this sense, they could be functionally dimorphic. Quite a different arrangement occurs in the peripheral zone. Here, about half the type II hair cells do not contact calyx endings. The appositions that do occur are narrow, they are typically interrupted by supporting-cell process (FIGURE 5C, cell 4, arrowheads), and they seldom show synaptic specializations (TABLE 4).

Ribbon synapses between type I hair cells and calyx inner faces are plentiful (FIGURE 5C, cell 3, and FIGURE 5D, cell 21). There are, on average, almost 20 such synapses per hair cell and the number is similar in all three zones (TABLE 4). Another

TABLE 4. Regional Variations in Synaptic Structures, Cristae Ampullares of Chinchilla[a]

Type I Hair Cells	Sample Size	Inner-Face Synaptic Ribbons	Calyceal Invaginations	Efferent Boutons	Outer-Face Synaptic Ribbons
Central	49 (40)	20 ± 1	46 ± 5	2.9 ± 0.4	2.9 ± 0.8
Intermediate	26 (22)	21 ± 3	24 ± 2	4.3 ± 1.0	1.6 ± 0.4
Peripheral	12 (11)	16 ± 1	16 ± 3	4.5 ± 0.5	0.1 ± 0.0
All zones	87 (73)	19 ± 2	36 ± 3	3.6 ± 0.5	1.9 ± 0.3

Type II Hair Cells	Sample Size	Synaptic Ribbons	Afferent Boutons	Efferent Boutons	Outer-Face Synaptic Ribbons
Central	39 (32)	19 ± 1	10 ± 1	4.6 ± 0.3	3.5 ± 0.8
Intermediate	22 (18)	26 ± 2	26 ± 3	4.8 ± 0.6	2.2 ± 0.7
Peripheral	18 (17)	20 ± 3	25 ± 3	3.4 ± 0.7	0.2 ± 0.1
All zones	79 (67)	21 ± 1	18 ± 2	4.4 ± 0.3	2.1 ± 0.5

[a]Dissector techniques[38] were used to estimate that, on average, each hair cell was present in $d = 133$ ultrathin sections. Sample size is expressed as the number of equivalent hair cells and was obtained from the ratio of the total number of hair-cell profiles for each of six classes (2 types × 3 zones) and d: there were 87 × 133 = 11,584 type I and 79 × 133 = 10,560 type II profiles in the entire sample. Counts of ribbon synapses required that serial sections be available, which neccessitated that some of the profiles not be used; the smaller numbers of equivalent hair cells (inside parentheses) are the sample sizes for the ribbon counts. Each of the last four columns was obtained from a regression between the number of synaptic features and the number of equivalent hair cells for each of nine samples. Entries are estimates of the mean ± standard error of the mean for the number of features per hair cell. In the upper half of the table, the values for efferent boutons and for outer-face calyx ribbons give the numbers of such structures on the calyx ending surrounding each type I hair cell. In the lower half, the values for synaptic ribbons include those synapsing on afferent boutons and those contacting the outer face of calyx endings.

kind of specialized contact is also seen. It consists of an invagination of the calyx ending into the type I hair cell. Because the invagination is associated with a reduction of the intercellular cleft, it was once thought to be a possible site of electrical transmission.[22,28] This is now considered unlikely since it does not have the ultrastructural features of a gap junction.[21] The function of the invagination is obscure. Despite this, it is of interest that invaginations are more numerous in central, as compared to peripheral, type I hair cells (TABLE 4).

Highly vesiculated efferent boutons are in contact with the base of type II hair cells, with the outside of calyx endings, with afferent boutons, and with unmyelinated nerve branches.[29,30] In our material, type II hair cells are each contacted by 2–6 efferent boutons and this is so both centrally and peripherally (TABLE 4). There are somewhat fewer efferent boutons contacting individual calyx endings in the central zone. Hence, the relatively large efferent responses of irregular units cannot be explained by the central zone receiving a particularly heavy efferent innervation.

STUDIES IN THE SQUIRREL MONKEY

Having completed our studies of the afferent innervation patterns in the chinchilla cristae,[3] we decided to do a comparable study in the monkey and expected that results would be similar in the two species. To our surprise, there are substantial differences. The most obvious of these concerns the proportion of type I and type II hair cells. In the chinchilla, the two kinds of hair cells occur in approximately equal numbers throughout the end organ (TABLE 1, upper). The situation is quite different in the monkey, as can be seen by an inspection of individual sections (FIGURE 2B). Type I hair cells in the monkey outnumber type II hair cells by a ratio that averages 3:1 for the entire crista and exceeds 5:1 in the central zone (TABLE 1, lower).

How are the differences in hair-cell populations reflected in the afferent innervation patterns for the two species? To answer this question, afferents in the monkey were labeled extracellularly. Results were qualitatively similar to those for the chinchilla.[3] Again, calyx units are concentrated in the central zone, bouton units are largely restricted to the peripheral zone, and dimorphic units are seen in all parts of the epithelium. Two quantitative differences were noted: (1) There are proportionally more calyx units and proportionately fewer bouton and dimorphic units in the monkey. (2) Dimorphic units in the central and intermediate zones of the monkey have relatively few bouton endings. Both differences serve to match the afferent innervation with the complement of hair cells found in the two species. An afferent reconstruction was done for the monkey cristae (TABLE 2, lower) and compared to similar calculations for the chinchilla cristae (TABLE 2, upper). In the central and intermediate zones, it is estimated that there are half as many bouton endings in the monkey, which almost precisely matches the smaller number of type II hair cells in this animal. As a result, the number of bouton endings per type II hair cell in the two zones is estimated to be similar in the two species. On the other hand, there would appear to be many more boutons on individual type II hair cells in the peripheral zone of the monkey. Another difference in the calculations is that there are proportionately more fibers innervating the central zone in the monkey. Most of these fibers should be calyx units.

What are the functional implications of this latter difference? Calyx units in the chinchilla cristae have a distinctive physiology: they are irregularly discharging, phasic afferents with relatively low rotational gains. Extracellular recordings show that there is a comparable population of afferents in the monkey (FIGURE 6). The identification of these as calyx units was confirmed by their antidromic conduction velocities (see Reference 31 for details), which showed them to be among the largest fibers in the vestibular nerve. Possible bouton units were identified by their low conduction velocities. They were regularly discharging, tonic fibers and thus resemble the peripheral dimorphs and the bouton units previously labeled in the chinchilla. From a functional perspective, the most obvious difference between the two species is the presence of a considerably larger proportion of low-gain irregular afferents in the monkey.

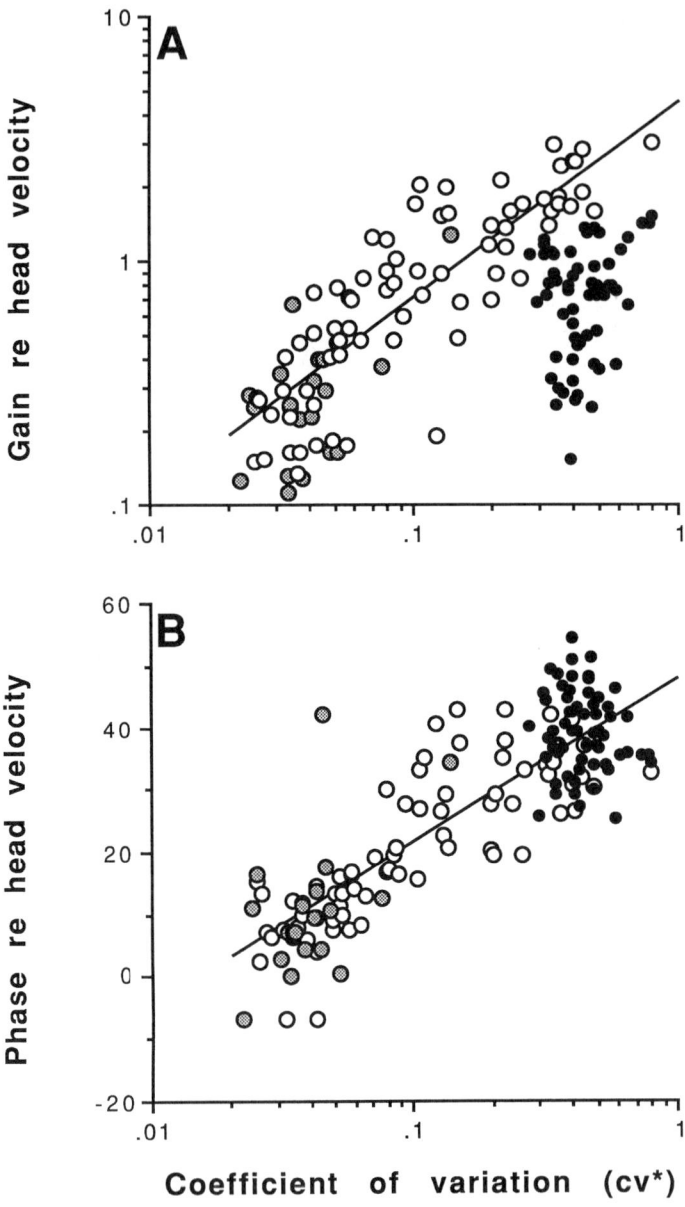

FIGURE 6. Gains and phases of the responses to 2-Hz sinusoidal head rotations plotted against a normalized coefficient of variation (cv*), extracellularly recorded semicircular canal units in squirrel monkey. Details as in FIGURE 3A and B. Units are separated into three categories by physiological criteria. Calyx units (black circles) are irregular units with low rotational gains. Bouton units (grey circles) had conduction velocities <18 m/second. All other units were classified as dimorphs (white circles). The lines are a power-law regression between gain and cv* for noncalyx units (A) and a semilogarithmic relation between phase and cv* for all units (B).

CONCLUDING REMARKS

The type I hair cell and its calyx ending are unusual structures, quite unlike the corresponding elements in any nonvestibular organ. They appeared late in evolution, being present in reptiles, birds, and mammals, but not in fish or amphibians.[10,32] They have a complicated ontogeny.[20] These facts suggest that the type I hair cell and the calyx ending play a distinctive role in vestibular processing. We had hoped to use morphophysiological techniques to discover this role. No clear answer has emerged. Here, it is well to emphasize the results from dimorphic units. Their discharge regularity and response dynamics are correlated with their locations in the neuroepithelium, but not with the types and numbers of hair cells they contact. It would thus appear that there are regional variations in transduction mechanisms and that these are shared by type I and type II hair-cell systems.

Three thoughts can be offered as to how progress can be made on this problem. First, a biophysical analysis of isolated type I hair cells should prove revealing. Work by Correia and Lang,[33] by Rennie and Ashmore,[34] and by Eatock and Hutzler[35] provides a convenient starting place. It will be important to correlate the biophysical properties of the elements with their epithelial locations and to work in end organs where the regional distribution of afferent types is known. Baird's work,[36] although done in a neuroepithelium not containing type I hair cells, is exemplary in this respect. Second, Sans and his colleagues have suggested that the calyx ending might be involved in a feedback regulation of the type I hair cell.[37] Regardless of the merits of the suggestion, it seems reasonable that to understand these structures requires that they be studied in their normal anatomical relation within the neuroepithelium. Third and finally, it may be important to choose an opportune species. In mammals, type I hair cells are found throughout the neuroepithelium and innervate afferents with a wide range of discharge properties. This may be contrasted with reptiles where type I hair cells are confined to central regions and where, one might guess, the afferents are more homogeneous and distinctive. One possibility is that the calyx ending and the type I hair cell have lost some of their distinctiveness during the course of mammalian evolution. In any case, the use of a reptilian species offers the possible advantage of being closer to the original idea behind these fascinating structures.

ACKNOWLEDGMENTS

The technical assistance of Ginger Tsien, Maureen Cummings, and Steven Price is gratefully acknowledged.

REFERENCES

1. LORENTE DE NÓ, R. 1926. Études sur l'anatomie et la physiologie du labyrinthe de l'oreille et du VIIIe nerf. Deuxième partie. Quelques donneés au sujet de l'anatomie des organes sensoriels du labyrinthe. Trab. Lab. Biol. Univ. Madrid **24**: 53–153.
2. POLJAK, S. 1927. Uber die Nervenendigungen in den vestibulären Sinnesendstellen bei den Säugetieren. Z. Anat. Entwicklungsgesch. **84**: 131–144.
3. FERNÁNDEZ, C., R. A. BAIRD & J. M. GOLDBERG. 1988. The vestibular nerve of the chinchilla. I. Peripheral innervation patterns in the horizontal and superior semicircular canals. J. Neurophysiol. **60**: 167–181.
4. BAIRD, R. A., G. DESMADRYL, C. FERNÁNDEZ & J. M. GOLDBERG. 1988. The vestibular

nerve of the chinchilla. II. Relation between afferent response properties and peripheral innervation patterns in the semicircular canals. J. Neurophysiol. **60:** 182–203.

5. FERNÁNDEZ, C., J. M. GOLDBERG & R. A. BAIRD. 1988. The vestibular nerve of the chinchilla. III. Peripheral innervation patterns in the utricular macula. J. Neurophysiol. **63:** 767–780.

6. GOLDBERG, J. M., G. DESMADRYL, R. A. BAIRD & C. FERNÁNDEZ. 1990. The vestibular nerve of the chinchilla. V. Relation between afferent discharge properties and peripheral innervation patterns in the utricular macula. J. Neurophysiol. **63:**791–804.

7. GOLDBERG, J. M., A. LYSAKOWSKI & C. FERNÁNDEZ. 1990. Morphophysiological and ultrastructural studies in the mammalian cristae ampullares. Hear. Res. **49:** 89–102.

8. LYSAKOWSKI, A., L. B. MINOR, C. FERNÁNDEZ & J. M. GOLDBERG. 1988. Physiological identification of calyx, dimorphic and bouton afferents in the vestibular nerve of the squirrel monkey. Soc. Neurosci. Abstr. **14:** 172.

9. FERNÁNDEZ, C., A. LYSAKOWSKI & J. M. GOLDBERG. 1991. Comparison of hair-cell populations and afferent innervation patterns in the cristae of the squirrel monkey and the chinchilla. Abstracts, 14th Midwinter Research Meeting of the Association for Research in Otolaryngology: 94.

10. WERSÄLL, J. & D. BAGGER-SJÖBÄCK. 1974. Morphology of the vestibular sense organ. *In* Handbook of Sensory Physiology. Vestibular System. Basic Mechanisms. H. H. Kornhuber, Ed. **6**(part 1): 123–170. Springer-Verlag. New York, N.Y.

11. LINDEMAN, H. H. 1969. Studies on the morphology of the sensory regions of the vestibular apparatus. Ergeb. Anat. Entwicklungsgesch. **42:** 1–113.

12. CARNEY, M. E., L. HOFFMAN & V. HONRUBIA. 1990. Quantitative analysis of canalicular nerves in the chinchilla. Abstracts 13th Midwinter Research Meeting of the Association for Research in Otolaryngology: 364.

13. GOLDBERG, J. M. & C. FERNÁNDEZ. 1971. Physiology of peripheral neurons innervating semicircular canals in the squirrel monkey. III. Variations among units in their discharge properties. J. Neurophysiol. **34:** 676–684.

14. FERNÁNDEZ, C. & J. M. GOLDBERG. 1976. Physiology of peripheral neurons innervating otolith organs of the squirrel monkey. III. Response dynamics. J. Neurophysiol. **39:** 996–1008.

15. SCHESSEL, D. A., R. GINZBERG & S. M. HIGHSTEIN. 1991. Morphophysiology of synaptic transmission between type I hair cells and vestibular primary afferents. An intracellular study employing horseradish peroxidase in the lizard, *Calotes versicolor.* Brain Res. **544:** 1–16.

16. GOLDBERG, J. M., C. E. SMITH & C. FERNÁNDEZ. 1984. Relation between discharge regularity and responses to externally applied galvanic currents in vestibular nerve afferents of the squirrel monkey. J. Neurophysiol. **51:** 1236–1256.

17. SMITH, C. E. & J. M. GOLDBERG. 1986. A stochastic afterhyperpolarization model of repetitive activity in vestibular afferents. Biol. Cybern. **54:** 41–51.

18. MCLAREN, J. W. & D. E. HILLMAN. 1979. Displacement of the semicircular canal cupula during sinusoidal rotation. Neuroscience **4:** 2001–2008.

19. GOLDBERG, J. M., C. FERNÁNDEZ & C. E. SMITH. 1982. Responses of vestibular-nerve afferents in the squirrel monkey to externally applied galvanic currents. Brain Res. **252:** 156–160.

20. FAVRE, D. & A. SANS. 1979. Morphological changes in afferent vestibular hair cell synapses during the postnatal development of the cat. J. Neurocytol. **8:** 765–775.

21. GULLEY, R. L. & D. BAGGER-SJÖBÄCK. 1979. Freeze-fracture studies on the synapse between the type I hair cell and the calyceal terminal in the guinea-pig vestibular system. J. Neurocytol. **8:** 591–603.

22. HAMILTON, D. W. 1968. The calyceal synapse of type I vestibular hair cells. J. Ultrastruct. Res. **23:** 98–114.

23. ENGSTRÖM, H. 1970. The first-order vestibular neuron. *In* Fourth Symposium on the Role of the Vestibular Organs in Space Exploration: 123–134. U.S. Government Printing Office (SP-187). Washington, D.C.

24. ROSS, M. D. 1985. Anatomic evidence for peripheral neural processing in mammalian graviceptors. Aviat. Space Environ. Med. **56:** 338–343.

25. Ross, M. D., C. M. Rogers & K. M. Donovan. 1986. Innervation patterns in rat saccular macula. A structural basis for complex sensory processing. Acta Otolaryngol. **102:** 75–86.

26. Dunn, R. F. 1980. Reciprocal synapses between hair cells and first order afferent dendrites in the crista ampullaris of the bullfrog. J. Comp. Neurol. **193:** 255–264.

27. Goldberg, J. M. & C. Fernández. 1980. Efferent vestibular system in the squirrel monkey: anatomical location and influence on afferent activity. J. Neurophysiol. **43:** 986–1025.

28. Spoendlin, H. 1966. Some morphofunctional and pathological aspects of the vestibular sensory epithelia. *In* Second Symposium on the Role of the Vestibular Organs in Space Exploration: 99–115. U.S. Government Printing Office (SP-115). Washington, D.C.

29. Hilding, D. & J. Wersäll. 1962. Cholinesterase and its relation to the nerve endings in the inner ear. Acta Otolaryngol. **55:** 205–217.

30. Smith, C. A. & G. L. Rasmussen. 1968. Nerve endings in the maculae and cristae of the chinchilla vestibule, with special reference to the efferents. *In* Third Symposium on the Role of the Vestibular Organs in Space Exploration: 183–201. U.S. Government Printing Office (SP-152). Washington, D.C.

31. Goldberg, J. M. & C. Fernández. 1977. Conduction times and background discharge of vestibular afferents. Brain Res. **122:** 545–550.

32. Jørgensen, J. M. 1974. The sensory epithelia of the inner ear of two turtles, *Testudo graeca* L. and *Pseudemys scripta* (Schoepff). Acta Zool. **55:** 289–298.

33. Correia, M. J. & D. J. Lang. 1990. An electrophysiological comparison of solitary type I and type II vestibular hair cells. Neurosci. Lett. **116:** 106–111.

34. Rennie, K. J. & J. F. Ashmore. 1991. Ionic currents in isolated vestibular hair cells from the guinea-pig crista ampullaris. Hear. Res. **51:** 279–292.

35. Eatock, R. A. & M. J. Hutzler. 1992. Ionic currents of mammalian vestibular hair cells. Ann. N.Y. Acad. Sci. (This volume.)

36. Baird, R. A. 1992. Morphological and electrophysiological properties of hair cells in the bullfrog utriculus. Ann. N.Y. Acad. Sci. (This volume.)

37. Scarfone, E., D. Demêmes, R. Jahn, P. De Camilli & A. Sans. 1988. Secretory function of the vestibular nerve calyx suggested by presence of vesicles, synapsin I, and synaptophysin. J. Neurosci. **8:** 4640–4645.

38. Gundersen, H. J. G. 1986. Stereology of arbitrary particles. A review of unbiased number and size estimators and the presentation of some new ones, in memory of William R. Thompson. J. Microsc. **143:** 3–45.

The Efferent Control of the Organs of Balance and Equilibrium in the Toadfish, *Opsanus tau*

STEPHEN M. HIGHSTEIN

Marine Biological Laboratory
Woods Hole, Massachusetts 02543

Department of Otolaryngology[a]
Box 8115
Washington University School of Medicine
4566 Scott Avenue
St. Louis, Missouri 63110

INTRODUCTION

Throughout phylogeny vertebrates have retained the organs of balance and equilibrium to detect and encode the position of the head with respect to gravity and the motion of the head upon the body and neck. Signals originating in these vestibular end organs combine in the brain stem with information about head and body orientation derived from other sensory systems to provide the organism with an internal representation of its position and motion. Vestibular and other sensory systems are under the control of the central nervous system being endowed with an efferent innervation with efferent cell bodies lying within the brain. This central efferent innervation or control must be in place to enhance the brain's perceptions under certain conditions rather than to degrade them.

The vestibular labyrinth and lateral line organs of the toadfish are examples of sense organs with an efferent innervation. Cell bodies in the medulla send axons to the labyrinth and lateral line organs to innervate hair cells and primary afferents. Individual vestibular and lateral line end organs contain a diverse spectrum of primary afferents that encode and report centrally different aspects of the adequate stimuli, e.g., in the case of the semicircular canals, head velocity versus head acceleration. Responses to efferent activation are not uniform across this spectrum of afferents, but are related to the dynamics of afferent response. Efferent neurons are tonically active and are recruited to a higher level of activity under certain conditions. When activated, efferents increase the resting discharge of primary afferents but decrease their responses to adequate stimulation. Below are reported the results of experiments concerning the efferent control of the acousticolateralis organs in the toadfish.

METHODS

Abbreviated methods follow: for a complete description of all methods please refer to References 1–6.

[a] Address for correspondence.

108

Adult toadfish of either sex weighing about one pound were utilized. These animals were collected from the Northeastern Atlantic tidal marshes from May until November by the Marine Biological Laboratory and housed in large aquaria supplied with running seawater.

Anatomical Experiments

For the demonstration of the location and organization of the efferent vestibular nuclei, fish were anesthetized by immersion in Finquel (Ayerst) and immobilized by an intramuscular injection of pancuronium bromide (Pavulon) 50 µg/kg. Animals were placed in a plastic experimental tank and perfused through the mouth with running seawater. A craniotomy was performed to visualize the nerves to each of the eight acousticolateralis end organs. Individual nerves were dissected free at their most distal intracranial locations and placed in contact with crystals of horseradish peroxidase (HRP). After two hours, nerves were rinsed with Ringers solution and the craniotomy closed in a layered, watertight fashion. Following survival of approximately two days, animals were anesthetized, heparinized, and perfused through the conus with saline followed by fixative. Composition of the fixative was 1% paraformaldehyde and 1% glutaraldehyde in phosphate buffer. Brains were removed and serially sectioned on a cryostat and processed to demonstrate the HRP reaction product.[7,8]

Intracellular Recording and Labeling

Animals were anesthetized and paralyzed as above. For antidromic identification of central neurons, individual end-organ nerves were dissected free and placed upon paired, silver wire stimulating electrodes for electric pulse stimulation. Efferent somata were penetrated with glass microelectrodes filled with standard recording solutions or with HRP. Individual neurons were injected utilizing iontophoresis. Efferent and/or primary afferent axons could also be penetrated with glass microelectrodes. For these experiments, paired stimulating electrodes were visually guided into the efferent vestibular nuclei within the medulla for electric pulse stimulation of the efferent system.

Characterization of the Responses of Primary Afferents to Adequate and Efferent Stimulation

Animals were placed into the experimental tank as noted above, and the efferent nuclei stimulated as above; the tank was centered on a rate table to deliver rotary stimulation. Recording of afferents was extracellular using glass microelectrodes. Responses were recorded by computer and subsequently analyzed off-line.

Characterization of Efferent Responses in Free-Swimming Animals

Animals were implanted with a miniature micromanipulator containing a tungsten microelectrode. A preamplifier was also mounted on the animal and signals led via fine wires to a computer for subsequent analysis. Animals were placed in a large tank, allowed to swim freely, and monitored via video. Visual, auditory, and visual stimuli were provided.

RESULTS

The efferent vestibular and lateral line nuclei are located in the posterior medulla spanning both sides of the midline. There are generally many fewer efferents than afferents to a given end organ. In the case of the semicircular canals, for example, there are 50 efferents and 350 afferents innervating each canal.

As noted in TABLE 1 (cf. also Highstein and Baker),[4] the majority of efferent somata are ipsilateral to the innervated end organ. Efferents to the horizontal canal are the exception to this rule, being distributed bilaterally. Efferent neurons surround the upper portions of the medial longitudinal fasiculus (MLF) and have been divided into dorsal and lateral subgroups for convenience. A dorsal-to-lateral ratio concerning the location of efferent somata was calculated. A high ratio indicates a predominance of dorsally located neurons, while a low ratio indicates the converse. The saccule, lagena, and anterior and posterior lateral line organs have high ratios while the anterior, posterior, and horizontal semicircular canals and the utricle have low ratios. Thus the dorsal subgroup tends to innervate the canals and utricle while the lateral subgroup tends to innervate the other end organs.

FIGURES 1 and 2 are reconstructions of intracellularly labeled dorsal and lateral subgroup neurons respectively. Dorsal neurons have horizontally oriented fusiform somata with a major dendrite issuing from each somatic pole. These dendrites travel laterally and ventrally to arborize extensively in the ventrolateral corners of the medullary brain stem. Cells in the lateral subgroup have large, pyramidal somata and exclusively ipsilaterally oriented dendritic trees that resemble half of the dendrites of the dorsal subgroup, except that they are more highly ramified than the latter. Both classes of cell often have dendritic processes that extend into the MLF and processes

TABLE 1. Distribution of Efferent Vestibular Neurons Projecting to Each Vestibular End Organ

End Organ	Total	Ipsilateral			Contralateral		Percent	Dorsal/ Lateral Ratio
		Dorsal	Lateral	Midline	Dorsal	Lateral		
Saccule	131	59	29	18	4	21	67	2.03
	119	89	11	6	6	7	84	8.09
Utricle	37	13	8	8	5	3	57	1.63
	65	25	30	8	2	0	85	0.83
Lagena	18	12	3	3	0	0	83	4.0
	30	12	11	6	0	1	77	1.09
Anterior L.L.	84	66	12	2	3	2	93	5.5
	69	60	5	0	4	0	94	12.0
Posterior L.L.	50	41	3	3	1	2	88	13.67
	39	38	0	1	0	0	97	>38.0
Posterior canal	51	4	27	13	0	7	61	0.1
	25	9	14	1	1	0	92	0.64
Horizontal canal	53	23	10	16	2	2	62	2.30
	44	9	14	15	1	5	52	0.64
Anterior canal	57	9	34	10	0	4	75	0.26
	36	20	10	4	0	2	83	2.0

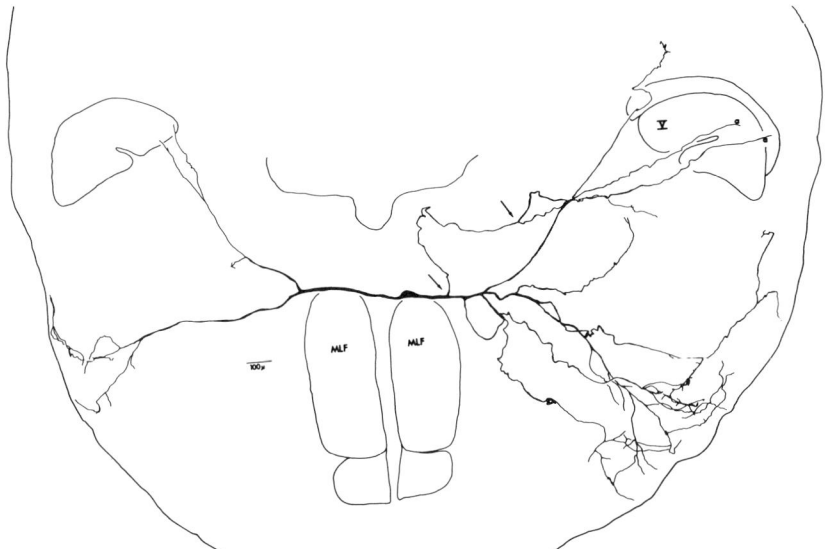

FIGURE 1. Reconstruction of a dorsal subgroup neuron injected with HRP. Note the bilateral dendritic tree with dendritic arbors similar to the cell in FIGURE 2. Note the dorsal dendrite traveling through the vestibular complex above the fifth nerve tract on the left. The arrows indicate the dendritic origin of the axon and its collateralization one-third of the distance between the somata and the edge of the medulla. Both axonal branches were followed to near the edge of the medulla and are indicated by the letters "a." (From Reference 4 with permission.)

that enter the dorsal vestibular nuclear complex. Labeled vestibular primary afferents are in a position to contact efferent dendrites either within the vestibular nuclei or within the ventrolateral corner of the medulla.[4,9,10]

Efferent neurons could be antidromically activated from more than one octavolateralis end organ and could be activated by multiple end organs via both electrical and chemical excitatory postsynaptic potentials.[4]

FIGURE 3 diagrams the peripheral organization of the afferent and efferent acousticolateralis nerve plexus to illustrate a major advantage of the toadfish for efferent study. Namely, the efferent nerves to several end organs can be visualized with a dissecting microscope *in vivo*. Efferent nerve branches are represented by blackened lines, and the arrow toward the right of the figure points to a small connecting nerve frequently present between the posterior canal and posterior lateral line nerves. Label of this connective demonstrated that it contained a pure population of efferent axons. Further, careful observation revealed that the efferent branches to other end organs could often be visualized as dense white bands lying on the surface of afferent nerves. That these bands were indeed efferent axons was verified by intracellular recording and injection of either HRP or biocytin.[4,11] The intense whiteness of these nerves may reflect a more heavy myelination than that of their afferent counterparts.

Recordings taken from hundreds of these efferent axons from "alert" fish in our experimental tank revealed a rather uniform pattern of resting discharge and activation under our experimental conditions. Efferent neurons are spontaneously

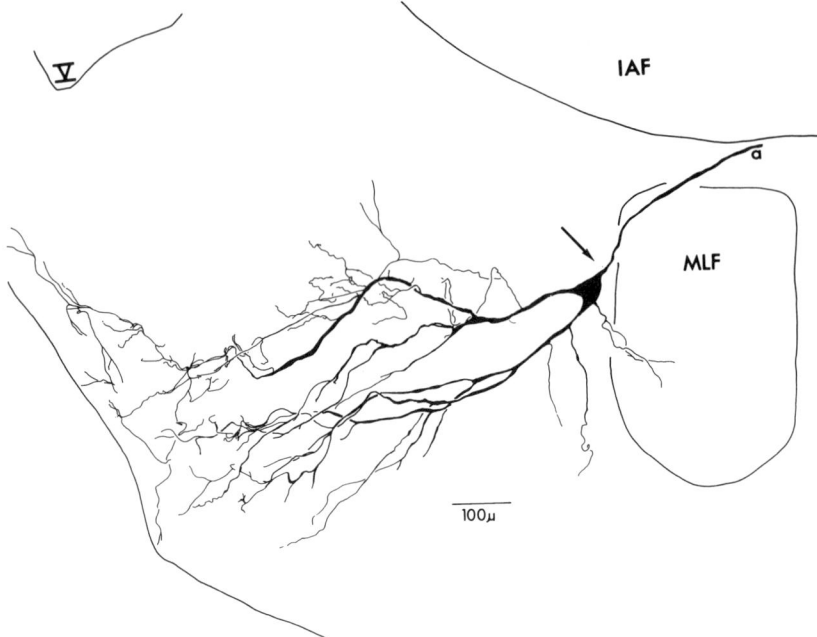

FIGURE 2. Reconstruction of a lateral subgroup neuron injected with HRP. The somata location and size are typical of lateral neurons. Note the highly branched dendritic tree with an extremely fine terminal arborization in the lateral and ventral medulla. The letter "a" indicates the axon and V is the fifth nerve complex. (From Reference 4 with permission.)

active when the fish is not moving, demonstrating a low discharge frequency (4–5 spikes/second) that is characteristically irregular.[3,6] Efferents continue to fire at this slow level if the fish is left unperturbed except for those cases when fish became spontaneously aroused. Light touch over the eye or snout results in an increase in the level of efferent activity. Single stimuli result in transient increases in firing rate, while sustained touching could result in prolonged elevations of efferent discharge. Multimodal sensory stimuli were also effective in increasing efferent neuronal firing. Not only light touch, but sound, vestibular sensation, and visual stimuli could activate the efferent system.[4] Efferent activation was always accompanied by an apparent arousal of the animal that appeared proportionate to the level of efferent activation. Full-blown arousal consisted of cessation of gilling, flaring of the pectoral and pelvic fins, and erection of the dorsal fin. As will be seen below, these same behavioral reactions were observed in free-swimming animals. The maximum sustained level of efferent activation was about 100 Hz, so this frequency of electrical stimulation of the efferent nuclei was adopted to investigate the efferent effects upon primary canal afferents (cf. below).

The response dynamics of the afferents in the horizontal semicircular canal nerve will now be described and classified as efferent actions on the population are related to these dynamics. In general, like other vertebrates, the population of afferents in toadfish demonstrates a continuum of response dynamics. We have utilized sinusoidal stimulation at various frequencies and amplitudes to group afferents into three broad classes.[1]

FIGURE 4 diagrams the Bode plots of the averaged phase (upper) and sensitivity (lower) for these three classes. One group was called acceleration afferents (diamonds; $n = 20\%$) because their phase of response leads stimulus velocity by more than 45 deg at all frequencies and amplitudes tested. A progressive high-frequency phase and gain enhancement is apparent (FIGURE 4). The 80% of afferents remaining showed a phase lead re velocity of less than 45 deg at one or more frequencies tested. These afferents have been divided into high- and low-gain velocity-sensitive afferents because of their sensitivity differences at stimulus frequencies greater than 0.1 Hz. High-gain afferents (triangles; $n = 42\%$) exhibit higher gains than low-gain afferents (circles; $n = 38\%$) and exhibit high-frequency phase and gain enhancements not seen in low-gain cells. Further, the sensitivity of low-gain afferents is similar for a 10-fold variation of stimulus amplitude and curves of sensitivity at fixed amplitudes, but with varying frequencies, superimpose, suggesting response linearity of this class of afferent. In contrast, high-gain afferents demonstrate a frequency-dependent increase in phase and changing sensitivities that can be best characterized as nonlinear, being both frequency and amplitude dependent. Further, low-gain

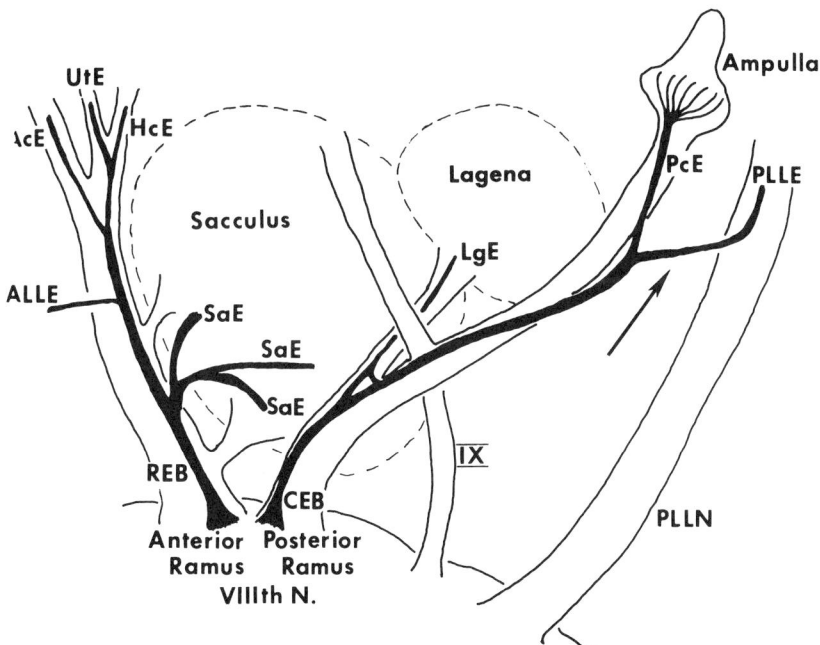

FIGURE 3. Diagram of the peripheral organization of the octavolateralis nerves. AcE is anterior canal efferent, Ut is utricle, Hc is horizontal canal, Sa is sacculus, Lg is lagena, Pc is posterior canal, PLL is posterior lateral line, ALL is anterior lateral line, REB is rostral efferent bundle, CEB is caudal efferent bundle, and IX is glossopharyngeal nerve. (From Reference 4 with permission.)

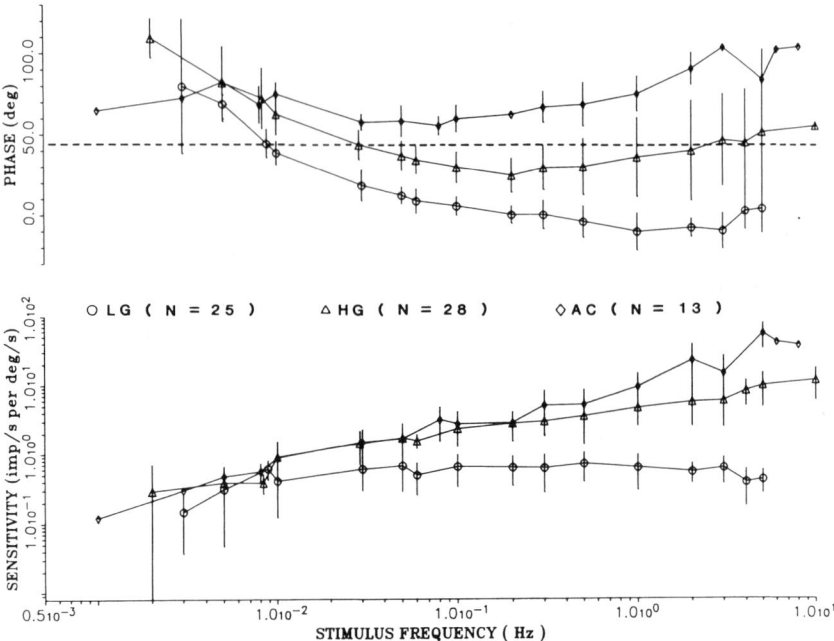

FIGURE 4. Bode plots of the averaged (+ standard deviation) responses of 25 low-gain (circles), 28 high-gain (triangles), and 13 acceleration (diamond) afferents to sinusoidal rotation between 0.001 and 10 Hz. Phase (deg) is expressed with respect to peak (0 deg) stimulus velocity; positive values correspond to a phase lead and negative values to a lag. A dashed line is drawn through the midpoint (45 deg) between peak stimulus acceleration and velocity. The three subgroups are distinguished by their differing phase and sensitivity relations across the tested bandwidth. (From Reference 1 with permission.)

afferents demonstrate a bidirectional modulation at all frequencies and amplitudes tested while high-gain cells can be driven into saturation by both inhibitory and excitatory stimuli. Acceleration afferents also demonstrated marked nonlinearities to rotational stimuli, similar to high-gain afferents except that the response distortion for acceleration cells began at lower frequencies than that for their high-gain counterparts.

Efferent action on these three broad classes of horizontal canal afferent was tested by electric pulse stimulation of the efferent nuclei at 100 Hz. Stimulus intensity was kept below the level that would have activated primary afferents via current spread. Electric pulse stimulation of the efferent nuclei could evoke all stages of the behavioral arousal described above. This arousal could be graded by varying the intensity of stimulation to the efferent nuclei. Mild reactions such as dorsal fin erection or full-blown arousal could be evoked.

Efferent effects upon resting discharge rates of horizontal canal afferents were related to the response dynamics of the particular afferent and are diagrammed in FIGURE 5. Low-gain afferents were only minimally affected by efferent stimulation, while high-gain and acceleration afferents demonstrated profound activation following efferent stimulation. In particular, afferents with low levels of spontaneous activity were the most profoundly affected. Afferent activation was generally observed within 10 mseconds of the onset of efferent stimulation, and maximum rates were reached in 2–4 seconds. TABLE 2 lists the averaged results of efferent stimulation upon the resting rates of afferents grouped by class.

Efferent stimulation was performed during rotation to evaluate its effects upon afferent responses. In no case was the phase of response changed by efferent stimulation. Bode plots of response of the three classes of afferent in combination with efferent stimulation are shown in FIGURE 6 and tabulated in TABLE 2.

The sensitivity of low-gain afferents is hardly affected by efferent stimulation. However, efferent stimulation reduced the sensitivity of high-gain and acceleration afferents by a proportionate amount at all frequencies tested without affecting the shapes of the input-output curves, e.g., the high-frequency gain increases were still present during efferent stimulation. In cases where an afferent was silenced during a part of a rotation cycle, efferent stimulation increased the fidelity of the response by allowing a bidirectional modulation of afferent discharge by raising the background discharge. However, the rise in background discharge is not matched by an equal rise in peak rate during rotation in the excitatory direction; thus the gain of the afferent is decreased. Utilizing caloric stimulation to increase the background discharge of afferents, it was demonstrated that the sensitivity decrease was not a function of excitatory saturation of afferents.[2]

To study some of the conditions under which the animal might utilize its efferent system, recording from peripheral lateral line nerves in alert, free-swimming fish was employed.[5,6] It was possible to record from putative efferent and primary afferent axons for prolonged periods of up to two weeks. Putative efferent axons demonstrated the same irregular, low-frequency spontaneous discharge observed in reduced preparations. In addition, efferents could be activated by the same set of sensory stimuli previously employed. Lateral-line primary afferents were classified into four groups by their patterns of activity, namely, silent, regular, irregular, and burster-type responses. Only one response type, the irregularly discharging fiber, was affected by the efferent system activated under our experimental conditions.

FIGURE 7 shows the recordings taken from the irregular afferent innervating an individual, superficial lateral-line neuromast located on the edge of the operculum or gill cover. This afferent was modulated by water movement across its hair cells as the animal respired, opening and closing its operculum about 24 times per minute. When

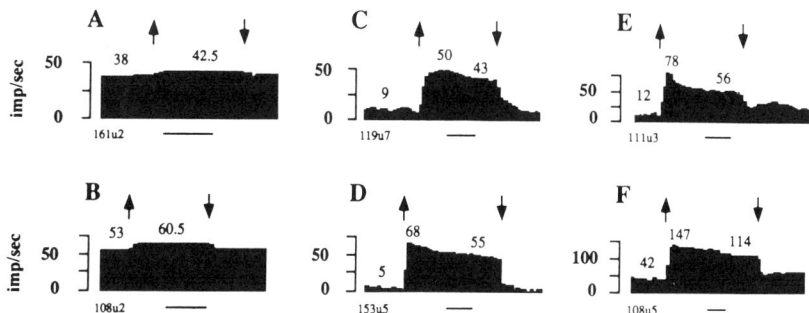

FIGURE 5. Facilitation of the background discharge of six horizontal canal afferents by electric pulse stimulation of the central, efferent vestibular nucleus. Downward and upward arrows indicate the beginning and end of efferent stimulation respectively. Horizontal calibration under each histogram is 10 seconds. (From Reference 1 with permission.)

TABLE 2. Action of Electrical EVS Stimulation on Background and Rotation-Induced Discharges of Horizontal Canal Afferents[a]

Cell Type	n	Condition	Background		Response to Rotation								
			CV	Rate (lps)	Stimulus Velocity	Stimulus (Hz)	DC (lps)	Rate (lps/cycle)	Sensitivity	Phase (deg)	Peak Rate (lps)	% Conduction	n Cutoff
Low gain	38 (27.1%)	Control	0.073 (0.02)	48.2 (14.7)	16.4 (7.6)	0.36 (0.16)	49.4 (22.3)	49.4 (22.3)	0.50 (0.23)	2.2 (10.3)	59.9 (23.9)	100 (0)	0
		Efferent	0.075 (0.03)	60.2 (19.4)			54.7 (25.3)	54.7 (25.3)	0.48 (0.25)	2.7 (8.8)	65.1 (26.1)	100 (0)	0
High gain R	21 (15%)	Control	0.074 (0.02)	68.3 (27.5)	12.8 (7.1)	0.39 (0.13)	67.9 (32.1)	67.9 (32.1)	3.49 (2.61)	20.1 (14.5)	105.1 (40.6)	100 (0)	0
		Efferent	0.09 (0.06)	85.1 (29.9)			81.9 (36.0)	81.9 (36.0)	2.89 (2.13)	22.0 (14.0)	112.7 (40.5)	100 (0)	0
M	23 (16.4%)	Control	0.22 (0.09)	34.5 (27.8)	11.7 (5.5)	0.43 (0.12)	31.3 (36.1)	39.3 (29.8)	2.95 (2.65)	19.5 (18.2)	68.4 (49.0)	82.0 (23.7)	10 (43.5%)
		Efferent	0.19 (0.14)	50.4 (33.8)			49.0 (36.8)	50.4 (36.7)	2.10 (2.01)	18.9 (16.9)	74.5 (52.7)	93.5 (12.7)	5 (21.7%)
I	26 (18.9%)	Control	0.68 (0.23)	6.4 (14.6)	13.5 (6.6)	0.42 (0.15)	-16.2 (36.4)	10.6 (9.6)	3.68 (3.28)	24.8 (12.2)	30.1 (20.2)	47.7 (22.2)	24 (92.3%)
		Efferent	0.38 (0.23)	25.5 (17.7)			23.8 (27.0)	34.7 (24.2)	1.99 (2.03)	28.0 (16.3)	48.6 (23.4)	84.6 (21.9)	12 (46.2%)

Acceleration		CV	rate	stimulus velocity	frequency	DC	average rate	sensitivity	phase	peak rate	percent	cutoff
R 4 (2.9%)	Control	0.09 (0.01)	85.8 (37.8)	16.7 (10.5)	0.30 (0.16)	98.6 (45.3)	98.6 (45.3)	4.77 (3.45)	60.4 (9.1)	154.5 (78.8)	100 (0)	0
	Efferent	0.09 (0.01)	109.8 (52.2)			108.0 (50.8)	108.0 (50.8)	4.11 (2.98)	61.6 (9.4)	155.4 (76.8)	100 (0)	0
M 9 (6.4%)	Control	0.20 (0.08)	52.3 (30.8)	16.8 (8.3)	0.38 (0.12)	51.2 (27.7)	61.2 (22.5)	5.59 (2.87)	65.6 (10.9)	130.4 (41.5)	75.8 (18.2)	7 (77.8%)
	Efferent	0.22 (0.07)	85.6 (29.5)			81.3 (28.3)	81.9 (28.1)	3.36 (1.48)	64.8 (9.6)	132.3 (38.4)	97.5 (7.4)	1 (11.1%)
I 19 (13.6%)	Control	0.60 (0.17)	13.3 (12.5)	14.4 (7.6)	0.42 (0.12)	−15.1 (36.3)	23.0 (34.3)	7.19 (4.89)	67.2 (13.0)	72.0 (40.0)	46.2 (19.1)	18 (94.7%)
	Efferent	0.53 (0.23)	66.2 (42.3)			31.7 (31.5)	65.8 (66.2)	4.46 (3.71)	69.6 (14.2)	88.5 (56.8)	77.8 (20.6)	14 (73.7%)
All 140 (100%)	Control	0.29 (0.19)	37.9 (34.1)	14.3 (7.3)	0.39 (0.14)	30.0 (46.2)	41.9 (36.5)	3.31 (3.78)	26.6 (31.8)	73.7 (47.3)	78.6 (27.5)	59 (42.1%)
	Efferent	0.22 (0.16)	59.8 (37.9)			52.2 (37.7)	59.1 (33.2)	2.22 (2.49)	27.7 (30.5)	80.8 (48.2)	92.9 (10.6)	32 (22.9%)

[a] For each cell type, n is the number of afferents (% of total) studied. For each condition (control and efferent) mean values (± standard deviation) are given of the CV and rate (imp/second) of the background discharge, stimulus velocity (°/second) and frequency (Hz), DC and average rate (imp/second), sensitivity (imp/second per °/second), phase (°/second), peak rate (imp/second), and percent (%) conduction of discharge over cycle time of the rotational responses. Number of afferents displaying a cutoff response is listed, with their percentage of total in that afferent group in parentheses. (From Reference 2 with permission.)

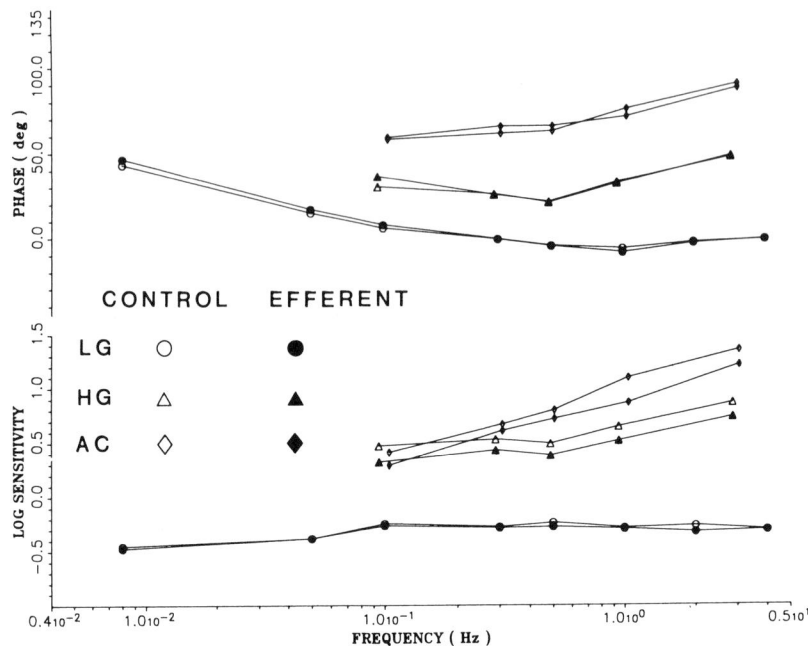

FIGURE 6. Bode plots of the averaged sensitivity and phase of responses to rotation alone (open symbols) and combined electric pulse stimulation of the efferent nuclei (filled symbols) for 19 low-gain (circles), 22 high-gain (triangles), and 11 acceleration (diamonds) afferents. (From Reference 1 with permission.)

stroboscopic visual stimulation was provided to a dark-adapted animal the unit apparently decreased its gain to adequate stimulation (bar above the top graph of FIGURE 7). Lines two and three of the figure demonstrate that neither the lateral nor the axial excursion of the operculum nor its frequency of opening and closing changed during this maneuver, thus confirming that the input to this neuromast did not change.

FIGURE 8 illustrates the effects of visually presenting a prey killifish on the primary afferent discharge of an irregular afferent. For these experiments the killifish was contained in a small, clear plastic box to eliminate all but visual clues. The killifish was hidden behind an opaque partition during the control portions of the experiment (FIGURE 8). When the partition was raised, afferent discharge was inhibited and returned to baseline values requiring more than one minute of elapsed time. These experiments were only possible in cases where toadfish did not lunge at prey because movement of the fish would affect the discharge of the lateral-line afferent. Thus experimental cycles wherein video monitoring of fish indicated that the animal remained stationary throughout the experiment were selected for analysis.

DISCUSSION

The location of the efferent vestibular nuclei has been studied in vertebrate species too numerous to mention. (For a recent review cf. Highstein.)[12] Are there any organizational or cytoarchitectural features of these nuclei that might be revealing of function? Firstly, the efferent nuclei generally lie in the medulla at the same coronal levels as the vestibular nuclei, are somewhat segmentally organized,[4] and tend to resemble facial motoneurons in the morphology of their axonal trajectories. These facts suggest a segmental origin of these neurons from the same segments as vestibular and facial neurons. The bilateral locations of neurons, or, in toadfish the bilateral dendritic trees of the dorsal subgroups with highly ramified dendritic trees, indicate that space for afferentation exists broadly on both sides of the brain stem. This is probably in keeping with the variety of inputs from multiple sensory modalities that have been shown to activate these cells. Even though we[3,4] and others[13] have shown that vestibular input to efferent neurons is monosynaptic, the shortest latency to activation of the efferent system in alert animals is about 160 mseconds,[3] probably indicating that a summation of multiple, small synaptic inputs is the adequate mode of bringing efferent neurons to threshold for firing an action potential. Further, physiological studies of efferent cells have indicated that although vestibular input to efferent neurons is monosynaptic, this input does not confer any spatial information to these cells (in the sense of responding in canal planes, for

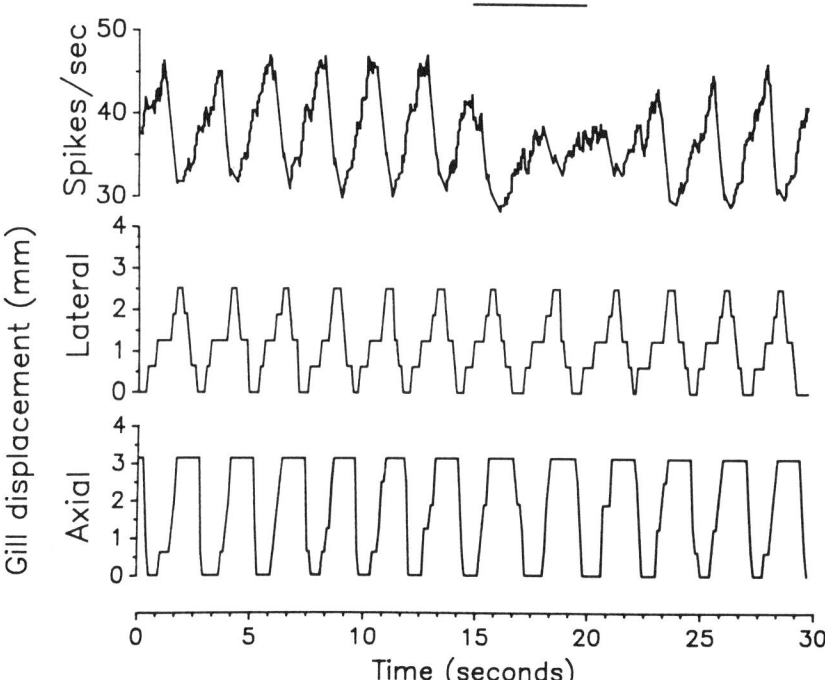

FIGURE 7. Stroboscopic activation of efferents inhibits lateral-line afferent activity evoked by movements of the operculum during respiration in an unrestrained toadfish. Upper trace shows mean firing rate of an irregular afferent neuron that innervated a neuromast on the operculum. Horizontal bar indicates period of stroboscopic stimulation (100 flashes/second for 5 seconds). Freeze frame analysis of the videotape shows that both lateral (center) and axial (lower trace) opercular displacements remain constant during the photostimulation period; thus decreased firing of the afferent was not due to changes in the mechanical stimulus generated by opercular movements. (From Reference 5 with permission.)

example). Additionally, all efferent neurons recorded to date respond in like fashion to multiple sensory stimuli, indicating a lack of specificity of inputs in a spatial sense. Thus we suppose that efferent neurons are in place to serve a more general function rather than changing the content of vestibular signals in a spatially related way.

Consideration of the factors that activate the efferent system in alert, free-swimming animals continues the above theme. These experiments were performed under quasi-natural conditions; the ultimate experiments must necessarily be per-

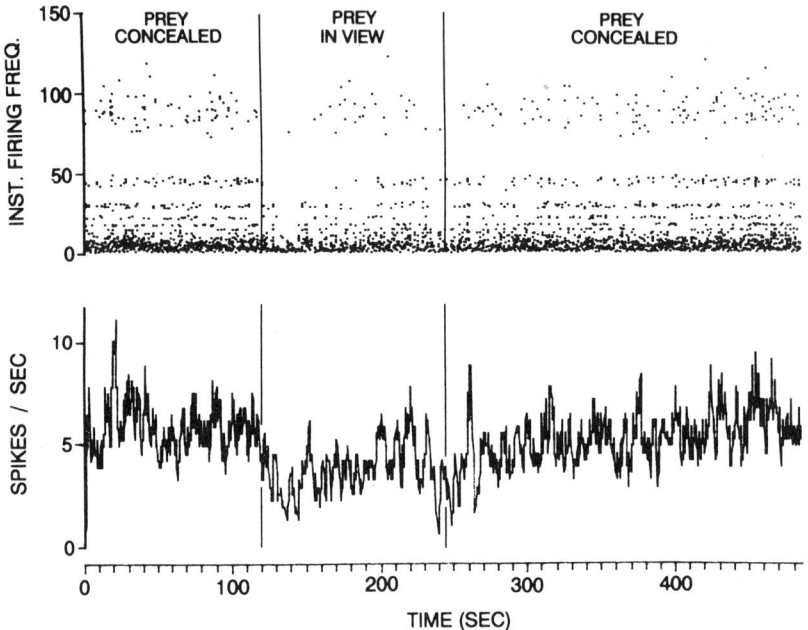

FIGURE 8. Inhibition of mechanically evoked lateral line activity in the behaving toadfish following visual presentation of a natural prey fish. Upper trace (instantaneous firing frequency) and lower trace (mean firing frequency) show activity of neuron during prestimulus (prey concealed), stimulus (prey in view), and poststimulus (prey concealed) periods of the experiment. Irregular-type afferent innervated superficial neuromast on the operculum. Mild vibration applied to the tank base biased the low spontaneous activity upwards. Firing decreased to lowest rates within 20 seconds after prey came into view, and returned to prestimulus rates approximately 60 seconds following prey removal from view. Fish did not approach or strike at prey during experiment. (From Reference 5 with permission.)

formed in the animals' natural habitat. Observation of toadfish indicates that they are ambush predators, at least in part, resting quietly and changing color to blend in with the background. Their tail and fins often wave in the current, resembling a piece of seaweed to this investigator. When a prey fish swims into view, toadfish will apparently become vigilant and alert, cease gilling, erect their dorsal fin, and wait until the prey comes near when they dart out and suddenly open their large mouths, sucking the prey and water into their oral cavity. This vigilant, alert, quiet behavior prior to darting out has also been observed in the laboratory and is associated with increased efferent vestibular activity. Consideration of the systemic results of this efferent activation below will further bolster the hypothesis for the role of this system that we currently favor.

As noted, effects of efferent activation are nonuniform across the spectrum of canal afferents (cf. also Reference 14). High-gain and acceleration afferents are the most profoundly affected, demonstrating a marked increase in their resting rates, a slight decrease in their sensitivity to adequate stimulation, and an increase in the fidelity of signals encoded due to bidirectional signaling. We presume that, in the case of the horizontal canal, for example, efferent effects are bilateral thus not signifying head movement to the brain. When the head actually moves, the movement is encoded by increases in activity in the afferents from one canal and parallel decreases in the activity of the contralateral, paired afferents from the paired canal. However, activation of a substantial proportion of canal afferents is not without consequence to the animal. Firstly, this increased barrage of activity might serve to "wake up" the brain stem and result in increased firing of neurons in all systems sequentially connected to the vestibular system. For example, high-gain and acceleration afferents terminate heavily in the nucleus magnocellularis, the counterpart of the mammalian dorsal Deiters' nucleus. N. magnocellularis gives rise to the ipsilateral descending lateral vestibulospinal tract that synapses monosynaptically upon extensor motoneurons.[9–11,15–17] This facilitation of extensor pathways might actually be the mechanism of the observed increased extensor tone that parallels efferent action. This increased extensor activity is reflected in the erecting of the dorsal fin for example. In this regard, it is interesting to note that graded electric pulse stimulation of the efferent nucleus leads to the sequence of behavioral responses seen during arousal in free-swimming toadfish, adding credence to this hypothesis. Increased extensor tone might be a necessary prelude to impending motor action, thus readying the animal to strike at prey or escape from danger. This hypothesis might be generalized to all vertebrates, as it has been demonstrated in squirrel monkeys that efferent action is most profound upon high-gain, irregular afferents[14] and these cells presumably project into pathways leading to input to extensor motoneurons.

Differential action on only a portion of the population of afferents from a given end organ has also been demonstrated for the lagena[18–20] where the most sensitive (to vibration), irregular, largest afferents are most profoundly affected by efferent stimulation. Regular, insensitive afferents are little affected by the efferents. In the lateral line, four classes of primary afferent have been described based upon the regularity of their resting, interspike intervals. These are silent, regular, burster, and irregular fibers. That these four classes of afferent innervate canal and superficial neuromasts in some ordered way and encode different aspects of water flow past the peripheral lateral line organs is under investigation. Presently it can be stated that only one class of afferent, the irregular cell, is affected by efferent stimulation.[5,6] This differential efferent action may function to enhance the central information transfer from individual lateral-line end organs. That is, by selectively decreasing the input from irregular-type afferents, the input from the other three classes may assume relatively more importance. Thus the animal may be utilizing its efferent system to enhance the peripheral processing of sensory information.

The results of these chronic recording experiments bear upon previous hypotheses of efferent function, namely, that efferent activation serves to prevent depletion of transmitter from hair cells.[21,22] In this regard it should be emphasized that neither locomotion, rapid movement, nor any other motor action invariably followed efferent activation in free-swimming animals. Furthermore, afferent discharge due to water motion past the skin during strikes at prey was most often robust. We thus favor the hypothesis that efferent action may function to decrease inputs from unwanted mechanical stimuli such as self-generated gill movements or other low-frequency background noise. If this is the case, then, as suggested above, higher frequency, more relevant signals, such as fin flutter of small prey fish, might be selectively enhanced. From all of the above evidence it may be again stated that the efferent control of sensory processes is most likely in place to enhance the perception of certain sensations in biologically relevant contexts in complex, behaving organisms.

SUMMARY

All vertebrates are endowed with a vestibular efferent system (EVS) consisting of somata within the central nervous system with long axons exiting the brain to innervate the labyrinth. Behaviorally relevant stimuli related to feeding and/or aggressive behaviors and conditions leading to enhanced attentional states or alerting activate the EVS. Increased EVS activity modifies the resting rate and response dynamics to motion of vestibular afferents. This modification is nonuniform across the fiber spectrum of the semicircular canals, for example, affecting the more-sensitive, low-spontaneous-activity cells more profoundly than their less-sensitive counterparts. The cellular bases for EVS effects are excitatory axoaxonic synapses upon primary afferents and axosomatic inhibitory synapses upon hair cells.

REFERENCES

1. BOYLE, R. & S. M. HIGHSTEIN. 1990. Resting discharge and response dynamics of horizontal semicircular canal afferents of the toadfish, *Opsanus tau.* J. Neurosci. **10:** 1570–1582.
2. BOYLE, R. & S. M. HIGHSTEIN. 1990. Efferent vestibular system in the toadfish: action upon horizontal semicircular canal afferents. J. Neurosci. **10:** 1570–1583.
3. HIGHSTEIN, S. M. & R. BAKER. 1985. The action of the efferent vestibular system upon primary afferents of the toadfish, *Opsanus tau.* J. Neurophysiol. **54:** 370–384.
4. HIGHSTEIN, S. M. & R. BAKER. 1986. Organization of the efferent vestibular nuclei and nerves of the toadfish, *Opsanus tau.* J. Comp. Neurol. **243:** 309–326.
5. TRICAS, T. C. & S. M. HIGHSTEIN. 1990. Visually mediated inhibition of lateral line primary afferent activity by the octavolateralis efferent system during predation in the free-swimming toadfish, *Opsanus tau.* Exp. Brain Res. **83:** 233–236.
6. TRICAS, T. C. & S. M. HIGHSTEIN. 1991. Inhibition of lateral line primary afferent activity by the octavolateralis efferent system during predation in the free-swimming toadfish, *Opsanus tau,* mediated by visual input. J. Comp. Physiol. **169:** 25–37.
7. CULLHEIM, S. & S. O. KELLERTH. 1976. Combined light and electron microscopic tracing of neurons including axons and synaptic terminals after intracellular injection of horseradish peroxidase. Neurosci. Lett. **2:** 307–313.
8. MESULAM, M. M. 1978. Tetramethylbenzidine for horseradish peroxidase neurochemistry: a non-carcinogenic blue reaction product with superior sensitivity for visualization of neural afferents and efferents. J. Histochem. Cytochem. **26:** 106–117.
9. CAREY, J., R. BOYLE & S. M. HIGHSTEIN. Differential central projections of physiologically characterized horizontal semicircular canal afferents in the toadfish, *Opsanus tau.* (In preparation.)
10. CAREY, J. P. & S. M. HIGHSTEIN. 1990. Physiology and brainstem morphology of single horizontal semicircular canal afferents in the toadfish, *Opsanus tau:* a biotin dye study. Neurosci. Abstr. **16:** 732.
11. HIGHSTEIN, S. M. Unpublished.
12. HIGHSTEIN, S. M. 1991. The central nervous system efferent control of the organs of balance and equilibrium. Neurosci. Res. **12:** 13–30.

13. WHITE, J. S. 1985. Fine structure of vestibular efferent neurons in the albino rat. Neurosci. Abstr. **11:** 322.
14. GOLDBERG, J. M. & C. FERNANDEZ. 1980. Efferent vestibular system in the squirrel monkey: anatomical location and influence on afferent activity. J. Neurophysiol. **43:** 986–1025.
15. GOLDBERG, J., S. M. HIGHSTEIN, A. K. MOSCHOVAKIS & C. FERNANDEZ. 1987. Inputs from regularly and irregularly discharging vestibular-nerve afferents to secondary neurons in the vestibular nuclei of the squirrel monkey. I. An electrophysiological analysis. J. Neurophysiol. **58:** 700–719.
16. GOLDBERG, J. M., L. B. MINOR, R. BOYLE & S. M. HIGHSTEIN. Vestibular nerve inputs to the vestibulo-ocular reflex in the squirrel monkey. Proceedings of the Barany Society, Tokyo, Japan. (In press.)
17. HIGHSTEIN, S. M., J. M. GOLDBERG, A. K. MOSCHOVAKIS & C. FERNANDEZ. 1987. Inputs from regularly and irregularly discharging vestibular-nerve afferents to secondary neurons in the vestibular nuclei of the squirrel monkey. II. Correlation with output pathways of secondary neurons. J. Neurophysiol. **58:** 719–739.
18. LOCKE, R. E. & S. M. HIGHSTEIN. 1989. Efferent modulation of synaptic noise and spike frequency in lagenar afferents of the toadfish. Assoc. Res. Otolaryngol. Abstr. 99.
19. LOCKE, R. E. & S. M. HIGHSTEIN. 1989. Efferent modulation of synaptic noise and spike frequency in lagenar afferents of the toadfish. Biol. Bull. Abstr. **177:** 324.
20. LOCKE, R. E. & S. M. HIGHSTEIN. 1990. Efferent modulation of synaptic noise in lagenar afferents of the toadfish. Neurosci. Abstr. **16:** 735.
21. RUSSELL, I. J. & B. L. ROBERTS. 1972. Inhibition of spontaneous lateral line activity by efferent nerve stimulation. J. Exp. Biol. **57:** 77–82.
22. FLOCK, A. & I. J. RUSSELL. 1973. The presynaptic action of efferent fibers in the lateral line organ of the burbot *Lota lota*. J. Physiol. **235:** 591–605.

Somatic versus Vestibular Gravity Reception in Man

H. MITTELSTAEDT

Max-Planck-Institut für Verhaltensphysiologie
D-W-8130 Seewiesen, Germany

INTRODUCTION

That the otoliths are not the only source of static information is evidenced by the residual faculties of bilaterally labyrinthectomized subjects. Common opinion appears to assume (but for exceptions see References 1 and 2) that the somatic gravity information is mediated by mechanoreceptors that measure forces acting on the limbs and/or the skin. It has been shown, however, that hitherto unknown gravity receptors must exist in the trunk.[3] After describing the methods, this existence proof shall first be reviewed briefly. Subsequently, I will attempt to localize these receptors and gain information on their mode of functioning.

METHODS

The subject (S), a paid or employed volunteer, lies on the side, the right-ear-down head position secured by means of a dental imprint, on a padded board, which S as well as the experimenter (E) can tilt via remote control about an axis parallel to the S's visual (x) axis. The E sets the initial tilt in total darkness to arbitrary, yet strictly up-down alternating, angles and then asks the S to rotate the board until she/he feels to be horizontal. The angular deviation ρ of the direction of gravity from the S's spinal (z) axis is measured by means of a potentiometer. The S has been instructed not to make cognitive inferences, but told, "Just act as you feel—as you are used to doing when sitting or standing upright in the dark." As a rule a session is terminated after 8 pairs of settings, but only the last 6 pairs are used to compute mean and variance of ρ as the "subjective horizontal position" (SHP). On average Ss are able to assume their personal SHP with a standard deviation (SD) of ±1 degree.

In a subsequent session the same procedure is followed with the S placed in the same way on a sled which can, by remote control, be moved radially over the horizontal surface of a human centrifuge (FIGURE 1). In all experiments the centrifuge is run with $\omega = 2\pi$ times 0.6 rotations per second and hence produces 0.0145 G per cm radius. This quantity is named G_r throughout this article (dimension: cm^{-1}). Sled velocity was chosen as 2.2cm/second, against 0.5°/second on the tiltable board, in order to compensate somewhat for the missing canal input. Sled position is measured as the distance d_v of the centrifuge axis (CA) from the binaural axis (BA), cranial positive, caudal negative.

Consider an S whose SHP on the tiltable board coincides with the objective horizontal position ($\rho = 90°$). If the otoliths were the only source of gravity information, this S would feel tilted upwards when the centrifuge axis is beyond the head, that is, if d_v is positive (upper pictograph of FIGURE 1). Correspondingly this S would feel tilted downwards if d_v is negative (lower pictograph in FIGURE 1), and accurately

124

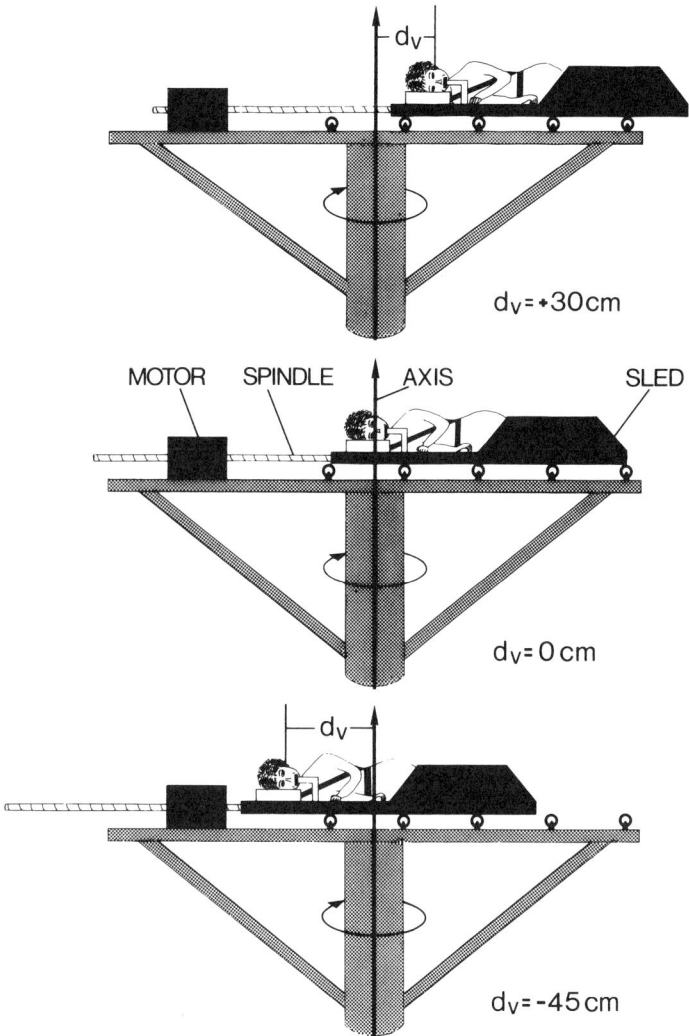

FIGURE 1. Exclusive variation of force component acting along S's spinal (z) axis by means of human centrifuge with vertical axis, horizontal platform, and radially mobile sled. The centrifuge is run with 0.6 rotations per second (angular velocity = 3.77 radian second^{-1}), thus produces $G_r = 0.0145\ G$ per cm radius, and hence a z-axis force component of $-d_v G_r$ at the site of the otoliths. d_v: distance of centrifuge axis from binaural axis, caudal negative. At $d_v = 0$ these two axes are collinear.

horizontal when CA coincides with BA ($d_v = 0$). In general, under exclusive vestibular control, the CA is expected to be set at $d_v = \cos\rho / G_r$ (in centimeters).

This prediction assumes that the SHP is based on information about the z-axis force component alone, gained essentially from saccular input, rather than (as is the subjective *visual* vertical) on a central representation, computed from otolithic input,

of the z- *and* the y-axis force component or of the angle ρ respectively.[4] In the latter
cases, cos ρ should be replaced by cotan ρ. Because of the small range of SHP
deviations from $\rho = 90°$ in all Ss, this difference is negligible here.

EXISTENCE PROOF

Basic Results

In the study to be reviewed all Ss except one set d_v up to 30 cm caudally of the
value to be expected under exclusive otolithic control. At present the interindividual
mean of this caudal deviation in 21 Ss, depending on conditions to be treated in the
section Normal Controls below, amounts to -19.9 and -27.8 cm, respectively, with
considerable differences between Ss [*inter*individual SD around 12 cm, against
*intra*individual (intrasession) SDs of around 2 cm]. In neurinomectomized Ss the
caudal deviation has been found to range between $d_v = -41$ and $d_v = -52$ cm. Hence
the centroid of the mass(es) governing the somatic gravity receptors is situated at the
height of the last ribs, that is, near the centroid of the body. In the normal Ss the
otolithic and the somatic inputs therefore counteract on the centrifuge. Yet obviously, they do not suppress one another, but reach a compromise. Under the
assumption of *additive superposition,* the effect of the somatic gravity receptors on the
z-axis component of the postural control system is thus, on average, equal to or even
larger than that of the otoliths.

Specifications

In a second series, the support forces on the centrifuge were altered: In
configuration A, with hips and knees both flexed by 90°, the centrifugal forces were
counteracted by supports under the thighs and soles; in B and C, with legs straight,
either by supports under the soles (B) or by mountaineer's straps which support the
pelvis (C). Consequently, the legs are subjected to pressure in B, whereas to tension
in C. Yet no difference of d_v between B and C was found, even though d_v varied from
-13 to -43 cm between Ss (cf. FIGURE 3, p. 29, in Reference 3). Hence force-
transducing receptors in the legs appeared to be excluded as source of the somatic
influence on the SHP.

There was, however, a clear-cut difference between d_v in configuration A and in
the two others, that is, between settings with legs flexed and legs straight, namely, a
caudad shift of CA by about 9 cm in B and C with respect to A in all those Ss. This
shift turned out to be correlated with the concomitant shift of the centroid of the
body ($r = 0.93$), raising the hypothesis that the somatosensory input might be caused
by shear and pressure forces at the skin. However, the measured shift was much
larger than expected under this hypothesis, the more so the smaller $|d_v|$ had been
with legs flexed ($r = -0.88$).

Taken together, then, the results point to the existence of gravity receptors *within*
the trunk.

EXPERIMENTS ON THE EFFECT OF TRANSVERSAL
LESIONS OF THE SPINAL CORD

In order to localize the somatic gravity input, paraplegic Ss with total bilateral
transversal sensory loss (TSL) at and caudad of a carefully diagnosed segment of the

spinal cord are recruited to be tested on the tiltable board and the centrifuge, both with legs flexed and with legs extended.

Selection Principles

Diagnosis is based on standard noninvasive neurological examinations on touch, pain, vibration sensitivity, and reflexes by a paraplegia specialist.[a] Although the sensory lesion is pertinent here, motor and vegetative dysfunctions were also checked. Only clear-cut complete lesions were included in the results given below. In several cases the diagnosis was corroborated by available anatomical evidence on complete local destruction or interruption of the cord. No differences in the settings of these Ss and the rest were found.

Normal Controls

In FIGURES 2 to 4 the settings with legs extended (dubbed "long") are plotted over those with legs flexed (dubbed "short"). To make comparison easy, the same measure is applied in all cases, namely, the z-axis component of the gravitoinertial force vector (apicad positive, caudad negative) at the binaural axis, in G, divided by G_r. On the tiltable board it is named $b_{long} = -\cos\rho_{long}/G_r$ and $b_{short} = -\cos\rho_{short}/G_r$; on the centrifuge it is named $c_{long} = -d_{v,long}$ and $c_{short} = -d_{v,short}$; in combination, the differences are named $\Delta_{long} = c_{long} - b_{long}$ and $\Delta_{short} = c_{short} - b_{short}$. The latter quantities are particularly instructive, because the subtraction of the b values from the respective c values eliminates the effect of the force-independent bias (contingent upon the assumptions of Methods, last paragraph) either completely or, because of the comparatively small deviations from $\rho = 90°$ on the board, almost completely. Hence the Δ-quantities show the force-dependent effect of the somatoreceptors in (almost) complete isolation.

As shown in FIGURE 2, on the board with legs extended the angle ρ (and hence the quantity b) tends to be increased. Mean SHP settings were 84.8° when legs were flexed and 88.6° when legs were extended. This increase is the larger, the smaller ρ had been with legs flexed. In 21 Ss this dependency, that is, the (negative) correlation between $b_{long} - b_{short}$ and b_{short}, is highly statistically significant ($p < 0.01\%$, two-tailed). The line of least-square distance (LSD) is $b_{long} = 0.66b_{short} + 2.47$ cm. The absolute parameter, the increase of 2.46 cm at $b_{short} = 0$ ($\rho_{short} = 90°$), is significantly different from zero, as well as the mean of the 21 differences ($b_{long} - b_{short}$) of 4.6 cm ($p < 0.01\%$).

On the centrifuge (FIGURE 3), the slope of the LSD of c_{long} over c_{short} is likewise smaller than unity. (The probability that the correlation of $c_{long} - c_{short}$ with c_{short} is by chance at least that much different from zero and negative is $p = 0.75\%$.) Furthermore, the mean of the differences ($c_{long} - c_{short}$) of 7.9 cm is significantly different from zero with $p = 0.015\%$. But the means of c_{long} and c_{short} (dashed lines) are far from the means of b_{long} and b_{short}, respectively, that is, far from where they should be under exclusive otolithic control.

If the settings on the board are subtracted from those on the centrifuge we find a noncorrelated relation between $\Delta_{long} - \Delta_{short}$ and Δ_{short} (FIGURE 4), hence an LSD of unity slope, and a comparatively small, not (yet) significant, mean increase ($\Delta_{long} - \Delta_{short}$)/$\Delta_{short}$ of 12.6%. However, mean Δ_{long} and Δ_{short} are quite large (29.4 and

[a]Dr. Erwin Weller, long-time neurologist at the *Berufsgenossenschaftliche Unfall-Klinik Murnau,* is gratefully thanked for devoting time and effort to this task.

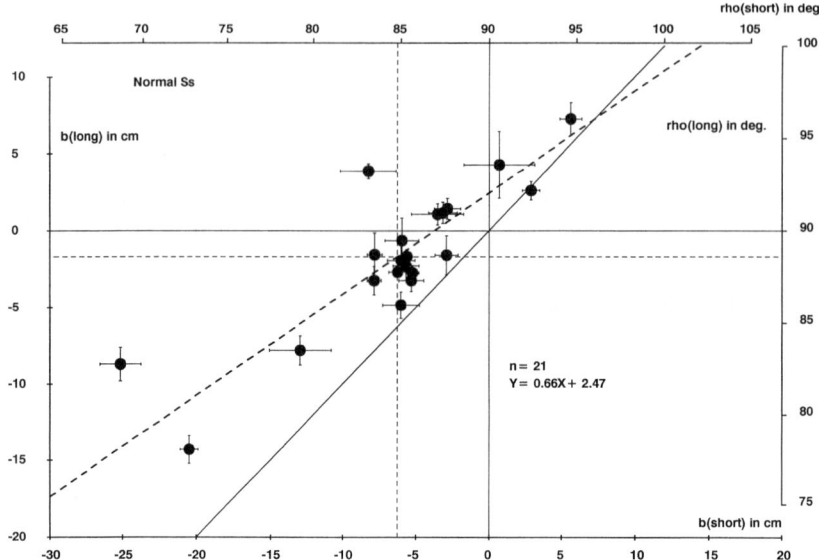

FIGURE 2. Subjective horizontal position (SHP) of normal Ss on the tiltable board, expressed as angle ρ and as distance (equivalent to $-d_v$) to be expected in the same S on the centrifuge of FIGURE 1 under exclusive otolithic control ($+b = -d_v$). b: expected distance, computed by dividing the z-axis force component of the gravity vector on the board by G_r. Thus $b = -\cos\rho/G_r$; where ρ is the deviation of the S's z-axis from the vertical and $G_r = \omega^2/(981 \text{ cm}$ second$^{-2}) = 0.0145 \text{ cm}^{-1}$. Ordinate: b_{long} with legs straight, in cm; abscissa: b_{short} with legs flexed (hip as well as knee by 90°), in cm. Brackets: SD. Dashed lines: means and LSD, respectively. Note the positive bias (2.47 cm) and the (highly significant) deviation of the slope of the LSD (0.66) from unity, indicating that the roll tilt increases with legs extended, and the more so the smaller it had been with legs flexed.

26.1 cm), even larger than mean c_{long} and c_{short} (27.8 and 19.9 cm), respectively. The respective differences are, by definition, equal to $-b_{\text{long}}$ and $-b_{\text{short}}$; and at least b_{short} is significantly different from zero ($p = 0.05\%$, two-tailed) and negative.

Paraplegics

FIGURES 5 to 7 give the same relations in 21 paraplegics with TSL from the 11th thoracic segment *inclusively* (named T11) to the 6th cervical segment inclusively (C6). On the board the increase shown by the normal Ss is missing, $b_{\text{long}} - b_{\text{short}} = -1.99 - (-2.36) = 0.38$ cm (FIGURE 5). However, on the centrifuge there is still a definite caudal deviation of the CA from that expected under exclusive otolithic control and a considerable effect of leg extension. Although mean c_{short} (14.8 cm) and Δ_{short} (17.2 cm) of these paraplegics are smaller than mean c_{short} (19.9 cm) and Δ_{short} (26.1 cm) of the normals [with at least the difference between the means of Δ_{short} of 9 cm significantly different from zero ($p = 0.9\%$, two-tailed)], the *relative* mean increase $(c_{\text{long}} - c_{\text{short}})/c_{\text{short}}$ is larger (59.8% than in the normals (39.6%), a relation that becomes even more pronounced in the respective differences Δ_{long} and Δ_{short} (49.3 and 12.6% respectively; FIGURES 6 and 7, compare also FIGURE 9).

The slopes of the LSD in FIGURES 5, 6, and 7 are consistently *larger* than the respective slopes in the normal Ss.

Also, most remarkably, *no correlation exists between any of these measures and the site of the lesion* (segment number of TSL). Hence this group, named TC-paraplegics, can be treated as a unit with, on average,

$$b_{long,TC} \approx b_{short,TC} \tag{1}$$

$$c_{long,TC} = K_c c_{short,TC} + A_c \tag{2}$$

$$\Delta_{long,TC} = K_\Delta \Delta_{short,TC} + A_\Delta \tag{3}$$

where A_c and A_Δ (which are not significantly different from 0) may in fact be near zero and K_c and K_Δ still larger than 1.3.

A very different picture emerges in paraplegics with TSL between the 5th and the 1st lumbar segment. These Ss, named L paraplegics, behave like normals on the tiltable board. Likewise on the centrifuge they show the same relative increase of leg extension as do the normal Ss $[(c_{long} - c_{short})/c_{short} = 44.6\%; (\Delta_{long} - \Delta_{short})/\Delta_{short} = 6.5\%]$, although the four means may be somewhat smaller than normally. The striking difference of this group to the rest of the paraplegics is illustrated by FIGURE 8, and means are graphically represented in FIGURE 9.

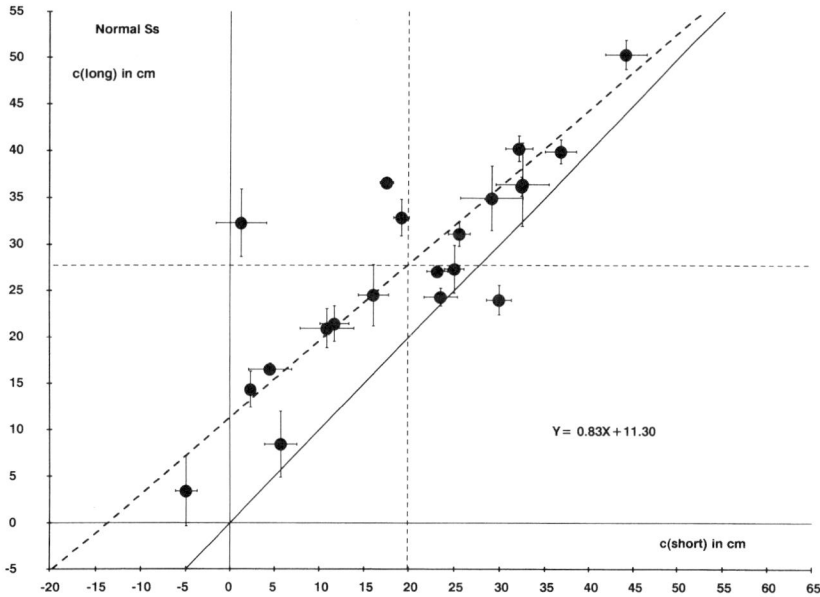

FIGURE 3. SHP of normal Ss on the sled centrifuge of FIGURE 1. c: z-axis component of force vector *at the binaural axis* (BA) divided by G_r (as in FIGURE 2). Hence $c = -d_v$, where d_v is the distance (in cm) of the centrifuge axis from the BA (see FIGURE 1). Index "long," "short:" legs straight, flexed, respectively. Brackets: SD. Note the large mean deviations of c_{long} and c_{short} from the respective b values in FIGURE 2, that is, from the values expected under exclusive otolithic control. The slope of the LSD is significantly different from unity and similar to that in FIGURE 2.

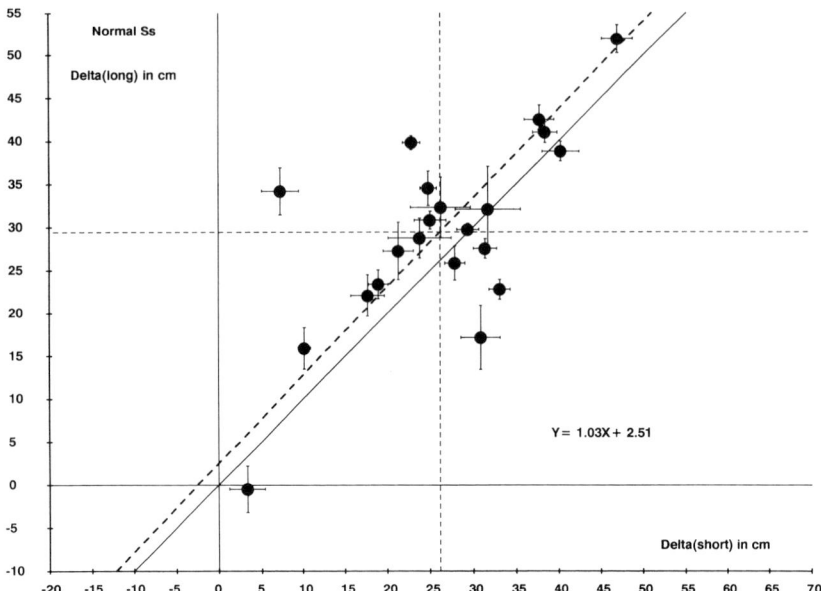

FIGURE 4. Difference $\Delta_{long} = c_{long} - b_{long}$ of normal Ss plotted over $\Delta_{short} = c_{short} - b_{short}$. It shows the effect of the force-dependent components of the rivaling gravity receptors in (almost) complete isolation from the force-*in*dependent components. Brackets: SD. Note that here, with Δ_{short} around 47 cm in labyrinthectomized Ss, the mean influence of the somatic graviceptors appears to be equal to or even larger than that of the otoliths.

CONCLUSIONS AND THEIR CONSEQUENCES

Because there is no correlation between the b, c, and Δ values on the one hand and the height of the lesion on the other hand within the L-paraplegic as well as in the TC-paraplegic group, even up to TSL at C6, shear and pressure forces at the skin, again (cf. Basic Results), are disproven as sources of somatic graviception in these experiments. By the same token the integrated inputs from mechanoreceptors between the vertebrae should be ruled out as possible candidates. Where, then, may the somatosensory information originate?

Consider the main findings of the preceding section. With TSL up to the first lumbar segments, the settings are close to normal. Between L1 and T10 a striking change occurs: the slopes of "long" over "short" grow to unity (at least); the mean increase of b_{long} over b_{short} disappears; the means of c_{long} and c_{short} become smaller, but not zero; the *relative* effect of leg extension on the centrifuge may even increase. These relations exist with TSL up to and including the 6th cervical segment.

Hence we must reckon with (at least) *two* different somatic graviceptive systems, one that is lost with destruction of the last two thoracic segments of the cord (it shall be named the "truncal" system) and another that is still operative with lesions up to C6 (inclusively). This residual second system shall be treated first.

The Residual Second System

Anatomy

With TSL at C6 the brain still receives information about a force acting on a mass with a centroid at least 23 cm, but rather more than twice as distant from the otoliths, and then still reacts to the extension of legs that are spinally deafferented.

Obviously, the messages transporting this information must somehow bypass an interruption of the cord between C6 and the segment where the relevant receptors are situated. Unless their spinal afference enters the cord above C6, these messages could then only reach the brain through the sympathetic trunk or through the vagal nerve. In the first case, because there is no correlation between the effect of leg extension and the height of the lesion, all of it would have to enter the cord via the first five *rami communicantes grisei*. But this seems unlikely since the main traffic between the sympathetic trunk and the cord is largely in*tra*- rather than in*ter*segmental. Thus the vagal nerve is the most likely carrier.

Function

Further insight is gained by considering Equations 1, 2, and 3. Such a relation results if the unknown sensory system would measure the z-axis force component of a (relatively denser) mass that *moves caudad if the legs are extended.* Let the centroid of

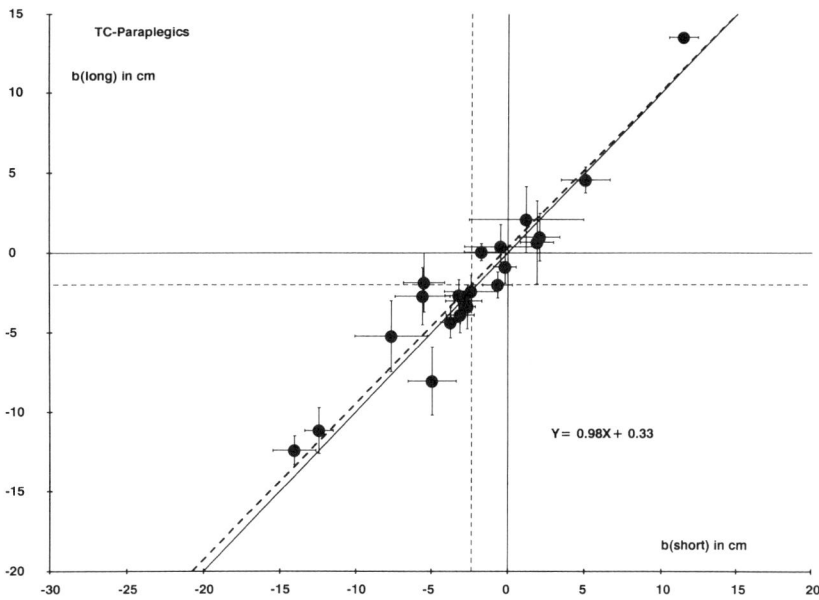

FIGURE 5. SHP in Ss with complete paraplegia at C6 to Th11 (TC paraplegics) on the tiltable board. All procedures and symbols as in FIGURE 2. Note that the difference between b_{long} and b_{short} is virtually zero.

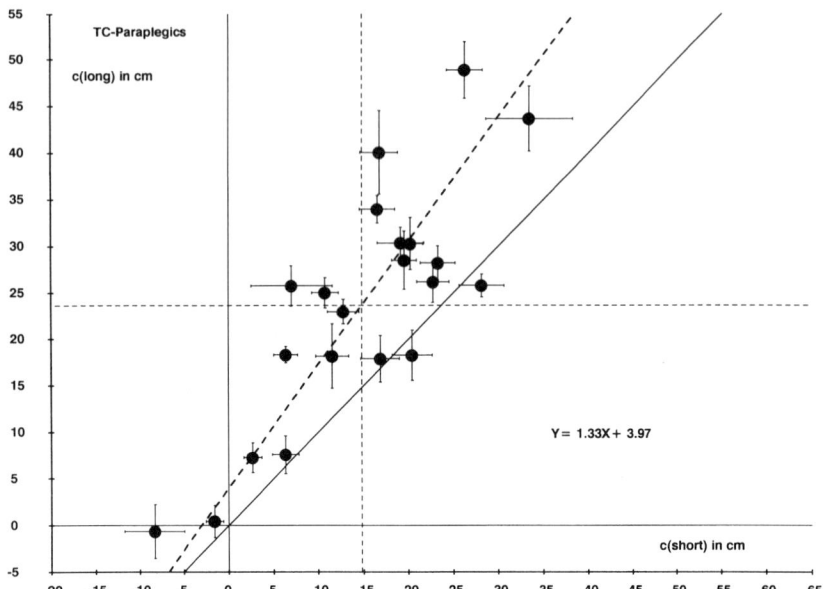

FIGURE 6. SHP in TC paraplegics on the centrifuge. All procedures and symbols as in FIGURE 3. Note that the means (dashed lines) are smaller than in FIGURE 3, yet far from the means of FIGURE 5, where they should be if the somatosensory influence had been totally abolished. The slope of the LSD is significantly larger than that in the normal Ss ($p/2 = 2.1\%$), and the absolute term not significantly different from zero ($p = 45.7\%$).

this mass then shift by $(\Delta_{long} - \Delta_{short})/\Delta_{short} \approx 50\%$ of its distance from the BA when the legs were flexed. But let the amplitude of this sensory system (i.e., its influence on postural control) remain unaltered. Consequently, leg extension could *not* affect the SHP on the tiltable board, because there the z-force component is independent of the position of the centroid. But on the centrifuge the z-force component necessarily increases with increasing distance of the centroid from the CA.

Only the centroid of a fluid within the body can obey these relations in TC paraplegics, notably the centroid of the mass of the blood. This may resolve the riddle of how information about the extension of spinally deafferent legs can reach the brain. On the other hand it raises the question of how the fluid shift can produce the required signal about the size of the z-axis force component at the centroid. It fits the emerging picture, however, that the vagal nerve is notably involved in cardiovascular control, that 80–90% of its fibers are afferent, and that the role of most of this input is still obscure. Moreover, body tilt is well known to affect heart rate and mean arterial blood pressure [5] (for review, see Reference 6). The second somatic graviceptive system shall here be provisionally—and perhaps prematurely—named the "vascular" graviceptive system.

The "Truncal" Graviceptive System

The function lost with TSL between L1 and T11 is very different from the one still operative in TC paraplegics. Therefore functional properties shall be treated first.

Function

On the truncal system leg extension has a twofold effect.

1. Because the slope of b_{long} over b_{short} and that of c_{long} over c_{short} are smaller than unity in normal Ss (see FIGURES 2 and 3), leg extension must enlarge the *amplitude* (R_1) of the unknown sensory system with respect to that of its otolithic (S_1) and vascular (V_1) counterparts.
2. Because on the tiltable board, in contrast to the TC paraplegics (FIGURE 5), the LSD in FIGURE 2 crosses the ordinate at a (significant) positive value, leg extension must produce or enlarge a positive force-independent bias (R_0). Yet the bias of the saccule (S_0) and of the vascular system (V_0) remains unaltered.

Let the S feel horizontal if

$$S_0 + V_0 + E_i R_0 + (S_1 + V_1 + E_i R_1) \cos \rho = 0 \tag{4}$$

where E is a weighting factor that is correlated with the amount of leg extension. Then, since $b = -\cos\rho/G_r$,

$$(S_0 + V_0 + E_{long}R_0)/G_r + (S_1 + V_1 + E_{long}R_1)b_{long}$$
$$= (S_0 + V_0 + E_{short}R_0)/G_r + (S_1 + V_1 + E_{short}R_1)b_{short} \tag{5}$$

Hence

$$b_{long} = \frac{S_1 + V_1 + E_{short}R_1}{S_1 + V_1 + E_{long}R_1} b_{short} + \frac{R_0(E_{long} - E_{short})/G_r}{S_1 + V_1 + E_{long}R_1} \tag{6}$$

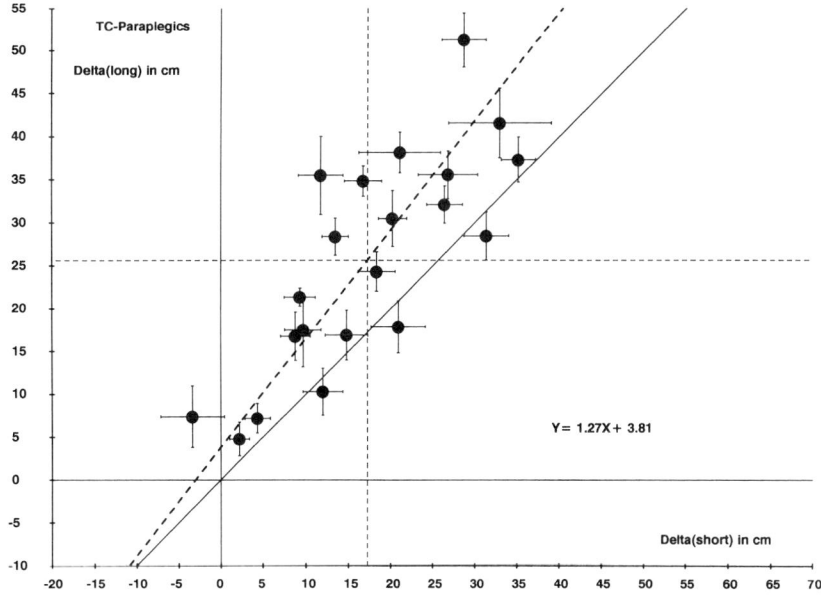

FIGURE 7. Δ_{long} over Δ_{short} in TC paraplegics, all symbols as in FIGURE 4. Note the comparatively large relative increase of mean Δ_{long} of 49.3% against 12.6% in normal Ss (FIGURE 4, see also FIGURE 9).

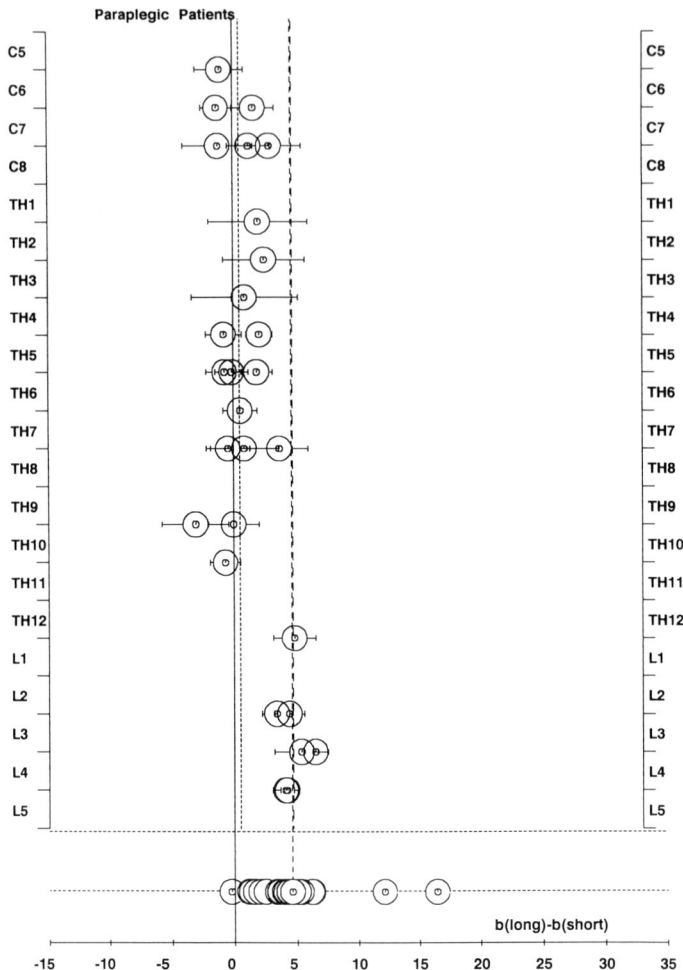

FIGURE 8. Dependence of the effect of leg extension on the height of the lesion in paraplegics, compared to that in normals (lowest line). Ordinate: site of the complete transversal loss of sensory input into the spinal cord (TSL). C: cervical segments. Th: thoracic segments. L: lumbar segments. Abscissa: $b_{long} - b_{short}$. Brackets: standard error of the mean. Note that this difference is virtually zero and independent of the height of TSL from C6 to Th11, yet indistinguishable from normal and highly significantly different from zero in Ss with TSL from L1 to L5.

Obviously, the experimentally found relations in normal Ss, namely, a slope of b_{long} over b_{short} smaller than unity and a positive bias, result if S_1, V_1, and R_1 vary stochastically between Ss and independently of leg extension, whereas in most or all Ss E_{long} is larger than E_{short}. In TC paraplegics, however, with $R_0 = R_1 = O$, it follows indeed that $b_{long} = b_{short}$.

The relations found on the centrifuge result in a just as straightforward way. A complete mathematical treatment of the data shall be published elsewhere.

Anatomy

The anatomical situation leads to the assumption that, in contrast to the vascular graviceptive system, the truncal system is locally concentrated and immobile. The last two thoracic segments (T11 and T12), where the truncal graviceptive afference presumably enters the cord, are connected via the *rami communicantes* and the sympathetic trunk with the lesser and the least splanchnic nerve, which is sometimes called the renal nerve. They contain among others relatively fast conducting (37 m/second) myelinated afferent fibers[7] which originate in the kidneys. Moreover, section of the respective *rami communicantes* is known to abolish kidney pain.[8]

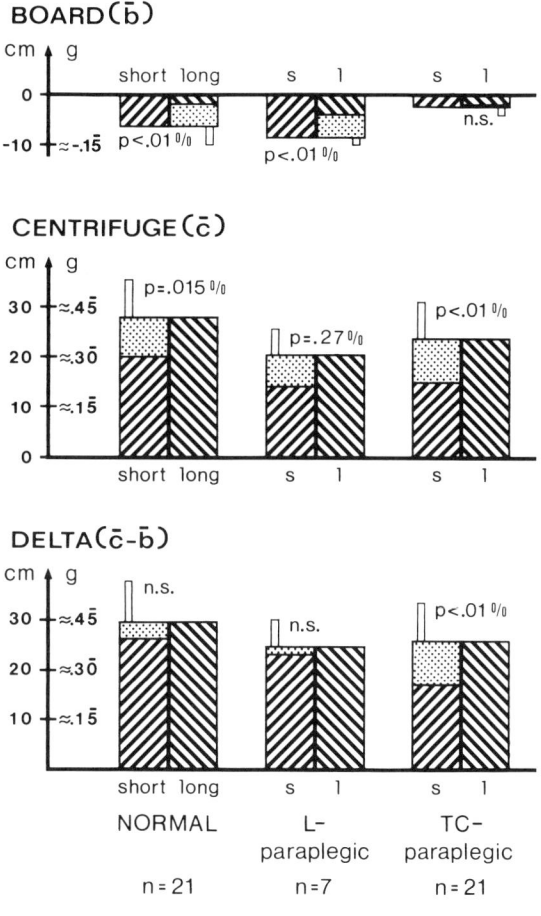

FIGURE 9. Comparison of mean values of 3 groups of Ss in the 2 experimental setups and their differences Δ, with legs straight (long, 1) and legs flexed (short, s). Ordinate: mean b, c, and Δ as well as G equivalents; n: number of subjects. Bars: SD of the n differences (long minus short). p: probability that these differences are by chance at least that much different from zero (two-tailed). Note the similarity of the b and the Δ values in the normal and the L-paraplegic Ss and the striking difference to those of the TC-paraplegic Ss.

Evaluation of all functional and anatomical findings thus centers on the question of whether and how the kidneys may partake in gravity reception. Kidneys are compact well-rounded structures filled with a constant amount of a fluid of greater density than their surrounds, notably the adipose capsules on which they rest. Furthermore, at any rate in one mammal, the rabbit, the kidney is densely covered by the receptive fields of pressure receptors of the slowly adapting kind subserved by relatively fast conducting (mean: 26 m/second) afferents in the renal nerve.[9]

Evidently, all these facts support our earlier conjecture of a supererogatory static function of the kidneys.

EXPERIMENTAL CROSS-EXAMINATION

Test of the Presumed Vascular Gravity Information

The method to be applied here rests on the result of Vaitl that an increase of mean arterial pressure, elicited by tilting the S from $\rho = 90°$ to $\rho = 45°$ can be nearly canceled by applying external pressure to the legs. [5]

Pilot experiments in 4 Ss showed no effect of pressure (30 to 45 mm Hg) to the (extended) legs on the board (mean $(b_{long} - b_{pressure}) = -0.41 \pm 1.04$ cm), but a mean *craniad* shift on the centrifuge (mean $c_{long} = 28.0 \pm 16.7$ cm and mean $c_{pressure} = 22.7 \pm 11.4$ cm, that is, close to c_{long} and c_{short} of the 21 normals of Normal Controls, last paragraph)—as expected according to the fluid-shift hypothesis of the Residual Second System—Function, first paragraph. It remains to be seen whether this result will be reproducible.

Test of the Presumed Role of the Kidneys

Here bilaterally nephrectomized Ss are recruited to be tested in the same four paradigms as were the paraplegic Ss. As of today, the result on 7 Ss has been as follows.

On the tiltable board the settings were distributed like those of the TC paraplegics (FIGURE 10). On the other hand their difference from those of normal Ss is clearly significant [the probability that the mean difference $(b_{long} - b_{short})$ in the normal Ss is by chance that large or larger than that found in the nephrectomized Ss *and* positive is $p = 0.24\%$), that is, the increase of b_{long} over b_{short} found in normal Ss is missing in nephrectomized Ss.

On the centrifuge, the slopes of the LSDs of c_{long} over c_{short} and Δ_{long} over Δ_{short} in the nephrectomized Ss are virtually equal to those of the TC paraplegics, and hence the former shows the same increase over that of the normal Ss. But the effect of leg extension is smaller than in the TC paraplegics, probably due to the considerable differences in cardiovascular conditions between the two groups.

Provided these results hold up in a larger sample, then hardly any doubt will remain on the conclusion that the kidneys in fact, in addition to their excretory function, partake in perception and control of posture.

FINAL REMARKS

In the light of this discovery, some findings in the literature may have to be reinterpreted or take on new meaning.

1. In a rather short (and apparently never followed up) study on thalamic and decorticated monkeys, Ito and Sanada have elicited tonic limb reflexes by pressing the viscera and by electrical stimulation of the splanchnic nerve close to the celiac ganglion as well as by tilt of the entire monkey.[10] Moreover, they reported the limb reflexes to be abolished after bilateral section of the splanchnic nerve.
2. A truncal graviceptive system that controls also pitch and roll is known in pigeons.[11-14] Remarkably, its influence becomes maximal when the pigeon stands on its legs.
3. The results of Do et al.,[1] mentioned in the Introduction, gain particular interest because they show a short latency response to sudden forward *pitch*

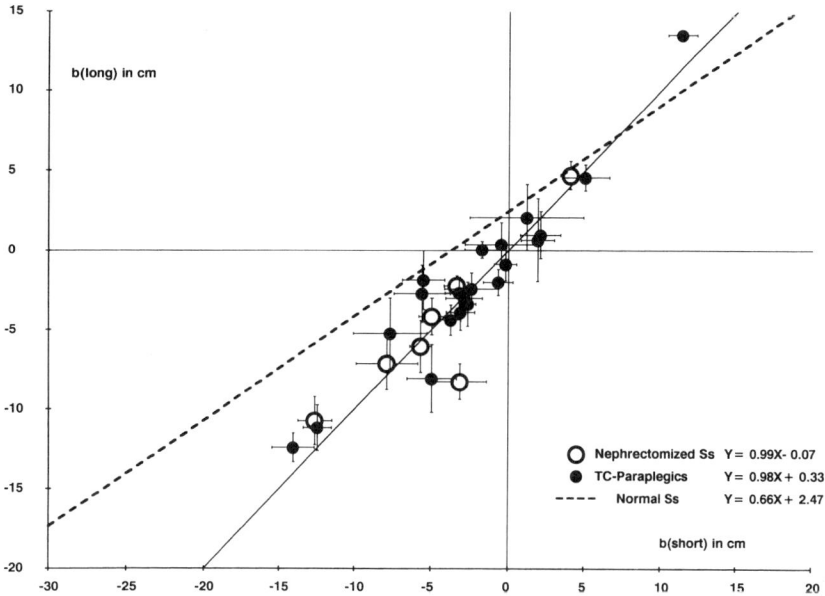

FIGURE 10. SHP of bilaterally nephrectomized Ss on the tiltable board. All procedures and symbols as in FIGURE 5. Open circles: nephrectomized Ss. Filled circles: TC paraplegics. Dashed line: LSD of normal Ss. Note that the difference $b_{long} - b_{short}$ is virtually zero, as in the TC paraplegics (see also FIGURE 5) but (significantly) different from that of the normal Ss (see also FIGURE 2).

which presumably cannot be attributed to the vestibulum or to leg proprioceptors, and hence supposedly originates in the trunk.
4. A somatosensory response to *roll* tilt in labyrinthine defective Ss has recently been found by Bles and De Graaf.[15] Their conclusion that it "must be due to somatosensory cues from the contact between body and chair"[15] should be revised in view of the present results, however.

Very likely, then, at least one of the two somatic graviceptive systems found here may turn out to be able to measure x- and y-axis as well as z-axis forces, and hence be a full-fledged three-dimensional processor of linear acceleration.

SUMMARY

In order to assess the effect of extravestibular gravity receptors on perception and control of body position against that of the otoliths, the subject (S) is exposed to gravitoinertial forces along the spinal (Z) axis on a tiltable board and on a sled centrifuge.

It turns out that (1) both effects, on average, are equally strong, although with considerable variance between Ss; (2) the centroid of the mass(es) governing the somatic receptors lies near the centroid of the body; and (3) somatic gravity reception contains two distinctly different systems.

Both appear unimpaired in paraplegic Ss with total bilateral sensory loss (TSL) from the 5th to the 1st lumbar spinal segment. One, the *truncal* system, is eliminated with TSL from the 11th thoracic segment upwards. Yet another is still functioning with TSL up to and including the 6th cervical segment, with the same effectiveness throughout this range. Hence it must be mediated by vagal or, less likely, sympathetic afference, that is, probably, by the influence of gravity on the cardiovascular system.

That the afference of the *truncal* system appears to enter the cord at the last two thoracic segments supports earlier conjectures about a supererogatory static function of the kidneys. In fact, on the tiltable board, 7 bilaterally nephrectomized Ss behaved like paraplegics with TSL between T11 and C6, yet differed significantly in the predicted direction from the normal controls.

ACKNOWLEDGMENTS

The present article marks the end point of a decade of work on human orientation by our group in Seewiesen. It would not have been possible without the endeavor and dedication of our staff, first and foremost of Evi Fricke, or without the extensive discussion in the group. Also from outside we were helped along by the leaders and the staff of neurological, nephrological, veterinarian, nuclear magnetic resonance–tomographical and aerospatial divisions of medical institutions in Murnau, Harlaching, the two Munich universities, and the Fliegerhorst Fürstenfeld-bruck, as well as by the crews promoting vestibular research at NASA, ESA, DARA, and DLR. We are also grateful for assistance and advice by Dr. E. Weller (Murnau; cf Selection Principles above), Prof. D. Vaitl (Giessen; cf. Test of the Presumed Vascular Gravity Information above), and Prof. W. Jänig (Kiel). Last but not least we thank our probands for devoting time and effort to what we had convinced them of being a worthwhile cause.

REFERENCES

1. Do, M. C., Y. Brenière & S. Bouisset. 1988. Compensatory reactions in forward fall: are they initiated by stretch receptors? Electroencephalogr. Clin. Neurophysiol. **69:** 448–452.
2. Horstmann, G. A. & V. Dietz. 1990. A basic posture control mechanism: the stabilization of the centre of gravity. Electroencephalogs. and Clin. Neurophysiol. **76:** 165–167.
3. Mittelstaedt, H. & E. Fricke. 1988. The relative effect of saccular and somatosensory information on spatial perception and control. Adv. Oto-Rhino-Laryngol. **42:** 24–30.
4. Mittelstaedt, H. 1988. The information processing structure of the subjective vertical. A cybernetic bridge between its psychophysics and its neurobiology. *In* Processing Structures for Perception and Action. H. Marko, G. Hauske & A. Struppler, Eds.: 217–263 VCH-Verlagsgesellschaft. Weinheim, Germany.

5. VAITL, D. & H. GRUPPE. 1990. Changes in hemodynamics modulate electrical brain activity. J. Psychophysiol. **4:** 41–49.
6. CIRIELLO, J., M. M. CARSON & C. POLOSA. 1989. The central neural organisation of cardiovascular control. Prog. Brain Res. **81**.
7. CALARESU, F. R., P. KIM, H. NAKAMURA & A. SATO. 1978. Electrophysiological characteristics of renorenal reflexes in the cat. J. Physiol. **283:** 141–154.
8. WHITE, J. C. & W. H. SWEET. 1969. Pain and the Neurosurgeon. C. C. Thomas, Publisher. Springfield, Ill. (Especially p. 578–582.)
9. NIIJIMA, A. 1975. Observation on the localization of mechanoreceptors in the kidney and afferent nerve fibers in the renal nerves in the rabbit. J. Physiol. **245:** 81–90.
10. ITO, T. & Y. SANADA. 1960. Location of receptors for righting reflexes acting upon the body in primates. Jpn. J. Physiol. **15:** 235–242.
11. MITTELSTAEDT, H. 1964. Basic control patterns of orientational homeostasis. Symp. Soc. Exp. Biol. **18:** 365–385.
12. BIEDERMAN-THORSON, M. & J. THORSON. 1973. Rotation-compensating reflexes independent of the labyrinth and the eye. Neuromuscular correlates in the pigeon. J. Comp. Physiol. **83:** 103–122.
13. VOLLRATH, F. W. & J. D. DELIUS. 1976. Vestibular projections to the thalamus of the pigeon. Brain Behav. Evol. **13:** 58–68.
14. BILO, D. & A. BILO. 1978. Influence of wind stimuli on vestibular and optokinetic wing and tail reflexes of the domestic pigeon (*Columbia livia*). Verh. Dtsch. Zool. Ges.: 263.
15. BLES, W. & B. DEGRAAF. 1991. Ocular rotation and perception of the horizontal under static tilt conditions in patients without labyrinthine function. Acta Otolaryngol. Stockh. **111:** 456–462.

Spatial Orientation of the Vestibular System[a]

THEODORE RAPHAN,[b,c] MINGJIA DAI,[c]
AND BERNARD COHEN[c,d]

[b]Department of Computer and Information Sciences
Brooklyn College
City University of New York
Bedford Avenue and Avenue H
Brooklyn, New York 11210

[c]Department of Neurology
[d]Department of Physiology and Biophysics
Mount Sinai School of Medicine
One Gustave Levy Place
New York, New York 10029

INTRODUCTION

By modeling the ocular response to head and surround rotations about a vertical axis when the body is aligned with gravity, we have identified a processing element in the vestibuloocular reflex (VOR) called the "velocity storage integrator".[1-3] It is responsible for the dominant time constant of the VOR, the slow dynamics of optokinetic nystagmus (OKN), and for optokinetic after-nystagmus (OKAN).[1,2,4,5]

Various architectures have been proposed to explain how the visual and vestibular systems utilize velocity storage in one dimension.[1,2,5-7] The Raphan model of the VOR and of OKN has been extensively tested and has been useful in predicting normal data on the VOR, OKN, and OKAN as well as the effects of lesions and drugs.[2,4,5,8] It models velocity storage as a nonideal integrator. The semicircular canals, the otoliths, and the visual system couple to the integrator to generate the slow components of compensatory eye velocity, and direct pathways around the integrator from both the visual and vestibular systems are responsible for the rapid dynamics of ocular compensation. This model also predicts differential effects of adaptation and habituation,[9] and an extension of the model to three dimensions has elucidated some of the relationships between velocity storage and the encoding of the spatial vertical during motion.[10-13]

The model of compensatory eye velocity for rotations about any axis in three-dimensional space was developed by representing velocity storage as a dynamical system in which the structure of the system matrix was determined by body orientation with regard to gravity.[11,13] It was assumed that the system matrix for a tilted position was the result of two linear transformations of the system matrix for the upright position. One modifies the eigenvalues of the system matrix while another is a similarity transformation which rotates the eigenvectors. A methodology was then developed to identify the eigenvalues and eigenvectors of the velocity storage system

[a]Supported by EY 04148, NS00294, EY01867, NASA-MVI project, PSC-CUNY award 662480, and NIH-NIGMS MRC 5T34 08078.

140

matrix from experimental data during OKAN for side-down tilts,[13] as well as for prone and supine tilts.[14]

Cross-coupling has also been noted during postrotatory nystagmus following off-vertical axis rotation (OVAR) in the cat[15] as well as in humans.[16] Whether the model structure obtained using OKAN data for side-down tilts would hold for supine or prone positions or for eye velocity generation during rotations that reorient the head and body with regard to gravity is not known.

The purpose of this paper is to review the model of the organization of velocity storage in three dimensions and determine whether the same organizational principles apply during OKAN which induces a roll component of eye velocity and during postrotatory nystagmus following OVAR which excites the semicircular canals.

EXPERIMENTAL METHODS

A complete description of the experimental methods, protocol and analysis techniques can be found in Raphan and Sturm[11] and Dai *et al.*[12] In brief, the horizontal, vertical, and roll position of one eye was detected with a scleral search coil.[17] Eye movements were induced by optokinetic stimulation or by angular acceleration about vertical or off-vertical axes using a vestibular stimulator that provides four axes of rotation (Contraves Goerz, Neurokinetics). The slow-phase eye velocity vector during the OKAN was related to the eigenvalues and eigenvectors of the velocity storage system matrix by fitting the data to weighted sums of exponentials using a Levenberg-Marquardt algorithm or a Chebychev optimization technique.[12-14]

Eigenvectors were also determined by putting the animal in various tilted positions and giving head-centered stimulation about axes that induced vertical or roll as well as horizontal components of eye velocity. We define this as "oblique head-centered" stimulation. By definition of an eigenvector,[18] the OKAN slow-phase velocity trajectory should stay along the axis of visual stimulation, with all of the velocity components decaying with the same time constant.[11] Thus, when pitch velocity was plotted against yaw velocity, the trajectory was linear.[13] The linearity was determined by fitting the data with a linear regression line. At each value of horizontal velocity, the difference between the vertical velocity and the straight-line fit was obtained. This represented a temporal sequence of the residuals relative to a linear trajectory. The power spectrum or Fourier transform of the autocorrelation of the temporal sequence of the residuals was then analyzed. For decays that were linear, there should be no signal component in the residuals, implying that the residuals were basically due to noise. A parameter called the spectral width was defined as the radius of gyration of the spectrum of the temporal sequence of the residuals. The spectral width should be large for linear declines, which had only noise components in the residuals. Temporal sequences further away from a straight line, due to cross-coupled components, should have a signal component in the residuals, and the spectral width should be substantially smaller. Thus, by using a spectral analysis of the temporal sequence of residuals, it was possible to evaluate the goodness of fit of linearity and proximity to eigenvector trajectories.[11]

During postrotatory nystagmus following off-vertical axis rotation, yaw and pitch eye velocities were fit with sums of exponentials. The eigenvalues and eigenvectors were then computed.[13] The trajectory was also evaluated as it approached the origin of the state space to determine the direction of the eigenvector.

RESULTS

Three-Dimensional Model of Visual-Vestibular Interaction

A simplified three-dimensional model of visual-vestibular interaction is shown in FIGURE 1. The external signals that drive the model are surround velocity, r_o, and head velocity, r_h, in space. These signals are transformed into canal-based coordinates for central processing.[19-22] Because the measurements of eye velocity are done in head coordinates, the transformations from head to canal and back to head coordinates have been left out for simplicity.

The head velocity signal is operated on by a dynamic three-dimensional transformation, D_{can}, due to the elastic properties of the cupula and the inertia and viscosity of the endolymph. The dynamic transformation can be represented by a diagonal system matrix whose output is the eighth nerve signal, r_v.[3] This signal activates the velocity storage integrator via a coupling matrix, G. It also activates the vestibular nuclei via a direct vestibular path.

The visual system also couples to the velocity storage integrator, as well as having a "direct" pathway around it which contributes to the generation of gaze in space. This is accomplished by subtracting the gaze velocity signal from surround velocity to give retinal slip, e, when the lights are on (LIGHT = ON). When the lights are off (LIGHT = OFF), the retinal slip is zero. The signal e is processed through a nonlinear network[5] $N(e)$ to generate e' which activates a multidimensional representation of velocity storage. In this manner, the central representation of slip contributes to the integrator state and the generation of v_n.

The velocity storage integrator state, x, represents a dynamic superposition of semicircular canal and visual signals. The integrator signal is summed with the direct vestibular pathway to generate a velocity command v_n, representing the signal found in "vestibular-only" neurons in the vestibular nuclei.[23-30] The signal in the vestibular nuclei, v_n, is summed with the direct optokinetic signal to generate eye velocity relative to the head, v_h. When v_h is mechanically summed with head velocity, gaze velocity in space, v_g, is obtained. The signal v_n also contributes to the generation of the direct optokinetic output signal that drives the oculomotor system.[5] It should be

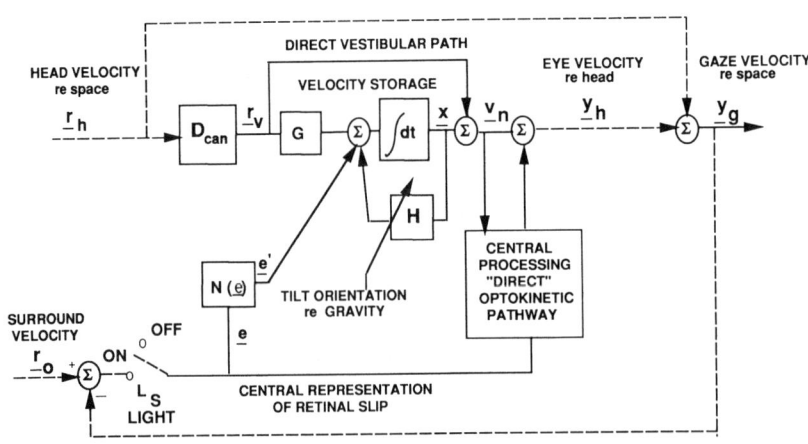

FIGURE 1. Three-dimensional model of visual vestibular interaction. See text for description.

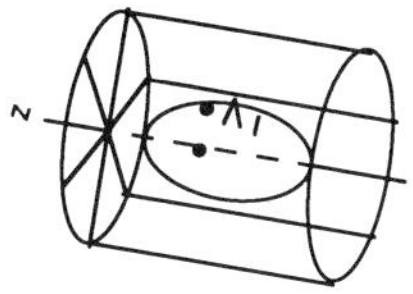

$$\begin{bmatrix} h_{11} & 0 & h_{13} \\ 0 & h_{22} & h_{23} \\ 0 & 0 & h_{33} \end{bmatrix}$$

EIGENVECTORS BASIS VECTORS

$$\underline{u}_1 = \begin{vmatrix} 1 \\ 0 \\ 0 \end{vmatrix} \qquad \underline{u}_2 = \begin{vmatrix} 0 \\ 1 \\ 0 \end{vmatrix} \qquad \underline{u}_3 = \begin{bmatrix} \dfrac{h_{13}}{h_{33}-h_{11}} \\ \dfrac{h_{23}}{h_{33}-h_{22}} \\ 1 \end{bmatrix}$$

RECIPROCAL BASIS VECTORS

$$\underline{r}_1 = \begin{vmatrix} 1 \\ 0 \\ \dfrac{-h_{13}}{h_{33}-h_{11}} \end{vmatrix} \qquad \underline{r}_2 = \begin{vmatrix} 0 \\ 1 \\ \dfrac{-h_{23}}{h_{33}-h_{22}} \end{vmatrix} \qquad \underline{r}_3 = \begin{vmatrix} 0 \\ 0 \\ 1 \end{vmatrix}$$

ZERO INPUT RESPONSE VECTOR

superposition of eigenvectors

$$\underline{x}(t) = \langle \underline{x}(0), \underline{r}_1 \rangle \underline{u}_1 e^{h_{11}t} + \langle \underline{x}(0), \underline{r}_2 \rangle \underline{u}_2 e^{h_{22}t} + \langle \underline{x}(0), \underline{r}_3 \rangle \underline{u}_3 e^{h_{33}t}$$

FIGURE 2. Mathematical structure of system matrix, H, for arbitrary tilted position of the head, assuming only cross-coupling from yaw to pitch and roll axes. Under this condition, the eigenvector basis vectors associated with pitch and roll move with the head, while the yaw axis eigenvector has a spatial component dependent on the parameters of the H matrix. The eigenvectors, and associated reciprocal basis vectors, determine the zero input response vector which is used to model OKAN.

noted that the summation of head velocity and eye velocity relative to the head ignores the processing through the oculomotor velocity to position integrator as well as the oculomotor plant. Only the velocity aspects of the signal flow have been modeled. The dotted lines represent mechanical variables, while the solid lines represent neural signals.

The dynamic response in three dimensions is determined by the velocity storage system matrix, H, whose mathematical structure is determined by the tilt orientation relative to gravity.

Mathematical Structure of the System Matrix as a Function of Tilt

The mathematical structure of the system matrix is determined by its elements, the **h** parameters (FIGURE 2, top right). They are functions of tilt angle and become modified as the head is reoriented with regard to gravity. In general, there would be nine elements, h_{ij}, associated with this 3×3 matrix, each designated by its row (i) and column (j), respectively. However, there are constraints that can be imposed on the parameters that restrict the degrees of freedom and simplify the modeling problem. The constraints also give insight into how velocity storage codes orientation information.

In previous studies,[13] it has been shown that over a wide range of side-down head positions, there is virtually no cross-coupling from the pitch to the yaw axis during OKAN. In addition, it is assumed that there is no roll to yaw axis coupling during OKAN. h_{31} and h_{32} represent cross-coupling to yaw from pitch and roll, respectively, and have been set to zero (FIGURE 2, top). It has also been postulated that there is no cross-coupling from pitch to roll and from roll to pitch, constraining h_{21} and h_{12} to be zero. Thus, for an arbitrary position of the head with regard to gravity, the system matrix (H) is upper triangular (FIGURE 2, top), and its eigenvalues are the diagonal elements.[18,31]

If the eigenvalues of the matrix are distinct, then the eigenvectors of the system matrix form a basis for describing eye velocity vectors in head coordinates.[11] It is this crucial insight that enables a definition of the code for spatial orientation embedded within the velocity storage integrator in terms of the eigenvectors of the system matrix (FIGURE 2). Note that the constraints on the **h** parameters derived from experimental observations demand that the pitch and roll eigenvectors, u_1 and u_2, respectively, move with the head (FIGURE 2), whereas the yaw eigenvector, u_3, has a spatial component induced by the parameters h_{13} and h_{23} (FIGURE 2). The zero input response vector is a superposition of eigenvectors (FIGURE 2, bottom). Therefore, the pitch and roll components of the state vector would contain a mode associated with the yaw eigenvalue (FIGURE 2, bottom).

In the upright position, there is no cross-coupling and the system matrix would be diagonal. Under these conditions, its eigenvectors are mutually orthogonal and are aligned with the body and spatial coordinate frame. Each component of the response state vector would contain only one mode, i.e., a single time constant corresponding to its eigenvalue.[18]

Thus, the model predicts that once the eigenvalues and eigenvectors of the H matrix are determined for a given head orientation, the cross-coupling of velocity storage can be characterized.

IDENTIFICATION OF H MATRIX PARAMETERS: MODEL SIMULATIONS AND DATA

To find the eigenvalues and eigenvectors of the velocity storage system matrix for a given head orientation, the solution to the state equations, $x(t)$, was compared with data of pitch, roll, and yaw components of slow-phase eye velocity obtained during OKAN, and an error sequence was generated. A modified Marquardt algorithm[11,13] or Chebychev approximation technique[12] was applied to search for the parameters that would minimize the mean square error.

The system matrix parameters found by the algorithm were incorporated into the proposed model, and simulations were compared to the data for OKN and OKAN (FIGURE 3A and B). For purposes of this simulation, the coupling matrix to the integrator [N (e) of FIGURE 1] was assumed to be linear and not modified by gravity.

Therefore, it was represented by a constant matrix G_0 and is given by:

$$G_0 = \begin{bmatrix} 0.25 & 0 & 0 \\ 0 & 0.25 & 0 \\ 0 & 0 & 0.25 \end{bmatrix} \tag{1}$$

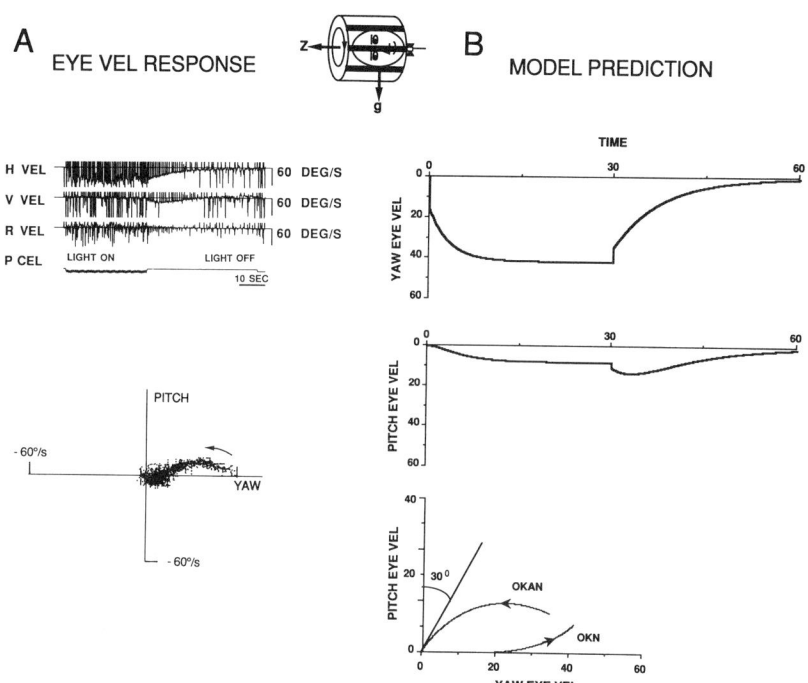

FIGURE 3. A: Three-dimensional eye velocity response during OKN and OKAN as well as the corresponding trajectory of eye velocity during OKAN when the monkey was right-side down (bottom). **B:** Model predictions for eye velocity as a function of time and the associated trajectory during OKN and OKAN. Note the close correspondence between the predicted and experimental data.

The *"direct" optokinetic pathway* is represented simply as a matrix, G_1, which is driven by *central representation of retinal slip*. G_1 is given by:

$$G_1 = \begin{bmatrix} 0.35 & 0 & 0 \\ 0 & 0.35 & 0 \\ 0 & 0 & 0.35 \end{bmatrix} \tag{2}$$

These matrices were chosen based on the coupling coefficient and direct pathway gain found for the one-dimensional model.[5,8]

Optokinetic Response to Yaw Axis Stimulation at Tilted Position

For a yaw axis optokinetic stimulus in the animal's frame of reference when at 90° right-side-down tilt, there was only horizontal eye velocity during OKN (FIGURE 3A). However, when the lights were turned off, a vertical component of OKAN developed with downward slow phases (FIGURE 3A). For this response, the **h** parameters of interest for the yaw to pitch cross-coupling were computed to be h_{11} = -0.2, $h_{33} = -0.133$, and $h_{13} = 0.12$. These correspond to yaw and pitch time constants of 7.5 and 5.0 seconds, respectively, and a yaw axis eigenvector of 30° relative to the spatial vertical.

The organization of the cross-coupling was such that the vertical component was oriented in the spatial frame as the horizontal component was in the body frame. That is, for right-side down, OKN slow-phase velocity to the left corresponds to an eye velocity vector directed upward along the z-axis according to a right-hand rule (FIGURE 3, inset). During OKAN, the vertical component of eye velocity is downward (FIGURE 3A) and corresponds to a velocity vector directed upward, opposite to gravity, g, in the spatial frame (FIGURE 3). This organizational principle is contained in the sign of the parameter h_{13}. The trajectories in eye velocity space were curved and were consistent with the response expected when the initial condition vector is not aligned with an eigenvector (FIGURE 3A, bottom).

Model simulations of OKN and OKAN when an animal is on its right side receiving an upward directed yaw stimulus velocity predicted the data shown in FIGURE 3A. During OKN the pitch component was small, and the velocity vector was close to the animal yaw axis (FIGURE 3B, first and second traces). The vertical velocity (V vel) of the data (FIGURE 3A) was closer to zero than the model prediction. This was probably due to additional "dumping" and "suppression" of the vertical component which have been left out of the model for simplicity.[1,2,4,5,32]

At the onset of OKAN, yaw eye velocity dropped rapidly and was followed by a slow decline (FIGURE 3B, first trace). There was a small jump in the pitch component of eye velocity, followed by a rounded peak which decayed to zero (FIGURE 3B, second trace). These compare favorably with the data (FIGURE 3A). The trajectory in velocity or state space during OKAN was a curved path which decayed toward zero (FIGURE 3B, third trace). The trajectory approached the origin of the state space at an angle of 30° which was the eigenvector chosen for the simulation (FIGURE 3B) and was the calculated eigenvector for this experiment.

Oblique Head-Centered Stimulation

The curved trajectory of the response in velocity state space considered above (FIGURE 3B, bottom) arose because the OKN stimulus velocity vector was not along an eigenvector. If the initial state vector is aligned with an eigenvector, then the inner products of the initial state vector with the reciprocal basis vectors associated with the other two eigenvectors are zero (FIGURE 2, bottom), and the trajectory of OKN and OKAN predicted by the model would be a straight line.[11] The model predictions show a jump and slow rise to a steady-state velocity in both yaw and pitch components of OKN which then decline to zero with the same time constant (FIGURE 4B, top two traces). The predicted trajectory in velocity state space is a straight line for both OKN and OKAN (FIGURE 4B, bottom).

When head-centered oblique optokinetic stimulation was given along the eigenvector direction, both the yaw and pitch axis eye velocity components (FIGURE 4A, H vel and V vel) were at an angle that was appropriate for the stimulus. When the lights were extinguished, both components of the OKAN decayed to zero with approximately the same time course (FIGURE 4A, top). The response in velocity state space was close to a straight line (FIGURE 4A, bottom). When a spectral analysis was done on the residual sequence (see Methods), it had narrow autocorrelation function and broad power spectrum (FIGURE 4A, bottom),[11,13] confirming the linearity. Thus, the model simulations (FIGURE 4B) predicted the data on the time course of the eye velocity as well as the trajectory accurately.

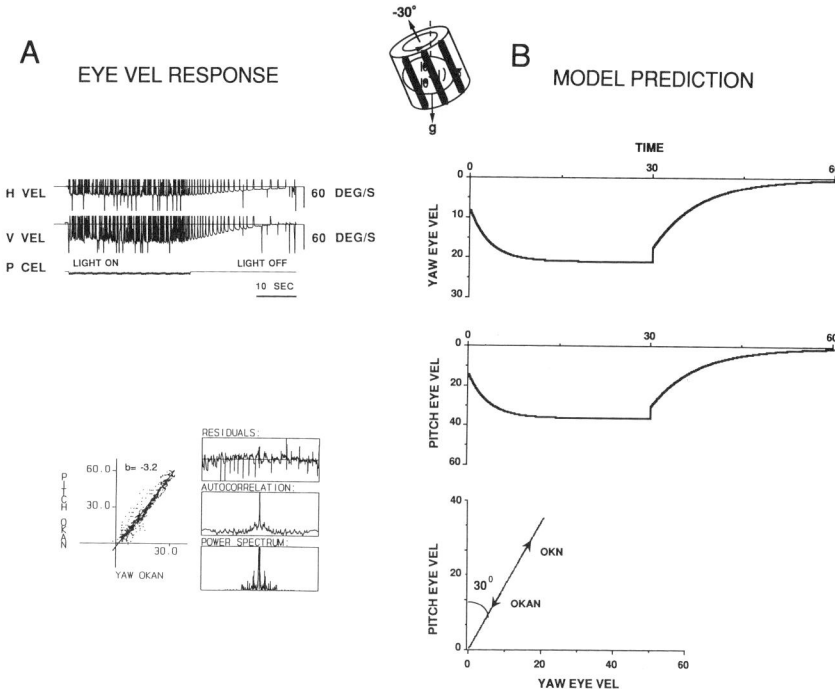

FIGURE 4. **A:** Eye velocity during OKN and OKAN (top) as well as the corresponding trajectory of eye velocity during OKAN when the monkey was right-side down and receiving oblique stimulation along an eigenvector direction (bottom). The narrow autocorrelation and the relatively wide power spectrum indicate that OKAN velocities were along the eigenvector. **B:** Model predictions for eye velocity as a function of time and the associated trajectory during OKN and OKAN. Note that the response predicted for OKN and OKAN is a straight line along the eigenvector in velocity space.

Cross-coupling also was present from horizontal or yaw axis OKN and OKAN to roll in the prone position (FIGURE 5). The roll eye velocity followed the right-hand rule for cross-coupling.[10,13] When the yaw eye velocity vector was directed toward the feet of the animal (inset), the cross-coupled vector of the roll component was downward in space (FIGURE 5A). When the eye velocity vector was fit by the model (FIGURE 5B), the computed eigenvector was approximately 20°. Oblique optokinetic

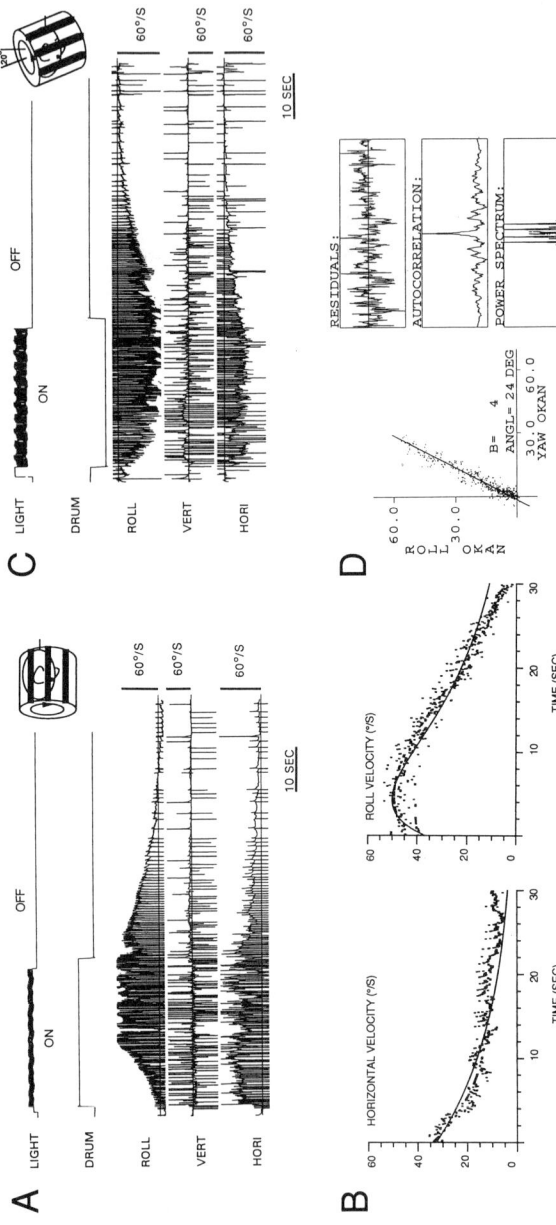

FIGURE 5. A: Cross-coupling from horizontal or yaw axis OKN and OKAN to roll when a monkey was in the prone position. The roll eye velocity followed the right-hand rule for cross-coupling.[10,13,14] When the yaw eye velocity vector was directed toward the feet of the animal (inset), the cross-coupled vector of the roll component was downward in space. **B:** When the eye velocity vector was fit by the model, the computed eigenvector was approximately 20°. **C:** Oblique optokinetic stimulation along the computed eigenvector direction resulted in oblique OKN along that direction followed by OKAN whose components decayed concurrently with approximately the same time constant, corresponding to a single eigenvalue. **D:** The trajectory during OKAN following oblique stimulation, as in C, resulted in a decay along a straight line. Evidence for this is the narrow autocorrelation and the wide spectrum of the residuals.

stimulation along the computed eigenvector direction resulted in oblique OKN, followed by OKAN whose components decayed concurrently with approximately the same time constant, corresponding to a single eigenvalue (FIGURE 5C). The trajectory during OKAN following oblique stimulation resulted in a decay along a straight line as evidenced by the narrow autocorrelation and the wide spectrum associated with the residuals (FIGURE 5D). Thus, the cross-coupling from yaw to roll during OKAN followed the same principles as for yaw to pitch.

Relationship of OKAN Cross-Coupling to the Perception of the Vertical

We have postulated that the yaw eigenvector, as determined by the h parameters of velocity storage,[11,13] codes the spatial vertical. It was of interest to compare the behavior of the yaw axis eigenvector (FIGURE 6A and B) to other representations of the spatial vertical, e.g., to data on the perceived spatial vertical in humans (FIGURE 6). Neither the eigenvectors for up nor down slow phases matched the human perceptual data in $1 \times g$. However, the shape of the curves was similar, and the up and down eigenvector data were symmetrically distributed around the perceptual data for tilt angles of 60° and above in $1 \times g$. This symmetry is shown by the finding that the mean of the eigenvector data corresponded almost exactly to the human perceptual data in this range (FIGURE 6C, monkey mean and human $1 g$).

One explanation for the difference between the eigenvector behavior and that of the perceived vertical could be that human spatial orientation is derived from static tilt information,[33-35] while the eigenvectors are determined by information about static position and motion. Thus, if direction of motion symmetrically modulated a static internal representation of the vertical by altering the h parameters of the velocity storage system matrix, it could account for the asymmetries observed in the eigenvector angles for upward and downward eye velocity.

A similar kind of modulation of the static representation of the spatial vertical occurs in an altered gravitational field. For example, perception of the spatial vertical is shifted up in a $2 \times g$ environment produced by a gondola centrifuge.[35-37] Data from this condition overlay the monkey eigenvector data with upward slow phase velocity (FIGURE 6C, human $2 g$ and monkey up). Thus, the altered gravitational field modulated the percept of the spatial vertical in a manner similar to that of upward motion.

Downward motion presumably modulates the static internal representation of the spatial vertical in the opposite direction, as evidenced by the fact that the downward eigenvectors (FIGURE 6C, monkey down) fell below the human $1 \times g$ data. By analogy, this would correspond to an environment with reduced g.[38] This suggests that in microgravity, where there is no reference for spatial orientation, the vertical reference should shift to the body frame.

One implication of the close association between the eigenvector behavior in the monkey and the perceptual data of human is that this methodology could be applied to human OKAN. Despite the fact that human OKAN is weak,[39-42] and the eigenvalues of velocity storage are large, as long as the relationship between yaw and pitch axis eigenvalues and the cross-coupled parameters was in the correct ratio,[11] cross-coupling of OKN and OKAN could play a similar role in sensing or determining the spatial vertical.

Cross-Coupling during Vestibular Activation

Cross-coupling of yaw to pitch during postrotatory nystagmus in tilted positions following OVAR has been noted in cat[15] and human.[16] In contrast to OKAN, there is input to velocity storage from the semicircular canals when the animal is stopped in the postrotatory period. It is not known, however, whether the organization of the eigenvectors of velocity storage is the same in this condition as during OKAN.

When a monkey was stopped in the right-side-down position after leftward yaw rotation about an axis tilted 90° from the vertical, it developed a downward eye velocity component in the postrotatory nystagmus (FIGURE 7A), consistent with the

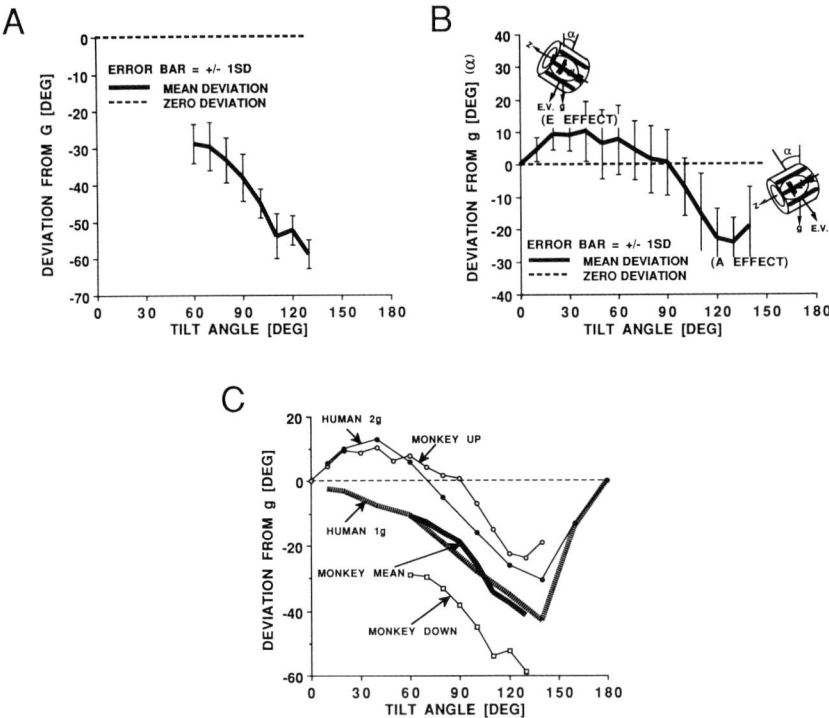

FIGURE 6. A: Deviation of the eigenvector angle with downward coupling from the spatial vertical (dotted line) as a function of tilt angle. The eigenvector angle was always below the spatial vertical, having solely Aubert-like effects. **B:** Deviation of the mean angle of the eigenvector with upward coupling as a function of tilt angle from the spatial vertical (dotted line). The angle of the eigenvector was close to the spatial vertical, although it changed from one side to the other (Müller and Aubert-like effects). **C:** Comparison of deviations from the spatial vertical of monkey eigenvectors and human perception (from Reference 35). For upward coupling (monkey up), the deviation of the eigenvector followed the human perceptual error of the vertical in a $2 \times g$ environment (human 2g). For downward coupling (monkey down), the deviation of the eigenvectors was greater than the human perceptual error in a $1 \times g$ environment (human 1g). However, the mean deviation of monkey up and monkey down eigenvectors overlay the error of human 1g.

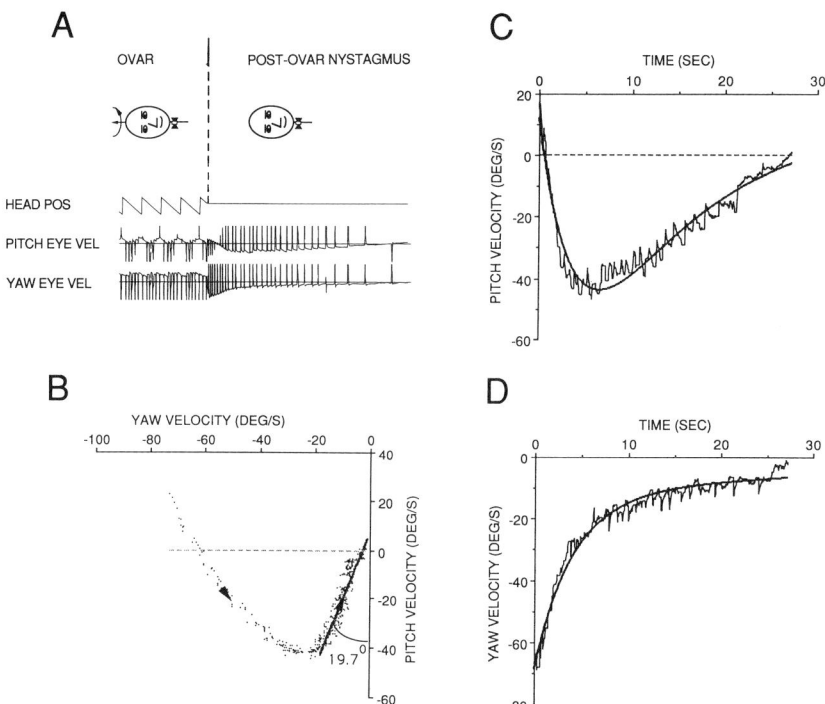

FIGURE 7. A: Cross-coupling following stopping right-side down after leftward OVAR stimulation. During OVAR, steady-state yaw eye velocity was to the right with an oscillating upward pitch eye velocity. Upon stopping, there was a reversal of the yaw eye velocity during postrotatory nystagmus. The cross-coupled component was in the downward direction. **B:** The trajectory of eye velocity during the post-OVAR postrotatory nystagmus approached a line that was approximately 20° relative to the spatial vertical. This corresponded to the computed eigenvector when the eye velocity vector was fit by exponentials (**C** and **D**).

organizational principle stated above. The trajectory of eye velocity in state space was curved and approached an angle of approximately 20° relative to the spatial vertical as eye velocity decayed to zero (FIGURE 7B). This was close to the eigenvector direction for OKAN in this and previous studies.[12,13] When eye velocity components during the postrotatory nystagmus were fit by sums of exponentials, the computed eigenvector was also close to 20° relative to the spatial vertical (FIGURE 7C and D).

DISCUSSION

The data indicate that the coding of the spatial vertical in the central nervous system is important for determining the temporal characteristics of compensatory eye movements in response to head or surround movement. This coding is accomplished by modifying the system matrix associated with velocity storage[2,11,13] such that its yaw axis eigenvector remains close to the spatial vertical while the pitch axis eigenvector rotates with the head. Since the eigenvectors form a basis for the state space that produces eye velocity, tilting the head skews this basis and generates a nonorthogonal coordinate frame.[11] Stimulation along the subject yaw axis, therefore, activates modes along both skewed axes giving rise to cross-coupling of eye velocity from the subject yaw to pitch axis.[3,10,11,13]

An important finding in this paper is that the yaw axis eigenvector for the right-side-down position with downward cross-coupling is approximately the same when the animal is receiving motion information from the semicircular canals about yaw rotation as during OKAN. This suggests that the eigenvectors established for velocity storage as a function of tilt angle during OKAN are not much altered by motion information input to velocity storage about the subject yaw axis detected by the semicircular canals. Thus, velocity storage could act as a "neural gyroscope" that maintains an estimate of the spatial vertical during a wide range of stimuli that generate motion signals. It could help maintain gaze compensation during circular locomotion about an axis aligned with gravity.[43] The data showing the close association of the angle of the yaw axis eigenvector with the angle of the perception of the vertical over a wide range of orientations with regard to gravity (FIGURE 6) support the link between velocity storage and postural orientation mechanisms during motion in a gravitational environment.

Spatial orientation has been widely studied as a static phenomenon with regard to gravity and to the external stimuli that affect it.[44] For example, pigeons trained to peck along a visual line will preferentially choose a line perpendicular to the floor in darkness. However, when the chamber in which they are housed is lit, they prefer the line parallel to the walls.[45,46] Human perception of the visual vertical is altered by body tilt[33,47,48] as well as by visual and somatosensory cues.[33] Image motion in a clockwise or counterclockwise direction also affects the perception of the spatial vertical.[49] Thus, perception of the spatial vertical is influenced by a wide variety of sensory stimuli and in many regards can be dissociated from reflex behavior that they induce.[50,51] This study shows that how we encode the spatial vertical determines oculomotor and gaze compensation and that velocity storage is an important link in that determination. The finding that orientation information is encoded in the velocity storage integrator, and that it controls the dynamics of compensatory eye movements, could be important for studying how spatial orientation is encoded in the central nervous system.

The demonstration that coupling between various states of the integrator in the matrix, H, is modified as a function of head tilt has physiological implications. If states of the integrator are represented in the activity of lateral and vertical canal related neurons in the vestibular nuclei,[29–30] then these neuron classes must be controlled during cross-coupling by central structures that are sensitive to otolith input, such as the nodulus and ventral uvula.[52] The methodology developed in this study could provide the theoretical basis for exploring the encoding of the spatial vertical in these neural networks.

There has been some controversy about how velocity storage is realized and activated by the visual and vestibular systems which has bearing on its representation in three dimensions. The Raphan model, which has been studied in detail and explains a wide range of data,[1,2,5,8,9] represents velocity storage as a nonideal integrator which is coupled to by the visual and vestibular systems as well as having direct paths around it. A model put forth by Robinson[6] suggested through teleology that the long time constant of the VOR and OKAN was due to low-pass filtering of the retinal slip signal (represented by an integrator with a small time constant), which was activated by a positive feedback of efference eye velocity when the semicircular canals were activated by head rotation. This architectural feature of visual-vestibular interaction can also simulate aspects of velocity storage in one dimension.

By examining parameter changes in the two models as a function of time constant and gain of vestibular nystagmus and OKAN of patients with labyrinthine disease, it was suggested that velocity storage was abolished after unilateral labyrinthine disease and that fewer parameter changes were necessary to simulate this in the Robinson model.[53] However, upon closer examination of the Robinson model, it was necessary to implicitly change the gain of the VOR to account for discrepancies between model predictions and data.[53] Thus, there were the same number of parameter changes in both models, vitiating the conclusion. In addition, the dynamics of the eye velocity trajectory were not examined, so it is not clear that velocity storage had actually disappeared after unilateral labyrinthectomy.

Early quantitative assessment of the Raphan model predicted that if gain and time constant were modified independently, the shape of the eye velocity trajectory would be altered due to differing plateau characteristics.[2] Recent experiments on adaptation and habituation of the VOR are consistent with these predictions[9] and suggest that characteristics of velocity storage are not controlled by a single feedback parameter but by distributed modifiable parameters.[9]

An attempt to generalize the Robinson model to three dimensions, using the positive feedback loop to model velocity storage, has also not given insight into the spatiotemporal aspects of the VOR.[54] Evidence that velocity storage contributes to OVAR[55] and pitch while rotating[3,56] and its close linkage to spatial orientation[11,12] and the somatosensory system[43,57,58] make it likely that positive efference feedback from eye velocity to the visual system[6] is not the dominant or significant aspect of velocity storage. As shown in this study, it is best represented by a feedback dynamical system with matrix, H, whose eigenvalues and eigenvectors are related to postural orientation. The peripheral vestibular system, the locomotor apparatus, and the visual system probably utilize the velocity storage integrator to help orient and stabilize gaze both temporally and spatially during motion in gravitational environments.

SUMMARY

1. A simplified three-dimensional state space model of visual vestibular interaction was formulated. Matrix and dynamical system operators representing coupling from the semicircular canals and the visual system to the velocity storage integrator were incorporated into the model.

2. It was postulated that the system matrix for a tilted position was a composition of two linear transformations of the system matrix for the upright position. One transformation modifies the eigenvalues of the system matrix while another rotates the pitch and roll eigenvectors with the head, while maintaining the yaw axis eigenvector approximately spatially invariant. Using this representation, the response characteristics of the pitch, roll, and yaw eye velocity were obtained in terms of the eigenvalues and associated eigenvectors.

3. Using OKAN data obtained from monkeys and comparing to the model predictions, the eigenvalues and eigenvectors of the system matrix were identified as a function of tilt to the side or of tilt to the prone positions, using a modification of the Marquardt algorithm. The yaw eigenvector for right-side-down tilt and for downward pitch cross-coupling was approximately 30° from the spatial vertical. For the prone position, the eigenvector was computed to be approximately 20° relative to the spatial vertical. For both side-down and prone positions, oblique OKN induced along eigenvector directions generated OKAN which decayed to zero along a

straight line with approximately a single time constant. This was verified by a spectral analysis of the residual sequence about the straight line fit to the decaying data. The residual sequence was associated with a narrow autocorrelation function and a wide power spectrum.

4. Parameters found using the Marquardt algorithm were incorporated into the model. Diagonal matrices in a head coordinate frame were introduced to represent the direct pathway and the coupling of the visual system to the integrator. Model simulations predicted the behavior of yaw and pitch OKN and OKAN when the animal was upright, as well as the cross-coupling in the tilted position. The trajectories in velocity space were also accurately simulated.

5. There were similarities between the monkey eigenvectors and human perception of the spatial vertical. For side-down tilts and downward eye velocity cross-coupling, there was only an Aubert (A) effect. For upward eye velocity cross-coupling there were both Müller (E) and Aubert (A) effects. The mean of the eigenvectors for upward and downward eye velocities overlay human $1 \times g$ perceptual data. The monkey upward eye velocity data overlay the data obtained in humans for $2 \times g$. The dynamics of OKAN may provide a model that can be utilized to study the neural basis for responding to gravity as well as to motion in a gravitational environment.

6. When monkeys were stopped in the right-side-down position after leftward yaw rotation about an axis tilted 90° from the vertical, they developed downward eye velocities during the postrotatory nystagmus, consistent with the organizational principle for converting the nystagmus along the body vertical to the same direction in the spatial frame of reference. The computed eigenvector was close to 20° relative to the spatial vertical. The trajectory of eye velocity in state space was curved as eye velocity decayed to zero. It approached an angle relative to the spatial vertical that was close to the eigenvector direction for OKAN in this and previous studies.

7. Thus, velocity storage is best represented by a feedback dynamical system, with matrix H, whose eigenvalues and eigenvectors are related to postural orientation relative to gravity. The peripheral vestibular system, the locomotor apparatus, and the visual system probably utilize the velocity storage integrator to help orient and stabilize gaze both temporally and spatially during motion in gravitational environments. This study lays the theoretical and experimental groundwork for studying gaze stabilization while moving in these environments.

ACKNOWLEDGMENTS

We thank Victor Rodriguez for technical assistance.

REFERENCES

1. RAPHAN, T., V. MATSUO & B. COHEN. 1977. A velocity storage mechanism responsible for optokinetic nystagmus (OKN), optokinetic after-nystagmus (OKAN), and vestibular nystagmus. *In* Control of Gaze by Brain Stem Neurons. R. Baker & A. Berthoz, Eds: 37–47. Elsevier/North Holland. Amsterdam, the Netherlands.
2. RAPHAN, T., V. MATSUO & B. COHEN. 1979. Velocity storage in the vestibulo-ocular reflex (VOR). Exp. Brain Res. **35:** 229–248.
3. RAPHAN, T. & B. COHEN. 1985. Velocity storage and the ocular response to multidimensional vestibular stimuli. *In* Adaptive Mechanisms in Gaze Control: Facts and Theories. A. Berthoz & G. Melvill-Jones, Eds.: 123–143. Elsevier. Amsterdam, the Netherlands.

4. COHEN, B., V. MATSUO & T. RAPHAN. 1977. Quantitative analysis of the velocity characteristics of optokinetic nystagmus (OKN) and optokinetic after nystagmus (OKAN). J. Physiol. London **270:** 321–344.
5. WAESPE, W., B. COHEN & T. RAPHAN. 1983. Role of the flocculus in optokinetic nystagmus and visual-vestibular interactions: effects of flocculectomy. Exp. Brain Res. **50:** 9–33.
6. ROBINSON, D. 1977. Vestibular and optokinetic symbiosis: an example of explaining by modeling. *In* Control of Gaze by Brain Stem Neurons. R. Baker & A. Berthoz, Eds.: 49–58. Elsevier/North Holland. Amsterdam, the Netherlands.
7. BUIZZA, A. & R. SCHMID. 1982. Visual vestibular interaction in the control of eye movement: mathematical modelling and computer simulation. Biol. Cybern. **43:** 209–223.
8. COHEN, B., D. HELWIG & T. RAPHAN. 1987. Baclofen and velocity storage: a model of the effects of the drug on the vestibulo-ocular reflex in the rhesus monkey. J. Physiol. London **393:** 703–725.
9. COHEN, H., B. COHEN, T. RAPHAN & W. WAESPE. Habituation and adaptation of the vestibulo-ocular reflex: a model of differential control by the vestibulo-cerebellum. Exp. Brain Res. (Submitted.)
10. RAPHAN, T. & B. COHEN. 1988. Organizational principles of velocity storage in three dimensions: the effect of gravity on cross-coupling of optokinetic after-nystagmus. Ann. N.Y. Acad. Sci. **545:** 74–92.
11. RAPHAN, T. & D. STURM. 1991. Modelling the spatio-temporal organization of velocity storage in the vestibulo-ocular reflex (VOR) by optokinetic studies. J. Neurophysiol. **66**(4): 1410–1421.
12. DAI, M. J., T. RAPHAN, B. COHEN & C. SCHNABOLK. 1991. Spatial orientation of velocity storage during post-rotatory nystagmus. Soc. Neurosci. Abstr. No. 314.
13. DAI, M. J., T. RAPHAN & B. COHEN. 1991. Spatial orientation of the vestibular system: dependence of optokinetic after nystagmus (OKAN) on gravity. J. Neurophysiol. **66:** 1422–1439.
14. DAI, M. J., T. RAPHAN & B. COHEN. 1992. Characterization of yaw to roll cross-coupling in the three-dimensional structure of the velocity storage integrator. Ann. N.Y. Acad. Sci. (This volume.)
15. HARRIS, L. R. 1987. Vestibular and optokinetic eye movements evoked in the cat by rotation about a tilted axis. Exp. Brain Res. **66:** 522–532.
16. HARRIS, L. R. & G. B. BARNES. 1987. Orientation of vestibular nystagmus is modified by head tilt. *In* The Vestibular System. M. D. Graham & J. L. Kemink, Eds.: 539–548. Raven Press. New York, N.Y.
17. HESS, B. J. M. 1990. Dual-search coil for measuring three-dimensional eye movements in experimental animals. Vision Res. **30:** 597–602.
18. ZADEH, L. A. & C. A. DESOER. 1963. Linear System Theory: The State Space Approach. McGraw Hill. New York, N.Y.
19. BLANKS, R. H. I., I. S. CURTHOYS & C. H. MARKHAM. 1972. Planar relationships of semicircular canals in the cat. Am. J. Physiol. **223:** 55–62.
20. BLANKS, R. H. I., I. S. CURTHOYS, M. BENNET & C. H. MARKHAM. 1985. Planar relationships of the semicircular canals in rhesus and squirrel monkeys. Brain Res. **340:** 315–324.
21. SIMPSON, J. I., W. GRAF & C. LEONARD. 1981. The coordinate system of visual climbing fibers to the flocculus. *In* Progress in Oculomotor research. F. A. F. & W. Becker, Eds.: 475–484. Elsevier. New York, N.Y.
22. SIMPSON, J. I. & W. GRAF. 1981. Eye muscle geometry and compensatory eye movements in lateral eyed and frontal eyed animals. Ann. N.Y. Acad. Sci. **374:** 20–30.
23. FUCHS, A. F. & J. KIMM. 1975. Unit activity in the vestibular nucleus of the alert monkey during horizontal angular acceleration and eye movement. J. Neurophysiol. **38:** 1140–1161.
24. WAESPE, W. & V. HENN. 1977. Neuronal activity in the vestibular nuclei of the alert monkey during vestibular and optokinetic stimulation. Exp. Brain Res. **27:** 523–538.
25. WAESPE, W. & V. HENN. 1977. Vestibular nuclei activity during optokinetic after-nystagmus (OKAN) in the alert monkey. Exp. Brain Res. **30:** 323–330.

26. WAESPE, W. & V. HENN. 1978. Conflicting visual vestibular stimulation and vestibular nucleus activity in alert monkeys. Exp. Brain Res. **33**: 203–211.
27. TOMLINSON, R. D. & D. A. ROBINSON. 1984. Signals in the vestibular nucleus mediating vertical eye movements in the monkey. J. Neurophysiol. **51**: 1121–1136.
28. REISINE, H., T. RAPHAN, B. COHEN & E. KATZ. 1988. Signal processing in the vestibular nuclei during off-vertical axis rotation (OVAR). Soc. Neurosci. Abstr. No. 173.
29. REISINE, H. & T. RAPHAN. 1992. Unit activity in the vestibular nuclei of monkeys during off-vertical axis rotation. Ann. N.Y. Acad. Sci. (This volume.)
30. REISINE, H. & T. RAPHAN. 1992. Neural basis for eye velocity generation in the vestibular nuclei during off-vertical axis rotation (OVAR). Exp. Brain Res. (Submitted.)
31. LANG, S. 1966. Linear Algebra. Addison-Wesley. Reading, Mass.
32. WAESPE, W., B. COHEN & T. RAPHAN. 1985. Dynamic modification of the vestibulo-ocular reflex by the nodulus and uvula. Science **228**: 199–202.
33. MITTELSTAEDT, H. 1983. Towards understanding the flow of information between objective and subjective space. In Neuroethology and Behavioral Physiology. F. Huber & H. Markl, Eds.: 382–402. Springer-Verlag. Berlin, Germany.
34. ORMSBY, C. C. & L. R. YOUNG. 1976. Perception of static orientation in a constant gravitational environment. Aviat. Space Environ. Med. **47**: 159–164.
35. SCHÖNE, H. 1964. On the role of gravity in human spatial orientation. Aerosp. Med. **35**: 764–772.
36. DAI, M. J., I. S. CURTHOYS & G. M. HALMAGYI. 1989. A model of otolith stimulation. Biol. Cybern. **60**: 185–194.
37. DAI, M. J., I. S. CURTHOYS & G. M. HALMAGYI. 1989. Linear acceleration perception in the roll plane before and after unilateral vestibular neurectomy. Exp. Brain Res. **77**: 315–328.
38. DIZIO, P. & J. R. LACKNER. 1988. The effects of gravitoinertial force level and head movements on post-rotational nystagmus and illusory after-rotation. Exp. Brain Res. **70**: 485–495.
39. COHEN, B., V. HENN, T. RAPHAN & D. DENNETT. 1981. Velocity storage, nystagmus, and visual vestibular interactions in humans. Ann. N.Y. Acad. Sci. **374**: 421–433.
40. DICHGANS, J. 1972. Circularvektion, optiche pseudo-coriolis effekte und optokinetischer nachystagmus: eine vergleichede untersuchung subjectiver und objectiver optokinetischer nacheffekte. Albrecht von Graefe's Arch. Klin. Exp. Ophthalmol. **184**: 42–57.
41. JELL, R. M., D. J. IRELAND & S. LAFORTUNE. 1984. Human optokinetic afternystagmus: slow-phase characteristics and analysis of the decay of slow-phase velocity. Acta Oto-Laryngol. **98**: 462–471.
42. ZASORIN, N. L., R. W. BALOH, D. R. YEE & V. HONRUBIA. 1983. Influence of vestibulo-ocular reflex gain in human optokinetic responses. Exp. Brain Res. **51**: 271–274.
43. SOLOMON, D. & B. COHEN. Stabilization of gaze during circular locomotion in darkness. II. Contribution of velocity storage to compensatory eye and head nystagmus in the running monkey. J. Neurophysiol. (In press.)
44. SCHÖNE, H. 1984. Spatial Orientation: The Spatial Control of Behavior in Animals and Man. Princeton Series in Neurobiology and Behavior. Princeton University Press. Princeton, N.J.
45. THOMAS, D. R. & J. LYONS. 1968. Visual field dependency in pigeons. Anim. Behav. **16**: 213–218.
46. THOMAS, D. R. & J. LYONS. 1968. Further evidence of sensory-tonic interaction in pigeons. J. Exp. Anal. Behav. **11**: 167–171.
47. AUBERT, H. 1861. Eine Scheibare bedeutende Drehung von Objecten bei Neigung des Kopfes nach rechts oder links. Virchows Arch. **20**: 381–393.
48. MÜLLER, G. E. 1916. Über das Aubersache Phänomenon. Z. Sinnesphysiol. **49**: 109–246.
49. BISCHOF, N. 1974. Optic-vestibular orientation to the vertical. In Handbook of Sensory Physiology: 155–192. Springer Verlag. Berlin, Heidelberg & New York.
50. DICHGANS, J. M., R. HELD & L. R. YOUNG. 1972. Moving visual scenes influence the apparent direction of gravity. Science **178**: 1217–1219.
51. YOUNG, L. R. 1984. Perception of the body in space mechanisms. In Handbook of Physiology. The Nervous System. Sensory Processes: 1023–1066. American Physiological Society Bethesda, Md.

52. MARINI, G., L. PROVINI & A. ROSINA. 1975. Macular input to the cerebellar nodulus. Brain Res. **99:** 367–371.
53. HAIN, T. C. & D. S. ZEE. 1992. Velocity storage in labyrinthine disorders. Ann. N.Y. Acad. Sci. (This volume.)
54. HAIN, T. 1986. A model of the nystagmus induced by off vertical axis rotation. Biol. Cybern. **54:** 337–350.
55. RAPHAN, T., B. COHEN & V. HENN. 1981. Effects of gravity on rotatory nystagmus in monkeys. Ann. N.Y. Acad. of Sci. **374:** 44–55.
56. RAPHAN, T., B. COHEN & V. HENN. 1983. Nystagmus generated by sinusoidal pitch while rotating. Brain Res. **276:** 165–172.
57. BRANDT, T., W. BUCHELE & F. ARNOLD. 1977. Arthrokinetic nystagmus and ego-motion sensation. Exp. Brain Res. **30:** 331–338.
58. BLES, W. & S. KOTAKA. 1986. Stepping around: nystagmus, self-motion perception and coriolis effects. *In* Adaptive Processes in Visual and Oculomotor Systems. E. L. Keller & D. S. Zee, Eds.: 465–471. Pergamon Press. Oxford, England.

Effect of Head Orientation and Position on Vestibuloocular Reflex Adaptation[a]

H. STEVIE TAN,[b] MARK SHELHAMER,[c]
AND DAVID S. ZEE[d,e]

[b]Department of Physiology I
Erasmus University
Rotterdam, the Netherlands

[c]Department of Ophthalmology
[d]Department of Neurology
The Johns Hopkins University
School of Medicine
Baltimore, Maryland 21205

INTRODUCTION

When the head and, to be more precise, the orbits move, one must generate compensatory rotations of the globes in order to maintain steady fixation upon objects of interest that are stationary within the external environment. The compensatory eye movements must be of the correct amplitude, direction, and timing in order to insure a stable visual image upon the retina. The requisite rotation of the globe will depend upon the degree to which the orbits are both rotated and translated, and upon the point of regard as reflected in the position of each eye in its orbit. The rotational and translational components of head motion are transduced by the semicircular canals and otoliths, respectively. The point of regard is derived from a variety of cues as to the location in depth of the object of interest. Finally, and critical for a correct calculation of the eye rotations that are needed to compensate for orbital motion, is an accurate internal model of the distances and the angles that relate the positions of the two labyrinths and the two orbits among each other. Thus, the vestibuloocular reflex (VOR) must do much more than simply generate an equal and oppositely directed slow-phase eye movement in response to a rotation of the head.

Since the VOR requires so much modification depending upon the particular circumstances in which it is elicited, questions arise about how the central nervous system can maintain an accurate VOR calibration for the many possible combinations of sensory inputs than can occur during head rotation. Thus, changes in the positions of the eyes in the orbit, which reflect the point of regard, and in the position and the orientation of the head with respect to the axis around which the body is rotating might be expected to influence adaptation of the rotation-induced VOR. Indeed, at this meeting, Shelhamer *et al* have shown that the vestibular system can elaborate two different *horizontal* VOR gains tied specifically to the *vertical* position of the eyes in the orbit.

[a]This research is supported by National Institutes of Health grants RO1-EY01849 and P60-DC00979 (D. S. Zee) and P30-EY01765 and T32-EY07047 (M. Shelhamer).
[e]Author to whom correspondence should be addressed.

Baker and colleagues have shown that short-term VOR adaptation in cats depends critically upon the attitude of the head, with respect to gravity, in which the learning has taken place.[1] In their experiment, the visual surround was oscillated around an axis 90 deg orthogonal to the axis of head rotation, inducing a change in the *direction* of the compensatory slow phase (cross-axis plasticity). The animals were trained in two different head orientations (left-ear down and right-ear down). In each case, the head was oscillated around an earth-vertical axis but the direction of motion of the visual surround was reversed so as to call for the slow-phase vector to cross couple in opposite directions with respect to the direction of head rotation. In response to this stimulus, VOR adaptation became *context specific.* The direction of the adapted slow-phase vector (measured in darkness) depended upon the static orientation of the head in which the learning had taken place, even though the pattern of canal stimulation was the same. Presumably, in these circumstances, otolith signals provide the contextual cue that guides the correct adaptive vestibular response. Baker and colleagues have also shown that if a cat is trained in just one head orientation, but then tested in another, there is some, but not complete, *transfer* of adaptation when the VOR is tested in the new head orientation, even though the pattern of canal stimulation is the same.[2] As one more example of transfer of vestibular learning, VOR adaptation has been shown to be frequency dependent.[3] With training at one frequency, adaptation is maximum when tested at that frequency.

We have begun investigations of similar phenomena in normal human subjects in order to study the general problem of *transfer* of vestibular learning. In particular, how much transfer is there when some, but not all, of the labyrinthine (and somatosensory) signals incurred during head rotation are similar to those experienced during training? We chose to examine the influence of head orientation and head position, relative to the axis around which the body is rotating, upon *short-term* vestibular learning.[4,5] In all circumstances, the horizontal semicircular canals were stimulated, but the pattern of otolith and vertical semicircular canal stimulation was varied. Adaptive modification of the horizontal VOR was induced during rotation of the body around an earth-vertical axis for one hour. The visual surround was artificially rotated to make visual and vestibular inputs discordant so as to call for an increase or a decrease in the amplitude of the horizontal VOR.

METHODS

Five subjects, ages 22–46, with no known vestibular or ocular motor abnormalities took part in this study. The horizontal position of the eyes in the orbit was measured with DC-coupled electrooculography (EOG). The amplitude of the VOR (gain) was measured using a modification of the method described by Gauthier and Robinson.[6] First, a light-emitting diode (LED) was presented at a distance of 120 cm. The target was extinguished, and the subject was then rotated in the dark through a fixed angle. The subjects were instructed to attempt to maintain gaze on the location of the previously visible LED. The waveform of the chair rotation was roughly a step of acceleration followed by a step of deceleration. Peak angular velocity of the chair was approximately 70–100 deg/second, and the amplitude of the chair movement was about 30 deg. The chair rotation either brought the target across the midline (midline steps), or to or away from primary position (centripetal or centrifugal steps). When the chair stopped rotating, the target light reappeared. Any refixation saccade that occurred after the target light was reilluminated allowed for a measure of the inaccuracy of the magnitude of the smooth compensatory movement in the dark. Consequently, a normalized gain of the VOR could be calculated from the ratio of the amplitude of the slow-phase movement to the size of the desired eye displacement (slow-phase eye movement ± corrective saccade). This method has the advantage of an internal calibration so that any fluctuation in the amplitude of the corneal-retinal potential or drift in the EOG signal would be unlikely to affect the results. The head of the subject was immobilized by a dental impression bite bar for the lateral head tilt experiments and in a chin rest when the head was displaced forward for eccentric rotations. For the lateral tilt experiments, the torso was shifted laterally in the chair so as to keep the head aligned along the axis of rotation. Eye and chair movements were displayed on a pen recorder as well as saved with a PDP 11/73 computer for later off-line analysis.

VOR adaptation was elicited by rotating the subject inside a 5-ft diameter optokinetic drum at 0.2 Hz, 30 deg/second peak velocity. The interior of the drum was covered with black squares, subtending angles of 2–4 deg, on a white background. To increase VOR gain, the chair was oscillated for one hour with the optokinetic drum moving in counter phase at an amplitude of 21 deg/second (×1.7 viewing). In this paradigm subjects were instructed to pick out a single feature on the optokinetic drum, and to follow it continuously from end to end as the *position* of the chair moved from one peak to the other. To decrease VOR gain, the chair and drum were oscillated in phase at the same amplitude (×0 viewing). In this paradigm, subjects were instructed to fixate continuously one single feature of the drum located directly in front of them. The subject was positioned with the head upright, so that the axis of rotation bisected the interaural line; or with the head tilted 45 deg laterally; or with the head upright but displaced about 17 cm forward from the center of rotation. VOR gain measurements were made in darkness, before adaptation and after one hour of adaptation.

RESULTS

Tilt Experiments

FIGURE 1 shows the change in VOR gain associated with training in the head-upright versus in the head-tilted position. After training with *head upright,* for both ×0 and ×1.7 viewing, there was an increase in VOR gain in all head positions tested. In other words, there was considerable transfer of adaptation from the trained to the nontrained positions. The degree of transfer, however, was greater with training after ×0 than after ×1.7 viewing. On the other hand, after training with the *head tilted,* for both ×0 and ×1.7 viewing, there was a considerable change in gain in the training position, but little transfer of adaptation to head positions tilted away from the training position.

We also compared the adaptive gain changes when testing with centripetal and centrifugal position steps (FIGURE 2). For ×0 training, there were only small differences between the two types of stimuli, but for ×1.7 training, the gain change was much larger for centripetal than for centrifugal motion of the eyes. For midline steps, the value of the gain change was intermediate between that for centripetal and centrifugal steps.

Eccentric Rotation Experiments

FIGURE 3 compares the changes in VOR gain after rotating with the head centered and with the head displaced forward (eccentric). (Recall that gain in our

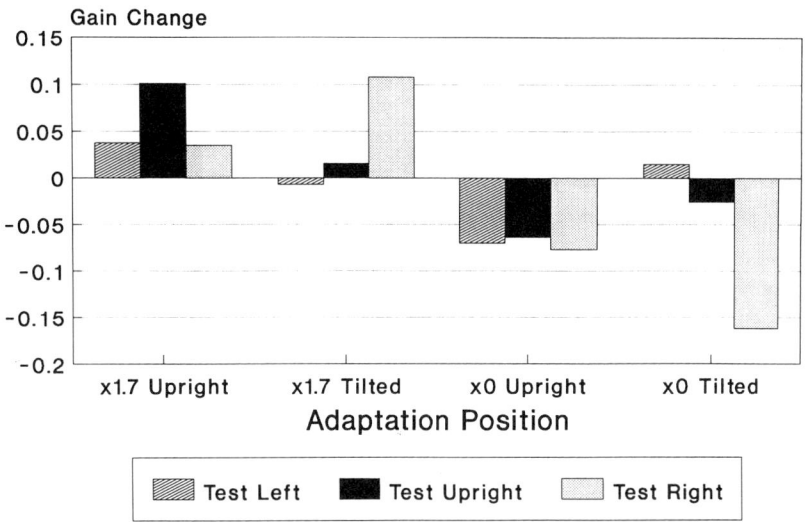

FIGURE 1. Comparison of VOR gain changes after head-tilted and head-upright training. For clarity, all head-tilted training is presented on the graph as tilt to the right. Note that after training in head-tilted positions, the gain change is larger than after head-up training, but there is less transfer to nontrained positions.

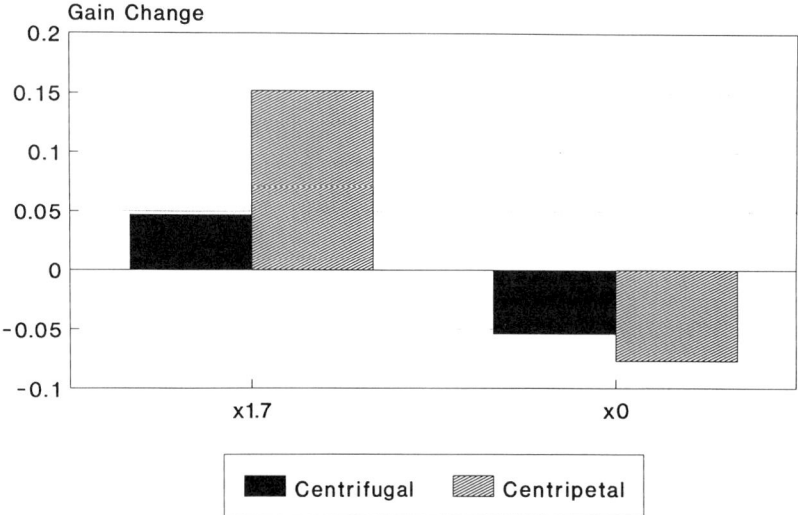

FIGURE 2. Comparison of VOR gain changes for centripetal and centrifugal motion of the globes during upright position-step rotations after head-upright training. Using a two-way analysis of variance, there was a statistically significant difference between centripetal and centrifugal steps after ×1.7 training, but not after ×0 training.

experiments is defined as actual slow-phase displacement/required slow-phase displacement, and that during rotation with the head forward, the *required* slow-phase displacement must be about 15% higher than during head-centered rotation.) After training in either position, for viewing at both ×0 and ×1.7 (×1.88 with the head forward since the optokinetic drum was closer), there was a significant change in VOR gain in both head positions tested. In other words, there was considerable transfer of adaptation from the trained to the nontrained position.

On the other hand, after training with the head-centered but not after training with the head eccentric, there was a significant difference between the amount of change in the two testing positions. In other words, there was more transfer of adaptation after eccentric training than after head-centered training. Finally, we

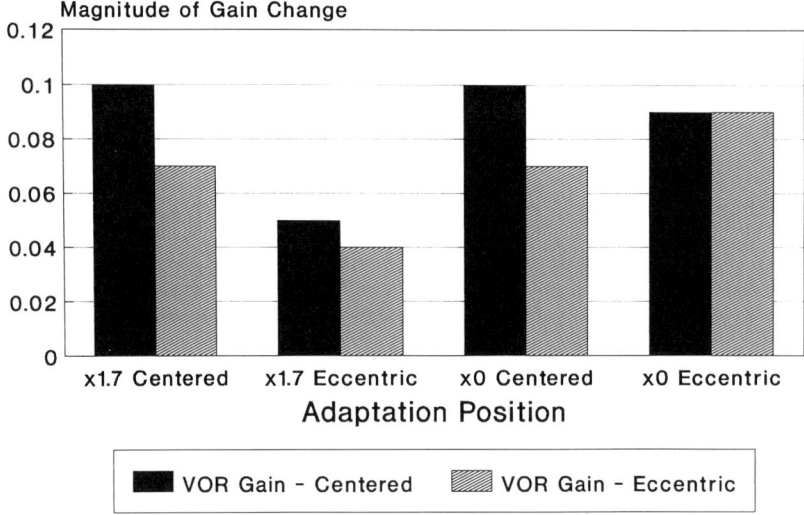

FIGURE 3. Comparison of the absolute value of the VOR gain changes after training with head eccentric and with head centered. Note the considerable transfer of VOR adaptation to the nontraining position. This was especially true after head-eccentric training.

noted that after ×1.7 training (actually ×1.88) with the head displaced forward, there was less adaptation than after training in any of the other conditions.

DISCUSSION

Lateral Tilt

Two main results emerged from the study of the effects of lateral head tilt on short-term adaptation of the rotational VOR. First, after head-upright training, there was considerable transfer of adaptation to the VOR when tested in head-tilted positions. After training in head-tilted positions, however, there was little transfer to other positions. Why this difference? During training in the upright, head-centered position, any adaptation that takes place would relate solely to an altered response from stimulation of the semicircular canals. Accordingly, one might expect considerable transfer of adaptation from upright to tilted positions since the central nervous system would have interpreted the VOR dysmetria experienced in the upright position as a problem primarily with horizontal semicircular-canal-induced reflexes. Therefore, any learning would be applicable to horizontal canal stimulation, no matter what the orientation of the head with respect to gravity. On the other hand, with training in a head-tilted position, both horizontal and vertical semicircular canals are stimulated so that the central nervous system might be less sure about the source of any retinal slip during head rotation. Looked at in another way, the oblique direction of retinal slip on the retina could not be readily attributed to dysfunction in a single pair of semicircular canals. These considerations, coupled with the extremely unnatural pattern of canal-otolith stimulation during prolonged training with the head tilted laterally, might make the adaptive mechanism tie its response to the specific and unusual circumstances in which the adaptation had been acquired.

Secondly, we found that after ×1.7 training there was considerable adaptive increase for head-position steps that moved the eyes centripetally in the orbit, but only a small amount for steps that moved the eyes centrifugally. This was not the case for ×0 viewing; there was only a small difference in adaptive change between centrifugal and centripetal steps. In the ×1.7 paradigm, subjects begin moving the eyes centripetally at the beginning of each half-cycle of head rotation, as head velocity (and presumably retinal slip) is increasing to a maximum. Consequently, it may be that adaptation is more potent when the head is accelerating and adaptive need is greatest. If one also accepts the idea that adaptive change can be both orbital-position and direction specific, one might expect adaptation when the eyes are moving centripetally but not centrifugally. Since, with ×0 viewing, the eyes are always kept near the primary position of gaze, one might expect little centripetal-centrifugal difference, as was the case.

An alternative explanation supposes that prolonged vestibular stimulation in the ×1.7 viewing condition leads to a decrease in the time constant of the brain stem velocity-to-position, gaze-holding integrator. Such a mechanism has been invoked to explain the dependence of the slow-phase velocity of a spontaneous nystagmus upon orbital position (Alexander's law).[7] If this were the case, compensatory eye rotations during head rotation would be enhanced by centripetally directed movement and retarded by centrifugally directed movement. Not clear, though, is why this effect would occur after ×1.7, but not after ×0, training. In these two conditions, however, the movements of the eyes during training are quite different. Further testing of gaze-holding capabilities during eccentric fixation in darkness after VOR stimulation, with and without adaptation, might help settle this issue.

Eccentric Forward Rotation

Two main results emerged from the comparison of VOR adaptation with centered and with eccentric positioning of the head. First, there was considerable transfer of learning to the nontraining position for both ×0 and ×1.7 viewing. This ready transfer of adaptation is not surprising since both the stimulus to the semicircular canals and the *static* otolith inputs are the same in the two training paradigms. Presumably, the changes in dynamic otolith inputs do not provide a compelling enough contextual cue to override any adaptive change in the canal-induced response.

An additional complication with interpretation of our results, however, is that the point of regard during training of the VOR was not the same as that during measurement of the VOR in darkness. (Viewing distance of 60 cm versus 103 cm.) Since any adjustment of VOR in response to changes in the *linear* component of head motion is critically dependent on viewing distance, the VOR should be assessed and trained at different viewing distances to determine the potency of depth of focus as a contextual cue for VOR adaptation.

We also noted that the amount of transfer of VOR adaptation was greater after head-eccentric than after head-centered training. This may reflect, in part, the fact that the stimulus to adaptation was greater in eccentric positions (more visual-vestibular conflict because of closer viewing of the optokinetic drum) and that the absolute value of the amplitude of the required slow phase relative to the amplitude of head rotation is lower with head-centered testing. Thus, any change in VOR gain acquired after eccentric rotation would have a relatively greater effect for head-centered rotation. We also found, however, that the amplitude of adaptation when trying to raise the VOR gain was less after head-eccentric than after head-centered rotations even though the adaptive stimulus was larger. We have no explanation for this difference.

Finally, it should be emphasized that the occurrence of any retinal slip during head motion—which is what drives VOR adaptation—can have a number of causes. Apart from an abnormality in the vestibular sensors themselves, beit the otoliths for translation or the canals for rotation, retinal slip during head movement can be due to a problem with the efferent limb of the VOR, in premotor structures or in the eye muscles themselves.[8] The brain has no easy way to distinguish among these possibilities, unless it has had additional experience in other contexts in which, for example, the otolith and canal contributions during rotation can be evaluated separately. For example, if there is no retinal slip during rotation with the head centered, an abnormal VOR during eccentric rotation can be attributed to an abnormal sensing of translation. Or, if retinal slip occurs when there is no vestibular stimulation, as during attempted smooth pursuit, a more central disorder, perhaps in premotor or motor structures, can be inferred. Or, if retinal slip occurs with the eyes in one orbital position, but not in another, a peripheral ocular motor problem might be deduced. Finally, when there is a vestibular lesion, the direction of retinal slip during head rotation will depend not only upon which semicircular canals are stimulated but also upon the positions of the eyes in the orbits.

The inherent ambiguity about the source of retinal slip during head motion—Is it specifically a vestibular problem? If so, is it an otolith or is it a canal problem? If a canal problem, which one(s)?—is perhaps an important reason why VOR adaptation becomes context specific. Experience in several different contexts is probably necessary for the adaptive mechanisms to be able to decipher the origin of VOR dysmetria, and to adjust the VOR more closely to the demands of particular circumstances. The relatively small changes in VOR gain that we elicited in our experimental paradigms were brought about by training in only one context, for only one hour. VOR adaptation after days and weeks of more varied experience and training almost certainly would be more precisely tailored to the specific needs of the organism.

CONCLUSIONS

Our results reinforce the importance of considering transfer and context dependency of motor learning when interpreting the adaptive performance of the VOR. While the VOR is ostensibly a relatively uncomplicated sensorimotor reflex—detection of rotation of the head followed by a compensatory rotation of the globe—its accuracy depends not only upon which semicircular canals are stimulated but upon the occurrence of any change in dynamic otolith inputs that signals a translation of the head, and upon a knowledge of the desired point of regard and the relative position of each eye in its orbit. Furthermore, the inherent ambiguity as to the cause of any retinal error signals that occur during head rotation may make it advantageous for VOR adaptation to become context specific. Thus, the orientation of the head with respect to gravity, the position of the head upon the trunk, the specific pattern in which the various canal and otolith signals are combined, the positions of the eyes in the orbits, and potentially a variety of other sensory and cognitive cues all may contribute to the judgment about whether one is moving in a circumstance known to require the use of a recalibrated vestibuloocular reflex. Finally, an understanding of the role of both context and transfer of learning in vestibular adaptation may be crucial when one prescribes programs of physical therapy to patients with disease or trauma. One needs to promote recovery of function that is optimal for activity experienced in normal life circumstances.

REFERENCES

1. BAKER, J. F., S. I. PERLMUTTER, B. W. PETERSON, S. A. RUDE & F. R. ROBINSON. 1987. Simultaneous opposing adaptive changes in cat vestibulo-ocular reflex direction for two body orientations. Exp. Brain Res. **69:** 220–224.

2. BAKER, J., C. WICKLAND & B. PETERSON. 1987. Dependence of cat vestibulo-ocular reflex direction adaptation on animal orientation during adaptation and rotation in darkness. Brain Res. **408:** 339–343.

3. LISBERGER, S. G., F. A. MILES & L. M. OPTICAN. 1983. Frequency-selective adaptation: evidence for channels in the vestibulo-ocular reflex? J. Neurosci. **3:** 1234–1244.

4. GONSHOR, A. & G. MELVILL JONES. 1976. Short-term adaptive changes in the human vestibulo-ocular reflex arc. J. Physiol. London **256:** 361–379.

5. KHATER, T. T., J. F. BAKER & B. W. PETERSON. 1990. Dynamics of adaptive human change in human vestibulo-ocular reflex direction. J. Vestib. Res. **1:** 23–29.

6. GAUTHIER, G. M. & D. A. ROBINSON. 1975. Adaptation of the human vestibuloocular reflex to magnifying glasses. Brain Res. **92:** 331–335.

7. ROBINSON, D. A., D. S. ZEE, T. C. HAIN, A. HOLMES & L. E. ROSENBERG. 1984. Alexander's law: its behavior and origin in the human vestibulo-ocular reflex. Ann. Neurol. **16:** 714–722.

8. VILIS, T. & D. TWEED. 1988. A matrix analysis for a conjugate vestibulo-ocular reflex. Biol. Cybern. **59:** 237–245.

Three-Dimensional Transformations from Vestibular and Visual Input to Oculomotor Output[a]

V. HENN, D. STRAUMANN, B. J. M. HESS,
TH. HASLWANTER, AND N. KAWACHI[b]

Neurology Department
University Hospital
University of Zürich
CH-8091 Zürich, Switzerland

INTRODUCTION

Eye muscles in primates are arranged such that the eyes can be moved with three rotational degrees of freedom. Yet there are motor programs that utilize only the horizontal and vertical degrees of rotational freedom while the torsion of the eye is kept constant. This has been described qualitatively by Donders[10] and quantitatively by Listing (cited as Listing's law by Helmholtz).[19]

Eye movements using three degrees of freedom can be induced by vestibular stimuli through canal activation. One important issue is the question of how well stimulus and eye rotation axes are aligned during and after velocity steps about different axes in darkness. Otolith input is also organized in three dimensions, and its input can take two forms. As static input it adds a bias to all eye positions, commonly described as counterrolling during roll tilt. We investigated how Listing's law is affected by static otolith input. As dynamic input it feeds into the velocity storage mechanism and can be tested during constant-velocity rotation about an off-vertical axis. This paradigm has been used to derive data to test hypotheses about central otolith information processing.

Eye movements with two degrees of freedom are observed in response to visual stimuli, while the head is erect and not moving. While earlier references in the literature mostly refer to fixation periods, and recently saccades, we could extend measurements to smooth pursuit movements.

METHODS

Representation of Eye Position in Three Dimensions

Eye positions will be described as rotation vectors in a head-fixed coordinate system.[18,20] Each eye position is defined in coordinates of a virtual rotation out of a reference position. The reference position is usually chosen as the position gaze

[a]The Vestibulooculomotor Laboratory is supported by grants from the Swiss National Foundation for Scientific Research (SNF 31-28008.89; SNF 3199-025239 ESPRIT Mucom 3149), EMDO Foundation Zürich, and SANDOZ Research Foundation Basel.
[b]Permanent affiliation: Otolaryngology Department, Tokyo Women's Medical College, Daini Hospital, Tokyo 116, Japan.

straight ahead with the subject upright. In many subjects, humans or monkeys, this position does not coincide with the primary position which is defined as the position out of which horizontal or vertical eye movements do not have a torsional component. FIGURE 1 defines the three axes and gives an example for the notation of an eye position vertical down. Positive directions are defined by the right-hand rule.

The quaternion description of eye positions is very similar, using four parameters to define an eye position.[34,36,42] The Fick[14] system and the Helmholtz[19] system use three rotation angles to define an eye position: horizontal, vertical, and torsional.

Eye Positions   Rotation Vectors

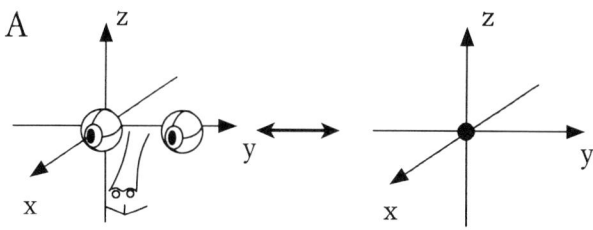

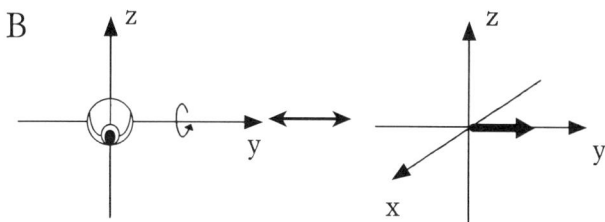

FIGURE 1. Definition of the right-handed coordinate system. Each eye position is described by a virtual rotation out of the reference position. A vector is defined by the direction of the rotation angle and axis. The angle of rotation α determines the length of the vector, $\tan(\tfrac{1}{2}\alpha)$. **A:** Reference eye position. **B:** A down position is described by a vector pointing along the positive y-axis.

Both systems describe ocular torsion as rotation of the eye about the line of sight. Because rotations are not commutative, a different sequence of rotations gives rise to different numerical values of the torsional angle. The amount by which the torsional angle differs from the value given by the rotation vector has been named "false torsion." The Fick or Helmholtz system does not give insight about the rotation components of the eye relative to the principal head axes (i.e., yaw, pitch, and roll) which is crucial for an understanding of the vestibuloocular reflex (VOR) as a compensatory ocular reflex.

Animal Preparation

Under intubation anesthesia, Rhesus monkeys were chronically prepared with the following implants: (1) a magnetic search coil described below to measure eye positions, (2) head bolts to fixate the head during experiments, (3) a stereotaxically placed receptacle over a trephine hole for a micromanipulator. Animals were trained to fixate a light spot according to the paradigm of Wurtz.[43]

Stimulation

For vestibular stimulation, the monkey was seated with its head in the center of a three-dimensional gimbal system, totally surrounded by an optokinetic sphere with a diameter of 170 cm (FIGURE 2; manufactured by Acutronic, Jona-Rapperswil, Switzerland). The whole superstructure can be tilted to any angle between upright and 90 deg to define the rotation axis of the sphere. Four axes are fitted with servo-controlled torque motors with a position resolution better than 0.1 degree. The acceleration of the inner axis reaches 1200 deg/second^2, the outer axis 400 deg/second^2. The velocity has been limited to 200 deg/second for all axes.

Eye Position Measurements

A dual search coil was implanted beneath the conjunctiva of one or both eyes.[22] It consists of a conventional coil to which two miniature coils are rigidly attached at an almost right angle. This allows us (1) to calibrate the coil before implantation, as the whole assembly is geometrically stable after embedding it in acrylic, (2) to implant it beneath the conjunctiva anterior to all muscle insertions so that a mechanical interference with eye movements is unlikely. Eye position is measured with two analog circuits (Skalar Co, Delft, the Netherlands). The calibration procedure is given elsewhere.[32]

The magnetic field generation and demodulation electronics are currently updated with a novel system which continuously rotates the magnetic field about two axes to allow phase angle measurements in three dimensions.[25]

Experimental Protocol

On each experimental day, eye positions were calibrated and compared to values prior to the implantation and to the preceding experimental days. Values usually remained stable, but could show DC offsets depending on whether a micromanipulator or other apparatus was introduced into the magnetic field.

Data Acquisition and Analysis

The signals of the two coils, turntable position, and light status (on or off) were continuously sampled at a rate of 833 Hz, and written onto the hard disk of a personal computer for off-line analysis. In a first step, eye positions were calculated and expressed as rotation vectors. In a second step, eye positions were related to turntable or other input parameters.

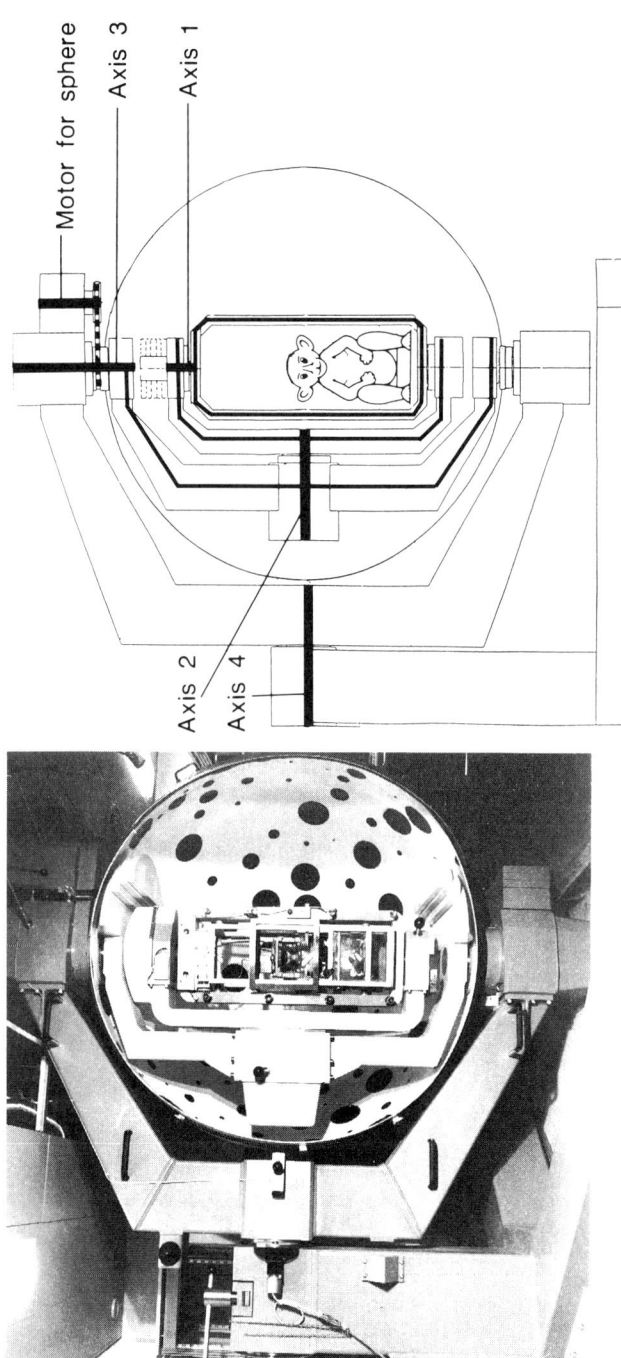

FIGURE 2. Photo and schematic drawing of the vestibular and optokinetic stimulation device. A three-dimensional gimbal system (axes 1–3) is totally surrounded by an optokinetic sphere with large sliding doors. This device can be rotated by a superstructure (axis 4) to align the rotation axis of the sphere which is collinear with rotation axis 3.

EXPERIMENTAL RESULTS: SLOW EYE MOVEMENT GENERATION

Vestibuloocular Reflex

The VOR usually displays a gain of less than unity, and therefore this reflex by itself does not lead to fully compensatory eye movements. We asked the question, how precisely are directions preserved, i.e. whether stimulus and eye rotation axes are always collinear. The animal was exposed to a step in velocity, always about an earth-vertical axis. Between experimental runs, the body position was systematically changed between upright and supine, or upright and right-ear down, or supine and right-ear down. FIGURE 3 shows an example. First, the body position was upright,

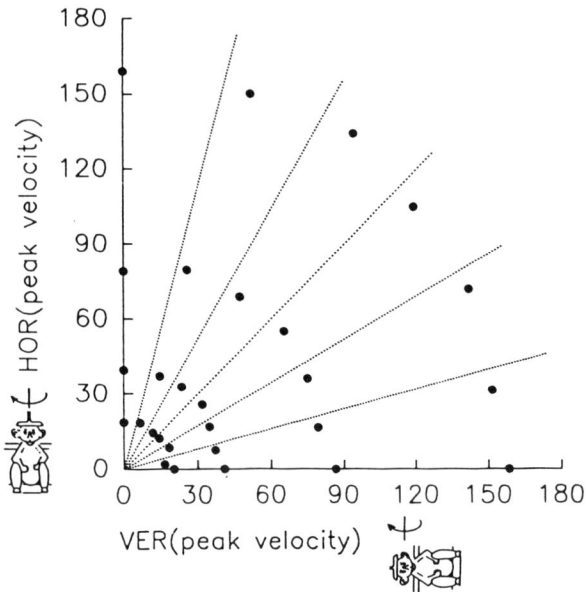

FIGURE 3. Vestibuloocular reflex in the dark (Cathy, mean value). Gain is plotted as eccentricity from the origin of the coordinate system, and direction of nystagmus is given in horizontal and vertical velocity components. The orientation of the animal was systematically shifted from upright to right-ear down in 15-deg steps as indicated by the dotted lines.

and the animal was accelerated in darkness with 100 deg/second^2 to a velocity of 22.5, 45, 90, and 180 deg/second. Peak slow-phase nystagmus velocity was measured, and the values obtained are plotted along the ordinate. Gain values are about 0.9 for all velocities. Then the body position of the animal was tilted by 15 deg, and the animal was again rotated with the same velocities about an earth-vertical axis. Gain data are plotted and lie next to the dotted line which forms a 15-deg angle with the ordinate. As expected, the horizontal gain is now attenuated, while a vertical velocity component emerges. However, the vertical component is relatively larger which has the consequence that the eye rotation axis is shifted towards the vertical. It is not collinear with the stimulation axis by the amount the data point moves away from the dotted 15-deg line. This procedure was repeated while the body axis was sequentially changed in 15-deg steps (30, 45, 60, 75, and 90 deg). Although VOR gain for horizontal or vertical nystagmus is similar, the eye rotation axis is systematically shifted to produce a stronger vertical nystagmus component.

FIGURE 4 gives further examples for different combinations of horizontal, vertical, and torsional rotations. FIGURE 4A shows data for the same paradigm as FIGURE 3. Each data point represents a single slow-phase velocity of nystagmus elicited by a velocity step of the turntable from zero to 180 deg/second. The data points farthest away from the origin of the coordinate system represent the highest velocity values and correspond to the gain values of FIGURE 3. While nystagmus declines during constant-velocity rotation, data points move towards the center of the coordinate system forming a line. The stronger the animal is tilted, the more curved these lines become. Seven different rotation axes were chosen, i.e., from upright to right-ear down in 15-deg steps, and every slow phase of nystagmus is shown as a data point.

FIGURE 4B shows nystagmus slow phase values for velocity steps while the animal's position was changed from upright to supine to elicit different combinations of horizontal and torsional nystagmus. Gain values are not symmetrical as in FIGURE 4A, as the torsional gain reaches values of only about 0.6, and lines that would connect data points become more curved while the animal reaches a supine position. This effect is partly due to the different time constants of horizontal and torsional nystagmus. In the horizontal plane, the velocity-storage mechanism prolongs the nystagmus time constant by a larger amount than in the torsional direction. Therefore, at peak acceleration, eye rotation and stimulus axes are much more closely aligned than during later phases when nystagmus declines.

FIGURE 4C shows data from rotations where the body axis was shifted from right-ear down to supine. Again, gain is asymmetric, and it is evident how the lines connecting the data points are curved.

Considering dynamic aspects of VOR, the response to a step in rotation is usually longer for horizontal than for vertical nystagmus. FIGURE 5 shows an example: the animal was accelerated in total darkness with 100 deg/second^2 to a constant velocity of 90 deg/second. The velocity of each nystagmus slow phase is plotted for the three directional components against time. When the animal was in an upright body position, horizontal nystagmus declined with a time constant of about 35 seconds which is quite typical for monkeys not yet habituated to vestibular stimuli. The corresponding time constant for vertical nystagmus has a value of only 6 seconds. For intermediate body positions that produce a combined horizontal-vertical nystagmus, the dynamics of nystagmus follow a time course neither of the horizontal nor of the vertical nystagmus curve. The time constant takes a value that lies in between, and it has a similar value for both the horizontal and the vertical components.

Optokinetic Nystagmus

In a way similar to that of the VOR, there are systematic deviations from collinearity of stimulus and eye rotation axes. FIGURE 6 shows a pertinent example: first, the monkey was upright, and the optokinetic sphere rotated about the animal. In 15-deg steps, the body position of the animal was tilted from upright to right-ear down, while the rotation axis of the sphere always remained vertical. The format of data presentation is the same as in FIGURE 3. Stimulus velocities were 22.5, 45, 90, and 180 deg/second. Gain values for these different velocities are connected. If stimulus and eye rotation axes were collinear, data points would have to lie on the dotted lines that radiate out of the origin of the coordinate system.

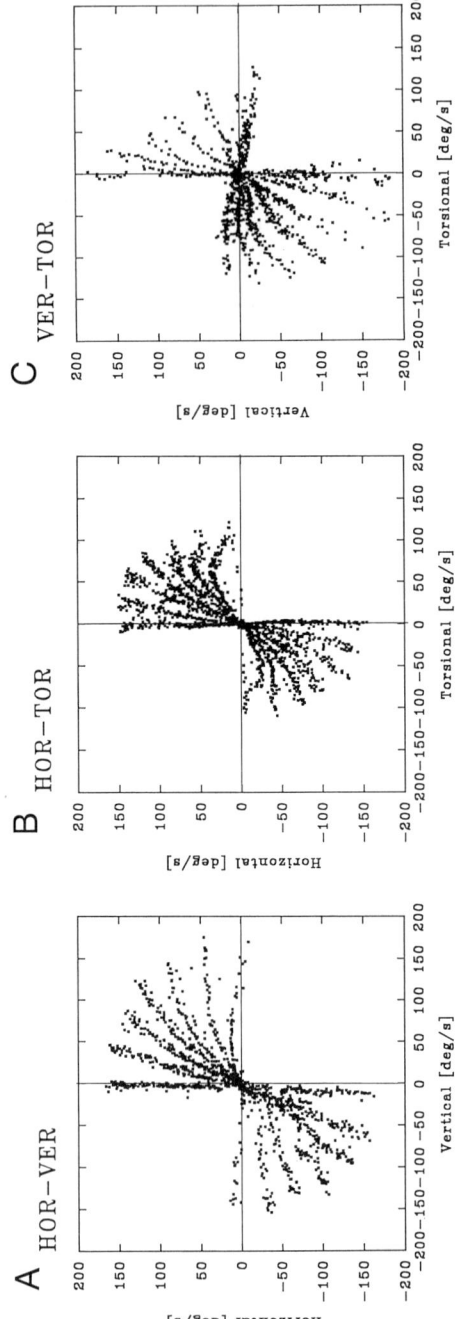

FIGURE 4. Nystagmus velocity for steps in darkness while the monkey was in different body positions. The stimulus axis was always earth vertical.

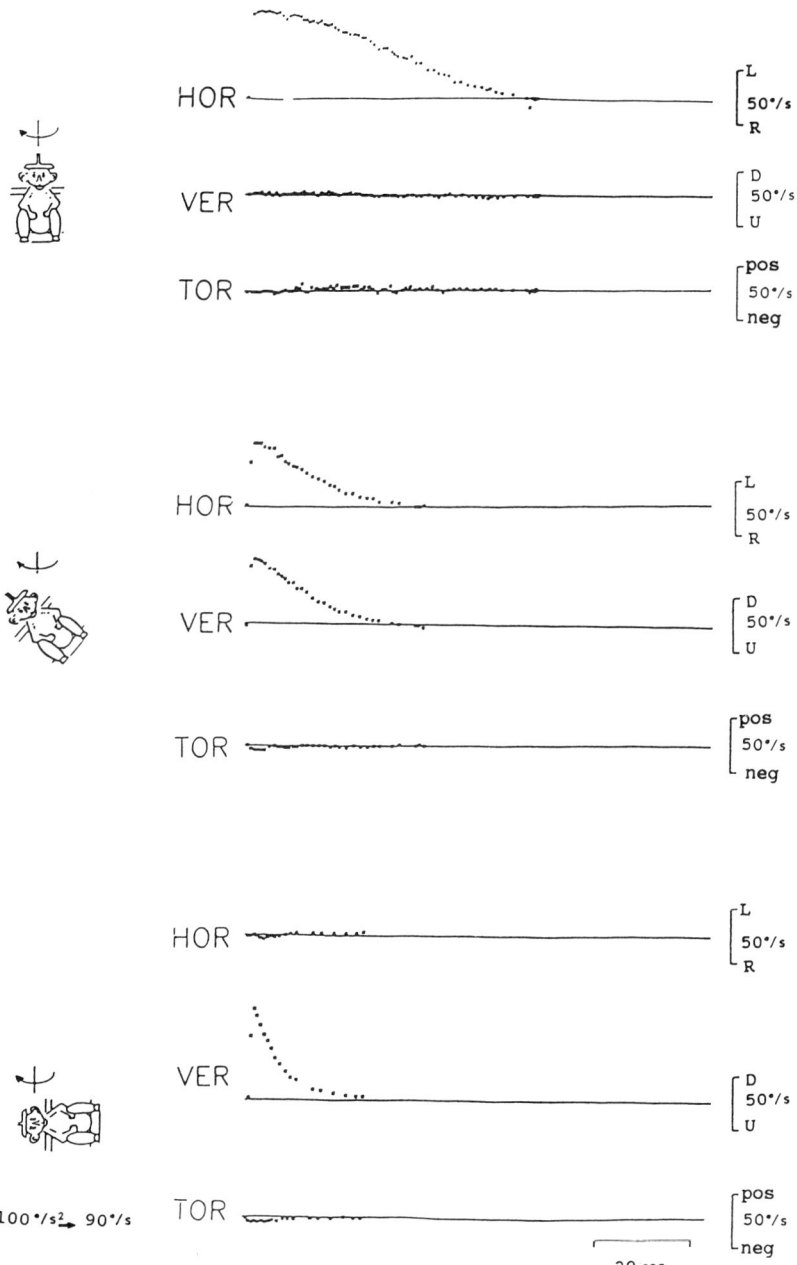

FIGURE 5. Dynamics of the different velocity components of vestibular nystagmus elicited by a step in different body positions.

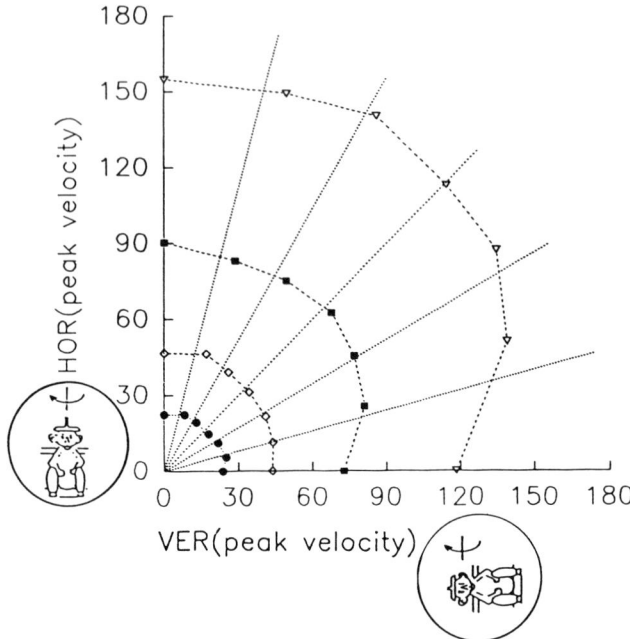

FIGURE 6. Optokinetic nystagmus (Cathy, mean value) as a function of different body positions, while the rotation axis of the sphere remained earth vertical. Same format as FIGURE 3.

In FIGURE 7, the animal's position remained upright, while the rotation axis of the optokinetic sphere was systematically changed from upright to horizontal. One sees even larger deviations from collinearity, and a marked gain attenuation for vertical nystagmus.

Static Otolith Input

Holding a monkey in different roll positions will produce counterrolling of the eyes.[6,9] We have extended these measurements by asking the question, how is Listing's plane affected, and found that it shifts in toto along the nasooccipital axis about an angle that corresponds to the typical amount of counterrolling for looking straight ahead. If the monkey is tilted about the pitch axis, Listing's plane is tilted. Further details are shown in Haslwanter *et al.*[17]

Dynamic Otolith Input

A convenient stimulus is off-vertical-axis rotation. When the rotation axis is horizontal, it is also known as barbecue spit rotation, first introduced as an experimental paradigm by Guedry[15] and Benson and Bodin.[3] Such a stimulus causes direction-specific nystagmus which usually consists of a bias component and a peak-to-peak modulation which is phase-locked to the stimulus. While this kind of stimulus had been widely investigated for rotation axes coinciding with the body longitudinal axis, and some experiments reported somersaulting,[44] we now tested rotation axes at different orientations relative to the principal body axis. Some implications for central otolith processing are given by Hess.[23]

Visual-Vestibular Interaction

Rotation in the light most closely mimics a natural situation save that all movements are passive instead of active. The visual pathways that couple through the vestibular system have been divided into an indirect, or optokinetic, system and a direct, or pursuit, system.[7,28] As pursuit movements are not possible in a torsional direction, one should not observe rapid changes of eye velocity in visual-vestibular interactions that are characteristic of the direct system. This has indeed been found, as there is virtually no visual suppression of the peak velocity of postrotatory torsional nystagmus; only the velocity decay of nystagmus is more pronounced in the light.[33]

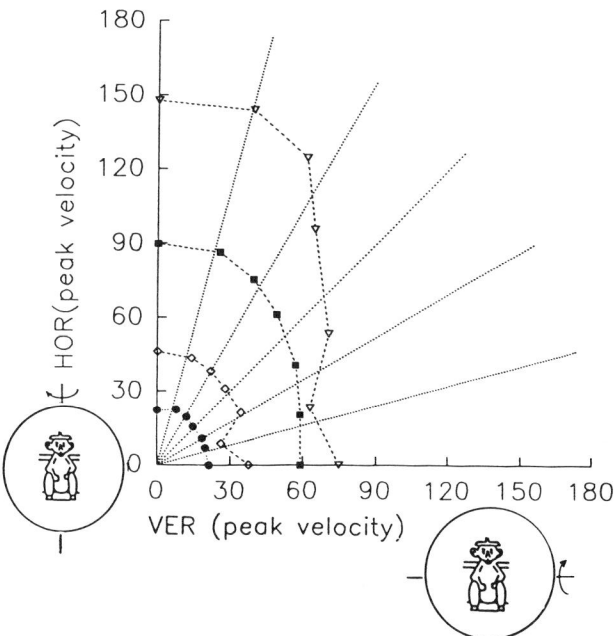

FIGURE 7. Optokinetic nystagmus (Cathy, mean value) as a function of different orientations of the axis of the optokinetic sphere. Note that the relative orientation between the animal's body and sphere rotation axis is the same in this figure and FIGURE 6, although nystagmus responses are quite different, which points towards a powerful otolith input.

DISCUSSION

Reduction of Degrees of Freedom for Motor Output

The possibility of measuring eye position in three dimensions with high temporal and spatial resolution initiated research into the problem of how the degrees of freedom for movement are reduced. Listing's law seems to represent a general organizational principle.[12,20,32] Originally put forward as a description of eye positions with the head stationary and upright, it can now be generalized to include saccades (man, References 37 and 38; monkey, Reference 40) as well as smooth pursuit movements (man, Reference 35; monkey, Reference 16). In an extension of the above-cited experiments, Straumann et al.[32] and Hore et al.[24] investigated eye-head and arm movements, and found similar reductions of degrees of freedom for gaze, and for certain classes of arm movements when pointing towards a target. Listing's law also remains valid for different head-roll positions that induce counterrolling, or in pitch positions that lead to an orientation change of Listing's plane. One can therefore put Listing's law within a broader context. It formalizes the observation that the three degrees of freedom of movement are not equivalent: for many movement programs only the horizontal and vertical components are independent parameters. But even during body tilt, when a third degree of freedom is added, Listing's plane is basically preserved, with an offset added or being tilted.

In a different kind of experimental approach Fetter et al. found that during optokinetic or vestibular stimulation, eye-rotation axes were tilted towards Listing's plane instead of rotating about an axis perpendicular to the line of sight.[13]

Collinearity of Stimulus and Eye-Rotation Axes

When talking about retinal slip and how it is minimized by the VOR or optokinetic nystagmus (OKN), most authors concentrated on the differences between stimulus and eye velocity. Indeed, when the stimulus direction is horizontal, vertical, or torsional, nystagmus direction is closely aligned with the stimulus. However, this cannot be generalized to arbitrary directions. The above-cited measurements, supported by a few reports from the literature,[5,39] show that deviations from collinearity can become quite large. These deviations are systematic, and can usually be shown already with low stimulus velocities where saturation effects are probably not the limiting factor. As a general trend, rotation axes tend to shift away from torsional towards the direction of vertical nystagmus. It remains speculative whether this effect can be linked to specific asymmetries in the three-dimensional organization of the velocity storage.[8,26] These authors found that the longitudinal and the two body axes orthogonal to it are not treated in a symmetrical way, either in perception or in motor, specifically, nystagmus, output.

Neuronal Mechanisms

We are far away from an understanding of neuronal mechanisms of how directions are coded. There were efforts to measure on-directions in neurons of the vestibuloocular reflex.[2,4,11,27] For many it came as a surprise how wide the standard deviation of on-directions was for neurons in central structures. In a similar way, on-directions of neurons in preoculomotor structures were usually not confined to on-directions of motoneurons[21,29,30,31,41] (for review, see Reference 21). From a theoret-

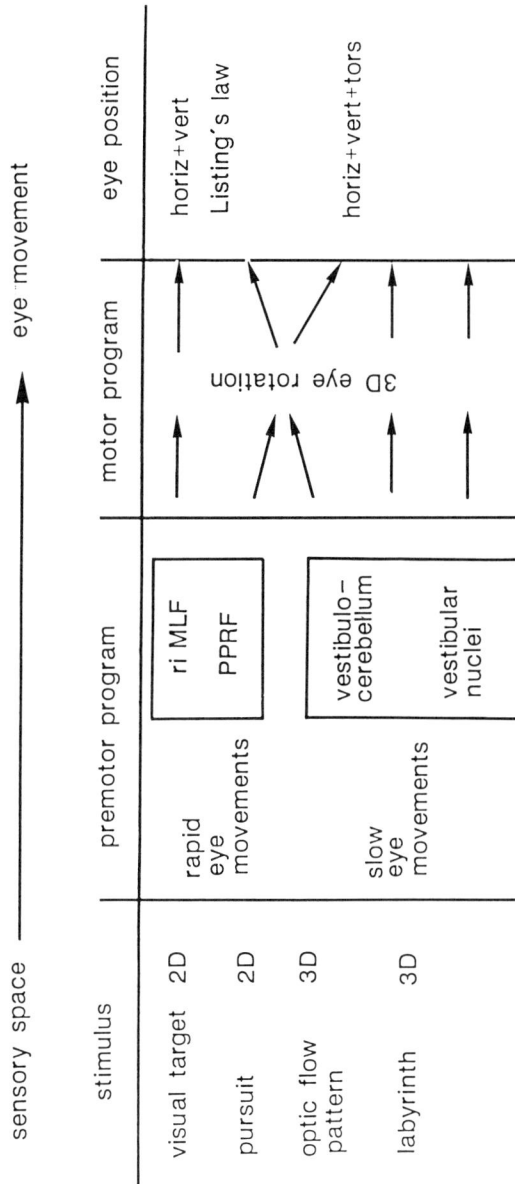

FIGURE 8. Scheme of the information flow from visual and vestibular inputs to generate conjugate rapid or slow eye movements.

ical standpoint, network models were developed that easily learned to provide specific directional information although single elements could have arbitrary on-directions.[1]

Information Flow

Based on the above considerations and experimental phenomena, one can outline a model of information flow for spatial transformations from visual and vestibular input to oculomotor output (FIGURE 8).

Stimuli on the sensory side are three-dimensional when originating in the vestibular system. When originating in the visual system, point targets displayed at optical infinity are two-dimensional, while optic flow patterns are usually three-dimensional. There is, of course, always the intermediate step of a two-dimensional projection on the retina, but the information extracted in central visual pathways for object or self-motion is that of motion in three-dimensional space.

Rapid eye movements have two degrees of freedom, if the head is stationary. The riMLF (rostral interstitial nucleus of the medial longitudinal fasciculus) and the PPRF (paramedian pontine reticular formation) contain the burst neurons which generate these movements. The same set of neurons can also generate rapid eye movements with three degrees of freedom during optokinetic or vestibular nystagmus.

Slow eye movements can be of the pursuit type which are two-dimensional, as we cannot make voluntary torsional movements. Compensatory movements in response to vestibular or optokinetic stimulation are three-dimensional. During optokinetic stimulation, both the "pursuit" and the "optokinetic" components superimpose to determine the rapid and slow changes in eye velocity, mediated by different neuronal pathways. There are neither sufficient experimental data nor detailed enough modeling to derive clear hypotheses about neural mechanisms of how the two- and three-dimensional inputs combine to generate eye movements.

Motor programs have to determine the actual rotation axis of the eye. Even when the possible degrees of freedom of eye movement are reduced to two during movements that obey Listing's law, the angular velocity axis for movements between two tertiary positions has to be described with three parameters.[38] Consider the example of horizontal eye movements at different elevations of the eye. If the movement goes through primary position, then by definition it has a zero torsional component and the angular rotation axis is strictly vertical. However with the eyes elevated, the angular rotation axis of the eye has to be tilted in a torsional direction to produce the same horizontal displacement. That has the consequence that although input to the eye movement generator can be two- or three-dimensional, motor output which determines the rotation axis of the eye is always three-dimensional.

REFERENCES

1. ANASTASIO, T. J. & D. A. ROBINSON. 1989. The distributed representation of vestibulo-oculomotor signals by brain-stem neurons. Biol. Cybern. **61:** 79–88.
2. BAKER, J., J. GOLDBERG, G. HERMANN & B. PETERSON. 1984. Spatial and temporal response properties of secondary neurons that receive convergent input in vestibular nuclei of alert cats. Brain Res. **294:** 138–143.
3. BENSON, A. J. & M. D. BODIN. 1966. Interaction of linear and angular accelerations on vestibular receptors in man. Aerosp. Med. **37:** 144–154.
4. BLANKS, R. H. I., M. S. ESTES & C. H. MARKHAM. 1975. Physiologic characteristics of

vestibular first-order canal neurons in the cat. II. Responses to constant angular acceleration. J. Neurophysiol. **38**: 1250–1268.

5. CRAWFORD, J. D. & T. VILIS. 1991. Axes of eye rotation and Listing's law during rotations of the head. J. Neurophysiol. **65**: 407–423.

6. COHEN, B. 1974. The vestibulo-ocular reflex arc. *In* Handbook of Sensory Physiology. Vestibular System. H. H. Kornhuber, Ed. **6**(1): 477–540. Springer. Heidelberg, Germany.

7. COHEN, B., V. MATSUO & T. RAPHAN. 1977. Quantitative analysis of the velocity characteristics of optokinetic nystagmus and optokinetic after-nystagmus. J. Physiol. London **270**: 321–344.

8. DAI, M., T. RAPHAN & B. COHEN. 1991. Spatial orientation of the vestibular system: dependence of optokinetic after nystagmus (OKAN) on gravity. J. Neurophysiol. **66**: 1422–1439.

9. DIAMOND, S. G., C. H. MARKHAM, N. E. SIMPSON & I. S. CURTHOYS. 1979. Binocular counterrolling in humans during dynamic rotation. Acta Otolaryngol. **87**: 490–498.

10. DONDERS, F. C. 1848. Beitrag zur Lehre von den Bewegungen des menschlichen Auges. Holländ Beitr. Anat. Physiol. Wiss. **1**: 104–145.

11. EZURE, K. & W. GRAF. 1984. A quantitative analysis of the spatial organization of the vestibulo-ocular reflexes in lateral- and frontal-eyed animals. I. Orientation of semicircular canals and extraocular muscles. Neuroscience **12**: 85–93.

12. FERMAN, L., H. COLLEWIJN & A. V. VAN DEN BERG. 1987. A direct test of Listing's law. II. Human ocular torsion measured under dynamic conditions. Vision Res. **27**: 939–951.

13. FETTER, M., D. TWEED, H. MISSLICH, W. HERMANN & E. KOENIG. 1992. The influence of head reorientation on the axis of eye rotation and the vestibular time constant during postrotatory nystagmus. Ann. N.Y. Acad. Sci. (This volume.)

14. FICK, A. 1854. Die Bewegungen des menschlichen Augapfels. Z rationelle Med. **4**: 101–128.

15. GUEDRY, F. E., JR. 1965. Orientation of the rotation-axis relative to gravity: its influence on nystagmus and the sensation of rotation. Acta Otolaryngol. **60**: 30–48.

16. HASLWANTER, TH., D. STRAUMANN, K. HEPP, B. J. M. HESS & V. HENN. 1991. Smooth pursuit eye movements obey Listing's law in the monkey. Exp. Brain Res. **87**: 470–472.

17. HASLWANTER, TH., D. STRAUMANN, B. J. M. HESS & V. HENN. 1992. Does counterrolling violate Listing's law? Ann. N.Y. Acad. Sci. (This volume.)

18. HAUSTEIN, W. 1989. Consideration on Listing's law and the primary position by means of a matrix description of eye position control. Biol. Cybern. **60**: 411–420.

19. HELMHOLTZ, H. V. 1866. Handbuch der physiologischen Optik. Voss. Hamburg, Germany. (English translation: 1962. Helmholtz' Treatise on Physiological Optics. Dover. New York, N.Y.)

20. HEPP, K. 1990. On Listing's law. Commun. Math. Phys. **132**: 285–292.

21. HEPP, K., V. HENN, T. VILIS & B. COHEN. 1989. Brainstem regions related to saccade generation. *In* The Neurobiology of Saccadic Eye Movements. R. H. Wurtz & M. E. Goldberg, Eds.: 105–212. Elsevier. Amsterdam, the Netherlands.

22. HESS, B. J. M. 1990. Dual-search coil for measuring 3-dimensional eye movements in experimental animals. Vision Res. **30**: 597–602.

23. HESS, B. J. M. 1992. How does the otolith system detect 3-dimensional head angular velocity? Ann. N.Y. Acad. Sci. (This volume.)

24. HORE, J., M. GOODALE & T. VILIS. 1990. The axis of rotation of the arm during pointing. Soc. Neurosci. Abstr. **16**: 1086.

25. KASPER, H. & B. J. M. HESS. 1991. Magnetic search coil system for linear detection of three-dimensional angular movements. IEEE Trans. Biomed. Eng. **38**: 466–475.

26. RAPHAN, T. & D. STURM. 1991. Modelling the spatio-temporal organization of velocity storage in the vestibulo-ocular reflex (VOR). J. Neurophysiol. **66**: 1410–1421.

27. REISINE, H., J. I. SIMPSON & V. HENN. 1988. A geometric analysis of semicircular canals and induced activity in their peripheral afferents in the rhesus monkey. Ann. N.Y. Acad. Sci. **545**: 10–20.

28. ROBINSON, D. A. 1981. The use of control systems analysis in the neurophysiology of eye movements. Annu. Rev. Neurosci. **4**: 463–503.

29. SCUDDER, C. A., A. F. FUCHS & T. P. LANGER. 1988. Characteristics and functional identification of inhibitory burst neurons in the alert monkey. J. Neurophysiol. **59:** 1430–1454.
30. STRASSMAN, A., S. M. HIGHSTEIN & R. A. MCCREA. 1986. Anatomy and physiology of saccadic burst neurons in the alert squirrel monkey. I. Excitatory burst neurons. J. Comp. Neurol. **249:** 337–357.
31. STRASSMAN, A., S. M. HIGHSTEIN & R. A. MCCREA. 1986. Anatomy and physiology of saccadic burst neurons in the alert squirrel monkey. II. Inhibitory burst neurons. J. Comp. Neurol. **249:** 358–380.
32. STRAUMANN, D., TH. HASLWANTER, M-C. HEPP-REYMOND & K. HEPP. 1991. Listing's law for eye, head and arm movements and their synergistic control. Exp. Brain Res. **86:** 209–215.
33. STRAUMANN, D., M. SUZUKI, V. HENN, B. J. M. HESS & TH. HASLWANTER. Visual suppression of torsional vestibular nystagmus in rhesus monkeys. Vision Res. (In press.)
34. TWEED, D., W. CADERA & T. VILIS. 1990. Computing three-dimensional eye position quaternions and eye velocity from search coil signals. Vision Res. **30:** 97–110.
35. TWEED, D., M. FETTER, S. ANDREADAKI, F. KOENIG & J. DICHGANS. Three-dimensional properties of human pursuit movement. Vision Res. (In press.)
36. TWEED, D. & T. VILIS. 1987. Implications of rotational kinematics for the oculomotor system in three dimensions. J. Neurophysiol. **58:** 832–849.
37. TWEED, D. & T. VILIS. 1988. Rotation axes of saccades. Ann. N.Y. Acad. Sci. **545:** 128–139.
38. TWEED, D. & T. VILIS. 1990. Geometric relations of eye position and velocity vectors during saccades. Vision Res. **30:** 111–127.
39. VAN DER STEEN, J., S. TAN & L. J. VAN RIJN. 1992. Three-dimensional optokinetic responses in man and rabbit: a comparison between a frontal and lateral eyed species. Ann. N.Y. Acad. Sci. (This volume.)
40. VAN OPSTAL, J., K. HEPP, B. J. M. HESS, D. STRAUMANN & V. HENN. 1991. Two- rather than three-dimensional representation of saccades in monkey superior colliculus. Science **252:** 1313–1315.
41. VILIS, T., K. HEPP, U. SCHWARZ & V. HENN. 1989. On the generation of vertical and torsional rapid eye movements in the monkey. Exp. Brain Res. **77:** 1–11.
42. WESTHEIMER, G. 1957. Kinematics of the eye. J. Optic Soc. Am. **47:** 967–974.
43. WURTZ, R. H. 1969. Response of striate cortex neurons to stimuli during rapid eye movements in the monkey. J. Neurophysiol. **32:** 975–986.
44. YOUNG, L. R. & V. HENN. 1975. Nystagmus produced by pitch and yaw rotation of monkeys about non-vertical axes. Fortschr. Zool. **23:** 235–246.

Floccular Contribution to Signal Processing in the Rabbit Vestibular Nucleus

JOHN S. STAHL AND JOHN I. SIMPSON[a]

Department of Physiology and Biophysics
New York University Medical Center
550 First Avenue
New York, New York 10016

INTRODUCTION

One approach to understanding a signal-processing device begins with describing the device's inputs and outputs, as well as any signals that are accessible within that device. Adopting this approach to studying the cerebellar flocculus would require recording from (at least) floccular mossy fibers and Purkinje cells. A variant of this approach—and the one adopted here—was to compare the signals carried by cells in the vestibular complex that do and do not receive direct floccular inhibition. Distinctive properties of "flocculus receiving neurons" (FRNs) might indicate the contribution of the flocculus to the control of eye movements. Whereas there have been some reports of recordings from identified FRNs in awake behaving primates[1,2] and anesthetized rabbits,[3] these studies do not provide the information necessary to contrast the dynamics of FRNs and non-FRNs.

In the present experiments, neurons of the vestibular complex were recorded in awake rabbits. The sample was restricted to neurons responding to rotational stimuli about the vertical axis. The sample was also constrained to include only cells receiving floccular inhibition and/or projecting toward the oculomotor nucleus, i.e., cells having a high probability of being second-order neurons of the three-neuron arc of the vestibuloocular reflex (VOR). Abducens neurons were recorded in the same animals to determine motoneuron dynamics. Primary afferents of the VIIIth cranial nerve were recorded in separate experiments in the ketamine anesthetized, paralyzed preparation.

METHODS

Four mongrel pigmented rabbits were prepared for chronic recording by aseptic surgery under ketamine/acepromazine/xylazine general anesthesia. The electrode arrangement is shown schematically in FIGURE 1. In each rabbit a scleral search coil was implanted on the left eye for monitoring eye movements. Two multistranded stainless steel wires (Cooner Sales AS-631) were placed in the left lateral rectus muscle for antidromic activation of abducens motoneurons. Bipolar stimulating electrodes were placed in the left oculomotor complex (just lateral to the midline and caudal to the medial rectus subdivision), in the left medial longitudinal fasciculus

---

[a] Author to whom correspondence should be addressed.

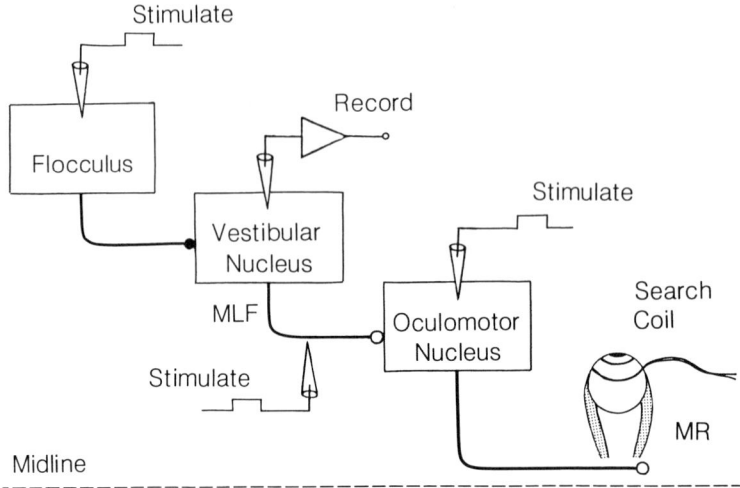

FIGURE 1. Schematic of experimental setup. All recordings and stimulations were performed on the left side. Stimulating electrodes were implanted in the flocculus, MLF, and III cranial nucleus. Eye position was recorded using a scleral search coil. A stimulating electrode was also placed in the lateral rectus muscle of the left eye (not shown). Extracellular recordings were made in the left vestibular complex.

(MLF), and left flocculus. Optimum electrode placement was determined by multiunit recordings and by observing the effects of electrical stimulation on eye movements. A stainless steel recording chamber was placed over the cerebellar vermis. Extracellular recordings were made using 0.005″ tungsten electrodes (Microprobe Inc. No. WE3003).

During recordings the animal was comfortably restrained on a servo-controlled turntable with its head fixed. The animal was surrounded by an optokinetic drum, which provided visual stimulation. The stimuli, restricted to rotations about a vertical axis, included sinusoidal table rotation in the dark and light and triangular rotation of the optokinetic drum. The stimulus parameters are listed in TABLE 1. Amplitudes are 0–peak.

TABLE 1

Device	Waveform	Frequency	Amplitude	Illumination
Table	Sine	0.05	5.0°	Light
		0.1	5.0	
		0.2	2.5	
		0.4	2.5	
		0.8	1.25	
Table	Sine	0.05	5.0°	Dark
		0.1	5.0	
		0.2	2.5	
		0.4	2.5	
		0.8	1.25	
Drum	Triangle	0.1	2.5	Light
Table	1° steps	—	—	Light

Isolated cells were classified by electrical stimulation. FRNs were inhibited following electrical stimulation of the ipsilateral flocculus. Midbrain projecting neurons (MPNs) were activated antidromically from stimulation at the MLF and/or oculomotor complex. FRNs and MPNs will be referred to together as "identified cells." Vestibular nucleus cells lacking a demonstrable relationship to at least one of these structures ("unidentified cells") were not studied.

Neuronal firing rate, eye position, head position, and optokinetic drum position were recorded as perstimulus histograms and stored on a computer. For quantitation, eye position and eye velocity sensitivities were measured by linear regression, with the model equation:

$$F = kE + rE' + c$$

where F is firing rate, E and E' are eye position and eye velocity, k is eye position sensitivity, and r is eye velocity sensitivity.

For analysis of sinusoidal data, the firing rate phase and "magnitude sensitivity" with respect to eye position were obtained by the relations:

$$\text{phase} = \arctan\left[2\pi f(r/k)\right]$$

$$\text{magnitude sens.} = \text{sqrt}\left[(r2\pi f)^2 + k^2\right]$$

where phase is firing rate phase re E, magnitude sens. is firing rate sensitivity re E, and f is stimulus frequency.

For the responses to nonsinusoidal stimuli, firing rate phase with respect to eye position was computed as the difference in the phases of the firing rate and eye position fundamentals, as obtained by Fourier decomposition.

Static eye position sensitivity (k_0) was determined from firing rates recorded during stable eye positions produced by stepwise rotation of the animal in the light. The firing rate from the last 0.6 second of each step was regressed against eye position to yield k_0. The regression equation included a dummy variable to compensate for a small hysteresis in the relationship of firing rate to eye position.[4] This variable appears as the bottom trace in FIGURE 3A, and the hysteresis component estimated from the regression is subtracted from firing rate to yield the "compensated firing rate" of FIGURE 3B.

At the conclusion of recording, the animals were sacrificed and their brains processed histologically. In two animals, reconstructions of the recording sites were made by registering the coordinates of the recording tracks with electrolytic lesions made in the final recording sessions.

RESULTS

MPNs and FRNs were located in the medial vestibular nucleus, in both the magnocellular (MVmc) and parvocellular (MVpc) divisions (nomenclature from Epema *et al.*),[5] at or posterior to the rostrocaudal level of the abducens nucleus. All identified cells were excited during eye movements to the contralateral side, whether produced spontaneously or in response to vestibular or optokinetic stimulation. Cells exhibited somewhat erratic saccadic activity, tending to burst weakly (in comparison to abducens neurons) for contralateral eye movements, and to pause weakly for ipsilateral movements.

For 0.2-Hz, 2.5° amplitude (0-peak), sinusoidal rotation in the light, the r and k values for FRN-MPNs and FRN-non-MPNs were similar, suggesting that other

things being equal, the existence of a midbrain projection was not a strong determinant of a neuron's dynamics. The similarity of FRN-MPNs and FRN-non-MPNs is
shown graphically in FIGURE 2, in which the r and k data are displayed in polar
coordinates after having been converted to phases and sensitivities by the equations
above. Note the similarity of the sectors representing the FRN-MPN and FRN-non-
MPN cell groups. Given their similarity, the two groups were lumped together as
"FRNs." The term "non-FRNs" will refer to the non-FRN-MPNs.

Static eye position sensitivity was determined for FRNs and non-FRNs. FIGURE 3
shows the perstimulus histogram and the regression analysis of the record. The k_0
values for FRNs (5.7 ± 2.8 spikes-sec^{-1}/deg, $n = 25$) and non-FRNs (5.6 ± 3.1,
$n = 13$) were similar ($p = 0.94$, two-tailed t-test).

In addition to demonstrating the similarity of FRN-MPNs and FRN-non-MPNs,
FIGURE 2 shows that firing rate modulation of cells receiving floccular input led

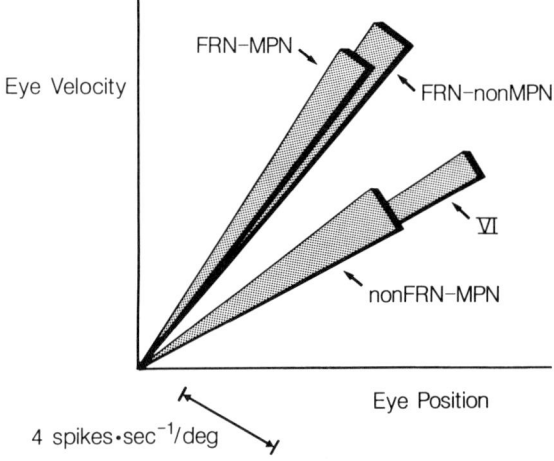

FIGURE 2. Polar plot of magnitude sensitivity and phase for response to rotation in light (0.2
Hz, 2.5° amplitude). Each sector corresponds to average values for a cell group. Shadowing is
added for graphic clarity; the data are represented by the stippled regions. Sector orientation
gives firing rate phase, ±2 standard errors of the mean. Sector length gives magnitude
sensitivity.

that of cells lacking the floccular input. FIGURE 4 demonstrates that this distinction
was present over a wide frequency band for sinusoidal rotation in the dark. Firing
rate phase (with respect to head position) is plotted for VIII primary afferents,
FRNs, non-FRNs, abducens neurons, and eye position. Since abducens (VI) neuron
data were only available for rotation in the light, the phase in the dark was predicted
by referencing the light data to the eye movements in the dark. This maneuver
is acceptable because at 0.1 Hz, the average abducens neuron phase lead re
eye position was minimally affected by illumination (light = $29.2 \pm 7.0°$, dark =
$33.7 \pm 6.1°$, $n = 16$). FRNs ($n = 22$) led non-FRNs ($n = 10$) for all frequencies in the
tested band (0.05–0.8 Hz). The phase differences were significant at $p < 0.05$ or
better, and they increased monotonically from 13.7° at 0.05 Hz to 22.1° at 0.8 Hz.
Non-FRNs were roughly in phase with abducens neurons at the highest frequencies,
and led slightly at lower frequencies.

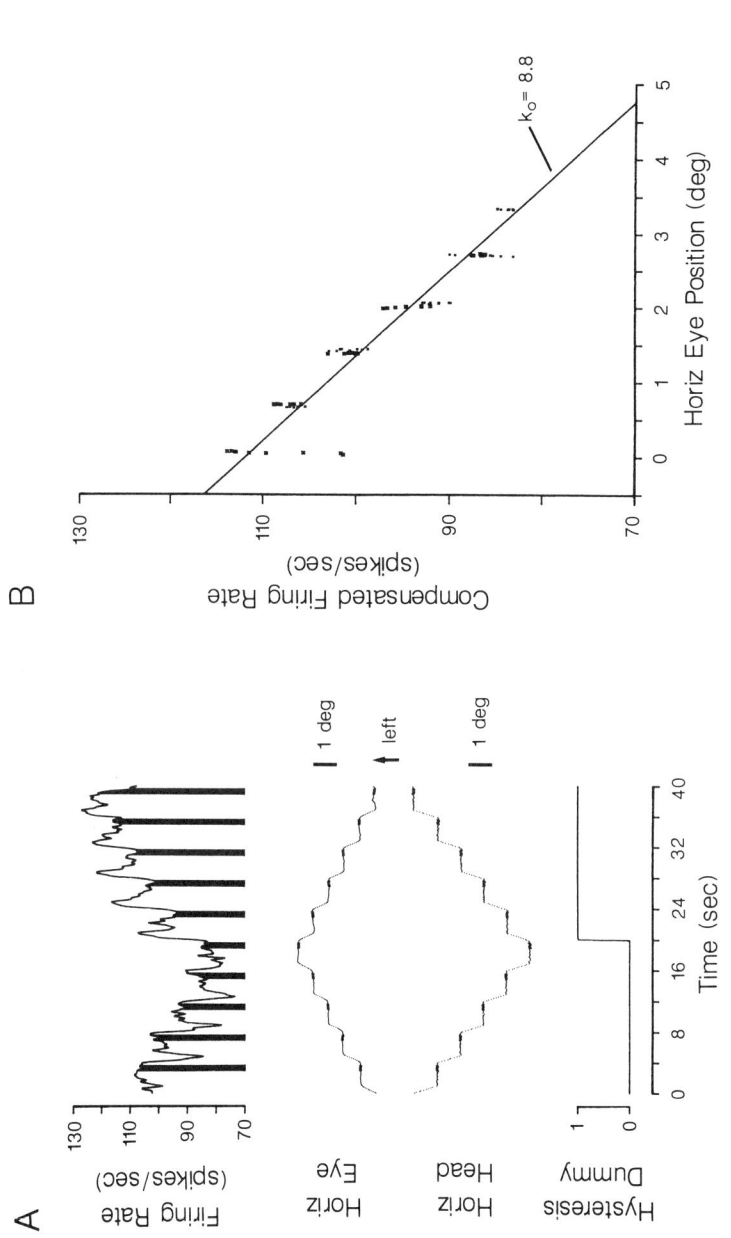

FIGURE 3. A: Response of an FRN-non-MPN to the static eye position sensitivity test. The perstimulus histogram (PSTH) is an average of two cycles of the stimulus, and has a binwidth of 100 mseconds. The turntable moves the rabbit through a series of 1-degree steps. Stationary periods are 3.5 seconds; ramp periods are 0.5 seconds. The stimulus is presented in the light. Quantitation is performed on the last 0.6 seconds of each stationary period (indicated by the filled sections of the firing rate trace and the heavy segments of the eye and head position traces). **B:** Regression analysis of the record in panel A. Data points obtained during nasally directed eye movements are indicated by heavy marks. Firing rate has been compensated for hysteresis.

FRN firing rate also led non-FRN firing rate during eye movements evoked by purely visual stimulation. Neuronal responses were recorded during 0.1 Hz oscillation of the optokinetic drum with a triangular position profile (velocity steps). The responses of representative neurons are shown in FIGURE 5. The responses were analyzed for firing rate phase by Fourier decomposition, as described in Methods.

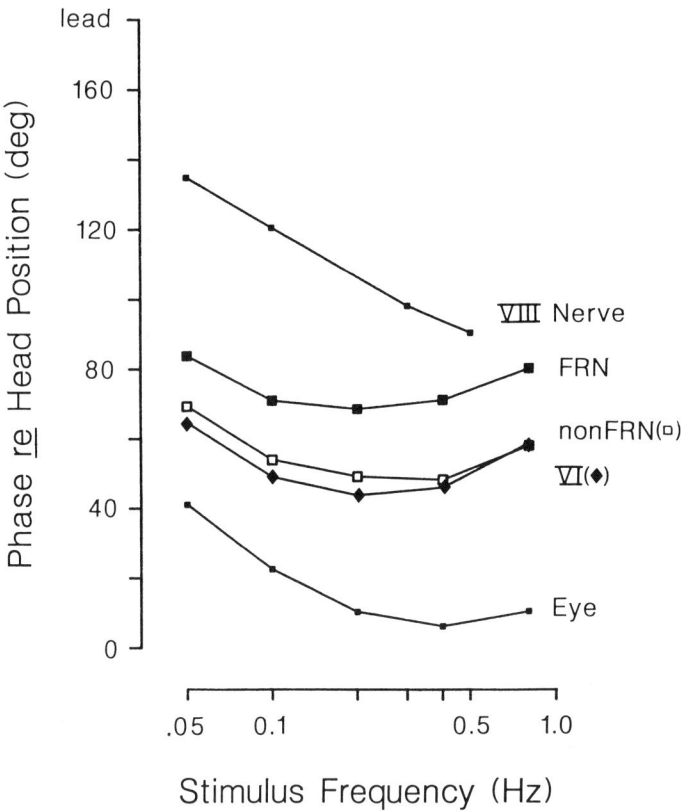

FIGURE 4. Summary of the phase relationships between the different elements of the VOR circuit for sinusoidal rotation in the dark. All phases are referenced to head position, with a positive sign denoting lead. FRN, non-FRN, and eye position data were obtained in the dark. Abducens (VI) neuron data were obtained in the light, then referenced to eye movements in the dark. VIIIth nerve data were obtained in the anesthetized animal.

Phase lead with respect to eye position was $41.6 \pm 10.3°$ for FRNs ($n = 52$) and $22.9 \pm 10.7°$ for non-FRNs ($n = 20$), a difference of $18.7°$ (significance $p < 0.0001$).

The phase advance might be interpreted to indicate that FRNs possess, on average, greater eye velocity sensitivities. In accord with this notion, FRNs frequently exhibited more marked changes in firing rate in response to changes in eye velocity. The brisker corner responses of the FRN in FIGURE 5 are an example of this difference.

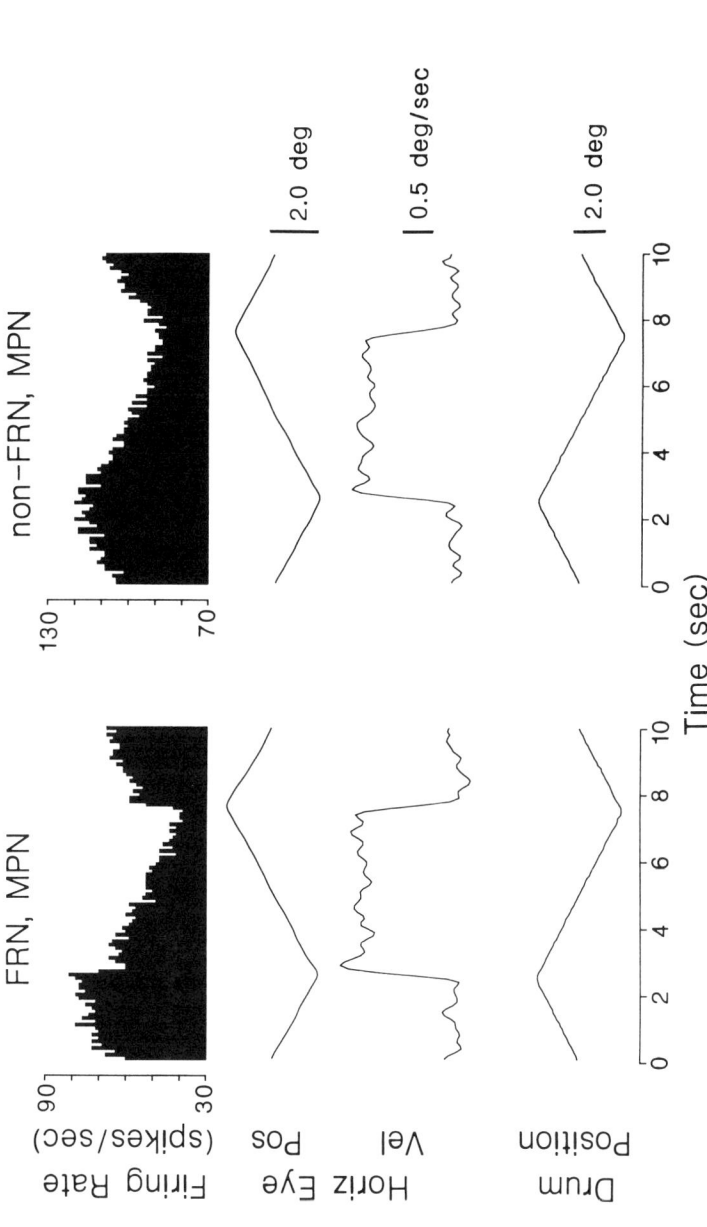

FIGURE 5. PST histograms for the response to triangular optokinetic stimuli at 0.1 Hz. **Left panel:** FRN. Average of ten cycles. **Right panel:** Non-FRN. Average of seven cycles. Note that the FRN evinces sharper transients at turnarounds than does the non-FRN.

DISCUSSION

The location of the FRNs recorded in this study accords well with the location reported in a study in the anesthetized rabbit,[3] as well as with the termination pattern found in our ongoing experiments in which floccular Purkinje cells related to rotation about the vertical axis were stained by extracellular placement of biocytin in the appropriate floccular zone.[6] The finding that the FRNs were excited during contralateral eye rotation (re the side of recording) also accords with past reports. FRNs of the anesthetized rabbit had the same sensitivity polarity[3] as did the majority of floccular target neurons (FTNs) of the awake primate, prior to modification of the VOR gain.[2]

The only attribute distinguishing FRNs from non-FRNs was that, on average, FRNs exhibited greater phase lead with respect to eye position. This phase difference was present at all stimulus frequencies investigated, for rotation in the dark, rotation in the light, and optokinetic stimulation. The two cell groups were similar in terms of their responses to stable eye positions. Their saccadic activity, compared qualitatively, was also similar. Responses to other conditions not discussed here, including VOR suppression by full-field optokinetic conflict, rapid transients of eye position, and tests of linearity, failed to demonstrate distinctive properties. The lack of differences militates against several of the roles that have been proposed for the flocculus. For instance, the proposal that the flocculus is a center for control of saccades[7] is not supported by the weakness of saccade-related modulation on FRNs, or by the qualitative similarity of that activity for FRNs and non-FRNs. The finding that k_0 of FRNs is no greater than that of non-FRNs makes less likely the hypothesis that the flocculus is an important pathway by which position signals enter the vestibular nucleus.[8,9] The similarity of the k_0 values is also significant because if a role of the flocculus were the immediate enhancement of VOR gain,[10] then FRNs might be expected to have greater values of k_0 than non-FRNs.

The one distinctive property of FRNs—their phase advancement—suggests that the flocculus may play a role in adjusting the phase of the premotor signal fed to the extraocular motoneurons. Presumably the difference comes about because the floccular Purkinje cells contribute a signal that is phase advanced in comparison to the other presynaptic influences on FRNs and non-FRNs. The phase angle by which FRNs lead non-FRNs is small, reflecting either the subtlety of the floccular contribution, or the existence of projections of FRNs and non-FRNs upon each other, which might reduce any differences in phase.

One way of looking at the floccular contribution is that it brings the phase of the net premotor signal closer to that present on the primary afferents. As such it would reduce the integration (in the mathematical sense) postulated to occur in the brain stem circuitry. This reduction might be necessary if the "phase-shifting" action of that brain stem circuitry is excessive. The flocculus would perform this function by sampling the vestibular nucleus signals, enhancing their velocity components (or equivalently, partially rejecting the position signals), and then returning the resultant phase-advanced signal to the premotor neurons of the vestibular nucleus. This idea is rendered plausible by suggestions that the vestibular nucleus is part of the neural integrator.[11] If there is a high degree of interconnectivity between cells within the vestibular nucleus, it may be difficult for that structure to preserve the original velocity signal; it may be virtually "integrated out of existence" on all cells. The preservation of the velocity component would have to be performed by a structure anatomically removed from the integrator, a criterion fulfilled by the flocculus.

The extent to which these conclusions can be generalized to other species is an open issue. In their recordings of primate FTNs, Lisberger and Pavelko found that FTNs did lead abducens neurons—by 23° during 0.2 Hz VOR and smooth pursuit.[1] However, primate FTNs and rabbit FRNs may not be entirely comparable, since primate FTNs may receive from floccular gaze velocity[12] cells, which are not present in the rabbit.[13] The idea that the flocculus rejects position signals has been pointed out in primate studies.[12,14] Additional recordings from primate "eye movement only" Purkinje cells[12] may determine whether they are similar to rabbit Purkinje cells, and so might clarify whether the phase-advancing role proposed for the rabbit is applicable to some portion of the primate flocculus.

REFERENCES

1. LISBERGER, S. G. & T. A. PAVELKO. 1984. Functional properties of brainstem cells inhibited from the cerebellar flocculus in monkey. Soc. Neurosci. Abstr. **10:** 290.7.
2. LISBERGER, S. G. & T. A. PAVELKO. 1988. Brain stem neurons in modified pathways for motor learning in the primate vestibulo-ocular reflex. Science **242:** 771–773.
3. KAWAGUCHI, Y. 1985. Two groups of secondary vestibular neurons mediating horizontal canal signals, probably to the ipsilateral medial rectus muscle, under inhibitory influences from the cerebellar flocculus in rabbits. Neurosci. Res. **2:** 434–446.
4. GOLDSTEIN, H. P. & D. A. ROBINSON. 1986. Hysteresis and slow drift in abducens unit activity. J. Neurophysiol. **43:** 1044–1056.
5. EPEMA, A. H., N. M. GERRITS & J. VOOGD. 1988. Commissural and intrinsic connections of the vestibular nuclei in the rabbit: a retrograde labeling study. Exp. Brain Res. **71:** 129–146.
6. VAN DER STEEN, J., J. I. SIMPSON & J. TAN. 1991. Representation of three-dimensional eye movements in the cerebellar flocculus of the rabbit. *In* Oculomotor Control and Cognitive Processes. R. Schmid & D. Zambarbieri, Eds.: 63–77. Elsevier Science Publishers, B.V. (North Holland). Amsterdam, the Netherlands.
7. NODA, H. & D. A. SUZUKI. 1979. The role of the flocculus of the monkey in saccadic eye movements. J. Physiol. **294:** 317–334.
8. NODA, H. & D. A. SUZUKI. 1979. The role of the flocculus of the monkey in fixation and smooth pursuit eye movements. J. Physiol. **294:** 335–348.
9. NODA, H. & T. WARABI. 1982. Eye position signals in the flocculus of the monkey during smooth-pursuit eye movements. J. Physiol. **324:** 187–202.
10. ITO, M. 1982. Cerebellar control of the vestibulo-ocular reflex—around the flocculus hypothesis. Annu. Rev. Neurosci. **301:** 275–296.
11. CHERON, G. & E. GODAUX. 1987. Disabling of the oculomotor neural integrator by kainic acid injections in the prepositus-vestibular complex of the cat. J. Physiol. **394:** 267–290.
12. MILES, F. A., J. H. FULLER, D. J. BRAITMAN & B. M. DOW. 1980. Long-term adaptive changes in primate vestibuloocular reflex. III. Electrophysiological observations in flocculus of normal monkeys. J. Neurophysiol. **43:** 1437–1476.
13. LEONARD, C. 1986. Signal characteristics of cerebellar Purkinje cells in the rabbit flocculus during compensatory eye movements. Ph.D. Thesis. New York University. New York, N.Y.
14. STONE, L. & S. G. LISBERGER. 1989. Processing oculomotor information within the monkey cerebellar flocculus. Soc. Neurosci. Abstr. **15:** 99.14.

The Algebra of Neural Response Vectors

ROBERT H. SCHOR[a] AND DORA E. ANGELAKI [b]

[a]Department of Otolaryngology
University of Pittsburgh
and
The Eye & Ear Institute of Pittsburgh
203 Lothrop Street
Pittsburgh, Pennsylvania 15213

[b]Department of Physiology
University of Minnesota
Minneapolis, Minnesota 55455

The concept of a "polarization vector" was introduced in two papers on the coding properties of otolith neurons in order to describe the spatial coding properties of these afferents.[1,2] The polarization vector was a spatial vector of unit length oriented in the direction of the stimulus which produced the maximal excitatory response in the afferent. This vector could be used to predict the response to a (constant) stimulus in an arbitrary direction by considering that only the stimulus component in the direction of the polarization vector was effective; this operation of taking a vector component can be realized by taking the scalar, or "dot," product between the polarization vector and a vector describing the stimulus orientation. Indeed, working backwards to the hair cell, which has long been known to have a morphological polarization vector given by the asymmetrically placed kinocilium, the depolarization of the hair cell is directly proportional to the component of bending of the hair bundle in the direction of the kinocilium.[3]

We will develop the concept of a neural response vector as a mathematical descriptor which will allow us to predict the response of a neuron or a reflex system to complex time-varying stimuli. We will also develop rules for combining these response vectors to enable predictions when, for example, multiple inputs are considered, or to describe responses expected from converging outputs. In the descriptions that follow, the main assumption is that we are dealing with linear systems; this allows us to describe the output to several inputs by considering the response to the inputs separately, then adding the responses. The other well-known consequence of linearity is that if we stimulate with a sinusoid, the response will be a sinusoid at the same frequency.

The concept of a "spatial response vector" arises directly out of the polarization vector concepts described above. For an afferent, A, we use the polarization vector $\hat{\mathbf{A}}$ to characterize the orientation in space, and use the sensitivity (or gain) g_A, the response when stimulated in this optimal orientation, as the vector length (FIGURE 1A). The response to an arbitrary constant stimulus $\hat{\mathbf{S}}$ can be expressed as the component of the stimulus along $\hat{\mathbf{A}}$, which is computed by taking the dot product of $\hat{\mathbf{A}}$ and $\hat{\mathbf{S}}$. If we consider the plane that contains both the $\hat{\mathbf{A}}$ and $\hat{\mathbf{S}}$ vectors, and express the orientations of $\hat{\mathbf{A}}$ and $\hat{\mathbf{S}}$ by ξ and θ, the response, as a function of stimulus orientation θ, can be written as

$$r_A(\hat{\mathbf{S}}) = g_A \, \hat{\mathbf{A}} \cdot \hat{\mathbf{S}} = g_A \cos (\theta - \xi_A). \tag{1}$$

Two natural questions are what happens to a neuron subjected to a combination

190

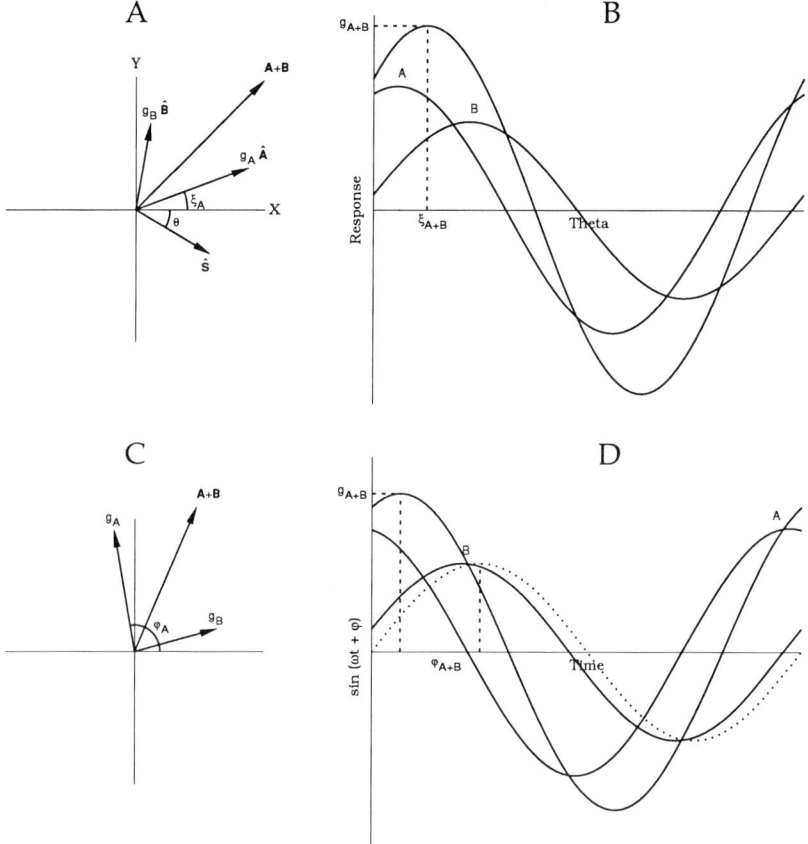

FIGURE 1. Sum of spatial and temporal response vectors. **A:** Spatial response vectors of neurons A and B, having gains g and orientation ξ, are acted upon by a stimulus $\hat{S}$ having orientation θ. The response resulting from adding the outputs of neurons A and B can be obtained by taking the vector sum of the two (spatial) response vectors. **B:** The response produced by a (constant) stimulus is a sinusoidal function of the orientation θ (it is actually a cosine function of the difference between stimulus and response vector orientations). **C:** The gain and phase of responses to sinusoidally time-varying stimuli can be plotted (in polar coordinates) in a "phase space"—the response to the sum of two temporal sinusoids can be obtained by a vector sum in this space. **D:** The sinusoidal stimulus (as a function of time) is plotted as a dotted line; the response sinusoids and their sum are also shown. Note that the phase is the difference between the peak stimulus and the peak response.

of stimuli, and how can the response of a collection of neurons to a single stimulus be analyzed. The first question is easily answered using the linearity assumption—the response to a (vector) sum of stimuli is simply the sum of the responses to each simple component of the stimulus. The second question, what is the response of a group of neurons to a (single) stimulus, is equivalent to asking how neural response vectors may be combined. We will again use linearity to operationally define response vector addition as follows: consider two neurons, A and B, and stimulate them both with $\hat{S}$ (FIGURE 1A). Plot the response of A and B as a function of stimulus orientation; as we've seen above, the response of each neuron (as a function of θ) will be a sinusoid (FIGURE 1B). We will now simply add the sinusoids together, as though these neurons converged on a summing neuron A + B—we can operationally define an orientation and gain of the sum by simply finding the peak of the summed sinusoids (which is simply another sinusoid). Expressing this in equation form gives

$$r_{A+B}(\hat{S}) = r_A(\hat{S}) + r_B(\hat{S}) = g_A\,\hat{A}\cdot\hat{S} + g_B\,\hat{B}\cdot\hat{S} = (g_A\,\hat{A} + g_B\,\hat{B})\cdot\hat{S} \qquad (2)$$

The last step above used the linearity of the dot product to show that this addition rule is equivalent to adding the A and B vectors together, and considering this (vector) sum as the new response vector.

Vectors have also been used to describe the response dynamics of a vestibular neuron. The vestibular system behaves, to a fairly good first approximation, as a linear system; if one stimulates it with a sinusoid of a given frequency (illustrated by the dotted line in FIGURE 1D), then the response is another sinusoid of the same frequency, having a particular gain and phase which may depend on the stimulus frequency,

$$r(\omega, t) = g(\omega) \sin (\omega t + \varphi(\omega)) \qquad (3)$$

This response to a sinusoidal stimulus can also be written as a "temporal response vector," where the gain and phase are the polar coordinates of this vector in some "phase space" (FIGURE 1C). Again, response vector addition can be operationally defined by considering the point-by-point sum of the two responses (FIGURE 1D),

$$r_{A+B}(\omega, t) = r_A(\omega, t) + r_B(\omega, t)$$

$$= g_A(\omega) \sin (\omega t + \varphi_A(\omega)) + g_B(\omega) \sin (\omega t + \varphi_B(\omega))$$

$$= g_{A+B}(\omega) \sin (\omega t + \varphi_{A+B}(\omega)) \qquad (4)$$

$$g_{A+B} = (g_A^2 + 2g_Ag_B \cos (\varphi_A - \varphi_B) + g_B^2)^{1/2}$$

$$\varphi_{A+B} = \arctan \frac{g_A \sin \varphi_A + g_B \sin \varphi_B}{g_A \cos \varphi_A + g_B \cos \varphi_B}$$

The sum of two sinusoids is again another sinusoid; we have written out the formulas for deriving the gain and phase of the sum from the input gains and phases; this same formula also expresses the rules for vector addition in polar coordinates, as illustrated by the vector sum $\hat{A} + \hat{B}$ in FIGURE 1C.

At this point, it is important to point out that gain and phase are not, themselves, directly the components of a mathematical "vector," which has the property that addition is defined as the addition of components. The addition rule for neither the spatial and temporal response vectors described above follows component-wise addition. One direct consequence of this is when describing "average" temporal properties of a collection of neurons, if we mean "what response would be expected if each neuron made an equal contribution to a higher-order neuron," we must use "vectorial" addition, as shown in FIGURE 1C, rather than averaging all the gains together and separately averaging all the phases.

Vestibular afferents have both spatial and temporal properties, but have the convenient "separation" that the spatial properties are independent of the temporal variable ω (except for a possible frequency dependence of the sensitivity parameter), and the temporal properties are independent of the spatial polarization vector (again, except for a simple cosine-rule dependence on the orientation of the stimulus). This may, or may not, be true for higher-order neurons. In trying to explain some peculiar response dynamics seen using a spatially fixed roll stimulus,[4,5] Schor and his colleagues realized that the spatial and temporal effects might be interacting. They accordingly examined the response of central vestibular neurons receiving otolith inputs using two-dimensional spatial stimuli (in place of a single fixed axis, which provides no spatial information) and varying sinusoidal frequencies (thus varying the temporal parameter as well).[4] They redefined the polarization vector to include orientation, gain, and phase parameters, and assumed the response would be expressed as

$$r(\hat{S}, \omega, t) = g(\omega) \, \hat{A}(\omega) \cdot \hat{S} \, \sin(\omega t + \varphi(\omega))$$

$$G(\hat{S}, \omega) = g(\omega) \, \hat{A}(\omega) \cdot \hat{S} \tag{5}$$

$$P(\hat{S}, \omega) = \varphi(\omega)$$

FIGURE 2 illustrates the method of determining the parameters of a response vector. First, record the (sinusoidal) response of a stimulus at some particular orientation, θ (FIGURE 2A). From the resulting sinusoidal function of time (FIGURE 2B), obtain the gain $G(\theta)$ and phase $P(\theta)$. Plot this gain (FIGURE 2C) and phase (FIGURE 2D) as a function of stimulus orientation. Determine ξ, the orientation parameter, as the stimulus orientation that produces the maximum response. Use the gain and phase at $\theta = \xi$ to define g and φ. Notice that when the orientation of the stimulus is perpendicular to the response vector orientation ξ, the response is zero. As the stimulus continues past perpendicular, the response appears reversed in phase (since the stimulus now starts in the "inhibitory" direction), illustrated in the phase plot (FIGURE 2D) by a flip in the phase of 180°. Furthermore, there are two maxima of the rectified cosine function (FIGURE 2C)—one can choose either of two opposing directions for the response vector orientation, provided one simultaneously flips the phase by 180°. The "cosine rule" is illustrated in polar form in FIGURE 2E—the double-ball figure illustrates the dual maximal response gains (and hence dual choices for the response vector orientation), as well as the null response expected for stimuli at right angles to this optimal direction. Finally, the response gain and phase have been plotted in polar form for varying orientations (FIGURE 2F), emphasizing that as orientation varies, response waxes and wanes, while the phase takes on one of two discrete values differing by 180°.

Central vestibular neurons in canal-plugged decerebrate cats, stimulated with sinusoidal tilts in varying vertical planes, produce responses that are consistent with this simple response vector model; the gain G could be fit with a cosine relationship as stimulus orientation varied, and the response phase P appeared to be independent of stimulus orientation. Furthermore, the orientation parameter $\hat{A}$ appeared to be a constant, independent of stimulus frequency; this greatly simplified the experimental determination of the frequency dependence of gain and phase, $g(\omega)$ and $\varphi(\omega)$.[5] A stimulus frequently employed in these experiments was the "wobble" stimulus, a stimulus related to off-vertical-axis rotation (OVAR), in which the orientation θ of the stimulus increased linearly with time, thus moving the direction of tilt around the animal. A consequence of the simple response vector model, with independence of spatial and temporal components, is that the amplitude of the sinusoidal modulation of the neuron's output was independent of the direction of the wobble.

At about the same time, Barry Peterson and his colleagues were describing the

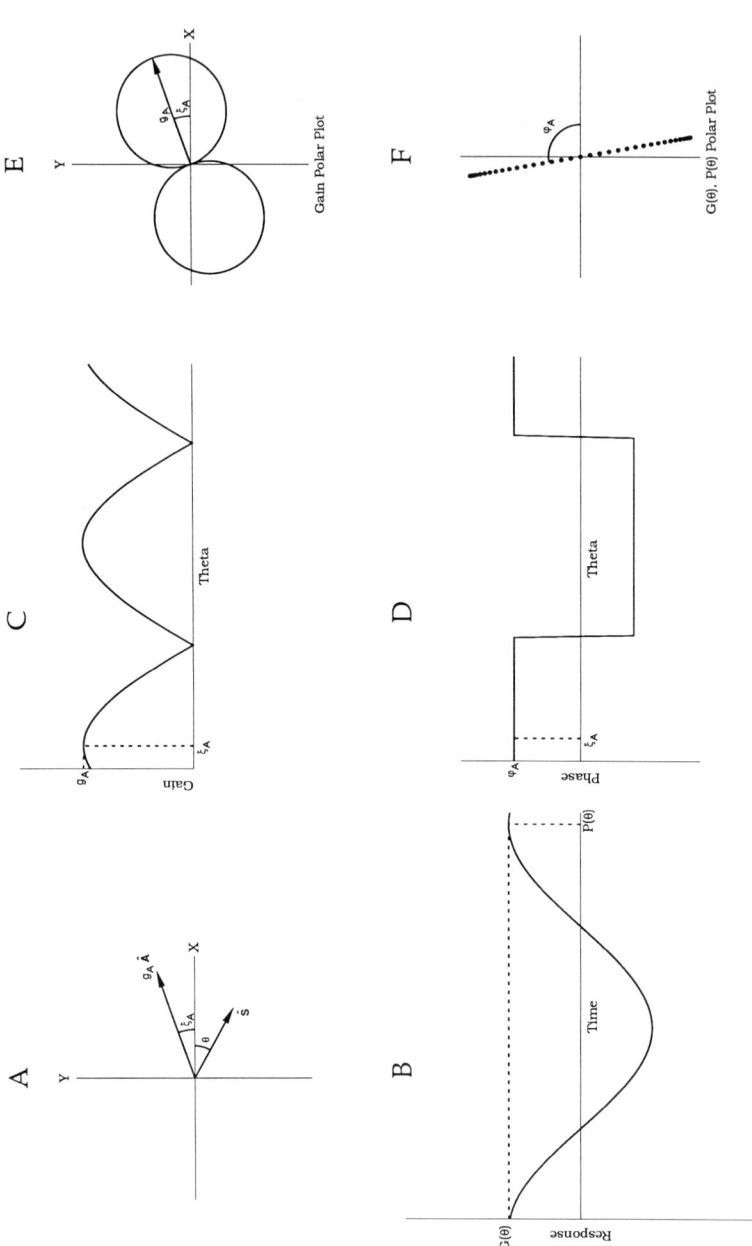

FIGURE 2. Simple response vector. **A:** The response vector has a spatial component ξ and gain g. The response phase φ is not illustrated. **B:** The response to a sinusoidal stimulus oriented at some angle θ will be a sinusoid, having a gain $G(\theta)$ and phase $P(\theta)$. **C:** Plotting response gain against stimulus orientation shows the expected rectified cosine. One of the two possible orientations (differing by $180°$) determined by the gain maximum is indicated. **D:** Phase for simple response vectors takes on two possible values, depending on whether the stimulus and response vectors point in the same or opposite directions. **E:** Plotting gain against stimulus orientation in polar coordinates shows the spatial direction producing the maximal gain. **F:** Each dot represents the gain and phase of the response at some fixed orientation, plotted in polar coordinates to illustrate phase constancy.

response pattern of the cervicocollic reflex, and of certain central vestibular neurons, also using sinusoidal stimuli with varying spatial orientations.[6,7] They described a new response called "spatial-temporal convergence" (STC) which specifically failed to show independence between the spatial and temporal aspects of the response. This behavior manifested itself in several ways. First, the "cosine rule" was clearly violated—the response amplitude not only did not vary according to a cosine function of stimulus orientation, it could even remain constant as the orientation varied over a full 360°. Under these same conditions, the phase (a "temporal" measure) varied directly with the stimulus orientation, instead of remaining constant. Finally, the orientation parameter, $\hat{A}$, could shift with frequency, though it often appeared to reach a stable orientation at 0.5 Hz and higher frequencies.

Such STC patterns could result from convergence between, say, semicircular canal afferents and otolith afferents having different spatial orientations (the afferents already have different inherent temporal properties).[7] As a simple example, suppose the neuron receives convergence from a "roll" otolith afferent and a "pitch" canal afferent (that is, the otolith afferent is best modulated by roll tilt, the canal afferent by pitch). Further assume that the responses of the two afferents in this preferred planes are equal at, say, 0.2 Hz. At frequencies much below this, the canal response will be minimal; hence the neuron's spatial and temporal properties should reflect the dominant otolith input, including having roll be the most effective stimulus. At very high frequencies, the canal response will dominate, and pitch will produce the most activation. Thus the response vector orientation will change from roll to pitch as a function of stimulus frequency. Note that the cosine rule is also violated; instead of zero modulation at right angles to the response vector, both roll and pitch produce responses (and, it can be shown, all intermediate vertical planes will produce a response). One final consequence of STC behavior is that when the wobble stimulus is employed, the amplitude of the sinusoidal responses to the two directions will vary; under certain conditions, one direction may produce no modulation at all.[8]

During discussions of these findings, Schor undertook an analysis of what he called "three-vectors" (reflecting the presence of orientation, gain, and phase aspects). Some of the mathematics to predict how STC responses could arise were developed at that time; some of these results were privately circulated, but never published. These results were used (and vaguely described) in studies on the convergence of neck and vestibular inputs.[9]

In an independent development, Angelaki began considering the problem of a linear system (such as the vestibular system) being stimulated in three orthogonal directions.[10,11] For each stimulus direction, the unit produced a response with a certain gain and phase, and Angelaki asked what predictions one could then make about the response to a stimulus in an arbitrary direction. In discussions prior to publication, Schor and Angelaki noted that many of the equations and conclusions in their works were identical, although the approaches were completely different. The present synthesis is an attempt to recognize this independent "convergence" of investigations, and to develop some of the material relating to the algebra of these spatial and temporal response vectors.

Let us begin by deriving the sum of two simple response vectors of the form given by Equation 5. Note that we are explicitly assuming that the gain and phase parameters we use to characterize the response vector can be considered as independent of the stimulus orientation. We will again use the operational definition that the sum of the response vectors of neurons A and B is simply the sum of the responses each would produce alone.

$$r_{A+B}(\hat{S}, \omega, t) = r_A(\hat{S}, \omega, t) + r_B(\hat{S}, \omega, t)$$

$$= g_A(\omega) \, \hat{A}(\omega) \cdot \hat{S} \sin (\omega t + \varphi_A(\omega)) \tag{6}$$

$$+ g_B(\omega) \, \hat{B}(\omega) \cdot \hat{S} \sin (\omega t + \varphi_B(\omega))$$

Although this expression looks fairly complicated, it is still a sum of two sinusoids, which we already know is simply another sinusoid. The question of interest is how the amplitude and phase of this response sum will vary as a function of $\hat{S}$, as that serves to define operationally the orientation, gain, and phase parameters of the sum.

Before tackling the full problem, consider some simpler special cases. To simplify the writing, the frequency dependence of the response parameters will be omitted. Thus we will consider sums of the form

$$r_{A+B}(\hat{S}, t) = g_A \, \hat{A} \cdot \hat{S} \sin (\omega t + \varphi_A) + g_B \, \hat{B} \cdot \hat{S} \sin (\omega t + \varphi_B) \tag{7}$$

Finally, note that $\hat{A}$ and $\hat{B}$ determine a plane (in general, unless $\hat{A}$ and $\hat{B}$ are colinear). We will, in this two-dimensional situation, assume that the stimulus vector $\hat{S}$ lies in the same plane ($\hat{S}$ can always be written as the sum of a vector lying in the plane and one perpendicular to the plane—this latter component can make no contribution to the sum in Equation 7 since it will be perpendicular to both $\hat{A}$ and $\hat{B}$, and hence both terms will vanish when the vector dot product is computed). If we express the orientation of $\hat{S}$ in this plane by θ, and the orientation of $\hat{A}$ by ξ_A, we can rewrite Equation 7 as

$$r_{A+B}(\theta, t) = g_A \cos (\xi_A - \theta) \sin (\omega t + \varphi_A) + g_B \cos (\xi_B - \theta) \sin (\omega t + \varphi_B) \tag{8}$$

For the first special case, consider two neurons with identical orientations, $\xi_A = \xi_B$. We can then factor out the cosine rule expression and are left with a simple sum of temporal sinusoids, whose addition was described in Equation 4. We can therefore state that the sum of two vectors with identical orientations is another response vector whose orientation is unchanged, and whose gain and phase are simply given by the vectorial addition of the two sinusoidal responses. Note that the only dependence of the response on stimulus orientation is the cosine-rule dependency which we factored out.

Now consider two neurons aligned temporally, that is, having identical phases, φ. We can write the response as

$$r_{A+B}(\hat{S}, t) = (g_A \, \hat{A} \cdot \hat{S} + g_B \, \hat{B} \cdot \hat{S}) \sin (\omega t + \varphi)$$

$$G_{A+B}(\hat{S}) = (g_A \, \hat{A} \cdot \hat{S} + g_B \, \hat{B} \cdot \hat{S}) = (g_A \, \hat{A} + g_B \, \hat{B}) \cdot \hat{S} \tag{9}$$

$$P_{A+B}(\hat{S}) = \varphi$$

Here we have written out the gain $G(\hat{S})$ and phase $P(\hat{S})$ of the sinusoid so we can examine how they vary with stimulus orientation. The constant phase allowed the sinusoid to be factored out, and the dot product sum was simplified to the dot product of a vector sum and $\hat{S}$ (as in Equation 2). If we express this vector sum as having magnitude g_{A+B} and orientation ξ_{A+B}, then the gain term can be further simplified to

$$G_{A+B}(\theta) = g_{A+B} \cos (\xi_{A+B} - \theta) \tag{10}$$

where ξ_{A+B} and g_{A+B} are the orientation and gain parameters of the sum response vector (the phase, of course, is simply φ). Again, the only dependency of the response on stimulus orientation is expressed by a cosine rule.

For the final example, consider the STC case discussed above, that is, two neurons whose orientation parameters are perpendicular, and whose phases differ by 90°. Without loss of generality, let the orientation of neuron A be 0°, that of B be 90°, and let the phase of neuron B lead that of A by 90°. If we let θ represent the orientation of the stimulus, Equation 7 now gives (after appropriate trigonometric substitutions for the 90° changes in orientation and phase)

$$r_{A+B}(\theta, t) = g_A \cos \theta \sin (\omega t + \varphi) - g_B \sin \theta \cos (\omega t + \varphi)$$

$$G_{A+B}(\theta) = (g_A^2 \cos^2 (\theta) + g_B^2 \sin^2 (\theta))^{1/2} \tag{11}$$

$$P_{A+B}(\theta) = \arctan \frac{g_A \cos \theta \sin \varphi - g_B \sin \theta \cos \varphi}{g_A \cos \theta \cos \varphi + g_B \sin \theta \sin \varphi}$$

Now we're in trouble—both gain and phase have a dependency on stimulus orientation, and a cosine-rule gain function is not evident. In fact, if we take the case where $g_A = g_B = g_0$, the response gain and phase become

$$G_{A+B}(\theta) = g_0$$

$$P_{A+B}(\theta) = \varphi - \theta \tag{12}$$

The gain loses all dependence on θ, while the phase appears to inherit the dependence. Thus the simple characterization of response vectors in terms of orientation, gain, and phase breaks down. Even if we stick with the operational definition of orientation as "the stimulus direction that produces the maximal response," how can we define it when the response is independent of stimulus orientation?

In order to begin a rescue of our definition, and of the response vector itself, consider again the expression for response gain in Equation 11. After a little rewriting,

$$G_{A+B}^2(\theta) = (g_A^2 - g_B^2) \cos^2 \theta + g_B^2 \tag{13}$$

Written in this form, it is easy to see that the gain varies between g_A and g_B, reaching the former when $\theta = 0°$, and the latter when $\theta = 90°$. Our previous characterization of response vector defined the vector orientation as the one that produced the maximal response gain; the "cosine rule" guaranteed that the minimal gain, when the stimulus was at right angles to the orientation parameter, would be zero. In the present case, the minimum response still occurs at right angles to the orientation producing the maximum response, but the minimum is nonzero. This suggests we modify our definition of the response vector to add a fourth parameter which expresses the ratio of minimum to maximum response. For now, let us call this the "deviant" parameter, as it expresses how far the response deviates from the cosine rule (this quantity has also been called the "tuning ratio").[11]

Afferentlike responses, our starting point in this derivation, follow the cosine rule, and can accordingly be characterized by an orientation, a deviant of zero, a gain, and a phase; let us call response vector having such a simple form a "simple response vector." We have previously demonstrated that in two special cases, namely, adding simple response vectors with identical orientation or phase parameters, the sum remains a simple response vector (with, respectively, the same orientation or phase).

In general, however, the sum of two simple vectors (FIGURE 3A) leads to responses with nonzero deviants, as the STC example demonstrated. To calculate the orientation, gain, deviant, and phase parameters for the general case, we start with the sum of two sinusoids shown in Equation 8 (FIGURE 3B), and write down the gain and phase of this sum as a function of θ.

$$G_{A+B}(\theta) = ((g_A \cos (\xi_A - \theta) \cos \varphi_A + g_B \cos (\xi_B - \theta) \cos \varphi_B)^2$$

$$+ (g_A \cos (\xi_A - \theta) \sin \varphi_A + g_B \cos (\xi_B - \theta) \sin \varphi_B)^2)^{1/2} \qquad (14)$$

$$P_{A+B}(\theta) = \arctan \frac{g_A \cos (\xi_A - \theta) \sin \varphi_A + g_B \cos (\xi_B - \theta) \sin \varphi_B}{g_A \cos (\xi_A - \theta) \cos \varphi_A + g_B \cos (\xi_B - \theta) \cos \varphi_B}$$

These expressions have been plotted in FIGURE 3C and D. Note that, again, the gain has a nonzero minimum, and the phase takes on all possible values as θ varies. By our convention, the orientation will be one of the two values of θ that maximize the gain (the resulting gain becomes the gain parameter, and the phase at that orientation is the phase parameter). We also want the value of θ that minimizes the gain to calculate the deviant. Setting the derivative of the gain equal to zero gives us the following equation:

$$\tan 2\theta = \frac{g_A^2 \sin 2\xi_A + g_B^2 \sin 2\xi_B + 2 g_A g_B \sin (\xi_A + \xi_B) \cos (\varphi_A - \varphi_B)}{g_A^2 \cos 2\xi_A + g_B^2 \cos 2\xi_B + 2 g_A g_B \cos (\xi_A + \xi_B) \cos (\varphi_A - \varphi_B)} \qquad (15)$$

This equation has two solutions at right angles to each other (there are two values of 2θ, differing by π, which have the same tangent). One value, when plugged back into Equation 14, will give the maximum gain, the other the minimum. The special case that arose in Equation 12, where the deviant is 1, results in the expression in Equation 15 being 0/0, which is indeterminate—in this degenerate case, one is free to choose the orientation parameter (since all directions are equally effective).

The expression of the response vector using the parameters of orientation (a unit vector), gain, deviant (which, when multiplied by the gain, gives the minimum response possible when stimulating in the plane of $\hat{A}$ and $\hat{B}$), and phase can be interpreted as the response to two different simple response vectors, one having the same orientation, gain, and phase parameters as the sum, the other having an orientation at right angles to the first, a gain given by the product of the gain and deviant parameters, and a phase 90° different from the first (dashed line in FIGURE 3E). Thus the same STC response can be produced by different simple response vectors (compare the vectors in FIGURE 3A with those in E). Finally, the ellipse resulting from plotting gain and phase of the responses as a function of stimulus parameter (FIGURE 3F) serves to emphasize that, by suitable choice of stimulus orientation θ, one can obtain any desired sinusoidal phase φ in the response.

Thus far, we have not worried much about the third spatial dimension, since when dealing with only two simple response vectors, the sum can be considered planar (since only the components of $\hat{S}$ in this plane can produce a response). What happens if we add more than two simple response vectors, or add two nonsimple vectors? Angelaki demonstrated that the general form of the three-dimensional response vector can always be written as the sum of two simple vectors, and can thereby be considered to lie in the plane defined by the orientation parameters of these two simple vectors.[10] In order to fully characterize the response, we must additionally include the orientation of this plane—we denote the unit vector normal (perpendicular) to this plane as $\hat{N}$. In the case of a simple response vector, with zero

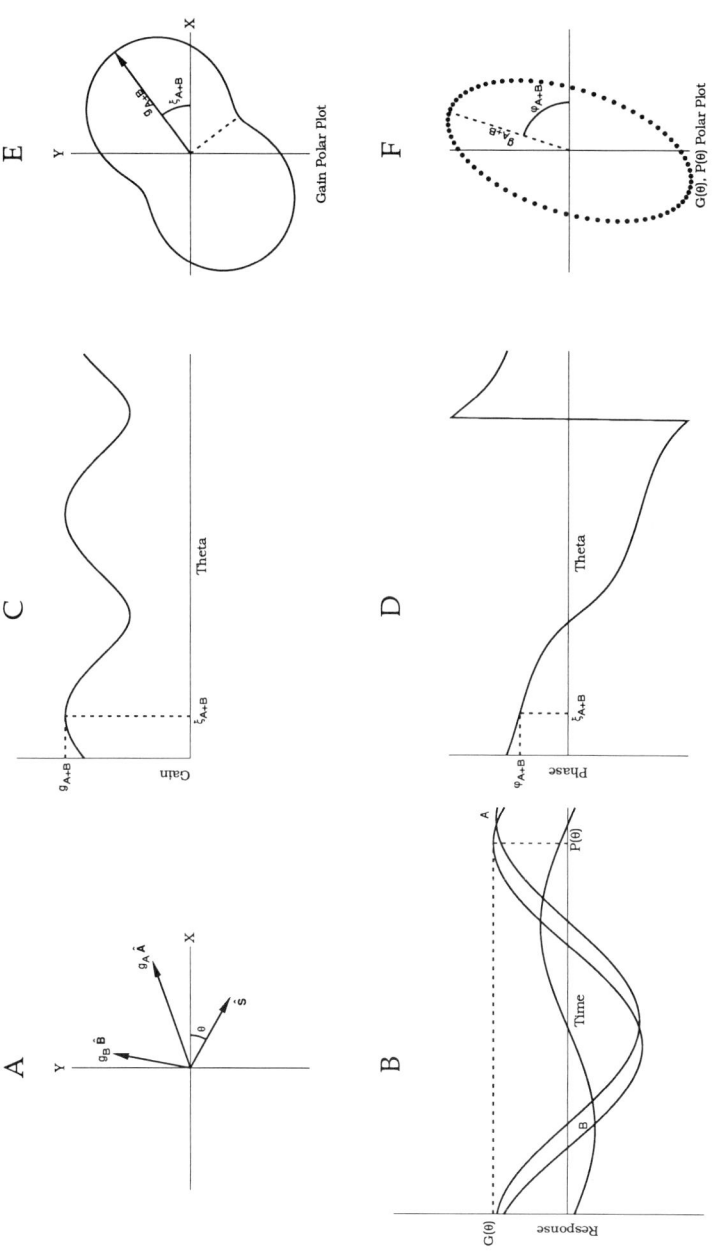

FIGURE 3. General sum of two simple response vectors. These vectors have the same spatial and temporal characteristics as those illustrated in FIGURE 1. The orientation, gain, and phase of A are 20°, 1.4, and 100°; those of B are 80°, 1, and 15°. **B:** This illustrates the sinusoidal responses of A, B, and their sum at a stimulus orientation of −30°. Note that the curves for A and B are scaled versions of the curves in FIGURE 1D, with the scaling given by the "cosine-rule" difference between stimulus and response orientations. **C:** Plotting gain as a function of orientation shows that the response does not have a cosine dependence; in particular, the minimum response is not zero. **D:** Phase constancy is similarly no longer observed. **E:** The polar plot of gain as a function of stimulus orientation is, in general, intermediate between the two extremes of the "two-ball" form characteristic of simple response vectors following a cosine rule (FIGURE 2E) and a circle or "one-ball" form described in Equation 12. **F:** The gain/phase polar plot has the general form of an ellipse, degenerating to a straight line for simple response vectors.

deviant, the normal $\hat{N}$ can be taken as any unit vector in the plane perpendicular to the orientation parameter. When considering a sum of two simple vectors, as above, $\hat{N}$ is perpendicular to the plane defined by the two orientation parameters. Note that an alternative interpretation of $\hat{N}$ is the direction of a null stimulus, one that produces no response.

In order to demonstrate the "planarity" of response vectors and the sufficiency of five parameters to characterize them, we will consider the effect of adding a third simple response vector to the sum of two others discussed above. If we can show that adding three such simple vectors still gives us a response vector that can be characterized by our five parameters (implying, among other things, that there is a null direction for the stimulus), then we are done, since (1) we can always express the resulting sum as a sum of two simple vectors (as described above), so adding further simple vectors reduces to the sum-of-three case, (2) if adding nonsimple vectors, we can first express them as a sum of two simple vectors, then add them up, as above.

The argument goes as follows. Let r_{A+B} be the sum of two simple vectors, as in Equations 8 and 14. Consider first a neuron C whose response vector orientation lies in the plane of $\hat{A}$ and $\hat{B}$. Stimuli in the direction $\hat{N}_{A+B}$ will produce no response from A + B, nor will they produce a response from C (since $\hat{N}$ is perpendicular to the plane of A and B, in which C lies). Therefore, we need only consider the case of C perpendicular to this plane.

Consider the orientation of the stimulus $\hat{S}$. In the two-dimensional case, we characterized it by means of a polar angle θ which measured its position in the A-B plane. Let θ continue to represent the orientation angle of the projection of $\hat{S}$ in the A-B plane, and let η be the angle between $\hat{S}$ and $\hat{N}$ (note that θ and η correspond to the angular components of spherical polar coordinates). Recalling that we are considering the case $\xi_C = \hat{N}$,

$$r_{A+B+C}(\theta, \eta, t) = \sin \eta \, r_{A+B}(\theta, t) + g_C \cos \eta \sin (\omega t + \varphi_C) \qquad (16)$$

Here we have explicitly written the sum of three vectors as the weighted sum of the planar response r_{A+B} we developed earlier, plus the response to C. We claim that at some value of θ and η, this response is 0, that is, there is some null plane, thereby defining $\hat{N}_{A+B+C}$. Recall that $\hat{N}$ defines the plane of the effective stimulus, that is, we only need to consider those components of $\hat{S}$ that lie perpendicular to $\hat{N}$. If we confine ourselves to this plane, we have effectively reduced the three-dimensional problem to the already-solved two-dimensional case. All that remains now is to let $\hat{S}$ vary in this plane, find the orientation of $\hat{S}$ that maximizes the response gain, and define orientation, gain, deviant, and phase as before.

Rather than attempt to solve Equation 16 for values of θ and η that make the response vanish, we will simply establish that such values exist. The response due to C is a sinusoid having phase φ_C and amplitude given by $g_C \cos \eta$. Consider the response r_{A+B}—if it represents a simple response vector, we are done, since we are just adding two simple response vectors, which we know how to do. Otherwise, $r_{A+B}(\theta)$ has the form of a sinusoid, with the phase of the response taking on all possible values as θ varies. In particular, for some value of θ, say θ_0, the phase of the sinusoid $r_{A+B}(\theta)$ is $\varphi_C + 180°$, i.e., the two sinusoids are exactly 180° out of phase, so they tend to cancel. Let G_0 be the gain G of r_{A+B} at that point. We now use stimuli whose projection in the X-Y plane is θ_0 (and thus whose phase is known to be $\varphi_C + 180°$), and let the parameter η vary (see FIGURE 4).

$$r_{A+B+C}(\theta_0, \eta, t) = -\sin \eta \, G_0 \sin (\omega t + \varphi_C) + g_C \cos \eta \sin (\omega t + \varphi_C)$$

$$\eta_0 = \arctan (g_C/G_0) \tag{17}$$

$$r_{A+B+C}(\theta_0, \eta_0, t) = 0$$

The first equation is simply due to the definition of θ_0 and G_0; the second defines the angle η_0 which makes the response vanish. Thus the coordinates (θ_0, η_0) provide the orientation of a stimulus that produces no response, hence are the coordinates for the $\hat{N}$ parameter of the sum.

We previously demonstrated that adding two simple response vectors yields a response vector that can be characterized by 5 parameters (two of which are spatial

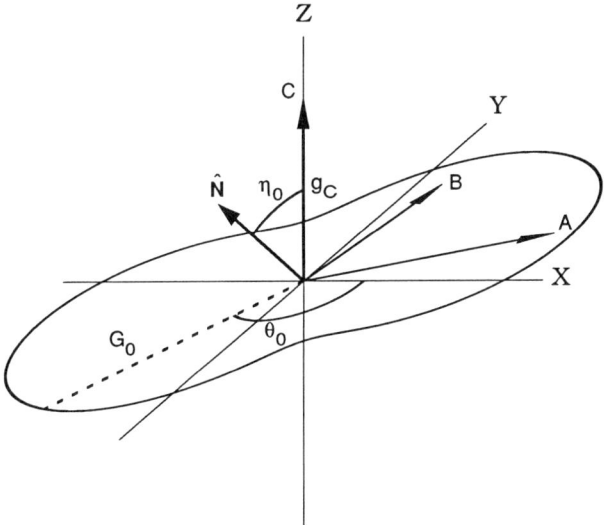

FIGURE 4. Existence of a null vector $\hat{N}$. The X-Y plane shows the response gain of $A + B$ (the plot is the same as FIGURE 3E); C (gain 1.2, phase 45°) is oriented perpendicular to A and B. At some stimulus orientation θ_0, the response from $A + B$ will be 180° out of phase with the stimulus from C. By adjusting the stimulus angle η (keeping θ_0 fixed), the response from C will exactly cancel that from $A + B$, showing that this is a null stimulus orientation $\hat{N}$.

unit vectors); for purposes of discussion, let us call this a "complete response vector." We further showed that any complete response vector can be decomposed into two simple response vectors at right angles to each other. Now, by demonstrating that the sum of three simple response vectors still yields a complete response vector (which, in turn, can be decomposed into two simple vectors), we have proved that the addition of complete response vectors is closed, that is, always produces other complete vectors. Our characterization of the vector is, in fact, "complete."

The development of the concept of a response vector started with two parameter vectors (orientation and sensitivity, if considering spatial tuning, or gain and phase, if considering responses to sinusoids). Three parameter vectors arose by considering both the spatial and temporal properties of neurons. The necessity for two additional parameters, the deviant and the normal, arose from a desire to combine responses of neurons in a "natural" way. Five parameters are necessary to describe the response in general; the closure of complete response vectors under addition demonstrates that five parameters are also sufficient.

The presence of two unit (spatial) vector parameters constrained to lie perpendicular to each other introduces three degrees of freedom to the vector; the other (scalar) parameters add three more, for a total of six degrees of freedom. Angelaki considered the response of a linear system to sinusoidal stimuli delivered in three orthogonal spatial directions. For each direction, a response gain and phase are obtained, for a total of six parameters. The response to an arbitrarily directed stimulus can be obtained by decomposing the stimulus into a linear combination of the three orthogonal axes, then adding together the appropriately weighted sinusoidal responses observed. She directly demonstrated that the gain and phase of the response could be derived for any stimulus orientation (thus six degrees of freedom, as above, are also necessary and sufficient). She also formulated the concept of a "response ellipse"—if the response vectors are considered as describing forces, or positions in space (that is, are considered "active"), the ellipse becomes the path traced out as the sinusoid varies as a function of time (FIGURE 5A). The parameters of the response vector obtained from this response ellipse are identical to those developed in this paper, considering the response vector as "passive" and acted upon by a stimulus $\hat{S}$ (FIGURE 5B). Note in particular that the inclination of the major axis of the ellipse is identical to the orientation parameter, its semimajor axis gives the gain parameter, and the ratio of the axes ("tuning ratio") gives the deviant. Finally, it is of interest to note the similarities between the response ellipse, developed using the "active" formulation of response vectors, and the gain/phase polar ellipse (FIGURE 5C), developed using the passive approach. To derive the response ellipse, one adds vectors and plots the vector sums as a function of time—the resulting ellipse expresses a gain and an orientation parameter. The gain/phase polar plot is derived by considering the summed sinusoidal responses as functions of orientation (a spatial parameter)—the resulting ellipse expresses the response phase, a timelike parameter. In some sense, the two approaches interchange spatial and temporal characteristics, but come to the same mathematical conclusions.

These results demonstrate that the neural response vector concept can be used to characterize the response of a linear system to combinations of spatial and temporal stimuli. These vectors themselves can be manipulated to further predict responses of interacting neural populations or systems, such as the reflex and neuron responses to vestibular stimulation that activates semicircular canals and otolith organs,[6,7,12] and the interaction of neck and vestibular responses.[9] Work in progress will describe STC-like responses resulting from pure linear accelerations,[13] and will examine how the central nervous system might use the encoded information to produce eye movements.[14]

Many neurons and reflexes appear to obey a cosinelike tuning curve, characteristic of simple response vectors, in two[4,8] and three[15,16] dimensions. To the extent that these spatial tuning properties arise from convergence of neural populations, this convergence must be between elements having similar spatial or temporal properties, or else we would observe the noncosine-response tuning characteristic of STC behavior. Otherwise, if simply combining arbitrary simple response vectors, spatial and temporal convergence is the rule, rather than the exception. We have demonstrated that STC behavior can alter and possibly reduce the spatial information content of the neural signal, but it cannot eliminate it entirely—there is always a null spatial direction (for linear systems) encoded by the neural response vector. Further theoretical and experimental studies may suggest how these more complex responses

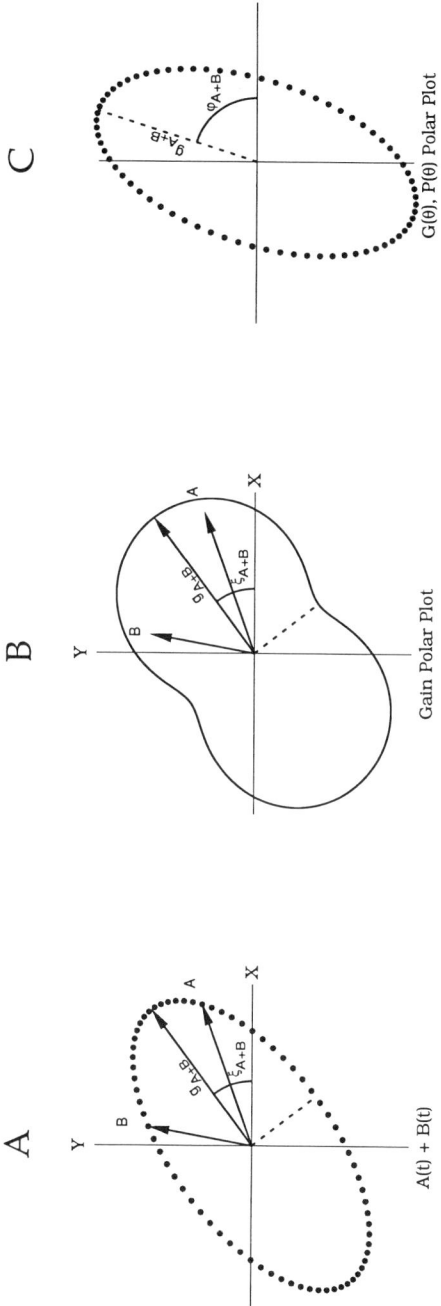

FIGURE 5. Relationship between "response ellipses" and response vector plots. **A:** Response ellipses result from considering vectors A and B as sinusoidally varying in length as a function of time, adding them, and plotting the resulting vector sum.[10] The inclination of the major axis gives a parameter identical to the orientation parameter developed in this paper; the length of the semimajor axis is the identical gain parameter. Similarly, the ratio of minor to major axis ("tuning ratio") gives a value identical to the deviant; the ratio of minimum-to-maximum gain. **B:** Plotting response gain as a function of stimulus orientation does not yield an ellipse. Note that the A and B vectors in this plot and that of FIGURE 5A do not, in general, touch the envelope. **C:** The gain/phase polar plot is closely related to the response ellipse, if spatial and temporal parameters are interchanged.

and interactions may be utilized by the central nervous system to analyze sensory information and produce appropriate motor responses.

REFERENCES

1. FERNANDEZ, C., J. M. GOLDBERG & W. K. ABEND. 1972. Response to static tilts of peripheral neurons innervating otolith organs of the squirrel monkey. J. Neurophysiol. **35:** 978–997.
2. LOE, P. R., D. L. TOMKO & G. WERNER. 1973. The neural signal of angular head position in primary afferent vestibular nerve axons. J. Physiol. **230:** 29–50.
3. SHOTWELL, S. L., R. JACOBS & A. J. HUDSPETH. 1981. Directional sensitivity of individual vertebrate hair cells to controlled deflection of their hair bundles. Ann. N.Y. Acad. Sci. **374:** 1–10.
4. SCHOR, R. H., A. D. MILLER & D. L. TOMKO. 1984. Responses to head tilt in cat central vestibular neurons. I. Direction of maximum sensitivity. J. Neurophysiol. **51:** 136–146.
5. SCHOR, R. H., A. D. MILLER, S. J. B. TIMERICK & D. L. TOMKO. 1985. Responses to head tilt in cat central vestibular neurons. II. Frequency dependence of neural response vectors. J. Neurophysiol. **53:** 1444–1452.
6. BAKER, J., J. GOLDBERG & B. PETERSON. 1985. Spatial and temporal response properties of the vestibulocollic reflex in decerebrate cats. J. Neurophysiol. **54:** 735–756.
7. BAKER, J., J. GOLDBERG, G. HERMANN & B. PETERSON. 1985. Spatial and temporal response properties of secondary neurons that receive convergent input in vestibular nuclei of alert cats. Brain Res. **294:** 138–143.
8. WILSON, V. J., R. H. SCHOR, I. SUZUKI & B. R. PARK. 1985. Spatial organization of neck and vestibular reflexes acting on the forelimbs of the decerebrate cat. J. Neurophysiol. **55:** 514–526.
9. KASPER, J., R. H. SCHOR & V. J. WILSON. 1988. Response of vestibular neurons to head rotations in vertical planes. II. Response to neck stimulation and vestibular-neck interaction. J. Neurophysiol. **60:** 1765–1778.
10. ANGELAKI, D. E. 1991. Dynamic polarization vector of spatially tuned neurons. IEEE Trans. Biomed. Eng. **38:** 1053–1060.
11. ANGELAKI, D. E., G. A. BUSH & A. A. PERACHIO. 1992. A model for the characterization of the spatial properties in vestibular neurons. Biol. Cybern. **66:** 231–240.
12. KASPER, J., R. H. SCHOR & V. J. WILSON. 1988. Response of vestibular neurons to head rotations in vertical planes. I. Response to vestibular stimulation. J. Neurophysiol. **60:** 1753–1764.
13. BUSH, G. A., A. A. PERACHIO & D. E. ANGELAKI. 1992. Quantification of different classes of canal-related vestibular nuclei responses to linear acceleration. Ann. N.Y. Acad. Sci. (This volume.)
14. ANGELAKI, D. E. 1992. Vestibular neurons encoding multidimensional linear acceleration assist in the estimation of rotational velocity during off-vertical-axis rotation. Ann. N.Y. Acad. Sci. (This volume.)
15. GEORGOPOULOS, A. P., A. B. SCHWARTZ & R. E. KETTNER. 1986. Neuronal population coding of movement direction. Science **233:** 1416–1419.
16. KASPER, J., R. H. SCHOR, B. J. YATES & V. J. WILSON. 1988. Three-dimensional sensitivity and caudal projection of neck spindle afferents. J. Neurophysiol. **59:** 1497–1509.

Neuronal Correlates of Optic Flow Stimulation

ROBERT H. WURTZ AND CHARLES J. DUFFY

Laboratory of Sensorimotor Research
National Eye Institute
Building 10, Room 10C101
Bethesda, Maryland 20892

INTRODUCTION

Optic flow fields are the patterns of visual stimulation generated by the movement of an observer through the environment. James Gibson emphasized that the structure of this flow field provides information to the observer regarding the observer's movement and the structure of the environment.[1] More recently, psychophysical experiments have demonstrated a remarkable precision in judgment based on optic flow information. For example, the work of Warren and his collaborators has demonstrated using simulated optic flow fields that the observer can determine the heading of simulated self-motion to within a few degrees of arc.[2,3] Such optic flow also contributes to stabilizing posture[4] and can produce powerful illusory movements.[5]

The neuronal basis for extracting information from optic flow stimuli has remained largely unknown. The critical characteristic of such neuronal mechanisms would seem to be sensitivity to motion over a large area of the visual field. In the monkey such visual areas as the primary visual cortex (striate cortex, or area V1) are responsive to motion but have minuscule receptive fields compared to the large-field stimulation of optic flow. However, several areas of extrastriate visual cortex have large receptive fields and are especially sensitive to visual motion. In particular, the middle temporal area (MT), which receives a direct projection from the striate cortex, has a high concentration of directionally selective cells,[6,7] the sine qua non of visual motion processing. MT, in turn, projects to another area, the medial superior temporal area (MST), that also has a high proportion of directionally selective cells.[8-10] For our present purposes, a key feature in this sequence along the pathway from V1 to MT to MST is the expansion in receptive field size. FIGURE 1 illustrates the approximate size of typical visual receptive fields found in these three areas of visual cortex with the receptive fields centered at 10° from the fovea. Superimposed on these mean receptive fields is an expanding optic flow field pattern with the center of expansion aligned with the point of fixation (FP). A V1 receptive field would subtend a negligible fraction of the optic flow field pattern, an MT receptive field substantially more but still a limited fraction, whereas the MST neuron would encompass a very large fraction of the optic flow stimulation.

Within the MST, Komatsu and Wurtz have identified functional subregions, one a lateral ventral region (MSTl), the other a dorsal medial division (MSTd).[10] MSTl has a large fraction of cells that respond to the motion of small spots and frequently have small receptive fields. Damage leads to a deficit in the maintenance of pursuit eye movements. The relationship of this area to pursuit eye movements has been summarized elsewhere.[11-13] The cells in the MSTd region have consistently large receptive fields (as illustrated in FIGURE 1) and in addition frequently prefer motion of large-field patterns rather than small spots of light. Receptive fields of MSTd

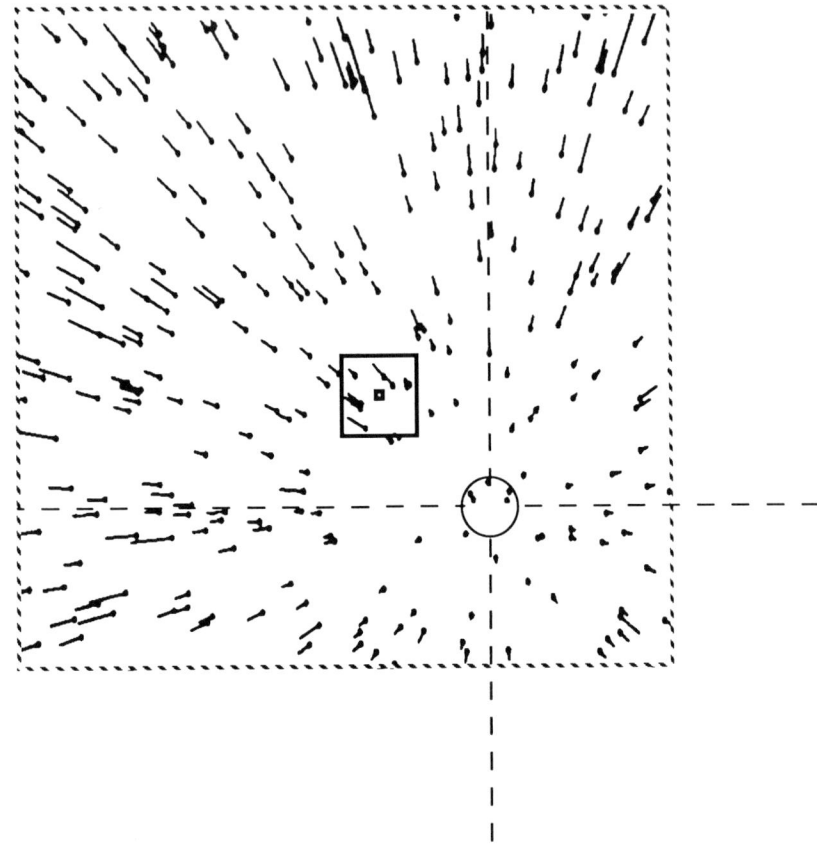

FIGURE 1. Increasing receptive field size for receptive fields moving from striate cortex (dark square—V1) to MT (small square) to MST (large square—MSTd) and their appropriateness for analysis of optic flow stimuli. Receptive field sizes taken from Albright and Desimone[20] and Komatsu and Wurtz.[10] The optic flow field illustrates radial-outward motion (after Warren and Hannon).[3]

neurons typically include or approach the fovea. Finally, MSTd cells not only respond to planar motion in front of the monkey, but also to rotation or looming of the visual stimulus.[14–16] These characteristics make the MSTd neurons ideal candidates for the analysis of optic flow patterns, and for this reason we have investigated the responses of MSTd neurons to simulated optic flow stimulation. In this discussion we summarize our investigations of the response of MSTd neurons to optic flow stimulation that have been recently published.[17,18] Work on the disparity sensitivity of MSTd neurons is considered elsewhere.[19]

RESPONSE OF MSTD NEURONS TO COMPONENTS OF OPTIC FLOW

In our experiments on MSTd neurons, we used a set of translational and rotational stimuli from which all optic flow fields can be derived. These stimuli were head centered and were designed to stimulate the components of optic flow that would be seen by subjects as they moved through the environment. The translational stimuli (FIGURE 2A) consisted not only of motion in the frontoplanar plane, such as left–right, and up–down, but of radial expansion and contraction of the full-field stimulation that would occur as one moved in depth along the axis labeled in–out.

A Translation along axes

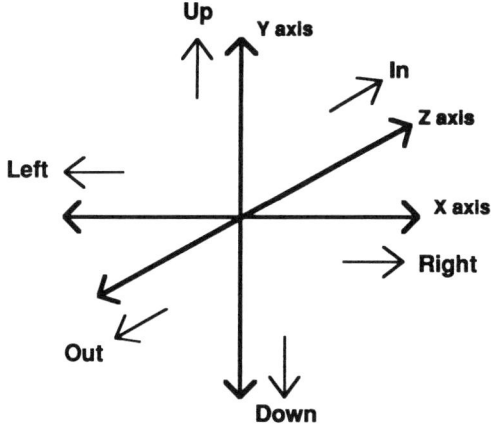

B Rotation around axes

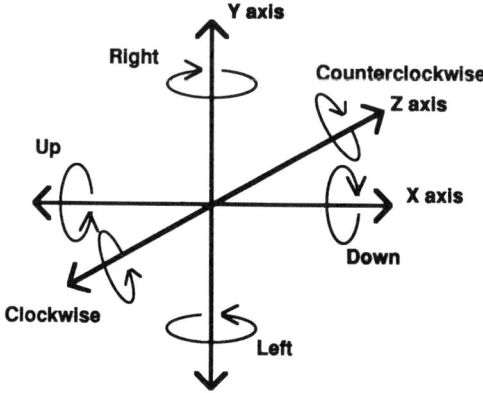

FIGURE 2. The basis set of motion components in head-centered coordinates. All motions in three-dimensional space (represented by the three axes, x, y, and z) consist of combinations of translation (as in A) and/or rotation (as in B) along or around these axes. In two cases rotation is very similar to translation: translation along the x-axis is similar to that for rotation around the y-axis; translation along the y-axis is similar to rotation around the x-axis and in these cases we tested only translation routinely. (From Duffy and Wurtz.)[17]

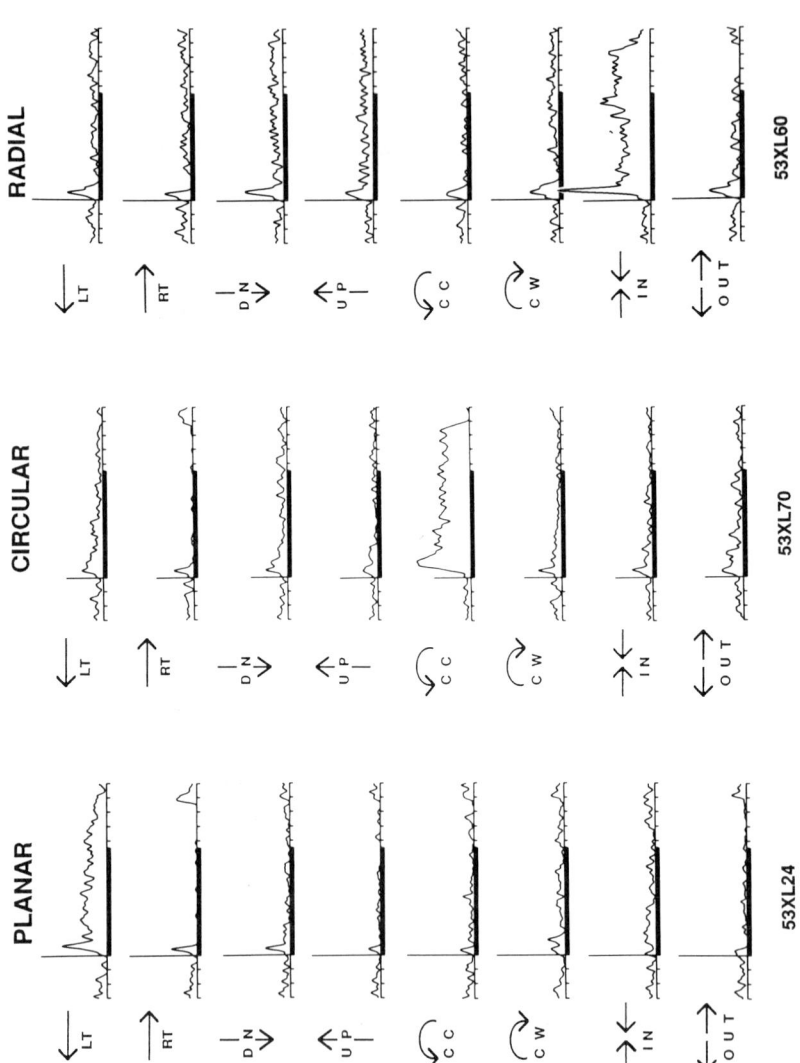

Rotation could occur along any of the same axes (FIGURE 2B), but in our experiments we used only those along the axis into the frontal plane (z-axis) that produce clockwise and counterclockwise motion. We dropped the other four rotatory motions because in preliminary experiments, we found that rotation around the axes labeled x and y approximated the response obtained with up–down or left–right motion, respectively. Our minimal test on all cells, therefore, consisted of four directions of frontal planar motion, in–out radial motion, and clockwise–counterclockwise rotatory motion. Applying the same stimulus set (frequently expanded to include eight directions of frontal planar motion) allowed us to systematically investigate and compare the responses of all MSTd neurons.

The monkey in our experiments faced a tangent screen 50 cm in front of it, onto which the $100° \times 100°$ stimuli were projected. It was rewarded for maintaining fixation on a spot, usually at the center of the screen, for the duration of the stimulus presentation. Since all receptive fields included or closely approached the fovea, and since that is the visual area of greatest responsiveness, we usually centered the stimuli over the fovea.

We studied over 200 single neurons in the MSTd and obtained a variety of response profiles to the optic flow stimuli.[17] Some neurons gave selective responses to either planar, circular, or radial motion, and FIGURE 3 shows an example of neurons responsive to each of these stimuli. In each column of FIGURE 3, the eight types of visual stimuli are shown down the edge of the column and the response of the cell to that stimulus is adjacent to it. We have used a spike-density display that shows the averaged response for between 6 and 12 presentations of each stimulus. The stimulus duration of 1500 mseconds is represented by the dark bar; the vertical line indicates the onset of the visual stimulus. We found that many MSTd neurons gave a relatively nonselective early phasic discharge followed by a sustained selective response, and we have therefore concentrated our analysis on the more selective tonic response. In FIGURE 3, the response of the cell in the left column is clear for leftward motion, that in the middle column is clearest for counterclockwise circular motion, and that in the right column clearest for inward radial motion. While these neurons responded primarily to just one component of the visual stimulus, most MSTd neurons responded to a combination of these motion components. Some neurons responded to planar and circular motion, or to planar and radial motion. In these cases, however, the motion selectivity was to one type of circular or radial motion or one direction of planar motion, not an indiscriminate response to all planar, circular, or radial motions. For example, a cell responding to planar and circular motion responded to counterclockwise motion only and to leftward motion only. Other neurons responded to planar, circular, or radial motion. We quantified the tonic response of

---

FIGURE 3. Examples of neurons that responded to planar, circular, or radial stimulation. The direction of motion of the random dot display is indicated to the left in each column: planar (leftward, LT; rightward, RT; downward, DN; or upward, UP), circular (counterclockwise, CC; or clockwise, CW) or radial (inward, IN; or outward, OUT) pattern. All stimuli contained 300 white dots rear projected on to a $100°$ square, tangent projection screen. The spike density display in each column shows responses over eight stimulus presentations. In the left column, the neuron responded to leftward planar motion; in the middle column, the neuron responded to clockwise circular motion; in the right column, the neuron responded to inward radial motion. The spike-density histograms represent the average of 6 to 8 trials with the location of the vertical bar indicating time of stimulus onset and its height indicating a discharge rate of 75 spikes/second for neurons 53XL24 and 53XL70, and 100 spikes/second for neuron 53XL60. The tick marks on the horizontal baseline indicate 200 mseconds. The horizontal bar indicates a stimulus period of 1.5 seconds. (From Duffy and Wurtz.)[17]

these neurons by requiring a statistically significant response of the neuron above the firing rate when the monkey fixated but no visual stimulation appeared. Using Student's t-test to make the distinction ($p < 0.01$), we found that 23% of the neurons responded to only a single component of optic flow (planar, circular, radial), 34% responded to two components (planocircular or planoradial), and 29% responded to three components. In our sample we did not find cells that gave a significant response to both circular and radial stimulation without having a significant response to planar motion. Further analysis showed that these categories were not discrete groupings of cell types, but rather points along a continuum ranging from responsiveness to a single component of the optic flow stimulation at one end of the continuum to responsiveness to all three components of the optic flow stimulation at the other end of the continuum.

This categorization of the cells was based on the excitatory response of the neurons, but as can be seen from FIGURE 3, some optic flow components produced an inhibitory response in these cells. We found that the frequency of these inhibitory responses also varied along the continuum of the neurons. Single-component neurons had an average of about 3 inhibitory responses per neuron for the set of eight stimuli that we tested, double-component neurons had an average of about 1 inhibitory component response per neuron, and triple-component neurons had only 0.4 inhibitory response per neuron. Thus, as we move along the continuum from single-component neurons to triple-component neurons, the frequency of both excitatory responses and inhibitory responses varied: the single-component neurons appeared to be the most selective; they had limited excitatory responses to one component of the stimuli we used and the clearest inhibition to other components.

In considering the relation of these neurons to optic flow, it is equally important to consider what stimulus factors did not influence these cells. On a smaller sample of cells, we found that most were responsive to a wide range of stimulus speeds over the range of 10° to 80°/second. The response of the cell was remarkably independent of dot density, responding not only to our usual density of 350 dots/100°,[2] but continuing to respond even when this density was reduced to 25 dots/100°.[2]

In summary, cells in MSTd do respond to the components of optic flow stimulation and their insensitivity to dot density and to speed make them ideal candidates for response to large-field optic flow stimulation. However, there is not one cell type in the MSTd, but rather a continuum ranging from highly selective single component neurons to less-selective multiple-component neurons.

MECHANISMS OF RESPONSE SELECTIVITY

The response of MSTd neurons to radial and circular motion over a large part of the visual field could indicate that these neurons are carrying out sophisticated higher-order analyses of visual motion. Alternatively, these responses might result from stimulation of parts of the receptive field of a cell that is essentially sensitive to planar motion. For example, FIGURE 4 shows how a circular response and a radial response might result from a neuron that had only sensitivity to planar motion. A neuron with sensitivity to downward planar motion (FIGURE 4, upper half) might respond to counterclockwise motion if it just overlapped the right side of the planar-sensitive field, or to clockwise motion if the stimulus overlapped the field on the left. Similarly, a receptive field sensitive to rightward planar motion (FIGURE 4, lower half) might respond preferentially to inward radial motion if that motion overlapped the receptive field of the cell on the right side, and to outward radial motion if it overlapped the receptive field on the left side.

We investigated this issue by determining first the response of a single neuron to the large-field stimulation, and then its response to the same stimuli placed in smaller subregions of the stimulus field.[18] If the responses of the cell were dependent upon the position of the stimulus, then we would expect that the stimulation in

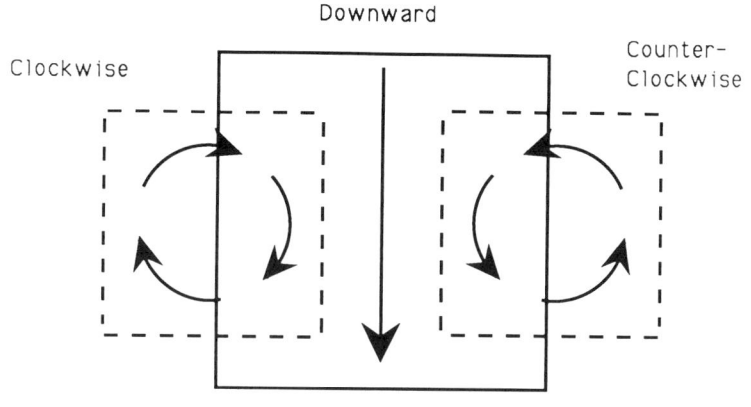

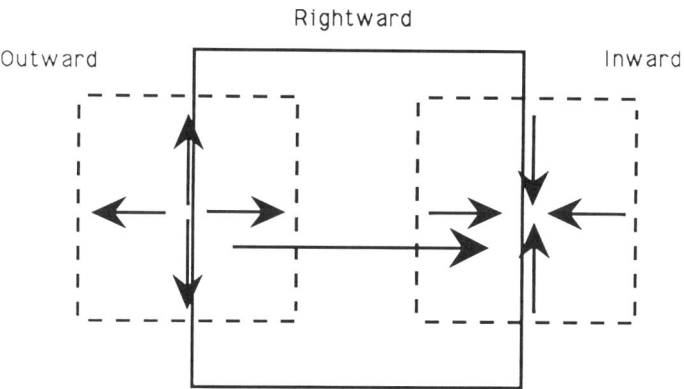

FIGURE 4. Response to circular (top) and radial (bottom) stimulation explained by the sensitivity of the hypothetical neuron to planar motion. See text for description.

different subregions of the field would produce different responses as in FIGURE 4. On the other hand, if the responses were position independent, then the same stimulus in different parts of the field would produce the same responses. FIGURE 5 shows the results of such an experiment. FIGURE 5A shows the response of the cell to full-field stimulation. In this figure we show the response to the planar, circular, and radial stimuli using a schematic vector summary diagram. Response to planar motion is indicated by the lines to the left, right, up, and down with the amplitude of the line

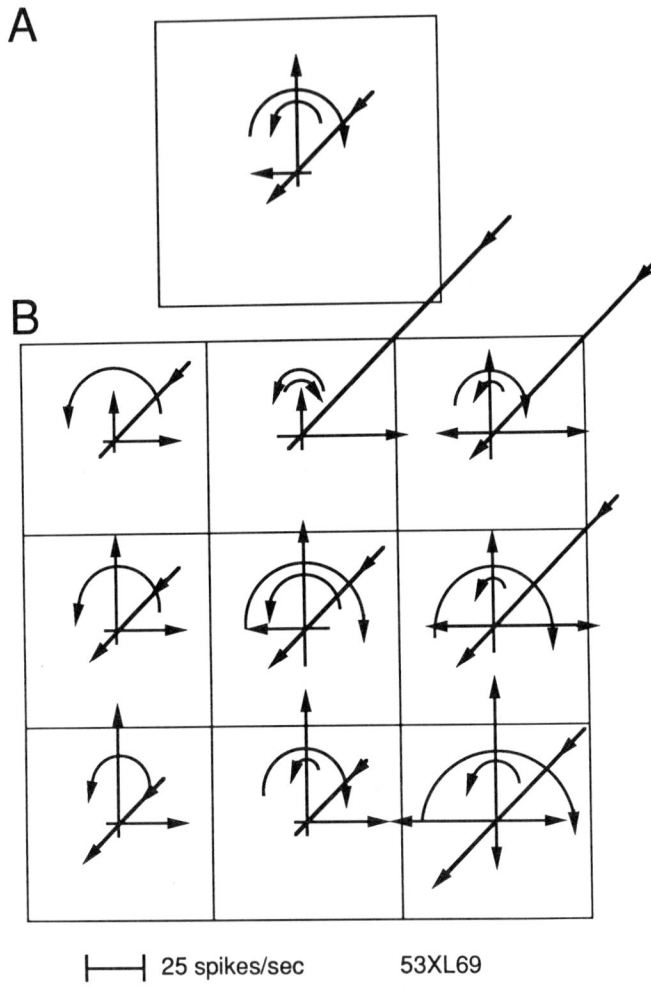

├──┤ 25 spikes/sec   53XL69

FIGURE 5. A triple-component, planocirculoradial neuron showing substantial response variability to small-field stimuli. **A:** The direction diagram shows responses to upward-planar, clockwise-circular, and inward-radial large-field stimulation (100° × 100°). **B:** Direction diagrams for the small-field studies (33° × 33°) showing variation in planar direction selectivity with strong upward selectivity in the right-lower segment and strong rightward selectivity in the middle-upper segment. Circular direction selectivity reverses from strong clockwise selectivity in the right-lower segment to strong counterclockwise selectivity in the left-upper segment. Radial selectivity is maintained but varies quantitatively, being strongest in the right-upper segment and weakest in the left-lower segment. In all cases the monkey fixated on a point at the center of the large field. (From Duffy and Wurtz.)[18]

indicating the magnitude of the response. Circular motion is indicated by the hemicircles with the diameter of the hemicircle indicating the size of the response. Inward radial motion is shown by the arrow running from upper right to center, and outward motion by the arrow running from the center to lower left with length again indicating amplitude of response. An arrowhead indicates that the response was statistically significant at the $p = 0.01$ level (Student's t-test). FIGURE 5B shows the same stimuli applied to a fraction of the $100° \times 100°$ field ($33° \times 33°$ stimuli). The upward planar sensitivity was present throughout the field; the rightward sensitivity varied considerably as did the amplitude of the inward radial response. The response to the circular stimulus varied dramatically, being clockwise for stimuli on the right side and counterclockwise for stimuli on the left side, and the response of this neuron to circular stimulation therefore could be explained on the basis of the upward planar motion using the logic illustrated in FIGURE 4. Thus, the responses are position dependent varying with the position of the stimulus within the field.

We obtained a very different result in a similar study on the single-component radial neuron illustrated in FIGURE 6. The response of this neuron to full-field stimulation (FIGURE 6A) indicated a clear response to radial-inward motion. The response to stimulation in the small fields continued to show this same inward-radial selectivity. While the amplitude of the response varied, the neuron basically showed a position-independent response to inward-radial motion. The response of this neuron is not easily explained on the basis of its minimal planar response, and its visual sensitivity is more likely to result from higher-order visual processing.

We compared the response of 160 neurons to the large-field stimulation and to stimulation within subregions of the field and found a variety of results ranging between those illustrated in FIGURES 5 and 6. The salient point was that neurons that had sensitivity to multiple components of optic flow, such as the triple-component neuron in FIGURE 5, were more likely to have a response that could be in large part explained by planar sensitivity. In contrast, neurons that were sensitive to single components of optic flow stimulation (such as the neurons sensitive primarily to inward-radial stimulation in FIGURE 6) were the least likely to be understood on the basis of any sensitivity to planar motion. The continuum of selectivity that we have observed in these neurons seems to be related to the structure of their visual receptive fields.[18]

RESPONSE TO COMBINATIONS OF OPTIC FLOW COMPONENTS

Using components of optic flow stimulation has been a useful first step in determining the response of the population of cells in MSTd, and has allowed initial experiments on the selectivity of these neurons. However, movement of an observer through the environment nearly always produces combinations of these optic flow components, and in the next step in our experiments we are investigating the effect of combining optic flow components on the single neuron's response.

One of the simplest combinations would be that of radial and planar stimulation. Such stimulation would occur as an observer moves forward but drifts to the side, as for example in the case of the observer in a row boat looking straight ahead while drifting sideways. This combination of radial and planar motion could also be taken as an approximation for the addition of radial and rotatory motion if the scene were at a substantial distance from the observer.

FIGURE 7 shows the stimulus when planar motion in 4 of the 8 directions tested was combined with radial motion (shown in the center). We applied these stimuli to neurons that were sensitive to radial motion, and in our initial experiments, we have

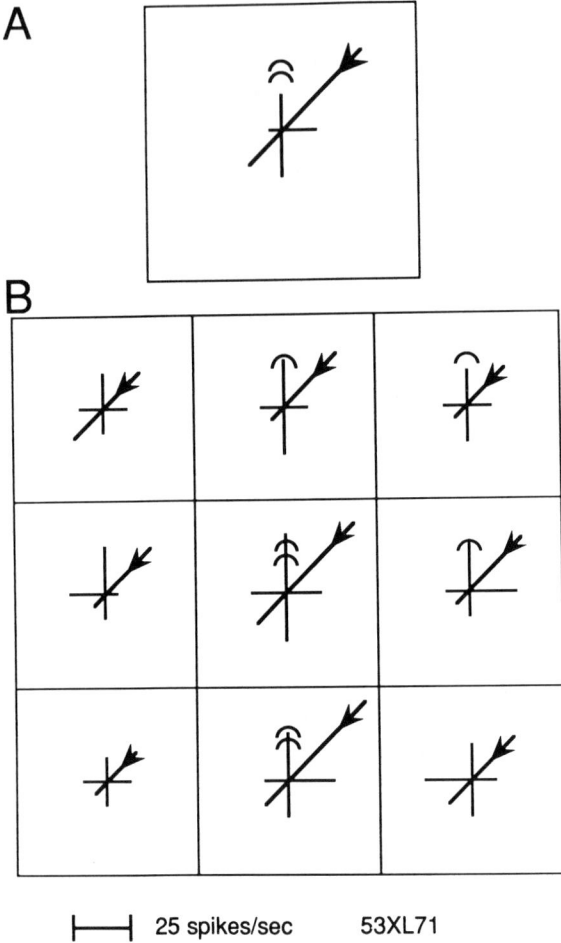

FIGURE 6. A single-component, radial neuron showing position-invariant selectivity for inward-radial motion. **A:** The direction diagram of responses evoked by large-field stimulation indicating a selective response to inward-radial motion. **B:** Direction diagrams for responses evoked by nine small-field stimulation experiments covering the same area stimulated in A. In the small-field experiments, selectivity for inward-radial motion is maintained although it varies quantitatively across the field. In all cases the monkey fixated on a point at the center of the large field. (From Duffy and Wurtz.)[18]

found once again that not all neurons respond to such combined stimulation in the same way. For example, the response of the single component neuron to full-field stimulation (FIGURE 8A) was substantial to radial-outward motion. In FIGURE 8B, the response was reduced when this radial motion was combined with one of the 8 directions of planar motion (indicated by the eight lines). In contrast, the triple-component neuron shown in FIGURE 9A had a large response to large-field outward-radial motion, to planar motion to the right and down, and counterclockwise circular motion. Combining radial motion with the eight directions of planar motion (FIGURE 9B) produced little effect for certain directions and an increase in the discharge rate for other directions. Our sample of neurons in this experiment is relatively limited, but there is again a tendency for single-component neurons and triple-component neurons, as two ends of the continuum within the population, to respond to the combinations of optic flow stimuli in different ways.

Clearly the combination of optic flow stimuli alters the response of some of the MSTd neurons. Exploring these combinations using stimuli that would actually occur with movement of the observer through the environment should yield a better understanding of the role of MSTd neurons in the cortical analysis of optic flow fields.

SUMMARY AND CONCLUSION

Neurons in a region of monkey extrastriate cortex, MSTd, respond to the components of optic flow stimulation. Some of these neurons (single-component

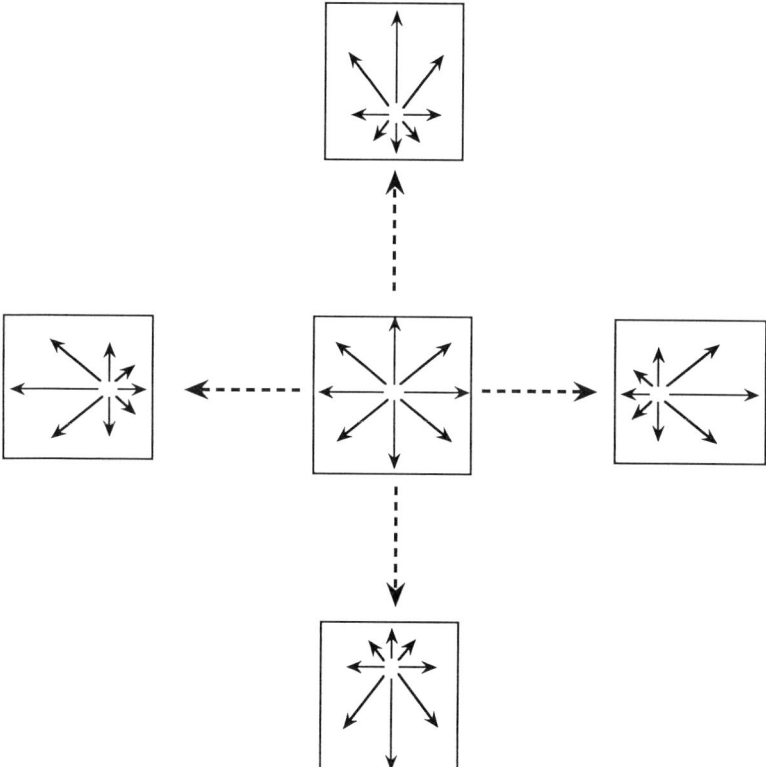

FIGURE 7. Effect of combining radial motion alone (center drawing) with each of four sample directions of the eight planar motion directions used (surrounding drawings). Note that the combination shifts the focus of expansion in each of the combination cases.

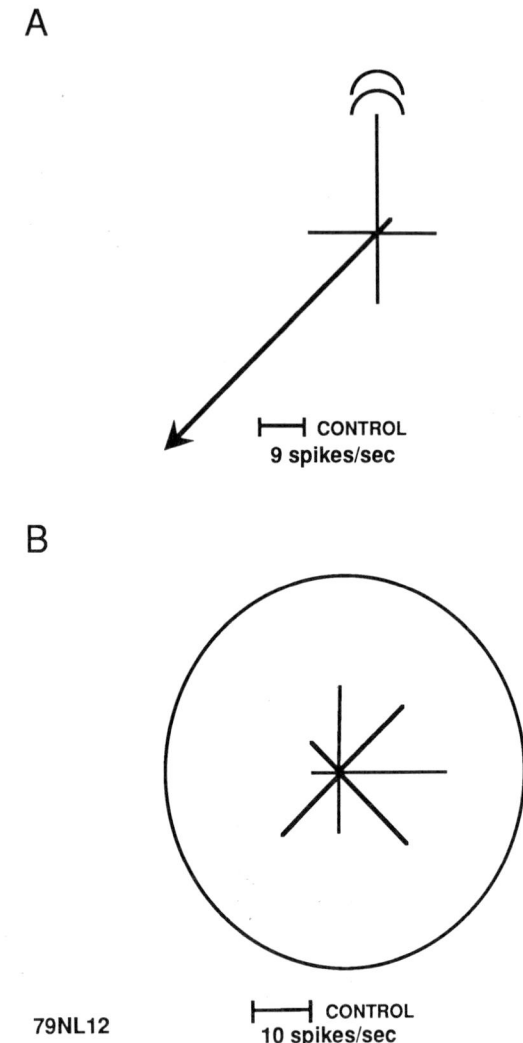

A

CONTROL
9 spikes/sec

B

79NL12

CONTROL
10 spikes/sec

FIGURE 8. Combined radial and planar motion reduces the response of a single-component neuron. **A:** Response to full-field stimulation showing primarily a response to radial-outward motion. Same response notation as in FIGURE 5. **B:** Response to the combined radial and planar motion for the same neuron. The response to the radial motion alone is indicated by the circle; that to the combination by the length of the lines that are in the direction of the planar motion.

neurons) are selective for a single type of motion such as inward- or outward-radial motion. Other neurons respond to multiple types of rotation, for example, rightward planar, clockwise circular, and inward radial. Rather than forming discrete groups, we think these neurons represent a continuum covering the range from single-component sensitivity to multiple-component sensitivity. By combining the optic flow stimuli, we have also been able to recognize that such combinations alter the response of cells in the continuum to varying degrees.

At this point, while our evidence is consistent with the hypothesis that cells in area MSTd contribute to the processing of optic flow stimuli, we do not know whether these neurons do in fact serve this function. As in all single-cell recording

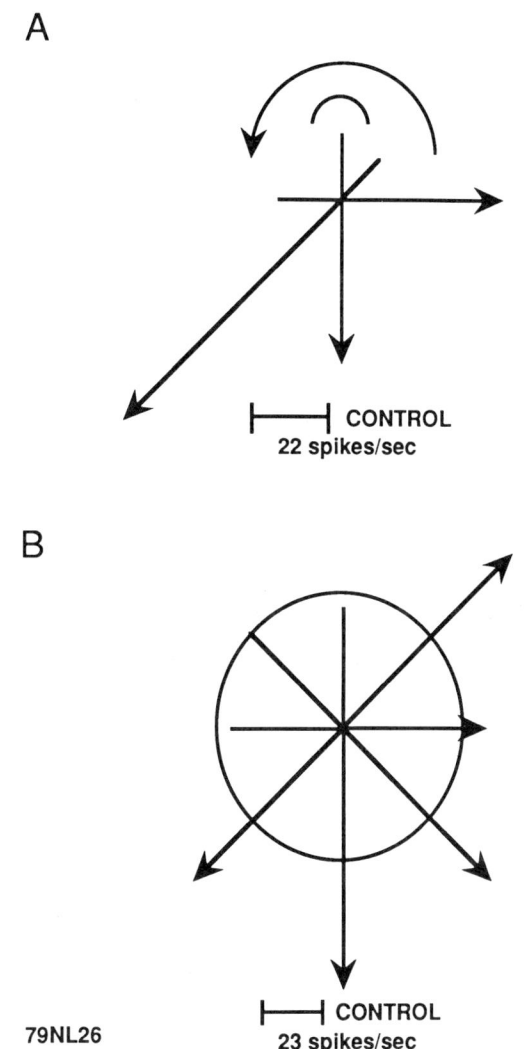

FIGURE 9. Combined radial and planar motion has little impact on the response of a triple-component neuron. **A:** Response to full-field stimulation showing a response to planar-circular and radial-outward motion. **B:** Response to the combined radial and planar motion for the same neuron. The response to the radial motion alone is indicated by the circle. Same notation as in FIGURE 8.

experiments, even those in awake animals performing tasks closer to real-world tasks than we have succeeded in emulating here, the activity of the cell in relationship to the visual stimulation is simply a correlate of the optic flow stimulation and may or may not contribute to the processing of optic flow stimulation upon which behavior depends. Further information on a number of characteristics of these cells might clarify their role. Information on such factors as whether heading in the environment is conveyed by individual neurons, or whether this property is more likely to be conveyed over a population of neurons, and the role of changes in the point of fixation of the eyes are critical points. Generation of behavior on the basis of the optic flow stimulation and determination that this behavior is modified by selective lesion of MSTd would also strengthen the argument that visual motion processing in this area is related to analyzing optic flow information.

REFERENCES

1. GIBSON, J. J. 1950. *In* The Perception of the Visual World. Houghton Mifflin. Boston, Mass.
2. WARREN, W. H., M. W. MORRIS & M. KALISH. 1988. Perception of translational heading from optic flow. J. Exp. Psychol. Hum. Percept. **14:** 646–660.
3. WARREN, W. H. & D. J. HANNON. 1990. Eye movements and optical flow. J. Opt. Soc. Am. A **7:** 160–169.
4. BERTHOZ, A., M. LACOUR, J. F. SOECHTING & P. P. VIDAL. 1979. The role of vision in the control of posture during linear motion. Prog. Brain Res. **50:** 197–209.
5. DICHGANS, J., R. HELD, L. R. YOUNG & T. BRANDT. 1972. Moving visual scenes influence the apparent direction of gravity. Science **178:** 1217–1219.
6. DUBNER, R. & S. M. ZEKI. 1971. Response properties and receptive fields of cells in an anatomically defined region of the superior temporal sulcus in the monkey. Brain Res. **35:** 528–532.
7. ZEKI, S. M. 1974. Functional organization of a visual area in the posterior bank of the superior temporal sulcus of the rhesus monkey. J. Physiol. London **236:** 549–573.
8. VAN ESSEN, D. C., J. H. R. MAUNSELL & J. L. BIXBY. 1981. The middle temporal visual area in the macaque: myeloarchitecture, connections, functional properties and topographic organization. J. Comp. Neurol. **199:** 293–326.
9. UNGERLEIDER, L. G. & R. DESIMONE. 1986. Cortical connections of visual area MT in the macaque. J. Comp. Neurol. **248:** 190–222.
10. KOMATSU, H. & R. H. WURTZ. 1988. Relation of cortical areas MT and MST to pursuit eye movements. I. Localization and visual properties of neurons. J. Neurophysiol. **60:** 580–603.
11. WURTZ, R. H., H. KOMATSU, M. R. DÜRSTELER & D. S. G. YAMASAKI. 1990. Motion to movement: cerebral cortical visual processing for pursuit eye movements. *In* Signal and Sense: Local and Global Order in Perceptual Maps. G. Edelman, W. E. Gall & W. M. Cowan, Eds.: 233–260. John Wiley. New York, N.Y.
12. WURTZ, R. H., H. KOMATSU, D. S. G. YAMASAKI & M. R. DÜRSTELER. 1990. Cortical visual motion processing for oculomotor control. *In* Vision and the Brain. B. Cohen & I. Bodis-Wollner, Eds.: 211–231. Raven Press. New York, N.Y.
13. WURTZ, R. H., D. S. YAMASAKI, C. J. DUFFY & J.-P. ROY. 1990. Functional specialization for visual motion processing in primate cerebral cortex. Cold Spring Harbor Symp. Quant. Biol. **55:** 717–727.
14. SAITO, H.-A., M. YUKIE, K. TANAKA, K. HIKOSAKA, Y. FUKADA & E. IWAI. 1986. Integration of direction signals of image motion in the superior temporal sulcus of the macaque monkey. J. Neurosci. **6:** 145–157.
15. TANAKA, K. & H. SAITO. 1989. Analysis of motion of the visual field by direction, expansion/contraction, and rotation cells clustered in the dorsal part of the medial superior temporal area of the macaque monkey. J. Neurophysiol. **62:** 626–641.
16. TANAKA, K., Y. FUKADA & H. SAITO. 1989. Underlying mechanisms of the response

specificity of expansion/contraction and rotation cells in the dorsal part of the medial superior temporal area of the macaque monkey. J. Neurophysiol. **62:** 642–656.

17. DUFFY, C. J. & R. H. WURTZ. 1991. Sensitivity of MST neurons to optic flow stimuli. I. A continuum of response selectivity to large-field stimuli. J. Neurophysiol. **65:** 1329–1345.

18. DUFFY, C. J. & R. H. WURTZ. 1991. Sensitivity of MST neurons to optic flow stimuli. II. Mechanisms of response selectivity revealed by small-field stimuli. J. Neurophysiol. **65:** 1346–1359.

19. ROY, J.-P. & R. H. WURTZ. 1990. The role of disparity-sensitive cortical neurons in signalling the direction of self-motion. Nature **348:** 160–162.

20. ALBRIGHT, T. D. & R. DESIMONE. 1987. Local precision of visuotopic organization in the middle temporal area (MT) of the macaque. Exp. Brain Res. **65:** 582–592.

Ocular Compensation for Self-Motion

Visual Mechanisms

F. A. MILES[a] AND C. BUSETTINI[a,b]

[a]Laboratory of Sensorimotor Research
National Eye Institute
National Institutes of Health
Building 10, Room 10C101
Bethesda, Maryland 20892

[b]Dipartimento di Elettrotecnica
Elettronica ed Informatica
University of Trieste
34100 Trieste, Italy

Motion of the observer threatens the stability of retinal images, a problem that is dealt with chiefly by labyrinthine and visual reflexes that combine to generate appropriate compensatory eye movements. In primates, the semicircular canals and the otolith organs, which respond selectively to angular and linear accelerations, respectively,[1] provide the input for two vestibuloocular reflexes (VOR) that compensate selectively for rotational and translational disturbances of the head: the RVOR and TVOR. Any residual shifts of gaze result in retinal image slip which activates visual tracking mechanisms that produce eye movements that reduce the slip. In studies of these visual backup systems, it has been usual to consider only rotational disturbances and to examine the associated visual compensatory mechanisms by rotating the visual surroundings around the stationary subject, the ensuing ocular following being termed *optokinetic nystagmus* (OKN). In the case of the monkey— our major concern here—if the surroundings are rotated rapidly enough, the development of OKN shows two distinct phases that reflect two distinct mechanisms: (1) An initial rapid rise in slow-phase eye speed that takes perhaps a few hundred milliseconds to reach a plateau and generally leaves the eyes somewhat short of the required speed. This has been termed the "direct" component of OKN,[2] but we shall refer to it as the *early* component (OKNe). (2) A subsequent gradual increase in slow-phase eye speed, during which the eyes reach an asymptotic speed more nearly approaching that of the surroundings. This has been termed the "indirect" component,[2] but we shall refer to it as the *delayed* component (OKNd). Cohen and his coworkers have likened OKNd to the gradual charging up of a velocity storage integrator.[2-5] We shall here review our recent suggestion that only OKNd actually evolved as a backup to the RVOR to help compensate for rotational disturbances of gaze, OKNe evolving as a backup to the TVOR to help compensate for translational problems.[6-9]

VISUAL AND VESTIBULAR MECHANISMS: SYNERGISTIC SYSTEMS

The primary support for this idea that the two components of the monkey's optokinetic response evolved independently as backups to the two VORs comes from the finding that selective changes in the gains of the two vestibular reflexes result in equally selective parallel changes in the amplitudes of the two components

220

of the optokinetic response. Changes in the gain of the RVOR can be induced with magnifying or minifying spectacles,[10,11] and such optical challenges have been shown to also result in changes in the amplitude of OKNd but, significantly, *not* of OKNe.[12] This is consistent with the idea that OKNd shares some central circuitry with the RVOR—indeed, that portion of the circuitry containing the variable gain element responsible for adaptive gain control of the RVOR—and supports the notion that the two systems are truly synergistic, combining to compensate selectively for rotational disturbances of the observer. The gain of the monkey's TVOR is a linear function of the inverse of the viewing distance—exactly as required by the optical geometry associated with lateral (side-to-side) translations of the observer—and this dependence on proximity is shared by OKNe.[13] (In the latter experiments, the visual scene was actually back projected onto a tangent screen in front of the stationary animal and laterally translated. We assume that the associated early ocular following is mediated by the same mechanisms as OKNe even though there are some significant differences in the stimuli.) Such a shared metric is consistent with the notion that the two systems, the TVOR and OKNe, share central pathways, are synergistic, and, in this case, compensate selectively for translational disturbances of the observer.

Our proposed schemes for explaining these findings are shown in block diagram form in FIGURE 1. In A of FIGURE 1, the RVOR and OKNd share the variable gain element (G) that mediates adaptive gain control in the RVOR. (Note also the shared velocity storage integrator.) In B of FIGURE 1, the TVOR and OKNe share two gain elements: a variable one $(k_1/d$, where k_1 is a constant and d is the estimate of target distance), which gives the dependence on proximity, and a fixed one (k_2), which accounts for the slight offset in our data. The variable gain element effectively allows the TVOR to receive inputs encoded in cartesian coordinates [translational velocity of the head (H_T)] and to respond with outputs encoded in polar coordinates [rotational velocity of the eyes (E_R)]. Clearly, the two models in FIGURE 1 depicting the RVOR and TVOR, with their respective visual backups, OKNd and OKNe, have a strong formal similarity.

THE OPTIC FLOW ASSOCIATED WITH MOTION OF THE OBSERVER

It is our thesis that OKNd and OKNe evolved to meet the very different visual challenges associated with rotational and translational disturbances of the observer. Neglecting for the moment any compensatory eye movements, pure rotation of the observer results in a rigid rotation of the entire retinal image, the flow pattern resembling the circles of latitude on a globe. From the viewpoint of the oculomotor system, the critical aspect is that the direction and speed of flow at all points are dictated entirely by the observer's own motion and, in principle, the retinal image motion can be entirely eliminated by appropriate compensatory eye movement: the angular speed of the observer's head in space must simply be offset by compensatory eye movements of matching angular speed. (Actually, the eyes lie some distance in front of the axis of rotation of the head and so undergo some translation during normal head turns. In fact, the RVOR takes account of this.)[14] Pure translation of the passive observer creates a flow pattern resembling the circles of longitude on a globe with images emerging from one pole ahead and disappearing into another behind. In this situation, the pattern (or direction) of flow again depends solely on the observer's motion, but the *speed* of the flow at any given point also depends on the *viewing distance* at that location. In fact, to a first approximation retinal image speed is inversely proportional to the viewing distance during lateral translation of the

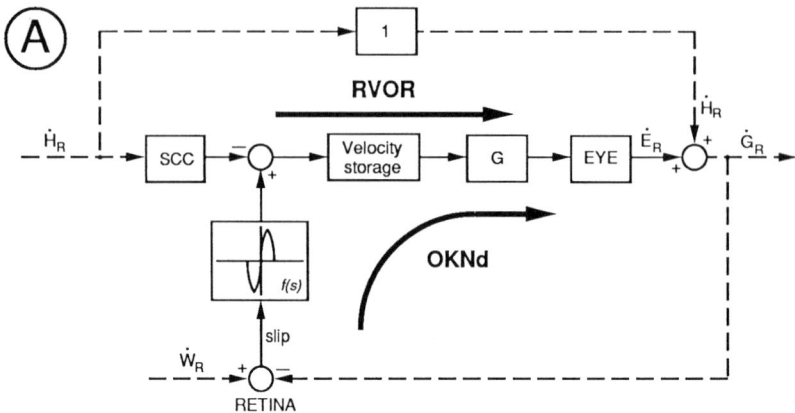

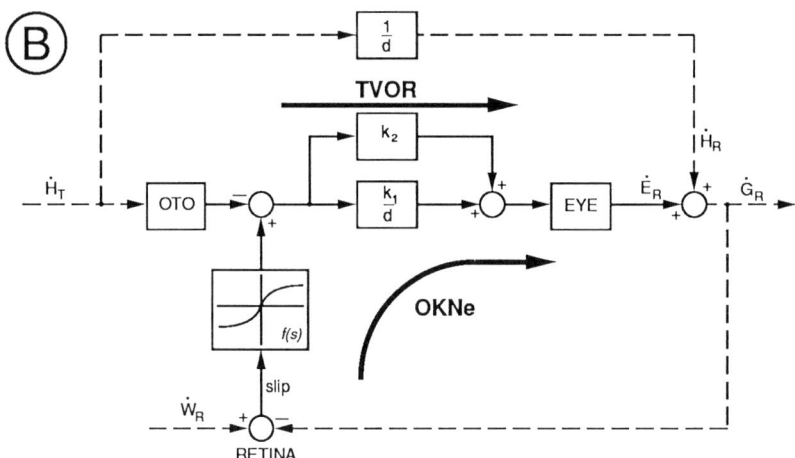

FIGURE 1. Block diagrams showing the proposed linkages between the visual and vestibular reflexes operating to stabilize gaze. **A:** The open-loop RVOR and the closed-loop OKNd generate eye movements, $\dot{E}_R$, that compensate for rotational disturbances of the head, $\dot{H}_R$. These reflexes share (1) a velocity storage element, which is responsible for the slow build-up in OKN and the gradual decay in RVOR with sustained rotational stimuli, and (2) a variable gain element, G, which mediates long-term regulation of RVOR gain. SCC, semicircular canals. The element $f(s)$ indicates that the visual input is sensitive to low slip speeds only. (From Reference 9.) **B:** The open-loop TVOR and the closed-loop OKNe generate eye movements that compensate for translational disturbances of the head, $\dot{H}_T$, which affect gaze in inverse proportion to the viewing distance, d. These reflexes share (1) a variable gain element, k_1/d, which gives them their dependence on proximity, and (2) a fixed gain element, k_2, which gives them an offset. We hypothesize that velocity saturation in the feedback path from the retina,[36] $f(s)$, tends to offset the influence of the variable gain element on the optokinetic response under normal viewing conditions. Since retinal slip speeds will tend to vary inversely with viewing distance, ocular following will tend to show increasing saturation with near viewing. Thus, under everyday conditions OKNe may show little dependence on viewing distance. OTO, otolith organs. Dashed lines represent physical links: $\dot{H}_T$, head velocity in linear coordinates; $\dot{H}_R$, $\dot{E}_R$, $\dot{G}_R$, and $\dot{W}_R$, velocity of head, eyes (in head), gaze, and visual surroundings, respectively, in angular coordinates. (From Reference 13.)

observer's eye, i.e., motion with a component orthogonal to the line of sight.[13] Thus, translation causes complex image shear as the nearby objects move across the field of view more rapidly than the distant ones (motion parallax) and a given compensatory eye movement can only eliminate the retinal image motion of selected objects.

Given these general characteristics of the optic flow, visual mechanisms involved with compensating for rotation of the observer might be expected to respond preferentially to en masse motion of the scene while those involved with translation must deal with complex differential motion. Clearly, normal head movements generally have both rotational and translational components, and we are suggesting that they are effectively dealt with separately by the two components of the monkey's optokinetic system. However, we do not think that this is achieved through a precise, mathematically complete decomposition of the optic flow into rotational and translational components. In our view, this is neither necessary nor possible, and the biological scheme that we have developed emerged from two key ideas. First, the two components of the primate optokinetic response are the product of *sequential*—rather than *parallel*—evolution, the compensation for rotational disturbances predating that for translational disturbances. Second, the evolutionary processes addressed the *specific* problems of normal everyday viewing conditions, not the abstract *general* problem, exploiting the fact that the layout of the visual environment and the observer's repertoire of interactions with it are highly constrained. Thus, our contention is that the optic flow normally provoked by the observer's own motion conforms to a relatively limited number of patterns—indeed, observers may unknowingly contrive to ensure that this is so—and that the nervous system has developed highly specific mechanisms for dealing with those patterns. This ecological view of the optical challenges facing the moving observer owes much to Gibson[15,16] and Lee,[17] who recognized the potential biological significance of the formal structuring of optic flow.

RVOR AND OKNd: REMNANTS OF A PRIMORDIAL SYSTEM?

Optokinetic reflexes with sluggish dynamics and canal-ocular reflexes with brisk dynamics appear to be ubiquitous among contemporary vertebrates with mobile eyes, suggesting that these visuovestibular mechanisms evolved early in a common ancestor. We suggest that these primordial mechanisms originated in a lateral-eyed progenitor but are manifest in contemporary monkeys as OKNd and the RVOR. We further suggest that the primordial visual tracking system compensated almost exclusively for rotational disturbances of the observer, developing special features that rendered it largely blind to the commonest translational disturbances of gaze—those associated with forward locomotion—and the primordial visual system actually relied on the motion parallax associated with locomotion to decode the 3-D structure of the surrounding world.

This view of the primordial optokinetic system comes in part from consideration of a contemporary lateral-eyed mammal, the rabbit, whose oculomotor system is largely insensitive to both the visual and the vestibular consequences of forward locomotion.[18,19] Presumably, this feature is desirable in a lateral-eyed animal since temporalward rotation of the two eyes (divergence) would disrupt the images of the scene ahead—the very region of the field where potentially interesting new images are being actively sought.[20] Recordings from single neurons assumed to mediate OKN in the rabbit suggest that in this species OKN has the same frame of reference as the RVOR, thereby facilitating the orderly summation of rotational information from the retina and the canals. Individual neurons have very large (binocular) visual receptive fields and respond best to global rotations of the entire visual scene about axes that coincide with the rotational axes that are optimal for activating particular semicircular canals.[21-24]

However, the mere fact that the visual receptive fields of the neurons mediating OKN are organized to respond optimally to rotations does not necessarily render those neurons totally insensitive to translation. Consider, for example, the situation during forward locomotion when each eye sees temporalward motion. Clearly, the visual input through one eye will tend to raise the activity of such a neuron while the input through the other will tend to depress the activity of that same neuron, i.e., the inputs from the two eyes will tend to cancel. However, differences in the three-dimensional arrangement of objects on either side of the animal would cause asymmetrical optic flow speeds at the two eyes so that the motion inputs to the eye seeing the nearer objects would be more potent than those to the other eye, the net result being activity that would tend to rotate the eyes towards the side with the nearer objects. Significantly, the rabbit's optokinetic system has at least three special features that operate specifically to reduce its sensitivity to the temporalward motion created by the animal's forward motion. Firstly, temporalward motion generally has only a suppressive effect on the activity of the neurons in the optokinetic pathway so that the major drive comes from the withdrawal of the resting maintained discharge, which is often low.[23-25] The effect here is akin to that of a rectifier. Secondly, these neurons are insensitive to motion in the lower visual field,[25-27] thereby excluding a major potential source of translational contamination since objects in the lower visual field are likely to be the ones most near and hence their retinal images most sensitive to translation. Thirdly, these neurons are insensitive to high accelerations/speeds, which is appropriate for a visual backup to the RVOR, since the latter functions sufficiently well that any retinal slip due to rotational disturbances of gaze will generally be minor and hence within the operating range; substantial distur-bances of gaze must therefore generally emanate from translational disturbances of gaze and will be ignored because they exceed the system's acceleration/speed range.[28] The net result is that the rabbit's optokinetic system responds selectively to the rotational component of optic flow, but this is clearly *not* achieved by a true mathematical decomposition. Thus, we envisage a system that was originally orga-nized to respond optimally to rotational flow patterns *and* developed a number of special features that exploit the layout of the environment and the observer's interactions with it to reduce the likelihood of a response to the commonest form of translation, forward locomotion. We further suggest that the rabbit's visual system may actually depend upon the optic flow associated with forward motion to map out the three-dimensional layout of the world to either side perhaps using relative-motion detectors such as those described in pigeon and primate tecta.[29-31]

Considerations such as the above lead us to suggest that the classical optokinetic system—manifest in primates as OKNd—originally evolved to deal selectively with rotational disturbances of gaze. However, in order to reveal the organization of the extensive receptive fields in the rabbit, it was necessary to use a planetarium projection system and such stimuli have yet to be tried in monkeys.

TVOR AND OKNe: A RECENT DEVELOPMENT IN PRIMATES?

Optokinetic responses with brisk dynamics (like OKNe) and otolith-ocular responses that compensate for translational accelerations (like the TVOR) have so far been found only in monkeys and humans, consistent with the idea that they evolved relatively recently. In the rabbit, for example, otolith-ocular reflexes have a very different function—maintaining the *orientation* of the eyes with respect to gravity by providing the tonic drive for sustained counterrolling during prolonged head tilts.[18] The suggestion is that this is the primordial otolith-ocular reflex. However, tonic counterrolling is vestigial in primates[32,33]—perhaps because the increasing dependence on manual feeding dispensed with one (the?) major reason for prolonged head tilting—and the *transient* counterrolling produced by canal-ocular reflexes seems to suffice. As foveal vision evolved—together with frontal placement of the eyes, vergence eye movements, and stereopsis—we suggest that the major new problem at hand was to stabilize binocular gaze on the depth plane containing the object of regard, the major threat to this being translational disturbances, especially lateral (side-to-side) ones. We propose that the TVOR and OKNe evolved primarily to meet this new translational challenge. Of course, the RVOR and OKNd were already in place so that rotational disturbances of gaze would be minor and so play little role in refining the visual input needed to support OKNe.

To be effective as a visual backup to the TVOR, OKNe must be able to single out the depth plane of interest and deal with the motion parallax associated with translation, which is a rich source of potentially conflicting motion cues for a visual stabilization mechanism. Significantly, OKNe can utilize relative depth cues, such as motion parallax, to aid in the decoding of translational optic flow and, thereby, help in stabilizing the eyes on the frontal depth plane of interest. Clearly, to use the standard optokinetic drum to assess this tracking system is to overlook such sophisticated properties and, to this degree, the drum is a less-than-adequate stimulus.

OKNe: SENSITIVITY TO RELATIVE DEPTH CUES

Optical geometry dictates that when his vantage point moves (i.e., translates) the passive observer experiences retinal image motion that is inversely related to viewing distance.[13] This optical lever effect is especially evident when looking out from a fast-moving train: the scene appears to pivot around the most distant objects. If the observer compensates for the motion by rotating his eyes, then the optical lever will now pivot about some intermediate point. This complicates the flow pattern yet further, the images of objects beyond this imaginary optical fulcrum now reversing their motion on the retina. If, as seems likely, the observer fails to compensate fully for his own motion, then the local image of interest—be it a tree or a house or whatever—will slowly drift back across his central retina while the image of the distant background will be swept forwards across his peripheral retina. Interestingly, there is evidence from studies of both monkeys and humans that the optokinetic system can utilize motion parallax cues such as these to improve its performance. Thus, concurrent motion in the central and surround regions that is opposite in direction actually improves the tracking of the central motion, an effect termed *antiphase enhancement;* conversely, surround motion that is in the same direction as that at the center degrades tracking performance, an effect termed *inphase suppression.*[34-38] Significantly, these effects due to concurrent motion in the peripheral retina are evident at short latency (< 100 mseconds) and hence we assume are characteristic of OKNe.[36] It is apparent from this that en masse motion is not the optimal stimulus for OKNe, and that antiphase motion in the surround (due to motion parallax) could actually help—rather than hinder—the moving observer attempting to stabilize an object off to one side.

THE SMOOTH PURSUIT SYSTEM: A SPATIAL FILTER
FOR STABILIZING LOCAL OPTIC FLOW?

There are many everyday situations in which it makes no sense for the oculomotor system to attempt a global analysis of the optic flow. For example, frontal-eyed animals such as ourselves and monkeys mainly see an expanding flow field during forward movement (centrifugal flow), and at any given moment eye movements can compensate for the flow only in a particular region of the field. If the selected region lies to the right of the direction of heading, for example, then the observer's compensatory eye movements should be rightward, while if the selected region lies to the left of the direction of heading then the eye movements should be leftward and so forth. (It is known that the eye movements generated by the TVOR in response to forward motion operate to increase the eccentricity of the eyes with respect to the direction of heading, as if the images in the region of the fovea take precedence.[39] It remains to be seen if extrafoveal locations can engage the system through selective attention of some sort.) Clearly, the only option for a visual stabilization mechanism here is to abandon the global analysis of flow fields and to concentrate solely on the local flow in the region of particular interest. Thus, appropriate ("compensatory") tracking in this situation first requires a decision as to which region should take precedence, presumably based on some assessment of the potential significance of the various features present, and then some spatial filtering to eliminate the visual inputs coming from other regions.

It is known that the primate smooth pursuit system can initiate ocular tracking with brisk dynamics in response to the motion of retinally eccentric targets *when the subject's attention is directed towards such targets.*[40–44] Indeed, we suggest that it is through this mechanism that human subjects can direct their optokinetic responses to accord with the flow in a restricted region of the visual field, regardless of whether that region is foveal or extrafoveal.[45–49] We assume that monkeys too are capable of this and further suggest that the need to concentrate from time to time upon selected elements of the shifting scene on the retina provided the major pressure to evolve the pursuit system. Thus, we question the general supposition that the pursuit system evolved to track small moving objects, useful though this ability is. In our scheme, the pursuit system evolved as part of an attentional focusing mechanism that spatially filters the visual motion inputs driving the oculomotor system. It is not difficult to imagine how, once evolved, such a mechanism could also be deployed to track small moving targets, even across textured backgrounds—the response property that has been regarded as the sine qua non of the pursuit system.

An idea that has gained some acceptance in recent years is that the OKNe of primates is generated by the smooth pursuit system,[2,12,50–53] one possibility being that pursuit results simply from the action of a spatial filter on the visual inputs driving OKNe. In this scheme, pursuit and OKNe would have much in common, differing only at the earlier stages when pursuit invokes some image-selection process. Optokinetic responses can sometimes be improved, at least transiently, by embedding small salient features in the rotating scene—exactly the kind of targets used to elicit pursuit tracking—suggesting that pursuit can contribute to OKN though this need not always be the case.[54] However, there are data that point to some significant differences between OKNe and pursuit. For example, some monkeys lack OKNe even though they can track small moving objects,[55] and lesions can result in monkeys with the reverse asymmetry—good OKNe but poor pursuit.[56] Further, we have recently obtained evidence that the differences extend beyond the visual processing stages: the monkey's initial pursuit responses—unlike OKNe—are *not* dependent on viewing distance.[57] Thus, it seems unlikely that pursuit simply results from filtering the inputs to OKNe. We suggest that the pursuit system must represent yet a third visual tracking mechanism that is substantially different from the other two and has evolved to stabilize the eyes on *local features* of interest in the often busy, swirling scene confronting the moving observer. It should not be forgotten that pursuit has several components—the initiation has at least two phases which are in turn quite distinct from the maintenance phase.[40,41,58]

Returning once more to the role of pursuit in stabilizing the eyes of the moving observer when the global flow cannot be resolved into a useful single vector. Once the feature selection has been made, appropriate motion signals must be extracted and, after pursuit has been enjoined, the system must be able to deal with the contrary motion of the background. Perhaps the system that sustains pursuit—in contrast to the one that initiates it—responds only to low retinal slip speeds and so is largely blind to the high-speed motion of the background images. Nonetheless, the background motion has the potential to induce optokinetic opposition, i.e., activate OKNe and OKNd. In fact, in both man and monkey the tracking of targets moving across a dark background is generally slightly better than the tracking of targets moving across a textured one.[59-64] However, the impact of the background can be quite small and it is not clear whether this simply reflects a linear interaction between pursuit and OKN or whether pursuit can override/veto OKN by some active suppression mechanism. It is known that fixation/pursuit only cancels the effects of OKNd (as measured during OKAN) and does not "dump" it—large-field stimuli appear to be necessary for that[12]—and we suspect that motion parallax may also achieve this under normal viewing conditions. Recovery from lesions that impair pursuit is sometimes less good for pursuit against a background than for pursuit in the dark, perhaps suggesting that there are special mechanisms that facilitate tracking across a background—perhaps by actively suppressing OKNe—and that these have been irrevocably damaged by the lesion.[65]

Even the initiation of pursuit is impaired by the presence of a (stationary) textured background,[61] and this cannot be due to optokinetic opposition since the initial pursuit results from target motion across the retina prior to any motion of the eyes. Also, the initiation of pursuit against a textured background recovers more slowly from some lesions than does pursuit in the dark,[66] perhaps indicating once more the presence of some special mechanism for dealing with the background. Recent studies have shown that the effect of the background on pursuit initiation is not due solely to the reduced physical salience (contrast) of the target spot, since selectively excluding the background texture from the path of the target reduces the impact of the background only slightly and the suppressive effect of the background shows interocular transfer: texture seen by one eye reduces the pursuit of a moving target seen only by the other eye.[67] The suggestion here is that the visual receptive fields of the neurons decoding the target motion are quite large, receive inputs from both eyes, and are sensitive to stationary texture. Thus, it seems that there is an object/ground discrimination problem of some kind, and other experiments indicate that the impact of the background on the initiation of pursuit is less when that background lies outside the plane of fixation/convergence and is not binocularly fused.[67] This supports the idea that the pursuit system can respond selectively to the motion of objects in the plane of fixation/convergence and ignore the motion of objects that are nearer or further. The implication is that the neurons mediating the initiation of pursuit are sensitive to disparity, responding best to the nondisparate images in the plane of convergence. Clearly, this will help to segregate the feature/object of interest from the background, though only at relatively near viewing (distances ranging up to a few meters at most). While the emphasis in these experiments was on pursuit of targets moving against a stationary background, we assume that the data are relevant to the object/ground discrimination that is imperative for effective visual stabilization of the eyes when the conflicting background cues emanate from the observer's own motion.

HUMANS

In sharp contrast with monkeys, humans show only weak optokinetic after-nystagmus (OKAN)[68,69] and their early ocular following does *not* show dependence

TABLE 1. Features of a Proposed Scheme for the Parsing of Optic Flow by the Monkey's Oculomotor System

Global Flow		Local Flow
Rotational Mechanism	Translational Mechanism	Feature-Tracking Mechanism
1. Delayed component of OKN:[2] <ul><li>Long time-constant.</li><li>Strong after-nystagmus.</li><li>Sensitive to low acceleration/speed.[56]</li></ul>	1. Early component of OKN:[2] <ul><li>Short time constant.</li><li>Weak after-nystagmus.</li><li>Sensitive to high acceleration/speed.[36,56]</li></ul>	1. Smooth pursuit system: <ul><li>Short time constant.[75]</li><li>Weak after-nystagmus.[12]</li><li>Sensitive to high acceleration/speed.[40,75,76]</li></ul>
2. Backup to canal-ocular reflex (RVOR)? <ul><li>Sensitive to gain of RVOR.[12]</li><li>Insensitive to gain of TVOR?</li></ul>	2. Backup to otolith-ocular reflex (TVOR)? <ul><li>Insensitive to gain of RVOR.[12]</li><li>Sensitive to gain of TVOR.[13]</li></ul>	2. Backup to spatial-attention mechanism? <ul><li>Insensitive to gain of RVOR.[77]</li><li>Insensitive to gain of TVOR.[57]</li></ul>
3. Organized in discrete (canal) planes?	3. Organized in (otolith) planes?	3. ?
4. Helps to stabilize gaze against *en masse global disturbances* (no obj/gnd discrim)? <ul><li>Dumped by motion parallax?</li><li>Insensitive to disparity?</li></ul>	4. Helps to stabilize gaze on *local depth plane of interest* (primitive obj/gnd discrim)? <ul><li>Can deal with some motion parallax.[36]</li><li>Sensitive to disparity?</li></ul>	4. Helps to stabilize gaze on the *feature of interest* (good obj/gnd discrim)? <ul><li>Can override motion parallax.[61,62]</li><li>Sensitive to disparity of background.[67]</li></ul>
5. Pretectum/accessory optic system.[78]	5. Corticopontocerebellar system.[79]	5. Corticopontocerebellar system.[42,50,53,66]

on viewing distance,[70] suggesting that OKNd and OKNe are vestigial in man. The implication is that man relies largely on pursuit. In fact, like the monkey's pursuit tracking, human OKN is better when the moving visual scene is binocularly fused than when disparate.[71] Further, the changes in nystagmus with disparity are quite abrupt, consistent with the brisk dynamics of pursuit. Human ocular following can be elicited by moving stereoscopic contours that lack monocular motion cues.[72,73] However, responses are not obligate and are observed only when the subject *perceives* the moving contour, perhaps representing an instance of "pursuing the perceptual rather than the retinal stimulus."[74]

SUMMARY

In monkeys, there are several reflexes that generate eye movements to compensate for the observer's own movements. Two vestibuloocular reflexes compensate selectively for rotational (RVOR) and translational (TVOR) disturbances of the head, receiving their inputs from the semicircular canals and otolith organs, respectively. Two independent visual tracking systems that deal with residual disturbances of gaze are manifest in the two components of the optokinetic response: the indirect or *delayed* component (OKNd) and the direct or *early* component (OKNe). We hypothesize that OKNd—like the RVOR—is phylogenetically old, being found in all animals with mobile eyes, and that it evolved as a backup to the RVOR to compensate for rotational disturbances of gaze. Indeed, optically induced changes in the gain of the RVOR result in parallel changes in the gain of OKNd, consistent with the idea of shared pathways as well as shared functions. In contrast, OKNe—like the TVOR—seems to have evolved much more recently in frontal-eyed animals and, we suggest, acts as a backup to the TVOR to deal primarily with translational disturbances of gaze. Frontal-eyed animals with good binocular vision must be able to keep both eyes directed at the object of regard irrespective of proximity, and in order to achieve this during translational disturbances, the output of the TVOR is modulated inversely with the viewing distance. OKNe shares this sensitivity to absolute depth, consistent with the idea that it is synergistic with the TVOR and shares some of its central pathways. There is evidence that OKNe is also sensitive to relative depth cues such as motion parallax, which we suggest helps the system to segregate the object of regard from other elements in the scene. However, there are occasions when the global optic flow cannot be resolved into a single vector useful to the oculomotor system (e.g., when the moving observer looks towards the direction of heading). We suggest that on such occasions a third independent tracking mechanism, the smooth pursuit system, is deployed to stabilize gaze on the local feature of interest. In this scheme, the pursuit system has an attentional focusing mechanism that spatially filters the visual motion inputs driving the oculomotor system. The major distinguishing features of the 3 visual tracking mechanisms are summarized in TABLE 1.

REFERENCES

1. GOLDBERG, J. M. & C. FERNANDEZ. 1975. Responses of peripheral vestibular neurons to angular and linear accelerations in the squirrel monkey. Acta Otolaryngol. **80:** 101–110.
2. COHEN, B., V. MATSUO & T. RAPHAN. 1977. Quantitative analysis of the velocity characteristics of optokinetic nystagmus and optokinetic after-nystagmus. J. Physiol. London **270:** 321–344.
3. MATSUO, V. & B. COHEN. 1984. Vertical optokinetic nystagmus and vestibular nystagmus in the monkey: up-down asymmetry and effects of gravity. Exp. Brain Res. **53:** 197–216.
4. RAPHAN, T., B. COHEN & V. MATSUO. 1977. A velocity-storage mechanism responsible for optokinetic nystagmus (OKN), optokinetic after-nystagmus (OKAN) and vestibular nystagmus. *In* Developments in Neuroscience. R. Baker & A. Berthoz, Eds. **1:** 37–47. Elsevier/North-Holland. Amsterdam, the Netherlands.
5. RAPHAN, T., V. MATSUO & B. COHEN. 1979. Velocity storage in the vestibulo-ocular reflex arc (VOR). Exp. Brain Res. **35:** 229–248.
6. MILES, F. A., U. SCHWARZ & C. BUSETTINI. 1989. Are the two components of the primate

optokinetic response concerned with translational and rotational disturbances of gaze? Soc. Neurosci. Abstr. **15:** 783.

7. MILES, F. A., U. SCHWARZ & C. BUSETTINI. 1991. The parsing of optic flow by the primate oculomotor system. *In* Representations of Vision: Trends and Tacit Assumptions in Vision Research. A. Gorea, Ed.: 185–199. Cambridge University Press. Cambridge, England.

8. MILES, F. A., U. SCHWARZ & C. BUSETTINI. The decoding of optic flow by the primate optokinetic system. *In* The Head-Neck Sensory-Motor System. A. Berthoz, P.-P. Vidal & W. Graf, Eds. Oxford University Press. New York, N.Y. (In press.)

9. MILES, F. A., C. BUSETTINI & U. SCHWARZ. Ocular responses to linear motion. *In* Vestibular and Brain Stem Control of Eye, Head and Body Movements. H. Shimazu & Y. Shinoda, Eds. Springer-Verlag/Japan Scientific Societies Press. Tokyo, Japan. (In press.)

10. MILES, F. A. & J. H. FULLER. 1974. Adaptive plasticity in the vestibulo-ocular responses of the rhesus monkey. Brain Res. **80:** 512–516.

11. MILES, F. A. & B. B. EIGHMY. 1980. Long-term adaptive changes in primate vestibuloocular reflex. I. Behavioral observations. J. Neurophysiol. **43:** 1406–1425.

12. LISBERGER, S. G., F. A. MILES, L. M. OPTICAN & B. B. EIGHMY. 1981. Optokinetic response in monkey: underlying mechanisms and their sensitivity to long-term adaptive changes in vestibuloocular reflex. J. Neurophysiol. **45:** 869–890.

13. SCHWARZ, U., C. BUSETTINI & F. A. MILES. 1989. Ocular responses to linear motion are inversely proportional to viewing distance. Science **245:** 1394–1396.

14. VIIRRE, E., D. TWEED, K. MILNER & T. VILIS. 1986. A reexamination of the gain of the vestibuloocular reflex. J. Neurophysiol. **56:** 439–450.

15. GIBSON, J. J. 1950. The Perception of the Visual World. Houghton Mifflin. Boston, Mass.

16. GIBSON, J. J. 1966. The Senses Considered as Perceptual Systems. Houghton Mifflin. Boston, Mass.

17. LEE, D. N. 1980. The optic flow field: the foundation of vision. Phil. Trans. R. Soc. London Ser. B **290:** 169–179.

18. BAARSMA, E. A. & H. COLLEWIJN. 1975. Eye movements due to linear accelerations in the rabbit. J. Physiol. **245:** 227–247.

19. COLLEWIJN, H. & H. NOORDUIN. 1972. Conjugate and disjunctive optokinetic eye movements in the rabbit, evoked by rotatory and translatory motion. Pflügers Arch. **335:** 173–185.

20. HOWARD, I. P. 1982. Human Visual Orientation. Wiley. London, England.

21. GRAF, W., J. I. SIMPSON & C. S. LEONARD. 1988. Spatial organization of visual messages of the rabbit's cerebellar flocculus. II. Complex and simple spike responses of Purkinje cells. J. Neurophysiol. **60:** 2091–2121.

22. LEONARD, C. S., J. I. SIMPSON & W. GRAF. 1988. Spatial organization of visual messages of the rabbit's cerebellar flocculus. I. Typology of inferior olive neurons of the dorsal cap of Kooy. J. Neurophysiol. **60:** 2073–2090.

23. SIMPSON, J. I., C. S. LEONARD & R. E. SOODAK. 1988. The accessory optic system of the rabbit. II. Spatial organization of direction selectivity. J. Neurophysiol. **60:** 2055–2072.

24. SOODAK, R. E. & J. I. SIMPSON. 1988. The accessory optic system of the rabbit. I. Basic visual response properties. J. Neurophysiol. **60:** 2037–2054.

25. COLLEWIJN, H. 1975. Direction-selective units in the rabbit's nucleus of the optic tract. Brain Res. **100:** 489–508.

26. SIMPSON, J. I., W. GRAF & C. LEONARD. 1981. The coordinate system of visual climbing fibers to the flocculus. *In* Progress in Oculomotor Research. A. Fuchs & W. Becker, Eds.: 475–484. Elsevier North Holland. Amsterdam, the Netherlands.

27. SIMPSON, J. I., R. E. SOODAK & R. HESS. 1979. The accessory optic system and its relation to the vestibulocerebellum. *In* Reflex Control of Posture and Movement. R. Granit & O. Pompeiano, Eds.: 715–724. Elsevier. Amsterdam, the Netherlands.

28. SIMPSON, J. I. 1991. Personal communication.

29. FROST, B. J. & K. NAKAYAMA. 1983. Single visual neurons code opposing motion independent of direction. Science **220:** 744–745.

30. FROST, B. J., P. L. SCILLEY & S. C. P. WONG. 1981. Moving background patterns reveal

double-opponency of directionally specific pigeon tectal neurons. Exp. Brain Res. **43:** 173–185.

31. DAVIDSON, R. M. & D. B. BENDER. 1991. Selectivity for relative motion in the monkey superior colliculus. J. Neurophysiol. **65:** 1115–1133.

32. COLLEWIJN, H., J. VAN DER STEEN, L. FERMAN & T. C. JANSEN. 1985. Human ocular counterroll: assessment of static and dynamic properties from electromagnetic scleral coil recordings. Exp. Brain Res. **59:** 185–196.

33. KREJCOVA, H., S. HIGHSTEIN & B. COHEN. 1971. Labyrinthine and extra-labyrinthine effects on ocular counter-rolling. Acta Otolaryngol. **72:** 165–171.

34. GUEDRY, F. E., JR., J. M. LENTZ, R. M. JELL & J. W. NORMAN. 1981. Visual-vestibular interactions: the directional component of visual background movement. Aviat. Space Environ. Med. **52:** 304–309.

35. HOOD, J. D. 1975. Observations upon the role of the peripheral retina in the execution of eye movements. J. Otorhinolaryngol. **37:** 65–73.

36. MILES, F. A., K. KAWANO & L. M. OPTICAN. 1986. Short-latency ocular following responses of monkey. I. Dependence on temporospatial properties of the visual input. J. Neurophysiol. **56:** 1321–1354.

37. TER BRAAK, J. W. G. 1957. "Ambivalent" optokinetic stimulation. Folia Psychiatr. Neurol. Neerl. **60:** 131–135.

38. TER BRAAK, J. W. G. 1962. Optokinetic control of eye movements, in particular optokinetic nystagmus. *In* Proceedings of the 22th International Congress on Physiological Science, Leiden **1:** 502–505.

39. PAIGE, G. D. & D. L. TOMKO. 1991. Eye movement responses to linear head motion in the squirrel monkey. II. Visual-vestibular interactions and kinematic considerations. J. Neurophysiol. **65:** 1183–1196.

40. LISBERGER, S. G. & L. E. WESTBROOK. 1985. Properties of visual inputs that initiate horizontal smooth pursuit eye movements in monkeys. J. Neurosci. **5:** 1662–1673.

41. LISBERGER, S. G. & T. A. PAVELKO. 1989. Topographic and directional organization of visual motion inputs for the initiation of horizontal and vertical smooth-pursuit eye movements in monkeys. J. Neurophysiol. **61:** 173–185.

42. NEWSOME, W. T., R. H. WURTZ, M. R. DÜRSTELER & A. MIKAMI. 1985. Deficits in visual motion processing following ibotenic acid lesions of the middle temporal visual area of the macaque monkey. J. Neurosci. **5:** 825–840.

43. RASHBASS, C. 1961. The relationship between saccadic and smooth tracking eye movements. J. Physiol. London **159:** 326–338.

44. TYCHSEN, L. & S. G. LISBERGER. 1986. Visual motion processing for the initiation of smooth-pursuit eye movements in humans. J. Neurophysiol. **56:** 953–968.

45. CHENG, M. & J. S. OUTERBRIDGE. 1975. Optokinetic nystagmus during selective retinal stimulation. Exp. Brain Res. **23:** 129–139.

46. VAN DEN BERG, A. V. & H. COLLEWIJN. 1987. Voluntary smooth eye movements with foveally stabilized targets. Exp. Brain Res. **68:** 195–204.

47. DUBOIS, M. F. W. & H. COLLEWIJN. 1979. Optokinetic reactions in man elicited by localized retinal stimuli. Vision Res. **19:** 1105–1115.

48. HOWARD, I. P., D. GIASCHI & C. M. MURASUGI. 1989. Suppression of OKN and VOR by afterimages and imaginary objects. Exp. Brain Res. **75:** 139–145.

49. MURASUGI, C. M., I. P. HOWARD & M. OHMI. 1986. Optokinetic nystagmus: the effects of stationary edges, alone and in combination with central occlusion. Vision Res. **26:** 1155–1162.

50. DÜRSTELER, M. R. & R. H. WURTZ. 1988. Pursuit and optokinetic deficits following chemical lesions of cortical areas MT and MST. J. Neurophysiol. **60:** 940–965.

51. ROBINSON, D. A. 1977. Linear addition of optokinetic and vestibular signals in the vestibular nucleus. Exp. Brain Res. **30:** 447–450.

52. WAESPE, W. & V. HENN. 1977. Neuronal activity in the vestibular nuclei of the alert monkey during vestibular and optokinetic stimulation. Exp. Brain Res. **27:** 523–538.

53. ZEE, D. S., A. YAMAZAKI, P. H. BUTLER & G. GÜÇER. 1981. Effects of ablation of flocculus and paraflocculus on eye movements in primate. J. Neurophysiol. **46:** 878–899.

54. MILES, F. A. 1991. Unpublished observations.

55. KATO, I., K. HARADA, T. HASEGAWA, T. IGARASHI, Y. KOIKE & T. KAWASAKI. 1986. Role of the nucleus of the optic tract in monkeys in relation to optokinetic nystagmus. Brain Res. **364:** 12–22.
56. ZEE, D. S., R. J. TUSA, S. J. HERDMAN, P. H. BUTLER & G. GÜÇER. 1987. Effects of occipital lobectomy upon eye movements in primate. J. Neurophysiol. **58:** 883–907.
57. BUSETTINI, C. & F. A. MILES. 1991. Unpublished observations.
58. MORRIS, E. J. & S. G. LISBERGER. 1987. Different responses to small visual errors during initiation and maintenance of smooth-pursuit eye movements in monkeys. J. Neurophysiol. **58:** 1351–1369.
59. BARNES, G. R. & J. W. CROMBIE. 1985. The interaction of conflicting retinal motion stimuli in oculomotor control. Exp. Brain Res. **59:** 548–558.
60. COLLEWIJN, H. & E. P. TAMMINGA. 1984. Human smooth and saccadic eye movements during voluntary pursuit of different target motions on different backgrounds. J. Physiol. London **351:** 217–250.
61. KELLER, E. L. & N. S. KHAN. 1986. Smooth-pursuit initiation in the presence of a textured background in monkey. Vision Res. **26:** 943–955.
62. MUSTARI, M. J., A. F. FUCHS & J. WALLMAN. 1988. Response properties of dorsolateral pontine units during smooth pursuit in the rhesus macaque. J. Neurophysiol. **60:** 664–686.
63. VAN DER STEEN, J., E. P. TAMMINGA & H. COLLEWIJN. 1983. A comparison of oculomotor pursuit of a target in circular real, beta or sigma motion. Vision Res. **23:** 1655–1661.
64. YEE, R. D., S. A. DANIELS, O. W. JONES, R. W. BALOH & V. HONRUBIA. 1983. Effects of an optokinetic background on pursuit eye movements. Invest. Ophthalmol. Visual Sci. **24:** 1115–1122.
65. HOOD, J. D. & E. WANIEWSKI. 1984. Influence of peripheral vision upon vestibulo-ocular reflex suppression. J. Neurol. Sci. **63:** 27–44.
66. MAY, J. G., E. L. KELLER & D. A. SUZUKI. 1988. Smooth-pursuit eye movement deficits with chemical lesions in the dorsolateral pontine nucleus of the monkey. J. Neurophysiol. **59:** 952–977.
67. KIMMIG, H., F. A. MILES & U. SCHWARZ. Effects of stationary textured backgrounds on initiation of pursuit eye movements in monkeys. (In preparation.)
68. COHEN, B., V. HENN, T. RAPHAN & D. DENNETT. 1981. Velocity storage, nystagmus and visual-vestibular interaction in humans. Ann. N.Y. Acad. Sci. **374:** 421–433.
69. KOENIG, E. & J. DICHGANS. 1981. Aftereffects of vestibular and optokinetic stimulation and their interaction. Ann. N.Y. Acad. Sci. **374:** 434–445.
70. BUSETTINI, C., F. A. MILES, U. SCHWARZ & J. R. CARL. 1991. Unpublished observations.
71. HOWARD, I. P. & E. G. GONZALEZ. 1987. Human optokinetic nystagmus in response to moving binocularly disparate stimuli. Vision Res. **27:** 1807–1816.
72. ARCHER, S. M., K. K. MILLER & E. M. HELVESTON. 1987. Stereoscopic contours and optokinetic nystagmus in normal and stereoblind subjects. Vision Res. **27:** 841–844.
73. FOX, R., S. LEHMKUHLE & L. E. LEGUIRE. 1978. Stereoscopic contours induce optokinetic nystagmus. Vision Res. **18:** 1189–1192.
74. STEINBACH, M. 1976. Pursuing the perceptual rather than the retinal stimulus. Vision Res. **16:** 1371–1376.
75. FUCHS, A. F. 1967. Saccadic and smooth pursuit eye movements in the monkey. J. Physiol. London **191:** 609–631.
76. LISBERGER, S. G., C. EVINGER, W. JOHANSON & A. F. FUCHS. 1981. Relationship between eye acceleration and retinal image velocity during foveal smooth pursuit in man and monkey. J. Neurophysiol. **46:** 229–249.
77. LISBERGER, S. G. 1991. Personal communication.
78. SCHIFF, D., B. COHEN, J. BÜTTNER-ENNEVER & V. MATSUO. 1990. Effects of lesions of the nucleus of the optic tract on optokinetic nystagmus and after-nystagmus in the monkey. Exp. Brain Res. **79:** 225–239.
79. KAWANO, K., M. SHIDARA, Y. WATANABE & S. YAMANE. Short-latency responses of neurons in dorsolateral pontine nucleus and cortical area MST of alert monkey to movement of large-field visual stimulus. *In* Vestibular and Brain Stem Control of Eye, Head and Body Movements. H. Shimazu & Y. Shinoda, Eds. Springer-Verlag/Japan Scientific Societies Press. Tokyo, Japan. (In press.)

Linear Vestibuloocular Reflex during Motion along Axes between Nasooccipital and Interaural[a]

DAVID L. TOMKO[b,c] AND GARY D. PAIGE[d]

bVestibular Research Facility
Life Sciences Division
National Aeronautics and Space Administration
Ames Research Center
Moffett Field, California 94035-1000

cDivision of Otolaryngology
Head and Neck Surgery
Stanford Medical Center
Stanford, California 94305

dDepartment of Neurology
University of Rochester
Rochester, New York 14642

INTRODUCTION

To maintain fixation on visual targets during head motion, eye movements must counteract and compensate for changes in eye and head position relative to the targets. This task is facilitated by the vestibuloocular reflexes (VORs), which use vestibular signals generated by angular and linear head motion to produce compensatory eye movements that help to maintain ocular fixation on targets in space.

In previous work,[1,2] we defined characteristics of compensatory eye movement responses to linear head motion along the nasooccipital (NO), interaural (IA), and dorsoventral (DV) axes in squirrel monkeys. Linear VORs (LVORs) are of two types: (1) responses to head tilt, which work primarily at low frequencies; and (2) responses to head translation, which act at higher frequencies. The present report concentrates on the latter, translational LVORs, which include horizontal eye movements in response to IA-axis motion, vertical responses to DV-axis motion, and horizontal/vertical responses to NO-axis motion.

The kinematics of eye movements required to achieve perfect ocular compensation during head translation are determined by the geometry of each eye's motion relative to the position of visual targets. The important variables that determine those kinematics include (1) target distance, and (2) target eccentricity relative to the axis of motion. Target distance is important for responses to all linear motions; fixation of closer targets requires larger eye movements (that is, higher LVOR sensitivity) than fixation of more distant ones. Target eccentricity is important primarily for head motion roughly parallel to the line of sight (near the NO axis). Fixation of targets that are more eccentric relative to the motion axis requires larger eye movements than does fixation of ones closer to the axis of motion. Further, the

[a]The financial support of NASA Space Medicine Tasks 199-16-12-17 and -24 is gratefully acknowledged.

233

direction of ocular response to NO-axis motion depends on the direction of gaze. For example, to maintain target fixation during *forward* (NO-axis) motion and *upward* gaze, upward eye movements are required. However, during the same forward motion but with *downward* gaze, downward eye movements must occur. Similarly, rightward responses must occur during rightward gaze, and leftward responses during leftward gaze. Thus, a 180° phase shift is required in the LVOR relative to head motion depending on whether the target (and gaze) is to the left or right, and up or down, relative to the axis of motion. Furthermore, vergence eye movements must occur during NO-axis motion as the head moves toward and away from the target. In *all* cases, each eye must move independently to maintain foveal vision.

Measured characteristics of the NO LVOR have confirmed that ocular responses in fact behave according to the above kinematic considerations.[2] However, an important experimental constraint in earlier studies was that head orientation was always fixed relative to the axis of linear motion, and was limited to either the NO, IA, or DV axes of the head. In contrast, normal behavior is not limited to these three orthogonal axes. This study was designed to determine how LVOR responses would change if head orientation were shifted to intermediate positions relative to the axis of motion. We hypothesized that reflexive eye movements would follow the same kinematics relative to the motion axis regardless of head orientation relative to linear motion. Gaze (gaze = eye position in space = head in space + eye in head) relative to the motion axis and binocular fixation distance are *both* required to define and execute ocular responses that maintain target fixation.

The hypothesis was tested by recording horizontal and vertical eye movements during linear oscillations at 5 Hz along the head's NO axis and along axes lying within ±30° of the NO axis. If the LVOR during motion along these axes maintains properties consistent with the kinematics required for proper compensation, then eye position information must be quickly and precisely integrated with otolith inputs to determine eye position relative to linear motion in space.

METHODS

Animals

Two adult male squirrel monkeys (*Saimiri sciureus*) weighing 0.9–1.0 kg were the subjects of this experiment. Detailed animal surgical preparations have been described before.[1] In brief, under general surgical anesthesia, a 1/4-inch stainless steel head-fixation bolt with a flat surface was aseptically fixed to the occiput so that the flat surface was parallel to the plane of the horizontal semicircular canals. Prefabricated, calibrated, 11-mm-diameter, three-turn search coils were sutured to the sclera concentric to the limbus in the frontal plane of both eyes.

Eye Movement Recordings

A detailed description of eye movement recording methods, calibration procedures, and sources of error have been described.[1,2] Briefly, horizontal and vertical eye positions were recorded using Neurodata Systems (EMD-82) coil position detection circuits. System calibration was performed prior to each recording session using reference coils identical to those implanted in the animals. Calibration was checked for each animal before each experiment by verifying that the gain of the angular VOR (0.5 Hz) in the light was close to unity. Vergence was calculated by subtracting the position of the right eye from that of the left.

Apparatus and Stimulus Characteristics

All experiments were conducted at NASA Ames Research Center's Vestibular Research Facility (VRF) using the linear sled illustrated in Paige and Tomko.[1] The monkey, seated in a primate chair containing the field coils, was held in a light-tight test container which was mounted to the sled by a three-axis gimbal that enabled repositioning the subject relative to the axis of motion. Earth-horizontal linear oscillations at 5 Hz were delivered to the subject with a peak acceleration of $0.36 \times g$ (11.4 cm/second peak velocity, 0.36 cm peak displacement).

Experimental Protocol

In each recording session, the monkey was seated in the primate chair with its head fixed and placed inside the test container on the sled. In the present experiments, linear oscillations of the upright subject were then carried out either along the animal's NO axis (FIGURE 1, center; 0°), or along 10 other head axes between the NO and IA axes. Three motion axes are illustrated in FIGURE 1. To achieve motion along intermediate axes, the head was repositioned to the right by angle $+\Theta_H$ (FIGURE 1, top) or to the left by angle $-\Theta_H$ (FIGURE 1, bottom). In the present experiments, values of Θ_H included 5, 10, 15, 20, and 30°. Ocular responses were recorded in darkness (LVOR) and with the visually rich interior surface of the test container illuminated (distance = 22 cm), which we have referred to as VSLVOR, since the head-fixed visual surround potentially allows visual "suppression" of the LVOR at lower frequencies.

Data Collection and Analysis

Horizontal and vertical eye position as well as head (sled) acceleration signals were recorded digitally on a PDP 11-83 minicomputer, and analyzed as described before,[2] to yield horizontal, vertical, and vergence sensitivity (degrees of eye movement per centimeter of head translation) and phase (phase angle of eye position minus phase angle of head position), as well as mean eye position values, for each cycle of head oscillation.

RESULTS AND DISCUSSION

The LVOR and VSLVOR during NO-Axis Motion

The ocular responses shown in FIGURE 2 demonstrate how the basic kinematic requirements described above are met during nasooccipital oscillation at 5 Hz. When horizontal gaze shifts from right to left, but vertical gaze remains down (FIGURE 2, see lines a and b), the right horizontal response shifts from roughly 0 to 180° in phase (thick horizontal eye velocity trace) while the vertical responses phase remains fixed (vertical velocity trace). When vertical gaze shifts from down to up, but horizontal gaze remains leftward (FIGURE 2, see lines b and c), the vertical response selectively shifts from around 180° to near 0°.

This behavior is characteristic of the directional properties of the NO VSLVOR and NO LVOR. With regard to response amplitude and left/right independence, ocular kinematics require that when the line of sight (gaze) and the axis of head

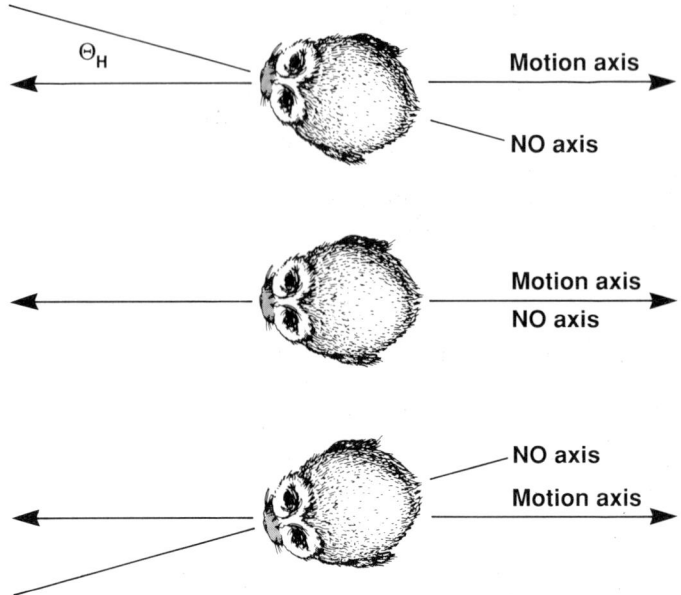

FIGURE 1. Experimental paradigm consists of linear oscillations (5 Hz, 0.36 × g peak) along the animal's nasooccipital (NO) axis (center), and along head axes between the NO and interaural (IA) axes. To achieve motion along intermediate axes, the head was repositioned to the right by angle $+\Theta_H$ as indicated in the top example, or to the left by angle $-\Theta_H$. In the present experiments, values of Θ_H varied from 0 to $\pm 30°$.

motion are parallel (in this case, when motion is along the NO axis; $\Theta_H = 0°$), a plot of response sensitivity as a function of horizontal eye position should be V shaped, with a minimum sensitivity at zero (NO-axis, or straight-ahead gaze), and increasing sensitivity as gaze becomes more eccentric. FIGURE 3C shows such a plot of averaged horizontal response sensitivity from a single animal (X33) during NO-axis head motion ($\Theta_H = 0°$) for the VSLVOR condition. FIGURE 3D illustrates (for the same response cycles shown in FIGURE 3C) the gaze-dependent 180° phase shifts as the eye crosses its NO axis of motion. Similar response characteristics were seen in the absence of visual targets (LVOR).

The LVOR and VSLVOR during Off-NO-Axis Motion

Examples are shown in FIGURE 3A and E of mean horizontal VSLVOR sensitivity when $\Theta_H = +5°$ and $-5°$, respectively. When the head is reoriented to the right with respect to the axis of motion ($\Theta_H = +5°$, FIGURE 3A), the NO axis is no longer parallel to the axis of motion. To achieve gaze parallel to the axis of motion, the eye must rotate to the left (negative in our convention) by $-5°$. Under these conditions, the eye is 5° to the left relative to the head, but is aligned with linear motion. This now becomes the point where the LVOR response should be minimal. In other words, the V-shaped sensitivity curve of FIGURE 3C and the phase shift curve of FIGURE 3D should shift by $-5°$ (to the left) if proper LVOR response kinematics are to be preserved. Indeed, FIGURE 3A and B show clearly that the "V-shaped" sensitivity function, as well as the phase plot, are shifted by roughly $-5°$ (to the left). Similarly, when the head is oriented $5°$ to the left ($\Theta_H = -5°$), the comparable sensitivity and phase curves (FIGURE 3E and F) are shifted by $+5°$ (to the right) relative to the head.

To compare results across all head orientations in this study ($\Theta_H = 0, \pm 5, \pm 10, \pm 15, \pm 20, \pm 30°$), horizontal gaze relative to the motion axis was calculated for each stimulus cycle by adding the value of head orientation with respect to motion (Θ_H) to the mean eye position in the head for each stimulus cycle, yielding eye position

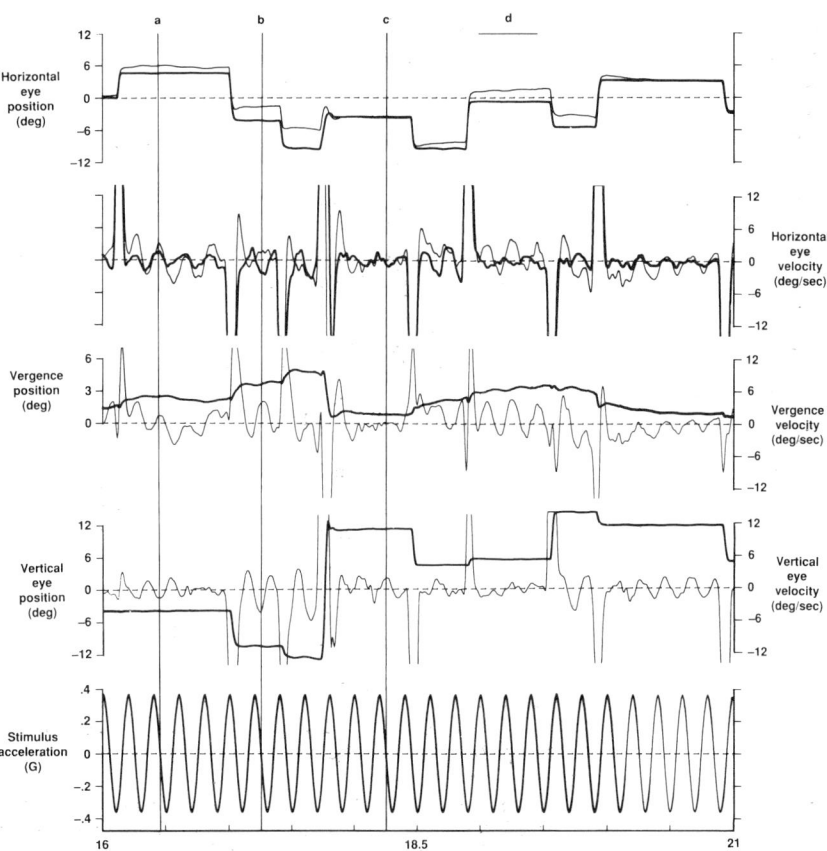

FIGURE 2. VSLVOR responses during 5 Hz NO-axis head oscillation are shown for monkey S12. Up corresponds to rightward (horizontal), upward (vertical), and increasing (vergence) eye position or velocity, and to nasal NO-axis acceleration. Horizontal left eye (thin) and right eye (thick) position (first traces) and velocity (second traces), vergence position and velocity (third traces), and vertical position and velocity (fourth traces) are plotted above the head acceleration (fifth trace). Vertical lines allow comparison of responses during rightward (a) and leftward (b) gaze while vertical gaze remained down, and during downward (b) and upward (c) gaze while horizontal gaze remained leftward. Region d illustrates disjunctive movements of the right and left eyes which result in a large vergence response.

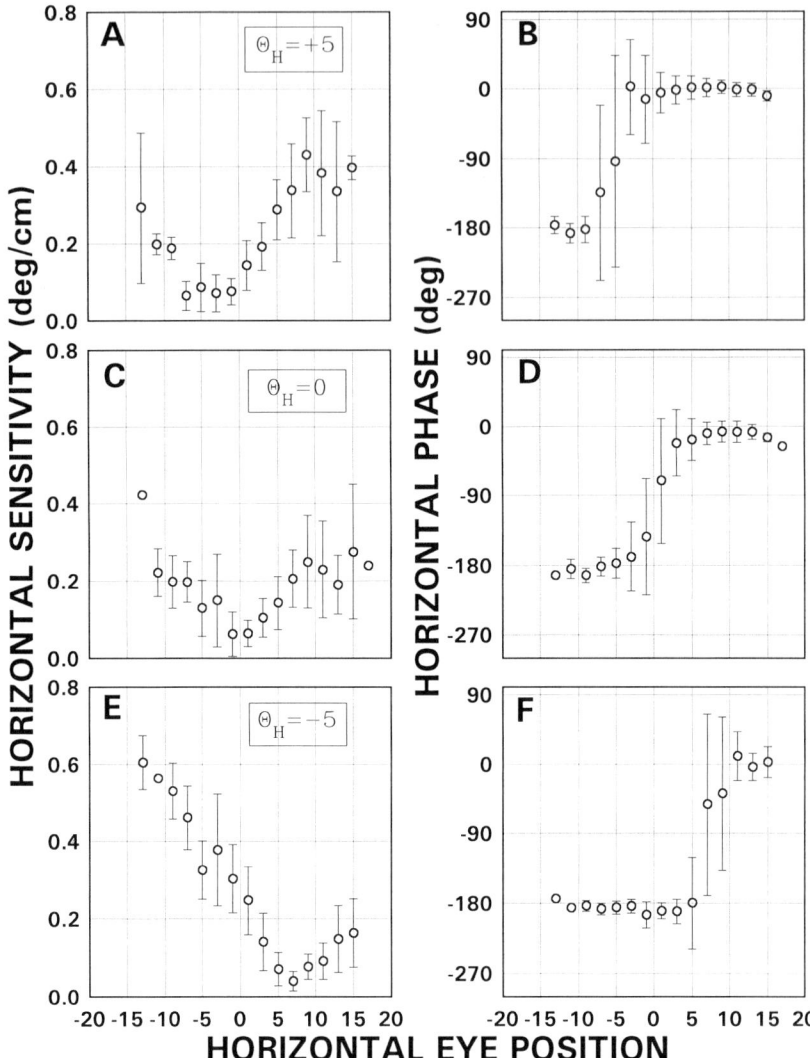

FIGURE 3. Mean VSLVOR horizontal sensitivity (plots A, C, and E) and phase (B, D, and F) for animal X33 during 5 Hz NO-axis linear oscillation (C and D; $\Theta_H = 0°$) and along axes in which head orientation was shifted left ($\Theta_H = -5°$; A and B) or right ($\Theta_H = +5°$; E and F). Gains and phases are plotted as a function of right eye position in the head (2° bins). Negative eye positions are to the left and positive values are to the right.

relative to the axis of motion. LVOR and VSLVOR data were then pooled separately, divided into 2° bins, and averaged. The results are plotted in FIGURE 4, where the abscissa now represents horizontal *gaze* relative to linear motion, not eye position in the head as in FIGURE 3. The superimposition of the data for both the LVOR and the VSLVOR conditions clearly demonstrates that regardless of the

direction of linear head motion in space, ocular responses depend upon the
orientation of the eye (gaze) relative to the axis of head motion in space. The roughly
twofold increase in response sensitivity between the LVOR and the VSLVOR
conditions is consistent with the measurement of a mean vergence value of 1.76° for
the LVOR and 3.44° for the VSLVOR in this data set (see Reference 2). Results
were the same for a second animal tested using this paradigm.

We conclude that the LVOR behaves according to the kinematic requirements of
compensatory eye movements during linear motion. Eye position information must
be integrated with otolith inputs rapidly and continuously to determine eye angle
relative to the axis of motion to account for the rapidity with which the reflex must

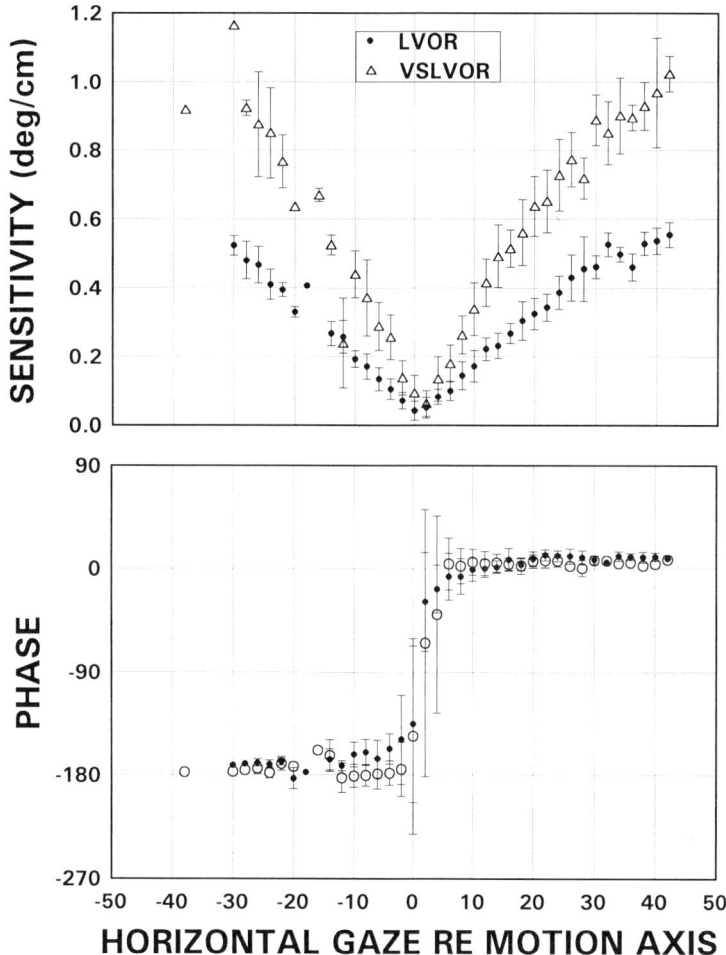

FIGURE 4. Mean LVOR and VSLVOR horizontal gain and phase (animal X33) during 5 Hz
($0.36 \times g$ peak) oscillations along axes where Θ_H equaled 0, 5, 10, 15, 20, or 30° right and left.
Responses are plotted as a function of gaze (2° bins) relative to (re) the axis of motion.

adjust its response characteristics as observed in the experiments. The neural mechanism that modifies the LVOR is surprisingly accurate (within 2° in our data). One potential mechanism is that LVOR responses to off-NO-axis motion simply represent the sum of component NO and IA LVOR characteristics.[1,2]

The LVOR characteristics described in this paper are an example of how otolith information is used by the brain to determine a subject's "heading"[3] in order to stabilize binocular fixation on targets during high-frequency translational motion. Otolith information is presumably important to other neural mechanisms that require heading information. This may be particularly critical during rapid linear motions which accompany the initiation or perturbations of locomotion. Heading from otolith inputs may be intimately related to visual heading information, or "optic flow."[4,5] For example, visual-vestibular interaction (VVI) during head rotation reflects synergistic influences of vestibular and visually guided eye movements, combining a high-pass-filtered angular VOR with low-pass-filtered visual-following mechanisms. The same may hold for VVI during linear head motion. Otolith responses to translation are high-pass filtered, and extend beyond the capacity of visual following mechanisms.[2] It seems logical, therefore, to hypothesize that the LVOR augments visually driven ocular following information derived from optic flow inputs which have been found to be important in specification of the perceptual heading in space.[4]

SUMMARY

Linear vestibuloocular reflexes (LVORs) stabilize retinal images by producing eye movements to compensate for linear head motion. LVOR response characteristics depend upon gaze relative to the motion axis and binocular fixation distance. LVOR sensitivity during NO-axis motion increases as gaze eccentricity relative to the motion axis increases and as binocular fixation distance decreases. To fixate targets during forward head motion and rightward gaze, eyes must move to the right, but when looking left, the eyes must move to the left. In this study, LVORs were measured (binocular search coils) during 5.0 Hz horizontal motion along axes between and including NO and IA. This reorients head and otolith inputs relative to linear motion. We found that LVORs follow the same kinematics regardless of eye position in the head or head orientation relative to motion. Eye position information must be quickly and accurately integrated with otolith inputs to determine eye position (gaze) relative to linear head motion in space. The LVOR provides a behaviorally useful reflex for maintaining ocular fixation on visual targets during translation along any axis.

ACKNOWLEDGMENTS

The authors thank the technical support staff of the Vestibular Research Facility and the Animal Care Facility at NASA Ames Research Center for their enthusiastic and expert technical assistance in all phases of these experiments.

REFERENCES

1. PAIGE, G. D. & D. L. TOMKO. 1991. Eye movement responses to linear head motion in the squirrel monkey. I. Basic characteristics. J. Neurophysiol. **65**(5): 1170–1182.

2. PAIGE, G. D. & D. L. TOMKO. 1991. Eye movement responses to linear head motion in the squirrel monkey. II. Visual-vestibular interactions and kinematic considerations. J. Neurophysiol. **65**(5): 1183–1196.
3. WARREN, R. 1976. The perception of egomotion. J. Exp. Psychol. **2**(3): 448–456.
4. WARREN, W. H. & D. J. HANNON. 1988. Direction of self-motion is perceived from optical flow. Nature **336**(6195): 162–163.
5. BUSETTINI, C., F. A. MILES & U. SCHWARZ. 1991. Ocular responses to translation and their dependence on viewing distance. II. Motion of the scene. J. Neurophysiol. **66**(3): 865–878.

Mislocalizations of Visual Elevation and Visual Vertical Induced by Visual Pitch: the Great Circle Model[a]

LEONARD MATIN AND WENXUN LI

Department of Psychology
Columbia University
Schermerhorn Hall
New York, New York 10027

EXPERIMENTS WITH CURARIZED HUMAN OBSERVERS

Somewhat more than 10 years ago the perceptual consequences of experimental reduction of extraocular muscle efficiency were examined in systemically curarized human observers.[1,2] The experiments were designed to learn some things about the way in which extraretinal eye position information contributed to where we see things. One of the things that was learned was that when the observer was in total darkness and foveated a single small, stationary visual target at true eye level, the target's elevation was badly misperceived. Its elevation could appear to lie either near the invisible floor or near the invisible ceiling. Through a series of experiments, it was discovered that whether it appeared near the floor or the ceiling depended systematically on the vertical orientation of the observer's eye in the orbit relative to the head, that its perceived elevation did not depend on the pitch of the head or body, and that the peculiar and very substantial deviations of visual direction were a consequence of the inappropriate extraretinal eye position information resulting from the curare-produced extraocular muscle paresis. By means of quantitative psychophysical measurements, it was determined that the physical elevation of a target that was set to appear at the eye level of the observer declined linearly with increase in the elevation of the eye in the orbit (FIGURE 1).

However, when illumination was turned on in the erect room, the errors in setting the elevation of the target to appear at visually perceived eye level (VPEL) by the curarized observer were essentially eliminated. This observation provided the basis for the work described in the remainder of this presentation.

EXPERIMENTS WITH THE ILLUMINATED PITCHROOM

Our analysis of the curare experiments led us to believe that the critical aspect of the illuminated room was its visual pitch—its orientation around a horizontal axis in the frontal plane of the observer.[3,4] The elimination of the errors in perceived elevation of the foveated target in the presence of an illuminated room was a consequence of the fact that the room was erect, whereas the body-referenced mechanism that made use of the incorrect extraretinal eye-position information in

[a]The research was supported by grants 91-0146 from the Air Force Office of Scientific Research, BNS 8617059 from the National Science Foundation, and EY 05929 from the National Eye Institute, National Institutes of Health.

242

conjunction with information regarding the direction of gravity was providing the curarized observer with information that was discordant with the information from the erect room. We reasoned that if we were able to change the pitch of the room itself, we would produce errors in the perceived elevation of objects, not only in curarized observers, but in normal observers as well. For, in the normal observer, changing the visual pitch of the room would introduce a discordance between the elevation of visually perceived eye level as determined by the orientation of the visual field and elevation of visually perceived eye level as determined by the normal input from the body-referenced mechanism.

FIGURE 2a shows how we went about pitching a normal environment.[3-5] The normal (noncurarized) observer sat on a stool on the level ground of the exterior room, and the pitch of the interior room (which did not contain a floor of its own) was systematically altered. The interior of the pitchroom constituted the entire visual field of the observer. With this arrangement we obtained the effects we expected. A

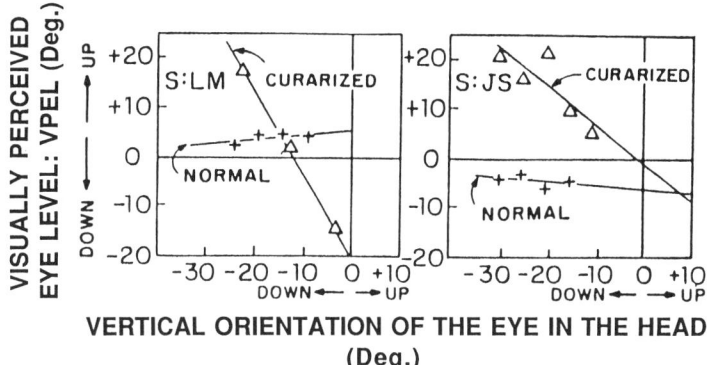

VERTICAL ORIENTATION OF THE EYE IN THE HEAD
(Deg.)

FIGURE 1. Physical elevation of a point in darkness that is visually perceived to lie at eye level (VPEL) for various vertical orientations of the eye in the head (abscissa). The orientation of the eye in the head was determined by the elevation of a second point employed for fixation. Results are displayed for each of two observers during systemic curarization and when in their (uncurarized) normal states. (Adapted from Matin et al.)[1]

single small red target was produced by a horizontal beam from a laser located behind the observer. The beam intersected the surface of the pitchroom that faced the observer. The physical elevation of the small visible target could be experimentally changed by raising or lowering the laser. When the laser target was truly at eye level against this background, its visually perceived elevation depended systematically on the angle at which the background was pitched. With room pitch in the direction shown in FIGURE 2a ("topbackward"), the single visual target at eye level appeared elevated from true eye level. When the pitchroom was rotated in the other direction—with the top toward the observer ("topforward")—the target at true eye level appeared displaced far below true eye level. The greater the pitch the greater the perceived deviation of the laser target from true eye level.

In order to measure these effects quantitatively we held constant the angle of room pitch and adjusted the physical elevation of the laser target until it appeared to lie at eye level, an adjustment that is quite easily and reliably made by the observer. With the pitchroom set 40° in the topbackward direction, the observer typically set

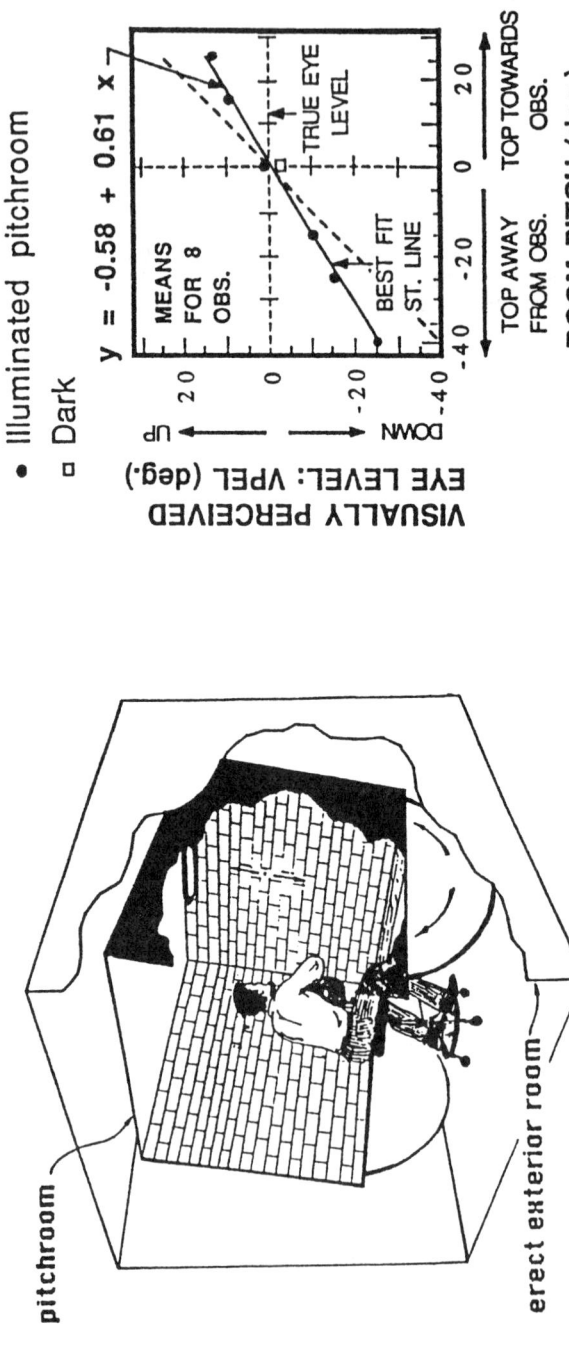

FIGURE 2. a: Cutaway drawing of the pitchroom inside the erect ("exterior") room of Schermerhorn Hall. The orientation of the pitchroom shown is "topbackward" (top away from observer). The pitchroom was illuminated by the lamp shown attached to the top of the interior wall facing the subject; a shield (not shown) prevented direct viewing of the lamp by the subject. The point on the interior wall facing the observer indicates the laser beam that was set by the observer's instruction to VPEL. The arrows indicate the possible directions of change of the target during psychophysical measurement. The black cloth was attached to the front of the chinrest and hung around the observer's shoulders extending forward to the front of the pitchroom's shelf where it was pinned across the shelf's horizontal extent. **b:** Average psychophysical measurements of VPEL for eight subjects in the illuminated pitchroom for each of six pitches of the room (filled circles); the average VPEL in darkness is shown by the unfilled rectangle at 0° pitch. The horizontal dashed line at ordinate zero represents the VPEL setting at true eye level for all angles of pitch. A VPEL setting at the level of the normal from the eye to the wall facing the subject would be equal to the pitch of the pitchroom (diagonal dashed line). The solid straight line is the line of best fit (least squares); slopes for the individual observers ranged between +0.45 and +0.81. (Adapted from Matin and Li.)[7]

the target down below his waist and stated that it appeared to lie at eye level—this corresponded to a deviation of the observer's setting from true eye level that averaged about 25°. When the observer took his chin off the chin rest, turned his head to face the erect exterior room, and viewed the laser source sending out the horizontal beam behind him, it appeared to lie below his waist as it was in fact. But when he turned his head to view the beam's projection on the wall of the pitchroom it appeared at eye level again. This could be done repeatedly, with the same effect each time. As we determined quantitatively later, the effect of the pitchroom takes place rapidly, reaching full magnitude in less than a minute; the change in VPEL elevation dark adapts exponentially to baseline levels with a time constant of 4–5 minutes. With topforward pitch, the observer sets the target far above true eye level when reporting that it appears at eye level; rise and decay of VPEL follow the same time course as with topbackward pitch (FIGURE 3).[6]

The quantitative influence of the fully illuminated pitchroom is shown in FIGURE 2b, which shows that when the pitch of the pitchroom was varied between 40° topbackward and 25° topforward (abscissa), the average value of VPEL (ordinate) changed from about 25.4° below to 13.5° above true eye level. The horizontal dashed line plots the locus that would be obtained if the observer set the target to true eye level at all pitches; the diagonal dashed line with a slope of +1.00 plots the locus that would be obtained if the observer's setting of the target deviated from true eye level by an angle that equaled the pitch of the pitchroom. The average effect shown for the eight observers is linear with pitch and has a slope of +0.61. Similar effects are manifested in the results of individual observers with the slope and intercept of the best-fit straight line differing among observers; individual slopes ranged from +0.45 to 0.81 and intercepts from 6.3° below true eye level to 5.1° above. These numbers are also characteristic of our subsequent measurements.[7]

The influence of visual pitch on VPEL that we measure is an indicator of a translation of the entire dimension of perceived elevation. This becomes most apparent when a familiar object—such as a person—is employed as a stimulus inside the pitchroom. The observer, outside of the pitchroom, sees the person standing inside the pitchroom as having grown enormously with the room turned topbackward, and as standing deep down in a hole with the room turned topforward.[3,4]

THE BODY-REFERENCED MECHANISM AND LINEAR WEIGHTING

The effects on VPEL produced by changing the angle of pitch of the illuminated pitchroom indicate the influence of a pitched visual field on VPEL. But they are not the only determinant of VPEL. An observer in complete darkness can set VPEL with a reliability of about 1° and a constant error that deviates from true eye level by an amount that typically does not exceed 5°. The open point in FIGURE 2b at zero on the abscissa indicates the average VPEL setting in total darkness, 2.9° below true eye level.

In order to set VPEL in darkness, the observer must make use of a combination of three kinds of information: (1) information about the orientation of the head relative to gravity, (2) extraretinal eye position information, and (3) information about the directional significance of stimulation at particular locations on the retina—retinal local sign. We refer to this combination as the "body-referenced mechanism." It is responsible for the ability of the observer to set VPEL reliably and reasonably accurately in total darkness. It is also responsible for the results with the pitchroom having the form that they do. Equation 1 shows that V, the influence of vision—the pitchroom in the present case, combines with B, the influence of the

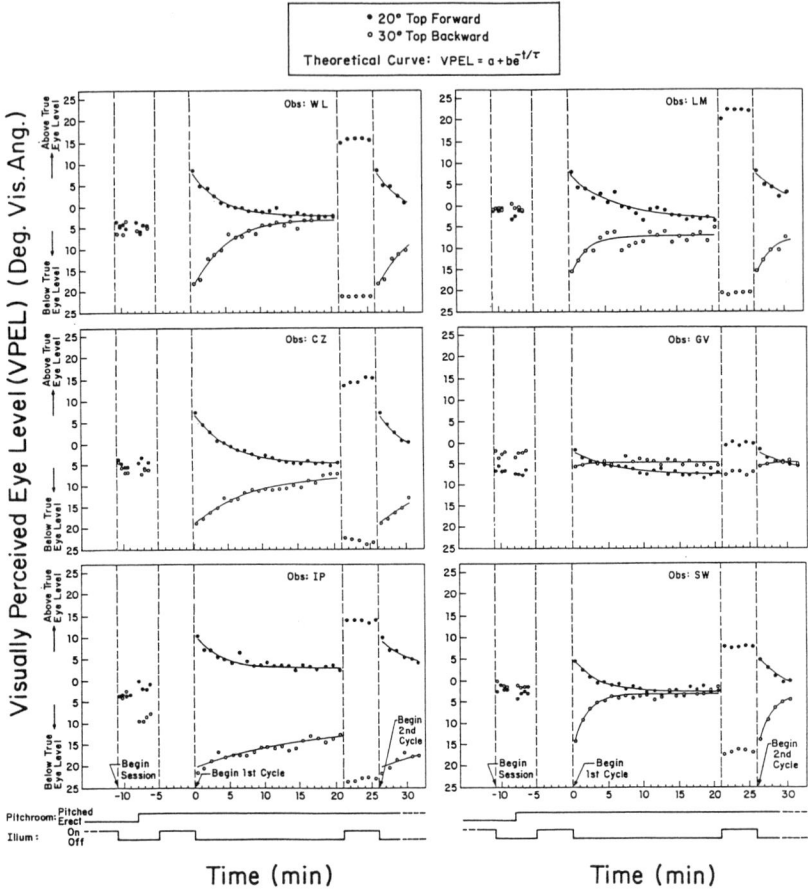

FIGURE 3. VPEL during light and dark adaptation is displayed separately for each of six observers in sessions involving 20° topforward pitch (filled circles) and 30° topbackward pitch (unfilled circles) of the fully illuminated pitchroom. All measurements were on the right eye of the observer; the left eye was occluded. Each displayed value in darkness and in the illuminated pitchroom is the mean of four settings (one in each of four cycles) except for IP where the values are means of eight measurements. The curves drawn through the points during the dark period are the best-fitting exponentials. The points between the first and second cycles were measurements against the illuminated pitchroom. During the first half of the 5½-minute period (approx.) from the beginning of the session to the beginning of the period of light adaptation, measurements were made in darkness against the erect pitchroom; during the second half of that period the room was pitched but remained dark and VPEL measurements were continued. (Adapted from Matin and Li.)[6]

body-referenced mechanism in a linear weighted average of effects to yield the resultant VPEL:

$$\text{VPEL} = k_v V + k_b B \tag{1}$$

In Equation 1, V and B are the influences of the visual field and the body-referenced

mechanism respectively, and k_v and k_b are constants reflecting the relative contributions of V and B to VPEL with $k_v + k_b = 1$. It is because of the linear weighting that
the slope of the results in FIGURE 2b falls between what would occur if the
body-referenced mechanism were the sole determinant of VPEL (the horizontal line
at 0.0 on the abscissa) and what would occur if the visual field were the sole
determinant (the diagonal line with slope of +1.00). Equation 1 predicts a linear
function in the normally illuminated environment to lie between the two, a prediction that is fulfilled.[3-5,7,8]

EXPERIMENTS WITH THE 2-LINE STIMULUS

What was there about the pitched visual field that generated the influence on
VPEL? Our attempt to answer this led us to an analysis in terms of retinal
perspective.[7,9] Consider a side view of the wall facing the observer from the pitched
visual field (FIGURE 4). The point that is closest to the observer is the normal from
the observer's eye to the plane of the visual field. When the field is pitched the
normal is turned by the angle of pitch. Consider the erect, stationary observer to be
looking from primary viewing position at a field made up of an erect grid of vertical

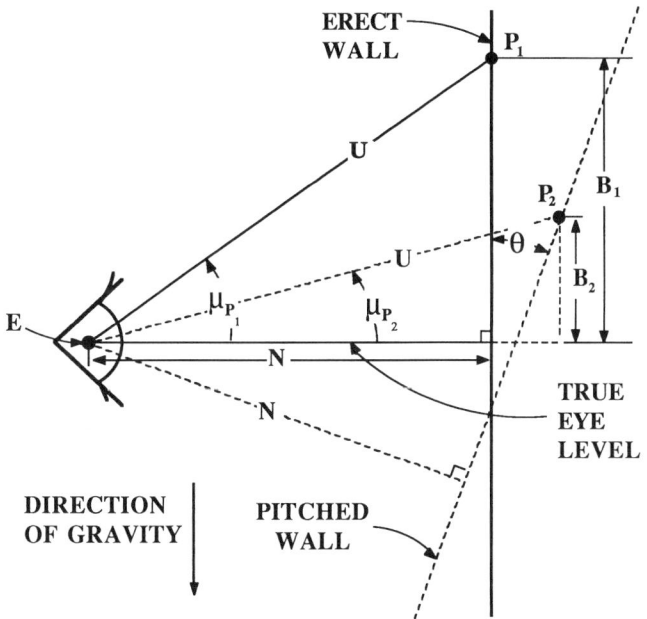

FIGURE 4. Vertical cross-section showing the observer at E viewing an erect wall (solid lines)
or a wall pitched topbackward by θ (dashed lines). Pitch is around the horizontal axis through E
with the normal from the wall to the eye at distance N at both orientations. The point P_1 on the
erect wall is elevated by angle μ_{P1} above true eye level; although the angular separation of the
point from the normal and its distance from the eye (U) remains unchanged on the pitched wall,
the angular elevation of the same point (now P_2) above true eye level is μ_{P2} ($= \mu_{P1} + \theta$). B_1 and
B_2 are the linear elevations of P_1 and P_2 above true eye level in the erect and pitched cases,
respectively. (Adapted from Matin and Li.)[7]

and horizontal lines (FIGURE 5a). Since the normal from the eye to the plane containing those lines is closest to the observer, the horizontal visual angle separation between the retinal images of any two vertical lines will be greatest between those points on the vertical line pair that lies at the normal. The visual angle separations between other points on the vertical line pair (e.g., at the level of point P in FIGURE 4) will decrease as they lie further from the normal. But, for an erect visual field, as the observer raises or lowers his body, no matter what his elevation, the maximum visual angle separation will lie at his true eye level since this will always be at the normal. This suggests an evolutionary conditioning basis for treating the maximum retinal separation between a pair of lines as a basis for setting the elevation of visually perceived eye level. Such will not be the case for a pair of horizontal lines.

FIGURE 5b displays the retinal image of a 2-line pitched-from-vertical stimulus, assuming central projection on a spherical approximation to the eye. Each of the functions in FIGURE 6 plots how the retinal separation between a pair of vertical lines varies with elevation on the lines; the different functions display this variation for different pitches. We have referred to this variation as MBP (for monocular biconvergence perspective).[7,9]

Thus, our first attempt at analysis of the visual stimulus in the pitchroom led us to examine the influence of a pair of vertical lines and separately a pair of horizontal lines; the displays are shown in FIGURE 7.

As predicted, the effect for vertical lines was substantial (FIGURES 8 and 9); the effect with horizontal lines was minimal (FIGURE 11).

FIGURE 8 is the same kind of plot displayed before, except that one of the sets of data was obtained with only two pitched-from-vertical lines. As shown in the figure the results are very similar to the results with the complexly structured visual field of the pitchroom.

FIGURE 9 shows the relative magnitude of the effects for the pitchroom and the 2-line pitched-from-vertical stimulus separately for each of the eight individual observers whose results are included in the averages in FIGURE 8. For each data point the pitch was equal in the two cases. That the results are very similar for the two cases is shown by the proximity of the points to the main diagonal.

Thus, a complex visual field is not necessary in order to obtain substantial influences on VPEL—a 2-line pitched-from-vertical stimulus is sufficient.

INSIGNIFICANCE OF BINOCULAR CUES

The results with the fully illuminated pitchroom in FIGURE 2b were obtained with binocular viewing; the results in FIGURE 8 were obtained with monocular viewing.[7,8] The slopes of the average VPEL-vs.-pitch functions were +0.61 and +0.63, respectively, for the two experiments. Although only two of the observers were common to the two groups of observers, the closeness of the two average values along with a great deal of stability of individual VPEL-vs.-pitch functions over time encourages the view that binocular cues play little if any role in determining the influence on VPEL.

INSIGNIFICANCE OF SEVERAL MONOCULAR GRADIENTS

As noted above in relation to FIGURE 4 different segments of a straight stimulus line lie at different distances from the eye. Three properties of the retinal image of

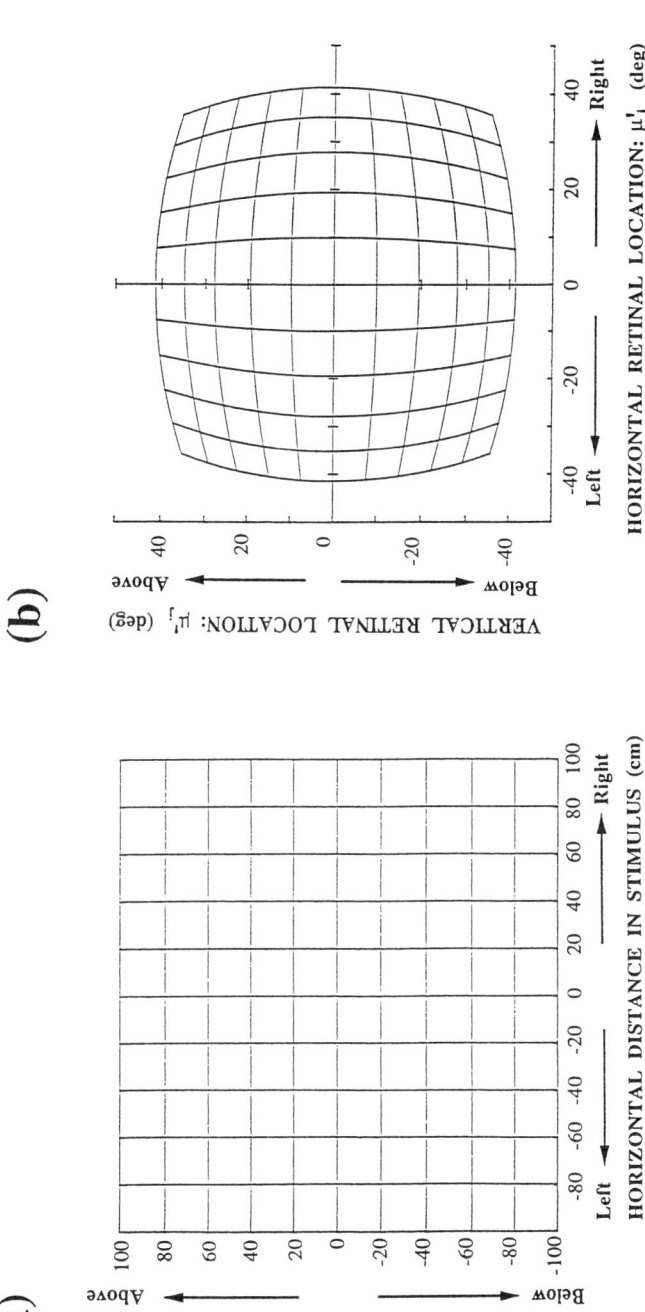

FIGURE 5. a: The grid stimulus. **b:** The retinal pattern of the grid imaged by the spherical approximation to the eye. The coordinates in panel a are in linear measure at the plane of the stimulus; coordinates in panel b are visual angles at a viewing distance of 1.13 meters. Pitching the plane of the grid in panel a topbackward or topforward results in a vertical translation (equal to the angle of pitch) of the retinal pattern in panel b upward or downward respectively in the figure on the page relative to the retinal coordinates. Upward directions in both object and image space are positive. (Adapted from Matin and Li.)[7]

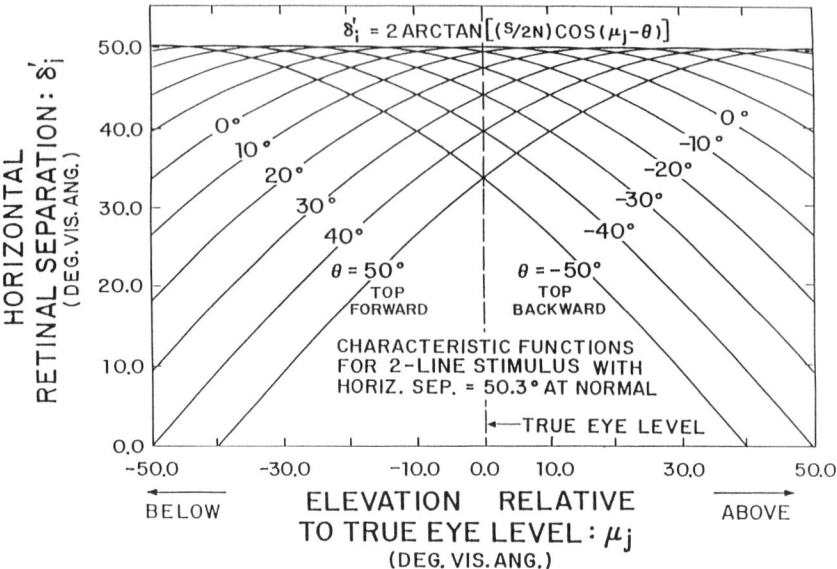

FIGURE 6. Horizontal separation (δ_i') within the retinal image of the pitched-from-vertical 2-line stimulus for different values of pitch (θ) vs. elevation (μ_j) in the stimulus plane. The values plotted are from the equation shown for which S is the linear horizontal distance between the two lines and N is the linear distance of the normal from the eye to the stimulus plane. (Adapted from Matin and Li.)[7]

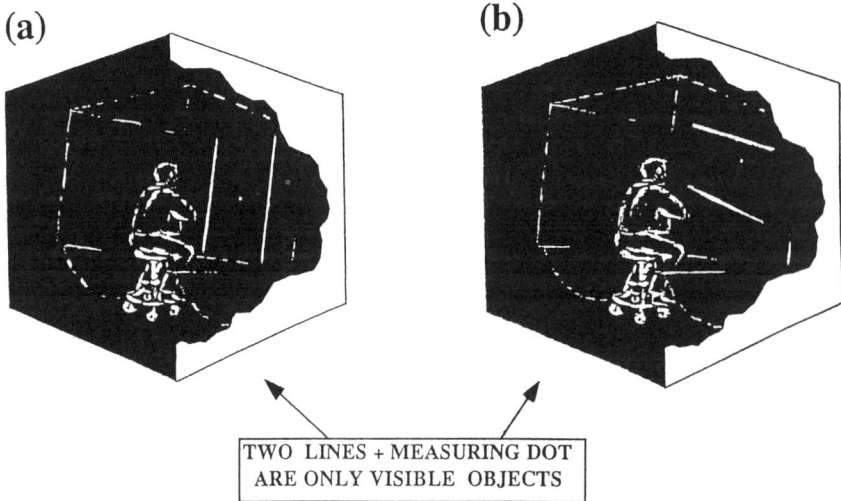

FIGURE 7. The 2-line stimuli pitched topbackward in the darkened pitchroom. **a:** Pitched-from-vertical stimulus. **b:** Horizontal stimulus. The VPEL measurement was made by the observer instructing the experimenter to set the laser beam ("measuring dot") so that it appeared at eye level. (Adapted from Matin and Li.)[7]

the line covary systematically with this differential distance and could conceivably be involved in the influence of the pitched-from-vertical line on VPEL: (1) width of the geometric image of the line; (2) luminous energy from each segment of the line; (3) sharpness of the image—image quality. However, it is clear that no one or any combination of them plays any substantial role.[7] In brief summary of the argument that leads to that conclusion we note that (1) the variation of width of the retinal image of the line is unrelated to the experimental VPEL-vs.-pitch function; (2) the VPEL-vs.-pitch function is unchanged by the topical instillation of pilocarpine which reduced pupil size to less than 2 mm with accommodative spasm and eliminated any reasonable possibility of the involvement of accommodative cues[8]; (3) the variation of VPEL with line orientation in a frontoparallel plane is identical to the variation of

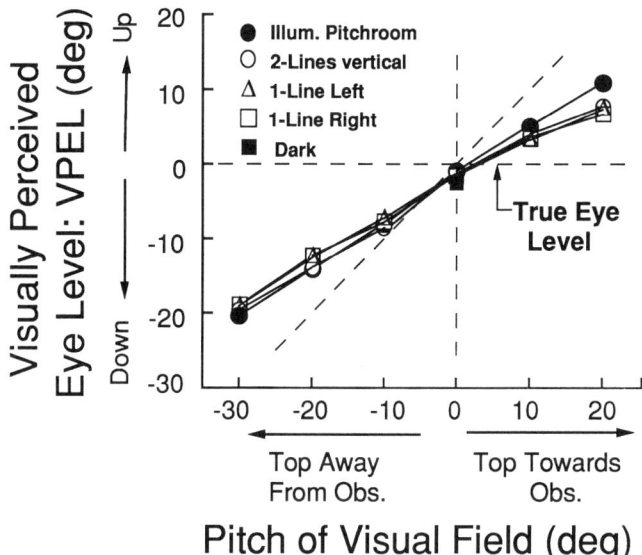

Pitch of Visual Field (deg)

FIGURE 8. VPEL vs. pitch in the fully illuminated pitchroom, when viewing the 2-line pitched-from-vertical stimulus, and when viewing one of the two lines alone as the 1-line stimulus with either the line on the right side of the median plane or on the left side of the median plane). (Adapted from Matin and Li.)[8]

VPEL with pitch for the parallel 2-line pitched-from-vertical stimulus when the retinal locations stimulated in the two cases are identical[10] (see below, FIGURE 16). However, the normal lines of visual direction to the eye from the pitched-only plane and from the erect plane fall at different elevations relative to the stimulus lines, and so both the retinal width and luminous flux gradients diverge from different points on the image of the lines in the two cases.

INFLUENCE ON VPEL OF TWO PARAMETERS OF THE 2-LINE STIMULUS

The experiments with the 2-line stimulus described above employed lines that were horizontally displaced by 25° from the median plane of the observer. Systematic variation of eccentricity from 1.25° to 35° (horizontal separation between the two lines from 2.5° to 70°) produced a threefold variation in the average slope of the VPEL-vs.-pitch function from 0.17 to 0.51.[11] But the relation of change in pitch to change in retinal orientation varies systematically with eccentricity, and, although the slope of the VPEL-vs.-pitch function increases with eccentricity (FIGURE 10a), following transformation of the angle of pitch to angle of retinal orientation, the slope of the VPEL-vs.-retinal orientation function decreases systematically with

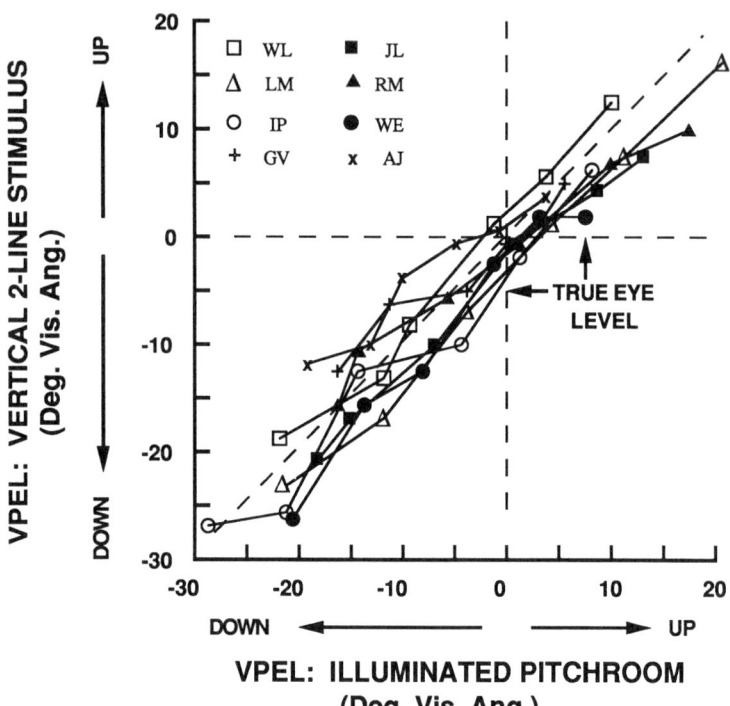

FIGURE 9. VPEL for the 2-line pitched-from-vertical stimulus plotted against the VPEL for the illuminated pitchroom. The results for each of eight observers are displayed for each of six different pitches. Each point plots the two VPEL values at a single pitch under the two conditions. For each observer the most negative pair of values are for the most topbackward pitch (−30°); increasing values of VPEL correspond to increasing topforward pitch. (Adapted from Matin and Li.)[7]

eccentricity along a curve that is similar to the decrease of visual acuity with eccentricity (FIGURE 10b).

Although horizontal eccentricity has a substantial influence on the slope of the VPEL-vs.-pitch function, change of the height of the 2-line stimulus in the visual field by 44° had no influence on the slope but did result in a change of the bias of the entire function by about 10%.[11]

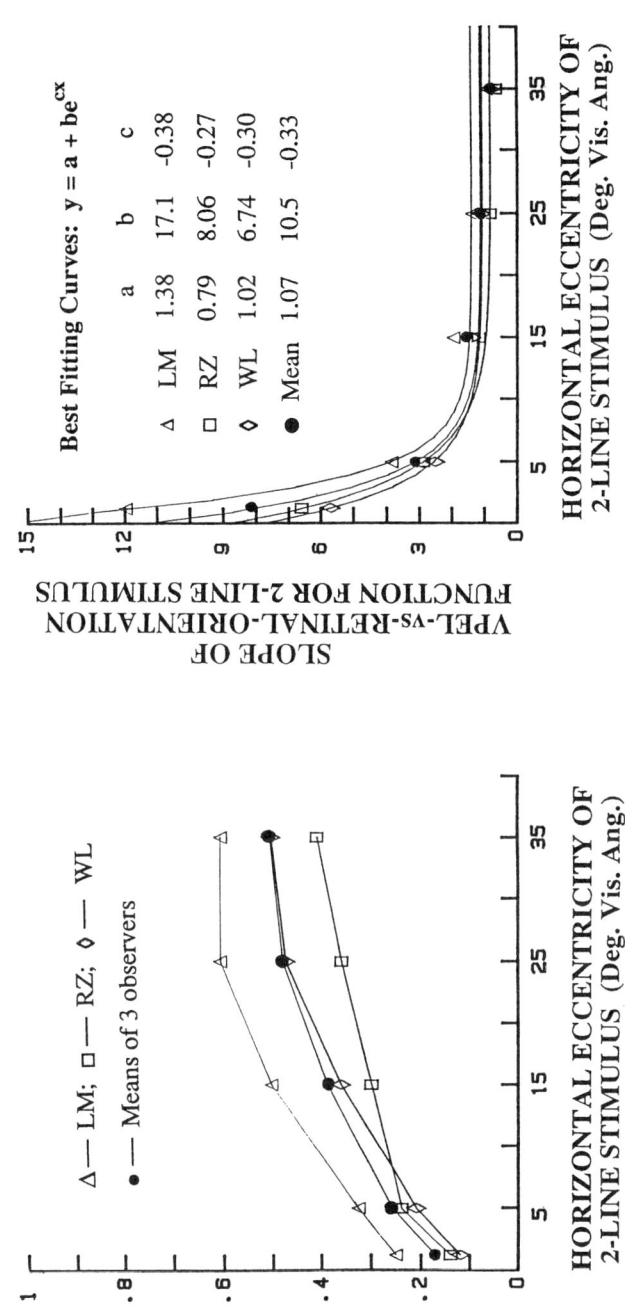

FIGURE 10. a: Slope of the best fitting VPEL-vs.-pitch function for the 2-line pitched-from-vertical stimulus for different horizontal eccentricities. **b:** The pitch variable was transformed to retinal orientation, and a VPEL-vs.-retinal-orientation function plotted for each horizontal eccentricity. Panel b displays the best fitting slope of the VPEL-vs.-retinal-orientation function for different horizontal eccentricities. The reversal of the direction of the function against the independent variable between panels a and b is a result of the differences in the pitch-to-retinal-orientation transform for the different horizontal eccentricities. (Adapted from Li and Matin.)[11]

EXPERIMENTS WITH THE 1-LINE STIMULUS

Our development of the Great Circle Model[7,8,10-16] (see below) led to a series of further discoveries. In order to examine predictions from this model, we first measured the influence of a single line against a dark field, and obtained effects very similar to those with the 2-line stimulus.[8,17] FIGURE 8 displays the average results for eight observers with both the 1-line and the 2-line pitched-from-vertical visual fields and with the pitchroom. The 1-line results are remarkably similar to those with the 2-line stimulus; the slopes of the VPEL-vs.-pitch functions were only a little shallower—slopes of +0.52 and +0.53 as compared to +0.56 for the 2-line stimulus. FIGURE 11 shows the results with the horizontal lines;[8,17] a small slope (+0.18) is obtained with the 2-line stimuli as noted above; the slope is even smaller with each of the 1-line stimuli (+0.07 and +0.09).

Thus, the effect of the 1-line pitched-from-vertical stimulus was large—almost as large as the effect with the full, complex visual field. But, of course, if a single point replaced the line we would not obtain any effect, and so, it was expected that variation of line length would result in variation of the magnitude of the effect. In the next experiment we repeated the previous experiment under conditions in which the length of the single line was systematically varied from 3° to 64°. Increase in line length resulted in a systematic increase in slope of the VPEL-vs.-pitch function (FIGURE 12). The curve relating the slope of the VPEL-vs.-pitch function to the length of the 1-line pitched-from-vertical visual field is a negatively accelerated exponential with space constant of 15.1°.[7,14,16]

THE GREAT CIRCLE MODEL

The reasons for the name given to the Great Circle Model will be immediately clear. FIGURE 13 shows a spherical approximation to an eye viewing a 4-line grid. In all three panels the eye is erect, stationary, and looking straight ahead. The equator and central vertical retinal meridian (CVRM) are shown as fixed in place in the three panels. A pinhole pupil at the center of the sphere provides for central projection on the interior rear surface of the sphere.

The central panel in FIGURE 13 shows the erect grid stimulus imaged on the spherical approximation to the eye. The images of the straight lines are projected on to great circles that intersect at the upper and lower poles of the eye.[b] When the plane containing the grid is pitched topforward around a horizontal axis through the center of the eye parallel to the plane of the grid as in the upper panel, the images move downward on the eye and the intersection point of the great circles containing the images of the pitched-from-vertical lines with the CVRM moves down by the angle of pitch. Similarly, when the plane containing the grid is pitched topbackward around a horizontal axis as in the lower panel, the intersection point of the great circles containing the images of the pitched-from-vertical lines with the CVRM move up by the angle of pitch.

[b]To refresh memories: (1) great circles are like lines of longitude on a globe of the earth; lines of latitude, except for the equator, are not great circles; (2) for central projection on the surface of a sphere, every straight line in object space is imaged on a great circle; (3) Any two great circles on a sphere intersect each other on the surface of the sphere at two points 180° apart; (4) the great circles containing the images of all lines that are parallel to a given line in object space intersect at the same two points at the surface of the sphere; the intersection points are different for parallel lines sets of different orientations.

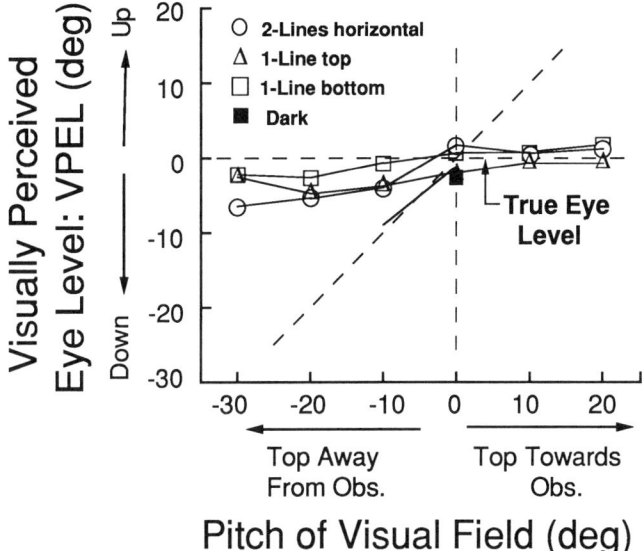

FIGURE 11. VPEL vs. pitch for the 1-line and 2-line horizontal stimuli. (Adapted from Matin and Li.)[8]

The three main aspects of the Great Circle Model that we emphasize here are the following:

1. We define ω' as the angular distance between the upper pole and the nearer point of intersection between the CVRM and the great circle containing the

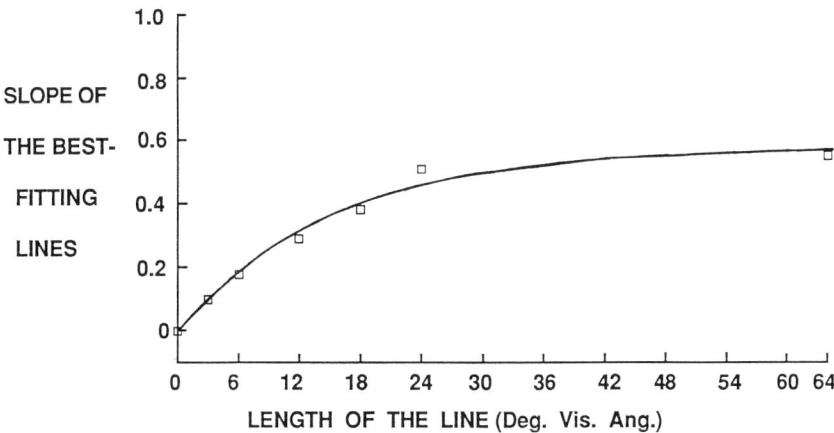

FIGURE 12. The slope of the VPEL-vs.-pitch function (ordinate) plotted as function of the length of the pitched-from-vertical 1-line stimulus (average results of five observers). The curve is a negatively accelerated exponential with a space constant of 15.1°. (Adapted from Li and Matin.)[16]

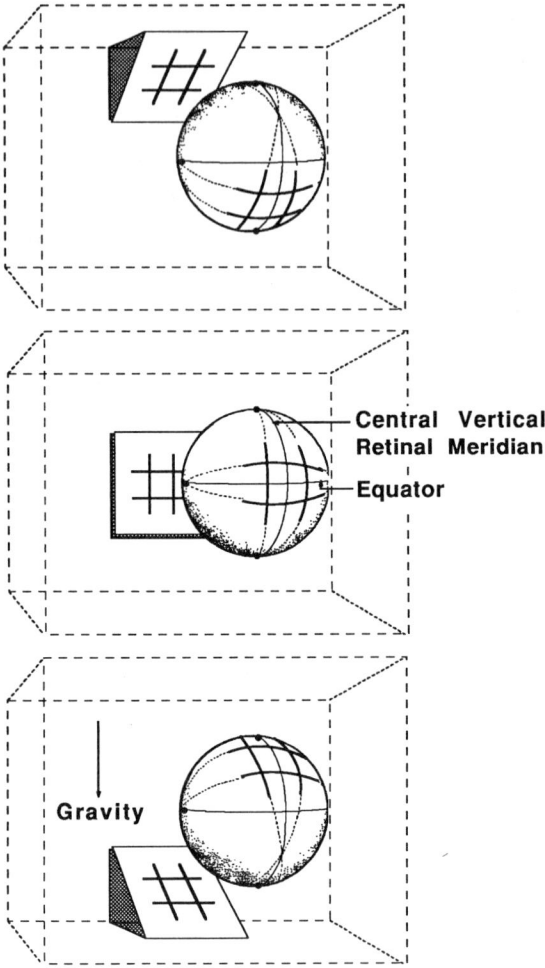

FIGURE 13. Three sketches of a spherical approximation of an eye viewing a grid consisting of crossed two-line pitched-from-vertical and horizontal stimuli. The image at the back of the eye results from central projection through the pinhole pupil at the center of the sphere. The figures are not drawn to scale. The eye is in the same fixed position within the room in the three panels. The front of the eye with its pinhole pupil faces the center of the stimulus in the center panel and remains in the same fixed orientation relative to the room (dashed lines) in the other two panels. The fixity of eye orientation is indicated by the three dark dots at the top, bottom, and left side of the sphere which represent the top and bottom poles on the vertical axis of the sphere and the left pole of the horizontal axis within a frontal plane through the center of the sphere, respectively. The equator and central vertical retinal meridian (CVRM) on the surface of the back of the sphere (the retina) are shown by light solid curves in the three panels. The figure shows the image of the stimulus as if it were visible from outside the sphere. The grid stimulus in the central panel is in a frontoparallel plane. In the bottom panel it is pitched topbackward around an axis coincident with the axis of the eye through the left and right poles; in the upper panel the stimulus is rotated topforward. Thus, the normal from the nodal point of the eye extends to the center of the plane containing the grid in the three panels and the distance along the normal is constant. (Adapted from Matin and Li.)[7]

image of a line.[c] For a line of fixed eccentricity and length, the value of V, the magnitude of the influence of a single straight line on VPEL, changes monotonically with ω'.

2. Define l' as the total length of the set of retinal images produced by all stimuli intersecting at a particular ω'. For a given value of ω', the magnitude of V changes monotonically with l'.

3. For the case in which each of n simultaneously presented sets of retinal images intersects the CVRM at a different ω', the magnitude of V is the weighted average of the n separate values that would be generated if each were presented alone; the weight contributed by each set increases with its associated l' value and with eccentricity.

NEUROPHYSIOLOGY OF THE GREAT CIRCLE MODEL

FIGURE 14a shows a neurophysiological embodiment of how the Great Circle Model processes the images of three long straight, parallel, pitched-from-vertical lines (long, thin rectangles with diagonal hatching) on a plane pitched at $-20°$ ($20°$ topbackward):

1. The leftmost line in FIGURE 14a is shown as being processed by five orientation-sensitive neural units, the shorter middle line by two neural units, and the rightmost line by four units. The outputs of all of these units are fed to a single node (all black dots on the horizontal line -20 represent the same point—the node—in the model) at a more central region where summation of the effects of all of the lines occurs. In FIGURE 14a this node is indicated as "-20." An output from this node, $V_{j,-20}$, carries the influence on VPEL.

2. FIGURE 14b shows that each point on the CVRM has its own output, $V_{j,\omega'}$.

3. All of the different outputs, $V_{j,\omega'}$, are treated together within a single weighted average to set the value of V in Equation 1.

Although the experiments with lines suggest control by orientation-sensitive units in visual cortex, the receptive fields of neural units in monkey primary visual cortex (V1) are no larger than about $3°$.[18] On the other hand, the space constant for summation of the influence on VPEL in the results shown here is $15.1°$, much too large to be set by neural units in V1 alone. This is compatible with the size of receptive fields of units in area 7a of the inferior temporal lobule of posterior parietal cortex, which have, in addition, several other properties that are essential for the VPEL discrimination.[7]

FOUR ADDITIONAL EXPERIMENTS SUPPORTING THE GREAT CIRCLE MODEL

The following four predictions were made directly from the Great Circle Model. One of them involves experiments described above; three of them involve additional sets of experiments (described below). The experimental results provide strong confirmation for the model.

---

[c]ω' is negative for topforward pitch; as the angle of topforward pitch decreases, the value of ω' increases monotonically from negative values to reach zero when the stimulus is erect; as the angle of pitch becomes increasingly topbackward ω' increases positively from zero.

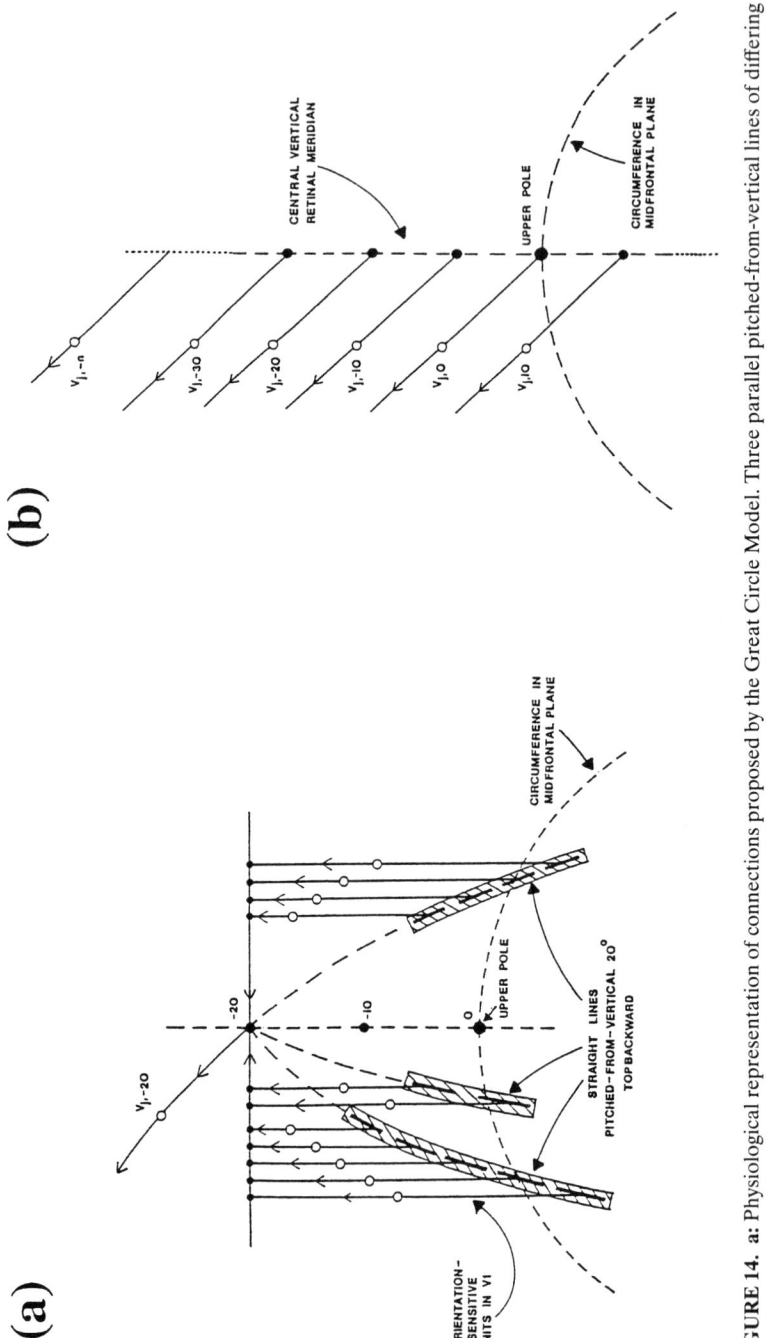

FIGURE 14. a: Physiological representation of connections proposed by the Great Circle Model. Three parallel pitched-from-vertical lines of differing length provide the stimulus; the three lines lie on a plane that is pitched topbackward by 20°. The leftmost one is sufficiently long so that it is processed by five different neural units in V1, the middle one is processed by two neural units, and the rightmost one is processed by four neural units. The curved representation of the images of the lines indicates the "curvature" of the great circle containing the images; each of the solid dashes within the image of each line represents processing by a different neural unit in V1. Since the three lines intersect the same point on the CVRM, their influences summate at the node designated −20 and the net influence on VPEL is designated as $V_{j,-20}$. **b:** Some of the output lines for influencing VPEL. Each collects inputs from a different location on the CVRM in the manner shown in a.

1. As expected from the significance given to the CVRM and the treatment of l' in the model above, the summation of the influences on VPEL of two short parallel pitched-from-vertical lines that are horizontally separated by 50.3° is as great as the summation of influences for the same two lines presented as a single coextensive pitched-from-vertical line. The magnitude of summation of the two horizontally separated lines is in line with the magnitude of the increase expected from the 1-line slope-vs.-length function in FIGURE 12.[12]

2. The effects of the long 2-line pitched-from-vertical stimulus—whose total length is 127.8°—and the pitchroom—which contains fragments of vertical lines adding up to 511.2° of length—fit on the same function as do the results from the 1-line variable-length experiment[7] (FIGURE 15). The theoretical

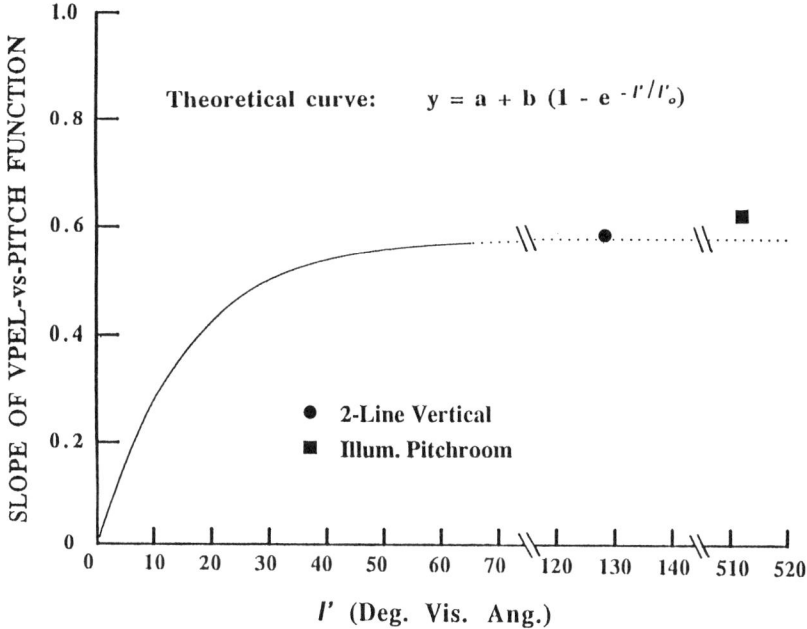

FIGURE 15. The results of experiments for the 2-line pitched-from-vertical stimulus (circles) and for the full pitchroom (squares) plotted as extrapolations on the exponential function that was fitted to the measurements relating the slope of the VPEL-vs.-pitch function and the length of the 1-line pitched-from-vertical stimulus (from Li and Matin)[16] under the assumptions of the Great Circle Model with l' equal to 127.8° and 511.2° respectively. The 1-line results were obtained with lengths between 3° and 63.9° and fitted ($p < 10^{-5}$) the negatively accelerated exponential with $a = -0.006$, $b = +0.585$, and $l'_0 = 15.1°$. (Adapted from Matin and Li.)[7]

curve is the curve from the 1-line variable-length experiment in FIGURE 12 above; the two data points displayed are the average slopes of the VPEL-vs.-pitch functions for the 2-line and the pitchroom experiments.

3. The third result predicted from the Great Circle Model can be understood with the assistance of FIGURE 16. In FIGURE 16 each of the two pitched-from-vertical lines—shown dashed in the pitched-only plane π'—and the nodal point of the eye determine a plane extending outward from the eye (as marked

by lines connecting the ends of the dashed lines with the eye). The intersections of those planes with the erect plane π—shown by the solid lines—are the two solid lines. Each dashed pitched-from-vertical line in the pitched plane and its corresponding solid oblique line in the erect plane strike identical retinal locations. Although the gradients of width and luminous flux of the retinal images of the pitched-from-vertical and oblique lines are different, the VPEL-vs.-pitch functions are indistinguishable. The results that demonstrate this identity for 1-line and for 2-line experiments are displayed in FIGURE 17. The results indicate the insignificance for VPEL of the difference in the depth plane from which the stimulus arises; they also demonstrate that a variety of details regarding variation of retinal illumination and width of the retinal

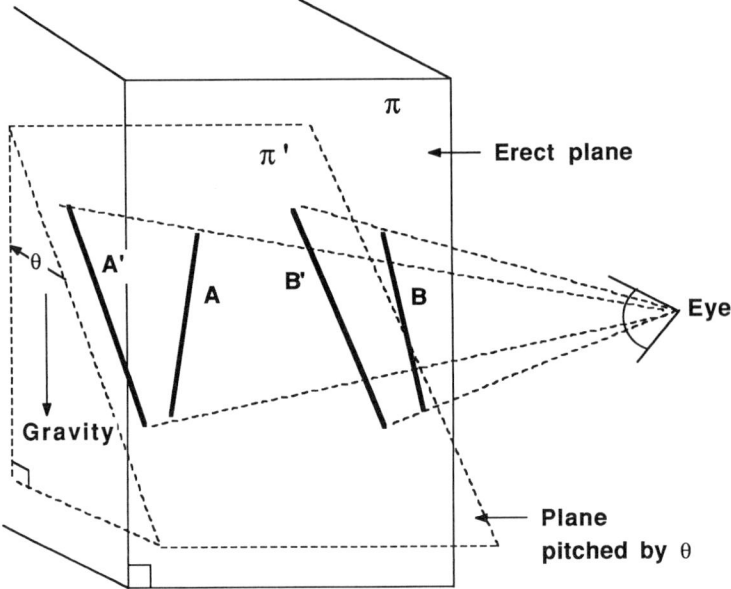

FIGURE 16. Figure shows relation of lines in plane π′ pitched topbackward by θ and erect plane π from the viewpoint of the eye of the observer. Lines A in π and A′ in π′ strike identical retinal loci as do B and B′. (Adapted from Matin and Li.)[10]

image of a line are irrelevant to the influence of the line on VPEL. What is essential is the retinal location and orientation of the image.

4. The fourth result is a bit more complex and, without the approach taken by the Great Circle Model, somewhat unexpected. However, making use of the spherical approximation, we are able to prove the following identity; the experimental prediction follows immediately from the identity. For central projection on the spherical approximation to the eye, stimulation by an infinitely thin parallel pair of rolled stimulus lines (FIGURE 18) is entirely equivalent to stimulation by two lines, one in each of two planes of numerically equal but opposite pitch. The experimental prediction based on this is that identical influences on VPEL should be obtained from a rolled line pair in an

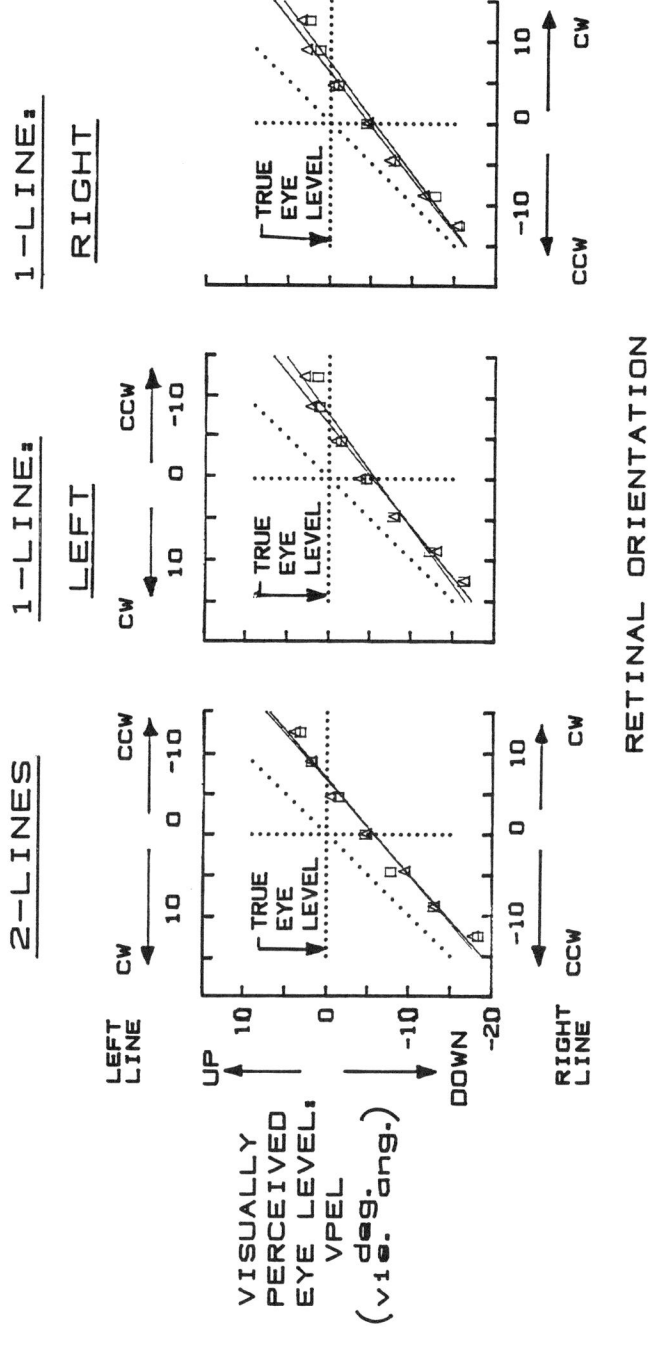

FIGURE 17. Results of experiments in which oblique lines in an erect plane and pitched-from-vertical lines in a pitched-only plane have identical retinal orientations. Average results of five observers: squares, real pitch; triangles, oblique line(s) on erect plane. Each graph plots the relation of VPEL to retinal orientation of the line. The left panel plots the relation for the 2-line stimulus, the other two panels plot relations for the 1-line stimuli. (Adapted from Matin and Li.)[10]

erect plane and the equivalent opposing-pitches line pair. The results in FIGURE 19 confirm the prediction.

We obtain some further understanding of these results by considering the fact that, not only are the results with the rolled and opposing-pitches line pairs indistinguishable, but neither variation of the pitch magnitude in the opposing-pitches case nor variation of the magnitude of roll brings about any variation of the net influence on VPEL—the results for both in FIGURE 19a fall along horizontal lines. This aspect of the results follows from the third aspect delineated for the Great Circle Model above and from the geometric identity between lines from pitched and rolled planes. Since the VPEL-vs.-pitch function for each individual line is linear, the sum of the influences from two lines of numerically equal but opposite pitch should be identical for all magnitudes of pitch. The fact that the VPEL for the two lines simultaneously presented does not change with the magnitude of pitch supports the linear weighting described in the Great Circle Model.

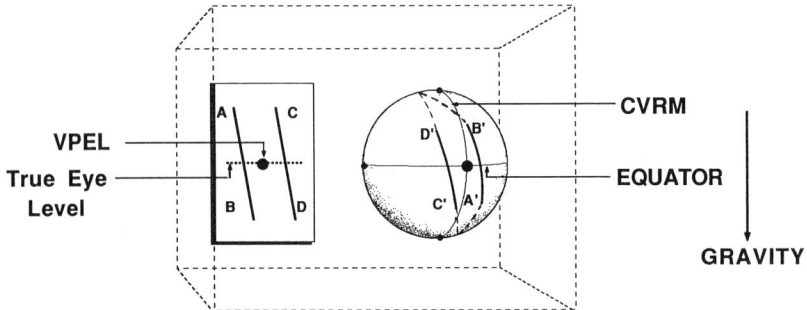

FIGURE 18. A rolled 2-line stimulus in an erect plane viewed by the erect stationary eye.

VISUALLY PERCEIVED VERTICAL, VISUALLY PERCEIVED EYE LEVEL, AND THE GREAT CIRCLE MODEL

A rolled stimulus (as in FIGURE 18) generates displacements of a line perceived as vertical in the frontal plane.[19-21] FIGURE 19 displays the average settings of a small line in the median plane to the visually perceived vertical (VPV) in the frontal plane of four observers. As shown in FIGURE 19a, indistinguishable results are obtained with the equivalent 2-line opposing-pitches stimulus.[13] Thus, both the rolled stimulus and the opposing-pitches stimulus produce no influence on VPEL but both result in systematic influences on VPV. Also shown in FIGURE 19b is the companion result described above in which both the same-pitch and opposing-roll 2-line stimuli systematically influence VPEL but produce no influence on VPV.

These additional sets of experimental results support the Great Circle Model presented above and extend it further: (1) The additional results support the implications we have drawn from the theorem relating the equivalence of rolled and opposing-pitches 2-line stimuli. (2) The results support the three main aspects of the Great Circle Model as presented earlier. (3) The results indicate that while the elevation of the intersection point of the great circle containing the image of a line with the CVRM is critical in determining the line's influence on VPEL, the location of the intersection point of the great circle containing the image of a line with the

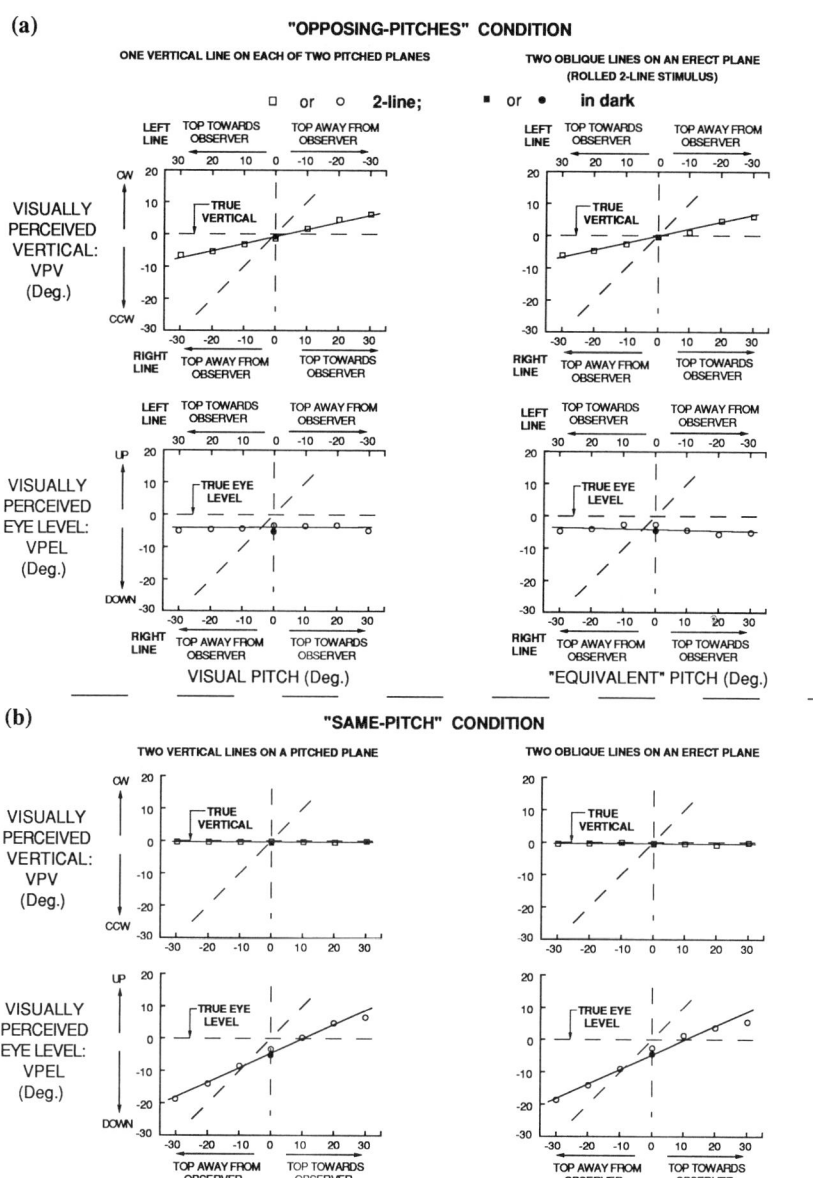

FIGURE 19. Results with VPEL and VPV discriminations which demonstrate identities between line stimuli in rolled and pitched planes. (Average results from 4 observers.) **a:** The four upper panels show identical results with two lines in the opposing pitches condition (two lefthand panels) and the two parallel lines that are rolled in the erect plane (two righthand panels). **b:** The four lower panels show identical results with the parallel 2-line pitched-from-vertical stimulus and two oblique lines in the erect plane. (Adapted from Matin and Li.)[13]

great circle circumscribing the midfrontal plane of the spherical approximation to the eye is critical for the determination of VPV and that a weighted average mechanism among a set of lines operates for VPV as well as for VPEL.

Above we have treated the eye as constrained to primary viewing position in a stationary, erect head that faces straight ahead relative to the body. The quantitative dependence of VPEL on extraretinal signals from movements of the head and eye is treated in a generalized version of the Great Circle Model in which the head and eye are free to move.

SUMMARY

The elevation visually perceived as eye level (VPEL) changes linearly with the pitch of an illuminated visual field. The magnitude of influence is only slightly less when the visual field contains only two dim vertical lines in darkness than when it is complexly structured and normally illuminated. Pitching a visual field consisting of only a single line in darkness produces an influence that is only slightly smaller than the 2-line stimulus. The slopes of the VPEL-vs.-pitch functions for the complex room, 2-line stimulus, and 1-line stimulus are +0.63, +0.56, and +0.52 respectively.

Although VPEL is systematically influenced by the pitch of the 2-line stimulus, the orientation of a small line within a frontal plane that is visually perceived as vertical is unaffected. However, when the two lines are pitched by equal amounts in opposite directions, the offset of VPV from true vertical changes linearly with pitch magnitude but VPEL is unaffected. These results are identical to those obtained when the two vertical lines are rolled within the frontal plane, a result that depends on some identities between roll and pitch: roll of two parallel lines in the same direction influences VPV but not VPEL; roll of the two lines in opposite directions influences VPEL but not VPV. The interaction between stimulus conditions and discriminations demonstrates that separate mechanisms are in control of VPEL and of VPV.

The slope of the VPEL-vs.-pitch function increases exponentially with line length for the 1-line stimulus (space constant = 15.1°). Summation of influences on VPEL for two lines horizontally separated by 50.3° is as great as for two coextensive lines.

The above results are predicted from the Great Circle Model which assumes (1) central projection on a spherical approximation to an erect stationary eye; (2) the sign and magnitude of influence of each line on VPEL and on VPV are determined by the direction and magnitude of the separation between the upper pole of the spherical eye and the intersection of the great circle containing the line's image with the central vertical retinal meridian and with the midfrontal retinal meridian, respectively; (3) the influence of individual nonparallel lines is determined by a weighted average of the influences of individual sets of parallel lines; (4) a generalized version of the Great Circle Model is indicated in which extraretinal signals from head and eye are taken into account.

REFERENCES

1. MATIN, L., E. PICOULT, J. K. STEVENS, M. W. EDWARDS, JR., D. YOUNG & R. MAC-ARTHUR. 1982. Oculoparalytic illusion: visual-field dependent mislocalizations by humans partially paralyzed with curare. Science **216:** 198–201.
2. MATIN, L., J. K. STEVENS & E. PICOULT. 1983. Perceptual consequences of experimental

extraocular muscle paralysis. *In* Spatially Oriented Behavior. A. Hein & M. Jeannerod, Eds. (Chapter 14): 243–262. Springer. New York, N.Y.

3. MATIN, L. & C. R. FOX. 1986. Perceived eye level: elevation jointly determined by visual field pitch, EEPI, and gravity. Invest. Ophthalmol. Visual Sci. Suppl. **27:** 333.

4. MATIN, L. & C. R. FOX. 1989. Visually perceived eye level and perceived elevation of objects: linearly additive influences from visual field pitch and from gravity. Vision Res. **29:** 315–324.

5. MATIN, L. & C. R. FOX. 1990. Visually perceived eye level and perceived elevation of objects: linearly additive influences from visual field pitch and from gravity. Vision Res. **30:** I.

6. MATIN, L. & W. LI. Light and dark adaptation of visually perceived eye level (VPEL) controlled by visual pitch. Percept. Psychophys. (In press.)

7. MATIN, L. & W. LI. 1992. Visually perceived eye level: changes induced by a pitched-from-vertical 2-line visual field. J. Exp. Psychol. Hum. Percept. Perform. **18:** 257–289.

8. MATIN, L. & W. LI. 1989. A single pitched line in darkness controls elevation of visually perceived eye level. Invest. Ophthalmol. Visual Sci. (Suppl.) **30:** 506.

9. MATIN, L., C. R. FOX & Y. DOKTORSKY. 1987. How high is up? Invest. Ophthalmol. Visual Sci. Suppl. **28:** 300.

10. MATIN, L. & W. LI. 1990. Identical effects on perceived eye level by oblique lines in erect planes and pitched-from-vertical lines in pitched planes. Invest. Ophthalmol. Visual Sci. Suppl. **31:** 328.

11. LI, W. & L. MATIN. 1990. Perceived eye level: sensitivity to pitch of a vertical 2-line stimulus grows with eccentricity but is biased by elevation. Invest. Ophthalmol. Visual Sci. Suppl. **31:** 84.

12. MATIN, L. & W. LI. 1989. Linear summation of visual influences on perceived eye level. J. Opt. Soc. Am. Tech. Digest Ser. **18:** 161.

13. MATIN, L. & W. LI. 1991. Separate mechanisms for perceived eye level and perceived vertical: dissection by pitch and roll of a 2-line stimulus. Invest. Ophthalmol. Visual Sci. Suppl. **32:** 900.

14. MATIN, L. & W. LI. 1991. The great circle model of spatial localization and visual perception of elevation. Soc. Neurosci. Abstr. **17:** 848.

15. MATIN, L. & W. LI. 1991. Visually perceived eye level, visually perceived vertical, and the great circle model. Bull. Psychonomic Soc. **29:** 526.

16. LI, W. & L. MATIN. 1991. Spatial summation of influences on visually perceived eye level from a single variably-pitched 1-line stimulus. Invest. Ophthalmol. Visual Sci. Suppl. **32:** 1272.

17. MATIN, L. & W. LI. Bilateral parity violation in visual processing of egocentric spatial localization: with implications for the evolution of partial decussation. (Submitted for publication.)

18. HUBEL, D. H. & T. N. WIESEL. 1974. Uniformity of monkey striate cortex: a parallel relationship between field size, scatter, and magnification factor. J. Comp. Neurol. **158:** 295–306.

19. ASCH, S. E. & H. A. WITKIN. 1948. Studies in space orientation. II. Perception of the upright with displaced visual fields and with body tilted. J. Exp. Psychol. **38:** 455–477.

20. WITKIN, H. A. & S. E. ASCH. 1948. Studies in space orientation. IV. Further experiments on perception of the upright with displaced visual fields. J. Exp. Psychol. **38:** 762–782.

21. WITKIN, H. A. 1949. Perception of body position and of the position of the visual field. Psychol. Monogr. **63**(7): 1–46.

Visual Signals in the Nucleus of the Optic Tract and Their Brain Stem Destinations[a]

ALBERT F. FUCHS, MICHAEL J. MUSTARI,[b]
FARREL R. ROBINSON, AND CHRIS R. S. KANEKO

Department of Physiology and Biophysics
and
Regional Primate Research Center, SJ50[c]
University of Washington
Seattle, Washington 98195

[b]*Department of Anatomy and Neuroscience*
University of Texas Medical Branch at Galveston
200 University Boulevard
Galveston, Texas 77550

INTRODUCTION

Since optokinetic nystagmus (OKN) was introduced as a clinical tool in the early 1900s, its neural substrate has been of intense interest to clinicians and basic scientists alike. Several lines of evidence suggest that the nucleus of the optic tract (NOT) of the pretectum constitutes the afferent limb of the pathway for horizontal OKN in mammals. Damage to the NOT severely impairs horizontal OKN slow phases towards the side of the lesion in both rabbits[1] and monkeys.[2,3] Electrical stimulation in the NOT of both rabbits[4] and monkeys[5] produces nystagmus with an ipsilateral slow phase, whose characteristics resemble those of OKN elicited in lower mammals. Finally, single units in the NOT prefer large-field stimuli moving in the ipsilateral direction. The preferred velocity sensitivities of NOT units increase from rabbits[1] to cats[6] to monkeys,[7] reflecting, in part, the progressively higher ranges of OKN velocity in those species.

Because NOT neurons receive a cortical input that might be influenced by alertness and because the NOT also might be involved in other types of slow eye movements, we recorded NOT unit activity in monkeys that were alert and trained to make eye movements. Since that study was published recently,[8] we will review its results only briefly. Then we will describe our more recent efforts to determine the destination of this NOT information and how it affects certain of its target neurons.

METHODS

Most of our methods have been described extensively in other publications. For the single-unit studies, eye movements were measured by the magnetic search coil

[a]These studies were supported by National Institutes of Health grants RR00166, EY00745, EY07991, and EY06558.
[c]Address for correspondence.

method, unit activity was recorded extracellularly with tungsten microelectrodes, and we correlated neuronal discharge patterns with ongoing eye, target, background, and head movement signals by means of interactive computer programs.[8] Neuronal sensitivity to background stimuli was evaluated by moving a large-field (70 × 50 deg) pattern of randomly sized and spaced dots in a variety of directions and with a variety of velocities on a screen while the monkey fixated a small, stationary (0.25-deg) spot of light. Smooth-pursuit sensitivity was determined by requiring the monkeys to follow the same small target spot while it moved sinusoidally in various directions with different frequencies in the dark. The possible interaction between visual and smooth-pursuit sensitivities was investigated by having the animal track the small spot as it moved across the dotted background. To determine whether neurons had a vestibular sensitivity, we oscillated one animal in the horizontal (yaw) plane at a variety of frequencies either in the dark or when the animal was suppressing the vestibuloocular reflex (VOR) by fixating a target rotating with it.

The locations of responsive neurons were marked by electrolytic lesions, which were later verified in frozen stained sections cut through the brain stem. On some electrode penetrations into the NOT, we passed trains of electrical pulses through the recording electrode to elicit eye movements. Typical stimulus parameters were 40 μA at 250 pulses/second for 10 to 30 seconds.

Connections of the NOT were determined in three monkeys by injecting horseradish peroxidase conjugated with wheat germ agglutinin (WGA-HRP) into areas first shown by unit recording to have neuronal activity characteristic of the NOT. One day after the injections, the monkeys were deeply anesthetized and given a lethal injection. Their brains were removed and the tissue was processed in the usual way, then reacted with tetramethyl benzidine (TMB).[9]

RESULTS

All neurons in the NOT responded either to large-field pattern motion, during smooth pursuit of a small spot in an otherwise dark room, or during both conditions. The range of discharge patterns obtained under these conditions is exemplified by the two representative NOT neurons in FIGURE 1. The neuron to the left was deeply modulated during motion of the background but displayed little, if any, modulation during smooth pursuit. Its modulation, therefore, was clearly a visual response. In contrast, the neuron to the right was well modulated during background movement, but showed an even greater modulation during smooth pursuit. The modulation during smooth pursuit also was of visual origin, however, as the firing rate returned to resting values if the target was briefly extinguished during pursuit, and eye movement continued without a visual target ("off" in second cycle). Consequently, the behavior of all NOT units can be attributed to a visual sensitivity per se. At the dorsal border of the NOT, we often encountered neurons that ceased firing for saccades in all directions but had no visual sensitivity; unlike similar neurons in the nucleus raphe interpositus, these pause neurons ceased firing only *after* saccade onset, and their pause was unrelated to saccade duration.

We expected that the visual NOT cells that were more deeply modulated during background movement might have large receptive fields, whereas those that were better modulated during smooth pursuit might have small receptive fields that included the fovea. The receptive fields, which were plotted in response to small, moving test spots, ranged in size from 2 × 2 deg to the entire tangent screen (70 × 50 deg). All but one of the cells that responded best during smooth pursuit had receptive fields that included only the foveal and perifoveal regions (e.g., the

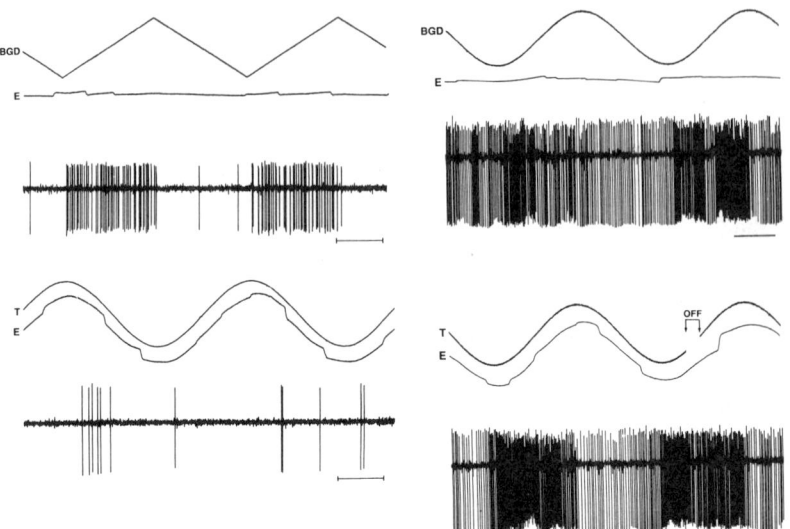

FIGURE 1. Discharge patterns showing the range of NOT neuronal behavior. The left-hand neuron has a strong response to ipsilateral background movement and little response during smooth pursuit, whereas the right-hand neuron responds better during ipsilateral smooth pursuit. The smooth-pursuit response is visual because it disappears when the target is briefly extinguished ("off" in second cycle).

right-hand unit in FIGURE 1). Seventy percent of the neurons with large receptive fields responded better during background movement than during smooth pursuit (e.g., the left-hand neuron in FIGURE 1). All of the units tested for ocular dominance were binocular.

The majority of NOT neurons preferred ipsilateral stimulus motion in the horizontal plane. Most were rather narrowly tuned (FIGURE 2A), whereas some responded well for stimulus directions as much as ±45 deg from their preferred directions (FIGURE 2B). The average width at half-maximum discharge was 127 deg, approximately that of the unit in FIGURE 2A. Although most NOT neurons preferred ipsilateral horizontal stimulus motion, occasional neurons had contralateral or even vertical preferred directions (FIGURE 2C).

The responses of all NOT neurons depended on the velocity of the visual stimulus. The vast majority (90%) of units discharged best for a particular velocity, and their responses declined for both higher and lower speeds (FIGURE 3A). The preferred velocities for these "tuned" neurons ranged from 4 to 180 deg/second (FIGURE 3C). About one-tenth of the neurons exhibited a monotonic increase in firing with velocity to 200 deg/second, the highest speed that we tested. In general, units that preferred the lower stimulus velocities tended to have smaller receptive fields, whereas those with very large receptive fields exhibited a variety of velocity preferences.

The predominantly horizontal preferred directions of NOT neurons and the wide range of their velocity preferences, which span the entire operating range of OKN, suggest that the NOT could play a pivotal role in the horizontal optokinetic response. Both stimulation and lesion studies support this suggestion. If a long train of electrical stimuli was delivered to a region where recording had identified NOT neurons, a brisk nystagmus was generated (FIGURE 4B). As first shown by Schiff *et al.,*[5] the slow-phase velocity was directed ipsilaterally and increased gradually over several seconds to a quasi steady state. After the stimulus was turned off, the eye velocity returned gradually to zero over 15–20 seconds. The time courses of the increase and decrease of electrically elicited slow-phase velocity closely resemble those of the charging of OKN and the discharge of optokinetic after-nystagmus (OKAN) that is elicited by surrounding a monkey with a rotating drum.[10] However, the initial rapid rise in velocity that is produced by drum rotation does not occur in response to electrical stimulation.

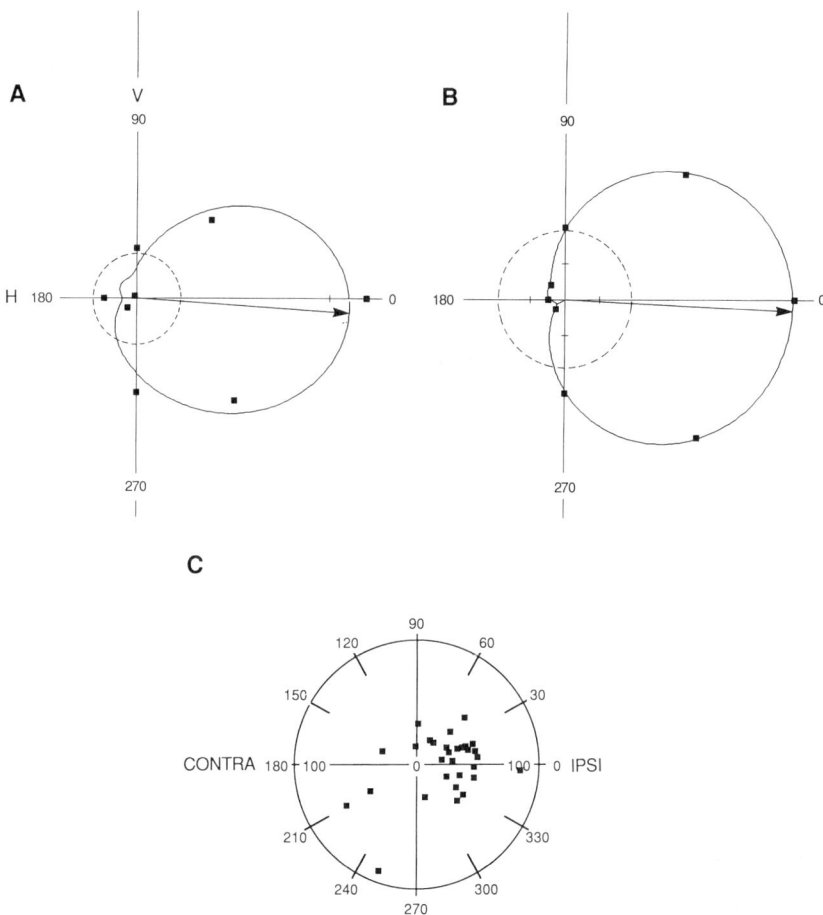

FIGURE 2. Directional tuning of NOT neurons. **A** and **B** show directional tuning curves in which the eight average responses obtained every 45 deg from the horizontal direction have been fitted by a fast Fourier transform algorithm to determine the best preferred directions (arrows) for the two units. Zero and 180 indicate ipsilateral and contralateral horizontal directions, respectively; 90 and 270 designate upward and downward, respectively. **C** shows the preferred directions with their associated average firing rates for all units. (After Mustari and Fuchs.)[8]

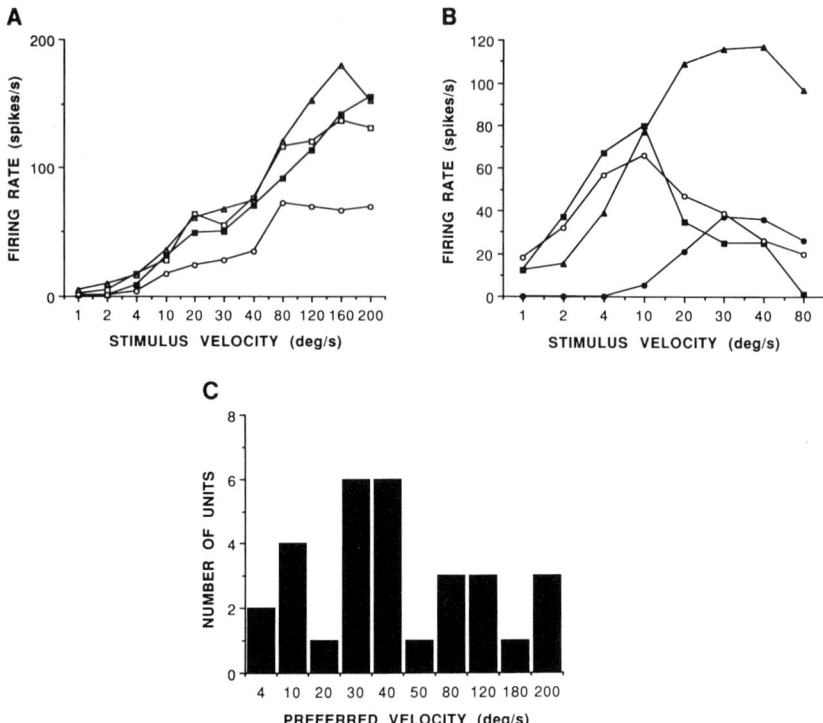

FIGURE 3. Velocity sensitivity of NOT neurons. The average firing rate as a function of stimulus velocity in the preferred direction is shown for neurons whose responses either increase monotonically with velocity to 180 deg/second (A) or are tuned for a particular velocity (B). C shows the distribution of preferred velocities with the three monotonically increasing units appearing at 200 deg/second, the highest velocity tested. (After Mustari and Fuchs.)[8]

The stimulation data suggest that the NOT is part of the so-called indirect pathway that feeds through the velocity storage mechanism but that the NOT is not involved in the rapid component of OKN. The deficits produced by damage of the NOT are consistent with this suggestion. FIGURE 4A shows that one week after kainic acid was delivered to the left NOT, the leftward slow phases of OKN could be elicited only sporadically and OKAN was completely abolished. In contrast, the initial rapid rise of OKN and the initial rapid fall in OKAN appeared to be essentially normal.

A velocity storage mechanism also has been suggested to explain the production of vestibular nystagmus, and some evidence indicates that the visual and vestibular systems share a common storage integrator. As to the question of whether vestibular and optokinetic signals already converge at the NOT, the answer is no. During suppression of the VOR, the representative NOT neuron shown in FIGURE 5A was deeply modulated in phase with peak head velocity. When the lights were extinguished and the animal was oscillated in the dark, the modulation disappeared (FIGURE 5B), indicating that the response was not the result of head rotation per se.

In fact, the monkey was not able to suppress the VOR completely and movement of the fixation spot over the retina elicited a visual response. Indeed, if the spot was extinguished briefly during suppression of the VOR, the modulation always disappeared. As can be seen in FIGURE 5, the two conditions—suppression of the VOR and occurrence of the VOR in the dark—provide a sensitive battery of tests for the kind of information that is relayed through the NOT.

Recently, we have begun to examine the routes by which the exquisite visual information available in the NOT might gain access to the oculomotor system.[11] Injections of WGA-HRP reveal that the NOT has heavy reciprocal connections to the contralateral NOT, the ipsilateral ventral lateral geniculate nucleus, and the ipsilateral peripeduncular nucleus. Strong reciprocal connections exist to the ipsilateral lateral terminal nucleus (LTN), the ipsilateral central mesencephalic reticular formation, and the contralateral Edinger-Westphal nucleus. Heavy efferent projections terminate in the ipsilateral parabigeminal nucleus, the ipsilateral zona incerta, and the ipsilateral thalamic reticular nuclei. With the exception of the projections to the contralateral NOT and LTN, none of these NOT destination sites have, as yet, been implicated in the generation of OKN.

Nuclei that *have* been implicated in the generation of horizontal OKN by studies in either primates or other mammals include the medial vestibular nucleus (MVN) and the nucleus prepositus hypoglossi (NPH) in the medulla and the nucleus reticularis tegmenti pontis (NRTP). In comparison to the NOT projections mentioned in the previous paragraph, those to the ipsilateral NRTP and NPH are rather sparse and those to the ipsilateral MVN are weaker still (FIGURE 6, upper). The magnitude of projections to the dorsal medulla contrasts markedly with the projection to the dorsal cap of Kooy in the inferior olive (FIGURE 6, lower), which was, by

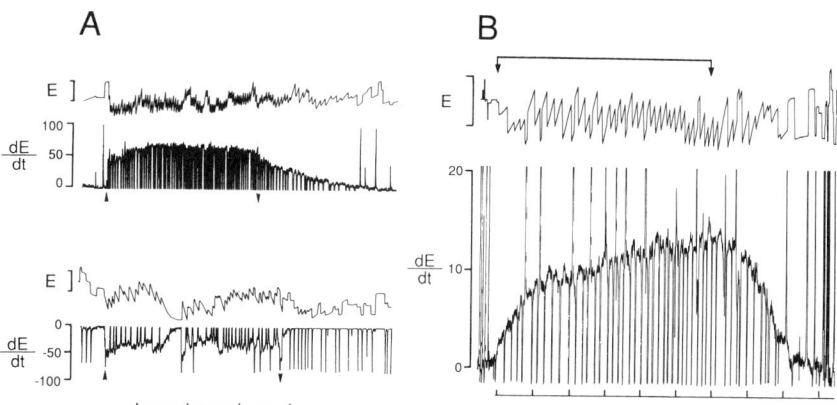

FIGURE 4. The effects on OKN and OKAN of lesions (A) and stimulation (B) in the monkey NOT. *E* and d*E*/d*t* are horizontal eye position and velocity, respectively; calibration for *E* is 30 deg, and each time mark is 10 seconds. Arrows indicate either the interval when the rotating drum was illuminated (A) or the duration of electrical stimulation (B). **A:** For a 90-deg/second drum velocity, slow phases of OKN and OKAN toward the side with the lesion (below) consisted of an initial rapid OKN component, an erratic "steady" OKN with fluctuating velocity, and no OKAN. The slow phases of both OKN and OKAN toward the intact side (above) were normal. (After Kato *et al.*,[2] with permission.) **B:** Electrical stimulation in the NOT produced a nystagmus that resembled OKN without an initial rapid component and an OKAN without an initial rapid fall.

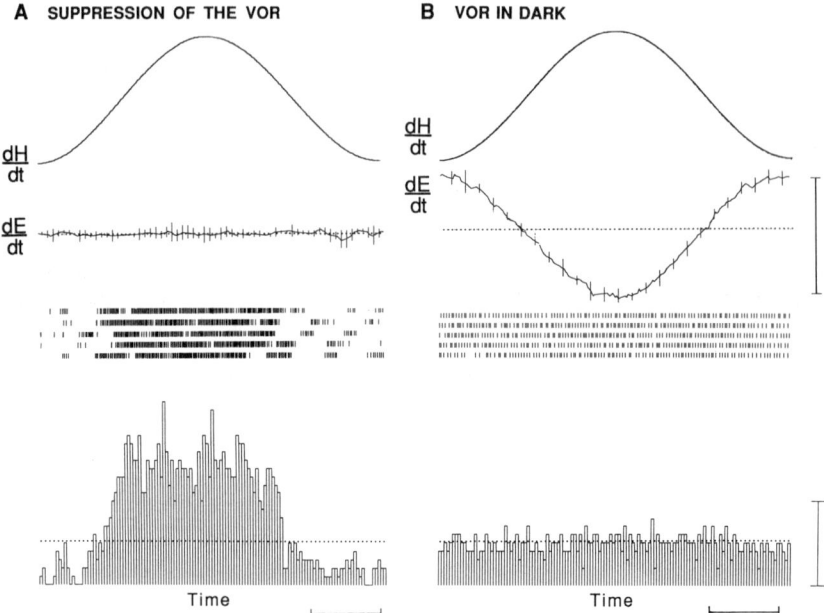

FIGURE 5. Discharge patterns of an NOT neuron during suppression of the VOR (A) and during the VOR in the dark (B). Traces from top to bottom are horizontal head velocity (dH/dt), average "desaccaded" eye velocity (dE/dt) with bars indicating standard deviations, neural response rasters of five individual trials, and frequency histograms of 10-cycle averages; dotted lines on histograms indicate the average resting rate. Sinusoidal head movement is ± 10 deg at 0.4 Hz. Horizontal scale bar = 0.5 second; vertical scale bar = 100 spikes/second or 50 deg/second for dE/dt.

far, the densest of any NOT efferent projection; indeed, it was possible to visualize the termination site in the dorsal cap without any magnification. Since the case shown in FIGURE 6 was deliberately chosen because it had the largest injection site, the relative magnitude of the projections to the dorsal medulla is somewhat disappointing. The ipsilateral dorsal lateral pontine nucleus also receives a strong input from the NOT; however, the dorsal lateral pontine nucleus is believed to project heavily to the cerebellum, which has been implicated only in the rapid component of OKN.[12,13]

We have attempted to confirm the destinations for the flow of information from the NOT by recording single-unit activity in both the NPH and the dorsal cap of Kooy under stimulus conditions that are effective at driving NOT neurons. We tested the possible visual sensitivity of NPH neurons by requiring a monkey to fixate a stationary spot while surrounded by a sinusoidally oscillating OKN drum. Under these conditions, which are very effective at driving NOT neurons, none of the NPH neurons tested showed any response to large-field moving stimuli.

Recently, we recorded from eight neurons in the left dorsal cap of Kooy while the monkey suppressed the VOR, another condition that is very effective at driving NOT neurons (recall FIGURE 5). FIGURE 7 summarizes our preliminary observations. All eight neurons responded during VOR suppression, with the greatest modulation occurring during leftward head velocity; cells no. 2 through 6 are particularly convincing examples. As the rasters demonstrate, the responses varied widely from cycle to cycle, with occasional stimulus cycles eliciting no spikes at all.

Since this monkey suppressed the VOR only poorly, leftward head movement produced a rightward eye movement with an average gain of 0.4. Consequently, the image of the spot was slipping leftward across the retina at a peak velocity of about 12.5 deg/second. Therefore, all of our dorsal cap neurons discharged the greatest number of spikes for ipsilateral image motion, consistent with an input from the ipsilateral NOT (recall FIGURE 6). However, some dorsal cap neurons (e.g., no. 1) continued to discharge weakly during contralateral image motion, a pattern that rarely occurred for NOT neurons. Because dorsal cap neurons have proven difficult to identify and isolate, we do not yet know anything about their velocity sensitivities. Two dorsal cap neurons also were tested when the animal was oscillated in complete darkness. Like NOT neurons, they did not respond (FIGURE 7, cell no. 4, lower histogram), indicating that they too receive no vestibular input.

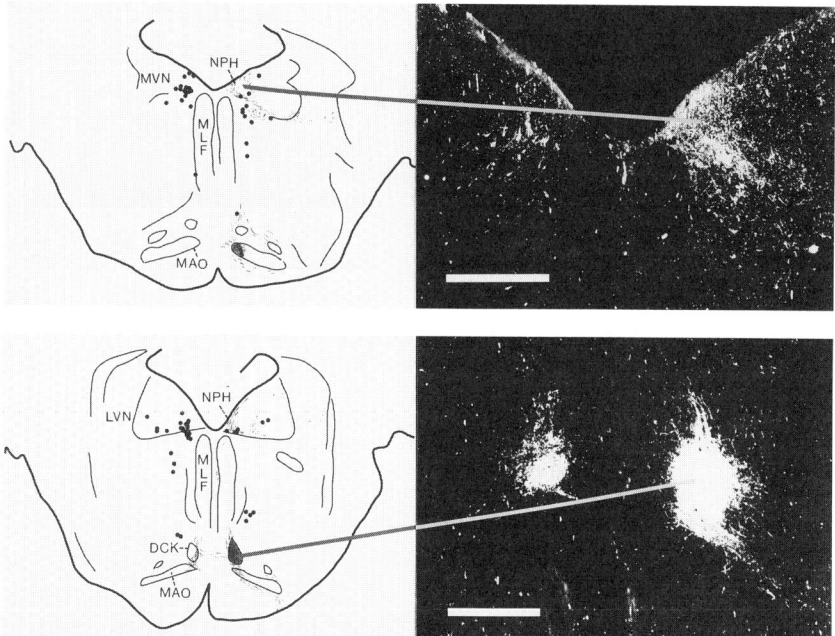

FIGURE 6. Anatomical projections from the right NOT to the medulla as revealed by WGA-HRP injections reacted with TMB. The sections to the left identify the locations of the high-power views of the light-field material to the right in the ipsilateral MVN/NPH region (above) and the ipsilateral and, to a lesser extent, the contralateral dorsal cap of Kooy (DCK) in the inferior olive (below). Other abbreviations: MAO, medial accessory olive; MLF, medial longitudinal fasciculus; LVN, lateral vestibular nucleus. The white bars are 1 mm. (After Mustari *et al.*)[11]

DISCUSSION

Our single-unit data in conjunction with the results of stimulation and lesion studies strongly suggest that the primate NOT, like that of lower species, is the afferent limb of the horizontal OKN pathway. The many NOT neurons with large

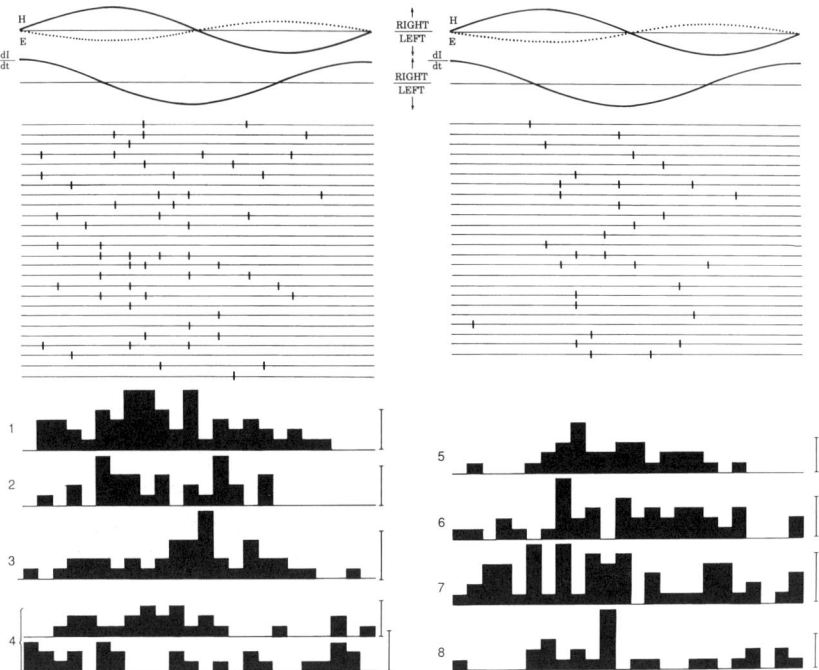

FIGURE 7. Unit activity in the dorsal cap of Kooy in the inferior olive during attempted suppression of the VOR. Traces from top to bottom indicate the schematic head (H) and eye (E) position, the schematic velocity of the visual image on the retina (dI/dt), rasters (spikes assigned to one of 24 bins for each cycle) that contribute to the first histogram below, and histograms for several additional units; all histograms are averages of from 20 to 26 responses and all calibration bars represent 2 spikes/second. The responses of the four cells in the left column were obtained while the oscillating animal attempted to fixate a small spot in an otherwise dark room; the last raster shows the response of cell no. 4 during the same oscillations without a fixation spot in the dark. The data of the four cells to the right were obtained during attempted suppression in a lighted room whose walls were covered with a "busy" design of assorted spots and edges. For all units, the sinusoidal head rotations were ±10 deg at 0.4 to 0.6 Hz.

receptive fields prefer exactly those visual stimulus conditions that are most effective at eliciting OKN. In particular, the total population of NOT neurons responds over the entire range of OKN velocities and therefore, in principle, could contribute to all parts of the optokinetic response. However, the NOT seems to be involved only with the slow component, or charging, of OKN because electrical stimulation of the NOT does not produce the initial rapid rise in nystagmic eye velocity and damage to the

NOT affects neither the initial rapid rise in OKN nor the initial rapid fall in OKAN. The neurons with small perifoveal receptive fields, which discharge during smooth pursuit and apparently are unique to primates, then pose a bit of a mystery. They apparently are not crucial either for smooth pursuit or for the rapid component of OKN, both of which survive NOT lesions. Perhaps they are involved in producing the OKN response to central retinal stimuli, which are so effective at eliciting OKN in human primates.[14]

Because of its clear involvement with the slow component of OKN, the NOT must send its information toward structures that have been implicated in velocity storage, such as the NPH and the vestibular nuclei.[15] We were surprised, therefore, that projections to the NPH, and particularly to the MVN, were relatively sparse when studied with HRP, even though we used the sensitive TMB reaction. In contrast, the nearby dorsal cap of Kooy was very densely labeled, indicating that HRP indeed could be transported over the requisite distance. Of course, it is possible that a terminal field that appears only lightly labeled may indeed have strong influences on its target neurons. However, unlike the neurons in the cat NPH, those in the monkey NPH showed no sensitivity to visual stimuli. Since some neurons in the vestibular nuclei of the monkey apparently have a frank visual response,[16] we next plan to examine those neurons under the stimulus conditions that are effective at driving NOT cells.

Our preliminary data indicate that neurons in the dorsal cap of Kooy, like climbing fibers in lower mammals,[17] respond to visual stimuli. Because the operating range over which they can be modulated is very limited, dorsal cap neurons probably cannot transmit the detailed velocity signals reaching them from the NOT. Instead, our data (FIGURE 7) suggest that, as for other sensory modalities,[18] inferior olive neurons report simply the occurrence of visual motion in a particular direction on the retina but probably provide little information as to its magnitude. Consequently, although connections have been reported from the inferior olive to the vestibular nuclei in lower mammals,[19] it seems unlikely that such a pathway, if it were to exist in the primate, could provide the high-fidelity signals that seem to be required for the generation of OKN.

REFERENCES

1. COLLEWIJN, H. 1975. Direction-selective units in the rabbit's nucleus of the optic tract. Brain Res. **100:** 489–508.
2. KATO, I., K. HARADA, T. HASEGAWA & T. IKARASHI. 1988. Role of the nucleus of the optic tract of monkeys in optokinetic nystagmus and optokinetic after-nystagmus. Brain Res. **474:** 16–26.
3. SCHIFF, D., B. COHEN, J. BÜTTNER-ENNEVER & V. MATSUO. 1990. Effects of lesions of the nucleus of the optic tract on optokinetic nystagmus and after-nystagmus in the monkey. Exp. Brain Res. **79:** 225–239.
4. COLLEWIJN, H. 1975. Oculomotor areas in the rabbit's midbrain and pretectum. J. Neurobiol. **6:** 3–22.
5. SCHIFF, D., B. COHEN & T. RAPHAN. 1988. Stimulation of the nucleus of the optic tract (NOT) induces nystagmus in the monkey. Exp. Brain Res. **70:** 1–14.
6. HOFFMANN, K. P. & A. SCHOPPMANN. 1981. A quantitative analysis of the direction-specific response of neurons in the cat's nucleus of the optic tract. Exp. Brain Res. **51:** 236–246.
7. HOFFMANN, K. P., C. DISTLER, R. G. ERICKSON & W. MADER. 1988. Physiological and anatomical identification of the nucleus of the optic tract and dorsal terminal nucleus of the accessory optic tract in monkeys. Exp. Brain Res. **69:** 635–644.

8. MUSTARI, M. J. & A. F. FUCHS. 1990. Discharge patterns of neurons in the pretectal nucleus of the optic tract (NOT) in the behaving primate. J. Neurophysiol. **64:** 77–90.

9. MESULAM, M.-M. 1978. Tetramethyl benzidine for horseradish peroxidase neurohistochemistry: a non-carcinogenic blue reaction-product with superior sensitivity for visualizing neural afferents and efferents. J. Histochem. Cytochem. **26:** 106–117.

10. COHEN, B., V. MATSUO & T. RAPHAN. 1977. Quantitative analysis of the velocity characteristics of optokinetic nystagmus and optokinetic afternystagmus. J. Physiol. London **270:** 321–344.

11. MUSTARI, M. J., A. F. FUCHS & C. R. S. KANEKO. 1990. Descending connections of the macaque nucleus of the optic tract. Soc. Neurosci. Abstr. **16:** 904.

12. ZEE, D. S., A. YAMAZAKI, P. H. BUTLER & G. GÜCER. 1981. Effects of ablation of flocculus and paraflocculus on eye movements in primate. J. Neurophysiol. **46:** 878–899.

13. WAESPE, W., D. RUDINGER & M. WOLFENSBERGER. 1985. Purkinje cell activity in the flocculus of vestibular neurectomized and normal monkeys during optokinetic nystagmus (OKN) and smooth pursuit eye movements. Exp. Brain Res. **60:** 243–262.

14. DUBOIS, M. F. W. & H. COLLEWIJN. 1979. Optokinetic reactions in man elicited by localized retinal motion stimuli. Vision Res. **19:** 1105–1115.

15. WAESPE, W. & V. HENN. 1987. Gaze stabilization in the primate. The interaction of the vestibulo-ocular reflex, optokinetic nystagmus, and smooth pursuit. Rev. Physiol. Biochem. Pharmacol. **106:** 33–125.

16. BUETTNER, U. W. & U. BÜTTNER. 1979. Vestibular nuclei activity in the alert monkey during suppression of vestibular and optokinetic nystagmus. Exp. Brain Res. **37:** 581–593.

17. MAEKAWA, K. & J. I. SIMPSON. 1973. Climbing fiber responses evoked in the vestibulocerebellum of the rabbit from the visual system. J. Neurophysiol. **36:** 649–666.

18. ROBINSON, F. R., M. O. FRASER, J. R. HOLLERMAN & D. L. TOMKO. 1988. Yaw direction neurons in the cat inferior olive. J. Neurophysiol. **60:** 1739–1752.

19. BALABAN, C. D., Y. KAWAGUCHI & E. WATANABE. 1981. Evidence of a collateralized climbing fiber projection from the inferior olive to the flocculus and vestibular nuclei in rabbits. Neurosci. Lett. **22:** 23–29.

20. FUCHS, A. F. & M. J. MUSTARI. The optokinetic response in primates and its possible neuronal substrate. In Reviews of Oculomotor Research. Visual Motion and Its Role in the Stabilization of Gaze. J. Wallman & F. A. Miles, Eds. **5.** Elsevier. Amsterdam, the Netherlands. (In press.)

The Nucleus of the Optic Tract[a]

Its Function in Gaze Stabilization and Control of Visual-Vestibular Interaction

BERNARD COHEN,[b,c,e] HARVEY REISINE,[b]
JUN-ICHI YOKOTA,[b] AND THEODORE RAPHAN[b,d]

[b]*Department of Neurology*
[c]*Department of Physiology and Biophysics*
Mount Sinai School of Medicine
New York, New York 10029

[d]*Computer and Information Sciences*
Brooklyn College
City University of New York
Brooklyn, New York 11210

INTRODUCTION

In a wide range of vertebrate species the nucleus of the optic tract (NOT) is an important link between the visual and the vestibular and oculomotor systems, contributing to the production of optokinetic nystagmus (OKN).[1-14] Until recently, however, it has been unclear whether NOT is functionally active in the primate. This now seems incontrovertible.[12,13,15-23] In this paper we will review recent research on NOT and consider how processing in NOT might be directed toward controlling eye movements that stabilize gaze, particularly in the primate.

VISUAL PROCESSING

Visual input to NOT is reviewed elsewhere in this volume,[23] and will not be considered in detail here. Several points should be emphasized, however.

In frontal-eyed animals, NOT receives direct input from both the visual cortex and the contralateral retina.[6,12,14,15,23–25] The retinal input in the cat is primarily from the slowly conducting axons[24] of cells morphologically identified as γ cells.[26,27] These cells are known as W cells in the cat, and a similar projection has been identified in the monkey.[13] The projection from X cells is minor.[27] Thus, retinal input to NOT is oriented predominantly toward detection of movement. The underlying influence of the retinal input is seen in the cat during development,[6,28] after visual deprivation,[29,30] or after the visual cortex is removed in cat and monkey.[6,20,31,32] After cortical removal, residual OKN remains, but it is weak. When elicited monocularly, OKN is asymmetric in the lesioned animals, with strong dominance of the contralateral eye, prominent temporal-nasal activation, and low acceleration levels.

[a]Supported by EY02296, EY04148, and Core Center Grant EY01867.
[e]Address correspondence to: Annenberg 21-74, Box 1135, Mount Sinai School of Medicine, 1 East 100th Street, New York, New York 10029.

Despite the direct retinal input, the major input to NOT in the cat comes from visual cortex,[12] via complex cells in Brodmann's areas 17 and 18 (V1 and V2), which are activated directly and indirectly by Y axons from the lateral geniculate nucleus.[25] This converts NOT to a binocular structure in the cat and monkey that detects relative movement of the visual field in the ipsilateral direction. The same is probably true in humans. The input information is directionally selective and head horizontal,[4,6,21,23,33] suggesting processing in planes parallel to the plane of the lateral semicircular canals.[10]

Collewijn's postulate that NOT has an important role in producing optokinetic nystagmus[3] has received wide support.[6,7,17–19,22,23] Hoffmann and colleagues[12,13,24,35] and Mustari and Fuchs[21,23] agree that the cells most widely encountered in NOT of cat and monkey are directionally selective neurons, responding to the retinal slip of large fields moving toward the ipsilateral side. An important characteristic of the NOT cells that project to the brain stem is that they respond over a broad range of stimulus velocities ($\geq 200°$/second),[8,13] with an average preferred velocity in the monkey of $64°$/second.[21,23] Therefore, NOT cells are also capable of sensing retinal slips produced by a wide range of eye velocities, and they have the requisite activity to produce ipsilateral slow phases of OKN.

EYE MOVEMENTS INDUCED BY NOT STIMULATION

In the monkey NOT lies interstitial to the fibers of the brachium of the superior colliculus at the junction between the pulvinar, superior colliculus, and pretectum (FIGURE 1A). The major downstream output from NOT is to the inferior olive on the ipsilateral side,[1,2,14,16,26,33,36–41] although there are also weaker projections to nucleus reticularis tegmenti pontis (NRTP)[5,7,16,33,41] and to the prepositus hypoglossi nuclei (PPH).[7,16,43–45] There is heavy GABAergic innervation of NOT.[40,46–49] The projection neurons from NOT to the inferior olive (IO) are not GABAergic and have relatively few GABAergic terminals, located mainly on the distal dendrites.[48] A large group of GABAergic neurons lies dorsal to the NOT-IO projection neurons,[40] in a region that projects to the NOT on the contralateral side. Their function is unknown.

When NOT was electrically stimulated at the sites shown in FIGURE 1B with monkeys in darkness, nystagmus was evoked, both during and after stimulation (FIGURE 2B).[f] A comparison of the slow-phase velocity of the induced responses with natural OKN and optokinetic after-nystagmus (OKAN) (FIGURE 2A) and with a model simulation of OKN and OKAN (FIGURE 2C; and References 50–52), shows that the induced nystagmus had the characteristics of the component of OKN produced by velocity storage (indirect pathway, FIGURE 2C). This included both the rising time constant or "charge" of the storage mechanism (FIGURE 2D), as well as the falling time constant of the stimulus-induced after-nystagmus, which resembled OKAN (FIGURE 2E). There was an inability to produce significant nystagmus by NOT stimulation in light.[19] This implies that there was visual suppression of velocity storage, a process that commonly occurs when there is reverse retinal slip created by a discrepancy between the direction of the movement of the eyes during the slow phases of nystagmus and the visual surround.[50,51,53,54]

---

[f]Methods used for inducing and recording eye movements in these experiments have been described in detail elsewhere.[19–22,50–52] Eye movements were recorded with a scleral search coil. Eye positions and eye velocities to the right and up are associated with upward trace deflections in the respective traces in the figures.

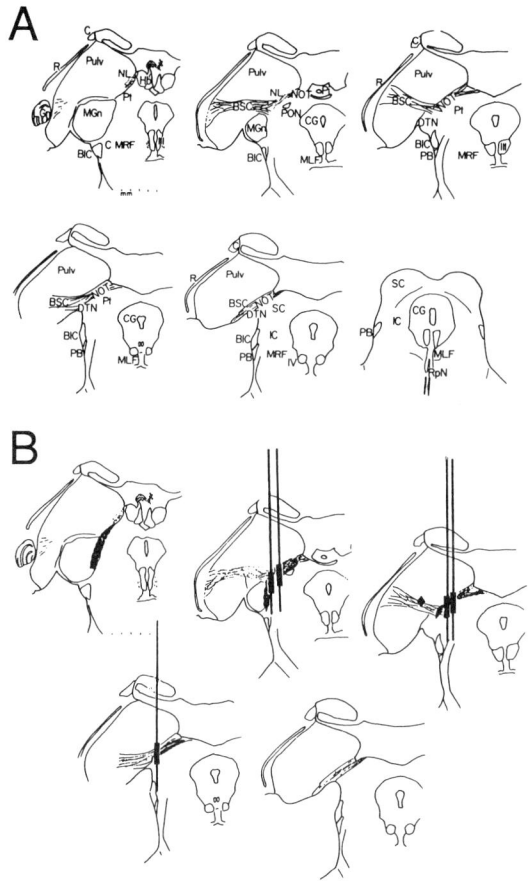

FIGURE 1. A: Upper brain stem, including NOT, DTN, and the pretectum. The sections are 300 μm apart and extend from rostral on the top left to caudal on the bottom right. The tick marks below the upper left section are 1 mm apart. Note location of NOT at the junction between the pulvinar (Pulv) and the pretectum (Pt). **B:** Location of 5 positive stimulation tracks in NOT and DTN in three animals from Reference 19. Regions from which nystagmus could be induced by stimulation are indicated by the solid bars. Positive sites lay within or adjacent to the brachium of the superior colliculus in NOT at the level of and posteriolateral to the pretectal olivary nucleus (PON). The darkened area, which overlies NOT, was a common region destroyed by kainic acid lesions in two animals. In both, the lesions caused a loss of OKAN and the slow component of OKN.[22] The diamond shows a stimulus site located in input fibers to NOT that caused similar nystagmus and after-nystagmus as NOT stimulation.[19] Presumably, it lay in the pathway from the visual cortex to NOT. A lesion at this site caused only transient changes in OKN and OKAN.[22]

Using NOT stimulation, it was possible to reproduce some of the key aspects of visual-vestibular interactions that are attributed to velocity storage, i.e., the ability to maintain eye velocity during constant-velocity rotation and to oppose inappropriate, "anticompensatory" eye velocities at the end of stimulation.[51] Thus, when animals are rotated in light and stopped in darkness, velocity storage maintains slow-phase velocity during rotation, and counters the postrotatory response at the end of stimulation (FIGURE 3B; and Reference 51). The per- and postrotatory response to rotation in darkness is shown in FIGURE 3C. By adding the slow component, elicited by NOT stimulation (FIGURE 3A), to the perrotatory response in darkness (FIGURE 3C), the nystagmus was maintained at a steady-state level during rotation in darkness and the after-response was reduced (FIGURE 3D).

The slow-phase velocities shown in FIGURES 2 and 3 were obtained with the monkeys upright, and were in one dimension, i.e., they were head and space horizontal. When the head is tilted with regard to gravity during OKN and OKAN, the vector of eye rotation tends to align with the spatial vertical according to a right-hand rule.[55–58] This also occurs during postrotatory nystagmus following off-vertical axis rotation (OVAR).[59] The nystagmus induced by stimulation of NOT had the same characteristics. Thus, when the animal was upright, the vector of the

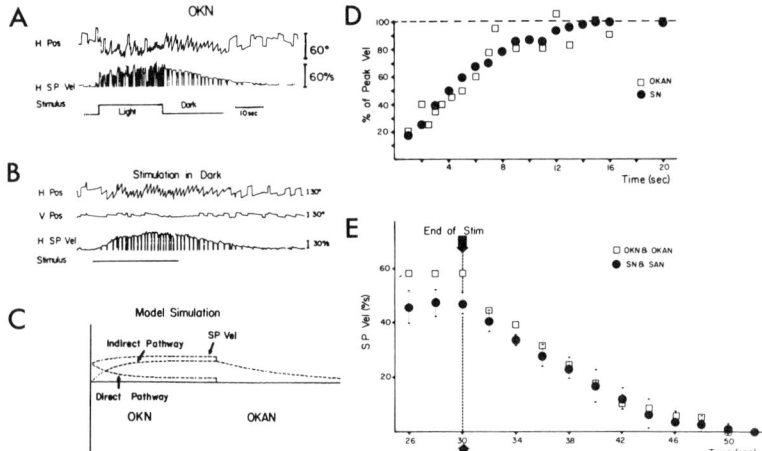

FIGURE 2. A: OKN and OKAN of a normal monkey responding to visual field motion of 60°/second. B: Stimulation of the right NOT of the same monkey in darkness. Eye movements were recorded with a scleral magnetic search coil. The top trace is horizontal eye position (H Pos), the second trace is vertical eye position (V Pos), and the third trace is horizontal slow-phase velocity (H SP Vel). The quick phases have been removed. The period of stimulation is shown by the solid bar under the slow-phase velocity trace. The stimulus frequency was 250 Hz. Movements to the right and up were associated with upward trace deflections in this and subsequent figures. Note that stimulation in darkness caused brisk nystagmus with ipsilateral (right) slow phases. This was followed by after-nystagmus in the same direction. C: Simulation of OKN and OKAN by the model of Raphan *et al.*[19,51] The initial jump in slow-phase velocity is due to activation of the direct pathway. The indirect pathway, which contains the velocity storage integrator, then becomes charged, causing a slow rise to a steady-state level. At the end of stimulation there is a rapid initial fall in slow-phase velocity due to the drop in activity of the direct pathway. This is followed by a slower fall to zero as the velocity storage integrator discharges. D: Comparison of the charge time of OKAN (open squares) with the slow rise in stimulus-induced nystagmus (filled circles). E: Comparison of the falling time course of OKN and OKAN (open squares) with that of the stimulus-induced nystagmus and after-nystagmus (filled circles). The time course and magnitude of the stimulus-induced nystagmus and after-nystagmus were the same as those of the slow component of OKN and of OKAN. (See Reference 19 for details.)

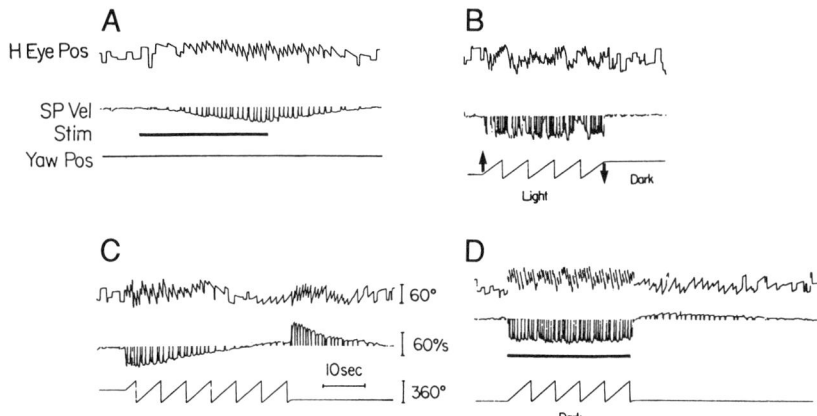

FIGURE 3. A: Nystagmus induced by stimulation at 250 Hz for the period shown by the black bar (Stim) under the eye velocity trace. The top trace in each panel is horizontal eye position (H Eye Pos); the second trace is slow-phase velocity (SP Vel). **B:** Nystagmus induced by rotation in light with a stop in darkness. The third trace is a potentiometer recording of the animal's position about the axis of rotation (Yaw Pos). It reset every 360° during rotation, creating the sawtooth appearance. The up and down arrows show the period in light. **C:** Per- and postrotatory nystagmus induced by rotation at a constant velocity around a vertical axis in darkness. Note the decline of slow-phase velocity to zero during constant-velocity rotation in darkness, and the symmetrical postrotatory response at the end of rotation. **D:** Summation of stimulation-induced nystagmus with perrotatory nystagmus. Activation of NOT caused eye velocity to be maintained for the duration of stimulation and reduced the postrotatory nystagmus. (See Reference 19 for details.)

nystagmus, determined according to the right-hand rule, was aligned along the animal-yaw and the spatial-yaw axis (FIGURE 4A). When the animal was tilted to the left and NOT was stimulated, slow-phase velocity to the left was accompanied by an upward component (FIGURE 4B, arrow), which tended to bring the vector of eye movement to the spatial vertical. When the tilt position was reversed, a downward component of slow-phase velocity did the same (FIGURE 4C, arrow; and Reference 19).

These experiments demonstrate that NOT activates velocity storage in the vestibular nuclei during angular motion in light to stabilize gaze through the slow component of OKN and to counter vestibular after-responses from the semicircular canals at the end of motion. If animals are upright, gaze stabilization is about a spatial vertical axis and the head yaw axis. When the head is tilted, gaze stabilization remains about a spatial vertical axis, maintaining spatial invariance.[58]

NOT LESIONS

While stimulation of NOT activated only the slow component of OKN, unilateral electrolytic lesions of NOT (FIGURE 5A–C) affected both the rapid and slow components. The most prominent effect of NOT lesions was a reduction of the slow component of OKN and of OKAN with ipsilateral slow phases (FIGURE 5E; and References 18 and 22). The loss of steady-state ipsilateral slow-phase velocities of OKN and of OKAN was most prominent at higher stimulus velocities (FIGURE 5G and H). There was also a reduction in the velocity of the initial jump (FIGURE 5F). There was no effect of NOT lesions on either the gain or time constant of the VOR, tested several weeks after lesion. OKN and OKAN were maintained normally to the contralateral side (FIGURE 5D). The loss of the slow component of OKN and of OKAN in only one direction after a unilateral NOT lesion suggests that the weak reciprocal activity in the intact NOT associated with field motion in the nonpreferred (contralateral) direction[13,21,23] is not adequate to maintain contralateral slow phases of OKN and OKAN.

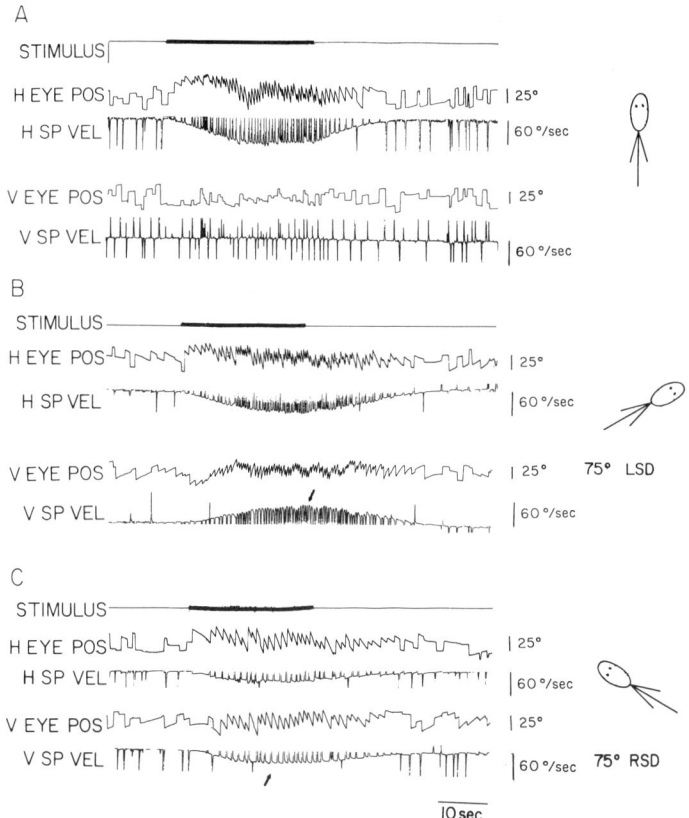

FIGURE 4. Effect of head position with regard to gravity on stimulus-induced nystagmus (200 Hz, 40 μA). The traces are, from top to bottom, period of stimulation, horizontal eye position (H EYE POS), horizontal slow-phase eye velocity (H SP VEL), vertical eye position (V EYE POS), and vertical slow-phase eye velocity (V SP VEL). The stick figures on the side of each panel show the position of the animal while receiving left NOT stimulation. The experiment was conducted in darkness. **A:** With the animal upright, stimulation induced pure leftward slow-phase velocity, and there was no vertical nystagmus. **B:** Stimulation with the animal tilted 75° left-side down (LSD) caused slow phases up and to the left. **C:** In the 75° right-side-down position (RSD), the induced slow phases were to the left and down. The velocity of the vertical components is marked with arrows in B and C. In both, the vertical components tended to bring the induced nystagmus toward a spatial horizontal plane (From Reference 19 with permission.)

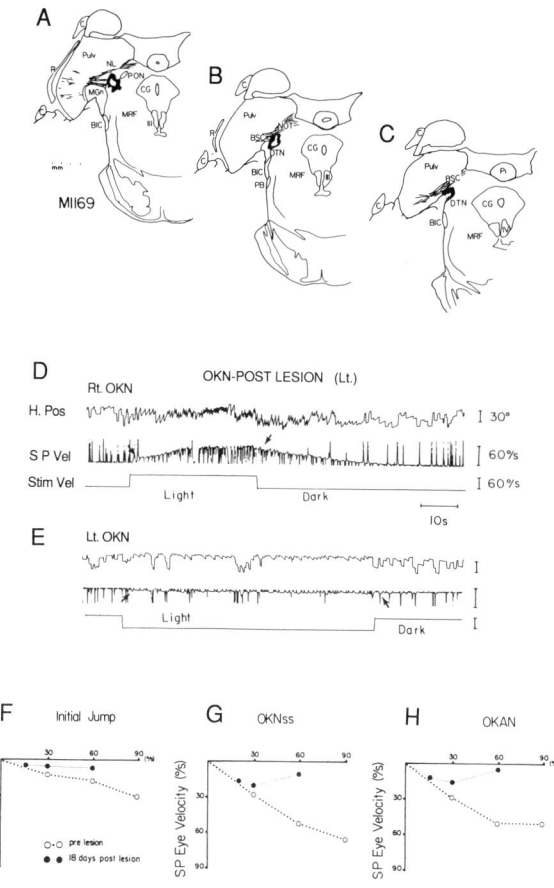

FIGURE 5. A–C: Diagrams of an electrolytic lesion of NOT and DTN in M1169. The sections were separated by approximately 300 μm. The lesion was approximately 1 mm in diameter. It lay in the caudal, lateral portion of NOT, in and under the brachium of the superior colliculus (BSC), ventral and lateral to the pretectal olivary nucleus (PON, A) and extended into DTN (C). **D** and **E:** OKN in response to movement of the visual surround at 60°/second 18 days after the left-sided NOT lesion shown in A–C. Arrows show the onset and end of optokinetic stimulation. OKAN and OKAN with contralateral slow phases were intact (D). The slow rise in OKN with ipsilateral (left) slow phases was lost (E), and there was little or no OKAN. Peak OKN velocities were less than 10°/second toward the lesioned side (E). **F–H:** Graphs of the initial jump (F) and steady-state velocity of OKN (G), and the OKAN peak velocity (H). Stimulus velocity is shown on the abscissa and slow-phase eye velocity on the ordinate. Open circles are prelesion values. The filled symbols are results of testing 18 days after lesion. Note the loss of peak steady state OKN and OKAN slow-phase velocities, which fell from 70°/second to 25°/second for OKN (G) and from 55°/s to 15°/s for OKAN (H). The initial jump was also affected in this animal (F). (Adapted from Reference 22.)

   In another animal, NOT was destroyed as a result of the spread of kainic acid into
the pretectum, from an injection into the mesencephalic reticular formation. Both
lateral canals had previously been plugged in this animal. The canal plugging had a
twofold effect on the OKN, presumably due to the loss of the horizontal VOR: there
was enhancement of the initial jump in slow-phase velocity and habituation of
OKAN (FIGURE 6A; and Reference 60). After the left NOT was destroyed in this
animal, slow-phase velocities of OKAN and the slow component of OKN were lost to
the left (FIGURE 6B). The initial jump of the OKN was also reduced to the ipsilateral

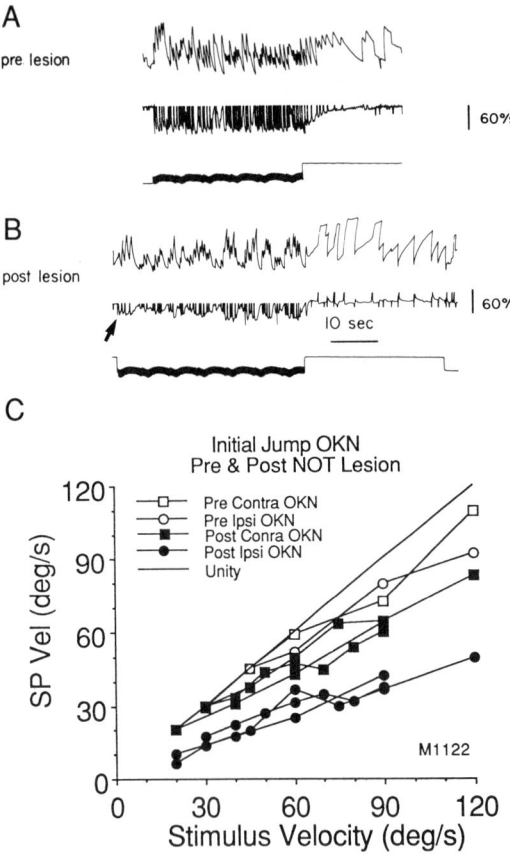

FIGURE 6. OKN of M1122 before (A) and after (B) a kainic acid lesion of the left MRF and
NOT (See Reference 22 for details.) **A:** Before lesion there was a large initial rapid rise in OKN,
which was maintained for the duration of stimulation. OKN was followed by OKAN. **B:** After
lesion the rapid rise in slow-phase velocity at the onset of stimulation was reduced (arrow), and
OKN velocities were not well maintained. There was no OKAN at the end of stimulation, the
OKN being replaced by spontaneous nystagmus in the opposite direction. **C:** Graph of the
initial jump of OKN to the ipsilateral (circles) and contralateral (squares) sides, before (open
symbols) and after (filled symbols) lesion. The results of several postlesion test sessions are
shown. After lesion there was a reduction in the initial jump to the ipsilateral side at all
velocities (filled circles). The contralateral initial jump was not much affected (filled squares).

(left) side (FIGURE 6B, arrow; C, filled circles). The reduction in the initial jump was to a level that was in the range of the initial jump in a large series of normal animals. The latter finding raises the possibility that adaptation of the direct visual pathway, secondary to the canal plugging, was lost as a result of the NOT lesion in this animal. Regardless, although the effect of NOT lesions on the direct visual pathway still must be elucidated, these studies support the conclusion that activity responsible for driving velocity storage from the visual system is mediated through NOT.

MUSCIMOL INJECTIONS INTO NOT

There is strong evidence that GABA (γ-aminobutyric acid) is utilized as an inhibitory transmitter in NOT.[40,46-49] Consequently, it was of interest to activate GABAergic receptors in NOT and study behavioral changes.

Muscimol, a GABA$_A$ agonist, was injected into the right NOT, at a site where stimulation induced typical nystagmus with ipsilateral (right) slow phases (FIGURE 7A).[g] After muscimol injection, there was spontaneous nystagmus with contralateral (left) slow phases (FIGURE 7B, arrow). There was also a reduction in slow-phase velocities of right OKN (FIGURE 7C, down arrow), and OKAN was absent in that direction (FIGURE 7C, up arrow). These effects would have been predicted from stimulation and lesions, presumably because muscimol inactivated a pathway through NOT, sensitive to GABA,[48] that drove velocity storage. Frequently, the deficit in ipsilateral OKN became greater as time after injection increased, and in some instances OKN and OKAN completely disappeared.

Surprisingly, muscimol injections into NOT also affected vestibular nystagmus (FIGURES 8 and 9). After an injection into the left NOT the envelope of ipsilateral (left) slow-phase velocity at the end of angular rotation (FIGURE 8C) was similar to that before injection (FIGURE 8A and B), and the peak velocities and time constants of the left slow phases before and after injection were approximately equal (FIGURE 9A and B, open and filled circles). In contrast, there were striking changes in contralateral (right) vestibular nystagmus. While the peak velocities were unaffected (FIGURE 8A–C; FIGURE 9A, open and filled squares), the time course became very long (FIGURE 8C, right slow-phase velocity), and all previous habituation of the contralateral VOR was lost. [Compare FIGURE 8C with FIGURE 8A and B; also FIGURE 9B, open squares (preinjection) and filled squares (postinjection).] OKAN habituation was also lost (FIGURE 8D). In accord with the increased time constant and unchanged gain, there was a considerable overshoot in the eye velocity response associated with the plateau,[51] 4–8 seconds after the onset of rotation (FIGURE 8C, arrow). Similar overshoots have been observed after adaptation of the gain of the VOR in unhabituated monkeys with long vestibular and OKAN time constants.[61]

The animals also lost the ability for rapid suppression of the storage component of the vestibular response to the contralateral side (FIGURE 8F, right slow-phase velocities). Therefore, the time constant of vestibular nystagmus, induced by a step of velocity in a lighted stationary surround, was substantially longer to the right (contralateral) than to the left (ipsilateral) side (FIGURE 9D, filled vs. open squares).

---

[g]The injection technique was similar to that utilized in anatomic studies of cMRF.[34,42] The region of interest was first identified by microstimulation. Then the stimulating microelectrode was withdrawn and replaced by a fine micropipette with a blunt tip, and small amounts (≤ 1 μl) of a 1 μg/μl solution of muscimol were injected over several minutes.

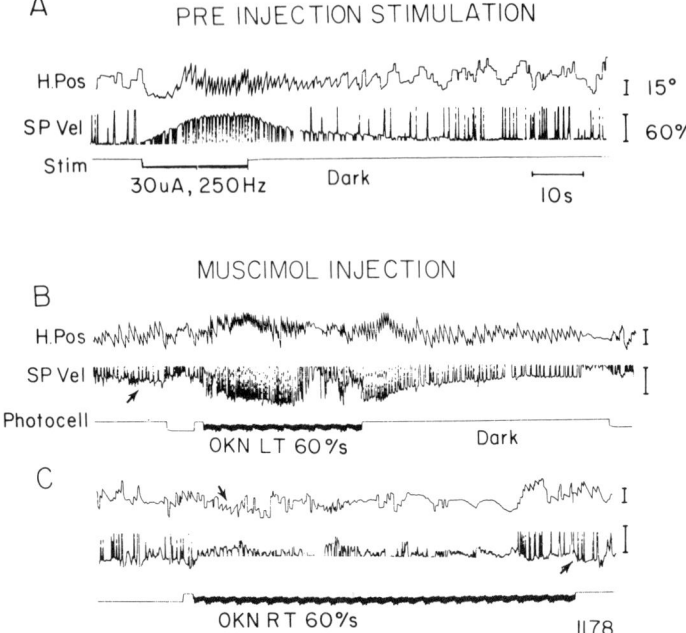

FIGURE 7. Effect of muscimol (0.5 μl of 1 μg/μl solution) injected over 1 minute into right NOT. **A:** Characteristic nystagmus with slow phases to the right induced by stimulation at site of injection in right NOT, with a slow rise to a steady-state level, followed by stimulus after-nystagmus. **B** and **C:** After muscimol there was spontaneous nystagmus with slow-phase velocity to the left (B, upward arrow). OKN with leftward slow phases could be induced by surround movement to the left (B), and was followed by OKAN. **C:** OKN could be induced to the right (down arrow), but it was irregular. At the end of stimulation, there was no OKAN (up arrow).

There was a variable effect of NOT injections on the initial jump in velocity at the onset or end of rotation in a relative stationary surround, i.e., in a conflict situation.[h] Normal animals will reduce the gain of the initial step by 40–60% if they are rotated in a relative stationary surround in light (FIGURE 8E; and Reference 54). Some injections caused a maximal effect, as in FIGURE 8F: visual suppression of the initial jump was lost, and there was little reduction in the initial jump in eye velocity at the onset of rotation. In other experiments the loss of suppression was present only at higher velocities (FIGURE 9C, filled squares).

Thus, after muscimol, animals lost previous habituation of nystagmus velocities to the contralateral side. They also lost the ability to suppress the slow component of the contralateral VOR through vision and reverse retinal slip. Finally, they frequently could not suppress the step in eye velocity to the contralateral side at the onset or end of rotation. These findings disappeared 4–6 hours after the muscimol

---

[h] A relative stationary surround is one that moves with the head. Experimentally, this was produced by rotating the optokinetic drum that surrounded the animal and the rate table on which it sat at the same velocity. Thus, during rotation in light the vestibular system signaled motion, but the visual system signaled that the subject was stationary, hence the conflict.

injections. Therefore, they were an acute effect of the muscimol inactivation of cells in and around NOT.

DISCUSSION

The experimental studies reviewed here show that NOT is an important link between the visual and vestibular systems in the monkey, carrying activity related to retinal slip that is responsible for producing slow changes in eye velocity during OKN. The stimulation and lesion studies demonstrate that it does this by producing

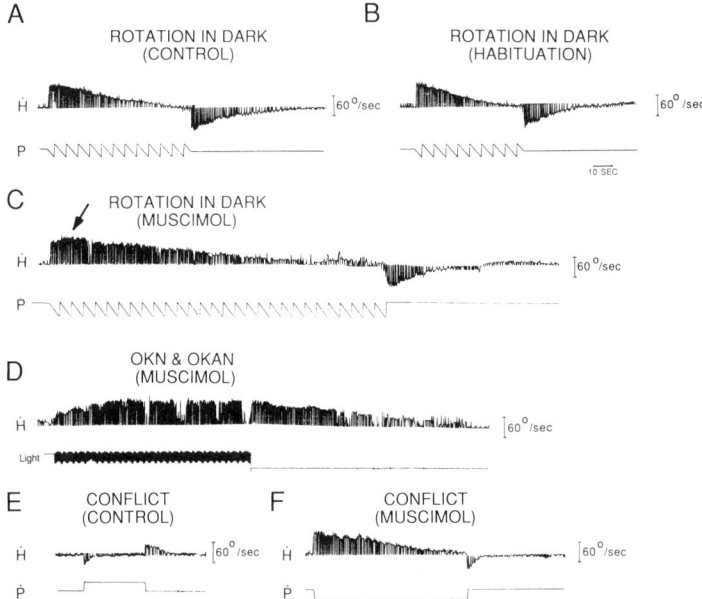

FIGURE 8. Effect of muscimol injected into left NOT on the velocity of vestibular nystagmus (first trace, P). The angular velocity of the animal in A– C and E, F, and of the visual surround in D, was 60°/second. In A–C the animal's position around the yaw axis (P) is shown in the second trace. In D the second trace is a photocell recording of light and the passage of 10° stripes of the OKN stimulus. The second trace in E and F shows yaw axis velocity (P̊). A–D were recorded in darkness, and E and F in a lighted subject-stationary surround (conflict; See Footnote g for explanation). **A:** Horizontal slow-phase velocity at the beginning and end of constant-velocity rotation. **B:** The time constant of the nystagmus, but not the gain were reduced by repeated testing during this test session (habituation). **C:** After injection habituation was maintained to the left, but was lost to the right. There was an overshoot in the plateau phase (arrow), 4–8 seconds after the onset of rotation, at a time when the velocity storage integrator would have been maximally charged. **D:** OKAN also had a long time course after muscimol. **E:** When the animal was rotated in a subject stationary surround before injection (conflict, control), the initial jump in velocity at the onset and end of rotation was small, and the time constant was short. After muscimol (conflict, muscimol) the animal could not suppress the initial jump of nystagmus to the right, and the time constant of the slow-phase velocity was substantially longer than to the left. Suppression of left vestibular nystagmus was unaffected by the left NOT muscimol injection.

activation of velocity storage in the vestibular system. The muscimol injection studies demonstrate that NOT is also responsible for carrying information that controls the time constant of velocity storage. Since NOT is closely linked to velocity storage, we will consider its function primarily in terms of how it contributes to visual-vestibular interaction, through its input and control of velocity storage.

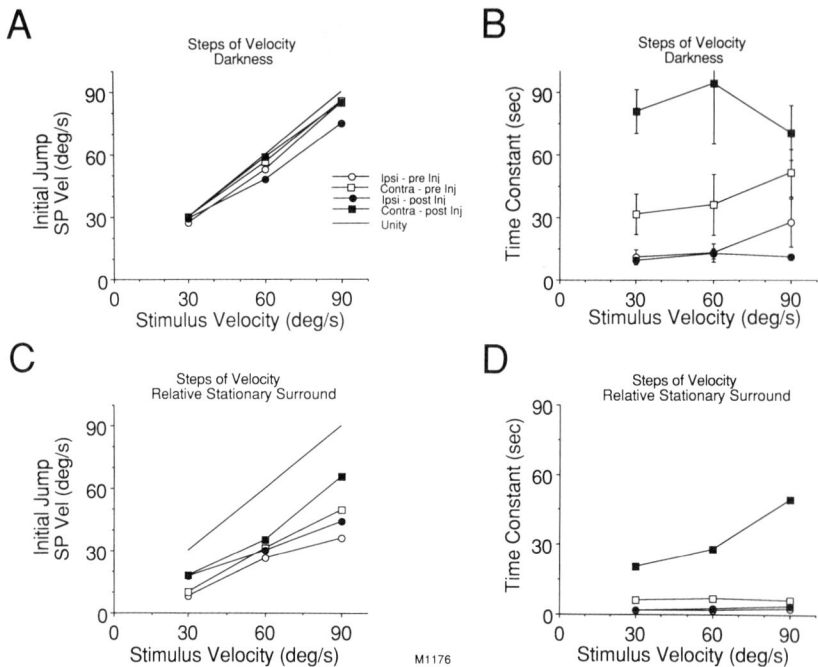

FIGURE 9. A and **B:** Graphs of effects of muscimol injection into NOT on the peak slow-phase velocity (A and C) and time constant of per- and postrotatory nystagmus (B and D), induced in darkness (A and B) and in a relative stationary surround (C and D). **A:** Slow-phase velocities were little affected by muscimol injection in NOT. There was a slight decrease in ipsilateral peak velocities induced by rotation at 90°/second. **B:** The time constant of decline of the ipsilateral slow-phase velocities during velocity steps was unaffected (open and filled circles). There was a striking increase in the time constant of the decline in slow-phase velocity to the contralateral side [compare values before (open squares) and after injection (filled squares)]. **C:** In a relative stationary surround, suppression of peak contralateral slow phase velocities was less at 90°/second. **D:** Suppression of contralateral slow-phase velocities was substantially poorer in a lighted subject stationary surround after injection, causing an increase in the time constant of per- and postrotatory nystagmus after injection (filled squares). Suppression of ipsilateral slow-phase velocities was unaffected by muscimol (filled circles).

Velocity storage is likely to be generated in the vestibular nuclei. It disappears after midline lesions of commissural fibers in the rostral medulla,[62–64] and can be evoked by electrical stimulation of the vestibular nuclei, at sites where neural activity related to velocity storage is encountered.[65] Two types of cells in the vestibular nuclei have activity whose dynamics are related to those of the slow component of OKN and of OKAN. These are the "vestibular-only" cells, which have no eye movement related activity, and the "vestibular plus saccade" cells.[66-68] The activity of the vestibular-only neurons, demonstrated during OKN and OKAN by Waespe and Henn,[69,70] is similar to the slow-phase velocity envelope of nystagmus induced by NOT stimulation (FIGURE 3A). Activity in the vestibular nuclei that was responsible for the type of visual-vestibular interaction shown in FIGURE 3D may have originated in NOT.

There are a number of gaps in our knowledge about NOT and its relation to velocity storage and the vestibular nuclei. As yet there has been no anatomical or physiological demonstration of a direct NOT-vestibular projection. Pathways have been demonstrated from NOT to NRTP, PPH, or the inferior olive,[1,2,5,7,10,14,15,16,26,33,39,40,43-45,47] but not to the vestibular nuclei. NRTP projects primarily to the cerebellum, and aside from a report of an NRTP-vestibular connection,[71] it is generally assumed that the vestibular nuclei do not receive input from NRTP (See Reference 41 for review). Specifically, the cellular activity in NRTP seems inappropriate for driving OKN.[72,73] Activity responsible for OKN or OKAN is not likely to be carried in the NOT-IO pathway, since the firing frequencies of IO neurons are too low[21] to produce the slow phase velocities of 60°/second–90°/second commonly encountered during OKN.[50] It has been proposed that the NOT-vestibular pathway synapses in PPH in the rat[5,7,43] and cat,[44] but a similar projection has not been explored in the monkey, and storage-related neural activity has not been found in PPH (see References 45 and 64 for review). Thus, while it is certain that a major target of NOT is the vestibular nuclei, the mode of transmission of activity from NOT to the vestibular nuclei in the primate is still uncertain.

Eye movements induced by NOT stimulation do not have a rapid component,[19,21] and OKAN and the slow component of OKN were mainly affected when NOT was lesioned.[18,22] This suggests that NOT primarily lies in the indirect pathway. Support for this idea comes from the finding that there is a characteristic nonlinearity in the firing of NOT neurons,[4,12,21] similar to the behavior of eye velocity in response to retinal slip, after flocculectomy has eliminated the direct pathway.[54] Increases in retinal slip past a threshold cause a decrement in the firing rates of NOT neurons, not an increment.[4,12,21] Similar shapes for the eye velocity nonlinearity were also found after the direct pathway was eliminated by occipital lobectomy.[32] The average peak firing rate of NOT neurons was $\approx$ 64°/second in the monkey,[21,23] and the turnaround point in the nonlinearity for the neural activity of many NOT neurons was in the range of 200°/second in both the cat and monkey.[12,21] This is much higher than would be expected from modeling eye velocity in the indirect pathway.[54] It implies that NOT may have some input or control over the direct pathway. This is consistent with the findings that the rapid component of OKN was smaller after NOT lesions (FIGURES 5F and 6C; and Reference 22). How NOT contributes to the rapid component is still unclear.

With regard to the initial question of whether NOT is functionally active in the primate, from our data and those presented by Fuchs *et al.*[21,23] and Kato *et al.*,[17,18] the postulate of Collewijn that NOT contributes heavily to production of OKN[4] is strongly supported in the monkey. From a functional point of view, full-field movement, which is the most effective way to induce OKN and OKAN,[50,74,75] occurs every time the head moves, if the gain of compensatory eye movements produced by the VOR is not unity. Therefore, OKN can be considered as the visual concomitant of angular head movement for stabilization of gaze in a lighted stationary environment, particularly in instances where the gain of the VOR is not unity, or where it is inadequate for stabilizing gaze because of the added demands of translation or vergence.[76-78] The data strongly suggest that NOT is one of the primary nuclei through which the retinal slip signal that generates the visual contribution to the VOR is processed in the monkey, and in man, as in other vertebrate species.[3,4,7,9,10-14]

Velocity storage probably makes little contribution to the VOR during normal gaze shifts produced by head-on-body movement. It is activated by frequencies of rotation below 0.1 Hz,[51,79] whereas most head-on-body movements occur in a range from 0.1 to 5–8 Hz.[76,80-82] However, there can be prolonged angular body movement during locomotion in light. This produces retinal slip that generates head and eye nystagmus and stabilizes gaze in space.[78] The compensatory gaze nystagmus is produced in part by activation of velocity storage through the visual, vestibular, and somatosensory systems. Thus, there can be compensatory gaze nystagmus and gaze stabilization relative to the surround through velocity storage even when monkeys run on an angular path in complete darkness.[83] Locomotion in light that produced retinal slip and OKN would generate excitation of velocity storage and gaze stabilization through NOT.

From these and other findings we have postulated that a major function of velocity storage is to stabilize gaze in space during passive motion or locomotion with an angular component.[51,83] It does this by producing low-frequency slow-phase head and eye movements through the vestibular system. The same would occur during locomotion along a straight path, when the head is turned to the side. Most previous work on NOT has considered the ocular response to visual surround movement in one dimension with the head fixed. However, the finding that velocity storage activates the vestibulocollic as well as the vestibuloocular reflex during circular locomotion[83] implies that NOT-vestibular connections must activate vestibulospinal pathways that move the head on the body, as well as vestibuloocular pathways.

As noted, velocity storage is determined by parameters that code the spatial vertical.[55-59] Moreover, the relationship of the yaw axis eigenvector of velocity storage to gravity can be characterized by the same Müller and Aubert effects[57-59,84] that characterize the percept of the spatial vertical.[85-87] Thus, NOT could play an important role in stabilizing gaze and posture in three dimensions, while one actively moves in light. This would be consistent with the data shown in FIGURE 4, which indicate that ocular compensation induced by activation of NOT was about an axis close to the spatial vertical, regardless of how the head or head and body were tilted.

In addition to exciting velocity storage, NOT appears to have an important role in controlling its characteristics. When activity in NOT was inhibited by the GABA agonist muscimol, all previous habituation of the VOR and of OKAN was lost (FIGURE 8). The NOT-olivary pathway is the major input from the visual system to the dorsal cap, carrying information about animal-horizontal surround motion; it dwarfs all other downstream visual projections to IO (See Simpson[10,14] for review). The visual input from the inferior olive is distributed by cells that have branching axons over climbing fibers to the nodulus-uvula and flocculus-paraflocculus of the vestibular-cerebellum.[88] Thus, it may be expected that NOT might participate in functions of both the flocculus/paraflocculus and nodulus/uvula.

It is known that habituation of the VOR and of OKAN disappears after nodulouvulectomy.[61,89] We assume that muscimol injections into NOT enhanced GABAergic inhibition and inactivated projection pathways from NOT to the brain stem. From this we speculate that the loss of habituation after muscimol was due to a functional nodulouvulectomy, secondary to inactivation of climbing fibers originating in the dorsal cap of Kooy and of mossy fibers entering the cerebellum through NRTP. The inability to suppress slow-phase eye velocity rapidly during the "storage" component of the response, which was lost after the NOT injections (FIGURE 8F), is also a function of the nodulus and ventral uvula, since it is lost after nodulouvulectomy.[89]

The flocculus and paraflocculus, activated by a pathway that originates in MT and MST[90,91] through the dorsolateral pontine nucleus,[92] is responsible for producing the rapid component of OKN, especially the initial jump in slow-phase eye velocity.[54,89,93,94] Suppression of the rapid component of the VOR is lost when the flocculus is lesioned.[53,54] The inability to suppress the initial jump in velocity can be viewed as dysfacilitation of the NOT-NRTP pathway carrying mossy fiber information and/or to inactivation of the NOT-IO pathway carrying climbing fiber activity to the flocculus and paraflocculus.

In summary, we postulate that muscimol caused a depression of NOT neurons that project to the dorsal cap of the inferior olive, dysfacilitating the climbing fiber input to the nodulus and flocculus, as well as mossy fiber input to these areas through NRTP. As the NOT-NRTP projection is small, and the NOT-olivary projection is massive, we assume that nodular and floccular inactivation was primarily due to dysfacilitation of climbing fibers. These results imply that the range of stimulus velocities that are processed in some neurons in NOT is broad because it must detect whether the gain of compensatory eye and head movements is appropriate for stabilizing visual targets during head and body movement.

An interesting aspect of the loss of habituation was its prompt return several hours later as the animals metabolized the muscimol. Habituation is a semipermanent phenomenon, lasting for months to years, and in some instances it can be permanent (See Reference 61 for review). Yet it was turned on and off simply by inactivating NOT. This indicates that the climbing fiber system and the nodulus and ventral uvula must be continuously active to maintain habituation.

Finally, the profound effects of inhibition of a visual structure, lying far in the central nervous system from the vestibular system on vestibular nystagmus, is of considerable interest clinically. It implies that vertigo, dizziness, a loss of gain adaptation, and a loss of habituation can occur on a transient or permanent basis from cerebral or brain stem lesions that do not directly involve vestibular structures.

SUMMARY

1. Electrical stimulation of the nucleus of the optic tract (NOT) induced nystagmus and after-nystagmus with ipsilateral slow phases. The velocity characteristics of the nystagmus were similar to those of the slow component of optokinetic nystagmus (OKN) and to optokinetic after-nystagmus (OKAN), both of which are produced by velocity storage in the vestibular system. When NOT was destroyed, these components disappeared. This indicates that velocity storage is activated from the visual system through NOT.

2. Velocity storage produces compensatory eye-in-head and head-on-body movements through the vestibular system. The association of NOT with velocity storage implies that NOT helps stabilize gaze in space during both passive motion and active locomotion in light with an angular component. It has been suggested that "vestibular-only" neurons in the vestibular nuclei play an important role in generation of velocity storage. Similarities between the rise and fall times of eye velocity during OKN and OKAN to firing rates of vestibular-only neurons suggest that these cells may receive their visual input through NOT.

3. One NOT was injected with muscimol, a GABA$_A$ agonist. *Ipsilateral* OKN and OKAN were lost, suggesting that GABA, which is an inhibitory transmitter in NOT, acts on projection pathways to the brain stem. A striking finding was that visual

suppression and habituation of *contralateral* slow phases of vestibular nystagmus were also abolished after muscimol injection. The latter implies that NOT plays an important role in producing visual suppression of the VOR and habituating its time constant.

　　4. Habituation is lost after nodulus and uvula lesions and visual suppression after lesions of the flocculus and paraflocculus. We postulate that the disappearance of vestibular habituation and of visual suppression of vestibular responses after muscimol injections was due to dysfacilitation of the prominent NOT-inferior olive pathway, inactivating climbing fibers from the dorsal cap to nodulouvular and flocculoparafloccular Purkinje cells. The prompt loss of habituation when NOT was inactivated, and its return when the GABAergic inhibition dissipated, suggests that although VOR habituation can be relatively permanent, it must be maintained continuously by activity of the vestibulocerebellum.

ACKNOWLEDGMENTS

We thank Dr. Martin Gizzi for critical comments.

REFERENCES

1.　MAEKAWA, K. & J. I. SIMPSON. 1973. Climbing fiber responses evoked in the vestibulo-cerebellum of rabbit from visual system. J. Neurophysiol. **36:** 649–666.
2.　MAEKAWA, K. & T. TAKEDA. 1979. Origin of descending afferents to the rostral part of the dorsal cap of inferior olive which transfers contralateral optic activities to the flocculus. An HRP study. Brain Res. **172:** 393–405.
3.　COLLEWIJN, H. 1975. Oculomotor areas in the rabbit's midbrain and pretectum. J. Neurobiol. **6:** 3–22.
4.　COLLEWIJN, H. 1975. Direction selective units in the rabbit's nucleus of the optic tract. Brain Res. **100:** 489–508.
5.　CAZIN, L., W. PRECHT & J. LANNOU. 1980. Firing characteristics of neurons mediating optokinetic responses to rat's vestibular neurons. Pfleugers Arch. **386:** 221–230.
6.　HOFFMANN, K. P. 1982. Cortical versus subcortical contributions to the optokinetic reflex in the cat. *In* Functional Basis of Ocular Motility Disorders. G. Lennerstrand, D. S. Zee & E. L. Keller, Eds.: 303–310. Pergamon Press. Oxford & New York.
7.　PRECHT, W., L. CAZIN, R. BLANKS & J. LANNOU. 1982. Anatomy and physiology of the optokinetic pathways to the vestibular nuclei in the rat. *In* Physiological and Pathological Aspects of Eye Movements. A. Roucoux & M. Crommelinck, Eds.: 153–172. Dr. Junk. The Hague, the Netherlands.
8.　MAEKAWA, K., T. TAKEDA & M. KIMUJRA. 1984. Responses of the nucleus of the optic tract neurons projecting to the nucleus reticularis tegmenti pontis upon optokinetic stimulation in the rabbit. Neurosci. Res. **2:** 1–25.
9.　MCKENNA, O. C. & J. WALLMAN. 1985. Accessory optic system and pretectum of birds: comparisons with those of other vertebrates. Brain Behav. Evol. **26:** 91–116.
10.　SIMPSON, J. I. 1984. The accessory optic system. Annu. Rev. Neurosci. **7:** 13–41.
11.　FITE, K. V. 1985. Pretectal and accessory-optic visual nuclei of fish, amphibia and reptiles: theme and variations. Brain Behav. Evol. **26:** 71–90.
12.　HOFFMANN, K. P. 1986. Visual inputs relevant for the optokinetic nystagmus in mammals. *In* The Oculomotor and Skeletalmotor Systems: Differences and Similarities. H.-J. Freund, U. Buettner, B. Cohen & J. Noth, Eds.: 75–84. Elsevier. Amsterdam, the Netherlands.
13.　HOFFMANN, K. P., C. DISTLER, R. G. ERICKSON & W. MADER. 1988. Physiological and anatomical identification of the nucleus of the optic tract and dorsal terminal nucleus of the accessory optic tract in monkeys. Exp. Brain Res. **69:** 635–644.

14. SIMPSON, J. I., R. A. GIOLLI & R. H. I. BLANKS. 1988. The pretectal nuclear complex and the accessory optic system. *In* Neuroanatomy of the Oculomotor System. J. A. Buettner-Ennever, Ed.: 335–364. Elsevier. Amsterdam, the Netherlands.

15. HUTCHINS, H. B. & J. T. WEBER. 1985. The pretectal complex in the monkey: a reinvestigation of the morphology and retinal termination. J. Comp. Neurol. **232:** 425–442.

16. SEKIJA, H. & K. KAWAMURA. 1985. An HRP study in the monkey of olivary projections from the mesodiencephalic structures with particular reference to pretecto-olivary neurons. Arch. Ital. Biol. **123:** 171–183.

17. KATO, I., K. HARADA, T. HASEGAWA, T. IGARASHI, Y. KOIKE & T. KAWASAKI. 1986. Role of the nucleus of the optic tract in monkeys in relation to optokinetic nystagmus. Brain Res. **364:** 12–22.

18. KATO, I., K. HARADA, K. HASEGAWA & Y. KOIKE. 1988. Role of the nucleus of the optic tract of monkeys in optokinetic nystagmus and optokinetic afternystagmus. Brain Res. **474:** 16–27.

19. SCHIFF, D., B. COHEN & R. RAPHAN. 1988. Nystagmus induced by stimulation of the nucleus of the optic tract in the monkey. Exp. Brain Res. **70:** 1–14.

20. COHEN, B., D. SCHIFF & J. A. BUETTNER-ENNEVER. 1990. Contribution of the nucleus of the optic tract to optokinetic nystagmus and optokinetic afternystagmus in the monkey: clinical implications. *In* Vision and the Brain; the Organization of the Central Visual System. B. Cohen & I. Bodis-Wollner, Eds.: 233–255. Raven Press. New York, N.Y.

21. MUSTARI, M. & A. F. FUCHS. 1990. Discharge patterns of neurons in the pretectal neucleus of the optic tract. J. Neurophysiol. **64:** 77–90.

22. SCHIFF, D., B. COHEN, J. A. BUETTNER-ENNEVER & V. MATSUO. 1990. Effects of lesions of the nucleus of the optic tract on optokinetic nystagmus and after-nystagmus in the monkey. Exp. Brain Res. **79:** 225–239.

23. FUCHS, A. F., M. J. MUSTARI, F. R. ROBINSON & C. R. S. KANEKO. 1992. Visual signals in the nucleus of the optic tract and their brain stem destinations. Ann. N.Y. Acad. Sci. (This volume.)

24. HOFFMANN, K. P. & A. SCHOPPMANN. 1975. Retinal input to direction selective cells in the nucleus tractus opticus of the cat. Brain Res. **99:** 359–366.

25. SCHOPPMANN, A. 1981. Projections from areas 17 and 18 of the visual cortex to the nucleus of the optic tract. Brain Res. **223:** 1–17.

26. BALLAS, I., K. P. HOFFMANN & H. J. WAGNER. 1981. Retinal projection to the nucleus of the optic tract in the cat as revealed by retrograde transport of horseradish peroxidase. Neurosci. Lett. **26:** 197–202.

27. SAWAI, H., Y. FUKUDA & K. WAKAKURA. 1985. Axonal projections of X-cells to the superior colliculus and to the nucleus of the optic tract in cats. Brain Res. **341:** 1–16.

28. MALACH, R., N. STRONG & R. C. VAN SLUYTERS. 1981. Analysis of monocular optokinetic nystagmus in normal and visually deprived cats. Brain Res. **210:** 367–372.

29. GRASSE, K. L. & M. S. CYNADER. 1986. Response properties of single units in the accessory optic system of the dark-reared cat. Dev. Brain Res. **27:** 199–210.

30. GRASSE, K. L. & M. S. CYNADER. 1987. The accessory optic system of the monocularly deprived cat. Dev. Brain Res. **31:** 229–241.

31. GRASSE, K. L., M. S. CYNADER & R. M. DOUGLAS. 1984. Alterations in response properties in the lateral and dorsal terminal nuclei of the cat accessory optic system following visual cortex lesions. Exp. Brain Res. **55:** 69–80.

32. ZEE, D. S., R. J. TUSA, P. H. BUTLER, S. J. HERMAN & C. GUCER. 1987. Effects of occipital lobectomy upon eye movements in primates. J. Neurophysiol. **58:** 883–901.

33. MAEKAWA, K., T. TAKEDA & M. KIMURA. 1984. Responses of the nucleus of the optic tract neurons projecting to the nucleus reticularis tegmenti pontis upon optokinetic stimulation in the rabbit. Neurosci. Res. **2:** 1–15.

34. BUETTNER-ENNEVER, J. A. & U. BUETTNER. 1978. A cell group associated with vertical eye movements in the rostral mesencephalic reticular formation of the monkey. Brain Res. **151:** 31–47.

35. HOFFMANN, K. P. 1983. Effects of early monocular deprivation on visual input to cat nucleus of the optic tract. Exp. Brain Res. **51:** 236–246.

36. TAKEDA, T. & K. MAEKAWA. 1976. The origin of the pretecto-olivary tract. A study using the horseradish peroxidase method. Brain Res. 117: 319–325.

37. HOFFMANN, K. P., K. BEHRENS & A. SCHOPPMANN. 1976. A direct afferent visual pathway from the nucleus of the optic tract to the inferior olive in the cat. Brain Res. 115: 150–153.

38. WEBER, J. T. & J. K. HARTING. 1980. The efferent projections of the pretectal complex: an autoradiographic and horseradish peroxidase analysis. Brain Res. 194: 1–28.

39. KAWAMURA, K. & S. ONODERA. 1984. Olivary projections from the pretectal region in the cat studies with horseradish peroxidase and tritiated amino axonal transport. Arch. Ital. Biol. 122: 155–168.

40. HORN, A. K. E. & K. P. HOFFMANN. 1987. Combined GABA-immunochemistry and TMB-HRP histochemistry of pretectal nuclei projecting to the inferior olive in rats, cats and monkeys. Brain Res. 409: 133–138.

41. BLANKS, R. H. I. 1988. Cerebellum. In Neuroanatomy of the Oculomotor System. J. A. Buettner-Ennever, Ed.: 225–272. Elsevier. Amsterdam, the Netherlands.

42. COHEN, B. & J. A. BUETTNER-ENNEVER. 1984. Projections from the superior colliculus to a region of the central mesencephalic reticular formation (cMRF) associated with horizontal eye movements. Exp. Brain Res. 57: 167–176.

43. CAZIN, L., M. MAGNIN & J. LANNOU. 1982. Non-cerebellar visual afferents to the vestibular nuclei involving the prepositus hypoglossal complex: an autoradiographic study in the rat. Exp. Brain Res. 48: 309–313.

44. MAGNIN, M., J. H. COURJON & J. M. FLANDRIN. 1983. Possible visual pathways to the cat vestibular nuclei involving the nucleus prepositus hypoglossi. Exp. Brain Res. 51: 198–203.

45. MCCREA, R. 1988. The nucleus prepositus. In Neuroanatomy of the Oculomotor System. J. A. Buettner-Ennever, Ed.: 203–223. Elsevier. Amsterdam, the Netherlands.

46. MUGNAINI, E. & W. H. OERTEL. 1978. An atlas of the distribution of GABAergic neurons and terminals in the rat CNS as revealed by GAD immunohistochemistry. In Handbook of Chemical Neuroanatomy, GABA and Neuropeptides in the CNS. A. Bjorklun & T. Hokfelt, Eds. 4: 436–605. Elsevier. Amsterdam, the Netherlands.

47. GIOLLI, R. A., G. M. PETERSON, C. E. RIBAK, H. M. MCDONALD, R. H. I. BLANKS & J. H. FALLON. 1985. GABAergic neurons comprise a major cell type in rodent relay visual nuclei: an immunocytochemical study of pretectal and accesory optic nuclei. Exp. Brain Res. 51: 194–203.

48. CARDOZO, B. N. & J. J. VAN DER WANT. 1990. Ultrastructural organization of the retino-pretecto-olivary pathway in the rabbit: a combined WGA-HRP tracing and GABA immunocytochemical study. J. Comp. Neurol. 291(2): 310–327.

49. CARDOZO, B. N., R. BUIJS & J. J. VAN DER WANT. 1991. Glutamate-like immunoreactivity in retinal terminals in the nucleus of the optic tract in rabbits. J. Comp. Neurol. 309(2): 261–270.

50. COHEN, B., V. MATSUO & T. RAPHAN. 1977. Quantitative analysis of the velocity characteristics of optokinetic nystagmus and optokinetic after-nystagmus. J. Physiol. London 270: 321–344.

51. RAPHAN, T., V. MATSUO & B. COHEN. 1979. Velocity storage in the vestibulo-ocular reflex arc (VOR). Exp. Brain Res. 35: 229–248.

52. COHEN, B., D. HELWIG & T. RAPHAN. 1987. Baclofen and velocity storage: a model of the effects of the drug on the vestibulo-ocular reflex. J. Physiol London 393: 703–725.

53. TAKEMORI, S. & B. COHEN. 1974. Loss of suppression of vestibular nystagmus after flocculus lesions. Brain Res. 72: 213–224.

54. WAESPE, W., B. COHEN & T. RAPHAN. 1983. Role of the flocculus in optokinetic nystagmus and visual-vestibular interaction: effects of flocculectomy. Exp. Brain Res. 50: 9–33.

55. RAPHAN, T. & B. COHEN. 1986. Multidimensional organization of the vestibulo-ocular reflex (VOR). In Adaptive Processes in Visual and Oculomotor Systems. E. L. Keller & D. S. Zee, Eds.: 285–292. Pergamon Press. Oxford & New York.

56. RAPHAN, T. & B. COHEN. 1988. Organizational principles of velocity storage in three dimensions: the effect of gravity on cross-coupling of optokinetic after-nystagmus. Ann. N.Y. Acad. Sci. 545: 74–92.

57. RAPHAN, T. & D. STURM. 1991. Modeling spatiotemporal organization of velocity storage in the vestibuloocular reflex by optokinetic studies. J. Neurophysiol. **66:** 1410–1421.

58. DAI, M., T. RAPHAN & B. COHEN. 1991. Spatial orientation of the vestibular system: dependence of optokinetic after-nystagmus on gravity. J. Neurophysiol. **66:** 1422–1439.

59. DAI, M. J., T. RAPHAN & B. COHEN. 1992. Characterization of yaw to roll cross-coupling in the three-dimensional structure of the velocity storage integrator. Ann. N.Y. Acad. Sci. (This volume.)

60. COHEN, B., D. HELWIG, T. RAPHAN, J. I. SUZUKI, K. KAGA & A. EDEN. 1988. Changes in visual and vestibular function after canal plugging in the monkey. Soc. Neurosci. Abstr. **14:** 172.

61. COHEN, H., B. COHEN, T. RAPHAN & W. WAESPE. Habituation and adaptation of the vestibulo-ocular reflex; a model of differential control by the vestibulo-cerebellum. Exp. Brain Res. (In press.)

62. DEJONG, J. M. B. V., B. COHEN, V. MATSUO & T. UEMURA. 1980. Midsagittal brainstem sections: effects on ocular adduction and nystagmus. Exp. Neurol. **68:** 420–442.

63. BLAIR, S. M. & M. GAVIN. 1981. Brainstem commissure and control of time constant of vestibular nystagmus. Acta Otolaryngol Stockh. **91:** 1–8.

64. KATZ, E., J. M. B. V. DEJONG, J. A. BUETTNER-ENNEVER & B. COHEN. 1991. Effects of midline medullary lesions on velocity storage and the vestibulo-ocular reflex. Exp. Brain Res. **87:** 505–520.

65. YOKOTA, J. I., H. REISINE & B. COHEN. 1990. Velocity storage and the ocular response to microstimulation of vestibular nuclei in alert monkey. Soc. Neurosci. Abstr. **16:** 733.

66. FUCHS, A. F. & J. KIMM. 1975. Unit activity in vestibular nucleus of the alert monkey during horizontal angular acceleration and eye movement. J. Neurophysiol. **38:** 1140–1161.

67. TOMLINSON, R. D. & D. A. ROBINSON. 1984. Signals in vestibular nucleus mediating vertical eye movements in the monkey. J. Neurophysiol. **51:** 1121–1137.

68. REISINE, H. & T. RAPHAN. 1992. Unit activity in the vestibular nuclei of monkeys during off-vertical axis rotation. Ann. N.Y. Acad. Sci. (This volume.)

69. WAESPE, W. & V. HENN. 1977. Neuronal activity in the vestibular nuclei of the alert monkey during vestibular and optokinetic stimulation. Exp. Brain Res. **27:** 523–538.

70. WAESPE, W. & V. HENN. 1977. Vestibular nuclei activity during optokinetic afternystagmus (OKAN) in the alert monkey. Exp. Brain Res. **30:** 323–330.

71. BALABAN, C. D. 1983. A projection from nucleus reticularis tegmenti pontis of Bechterew to the medial vestibular nucleus in rabbits. Exp. Brain Res. **51:** 304–309.

72. KELLER, E. L. & W. F. CRANDALL. 1981. Neural activity in the nucleus reticularis tegmenti pontis in the monkey related to eye movements and visual stimulation. Ann. N.Y. Acad. Sci. **374:** 249–261.

73. KELLER, E. L. & W. F. CRANDALL. 1983. Neuronal responses to optokinetic stimuli in pontine nuclei of behaving monkey. J. Neurophysiol. **49:** 169–187.

74. HOOD, J. D. 1967. Observations upon the neurological mechanism of optokinetic nystagmus with special reference to the contribution of peripheral vision. Acta Otolaryngol Stockh. **63:** 208–215.

75. DICHGANS, J., B. NAUCK & E. WOLPERT. 1973. The influence of attention, vigilance and stimulus area on optokinetic and vestibular nystagmus and voluntary saccades. *In* The Oculomotor System and Brain Functions: Proceedings of the International Colloquium. B. Zikmund, Ed.: 280–294. Butterworths. London, England.

76. BRONSTEIN, A. M. & M. A. GRESTY. 1991. Compensatory eye movements in the presence of conflicting canal and otolith signals. Exp. Brain Res. **86:** 697–700.

77. VIIRRE, E., D. TWEED, K. MILNER & T. VILIS. 1986. A reexamination of the gain of the vestibuloocular reflex. J. Neurophysiol. **56:** 439–450.

78. SOLOMON, D. & B. COHEN. Stabilization of gaze during circular locomotion in light. I. Compensatory head and eye nystagmus in the running monkey. J. Neurophysiol. (In press.)

79. RAPHAN, T. & B. COHEN. 1985. Velocity storage and the ocular response to multidimensional vestibular stimuli. *In* Adaptive Mechanisms in Gaze Control; Facts and Theories. A. Berthoz & G. Melvill Jones, Eds.: 123–143. Elsevier. Amsterdam, the Netherlands.

80. TOMLINSON, D. & P. S. BAHRA. 1986. Combined eye-head shifts in the primate. I. Metrics. J. Neurophysiol. **56:** 1542–1557.
81. GROSSMAN, G. E., R. J. LEIGH, L. A. ABEL, D. J. LANSKA & S. E. THURSTON. 1988. Frequency and velocity of rotational head perturbations during locomotion. Exp. Brain Res. **70:** 470–476.
82. GROSSMAN, G. E., R. J. LEIGH, E. N. BRUCE, W. P. HUEBNER & D. LANSKA. 1989. Performance of the human vestibuloocular reflex during locomotion. J. Neurophysiol. **62:** 264–272.
83. SOLOMON, D. & B. COHEN. Stabilization of gaze during circular locomotion in darkness. II. Contribution of velocity storage to compensatory eye and head nystagmus in the running monkey. J. Neurophysiol. (In press.)
84. RAPHAN, T., M. DAI & B. COHEN. 1992. Spatial orientation of the vestibular system. Ann. N.Y. Acad. Sci. (This volume.)
85. SCHÖNE, H. 1964. On the role of gravity in human spatial orientation. Aerosp. Med. **35:** 764–772.
86. MITTELSTAEDT, H. 1983. Towards understanding the flow of information between objective and subjective space. *In* Neuroethology and Behavioral Physiology. F. Huber & H. Markl, Eds.: 382–402. Springer-Verlag. Berlin, Germany.
87. SCHÖNE, H. 1984. Spatial Orientation: the Spatial Control of Behavior in Animals and Man. Princeton Series in Neurobiology and Behavior. Princeton University Press. Princeton, N.J.
88. TAKEDA, T. & K. MAEKAWA. 1984. Collateralized projection of visual climbing fibers to the flocculus and nodulus of the rabbit. Neurosci. Res. **2:** 125–132.
89. WAESPE, W., B. COHEN & T. RAPHAN. 1985. Dynamic modification of the vestibulo-ocular reflex by the nodulus and uvula. Science **228:** 199–202.
90. NEWSOME, W. T., R. H. WURTZ, M. DURSTELER & A. MIKAMI. 1985. Deficits in visual motion processing following ibotenic acid lesion of the middle temporal visual area of the macaqued monkey. J. Neurosci. **5:** 825–840.
91. WURTZ, R. H., H. KOMATSU, D. S. G. YAMASAKI & M. R. DUERSTELER. 1990. Cortical visual motion processing for oculomotor control. *In* Vision and the Brain; the Organization of the Central Visual System. B. Cohen & I. Bodis-Wollner, Eds.: 211–231. Raven Press. New York, N.Y.
92. MAY, J. G., E. L. KELLER & D. A. SUZUKI. 1988. Smooth pursuit eye movement deficits with chemical lesions in the dorsolateral pontine nucleus of the monkey. J. Neurophysiol. **59:** 952–977.
93. ZEE, D. S., A. YAMAZAKI, P. H. BUTLER & G. GUCER. 1981. Effects of ablation of flocculus and paraflocculus on eye movements in primate. J. Neurophysiol. **46:** 878–899.
94. WAESPE, W., D. RUDINGER & M. WOLFENSBERGER. 1985. Purkinje cell activity in the flocculus of vestibular neurectomized and normal monkeys during optokinetic nystagmus and smooth pursuit eye movements. Exp. Brain Res. **60:** 243–262.

Velocity Storage in Labyrinthine Disorders[a]

TIMOTHY C. HAIN[b,d] AND DAVID S. ZEE[c]

[b]Departments of Neurology and Otolaryngology
Northwestern University School of Medicine
Chicago, Illinois 60616

[c]Departments of Neurology, Ophthalmology,
Otolaryngology, and Neuroscience
The Johns Hopkins University School of Medicine
Baltimore, Maryland 21205

INTRODUCTION

It is generally thought that a single central velocity storage mechanism is responsible for perseverating both canal-ocular responses and optokinetic responses.[2,6–8] The velocity storage mechanism transforms the signal from the cupula of the semicircular canal, which has a time constant of about 7 seconds, into the vestibular ocular reflex (VOR), which has a time constant of about 20 seconds. It is also responsible for optokinetic after-nystagmus (OKAN), that is, the nystagmus that continues with a time constant of 10 seconds in the dark following elicitation of an optokinetic nystagmus by a constant-velocity drum rotation.

The neural circuitry underlying the velocity storage mechanism is uncertain. However, two mathematical models of vestibular central processing have been proposed, one by Robinson[8] and the other by Raphan, Cohen and Matsuo.[7] FIGURE 1 shows the model by Robinson. Velocity storage is implemented via a positive feedback loop, which incorporates a gain parameter, "k," and a storage integrator element with a time constant, T_0. The transfer functions of this model for the VOR and OKAN are shown in Equation 1 below. The time constant of the cupula, T_c, is assumed to be 7 seconds. When the time constant of the internal storage element, T_0, is made equal to T_c, the "$sT_c + 1$" and "$sT_0 + 1$" terms cancel and the transfer function is simplified, providing a vestibuloocular reflex and optokinetic after-nystagmus with the same time constant, namely ($T_0/(1 - k)$). While OKAN and the VOR do not have identical time constants in humans, we will work with this assumption for the moment for clarity as the modifications in parameters needed to simulate differing time constants for OKAN and the VOR have been discussed elsewhere.[3] The simulated gain of the high-frequency response of the VOR is 1.0. In normal humans, the gain of OKAN is much less than 1.0 and a gain element prior to velocity storage (not shown on FIGURE 1) is used to adjust it appropriately in the

[a]This research was supported by research grants NS00914 (T.C.H.) and EY001849 (D.S.Z.) from the National Institutes of Health, U.S. Public Health Service.

[d]Address correspondence to: Northwestern Memorial Hospital, Department of Neurology, Passavant 10, 303 East Superior, Chicago, Illinois 60611.

297

model. A value of k of 0.66 produces a time constant of the VOR (T_{vor}) and OKAN (T_{okan}) of 21 seconds.

$$VOR = \frac{sT_c}{sT_c + 1} \cdot \frac{1}{1 - k} \cdot \frac{sT_0 + 1}{s\dfrac{T_0}{1 - k} + 1}$$

with $T_0 = T_c$

$$VOR = \frac{1}{1 - k} \cdot \frac{sT_c}{s\dfrac{T_c}{1 - k} + 1}$$

$$OKAN = \frac{1}{1 - k} \cdot \frac{1}{sT_0 + 1} \tag{1}$$

FIGURE 2 shows the model of Raphan, Cohen and Matsuo.[6] We will subsequently call this the "Raphan" model. In contrast to Robinson's model, velocity storage is provided by a feedforward pathway. Canal input is connected via a gain element, g_0, to an integrating element, the time constant and gain of which are controlled with parameter h_0. Equation 2 gives the transfer functions of this model. As in Robinson's model, by appropriate adjustment of two parameters [by letting $1/(g_0 + h_0) = T_c$ and $1/h_0 = T_{vor}$] this model can also be made to produce output with a single time

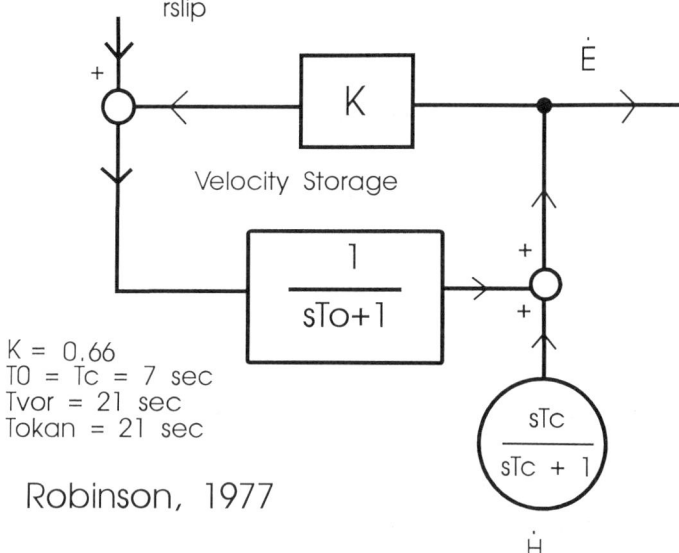

FIGURE 1. Robinson's model of velocity storage.[8] Laplace notation is used. Head velocity and eye velocity are designated by $\dot{H}$ and $\dot{E}$ respectively. Canal dynamics are designated by the operator ($sT_c/sT_c + 1$). Parameter "k" controls the amount of positive feedback within the velocity storage mechanism. T_0 is a lag element with a time constant matched to that of the cupula in order to obtain single exponential decay of the VOR, and a similar time constant for both OKAN and the VOR.

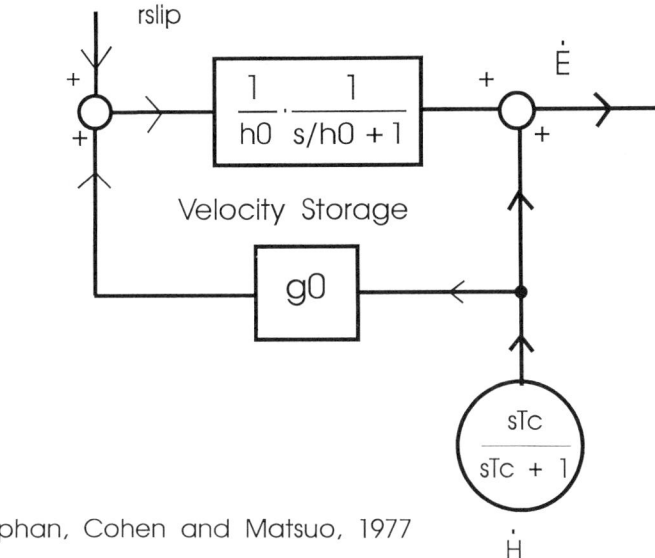

FIGURE 2. The Raphan model of velocity storage,[6] rearranged to be more easily compared to Robinson's model. Canal dynamics are the same as for Robinson's model. A pure integrator with negative feedback is used to create a velocity storage lag element, similar to that found in Robinson's model. Parameter h_0 adjusts the time constant and gain of the lag element and directly determines the time constant of OKAN. Parameter g_0 is adjusted in tandem with parameter h_0 in order to obtain single exponential decay of the VOR, and a similar time constant in both OKAN and the VOR.

constant common to both the VOR and OKAN. Values of T_c, g_0, and h_0 used to simulate responses in monkeys are 4 seconds, 0.25, and 0.085, respectively.[7] Given that for humans, $T_c = 7$ seconds, and $T_{vor} = 21$ seconds, g_0 should be 0.095; and h_0, 0.047. As in the Robinson model, OKAN gain is adjusted via a constant external to the velocity storage mechanism.

$$ \text{VOR} = \frac{sT_c}{sT_c + 1} \cdot \frac{g_0 + h_0}{h_0} \cdot \frac{s\dfrac{1}{g_0 + h_0} + 1}{s\dfrac{1}{h_0} + 1} $$

$$ \text{with } \frac{1}{h_0} = T_{vor} \quad \text{and} \quad \frac{1}{g_0 + h_0} = T_c $$

$$ \text{VOR} = \frac{g_0 + h_0}{h_0} \cdot \frac{sT_c}{s\dfrac{1}{h_0} + 1} $$

$$ \text{OKAN} = \frac{1}{h_0} \cdot \frac{1}{s\dfrac{1}{h_0} + 1} \tag{2} $$

Unilateral loss of vestibular function reduces the time constant of the VOR to about 7 seconds,[4] which is also thought to be the time constant of the cupula of the semicircular canal. In other words, velocity storage for canal input is abolished. We will now consider how one might accomplish this in the context of the two models discussed above, and how this might interact with optokinetic after-nystagmus. First, as a general observation, the reduction of velocity storage caused by unilateral loss of vestibular input must be an *active process,* since linear models such as we are discussing are by definition constrained by the scaling property, so that dynamics cannot change when input is scaled.[5] In order to change the time constant of processes related to velocity storage, one must change internal parameters.

In the model of Robinson, there are two methods of eliminating velocity storage for canal input. Equation 3 gives the transfer functions for the two cases. For case 1, parameter "k" is reduced to zero and canal input is no longer connected to the velocity storage mechanism. VOR gain at high frequencies is unaffected. OKAN persists, with a time constant of T_0 (7 seconds), but with a gain reduced by a factor of $1 - k$ (0.33). A second way to reduce velocity storage is to reduce T_0 (case 2). In the limit, where T_0 is zero, velocity storage is abolished, and OKAN is abolished along with it [as $1/(1 - k)$ in Equation 3 has no "s" term, OKAN vanishes when the lights are turned off]. Interestingly, for case 2, the VOR gain for high frequencies should increase by a factor of $1/(1 - k)$ or 3. When T_0 is intermediate between T_c and 0, double-exponential decay occurs in the VOR,[3] and the time constant of OKAN appears less than that of the VOR (although one cannot really assign a single time constant to a double-exponential process).

$$\text{Case 1 } (k = 0) \qquad \text{Case 2 } (T_0 = 0)$$

$$\text{VOR} = \frac{sT_c}{sT_c + 1} \qquad \text{VOR} = \frac{1}{1 - k} \cdot \frac{sT_c}{sT_c + 1}$$

$$\text{OKAN} = \frac{1}{sT_0 + 1} \qquad \text{OKAN} = \frac{1}{1 - k} \qquad (3)$$

In the Raphan model, again there are two methods of reducing velocity storage and Equation 4 gives the transfer functions for the two limiting cases. For case 3, canal velocity storage is eliminated without affecting the initial velocity or time constant of OKAN by reducing parameter g_0 to zero. Alternatively in case 4, by increasing parameter h_0 to infinity, both canal storage and optokinetic storage are eliminated. No change in VOR gain occurs in either subcase. Again, in intermediate cases, two-time-constant decay of the VOR should be present.

$$\text{Case 3 } (g_0 = 0) \qquad \text{Case 4 } (h_0 = \infty)$$

$$\text{VOR} = \frac{sT_c}{sT_c + 1} \qquad \text{VOR} = \frac{sT_c}{sT_c + 1}$$

$$\text{OKAN} = \frac{1}{h_0} \cdot \frac{1}{s\dfrac{1}{h_0} + 1} \qquad \text{OKAN} = 0 \qquad (4)$$

To summarize, limiting cases of the Robinson and Raphan models that eliminate velocity storage for canal input can be used to generate four predictions about the effects of labyrinthine lesions on visual velocity storage (see TABLE 1). We studied 13 patients with unilateral vestibular loss in order to simultaneously measure the time constant of the VOR and of OKAN so as to be able to evaluate which model and limiting case most accurately predicts the effect of labyrinthine loss upon velocity storage in human beings.

METHODS

Thirteen subjects were studied after surgical resection of an acoustic neurinoma. All but one patient was studied within six months of their surgery. The surgery required sectioning of the vestibular portion of the eighth nerve. The subjects were seated upright inside a five-foot diameter optokinetic drum which extended three feet above and below the line of sight. The head was stabilized by a chin rest. Optokinetic nystagmus (OKN) was induced by rotating the drum at 60 deg/second. Vestibular responses were measured using steps of chair rotation at a rate of 60 deg/second. The gain and time constant were calculated by semilogarithmic regression of eye velocity versus time, weighted by slow-phase duration.

Horizontal eye movements were recorded using DC electrooculography (EOG) with miniature silver–silver chloride electrodes placed lateral and medial to the right eye. Vertical eye movements and blinks were registered by electrodes placed above

TABLE 1. Summary of Limiting Cases

Case	Parameter	T_{vor}	T_{okan}	G_{okan}	G_{vor}
Robinson Model					
1	$k = 0$	T_c	T_o	*0.33	—[a]
2	$T_o = 0$	T_c	0	—	*3.0
Raphan Model					
3	$g_0 = 0$	T_c	—	—	—
4	$h_0 = \infty$	T_c	0	0	—

[a] "—" indicates no change from normal value.

and below the right eye. Eye position signals were amplified differentially (Tektronix 5A22 amplifier), low-pass filtered (3 dB point at 40 Hz), and then sampled at 100 Hz with 12-bit resolution by a laboratory computer. Calibration of eye position was performed at the beginning of each test by alternately viewing one of two light-emitting diodes (LEDs) located 40 deg apart. Spontaneous nystagmus in the dark was recorded prior to optokinetic stimulation.

Optokinetic stimulation lasted 87 seconds. During the stimulation the lights were switched off for two seconds at 12, 26, 40, and 60 seconds in order to "sample" the OKAN and produce a reliable estimate of initial velocity using averaging.[9] At 87 seconds, the lights were switched off and remained off for 1 minute. The lights were then turned back on for 1 minute prior to the next trial. The subjects were instructed to "look at the squares as they pass directly in front of you" during the optokinetic stimulation and to "look straight ahead" during periods of darkness. Subjects were encouraged at the beginning of each trial, but otherwise no other alerting maneuvers were utilized. Two sets of trials were obtained in six patients and one set in the remainder.

Spontaneous nystagmus was subtracted from slow-phase velocities and then OKAN was analyzed using the methods described elsewhere for normal subjects.[10] Three parameters were calculated from each trial. *Initial velocity* (E_0) was taken from the mean slow-phase eye velocity during the second second after the lights were extinguished. This occurred during the four samples and at the start of the one-minute period of darkness. *Slow-phase cumulative eye position* (SCEP) was calculated by numerically integrating the area under the first 30 seconds of the slow-phase velocity curve, excluding the first 2 seconds, and considering eye velocity going contrary to the drum as zero. To this quantity was added the mean initial velocity calculated from the samples, multiplied by 2 seconds. The *time constant* of OKAN was calculated by dividing the SCEP by the initial velocity, and adjusting the result by a correction formula. It has been shown for normal subjects that this method of deriving the time constant produces values that are identical to other methods.[12] Normal comparison data for OKAN were obtained from a study of 30 normal subjects reported previously from this laboratory.[10,12] Comparison data for the VOR were obtained from 40 normal subjects.

RESULTS

FIGURE 3A shows the distribution of the values of OKAN and VOR time constants in normal subjects; and FIGURE 3B, the same parameters in the patients. For the normal subjects, neither the OKAN nor the VOR distribution was statistically normal, but rather both are skewed towards larger values. Also, the modal bins for the VOR (14–20) and OKAN (6–8) are different, contrary to the assumptions necessary to produce single exponential decay of the VOR for both the Robinson and Raphan models. After unilateral vestibular loss, the range of the VOR and OKAN time constants was smaller, eliminating the skew. The mean values of the OKAN time constant (7.2 ± 1.8) and VOR time constant (6.4 ± 2.6) are not significantly different. The mean initial velocity of OKAN in patients was 9.7 ± 2.4 deg/second, which is reduced by a factor of 0.82 from that of normal subjects (11.7 ± 5.9 deg/second).

DISCUSSION

Our data document that following surgery for removal of an acoustic neuroma, the time constants both of the VOR and of OKAN are reduced to about 7 seconds. Quantitatively, our results closely resemble those of Blakely and associates, who studied 7 patients who underwent either vestibular neurectomy or labyrinthectomy for Meniere's disease.[1] They also closely resemble findings following unilateral labyrinthectomy in monkeys.[13] Teleologically, it is not surprising that the time constants of OKAN and the VOR remain similar even when velocity storage in the VOR has been lost. In this way, after rotation in the light, OKAN can still serve to null the postrotatory VOR.

Returning to the models, our results for the change in time constants are identical to case 1 of the Robinson model. In this case, reduction of the feedback gain parameter, k, to zero explains the pattern of behavior of the time constant for both the VOR and OKAN. While OKAN gain does not drop by the factor of 0.3 as predicted, after unilateral labyrinthine loss, one would expect an independent increase of gain by 2.0 in the VOR gain pathway, driven by the need to restore the overall VOR gain to 1.0. Furthermore, as it is thought that OKAN and the VOR share a common gain element,[11] the overall gain decrease predicted by combining these two considerations would be 0.6. This prediction is close to what was observed (0.82).

The Raphan model can also be adjusted to produce the same behavior, but it requires manipulation of two parameters in tandem. Parameter g_0 must be reduced

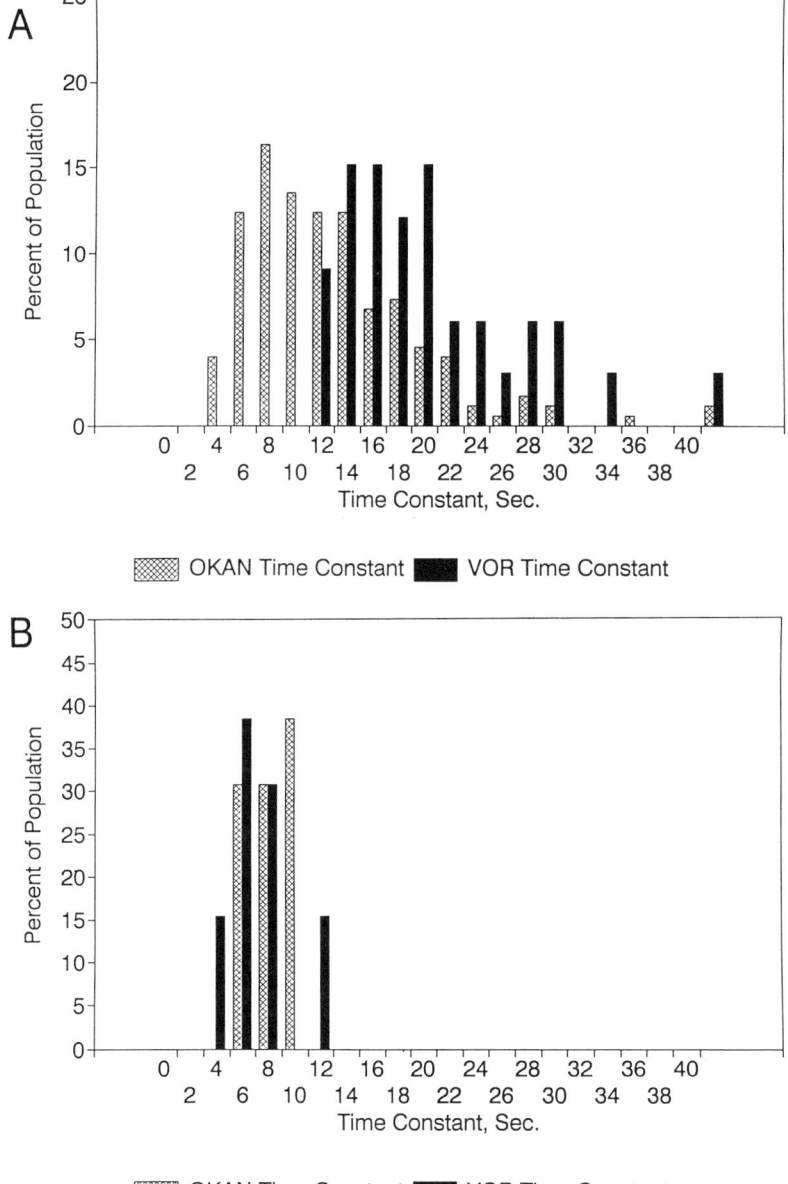

FIGURE 3. The distribution of time constants of the VOR and OKAN in normal subjects (A) and patients with unilateral loss of vestibular function (B). Dark bars indicate OKAN; and hatched bars, VOR.

to zero, to eliminate vestibular storage, and parameter h_0 must be tripled in order to decrease the time constant of OKAN to 7. OKAN gain is affected by these manipulations exactly as noted in Robinson's model.

Which model is better? Both can be used to fit the data. Robinson's model requires only a change of one parameter (k), while Raphan's model requires a change of two parameters $(h_0$ and $g_0)$. For reasons of parsimony, the Robinson model might be preferred although there are no obligatory constraints preventing multiple parameter adjustments.

SUMMARY

We studied 13 patients with unilateral peripheral vestibular lesions following removal of acoustic neurinomas. The time constant of the VOR after surgery was 6.4 ± 2.6 seconds (normal is 18.5 ± 7.7 seconds). The time constant of OKAN after surgery was 7.2 ± 1.8 seconds (normal is 11.3 ± 3.2 seconds). The mean initial velocity of OKAN after surgery was 9.7 ± 2.4 deg/second (normal is 11.7 ± 5.9 deg/second). These data suggest that unilateral peripheral vestibular loss is associated with a *complete* loss of velocity storage for canal input but only a *partial* loss of velocity storage for visual input. These results can be accounted for by current mathematical models of the velocity storage mechanism.

REFERENCES

1. BLAKELY, B. W., H. O. BARBER, R. D. TOMLINSON & L. MCILMOYL. 1989. Changes in the time constants of the vestibulo-ocular reflex and optokinetic after-nystagmus following unilateral ablative vestibular surgery. J. Otolaryngol. **18:** 210–217.
2. COHEN, B., V. HENN, T. RAPHAN & D. DENNETT. 1981. Velocity storage, nystagmus, and visual-vestibular interactions in humans. Ann. N.Y. Acad. Sci. **374:** 421–433.
3. DEMER, J. L. & D. A. ROBINSON. 1983. Different time constants for optokinetic and vestibular nystagmus with a single velocity storage element. Brain Res. **276:** 173–177.
4. HONRUBIA, V., H. A. JENKINS, R. W. BALOH, R. D. YEE & C. G. LAU. 1984. Vestibulo-ocular reflexes in peripheral labyrinthine lesions: I. unilateral dysfunction. Am. J. Otol. **5:** 15–26.
5. OPPENHEIM, A. V., A. S. WILLSKY & I. T. YOUNG. 1983. Signals and Systems. Prentice-Hall. Englewood Cliffs, N.J.
6. RAPHAN, T., B. COHEN & V. MATSUO. 1977. A velocity storage mechanism responsible for optokinetic nystagmus (OKN), optokinetic after-nystagmus (OKAN), and vestibular nystagmus. *In* Control of Gaze by Brainstem Neurons. R. Baker & A. Berthoz, Eds.: 37–47. Elsevier/North-Holland. Amsterdam, the Netherlands.
7. RAPHAN, T., V. MATSUO & B. COHEN. 1979. Velocity storage in the vestibulo-ocular reflex arc (VOR). Exp. Brain Res. **35:** 229–248.
8. ROBINSON, D. A. 1977. Linear addition of the optokinetic and vestibular signals in the vestibular nucleus. Exp. Brain Res. **30:** 447–450.
9. SEGAL, B. N. & S. LIBEN. 1985. Modulation of human velocity storage sampled during intermittently-illuminated optokinetic stimulation. Exp. Brain Res. **59:** 515–523.
10. TIJSSEN, M. A., C. S. STRAATHOF, T. C. HAIN & D. S. ZEE. 1989. Optokinetic after-nystagmus in humans: normal values of amplitude, time constant and asymmetry. Ann. Otol. Rhinol. Laryngol. **98:** 741–746.
11. ZASORIN, N. L., R. W. BALOH, R. D. YEE & V. HONRUBIA. 1983. Influence of vestibulo-ocular reflex gain on human optokinetic responses. Exp. Brain Res. **51:** 271–274.
12. HAIN, P. C. & G. PATEL. Use of slow-cumulative eye position to evaluate optokinetic after-nystagmus. Ann. Otol. (In press.)
13. FETTER, M. & D. S. ZEE. 1988. Recovery from unilateral labyrinthectomy in rhesus monkey. J. Neurophysiol. **59:** 394–407.

High-Frequency Vestibuloocular Reflex as a Diagnostic Tool[a]

R. JOHN LEIGH,[b] ROBERT N. SAWYER,
MICHAEL P. GRANT, AND SCOTT H. SEIDMAN

Departments of Neurology, Neuroscience, Otolaryngology, and
Biomedical Engineering
Case Western Reserve University
Department of Veterans Affairs Medical Center,
and University Hospitals
Cleveland, Ohio

INTRODUCTION: NATURAL DEMANDS ON THE VESTIBULOOCULAR REFLEX

The purpose of the vestibuloocular reflex (VOR) is to guarantee a clear and stable view of the environment during activities that entail head movements. The VOR generates eye movements that hold images of stationary objects relatively still upon the retina. When we view distant objects, only rotational perturbations require vestibular compensation; however, if we fix upon near objects, translations (linear displacements) of the head also become important.[1,2]

Recently, the existence of the VOR as an independent subsystem has been questioned.[3,4] Nonetheless, strong evidence for the indispensable role of the VOR in everyday life comes from reports of patients who have lost the function of their vestibular labyrinths.[5,6] Acutely, all head movements (even transmitted cardiac pulsations) impair vision and cause oscillopsia—illusory movement of the environment. Although some abatement of these visual symptoms eventually occurs while such patients are stationary, they continue to complain of blurred vision during locomotion. During walking or running in place, rotational head perturbations with predominant frequencies in the range 0.5–5.0 Hz occur; the highest frequency head rotations occur in the pitch (vertical) plane.[7,8] Thus the predominant frequency of these natural head movements greatly exceeds the frequencies of chair rotations used to test the VOR in many clinical laboratories.

In this paper we propose that (1) an important function of the VOR is to guarantee clear and stable vision during locomotion, and that this reflex is less important during stationary activities; (2) patients with vestibular disturbances complain of impaired vision during locomotion because of instability of gaze, not instability of the head; and (3) nonpredictable, transient head rotations, of similar frequency and velocity to those occurring during locomotion, are useful stimuli to test patients with vestibular symptoms. To support these propositions, we present preliminary results from studies of head stability during locomotion and gaze stability during transient head rotations in pitch. We have focused principally on head rotations in pitch because the most strenuous demands are made of the VOR in

[a] Supported by U.S. Public Health Service grant EY06717 (to Dr. Leigh), the Department of Veterans Affairs, and the Evenor Armington Fund.
[b] Address Correspondence to the Department of Neurology, University Hospitals, 2074 Abington Road, Cleveland, Ohio 44106.

this plane, and because the vertical VOR has seldom been tested with higher-frequency stimuli.

SUBJECTS AND METHODS

Subjects

We studied 2 patients who had lost most function of their vestibular labyrinths; both had absent responses to ice-water caloric stimulation. Patient 1 was a 68-year-old man who had lost right ear function from barotrauma during World War II, and had lost left ear function, presumably due to labyrinthine artery occlusion in 1990; he was totally deaf. Patient 2 was a 62-year-old woman who had lost vestibular function secondary to aminoglycoside antibiotics that were given for severe sepsis; her hearing was preserved. We compared the yaw and pitch head rotations of these two patients during locomotion with those of 20 normal subjects (age range 19–64 years), previously reported.[7,8] We also studied gaze stability in the pitch plane, during transient head rotations, in these 2 patients and in 4 normal male subjects (age range 29–44 years). All patients and subjects gave informed consent.

Recording Techniques

We used an angular rate sensor (Watson Industries, Eau Claire, Wis.) to measure yaw and pitch rotations while the patients (1) walked briskly in place, and (2) ran in place, at a gentle pace, for 21-second epochs. Because of a cardiac condition in patient 1, studies of head rotation during running in place were limited to the pitch plane. Note that the measurements of each plane of head movements of patients 1 and 2 were not made synchronously. Patients wore their shoes and viewed, through a window, a group of trees approximately 150 meters away.

To study gaze stability during transient pitch head rotations, we used the magnetic search coil technique.[9] Subjects wore a scleral search coil on their left eye; the other was patched. Head rotations were measured using a search coil that was attached to a bite plate. Search coils were calibrated prior to each experimental session. Subjects and patients wore a modified rubber diving helmet, to which was attached a square wooden frame. One of the investigators held the frame and delivered brief head rotations. The subject was instructed to visually fixate a small spot of light located at a distance of 1.3 meters. The diving mask allowed us to deliver head rotations of predominant frequency 1.0–3.0 Hz, and prevented the subjects from detecting cues about the timing of stimuli, which were applied in a nonpredictable sequence. Sixteen pitch rotations, intermixed with control rotations in the roll plane, were applied to each subject and patient. At the end of the recording session, the eye coil was removed and firmly taped to the head coil. Then 2–3 control trials were carried out to verify that there was no phase shift between eye and head coil signals.

Data Collection and Analysis

For the measurements of head rotation during locomotion, signals from the angular rate sensor were passed through Butterworth analog filters (bandwidth 0–40 Hz) prior to digitization, with 16-bit resolution, at 100 Hz. For each trial, a

2048-point fast Fourier transform was performed to determine the predominant frequency of the head rotations. Using an interactive program,[10] mean peak head velocity was calculated by averaging the 20 highest values of head velocity for each record.

For the measurements of gaze stability during transient head rotations, coil signals of gaze and head position in the vertical, torsional, and horizontal planes were filtered (bandwidth 0–50 Hz) prior to digitization at approximately 1 kHz. The digital signals were filtered (bandwidth 0–50 Hz), using a Hamming window,[11] and then differentiated to obtain gaze and head velocity in all three planes. Each response was then analyzed interactively; responses contaminated by blinks or early saccades were rejected. We estimated the peak velocity, acceleration, and predominant frequency of the head rotation; the peak velocity of the gaze perturbation; and the latency of the VOR, as previously reported.[12] Because a saccade or quick phase of nystagmus frequently followed the head rotation, it was often difficult to reliably measure peak eye-in-orbit velocity; nevertheless, the peak gaze velocity measure-

TABLE 1. Velocity and Frequency of Head Rotations during Locomotion in Patients with Deficient Vestibular Function

Patient/ Activity[a]	Mean Peak Velocity[b]		Predominant Frequency[c]	
	Yaw	Pitch	Yaw	Pitch
1/W	24	38	0.8	3.2
1/R	—	97	—	2.7
2/W	21	16	1.0	0.6
2/R	32	31	1.6	2.9
3/W	29	22	1.2	1.6
4/W	41	53	0.8	0.8
N/W	16–89	14–81	0.6–1.0	0.6–4.2
N/R	26–590	23–240	1.1–2.7	2.2–8.2

[a]Patients 1 and 2 from this study; patients 3 and 4 from Grossman and Leigh;[6] N: ranges for 20 normal subjects.[8,9] W: walk; R: run.
[b]Peak velocity values expressed in degrees/second.
[c]Frequency values expressed in Hz.

ment provided an index of the reliability of the VOR in holding images of stationary objects steady upon the retina.

RESULTS

Head Stability during Locomotion

The results of analysis of head rotations of the 2 patients during walking and running in place are summarized in TABLE 1; these data are compared with corresponding results from 20 normal subjects. Also included in TABLE 1 are data from a study of 2 other labyrinthine-deficient patients,[6] using the magnetic search coil technique to measure yaw and pitch head rotations. Taken together, none of these patients showed any degree of "head instability" during walking or running in place. As an example, a record of pitch rotations during running in place from patient 1 is shown in FIGURE 1; peak head velocity did not exceed 150 degrees/

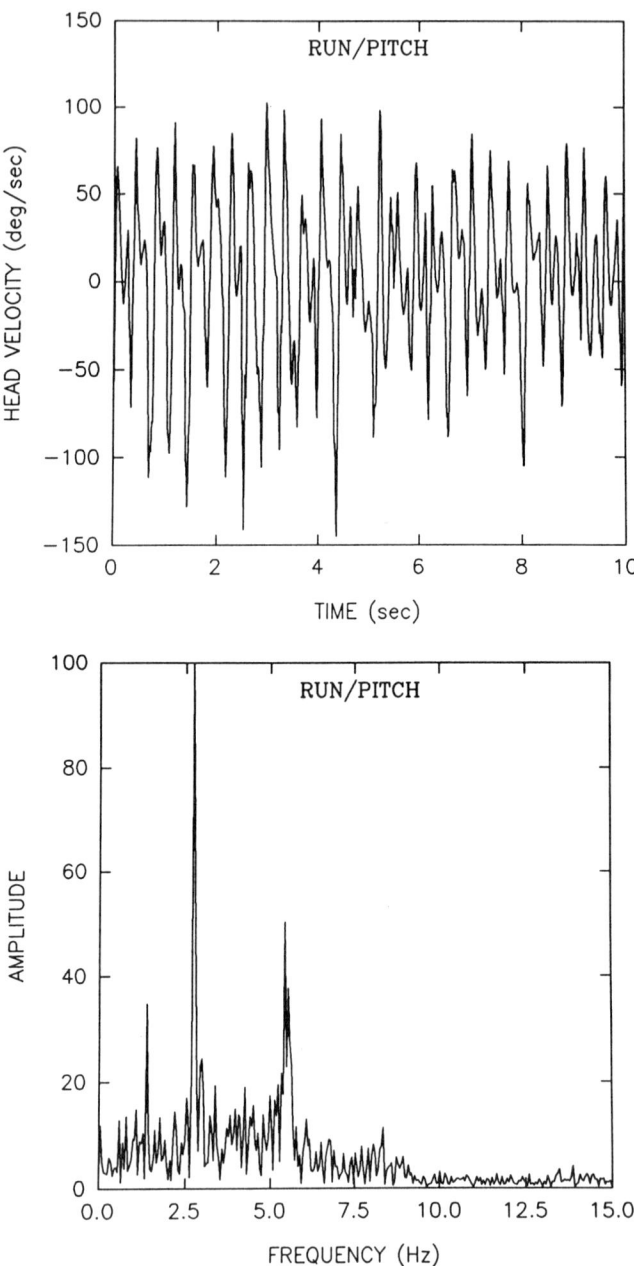

FIGURE 1. Head stability in patient 1, who had deficient vestibular function (see text). **Top:** Record of pitch head velocity as he ran in place. Note that peak head velocity did not exceed 150 deg/second. **Bottom:** Fourier transform based on head pitch data (2048-point). Amplitude (relative scale) indicates a predominant frequency of 2.7 Hz.

second, and the predominant frequency was 2.7 Hz with a harmonic at 5.4 Hz. Thus, the head perturbations of these four patients with deficient labyrinths were within the peak velocity and frequency ranges of normal subjects.

Gaze Stability during Transient Head Rotations

Typical head rotations and corresponding gaze perturbations from one of the *normal subjects* are shown in FIGURE 2. Analyses are preliminary, and will be reported in detail elsewhere.[13] Note that the peak head velocities and predominant frequencies of the stimuli shown in FIGURE 2 are within the range of values of pitch head movements occurring during walking and running in place (see TABLE 1). In the two examples shown, peak gaze velocity, in the direction of the head movement, was less than 15% of peak head velocity. Nonetheless, peak gaze velocity exceeded 10 deg/second for both stimuli. The latency of the vertical VOR was similar to the horizontal VOR (i.e., less than 16 mseconds).

Typical head rotations and corresponding gaze perturbations from patient 2 are shown in FIGURE 3. Although the velocity and frequency of the stimuli are less than those shown for the normal subject, they are within the range of values of pitch head movements occurring during locomotion, and a comparison of responses is possible. Note how the patient showed peak gaze velocities that were over 60% of the peak velocity of the pitch head rotation. The latency to onset of the compensatory eye movement was 40–50 mseconds, suggesting a nonvestibular origin.

DISCUSSION

The head perturbations that occur during locomotion constitute a major threat to clear and stable vision. Because of transmitted vibrations during heel strike, the head is subjected to rotations, particularly in pitch, that have fundamental frequencies of up to 5 Hz, or even higher during running.[7,8,14] The highest frequency head rotations during locomotion occur in the pitch plane. Although head perturbations during locomotion are of relatively high frequency (higher frequencies than are usually tested in most clinical laboratories), the peak velocities do not usually exceed about 200 deg/second, which is within the operating range of the vestibuloocular reflex.[15]

We have found that the peak velocities and predominant frequencies of head rotations in patients who have lost vestibular function do not differ from those of normal subjects. This finding is consistent with prior studies, which have indicated that the main factor determining stability of the head for stimuli of similar frequencies to those occurring during locomotion is mechanical rather than neurogenic,[16,17] although vision and prediction may both contribute to head stability.[17,18] It remains possible, however, that patients who have lost labyrinthine function may show a change in mean head position, such as occurs in normal subjects when they walk in darkness.[14]

Since the frequencies of the head perturbations occurring during locomotion are much above the ability of visually mediated eye movements to compensate, then it is not surprising that patients who have lost vestibular function are persistently troubled by blurred vision and oscillopsia whenever they walk. These symptoms can be ascribed to instability of gaze and, consequently, excessive motion of images upon their retina.[6]

Since normal subjects do not experience such symptoms during locomotion, and

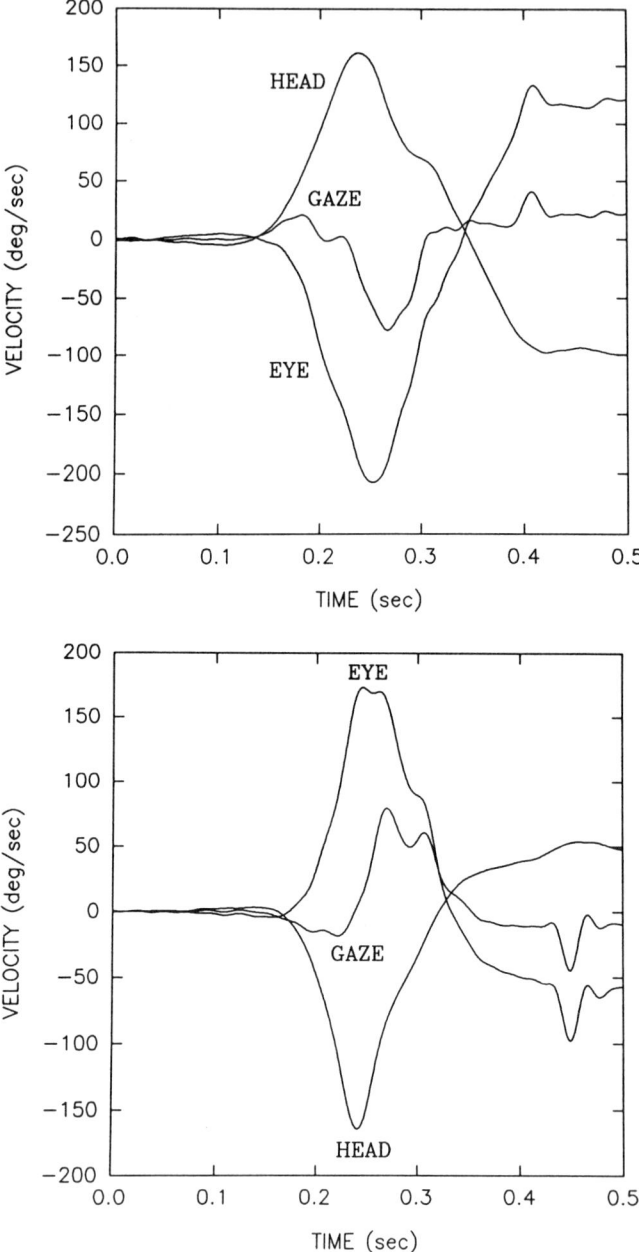

FIGURE 2. Perturbations of gaze produced by transient head rotations in the pitch plane in a normal subject. Head, head velocity; gaze, gaze velocity; eye, eye-in-orbit velocity. Upward deflections indicate upward rotations.

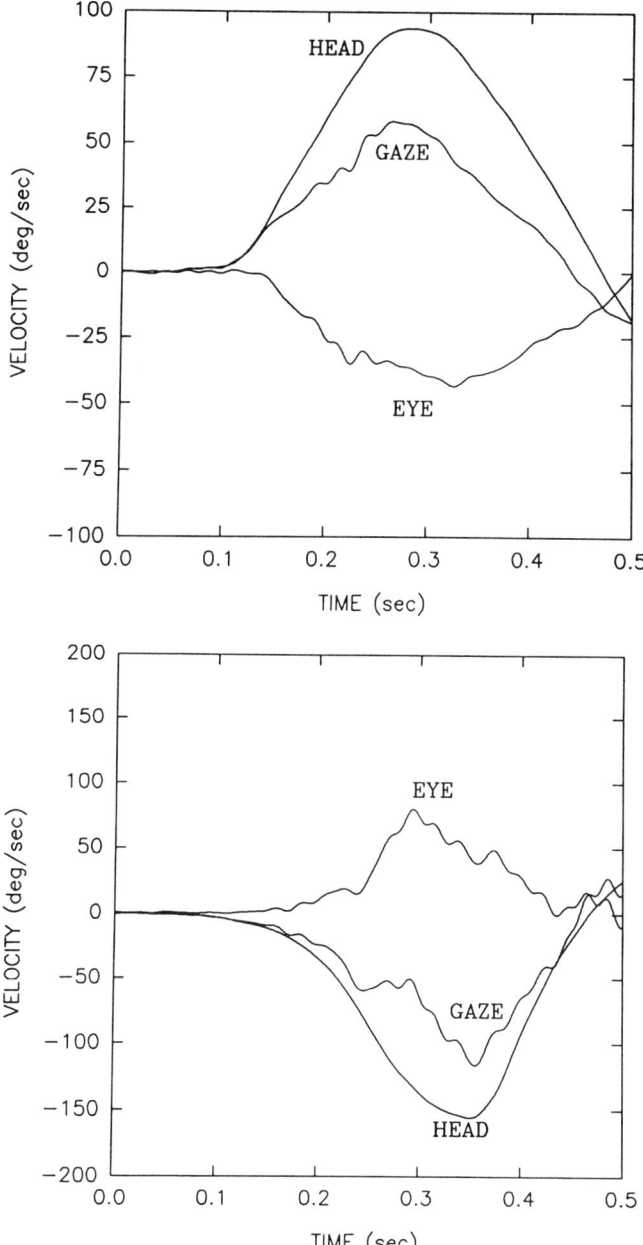

FIGURE 3. Perturbations of gaze produced by transient head rotations in the pitch plane in patient 2, who had deficient vestibular function. Note that although the head velocities are not as great as those shown in FIGURE 2, the perturbations of gaze produced are proportionally much larger. Same abbreviations and conventions as for FIGURE 2.

their gaze remains relatively stable,[19] why is it that the gain of the VOR has been frequently reported to be much less than 1.0? One reason concerns the conditions of testing. For example, in many studies of the VOR, subjects were rotated in darkness, an artificial condition that is known to reduce the gain of the VOR, even if the subject imagines a stationary target.[20,21] Furthermore, in many of these studies, the rotational stimuli were at much lower frequencies than those occurring during locomotion. Recent studies suggest that, during stationary or sedentary activities, we may not need a VOR. So, for example, vision is not appreciably impaired in patients with deficient vestibular function, compared to normal subjects, while they sit or stand stationary, and their gaze stability is similar to normals.[6] In normal subjects, it has been recently shown that the gain of the VOR while at rest is approximately 0.75 but, after the onset of head rotation, rapidly rises to about 0.95.[22,23] This finding may explain the perturbations of gaze shown by normal subjects soon after the onset of head rotations, such as is shown in FIGURE 2. It seems possible that humans show a lower VOR gain while at rest because of the increased importance of performing fine motor tasks while stationary. The head perturbations that occur during locomotion require the gain of the VOR to be close to 1.0, but the rotations that occur during stationary activities do not pose such a threat to clear vision. During the latter, certain tasks entailing fine coordination of head, eyes, and hands require that the VOR be negated, and this may be more easily achieved if VOR gain is set at a lower level.

During locomotion, head rotations not only contain high-frequency components, but they also have a randomness associated with them.[7,19] It has been shown that normal subjects can generate anticipatory eye movements to compensate for active head rotation.[24] It is also well established that patients who have lost vestibular function can use predictive mechanisms to generate compensatory eye movements.[6,25,26] During locomotion, however, such predictive mechanisms are unable to compensate for an absent VOR. These findings raise questions about the diagnostic value of studies using either rotational stimuli applied in a predictable manner, or active (self-generated) head movements.[27]

All in all, extrapolation of the results of conventional vestibular testing to account for patients' symptoms during natural activities requires caution, and it seems wiser to use stimuli that correspond to the head perturbations that occur during locomotion; such head rotations should have nonpredictable characteristics and a frequency range of 0.5 to 5.0 Hz. If transient head rotations are used, then a perturbation of gaze should be expected at the onset, but not the offset, of the stimulus.

SUMMARY

During locomotion, the head is subject to rotational perturbations with fundamental frequencies in the range 0.5–5.0 Hz, and significant harmonics up to 20 Hz. Patients who have lost labyrinthine function complain of oscillopsia and visual impairment during locomotion. Measurements of head movements during walking and running in place in such patients indicate that head stability is similar to that in normal subjects. Therefore, head stability is mainly guaranteed by mechanical, not neurogenic, factors. On the other hand, the visual symptoms of such patients can be ascribed to instability of gaze. Thus, it seems that other mechanisms such as visual following, the cervicoocular reflex, or anticipatory eye movements cannot compensate for loss of the VOR during locomotion (though they may do so for lower-frequency or active head rotations). The indispensable role of the VOR during locomotion is probably a reflection of its short latency (16 mseconds or less in the horizontal and vertical planes), which guarantees short phase lags during high-frequency head rotations. Our results indicate that laboratory testing of patients with vestibular symptoms should employ stimuli that correspond to those occurring during locomotion.

ACKNOWLEDGMENTS

We are grateful to Dr. U. P. K. Kumar for referring patient 2, and to A. O. DiScenna and L. F. Dell'Osso for technical assistance.

REFERENCES

1. SCHWARZ, U., C. BUSETTINI & F. A. MILES. 1989. Ocular responses to linear motion are inversely proportional to viewing distance. Science **245:** 1394–1396.
2. PAIGE, G. D. 1989. The influence of target distance on eye movement responses during vertical linear motion. Exp. Brain Res. **77:** 585–593.
3. COLLEWIJN, H. 1989. The vestibulo-ocular reflex: is it an independent subsystem? Rev. Neurol. Paris **145:** 502–512.
4. COLLEWIJN, H. 1989. The vestibulo-ocular reflex: an outdated concept? Prog. Brain Res. **80:** 197–209.
5. J. C. 1952. Living without a balancing mechanism. N. Engl. J. Med. **246:** 458–460.
6. GROSSMAN, G. E. & R. J. LEIGH. 1990. Instability of gaze during locomotion in patients with deficient vestibular function. Ann. Neurol. **27:** 528–532.
7. GROSSMAN, G. E., R. J. LEIGH, L. A. ABEL, D. J. LANSKA & S. E. THURSTON. 1988. Frequency and velocity of rotational head perturbations during locomotion. Exp. Brain Res. **70:** 470–476.
8. KING, O. S., S. H. SEIDMAN & R. J. LEIGH. 1991. Control of head stability and gaze during locomotion in normal subjects and patients with deficient vestibular function. *In* Second Symposium on Head-Neck Sensory-Motor System. A. Berthoz, W. Graf & P. P. Vidal, Eds. Oxford University Press. New York, N.Y.
9. FERMAN, L., H. COLLEWIJN, T. C. JANSEN & A. V. VAN DEN BERG. 1987. Human gaze stability in the horizontal, vertical and torsional direction during voluntary head movements, evaluated with a three-dimensional scleral induction coil technique. Vision Res. **27:** 811–828.
10. HARY, D., K. OSHIO & S. D. FLANAGAN. 1987. The ASYST software for scientific computing. Science **236:** 1128–1132.
11. OPPENHEIM, A. V. & R. W. SCHAFER. 1975. Digital *Signal Processing.* Prentice Hall. Englewood Cliffs, N.J.
12. MAAS, E. F., W. P. HUEBNER, S. H. SEIDMAN & R. J. LEIGH. 1989. Behavior of human horizontal vestibulo-ocular reflex in response to high-acceleration stimuli. Brain Res. **499:** 153–156.
13. SAWYER, R. N., M. P. GRANT, S. H. SEIDMAN, R. L. RUFF & R. J. LEIGH. A comparison of the vertical and torsional vestibulo-ocular reflexes using transient stimuli. (Submitted.)
14. POZZA, T., A. BERTHOZ & L. LEFORT. 1990. Head stabilization during various locomotor tasks in humans. I. Normal subjects. Exp. Brain Res. **82:** 97–106.
15. PULASKI, P. D., D. S. ZEE & D. A. ROBINSON. 1981. The behavior of the vestibulo-ocular reflex at high velocities of head rotation. Brain Res. **222:** 159–165.
16. GRESTY, M. 1987. Stability of the head: studies in normal subjects and in patients with labyrinthine disease, head tremor, and dystonia. Movement Dis. **2:** 165–185.
17. GUITTON, D., R. E. KEARNEY, N. WERELEY & B. W. PETERSON. 1986. Visual, vestibular and voluntary contributions to human head stabilization. Exp. Brain Res. **64:**59–69.
18. DEMER, J. L., J. GOLDBERG & F. I. PORTER. 1991. Effect of telescopic spectacles on head stability in normal and low vision. J. Vestibular Res. **1:** 109–122.
19. GROSSMAN, G. E., R. J. LEIGH, E. N. BRUCE, W. P. HUEBNER & D. J. LANSKA. 1989.

Performance of the human vestibuloocular reflex during locomotion. J. Neurophysiol. **62:** 264–272.

20. BARR, C. C., L. W. SCHULTHEIS & D. A. ROBINSON. 1976. Voluntary, non-visual control of the human vestibulo-ocular reflex. Acta Otolaryngol. Stockh. **81:** 365–375.

21. COLLEWIJN, H., A. J. MARTINS & R. M. STEINMAN. 1983. Compensatory eye movements during active and passive head movements: fast adaptation to changes in visual magnification. J. Physiol. London **340:** 259–286.

22. HUEBNER, W. P. & R. J. LEIGH. 1992. Dynamic modulation of VOR gain during passive head rotation. Ann. New York Acad. Sci. (This volume.)

23. GAUTHIER, G. M. & J.-L. VERCHER. 1990. Visual vestibular interaction: vestibulo-ocular reflex suppression with head-fixed target fixation. Exp. Brain Res. **81:** 150–160.

24. VERCHER, J.-L. & G. M. GAUTHIER. 1991. Eye-head movement coordination: vestibulo-ocular reflex suppression with head-fixed target fixation. J. Vestibular Res. **1:** 161–170.

25. HALMAGYI, G. M., I. S. CURTHOYS, P. D. CREMER, C. J. HENDERSON & M. STAPLES. 1991. J. Vestibular Res. **1:** 187–197.

26. KASAI, T. & D. S. ZEE. 1978. Eye-head coordination in labyrinthine-defective human beings. Brain Res. **144:** 123–141.

27. FINEBERG, R., D. P. O'LEARY & L. L. DAVIS. 1987. Use of active head movements for computerized vestibular testing. Arch. Otolaryngol. Head Neck Surg. **113:** 1063–1065.

Perception of Motion and Position Relative to the Earth

An Overview

FRED E. GUEDRY

Naval Aerospace Medical Research Laboratory
Naval Air Station
Pensacola, Florida 32508-5700

University of West Florida
11000 University Parkway
Pensacola, Florida 32514-5732

INTRODUCTION

Knowledge of the statics and dynamics of spatial orientation perception is an essential factor in the pursuit of solutions to three major problems, each involving systems that mediate skilled control of motion and angular position of the head and body (in three dimensions) relative to the earth. (1) Ninety million Americans have a medically significant dizziness or balance problem. Each year, an estimated 97,000 new cases of Ménière's disease present for medical attention in the United States alone.[1] (2) The motion sickness syndrome is significant in most modern forms of transportation for many healthy citizens[2] and is a costly operational problem in the aerospace community;[3,4] more than 60% of U.S. astronauts curtail unnecessary movement for several days into orbital missions to minimize symptoms of the "space-adaptation syndrome." (3) The control of spacecraft, aircraft, and aerospace craft like the shuttle is directly jeopardized when the pilot is disoriented concerning position (angular and linear) and motion of the craft relative to the earth or other significant object. Pilot disorientation at a critical moment in a mission is too often catastrophic. Each year the armed forces lose about 30 aircraft and a number of excellent pilots in disorientation-error accidents. Dollar cost estimates sum to hundreds of millions of dollars each year[5] for the combined annual dollar cost of disorientation-error aircraft accidents in the Army, Air Force, and Navy.

Perceptual events are central to each problem area. Disturbance by abnormality in spatial orientation perception brings patients to the medical clinic. In current concepts of the mechanisms of motion sickness, revision of an internal spatial orientation model is central to the "motion adaptation syndrome."[3,6] Spatial disorientation perceptions are the primary, if not sole, cause of spatial-orientation-error accidents in aviation.

Every approach represented in this conference is necessary for the level of understanding required for real solutions in the three problem areas, and any comprehensive model developed must at least be consistent with data from each approach. Models are a necessary way to handle the range of data being generated. It is my belief that no approach is more germane to testing and developing of comprehensive models than the approach implied by the title of this session, Perception of Motion and Position in Hypergravity and Hypogravity.

A brief overview of spatial orientation research is that we have substantial information on the perception of static (or very slowly changing) tilt position of the head (and body) relative to the earth but very little information on dynamics. We also have substantial information on the perception of turning (angular velocity) during and after rotation about an earth-vertical axis when the head-to-seat body axis is located on or very near the center of rotation. Even within this scope, research has far from exhausted our quest for knowledge. Static tilt is curiously different when reference is to the body as opposed to a visual line, and it remains to be determined whether hypo-G results can be extrapolated from results obtained during different levels of hyper-G above 1.0 G. With respect to on-axis rotation, we still have much to learn. Consider for example adaptation to complex interactions between the visual, proprioceptor, and vestibular systems, or voluntary effort to control a motor response superimposed on reflexive control, or differences in reactions to "active" and "passive" stimuli, or any of these studied in a strange environment such as earth-orbit trajectory.

When we move on to dynamics of attitude perception during rapidly changing magnitude and direction of resultant force vectors (of which gravity is a component), or to perceived movement trajectories in the earth-horizontal plane or out of it, or to the concomitant perception of attitude position coupled with perceived linear and angular velocity, then we have very little information. Dynamic spatial orientation perception is not a simple inferential extension of static perception although well-conceived theory on static perception is a necessary component in the modeling of dynamics. Herein I will describe results of experiments that suggest that the same mechanism mediates the dynamics of spatial orientation perception in earth-horizontal and earth-vertical planes. Some of the experiments are firmly based, others are more preliminary in nature, but a common thread connects the observations: subtle interactions between canal and otolith systems influence perceptual dynamics.

GRAVITY VECTOR DIRECTIONALLY FIXED IN PLANE
OF CANAL RESPONSE

The first experiment involves simple measurement of the duration of postrotation turning sensation following sustained whole-body rotation, i.e., a repetition of one of the oldest experiments in vestibular neuroscience.[7,8] However in this series of observations,[9] the axis of rotation is fixed at different angles of tilt relative to gravity.

The results are partially represented in FIGURE 1 where it is apparent that the response duration is reduced as the plane of the responding canal approaches alignment with gravity. The postrotation mean response durations are shown here to illustrate that a definite response change occurred, but the perception is quite different at different tilt angles. The postrotation perception changes from a simple clear sensation of whole-body rotation at the zero-tilt position to a more complex perception when the axis is tilted. The sense of rotation seems to be opposed by an "alien force," a phrase used by Purkinje in about 1820 to describe a similar perception (Reference 9, p. 137).

Perceptions induced by deceleration from a substantially tilted axis are unpleasant. The otolithic system senses the angular position of the rotation axis relative to gravity (which the canals cannot). Considering only vestibular information, combined information from canal and otolith systems indicates the nose-up position is changing relative to gravity but the otolith system is simultaneously indicating static angular position, i.e., that the nose-up position is not changing. A "vestibular" interpretation of the unpleasant and short response is that the otolith position signal lying fixed in the plane of the canal response is the "alien force" that opposes the sense of rotation.

Another possible interpretation of these data is that velocity stored during the sustained off-vertical rotation persists and counteracts the response to deceleration, greater storage effects being related to slightly greater per-rotational reaction with greater tilts of the rotation axis. However velocity storage seems inappropriate for the following observations.

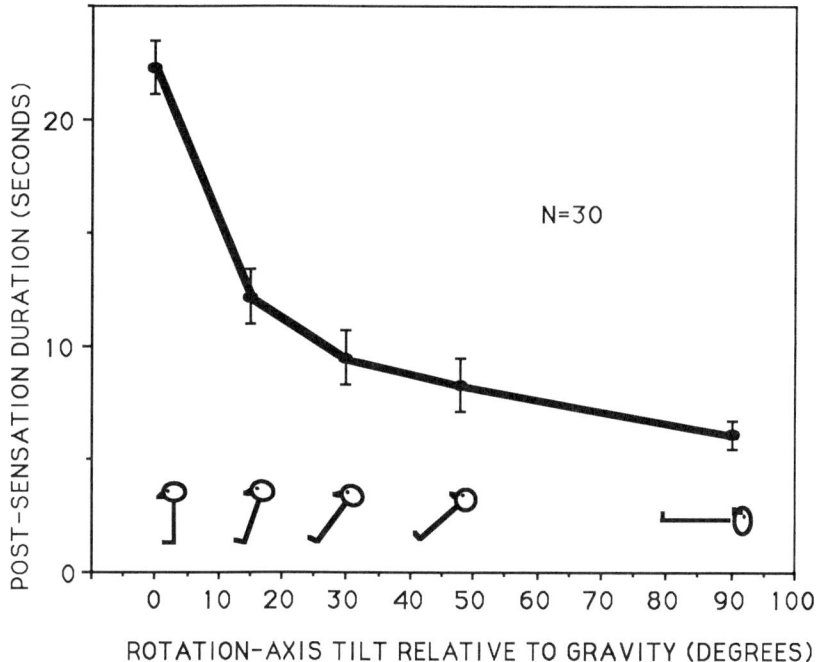

FIGURE 1. Duration of postrotation turning sensation when rotation axis was vertical and when it was tilted. When axis was tilted, stopping position was "nose up." Error bars indicate mean plus or minus standard error of the mean. Rotation at 22 rpm was sustained for 60 seconds before $20°/\text{second}^2$ deceleration was commenced.

LARGE LINEAR VECTOR DIRECTIONALLY FIXED
IN PLANE OF CANAL RESPONSE

The second experiment was carried out on a very large centrifuge.[10] Subjects were in fixed upright position relative to gravity, 50 feet from rotation center, as the centrifuge turned rapidly through about 330 degrees. The stimulus was complete in 3.2 seconds. During the turn, the subject was turned about a secondary axis parallel to but 50 feet displaced from the main axis so as to maintain subject's nose (x-axis) perfectly on the earth-horizontal resultant vector (which can be calculated from the centripetal and tangential acceleration components) as illustrated in FIGURE 2. The change in heading of the subject to maintain alignment with the direction of the horizontal resultant vector is illustrated in FIGURE 3.

The results of these observations, at least that part of the results of interest at the moment, is that subjects felt little or no turn in spite of the fact that they had actually turned through 330 degrees in 3.2 seconds (approximately 147 degrees of the turn was centrifuge and 180 degrees of the turn was added by the secondary axis). The absence of turn perception is impressive because fast turns through 330 degrees are, on average, accurately perceived when subjects are at the center of rotation.[11] As indicated by ability to redirect gaze in darkness to imagined starting position, short

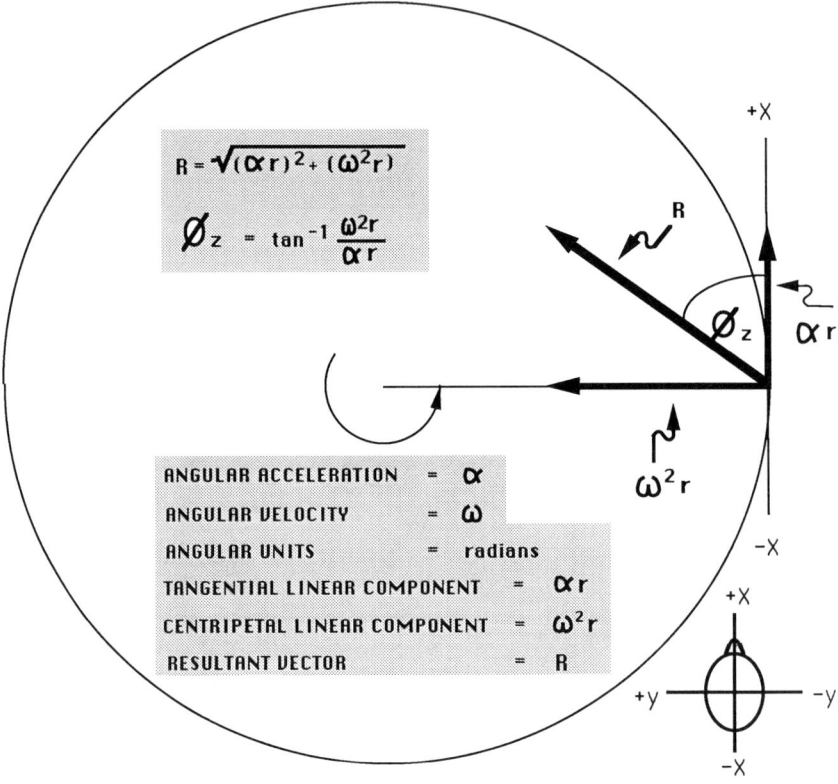

FIGURE 2. During angular acceleration of a centrifuge, when the head is upright and fixed in position away from the center of the centrifuge, a horizontal linear acceleration, which is the resultant of centripetal and tangential vectors, rotates in the horizontal head plane; it changes magnitude and direction relative to the x-axis.

turns are very accurately judged,[12] and, with similar procedure, average accuracy of perceived 330 degree turns has been confirmed.[13] Something suppressed the canal information or the canal input itself was suppressed.

A velocity storage explanation appears inappropriate for the centrifuge, where a strong angular velocity semicircular canal signal peaked in 1.6 seconds and was complete in 3.2 seconds. The directionally fixed linear resultant vector in the plane of the responding horizontal canals seems the most viable explanation. The direction-

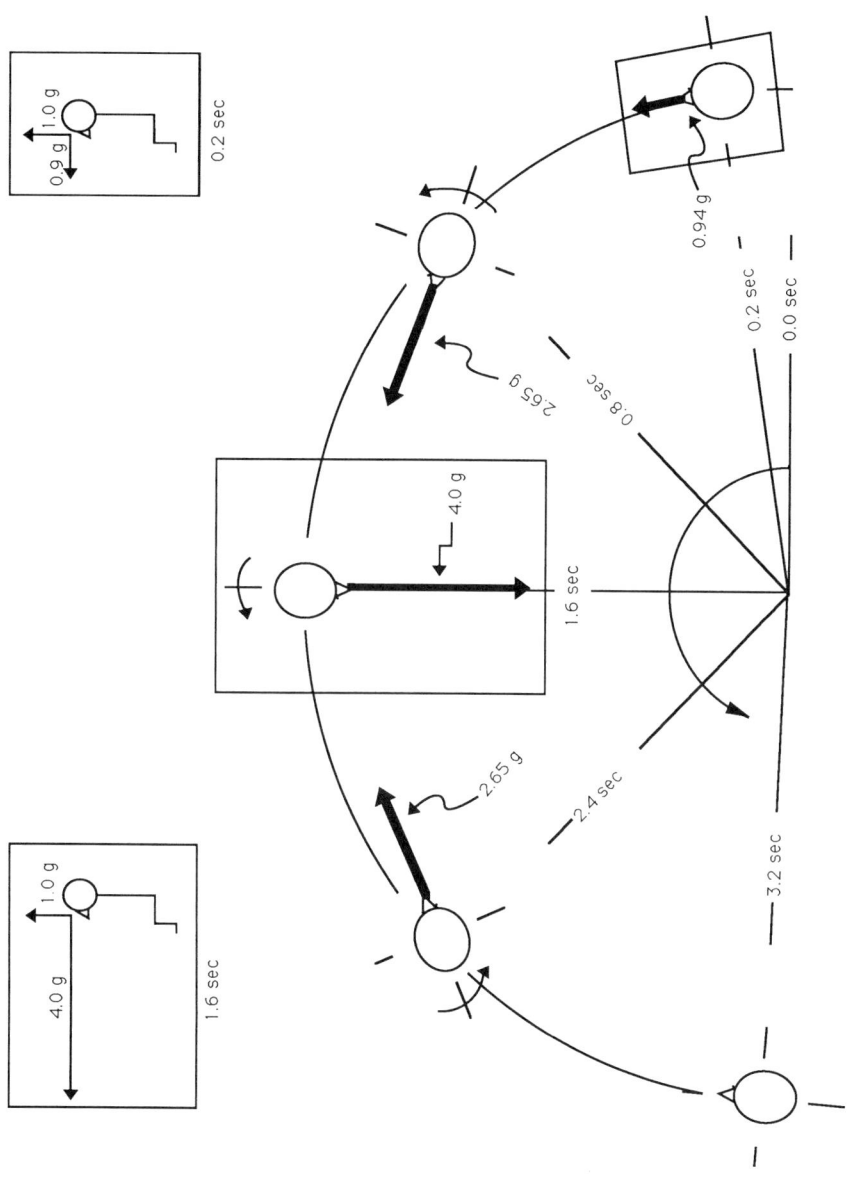

FIGURE 3. Subjects were seated upright relative to gravity and turned to maintain x-axis of head aligned with the resultant horizontal linear acceleration. The resultant vector peaked at 4.0 g-units in 1.6 seconds from start; the horizontal linear acceleration profile simulated the linear acceleration experienced by pilots during catapult take off from aircraft carriers.

ally fixed otolith position signal denies angular position change in the plane that the canals signal angular velocity.

MINOR ROTATION OF LARGE LINEAR VECTOR
IN PLANE OF HORIZONTAL CANAL

Additional runs with the large centrifuge were made.[14] An A-7 cockpit was installed on the secondary turn axis of the centrifuge in order to achieve realism in simulating a catapult takeoff from an aircraft carrier. The peak linear acceleration along the subject's nasal-occipital axis (x-axis) was 4.0 g-units, equivalent to that experienced during catapult takeoff. But the A-4 cockpit configuration displaced the subject's head 2 feet from the secondary axis, which means that the resultant horizontal linear vector rotated slightly in the horizontal plane of the head, but only about six degrees. Rotation of the vector was in the same direction as the direction of turn signaled by the horizontal canals. Subjects now experienced a curvilinear path, distorting the realism of the simulated catapult takeoff straight down a carrier deck. However keep in mind that if the canals had dominated the perception, subjects would have perceived that they were moving backward down the carrier deck about midway in the profile.

SMALL LINEAR VECTOR ROTATING IN HORIZONTAL CANAL PLANE

That a rotating low-magnitude linear acceleration vector can alter the perception of canal-mediated turn is suggested by the following experiment. When a subject on a centrifuge is fixed upright (relative to gravity) in forward-facing position as shown in FIGURE 4, angular acceleration of the centrifuge produces a linear acceleration vector that rotates in the horizontal plane of the head, i.e., approximately in the plane of the horizontal canals and the utricular otolithic membrane. In this experiment, the subject was not turned about a secondary axis.

A 20-foot-radius centrifuge accelerated to 7 rpm in 6 seconds yields a maximum linear acceleration in the horizontal plane of about 0.33 g-units. During acceleration of the centrifuge up to 7 rpm, subjects reported linear motion with only a little turn—for example, turn at a radius or curvilinear motion. After several minutes at constant angular velocity, deceleration produced a sensation of true rotation that persisted for awhile. When asked where the axis of turn was located, all of seven subjects indicated that the axis of turn was displaced from the body during acceleration but during deceleration the perceived axis of turn was near or through the body.

FIGURE 5 shows that during acceleration, the horizontal linear vector rotated in the same direction as the concurrent signal from the semicircular canals, whereas during deceleration the linear vector rotated in opposite sense to the canal signal. Note the similarity of the latter combination to what occurs during any natural head turn around an earth-horizontal axis; the canals signal head rotation in one direction, while the gravity vector rotates relative to the head in the opposite direction. A weak linear vector rotating in the same direction as the canal signal seems to reduce the perception of turn and to yield a perceived curved path of motion, whereas turn perception is enhanced when canal signals and rotating linear vectors are opposite in direction.

The acceleration/deceleration difference in turning sensation is partially attributable to the effect described in experiment 1. At the end of the acceleration, a

0.33-g-unit linear vector was fixed in the plane of the responding horizontal canals (equivalent to an 18-degree tilt in the roll plane). An 18-degree tilt in pitch reduced the duration of turn sensation by about 50% (FIGURE 1). At the end of deceleration, the resultant linear vector is gravity, tangential and centripetal components having gone to zero, so that the turn sensation from responding horizontal canals would be unsuppressed as in the zero-tilt position of FIGURE 1. This does not, however, account for the linear component of the motion perception in experiment 4. Resultant linear and angular vectors must be assessed in three dimensions to appreciate the perceptual and perceptual-motor effects of vestibular stimuli.

The linear vectors in the horizontal plane were small; maximum centripetal acceleration was 0.33 g-units and maximum tangential was 0.076 g-units. More data

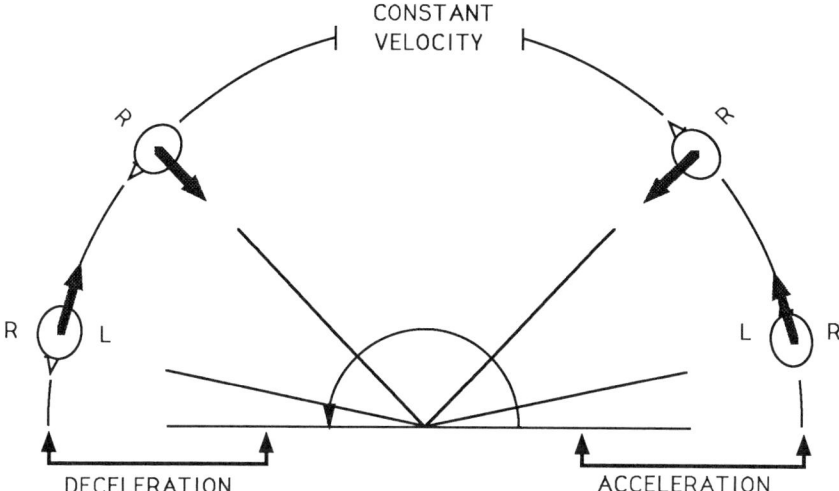

FIGURE 4. When subjects are maintained in initial position on the centrifuge, the resultant horizontal linear acceleration vector rotates relative to the head during acceleration and deceleration of the centrifuge. Direction of rotation of the linear vector relative to the head is the same during acceleration and deceleration, but signals from the semicircular canals are opposite when a sufficient period of constant velocity intervenes between centrifuge acceleration and deceleration.

are needed to test our interpretation and reliability of results when subjects are naive concerning angular and linear position on the centrifuge. Similarly, the turn perception, or lack thereof, during the catapult simulation needs confirmation. The investigators were focused on a different component of the perception, viz., pitch, which is discussed in a later section of this paper.

LARGE LINEAR VECTOR FIXED IN THE PITCH PLANE

The centrifuge run used in traditional G-tolerance training of pilots maintains the subject's z-axis alignment with the vector resultant of the centripetal and gravity components. Here, canal responses are governed by cross-coupled angular velocity effects plus the effects of change in canal planes relative to the centrifuge angular acceleration.[15] In the run we consider first, the subject was placed in forward-facing tangential heading. The subject's head-to-seat axis (z-axis) remained aligned with the resultant linear vector as it rolled 70 degrees, relative to gravity (FIGURE 6). Because we used a low-onset acceleration, the tangential acceleration displaced the total resultant linear acceleration relative to the z-axis less than four degrees. Thus, as head-to-seat G increased from 1.0 to 3.0 g-units, the direction of the total resultant linear acceleration vector remained nearly fixed relative to the otolithic membranes (despite 70-degree roll relative to gravity).

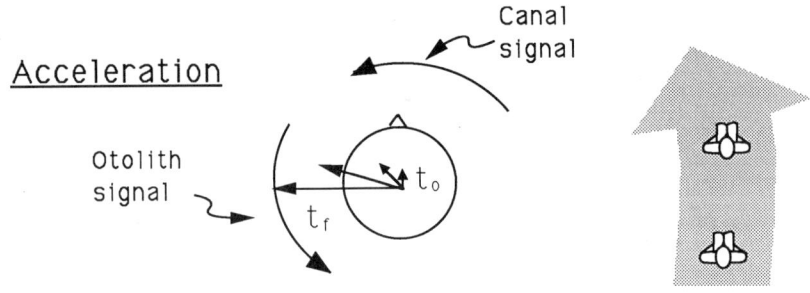

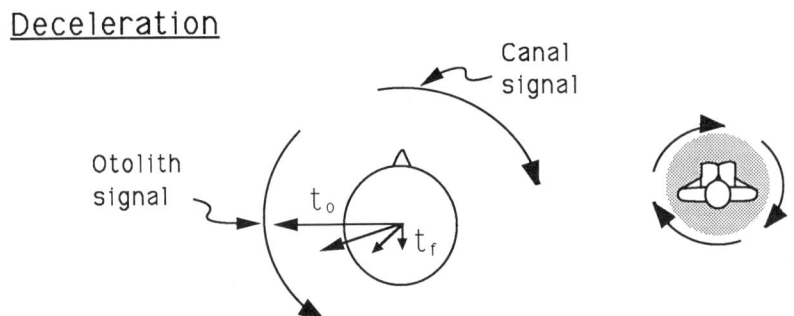

FIGURE 5. Illustrating the direction of rotation of the linear acceleration vector in the horizontal head plane relative to the direction of rotation signaled by the semicircular canals. Perceptions indicated by stippled figures on right differed approximately as illustrated.

During acceleration, subjects perceived an average maximum pitch-up position of about 15 degrees. This much pitch-position change may be solely otolith mediated; the very slight tangential effect (4.0 degrees in pitch) coupled with the increased utricular-shear component from the 3-g-unit z-axis vector could account for all of the pitch change.[16,17] In other words, the pitch-plane angular velocity signal from the canals may have contributed little toward the 15 degree pitch-up attitude perception. Moreover very little pitch plane angular velocity was perceived. The linear vector, which increased to 3 g-units and remained almost directionally fixed in the pitch plane, suppressed the perceptual effects of the strong pitch-plane canal stimulus as it accumulated during the acceleration.

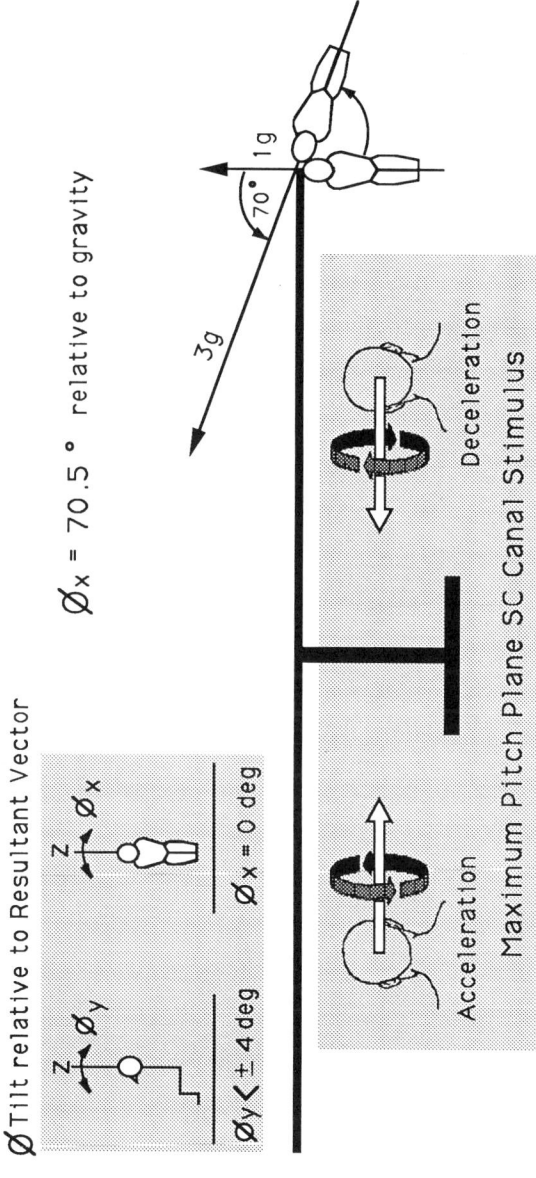

FIGURE 6. In a conventional centrifuge run, the subject rolls through 70.5 degrees relative to gravity but remains upright in the roll plane relative to the resultant force vector while the head-to-seat (z-axis) acceleration increases to 3 g-units. A low angular acceleration (0.112 rad/second²) produced a less than 4-degree shift of the total resultant vector in the subject's pitch plane. Angular acceleration vectors in the lower stippled area show that the total pitch plane semicircular canal stimulation was equal in magnitude but opposite in direction during acceleration and deceleration.

Spatial orientation dynamics perceived during deceleration differed markedly from the perceptual dynamics during acceleration, as illustrated in FIGURE 7. During deceleration, the average response was forward pitch displacement of about 90 degrees, so that subjects reported feeling face down as though diving directly toward the earth. Superimposed on the pitch displacement position was a pitch plane angular velocity, paradoxically much too great for the pitch *position* attained; in some subjects pitch-position distortion persisted for several minutes after the run terminated. During deceleration, a major pitch-plane attitude shift was perceived despite the presence of a linear acceleration vector almost fixed in the resultant plane of the semicircular canal response.

A linear vector in the pitch plane that *increased* in magnitude (1 to 3 *g*-units) suppressed pitch-plane angular *velocity perception* during acceleration, whereas

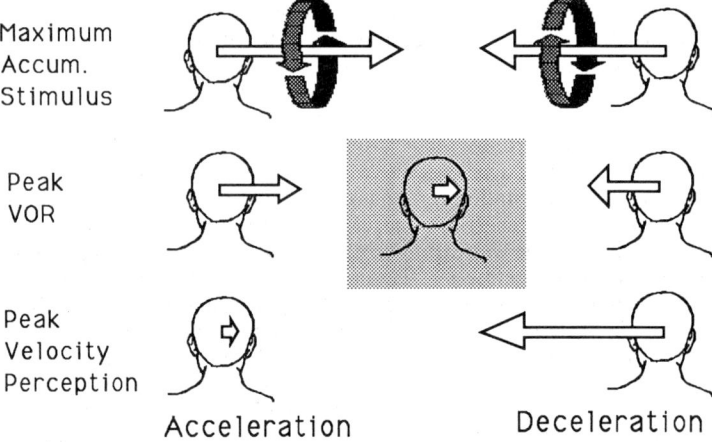

Maximum
Accum.
Stimulus

Peak
VOR

Peak
Velocity
Perception

Acceleration   Deceleration

FIGURE 7. Angular vectors illustrate the pitch plane acceleration and deceleration stimuli and responses. The maximum pitch plane vestibuloocular reflex (VOR) slow-phase velocity was slightly greater during acceleration than during deceleration. By contrast, the deceleration pitch velocity perceived was many times greater than the pitch velocity perceived during acceleration. (VOR fast-phase direction was used to facilitate pictorial comparisons of response magnitudes.) Perceived pitch velocity during deceleration is illustrated with a gain of 1.0. The gain of pitch plane VOR was 0.44 and 0.40 for the acceleration and deceleration VOR respectively. The stippled head in the center illustrates a mean 8 degrees/second sustained up-beating nystagmus during constant velocity at 3 G_z.

during deceleration a linear vector in the pitch plane that *decreased* in magnitude (3 to 1 *g*-unit) *did not suppress* pitch-plane angular *velocity perception*. And the angular *velocity* canal signal asserted strong influence on otolith-mediated angular *position* (attitude) *perception* only when the linear vector was decreasing in magnitude.

The accumulative pitch-plane canal stimuli were identical in magnitude at the end of centrifuge acceleration and deceleration, but the total resultant linear vector did rotate through a small angle in the pitch plane. The timing and direction of the pitch-plane vector rotation (Φ_y in FIGURE 6) may be important despite its small size, when considered in relation to concurrent canal activity. The initial Φ_y shift occurred just as angular acceleration commenced, but at this moment, only the horizontal canals were being stimulated. As centrifuge acceleration continued, the resultant vector rotated gradually back toward the z-axis, owing to the increasing magnitude of the G_z vector, and Φ_y rotation ended with a step change to z-axis alignment when centrifuge angular acceleration stopped. The direction of linear vector rotation (Φ_y) was the same as the pitch plane direction being signaled by the canals which, according to rationale presented in experiment 4, would diminish canal-mediated pitch angular velocity and add a slight linear velocity component to the perception. Conversely, as deceleration commenced, linear vector rotation (Φ_y) in the pitch plane was directionally opposite the pitch plane angular velocity signaled by the canals; moreover the pitch plane canal stimulus was fairly strong as this vector rotation commenced. The gradual reduction in the G_z force as the centrifuge slowed produced a very gradual rotation of the linear vector in a direction that would continue to enhance the canal-mediated pitch-plane velocity perception throughout the deceleration. At the end of the deceleration, a step Φ_y shift in the same direction as the canal response occurred but, by this time, a very strong pitch angular velocity perception was in progress. If the great differences in perceptual dynamics during centrifuge acceleration and deceleration are attributable to the relationship of concurrent canal activity to these very small vector rotations, then these observations have important far-reaching implications for model development and to heretofore unappreciated disorienting conditions in flight.

This series included runs in which subjects were placed in different positions relative to the developing acceleration vectors, e.g., upright but backward facing or supine with feet toward rotation center as centrifuge rotation commenced. Irrespective of body positioning, centrifuge deceleration produced the "strong" response, i.e., major position change with strong paradoxical angular velocity superimposed.

The vestibuloocular reflex (VOR) does not reflect the drastic acceleration/deceleration difference[18] that characterizes the perceptual dynamics although the VOR deviates in some respects from classical theory; a "linear nystagmus" component becomes apparent during constant velocity.[18] Both aspects of the response, the VOR and the perceptual dynamics, in these hyper-g runs must be subsumed by any comprehensive model that is meant to be predictive in the aerospace environment or in the world of dynamic flight simulation or in clinical medicine where unusual combinations of vestibular signals are generated by changes in sensory and central processes of the body.

We now consider briefly our five observation sets in relation to some ideas about the dynamics of responses sometimes viewed as indicative of otolith transduction dynamics. In the first catapult simulation (experiment 2), the direction of the resultant linear vector in the subject's pitch plane, initially gravity, rotated forward from the subject's z-axis through almost 76 degrees toward the x-axis of the head as the 4-g-unit maximum acceleration was achieved. At this moment the subject was, in effect, almost on his back in the $4G_x$ force field. If this condition were to be maintained ("steady state"), a pitch-up attitude of at least 76 degrees would be perceived. But subjects perceived little change in pitch attitude. The underperception of attitude change (pitch, roll, or yaw) during fast attitude change could be attributed to sluggish otolith system transduction or to slow central nervous system dynamics or to both, but an important additional consideration is the influence on perceptual dynamics of canal/otolith systems interaction. Otolith "change" information "on the x-axis" coupled with absence of pitch plane canal input seems to have generated more forward linear motion perception than pitch perception. Curiously the pitch-up attitudes recorded, about 15 degrees, persisted beyond the end of the stimulus, operationally very significant because the "error" could account for aircraft that flew into the water after catapult launch.

We project that pitch perception dynamics will be improved under stimulus conditions that generate concordant pitch plane canal/otolith information: specifically required would be a linear vector rotating in the same head plane but in opposite sense to the canal signal. In the five experiments described, perceived change in pitch attitude was produced in two different ways: (1) In the catapult simulation, the direction of the force field relative to otolith system changed quickly through a large angle without concordant pitch plane stimulation of the semicircular canals. A small (relative to the 76-degree pitch change) pitch-up attitude was perceived—which persisted beyond the end of the stimulus. (2) In the conventional centrifuge run (experiment 5), a large pitch plane attitude shift occurred during deceleration despite a stimulus to the otolith system that was directionally fixed except for a 4-degree rotation in the pitch plane. The attitude shift persisted beyond the end of the stimulus. The relation of otolith transduction dynamics to attitude perception dynamics and to dynamics of perceived linear translation is a significant unknown in our field. When a linear vector changes direction relative to the head and body without concordant stimulation of the semicircular canals, how slow is slow enough for the attitude perception to be equivalent to that of the static tilt condition? At what frequency of change is the perception transformed from tilt perception to perceived linear translation, a change that seems to be influenced by adaptation to prolonged weightlessness.[19]

SUMMARY

Results of the five experiments are consistent with the following generalizations. Canal-mediated turn perception (pitch, roll, or yaw) in earth-horizontal or earth-vertical plane, is suppressed in direct relationship to the magnitude of a linear acceleration vector lying in the plane of a responding canal when the magnitude of the linear vector is constant or increasing and when its direction is either fixed or rotating in the same direction as the concomitant canal signal. Canal-mediated turn perception (pitch, roll, or yaw) is not suppressed by a coplanar linear vector that is counterrotating relative to the canal signal. *Change* in perceived attitude (pitch, roll, or yaw) is very sluggish in the absence of concordant canal information; attitude change may not be an immediate otolith-mediated perceptual event but a slowly developing perception dependent upon cognitive appreciation of an immediate otolith angular position signal. Otolith phasic neural units, unreinforced by appropriate canal signals, may contribute more to a brief linear velocity component in perception than to rate of attitude change. Otolith-mediated attitude perception within a given earth-vertical plane can be distorted by strong coplanar angular velocity canal information. Once distorted, return to veridical attitude perception can be gradual because, in the absence of complimentary canal or visual information, recovery is dependent upon relatively slow cognitive appreciation of a prevailing otolith position signal. Several attractive hypotheses relating to the dynamics of attitude perception can only be tested by substantially more data on the dynamics of spatial orientation perception. Most of our objectives cannot be achieved without models that yield valid prediction of the dynamics of spatial orientation perception.

All of the observations in these experiments were carried out in darkness, or, in the simulated catapult experiment, without external visual reference. Various forms of visual information will change the dynamics of spatial orientation perception. My discussion has been limited to consideration of the vestibular system, as though the canal and otolith systems completely controlled the dynamics of spatial orientation perceptions. Obviously other partners in the dynamics of postural control, including vision, proprioception, and expectation, must be included in this challenging field of research.

Dedication to stereotyped ideas about objectivity in the 20th century has hindered advancement of knowledge on the dynamics of spatial orientation perception relative to rate of progress achieved by several scientists of the 18th and 19th centuries,[7,8,20,21] who provided word pictures of perceived motions and tilts along with descriptions of the motions that engendered the pictures. Word pictures coupled with sufficient description of accelerative stimuli are a productive first step in obtaining information on the dynamics of spatial orientation perception. Initial exploration can reveal trends and conditions that can be selected, refined, and tested for reliability by independent observation. When gross but reliable spatial orientation dynamics are discovered, innovative procedures will be found to add quantitative description. Results are likely to be unreliable when subjects are required to provide multidimension description of perception during single runs. Use of a "joy stick" to track concomitant rapidly changing pitch, roll, or yaw perception can be very misleading[11] and is nonsense in some responses like the paradoxical perceptions of centrifuge deceleration. A few simple rules can serve to provide descriptions of perceptual events that are as reliable as any response event in vestibular neurosciences. Witness for example the postrotation sensations of active and passive turning described by Purkinje in 1820, independently rediscovered recently.[22] Careful description of the statics and dynamics of perceptual events should be a major part of our effort to develop the models that are intended to represent our levels of understanding and that will eventually predict outcomes in complex motion environments.

REFERENCES

1. January 1991. NATIONAL STRATEGIC RESEARCH PLAN. NIDCD working document. National Institute of Defense and Communication Disorders, Balance/Vestibular section. Bethesda, Md.
2. REASON, J. T. & J. J. BRAND. 1975. Motion Sickness. Academic Press. London, England.
3. BENSON, A. J. 1988. Motion Sickness. In Aviation Medicine. J. Ernsting & P. King, Eds. Buttersworth. London, England.
4. LEGER, A. 1991. Operational significance of motion sickness in aerospace operations and sea survival. In AGARD Lecture Series No. 175. AGARD, NATO-OTAN. Neuilly-sur-Seine, France.
5. NEAL, W. 1990. Summer study, July. Aviator Physical Stress Panel (W. Neal, Panel Chairman). Naval Research Advisory Committee. San Diego, Calif.
6. GUEDRY, F. E. 1991. Motion sickness and its relation to some forms of spatial orientation: mechanisms and theory. In AGARD Lecture Series No. 175. AGARD, NATO-OTAN. Neuilly-sur-Seine, France.
7. COHEN, B. 1984. The roots of vestibular and oculomotor research. Introduction. Hum. Neurobiol. 3: 121–128.
8. GRÜSSER, O. J. 1984. J. E. Purkyne's contribution to the physiology of the visual, the vestibular and the oculomotor systems. Hum. Neurobiol. 3: 129–144.
9. GRISSETT, J. D. & F. E. GUEDRY. Variation in post rotation turning sensation with different angles of axis tilt. (In preparation.)
10. COHEN, M. M., R. J. CROSBIE & L. H. BLACKBURN. 1973. Disorienting effects of aircraft catapult launchings. Aerosp. Med. 44(1): 37–39.
11. GUEDRY, F. E. 1974. Psychophysics of vestibular sensation. In Handbook of Sensory Physiology. H. H. Kornhuber, Ed. 6: 1–154. Springer-Verlag. New York, Heidleberg & Berlin.
12. BLOOMBERG, J., G. MELVILL JONES & B. SEGAL. 1991. Adaptive modification of vestibularly perceived rotation. Exp. Brain Res. 84: 47–56.

13. CARGILE, R. M., III. 1991. Attenuation of the vestibular memory contingent saccade. Honors Thesis. Department of Biomedical Engineering. Tulane University. New Orleans, La.
14. COHEN, M. M. 1991. Personal communication.
15. GUEDRY, F. E. & C. M. OMAN. 1990. Vestibular stimulation during a simple centrifuge run. Report 1353. Naval Aerospace Medical Research Laboratory. Pensacola, Fla.
16. CORREIA, M. J., W. C. HIXSON & J. I. NIVEN. 1968. On predictive equations for subjective judgments of vertical and horizontal in a force field. Acta Otolaryngol. Stockh. (Suppl. 230).
17. SCHÖNE, H. 1964. On the role of gravity in human spatial orientation. Aerosp. Med. **35:** 764–772.
18. MCGRATH, B. J. 1990. Human vestibular response during 3 G_z centrifuge stimulation. Master of Science Thesis. Massachusetts Institute of Technology. Cambridge, Mass.
19. PARKER, D. E., M. F. RESCHKE, A. P. ARROTT, J. L. HOMICH & B. K. LICHTENBURG. 1985. Otolith tilt-translation reinterpretation following prolonged weightlessness: implications for preflight training. Aviat. Space Environ. Med. **56:** 601.
20. HENN, V. 1984. E. Mach on the analysis of motion sensation. Hum. Neurobiol. **3:** 145–148.
21. BORING, E. G. 1942. Sensation and perception in the history of experimental psychology. D. Appleton–Century Co. New York, N.Y.
22. GUEDRY, F. E., C. E. MORTENSEN, J. B. NELSON & M. J. CORREIA. 1978. A comparison of nystagmus and turning sensation, generated by active and passive turning. *In* Vestibular Mechanisms in Health and Disease. J. Hood, Ed. Plenum. New York, N.Y.

Sense of Body Position in
Parabolic Flight[a]

JAMES R. LACKNER

Ashton Graybiel Spatial Orientation Laboratory
Brandeis University
Waltham, Massachusetts 02254-9110

Under terrestrial conditions we normally take for granted the perceptual continuity and stability of our own orientation and that of our surroundings. For example, if we close our eyes, we expect to continue to experience the same orientations of our bodies. Not only do we expect our sense of the direction of up and down not to be influenced by whether our eyes are open, we also expect to continue to have a sense of up and down if we close our eyes. However, some years ago in the course of studying gravitoinertial force influences on otolithic function, we became aware of the complexity of the sensory and motor factors that determine perceived orientation in a weightless environment.[1] This made us realize that perceived orientation under terrestrial conditions which we so take for granted is also the result of quite complex processes.

In these early experiments, subjects were being rotated in barbecue spit fashion (see FIGURE 1) at constant velocity and their eye movements and their sense of position within a rotary cycle were being continuously monitored. At constant velocity after the semicircular canal response to acceleration had decayed, their otolith organs were continuously reoriented in relation to the force of gravity. In addition, there was a continuously changing pattern of touch and pressure stimulation of the body surface, with the "down" side of the body being stimulated by the contact forces of support provided by the restraint harness and body mold. If a subject's eyes were open, he also received visual information about his changing spatial orientation in relation to the test chamber.

At constant velocities of rotation above 10–12 rpm (sometimes as low as 6 rpm), subjects did not experience rotation about their z-axis if their eyes were closed.[2,3] Instead, most subjects experienced themselves to be going through an orbital motion while always facing in the same direction, up or down. The direction of orbital motion was opposite to that of the actual direction of rotation. With full vision, subjects correctly perceived themselves to be turning in the true direction of rotation. FIGURE 2 illustrates the eyes-closed situation.

We had the opportunity to test subjects in this situation in parabolic flight maneuvers during which there are periods of increased and decreased force level. The flight profile of the Boeing KC-135 aircraft used in our experiments is illustrated in FIGURE 3. Periods of high force, approximately 1.8 G peak, and free fall, 0 G, each lasting about 25 seconds alternate. During the 1.8-G period, the forces acting on the otolith organs nearly double as do the contact forces of support acting on the subject's body. In free fall, the net force on the otolith organs is effectively zero, and the contact forces of support are virtually zero as well.

Subjects rotating at constant velocity in straight and level flight experienced the

---

[a]Support was provided by National Aeronautics and Space Administration Grant NAG9-515.

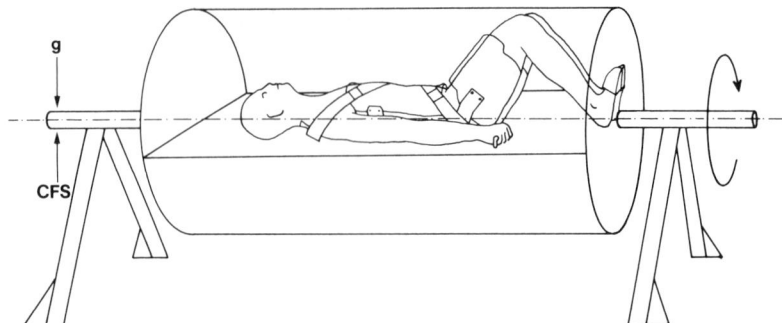

FIGURE 1. Schematic illustration of the forces acting on a subject's body during constant-velocity rotation about his horizontal z-axis. The force of gravity, G, is opposed by the contact forces of support, CFS, provided by the body mold (not illustrated) which encloses and restrains the subject.

same pattern of self-motion as in the laboratory. With the onset of parabolic maneuvers, however, the subjects experienced dramatic changes in their experienced orientations.[1] On transition to 1.8 G with eyes blindfolded, they experienced their "orbits" to spiral out to a greatly increased diameter and their orbital velocity to have greatly increased as well (they "completed" one orbit each time they physically rotated through 360°). When permitted normal vision in 1.8 G, they experienced rotation of their bodies in the direction of actual rotation but simultaneously felt themselves to be undergoing an orbital motion (15–30 inches in diameter) in the opposite direction.

In 0 G, by contrast, with eyes blindfolded the subjects experienced no motion whatsoever; they felt absolutely stationary and lost a sense of orientation to the aircraft. That is, they knew where they were in the aircraft and that they were rotating, but they did not feel in any orientation vis-à-vis the aircraft and they did not

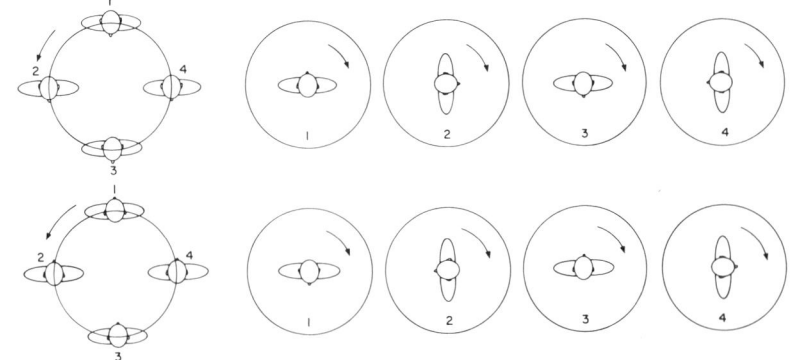

FIGURE 2. The leftmost figures illustrate the face-down and face-up orbital motions experienced by blindfolded subjects rotating at constant velocity. The numbered figures to the right show the true body orientations for the corresponding, numbered felt positions in the orbits.

have a sense of up and down. Although they had lost their spatial anchoring to their surroundings and had lost perception of the vertical, they retained spatial awareness of their relative body configuration. When permitted normal vision in 0 G, they saw themselves as turning in relation to the aircraft and saw it as stable, but they did not experience a compelling feeling of self-rotation. All of the subjects while blindfolded in 0 G found that if they pushed with their head or feet against the ends of the apparatus, this would restore their sense of orientation. Foot pressure made them feel as if they were upright standing on their feet, pressure on top of their head as if they were upside down resting on their head.

These observations made us interested in the general issue of what factors contribute to the sense of orientation in free fall as well as in other force backgrounds. Since the earliest space flights, astronauts have reported a variety of orientation illusions.[4] One of the most interesting of these is the so-called inversion illusion in which an astronaut who is free floating and normally oriented in relation to the visual layout of the spacecraft reports that both he and the spacecraft are

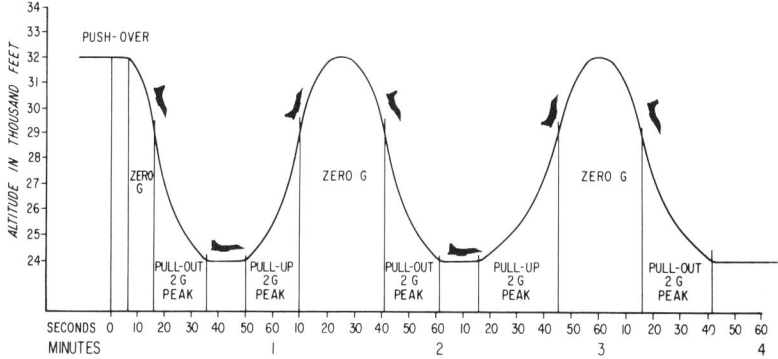

FIGURE 3. Schematic illustration of the flight profile of the Boeing KC-135 aircraft used in our experiments.

inverted.[5] This is a situation that Mittelstaedt has referred to as cue-free inversion because the visual cues are concordant with being upright in a normal environment.[6] In the course of experiments on space motion sickness, we have had the opportunity to evaluate sensory and motor influences on orientation in parabolic flight under a variety of conditions including (1) with subjects strapped in flight seats, (2) with subjects free floating, and (3) with subjects exposed to optical inversion of the visual array.

METHODS

Subjects

Sixty-eight individuals participated. Each had normal vestibular function as demonstrated by systematic assessments of the vestibuloocular reflex, thresholds for rotation, and ocular counterrolling. All had passed an FAA class II flight physical.

Flight Path

FIGURE 3 illustrates the flight path: typically 40 parabolas were flown on a given test day, usually in four sets of 10 consecutive parabolas. During a test week, each subject participated in four consecutive days of flights. Most of the subjects participated in several flight weeks.

Procedure

Initial Baseline Observations

Each of the 68 subjects was restrained in an aircraft seat during his first two, 40-parabola flight days. The subjects were restrained in place by tightly drawn lap belts. On one of the test days a subject was blindfolded, on the other permitted normal vision, with the order balanced across subjects. Each subject's task during these flights was to report periodically on his apparent orientation and that of the aircraft, and whether rising or falling movements of his body were experienced either during transitions in force level or at steady state, low or high force levels.

Free Floating

Thirty-one subjects were tested while free floating. They were maneuvered into different orientations in relation to the aircraft by the experimenters, then released. The subject then reported on his own experienced orientation and that of the aircraft as he looked straight ahead of his body, toward his feet, and beyond his forehead. In some parabolas, the free-floating subject would close his eyes and then describe his orientation as he directed his gaze (head and eyes) toward his feet, straight ahead, or toward his forehead.

Visual Inversion

Eighteen of the subjects were tested wearing a head-mounted prism that inverted the optical array. The device provided a field of view of about 115° horizontally and 45° vertically. The subjects wore the prism while seated in position and restrained by a lap belt first in straight and level flight and then for at least five parabolas. Subjects reported on how the optical inversion affected their sense of self-orientation.

RESULTS

Baseline Observations

Eyes Blindfolded

Virtually all of the subjects (66 of 68) felt upside down in at least the first several parabolas when they were blindfolded; it felt to them as if they were suspended from their seat belts. Some felt upside down in an upright aircraft, others felt the aircraft was also upside down. During the transition into 0 G when the resultant force dropped to between 0.1 G and 0 G, subjects felt as if their bodies had done a backward flip to become inverted, or an internal telescoping such that their head moved down and their feet up and became inverted without any rotation. On transition out of 0 G, at approximately 0.05–0.2 G subjects again felt upright, but did not feel their bodies rotate through space to become rightside up. Thus, the perceptual responses to transitions in force level are asymmetric in character, not mirror images, for going to 0 G versus going to greater than 0 G.

Normal Vision

During their first several exposures to 0 G while strapped in their aircraft seats, 35 of the 68 subjects reported feeling that both they and the aircraft were inverted. Eight others reported themselves to be inverted but not the aircraft, and two said the aircraft was inverted in relation to them but that they felt upright. The remaining subjects always felt themselves and the aircraft to be upright. The subjects who experienced the different variants of inversion never experienced a physical rotation

TABLE 1. Apparent Orientations of Self and Aircraft during Exposure to Weightlessness in Parabolic Flight Maneuvers[a]

	Self Upright Plane Upright	Self Upright Plane Inverted	Self Inverted Plane Upright	Self Inverted Plane Inverted
Parabolas 1–5				
Blindfolded	2	0	52	14
Normal vision	23	2	8	35
Parabolas > 80				
Blindfolded	5	0	51	12
Normal vision	62	0	1	5

[a]The subjects were strapped in aircraft seats at the time of testing. $n = 68$.

of themselves and the aircraft to become inverted or a visual rotation. Instead as the G force went to 0, they experienced themselves or the aircraft to be upside down. The experience was of a fading-out, fading-in character: the subjects would feel less and less vividly upright then more and more inverted until the sensation was completely compelling. When about 0.1 G was attained in the pull-up phase of a parabola, they again felt themselves and the aircraft to be normally oriented without experiencing any body or aircraft rotation through space. TABLE 1 summarizes the subjects' perceptual reports.

Sensations of Falling and Rising

Only three subjects reported sensations of downward motion in the 0-G phases of parabolic flight and only during their first five or six parabolas. Afterwards the feelings were gone. The sensations reported were those of an apparent downward motion of themselves and the aircraft without a physical displacement through space. With exposure to high force levels, three subjects during their first four or five parabolas felt a sense of downward displacement of themselves and the aircraft but not a sense of falling. The others felt as if they were being pulled or sucked toward the floor of the aircraft but did not feel physical displacement.

Sensations of rising upward without displacing through space were reported by four subjects in 0 G during their first two or three parabolas while blindfolded. None of the subjects had any sensations of upward motion in the high force phases of the parabolas.

Free Floating

Thirty-one subjects were tested as they free floated in 0 G. In the absence of any contact cues (except from their clothing), 23 of them lost all sense of spatial anchoring to the aircraft and to external space if they closed their eyes, although they retained cognitive awareness of their actual position in relation to the aircraft. The other subjects felt their orientation to the aircraft had not changed but was less vivid. Seven of the 23 subjects found that "looking" toward their feet made them feel upright as did "directing" their closed eyes toward their foreheads, but they still had no firm sense of position vis-à-vis the aircraft.

If a subject's eyes were open as he free floated, his experience of self and aircraft orientation depended on his physical position in the aircraft and whether parts of his body were visible in relation to architectural features of the aircraft. The relationships between a subject's actual and experienced orientations are presented in

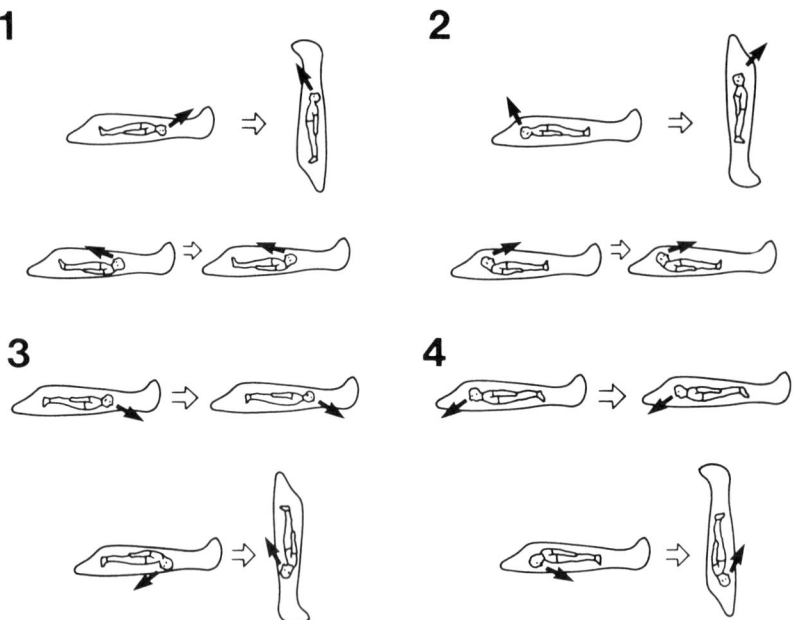

FIGURE 4. The leftmost figure of each pair shows the actual orientation of the subject to the aircraft; the rightmost figure depicts what the subject perceived his orientation and that of the aircraft to be.

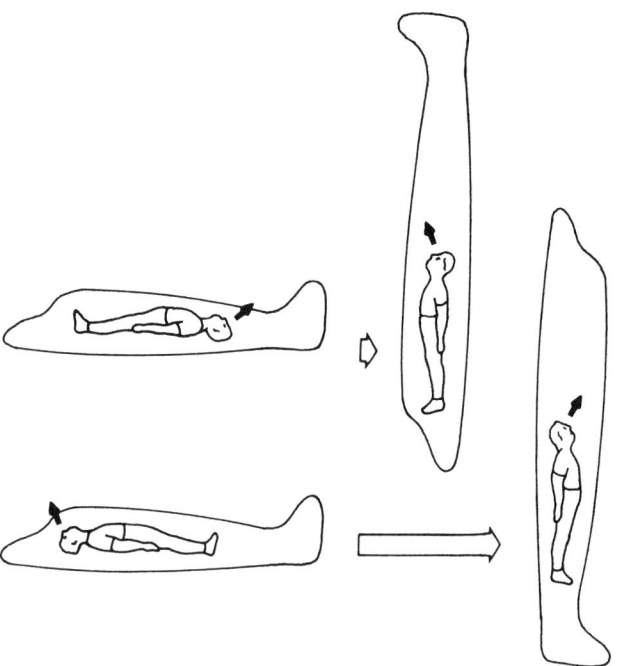

FIGURE 5. Tunnel illusions of aircraft elongation associated with perceptual remappings of self and aircraft orientation.

FIGURE 4 for several test configurations. The changes in experienced orientation associated with changing gaze were always of the fading out–fading in variety. The subject would feel less and less compellingly in the old orientation and then more and more vividly in the new. Experienced rotations through space were never reported.

Many of the subjects reported that a striking aspect of the changes in orientation associated with gaze shifts was a severalfold visual elongation of the fuselage. This is illustrated in FIGURE 5.

Optical Inversion

Thirteen of the 18 subjects tested found that looking through the inverting prism while strapped in their aircraft seats affected their apparent orientation in 0 G. Nine of these subjects no longer experienced inversion illusions when seated and looking about the aircraft normally, but looking through the prism they perceived both themselves and the aircraft to be upside down. Of the other four subjects, three of them when they were not looking through the prism felt both themselves and the aircraft to be inverted, the other felt just himself to be inverted. All four felt rightside up looking through the prism. The changes in apparent self-orientation evoked by the optical inversion were always of the fading out–fading in character and took several seconds to be completed. Looking through the prism had no influence on apparent self-orientation in 1-G straight-and-level flight or in the high-force phases of the parabolas.

A number of the subjects were experiencing symptoms of motion sickness prior to looking through the prism. Most of them felt that the resulting change in experienced orientation made their symptoms worse.

Abatement of Inversion Illusions with Repeated Exposure to Parabolas

Sixty-two of the 68 subjects after repeated exposure to parabolic maneuvers came to feel rightside up in a rightside up aircraft when strapped in their flight seats and permitted normal vision. By contrast, when strapped in and blindfolded 63 of the 68 still felt inverted in 0 G, hanging from their seat belts. Six subjects always felt inverted in free fall when strapped in their seats with normal vision, or when blindfolded without contact cues on their bodies. These subjects had had many exposures to free fall in parabolic flight, having participated in 16 to 24 40-parabola flights.

DISCUSSION

A key feature of our observations is that in the absence of touch and pressure cues, individuals who were blindfolded and free floating generally lost any sense of firm, spatial anchoring to their surroundings. They knew where they were but did not feel in that location or in any particular location at all. Simply by opening and closing their eyes, they could switch back and forth from feeling a sense of orientation to not. On earth, opening and closing our eyes does not affect our experienced orientation in relation to the environment. In this context, it is worth noting that cells have been identified in the hippocampus of experimental animals that code "place."[7] These cells seem to code locations in the environment with different cells responsive with the animal in different locations and orientations in the environment. Such "place cells" continue to fire if the room lights are extinguished, which has led investigators to assume they are also coding memory of position. Our observations lead to the interesting and testable speculation that perhaps such place cells, or their equivalent for perception, are not activated in the absence of vision if vestibular and touch and pressure cues about orientation to gravity are absent, e.g., in weightlessness, or are activated differently if vestibular and contact cues are abnormal, e.g., the animal is suspended inverted.

Regardless of the value of this speculation, it is clear that in the absence of vision, touch and pressure cues have a dominant influence on apparent spatial orientation in 0 G. Nearly all of our subjects continued to feel inverted in 0 G when they were strapped tightly in their seats with a lap belt. It felt to them as if they were suspended from the belt. Similarly, in our barbecue spit rotation studies, if blindfolded subjects pushed in the apparatus with their feet, they often felt standing upright; pressure on the top of their heads could make them feel upside down.[1-3] Accordingly, in the absence of visual and vestibular cues to the gravitational vertical, the sense of body orientation is determined by touch and pressure stimulation of the body surface, with the region of greatest overall force or pressure indicating "down."[8]

Seated, strapped in position, and permitted normal vision, many of our subjects experienced sensations of inversion during their initial exposures to 0 G. Interestingly, all possible combinations of self and aircraft inversion were reported, although feeling upside down in an upside down aircraft was most commonly reported. The unloading of the otolith organs in 0 G has often been evoked as the basis of the "inversion illusion."[5] Our observations show that actually there are a variety of inversion illusions. With repeated exposure while seated and permitted normal vision, most subjects came to feel upright in an upright aircraft. However, some subjects continued to feel inverted regardless of their number of exposures to 0 G. Mittelstaedt has developed a model of otolith function[6] which predicts that subjects with a negative saccular bias and an "ideotropic vector" less than unity will feel inverted in 0 G. It may well be that these subjects met these criteria. An advantage of the model is that the ideotropic vector and saccular bias are operationally defined so that ground-based assessments of their magnitude can be made.

The unloading of the otolith organs in 0 G did not, with only a few minor exceptions, elicit sensations of falling, nor were sensations of upward displacement and motion triggered in the high-force periods. The saccules are likely key receptors in the otolith spinal reflex elicited by falling on earth.[9,10] Nevertheless, sensations of falling were not elicited by actual free fall. The presence of a subject-stable, visual surround could, on the basis of the observations of Nashner and Berthoz,[11] Soechting, Berthoz, and Nashner,[12] and Lacour, Vidal, and Xerri,[13] be expected to depress both vestibulospinal and perceptual responses to free fall. However, even when blindfolded the subjects did not experience themselves to be falling. Accordingly, it is likely that cognitive factors, such as awareness that the aircraft could not move straight up or down, influenced the perceptual responses to 0 G. The relatively gradual transitions to 0 G may also have been a factor (see FIGURE 3).

The importance of visual and cognitive factors to experienced orientation in 0 G is underscored by the patterns perceived by subjects free floating without touch and pressure cues. Their apparent orientation was influenced by their physical orientation in relation to the architectural layout of the aircraft and whether parts of their bodies were visible in relation to the aircraft. For example, subjects free floating, facing the deck of the aircraft, felt upside down in a vertical aircraft if they looked toward their feet. Under terrestrial conditions it is impossible, except momentarily, to see one's body above the floor and not in contact with it unless one is artificially suspended. A perceptual reinterpretation occurs during transient exposure to 0 G that makes the visual input compatible with terrestrial conditions, e.g., one can see the front of one's body and feet if one is upside down looking upward. It remains to be determined if "terrestrial remappings" would occur during prolonged microgravity in spaceflight.

Interestingly, none of the situations involving physical and visual separation of the subject from the support surface evoked sensations of falling or of startle. Under terrestrial conditions, placing an animal in a visual cliff apparatus in which it stands on transparent glass above the visually identified support surface (ground) readily evokes startle or avoidance behavior.[14] This is true both for many species of animals and for human infants. No subject reported anything comparable in our experiments. Inverting the optical array also evoked sensations of inversion (and in some cases "un-inversion") in many of our subjects, again emphasizing the importance of the visually specified, architectural up and down of the environment in affecting apparent self-orientation.

Experienced orientation in weightlessness is under multimodal control and cognitive influence. This is clear from the nature of the changes in orientation experienced under different circumstances: (1) the backward flips and telescoping during transitions into 0 G, (2) the telescoping and fading-out, fading-in reorientations associated with closing and opening the eyes and redirecting gaze. Changes in orientation not associated with force transitions are of the fading-out, fading-in variety probably because the absence of canal signals and of appropriate phasic touch and pressure cues indicates that the body has not physically rotated. During transition to 0 G, subjects facing forward in the aircraft may experience backward flips as the gravitoinertial force vector diminishes in magnitude and tilts slightly in relation to them. By contrast, in other experiments we have found that seated subjects who are rotating experience a telescoping or fading-out, fading-in change in orientation. At present, we have no way of explaining the airplane elongation illusion (FIGURE 4).

Thus, in free fall, the nervous system is generating a sense of the spatial orientation of the body in relation to the environment on the basis of available patterns of sensory stimulation. In the absence of any exteroceptive stimulation, all sense of spatial orientation of self and of self to the environment can be lost, while sense of relative body configuration and cognitive awareness of body position in the aircraft are retained. If only touch and pressure cues are present, they can serve to specify the direction of "down." With vision permitted, there is a complex interplay of the architectural features of the vehicle, body position in relation to these features, and whether the body can be seen separated from the floor of the aircraft in determining orientation. Orientations tend to be generated that are compatible with terrestrial conditions, that is, with being in a 1-G environment. The absence of sensations of falling is likely due to cognitive influences. It should not be surprising that orientation in microgravity is under multimodal and cognitive influence because of course it is on earth as well. Moreover, just as on earth, exposure history is a key factor affecting the adaptive control of orientation and movement.

ACKNOWLEDGMENTS

I thank Charles Diamond, G. E. Tanner, and L. B. Lamolinara for technical assistance and Dr. H. Dolezal who provided the inverting prism we used.

REFERENCES

1. LACKNER, J. R. & A. GRAYBIEL. 1979. Parabolic flight: loss of sense of orientation. Science 206: 1105–1108.
2. LACKNER, J. R. & A. GRAYBIEL. 1978. Postural illusions experienced during z-axis recumbent rotation and their dependence on somatosensory stimulation of the body surface. Aviat. Space Environ. Med. 49: 484–488.
3. LACKNER, J. R. & A. GRAYBIEL. 1978. Some influences of touch and pressure cues on human spatial orientation. Aviat. Space Environ. Med. 49: 798–804.
4. TITOV, G. & M. CAIDIN. 1962. I Am Eagle. Bobbs Merrill. Indianapolis, Ind.
5. GRAYBIEL, A. & R. KELLOGG. 1967. The inversion illusion in parabolic flight: its probable dependence on otolith function. Aerospace Med. 38: 1099–1103.
6. MITTELSTAEDT, H. 1987. Inflight and post-flight results on the causation of inversion illusions and space sickness. In Scientific Results of the German Spacelab Mission D1. P. Sahm, R. Jansen & M. Keller, Eds.: 525–536. DFVLR. Koln, Germany.
7. EICHENBAUM, H. & N. COHEN. 1988. Representation in the hippocampus: what do hippocampal neurons code? TINS 11: 244–248.
8. LACKNER, J. R. & A. GRAYBIEL. 1983. Perceived orientation in free fall depends on visual, postural, and architectural factors. Aviat. Space Environ. Med. 54: 47–51.
9. MELVILL JONES, G. & D. WATT. 1971. Muscular control of landing from unexpected falls in man. J. Physiol. 219: 729–737.

10. GREENWOOD, R. & A. HOPKINS. 1976. Muscle responses during sudden falls in man. J. Physiol. **254:** 507–518.
11. NASHNER, L. & A. BERTHOZ. 1978. Visual contribution to rapid motor responses during postural control. Brain Res. **150:** 403–407.
12. SOECHTING, J. & A. BERTHOZ. 1979. Dynamic role of vision in the control of posture in man. Exp. Brain Res. **36:** 551–561.
13. LACOUR, M., P. VIDAL & C. XERRI. 1981. Visual influences on vestibulospinal reflexes during vertical linear motion in normal and hemilabyrinthectomized monkeys. Exp. Brain Res. **43:** 383–394.
14. WALK, R. & E. GIBSON. 1961. A comparative and analytical study of visual depth perception. Psychol. Monogr. **75:** 2–34.

Multisensory Integration in Microgravity[a]

LAURENCE R. YOUNG, D. KEOKI JACKSON, NICOLAS
GROLEAU, AND SHERWOOD MODESTINO

Man-Vehicle Laboratory
Department of Aeronautics and Astronautics
Massachusetts Institute of Technology
77 Massachusetts Avenue
Cambridge, Massachusetts 02139-4307

INTRODUCTION

It is known that visually induced feelings of self-motion are normally inhibited if vestibular signals fail to confirm self-motion. It is hypothesized that, with increasing exposure to weightlessness, the strength of visually induced self-motion will increase due to the absence of confirming vestibular signals. It is expected that crew members will become increasingly dependent on visual and tactile (touch) cues, as opposed to vestibular cues, for spatial orientation. The alteration of the visual-vestibular interaction by loss of gravity is considered one of the possible causes of space motion sickness.[1]

METHODS

To investigate the interaction between conflicting visual, vestibular and tactile information, the crew member places his head inside a "rotating dome" hemispherical display. The moving visual field rotates at a constant velocity of 30, 45, or 60 deg/second. The stimulus and recording apparatus has been used twice before in space[2,3] and is described in earlier papers. For the SLS-1 experiments, the exposure time was shortened to 20 seconds of constant-velocity rotation, separated by a 10-second pause. In addition, in order to pursue the possibility that neck proprioception was enhanced in microgravity, each subject was tested in a free-floating condition, held in place only by the bite board, in which the initial body position was offset by approximately 45 degrees laterally, to induce a neck twist. The remaining "free float" and "tactile" (restrained by bungee cords pulling the subject footward) conditions were retained as earlier. A sensation of self-rotation in the direction opposite to that of dome rotation is felt by the crew member. Before entering the dome, a soft contact lens, marked with a "starburst" pattern to serve as landmarks for postflight data analysis, is inserted and wetted with distilled water causing temporary adherence of the lens to the cornea. Crew members move a joystick to indicate their perception of self-motion. Torsional (rolling) eye movements and body angle are recorded by split-screen, closed-circuit television. Neck torque, produced when the crew member attempts to "correct" for the perceived roll, is recorded using

[a]This research was supported under National Aeronautics and Space Administration contract NAS9-15343.

strain gauges attached to the rotating dome bite board. For some of the pre- and postflight tests the subject stood in front of the rotating dome with headfree of the bite board, to reduce the localized nondirectional tactile cues. In some of those tests additional nondirectional localized tactile cues were applied to both shoulders, producing a light force that was independent of sway. For those tests, lateral sway of the subject was recorded.

Electrodes are placed on the skin above the neck muscles to record the electromyogram (EMG) representing muscle activity. Tactile and proprioceptive cues (those having to do with the physical state of the body, tendon, muscle, pressure sensations) simulating gravity are applied as a control using elastic cords to create an upward force on the bottom of the foot.

Visual field dependence is measured pre- and postflight by the "rod and frame" test, whereby subjects, seated in a darkened room, cause a rod to be aligned to the subjective vertical. A surrounding lighted "frame," which can be tilted 28 degrees clockwise or counterclockwise, may produce a static visual influence on the perceived vertical. The magnitude of this influence represents the contribution of visual cues relative to inertial signals in static orientation.

RESULTS

This reserve in-flight experiment was scheduled for one performance by 2 payload crew members, and anticipated further participation by the orbiter crew. Although the orbiter crew never did participate in the flight tests, the payload crew set up the "rotating dome" on the fifth day of the flight (MD5), despite two major delays and a failure in the connector to one EMG amplifier. The crew repaired the broken connector on their own initiative, and ran the dome for a full set of trials on all four payload specialists on the sixth day of flight (MD6).

Subject P completed a full set of dome runs on MD5, with 6 trials each for the "free-float" case (restrained only by the bite board and a hand on the joy stick), followed by the "neck twist" condition, in which the subject began each trial with significant head roll toward the right shoulder, and finally the "bungee" run, in which the feet were forced to an adjustable grid plate by stretched elastic cords. Since no differences were seen between the two conditions, "neck twist" trials were combined with free-float trials for the analyses. The measure of perceived rotation rate, obtained from the potentiometer controlled by the subject, is maximum vection strength. The vection magnitude is scaled for each run, by the subject, so that a self-induced motion rate equal to the speed of the rotating dome, or "saturated vection," is equated to 100%. The first day results for P (FIGURE 1 top) were partially in conformity with earlier tests; the average vection for all runs was significantly stronger than the average preflight erect or preflight supine tests with the dome. (This is in conformity with the hypothesis of shifting emphasis from vestibular to visual cues in space.) However, an unexpected finding was the *stronger* vection experienced by P for the bungee runs, with artificial localized tactile cues, than for the free floating or neck twist conditions.

When P repeated the full set of tests on the following day (FIGURE 1 bottom), the average vection was reduced, and now the free-float vection was not significantly higher than preflight. However, the body sway observed during some of the free-float trials was quite large, as the subject apparently attempted to maintain head orientation against apparent falling opposite to the dome direction. The subjective report for the free-float trials was of even greater sway, and surprise at the limit of actual vestibulocollic reflex. Postflight tests showed a similar unusual reaction to tactile

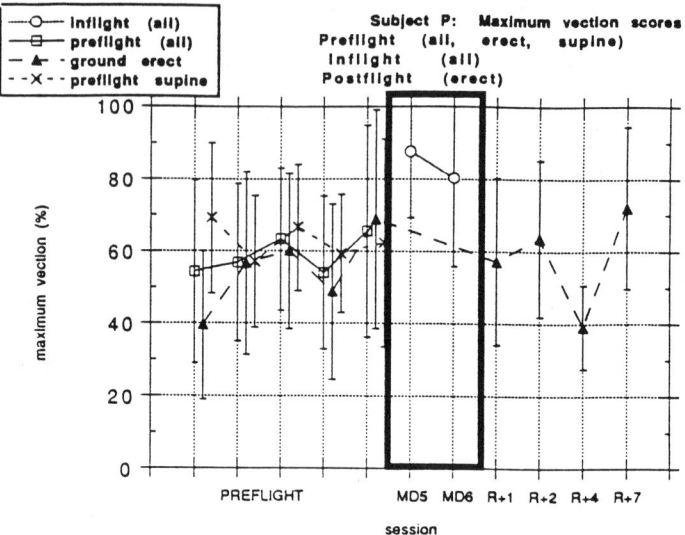

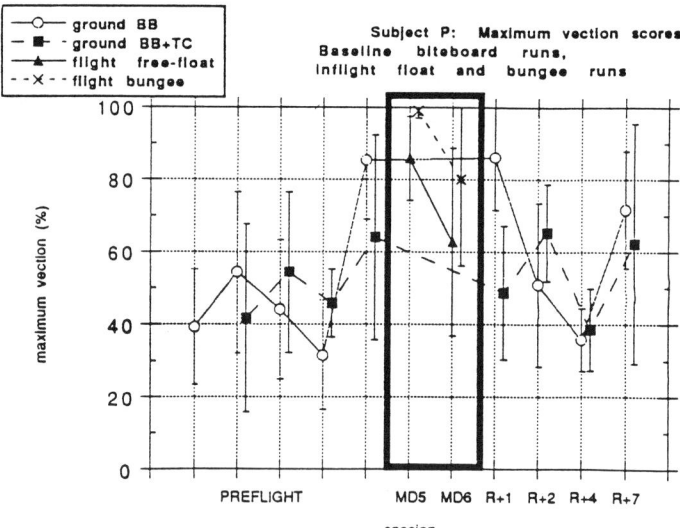

FIGURE 1. Visual-vestibular interaction (rotating dome) experiment. **Top:** Maximum vection scores, all conditions, subject P. **Bottom:** Maximum vection scores, bite board, with and without tactile cues, subject P.

cues. Whereas the early postflight tests with a bite board for head stabilization showed elevated vection, consistent with a carryover of flight adaptation, the addition of further nondirectional tactile cues by applying light force to the shoulders had little influence on vection. P reacted to weightlessness by increasing attention to visual cues, but reacted unexpectedly to tactile cues, which seemed to reinforce a sense of vection rather than to inhibit it.

Subject M, who never experienced strong or saturated vection except during the pre-flight parabolic flights, showed the expected influence of microgravity and of tactile cues during the MD6 tests. Free-float vection was significantly stronger and

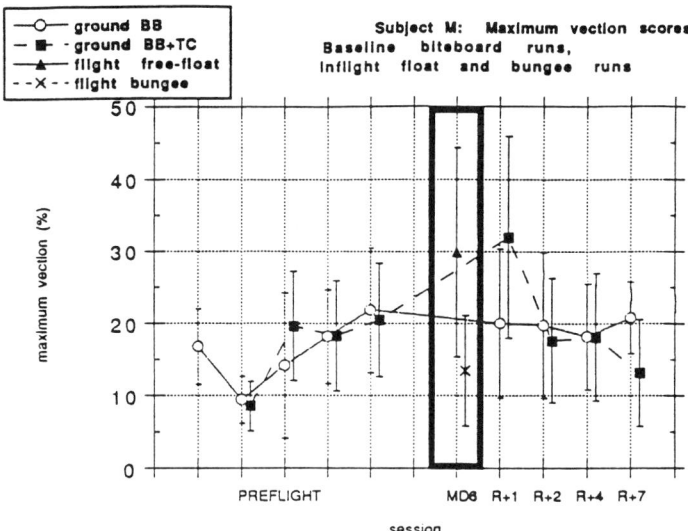

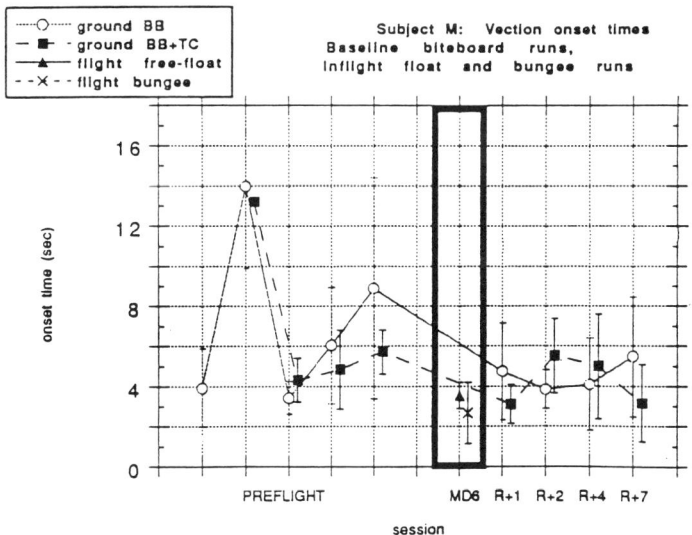

FIGURE 2. Visual-vestibular interaction, subject M. **Top:** Maximum vection scores, all conditions. **Bottom:** Onset latency for vection, biteboard, with and without tactile cues.

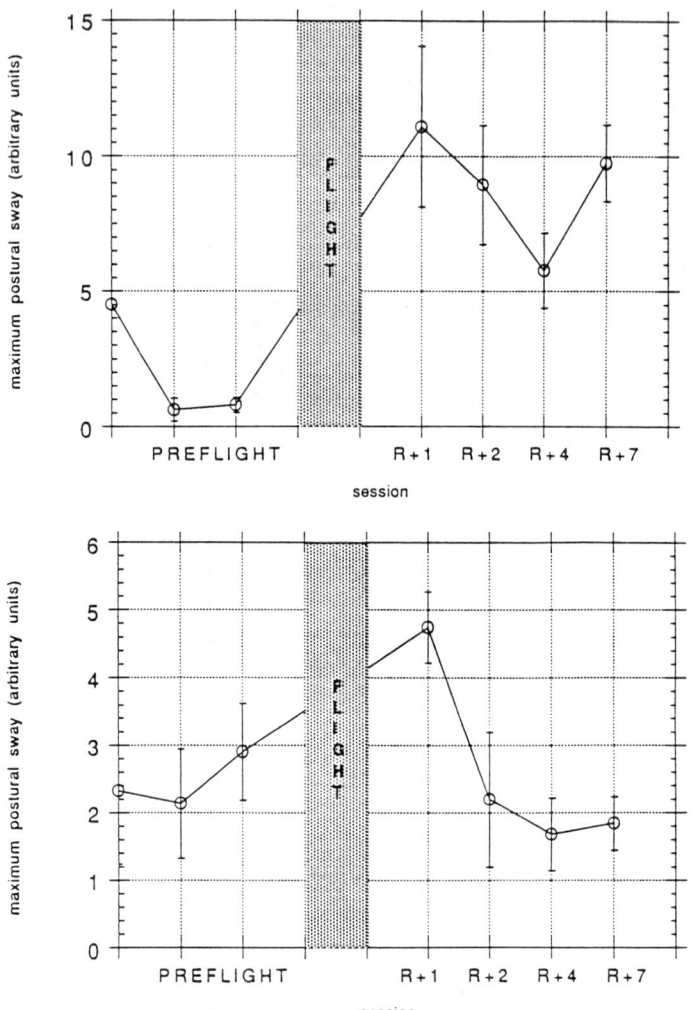

FIGURE 3. Maximum postural sway, pre- and postflight. **Top:** Subject M. **Bottom:** Subject T.

came on earlier than preflight, and bungee runs produced significantly less vection than floating or preflight testing (FIGURE 2). Postflight, vection remained elevated and latencies short until four days after the return. Postflight postural instability, in response to the rotating dome without bite board stabilization, was evident for the entire postflight week of testing (FIGURE 3 top).

Subject N was the most unusual. Alone among the 13 subjects tested in flight, N indicated *reduced* vection and longer latencies in flight, for both free-float and bungee conditions, than preflight or postflight. (FIGURE 4).

Subject T reported a subjective sense of substantial body sway (approaching 180 degrees) during free-float runs on MD6, but failed to show any differences in joystick indications of sway, for either bungee or free-float conditions, relative to pre- and postflight measures (FIGURE 5). The evidence of altered processing of spatial information was most clearly seen in the postflight body sway during erect dome tests on the day after return (FIGURE 3 bottom).

Measurements of bite board torque for erect dome runs demonstrated a significant and continuing instability in posture and dependence on visual cues for posture

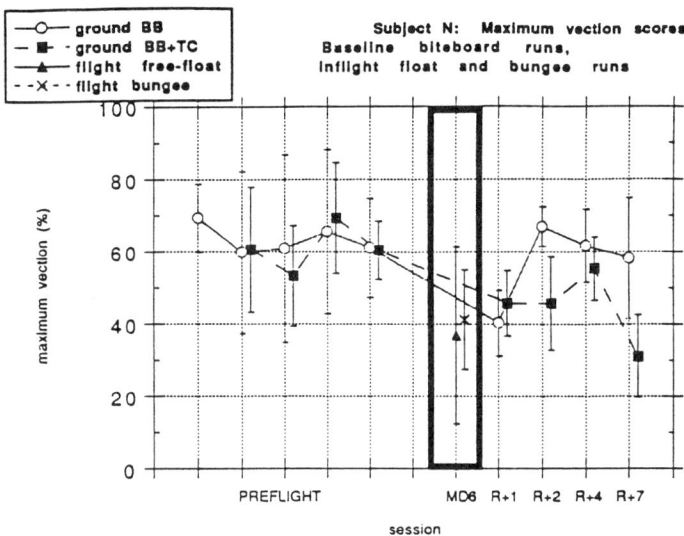

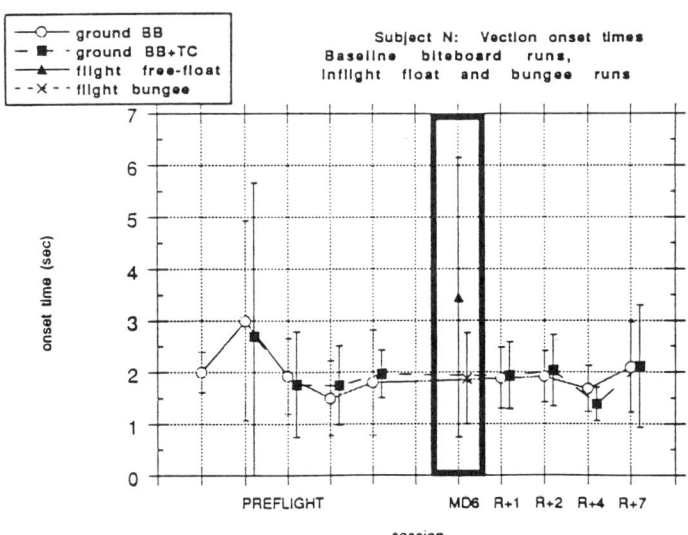

FIGURE 4. Visual-vestibular interaction, subject N. **Top:** Maximum vection scores, all conditions. **Bottom:** Onset latency for vection, bite board, with and without tactile cues.

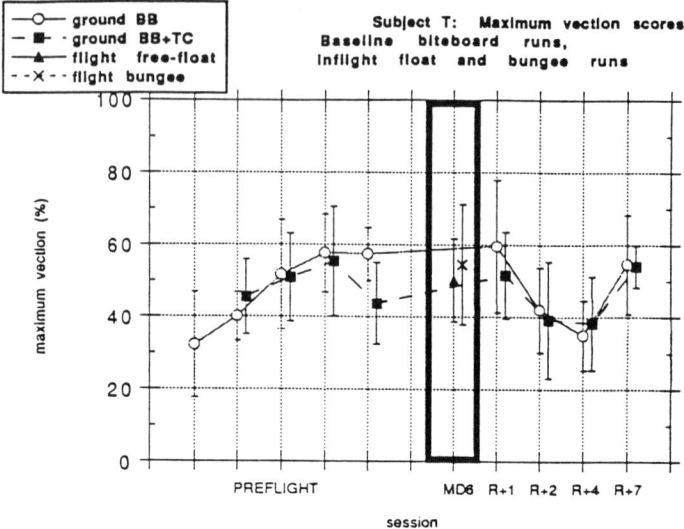

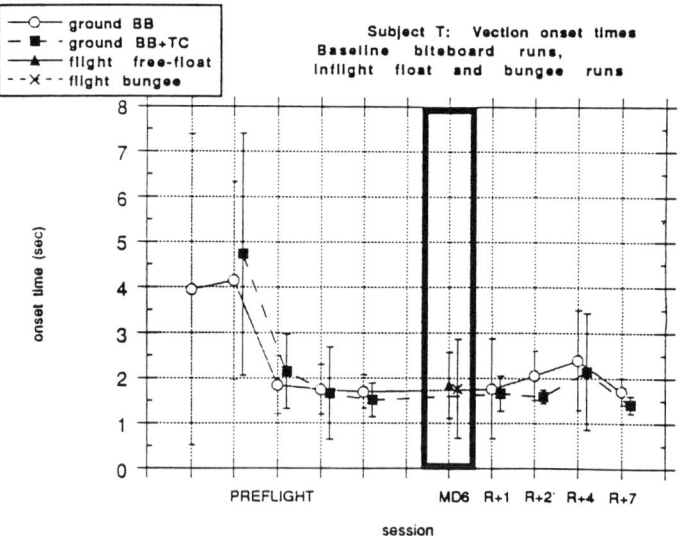

FIGURE 5. Visual-vestibular interaction, subject T. **Top:** Maximum vection scores, all conditions. **Bottom:** Onset latency for vection, bite board, with and without tactile cues.

for periods from 1 to at least 7 days after the flight. Neck muscle activity frequently corroborates body sway and vection, but quantitative relationships have yet to be determined. Ocular torsion is still being analyzed. Only pre- and postflight comparisons were available for the three members of the orbiter crew, two of whom were tested shortly after landing on R + O, and all three of whom were tested on R + 10.

Subject R showed the carryover of in-flight dependence on visual cues, with or without a bite board, or external tactile cues, with strengthened vection and reduced latency on R + O, still evident 10 days later (FIGURE 6A–C). This subject experienced residual motion after the rotating field stopped, and showed more postural sway when viewing the rotating field postflight.

Subject O showed unexpected results for the postflight trials. The strength of vection decreased with respect to the preflight sessions. (An opposite behavior was expected due to a carryover of the in-flight adaptation.) The presence of additional cues (bite board and shoulder pressure) did not have significant effects on the vection sensations. There was not a significant difference in postural stability after flight.

Subject Q showed an increase in vection when tested on R + 10 (the only day in which this subject was available for testing).

All of the flight subjects were relatively field independent, as measured by the rod and frame test (maximum preflight average deviation less than 4 degrees). Each subject, however, showed a change in field dependence postflight, beginning on the first day of testing (1 day after landing) and lasting from 2 to 6 days. M, who was the most typical subject on the dome tests, showed 1 degree of field dependence on R + 1, and returned to independence by the next test on R + 4 (FIGURE 7A top). N, who showed the opposite influence of vection stimuli on orbit, also developed an independence to visual tilt cues on R + 1 in the rod and frame test, and recovered to preflight visual tilt bias by R + 4 (FIGURE 7A bottom). Subject P, who was field independent preflight, demonstrated a tendency to be biased toward the counterclockwise direction when exposed to a tilted frame in either direction postflight, and retained this characteristic for the postflight week (FIGURE 7B top). Finally, T, who was apparently unaffected in judgment of self-motion during weightlessness, showed no consistent postflight change in field influence, and remained field independent (FIGURE 7B bottom).

DISCUSSION

Beyond the previously reported tendency to increased use of visual cues for spatial orientation during and following spaceflight, the SLS-1 results uncovered interesting variability in individual perceptual styles. Some subjects were strongly tied to a seemingly solid anchor by the localized tactile cues from bungee cord induced foot pressure, whereas another felt a strong contribution toward self-rotation under the same circumstance. Similarly, some subjects indicated a strong inhibition of visually induced motion because of the tight grip on the bite board, although it provided no directional cues. (In reaction to these findings a protocol modification for SLS-2 will include some tests with the subject loosely tethered in front of the dome, to eliminate the definite cues from the bite board.) The differences in perceptual styles may be related to the contributions of sensory and body-centered vectors in the normal and adaptive calculation of spatial orientation.

ACKNOWLEDGMENTS

We gratefully acknowledge the outstanding performance of the flight crew, Rhea Seddon, Millie Hughes-Fulford, Drew Gaffney and James Bagian, both for their diligence in performing the tests and for their ingenuity and persistence in accomplishing "in-flight maintenance." We also thank the supporting personnel at the NASA

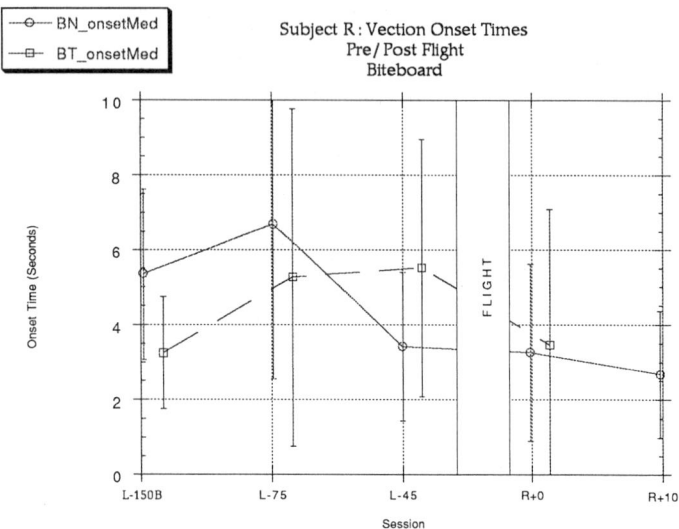

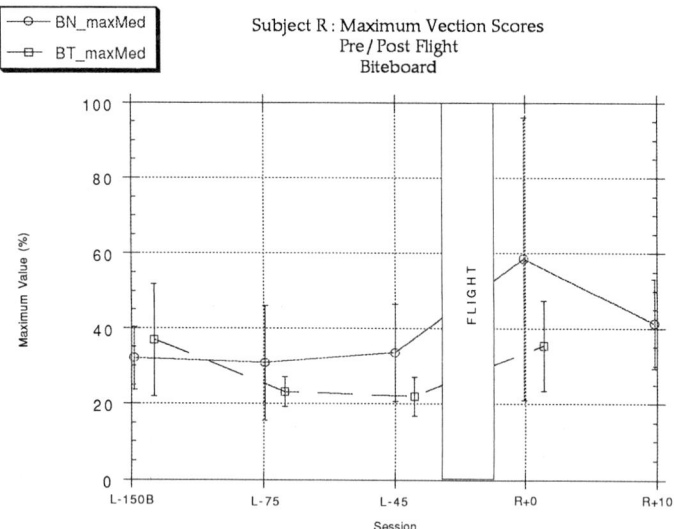

FIGURE 6A. Subject R, showed an elevated degree of vection after flight. Additional cues, in this case, shoulder pressure, reduced the increase of vection sensation. BN: Use of bite board, no shoulder tactile cues. BT: Bite board, no tactile cues.

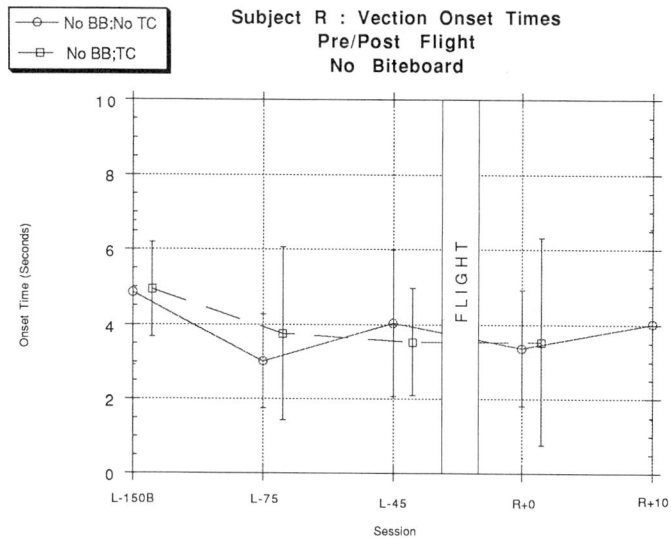

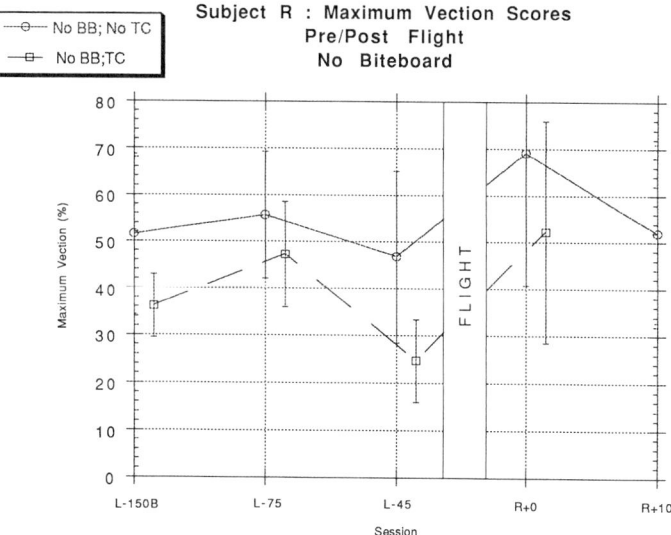

FIGURE 6B. Once again, the in-flight adaptation causes an elevation of vection after flight. This rate of increase is independent of the inclusion of additional cues, though as seen in the other subjects, the degree of vection is higher when no additional cues are present.

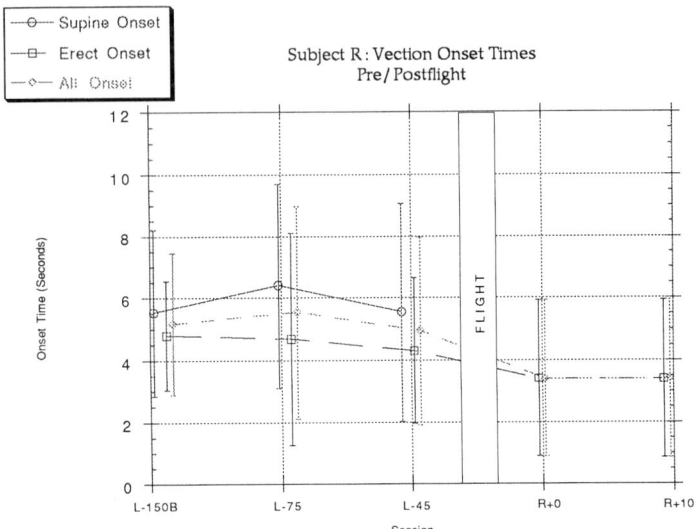

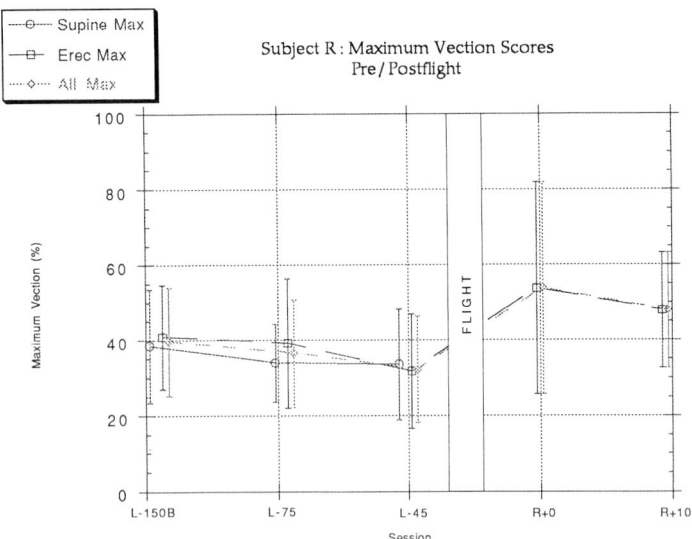

FIGURE 6C. As in FIGURE 6A, but subject not using bite board. Comparison between all preflight and postflight trials. The increase in postflight vection can be seen as late as 10 days after landing.

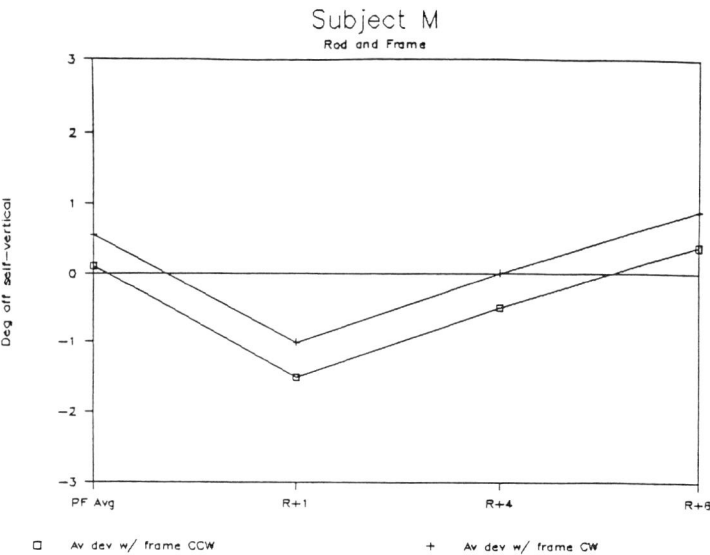

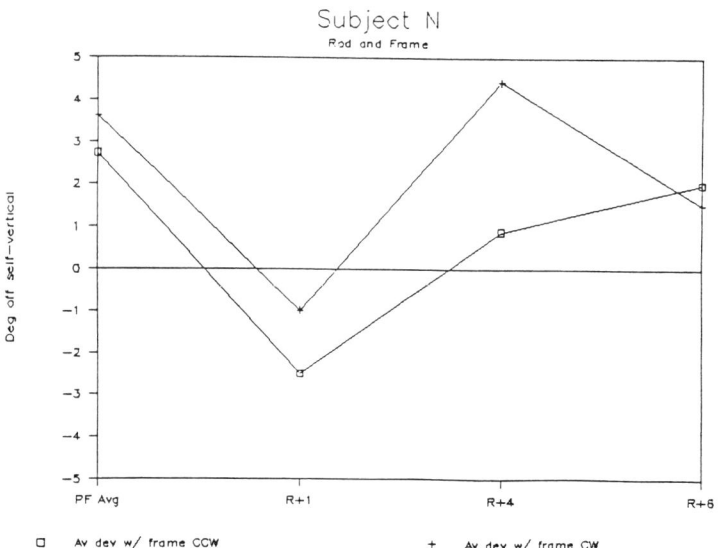

FIGURE 7. Field dependence based on rod and frame test. Preflight average (PF) and 3 days of postflight deviation of the rod in the direction of the frame (positive) or opposite (negative) in degrees, with respect to the setting in the absence of a frame. FIGURE 7A (above). **Top:** Subject M. **Bottom:** Subject N.

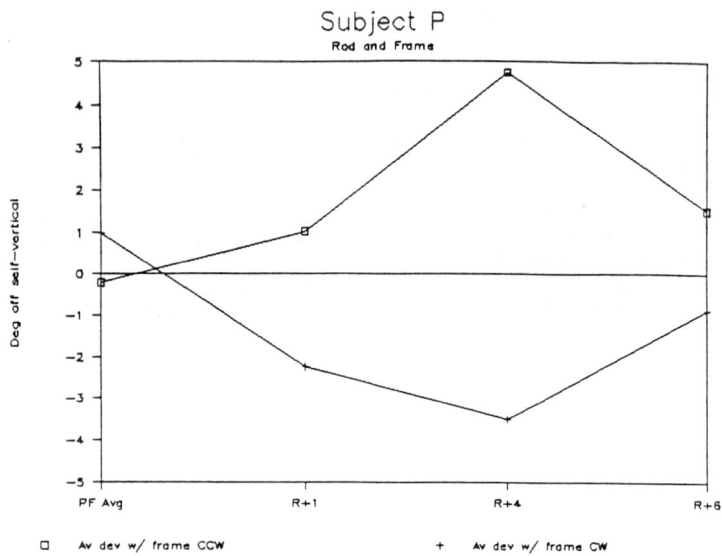

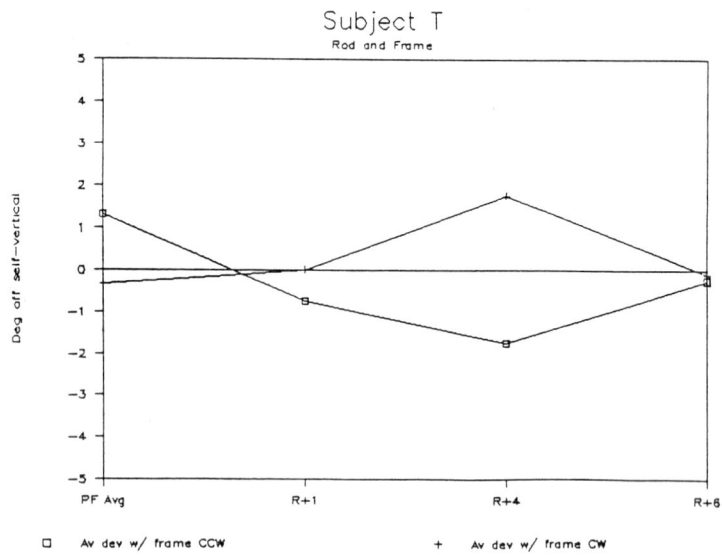

FIGURE 7B. Top: Subject P. **Bottom:** Subject T.

Centers (KSC, JSC, ARC, and DFRF and Headquarters) who contributed to the success of the experiments, particularly Stuart Johnston, Gloria Salinas, Vanessa Ellerbe, Mel Buderer, and Ron White.

REFERENCES

1. YOUNG, L. R., C. M. OMAN, D. G. D. WATT, K. E. MONEY & B. K. LICHTENBERG. 1984. Spatial orientation in weightlessness and readaptation to earth's gravity. Science **225:** 205–208.
2. YOUNG, L. R., M. SHELHAMER & S. MODESTINO. 1986. MIT/Canadian vestibular experiments on the Spacelab-1 mission. II. Visual vestibular tilt interaction in weightlessness. Exp. Brain Res. **64:** 299–307.
3. YOUNG, L. R. & M. SHELHAMER. 1990. Microgravity enhances the relative contribution of visually-induced motion sensation. Aviat. Space Environ. Med. **61:** 525–530.

Perception and Action in Altered Gravity

MALCOLM M. COHEN

National Aeronautics and Space Administration
Ames Research Center
Moffett Field, California 94035

INTRODUCTION

Human perception and action are dramatically influenced by the gravitational-inertial forces (GIFs) in the surrounding environment. Forces due to acceleration and those due to gravity interact with one another, and have virtually identical physical effects. These forces cause novel mechanical loadings of the entire organism, and provide systematically altered stimulation to the otolith organs and the somesthetic receptors.

Over the past several years, studies of human spatial orientation, visual localization, force estimation, and hand-eye coordination under a variety of conditions that involve exposure to altered GIFs have revealed how these various functions are influenced by the force environment.

This presentation reviews some of our earlier studies, and presents a general framework for examining the effects of altered GIFs on both perception and action. The critical role of motor-sensory feedback in adaptation to altered GIFs is highlighted from the results of these studies.

PERCEPTUAL AND MECHANICAL EFFECTS OF ALTERED GIFS

Two phenomena of particular interest that influence human perception and perceptual-motor activity in altered GIFs are the elevator illusion and muscle loading.

In the elevator illusion, a visual target appears to move and be displaced from its objective physical position when it is viewed under altered GIFs; the apparent location of the target is displaced upward in hypergravity, and downward in hypogravity.[1-3] The illusion is most effective when a single visual target is viewed against an otherwise featureless background, most commonly a spot of light in the dark; in a well-structured visual environment, the illusion is virtually eliminated.[4]

Altered muscle loading is the direct mechanical result of any change in the strength of the gravitoinertial environment; its influence on motor control can be both immediate and dramatic. Under hypergravity, the increased weight of the limbs requires greater muscle force to move the limbs to a specific location at a given velocity than under normal terrestrial conditions. Under hypogravity, the reduced weight of the limbs requires less force than under normal terrestrial conditions. As a result of these mechanical influences, the initial location of a directed reaching movement would be expected to be too low in hypergravity, and too high in hypogravity, particularly for ballistic pointing responses.

We will next examine in detail how these two phenomena, the elevator illusion and muscle loading, working in opposite directions, come to influence perception and action in altered gravity.

EXPERIMENTAL FINDINGS

Visual Localization under Altered GIFs

Elevator Illusion Study

It is well known that the localization of a visual target depends on both retinal and extraretinal signals.[5,6] Although retinal signals are generally unaffected by exposure to altered GIFs, extraretinal signals may be altered dramatically by even relatively minor perturbations in the magnitude of GIFs; changes in these signals are probably responsible for the elevator illusion.

Specifically, it has been proposed that normal otolith-oculomotor reflexes are calibrated to function appropriately in the terrestrial environment, and that altered GIFs change their gain.[7,8] These changes, in turn, can lead to illusory displacements of visual targets. According to this explanation, exposure to hypergravity stimulates the otolith organs to drive the natural resting position of the eyes downward. As a result, light from a visual target that was at the normal line of regard under terrestrial conditions now strikes the retina at a point below the fovea. This point was formerly associated with an object in the upper portion of the visual field. Thus, when viewed in hypergravity, the target appears to be elevated.

Further, to achieve this illusion, no net physical displacement of the eyes is necessary. With only a single spot of light as a visual target, increased stimulation of the otolith organs could initiate a vertical nystagmus, with slow phase downward.[9] To maintain foveal vision of a target that is objectively at the level of the eyes during exposure to hypergravity, an additional "upward" command to the oculomotor muscles could be initiated. Efferent monitoring of this "upward" command could be sufficient to generate an apparent upward displacement of the target. The same mechanism, but with directionally opposite effects, is suggested for the events that would occur during exposure to hypogravity.

The elevator illusion has been studied by several investigators, and its magnitude has been shown to depend on the orientation of the subject's head with respect to the direction of the altered GIFs as well as on the intensity of the GIFs themselves.[4,7,10] When the head is pitched forward by only 15°, the magnitude of the elevator illusion is reduced to about half of that produced when the head is erect; when the head is pitched forward by 30°, the elevator illusion is eliminated.[7] (See FIGURES 1 and 2.)

Based on these results, it is clear that the orientation of the subject's head should be considered in any studies involving hand-eye coordination where the task involves reaching for a visual target whose apparent location can be changed by the elevator illusion.

Studies of Hand-Eye Coordination under Altered GIFs

Centrifuge Study

The effects of increased GIFs on hand-eye coordination were investigated in a human centrifuge. Eight volunteer subjects sat with head erect, and were exposed to a GIF of 2.0 Gz, while they pointed repeatedly at intervals of five seconds to a mirror-viewed target, as shown in FIGURE 3.

Because the mirror precluded direct vision of the reaching hand and prevented the subjects from seeing any errors in their reaching attempts, the location of each

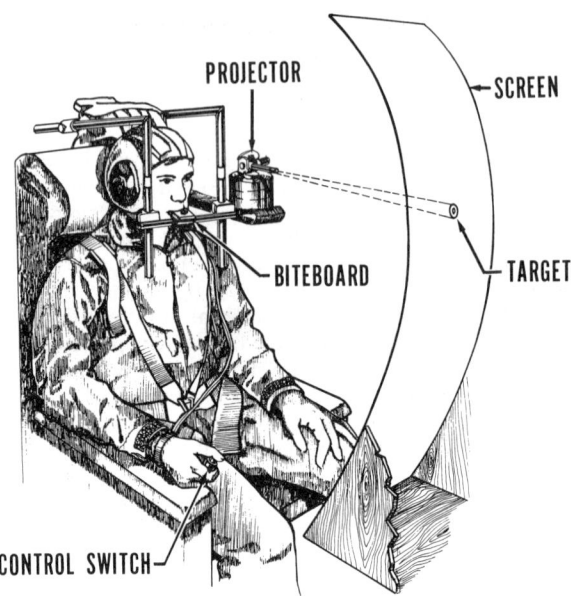

FIGURE 1. Equipment used to study the elevator illusion in a human centrifuge. The biteboard rigidly keeps the position of the subject's head at a fixed orientation relative to the applied GIFs, and the control switch allows the subject continuously to adjust the elevation of the target to keep it at the apparent horizon.

reaching movement was primarily determined by where the target appeared to be located and by any influences of the altered GIFs on the reaching movements themselves. As shown in FIGURE 4, the subjects initially reached below the target for the first few reaching attempts, but soon they consistently reached above the target.

These data suggest that the increased GIFs initially load the musculoskeletal system so that, for the specific trajectories used, preexisting motor programs are not sufficient to raise the arm to the target. Because of proprioceptive/kinesthetic feedback from the arm, the subjects become aware of these errors in reaching attempts, and correct for this underreaching in a few trials. The remainder of the reaching movements made under the hypergravity condition are above the target; these errors are attributed to the elevator illusion, and suggest that the subjects are now pointing to the apparently elevated (but incorrectly localized) target.

When the centrifuge stops, and the GIFs are returned to normal terrestrial values, the subjects initially reach *above* the target by about the same amount as they had initially reached *below* the target when they first responded under the 2.0 Gz condition. Many of the subjects spontaneously expressed surprise at the results of these initial reaching movements, and rapidly corrected them to preexposure values.[11]

Parabolic Flight Studies

Expanding on the earlier efforts by Gerathewohl, Strughold, and Stallings,[2] a series of studies in a specially equipped Learjet aircraft was directed to evaluating the specificity of activity in altered GIFs, and the resulting adaptation.[12] As shown in FIGURE 5, a computerized hand-eye coordination test apparatus, similar to that depicted in FIGURE 3, was fitted into the Learjet.

The aircraft was flown in a series of five parabolas, with each parabola of the series having the characteristics shown in FIGURE 6.

During each parabola, the subjects attempted to reach out and touch the apparent position of the mirror-viewed target. When not making reaching movements, the limbs were supported on the subject's knees or thighs.

FIGURES 7–9 demonstrate the effects of exposure to alternating hypogravity and hypergravity in parabolic flight on hand-eye coordination. FIGURE 7 depicts the effects of reaching movements made during *both* the hypogravity and hypergravity phases of all parabolas.

The initial reaching movements during the hypogravity phase of the first parabola tended to be slightly above the target, an effect that may have resulted from the reduced weight of the limbs. The first reaching movement during the first episode of hypergravity demonstrates a dramatic downward shift in the reaching movement, followed by a rapid upward trend, quite similar to that observed in the studies on the human centrifuge. In the next episode of hypogravity, the initial reaching movement was considerably higher, but it was followed by a return to about the same level as that obtained during the previous episode of hypergravity. Throughout the remaining parabolas, the alternating periods of hypogravity and hypergravity did not appear to produce dramatic differences in reaching movements, and the hypogravity and hypergravity responses did not separate from one another. Upon return to level flight

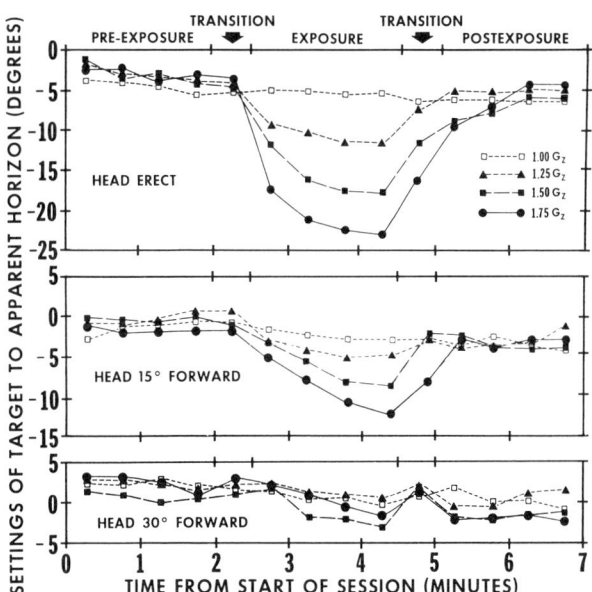

FIGURE 2. Settings of a visual target to the apparent horizon. The position at which the subjects set the target so that it appears to remain at the apparent horizon is progressively lower for increased intensities of Gz; the effects of increased Gz also depend on head orientation, with the effects being attenuated, and then, eliminated, as the head is pitched forward by 30 degrees.

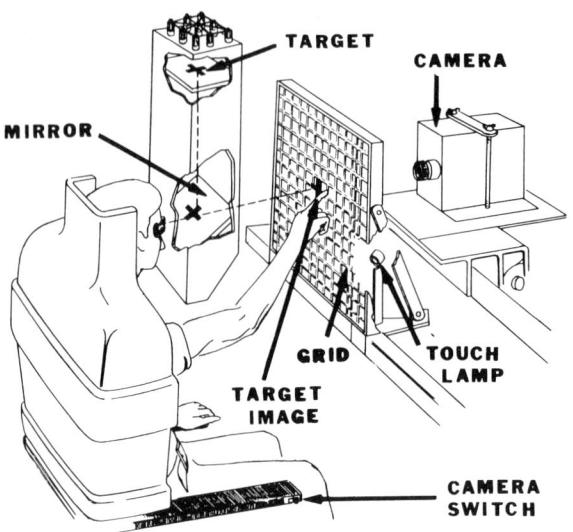

FIGURE 3. Equipment used to study hand-eye coordination in a human centrifuge. The mirror precludes direct vision of the hand, and the target image, a backlighted illuminated cross-hair, appears on the grid ahead of the subject. Each reaching movement is recorded by the camera located behind the transparent grid.

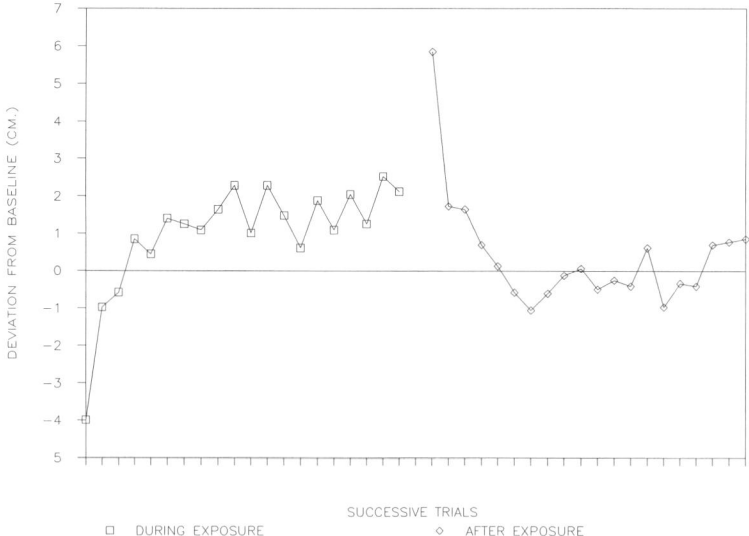

FIGURE 4. Reaching movements made during and after exposure to 2.0 Gz. Upon introduction of the increased GIFs, subjects initially reach below the target; after a few reaching responses, subjects consistently reach above the target; when returned to normal terrestrial conditions, subjects manifest transient aftereffects of overreaching.

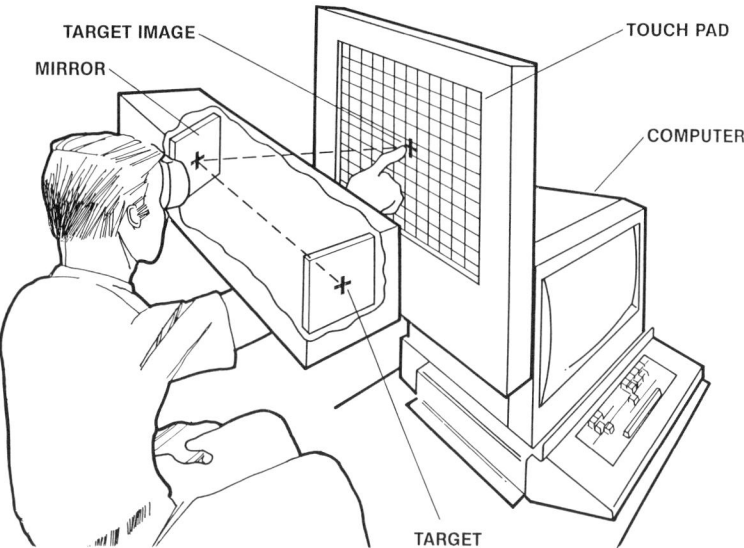

FIGURE 5. Equipment used to study hand-eye coordination in a Learjet aircraft. This device is similar to that depicted in Figure 3, but is an updated version that uses a computer, rather than photographic techniques.

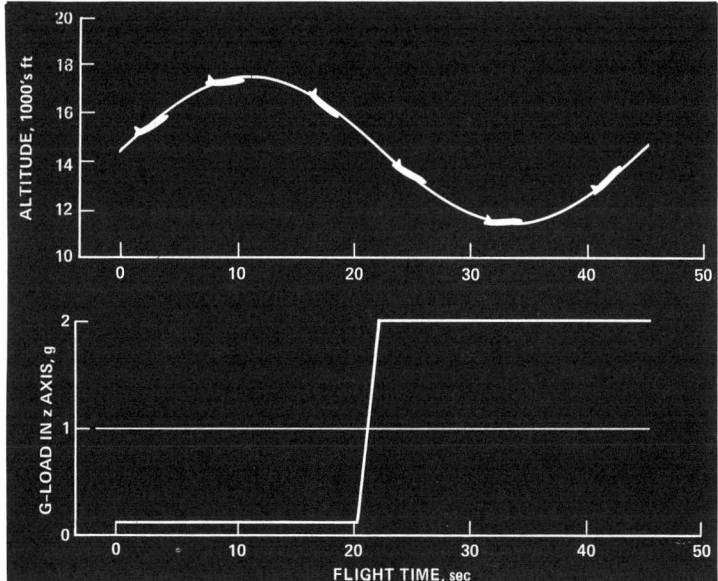

FIGURE 6. Idealized flight and gravitational profiles obtained in the Learjet aircraft during parabolic flight. The profile depicted was repeated five times for each subject to comprise a single data collection session; each session was preceded and followed by prolonged periods of straight and level flight.

after the final episode of hypergravity, there was a transient aftereffect of reaching above the target, similar to that obtained in the studies on the human centrifuge.

FIGURE 8 illustrates the effects of reaching movements made *only* during the hypogravity phase of each parabola. In the first parabola, the initial reaching movement was below the target, and this pattern was repeated throughout the remaining parabolas. Unlike the effects obtained in hypogravity when the subjects responded during the hypergravity phase as well, they tended to reach fairly consistently *below* the target. Upon return to level flight after the fifth parabola, they continue to reach below the target, possibly reflecting an aftereffect of the formerly reduced loading of the limbs in hypogravity.

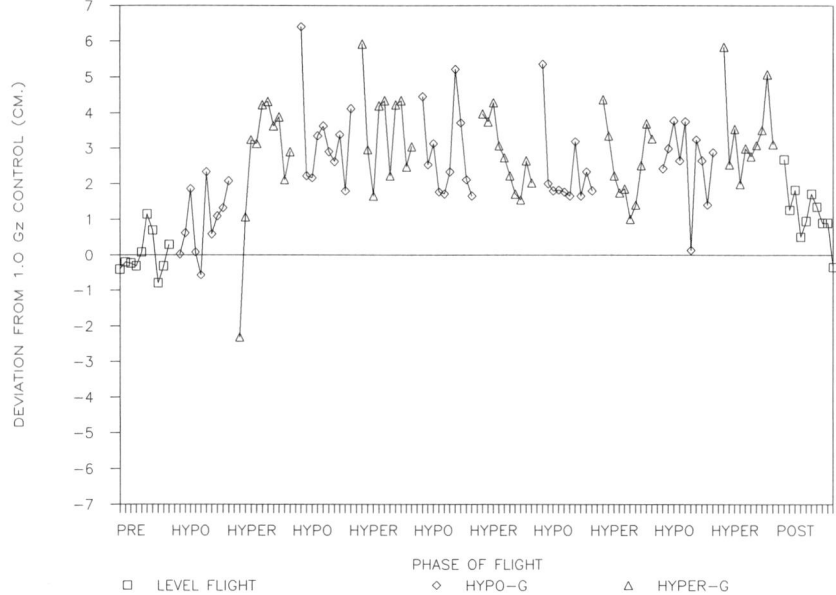

FIGURE 7. Measures of pointing responses obtained before, during, and after parabolic flight. These data were obtained when subjects reached for a mirror-viewed target under both hypogravic and hypergravic conditions.

As shown in FIGURE 9, the effects of reaching movements made *only* during the hypergravity phase of each parabola differ dramatically from those made *only* during the microgravity phase. The initial reaching movement was below the target, but this was rapidly replaced by a consistent pattern of reaching above the target in this, and subsequent parabolas. Upon return to level flight after the fifth parabola, there was a transient overreaching, which rapidly returned to the preparabola baseline.

Collectively, these results suggest that the initial effects of even brief exposures to altered gravity involve both the effects of the elevator illusion and those of muscle loading/unloading. With reliable and consistent feedback from motor activity, the organism apparently can learn to correct for the effects of the altered muscle loading; the direction of the errors in reaching movements is appropriate for the apparent location of the target (which is displaced by the elevator illusion). Further, the motor corrections persist briefly upon return to normal terrestrial conditions, and afteref-

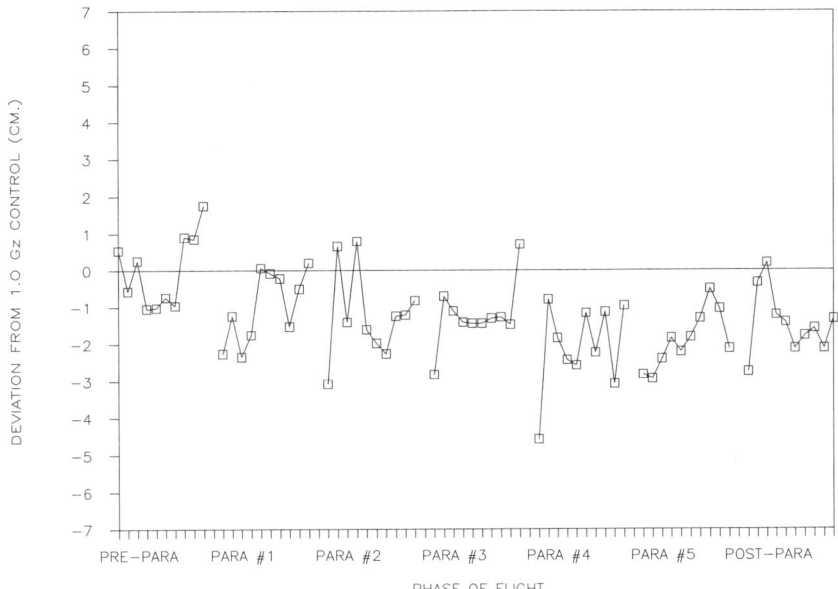

FIGURE 8. Measures of pointing responses obtained before, during, and after parabolic flight. These data were obtained when subjects reached for a mirror-viewed target only during hypogravic conditions.

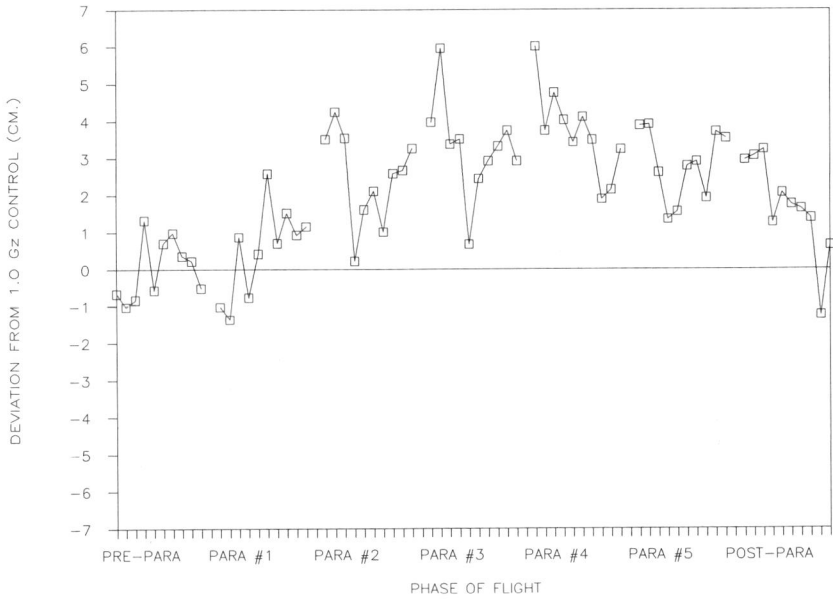

FIGURE 9. Measures of pointing responses obtained before, during, and after parabolic flight. These data were obtained when subjects reached for a mirror-viewed target only during hypergravic conditions.

fects of hypogravity or hypergravity, depending on the specific phase of flight in which the subjects responded, may be observed.

Thus, motor-generated sensory feedback, or "reafferent" stimulation,[13] may be essential in allowing the organism to adapt to specifically altered gravitoinertial conditions. Clearly, spatially directed responses that depend on precise visual-motor coordination have both motor and sensory components. The spatial direction of a response involves the localization of stimulus sources in the physical environment, as well as the precise specification of motor programs to direct movements of the relevant body parts. Because motor outputs have sensory consequences (e.g., the sight or the "feel" of the moving hand), they can serve as "probes" of the environment, allowing organisms to evaluate the relationships between actions and their spatial effects. Thus, sensory consequences of motor acts (motor-generated sensory feedback) can provide needed information to calibrate subsequent motor activity in altered gravity.

REFERENCES

1. GERATHEWOHL, S. J. 1952. Physics and psychophysics of weightlessness visual perception. J. Aviat. Med. **23:** 373–395.
2. GERATHEWOHL, S. J., H. STRUGHOLD & H. D. STALLINGS. 1957. Sensomotor performance during weightlessness. J. Aviat. Med. **28:** 7–12.
3. GRAYBIEL, A., B. CLARK & K. MacCORQUODALE. 1947. The illusory perception of movement caused by angular acceleration and by centrifugal force during flight. I. Methodology and preliminary results. J. Exp. Psychol. **37:** 170–177.
4. SCHONE, H. 1964. On the role of gravity in human spatial orientation. Aerosp. Med. **35:** 764–772.
5. HELMHOLTZ, H. VON. 1866. Handbuch der Physiologischen Optik (Leipzig: Voss). English translation from 3rd edition, 1924. 1962, Helmholtz's Treatise on Physiological Optics, J. P. C. Southall, Ed. **3:** 242–281. Dover. New York, N.Y.
6. MATIN, L. 1976. A possible hybrid mechanism for modification of visual direction associated with eye movements—the paralyzed-eye experiment reconsidered. Perception **5:** 233–239.
7. COHEN, M. M. 1973. Elevator illusion: influences of otolith organ activity and neck proprioception. Percept. Psychophys. **14:** 401–406.
8. WHITESIDE, T. C. D., A. A. GRAYBIEL & J. I. NIVEN. 1965. Visual illusions of movement. Brain **88:** 193–210.
9. MARCUS, J. T. & C. R. VAN HOLTEN. 1990. Vestibular-ocular responses in man to +Gz Hypergravity. Aviat. Space Environ. Med. **61:** 631–635.
10. CORREIA, M. J., W. C. HIXSON & J. I. NIVEN. 1968. On predictive equations for subjective judgments of vertical and horizon in a force field. Acta Oto-laryngol. Monogr. Suppl. No. 230.
11. COHEN, M. M. 1970. Sensory-motor adaptation and after-effects of exposure to increased gravitational forces. Aerosp. Med. **41:** 318–322.
12. COHEN, M. M. & R. B. WELCH. 1988. Hand-eye coordination during and after parabolic flight. Aviat. Space Environ. Med. **59:** 68.
13. HELD, R. & A. HEIN. 1958. Adaptation to disarranged hand-eye coordination contingent upon reafferent stimulation. Percept. Motor Skills **8:** 87–90.

Patterns of Connectivity in the
Vestibular Nuclei[a]

J. A. BÜTTNER-ENNEVER[b]

Institute of Neuropathology
University of Munich
Thalkirchnerstrasse 36
D-8000 Munich, Germany

This article is dedicated with great respect to Parviz Mehraein, Professor of Neuropathology, University of Munich, on the occasion of his 60th birthday.

INTRODUCTION

There have been several recent reviews on the neuroanatomy of the vestibular nuclei, describing their cytoarchitecture, location, nomenclature, and connections.[1-4] But it is still very difficult to relate the anatomy to the functions of the vestibular nuclei, such as their role in smooth pursuit eye movements, the velocity storage mechanism, adaptation of the vestibuloocular reflex (VOR), the common neural integrator, and the role in optokinetic responses.[5] The reason is that the information available to do this is meager, and widely scattered. Any attempt to localize function reveals only a few consistent patterns, and these unfortunately do not conform to any neuroanatomical borders. In this review the more basic patterns of connectivity in the vestibular nuclei will be described. These relate almost exclusively to labyrinthine canal sinals since there is very little information on otolith pathways.

The diagramatic outlines of the vestibular nuclei in the horizontal plane that are used in many of the following diagrams roughly represent the arrangement in mammals. The ventromedial border of the lateral vestibular nucleus (lv) has been drawn so that the smaller-celled region containing many secondary vestibular neurons [the ventrolateral nucleus, or the magnocellular subdivision of the medial vestibular nucleus (mv)] lies within the boundaries of the mv.

VESTIBULAR NERVE AFFERENTS

The primary (1°) vestibular afferents in the VIIIth nerve enter the medulla at the level of the lateral vestibular nucleus. Nearly every fiber divides at this point and sends one descending branch to terminate in the medial and inferior vestibular nuclei (mv and iv), and an ascending branch to the superior vestibular nucleus (sv) where it gives off branches before continuing on to the cerebellum. These fibers project to the anterior and posterior vermis bilaterally, the major input being to the nodulus (lobe X) and the lobule IXd of the uvula. There are conflicting reports of the input to the flocculus; if it is present it is very small.[6]

FIGURE 1 represents a generalized summary of the termination sites of primary afferents found in several studies,[7-9] including the terminal branches of a single

[a]This work was supported by the Deutsche Forschungsgemeinschaft SFB 220/D8.
[b]Address all correspondence to the Institute of Physiology, University of Munich, Pettenkoferstr. 12, D-8000 Munich 2, Germany.

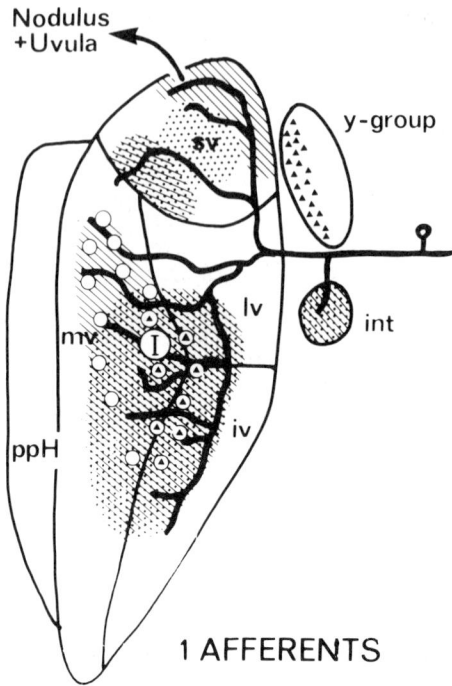

FIGURE 1. The pattern of vestibular nerve termination in the vestibular nuclei, generalized from studies that disagree on several points.[7-9] The canal projections are represented by stripes (HC and AC) and dots (PC), the utricular input by open circles, and sacculus by triangles. Only one area (zone I) receives inputs from all canals and otoliths. In sv the canal afferents each have individual sites of termination, and project to the interstitial nucleus of the vestibular nerve (int). The fibers supplying peripheral sv continue on to the nodulus and uvula. The utricle terminates mainly in mv, and the sacculus in the ventral y-group.

horizontal canal (HC) afferent.[10] There appears to be only one region, marked I in FIGURE 1, that receives all types of vestibular inputs,[7] i.e., all three canals [HC, anterior canal (AC), and posterior canal (PC)] and both otolith (utriclar and saccular) inputs. The canal inputs to sv converge on a small patch ventromedially, but in the central and dorsolateral regions of sv they terminate differentially. Another point of convergence of canal afferents is the interstitial nucleus of the vestibular nerve (a known source of input to the flocculus),[52] while the saccule innervates the ventral y-group and rostral mv receives utricular inputs. In some studies these 1° projections are more extensive, but most agree that there is very little input to lv and prepositus hypoglossi (ppH).

VESTIBULOOCULAR PATHWAYS

Secondary Vestibuloocular Neurons

There are many direct pathways that run in parallel from the vestibular nuclei to the oculomotor nuclei.[11,12] Of these the most thoroughly investigated are those arising from vestibular neurons with a direct, monosynaptic, input from a canal nerve, i.e., secondary vestibular neurons (2°). Each of these has a dominant input from one canal, which is clear under anesthesia, but in the awake state several additional inputs from other canals or otoliths converge on these cells.[13,14] The pathways by which otoliths induce ocular responses are poorly understood and not included here.[15] FIGURES 2 and 3 show the location of 2° canal vestibular neurons which project directly to either the abducens nucleus (nVI), the trochlear nucleus (nIV), or the oculomotor nucleus (nIII), and not to the spinal cord. They are called vestibuloocular neurons (VO). The figures were complied from recent studies.[11,16–27] The pattern of distribution varies very little between species and consists of two main areas: zone

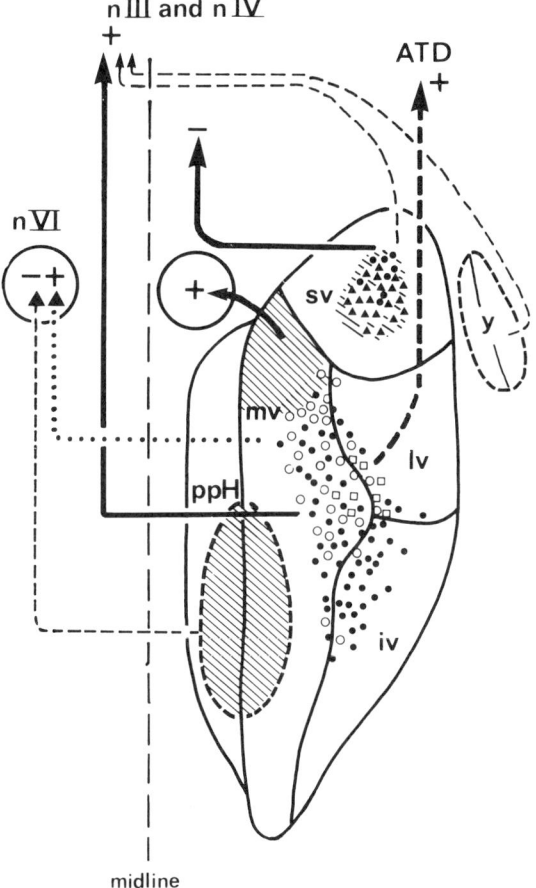

2° VESTIBULO-OCULAR CELLS

FIGURE 2. The location of secondary vestibular neurons with monosynaptic projections to the extraocular motoneurons are indicated by dots, circles, squares, and triangles. The hatched regions indicate the location of eye-position-related, burst-tonic neurons that are not secondary. The y-group projection to nIII also arises from nonsecondary neurons.

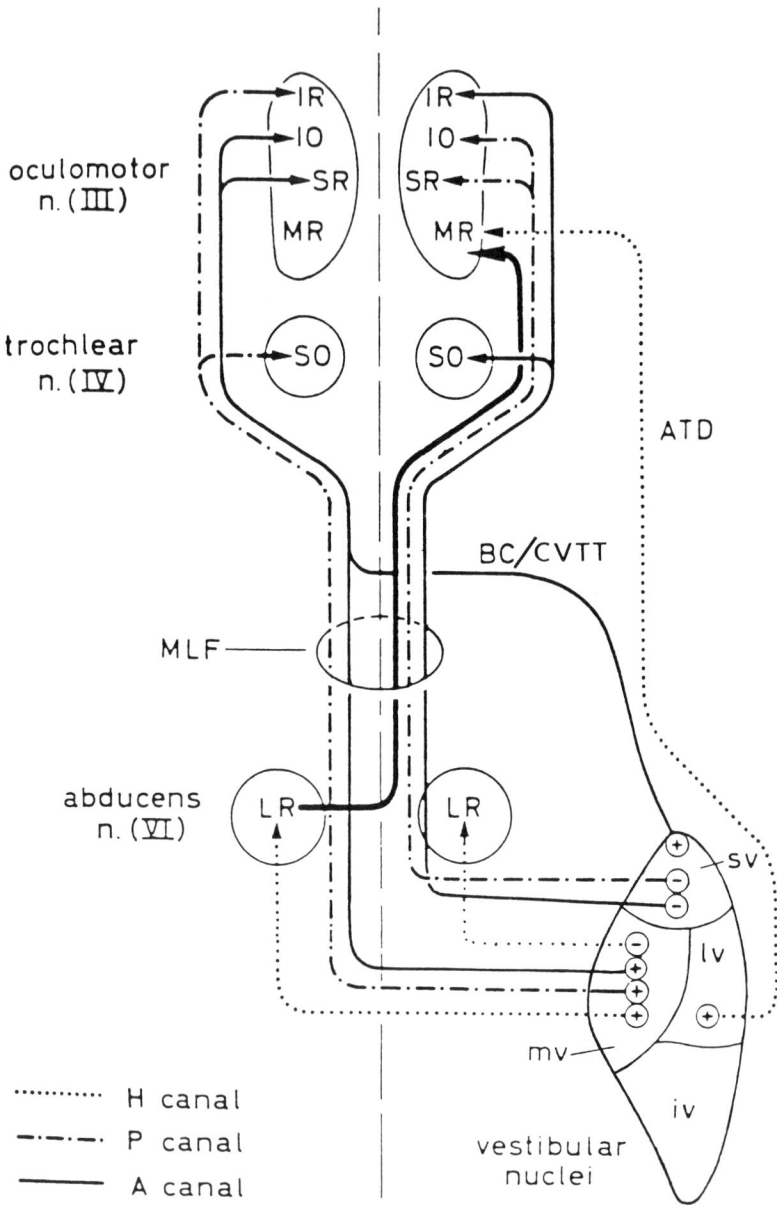

FIGURE 3. Direct pathways from the vestibular nuclei with dominant canal inputs, and their projections to individual eye muscle motoneurons. In monkey and cat there are collaterals to other motoneurons in addition.[23] The crossing pathways are excitatory and the ipsilateral projections, with the exception of ATD, are inhibitory.

I described above, which lies around the borders of ventromedial lv, dorsomedial iv, and medial mv; and the center of sv.

1. Efferents from zone I cross the midline and provide excitatory inputs related to the appropriate muscle motoneurons (FIGURE 3)[11,28] to nVI, or to nIV and nIII via an ascending axon in the medial longitudinal fasciculus (MLF). The activity of these cells generally encodes eye position, head velocity, and pauses with saccades (PVP cells). In addition zone I is the origin of the ascending tract of Deiters (ATD) which runs rostrally lateral to the MLF and terminates in an excitatory fashion on medial rectus neurons in nIII. It also carries PVP activity.[24,29] In the rostral mv there are secondary vestibular neurons projecting to the ipsilateral nVI and carrying a PVP inhibitory signal.[30] These are not included in FIGURE 2 because they also project to the spinal cord, i.e., vestibulooculospinal neurons (VOC) not VO neurons. In general the horizontal secondary neurons lie more rostrally than do the vertical.

2. The large-celled central part of sv gives rise to several different types of projections. Ipsilateral inhibitory pathways ascend in MLF, from secondary neurons with PC and AC dominant inputs.[16,31,32] Only vertical canal-related neurons have been found in this region, and unlike the cells of zone I, which are mainly of the PVP type, these carry a burst-tonic (BT), or eye-position-related, signal.[25] Those with an anterior canal input tend to receive a commissural facilitation rather than the usual commissural inhibition.[33] Another unusual feature of sv is the presence of only AC-related neurons that send an excitatory projection to the inferior oblique motoneurons of nIII, via a crossing ventral tegmental tract (CVTT) in the cat,[22] or the brachium conjunctivum in the rabbit.[34] There are therefore two separate pathways carrying AC to nIII and making direct excitatory connections with inferior oblique muscle (IO) motoneurones: one lies in the PVP "zone I" of FIGURE 1, and the other in the burst tonic neuron (BT) area of central sv (FIGURES 2 and 3).

Nonsecondary Vestibuloocular Cell Groups

There is very little information on cells projecting to the extraoculomotor nuclei other than from the 2° neurons described above. Most of the reports on nonsecondary neurons find that they have a burst-tonic type of activity (see hatched areas in FIGURE 2). This is expected from neurons whose input is the common neural integrator, i.e., neurons whose velocity signal has been integrated to code position. An inhibitory signal reaching nVI arises from nonsecondary cells in the marginal zone between ppH and mv.[35,36] This region is far less prominent in cats than in monkeys, where 70% of the units were related only to eye position, the rest carried a gaze velocity signal.

The dorsal y-group was shown by Sato and Kawakasi[37] to project to the upward moving eye muscle motoneurons [inferior oblique muscle (io) and superior rectus muscle (sr)] in contralateral nIII, by the same pathway, CVTT, as the excitatory 2° sv target neurons activated by AC. However these y-group neurons, which were target cells for a flocculus input, were *not* secondary vestibular neurons (and therefore not on FIGURE 2). It is interesting that in experiments using tetanus toxin as the tracer, they were found to be the strongest projection from the vestibular complex to the nIII motoneurons.[38] Chubb and Fuchs showed that the y-group neurons responded to upward eye position, head velocity, and often to saccades.[39]

Uchino and Hirai reported that the majority of VO neurons, with no spinal projection, were excited polysynaptically, and not monsynaptically, from the anterior canal nerve.[33] These neurons lay in zone I mostly in iv, with many secondary vestibular cells, but their activity type is not known. Another example of nonsecondary vestibular neurons is reported by Keller and Kamath[40]: they lay in rostral mv-sv region, carried an eye-position-related signal, and were disynaptically activated from the vestibular nerve.[41] It is possible that their units are the physiological counterpart of the large group of neurons in rostral mv, which are filled with horseradish peroxidase (HRP) after injections in the ipsilateral abducens nucleus.[3,35,42]

In summary there are two main areas of secondary vestibular neurons that project directly to extraocular motoneurons. Those lying around the mv, lv, and iv border (zone I), whose axons cross the midline and send, via MLF, excitatory (PVP) signals to the appropriate muscles to move the eyes in a compensatory fashion (FIGURE 3). Second, those in the center of sv: some of these are related to vertical canals, whose axons carry an eye-position-related (BT) signal, and ascend in the ipsilateral MLF to inhibit the antagonist muscles (FIGURE 3). In addition there is an excitatory AC pathway to the contralateral motoneurons. All other nonsecondary vestibuloocular inputs, and there appear to very many of them, receive polysynaptic inputs from the vestibular nerve, and many carry an eye-position signal. These include the marginal zone and a less-well-defined area in rostral mv.

VESTIBULOSPINAL PROJECTIONS

The vestibular nuclei project to the spinal cord through the "lateral vestibulospinal tract" (LVST) and the "medial vestibulospinal tract" (MVST). LVST has its origin mainly in ipsilateral lv, excites neck, axial and extensor muscle motoneurons, and inhibits flexors. MVST arises mainly contralaterally from mv and to a lesser extent in iv and lv; it regulates the activity of neck and axial motoneurons.[43,44] Sv, ppH, and the y-group have no spinal projections. Many of the neurons with monosynaptic inputs onto spinal motoneurons also project directly to extraocular motoneurons. The location of such units which also have a monosynaptic input from the vestibular nerve (termed VOC units) has been studied and is shown on FIGURE 4.[18,27,30,33,45,46] They are scattered throughout zone I and perhaps even beyond its borders, with the vertical canal-related cells further caudal than the horizontal units. They project predominantly to the contralateral side (but there is a small ipsilateral MVST projection),[30,2] and they receive no input from the flocculus.[47] The lack of flocculus projections implies that these pathways take no part in adaptation of the vestibuloocular reflex, which is abolished by flocculectomy.[12] The two neurons with only an ipsilateral spinal projection lie in ventral lv, and those with only a contralateral descending axon lay further caudally in rostral iv.

CEREBELLAR AFFERENTS AND EFFERENTS

Several studies have investigated the mossy fiber input to the cerebellum from the vestibular nuclei.[2,4] With the exception of lv, all vestibular nuclei projected bilaterally to the nodulus and uvula (mainly lobules X and IXd). Most terminals arose from sv and caudal mv and iv[48,49] (see FIGURE 5). The projections to the floccular region arose bilaterally from caudal ppH, central mv, central iv, central sv, and ventral y-group.[50,51,37] The basal interstitial nucleus of the cerebellum provides the flocculus with a reciprocal projection, but as yet nothing else is known about this nucleus.[52] An additional group of neurons that project to the flocculus, and are described by Blanks *et al.*[53] and Langer *et al.*,[50] lie in groups scattered along the pontine and medullary midline. They have been called collectively nuclei of the paramedian tracts (PMTn) by Büttner-Ennever *et al.*,[54] who show that they also receive afferents from all premotor regions of the oculomotor system. This indicates that PMTn provide the flocculus with the eye movement feedback signal necessary for its control of eye position.[12,54]

SPINAL PROJECTIONS

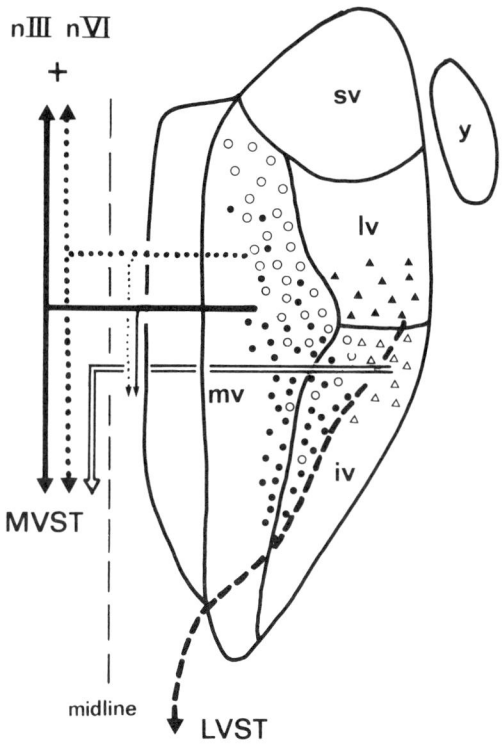

FIGURE 4. The location of secondary vestibular neurons that project directly to both oculomotor and spinal motoneurons (VOC neurons: circles, HC; and dots, AC and PC), and that project only to the spinal cord motoneurons (VC neurons: triangles).

The cerebellar efferents supply all vestibular nuclei: the lv receiving input from the anterior lobe, zone B. The flocculus projects mainly to dorsal y-group, central sv, and rostral mv,[55] while the nodulus terminates in peripheral sv and caudal mv. This is slightly different from the uvula efferents, which supply iv and dorsal sv,[56] but they are treated together in FIGURE 5.

In summary, the flocculus receives information from the secondary neurons via the PMTn pathways, and saccular information from the ventral y-group: it has

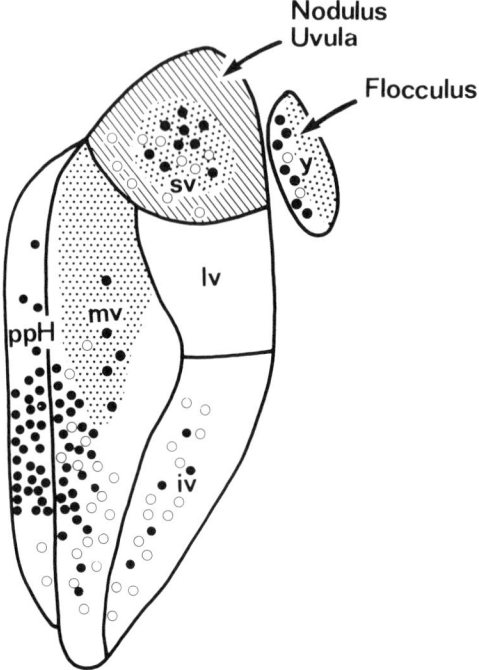

FIGURE 5. Interconnections between the vestibulocerebellum and the vestibular nuclei. Cells projecting to the flocculus are represented by filled circles, to the nodulus and uvula by open circles. The locations of projections from the flocculus are shown by small dots and from the nodulus and uvula by hatching. The floccular region controls the rostral mv, central sv, and the dorsal y-group, whereas the nodulus and uvula project to the peripheral sv and the caudal mv and iv (not shown). The location of the floccular projecting neurons in the marginal zone do not overlap with the efferent areas.

reciprocal connections with central sv, and afferents from ppH, mv, and iv. Its efferents are confined to central sv, dorsal y-group, and rostral mv, all of which, apart from the y-group, are areas containing burst-tonic neurons that project to extraocular motoneurons (see above and FIGURE 2). In contrast, the nodulus and uvula project to the more peripheral regions—in particular the peripheral sv, which is interconnected with the contralateral mv and sv by commissural fibers.

COMMISSURAL AND INTRINSIC CONNECTIONS

Although all primary vestibular and cerebellar inputs to the vestibular nuclei are strictly ipsilateral, the vestibular nuclei of each side operate together in an integrated manner. Clinically, this is seen by the compensation of the effects of unilateral lesions, and theoretically, it is thought to provide the substrate for the function of the common neural integrator, and the velocity storage mechanism. The basis for this coherence is a massive commissural system interconnecting the vestibular nuclei of both sides, and prominent intrinsic interconnections on each side.[57-60] The most prominent commissural connections are between mv and mv throughout its length; next are the reciprocal projections from the periphery of sv to the contralateral sv and mv. Gacek found the ventral y-group responsible for strong commissural connections using the HRP technique,[61] but other studies could not confirm this projection using amino acid tracers.[62] Several pathways described in the above studies are not in agreement with each other, but Epema and collegues re–interpreted and summarized the main findings to reveal a consistent pattern[59] (FIGURE 6). The peripheral regions of the vestibular complex (comprised of periph-eral sv and iv, rostral and caudal mv, and perhaps also ppH) are linked through reciprocal intrinsic connections, and are also the origin of the major commissural connections. The more central vestibuloocular relay regions (zone I and central sv) only received intrinsic connections from the more peripheral regions, and played a very minor role in commissural pathways.

Velocity Storage

There is a network of neurons that prolongs the generation of eye movements evoked by a vestibular or optokinetic stimulus beyond the stimulation period. This

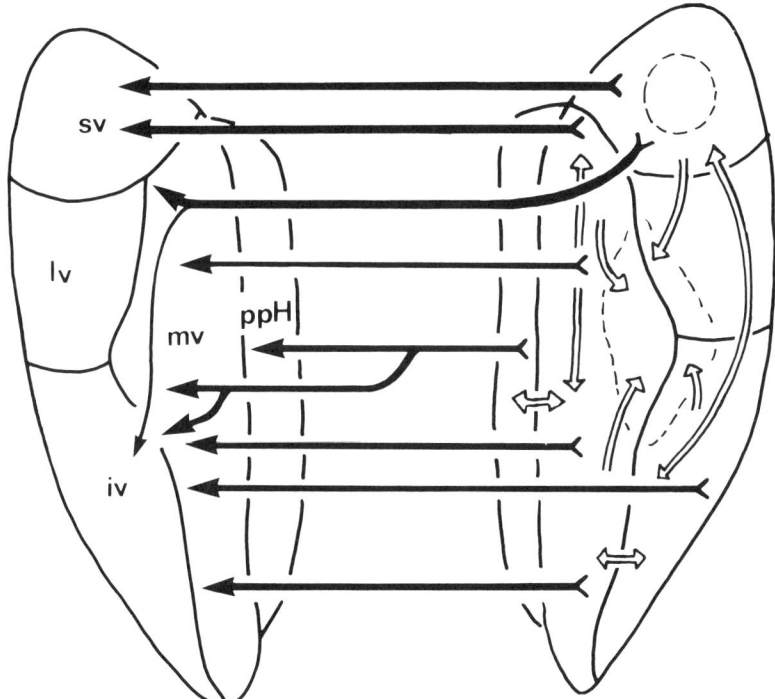

FIGURE 6. The major intrinsic (open arrows) and commissural pathways of the vestibular nuclei arise from the more peripheral regions, especially mv and sv. The oculomotor relay areas (dashed lines) receive from these areas but do not participate in intrinsic or commissural connections as intensely as the more peripheral regions.

mechanism, called the "velocity storage integrator," is highly developed in monkey.[5] Several lesion experiments have been designed to locate the network generating velocity storage, using the duration of optokinetic after-nystagmus as a simple measure of its activity. Midline cuts in the medulla shown in FIGURE 7, to a depth of at least 2 mm below the floor of the 4th ventricle, and caudal to the abducens nucleus, abolished velocity storage but left gaze-holding (the common neural integrator) intact: in addition saccades were normal, and adaptation and gain of the VOR remained intact.[63–65] The results indicate that fibers crossing the midline between 1 and 2 mm below the ventricle floor are essential for the velocity storage mechanism. The fiber bundles that cross at this point are shown in FIGURE 7: (1) from the periphery of sv crossing to its contralateral counterpart and mainly rostral

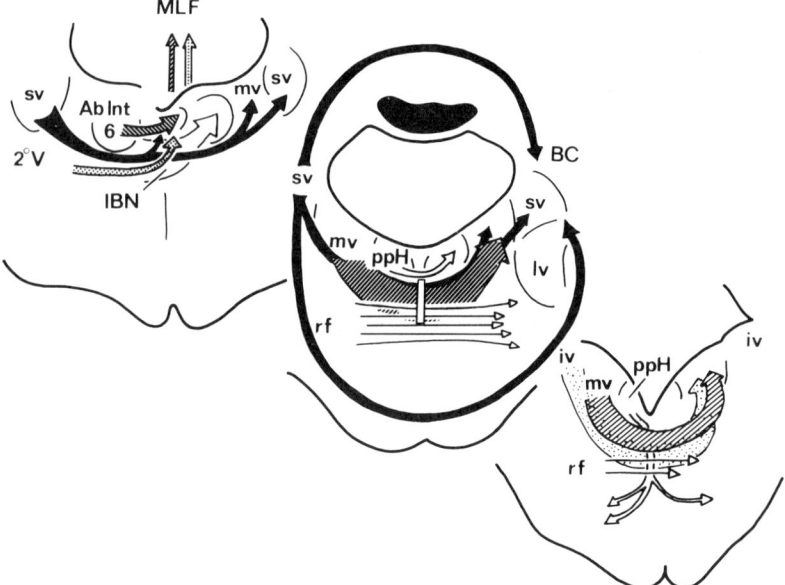

FIGURE 7. The major pathways crossing the midline in the rostral medulla. The midline cut which abolishes the velocity storage mechanism is shown in the middle figure.

mv; (2) from rostral mv crossing to the contralateral mv; (3) from principle cells of the prepositus hypoglossi[60]; and (4) from iv crossing to mv. Of these possibilities the last seems less likely because the iv fibers cross somewhat ventral and caudal to the lesions.[8,59] The role of prepositus is unclear; it certainly has fibers that cross at this level, but its connectivity is so ubiquitous (including however contralateral ppH, mv, and iv) that it is hard to relate them to a specific function such as the velocity storage mechanism. The most likely commissures for this are those from sv and from mv, since they both cross at the level of the effective lesions; they are also the two main commissural systems. In addition sv receives a major input from the nodulus, a structure whose ablation leads to "dumping" or abolition of velocity storage.[66] Since the peripheral sv projection terminates in both rostral mv and peripheral sv, an involvement of both these regions in the velocity storage mechanism seems probable, and lesions in both these structures affect velocity storage.[67]

OPTOKINETIC INPUTS

The pathways through which optokinetic signals reach the vestibular nuclei are not well understood.[68,69] The visual afferents are known to enter via the superior fascicle of the accessory optic system, and terminate in the accessory optic relay nuclei [medial terminal nucleus (MTN), dorsal terminal nucleus (DTN), lateral terminal nucleus (LTN), and interstitial nucleus of the superior fascicle, posterior fibers (InSFp)]. The visual information also reaches the nucleus of the optic tract (NOT). Lesions in these areas can affect the optokinetic response (OKR),[70] but it is unclear through which efferent pathways. The relay nuclei all project to the inferior olive but this pathway is not essential for the OKR and neither is the cerebellum.[71] The MTN projects directly to the contralateral vestibular nuclei, to the periphery of sv.[72] This is a region that is involved in commissural pathways and receives a projection from the nodulus (see above, FIGURE 8). For these reasons it is a good candidate for involvement in vestibular functions with access to the velocity storage mechanism. DTN has also been shown to have similar projections, but NOT and LTN do not.[73] NOT does send afferents to ppH, but the functions of ppH have not been specifically related to optokinetic inputs.[74] However all the relay nuclei—MTN, DTN, LTN, and NOT—project to the visual tegmental relay zone (VTRZ) in the mesencephalon which in turn projects to the vestibular nuclei, in the same way as MTN.[75] It is not clear what role VTRZ plays in the vestibular response to optokinetic stimuli. But it could provide a route for some optokinetic information to reach the vestibular nuclei.

OTHER VESTIBULAR CONNECTIONS

There are several other structures close to MTN and VTRZ in the mesencephalon that are also interconnected with the vestibular complex, albeit mainly ipsilaterally. Most important is the interstitial nucleus of Cajal (iC). It receives afferents from all secondary vestibular neurons that supply the oculomotor nucleus. Only one exception to this is known, and that is from the y-group floccular target neurons (which are not secondary), where no projection to the iC from these y-group neurons that innervated upward moving motoneurons in nIII could be found.[37] The iC sends a reciprocal projection back to the vestibular nuclei, mainly ipsilaterally. These pathways are known to modulate vestibulospinal activity,[76] and in addition are implicated in the role of iC as part of the vertical common neural integrator.[77] The iC projections to vestibular nuclei are not well delineated, but they do involve the mv/iv border, a region interconnected with the central sv.[62] Rostral iC, caudal iC, and the reticular formation around iC appear to have different connections to the lower brain stem.[78] A separate vestibular projection arises from small fusiform cells clustered around the tractus retroflexus. It was labeled by HRP injections into the caudal mv.[78,79] These studies also demonstrated the prominent connection from midline cells of nIII to the vestibular complex, whose function is completely unknown.

SUMMARY

In FIGURE 8 the peripheral areas of the vestibular complex are all hatched to indicate that they contain the majority of intrinsic interconnections, in particular mv and peripheral sv. They are also the origin of commissural connections, in contrast to

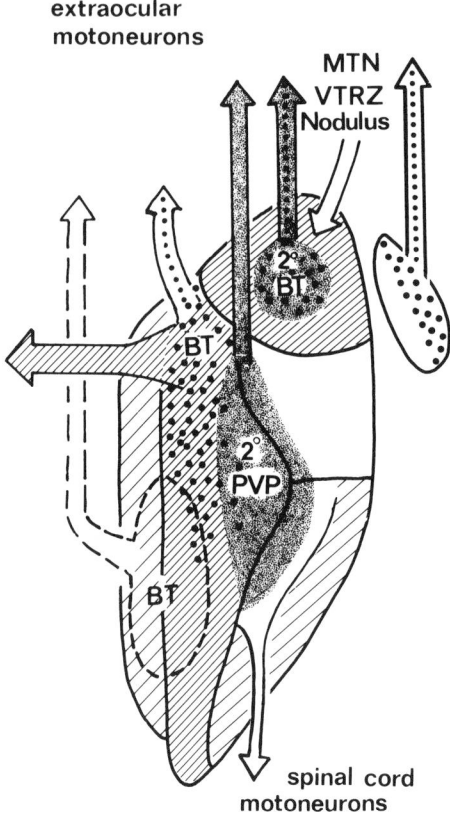

FIGURE 8. Summary diagram of the pattern of connections of the vestibular nuclei and the areas projecting to eye muscle and spinal cord motoneurons, shown by arrows. The dotted regions receive a strong floccular input, and the hatched areas (in particular mv and peripheral sv) subtain intrinsic and commissural projections. Note that inputs from the accessory optic system (MTN and VTRZ) and nodulus project to the peripheral (hatched) sv. 2° indicates the location of secondary vestibular neurons, BT of the burst tonic (eye-position-related) neurons, and PVP the position-velocity-pause cell types.

the more central zone I and central sv. Peripheral sv also receives a major projection from the nodulus, and afferents from some accessory optic nuclei.

There are many parallel output pathways from the vestibular nuclei directly to eye, head and axial muscle motoneurons, but only those receiving a monosynaptic input from the vestibular nerve (2°) have been well studied. Zone I contains secondary excitatory neurons (VO, VOC, and VC) modulated by canals (and probably the otoliths), the region being under very little cerebellar influence. The dorsal y-group, marginal zone, rostral mv, and central sv are under the influence of the floccular complex (dotted areas in FIGURE 8), and project only to extraocular motoneurons, and not the spinal cord. Of these regions only the central sv, and possibly the rostral mv have secondary neurons, while central sv, rostral mv, and the marginal zone contain neurons carrying predominantly an eye-position signal.

These patterns of connectivity support the hypotheses that several areas contain the floccular target neurons, involved in the cerebellar control of eye position, and that the zone I of FIGURE 1 relays the fast (unmodulated) responses generated by vestibular or optokinetic signals. Furthermore the periphery of sv, and mv, are implicated as relays in the velocity storage mechanism.

ACKNOWLEDGMENTS

The author is grateful for the excellent assistance of Lydia Sacher.

REFERENCES

1. BRODAL, A. 1984. The vestibular nuclei in the Macaque monkey. J. Comp. Neurol. **227:** 252–266.
2. CARPENTER, M. B. 1988. Vestibular nuclei: afferent and efferent projections. Prog. Brain Res. **76:** 5–15.
3. HIGHSTEIN, S. M. & R. A. McCREA. 1988. The anatomy of the vestibular nuclei. *In* Neuroanatomy of the Oculomotor System (Reviews of Oculomotor Research 2). J. A. Büttner-Ennever, Ed.: 177–202. Elsevier. Amsterdam, the Netherlands.
4. GERRITS, N. M. 1990. Vestibular nuclear complex. *In* The Human Nervous System. G. Paxinos, Ed.: 863–888. Academic Press. San Diego, Calif.
5. BÜTTNER, U. & J. A. BÜTTNER-ENNEVER. 1988. Present concepts of oculomotor organization. *In* Büttner-Ennever JA (Ed.) Neuroanatomy of the Oculomotor System (Reviews of Oculomotor Research 2). J. A. Büttner-Ennever, Ed.: 3– 32. Elsevier. Amsterdam, the Netherlands.
6. GERRITS, N. M., A. H. EPEMA, A. VAN LINGE & E. DALM. 1989. The primary vestibulocerebellar projection in the rabbit: absence of primary afferents in the flocculus. Neurosci. Lett. **105:** 27–33.
7. GACEK, R. R. 1969. The course and central termination of first order neurons supplying vestibular end organs in the cat. Acta Oto-Laryngol. Suppl. No. **254:** 1–66.
8. CARLETON, S. C. & M. B. CARPENTER. 1984. Distribution of primary vestibular fibers in the brainstem and cerebellum of the monkey. Brain Res. **294:** 281–298.
9. KORTE, G. E. 1979. The brainstem projection of the vestibular nerve in the cat. J. Comp. Neurol. **184:** 279–292.
10. ISHIZUKA, N., H. MANNEN, S. SASAKI & H. SHIMAZU. 1980. Axonal branches and terminations in the cat abducens nucleus of secondary vestibular neurons in the horizontal canal system. Neurosci. Lett. **16:** 143–148.
11. BÜTTNER-ENNEVER, J. A. 1981. Vestibular-oculomotor organization. *In* Progress in Oculomotor Research. A. F. Fuchs & W. Becker, Eds.: 361–370. Elsevier. Amsterdam,
12. LISBERGER, S. G. 1988. The neural basis for learning of simple motor skills. Science **242:** 728–735.
13. PRECHT, W. 1979. Vestibular mechanisms. Annu. Rev. Neurosci. **2:** 265–289.
14. FUKUSHIMA, K., S. I. PERLMUTTER, J. F. BAKER & B. W. PETERSON. 1990. Spatial properties of second-order vestibulo-ocular relay neurons in the alert cat. Exp. Brain Res. **81:** 462–478.

15. MARKHAM, C. H. 1989. Anatomy and physiology of otolith-controlled ocular counterrolling. Acta Otolaryngol. Suppl. No. **468:** 263–266.
16. UCHINO, Y., N. HIRAI, S. SUZUKI & S. WATANABE. 1981. Properties of secondary vestibular neurons fired by stimulation of ampullary nerve of the vertical, anterior or posterior, semicircular canals in the cat. Brain Res. **223:** 273–286.
17. UCHINO, Y., N. HIRAI & S. SUZUKI. 1982. Branching pattern and properties of vertical- and horizontal-related excitatory vestibuloocular neurons in the cat. J. Neurophysiol. **48:** 891–903.
18. ISU, N. & J. YOKOTA. 1983. Morphophysiological study on the divergent projection of axon collaterals of medial vestibular nucleus neurons in the cat. Exp. Brain Res. **53:** 151–162.
19. MITSACOS, A., H. REISINE & S. M. HIGHSTEIN. 1983. The superior vestibular nucleus: an intracellular HRP study in the cat. I. Vestibulo-ocular neurons. J. Comp. Neurol. **215:** 78–91.
20. MITSACOS, A., H. REISINE & S. M. HIGHSTEIN. 1983. The superior vestibular nucleus: an intracellular HRP study in the cat. II. Non-vestibulo-ocular neurons. J. Comp. Neurol. **215:** 92–107.
21. GRAF, W., R. A. McCREA & R. BAKER. 1983. Morphology of posterior canal related second order vestibular neurons in rabbit and cat. Exp. Brain Res. **52:** 125–138.
22. HIRAI, N. & Y. UCHINO. 1984. Superior vestibular nucleus neurones related to the excitatory vestibulo-ocular reflex of anterior canal origin and their ascending course in the cat. Neurosci. Res. **1:** 73–79.
23. GRAF, W. & K. EZURE. 1986. Morphology of vertical canal related second order vestibular neurons in the cat. Exp. Brain Res. **63:** 35–48.
24. McCREA, R. A., A. STRASSMANN, E. MAY & S. M. HIGHSTEIN. 1987. Anatomical and physiological characteristics of vestibular neurons mediating the horizontal vestibuloocular reflex in the squirrel monkey. J. Comp. Neurol. **264:** 547–570.
25. McCREA, R. A., A. STRASSMANN & S. M. HIGHSTEIN. 1987. Anatomical and physiological characteristics of vestibular neurons mediating the vertical vestibulo-ocular reflexes in the squirrel monkey. J. Comp. Neurol. **264:** 571–594.
26. OHGAKI, T., I. S. CURTHOYS & C. H. MARKHAM. 1988. Morphology of physiologically identified second-order vestibular neurons in cat, with intracellularly injected HRP. J. Comp. Neurol. **276:** 387–411.
27. ISU, N., N. I. UCHINO, H. NAKASHIMA, S. SATOH, T. ICHIKAWA & S. WATANABE. 1988. Axonal trajectories of posterior canal-activated secondary vestibular neurons and their coactivation of extraocular and neck flexor motoneurons in the cat. Exp. Brain Res. **70:** 181–191.
28. COHEN, B. 1974. The vestibulo-ocular reflex arc. In Handbook of Sensory Physiology. H. H. Kornhuber, Ed. **6**(1): 477–540. Vestibular System. Springer. New York, N.Y.
29. REISINE, H., A. STRASSMANN & S. M. HIGHSTEIN. 1981. Eye position and head velocity signals are conveyed to medial rectus motoneurons in the alert cat by the ascending tract of Deiter's. Brain Res. **211:** 153–157.
30. McCREA, R. A., K. YOSHIDA, A. BERTHOZ & R. BAKER. 1980. Eye movement related activity and morphology of second order vestibular neurons terminating in the cat abducens nucleus. Exp. Brain Res. **40:** 468–473.
31. HIGHSTEIN, S. M. 1973. Organization of the vestibulo-oculomotor and trochlear reflex pathways in the rabbit. Exp. Brain Res. **17:** 285–300.
32. ITO, M., N. NISIMARU & M. YAMAMOTO. 1973. The neuronal pathways relaying reflex inhibition from semicircular canals to extraocular muscles of rabbits. Brain Res. **55:** 189–193.
33. UCHINO, Y. & N. HIRAI. 1984. Axon collateral of anterior semicircular canal–activated vestibular neurons and their coactivation of extraocular and neck motoneurons in the cat. Neurosci. Res. **1:** 309–325.
34. YAMAMOTO, M., I. SHIMOYAMA & S. M. HIGHSTEIN. 1978. Vestibular nucleus neurons relaying excitation from the anterior canal to the oculomotor nucleus. Brain. Res. **148:** 31–42.
35. LANGER, T., C. R. S. KANEKO, C. A. SCUDDER & A. F. FUCHS. 1986. Afferents to the abducens nucleus in the monkey and cat. J. Comp. Neurol. **245:** 379–400.

36. McFARLAND, J. L., A. F. FUCHS & C. R. S. KANEKO. The nucleus prepositus and nearby medial vestibular nucleus and the control of simian eye movements. *In* Vestibular and Brain Stem Control of Eye, Head and Body Movements. Y. Shinoda & H. Shimazu, Eds. Japan Scientific Societies Press. Tokyo, Japan. (In press.)

37. SATO, Y. & T. KAWASAKI. 1987. Target neurons of floccular caudal zone inhibition in Y-group nucleus of vestibular nuclear complex. J. Neurophysiol. **57:** 460–480.

38. HORN, A. K. I. & J. A. BÜTTNER-ENNEVER. 1990. The time course of retrograde transsynaptic transport of tetanus toxin fragment C in the oculomotor system of the rabbit after injection into extraocular eye muscles. Exp. Brain Res. **81:** 353–362.

39. CHUBB, M. C. & A. F. FUCHS. 1982. Contribution of y group of vestibular nuclei and dentate nucleus of cerebellum to generation of vertical smooth eye movements. J. Neurophysiol. **48:** 75–99.

40. KELLER, E. L. & B. Y. KAMATH. 1975. Characteristics of head rotation and eye movement related neurons in alert monkey vestibular nucleus. Brain Res. **100:** 182–187.

41. KELLER, E. L. & P. D. DANIELS. 1975. Oculomotor related interaction of vestibular and visual stimulation in vestibular nucleus cells in alert monkey. Exp. Neurol. **46:** 187–198.

42. MACIEWICZ, R. Z., K. EAGEN, C. R. S. KANEKO, & S. M. HIGHSTEIN. 1977. Vestibular and medullary brainstem afferents to the abducens nucleus in the cat. Brain Res. **123:** 229–240.

43. HOLSTEGE, G. 1988. Brainstem–spinal cord projections in the cat related to control of head and axial movements. *In* Neuroanatomy of the Oculomotor System (Rev. Oculomotor Res.). J. A. Büttner-Ennever, Ed.: 431–470. Elsevier. Amsterdam, the Netherlands.

44. SHINODA, Y., T. OHGAKI, T. FUTAMI & Y. SUGIUCHI. 1988. Vestibular projections to the spinal cord: the morphology of single vestibulospinal axons. Prog. Brain Res. **76:** 17–27.

45. IWAMOTO, Y., T. KITAMA & K. YOSHIDA. 1990. Vertical eye movement–related secondary vestibular neurons ascending in medial longitudinal fasciculus in cat. II. Direct connections with extraocular motoneurons. J. Neurophysiol. **63:** 918–935.

46. MINOR, L. B., R. A. MCCREA & J. M. GOLDBERG. 1990. Dual projections of secondary vestibular axons in the medial longitudinal fasciculus to extraocular motor nuclei and the spinal cord of the squirrel monkey. Exp. Brain Res. **83:** 9–21.

47. HIRAI, N. & Y. UCHINO. 1984. Floccular influence on excitatory relay neurons of vestibular reflexes of anterior semicircular canal origin in the cat. Neurosci. Res. **1:** 327–340.

48. BRODAL, A. & P. BRODAL. 1985. Oberservations on the secondary vestibulo-cerebellar projections in the Macaque monkey. Exp. Brain Res. **58:** 62–74.

49. THUNNISSEN, I. E., A. H. EPEMA & N. M. GERRITS. 1989. Secondary vestibulocerebellar mossy fiber projection to the caudal vermis in the rabbit. J. Comp. Neurol. **290:** 262–277.

50. LANGER, T., A. F. FUCHS, C. A. SCUDDER & M. C. CHUBB. 1985. Afferents to the flocculus of the cerebellum in the rhesus macaque as revealed by retrograde transport of horseradish peroxidase. J. Comp. Neurol. **235:** 1–25.

51. EPEMA, A. H., N. M. GERRITS & J. VOOGD. 1990. Secondary vestibulocerebellar projections to the flocculus and uvulonodular lobule of the rabbit: a study using HRP and double fluroescent tracer techniques. Exp. Brain Res. **80:** 72–82.

52. LANGER, T. 1985. Basal interstitial nucleus of the cerebellum: a deep cerebellar nucleus related to the flocculus. J. Comp. Neurol. **235:** 38–47.

53. BLANKS, R. H. I. & Y. TORIGOE. 1983. Two neuronal groups associated with the medial longitudinal fasciculus (MLF) of cat and rat that convey vestibular impulses to the cerebellar flocculus. Soc. Neurosci. Abstr. **9:** 608.

54. BÜTTNER-ENNEVER, J. A. Cell groups of the paramedian tracts. *In* Vestibular and Brain Stem Control of Eye, Head and Body Movements. Y. Shinoda & H. Shimazu, Eds. Japan Scientific Societies Press. Tokyo, Japan. (In press.)

55. LANGER, T., A. F. FUCHS, M. C. CHUBB, C. A. SCUDDER & S. G. LISBERGER. Floccular efferents in the rhesus macaque as revealed by autoradiography and horseradish peroxidase. J. Comp. Neurol. **235:** 26–37.

56. HAINES, D. E. 1975. Cerebellar corticovestibular fibers of the posterior lobe in a prosimian primate, the lesser bushbaby (*Galago senegalensis*). J. Comp. Neurol. **160:** 363–398.

57. POMPEIANO, O., T. MERGNER & N. CORVAJA. 1978. Commissural, perihypoglossal and reticular afferent projections to the vestibular nuclei in the cat: an experimental anatomical study with horseradish peroxidase. Arch. Ital. Biol. **116:** 130–172.
58. CARLETON, S. C. & M. B. CARPENTER. 1983. Afferent and efferent connections of the medial, inferior and lateral vestibular nuclei in the cat and monkey. Brain Res. **278:** 29–51.
59. EPEMA, A. H., N. M. GERRITS & J. VOOGD. 1988. Commissural and intrinsic connections of the vestibular nuclei in the rabbit: a retrograde labeling study. Exp. Brain Res. **71:** 129–146.
60. MCCREA, R. A. & R. BAKER. 1985. Anatomical connections of the nucleus prepositus of the cat. J. Comp. Neurol. **237:** 377–407.
61. GACEK, R. R. 1978. Location of commissural neurons in the vestibular nuclei of the cat. Exp. Neurol. **59:** 479–491.
62. CARPENTER, M. B. & R. J. COWIE. 1985. Connections and oculomotor projections of the superior vestibular nucleus and cell group "y." Brain Res. **336:** 265–287.
63. DEJONG, J. M. B. V., B. COHEN, V. MATSUO & T. UEMURA. 1980. Midsagittal pontomedullary brain stem section: effects on ocular adduction and nystagmus. Exp. Neurol. **68:** 420–442.
64. CHERON, G., P. GILLIS & E. GODAUX. 1986. Lesions in the cat prepositus complex: effects on the optokinetic system. J. Physiol. **372:** 95–111.
65. KATZ, E., J. M. B. V. DEJONG, J. A. BÜTTNER-ENNEVER & B. COHEN. 1991. Effects of midline medullary lesions on velocity storage and the vestibulo-ocular reflex. Exp. Brain Res. **87:** 505–520.
66. WAESPE, W., B. COHEN & TH. RAPHAN. 1985. Dynamic modification of the vestibulo-ocular reflex by the nodulus and uvula. Science **228:** 199–202.
67. UEMURA, T. & B. COHEN. 1973. Effects of vestibular nuclei lesions on vestibulo-ocular reflexes and posture in monkeys. Acta Oto-Laryngol. Stockh. Suppl. **315:** 1–71.
68. MAGNIN, M., J. H. COURJON & J. M. FLANDRIN. 1983. Possible visual pathways in the cat vestibular nuclei involving the nucleus prepositus hypoglossi. Exp. Brain Res. **51:** 298–303.
69. SIMPSON, J. I. 1984. The accessory optic system. Annu. Rev. Neurosci. **7:** 13–41.
70. SCHIFF, D., B. COHEN, J. BÜTTNER-ENNEVER & V. MATSUO. 1990. Effects of lesions of the nucleus of the optic tract on optokinetic nystagmus and after-nystagmus in the monkey. Exp. Brain Res. **79:** 225–239.
71. KELLER, E. L. & W. PRECHT. 1979. Visual-vestibular responses in vestibular nuclear neurons in the intact and cerebellectomized, alert cat. Neuroscience **4:** 1599–1613.
72. GIOLLI, R. A., R. H. I. BLANKS & Y. TORIGOE. 1984. Pretectal and brainstem projections of the medial terminal nucleus of the accessory optic system of the rabbit and rat as studied by anterograde and retrograde neuronal tracing methods. J. Comp. Neurol. **227:** 228–251.
73. GIOLLI, R. A., R. H. I. BLANKS, Y. TORIGOE & H. M. MCDONALD. 1988. Projections of the dorsal and lateral accessory optic nuclei and the interstitial nucleus of the superior fascicle (posterior fibres) in the rabbit and rat. J. Comp. Neurol. **277:** 608–620.
74. MCCREA, R. A. 1988. The nucleus prepositus. *In* Neuroanatomy of the Oculomotor System. J. A. Büttner-Ennever, Ed.: 203–223. Elsevier. Amsterdam, the Netherlands.
75. GIOLLI, R. A., R. H. I. BLANKS, Y. TORIGOE & D. D. WILLIAMS. 1985. Projections of medial terminal accessory optic nucleus, ventral terminal nuclei, and substantia nigra of rabbit and rat as studied by retrograde axonal transport of horseradish peroxidase. J. Comp. Neurol. **232:** 99–116.
76. FUKUSHIMA, K. 1987. The interstitial nucleus of Cajal and its role in the control of movements of head and eyes. Prog. Neurobiol. **29:** 107–192.
77. FUKUSHIMA, K. 1991. The interstitial nucleus of Cajal in the midbrain reticular formation and vertical eye movement. Neurosci. Res. **10:** 159–187.
78. SPENCE, S. J. & J. A. SAINT-CYR. 1988. Mesodiencephalic projections to the vestibular complex in the cat. J. Comp. Neurol. **268:** 375–388.
79. BARMACK, N. H., C. K. HENKEL & V. E. PETTOROSSI. 1979. A subparafascicular projection to the medial vestibular nucleus of the rabbit. Brain Res. **172:** 339–343.

Responses of Vestibular and Prepositus Neurons to Head Movements during Voluntary Suppression of the Vestibuloocular Reflex

ROBERT A. McCREA AND KATHLEEN E. CULLEN

Committee on Neurobiology
Department of Pharmacological and Physiological Sciences
University of Chicago
947 East 58th Street
Chicago, Illinois 60637

INTRODUCTION

In order for a smooth head movement to contribute to visual tracking or smooth gaze pursuit, the eye movements generated by the vestibuloocular reflex (VOR) must be suppressed or canceled in some way, since the vestibuloocular reflex is designed to stabilize gaze in space by generating an eye movement that is in the opposite direction to an ongoing head movement. The results of our behavioral studies in squirrel monkeys suggest that there are at least two mechanisms by which the VOR is canceled during gaze pursuit.[1,2] One mechanism utilizes visual feedback regarding the relative motion of the visual target on the retina and produces a smooth pursuit eye movement in the opposite direction from the VOR.[3–5] This visual feedback mechanism is limited by a central delay of approximately 100 mseconds in squirrel monkeys. A second, nonvisual mechanism for VOR cancellation is also available that can supplement the smooth pursuit mechanism when the head velocity generated during gaze pursuit lags target velocity. The nonvisual mechanism operates at a much shorter latency than visual feedback following unexpected perturbations of the head during gaze pursuit.[6,2]

In the present study we recorded the responses of several types of neurons in the vestibular nuclei and the prepositus nucleus that are likely to be involved in controlling the horizontal canal evoked VOR during smooth pursuit eye movements and during horizontal VOR cancellation (VORC), and compared their responses to those generated by abducens motor neurons. We found that the visual and nonvisual mechanisms for VORC are mediated by largely separate brain stem pathways.

METHODS

Five squirrel monkeys were prepared for chronic extracellular recording from the brain stem. A head-restraining bolt was implanted on the occipital bone so that the head would be held with the horizontal canals in the plane of table rotation (15 deg nose down), and an eye coil was attached to the sclera with small sutures. Pairs of Teflon-insulated electrodes (250 μm in diameter of which 1 mm of the tip was exposed and coated with AgCl$_2$) were implanted in the middle ear bilaterally for electrical stimulation of the vestibular nerves. A small plastic chamber with a removable lid was attached to the parietal bone with dental cement to allow the insertion of recording microelectrodes.

During experiments, a monkey was seated in a primate chair, and both the chair and the bolt on the animal's head were fixed to the suprastructure of a vestibular turntable. Eye movements were recorded using the magnetic search coil technique. The monkeys were trained to fixate a small target light. The target light was generated by an HeNe laser mounted on the turntable whose beam was projected onto a cylindrical screen that surrounded the turntable. The target could be moved horizontally or vertically by rotating mirrors attached to a pair of galvanometers (General Scanning) that were controlled by a position feedback servo control system. The water-deprived monkey was given a vanilla-flavored milk reward during experimental trials for fixating the target light for variable periods of time (1–3 seconds).

Computer-generated waveforms (digitized at 500 Hz) were used to control the movement of the vestibular turntable, which was capable of > 360 degrees of angular rotation. The actual angular head velocity was monitored with an angular velocity sensor (Watson Industries, bandwidth: DC-200Hz) mounted below the primate chair and centered on the axis of table rotation.

Single-unit recordings were made with enamel-insulated tungsten microelectrodes that were inserted into the brain stem through a 22-gauge guide tube. The anatomical location of recorded cells was determined by their location relative to the abducens nucleus, recognized by its characteristic burst tonic activity, and monosynaptic latency of spike-triggered averages of lateral rectus activity, and through analysis of the field potential elicited by stimulation of the vestibular nerve. Spike potentials were conventionally amplified and filtered (bandpass 300 Hz–8 kHz), and unit spikes were converted into trigger pulses by setting the appropriate level and slope of an oscilloscope discriminator circuit. The pulses were input to an event channel of intelligent peripheral (CED1401, Cambridge Electronic Design) for further analysis.

The response of most units was recorded following electrical stimulation of the vestibular nerve. Brief monophasic pulses (0.1 msecond, 200–400 μA) were used to electrically stimulate the vestibular nerve at 2 Hz. Units that were activated at latencies of 0.7–1.3 mseconds were considered to be second-order vestibular neurons.

Visual target position, vestibular turntable velocity, data acquisition, on-line display, and on-line analysis were preformed using the Compaq Deskpro 60 80386 microcomputer equipped with a Data Translation DT2825 data acquisition board, a DATAQ waveform scroller board, and a CED1401 peripheral device.

Experimental Paradigms

The responses of a unit were recorded during sinusoidal vestibular and visual stimulation and during unpredictable steps in head or target acceleration.

Sinusoidal Stimulus Paradigms

The responses of a unit were recorded during several sinusoidal tracking tasks: (1) First the monkey tracked both horizontal and vertical sinusoidal target motion at 0.5 Hz, 20 deg/second peak velocity (smooth pursuit, SP$_s$). (2) The monkey canceled its VOR during vestibular turntable rotation at 0.5 Hz, 40 deg/second peak velocity, by fixating a target that remained stationary with respect to the head (VORC$_s$). (3) The monkey was rotated sinusoidally in the light and the dark at 0.5 Hz, 40 deg/second peak velocity in the absence of a fixation target (VOR$_s$). Typical eye movements evoked in the VOR$_s$ and VORC$_s$ behavioral paradigms are illustrated in FIGURE 1A.

A

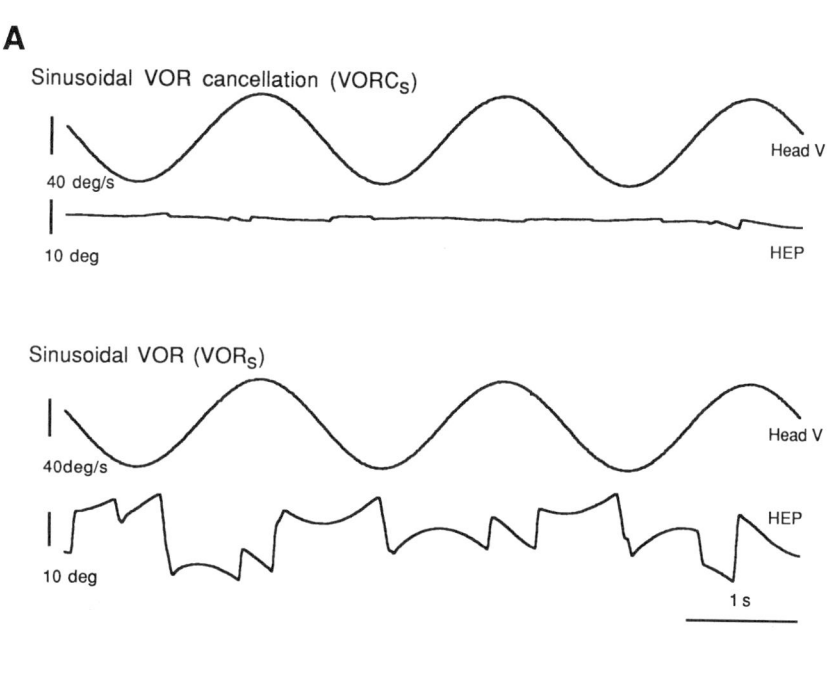

B

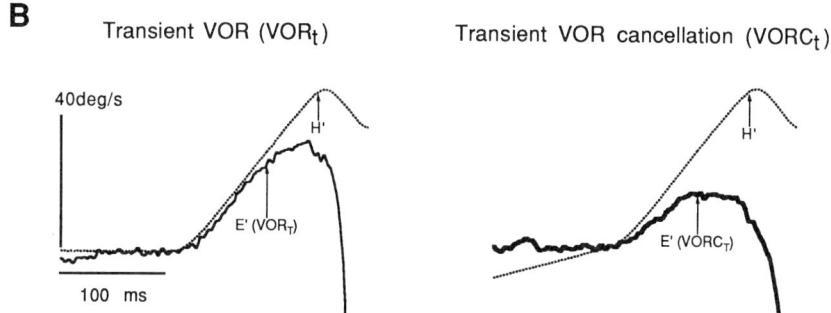

FIGURE 1. Examples of the eye movements evoked in four of the behavioral paradigms used in this study. A: Sinusoidal VOR$_s$ and VORC$_s$. Horizontal eye position (HEP) and head velocity (Head V) traces are shown. Upward eye movements are ipsilaterally directed, upward head movements are contralaterally directed. B: Head acceleration step paradigms. Upward eye velocity is contralateral, upward head velocity is ipsilateral.

Unpredictable Acceleration Step Paradigms

In the one paradigm (VORC$_t$) a head acceleration step was triggered while the monkey was canceling its VOR during a background table movement. The acceleration step was triggered once the monkey fixated a head stationary target for a variable time period (between 1 and 10 seconds). The background table movement was usually a cyclic velocity trapezoid whose plateau velocities were 40 deg/second counterclockwise and 40 deg/second clockwise. The acceleration of the background table movement was 80 deg/second and the background table movement at the time of the step was always 0 deg/second in these studies. The steps in head acceleration were made in the same direction as the current background table acceleration.

In a second head acceleration step paradigm (VOR$_t$), the animal was initially stationary. An acceleration step was unexpectedly triggered while the monkey fixated a head stationary target. Although the monkey was attempting to fixate a head stationary target in this paradigm, the eye movement response to the head acceleration step was identical to the eye movement evoked in the absence of a target light for ≈ 100 mseconds. We used this target fixation paradigm as an index of the response of a neuron during the VOR instead of turntable rotation in the absence of a target, since in the absence of a target the initial position of the eye prior to a step in head acceleration was impossible to control. The eye movements typically evoked during the VOR$_t$ and VORC$_t$ paradigms are illustrated in FIGURE 1B.

Data Analysis

The data were analyzed with the aid of the IGOR analysis package (WaveMetrics) using a Macintosh IIx computer. The horizontal and vertical eye position sensitivity of a neuron was determined by selecting periods of steady fixation 100 mseconds or more in length that did not include a period following or proceeding saccadic eye movements by 40 mseconds. A multiple regression analysis was done for each unit, in order to determine its best position vector.

Neuronal sensitivity to smooth pursuit eye movements was analyzed by using a multiple regression analysis of desaccaded records and by fitting the desaccaded eye velocity and eye position corrected firing rate with a sine wave of the stimulus frequency, using a least-squares algorithm. The eye velocity and the eye position coefficients and intercept (which corresponded to the neuron's spontaneous firing rate) were determined over 5–15 cycles of the smooth tracking behavior.

In determining the sensitivity of the neurons to head velocity during the VOR in the dark and the light (VOR$_s$), 5–15 cycles of the firing rate and head velocity were fit with a sine wave of the stimulus frequency, using a least-squares algorithm. The firing rate was corrected for its dc level, and the values in an interval around a saccade were not included in the determination of best fit. In cells that had an eye position sensitivity, the eye position dependent portion of the firing rate was subtracted out before analysis. A similar procedure was used to determine the sensitivity of the neurons to head velocity during cancellation of the VOR (VORC$_s$). Only cycles during which the monkey successfully suppressed its eye velocity during head rotation were included in this analysis.

In order to analyze data from the acceleration step paradigm, data from between 10 and 30 trials were averaged. The selection of the trials that were included in each average was based on both behavioral performance and on a relative lack of saccades in the record. Trials were only used when a saccade did not occur 150 mseconds before the step and when the monkey successfully reacquired the target within 200 mseconds after the onset of the stimulus.

RESULTS

Classification and Location of the Units Included in this Study

Neurons were classified on the basis of their firing behavior during spontaneous eye movements, sinusoidal smooth pursuit eye movements (SP_s), cancellation of the horizontal sinusoidal VOR ($VORC_s$), and horizontal sinusoidal VOR in the light (VOR_s) and on the basis of their location in the brain stem. Six classes of cells in the vestibular nuclei, the prepositus hypoglossi nucleus (PH), and the abducens nucleus (ABD) were studied: (1) type I "pure vestibular neurons"; (2) type I position vestibular pause neurons; (3) type II vestibular neurons; (4) abducens neurons; (5) burst tonic neurons in the rostral medial vestibular nucleus (MVN) and the PH, and (6) smooth pursuit neurons (SP neurons) in the rostral MVN and PH.

All of the type I vestibular neurons increased their firing rate during ipsilateral turntable rotation. Twenty-five of these units were identified as horizontal position vestibular pause neurons (PVP neurons). All of the PVP neurons increased their firing rate as a function of contralateral eye position, and the firing rate of all but two was correlated with eye velocity during horizontal smooth pursuit eye movements. The spiking activity of 20 type I vestibular units was not well correlated ($r^2 < 0.5$) with eye position during spontaneous saccades or horizontal smooth pursuit eye movements. These cells will be referred to as "pure" vestibular neurons, since we were not able to detect any other signals with the paradigms we used. Electrical stimulation of the vestibular nerve evoked an action potential in 14 of the 16 PVP neurons tested, and in 5 of the 14 pure vestibular neurons tested. The latency of the electrically evoked action potentials was always less than 1.3 mseconds (range = 0.7–1.3 mseconds), which suggests that those neurons received monosynaptic inputs from the vestibular nerve.

Behavior of Type I Pure Vestibular Neurons during VOR_s and $VORC_s$

The head movement sensitivity of most pure vestibular neurons did not change when the monkeys canceled their VOR during sinusoidal turntable rotation. FIGURE 2 shows the firing behavior of a typical pure vestibular neuron during cancellation of the horizontal VOR (FIGURE 2A), and during the VOR (FIGURE 2B). The gain and phase of each cell's response to 0.5 Hz sinusoidal vestibular stimuli were calculated by fitting a sine wave to its response during $VORC_s$ and VOR_s. The response gain of the population of pure vestibular units as a whole was 0.8 (spikes/second)/(deg/second) both during VOR_s and $VORC_s$. None of the pure vestibular neurons had firing behavior that was well correlated with eye movements. This lack of correlation included saccadic eye movements. Although some cells exhibited highly irregular activity during spontaneous eye movements, none of them consistently paused during saccadic eye movements.

Firing Behavior of PVPs during VOR and VORC

The firing rate of PVP neurons was modulated during spontaneous eye movements and during horizontal smooth pursuit. As noted above, during spontaneous

A   VOR cancellation

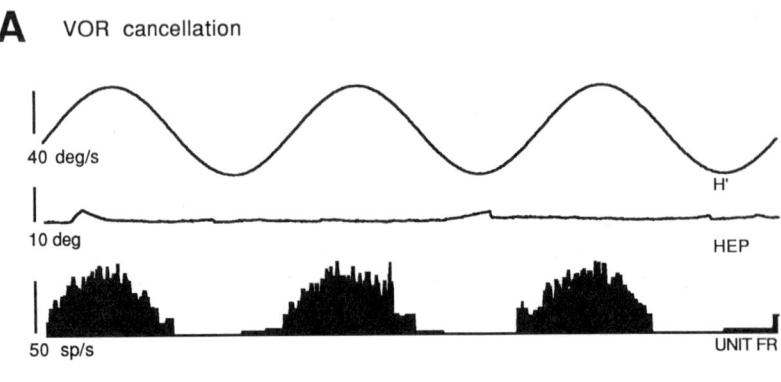

B   VOR

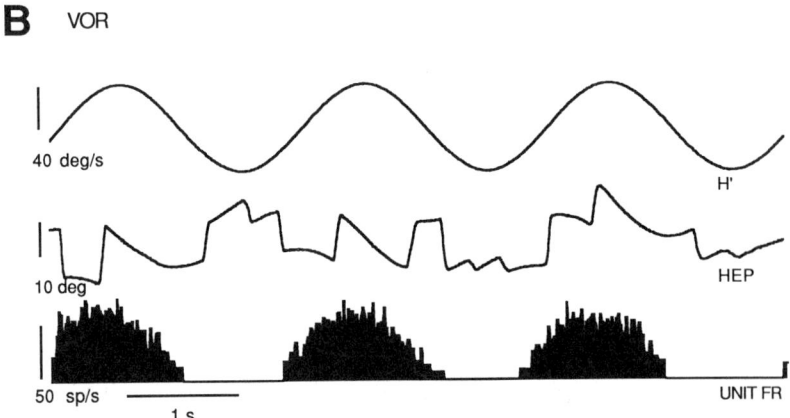

FIGURE 2. Firing behavior of a pure vestibular neuron during VOR cancellation (A) and during the VOR (B). H' = head velocity, HEP = horizontal eye position, unit FR = unit firing rate.

eye movements, all of the PVP neurons recorded in this study increased their firing rate when the eye was fixated in the contralateral direction and decreased during ipsilateral fixations. The eye position sensitivity ranged from 1.1 to 7.9 spikes/second per deg. (mean = 3.4). Most PVP neurons paused during most saccades, although the pause in firing rate was less profound during contralaterally directed horizontal saccades. When a saccade was near, or in, the preferred direction of a PVP cell, the pause was frequently absent. During 0.5 Hz sinusoidal smooth pursuit the firing rate of most PVPs phase led eye position and lagged eye velocity. The sensitivity to eye velocity during smooth pursuit varied considerably from neuron to neuron, ranging from 0.0 to 1.8 (spikes/second)/(deg/second) (mean = 0.7).

FIGURE 3 shows the response of a PVP neuron during 0.5 Hz sinusoidal VOR cancellation (A), and VOR (B). Since PVPs were sensitive to eye position, and tended to pause during saccades, it was necessary to analyze desaccaded records and to correct the firing rate of PVPs for their eye position sensitivity, in order to compare the head movement sensitivity of PVPs during $VORC_s$ and VOR_s. The results of this type of analysis are illustrated in FIGURE 6A. The upper traces in FIGURE 6A are the average eye position corrected response of a PVP during 0.5 Hz VOR$_s$ and VORC$_s$ superimposed on the vestibular stimulus. The shaded area indicates the difference in the neuron's response during the two behavioral conditions. The lower traces in FIGURE 6A are sine wave fits to the unit's response in each condition. The results of this analysis show that in the case of the illustrated PVP the sensitivity to head velocity during VORC$_s$ [0.7 (spikes/second)/(deg/second)] was less than half of its sensitivity during VOR$_s$. The head velocity sensitivity of 19 of the 24 PVPs tested was less during VORC$_s$ than during VOR$_s$. The mean difference in sensitivity for those 19 cells was 0.4 (spikes/second)/(deg/second), or a reduction in sensitivity of approximately 31%. The remaining five PVP cells did not show a change in their head velocity sensitivity during VORC, either during VORC$_s$ or during VORC$_t$.

Response of Type II Vestibular Neurons during VORC and during the VOR

The firing rates of all of the type II vestibular neurons we recorded from were related to ipsilateral eye movements and contralateral head movements. The head

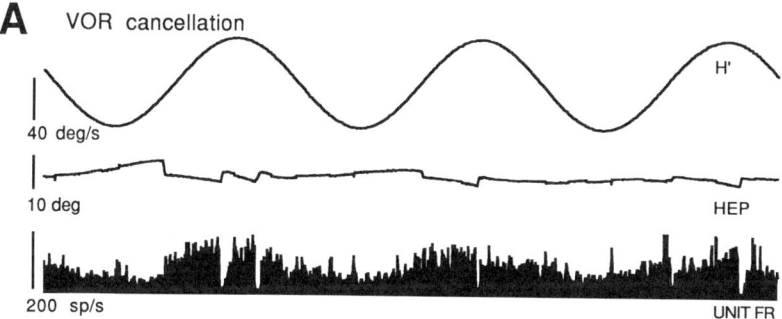

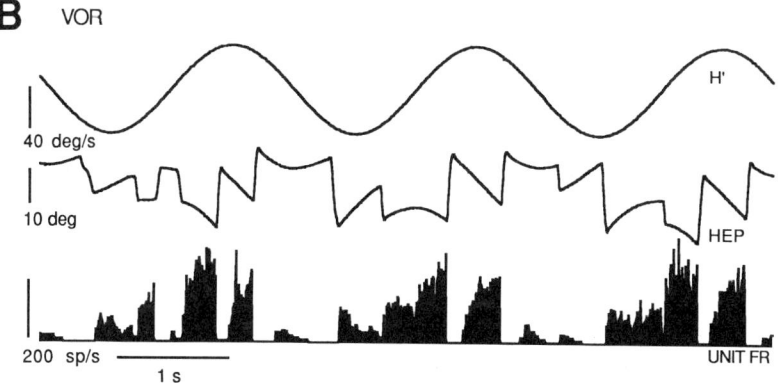

FIGURE 3. Firing behavior of a PVP neuron during VORC and VOR.

movement sensitivity of most type II vestibular neurons was significantly larger when the monkeys canceled their VOR by fixating a head stationary target. The increase in head velocity sensitivity of the entire population of 12 type II units was 1.2 (spikes/second)/(deg/second) during $VORC_s$ compared to 0.7 (spikes/second)/(deg/second) during VOR_s (an increase in sensitivity of 58%), even though two of the cells had no net change in their sensitivity to head velocity in the two behavioral conditions.

Firing Behavior of Abducens Neurons and Burst Tonic Neurons during VORC and VOR

FIGURE 4 shows the firing behavior of an abducens motor neuron during sinusoidal smooth pursuit, VORC, and during the VOR. The firing behavior of 15 burst tonic neurons in the PH and rostral MVN, like that of 12 ABD neurons, was related to the horizontal eye position and eye velocity during VORC and during the VOR. Burst tonic neurons were, on average, more sensitive to eye position during spontaneous eye movements than were motor neurons. The firing rate of most burst

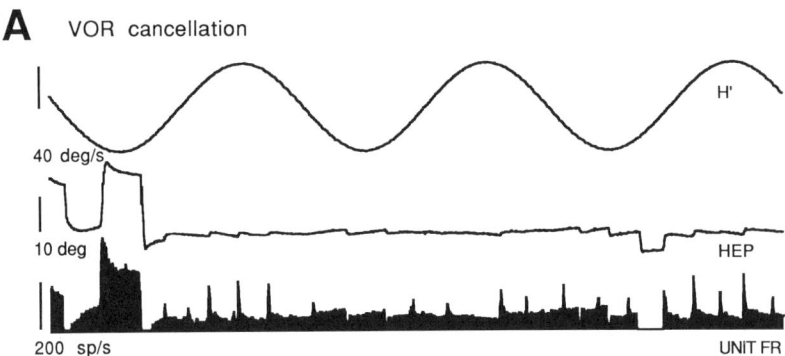

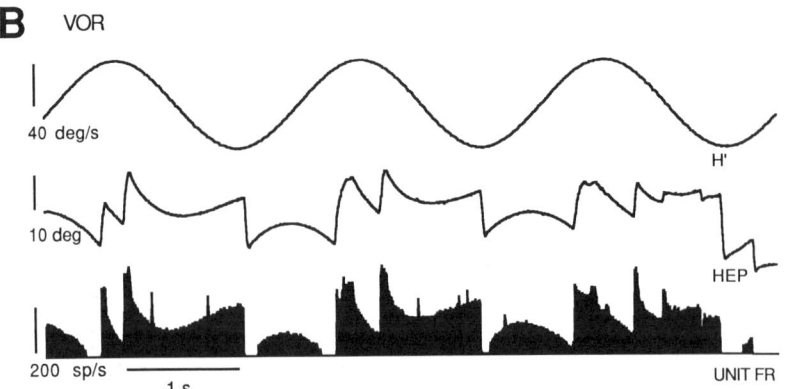

FIGURE 4. Firing behavior of an abducens motor neuron during VORC and VOR.

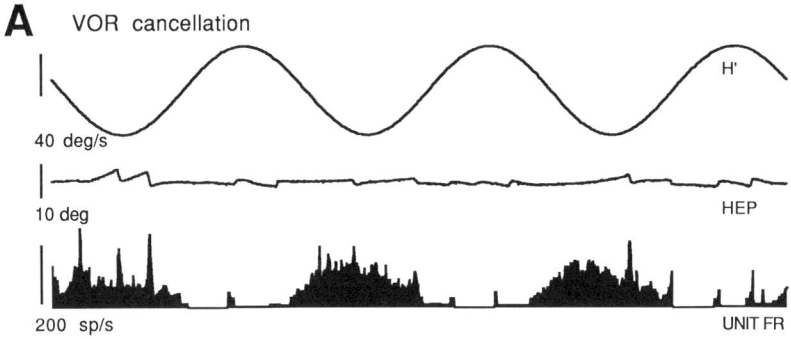

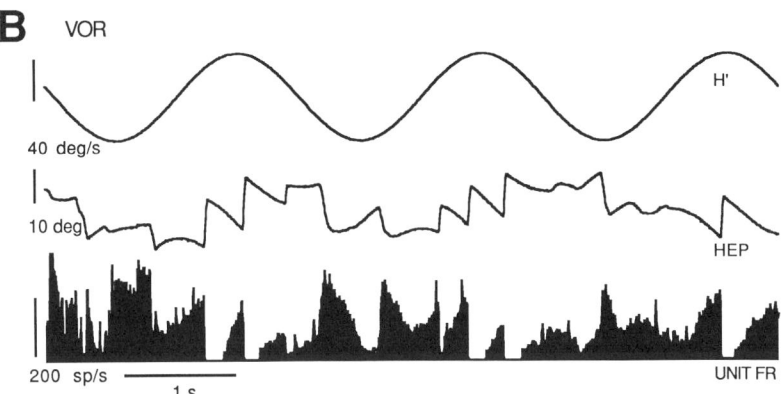

FIGURE 5. Firing behavior of an SP neuron during VORC and VOR.

tonic neurons, and virtually all ABD neurons, was not significantly related to head velocity during VORC, if their responses were corrected for the residual eye movements that were made during VORC.

Firing Behavior of SP Neurons during VOR and VORC

The firing behavior of 15 units in the MVN and PH was related to eye movements during smooth pursuit and to head movements during VORC. Six of these units fired in phase with ipsilateral eye velocity during smooth pursuit, while the preferred direction of the other nine was in the opposite direction. All but four of these smooth pursuit cells were sensitive to eye position during spontaneous saccadic eye movements and generated a burst of spikes during ipsilateral saccades. Many (8/15) of them generated bursts of spikes during all saccades, except those in the off direction of the cell.

All of the SP cells exhibited dramatically different sensitivities to head velocity during the VOR_s and $VORC_s$. FIGURE 5 illustrates the behavior of an SP neuron that fired in phase with contralateral eye velocity during sinusoidal smooth pursuit and with contralateral head velocity during VORC (FIGURE 5A). During the VOR (FIGURE 5B), the unit's firing rate was correlated with ipsilateral head velocity and contralateral eye position. The cell's response during the VOR was complicated by the fact that it generated bursts of spikes during most saccades. In FIGURE 6B the responses of the same SP cell during saccade free periods of VOR_s and $VORC_s$ are illustrated after the firing rate was corrected for the cell's contralateral eye position sensitivity. Note that the unit's response was phase shifted nearly 180° (159°) in respect to head velocity during VOR. The head velocity sensitivity of a majority (9/12) of the SP cells was essentially reversed, and all of the cells increased the gain of their response to head velocity in the direction of their pursuit eye movement sensitivity during VORC.

The signals generated by MVN and PH SP neurons were qualitatively similar to those generated by flocculus Purkinje cells,[7] but only three of the brain stem cells generated signals that could be characterized as being related to gaze velocity, since the head velocity sensitivity of most cells during VORC was usually significantly less than their eye velocity sensitivity during smooth pursuit. It is important to note that the response of neurons during the VOR_s was studied in an illuminated room. This was done because the VOR has no obvious function in the dark, and our drug-free squirrel monkeys typically behaved accordingly: the VOR gain at 0.5 Hz usually decreased significantly when it was tested in total darkness. If one assumes that smooth pursuit contributes to gaze stabilization during head rotations in an illuminated environment by generating an eye movement in the direction opposite to the ongoing head movement, then the "head velocity" sensitivity of SP neurons during VOR that we recorded could simply reflect the contribution of smooth pursuit to gaze stabilization during head rotations in the light.

Response of Vestibular and PH Neurons to Steps in Head Acceleration during VOR and VORC

In FIGURE 7, the responses of pure vestibular neurons, PVP neurons, SP neurons, and burst tonic neurons during unpredictable steps in head acceleration that were generated while the monkey was canceling its VOR ($VORC_t$, thick lines) and when it was not (VOR_t, thin lines) are superimposed. The figure illustrates the average responses of all of the cells in each class in the two behavioral paradigms. All of the responses have been corrected for the eye position sensitivity of the cells. The top trace in FIGURE 7 is the head velocity generated by the acceleration step. Most pure vestibular neurons generated the same response to the acceleration step whether the monkey was canceling its VOR or not. The firing rate of burst tonic neurons (like abducens motorneurons) reflected the eye velocity generated by the acceleration step (see FIGURE 1B).

The reduced head movement sensitivity of PVP neurons was apparent almost immediately after the initiation of the $400°/second^2$ step in head acceleration. The reduction in head movement sensitivity observed with these transient vestibular stimuli during VORC was similar to that observed during $VORC_s$. The short latency at which the response of PVP neurons was affected during $VORC_t$ suggests that the reduction in head movement sensitivity was not related to visual feedback.

The head acceleration steps generated in the VOR_t and $VORC_t$ paradigms did not evoke a change in the firing rate of SP neurons until 80–100 mseconds after the

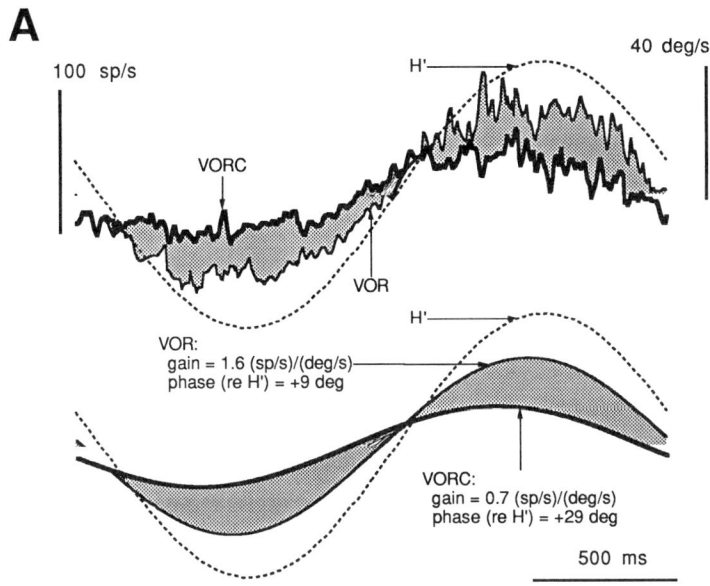

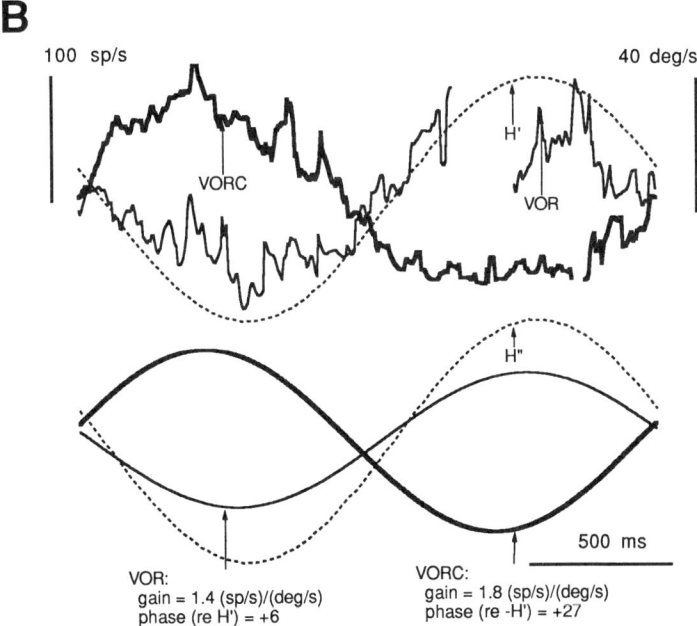

FIGURE 6. **A:** Averaged desaccaded response of a PVP neuron during $VORC_s$ and VOR_s. The lower traces show the best-fit sinusoids during each behavior. **B:** Averaged desaccaded response of an SP neuron during $VORC_s$ and VOR_s.

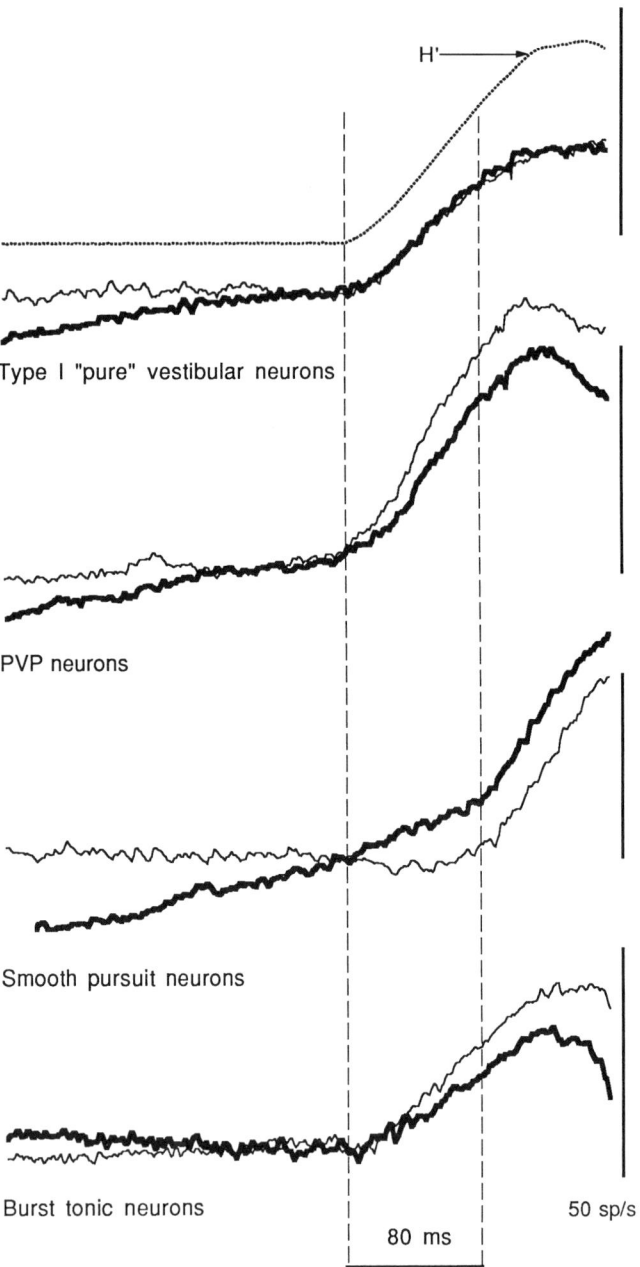

FIGURE 7. Mean response of different classes of neurons to unexpected steps in head acceleration ($400°/\text{second}^2$) generated while the monkey was canceling its VOR (thick lines) and while it was not (thin lines). Dotted lines indicate onset of step and 80 mseconds after the initiation of the step. The top trace shows the head velocity change during the step.

onset of the step (second dotted line in FIGURE 7). Thereafter, the firing rate of the cell increased dramatically by 50–100 spikes/second, if the step was in its on direction. Since the head acceleration steps evoked an eye movement in the compensatory direction in both behavioral paradigms, the lack of a short latency change in the firing rate of SP neurons suggests that their apparent sensitivity to eye and head movements during VOR_s, $VORC_s$ was not directly related to eye or head movements per se. The large transient response that began ≈ 80 mseconds after the acceleration step was not significantly different in $VORC_t$ trials and VOR_t trials. However, the eye movement evoked by the vestibular stimulus caused the head stationary visual target to move in respect to the retina in both behavioral paradigms. Since a smooth pursuit eye movement was evoked by a comparable acceleration of the target at a latency of 90–100 mseconds,[2] it is likely that smooth pursuit eye movements were generated 90–100 mseconds after the initiation of the step in both paradigms. Thus the delay in the response of SP cells during head acceleration steps was most likely due to the delays involved in the generation of visual smooth pursuit eye movements.

DISCUSSION

When head-restrained squirrel monkeys voluntarily suppress or cancel their horizontal VOR, the head movement sensitivity of secondary VOR pathways is reduced. Most of the secondary vestibular neurons that exhibited reduced head movement sensitivity during VOR cancellation were position-vestibular-pause neurons (PVP neurons), which are the secondary vestibular neurons that project to extraocular motor neurons and presumably mediate the vestibuloocular reflex.[8-16] The mechanism by which the head velocity sensitivity of PVP neurons was reduced did not appear to require visual feedback or smooth pursuit, since the reduction was apparent at a short latency when the head was transiently accelerated while the monkey was canceling its VOR. Taken together, these observations lead us to conclude that the nonvisual mechanism for VOR cancellation that we described in previous behavioral studies[2] involves a reduction in the head movement sensitivity of secondary vestibuloocular reflex pathways.

The firing rate of all of the horizontal canal related secondary vestibular neurons that we recorded from, including PVP neurons, was sensitive to head movements regardless of whether the monkey was canceling its VOR or not. During voluntary cancellation of the VOR evoked by a 0.5-Hz stimulus, the head movement sensitivity of many cells was reduced 20–60%, but it was never abolished. Yet monkeys are capable of completely suppressing their VOR under these stimulus conditions. Obviously, the reduction in head movement sensitivity of secondary VOR pathways is not the only mechanism that monkeys utilize for canceling their VOR. The signals that MVN and PH SP neurons generate could be used to cancel the residual head movement signals in VOR pathways at the level of the motor neuron.

The Role of PH and MVN Smooth Pursuit Neurons in VOR Cancellation

We referred to the PH and MVN neurons whose firing rates were related to eye and head velocity in the same direction during smooth pursuit and VORC as SP neurons because their firing behavior was best explained in terms of smooth pursuit. The lack of an early response during steps in head acceleration suggests that the head velocity sensitivity of those cells was very indirectly related to vestibular nerve activity. It seems more likely that the apparent head velocity sensitivity of the cells during the VOR in the light reflects the fact that smooth pursuit contributes to gaze stabilization during head rotations in the light. The change in their head movement response during VORC can be explained if one assumes that the smooth pursuit eye movement commands are oppositely directed during that behavior. The 80-msecond latency of their response to head acceleration steps generated while the monkeys were fixating a head stationary target was reasonably close to the latency at which smooth pursuit eye movements are initiated (90–100 mseconds) when a visual target is accelerated in a similar manner.[2]

SP cells could play an important role in canceling the residual head movement signals in VOR pathways during VORC if they project to the abducens nucleus. In order to cancel the activity of PVP neurons during VORC, SP neurons whose firing behavior is related to contralateral head movements during VORC should inhibit ipsilateral abducens neurons or excite contralateral abducens neurons. Since the rostral MVN and the PH project bilaterally to the abducens nucleus,[17-19] either or both connections are feasible, albeit unproven.

The Contribution of Burst Tonic Neurons to VORC

The firing rate of burst tonic neurons in the MVN and PH was primarily related to the position of the eye during VORC. The response of those neurons during unpredictable steps in head acceleration that were generated during VORC was also related to the eye movements that were evoked. This suggests that the nonvisual mechanism for VOR cancellation affects the firing behavior of these neurons as well as abducens neurons, and that horizontal burst tonic neurons receive inputs from the same vestibular pathways that project to the abducens nucleus. Since horizontal PVP neurons send collateral projections to the PH and rostral MVN[12] as well as the abducens nucleus, the short latency reduction in the head movement sensitivity of burst tonic neurons is reasonable if one assumes that their vestibular inputs are derived primarily from PVPs.

There is both anatomical and physiological evidence that burst tonic neurons in the rostral MVN and PH project to the abducens nucleus,[12,14] although it is not clear how influential their inputs to abducens motoneurons are. Since the firing behavior of most burst tonic neurons was also strongly related to smooth pursuit eye movements, it is clear that their inputs to the abducens nucleus would contribute to both the nonvisual and smooth pursuit dependent mechanisms for VORC. The importance of the role they play in shaping the response properties of abducens motor neurons during VORC depends in large part on how powerful a premotor input they are.

Hypothesized Organization of the Central Pathways that Mediate VOR Cancellation

The mean firing behavior of five of the classes of cells recorded in those nuclei during VOR_s and $VORC_s$ is illustrated in FIGURE 8. The firing rate of all of the neurons, except pure vestibular neurons, was modulated during smooth pursuit. For the sake of simplicity, the responses of SP neurons whose on direction was ipsilateral were inverted and combined with those whose firing rate was related to contralateral eye velocity. The behavior of most pure vestibular neurons was the same during

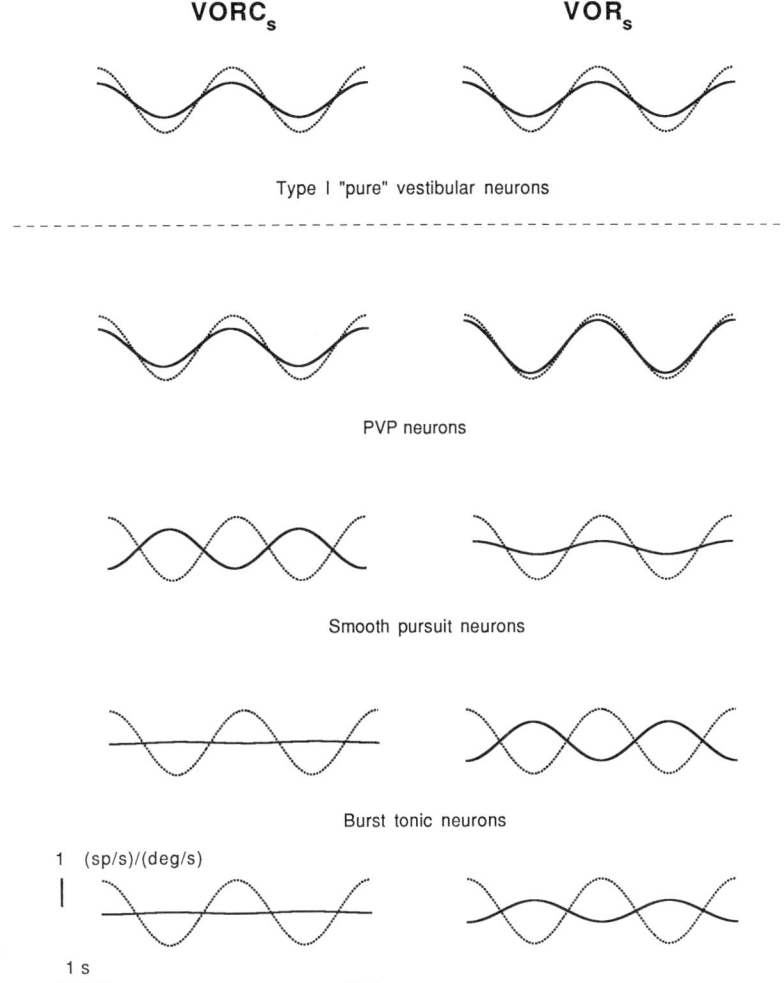

FIGURE 8. The averaged responses of "pure" vestibular neurons, PVP neurons, smooth pursuit neurons, and burst tonic neurons during $VORC_s$ and VOR_s. The eye position sensitivities of the neurons were subtracted from the responses. The signs of the responses of SP neurons whose on direction was ipsilateral were inverted so they could be grouped with the SP neurons with contralateral on directions. The dotted lines represent head velocity during sinusoidal vestibular stimulation (0.5 Hz, peak velocity = 40 deg/second).

VOR_s and $VORC_s$, but the other classes of cells showed significant changes in their firing behavior in the two behavioral conditions.

It is possible to explain the behavior of abducens neurons during VORC and during the VOR if one assumes that they receive inputs from PVP neurons, SP neurons, and burst tonic neurons. All three inputs would contribute to the generation of the eye position and eye velocity signals generated by motorneurons during the VOR. When the VOR is voluntarily cancelled, the head movement sensitivity of PVP neurons and burst tonic neurons is reduced by a nonvisual mechanism. The residual head movement related signals generated by PVPs during VORC could be cancelled by SP inputs to burst tonic neurons and abducens neurons.

SUMMARY AND CONCLUSION

Neurons in the vestibular nuclei and the prepositus nucleus exhibited several different types of changes in their firing behavior during voluntary cancellation of the horizontal VOR. The head velocity sensitivity of type I position-vestibular-pause neurons was reduced during cancellation, while type II vestibular neurons exhibit an increase in their sensitivity. The firing behavior of burst tonic neurons in the medial vestibular nucleus, the prepositus nucleus, like the cells in the abducens nucleus, was closely related to the eye movements generated when the VOR is cancelled. Other cells in the PH and MVN respond primarily to smooth pursuit eye movements. We suggest that the behavior of abducens neurons during the VOR and during VOR cancellation can be explained if they receive inputs from PVP neurons, burst tonic neurons, and smooth pursuit neurons.

REFERENCES

1. CULLEN, K. E. & R. A. McCREA. 1990. Gaze pursuit of predictable and unpredictable moving visual targets compared to smooth pursuit and vestibulo-ocular reflex cancellation. Abst. Soc. Neurosci. **16**: 903.
2. CULLEN, K. E., T. BELTON & R. A. McCREA. 1991. A non-visual mechanism for voluntary cancellation of the vestibulo-ocular reflex. Exp. Brain Res. **83**: 237–252.
3. LANMAN, J., E. BIZZI & J. ALLUM. 1978. The coordination of eye and head movement during smooth pursuit. Brain Res. **153**(1): 39–53.
4. LISBERGER, S. G., C. EVINGER, G. W. JOHANSON & A. F. FUCHS. 1981. Relationship between eye acceleration and retinal image velocity during foveal smooth pursuit in man and monkey. J. Neurophysiol. **46**: 229–249.
5. GAUTHIER, G. M. & J. L. VERCHER. 1990. Visual vestibular interaction: vestibulo-ocular reflex suppression with head-fixed target fixation. Exp. Brain Res. **81**: 150–160.
6. LISBERGER, S. G. 1990. Visual tracking in monkeys: evidence for short-latency suppression of the vestibulo-ocular reflex. J. Neurophysiol. **63**: 676–688.
7. LISBERGER, S. G. & A. F. FUCHS. 1978. Role of primate flocculus during rapid behavioral modification of the vestibulo-ocular reflex. I. Purkinje cell activity during visually guided horizontal smooth-pursuit eye movements and passive head rotation. J. Neurophysiol. **41**: 733–763.
8. McCREA, R. A., K. YOSHIDA, A. BERTHOZ & R. BAKER. 1980. Eye movement related activity and morphology of second order vestibular neurons terminating in the cat abducens nucleus. Exp. Brain Res. **40**: 468–473.
9. HIGHSTEIN, S. M. & H. REISINE. 1981. The ascending track of Deiters and horizontal gaze. Ann. N.Y. Acad. Sci. **374**: 102–111.
10. McCREA, R. A., K. YOSHIDA, C. EVINGER & A. BERTHOZ. 1981. The location, axonal arborization and termination sites of eye-movement-related secondary vestibular neurons demonstrated by intra-axonal HRP injection in the alert cat. *In* Progress in Oculomotor Research. A. F. Fuchs & W. Becker, Eds.: 379–386. Elsevier/North Holland. New York, Amsterdam & Oxford.
11. YOSHIDA, K., A. BERTHOZ, P. P. VIDAL & R. McCREA. 1981. Eye movement-related activity of identified second order vestibular neurons in the cat. *In* Progress in Oculomotor Research. A. F. Fuchs & W. Becker, Eds.: 371–378. Elsevier/North Holland. New York, Amsterdam & Oxford.

12. McCREA, R. A., A. STRASSMAN, E. MAY & S. M. HIGHSTEIN. 1987. Anatomical and physiological characteristics of vestibular neurons mediating the horizontal vestibulo-ocular reflexes in the squirrel monkey. J. Comp. Neurol. **264:** 547–570.

13. McCREA, R. A., A. STRASSMAN & S. M. HIGHSTEIN. 1987. Anatomical and physiological characteristics of vestibular neurons mediating the vertical vestibulo-ocular reflexes in the squirrel monkey. J. Comp. Neurol. **264:** 571–594.

14. ESCUDERO, M. & J. M. DELGADO-GARCIA. 1988. Behavior of reticular, vestibular and prepositus neurons terminating in the abducens nucleus of the alert cat. Exp. Brain Res. **71:** 218–222.

15. IWAMOTO, Y., T. KITAMA & K. YOSHIDA. 1990. Vertical eye movement–related secondary vestibular neurons ascending in medial longitudinal fasciculus in cat. I. Firing properties and projection pathways. J. Neurophysiol. **63:** 902–917.

16. IWAMOTO, Y., T. KITAMA & K. YOSHIDA. 1990. Vertical eye movement-related secondary vestibular neurons ascending in medial longitudinal fasciculus in cat. II. Direct connections with extraocular motorneurons. J. Neurophysiol. **63:** 918–935.

17. MACIEWICZ, R. J., K. EAGEN, C. R. S. KANEKO & S. M. HIGHSTEIN. 1977. Vestibular and medullary brain stem afferents to the abducens nucleus in the cat. Brain Res. **123:** 229–240.

18. GACEK, R. R. 1979. Location of abducens afferent neurons in the cat. Exp. Neurol. **64:** 342–353.

19. BELKNAP, D. & R. A. McCREA. 1988. A comparison of the afferent and efferent connections of the abducens nucleus and prepositus nucleus in the squirrel monkey. J. Comp. Neurol. **268:** 13–28.

Vestibular and Visual Interaction in Generation of Rapid Eye Movements[a]

TOSHIHIRO KITAMA, HIROSHI SHIMAZU,[b] MAKI
TANAKA, AND KAORU YOSHIDA[c]

Department of Physiology
Institute of Basic Medical Sciences
University of Tsukuba
Tsukuba
Ibaraki 305, Japan

[b]*Tokyo Metropolitan Institute for Neurosciences*
2-6 Musashidai
Fuchu
Tokyo 183, Japan

It is generally accepted that saccades and quick phases of nystagmus are generated by a common premotor neuronal mechanism. To make rapid eye movements, ocular motoneurons innervating agonists exhibit a burst of spikes and those innervating antagonists cease or decrease firing.[1,2] There is good evidence that these changes in motoneuron activity are produced by high-frequency bursts of spikes from premotor burst neurons.[3-5] For horizontal eye movements, both excitatory[6-8] and inhibitory burst neurons[9-12] have been identified, and their locations in the pontomedullary reticular formation and their monosynaptic connections with abducens motoneurons have been verified electrophysiologically and morphologically. These premotor burst neurons participate in any type of rapid eye movement and exhibit a burst of spikes for saccades and quick phases of vestibular and optokinetic nystagmus with an ipsilateral horizontal component.[5,7,9-11] The number of spikes in the burst is linearly related to the amplitude of the ipsilateral component of eye movement, and the intraburst firing rate to the ipsilateral component of eye velocity.[11,13-15]

Another neuronal element in the premotor circuitry, the omnipause neurons, plays an important role in generation of burst activity in burst neurons. Omnipause neurons fire at high rates during fixation and the slow phase of nystagmus, and cease firing during all types of rapid eye movement in any direction.[3-5,16] It is believed that they tonically inhibit burst neurons[5,17,18] and that the pause in their activity causes disinhibition of burst neurons and thereby controls burst duration.[19] However, such disinhibition apparently does not contain information on the direction, amplitude, and velocity of rapid eye movements. Together with disinhibition, excitatory signals coding this information must be fed to burst neurons to control the metrics of saccades or quick phases.

A recent paper by Ohki *et al.* has provided evidence for the existence of candidate neurons for an excitatory source generating spike bursts associated with the quick phase of vestibular nystagmus.[20] These neurons are located in the prepositus hypoglossi nucleus and the underlying medullary reticular formation, and project contralaterally to excite burst neurons monosynaptically. They display type II

[a]Supported in part by the Human Frontier Science Program.
[c]Author to whom correspondence should be addressed.

response to horizontal head rotation, receive short latency activation from the contralateral labyrinth, and exhibit a burst of spikes in association with the quick phase directed to the contralateral side. On the basis of these findings, these neurons have been suggested to be an excitatory source that drives bursters during the quick phase, thus being called burster-driving neurons (BDNs).

The present paper will describe the activity of BDNs during saccades and quick phases of nystagmus to clarify what information on eye movements, such as the direction, amplitude, or velocity, is transmitted by BDNs to burst neurons. We will also deal with convergence of vestibular and visual inputs to BDNs and their responses to these inputs to characterize the afferent organization and vestibular-visual interactions in the generation of rapid eye movements.

DISCHARGE CHARACTERISTICS OF BDNs ASSOCIATED WITH NYSTAGMUS AND SACCADES

Spikes of BDNs were recorded in alert cats mostly in the area 0.5–1.5 mm below the surface of the fourth ventricle close to the border between the prepositus hypoglossi nucleus and the underlying reticular formation. As in previous experiments,[20] BDNs were identified by their activation following stimulation of the contralateral vestibular nerve with short latencies and their characteristic discharge pattern during vestibular nystagmus, which clearly distinguish them from other prepositus neurons.[21] FIGURE 1 illustrates for a BDN the typical responses to contralateral and ipsilateral horizontal head rotations. When nystagmus was induced by contralateral head rotation (FIGURE 1A), BDNs increased their firing rate during the slow phase (type II response) and exhibited a high-frequency burst of spikes in association with the quick phase directed to the contralateral side. With ipsilateral head rotation (FIGURE 1B), BDNs decreased their firing rate during the slow phase (type II response). BDNs showed little change or a slight decrease in firing rate during the quick phase directed to the ipsilateral side.

The activity of BDNs associated with saccades was similar to those associated with quick phases of nystagmus. Thus, all BDNs exhibited a burst of spikes before and during saccades with a contralateral horizontal component as well (FIGURE 2). For saccades with an ipsilateral component, the firing rate occasionally showed a slight decrease. No relationship was observed between the activity of BDNs and the vertical component of saccades. All BDNs showed fairly high frequency of tonic, though irregular, discharges during fixation or slow eye movements. This contrasts with burst neurons which usually exhibit no spike activity except for the period of fast eye movement. The tonic firing rate of BDNs during fixation was not related to eye position, as shown in FIGURE 2.

Firing rate of BDNs began to gradually increase 100–150 mseconds, and exhibited an abrupt increase 30–50 mseconds, prior to the onset of quick phase or saccade. The spike activity in a burst increased with an increase in amplitude of horizontal component of rapid eye movement. FIGURE 3 exemplifies, for a BDN, the relationship between the number of spikes in the burst and the amplitude of the contralateral horizontal component of rapid eye movement (horizontal amplitude). The number of spikes in the burst increased linearly with an increase in horizontal amplitude. In all BDNs investigated, the correlation between the number of spikes in the burst and the horizontal amplitude was statistically significant ($p < 0.01$). The relation shown in FIGURE 3 is similar to that for burst neurons.[11,13–15]

It is also noted that, for both saccade and quick phase, the number of spikes in the burst was well correlated with horizontal amplitude with similar slopes of regression lines. This suggests that the functional role of BDNs is similar for the both kinds of rapid eye movements with different origins.

Since the amplitude and the duration of the horizontal component tended to be correlated with each other, the number of spikes in the burst might carry information

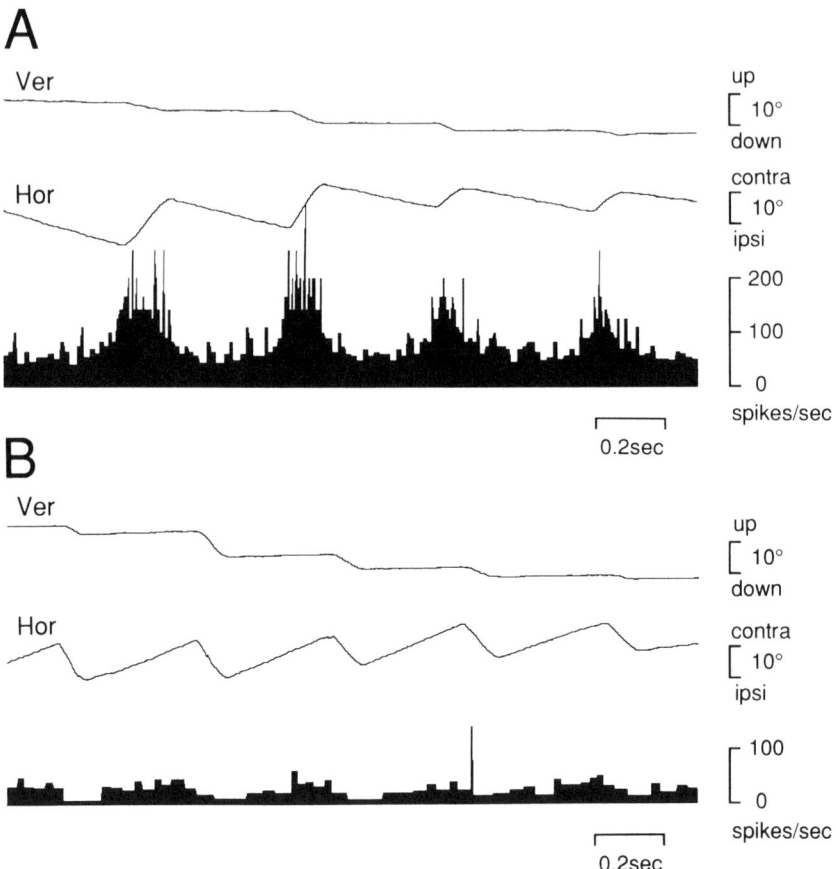

FIGURE 1. Firing pattern of a BDN during vestibular nystagmus. In A and B, traces indicate from top to bottom, vertical eye position (Ver) horizontal eye position (Hor), and instantaneous firing rate of the BDN. **A:** Activity during nystagmus induced by horizontal head rotation to the contralateral side. Firing activity of the BDN increased in a burst fashion before and during quick phases directed to the contralateral side. **B:** Activity during rotation to the ipsilateral side. Note that the firing rate during slow phase was higher during rotation to the contralateral side than during rotation to the ipsilateral side (type II response).

about the duration. However, for fast eye movements with similar horizontal amplitudes and different durations, the number of spikes in the burst was not correlated with the duration. In contrast, the correlation between the number of spikes and horizontal amplitude was highly significant even for eye movements with

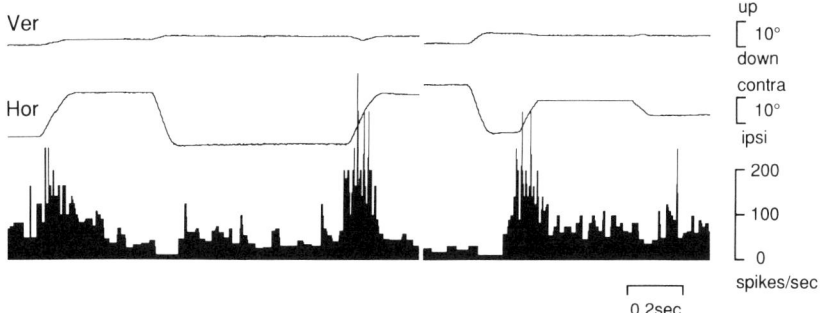

FIGURE 2. Firing pattern of the same BDN as in FIGURE 1 during spontaneous saccadic eye movements. Same display as in FIGURE 1. Note that the cell exhibited a burst of activity in association with saccades with a contralateral horizontal component.

almost the same duration. This indicates that the number of spikes in the burst essentially provides information about the amplitude, not the duration, of fast eye movements. It may therefore be reasonable to suggest that the firing rate of BDNs in the burst carries information about the velocity of fast eye movements, similar to coding of velocity signal by premotor burst neurons.

In summary, the above results demonstrate a great similarity between BDNs and

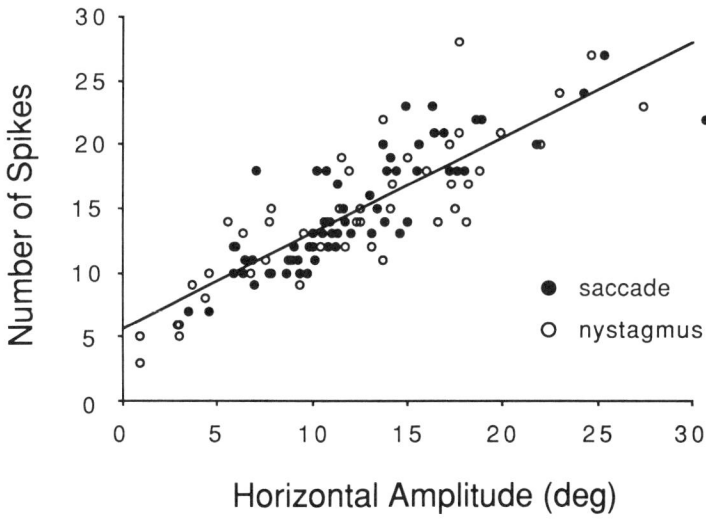

FIGURE 3. Relationship between the number of spikes in the burst (ordinate) and amplitude of horizontal component of rapid eye movement (abscissa) for a BDN. Data for saccades (filled circles) and quick phases of nystagmus (open circles) with a contralateral horizontal component. The number of spikes in the burst was counted over a period equal to the duration of saccade or quick phase, starting from 30 mseconds before the beginning of eye movement. A solid line indicates the least-squares linear regression line fitted to the data points. The correlation coefficient was 0.84, and was highly significant ($p < 0.001$).

premotor burst neurons in their burst discharges. It appears that the burst of BDNs contains appropriate information to determine the direction, amplitude, and velocity of the horizontal component of rapid eye movements. As in the case of premotor burst neurons, BDNs participate in all types of rapid eye movements, i.e., saccades and quick phases of vestibular and optokinetic nystagmus. They are clearly direction specific and exhibit a burst of spikes for rapid eye movements with a contralateral component. Thus, the burst discharges of BDNs on one side resemble those of burst neurons on the opposite side, to which BDNs project to make excitatory connections. These results are consistent with the suggestion that excitatory input from BDNs, together with disinhibition from omnipause neurons, plays an important role in generation of burst activity in premotor burst neurons and consequently rapid eye movement. On the other hand, the behavior of BDNs during fixation and slow phases is quite different from that of premotor burst neurons. While burst neurons are silent except during rapid eye movements, BDNs display tonic firing during fixations. This may suggest that, when omnipause neurons are active, their inhibitory effect predominates over the excitation from the BDNs and prevents burst neurons from firing.

RESPONSES OF BDNs TO INPUT FROM THE VESTIBULAR LABYRINTH

The tonic activity of BDNs was unrelated to eye position, but was modulated in response to vestibular inputs. As mentioned above, BDNs increased their firing rate with head rotation to the contralateral side and decreased it with rotation to the ipsilateral side. This type II behavior was quantified by analyzing responses of BDNs to sinusoidal horizontal head rotation. When nystagmus was absent, the firing rate was modulated approximately sinusoidally around the resting level (FIGURE 4). The phase and gain of response relative to the angular acceleration were determined at various stimulus frequencies using acute preparations.[22] The mean phase lag varied from 80 to 110 degrees over the frequency range of 0.05–0.5 Hz (FIGURE 4). Thus, the firing rate of BDNs was approximately in phase with the contralateral head velocity. The phase lag in BDNs was, however, somewhat larger than the mean phase lag in horizontal canal afferents (Ezure et al.,[23] open circles in FIGURE 4). Since some secondary vestibular neurons have also been shown to exhibit a phase lag larger than that of primary afferents,[23,24] BDNs might receive canal input through these secondary neurons. Similar phase lag of BDNs was observed in alert cats as well.

In confirmation of previous findings by Ohki et al.,[20] all BDNs were excited by electrical stimulation of the contralateral vestibular nerve at short latencies (FIGURE 5B). The mean latency of activation was 1.8 mseconds, suggesting disynaptic excitatory connections from the vestibular nerve. Stimulation of the ipsilateral vestibular nerve had no excitatory effect on BDNs (FIGURE 5A).

RESPONSES OF BDNs TO VISUAL AND SUPERIOR
COLLICULUS STIMULATION

It was noted that BDNs sometimes showed a burst of spikes even during fixation periods. Subsequent tests with visual stimuli revealed that presentation of an object in the contralateral visual field induced a brief burst of spikes in BDNs even in the absence of any eye movement. An example of such responses evoked by visual stimulation is illustrated in FIGURE 6A. The latency of the burst was approximately 30–40 mseconds after the onset of visual stimulation. When the cat made a saccade in response to a visual stimulus, this BDN exhibited a burst consisting of early and late components (FIGURE 6B). The early component had a latency of about 30 to 40 mseconds, and was considered to be visual response. The late component appeared to correspond to the burst associated with saccadic eye movement. Thus it seems that brief visual responses of BDNs alone are not sufficient to induce eye movement, but their relatively long lasting, intense burst activity would recruit neural elements

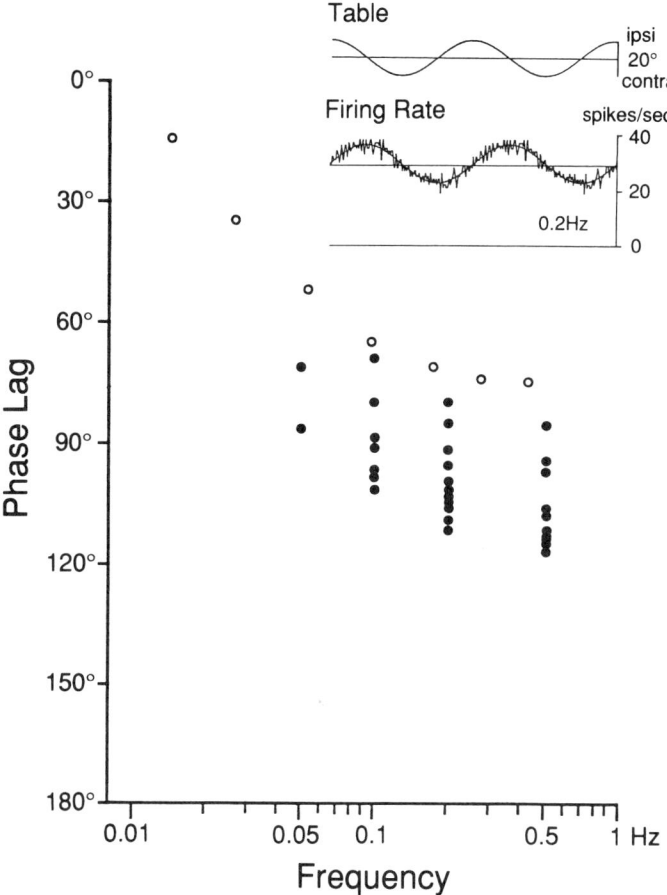

FIGURE 4. Phase characteristics of BDN responses to sinusoidal rotation. Filled circles represent phase lags with respect to contralateral angular acceleration for 11 BDNs at different stimulus frequencies. Open circles indicate mean phase lags for horizontal canal afferents (Ezure *et al.*, 1987).[23] **Inset:** Averaged response of a BDN at 0.2 Hz. Superimposed sine wave is best-fitting response fundamental. A single averaged cycle is repeated for clarity.

involved in the generation of saccade. Visual stimuli presented on the ipsilateral visual field had little, if any, effect on the BDN shown in FIGURE 6.

Single shock stimulation of the ipsilateral superior colliculus consistently evoked spike potentials in all BDNs tested (FIGURE 5C). Spike potentials were induced transsynaptically at variable latencies, and threshold current for activation was as low

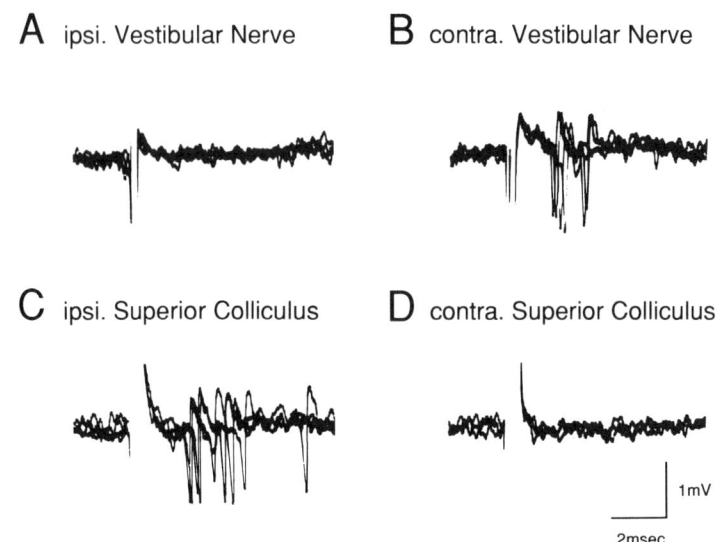

A ipsi. Vestibular Nerve   **B** contra. Vestibular Nerve

C ipsi. Superior Colliculus   **D** contra. Superior Colliculus

1mV

2msec

FIGURE 5. Responses of a BDN to electrical stimulation of the vestibular nerve and the superior colliculus. All traces were recorded extracellularly from the same BDN. Each record is composed of 4 superimposed traces. **A** and **B**: Responses to single shock stimulation of the ipsilateral vestibular nerve (A) and the contralateral vestibular nerve (B). **C** and **D**: Responses to single shock stimulation of the ipsilateral superior colliculus (C) and the contralateral superior colliculus (D).

as 20 μA and was below 100 μA for most units. The latency ranged from 1.7 to 3.5 mseconds with a mean of 2.3 mseconds. Thus, the shortest excitatory pathway from the ipsilateral colliculus was most likely di- or trisynaptic. In contrast to consistent activation from the ipsilateral superior colliculus, BDNs showed no short-latency excitatory response to single shock stimulation of the contralateral superior colliculus at the intensity up to 200 μA.

The response of BDNs to stimulation of the ipsilateral superior colliculus was greatly affected by horizontal head rotation. When collicular stimulation was combined with contralateral head rotation, the excitation from the superior colliculus was clearly facilitated, and 3–4 spike potentials could be generated in most cases. A significant decrease in latency of first spikes was also observed during contralateral rotation. Ipsilateral head rotation was shown to have an opposite action; it effectively reduced the response to collicular stimulation. These results indicate that visual and vestibular interactions occur at the level of BDNs. Convergence of afferent inputs from the contralateral horizontal canal and the ipsilateral superior colliculus makes sense in terms of the generation of spike bursts in BDNs, since activation of both afferents leads to induction of rapid eye movements to the contralateral side.

EFFECTS OF HEAD ROTATION ON SACCADES EVOKED FROM THE SUPERIOR COLLICULUS

In view of the convergence and interaction of vestibular and tectal inputs at BDNs, it seems very likely that their burst discharges associated with saccades are also affected by inputs from the horizontal canal, resulting in modulation of amplitude of colliculus-induced saccades. In agreement with previous findings,[25–28] the direction and amplitude of saccades evoked from the superior colliculus were determined by the site of stimulation in the colliculus. In FIGURE 7 saccades were evoked by a 25-msecond pulse train (0.2-msecond pulses, 400/second, 35 μA) applied to the left superior colliculus. Without vestibular influence, collicular stimulation evoked saccades with contralateral and upward components in this case (FIGURE 7A). When the same collicular stimulation was combined with horizontal head rotation, the amplitude of the horizontal component of evoked saccades was significantly affected. In FIGURE 7B and C, the stimulus was delivered at different phases of sinusoidal horizontal head rotation. In association with ipsilaterally directed head velocity, the colliculus-induced saccades had a small horizontal component (FIGURE 7B) compared to the control (FIGURE 7A). During contralateral head rotation, the same stimulation produced saccades with a larger horizontal component (FIGURE 7C). Comparisons of colliculus-induced saccades at various phases of sinusoidal head rotation showed that the modulation of the horizontal amplitude was approximately in phase with contralateral head velocity, i.e., evoked

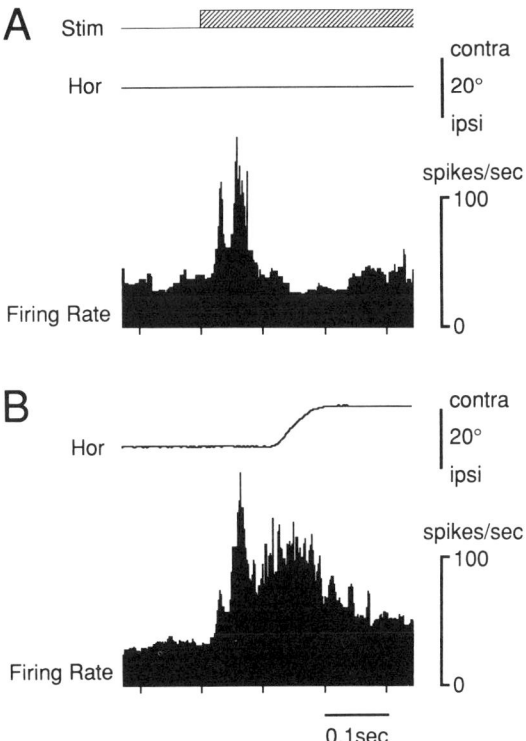

FIGURE 6. Responses of a BDN to visual stimulation. **A:** Response trials in which the cat made no eye movement after the appearance of a visual target (Stim). The visual target (a 5° × 5° square board) was presented for 1.2 seconds on the contralateral side at 30° from the midsagittal plane. Histogram shows the average instantaneous firing rate for 7 trials aligned on the appearance of the visual target (Stim). Horizontal eye position trace (Hor) is shown for one trial. **B:** Response trials in which the cat made saccades toward the visual target.

saccades with the largest horizontal components were associated with maximal contralateral head velocity, and the smallest with maximal ipsilateral head velocity. In contrast to clear changes in the horizontal component of responses, the amplitude of the vertical component was usually affected very little during horizontal head rotation (FIGURE 7, bottom traces). This suggested that the vestibular influence on colliculus-induced responses was direction specific and caused modulations in a component parallel to the plane of rotation.

In agreement with previous findings,[27,28] the amplitude of saccades evoked by collicular stimulation depended on the eye position prior to stimulation. The predicted modulation due to position dependency was, however, generally much smaller than the modulation actually observed and approximately in phase with ipsilateral head position rather than with contralateral head velocity.

The finding that the modulation of the horizontal amplitude of colliculus-induced saccade was in phase with contralateral head velocity is well explained by the

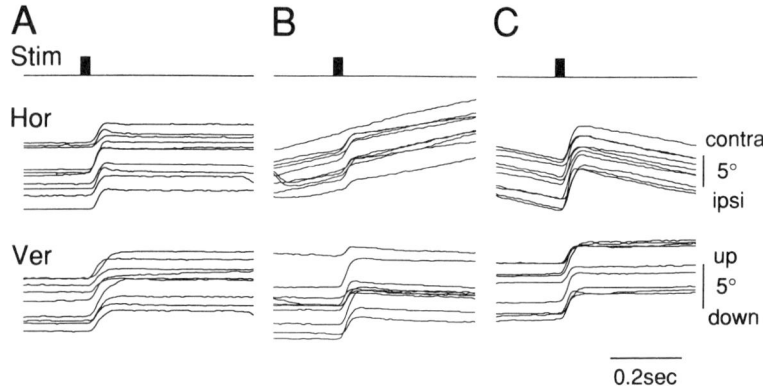

FIGURE 7. Effects of head rotation on colliculus-induced saccades. **A:** The horizontal (Hor) and vertical (Ver) component of saccades induced by stimulation of the superior colliculus (Stim) with 25-msecond pulse trains (35 μA, 0.2-msecond pulses, 400/second) with the head stationary. In B and C, the stimulus was delivered during sinusoidal horizontal head rotation at 0.1 Hz, ±13.5°/second. **B:** Evoked saccades during maximal ipsilateral head velocity. **C:** Evoked saccades during maximal contralateral head velocity.

fact that the firing rate of BDNs, hence facilitatory effects from the contralateral horizontal canal on BDNs, was in phase with contralateral head velocity (FIGURE 4). This parallelism is also consistent with the notion that the colliculus-induced saccade is modulated, at least in part, through BDNs.

POSSIBLE PATHWAYS FROM THE SUPERIOR COLLICULUS
TO PREMOTOR BURST NEURONS

It is well known that the intermediate and deep layers of the superior colliculus contain many neurons that discharge before saccades.[29-31] These saccade-related neurons are organized topographically according to their movement fields. That is, neurons located in a particular site discharge most vigorously for saccades with a particular direction and amplitude. It is considered that the metrics (direction, amplitude, velocity) of a saccade are specified by the location of the active neurons in the superior colliculus, and not by the firing rate or the number of spikes in the burst of individual neurons. In contrast to the discharge of premotor burst neurons in the reticular formation, in which the firing rate is linearly correlated with the component velocity and the number of spikes in the burst with the component amplitude, the discharge of a single collicular neurons does not contain information required to determine the metrics of saccade. Therefore, it has been suggested that neuronal mechanisms intercalated between the superior colliculus and the burst neurons transform the spatially organized collicular signal into the temporally encoded premotor signals proportional to the amplitude and velocity of saccade (see review by Sparks and Mays).[32,33] Our results demonstrate that the burst discharge of BDNs resembles that of burst neurons and contains information required to control the metrics of all types of rapid eye movements. Since the pathway from the ipsilateral superior colliculus to BDNs is polysynaptic, it might be suggested that the spatial-temporal transformation occurs at the level of intermediate neurons in the pathway. It should be pointed out that activity of BDNs could be dissociated from eye movements. BDNs discharge in a burst fashion in response to visual stimuli even when no saccades are induced. Similar behavior of neurons is observed in the deep layers of the superior colliculus.[31,34] This may suggest that the collicular efferent signals impinging on BDNs arise from the visually responsive neurons in the deep layers, many of which are not only responsive to sensory stimuli but also discharge before saccade. It is also worth mentioning that BDNs modulate their activity not only in relation to eye movements during quick phases, but also in relation to vestibular sensory signals during slow phases. This may suggest that BDNs are involved in neural mechanisms of sensorimotor transformation in vestibulooculomotor function.

For the present, the location and response properties of neurons that mediate signals from the superior colliculus to BDNs are unknown. A candidate structure containing these intermediate neurons could be the central mesencephalic reticular formation (cMRF).[35,36] Neurons in the intermediate and deep layers of the superior colliculus project to the ipsilateral cMRF, and in turn, neurons in cMRF project to the pontine reticular formation. Neurons in cMRF show bursts of activity in association with contralateral horizontal saccades and quick phases of nystagmus. The number of spikes in their burst increases as the amplitude of the horizontal component of saccade increases. Thus, the discharge characteristics of cMRF neurons are similar to those of BDNs. However, differences appear to exist between cMRF neurons and BDNs with respect to responses to slow eye movements and visual input. In contrast to BDNs, cMRF neurons show no overt visual response and do not change their firing rates during slow phase.

BDNs are, however, not a single source that drives burst neurons. There seem to be multiple parallel pathways from the superior colliculus to burst neurons,[37–39] i.e. a direct pathway, a pathway through cMRF neurons, and a pathway through BDNs. The functional role of each pathway may be different, and a possible interaction between cMRF neurons and BDNs remains to be studied. It should be stressed that the pathway through BDNs is susceptible to vestibular input. BDNs receive strong disynaptic input from the contralateral horizontal canal, and their response to collicular stimulation is facilitated during contralateral head rotation and suppressed during ipsilateral rotation. Therefore, BDNs may serve as an important locus for visual-vestibular interactions. Our results demonstrate that the amplitude of saccades evoked by collicular stimulation is also significantly modulated by the vestibular inputs. Since the modulation is approximately in phase with contralateral head velocity, as is the modulation of the response in BDNs, it seems likely that the effect may be in part due to interaction of vestibular and visual inputs converging on BDNs. It is also suggested that such interaction might be related to the phenomenon that, during head rotation, saccades in the opposite direction seldom occur.

REFERENCES

1. FUCHS, A. F. & E. S. LUSCHEI. 1970. Firing patterns of abducens neurons of alert monkeys in relationship to horizontal eye movement. J. Neurophysiol. **33:** 382–392.
2. ROBINSON, D. A. 1970. Oculomotor unit behavior in the monkey. J. Neurophysiol. **33:** 393–404.
3. COHEN, B. & V. HENN. 1972. Unit activity in the pontine reticular formation associated with eye movements. Brain Res. **46:** 403–410.
4. LUSCHEI, E. S. & A. F. FUCHS. 1972. Activity of brain stem neurons during eye movements of alert monkeys. J. Neurophysiol. **35:** 445–461.
5. KELLER, E. L. 1974. Participation of the medial pontine reticular formation in eye movement generation in monkey. J. Neurophysiol. **37:** 316–332.
6. IGUSA, Y., S. SASAKI & H. SHIMAZU. 1980. Excitatory premotor burst neurons in the cat pontine reticular formation related to the quick phase of vestibular nystagmus. Brain Res. **182:** 451–456.
7. SASAKI, S. & H. SHIMAZU. 1981. Reticulovestibular organization participating in generation of horizontal fast eye movement. Ann. N.Y. Acad. Sci. **374:** 130–143.
8. STRASSMAN, A., S. M. HIGHSTEIN & R. A. MCCREA. 1986. Anatomy and physiology of saccadic burst neurons in the alert squirrel monkey. I. Excitatory burst neurons. J. Comp. Neurol. **249:** 337–357.
9. HIKOSAKA, O. & T. KAWAKAMI. 1977. Inhibitory reticular neurons related to the quick phase of vestibular nystagmus—their location and projection. Exp. Brain Res. **27:** 377–396.
10. HIKOSAKA, O., Y. IGUSA, S. NAKAO & H. SHIMAZU. 1978. Direct inhibitory synaptic linkage of pontomedullary reticular burst neurons with abducens motoneurons in the cat. Exp. Brain Res. **33:** 337–352.
11. YOSHIDA, K., R. MCCREA, A. BERTHOZ & P. P. VIDAL. 1982. Morphological and physiological characteristics of inhibitory burst neurons controlling horizontal rapid eye movements in the alert cat. J. Neurophysiol. **48:** 761–784.
12. STRASSMAN, A., S. M. HIGHSTEIN & R. A. MCCREA. 1986. Anatomy and physiology of saccadic burst neurons in the alert squirrel monkey. II. Inhibitory burst neurons. J. Comp. Neurol. **249:** 358–380.
13. KING, W. M. & A. F. FUCHS. 1979. Reticular control of vertical saccadic eye movements by mesencephalic burst neurons. J. Neurophysiol. **42:** 861–876.
14. KANEKO, C. R. S., C. EVINGER & A. F. FUCHS. 1981. Role of cat pontine burst neurons in generation of saccadic eye movements. J. Neurophysiol. **46:** 387–408.
15. SCUDDER, C. A., A. F. FUCHS & T. P. LANGER. 1988. Characteristics and functional identification of saccadic inhibitory burst neurons in the alert monkey. J. Neurophysiol. **59:** 1430–1454.
16. EVINGER, C., C. R. S. KANEKO & A. FUCHS. 1982. Activity of omnipause neurons in alert cats during saccadic eye movements and visual stimuli. J. Neurophysiol. **47:** 827–844.
17. STRASSMAN, A., C. EVINGER, R. A. MCCREA, R. G. BAKER & S. M. HIGHSTEIN. 1987. Anatomy and physiology of intracellularly labelled omnipause neurons in the cat and squirrel monkey. Exp. Brain Res. **67:** 436–440.
18. NAKAO, S., I. S. CURTHOYS & C. H. MARKHAM. 1980. Direct inhibitory projection of pause neurons to nystagmus-related pontomedullary reticular burst neurons in the cat. Exp. Brain Res. **40:** 283–293.
19. ROBINSON, D. A. 1975. Oculomotor control signals. *In* Basic Mechanisms of Ocular Motility and Their Clinical Implications. P. Bach y Rita & G. Lennerstrand, Eds.: 337–374. Pergamon. Oxford, England.
20. OHKI, Y., H. SHIMAZU & I. SUZUKI. 1988. Excitatory input to burst neurons from the

labyrinth and its mediating pathway in the cat: location and functional characteristics of burster-driving neurons. Exp. Brain Res. **72:** 457–472.

21. LOPEZ-BARNEO, J., C. DARLOT, A. BERTHOZ & R. BAKER. 1982. Neuronal activity in prepositus nucleus correlated with eye movement in the alert cat. J. Neurophysiol. **47:** 329–352.

22. KITAMA, T., Y. OHKI, H. SHIMAZU & K. YOSHIDA. 1988. Physiological and morphological properties of brain stem neurons related to vestibular and saccadic eye movements in the cat. *In* Basic and Applied Aspects of Vestibular Function. J. C. Hwang, N. G. Daunton & V. J. Wilson, Eds.: 45–53. Hong Kong University Press. Hong Kong.

23. EZURE, K., R. H. SCHOR & K. YOSHIDA. 1978. The response of horizontal semicircular canal afferents to sinusoidal rotation in the cat. Exp. Brain Res. **33:** 27–39.

24. SHINODA, Y. & K. YOSHIDA. 1974. Dynamic characteristics of responses to horizontal head angular acceleration in the vestibuloocular pathway in the cat. J. Neurophysiol. **37:** 653–673.

25. ROBINSON, D. A. 1972. Eye movements evoked by collicular stimulation in the alert monkey. Vision Res. **12:** 1795–1808.

26. SHILLER, P. H. & M. STRYKER. 1972. Single-unit recording and stimulation in superior colliculus of the alert rhesus monkey. J. Neurophysiol. **35:** 915–924.

27. ROUCOUX, A. & M. CROMELINCK. 1976. Eye movements evoked by superior colliculus stimulation in the alert cat. Brain Res. **106:** 349–363.

28. GUITTON, D., M. CROMMELINK & A. ROUCOUX. 1980. Stimulation of the superior colliculus in the alert cat. Exp. Brain Res. **39:** 63–73.

29. WURTZ, R. H. & M. E. GOLDBERG. 1971. Superior colliculus cell responses related to eye movements in awake monkeys. Science **171:** 82–84.

30. SHILLER, P. H. & F. KOERNER. 1971. Discharge characteristics of single units in the superior colliculus of the alert rhesus monkey. J. Neurophysiol. **34:** 920–936.

31. WURTS, R. H. & M. E. GOLDBERG. 1972. Activity of superior colliculus cell in behaving monkey. III. Cells discharging before eye movements. J. Neurophysiol. **35:** 575–586.

32. SPARKS, D. L. & L. E. MAYS. 1980. Movement fields of saccade-related burst neurons in the monkey superior colliculus. Brain Res. **190:** 39–50.

33. SPARKS, D. L. & L. E. MAYS. 1990. Signal transformations required for the generation of saccadic eye movements. Annu. Rev. Neurosci. **13:** 309–336.

34. MAYS, L. E. & D. L. SPARKS. 1980. Dissociation of visual and saccade related responses in superior colliculus neurons. J. Neurophysiol. **43:** 207–232.

35. COHEN, B. & J. A. BÜTTNER-ENNEVER. 1984. Projections from the superior colliculus to a region of the central mesencephalic reticular formation (cMRF) associated with horizontal saccadic eye movements. Exp. Brain Res. **57:** 167–176.

36. COHEN, B., D. M. WAITZMAN, J. A. BÜTTNER-ENNEVER & V. MATSUO. 1986. Horizontal saccade and the central mesencephalic reticular formation. Prog. Brain Res. **64:** 243–256.

37. KAWAMURA, K., A. BRODAL & G. HODDEVIK. 1974. The projection of the superior colliculus onto the reticular formation of the brainstem. An experimental anatomical study in the cat. Exp. Brain Res. **19:** 1–19.

38. RAYBOURN, M. S. & E. L. KELLER. 1977. Colliculoreticular organization in primate oculomotor system. J. Neurophysiol. **40:** 861–878.

39. GRANTYN, A. & R. GRANTYN. 1982. Axonal patterns and sites of termination of cat superior colliculus neurons projecting in the tecto-bulbar-spinal tract. Exp. Brain Res. **46:** 243–256.

Effects of Ibotenic Acid Lesions of Nucleus Prepositus Hypoglossi on Optokinetic and Vestibular Eye Movements in the Alert, Trained Monkey[a]

CHRIS R. S. KANEKO

Department of Physiology and Biophysics and
Regional Primate Research Center
University of Washington
Seattle, Washington 98195

INTRODUCTION

In the last 15 years, a number of studies have suggested that the nucleus prepositus hypoglossi (nph) may function as a neural integrator (in the mathematical sense) for vestibular and oculomotor signals to convert velocity-coded input into position-coded commands that drive the ocular motor neurons. The anatomical study by Graybiel and Hartweig suggested,[1] and the physiological work of Baker and Berthoz confirmed,[2] that the classical notion that the nph serves a tongue motor function is incorrect. Instead, the connections of the nph[1–4] and neural recordings in alert cats showed that the nph probably subserves an oculomotor function. Later, Baker, Evinger, and McCrea deduced that the nph was the location of the neural integrator.[5] They reasoned that the only major afferent to the abducens nucleus without an established function was the nph input and the only function without an anatomical location was the neural integrator; therefore, the nph had to be the site of the integrator. More recent studies have corroborated this suggestion. The discharge of nph neurons seems appropriate in both cats[6] and monkeys.[7,8] The early anatomical connections have been confirmed by a variety of techniques in a number of different laboratories (see Belknap and McCrea[9] for review). Finally, recent studies have correlated damage of the nph with loss of integrator function.[10–13]

Despite this body of evidence, there are still many questions remaining regarding the anatomical extent of the integrator as well as the detailed contribution of the nph to integrative function. For example, Cheron, Godaux, and colleagues made electrolytic lesions of the nph region in cats and showed a correlated deficit in oculomotor integrative function.[10,12] Unfortunately, those lesions encroached on the medial vestibular and abducens nucleus and also interrupted portions of the medial longitudinal fasciculus and the numerous fibers that pass through the region carrying the signals for the vestibular ocular reflex (e.g., McCrea *et al.*)[14] as well as saccades.[15] Recognizing the problems for interpreting their results created by the damage to fibers of passage, Cheron and Godaux repeated their lesion study using kainic acid,[11] which spares axons. Unfortunately, neither the lesions nor the associated oculomotor

[a]This study was supported by grants EY06558 and RR00166 from the National Institutes of Health.

deficits were permanent. Thus, it was impossible to conclude whether the transient oculomotor deficits were actually due to nph damage or to some other effect such as the well-known transmitter mimetic excitatory effects of the neurotoxins. Using both kainic and ibotenic acid, Cannon and Robinson attempted similar studies in monkey, with similar results and similar problems in interpretation.[13]

Our own studies of the nph began with two findings in our laboratory. First, we discovered that the projection of the nph to the abducens is not uniform in monkeys[16] as it is in cats. Furthermore, a histologically distinct subregion that lies in the rostromedial margin of the nucleus apparently supplies the majority of the abducens input[16] (see also Belknap and McCrea).[9] We have called this area the marginal zone because it borders on the nph rostromedially and the medial vestibular nucleus caudolaterally, blends with each at its borders, and is distinct over a portion of its rostrocaudal extent. Second, electrophysiological recordings from alert, trained monkeys in our laboratory suggested that this region contained neurons that exhibit discharge appropriate for integrator function.[7,8] We decided to reinvestigate the marginal zone using ibotenic acid lesions since we could distinguish the area in normal anatomical material, thus making it possible to correlate the lesion damage with a distinct subnuclear group that was relatively restricted in size and contained cells with appropriate discharge. We were interested in saccade-related integrative function, since the previous studies[10-13] had not looked particularly at saccadic eye movements. Our previous lesion experiments[17] convinced us that we could produce permanent axon-sparing lesions using the neurotoxin ibotenate. In the course of investigating the role of the nph neural integrator in producing saccades,[18,19] we discovered a variety of long-lasting optokinetic and vestibular deficits that had not been reported previously. Here we report preliminary analyses of some of those deficits.

MATERIALS AND METHODS

We trained three juvenile male rhesus macaques (*Macaca mulatta*) to track a laser spot for food reward during a variety of visual and vestibular stimulus situations. Then we used standard extracellular electrophysiological recording techniques to map the area of interest.[18] Electrolytic marking lesions were placed around the physiologically identified regions for later histological recovery. We located the marginal zone in three animals by recording the neural activity in relation to eye movements and visual and vestibular stimulation. We then made lesions in the area using pressure injections of ibotenic acid (15 μg/μl ibotenate in phosphate buffered saline, pH 7.4). Monkey 1 received a large (~1.5 μl), unilateral injection so that we could ascertain the feasibility of the technique in light of previous reports that excitatory neurotoxins in that region were ineffective.[10-13] Monkey 2 received a series of more punctate (200–500 nL), sequential injections so that we could titrate the amount of damage against the saccadic deficits. To date, monkey 3 has received a single 200 nL injection.

Eye movements were measured (DC to 320 Hz, −3 dB) with the magnetic search coil technique before and after each lesion. Following the fourth lesion in monkey 2, we used amphetamine [0.25 mg/kg, intramuscularly (im)] during behavioral testing because it made the vestibular and optokinetic responses much more consistent. A large battery of data was obtained immediately following each injection (except for the third which was followed after two days by the fourth) and for a period lasting until there were no further changes in the vestibular and optokinetic responses. The battery included responses to sinusoidal and ramp vestibular stimuli presented in the dark, smooth pursuit and saccadic tracking of a small laser diode target projected against a white background, fixation of the target spot in the light and in the dark, saccades and drift in the dark, and optokinetic responses. This report presents only the vestibular and optokinetic responses to ramp stimuli that we have analyzed thus far.

Vestibular stimulation consisted of whole-body rotation in the dark from a stationary position via a short period of acceleration at $180°/\text{second}^2$ to a constant $40°/\text{second}$ or $80°/\text{second}$. When perrotatory nystagmus had subsided, the animal was immediately decelerated to resting and the course of the postrotatory and after-nystagmus was followed until peak after-nystagmus slow eye velocity had been reached. The lights were then turned on for at least 10 seconds, and a new direction or velocity of stimulation was tested. Optokinetic stimulation was produced by an encircling white drum whose surface was overlaid with a random pattern of black, 2° squares on a white background such that the overall surface was 50% black and 50% white. The drum was diffusely but intensely illuminated, and the effect of the stimulation was augmented by mirrors placed above and below the animal's head so that there were no available stationary fixation points and so that the image movement subtended virtually the entire visual field.

A trial consisted of accelerating the drum in the dark to a constant rightward or leftward velocity of 30, 60, 90, or 120°/second, then turning on the drum lights and following the course of the optokinetic nystagmus (OKN) until peak slow eye velocity had been sustained for a minimum of 60 seconds, then extinguishing the lights and tracking the course of optokinetic after-nystagmus (OKAN) and after-after-nystagmus until peak after-after-nystagmus slow eye velocity had been reached. Each test was separated by a period of at least 10 seconds in the light with the drum and chair stationary. Stimulus conditions were presented in random order.

Slow eye velocity was produced by a five-pole, linear-phase, low-pass analog differentiator (DC to 200 Hz) and recorded directly onto an eight-channel thermal chart recorder (Astro-Med MT 95000) with a resolution of 300 dots per inch and a bandwidth from DC to 20 kHz. In monkey 1 and for the first lesion in monkey 2, the velocity signal was reconstructed from taped recordings of eye position. After that, the velocity signal was recorded on-line which resulted in improved signal-to-noise ratio. Eye velocities and time course were measured directly from these records by hand.

After testing was completed in monkeys 1 and 2, the animals were given a lethal injection of nembutal and perfused with phosphate-buffered saline followed by buffered formalin. The brains were blocked stereotaxically in the skull, fixed for several more days in formalin, and then allowed to sink in formol sucrose before being cut in 60-μm sections on a freezing microtome. Sections were mounted, stained for Nissl, coverslipped, and examined at 2.7× and 12× with a projection microscope. Lesions were confirmed histologically as damage to the distinct nucleus and by the previously placed marking lesions that bracketed the nucleus. Monkey 3 is still in progress.

RESULTS

In monkey 1, we destroyed the marginal zone of the nph (FIGURE 1, M on right side of the brain stem). That is, we damaged the putative neural integrator by creating a lesion in the rostrolateral edge of the nph where it abuts the medial vestibular nucleus. We first mapped the area by recording the distinctive activity of the neurons of the region,[7,8] and then placed a large (1.5- to 2.0-μL) injection at the site of those recordings (FIGURE 1, star). Because previous investigators had been unable to permanently damage neurons in the region using excitatory neurotoxins,[10-13] we used a very large injection and thus were able to produce a permanent marginal zone lesion. The distinctive structure of the marginal zone seen on the right in FIGURE 1 (M) has been obliterated on the left. The area pictured shows the region of maximum dorsoventral (~4.0 mm) and mediolateral (2.5 mm) extent of our estimate of cell death (open arrows on left). The rostrocaudal destruction extended from just caudal to the abducens about 1.4 mm. Thus, virtually the entire marginal

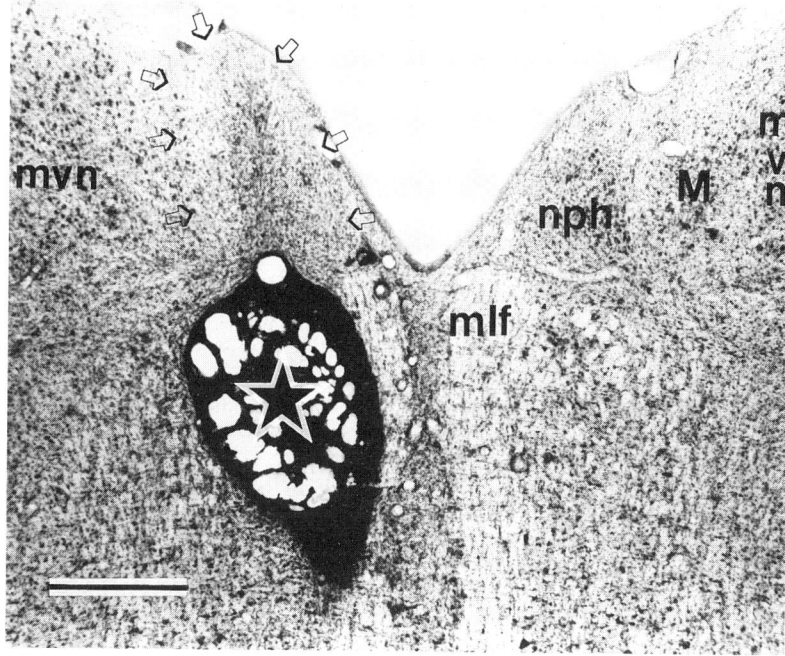

FIGURE 1. Ibotenic acid lesion of the marginal zone (M) of the nph in monkey 1. Nissl stained, transverse section taken about 1 mm caudal to the end of the abducens nucleus. The nph/marginal zone on the left (between open arrows) was completely disrupted; obvious cell death diminished with distance from the injection center. Note the large, frank, mechanical damage (star) caused by the large volume used and that most of the cell death occurred dorsal to the injection center. Abbreviations for FIGURES 1 and 2: mlf, medial longitudinal fasciculus; nph, nucleus prepositus hypoglossi; M, marginal zone; mvn, medial vestibular nucleus. Calibration = 1.0 mm in FIGURES 1 and 2.

zone on the left side was destroyed but the caudal nph and the medial vestibular nucleus (mvn) were almost completely intact. This measure is necessarily somewhat subjective since the borders of the destroyed area are not distinct. It is based on visual comparison of the cell density on the two sides of the brain. Note that there is an area of mechanical damage due to the volume of injected fluid (FIGURE 1, star); this area is a much smaller portion (2.5 mm × 1.2 mm) of the estimated total region of cell death.

Having demonstrated an ability to both damage the marginal zone and create the permanent deficits described below, in monkey 2 we attempted to titrate the amount of damage against the amount of deficit by producing a series of much smaller lesions that would also eliminate the problem of mechanical damage due to larger volumes. We therefore mapped the extent of the nph region and produced a series of lesions using smaller injections that were more broadly spaced around the nph. Unfortunately, but as expected, the damage from any one of these single small injections, and especially the smallest ones, was more difficult to see and its extent was correspondingly difficult to assess. The first injection (180 nL, left side) and the second (200 nL, right side) produced barely discernible cell death (maximum extent ≤ 300 μm). Because these original lesions were associated with minimal behavioral deficits, we used larger injections for lesions 3 to 5. Lesions 3 (500 nL) and 4 (250 nL) were placed rostrocaudally adjacent to each other on the right side (FIGURE 2A) on successive days; lesion 5 (360 nL, left side) is shown in FIGURE 2B. We estimate that the rostrocaudal extent of lesions 3 and 4 was about 600 μm and that of lesion 5 was about 400 μm. Although the last lesions produced obvious cell loss (FIGURE 2, open arrows on right in A and left in B), especially the overlapping lesions 3 and 4, none were centered on the marginal zone. Lesions 3 and 4 were at the very caudal end and lesion 5 was placed at the rostral tip so that the center of the marginal zone remained distinct and intact.

The damage to the marginal zone and the underlying reticular formation was associated with corresponding long-lasting deficits in eye movement. Monkey 1 (lesion in FIGURE 1) did not recover the ability to hold ipsilateral fixation or to produce normal ipsilateral vestibular nystagmic or optokinetic responses over the 5 weeks before it was sacrificed. FIGURE 3 shows the response to ramps of vestibular velocity of 50°/second (top trace) and 150°/second (middle trace) as well as the optokinetic response to constant-velocity drum rotation of 60°/second (bottom trace). Rotation at 50°/second in the contralateral direction in the dark resulted in a relatively normal perrotatory response (top trace, first down arrowhead) that consisted of an initial increase in slow eye velocity followed almost immediately by a gradual decay. When the chair was decelerated to rest (first up arrowhead), there was a nearly symmetric postrotatory response. In sharp contrast, rotation at the same velocity but toward the side with the lesion (ipsi) resulted in a normal jump in slow eye velocity (second up arrowhead) but in very minimal decay. Instead, a constant low-velocity perrotatory response and virtually no postrotatory response (second down arrowhead) followed. In other words, a leftward step of head velocity generated a rightward step of eye velocity. This asymmetrical prolongation is also seen at 150°/second (middle trace) but is not as pronounced (note the lower velocity gain for this record). Full-field optokinetic drum rotation in the contralateral direction (bottom trace, first up arrowhead) resulted in OKN and OKAN (second up arrowhead) that were relatively intact. Rotation of the drum toward the side of the lesion (bottom trace, first down arrowhead) resulted in little, if any, OKN or OKAN.

In an attempt to titrate the amount of damage of particular rostral medullary structures against the eye movement deficits, we measured the vestibular and optokinetic response following smaller lesions in monkey 2. The vestibular changes were consistent with the results from monkey 1 except that there was no change in the peak slow eye velocity with these smaller lesions (FIGURES 4–6). FIGURE 4 shows a series of eye velocity responses to rotation at 40°/second following each of the lesions in monkey 2 (rows). As can be seen, especially by comparing the upper and lower traces in each column, there is a prolongation of the initial slow eye velocity, particularly following lesions 3–5 (bottom rows). For example, the dashed line in the first column shows the estimated point at which the response began to diminish

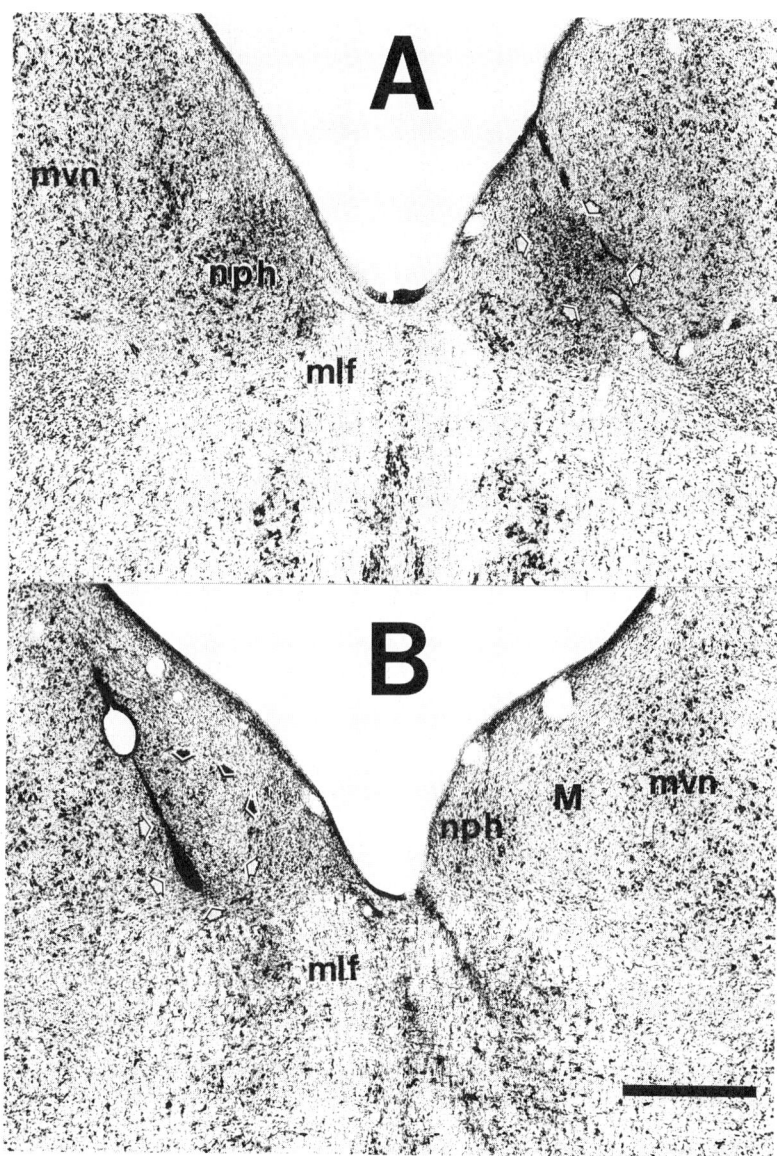

FIGURE 2. Lesions 3–5 in monkey 2. Nissl stained, transverse sections from the rostral medulla. **A:** Lesions 3 and 4. Section is from about the middle of the region of cell death at the level of the caudal end of the marginal zone, where the zone is no longer distinct. Damage (arrows) includes the lateral portion of nph, the edge of the medial mvn, and the border between them. **B:** Lesion 5. Section from rostral edge of marginal zone at the caudal tip of the abducens nucleus showing damage to medial mvn, sparing of the nph, and slight damage to the rostral tip of the marginal zone (M).

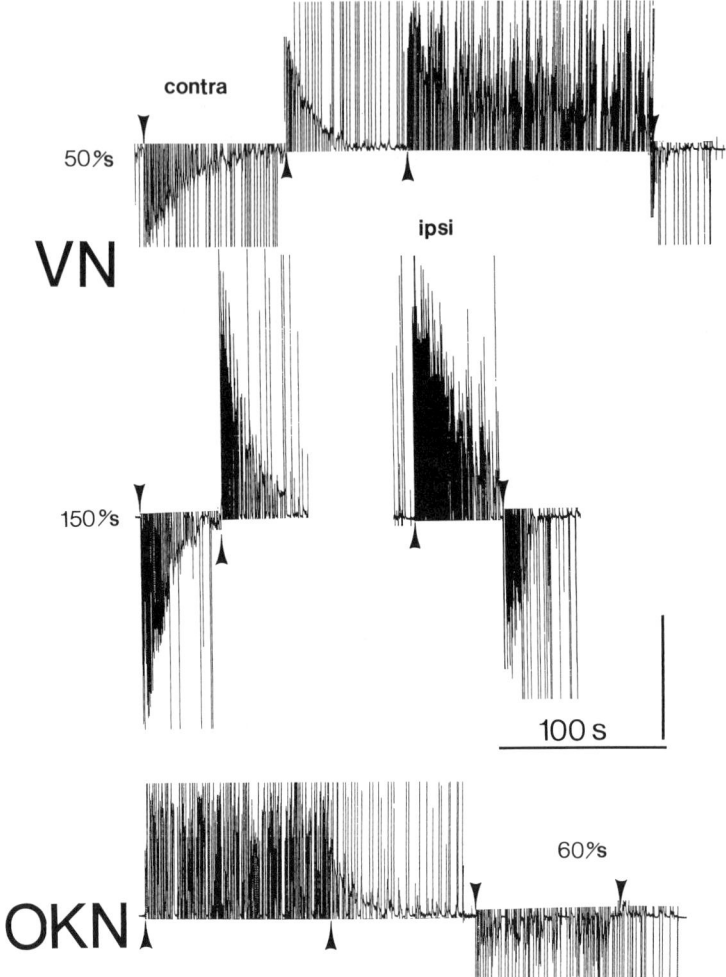

FIGURE 3. Vestibular (top traces) and optokinetic (bottom trace) deficits following large lesion in monkey 1 (FIGURE 1). Horizontal eye velocity records (rightward is represented as upward deflections in this and all other figures); quick phases masked photographically. Contra indicates the response to rightward stimulus rotation for the first series in each trace; ipsi for leftward. Arrowheads indicate moment of stimulus onset and offset, chair for VN and lights for OKN. See text for description. Calibration: 100-second bar applies to all; vertical bar is 125°/second for top and bottom, 100°/second for middle. Note, apparent episodes of zero eye velocity during nystagmus are due to saturation of the movement transducer because of the large offset of the central gaze position in this animal.[17–19]

following a rightward step after the last lesion (bottom). In contrast, the prelesion response or that following the first two lesions (top three traces) has decayed about halfway to baseline in a similar amount of time.

   To assess the effects of nph lesions more accurately and quantitatively, we measured the peak slow eye velocity in response to steps of rotation (FIGURE 5) and the time it took to decline from peak and reach baseline following each lesion and plotted these measures against postlesion time. In FIGURE 5 and the following figures, the connected symbols are measurements that were taken following each lesion so that they form four clusters. The initial symbols, near the ordinate, were prelesion measurements that were taken *after* many recording tracks had been run[8] on the left side. The first cluster immediately adjacent to the ordinate indicates the measurements taken after the initial small lesion, which was well centered in the marginal zone but did very little damage. The second cluster, recorded at around 90 days, indicates the measurements following the second small rostral lesion (FIGURE 2). Since other eye movement deficits were minimal,[18] we unfortunately did not take extensive vestibular and optokinetic measurements subsequent to this lesion. The third cluster, recorded at about 200 days, indicates measurements subsequent to the closely spaced third and fourth injections (FIGURE 3A) on the right side, and the last cluster at 250–300 days indicates measurements after lesion 5 (FIGURE 3B). FIGURE 5 shows that there was no change in the peak slow eye velocity response to vestibular stimulation at 40 or 80°/second (squares and circles, respectively) for either per (open symbols) or postrotatory (filled symbols) stimulation.

FIGURE 6 quantifies the prolongation of the slow eye velocity decay for the per

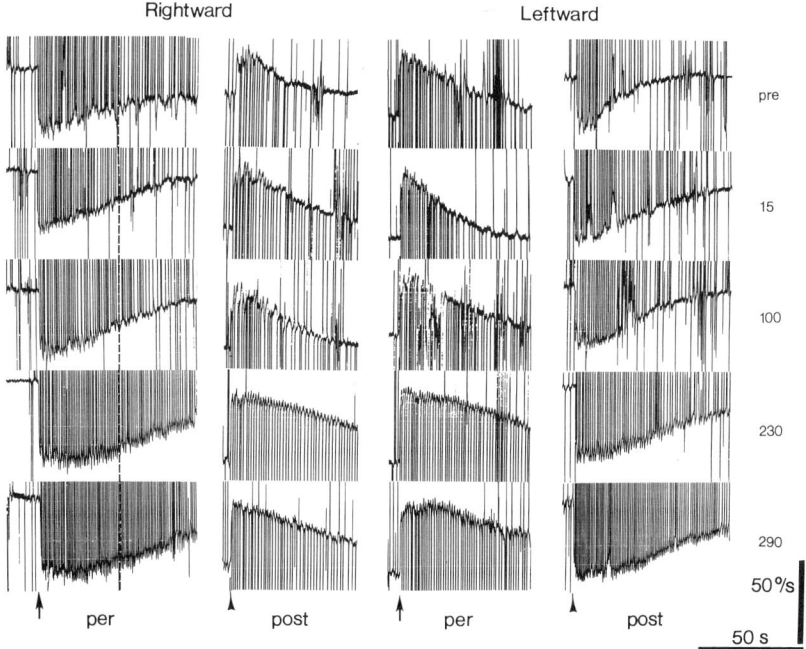

FIGURE 4. Progressive prolongation of perrotatory responses in monkey 2. Perrotatory (columns 1 and 3) and postrotatory (columns 2 and 4) responses to velocity steps of 40°/second for rightward (columns 1 and 2) and leftward (columns 3 and 4) whole-body rotation. Measurements were taken before the first lesion (pre, top row) and several weeks after each of the lesions (rows 2–4). Days postlesion are indicated to the right of the rows. Upward arrows at bottom indicate stimulus onset. Calibration is 50°/second and 50 seconds.

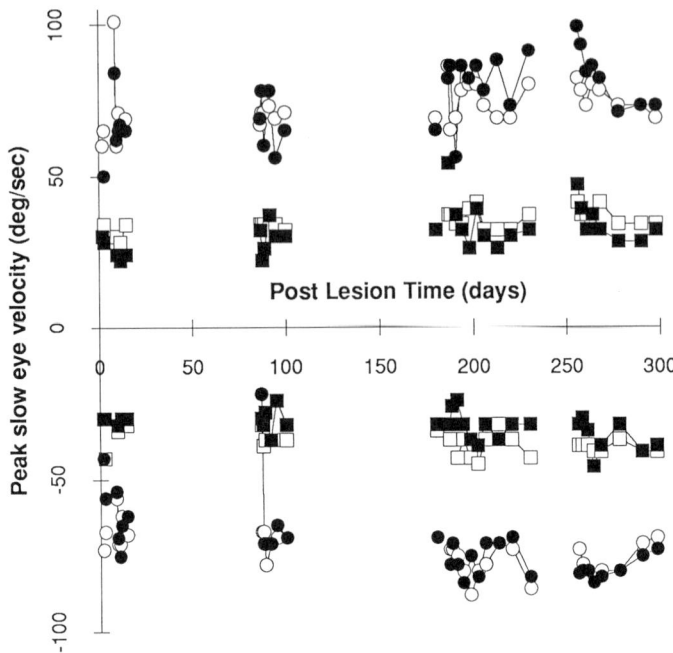

FIGURE 5. Changes in peak slow eye velocity of per- (open symbols) and postrotatory (filled symbols) nystagmus over the course of the five injections in monkey 2. Clockwise rotation (to the right) resulted in leftward eye velocity (negative numbers, bottom), while counterclockwise rotation yielded rightward eye movements (top). For this and subsequent figures, the initial measurements (near ordinate) are control measurements; measurements after each injection are connected by thin lines (see text for details). Abscissa is time (days) following the initial control measurements made after surveying recording tracks. Different symbols in this and following graphs represent responses to diferent velocities of stimulation (e.g., in this figure, squares represent 40°/second and circles, 80°/second).

(FIGURE 6A) and postrotatory (FIGURE 6B) responses. The increase in decay time at all velocities is most readily seen during postrotatory nystagmus (FIGURE 6B) and is obvious at 40°/second (squares) and 80°/second (circles) for both rightward (filled symbols) and leftward (open symbols) rotation, particularly following the first four lesions (clusters 1–3). There is a hint of compensation for right-side lesions (third cluster at about 200 days) following the last, left-side lesion (last cluster). This possible compensation is also seen for perrotatory responses (FIGURE 6A). Note that the smaller increases in decay time seen for perrotatory responses (compare FIGURE 6A and B) are associated with longer initial decay times for perrotatory responses.

The most unexpected finding from the series of lesions in monkey 2 was the fractionation of OKN into subcomponents. FIGURE 7 shows a series of OKN responses to 30°/second drum rotation and is typical of the fractionation we saw to a greater or lesser extent at all velocities over the course of these lesions. In between the well-known initial jump in slow eye velocity (FIGURE 7, open arrows) and the more gradual buildup (i.e., charging) to steady-state slow eye velocity, an interposed, previously undescribed delay can be distinguished readily (FIGURE 7, bottom traces). It was exacerbated selectively by each small nph lesion (FIGURE 7) so that eye velocity often fell back to zero for several seconds (FIGURE 7, curved arrows) and charging was greatly prolonged (compare top and bottom traces). The slow buildup, although greatly prolonged (note the decreased slope in lower traces), was not abolished by these lesions. In addition, the first component, the fast jump, was diminished progressively but the steady-state plateau seemed unaffected.

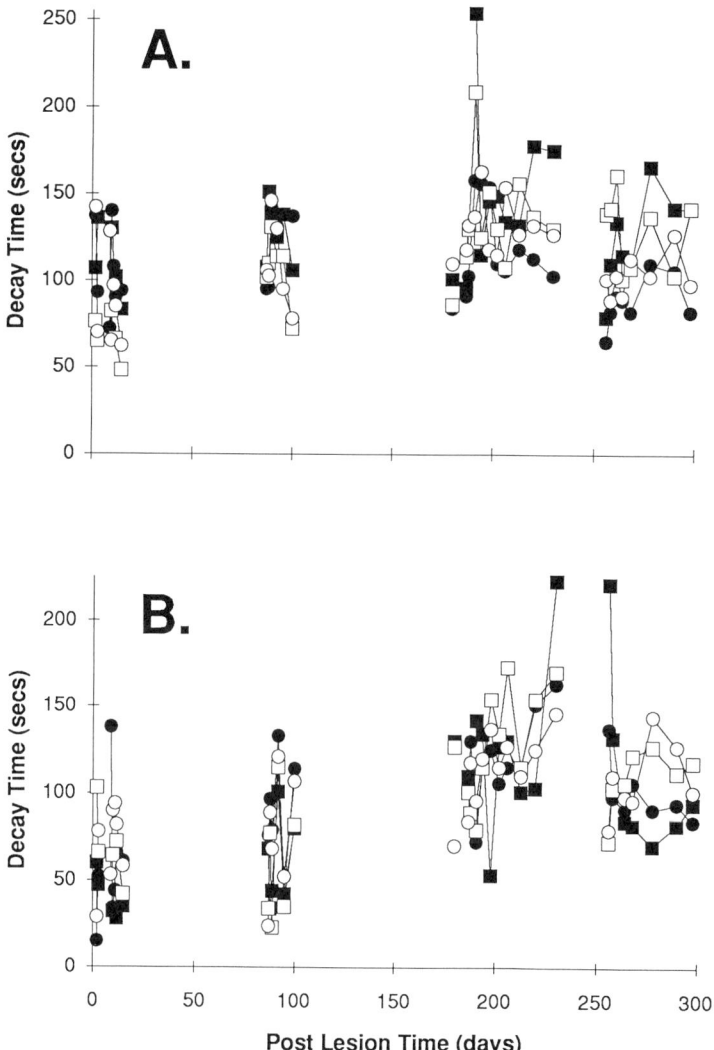

FIGURE 6. Decay time from peak slow eye velocity to baseline during per- (A) and postrotatory (B) nystagmus in monkey 2. **A:** Note the slight increase in decay time at 40°/second (squares) for both clockwise (filled symbols) and counterclockwise (open symbols) rotation, particularly following the first four lesions (clusters 1–3). There is a hint of compensation for right-side lesions (third cluster at about 200 days) following the last, left-side lesion (last cluster), especially at 80°/second (circles). **B:** The increase in decay time at all velocities and the hint of compensation are more readily seen during postrotatory nystagmus.

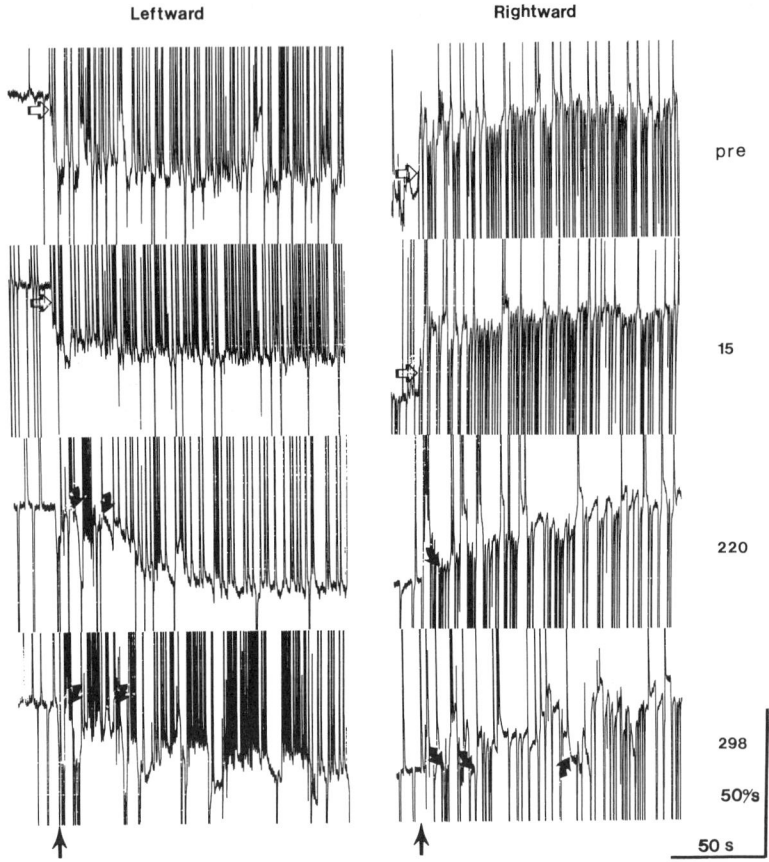

FIGURE 7. Progressive fractionation of OKN following nph lesions. Series of OKN recordings taken after several weeks recovery following each lesion. Drum velocity was 30°/second in all cases, left column is leftward rotation. Numbers (to right) represent days after the first lesion and correspond to those of FIGURES 5–12. Note the decrease in and abolition of the initial jump (open arrows) in slow eye velocity and the increase in the delay between stimulus onset (light-on indicated by upward arrows below bottom trace) and steady-state eye velocity. Also note the drops in eye velocity (curved arrows) and the increase time to steady state reflected in the slower charging (lower slope).

In order to quantify some of these changes, we measured the various parameters of the OKN response. Measuring the peak slow eye velocity achieved by the initial jump following the illumination of the drum, we found a progressive decrease in this response at all drum speeds, in both directions, following each lesion (FIGURE 8). The decrease in initial jump was obviously symmetric and amounted to about a 50% decrease at all velocities. We also measured the delay between the initial jump and the point at which slow eye velocity reached steady-state levels and plotted those delays as a function of postlesion time (FIGURE 9). Several points could be discerned.

First, the delay time increased with stimulus velocity and that relation was not affected by the lesions. That is, the higher the drum velocity, the longer it took to reach steady-state levels. Second, the small lesions were associated with only slightly increased delays. These delays were enhanced by the larger third and fourth lesions that damaged more of the nph. Third, while the delays for the lowest velocity (30°/second) were increased, the higher velocities were less affected, resulting in a convergence toward a delay time of about 35 seconds. Finally, the suggestive increases in delay following lesions 3 and 4 (on the right side; third cluster) were apparently partially compensated for by lesion 5 (on the left side; last cluster). Despite these changes, the steady-state velocity was not affected, as documented in FIGURE 10. Slow eye velocity was initially proportional to drum velocity and changed very little over the course of the lesions except for a slight increase in variance.

OKAN was also little affected in the series of small lesions. FIGURE 11 shows that the initial drop in slow eye velocity remained proportional to stimulus velocity and changed very little although there was some hint, again, that lesion 5 may have compensated for the effects of lesions 2–4. Likewise, the time it took to reach baseline following the initial drop in eye velocity (FIGURE 12) stayed about the same

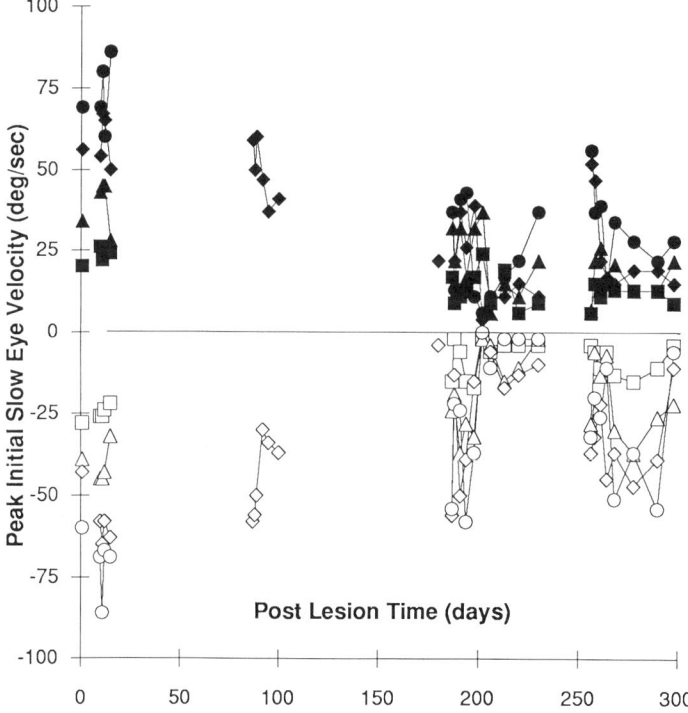

FIGURE 8. Decrease in the initial OKN response with nph damage in monkey 2. In this and subsequent figures, filled symbols indicate responses to clockwise drum rotation, open symbols indicate responses to counterclockwise rotation. Responses to drum velocities are indicated by circles (120°/second), diamonds (90°/second), triangles (60°/second), and squares (30°/second).

for all drum velocities in both directions. Again there was a hint of an increase in decay time following lesions 2–4 (right side), but values returned to about initial levels (~50 seconds) after lesion 5. With the increase in variability, it seems unlikely that these values are significantly different from each other.

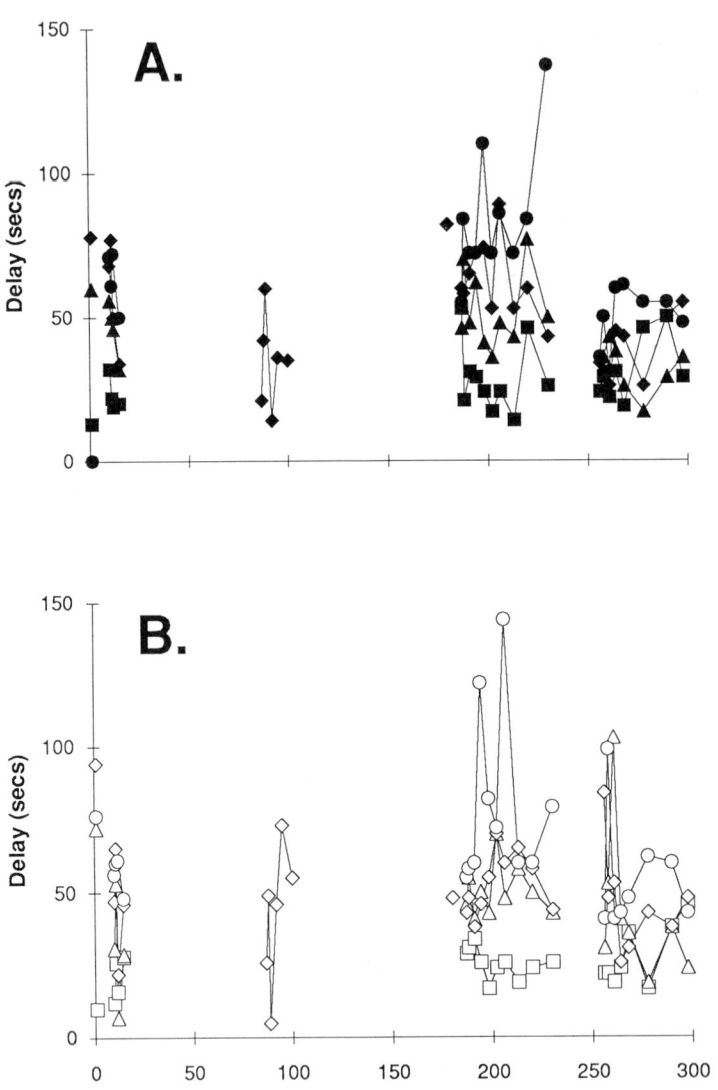

FIGURE 9. Increased delay between initial and final OKN response to rightward (A, clockwise) and leftward (B, counterclockwise) drum rotation. Increased delay is most obvious following rightward stimulation (B), is exacerbated by right-side lesions (second and third cluster), and may be compensated by the final, left-side lesion. Symbols as in FIGURE 8.

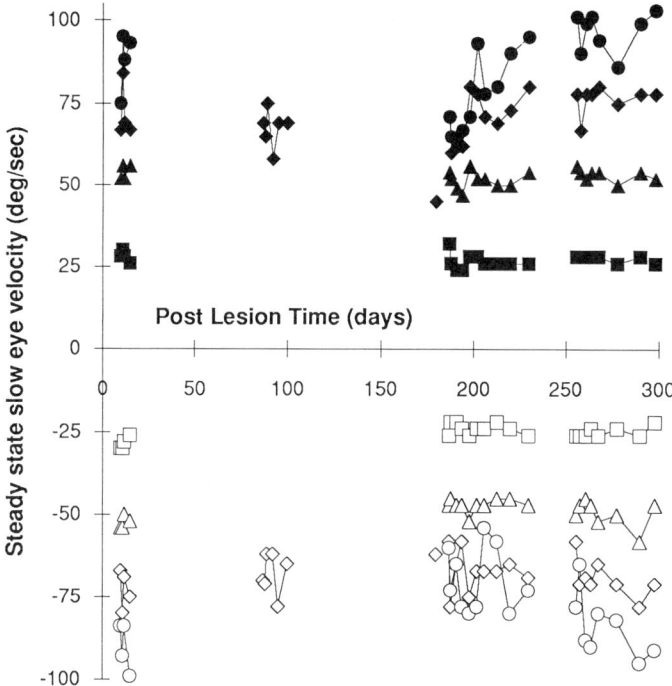

FIGURE 10. Steady-state slow eye velocity is unaffected by small nph lesions. Symbols as in FIGURE 8.

DISCUSSION

Our results differ significantly from those of similar experiments.[10–13] First, and in our opinion most important, we were able to produce enduring lesions as well as long-lasting deficits. This is important because the transient behavioral results reported previously are open to many interpretations. While electrolytic lesions of the cat prepositus resulted in permanent lesions, much of the deficit observed could be explained by damage to fibers of passage.[15,20] In addition, although it is probable that the neurotoxins used in other studies[11,13] spared axons of passage and destroyed only somata, the fleeting effects that were observed might also have been due to the known transmitter-mimetic properties of the toxins. For example, some of the same behaviors have been observed following injections of transmitter agonists into the same region.[21] It is also important to remember that the earlier lesions were produced with kainic acid, which has a tenfold greater potency than ibotenic acid.[22] They were also created with much larger quantities of neurotoxin, which we have found to result in frank mechanical damage to the tissue (see FIGURE 1). Finally, kainic acid has distant as well as local effects.

Another difference is the response to vestibular stimulation. Cheron and Godaux found that sinusoidal rotation of cats produced sinusoidal, symmetric eye movements with a very low gain following unilateral injections.[11] Although we did not use sinusoidal stimulation in monkeys 1 and 2, preliminary data from monkey 3 are

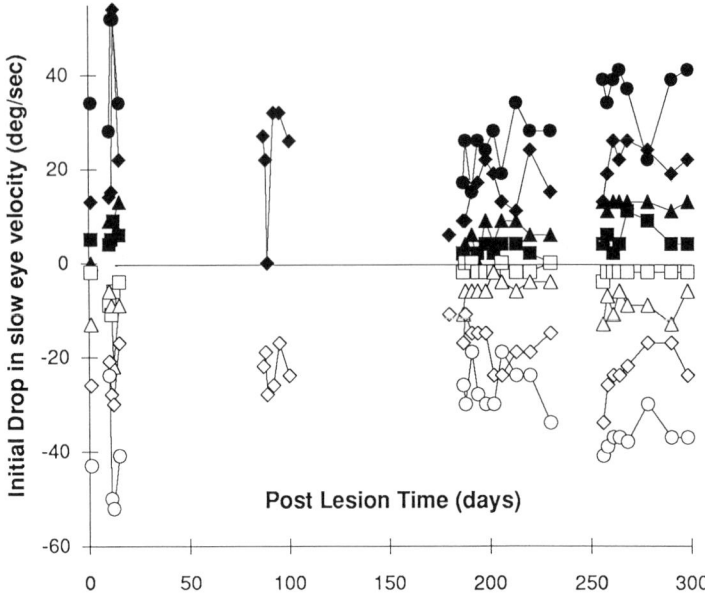

FIGURE 11. Initial optokinetic after-nystagmus responses after small nph lesions. Symbols as in FIGURE 8. Unlike initial jump in OKN, initial drop of OKAN is unaltered by lesions.

consistent with the continuous rotation response of the first two animals in showing an *asymmetric* response following unilateral lesions. In monkeys, Cannon and Robinson reported that a step in head velocity produced a phasic step in eye position.[13] We have consistently found that the response is a prolongation of the eye velocity response, i.e., the greater the damage, the more nearly the response to a step in head velocity approaches a step in eye velocity.

The optokinetic deficits were not as dissimilar. Although Cheron and Godaux injected unilaterally, they devastated OKN in both directions.[11] Cannon and Robinson injected bilaterally[13] so it is more difficult to compare findings directly; however, they also showed that the steady-state slow eye velocity all but disappeared. Similarly, we found that a large lesion that destroyed the entire marginal zone greatly decreased the peak steady-state slow eye velocity. However, the effect of our unilateral lesions was asymmetric and smaller lesions showed that the steady-state slow eye velocity was only modestly affected. When the response was fractionated into its several subcomponents by smaller lesions, we also found that those components were affected symmetrically. We do not think the differences in symmetry and decrease in the steady-state slow eye velocity were due to placement of the lesions, although that is a possibility that must be investigated more closely. Rather, the results suggest to us that there is a threshold level of destruction of the marginal zone that must be exceeded before the steady-state slow eye velocity is diminished and that the symmetrical responses observed in our laboratory and elsewhere are caused by damage to both sides of the brain stem.

We do not think the results of our injections can be explained by damage to neighboring regions. The first injection was overlarge and affected a region ventral to the marginal zone. In addition, the frank mechanical damage owing to the volume displaced by the injected fluid may have affected axons in the region as well as cell bodies. If the devastating effects of our larger lesion were due to damage to crossing fibers, then we should have observed results similar to those obtained with midline surgical lesions of the crossing fibers in this area. Katz *et al.* showed that the time constant of per and postrotatory nystagmus decreased and that the slow rises in OKN and OKAN were both lost but that steady-state slow eye velocity remained the same.[20] In contrast, our deficits showed an *increase* in the time constant of per and postrotatory nystagmus for ipsilateral rotation. In monkey 1, we also abolished OKN including the initial fast rise *and* the steady state, but a rudimentary slow rise may

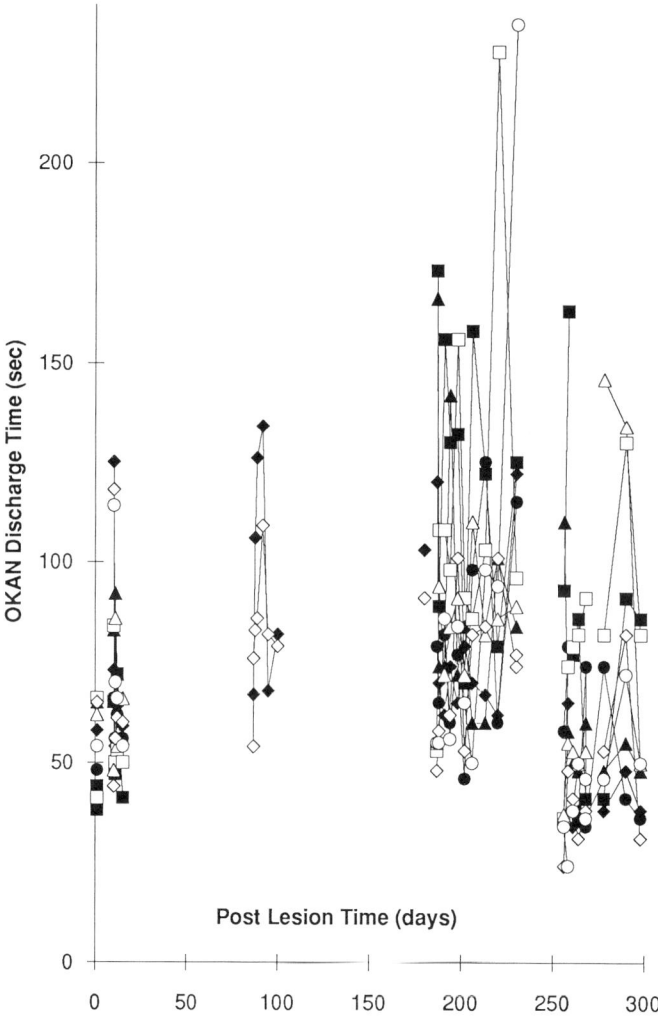

FIGURE 12. Optokinetic after-nystagmus slow eye velocity decay time changes little with small nph lesions. Symbols as in FIGURE 8.

have been spared. Even though there was no OKAN, the interpretation is problematic owing to the lack of OKN.

The series of lesions in monkey 2 affected a much larger portion of the nph, particularly its caudal end, but spared the marginal zone. Even so, the deficits were consistent with those in monkey 1. There was a prolongation of the vestibular per and postrotatory nystagmic time constants. There was also a fractionation of OKN, but it was clear that the fast jump in slow-phase eye velocity was decreased and the slow buildup was prolonged but OKAN was preserved. Despite their purposely restricted size, these lesions were more numerous and together destroyed a larger region of the nph but spared most of the marginal zone. The total effect of the five small lesions was not as devastating as that of the single lesion. It is possible that their distributed pattern and staggered administration allowed for more compensation, but we think the most likely interpretation is that the marginal zone is the critical region for producing the prolongation of the vestibular response, the decrease in the initial OKN response, and the delay in OKN charging reported here. These effects cannot be attributed to the neighboring mvn or abducens nucleus or any of the fiber pathways in the region.[14,15]

Our data do not provide a definitive affirmation of the suggestion that the marginal zone may be part of the neural integrator that converts vestibular and optokinetic velocity afferent information into eye position output. The large lesion in monkey 1 completely destroyed the marginal zone while sparing the caudal nph and mvn. Yet the vestibular responses for ipsilateral rotation were prolonged, not shortened. While the smaller, more distributed lesions in monkey 2 did not have as large an effect, the results were similar. This most puzzling result is *not* consistent with the loss of neural integrator function. Neither is the lack of changes in the OKN and OKAN responses in monkey 2, particularly steady-state slow eye velocity. It is possible that the OKN deficits seen in monkey 1 may have been due to not allowing sufficient time for OKN to build, since we did not understand the delay in the OKN response until we did the series in monkey 2. We hope our studies in monkey 3 will clarify the OKN results.

In addition to this conundrum, there were also a number of lesser enigmas. The large lesion in monkey 1 affected vestibular responses most severely at 50°/second and less at 150°/second, and the smaller lesions in monkey 2 had markedly less effect, especially at the higher velocity. While there was a prolongation of the per and postrotatory nystagmus at 40°/second and possibly at 80°/second as expected, there was much more symmetry in the response and virtually no change in the peak slow eye velocity. We do not understand why the vestibular deficits of monkey 1 were not symmetric. Normally, postrotatory nystagmus in one direction is expected to mirror perrotatory nystagmus in the opposite direction. Nevertheless, the results from our series of small injections in monkey 2 corroborate the findings in monkey 1 since perrotatory responses were more affected than postrotatory and since there was some hint that the side with the lesion was most severely affected. We do not have a ready explanation for these minor discrepancies. We did use monkey 2 for a recording experiment[8] before initiating the lesion series. It is possible that damage from those recordings altered the baseline values that we used for comparison, making the deficits that were created seem less dramatic. Alternatively, the variability of the responses before we began to use amphetamine might have masked some true differences, especially asymmetry. In monkey 3, which is currently being tested, we are using sinusoidal vestibular stimuli in the hope that phase shifts produced by the small lesions will be a more sensitive measure of loss of neural integration in the vestibular system.

Another confusing detail is that while the deficit in ipsilateral vestibular response makes sense, the optokinetic deficit seen in monkey 1 is puzzling because it is asymmetric and *ipsilateral.* Very few vestibular neurons show both ipsilateral vestibular and optokinetic activation,[23] and the normal vestibular and optokinetic nystagmi are synergistic. Since nph neurons do not show optokinetic or visual slip responses,[8] it is surprising that our lesions had such a significant effect in monkey 1. At present we have no explanation for these puzzling results except to suggest that the marginal zone damage in monkey 1 was so complete that it might have affected other related systems including the optokinetic system. Such an explanation is consistent with the results from monkey 2 which did not affect steady-state slow eye velocity and which were more symmetric. The balanced effects are more consistent with the known input from the marginal zone to abducens that is bilateral though predominantly contralateral.[16]

Although some of our findings remain mysterious, it is clear that our results offer some insight into the oculomotor system and are germane to a more general understanding of its function. First, we began this series of experiments to test whether the integrator circuits hypothesized for saccades consisted of a single group of neurons in the marginal zone. Our results have revealed a surprising spectrum of effects suggesting that instead of a single integrator, there may be several for horizontal eye movements, as posited some time ago on the basis of clinical findings.[24] The marginal zone neurons apparently are not involved in neurally integrating oculomotor velocity signals from vestibular and optokinetic sources, but they do maintain steady fixation following saccades to eccentric horizontal positions.[6,7,10–13,18,19] The internal saccadic integrator apparently resides elsewhere,[18,19] while the velocity storage integrator is yet another distinct circuit.[20] Second, the OKN and vestibular responses are apparently even more complicated than previously realized and the various components may be associated, at least in part, with different neural structures. This parcellation may permit a more careful study of the individual components of the more complex coordinated responses. For example, the specific decrement in the initial OKN response suggests the possible disruption of that "ocular-following" response associated with the marginal zone damage. Thus the marginal zone may act either as a site for generating that response or as a relay for visual information that drives the response, possibly from the flocculus. Third, the slip of visual images across the retina cannot be the input for the velocity storage mechanism as posited by most models of that system,[25,26] since the fractionation of the OKN response that we produced in monkey 2 markedly increased the initial retinal slip while increasing the time it took to reach steady-state slow eye velocity and to charge the velocity storage mechanism.

ACKNOWLEDGMENTS

The superb technical assistance of Ms. Sue Usher and the journalistic savoir faire of Ms. Kate Elias are gratefully acknowledged.

REFERENCES

1. GRAYBIEL, A. M. & E. A. HARTWIEG. 1974. Some afferent connections of the oculomotor complex in the cat: an experimental study with tracer techniques. Brain Res. **81:** 543–551.
2. BAKER, R. & A. BERTHOZ. 1975. Is the prepositus hypoglossi nucleus the source of another vestibulo-ocular pathway. Brain Res. **86:** 121–127.

3. BAKER, R., M. GRESTY & A. BERTHOZ. 1975. Neuronal activity in the prepositus hypoglossi nucleus correlated with vertical and horizontal eye movement in the cat. Brain Res. **101:** 366–371.
4. MACIEWICZ, R. J., K. EAGEN, C. R. S. KANEKO & S. M. HIGHSTEIN. 1977. Vestibular and medullary brainstem afferents to the abducens nucleus in the cat. Brain Res. **123:** 229–240.
5. BAKER, R., C. EVINGER & R. A. McCREA. 1981. Some thoughts about the three neurons in the vestibular ocular reflex. Ann. N.Y. Acad. Sci. **374:** 171–188.
6. LOPEZ-BARNEO, J., C. DARLOT, A. BERTHOZ & R. BAKER. 1982. Neuronal activity in prepositus nucleus correlated with eye movement in the alert cat. J. Neurophysiol. **47:** 329–352.
7. McFARLAND, J. L. 1988. The role of the nucleus prepositus hypoglossi and the adjacent medial vestibular nucleus in the control of horizontal eye movement in the behaving monkey. Ph.D. Thesis. University of Washington. Seattle, Wash.
8. McFARLAND, J. L., A. F. FUCHS & C. R. S. KANEKO. The nucleus prepositus and nearby medial vestibular nucleus and the control of simian eye movements. In Vestibular and Brain Stem Control of Eye, Head and Body Movements. H. Shimazu & Y. Shinoda, Eds. Springer-Verlag. New York, N.Y. (In press.)
9. BELKNAP, D. E. & R. A. McCREA. 1988. Anatomical connections of the prepositus and abducens nuclei in the squirrel monkey. J. Comp. Neurol. **268:** 13–28.
10. CHERON, G., P. GILLIS & E. GODAUX. 1986. Lesions in the cat prepositus complex: effects on the optokinetic system. J. Physiol. **372:** 95–111.
11. CHERON, G. & E. GODAUX. 1987. Disabling of the oculomotor neural integrator by kainic acid injections in the prepositus-vestibular complex of the cat. J. Physiol. **394:** 267–290.
12. CHERON, G., E. GODAUX, J. M. LAUNE & B. VANERKELEN. 1986. Lesions in the cat prepositus complex: effects on the vestibulo-ocular reflex and saccades. J. Physiol. **372:** 75–94.
13. CANNON, S. C. & D. A. ROBINSON. 1987. Loss of the neural integrator of the oculomotor system from brain stem lesions in monkey. J. Neurophysiol. **57:** 1383–1409.
14. McCREA, R. A., A. STRASSMAN, E. MAY & S. M. HIGHSTEIN. 1987. Anatomical and physiological characteristics of vestibular neurons mediating the horizontal vestibuloocular reflex in the squirrel monkey. J. Comp. Neurol. **264:** 547–570.
15. HIKOSAKA, O. & T. KAWAKAMI. 1977. Inhibitory reticular neurons related to the quick phase of vestibular nystagmus—their location and projection. Exp. Brain Res. **27:** 377–396.
16. LANGER, T. P., C. R. S. KANEKO, C. A. SCUDDER & A. F. FUCHS. 1986. Afferents to the abducens nucleus in the monkey and cat. J. Comp. Neurol. **245:** 379–400.
17. KANEKO, C. R. S. & A. F. FUCHS. 1987. The effect of ibotenic acid lesions of the omnipause neurons on saccadic eye movements in the monkey. Soc. Neurosci. Abstr. **13:** 393.
18. KANEKO, C. R. S. Tests of two models of the neural saccade generator: saccadic eye movement deficits following lesions of the nuclei interpositus and prepositus hypoglossi in monkey. In Vestibular and Brain Stem Control of Eye, Head and Body Movements. H. Shimazu & Y. Shinoda, Eds. Springer-Verlag. New York, N.Y. (In press.)
19. KANEKO, C. R. S. & A. F. FUCHS. 1991. Saccadic eye movement deficits following ibotenic acid lesions of the nuclei raphe interpositus and prepositus hypoglossi in monkey. Acta Otolaryngol. Suppl. **481:** 213–216.
20. KATZ, E., J. M. B. VIANNEY DE JONG, J. BÜTTNER-ENNEVER & B. COHEN. Commissural contributions to velocity storage and the vestibulo-ocular reflex. Exp. Brain Res. (In press.)
21. BÜTTNER, U. 1992. Oculomotor effects of GABA injections into the vestibular nuclei. Ann. N.Y. Acad. Sci. (This volume.)
22. KÖHLER, C. 1983. Neuronal degeneration after intracerebral injections of excitotoxins. A histological analysis of kainic acid, ibotenic acid and quinolinic acid lesions in the rat brain. Werner-Gren Symp. **39:** 99–111.
23. WAESPE, W. & V. HENN. 1977. Neuronal activity in the vestibular nuclei of the alert monkey during vestibular and optokinetic stimulation. Exp. Brain Res. **27:** 523–538.

24. ABEL, L. A., L. F. DELL'OSSO & R. B. DAROFF. 1978. Analog model for gaze-evoked nystagmus. IEEE Trans. Biomed. Eng. **25:** 71–75.
25. RAPHAN, TH., B. COHEN & V. MATSUO. 1979. The velocity storage mechanism responsible for optokinetic nystagmus (OKN), optokinetic after-nystagmus (OKAN) and vestibular nystagmus. *In* Control of Gaze by Brain Stem Neurons. R. Baker & A. Berthoz, Eds. Dev. Neurosci. **1:** 37–47. Elsevier/North-Holland Biomedical Press. New York, N.Y.
26. ROBINSON, D. A. 1979. Vestibular and optokinetic symbiosis: an example of explaining by modelling. *In* Control of Gaze by Brain Stem Neurons. R. Baker & A. Berthoz, Eds. Dev. Neurosci. **1:** 49–58. Elsevier/North-Holland Biomedical Press. New York, N.Y.

Climbing Fiber Intervention Blocks Plasticity of the Vestibuloocular Reflex[a]

A. E. LUEBKE[b] AND D. A. ROBINSON[c]

Departments of Ophthalmology
and Biomedical Engineering
The Johns Hopkins University
School of Medicine
Baltimore, Maryland

INTRODUCTION

It has been demonstrated in a variety of circumstances that there exists a reciprocal relationship between the discharge rates of simple and complex spikes in Purkinje cell firing rate activity. Demer *et al.* demonstrated this again in the cat flocculus.[1] They stimulated climbing fibers near the inferior olive in the lightly anesthetized cat while recording from the flocculus and showed that as they evoked complex spikes, at an increasing rate from 1 to 10 Hz, the simple-spike rate of Purkinje cells, normally about 23 Hz, decreased steadily to zero. At a stimulus rate of 7 Hz, 87% of the Purkinje cells, in their study, showed no simple-spike discharges at all.

This situation presents an interesting opportunity. With no simple spikes and the complex spikes artifically clamped at, say, 7 Hz, no useful signal can possibly emerge from the flocculus. This situation should constitute a functional flocculectomy. If motor learning in the vestibuloocular reflex (VOR) is stored by long-term depression of synapses between parallel T fibers and Purkinje cell dendrites, such learning should be abolished if the Purkinje cells are prevented from sending any meaningful signal to any motor target. We tested this idea.

METHODS

Alert cats were studied. Movements of one eye were recorded by the eye-coil/magnetic-field method. A skull implant was used to immobilize the head. Two microelectrode chambers were mounted on the skull, one aimed at the midline between the inferior olives, the other at the right flocculus. All surgery was done under pentothal anesthesia and aseptic conditions.

The gain of the VOR was altered by rotating the cats for four hours daily at 0.15 Hz, 30 deg/second peak velocity, in an optokinetic drum that either rotated with the vestibular platform ($\times 0$ viewing) or in the opposite direction ($\times 2$ viewing). Between optokinetic training sessions, the cats were fitted with fixed-field goggles (a focussed

[a]This research was supported by a grant from the National Eye Institute (EY00598). Anne Luebke has been supported by the Johns Hopkins University Whiting School of Engineering.

[b]Present affiliation: Department of Physiology & Biophysics, University of Miami School of Medicine, Miami, Florida 33101.

[c]Author to whom correspondence should be addressed at Room 355 Woods Research Building, the Wilmer Institute, the Johns Hopkins Hospital, 600 N. Wolfe St., Baltimore, Maryland 21205.

scene fastened to the head) or ×2 magnifying goggles to help retain the induced changes. After three days, gains of the VOR had approximately doubled with ×2 viewing, or halved with ×0 viewing.

After such changes, cats were rotated in the optokinetic drum with the drum moving in the opposite direction to that used for adaptation (×0 viewing from high gains, ×2 from low gains). This procedure causes the gain of the VOR to return close to its normal value in 30 minutes. The effect of climbing fiber stimulation on this rapid deadaptation was investigated.

With a recording electrode in the flocculus, the medullary electrode was moved about until a minimum threshold was found for evoking a field potential and complex spikes in individual cells with a biphasic, constant-current pulse with pulse durations of 0.2 msecond. Cats were kept alert by the administration of dextroamphetamine sulfate. Locations of stimulating and recording electrodes were verified histologically at the end of all experiments.

RESULTS

A total of 60 Purkinje cells were isolated (about 15 cells in each of 4 cats), and the climbing fibers were stimulated at 7 Hz. The simple-spike rate shortly began to decrease and went to zero within two minutes. At this time, the extracellular recordings consisted of seven complex spikes/second and nothing else. When stimulation stopped, the simple spikes began to return and resumed their former rate after two minutes. Of these cells, 57 or 95% were totally silenced by 7 Hz stimulation.

We moved our recording electrode by several millimeters in all directions from the center of the floccular region, and feel that we sampled cells in all regions including the H (horizontal) zone. We conclude that stimulating the climbing fibers to the flocculus turned the flocculus off by eliminating all simple-spike firing and preventing any meaningful neural output.

We stimulated the climbing fibers and shut down the flocculus during the 30 minutes when deadaptation, from high or low VOR gains toward normal gain, would normally occur by reversing the drum's motion. We measured the gain of the VOR every 5 minutes. We measured the initial gain (usually around twice or half normal gain) and then began to stimulate the climbing fibers at 7 Hz. During the next 30 minutes, no change occurred in the gain of the VOR, although in control experiments without electrical stimulation, the gain always changed by a factor of 2:1, up or down, with such visual-vestibular experience. After 30 minutes, the 7-Hz stimulation was stopped, the cat and drum continued rotating, and the gain immediately began to change toward normal, completing this change, as usual, within the next 30 minutes. In summary, so long as the flocculus was functionally abolished by 7-Hz, climbing-fiber stimulation, motor learning was blocked.

Of more interest is that after the three-day adaptation to an altered VOR gain, blocking floccular output by 7-Hz stimulation, by itself, did not change the gain of the VOR. If the gain was high, it stayed high. If low, it stayed low. If normal, it stayed normal. This result was obtained in 3 trials at each of the three gains in all three cats tested (27 trials total). The gains before and two minutes after climbing fiber stimulation began did not show a statistically significant difference.

DISCUSSION

That shutting down the flocculus by 7-Hz stimulation prevented motor learning is not surprising. It simply confirms earlier reports that flocculectomy abolishes VOR plasticity.[2,3] Seven-hertz stimulation also clamped the complex spike rate (no sponta-neous complex spikes ever appeared), approximating a lesion of the inferior olive (in the sense that no meaningful signals could get through) which also blocks motor learning.[4,5] Consequently, like previous lesion studies, these results alone do not allow one to say which synapses are modified when the gain of the VOR is altered.

That floccular shut down by 7-Hz stimulation does not cause changes in the modified gains of the VOR is, to us, surprising. It is difficult, we think impossible, to reconcile our results with learning stored by any sort of synaptic changes in the cerebellum. If the modified gains are created by a synaptic change that modifies the contribution of Purkinje cells to the VOR, how could such modifications survive when these Purkinje cells are effectively silenced?

We realize that stimulating in the medulla can excite other pathways and that axon collaterals of climbing fibers to floccular target neurons in the vestibular nuclei are also excited at 7 Hz by our stimulation, but we are unable to conjure up a scheme that leaves the VOR gain invariant of a complete loss of a meaningful, cerebellar contribution.

Miles *et al.*[6] and Lisberger[7] have also found results, in the monkey, that cannot be explained by synaptic modifications in the flocculus alone, and have proposed modifiable synapses on cells in the vestibular nucleus that receive Purkinje cell axons. The surprising (to us) results of our experiments do not agree with the hypothesis of Ito,[8] but are consistent with those of Lisberger: that motor learning may occur in the deep cerebellar nuclei, in this case in the vestibular nuclei.

ACKNOWLEDGMENTS

We thank A. McCracken for preparing the manuscript.

REFERENCES

1. DEMER, J. L., D. A. ECHELMAN & D. A. ROBINSON. 1985. Effects of electrical stimulation and reversible lesions of the olivocerebellar pathway on Purkinje cell activity in the flocculus of the cat. Brain Res. **346:** 22–31.
2. ITO, M., T. SHIIDA, N. YAGI & M. YAMAMOTO. 1974. The cerebellar modification of rabbit's horizontal vestibulo-ocular reflex induced by sustained head rotation combined with visual stimulation. Proc. Jpn Acad. **50:** 85–89.
3. ROBINSON, D. A. 1976. Adaptive gain control of vestibuloocular reflex by the cerebellum. J. Neurophysiol. **39:** 954–969.
4. ITO, M. & Y. MIYASHITA. 1975. The effects of chronic destruction of the inferior olive upon visual modification of the horizontal vestibulo-ocular reflex of rabbits. Proc. Jpn Acad. **51:** 716–720.
5. HADDAD, G. M., J. L. DEMER & D. A. ROBINSON. 1980. The effect of lesions of the dorsal cap of the inferior olive on the vestibulo-ocular and optokinetic systems of the cat. Brain Res. **185:** 265–275.
6. MILES, F. A., D. J. BRAITMAN & D. M. DOW. 1980. Long-term adaptive changes in primate vestibuloocular reflex. IV. Electrophysiological observations in flocculus of adapted monkeys. J. Neurophysiol. **43:** 1477–1493.
7. LISBERGER, S. G. 1988. The neural basis for learning of simple motor skills. Science. **242:** 728–735.
8. ITO, M. 1982. Cerebellar control of the vestibulo-ocular reflex—around the flocculus hypothesis. Annu. Rev. Neurosci. **5:** 275–296.

Otolith-Ocular Testing in Human Subjects[a]

JOSEPH M. R. FURMAN[b] AND ROBERT W. BALOH[c]

[b]*Department of Otolaryngology*
The Eye and Ear Institute of Pittsburgh
203 Lothrop Street
Pittsburgh, Pennsylvania 15213

[c]*Department of Neurology*
University of California at Los Angeles
Medical Center
10833 LeConte Avenue
Los Angeles, California 90024

INTRODUCTION

This paper will review the current status of otolith-ocular testing in human subjects with specific emphasis on methods that have potential for becoming useful in a clinical setting. Most of the methods to be discussed can use conventional electrooculography (EOG) for eye movement recording, although some stimuli discussed require more sophisticated eye movement recording techniques such as the magnetic scleral search coil.

Each of the paired labyrinths contains five suborgans (three semicircular canals and two otolith organs). The natural stimulus to the semicircular canals is angular acceleration in the plane of the canal, and the natural stimulus to the otolith organs is linear acceleration caused either by translation or changes in orientation with respect to gravity. The utricular macula is oriented in about the same plane as the horizontal semicircular canals, and the saccular macula is oriented in the sagittal plane of the head.

Most head movements stimulate more than one of the five suborgans of the labyrinth. The determination of which suborgans are stimulated requires a knowledge of three factors: (1) whether the stimulus is an angular or a linear acceleration, (2) the orientation of the skull (and thus the labyrinths) with respect to the movement, and (3) the orientation of the movement with respect to gravity. TABLE 1 lists the tests that are to be described in this paper and indicates which suborgans the various tests are designed to evaluate. Note that for purposes of simplification, the two vertical semicircular canals have been grouped and the two otolith organs have been grouped. In general, otolith-ocular testing requires equipment that is more complex than simple caloric irrigation or earth-vertical axis yaw rotation, which together form the basis for contemporary clinical vestibular laboratory testing.

FIGURE 1A, B, and C depict three different types of yaw rotation, i.e., the axis of rotation is collinear with the rostral-caudal body axis. Two of these, off-vertical axis rotation (OVAR) and eccentric rotation, stimulate the otolith organs as well as the

[a]This work was supported by a grant from the Whitaker Foundation (J.F.) and by National Institutes of Health grants DC01317 (J.F.) and DC00097 (R.B.).

TABLE 1[a]

| | Semicircular Canals | | |
Test	Horizontal	Vertical	Otoliths
Conventional rotatory chair	x	—	—
Pitch rotation			
Upright	—	x	x
On-side	—	x	—
Ocular counterrolling			
Static	—	—	x
Dynamic	—	x	x
Eccentric rotation	x	—	x
Off-vertical rotation	x	—	x
Linear track	—	—	x
Parallel swing	—	—	x

[a]x denotes stimulation. Adapted from Baloh and Furman, 1989.[51]

horizontal semicircular canals. FIGURE 1A depicts upright earth-vertical axis (EVA) yaw rotation wherein only the horizontal semicircular canals are stimulated. This stimulus is widely used in clinical vestibular laboratories. FIGURE 1B depicts OVAR wherein the axis of rotation is tilted with respect to gravity; if the axis is tilted by 90 degrees, the stimulus is called earth-horizontal axis (EHA) rotation. It is the changing orientation with respect to gravity during OVAR that stimulates the otolith organs. FIGURE 1C depicts off-axis (to be distinguished from off-vertical) rotation, also known as *eccentric* rotation, wherein the subject is rotated about an earth-vertical axis that is parallel *but not collinear* with the rostral-caudal body axis. It is the tangential and centripetal accelerations during eccentric rotation that stimulate the otolith organs.

FIGURES 1D, E, and F depict three other types of otolith stimuli. FIGURE 1D shows upright (earth-horizontal axis) roll wherein the axis of rotation is collinear with the subject's nasal-occipital axis. FIGURE 1E depicts upright (earth-horizontal axis) pitch wherein the axis of rotation is collinear with the subject's interaural axis. Both stimuli activate the vertical semicircular canals and the otolith organs as a result of the combined rotational acceleration and changing orientation of the subject with respect to gravity. FIGURE 1F depicts upright interaural (earth-horizontal axis) linear acceleration. This stimulus is purely otolithic and, based on the orientation of the maculae, is thought to activate primarily the utriculus.

For each topic, the discussion will begin with a brief review of basic principles and pertinent animal studies. Then, results from humans will be described including both normative and patient responses. For ease of discussion, the interaural, nasooccipital, and rostral-caudal axes will be denoted as the x, y, and z axes respectively.

OFF-VERTICAL AXIS ROTATION

Basic Principles

Off-vertical axis rotation stimulates the otolith organs by constantly changing a subject's orientation with respect to gravity. *Constant-velocity* OVAR induces a continuous nystagmus whose slow-component velocity is in the direction opposite to that of the direction of rotation.[1] The slow-component velocity of this continuous nystagmus contains a periodic component, the modulation, and a nonzero component, the bias. It is these two response components, which may be seen in isolation when canal-ocular responses have decayed to zero, that form the basis for OVAR being an otolith-ocular reflex assessment tool. *Sinusoidal* OVAR is a means of assessing semicircular canal–otolith interaction.

Animal studies have shown that the modulation component depends on stimulation of the otolith organs.[2] The bias component depends on stimulation of the otolith organs and the integrity of semicircular canal afferents, suggesting that the bias component is due, in part, to activity in the so-called velocity-storage element of the central vestibular system.[3,4]

As noted above, OVAR may be performed with the axis of rotation tilted by as much as 90 degrees, so-called EHA rotation. EHA rotation is quite cumbersome to perform and removing subjects from the test apparatus quickly is difficult, thereby reducing the potential clinical usefulness of such a technique. However, EHA rotation has been performed in both normal subjects and patient populations.[5–8] Such studies have indicated that a consistent, though somewhat variable, response can be obtained from normal subjects.

More recently, OVAR using smaller angles of tilt (30 degrees or less) has been employed at the University of Pittsburgh in that testing is relatively simple to perform, the apparatus can also be used for earth-vertical axis rotation, and subjects can be removed from the apparatus with relative ease. This last feature is important

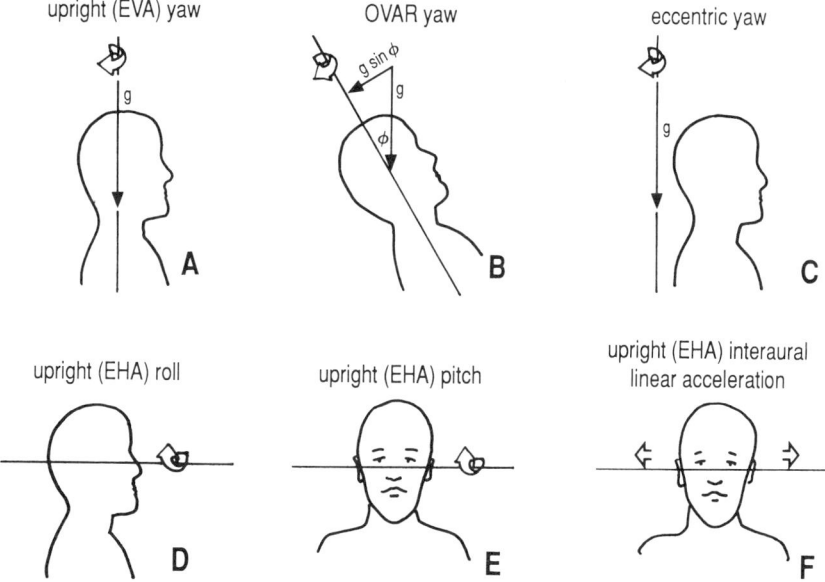

FIGURE 1. Schematic diagram of several otolith-ocular assessment methods. **A:** "Conventional" upright earth-vertical axis yaw rotation, which is solely a horizontal semicircular canal stimulus. **B:** Off-vertical axis rotation in yaw, which stimulates both the horizontal semicircular canals and the otolith organs as a result of a changing orientation with respect to gravity. **C:** Eccentric yaw rotation wherein tangential and/or centripetal forces are used to stimulate the otolith organs. **D:** Upright earth-horizontal axis roll, and **E:** Upright earth-horizontal axis pitch wherein the vertical semicircular canals are stimulated as a result of the rotation and the otolith organs are stimulated as a result of the changing orientation with respect to gravity. **F:** Upright earth horizontal axis interaural linear acceleration, which is a pure otolith stimulus.

because OVAR is nausea producing and, on occasion, patients may need to be removed from the device quickly. OVAR has been performed using both constant velocity and sinusoids. Testing has been performed either by first tilting the subject off-vertical and then rotating the individual about this tilted axis, the so-called T,R paradigm and, alternatively, by rotating subjects about an earth-vertical axis until canal-ocular responses have decayed and then tilting the axis of rotation. This latter paradigm is called the R,T paradigm and is useful for constant-velocity rotations in that the semicircular canal–ocular reflex can be allowed to decay to zero and then the otolith-ocular reflex can be studied in isolation.

Normative Data

FIGURE 2 top illustrates the slow-component eye velocity response to constant-velocity OVAR using the "T,R" paradigm. Note the presence of a modulation and a bias component. Our data, in agreement with those of Darlot et al.,[9] have indicated that for constant-velocity OVAR to produce reliable modulation and bias components, a tilt angle of about 30 degrees and a constant velocity of about 60 degrees per second are required. Stimuli of lower magnitude are not sufficient to provide clearly identifiable otolith-ocular responses, at least when using electrooculography. Possibly, with more sensitive eye movement recording techniques, such as the magnetic scleral search coil, lower intensity stimuli could be used. Our data have also indicated that responses using the "T,R" paradigm are the same as those obtained using the "R,T" paradigm. Interestingly, however, although the "R,T" paradigm may be extremely disorienting and may produce unusual illusions, the stimulus is not as nausea-producing as the "T,R" paradigm. Studies are still under way to determine the optimal stimulus characteristics.

Normative studies using sinusoidal OVAR have indicated, surprisingly, that such a stimulus is *less* nausea producing than constant-velocity OVAR. Sinusoidal OVAR produces responses that contain both a sinusoidal component at the frequency of stimulation and higher frequency components presumably of otolithic origin (see FIGURE 2 bottom). Sinusoidal OVAR produces responses of lower gain and smaller phase lead than sinusoidal earth-vertical axis rotation presumably as a result of interaction between responses to semicircular canal and otolith stimulation.

Patient Data

Studies of patients using OVAR have included two studies of patients with surgically confirmed unilateral peripheral vestibular lesions[8,10] and two studies of patients with cerebellar lesions.[11,12]

FIGURE 3 shows the response to EHA rotation at a constant velocity of 60 degrees per second of a patient who underwent resection of an acoustic neuroma on the left. It can be seen that the semicircular canal–ocular reflex component decays exponentially and is succeeded by a nonzero baseline. A modulation is superimposed on these response components. Note that the bias component is of larger amplitude when rotating away from the lesioned ear. This asymmetry in the bias component has been confirmed using OVAR[10,13] and may reflect the asymmetry in peripheral otolith function per se or may reflect asymmetric velocity storage caused by unilateral peripheral vestibular abnormalities. Remarkably, the modulation component was normal, suggesting that for this response component, a single labyrinth is sufficient. These results are disappointing in terms of OVAR becoming a routine clinical

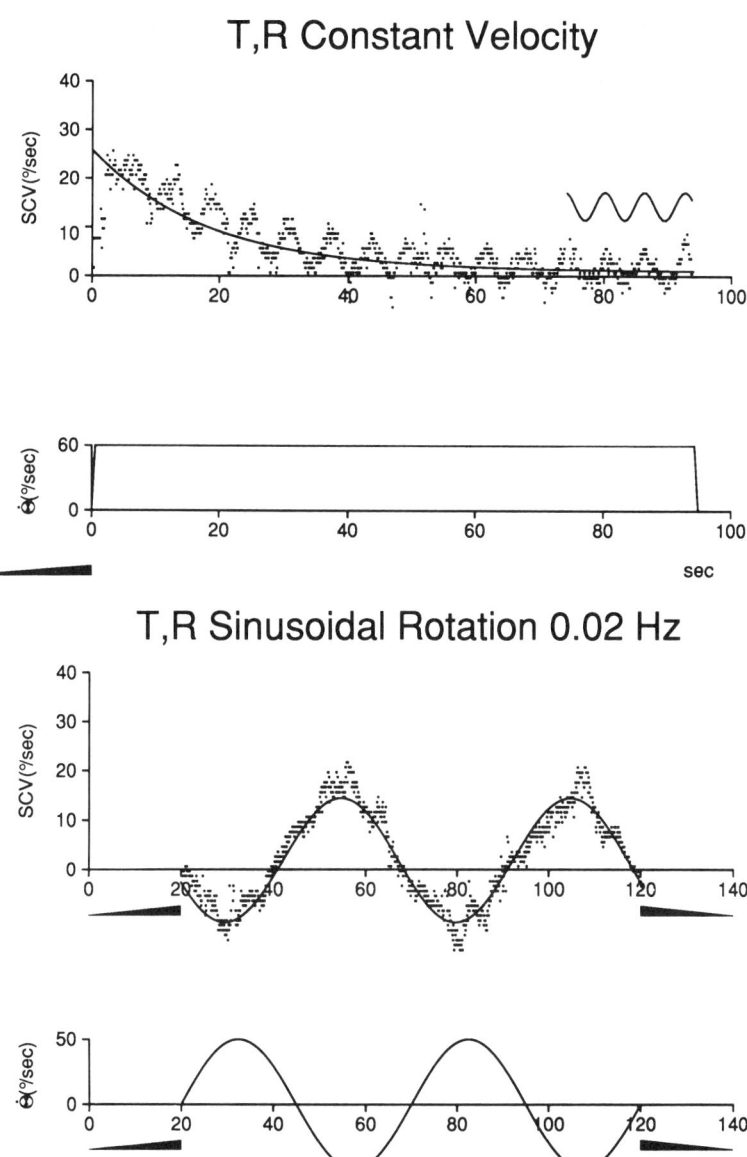

FIGURE 2. Slow component eye velocity induced by off-vertical axis rotation (OVAR) of a normal subject. Eye movements were recorded using the magnetic scleral search technique. **Top:** Responses to constant velocity OVAR with a tilt angle of 30 degrees and a velocity of 60 degrees per second. Note the presence of bias and modulation components. **Bottom:** Responses to sinusoidal OVAR with a tilt angle of 30 degrees at a frequency of 0.02 Hz with a peak velocity of 60 degrees per second. Note the higher frequency components as a result of the otolith stimulation.

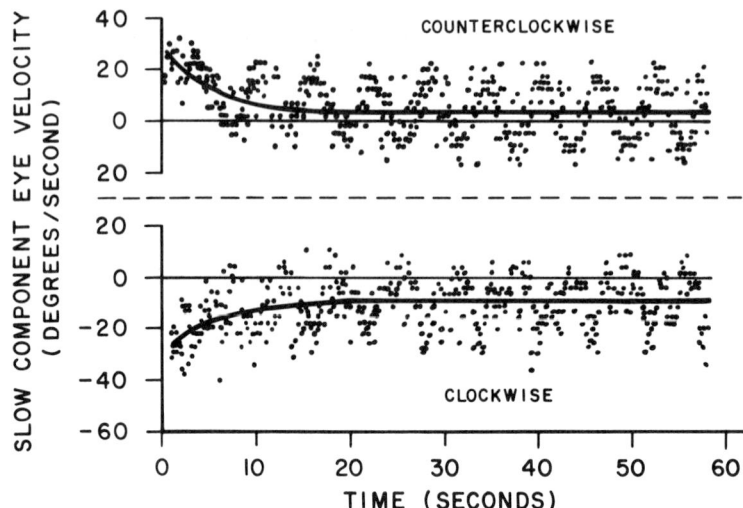

FIGURE 3. Earth-horizontal axis rotation responses of patient two months following a resection of a left acoustic neuroma. Note the asymmetry of the bias components and normal modulation components. (From Furman and Wall[8] with permission.)

vestibular assessment tool in that alterations in OVAR responses as a result of a complete unilateral peripheral vestibular asymmetry are meager. However, further studies will be performed in an attempt to determine the clinical usefulness of OVAR.

Studies of patients with cerebellar abnormalities using OVAR have consisted of (1) assessing a group of patients with periodic alternating nystagmus (PAN),[11] a disorder thought to be localized to the cerebellar nodulus or uvula or their connections with the brain stem vestibular nuclei, and (2) assessing several patients with olivopontocerebellar atrophy.[12]

The study of patients with PAN revealed that their modulation component during OVAR was several times larger than that seen in normal subjects (see FIGURE 4). This increase in otolith-ocular reflex responses is analogous to the increase in semicircular canal–ocular reflex gain seen in patients with vestibulocerebellar lesions.[14,15] Apparently, the cerebellar nodulus and uvula are important for the regulation of otolith-ocular reflexes as well as canal-ocular reflexes. Our study of patients with olivopontocerebellar atrophy revealed that the slow saccades seen during ocular motor testing and during vestibular nystagmus induced by canal stimulation were also seen during the nystagmus induced by OVAR.[12] Also, during OVAR, there was a dearth of quick components and the fluctuation in mean eye position, which is usually minimal in normal subjects, was large, sometimes causing the eyes to be pinned to the right or left.

Related to OVAR is the study of the effect of postrotatory head tilt on postrotatory nystagmus. This stimulus can be delivered with earth-vertical axis rotation by having subjects tilt their heads immediately upon cessation of rotation or by using OVAR and stopping rotation while the axis of rotation remains tilted. In either case, the subject is exposed to the combination of a semicircular canal stimulus from the rotational deceleration and a static otolith stimulus from the tilt with respect to gravity. Normal subjects are known to show a shortened time constant with postrotatory head tilt[16] (see FIGURE 5). Both animal studies[17] and human studies[10] have shown that unilateral peripheral vestibular injury does not abolish this shortened time constant with postrotatory head tilt. However, patients with caudal midline cerebellar lesions have been shown to lose this effect.[18] Our patients with PAN also showed a loss of the effect of postrotatory head tilt (FIGURE 5).

These studies suggest that although OVAR appears to be worthy of further study, it cannot be considered to be an established method of assessing the otolith-ocular reflex.

ECCENTRIC ROTATION

Basic Principles

When a subject is rotated about an axis parallel to but at a distance from the z (rostral-caudal) axis, they experience an off-axis or eccentric rotation, which consists of a combination of angular acceleration and two kinds of linear acceleration, namely, tangential and centripetal. The tangential acceleration is a function of the rate of change of the angular velocity while the centripetal acceleration is a function of the instantaneous rotational velocity. Thus, at constant rotational velocity there is no angular acceleration and no tangential acceleration, only centripetal acceleration comparable to that achieved by rolling the head into a fixed tilt. However, during nonconstant angular velocity, such as sinusoidal rotations, in addition to the centripetal acceleration, both the angular acceleration and the tangential linear accelerations are nonzero. Studies of eccentric rotation in monkeys[19,20] indicate that eccentric rotation at a distance of 23 cm causes an increase in vestibuloocular reflex (VOR)

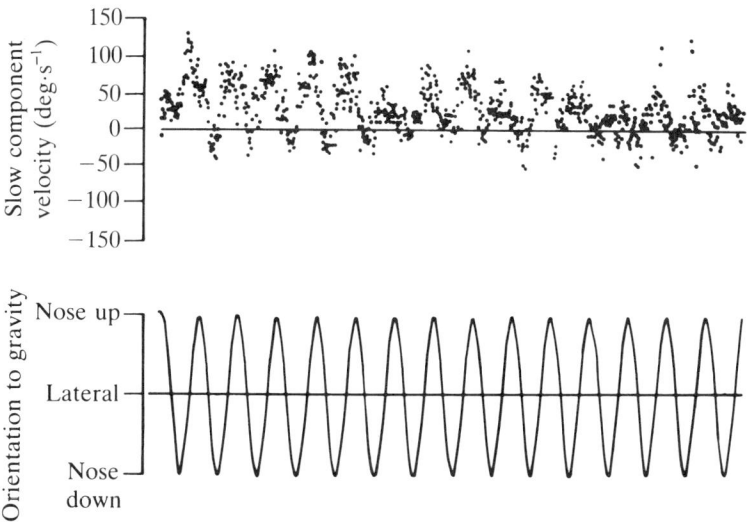

FIGURE 4. Earth-horizontal axis rotation responses from a patient with periodic alternating nystagmus. Rotation was at a constant velocity of 60°/second. Note that the modulation component amplitude is several times normal. (From Furman *et al.*[11] with permission).

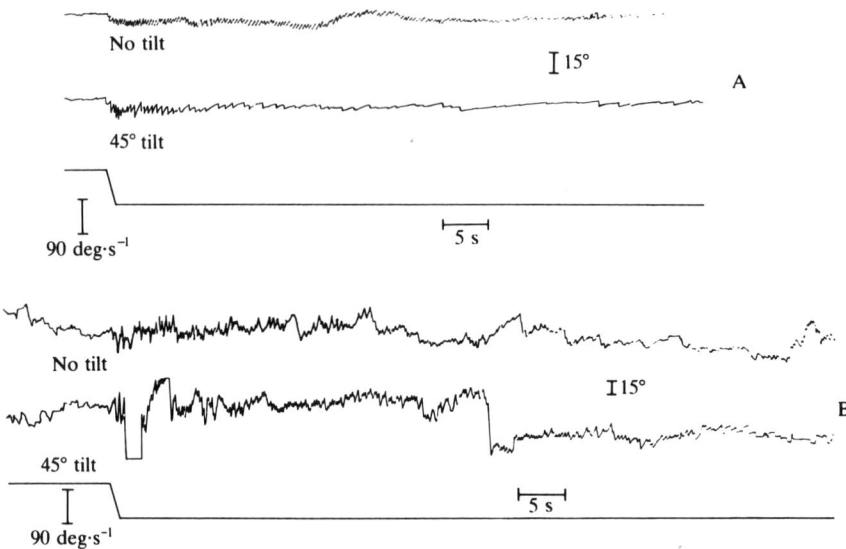

FIGURE 5. Effects of head tilt on postrotatory responses following earth-vertical axis rotation trapezoids. **A:** Responses from a normal subject. **B:** Responses from a patient with periodic alternating nystagmus. Note that whereas the postrotatory nystagmus in the normal subject is markedly diminished by head tilt, the response of the patient is not altered by head tilt. (From Furman et al.[11] with permission.)

gain when animals are oriented such that the tangential linear acceleration is directed along the animal's x-axis. Specifically, gain was noted to increase from 0.67 to 1.0 at 1.0 Hz in the dark. No gain enhancement was noted when animals were oriented such that tangential forces were oriented along the y-axis or at 0.5 Hz regardless of orientation. These same investigators also observed that after unilateral deafferentation of the otolith organs, the enhancement of gain during eccentric rotation decreased and then recovered completely within eight weeks.[19] Following bilateral otolith ablation no gain enhancement was seen during eccentric rotation. These studies suggest that it is the tangential acceleration, sensed by the otolith organs, that underlies changes in VOR gain during eccentric rotation. It is this gain enhancement that forms the basis of eccentric rotation as a tool for assessment of the otolith-ocular reflex.

Normative Data

Several normative studies of eccentric rotation have been performed in humans.[21-24] Such studies have provided somewhat conflicting results. Benson and Th. Vieville failed to show any significant difference in the response evoked by on-center and by eccentric oscillation at 0.3 and 1.0 Hz.[21] However, reliable data were obtained only from one subject. Gresty and Bronstein,[22] Gresty et al.,[23] and Koizuka et al.[24] observed an increased gain in normal humans during eccentric rotation with tangential forces acting along the x-axis as compared to on-center rotation (see FIGURE 6).

As in animal studies, such gain increases have been most evident at frequencies of about 1 Hz. Gresty *et al.*[23] observed that VOR gain increased with the head positioned 30 centimeters eccentric to the axis of rotation by extension of the neck and leaning forward. Gain increases were particularly evident at frequencies of 0.5 and 1.2 Hz with peak angular velocities of 60 degrees per second, which corresponds to peak tangential accelerations of 0.1 and 0.24 *g* peak linear accelerations. Gain increases were particularly large when subjects imagined near targets. This latter finding is in accord with the idea that eye movement compensating for linear head translation is most important when viewing near targets (see below). No changes were noted in the phase of the responses as a function of the head being on-center or eccentric from the axis of rotation. Gresty *et al.* listed several possibilities to account for the increased gain seen during eccentric rotation: (1) voluntary nonvisual enhancement of the VOR (although gains were increased beyond the level that would be expected from such an effect), (2) an effect of the position of the neck, and (3) otolithic stimulation.[23] Koizuka *et al.* used an eccentric position of 90 centimeters with head oriented so that tangential forces acted along the *x*-axis.[24] Head and neck posture were identical for both on-center and eccentric rotations. They found that normal subjects showed an increase in the VOR gain at 0.64 Hz with eccentric rotation. They ascribed this gain enhancement to stimulation of the utricle.

Patient Data

Studies of patients using eccentric rotation[25,26] have suggested that such testing, although promising, may not be a particularly useful clinical test of the otolith-ocular reflex. Barratt *et al.* have observed that surgically confirmed unilateral peripheral vestibular lesions lead to a greater asymmetry in vestibuloocular responses during eccentric rotation as compared to on-center rotation.[25] These authors also showed

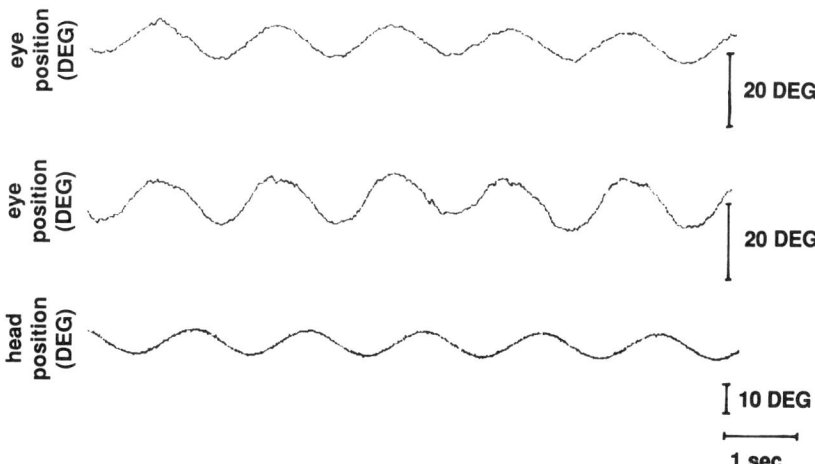

FIGURE 6. Eye movement responses to on-center (upper trace) and to eccentric rotation (lower trace) in a normal subject. Rotations were at 0.64 Hz with a peak velocity of 20 degrees per second. For eccentric rotation, the subject was placed at 90 cm from the axis of rotation. (Data donated by I. Koizuka.)

that cerebellar lesions may cause an increase or a decrease in the asymmetry of vestibular-ocular responses. Further case studies indicated a significant association between abnormal responses during eccentric rotation and static positional nystagmus. However, the direction of the positional nystagmus was not necessarily the same as the direction of the effect on eye movements seen during eccentric rotation. Patients with benign paroxysmal positional nystagmus and vertigo, and other patients in whom otolithic abnormalities were suspected, were not found to have abnormal responses during eccentric rotation. Koizuka et al. studied patients with Meniere's disease and patients with "vestibular Meniere's disease."[26] They found that patients with Meniere's disease and patients with vestibular Meniere's disease tested within one month of an exacerbation showed no gain enhancement during eccentric rotation. These investigators concluded that the utricular function of patients with Meniere's disease and of those with "vestibular Meniere's disease" was not different.

These studies of eccentric rotation suggest that although the otolith ocular reflex can be tested in this manner, otolith function can be assessed only by comparing combined canal-otolithic responses with responses to pure semicircular canal stimulation; moreover, the affect of near fixation cannot be underestimated. At this time, although eccentric rotation appears to be worthy of further study, it cannot be considered to be an established method of assessing the otolith-ocular reflex.

PITCH AND ROLL

Basic Principles

As suggested earlier, pitch and roll movements in the upright position stimulate both canal and otolith afferents, and the resulting compensatory eye movements represent a combination of canal-ocular and otolith-ocular reflexes. On the other hand, pitch in the onside position or roll in the supine position activate only the canal-ocular reflexes since the orientation of the subject with respect to gravity does not change during rotation. Therefore, subtracting the response to pitch or roll in the onside or supine positions from the response in the upright position is a measure of otolith-ocular function. Also, if the velocity of pitch or roll in the upright position is low, below the threshold for the semicircular canals, only the otolith-ocular reflex is activated. Ocular counterrolling induced by either static tilt or low-velocity roll has long been championed as a clinical test of otolith-ocular function.

Matsuo and Cohen found an asymmetry in the time constant of the vertical vestibuloocular reflex (VOR) to step stimuli when monkeys were tested in the onside position (vertical interaural axis) but it was not present when they were tested in the upright position (horizontal interaural axis).[27] The time constant of upward slow-phase velocity was much greater than that of downward slow-phase velocity when animals were tested in the onside position. Vertical optokinetic after-nystagmus (OKAN) (another measure of velocity storage) showed the same asymmetry when the animals were tested in the onside position. These investigators concluded that otolith stimulation in the onside position resulted in asymmetric central velocity storage (up greater than down) that enhanced upward compensatory eye movements induced by vestibular and optokinetic stimulation. In the cat and monkey, upright pitch elicits a vertical VOR which is more symmetric and has a more compensatory gain than VOR induced by onside pitch.[28,29] Presumably, the vertical VOR requires both otolith and canal signals for normal functioning. This could have important implications regarding space motion sickness in humans since otolith signals are not generated with pitch in microgravity and an inadequate vertical VOR in space could lead to visual-vestibular mismatches.[30]

Normative Data

Pitch

As in the monkey, the time constant of the human vertical VOR to pitch in the onside position is asymmetric (up slow phases > down slow phases).[31] However, the time constant is short in both vertical directions compared to that of the horizontal VOR and the amount of asymmetry in humans is much less than that found in monkey. There have been no direct comparisons of the time constant of the vertical VOR to rotational stimuli in the onside and upright position in humans. However, studies of vertical OKAN using the scleral search coil technique show the same up-down asymmetry (up > down) when subjects are tested in the onside and upright position (FIGURE 7).[32] Most subjects had an immediate reversal in OKAN slow-phase velocity after downward OKN stimuli in the sitting and onside positions. These data indicate that there is some upward velocity storage (but much less than that in the horizontal system) and minimal downward velocity storage in either position. Although upward velocity storage tended to be slightly greater in the onside position compared to the upright position, the mean OKAN upward slow-phase velocity was not significantly different in these two positions. Thus, vertical velocity storage in humans is less dependent on otolith influence than vertical velocity storage in the monkey and cat.

Also, unlike the cat, studies of low-frequency onside pitch in humans have not found a consistent gain asymmetry in the vertical VOR.[31,33] We compared the vertical vestibuloocular reflex and the vertical visual-vestibular ocular reflex induced by voluntary pitch in the upright and onside positions (FIGURE 8).[34] Subjects were trained to produce sinusoidal (0.4–1.6 Hz) pitch head movements guided by frequency-modulated sound signals. Eye and head movements were recorded with a magnetic scleral search coil. Vertical VOR gain in any position was often less than 1 for individual subjects, whereas the visual-vestibular ocular reflex gain with an earth-fixed visual target was always near 1. Asymmetries in the gain of upward and downward VOR occurred in individual subjects, but in the group of normal subjects there was no significant difference between the gain of up and down eye movements induced by vestibular or visual-vestibular stimulation in any position. From these data, one can conclude that during voluntary pitch otolith signals are not critical for normal functioning of the vertical VOR in humans. Presumably, sensory signals and efferent "motor command" signals are available to correct any deficiencies in the vertical VOR.

Roll

Studies of normative data on ocular counterrolling are limited, primarily because of the difficulty in accurately measuring torsional eye movements. There is general agreement that the counterrolling response to static tilt or low-velocity roll in normal human subjects is small.[35,36] For example, the maximum ocular counterroll for a lateral tilt of 20 degrees is about 4 degrees and for a tilt of 90 degrees about 9 degrees (gains of 0.2 and 0.1 respectively). More recently, dynamic ocular counterrolling, i.e., torsional responses to roll stimuli that probably activate the semicircular canals, has

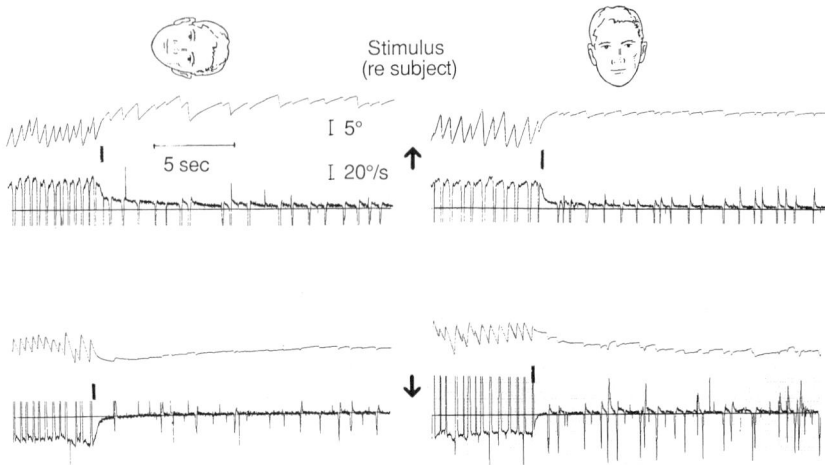

FIGURE 7. Eye position (top) and eye velocity (bottom) curves of vertical optokinetic nystagmus and vertical optokinetic after-nystagmus elicited with 45 deg/second optokinetic stimulation in head onside (left) and upright position (right) in a normal human subject. Eye movements recorded with a scleral search coil. Vertical bars indicate "lights off."

been measured using scleral search coil and video eye movement recording techniques.[36–39] These studies show that counterrolling is regularly interrupted by saccades in the opposite direction leaving only a comparatively small residual torsion which is maintained during static tilt (the 0.1 to 0.2 gain mentioned above). The gain of dynamic ocular counterrolling is in the range of 0.3 to 0.8, increasing gradually with frequency and only modestly enhanced by visual-vestibular interaction. Morrow compared the gain of dynamic ocular counterrolling in the sitting and supine positions and found that the gain at multiple frequencies was significantly greater in the sitting versus supine position (FIGURE 9).[39] Vieville and Masse found a similar difference between the gain of dynamic counterrolling in a single subject tested while performing active roll in the sitting and head-tilted-forward position.[37] Thus, comparing ocular counterrolling in the sitting versus supine positions might be a useful test of otolith-ocular function. Surprisingly, the difference in dynamic counterrolling gain from the supine to sitting position is greater than the average gain typically seen with static tilt measurements. As with pitch rotations, the gain associated with active roll was consistently higher than that seen with passive roll suggesting that neck proprioceptive and efferent copy signals can enhance the VOR.

Patient Data

Of the different stimuli mentioned above, only static tilt and low-velocity roll stimuli have been used in patients. Normal subjects typically exhibit symmetrical conjugate ocular counterrolling responses with a maximum of from 6–9 degrees of counterrolling for a 60–90-degree tilt (FIGURE 10).[35,40] Patients with unilateral peripheral vestibular lesions show variable responses, some exhibiting reduced ocular counterrolling with tilt towards the lesion side,[41] while others show reduced responses with tilt toward the normal side.[40] Some patients with central lesions exhibit roll rather than counterroll responses.[42] Clearly, more studies are needed in patients with well-defined localized lesions (e.g., VIII nerve section).

LINEAR ACCELERATION

Basic Principles

Each macule is sensitive to the resultant linear acceleration acting on it, and gravity (g) is usually the major component of this resultant acceleration. Purely vertical linear acceleration changes only the magnitude of the experienced gravity, but horizontal acceleration changes both the magnitude and direction. For a strictly

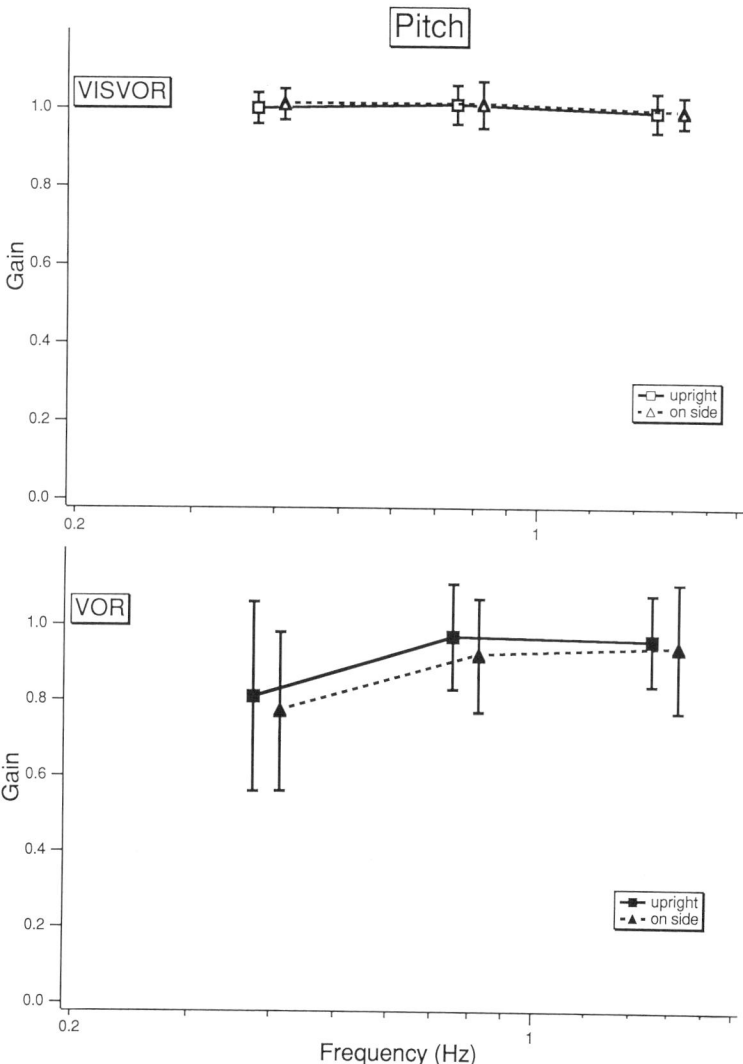

FIGURE 8. Mean gain (± 1 standard deviation) of the vertical VOR to voluntary pitch at multiple frequencies in the upright and onside position in 10 normal human subjects.

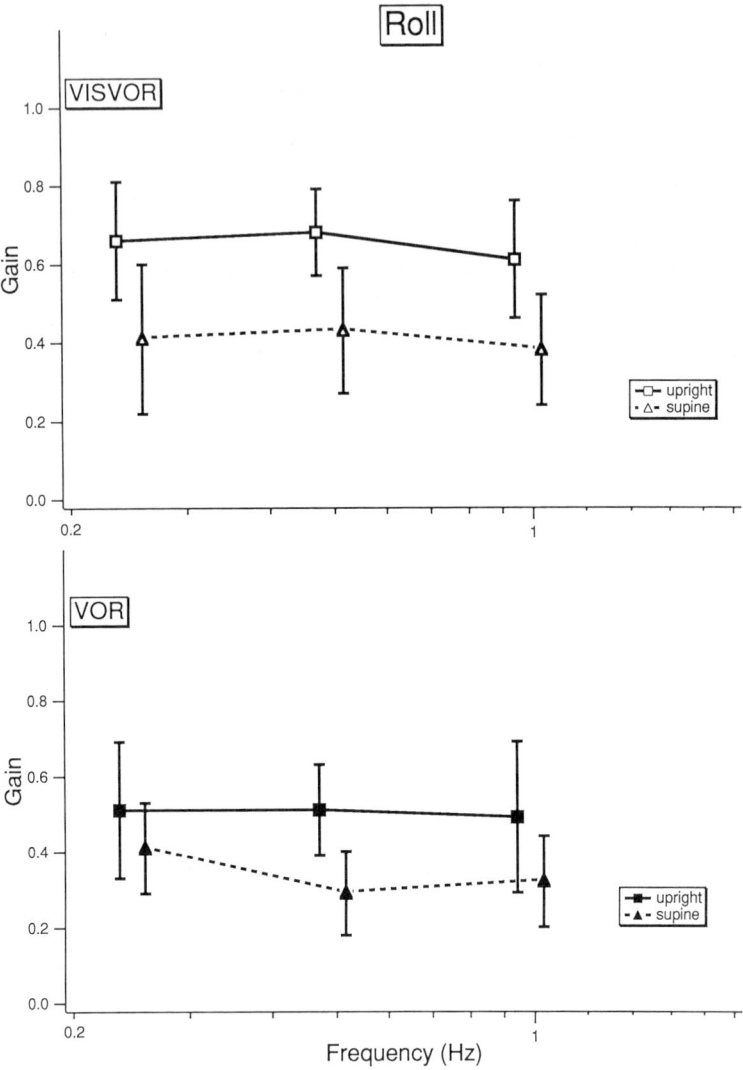

FIGURE 9. Mean gain (±1 standard deviation) of the torsional VOR to voluntary roll at multiple frequencies in the upright and supine position in five normal human subjects. (Courtesy, M. Morrow, UCLA School of Medicine.)

horizontal acceleration, h, the angle θ between the apparent and true directions of gravity is defined by $\theta = \arctan h/g$. If eye movements are compensatory, the angle of eye deviation should be equal to θ. This simple model predicts that torsional eye movements should be induced by horizontal linear acceleration along the interaural axis and vertical eye movements by horizontal linear acceleration along the occipito-nasal axis. However, monkeys[43,44] and humans[45,46] have predominantly horizontal eye movements induced by horizontal linear acceleration along the interaural axis and minimal or no vertical eye movements induced by horizontal linear acceleration along the occipitonasal axis.

Apparently, the otolith-ocular reflex in primates responds to horizontal and vertical linear accelerations with horizontal and vertical eye movements that are predominantly related to the horizontal and vertical displacement and only minimally related to the angle of a resultant vector combining linear accelerations (as predicted by the above equation). The brain must be able to distinguish between gravity and other linear acceleration components of the otolith signal. This makes sense from a functional point of view inasmuch as the logical function of the reflex is to augment visual pursuit during linear displacement of the head (analogous to the role of canal-ocular reflexes during angular displacement of the head). In primates, lateral head movements require horizontal, not torsional, eye movements to maintain fixation on an earth-fixed target. Similarly, with fore-aft movements, vertical eye movements would impair rather than improve fixation on an earth-fixed target in the axis of movement.

One technique used to study the otolith-ocular reflex in the research laboratory is a roller-coaster-like sled that runs along a linear track. For the relatively simple case in which the subject is placed on the sled facing the side, as if looking out the side window of an automobile moving forward, a predominant horizontal eye movement (including nystagmus) can be recorded with standard EOG or infrared reflective

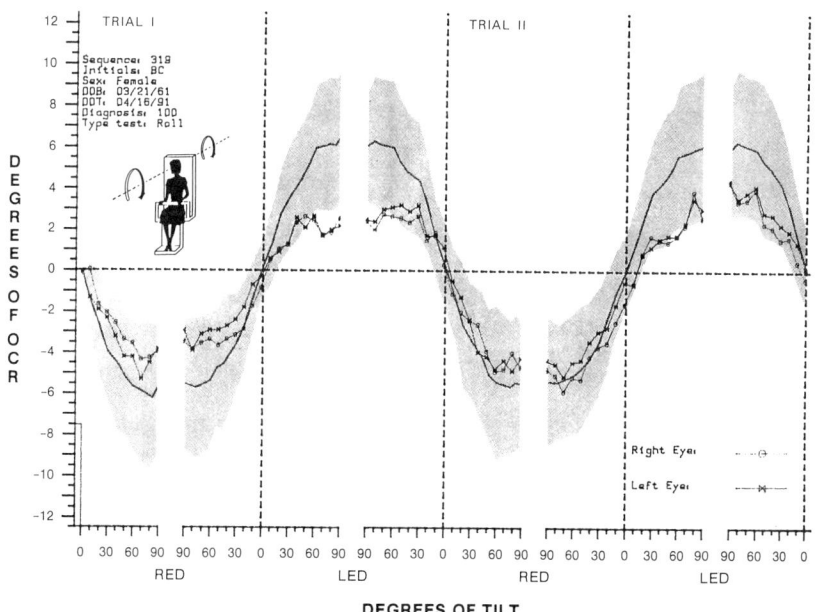

FIGURE 10. Normal ocular counterrolling response to two consecutive trials. The subject was rotated at 3 deg/second to 90 deg right-ear down, held for 30 seconds, rolled in the opposite direction to 90 deg left-ear down, held for 30 seconds for each trial. Each eye was measured independently and is shown separately. Shaded areas—normal range. (Courtesy, S. Diamond and C. H. Markham, UCLA School of Medicine.)

techniques. For other head orientations, vertical or torsional eye movements are induced requiring other eye movement recording techniques such as a magnetic scleral search coil or a video system. The use of linear sleds is severely limited for clinical testing, however, due to the expense and size of the equipment required.

The parallel swing is a simple method for inducing linear acceleration that may be practical in the clinical laboratory. It consists of a platform suspended from the ceiling by supporting cables at each of its four corners. For small-amplitude displacements, almost pure horizontal linear acceleration is experienced by the subject seated on the platform. As with the linear sled, the eye movement response on a parallel swing depends on the orientation of the subject's head relative to the linear acceleration of the swing and gravity. Combinations of horizontal, vertical, and torsional eye movements may be induced. So far, parallel swings have been used only at their natural frequency (usually in the range of 0.3 Hz), providing limited information about the dynamic range of the otolith-ocular reflex. Spring- or motor-driven devices would allow a more extended evaluation of the reflex.

Normative Data

Niven et al. used a linear track to produce periodic linear accelerations at different frequencies and in different head orientations.[45] Linear acceleration along the interaural axis induced compensatory horizontal eye movements (including nystagmus), but acceleration in the z (lying) or y (sitting) axis did not induce vertical eye movements. Surprisingly, the horizontal eye movements induced by linear acceleration along the interaural axis were about the same whether the subjects were lying or sitting. The magnitude and phase of the horizontal nystagmic eye movements induced by linear acceleration (so-called L-nystagmus) were different from those associated with periodic angular acceleration of the canals in a comparable frequency range, so the authors considered it unlikely that these eye movements resulted from unanticipated stimulation of the horizontal canals. Eye movements were recorded with electrooculography so that vertical eye movements could not be precisely recorded and torsional eye movements were not recorded at all. Buizza et al.[47] and Skipper and Barnes[48] confirmed the findings of Niven et al. by producing horizontal L-nystagmus in seated normal subjects during horizontal acceleration along the interaural axis in the frequency range of 0.2–0.8 Hz (also, on a linear track).

Paige studied the vertical linear vestibuloocular reflex in the frequency range of natural head movements by seating subjects on a spring-suspended stool and instructing them to oscillate up and down using their feet as drivers while holding their head stable and parallel to the ground.[49] The mean frequency obtained was 2.7 Hz with a mean peak excursion of 3.2 cm (corresponding to 0.94 g). Vertical eye movements were recorded uniocularly using the scleral search coil technique, and linear vertical head movements were recorded using a low-inertia potentiometer connected by a rod to a headband firmly secured to the head. The two normal subjects tested exhibited robust compensatory vertical eye movements whose gain was markedly influenced by the viewing distance of a visual target in the light or of the distance of an imagined target in the dark (FIGURE 11). Studies using optical manipulations, including spherical lenses to modify accommodation and accommodative convergence, and prisms to modify fusional vergence without altering accommodation, indicated that the amount of vergence was the most important variable in determining the vertical linear VOR gain. Furthermore, trials in darkness and with head-fixed targets indicated that, while visual following and perhaps "mental set" influenced the results, the major portion of the vertical linear VOR was driven by otolith inputs.

We studied horizontal and vertical eye movements induced in normal human subjects by sinusoidal linear acceleration on a parallel swing.[46] The swing frequency was 0.3 Hz, and the peak horizontal and vertical acceleration range from 0.17 to 0.48 and 0.03 to 0.34 × g, respectively. Eye movements were recorded with the scleral search coil technique. With subjects seated in the dark, swing displacement along the interaural axis induced robust compensatory horizontal eye movements with both slow and fast components (FIGURE 12A). The magnitude and form of these compensatory eye movements varied with the mental set of the subject. If the subject was instructed to perform mental arithmetic, nystagmus was induced (FIGURE 12A, top traces), whereas if the subject was instructed to imagine an earth-fixed target, saccades occurred in the same direction as the slow eye movements as the subjects made a voluntary effort to track the imagined target (Fig. 12A, lower traces). As

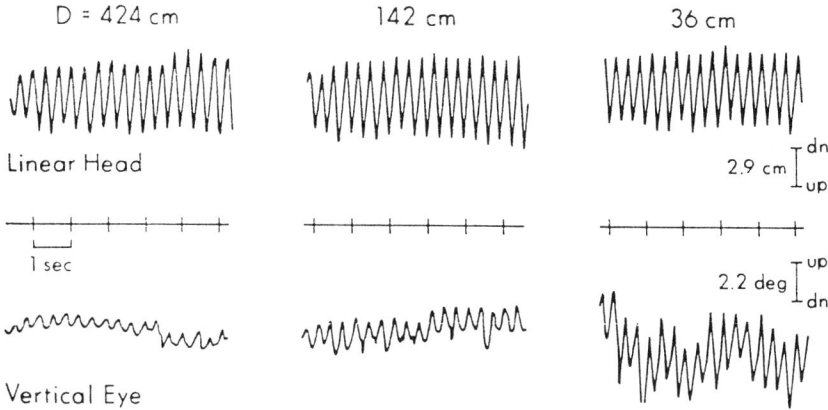

FIGURE 11. Vertical eye movements recorded with a scleral search coil (bottom traces) induced by vertical linear head oscillation (top traces) while a normal human subject views targets at different distances (D). Sensitivity to translation (ST) (peak eye velocity/peak stool velocity) increased by a factor of over 8 in this subject as D decreased from 424 cm to 36 cm. (From Reference 49, with permission.)

Paige observed,[49] we found that the distance of a real or imagined visual target markedly affected the eye movement sensitivity to translation (ST) (peak eye velocity/peak swing velocity). Halving the distance of an imagined target approximately doubled the ST. We concluded that the otolith-ocular reflex interacts with visual pursuit to improve ocular stability during translational head movements.

In TABLE 2, we compare normative data on the linear VOR obtained under different experimental circumstances in the studies mentioned above. Variability in ST measurements can largely be attributed to the frequency of stimulation and to the mental set of the normal subjects. If subjects are given no specific instructions or if they are asked to perform mental arithmetic, mean ST values are low and variable. This variability in ST is presumably explained by variable default values for the internal estimate of target distance. If the default value is large, ST measured in the dark is low, whereas if the default value is small, the ST is high. This setting of a default value, no doubt, depends on prior experience and on the combined sensory information available at the time of testing. However, when subjects are asked to imagine a specific target distance, ST values are less variable and much higher (as much as an order of magnitude higher than when subjects are not given specific instructions).

Patient Data

Jongkees and Philipszoon first recorded compensatory eye movements with electrooculography as patients received linear acceleration in the head-to-foot axis (vertical eye movements) or interaural axis (horizontal eye movements) while lying on a parallel swing.[50] No quantitative measurements were reported, but the authors suggested that the eye movements were easily induced in normal human subjects and were absent in patients without labyrinthine function. We have measured the linear VOR on a parallel swing in a small number of patients with absent or markedly decreased horizontal canal ocular reflex function (absent response to caloric stimulation bilaterally including iced water). Most subjects also exhibit a diminished or absent LVOR (e.g., FIGURE 12B, upper traces). When such subjects attempt to imagine an earth-fixed target, they generate inaccurate and poorly timed saccadic eye movements with minimal or no slow eye movements (FIGURE 12B, lower traces). They probably use somatosensory clues to generate these inaccurate saccades. Such patient data clearly indicate that an otolith signal is required to generate the internal reference signal used to augment the slow phase of the otolith-ocular reflex with an

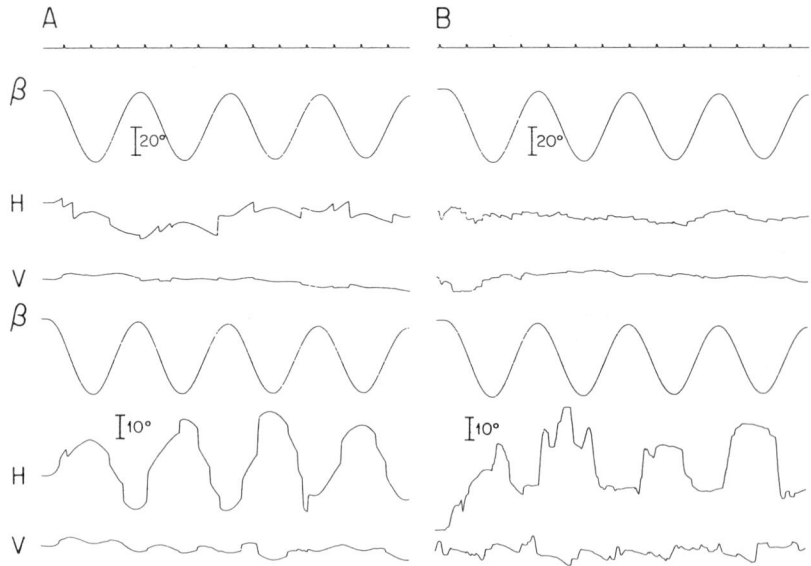

FIGURE 12. Horizontal (H) and vertical (V) eye movements induced on a parallel swing with the subject facing the side in the dark (horizontal linear acceleration along the interaural axis) while responding to questions (upper 3 traces) and while imaging a target at 2.3 meters (lower 3 traces). **A:** Normal subject. **B:** Patient with bilateral Meniere's syndrome and absent response to caloric stimulation. Tick marks—1 second. (From reference 46, with permission.)

TABLE 2. Normative Data for Linear Vestibuloocular Reflex in Human Subjects

Frequency (Hz)	Peak Linear Acceleration (g)	Sensitivity to Translation (ST) (deg/m)	Mental Set[a]	References
0.2	0.58	2.1 ± 0.7	NI	Niven et al.,[45] 1965
0.4	0.58	2.2 ± 0.6	NI	
0.8	0.58	2.1 ± 0.7	NI	
0.2	0.10	3.2 ± 1.5	NI	Buizza et al.,[47] 1980
0.2	0.15	2.8[b]	MA	Skipper & Barnes,[48] 1989
0.4	0.15	4.6	MA	
0.8	0.15	10.5	MA	
0.3	0.44	4.8 ± 2.8	MA	Baloh et al.,[46] 1988
0.3	0.44	8.8 ± 2.9	2.3 m[d]	
0.3	0.44	16.7 ± 7.4	1.1 m	
2.7	0.94	11[c]	4.2 m	Paige,[49] 1989
2.7	0.94	17	1.4 m	
2.7	0.94	31	0.4 m	

[a] NI, no specific instructions; MA, mental arithmetic.
[b] Standard deviations not given.
[c] Mean of two subjects.
[d] Distance of imagined target.

imagined target. So far, we have identified one patient with absent caloric responses and minimal response to horizontal angular acceleration who exhibited high normal otolith-ocular responses on the parallel swing. Presumably, the lesion in this patient selectively damaged the canal ocular system, leaving the otolith ocular pathways intact. Studies are currently under way to systematically evaluate patients with well-defined bilateral and unilateral peripheral vestibular lesions.

SUMMARY

Assessment of the otolith-ocular reflex of human subjects involves linear acceleration and/or changes in the orientation of the head with respect to gravity. Several such stimuli are currently under investigation regarding their applicability to the evaluation of patients with dizziness and balance disorders. Discussed in this paper are off-vertical axis rotation, eccentric rotation, pitch and roll rotation, and linear acceleration. For each of these stimuli, basic principles, normative human data, and patient data are described. Although none of these methods are currently established for clinical use, each of them, especially off-vertical axis rotation and linear acceleration, have the potential for developing into a clinically useful method for assessing otolith function in man.

REFERENCES

1. STOCKWELL, C. W., G. T. TURNIPSEED & F. E. GUEDRY. 1971. Nystagmus responses during rotation about a tilted axis. Naval Aerospace Medical Research Laboratory-1129.
2. GOLDBERG, J. M. & C. FERNANDEZ. 1981. Physiological mechanisms of the nystagmus produced by rotations about an earth-horizontal axis. Ann. N.Y. Acad. Sci. 374: 40–43.
3. COHEN, B., J. SUZUKI & T. RAPHAN. 1983. Role of the otolith organs in generation of horizontal nystagmus: effects of selective labyrinthine lesions. Brain Res. 276: 159–164.

4. RAPHAN, T., W. WAESPE & B. COHEN. 1983. Labyrinthine activation during rotation about axes tilted from the vertical. Adv. Oto-Rhino-Laryngol. **30:** 50–53.
5. BENSON, A. J. & M. A. BODIN. 1966. Interaction of linear and angular accelerations on vestibular receptors in man. Aerosp. Med. **37:** 144–154.
6. GUEDRY, F. E. 1965. Orientation of the rotation-axis relative to gravity: its influence on nystagmus and the sensation of rotation. Acta Otolaryngol. Stockh. **60:** 30–49.
7. WALL, C., III & J. M. R. FURMAN. 1989. Nystagmus responses in a group of normal humans during earth horizontal axis rotation. Acta Otolaryngol. Stockh. **108:** 327–335.
8. FURMAN, J. M. R., C. WALL III & D. B. K. KAMERER. 1989. Earth horizontal axis rotational responses in patients with unilateral peripheral vestibular deficits. Ann. Otol., Rhinol. Laryngol. **98**(7): 551–555.
9. DARLOT, C., P. DENISE, J. DROULEZ, B. COHEN & A. BERTHOZ. 1988. Eye movements induced by off-vertical axis rotation (OVAR) at small angles of tilt. Exp. Brain Res **73:** 91–105.
10. KAMERER, D. B. K. & J. M. R. FURMAN. 1990. Off-vertical axis rotational responses in patients with unilateral peripheral vestibular lesions. Bárány Society Meeting, Tokyo, Japan.
11. FURMAN, J. M. R., C. WALL III & D. PANG. 1990. Vestibular function in periodic alternating nystagmus. Brain **113:** 1425–1439.
12. GIAMPOLO, A. J., M. EL-AWAR & J. M. R. FURMAN. 1990. Vestibular function in olivopontocerebellar atrophy. American Academy of Neurology. Miami Beach, Fla.
13. DENISE, P. 1986. Rotation d'axe incline par rapport a la gravite. Effets des petits angles et interet clinique. Doctoral Thesis. Universite Pierre et Marie Curie, Paris VI Faculte de Medecine Pitie, Salpetriere, Dacylosorbonne, 8 rue, Casimir-Delavigne, France.
14. ZEE, D. S., A. R. FRIENDLICH & D. A. ROBINSON. 1974. The mechanism of downbeat nystagmus. Arch. Neurol. Chicago **30:** 227–237.
15. BALOH, R. W., R. D. YEE, J. KIMM & V. HONRUBIA. 1981. Vestibular-ocular reflex in patients with lesions involving the vestibulocerebellum. Exp. Neurol. **72:** 141–152.
16. SCHRADER, V., E. KOENIG & J. DICHGANS. 1985. Direction and angle of active head tilts influencing the purkinje effect and the inhibition of postrotatory nystagmus I and II. Acta Otolaryngol. Stockh. **100:** 337–343.
17. IGARASHI, M., M. TAKAHASHI, T. KUBO, B. R. ALFORD & W. K. WRIGHT. 1980. Effect of off-vertical tilt and macular ablation on postrotatory nystagmus in the squirrel monkey. Acta Otolaryngol. **90:** 93–99.
18. HAIN, T. C., D. S. ZEE & B. MARIA. 1988. Modification of the dynamics of the vestibulo-ocular reflex by head tilt in patients with cerebellar lesions. Acta Otolaryngol. **105:** 13–20.
19. TAKEDA, N., M. IGARASHI, I. KOIZUKA, S. Y. CHAE & T. MATSUNAGA. 1990. Recovery of the otolith-ocular reflex after unilateral deafferentation of the otolith organs in squirrel monkeys. Acta Otolaryngol. Stockh. **110:** 25–30.
20. TAKEDA, N., M. IGARASHI, I. KOIZUKA, S. Y. CHAE & T. MATSUNAGA. 1991. Effects of otolith stimulation in eccentric rotation on the vestibulo-ocular reflex in squirrel monkeys. Acta Otolaryngol. Stockh. Suppl. **481:** 27–30.
21. BENSON, A. J. & TH. VIEVILLE. 1986. European vestibular experiments on the Spacelab-1 mission. VI. Yaw axis VOR. Exp. Brain Res. **64:** 279–283.
22. GRESTY, M. A. & A. M. BRONSTEIN. 1986. Otolith stimulation evokes compensatory reflex eye movements of high velocity when linear motion of the head is combined with concurrent angular motion. Neurosci. Lett. **65:** 149–154.
23. GRESTY, M. A., A. M. BRONSTEIN & H. BARRATT. 1987. Eye movement responses to combined linear and angular head movement. Exp. Brain Res **65:** 377–384.
24. KOIZUKA, I., N. TAKEDA, H. OGINO, T. KUBO & T. MATSUNAGA. 1988. Centric and eccentric pendular rotation test by use of MVM-C2. Equilibrium Res. Suppl. 4.
25. BARRATT, H., A. M. BRONSTEIN & M. A. GRESTY. 1987. Testing the vestibular-ocular reflexes: abnormalities of the otolith contribution in patients with neuro-otological disease. J. of Neurol. Neurosurg. Psychiatry **50:** 1029–1035.
26. KOIZUKA, N., N. TAKEDA, S. SATO, M. SAKAGAMI & T. MATSUNAGA. 1991. Centric and eccentric VOR tests in the patient's with Meniere's disease and vestibular Meniere's disease. Acta Otolaryngol. Stockh. Suppl.

27. MATSUO, V. & B. COHEN. 1984. Vertical optokinetic nystagmus and vestibular nystagmus in the monkey: up-down asymmetry and effects of gravity. Exp. Brain Res. **53:** 197–216.

28. TOMKO, D. L., C. WALL III, F. R. ROBINSON & J. P. STAAB. 1988. Influence of gravity on cat vertical vestibulo-ocular reflex. **69:** 307–314.

29. PAIGE, G. D. & D. L. TOMKO. 1987. Canal-otolith interactions in the vestibulo-ocular reflex (VOR). ARVO Abstr.: 332.

30. LACKNER, J. R. & A. GRAYBIEL. 1981. Variations in gravito-inertial force level affect the gain of the vestibulo-ocular reflex: implications for the etiology of space motion sickness. Aviat. Space Environ. Med. **52:** 154–158.

31. BALOH, R. W., L. RICHMAN & V. HONRUBIA. 1983. The dynamics of vertical eye movements in normal human subjects. Aviat. Space Environ. Med. **54:** 32–38.

32. BOHMER, A. & R. W. BALOH. 1991. Vertical optokinetic nystagmus and optokinetic afternystagmus in humans. J. Vestibular Res. **1:** 309–315.

33. BENSON, A. J. & F. E. GUEDRY. 1965. Comparison of tracking-task performance and nystagmus during sinusoidal oscillation in yaw and pitch. Aerospace Med. **42:** 593–601.

34. BALOH, R. W. & J. DEMER. 1191. Gravity and the vertical vestibulo-ocular reflex. Exp. Brain Res. **83:** 427–433.

35. DIAMOND, S. G., C. H. MARKHAM, N. E. SIMPSON & I. S. CURTHOYS. 1979. Binocular counterrolling in humans during dynamic rotation. Acta Otolaryngol. **87:** 490–498.

36. COLLEWIJN, H., J. VAN DER STEEN, L. FERMAN & T. C. JANSEN. 1985. Human ocular counterroll: assessment of static and dynamic properties from electromagnetic scleral coil recordings. Exp. Brain Res. **59:** 185–196.

37. VIEVILLE, T. & D. MASSE. 1987. Ocular counter-rolling during active head tilting in humans. Acta Otolaryngol. **103:** 280–290.

38. LEIGH, R. J., E. F. MAAS, G. E. GROSSMAN & D. A. ROBINSON. 1989. Visual cancellation of the torsional vestibulo-ocular reflex in humans. Exp. Brain Res. **75:** 221–226.

39. MORROW, M. J. 1989. Effects of head and body position on torsional optokinetic and vestibular eye movements in humans. Soc. Neurosci. Abstr.: 514.

40. DIAMOND, S. G. & C. H. MARKHAM. 1981. Binocular counterrolling in humans with unilateral labyrinthectomy and in normal controls. Ann. N.Y. Acad. Sci. **374:** 69–79.

41. NELSON, J. R. & W. F. HOUSE. 1971. Ocular countertorsion as an indicator of otolith function: effects of unilateral vestibular lesions. Trans. Am. Acad. Ophthalmol. Otolaryngol. **75:** 1313–1315.

42. DIAMOND, S. G., C. H. MARKHAM & R. W. BALOH. 1988. Ocular counterrolling abnormalities in spasmodic torticollis. Arch. Neurol. **45:** 164–169.

43. PAIGE, J. G. & D. D. L. TOMKO. 1991. Eye movement responses to linear head motion in the squirrel monkey. I. Basic characteristics. J. Neurophysiol. **65:** 1170–1182.

44. PAIGE, G. D. & D. L. TOMKO. 1991. Eye movement responses to linear head motion in the squirrel monkey. II. Visual-vestibular interactions and kinematic considerations. J. Neurophysiol. **65:** 1183–1196.

45. NIVEN, J. I., W. C. HIXSON & M. J. CORREIA. 1965. Elicitation of horizontal nystagmus by periodic linear acceleration. Acta Oto-Laryngol. **62:** 429–262.

46. BALOH, R. W., K. BEYKIRCH, V. HONRUBIA & R. D. YEE. 1988. Eye movements induced by linear acceleration on a parallel swing. J. Neurophysiol. **60:** 2000–2013.

47. BUIZZA, A., A. LEGER, J. DROULEZ, A. BERTHOZ & R. SCHMID. 1980. Influence of otolithic stimulation by horizontal linear acceleration on optokinetic nystagmus and visual motion perception. Exp. Brain Res. **39:** 165–176.

48. SKIPPER, J. J. & G. R. BARNES. 1989. Eye movements induced by linear acceleration are modified by visualization of imaginary targets. Acta Otolaryngol. Suppl. **468:** 289–293.

49. PAIGE, G. D. 1989. The influence of target distance on eye movement responses during vertical linear motion. Exp. Brain Res. **77:** 585–593.

50. JONGKEES, L. B. W. & A. J. PHILIPSZOON. 1962. Nystagmus provoked by linear acceleration. Acta Physiol. Pharmacol. Neerl. **10:** 239–247.

51. BALOH, R. W. & J. M. R. FURMAN. 1989. Modern vestibular function testing. West. J. Med. **150:** 59–67.

Central Organization and Modeling of Eye-Head Coordination during Orienting Gaze Shifts[a]

H. L. GALIANA[b] AND D. GUITTON[c]

[b]Department of Biomedical Engineering
[c]Department of Neurology and Neurosurgery
and Montreal Neurological Institute
McGill University
3801 University Street
Montreal, Quebec, Canada H3A 2B4

An important function of motor control is to coordinate the trajectories of many simultaneously moving and coupled body segments. Perhaps the best understood of such neural systems is the one, reviewed here, that coordinates the eyes and head during a rapid saccadelike orienting movement of the visual axis, referred to as gaze. In this text the main characteristics of normal gaze shifts in alert subjects will be reviewed, and a model control scheme will be proposed which can duplicate the experimental data with several important implications.

EYE, HEAD, AND GAZE TRAJECTORIES

FIGURE 1A–D shows examples of rapid gaze shifts produced respectively by the human,[1] monkey,[2] and cat.[3,4] The shapes of the eye (E), head (H), and gaze (G) trajectories are qualitatively similar in the three species. The initial rapid rise in the gaze saccade is caused by a saccadic eye movement, the initial acceleration of the head being comparatively much slower than that of the eye. At the end of the gaze shift the visual axis is immobile in space (except for the 120° human gaze shift which has a small drift) in spite of continuing head motion. The vestibuloocular reflex (VOR) is responsible for this gaze stability. The semicircular canals sense head rotation and send signals to brain stem oculomotor centers that drive the eyes in the orbit in a trajectory equal and opposite to that of the head (the "compensatory phase").

To explain the operation of this gaze-control system, Bizzi and collaborators proposed that (1) the same saccadic eye movement is programmed to acquire a target irrespective of whether or not the head moves; and (2) a head movement occurring during a saccade attenuates, by the action of the VOR, the saccade amplitude by an amount equal to the head displacement during the saccade (to be called VOR-saccade interaction).[5–7] In this scheme, the VOR acts throughout the eye saccade: head-in-space motion is added to that of the eye-in-head (because the eye is being carried by the head) and subtracted out by the VOR. This view of eye-head coordination is "oculocentric" in that the gaze trajectory produced when the head is restrained (a saccadic eye movement) is not altered when the head is unrestrained.

[a]This research was supported by the Medical Research Council of Canada and Le Fonds de la Recherche en Santé du Québec.

452

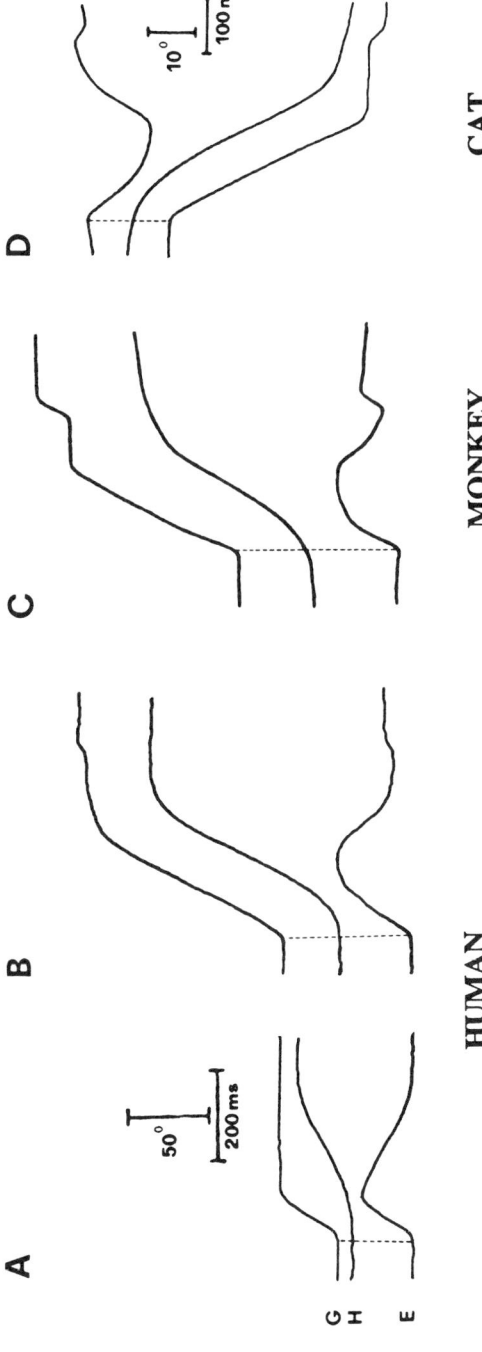

FIGURE 1. Examples, in different species, of horizontal saccadic gaze shifts composed of coordinated eye and head movements. Gaze (G) = eye-in-space = eye-in-head (E) + head-in-space (H). **A** and **B:** Human gaze shifts of amplitudes 35° and 120°, respectively. **C:** Monkey, 100°. A small "corrective" gaze shift follows. **D:** Cat, 40°. Vertical dashed line indicates onset of gaze shift. In each example, there are two distinct directions of eye motion: the initial saccadic eye movement that is responsible for the rapid onset of the gaze shift, and the final "compensatory" phase wherein the eye rotates in a direction equal and opposite to head rotation thereby assuring gaze stability. In large gaze shifts the two modes of eye motion are separated by a third, the "plateau" phase, wherein the eye remains about motionless (e.g., panel C) in the orbit while gaze is being carried by the head.

According to the oculocentric view, if head motion is suddenly stopped during an eye saccade, the attenuating influence of the VOR on the saccade pulse generating motor circuitry is removed and the eye should suddenly accelerate after a VOR latency of 10–20 mseconds. Conversely, if the head is suddenly speeded up during the saccade, the eye should decelerate with the same latency. In either case the *gaze* trajectory should not be affected. (In these conditions, neck proprioceptive influences are negligible).[6,8]

Experiments in which head motion was mechanically perturbed, during an eye saccade, have been performed in humans,[1,9] monkeys,[10,11] and cats.[3,12,13] The following points summarize what is generally agreed upon for human and monkey data: (1) The oculocentric strategy—involving VOR-saccade interaction and the coding of an eye saccade of amplitude equal to target offset angle—is strictly valid only for gaze shifts of about < 10°. (2) For large gaze shifts > 50°, the strategy is different and the VOR is essentially off during an eye saccade. (Except in the occasional human subject.)[1] The VOR gain (change in eye velocity/change in head velocity) was estimated in humans[14,15] and monkeys[11] to decrease about linearly between gaze amplitudes of 10° and 50°, from a high value (0.5–0.8) to a low value (0.1–0.2) respectively. (3) In large gaze shifts in which VOR-saccade interaction is initially absent, there is clearly a time, within the gaze shift, at which the VOR signal again becomes functional and stabilizes the visual axis in space. This reactivation of the VOR occurs when gaze error approaches or equals zero, i.e., when gaze is on or near the target.

In cats, similar experiments produced variable results, suggesting that VOR and saccades either partially or fully interact at all gaze shift amplitudes.[3,12,13] This differs from monkey but resembles some human data.[1] Cat ocular saccades are small (< 15° usually) and slow, suggesting in general that VOR-saccade interaction may depend on both gaze shift amplitude and eye velocity.

Testing for VOR-saccade interaction is not the only means of determining whether or not the oculocentric view is applicable; the size of the programmed eye saccade is also relevant. When cat and human head motion is prevented just *before* gaze shifts of different amplitudes, the maximum saccade amplitude is limited neurally, not mechanically, to about 15° and 45° respectively,[1,3] well short of the oculomotor range. This neural limit accounts for the "plateau" phase of eye motion, seen in large gaze shifts (e.g., FIGURE 1C), wherein the eye remains relatively immobile while gaze is being displaced by the head. For target offsets beyond this plateau value, an eye saccade is insufficient to reach the target and the oculocentric view is incorrect even if VOR and eye saccades do interact.

THE GAZE FEEDBACK HYPOTHESIS

An alternative to the oculocentric model must be provided to explain the control of large gaze shifts. In recent models of the saccadic oculomotor system in the head-fixed animal, the eye saccade is driven by a burst generator in turn responding to eye motor error [desired change in eye position–actual change in eye position obtained from a corollary discharge (FIGURE 2A)].[16,17] An extension of this model to gaze control in the head-free condition has been proposed, namely, that brain stem eye and head motor circuits are driven by a gaze-motor-error signal.[1,3,4,9–11,13–15,18]

FIGURE 2B shows a schematic model of a gaze-control system similar in structure to that shown in FIGURE 2A but now incorporating the head motor system. In this model an estimate of the actual change in gaze position is obtained by integrating the sum of eye and head velocities. The head-velocity signal is assumed to be derived

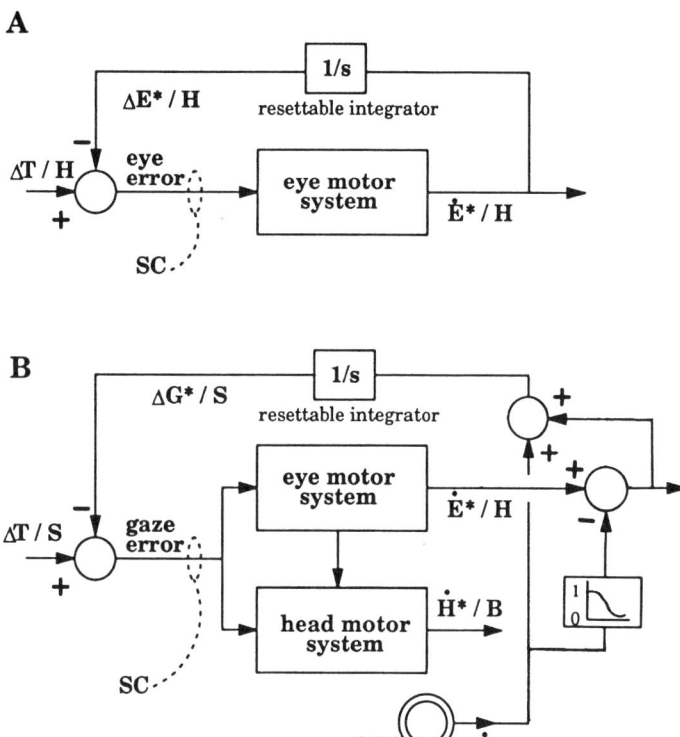

FIGURE 2. Simplified feedback circuits for controlling movements. **A:** Classic schema of saccadic eye movement control system. Initially, a subject is assumed to fixate a target and a new target appears elsewhere. The input to the system is change in target position relative to the head ($\Delta T/H$). An eye error signal drives the saccadic eye movement generator and is constructed by subtracting change in eye position relative to the head ($\Delta E^*/H$) from $\Delta T/H$. When eye error $= 0$, the eye saccade stops. $\Delta E^*/H$ is a corollary discharge obtained by integrating (Laplacian symbol, $1/s$) an eye velocity signal given, say, by brain stem burst generator circuits. After a saccade the integrator output is reset to zero. **B:** Schema of saccadic gaze control system. The input to the system is now change in target position relative to space ($\Delta T/S$). A gaze error signal drives the eye saccade and head movement generators. It is constructed by subtracting change in gaze position relative to space ($\Delta G^*/S$) from $\Delta T/S$. When gaze error $= 0$, the saccadic gaze shift stops and the compensatory phase of eye rotation (not shown here, see text) begins. $\Delta G^*/S$ is a corollary discharge obtained by integrating the sum of eye velocity relative to head ($\dot{E}^*/H$) and head velocity relative to space (d] p;$\dot{H}^*/S$). The latter signal is assumed to emanate from the semicircular canals (SCC). Before being added to $\dot{E}^*/H$, it is attenuated as a function of gaze error. Horizontal axis on gain function in box downstream of SCC is gaze error; vertical axis is gain of canal signal. Gaze and eye errors may be provided by SC.

from the semicircular canal output[19] but alternatively could be generated by a "corollary" discharge provided by brain stem circuits. The eye and head motor systems are driven until the gaze error signal is zero. At that point the saccadic gaze-generating system of FIGURE 2B is disengaged and the compensatory phase of eye rotation, driven by the VOR, commences. This gaze-stabilization system is not shown in FIGURE 2B.

In FIGURE 2B the head motor system is also driven by a command emanating from the eye motor system (a shared burst generator). The basis for this assumption is that in head-restrained humans, monkeys, and cats, the tonic level of electromyographic activity in many dorsal neck muscles is modulated in relation to the position of the eye in the orbit.[20-22] The eye-motor to head-motor link helps accelerate the head during the saccadic gaze shift.[4] Although not shown in FIGURE 2B, the eye-head coupling also assures, during the compensatory phase, that head motion continues in spite of zero gaze error.[4]

With the scheme shown in FIGURE 2B, it is clear why excellent gaze accuracy is obtained in humans, monkeys, and cats, independent of whether the movement is made in the light or dark or whether head motion is normal, perturbed, or passively generated.[1,3,9-11,13,15,23] The system attempts to null gaze error no matter what the trajectories of the eye and head may be.

A further test of the gaze feedback concept was made by Pélisson et al.[24] Cats were trained to orient in the dark to a target that had flashed but was extinguished when the gaze shift began. FIGURE 3 shows results for one cat. During control trials (FIGURE 3A), the cat produced leftward single-step gaze shifts towards the target. In randomly selected trials, the left superior colliculus (SC) was stimulated during the dark period to perturb eye and head positions by generating a rightward horizontal gaze shift of about 20° ("test" condition in FIGURE 3A). Despite this gaze perturbation and the absence of visual feedback, the leftward corrective saccadic gaze shift terminated very close to the control final position shown by the horizontal dashed line. Gaze and head planar trajectories (FIGURE 3B) nicely illustrate the accuracy of this compensation. The control gaze shift had a normal straight trajectory between the start "S" and target "T" positions. In the test trial, the stimulation-induced gaze perturbation was fully compensated by a twofold increase in response amplitude.

NEURAL CONTROL OF GAZE

Superior Colliculus and Feedback Control of Gaze Shifts

The mammalian SC is a laminated structure consisting of seven alternating fiber and cell body layers which are oriented parallel to its surface. The deeper layers are organized into a motor map that traditionally in monkey has been thought to provide the brain stem oculomotor circuitry with a topographically coded vector command that specifies, depending on which site on the map is active, the amplitude and direction of the desired eye saccade (the $\Delta T/H$ signal in FIGURE 2A; see reviews in References 25–28).

It transpires that this "oculomotor" view of the motor map is species specific.[29] For example, in the cat—whose small oculomotor range[30] requires that head movements be used to attain targets situated in a large portion of visual space—microstimulation of the motor map evokes gaze shifts composed of coordinated eye-head movements. In the barn owl, an animal with almost no ocular motility, microstimulation evokes orienting gaze shifts composed of head movements.[31]

Munoz and colleagues have investigated, in the cat whose head is unrestrained, the nature of the SC's gaze command signals by recording—from output cells of the deep layers—the tectoreticular (TRN) neurons (reviewed in Reference 37) that project to contralateral brain stem eye and head motor centers.[32-36] A given TRN in the motor map has a burst discharge that precedes *gaze* saccades of a specific amplitude and direction (its "preferred" vector) irrespective of whether the head is fixed or unrestrained. Weaker discharge is associated with gaze saccades whose directions deviate from the cell's preferred vector. A saccadic gaze movement is preceded by activity in a large ensemble of TRNs centered at whatever location, on the retinotopically coded motor map, is appropriate for the required gaze vector.

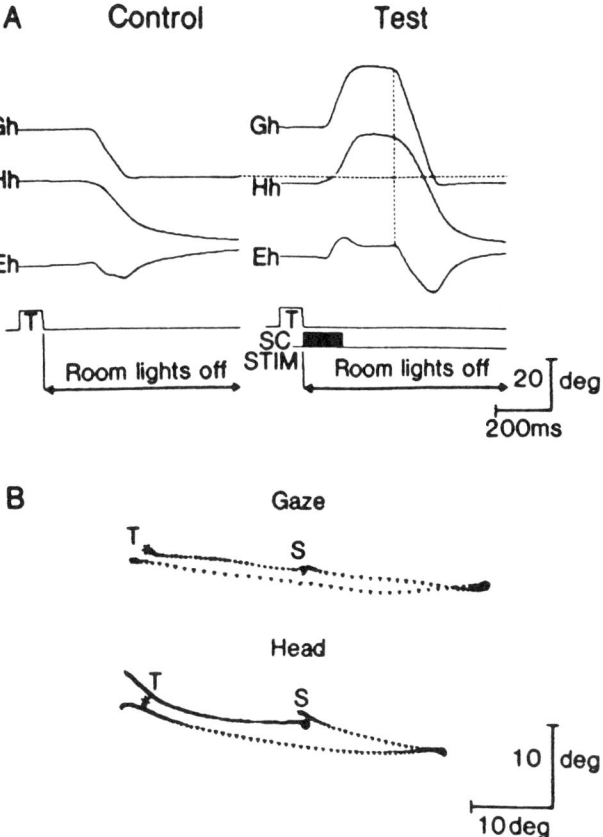

FIGURE 3. Compensation, by the cat, for perturbation of gaze position. **A:** Horizontal components of gaze (Gh), head (Hh), and eye (Eh) movements in response to a target flash (T) to the left of the start position. The room lights were turned off for 1 second at target offset. In the control condition (left), the combined eye-head movements brought gaze to the position shown by the horizontal dashed line. In the test condition (right), the eyes and head were deviated to the right, in the dark, by electrically stimulating the left SC (trace labeled SC STIM). About 100 mseconds after the deviated gaze shift ended, the cat made a corrective saccadic gaze shift, in the dark, to the correct position of the target. **B:** Gaze and head trajectories, projected onto a vertical frontal plane, of responses shown in A. S, start position; T, target position. (Adapted from Reference 41.)

The average burst frequency of the ensemble of active TRNs determines how fast gaze gets on target. An example for one TRN is shown in FIGURE 4. In FIGURE 4A the left vertical line marks the disappearance of a fixation target, telling the trained cat to orient to another location in space where it predicts a target will appear. The

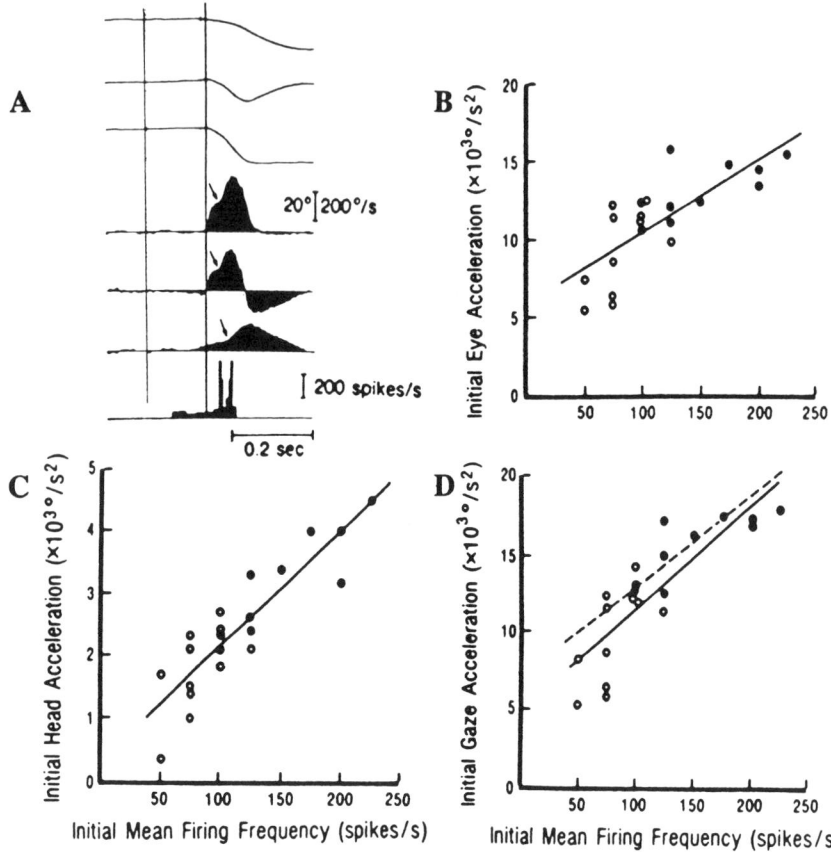

FIGURE 4. Dependence of eye, head, and gaze acceleration on TRN burst frequency. **A:** A fixation target was extinguished at the left vertical line. This signaled, to the trained cat, that a new target would reappear at a location on its left. The cat generated a leftward gaze shift that began (right vertical solid line) about synchronously with the appearance of the new target. This gaze shift was triggered by "predictive" cues. About 50 mseconds after target appearance the TRN produced a burst discharge followed, shortly after, by sharp accelerations (marked by arrows) in the eye, head, and gaze trajectories. Traces represent from top to bottom: horizontal head (Hh), eye (Eh), and gaze (Gh) position traces; horizontal gaze (Gh), eye (Eh), and head (Hh), velocity traces; cell firing frequency. **B–D:** Relation, for cell shown in A, between initial eye, head, and gaze accelerations, respectively, and initial mean discharge frequency measured in the interval 20 mseconds before to 20 mseconds after initiation of gaze saccade acceleration. **Filled circles:** Movements, head-free, to a visible target. **Open circles:** Movements, head-free, to a predicted target; **solid lines,** Linear regression fit through data points. **Dashed line:** Head-fixed (data points not shown). (Reproduced with permission, from Reference 42.)

appearance, just before the gaze shift begins, of this target in the cell's visual receptive field generates a burst discharge, followed about 10 mseconds and 30 mseconds later by an abrupt acceleration of the eye and head respectively. FIGURE 4B–D show that the initial eye, head, and therefore gaze accelerations all increase with burst frequency.[35] The dependence of eye and/or head saccade speed on the level of SC output activity in barn owl, monkey, and cat is reviewed in Reference 31.

Traditional models of the collicular control of saccadic eye movements postulate (FIGURE 2A) that the SC provides the $\Delta T/H$ command specifying only the initial vector of the intended movement. It is known that the SC drives a brain stem burst generator whose summed output, total number of spikes, is proportional to saccade amplitude. The models assume that the feedback control system is downstream of the SC around the burst generator (see Reference 28 for review). Observations that gaze shifts are accelerated by a sudden increase in TRN discharge frequency (FIGURE 4A) without affecting gaze accuracy are difficult to incorporate in such a scheme: they imply that the SC "knows" a priori just how to modulate the number of spikes it sends to the eye and head movement generators to drive a given amplitude gaze saccade irrespective of the movement's velocity profile.

Recently, new data have emerged that suggest the SC provides the instantaneous gaze motor error signal (FIGURE 2B). Waitzman *et al.* suggest that a hill of neural activity (height = frequency), appropriately located on the monkey SC's motor map, to code the desired eye saccade (FIGURE 5A), is "flattened" by feedback, the movement stopping when activity ceases.[38] Alternatively, Munoz *et al.* reported that in cat the hill of activity moves continuously (FIGURE 5A–D) on the motor map from an initial location,[36] defining the vector of the desired gaze shift, to a final "zero" position where neurons active during fixation are located.[32–36] This result clarifies how the spatial representation of activity on the motor map is transformed into the temporal code (frequency and duration of discharge) required by motor neurons. The spatial-temporal transformation is accomplished by controlling gaze throughout the spatial trajectory of activity on the motor map, implying that the SC itself spatially encodes the instantaneous gaze motor error signal shown in FIGURE 2B.

The monkey and cat data, although different specifically, both suggest the SC resides within a gaze feedback loop. FIGURE 5E is an extension of the gaze feedback circuit (FIGURE 2B) proposed by Munoz and Guitton.[33,35] A dorsal view of the TRN layer is schematically represented for the left and right SCs. Output neurons project to brain stem oculomotor and head motor circuitry. Axons from two hypothetical TRNs leave the colliculus, one emanating from the rostral fixation zone (fTRN, area centralis representation, where the two solid axes intersect), the other originating from a more caudal site whose activation is appropriate for driving the required gaze shift. A feedback signal related to the change in gaze position ($\Delta G'$) is returned to the SC and is subtracted from the original retinal error signal specifying desired change in gaze position (ΔG) to yield instantaneous gaze motor error (gme). This signal moves the neural activity along the motor map. At each location, TRNs appropriate for the current gaze motor error vector are active and drive the brain stem saccadic eye movement burst generator (oculomotor system in FIGURE 5E) and, by the way of reticulospinal (RS) projections, the head premotor circuits. (By comparison, in the Waitzman *et al.* approach,[38] the gme signal decreases activity at the caudal site.)

FIGURE 5E shows premotor elements that are well known to oculomotor physiologists:[25] long-lead burst neurons (LLBNs), excitatory burst neurons (EBNs), and omnipause neurons (OPNs). Traditionally, LLBNs and EBNs are thought to constitute the brain stem burst generator that drives saccades. OPNs gate saccades, by tonically inhibiting EBNs. In the head-fixed animal they are silent for the duration of a saccade. An important implication of the schema in FIGURE 5E is that these brain stem neurons, hitherto considered to have strictly oculomotor functions, may in fact be controlled by gaze signals. It has now been verified that OPNs, in the cat whose

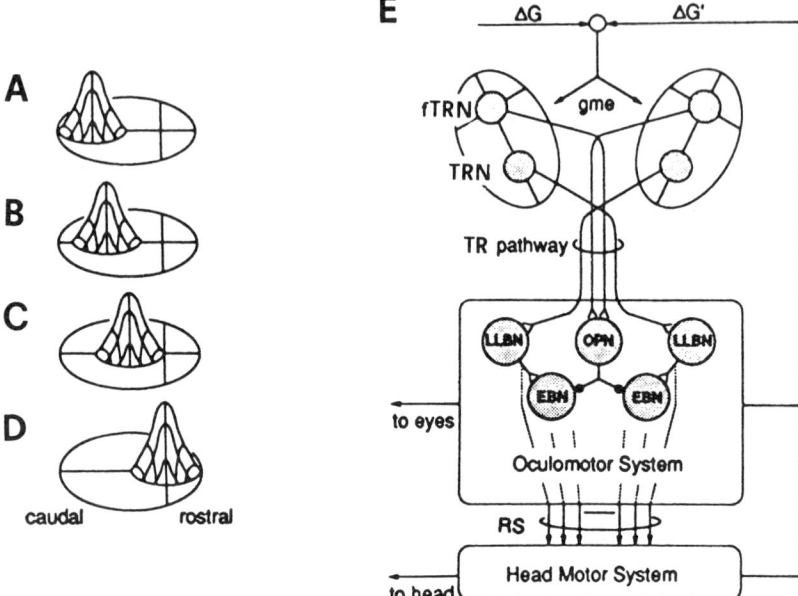

FIGURE 5. Proposed activity patterns (A–D) in the TRN layer of the SC during the execution of an orienting gaze shift.[36] Neural activity (firing frequency) is represented schematically as a hill protruding from the two-dimensional, oval surface of the SC motor map. The "zero" of the map (or fixation area) is at the intersection, in the rostral pole, of the horizontal and vertical meridians. Each section shows the TRN layer at different times during the gaze shift. The movement begins with activity at A and terminates with activity at the fixation site (D). **E:** Proposed links between SC and brain stem oculomotor circuitry. Schematized dorsal view of the left and right SC is shown at top. Two axons leave each colliculus. The cell body of one, an fTRN, is situated in the fixation area and projects to omnipause neurons (OPNs). The other, TRN, originates from a more caudal site and projects to long-lead burst neurons (LLBNs). LLBNs and OPNs then terminate monosynaptically onto excitatory burst neurons (EBNs). LLBNs and EBNs form part of the burst generator that drives saccades. Excitatory connections are drawn as open triangles, while inhibitory connections are denoted by filled circles. The reticulospinal (RS) system projects to the head motor system whose contents are left blank. The outputs of the eye and head motor systems sum to yield an internal representation of change in gaze position ($\Delta G'$). This signal is subtracted from desired change in gaze position (ΔG) to yield instantaneous gaze motor error (gme), which is topographically represented upon the SC motor map as shown in A–D.

head is unrestrained, cease discharging for the duration of a gaze shift and not, as the traditional view implies, for the duration of the saccadic eye movement component.[39]

Gaze-Related Discharges of Omnipause Neurons

The recording of a typical OPN (cell P1) is shown in FIGURE 6A. The cell responded with a complete cessation of its relatively regular tonic activity during a saccadic eye movement made by a head-restrained animal.

In the head-unrestrained condition, the cessation of OPN activity was associated with shifts of gaze, i.e., the displacements of the visual axis relative to space (FIGURE 6B, C, and D). In gaze shifts accomplished with a steplike sequence of small gaze displacements (FIGURE 6B), the durations of the eye saccades and the gaze saccades were indistinguishable, and the pause in OPN firing could not be linked preferentially to either eye or gaze saccades. However, in large single-step gaze shifts, the ocular saccade duration was often shorter than the gaze shift duration. In these cases the pause duration was closely time locked to the duration of the gaze displacement. Indeed, OPNs paused for the duration of the gaze shift even when the rapid eye saccade was already terminated and the eye was moving back in a compensatory direction (FIGURE 6C). This relationship between gaze saccade duration and OPN silent period was more impressive for large eye-head coordinated movements in which a "plateau" phase appeared in the eye trace (FIGURE 6D). In such gaze shifts, OPNs were found to resume discharging not when the eye had achieved a stationary position in the orbit but only when gaze was on target, or dynamic gaze error was null.

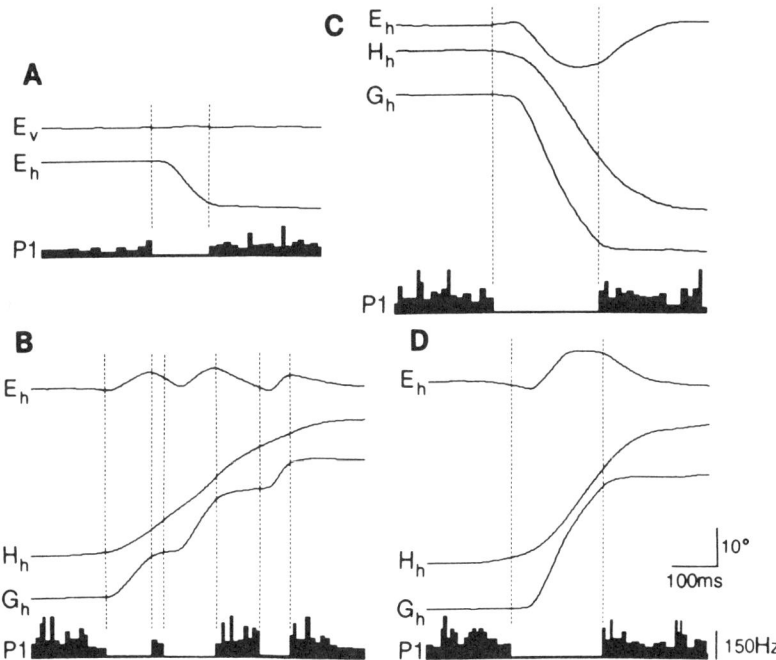

FIGURE 6. **A:** Discharge pattern of a representative OPN during a horizontal saccadic eye movement in the alert head-restrained cat. **B–D:** Examples of the activity of the same OPN during combined eye-head gaze shifts in the head-unrestrained condition. E_v: vertical eye position; E_h: horizontal eye position; H_h: horizontal head position; G_h: horizontal gaze position. Upward deflections are rightward movements. Lower traces show instantaneous firing rate of the neuron. Dotted lines indicate beginning and end of the cessation of activity.

MODEL OF GAZE CONTROL

The simple schemas shown in FIGURES 2 and 5 have been recast, recently, into a more complete network and control systems model that incorporates realistic neural connectivity and signals.[4,40] The model relies on the calculation of gaze motor error during a gaze shift, and incorporates some key innovations which allow for the first time the coordinated control of *both* eye and head platforms. In the following sections we present a revised and improved version of this model.

The Model Structure

Gaze motor error can be extracted by comparing desired target angle in space with efference copies of the sum of eye-in-head and head-in-space positions.[4] The model in FIGURE 7 calculates gaze motor error, but in a manner that corresponds more realistically to the physiology and anatomy of the system. First, the SC only provides retinotopic error at the beginning of a gaze shift, as transmitted, say, by the retina and visual cortex; hence, the SC in our model now compares *initial* gaze error to efference feedback signals. Second, several experimental observations indicate that the final eye and head contributions to a gaze shift are better correlated with the *size* of the required head or gaze shift[43] (Guitton and Volle, personal communication) rather than with the gaze position in space; this implies relative movement feedback similar to that hypothesized by Scudder,[17] which is compatible with the use of efference velocity signals.[44,45]

As a result, the so-called resettable integrator (FIGURE 2) proposed for the control of eye saccades is now placed at the level of the SC to control gaze saccades in the head-free condition. Postulating that the SC serves as a resettable spatial and temporal integrator is compatible with its position in the gaze premotor pathways, and is easily implemented in a two-dimensional sheet of neurons.[44,46,47] Note that comparison of a spatial target to the sum of eye and head positions in a summing junction is mathematically equivalent to the now-proposed integration of the difference between an initial (impulsive) gaze error and an efference copy of gaze velocity during the movement. Both approaches provide a dynamic measure of gaze error represented in the position coordinates of activity of TRNs.

The schematic in FIGURE 7 incorporates a bilateral model of the VOR, collapsed into its linear (push-pull) form,[48] and extended to include the control of head movements and the role of a collicular feedback loop. On the left of the figure, initial gaze error (ΔG_D) is continuously compared to the sum of an efference copy of eye velocity (E^*) in the head, and head velocity (H^*) in space. The result is spatially and temporally integrated on the collicular (SC) map to provide an estimate of gaze error. The velocity efference copies could be provided by cells with appropriate firing patterns in the prepositus hypoglossi nucleus, for example, or by the activity on primary vestibular fibers: both options are compatible with known anatomical projections.[49] As a consequence, the activity on the SC map moves spatially towards the foveation zone during the gaze shift, and the activity on TRN axons at the initial site of activation on the motor map decreases with the dynamic gaze error.

The gaze error signal projects directly, or indirectly via reticulospinal neurons, to various oculomotor centers such as the burst-pause area in the pontine reticular formation, and the vestibular nuclei (VN)–prepositus hypoglossi (PH) complex; and directly to head motor centers in the spinal cord. Note that each platform to be controlled has an internal representation of its dynamics (eye and head models) which is used to calculate efference copies as needed; in addition, afferent signals from the semicircular canals (SCC) to VN and SC provide another potential feedback loop carrying estimates of head velocity. In this *gaze*-control model, there are therefore two mechanisms for the interaction of eye and head trajectories: first, gaze error and canal signals can interact at the level of the VN to enhance or suppress the VOR, depending on the sign of the gaze error and the direction of head rotation; second, all ocular premotor signals (from VN or B cells) are transmitted via

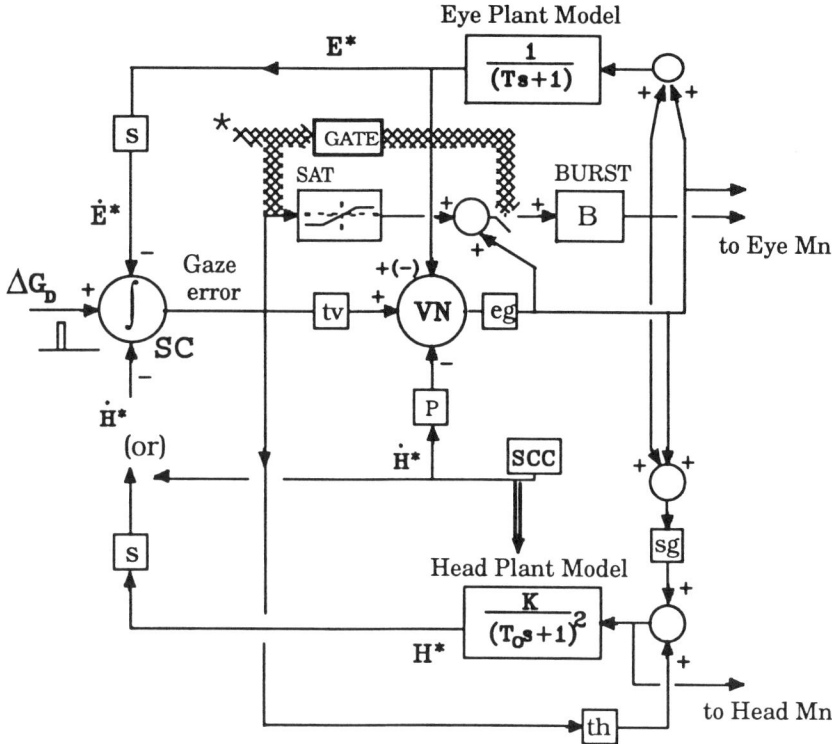

FIGURE 7. A: Model for coordinated eye and head gaze shifts. ΔG_D, initial gaze error presented to the colliculus as an impulse; SC, superior colliculus which "integrates" the difference between initial gaze error and the sum of efference copies of eye ($\dot{E}^*$) and head ($\dot{H}^*$) velocities; SCC, semicircular canals, or head plant model provide estimate of head velocity; Mn, motoneurons; VN, vestibular nuclei. **B:** Short-lead reticular burster neurons active only during saccades. GATE, saccade release and termination strategy, corresponding to silent period on brain stem OPNs, is controlled by SC gaze error, or a brain stem mechanism (*). Other symbols are scalar parameters, and "s" is Laplace operator. See text for more details.

collaterals as head premotor signals to allow the modulation of neck electromyographic (EMG) signals with eye position.

The VN summing point in FIGURE 7 represents the difference between left and right vestibular nuclei activity during fixation or VOR slow-phase compensation: a positive difference drives the eyes to the right as a result of axons crossing the midline. In contrast, during saccades, VN cells on the *left* and burst cells on the *right* will now contribute to rightward eye and head movements: because of interconnections between these two areas, VN cells on one side will also carry mild bursts during saccades. Anatomical and physiological data supporting saccadic interactions between contralateral VN and B cells are reviewed in detail elsewhere.[50]

Modeling Initiation and Termination of Eye and Gaze Saccades

The final point to be addressed is a mechanism for the initiation and termination of eye and gaze saccades. Recall that the initiation of eye (or gaze) saccades must be associated with the silencing of OPNs (gate in FIGURE 7) which normally inhibit reticular bursters. Several strategies for the control of OPNs have been proposed, normally based on a calculation of eye error as in FIGURE 2A.[16,25]

A simple alternative to controlling OPNs with a collicular gaze error signal has been proposed earlier, which relies only on brain stem firing activities,[50] to permit the generation of vestibular quick phases even in the absence of superior colliculi or frontal eye fields.[51] In this VOR-nystagmus strategy, it is postulated that the known bilateral projections of VN efferents onto OPN areas (Ret) could intermittently silence them and allow the generation of rapid eye movements driven by unilateral burst and VN cell activity; the fast phase would then end when ocular premotor signals (Mn of eye in FIGURE 7) are closely matched by the efference copy of the response (E^*). This again is compatible with the known projection of PH and VN cells onto burst and pause areas.

In the head-free case, we propose that this strategy be extended by comparing the sum of gaze premotor signals (Mn eye and head) to the sum of platform efference copies ($E^* + H^*$). Hence, if the SC is absent, OPN pause duration, acting as the "gate" controlling the length of the saccadic mode interval, would be correlated with the duration of a saccadic gaze shift when the head is free, and of a saccadic eye movement when the head is fixed. When the SC is intact, this simple strategy could be overridden by collicular projections onto OPNs, via a postulated direct excitatory effect exerted by activity in the SC's foveation zone, and an indirect inhibitory effect (say via long-lead bursters) of activity in caudal SC areas.

Simulation Parameters and Methods

Simulations were performed on an IBM Risc/6000, using MATLAB and its simulation toolbox, SIMULAB, with a fixed step interval of 0.5 msecond. The eye plant was represented as a first-order low-pass filter, with time constant 0.2 second[52]; the cat's head plant, with higher inertia, was approximated as a critically damped second-order system with time constant 0.3 second, and $K = 1$ (adapted from Reference 53).

The remaining parameters were selected to qualitatively reproduce the characteristics of head-free gaze shifts in the cat: at the end of a gaze shift, the head is nearly centered on the target, while the eye has returned near the central rest position in the orbit; also maximal eye deviation during very large gaze shifts appears to be neurally controlled to a limit near 15–20 degrees. The parameter set below incorporates these observations and provides a gaze holding time constant of 12 seconds, together with a slow-phase VOR gain of 1 during passive head perturbations. (The "VOR" gain during saccades can vary with the value of P).

B = 2   SAT = gain of 4.0 with output saturation of 20

tv = 0.5   eg = 0.98, slow-phase or (0.5, saccade)

th = 0.5   P = 0.2, for slow-phase VOR gain = 1

sg = 6.0   = 0, for no VOR-saccade interaction

During saccadic segments, the silencing of OPNs (gate) closes the burst pathway and it responds to VN and gaze error signals; during slow phases this pathway is silent. Signal interactions at the VN level are modified during saccades to reflect the structural changes associated with sudden silencing of various premotor cell groups. This involves modifying the primary vestibular projection, P, reducing the VN efferent gain, e.g., for the unilateral saccadic drive, and reversing the sign of a projection from E^* to VN from excitation to inhibition (see Reference 50 for more details). The strategy used for the initiation and termination of saccades consisted of: (1) Start saccade mode if Gaze error > ON threshold of 2 deg. Burst cells turn on; Pause cells off. (2) Stop saccade mode (return to slow-phase) if, Gaze error OR ($Mn_{eye} - E^* + Mn_{head} - H^*$) < OFF threshold of 0.5 deg. Burst cells turn off; Pause cells on. The Mn above need not be collaterals of the actual motoneural signals; collaterals or interneurons from the premotor contributors (B, VN, TRSN) would be just as effective.

The parameter set above was used for the simulation of the slower gaze trajectories, and for the initial part of the accelerated trajectories in FIGURE 9B. The faster trajectories were achieved by simply doubling the parameters receiving the collicular (SC) gaze error (tv = 1, th = 1, SAT gain = 8 with same saturation), to simulate the greater cell recruitment expected with more intense collicular activity (see below).

SIMULATION RESULTS AND MODEL IMPLICATIONS

The Effect of Target Eccentricity

The upper panels in FIGURE 8 compare eye, head, and gaze trajectories for targets within and beyond the oculomotor range, with the same "slow" parameter set listed above. Small and large gaze shifts exhibit the expected coordinated contributions from both eye and head platforms (FIGURE 1). Specifically, during large gaze shifts, the eye hovers near some plateau position in the orbit, until the head sufficiently reduces gaze error to allow refixation. In the model, this is caused by the saturation limit which is applied to the gaze error before being transmitted to the burst circuit: this imposes a neural limit on the allowed eye deviation, well within its mechanical limit in the orbit, which we described earlier in this chapter.

The Duration of the Saccadic Mode

A simple strategy was proposed above for the control of the saccadic mode during a gaze shift. It forces the silent period on OPNs and active period on EBNs to be correlated with the *gaze* parameters (horizontal bars and lower panels in FIGURE 8). In particular, the fixation mode is reactivated when gaze is near target, *not* when the eye reaches its maximal deviation in the orbit as previously modeled. Both the

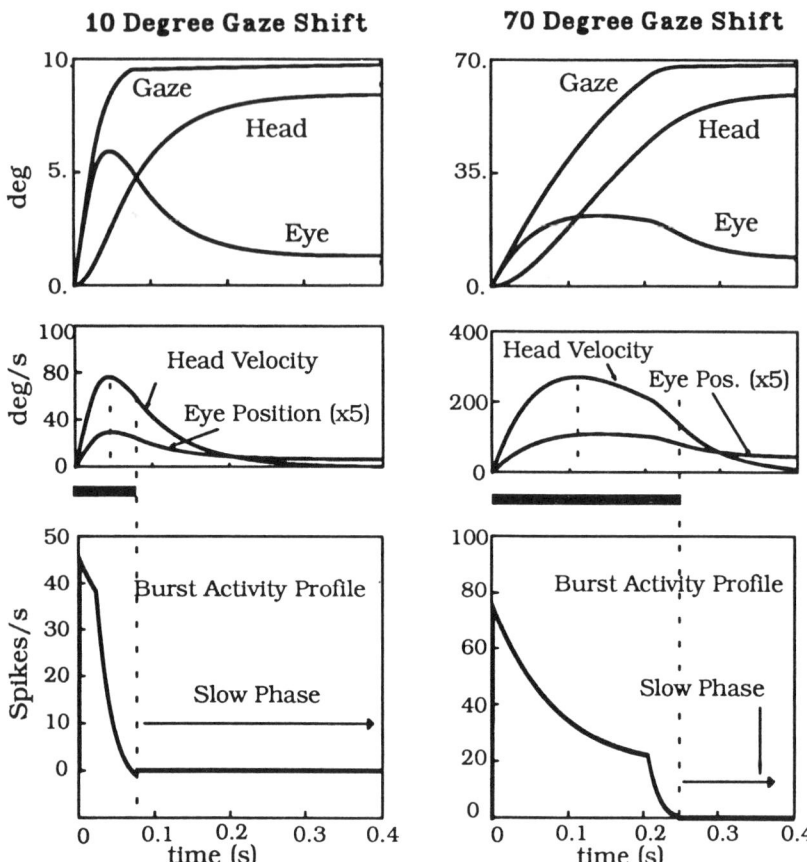

FIGURE 8. Ten- and 70-degree gaze shifts with the slow parameter set (see text). Top panels show eye, head, and gaze profiles, with saturation of eye's contribution during large gaze shifts. Middle panel emphasizes how head velocity (or acceleration) and eye position (or velocity) can covary in time, especially in the synchrony of peak levels. Heavy horizontal bars below middle panels indicate OPN silent period. Burst profiles are in lower panels. Both OPN and burst discharges, and by extension the "saccade mode," are controlled by *gaze* parameters (see text for details).

collicular and brain stem threshold comparisons produce this effect, when the head is free. In addition, for larger gaze shifts both mechanisms reactivate fixation earlier in the gaze shift, resulting in longer lead times of OPN reactivation with respect to the time of target acquisition (observed experimentally; Paré and Guitton, unpublished): this would prevent overshooting a target when the head platform is moving faster. In any case, any residual gaze error can still be reduced since some collicular pathways project onto premotor areas even in the slow-phase mode.

There are two other aspects to the proposed pause-burst strategy. First, it allows the termination of saccades, even when the collicular error signal does not drop below threshold levels: this can occur, for example, during attempted head-fixed eye saccades to targets beyond the linear oculomotor range (mimicked by electrical stimulation of the colliculus). In the head-fixed case, the pause cell duration is correlated with the duration of the eye movement. Second, the strategy allows the model to duplicate observations during electrical stimulation of the SC in the head-fixed condition: retinotopic eye saccades for stimulation near the foveal zone (inside the oculomotor range), and craniocentric saccades for stimulation in the caudal zone beyond the oculomotor range.[18]

Coupling of Eye and Head Trajectories

In the middle panels of FIGURE 8, we show that eye position and head velocity profiles can be very similar, regardless of the dynamics of the gaze trajectory. For example, peak eye deviations and head velocities are coincident in this model because of the shared premotor drive. This is reinforced in FIGURE 9, which represents task-dependent modifications in the gaze dynamics.

It was stated above that increased intensity in the activity of SC efferent cells would be translated into parallel, increased gains in the tectal–brain-stem projections of the model; this would represent the postulate that more intense activity on the SC could result in stronger responses on cells in the brain stem participating in the mapping from SC spatial coordinates to temporal premotor activity. FIGURE 9A presents the results of simulated gaze shifts using the slow and fast parameter set, showing that both eye and head contributions are faster, so that target acquisition is

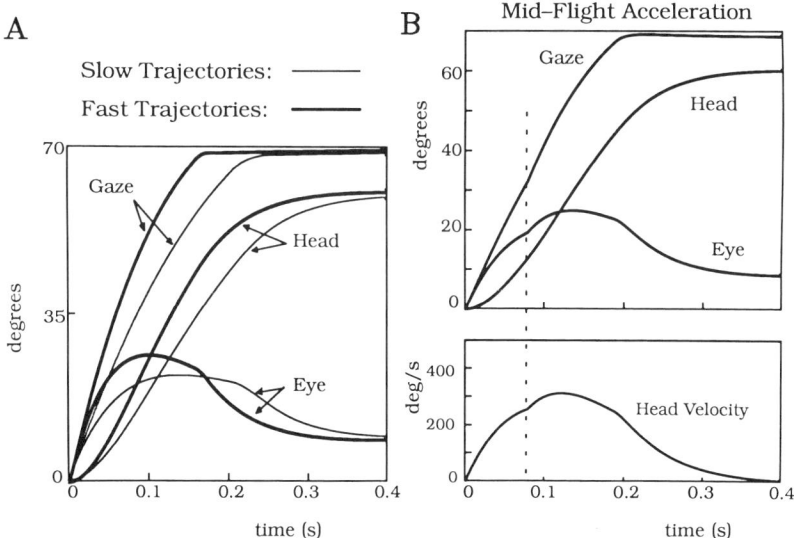

FIGURE 9. Seventy-degree gaze shifts with context-dependent dynamics. **A:** Faster gaze shifts to the same target involve faster movements for both eye and head platforms, when caused by increased activity in SC circuits. **B:** Head and eye platforms show co-acceleration in their profiles, if the SC activity intensifies during the course of a saccade, to simulate the sudden appearance of a target during a saccade to its remembered location (see data in FIGURE 4).

achieved much earlier. As reported experimentally, the duration and shape of a gaze shift to large target eccentricities can be very variable, without loss of accuracy.

We discussed above how the appearance of a target just prior to a saccade to its remembered location is associated with a sudden increase in the intensity of the TRN activity, and a concurrent increase in the speed of the gaze trajectory after a short delay (FIGURE 4). The panel in FIGURE 9B illustrates that a sudden increase in collicular burst intensity during a saccade will also cause a midflight acceleration of the eye and head platforms in the model. In all cases, regardless of the intensity of the SC activity, its location inside the feedback loop guarantees that the saccade will terminate on the appropriate target, albeit with various trajectory details. The profile of TRN activity at the initial site of SC activation will decay with gaze error, in a way that varies with the dynamics of the movements.

The Nature of Central Burst Profiles

According to the pulse-step model for the control of head-fixed eye saccades, it is presumed that the instantaneous activity on premotor burst cells codes desired eye velocity.[25] The lower panel in FIGURE 8 implies that this hypothesis may still hold for small saccades in the head-free condition, but can fail in the case of large saccades. Note, for example, that the predicted burst profile in the lower right panel of FIGURE 8 continues unabated, despite the fact that eye position has reached a maximum and is even decreasing. The burst profile in this case seems to have taken on the properties of a tonic oculomotor discharge! Tomlinson (private communication) has reported such behavior in monkey bursters during very large gaze saccades beyond the oculomotor range.

Through simulations, we reexamined (FIGURE 10) the eye velocity hypothesis in the context of 10- and 70-degree head-free gaze saccades (full lines) and 10–20 degree head-fixed eye saccades (dashed lines). All trajectories are superimposed and related to various eye and gaze parameters, in order to visualize the most consistent relationship. Fast (F) and slow (S) movements are shown for the head-free condition. For small gaze shifts well within the *linear* oculomotor range (about 10 deg for this cat model), instantaneous burst cell discharge frequency appears related to eye velocity head fixed and head free (FIGURE 10C). Equivalently, integrated burst profiles (number of spikes in pulse) superimpose closely onto a single line related to saccadic eye deviation in the orbit (FIGURE 10E). Here burst activity would appear related to eye movement, for both head-free and head-fixed saccades less than approximately 10 degrees. This limit would change with the model parameters used to represent a different species.

For larger gaze saccades, most x-y curves related to either eye or gaze parameters become progressively more nonlinear and dispersed. Note, for example, the changing slope of burst activity versus gaze error (FIGURE 10B) during a movement, and the offset of trajectories depending on saccade speed, or saccade size. This behavior has been reported experimentally by Van Gisbergen et al. in the head-fixed condition,[54] and by Tomlinson in the head-free monkey (personal communication). When related to gaze error, the burst profiles are only similar when the error is again within the linear oculomotor range. Burst activity is very variable when plotted against gaze velocity (FIGURE 10D); this is not surprising given the mixed dynamics (first and second order) of the eye and head platforms.

A notable exception is found in FIGURE 10F. Here, all trajectories are very closely superimposed when integrated burst activity is plotted against concurrent gaze deviation. Unless a large head-fixed saccade is requested beyond the plateau or

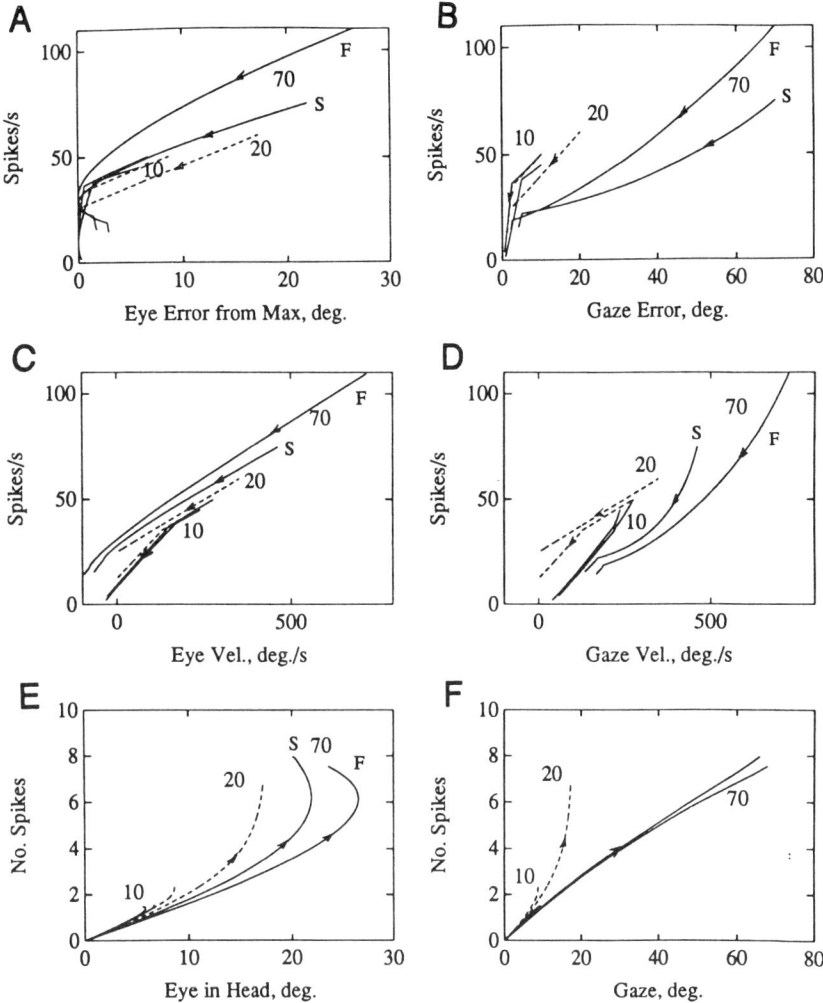

FIGURE 10. Burst cell activity is best correlated with gaze deviation in the model. Burst cell profiles (Spikes/s) or their integral (No. Spikes) are plotted against eye parameters on the left (A, C, E) and gaze parameters on the right (B, D, F). Simulation data for fast (F) and slow (S) trajectories are superimposed for 10-, 20-, and 70-degree target deviations, with dashed lines for head-fixed and solid lines for head-free saccades. Note that the best correlation is obtained by relating integrated burst activity to the concurrent gaze deviation achieved (F), unless large head-fixed saccades are requested beyond the linear oculomotor range.

linear oculomotor range (e.g., 20 deg dashed curve), all trajectories correspond to a nearly linear curve for both head-free and head-fixed saccades. Hence the model predicts that burst activity should in general be related to gaze parameters, and that the number of spikes in the burst is the most parsimonious predictor of gaze angle for all contexts.

Collicular vs. VN Contributions to VOR-like Responses

In many of the simulated trajectories, but in particular during large gaze shifts, the eye may turn around in the orbit in the "compensatory" direction before the target is acquired. This VOR-like response is obtained, despite the fact that the classical VOR in the model was set to zero (P = 0) during the saccadic simulations; the saccade mode remained enabled (OPNs off) until gaze error achieved the lower threshold. Such behavior has been experimentally observed in the cat[39] where the silent period on pause cells was sustained during eye reversals, until target acquisition. Thus, it is clear that the beginning of the classical VOR slow phase in a gaze shift must be evaluated from the gaze error profile, and not simply from the direction of the eye movement in the orbit.

Indeed, the question now arises as to what constitutes a VOR-saccade interaction. If one simply refers to eye trajectories compensating for head movements, the strong coupling between eye and head platforms will confuse the issue. For example, passive acceleration of the head trajectory might cause a reduction in the eye contribution, but head acceleration due to more intense collicular activity may instead accelerate the eye (FIGURE 9B) and result in a shorter gaze trajectory. Furthermore the collicular loop can provide a form of compensation for head perturbations, by updating gaze error from both eye and head contributions, even in the absence of vestibular projections to the VN. Collicular contributions to the compensation for head perturbations should not be considered VOR-saccade interaction.

Our simulation studies indicate that an active VOR (in terms of VN projections) is not required to preserve saccade accuracy with head perturbations. So long as the SC is properly updated (by efference copy or canal projections), the eye will indeed partially correct for the head trajectory, and gaze saccade duration will simply be extended until target acquisition is achieved.[50] In fact, it appears that an active VOR at VN levels would severely degrade the speed of large gaze saccades. This may be an explanation for the observation that VOR gain appears to be restored only when the gaze trajectories are near the target. Indeed, during small gaze saccades within the range of the eyes alone, model gaze trajectories are not significantly affected by the VOR gain level, so the system could behave as if the oculocentric mechanism, discussed above, were operating.

REFERENCES

1. GUITTON, D. & M. VOLLE. 1987. J. Neurophysiol. 58: 427–459.
2. TOMLINSON, R. D. & P. S. BAHRA. 1986. J. Neurophysiol. 56: 1542–1557.
3. GUITTON, D., R. M. DOUGLAS & M. VOLLE. 1984. J. Neurophysiol. 52: 1030–1050.
4. GUITTON, D., D. P. MUNOZ & H. L. GALIANA. 1990. J. Neurophysiol. 64: 509–531.
5. BIZZI, E., R. E. KALIL & V. TAGLIASCO. 1971. Science 173: 452–454.
6. DICHGANS, J., E. BIZZI, P. MORASSO & V. TAGLIASCO. 1973. Exp. Brain Res. 18: 548–562.
7. MORASSO, P., E. BIZZI & J. DICHGANS. 1973. Exp. Brain Res. 16: 492–500.
8. BIZZI, E. 1981. In Handbook of Physiology. The Nervous System. Motor Control: 1321–1336. American Physiological Society. Bethesda, Md.
9. LAURUTIS, V. P. & D. A. ROBINSON. 1981. J. Neurol. Sci. 373: 209–233.
10. TOMLINSON, R. D. & P. S. BAHRA. 1986b. J. Neurophysiol. 56: 1558–1570.
11. TOMLINSON, R. D. 1990. J. Neurophysiol. 64: 1873–1891.
12. BLAKEMORE, C. & M. DONAGHY. 1980. J. Physiol. London 300: 317–335.
13. FULLER, J. H., H. MALDONADO & J. SCHLAG. 1983. Brain Res. 271: 241–250.
14. PÉLISSON, D. & C. PRABLANC. 1986. Brain Res. 380: 397–400.
15. PÉLISSON, D., C. PRABLANC & C. URQUIZAR. 1988. J. Neurophysiol. 59: 997–1013.

16. ROBINSON, D. A. 1975. *In* Basic Mechanisms of Ocular Motility and Their Clinical Implications. G. Lennerstand & P. Bach-y-Rita, Eds.: 337–374. Pergamon. Oxford, England.
17. SCUDDER, C. A. 1988. J. Neurophysiol. **59:** 1455–1575.
18. ROUCOUX, A., D. GUITTON & M. CROMMELINCK. 1980. Exp. Brain Res. **39:** 75–85.
19. WILSON, V. J. & G. MELVILL JONES. 1979. Mammalian Vestibular Physiology. Plenum Press. New York, N.Y.
20. ANDRE-DESHAYS, C., A. BERTHOZ & M. REVEL. 1988. Exp. Brain Res. **69:** 399–406.
21. LESTIENNE, F., P. P. VIDAL & A. BERTHOZ. 1984. Exp. Brain Res. **53:** 349–356.
22. VIDAL, P. P., A. ROUCOUX & A. BERTHOZ. 1982. Exp. Brain Res. **46:** 448–453.
23. GRESTY, M. A. 1974. Vision Res. **14:** 395–403.
24. PÉLISSON, D., D. GUITTON & D. P. MUNOZ. 1989. Exp. Brain Res. **78:** 654–658.
25. FUCHS, A. F., C. R. S. KANEKO & C. A. SCUDDER. 1985. Annu. Rev. Neurosci. **8:** 307–337.
26. SPARKS, D. L. 1986. Physiol. Rev. **66:** 118–171.
27. SPARKS, D. L. & L. E. MAYS. 1990. Annu. Rev. Neurosci. **13:** 309–336.
28. GUITTON, D. 1991. *In* Vision and Visual Dysfunction. Eye Movements. R. H. S. Carpenter, Ed. **8:** 244–276. MacMillan Press. London, England.
29. DEAN, P., P. REDGRAVE & W. M. WESTBY. 1990. TINS **12:** 137–147.
30. GUITTON, D., M. CROMMELINCK & A. ROUCOUX. 1980. Exp. Brain Res. **39:** 63–73.
31. DU LAC, S. & E. I. KNUDSEN. 1990. J. Neurophysiol. **63:** 131–146.
32. GUITTON, D. & D. P. MUNOZ. 1991. J. Neurophysiol. **66:** 1605–1623.
33. MUNOZ, D. P. & D. GUITTON. 1989. Rev. Neurol. Paris. **145:** 567–579.
34. MUNOZ, D. P. & D. GUITTON. 1991. J. Neurophysiol. **66:** 1624–1641.
35. MUNOZ, D. P., D. GUITTON & D. PÉLISSON. 1991. J. Neurophysiol. **66:** 1642–1666.
36. MUNOZ, D. P., D. PÉLISSON & D. GUITTON. 1991. Science **251:** 1358–1360.
37. GRANTYN, A. & A. BERTHOZ. 1988. *In* Control of Head Movement. B. W. Peterson & F. J. Richmond, Eds.: 224–244. Oxford University Press. Oxford, England.
38. WAITZMAN, D. M., T. P. MA, L. M. OPTICAN & R. H. WURTZ. 1988. Exp. Brain Res. **72:** 649–652.
39. PARÉ, M. & D. GUITTON. 1990. Exp. Brain Res. **83:** 210–214.
40. GALIANA, H. L., D. GUITTON & D. P. MUNOZ. 1991. *In* Head-Neck Sensory Motor System. A. Berthoz, P. P. Vidal & W. Graf, Eds. Oxford University Press. Oxford, England.
41. PÉLISSON, D., D. GUITTON & D. P. MUNOZ. 1991. *In* Oculomotor Control and Cognitive Processes. R. Schmid & D. Zambarbieri, Eds.: 213–228. Elsevier/North Holland. Amsterdam, the Netherlands.
42. GUITTON, D., D. P. MUNOZ & D. PÉLISSON. 1991. *In* Brain and Space. J. Paillard, Ed.: 20–37. Oxford University Press. Oxford, England.
43. DELREUX, V., S. VANDEN ABEELE, P. LEFEVRE & A. ROUCOUX. 1991. *In* Brain and Space J. Paillard, Ed.: 38–48. Oxford University Press. Oxford, England.
44. DROULEZ, J. & A. BERTHOZ. 1991. *In* Motor Control: Concepts and Issues. D. R. Humphrey & J. R. Freund, Eds.: 137–161. Wiley and Sons. New York, N.Y.
45. DROULEZ, J. & A. BERTHOZ. 1988. *In* Neural Computers. R. Eckmiller & Cvd. Malsburg, Eds.: 345–358. Springer-Verlag. Berlin, Germany.
46. LEFEVRE, P. & H. L. GALIANA. Neural Networks. (In press.)
47. LEFEVRE, P. & H. L. GALIANA. 1990. Soc. Neurosci. Abstr. **16:** 1084.
48. GALIANA, H. L. & J. S. OUTERBRIDGE. 1984. J. Neurophysiol. **51:** 210–241.
49. MCCREA, R. A. & R. BAKER. 1985. J. Comp. Neurol. **237:** 377–407.
50. GALIANA, H. L. 1991. IEEE Trans. Biomed. Eng. **38:** 532–543.
51. SCHILLER, P. H., S. D. TRUE & J. L. CONWAY. 1980. J. Neurophysiol. **44:** 1175–1189.
52. ROBINSON, D. A. 1981. *In* Models of Oculomotor Behavior and Control. B. L. Zuber, Ed.: 21–41. CRC Press. Boca Raton, Fla.
53. GOLDBERG, J. & B. W. PETERSON. 1986. J. Neurophysiol. **56:** 857–875.
54. VAN GISBERGEN, J. & D. A. ROBINSON. 1981. J. Neurophysiol. **45:** 417–442.

Cortical Control of Vestibular Memory-Guided Saccades[a]

I. ISRAËL,[b] S. RIVAUD,[c] A. BERTHOZ,[b] AND
C. PIERROT-DESEILLIGNY[c]

[b]CNRS—Laboratory of Neurosensory Physiology
15, rue de l'Ecole de Médecine
75006 Paris, France

[c]INSERM—U. 289
Salpêtrière Hospital
47, Boulevard de l'Hôpital
75013 Paris, France

INTRODUCTION

By measuring the magnitude of subjective angular velocity in humans, Guedry had shown that human subjects can correctly evaluate the amplitude of a passive angular whole-body displacement,[1] from the angular acceleration signal transduced into angular velocity signal by the mechanical properties of the semicircular canals.[2,3] The exact mechanism of this evaluation of displacement is not known. However, it confirmed the old proposal of Beritoff[4] and the more recent ideas of Potegal[5] concerning an involvement of the vestibular system in "path integration."

Recently, quantitative studies on the reproduction, with eye movements, of passive horizontal whole-body angular displacements in darkness have been performed in humans, for angular movements by Bloomberg et al.[6–8] (who have proposed a "vestibular memory contingent saccade" task) and Israël et al.,[9] and for linear displacements by Israël et al.[10,11] The results showed that humans can correctly match the amplitude of a preceding head displacement with a voluntary ocular saccade of equal but opposing amplitude. This oculomotor return after passive transport is possible only if the vestibular system is intact.[11] These authors not only showed that the amplitude of a passive body displacement in darkness is correctly estimated by the brain, but also that this information can be adequately stored, retrieved, and used by the oculomotor (saccadic) system.

It has also recently been suggested[12–14] that the vestibuloocular reflex (VOR) is not a simple vestibular-induced reflex, but proceeds from an internal reconstruction of the spatial relations between subject and target, or of a mental "reference." Such a hypothesis is supported by the recent finding by Ventre and Faugier-Grimaud of cortical afferences on the vestibular nuclei and VOR gain modification after parietal lesions,[15,16] and also by the discovery of a vestibular area in the parietoinsular cortex.[17] Segal and Katsarkas have furthermore suggested that the VOR is goal-directed, and that it cooperates with the saccadic system.[18] This vestibulosaccadic cooperation, or functional coupling which had been suggested by Berthoz,[19] has been confirmed by Bloomberg et al.[7,8] The latter further suggested that this synergistic gaze compensation of slow-phase and saccadic components could arise from vestibular perception of body rotation in space.

[a]I. Israël was supported by a grant from the "Centre National d'Etudes Spatiales," France.

This delayed saccade task following vestibular stimulation is actually a spatial delayed-response task different from the ones in which the monkey has to make a saccade toward a memorized visual target.[20] A fundamental difference between the classical spatial delayed-response tasks and those "vestibular memory-contingent saccades" is that in the former, the subject has been presented with a model of the response to execute, since he has seen where the target is to point to. In the latter, the subject has to elaborate the adequate response, also toward a disappeared visual target, but after his own body displacement with respect to this target. The subject has to memorize the starting point using only the vestibular information about his own passive displacement, instead of a retinal error.

To study this question, a vestibular memory-guided saccades task has been performed by 15 patients with limited unilateral cortical infarctions. The lesions affected those cortical areas that would appear to be involved in saccade control. It has been shown that, in humans, the different types of saccades are triggered by the frontal eye field (FEF), or the posterior parietal cortex (area 39 of Brodmann).[21,22] In addition, the frontal cortex is involved in preventing unwanted saccades,[22,23] and the prefrontal cortex appears to be involved in memory-guided saccades.[24]

PATIENTS AND METHODS

Fifteen patients have been tested, together with seven age-matched healthy (control) subjects. All patients had relatively recently developed a unilateral limited cortical infarction:

Eight patients had lesions affecting the dorsolateral frontal cortex, with the prefrontal cortex (PFC) or area 46 of Brodmann as area of common damage (FIGURE 1). Six patients had a lesion on the right side (RPFC group), and two on the left side (LPFC group).

Seven patients had lesions affecting the posterior parietal cortex (PPC), in each case with damage to the superior part of the angular gyrus, which was chosen as the area of common damage (FIGURE 1). Three patients had lesions on the right side (RPPC group) and four on the left side (LPPC group).

The PPC group was slightly older than the PFC group and than the control group. Ages were respectively 58 (RPFC), 49 (LPFC), 68 (RPPC), 66 (LPPC), and 60 (control) years. All subjects were right-handed. None of the patients had any visual defect using Goldman's perimetry. Visual neglect was always contralateral to the lesion. In the RPPC group, visual neglect was severe in two patients and absent in one patient. In the LPPC group, neglect was moderate in one patient and absent in the other three. In the RPFC group, visual neglect was severe in three patients and absent in the other three. It was absent in the two LPFC patients.

Experimental Setup

The whole experiment was performed in complete darkness. The subject sat on a chair that could be manually rotated around the vertical axis. The head was fixed to the chair by two large soft pads, adjusted for each subject to hold tight against the ears. In order to control rotation amplitude, a fluorescent angular scale of 5-deg resolution was drawn on the floor, at the base of the rotating chair, and a fluorescent arrow fixed to the pivoting axis.

Horizontal eye position was measured by DC-electrooculographic (EOG) technique. Vertical eye position was also monitored with the EOG, and trials with too many eye blinks or with eye closure were repeated. Head position was measured with a potentiometer attached to the chair and aligned with its rotation axis.

The earth-fixed target (EFT) was a red spot about 2 cm diameter, projected on the center of a screen 80 cm away, thus straight ahead of the subject at the initial position. The chair-fixed target (CFT) was a small faint red light-emitting diode (LED) placed on an L-shaped bracket fixed to the chair. The L extremity, i.e., the

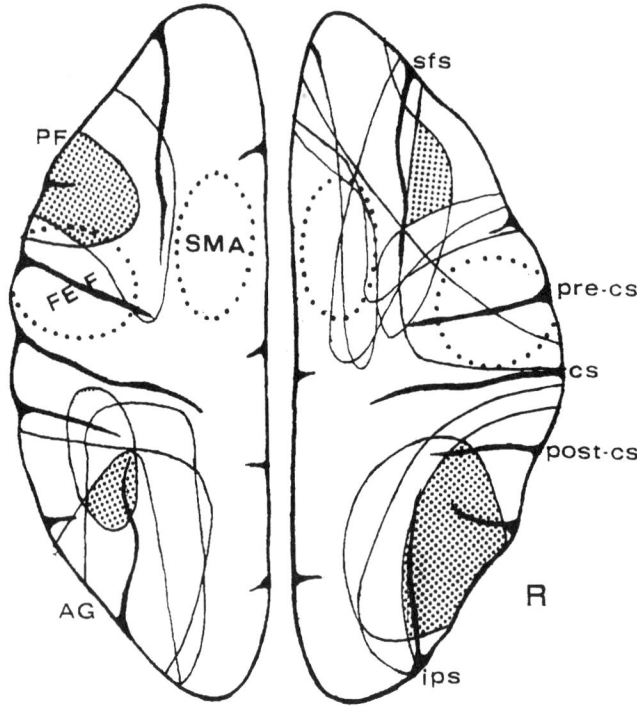

FIGURE 1. Dorsal view on reconstructed lesions of the patients. Areas where infarctions were localized are drawn on a horizontal section passing 30 mm below the upper limit of the cerebral hemisphere. Each thin line corresponds to a different lesion. In each group, the dotted zone is the common area of damage. The dotted lines indicate the approximate limits of the FEF and the SMA. AG, angular gyrus; cs, central sulcus; FEF, frontal eye field; ips, intraparietal sulcus; post-cs, postcentral sulcus; PF, prefrontal cortex; sfs, superior frontal sulcus; SMA, supplementary motor area.

CFT position, was 75 cm away from the subject, straight ahead. The vertical position of both targets was adjusted to each subject's eye level.

Horizontal eye position, head position, as well as both targets on/off timing signals were filtered (0–35 Hz range) and continuously plotted during the experiment on a paper chart, while recorded on analog tape. Those signals were thereafter digitized (100-Hz sampling rate) onto a PC microcomputer for off-line analysis.

Experimental Protocol

VOR integrity and VOR suppression ability were qualitatively assessed before each experimental session. The subject viewed the EFT in front of him for about 5 seconds. After the EFT was turned off, the chair was manually rotated in complete darkness, with a quasi-sinusoidal motion at a frequency of about 0.5 Hz. The subject was instructed to keep his eyes on the imagined/memorized EFT, i.e., to stabilize his gaze in space, during a few oscillatory cycles (VOR). Then, the EFT was turned off and replaced by the CFT, which the subject was required to fixate during chair oscillation (VOR suppression).

VOR function was considered as normal if the vestibular-induced eye movements' peak to peak amplitude was about equal to the head (the chair) oscillations' amplitude. Similarly, VOR suppression was judged as adequate if no slow compensatory eye movements were seen on the paper chart.

The experimental protocol for testing oculomotor return from passive body rotation was as follows (FIGURE 2):

1. The visual EFT was presented straight ahead to the subject during about 5 seconds; the subject was required to fixate it. The EFT was then turned off and replaced by the CFT, which was superimposed to the EFT location when the chair was at the initial position.
2. The chair was then manually rotated clockwise or counterclockwise, with a nonpredictible angular amplitude (H): 10, 20, 30, or 40 deg. During the rotation, the subject was required to keep his eyes on the CFT (VOR suppression).
3. After the rotation, when the chair had stopped moving, the subject was asked to keep maintaining his eyes on the CFT during a delay (D) of 5 seconds.
4. After this delay, the CFT was turned off. This was the signal for the subject to move his eyes to the memorized position of the EFT previously shown, and to keep his eyes on that position (E) in complete darkness.
5. The EFT was turned on again, about 2 to 3 seconds after the CFT had been turned off. The subject made, if necessary, a corrective saccade to foveate the visual EFT.
6. The chair was then rotated back to its initial position, with the subject still foveating the visual EFT.

In order to familiarize the subjects with the equipment and procedures, about 5 test trials were performed at the beginning of each experiment.

Data analysis

We have measured both latency and accuracy of memory-guided saccades. This analysis was performed with a specially designed interactive graphical software of events marking and storage. In order to obtain objective and reproducible data for all subjects, a histogram of eye positions has been computed for each trial, from the first saccade after CFT off, up to the last final correction saccade, i.e., when the subject was in complete darkness. Each bin is the eye position average on 200 mseconds, (FIGURE 3).

This histogram provided eye position accuracy at the first saccade after CFT off, last saccade before EFT on, as well as the duration spent at different levels of eye position relative error ($|E - H/H|$), and also first saccade latency or eye position

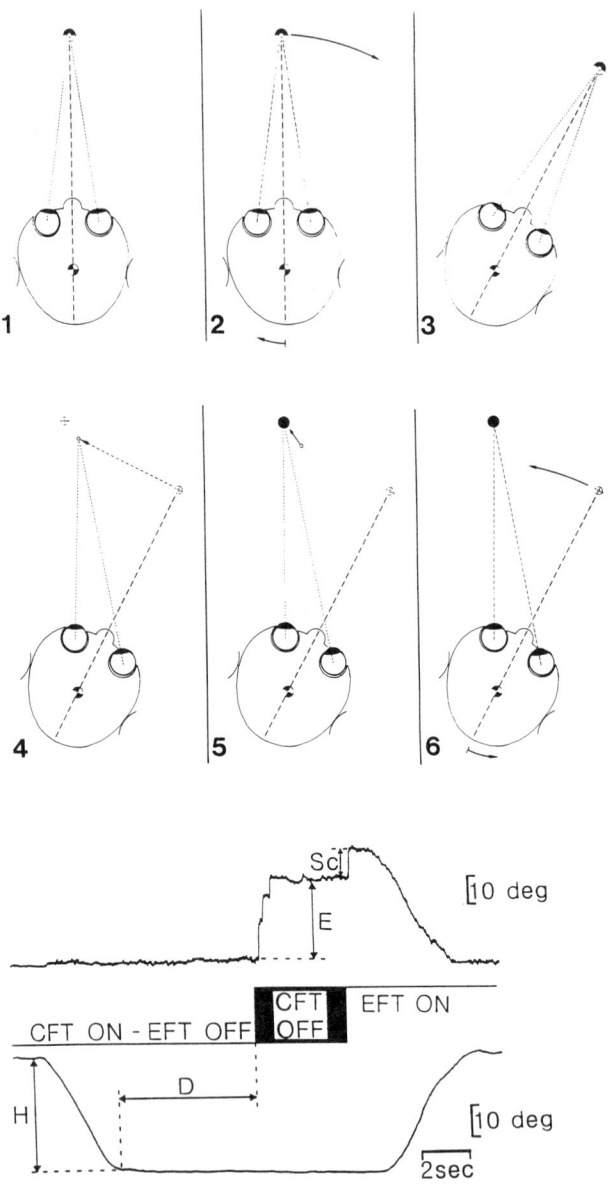

FIGURE 2. Panels 1–6 describe the experimental procedure (see text). Lower panel shows the signals recorded during one trial, with eye position at the top, head position at the bottom, and both targets' on-off timing signals at the middle.

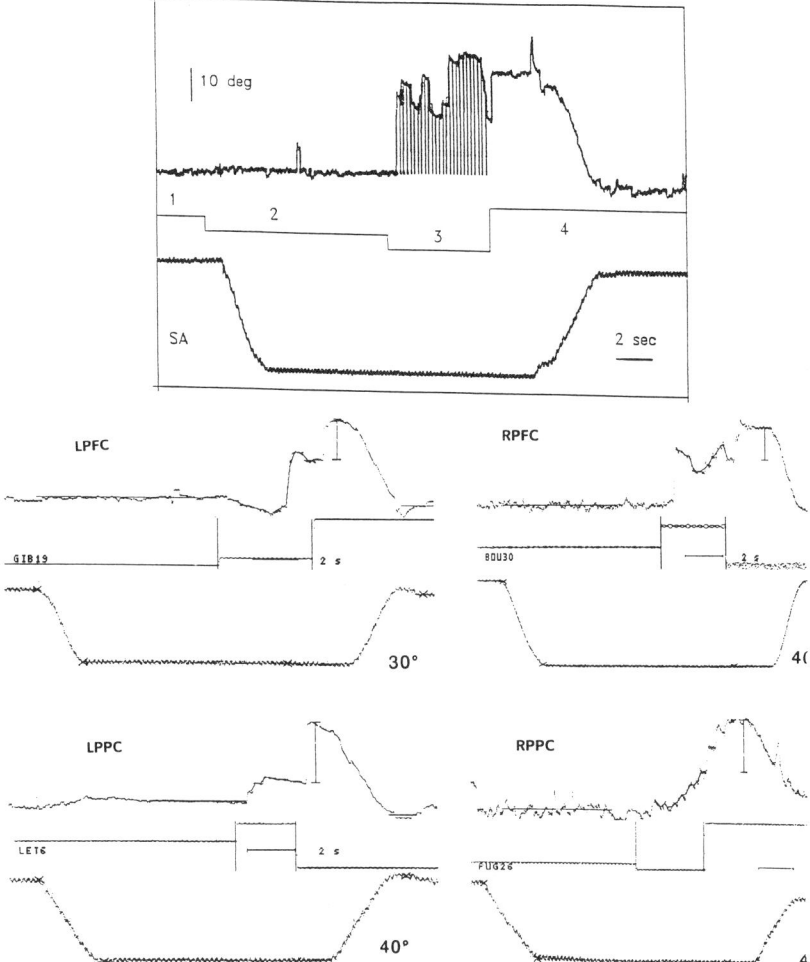

FIGURE 3. Upper central panel shows an example of the data analysis histogram, superimposed on raw eye position trace. The lower four panels show examples of trials performed by patients, with the histogram values represented with discontinuous lines (smoothed eye position) instead of columns, and also superimposed on raw eye position trace.

stability in complete darkness. Besides, it also provided "smoothed" views of eye position during the dark period of each trial, as can be seen in FIGURE 4.

RESULTS

Qualitative Analysis of Memory-Guided Saccades

A first observation is brought about by those smoothed views: it is the apparent difficulty for all subjects to fixate an eccentric position in complete darkness. All

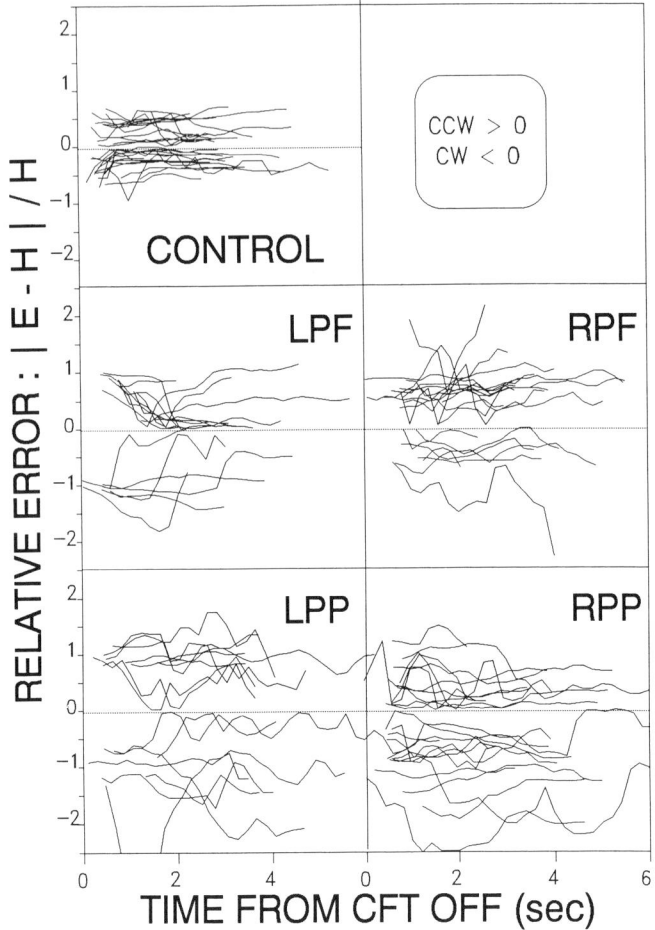

FIGURE 4. Histogram values (smoothed eye position) converted into relative error, for one subject in each group.

patients exhibited important and quasi-continuous variations of eye position, i.e., multiple saccades. This can be seen in FIGURE 4, showing analysis histogram values in lines instead of columns, for each trial of several subjects. This inability to fixate in the dark, independent of the desired position, was observed in all patients. Although this deviation from fixation was considerably larger with patients than with control subjects, the difference is not statistically significant.

Latency of Memory-Guided Saccades

Before we analyze latency of saccades toward memorized EFT, namely, duration between CFT off and saccade, it was necessary to examine the cases in which the subjects anticipated the timing for the memory-guided saccade to the remembered stationary target location. Those trials in which the first memory-guided saccade was triggered before the CFT went off have been counted for each group. Although all patient groups in general anticipated more often than the control group (except the LPF group who made no anticipation for CCW rotations), this difference was statistically significant only for the LPF group and for CW rotations. On average, the PFC groups had more difficulties in waiting for the extinction of the CFT before making a saccade (FIGURE 5). These anticipated saccades have been omitted from all the following average calculations, including latency averages.

Latency of the first saccade was greater for all patient groups compared to the control group, except LPF for CW rotations, and LPP for CCW rotations (FIGURE 5). In general the latency increased with the frequency of anticipation. Latency was also significatively longer for R-groups than for L-groups.

Saccade Accuracy

As was seen above, patients had difficulty stabilizing their eyes on an eccentric position in complete darkness (FIGURE 4). For this reason, we chose the accuracy of the first saccade as the main criterion for the assessment of spatial performance.

We first measured the direction errors occurring for this first saccade. Those direction errors were detected by the computer when the amplitude ratio E/H was negative, and counted for each subject. Direction errors are clearly more frequent in

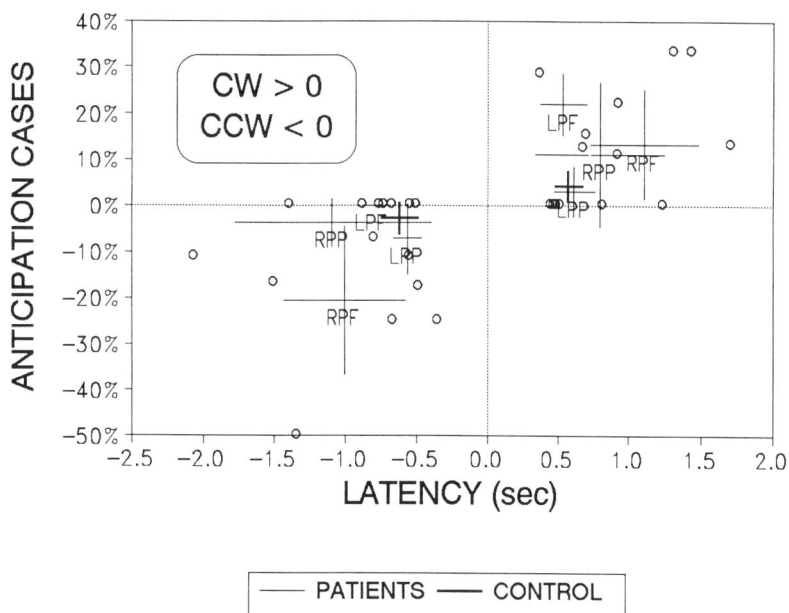

FIGURE 5. Frequency of anticipation with respect to latency of the first memory-guided saccade. Average ± standard deviation is shown for both parameters (axes), and small circles represent individual data for all patients.

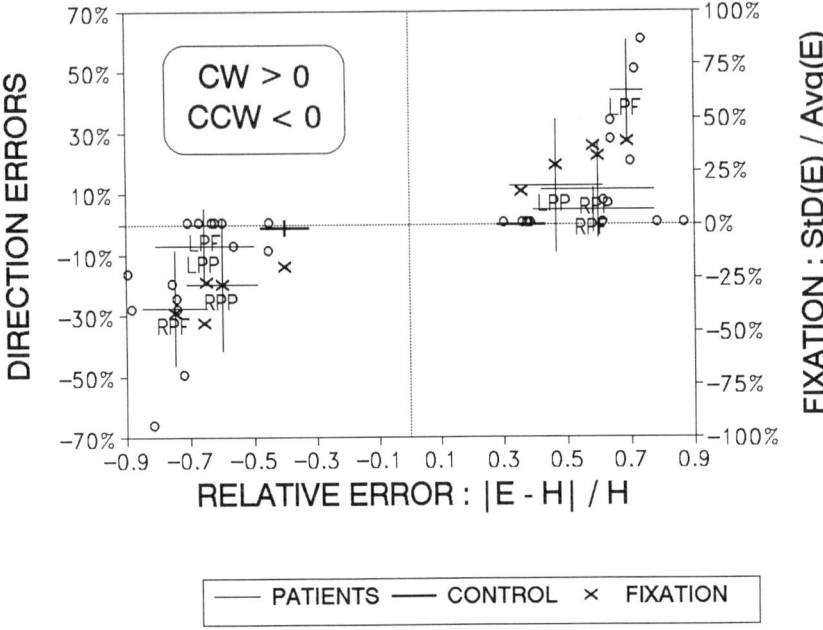

———— PATIENTS ———— CONTROL   ×   FIXATION

FIGURE 6. Direction error occurrence (left *y*-axis) and deviation from fixation (right *y*-axis), with respect to relative error of the first saccade. Same notation as in FIGURE 5, with × symbols related to right *y*-axis, and showing average of deviation from fixation, or coefficient of variation [StD(E)/Avg(E) × 100] for each group.

all patient groups than in the control group (FIGURE 6). The difference is significant (*t*-test) only for the PFC groups: $p < 0.001$ in the LPFC group for CW rotations, and $p < 0.01$ in the RPFC group for CCW rotations, showing that the PFC groups made direction errors mostly for saccades that should be aimed toward the hemifield ipsilateral to the lesion. This hemifield contingency is not observed for PPC groups, who made fewer direction errors.

Some correlation trends can be seen in FIGURE 6 between relative error and direction error occurrence, as well as between relative error and deviation from fixation. Furthermore, we found a positive correlation between anticipation and direction error occurrence ($p < 0.01$), at the group level (not at the individual level because of the cases without anticipation or without direction errors). This suggests a common impairment of both timing and amplitude control of those vestibular memory-guided saccades.

The trials in which there was a direction error have been omitted from the following saccade accuracy averages. First saccade end position with respect to head rotation amplitude for all subjects and all trials (including direction errors) is plotted in FIGURE 7, and averages are shown in FIGURE 6. It can be seen in FIGURES 6 and 7 that the first saccade relative error is larger for all patients groups than for the control group. Although this is not always significant, this is mainly due to the small number of subjects. Here again, the PFC groups are most deficient, for both rotation directions, particularly the LPFC group for CW rotations (saccades to the left, i.e., the contralateral hemifield). However, the difference in saccade latency that has been found between R-groups and L-groups does not correspond to an equivalent accuracy difference.

It can also be noted in FIGURE 7 that whereas the majority of the first saccades of the control group are hypometric, as well as those of both PPC groups, saccades of the PFC groups are quasi-exclusively hypometric.

Although the most relevant parameter seemed to be the first saccade, we checked whether there was some improvement at the last saccade end position, i.e., eye position just before the correction saccade, when the EFT had been turned back on (see FIGURE 3). This improvement was actually observed for all groups, patients and control, considering either the average relative error decrease ($|E - H|/H$), or the number of cases where the last saccade was closer to the desired position than the first one. Furthermore, the last saccade relative error was closer to the control group

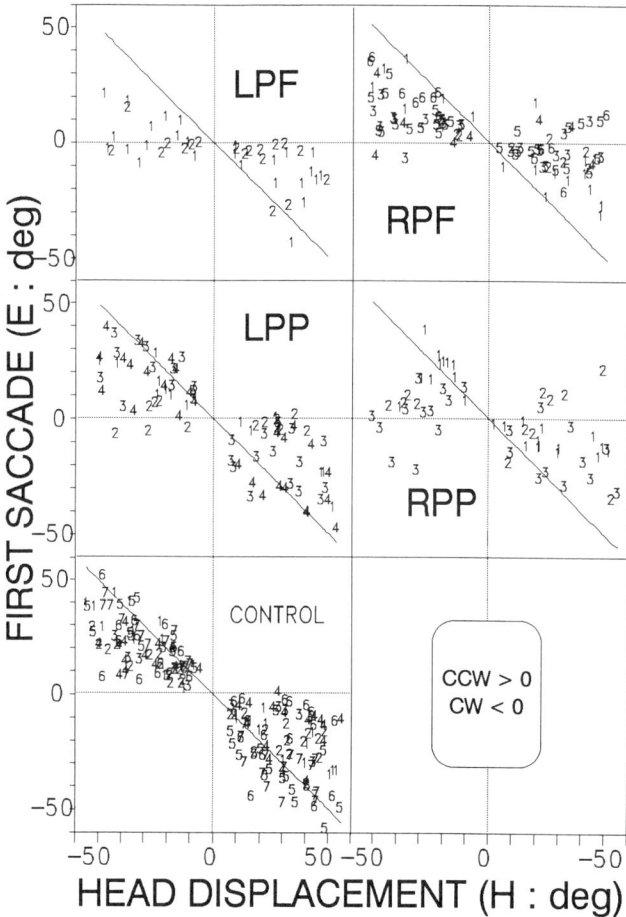

FIGURE 7. First saccade end position with respect to head rotation amplitude. Numerals are subjects' identification, and replace drawing symbols in the figure.

saccade relative error than the first saccade, for all patient groups. Relative error was always larger in patients groups than in controls, however.

DISCUSSION

This study shows that patients with parietal lesions have deficits in the performance of a vestibular memory-contingent saccade task. Therefore, we confirm the findings reported by Bloomberg et al.[7,8] In addition, the performance is greatly impaired in patients with prefrontal lesions.

The measurements made on the control age-matched group have in fact revealed a rather new and unexpected result which may deserve a follow-up study: their performance in this vestibular memory task was very low in many instances. It may be interesting to study in detail the possible deficits of vestibular representational memory. We could propose the hypothesis that if this finding were confirmed, a number of accidents occurring to old persons would not be due to mechanical or muscular factors, but to deficits in spatial cognition of the kind we have explored in this work.

An interesting aspect of our results is that this deficit cannot be due to a spatial neglect, because neglect is generally contralateral, and the deficits we observed were generally ipsilateral. In addition there was no relation between the amount of neglect and the error of the final eye position.

Lastly, the particular paradigm that we used requires a vestibulospatial integration, which could be controlled as visuospatial integration by the posterior parietal cortex, and a contribution of vestibular memory which could depend, as spatial memory, on prefrontal cortex.

The overall conclusion for this work is that the prefrontal cortex seems to be involved not only in visual representational memory[20] and in the organization of memorized saccades,[24-27] but may also be involved in "vestibular representational memory" of the displacement and orientation of the body in space.

Further work will allow us to understand the respective contribution of the parietal and prefrontal cortex in this mechanism. The same patients had also participated in experiments of visual memory-guided saccades,[24] so that a comparison between those two paradigms will possibly be presented in a future paper.

ACKNOWLEDGMENTS

We thank M. Ehrette and M. Loiron for their contribution to the experimental setup.

REFERENCES

1. GUEDRY, F. E. 1974. Psychophysics of vestibular sensation. *In* Handbook of Sensory Physiology. Vestibular System. Psychophysics, Applied Aspects and General Interpretations. H. H. Kornhuber, Ed. **6**(2, Part 2): 3–154. Springer. Berlin, Heidelberg & New York.
2. GOLDBERG, J. M. & C. FERNANDEZ. 1975. Responses of peripheral vestibular neurons to angular and linear accelerations in the squirrel monkey. Acta Otolaryngol. Stockh. **80**: 101–110.
3. YOUNG, L. R. 1984. Perception of body in space: mechanisms. *In* Handbook of Physiology—

the Nervous System III. I. Darian-Smith, Ed.: 978–1023. American Physiological Society. Bethesda, Md.

4. BERITOFF, J. S. 1965. Neural Mechanisms of Higher Vertebrates. Little Brown & Co. Boston, Mass.

5. POTEGAL, M. 1982. Vestibular and neostriatal contributions to spatial orientation. *In* Spatial Abilities. Development and Physiological Foundations. M. Potegal, Ed.: 361–387. Academic Press. New York, N.Y.

6. BLOOMBERG, J., G. MELVILL JONES, B. N. SEGAL, S. MCFARLANE & J. SOUL. 1988. Vestibular-contingent voluntary saccades based on cognitive estimates of remembered vestibular information. Adv. Oto-Rhino-Laryngol. **40:** 71–75.

7. BLOOMBERG, J., G. MELVILL JONES & B. SEGAL. 1991. Adaptive plasticity in the gaze stabilizing synergy of slow and saccadic eye movements. Exp. Brain Res. **84:** 35–46.

8. BLOOMBERG, J., G. MELVILL JONES & B. SEGAL. 1991. Adaptive modification of vestibularly perceived rotation. Exp. Brain Res. **84:** 47–56.

9. ISRAËL, I., S. RIVAUD, C. PIERROT-DESEILLIGNY & A. BERTHOZ. "Delayed VOR:" an assessment of vestibular memory for self motion. *In* Tutorials in Motor Neuroscience. J. Requin & G. E. Stelmach, Eds. Kluwer Academic Pub. The Hague, the Netherlands. (In press.)

10. BERTHOZ, A., I. ISRAËL, T. VIEVILLE & D. S. ZEE. 1987. Linear head displacement measured by the otoliths can be reproduced though the saccadic system. Neurosci. Lett. **82:** 285–290.

11. ISRAËL, I. & A. BERTHOZ. 1989. Contribution of the otoliths to the calculation of linear displacement. J. Neurophysiol. **62**(1): 247–263.

12. BERTHOZ, A. 1989. Coopération et substitution entre le système saccadique et les réflexes d'origine vestibulaires: faut-il réviser la notion de réflexe? Rev. Neurol. Paris **145:** 513–526.

13. MELVILL JONES, G. & A. BERTHOZ. 1985. Mental control of the adaptative process. Rev. Oculomotor Res. **1:** 203–208.

14. MELVILL JONES, G., A. BERTHOZ & B. N. SEGAL. 1984. Adaptative modification of the vestibulo-ocular reflex by mental effort in darkness. Exp. Brain Res. **56:** 149–153.

15. VENTRE, J. & S. FAUGIER-GRIMAUD. 1988. Projections of the temporo-parietal cortex on vestibular complex in the macaque monkey (*Macaca fascicularis*). Exp. Brain Res. **72:** 653–658.

16. FAUGIER-GRIMAUD, S. & J. VENTRE. 1989. Anatomic connections of inferior parietal cortex (area 7) with subcortical structures related to vestibular-ocular function in a monkey (*Macaca fascicularis*). J. Comp. Neurol. **280:** 1–14.

17. GRÜSSER, O. J., M. PAUSE & U. SCHREITER. 1982. Neuronal responses in the parieto-insular vestibular cortex of alert Java monkeys (*Macaca fascicularis*). *In* Physiological and Pathological Aspects of Eye Movements. A. Roucoux & M. Crommelinck, Eds.: 251–270. Junk Publishers. The Hague, the Netherlands.

18. SEGAL, B. N. & A. KATSARKAS. 1988. Goal-directed vestibulo-ocular function in man: gaze stabilization by slow-phase and saccadic eye movements. Exp. Brain Res. **70:** 26–32.

19. BERTHOZ, A. 1985. Adaptive mechanisms in eye-head coordination. Rev. Oculomotor Res. **1:** 177–201.

20. GOLDMAN-RAKIC, P. S. 1987. Circuitry of primate prefrontal cortex and regulation of behavior by representational memory. *In* Handbook of Physiology—the Nervous System V. V. B. Mountcastle, F. Plum & S. R. Geiger, Eds.: 373–417. American Physiological Society. Bethesda, Md.

21. PIERROT-DESEILLIGNY, C. 1991. Cortical control of saccades. Neuro-ophthalmology **11:** 63–75.

22. PIERROT-DESEILLIGNY, C., S. RIVAUD, B. GAYMARD & Y. AGID. 1991. Cortical control of reflexive visually-guided saccades. Brain **114:** 1473–1485.

23. GUITTON, D., H. A. BUCHTEL & R. M. DOUGLAS. 1985. Frontal lobe lesions in man cause difficulties in suppressing reflexive glances and in generating goal-directed saccades. Exp. Brain Res. **58:** 455–472.

24. PIERROT-DESEILLIGNY, C., S. RIVAUD, B. GAYMARD & Y. AGID. 1991. Cortical control of memory-guided saccades in man. Exp. Brain Res. **83:** 607–617.

25. JOSEPH, J. P. & P. BARONE. 1987. Prefrontal unit activity during a delayed oculomotor task in the monkey. Exp. Brain Res. **67:** 460–468.
26. FUNAHASHI, S. C., J. BRUCE & P. S. GOLDMAN-RAKIC. 1989. Mnemonic coding of visual space in the monkey's dorsolateral prefrontal cortex. J. Neurophysiol. **61**(2): 331–349.
27. FUNAHASHI, S. C., J. BRUCE & P. S. GOLDMAN-RAKIC. 1990. Visuospatial coding in primate prefrontal neurons revealed by oculomotor paradigms. J. Neurophysiol. **63**(4): 814–831.

Neuronal Substrates of Spatial Transformations in Vestibuloocular and Vestibulocollic Reflexes[a]

BARRY W. PETERSON, JAMES F. BAKER, STEVE I.
PERLMUTTER, AND YOSHIKI IWAMOTO

Northwestern University Medical School
303 East Chicago Avenue
Chicago, Illinois 60611

INTRODUCTION

This paper will examine the neuronal mechanisms that allow vestibular reflex systems to produce eye or head rotations that are in the appropriate direction to compensate for perturbations that disturb the angular orientation of the head. After a decade of exploring the dynamic transformations that occur in vestibular and optokinetic reflexes, a number of investigators became interested in analysis of the three-dimensional spatial properties of these reflexes. Robinson[1] and Pellionisz and Graf[2] proposed models of spatial transformations in the vestibuloocular reflex (VOR). Graf,[3,4] McCrea,[5-7] Uchino,[8-11] Wilson,[12,13] and their colleagues used electrophysiological and morphophysiological techniques to reveal circuitry involved in these transformations. Simpson and colleagues studied spatial transformations in the rabbit's optokinetic reflex,[14] while we[15-17] and others[18,19] undertook parallel studies of the cat's VOR and vestibulocollic reflex (VCR). The latter studies have provided information on vestibular reflexes elicited by both semicircular canal and otolith receptors of the vestibular labyrinth. We are closer to understanding the neuronal substrates of the canal-ocular and canal-neck reflexes, which will therefore be the focus of this review.

The spatial properties of the input to the canal-ocular and canal-neck reflexes are simple since all afferent fibers of a given canal have the same spatial tuning, which is determined by the geometry of the canal. The directional sensitivity of each afferent of a given canal can thus be represented by a unit vector perpendicular to the plane of that canal extending in the direction corresponding to the rotation that activates that afferent according to the right-hand rule. Throughout this paper we will refer to such a vector as the maximal activation direction (MAD) vector of an afferent, neuron, or muscle. The response of a canal afferent to rotation about other axes is simply the product of the maximal response it exhibits for rotation around its MAD vector multiplied by the cosine of the angle between the axis and its MAD vector. The MAD vectors of the three canals in the cat are well characterized by the work of Curthoys and colleagues.[20] The remainder of this review will examine first what spatial transformations take place between the incoming canal signal and muscle output in the VOR and VCR and second how those transformations may be implemented by vestibuloocular and vestibulospinal neurons.

[a]This work was supported by grants EY 05289, EY 06485, EY 07342 and NS 17489.

485

SPATIAL TRANSFORMATIONS BETWEEN CANAL AND MUSCLE SIGNALS

The patterns of muscle activation that are required to generate a compensatory VOR or VCR response to a three-dimensional rotation are constrained by the pulling actions of the eye or neck muscles, which have been determined from their geometry in the cat.[21,22] However, because the number of muscles (6 for eye, 30 for neck) exceeds the three rotational degrees of freedom, a unique reflex muscle pattern cannot be determined from such pulling directions. Rather, there exist an infinite number of possible patterns that could generate any particular eye or head rotation. We have therefore employed electromyographic (EMG) recording to determine experimentally the muscle activation patterns underlying the canal-ocular and canal-neck reflexes.

Two general features of EMG responses elicited by the VOR and VCR are: (1) muscles respond to sinusoidal rotations with sinusoidal modulations of firing rate, and (2) like primary canal afferents, each muscle has a plane and direction of rotation for which the response was maximal, and a null-response for rotations orthogonal to that plane. A scaled cosine function of the angle between the maximal response axis and the axis of rotation describes the response between these extremes. As described above, this behavior can be summarized by calculating for each muscle a MAD vector, representing the three-dimensional axis and direction of rotation which maximally excites it. MAD vectors of each muscle were consistent across cats as illustrated for the VOR in Fig. 1B.

Vestibuloocular Reflex

To define the spatial transformations in the VOR, we measured responses of the 12 extraocular muscles in decerebrate cats to rotation in many planes.[23] A striking finding, illustrated in FIGURE 1B, was that the 6 muscles of each eye operated as antagonistic pairs: paired muscles had colinear axes, but opposite directions, of maximal activation. The muscles could thus be treated as three yoked pairs as postulated by Robinson[1] in his matrix model of the VOR. Assuming that the six canals also act as pairs, which is reasonable given their commissural connections,[24] the overall transformation in the cat's VOR can be expressed as a 3×3 matrix, labeled B in FIGURE 1C.

FIGURE 1A shows that MAD vectors of horizontal rectus muscles are closely aligned with those of the paired horizontal semicircular canals. Thus activation of horizontal eye movers in the cat during normal VOR can be explained by connections of the horizontal, but not vertical, semicircular canals to the medial and lateral rectus. Correspondingly the first two terms in the third row of B are ~ 0. Similarly horizontal canal input to vertical rectus and oblique muscles is negligible (first two terms in third column of B). However, the vertical rectus and oblique muscle vectors are misaligned with those of the nearest vertical canal pair so that appropriate activation of these muscles requires a combination of signals from both vertical canal pairs onto vertical rectus and oblique muscles. As indicated for muscles of the right eye in FIGURE 1C, the obliques require a 0.17 input from the ipsilateral RALP canal pair in addition to their primary 0.98 input from the RPLA pair while the vertical recti require a 0.42 input from the contralateral left anterior-right posterior (equivalent to a -0.42 input from RPLA) canal pair in addition to their primary input from the RALP pair. The connection strengths within the B matrix provide a quantitative framework for neuronal analysis of VOR pathways, suggesting what kind of convergent canal signals must be present in reflex circuits to execute the observed VOR.

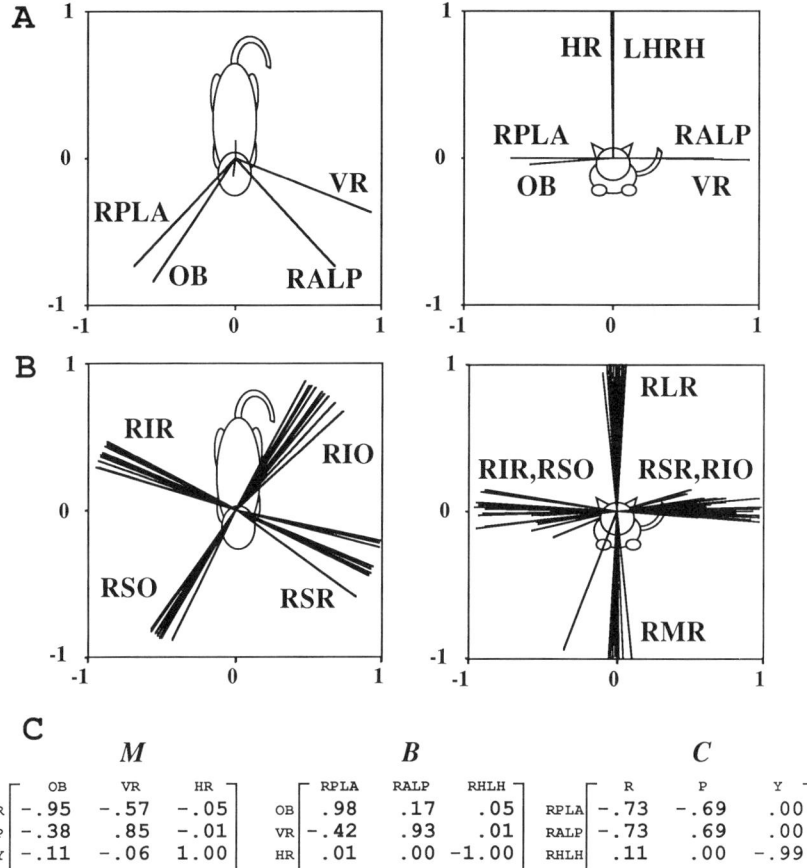

FIGURE 1. Spatial transformation between canals and eye muscles in VOR (from Baker and Peterson).[23] **A** shows MAD vectors of the three pairs of canals (left and right horizontal canals, LHRH; right posterior and left anterior canals, RPLA; right anterior and left posterior canals, RALP) and of the three eye muscle pairs (lateral and medial rectus, HR; superior and inferior oblique, OB; superior and inferior rectus, VR) of the right eye. Note alignment of MADs in front view at right, which indicates that interaction of vertical and horizontal canals is not required in cat's VOR. **B** shows populations of MAD vectors of 79 muscles from 13 cats, which indicate that the MAD vectors of the two muscles in each pair are opposite to one another. **C** contains matrices that describe the spatial transformations in the VOR based on data shown in A. Matrix *C* indicates the transformation of Roll, Pitch and Yaw components of head movements into activation of the three canal pairs. Matrix *M* indicates how activity of three muscle pairs generates roll, pitch, and yaw components of head movements. Matrix *B* summarizes the brain transformations that convert canal signals into activity of motoneurons. The product of the three matrices is assumed to generate a perfectly compensatory VOR equal to $-I$, where I is the identity matrix. Translating the numbers in B into actual neural connections requires knowledge of the anatomy of the vestibuloocular connections. For instance, the 0.98 connection between RPLA and OB involves second-order vestibular neurons that receive input from the right posterior canal, some of which project to and excite motoneurons of the right superior oblique muscle that are located in the left trochlear nucleus, others of which project to and inhibit motoneurons of the right inferior oblique that are located in the right oculomotor nucleus.

Vestibulocollic Reflex

In the VCR the situation is much more complex since the neck is a multijoint system with far more muscles than the eye. Once again we first addressed this problem by recording the activation of seven different neck muscles during rotation in multiple planes in decerebrate[25] and alert[26] cats. The VCR differs from the VOR in two ways. First, at low frequencies nonaligned canal and otolith inputs often interact to give a VCR with a significant gain and varying phase for rotations in all directions. This behavior may match VCR output to orientation-dependent loads imposed by the head-neck system. At 0.25 Hz or above, where canal input dominates, it was possible to determine neck muscle MAD vectors defining the spatial transformation occurring in the canal-neck reflex. The figurines at the bottom of FIGURE 2 show such MAD vectors for 7 of the 15 neck muscles, which indicate that each muscle receives highly convergent canal input.

The second difference is that there is no obvious grouping of muscle vectors along three axes. Thus from the point of view of canal-induced stabilization of head angular position, the muscle pattern is nondeterministic because the motor plant is overcomplete: 30 neck muscles are used to generate three degrees of freedom of rotational movement. The central nervous system (CNS) must therefore choose a muscle pattern from an infinite number of possibilities. As can be seen by comparing the MAD and pulling vectors at the bottom and top of FIGURE 2, the chosen MAD vectors are not aligned with muscle pulling directions. However, they were well approximated by predictions of a tensorial model[27] that treats the VCR as three successive transformations. The first two transformations, which convert the incoming canal signal from a covariant (projection) form to a contravariant (parallelogram) form and then project those signals onto the coordinate frame of the neck muscles, are uniquely determined by the canal and neck muscle geometry and therefore are not the site of the optimizing process. They generate a covariant representation of the intended movement in muscle space. However, the system requires as its output a contravariant representation in which torque forces generated by muscle activation sum in parallelogram fashion to generate the desired head movement. Because the number of muscles exceeds three, there are an infinite number of transformations between the covariant and contravariant signal in neck muscle space: it is here that an optimizing choice is required. The model chooses the Moore Penrose generalized inverse of the covariant motor metric[28] and arrives at a good approximation of the output pattern that is consistently chosen by the cat's CNS. It is relevant to the examination of neural behavior that follows since it also predicts the types of signals that should exist within VCR neural circuits. Among these are signals aligned with neck muscle pulling and activation directions.

NEURONAL SUBSTRATES OF SPATIAL TRANSFORMATIONS IN THE VOR

Because it makes a major contribution to the VOR and is accessible for study, we have focused on the role of the three-neuron VOR arc in implementing the spatial transformations that were discussed in the previous section. Our approach has been to quantify the spatial response properties of second-order vestibuloocular relay neurons (VOR relay neurons) whose projections to extraocular motor nuclei are identified either electrophysiologically or morphologically. VOR relay neurons can be classified by the canal from which they receive their direct, monosynaptic input. Morphologically each canal group can be subdivided into ipsilaterally projecting (inhibitory) and contralaterally projecting (excitatory) neurons.[10,11,29,30] Relay neurons

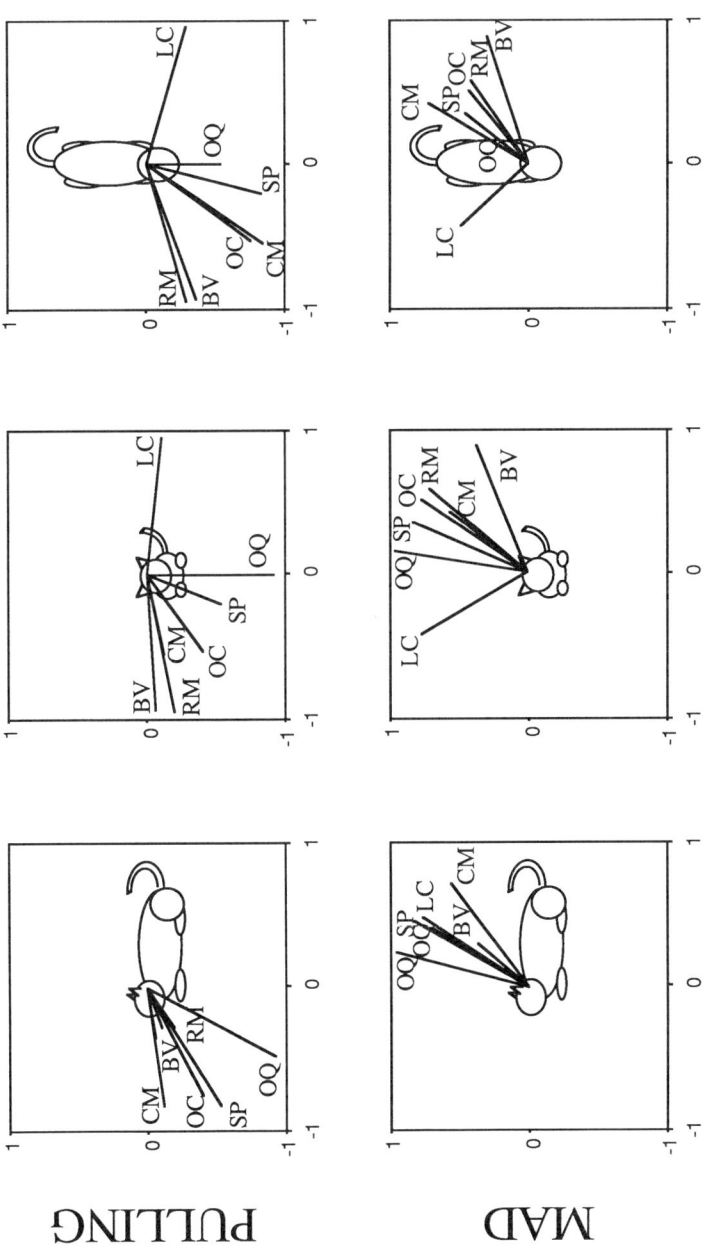

FIGURE 2. Spatial transformation between canals and 7 neck muscles in VCR (from Wickland *et al.,*[22] 1991). Top row of figurines shows the head torque vectors produced by seven major neck muscles. These were estimated from anatomical measurements of the origins and insertions of each muscle and of the rotation axes of the C1-skull and C1-C2 joints. The bottom row shows the MAD vectors of the same neck muscles measured in the decerebrate cat by Baker *et al.*[25] Note that pulling and MAD vectors are typically not colinear.

also may or may not send branches to the spinal cord, making four subclasses. Vertical canal VOR relay neurons are further divided according to whether they terminate uni- or bilaterally in the oculomotor nucleus thus yielding eight subclasses per canal.[3,4,7,10,31] To determine what roles these subclasses play in spatial transformations in the VOR, we must realize that a VOR relay neuron can contribute to such transformations by receiving convergent input from multiple canals and/or by sending divergent projections to multiple motor pools. Robinson has theorized that divergent projections implement the transformation,[1] whereas the model of Pellionisz and Graf implies that most of the transformation occurs at the second-order neuron.[2] Our data, which are described below, indicate that both mechanisms play a crucial role.

Fukushima et al. studied second-order vestibular nucleus neurons which were antidromically activated from the region of the oculomotor nucleus (VOR relay neurons) in alert cats during whole-body rotations in many horizontal and vertical planes.[15] Results are similar to those obtained in collaboration with W. Graf from intraaxonal recordings of neurons driven monosynaptically with labyrinthine shock in decerebrate animals.[17] FIGURE 3 illustrates the properties of three typical neurons. Like primary afferents, VOR relay neurons responded to sinusoidal rotations with sinusoidal modulation of firing rates, response gains (spikes/second per deg/second) varied as a cosine function of the angle between the rotation axis and a MAD axis, and phases were near that of head velocity, suggesting linear summation of canal inputs. Plots of response gain versus the angle between the rotation axis and the pitch axis in FIGURE 3B indicated that the three neurons from left to right received their vertical input from the posterior canal only, both anterior and posterior canals, and the anterior canal only. In addition the neuron at the right received significant convergent input from the ipsilateral horizontal canal. These canal inputs give rise to the MAD vectors shown in FIGURE 3C. FIGURE 4 shows the MAD vectors of our entire population of VOR relay neurons, which divided into three distinct populations, indicating primary canal input from either anterior, posterior, or horizontal semicircular canal (AC, PC, HC cells). Of 80 AC and PC neurons studied by Fukushima et al.,[15] at least 6 received input from more than one vertical canal, indicated by MAD azimuths which were ≥ 10° from their primary canal plane (either the canal pair plane shown in FIGURE 4 or the plane of the single ipsilateral canal). In addition, 21/80 received convergent input from a horizontal canal, with about equal number of type I and type II yaw responses. In alert cats activity of many vertical VOR relay neurons was also related to vertical and/or horizontal eye position. All AC and PC cells that had vertical eye position sensitivity had upward and downward on-directions, respectively.

A number of VOR relay neurons in FIGURE 4 had MAD vectors centered around the MAD vectors of the oblique eye muscles (PC cells aligned with superior oblique, AC cells with inferior oblique). Thus, signals with spatial properties appropriate to activate these muscles are present at the VOR relay neuron level. FIGURE 4C indicates that such spatial properties are not common in irregularly discharging VOR relay neurons, which send branches to the spinal cord.[31] We therefore hypothesize that the oblique signal is carried by a specific group of contralaterally projecting vestibuloocular neurons that terminate only unilaterally in trochlear and oculomotor nuclei and end especially heavily in the former (cf. Reference 3, Figure 4).

In contrast to the presence of VOR relay neurons that had MAD vectors resembling those of oblique motoneurons, no VOR relay neuron had a MAD vector near those of superior or inferior rectus muscles. Signals appropriate to activate these muscles might be produced by combining excitatory signals from ipsi- and

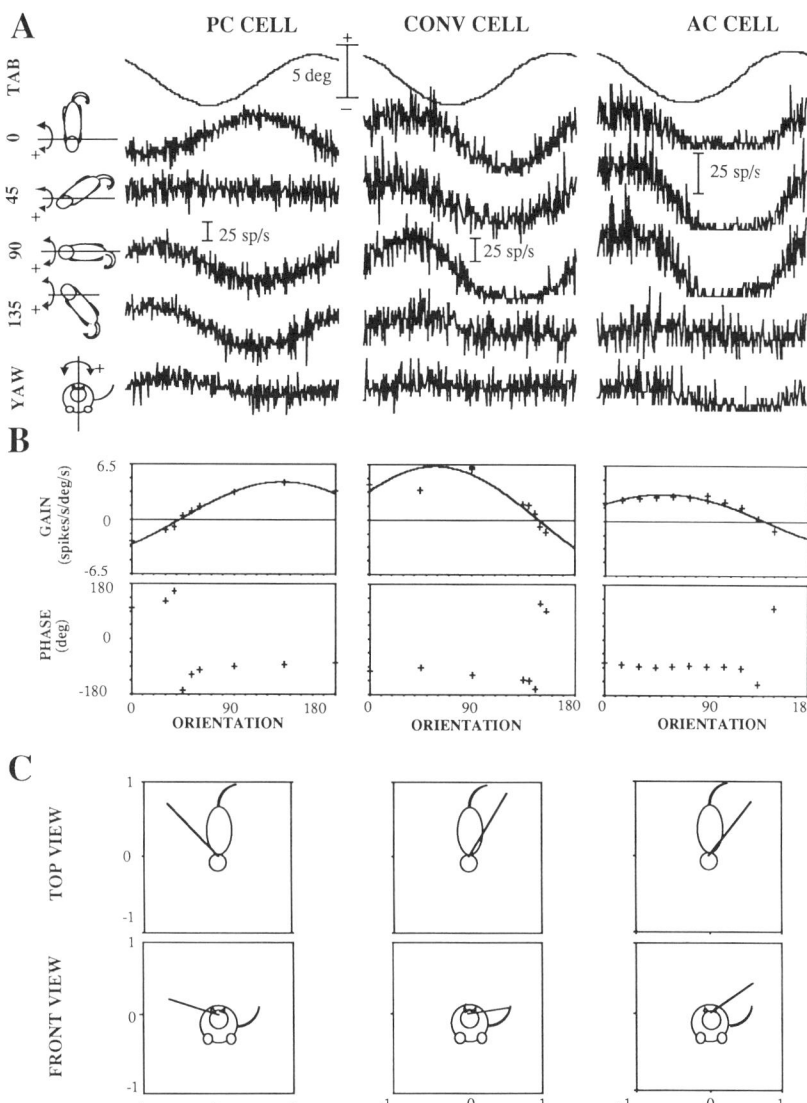

FIGURE 3. Responses of three second-order vestibuloocular neurons during 0.5 Hz sinusoidal rotation in multiple planes. In **A** figurines show four vertical rotations, specified by the angle between their axis and the pitch axis plus horizontal (yaw) rotation. Histograms of the spikes of each neuron during the rotation cycle are shown in three columns at right. **B** plots gain and phase of each neuron's response in a number of different vertical planes versus the orientation of the rotation axis. Curves are least-squares sinusoids fitted to the data. **C** shows normalized MAD vectors for each neuron determined by its response to vertical plane rotations and to yaw. Use the right-hand rule to determine the rotation represented by each vector.

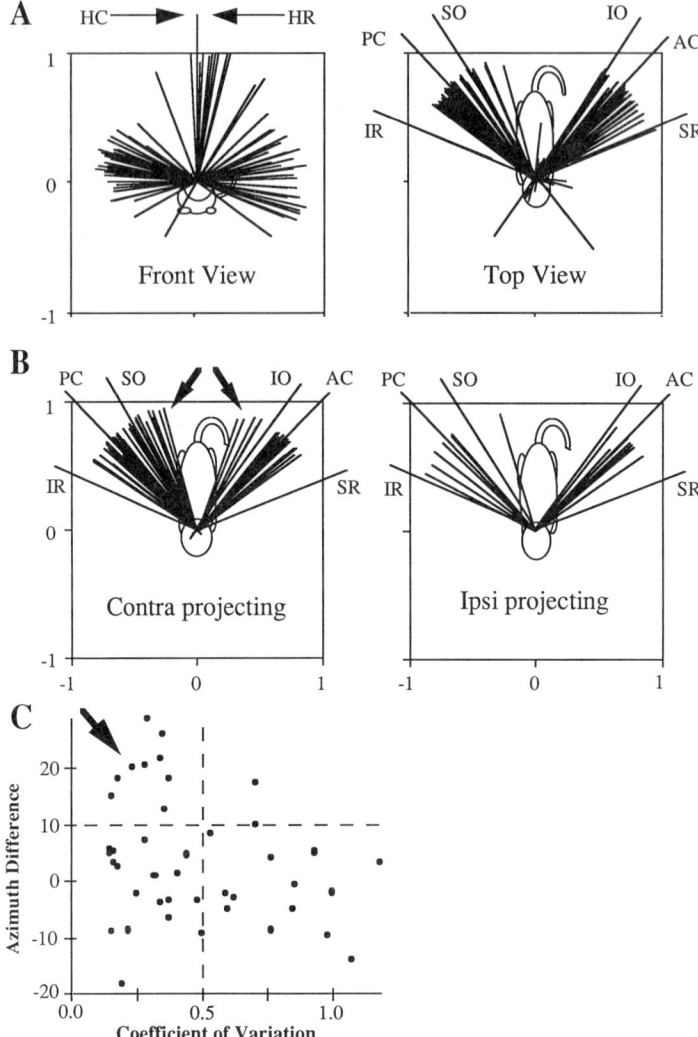

FIGURE 4. Maximal activation directions of second-order vestibuloocular neurons in alert and decerebrate cats. **A:** Front view at left contains MAD vectors of second order neurons antidromically activated from oculomotor nucleus in alert cats from Fukushima et al.[15] and recorded intraaxonally in ascending medial longitudinal fasaculus (MLF) in decerebrate cats (Graf, Baker, Perlmutter and Peterson, in preparation). Top view at right contains MAD vectors only from neurons in alert cats, vectors of the remaining neurons from decerebrate cats are shown in B. These neurons were classified as contra- or ipsilaterally projecting depending upon whether they were monosynaptically activated from the vestibular nerve on the side opposite or adjacent to the recording site. Arrows highlight the fact that most neurons with MAD vectors shifted in direction of oblique muscles were contralaterally projecting (excitatory) neurons. Vectors extending beyond the frames in A and B indicate the MAD vectors of horizontal, anterior, and posterior semicircular canals (HC, AC, PC) and horizontal rectus (HR), superior and inferior oblique (SO, IO) and superior and inferior rectus (SR, IR) eye muscles. **C:** Relation between firing regularity (measured as coefficient of variation of resting discharge) and direction of MAD vector for the population of neurons whose vectors are plotted in B. Ordinate plots azimuth angles between neuron and anterior or posterior canal MAD vectors measured so that a positive difference angle indicates a shift towards the posteriorly extending roll axis. (Azimuth angle refers to the angle between the projections of two vectors onto the pitch-roll plane.) Note that all but one neuron with azimuth difference $> 10°$ had a regular firing pattern (coefficient of variation < 0.5).

contralateral AC neurons (for superior rectus) or PC neurons (for inferior rectus). However, even assuming that these projections arise selectively from the nonconvergent VOR relay neurons of FIGURE 4, the recrossing projections must have 40% of the weight of the main projections to generate the required vertical rectus response. This value seems large given the morphological data[3,4] and must be confirmed experimentally for the proposed mechanism to be accepted. Alternatively, less direct pathways such as those involving higher-order vestibular, prepositus, or interstitial nucleus of Cajal neurons might play a crucial role in generating vertical rectus signals. Our data suggest that higher-order neurons within the vestibular nucleus are an unlikely source of these signals. Such neurons constitute less than 25% of oculomotor-projecting vestibular neurons, and very few of them had MAD vectors aligned with those of vertical rectus muscles. Further studies on vestibular responses outside the vestibular nuclei (and on responses of superior vestibular nucleus which we did not extensively sample) are required to determine if premotor signals appropriate to drive vertical recti are present outside the oculomotor nucleus.

As shown in FIGURE 4B, our intracellular experiments have revealed that VOR relay neurons with ipsi- and contralaterally projecting axons have different spatial properties. Most neurons with MAD vectors shifted $\geq 10°$ in the direction of oblique muscle MAD vectors were contralaterally projecting (excitatory) VOR relay neurons. They also had regular discharge patterns, thus are less likely to send branches to spinal cord than are neurons with more irregular discharge.[31] Because of the high degree of canal convergence on neck motoneurons, we had expected VOR relay neurons with branches descending to spinal cord, most of which discharge irregularly,[31] to exhibit more convergent input than those projecting only to oculomotor nuclei—exactly the opposite of the result shown in FIGURE 4B. In retrospect the lack of convergence in branching neurons is understandable if we assume that convergence is shifting the spatial tuning of vestibular neurons into alignment with specific target motoneurons. Such tuning would be most appropriate in neurons with projections to a limited number of motor pools, such as neurons projecting only to oculomotor or neck motor nuclei. Branching neurons might then serve as a "party line" to convey unmodified canal signals to multiple sites, where they could be modified by local, intrinsic circuitry or by patterns of terminations of the branching neuron. Further evidence for such a distinction between party line and "private line" neurons will be presented in the next section.

Horizontal-vertical convergence was common in both ipsi- and contralaterally projecting VOR relay neurons. However, type II horizontal responses tended to occur more frequently in AC and PC type VOR relay neurons with a regular discharge pattern. This may be a special response pattern that helps to balance type I and II inputs to vertical extraocular motoneurons.

In conclusion, the spatial properties of VOR relay neurons are sufficient to explain the spatial transformation that occurs in the VOR provided that neurons with bilateral terminations in vertical rectus motor nuclei have a rather strong recrossing action on ipsilateral vertical rectus motoneurons. If this were so, we would conclude that the three-neuron arcs and more complex VOR pathways form a parallel system both of whose parts are adapted to generate spatially accurate eye movements. This is the alternative we favor since we have shown that the directional properties of VOR relay neurons are plastic,[32] so that adaptive adjustments should lead them to develop fully compensatory spatial properties. If the ipsilateral terminations do not have the requisite strength, the higher-order pathways will have to be sought to explain those aspects of the transformation not implemented by the three-neuron arcs.

NEURONAL SUBSTRATES OF SPATIAL TRANSFORMATIONS IN THE VCR

Circuitry of the vestibulocollic system is more complex because there are several populations of vestibulospinal neurons that project to upper cervical grey matter and that could be involved in generating the VCR. We have utilized a combination of electrophysiological and morphophysiological techniques to categorize these neurons and to determine the signals they carry. The projection pattern of each neuron was identified by antidromic activation from, or by recording intraaxonally in, the ipsi- or contralateral vestibulospinal tracts at the C1 level, and by responses to electrical stimulation of IIIrd nucleus and C6 segment. The spatial properties of the signal carried by the neuron was then determined from its responses to 0.5 Hz sinusoidal rotations in many planes as described above. Finally, some axons were filled with horseradish peroxidase or neurobiotin to allow their terminal arbors in the C1 segment to be visualized and reconstructed.

MAD vectors of four classes of vestibulospinal neurons are shown in FIGURE 5. Neurons with different projections responded differently. Vestibulooculocollic (VOC) neurons, which had axon branches ascending to oculomotor nucleus, typically terminated rostral to the C6 segment and had spatial properties that closely resembled VOR relay neurons, with MAD vectors approximately aligned with those of an ipsilateral semicircular canal. VC neurons, which projected only to the neck and not to C6 or oculomotor nucleus, received more convergent input. While some had MAD vectors that aligned with those of ipsilateral canals, others exhibited strong vertical-horizontal canal convergence, in some cases sufficient to preclude any classification as a vertical canal or HC neuron. A few VC neurons, indicated by an asterisk in FIGURE 5, received their primary input from contralateral canals. The greater degree of convergence in VC neurons versus VOC neurons is consistent with our hypothesis that branching neurons will show little convergence. The diversely distributed MAD vectors of VC neurons may reflect the diverse MAD vectors of neck muscles shown in FIGURE 2.

MAD vectors of lateral vestibulospinal neurons and medial vestibulospinal neurons activated antidromically from the C6 segment were very different from those of VOR relay neurons and VC neurons. Neurons excited strongly by contralateral yaw rotation (type II response indicated by asterisk) or receiving convergent input from two vertical canals were common. Thus, whereas three-neuron VOR and VCR arcs relay primarily single ipsilateral canal signals to extraocular and neck motor nuclei, projections beyond the neck involve extensive processing of canal information at the second-order level.

Following characterization of their responses to rotations in many planes, 21 neurons studied using intraaxonal recording were stained by passing current pulses to inject a tracer into the axon. Neurobiotin staining enabled reconstruction of axonal collaterals and terminal arbors over long distances. The projection pattern of a neuron's terminal arbor in spinal gray matter varied according to the neuron's physiological response properties. Neurons that received their primary input from the ipsilateral posterior canal ended most heavily in motor pools of neck flexors that produce head movement in a direction opposite to the upward head rotation that activates the posterior canals. Two neurons with primary input from the ipsilateral anterior canal ended most heavily in neck extensor motor pools, but the third, whose terminations are shown at the bottom of Fig. 6, ended in the vicinity of neck flexor motoneurons. If we assume that the first two neurons were excitatory and the third inhibitory, these neurons would help generate neck extension required to counter the downward head rotation that activates the anterior canals. Horizontal-canal-

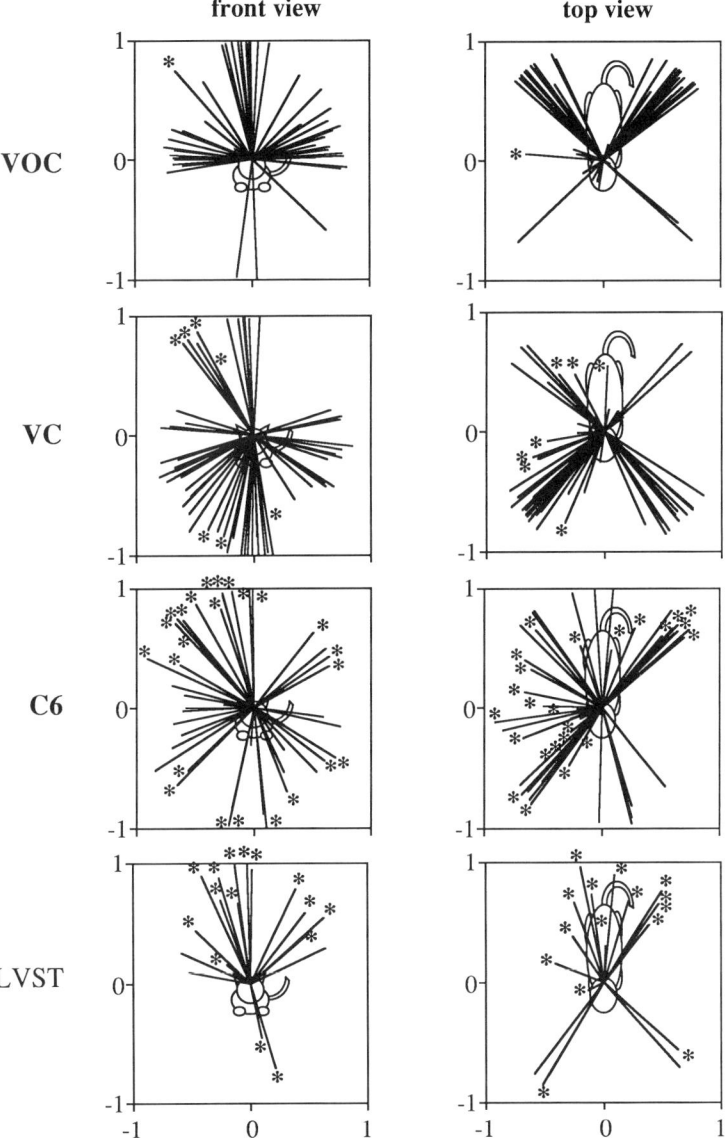

FIGURE 5. Maximal activation directions of second-order medial and lateral vestibulospinal neurons. Ipsilaterally and contralaterally projecting medial vestibulospinal tract (MVST) neurons have been merged so as to indicate the combined population of spatial responses that reach the right neck segments. Asterisks indicate neurons that were primarily activated by contralateral canals. **A:** VOC-type second-order MVST neurons. The majority of these neurons were contralaterally projecting. **B:** VC-type second-order MVST neurons. The majority of these neurons were ipsilaterally projecting. **C:** Second-order MVST neurons whose axons continued beyond C6. **D:** Lateral vestibulospinal tract neurons. Most of these projected beyond C6.

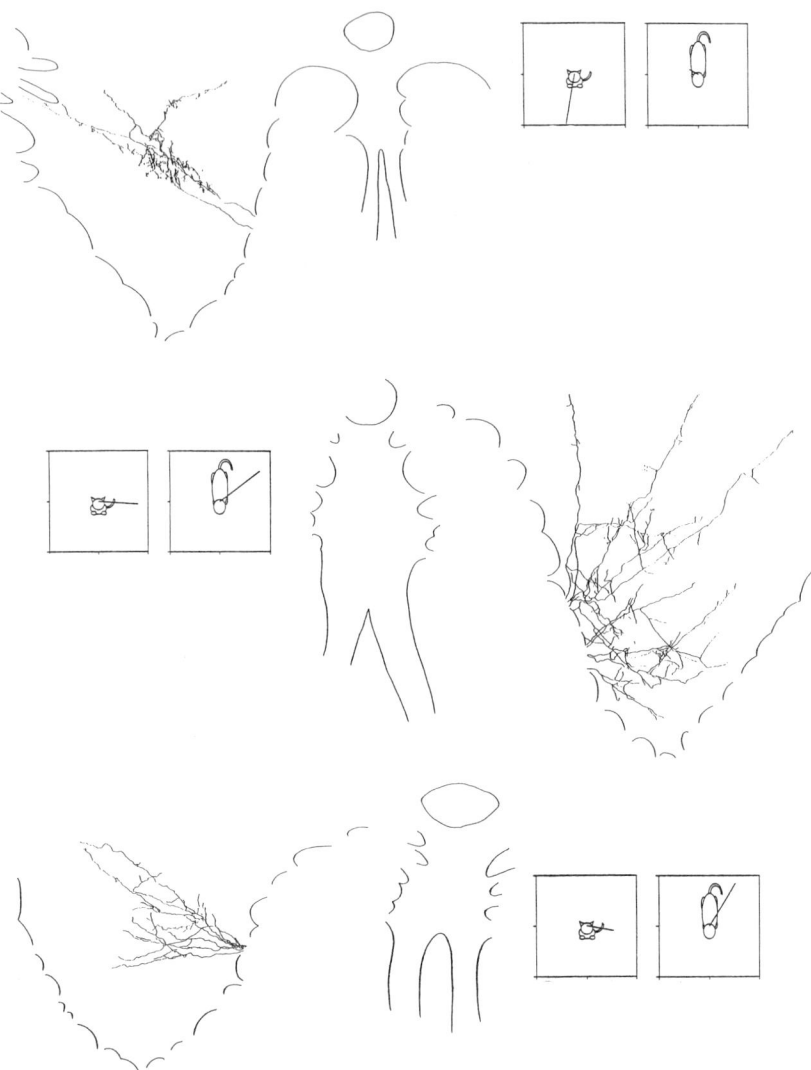

FIGURE 6. Terminations of different classes of vestibulospinal neurons in C1 ventral horn. Each cross-sectional diagram of the C1 ventral horn superimposes several collaterals of an intracellularly labeled second-order medial vestibulospinal neuron. Adjacent figurines depict the MAD vector of that neuron. At the top is a vestibulospinal neuron that projected ipsilaterally beyond the C6 level whose response was dominated by input from the contralateral horizontal canal (even though the neuron responded monosynaptically to stimulation of the ipsilateral vestibular nerve). In the middle is a contralaterally projecting VOC neuron that responded only to the ipsilateral anterior canal. At the bottom is an ipsilaterally projecting VC neuron that received convergent input from ipsilateral anterior and posterior canals. Its termination in the neck flexor motor pools suggests that it was an inhibitory neuron.

related neurons, on the other hand, had more widespread arbors that included both flexor and extensor motor pools, presumably because flexors and extensors are coactivated or coinhibited during a horizontal head movement.

FIGURE 6 illustrates another anatomical correlate of spatial activity. Neurons, like the VOC neuron shown in the middle of FIGURE 6, that received input from a single canal arborized extensively in the C1 grey matter and often projected to both spinal and oculomotor nuclei, whereas neurons that received convergent canal input often projected only to the neck and terminated in a more focal fashion as illustrated by the VC neuron at the bottom of FIGURE 6. We therefore view the latter as "private line" neurons that carry a specialized signal to a select group of spinal neurons and the former as "party line" neurons that carry an unmodified afferent signal to many different target sites. The party line signal could either be transformed into more specific motor patterns in each target zone or could serve some other motor priming function. The final neuron shown at the top of FIGURE 6 is a C6 projecting neuron that received a convergent input dominated by contralateral canals. As was typical of such neurons, its terminal arbor appeared to avoid the motor pools and terminate in interneuron areas. It could be providing information needed to coordinate neck and limb movements. Such neurons had significantly fewer terminal branches in the upper cervical segments than VOC or VC neurons.

If C6 projecting neurons are excluded, the spatial transformations mediated by the medial vestibulospinal tract can be appreciated by comparing FIGURE 5A and B with the MAD vectors of neck muscles shown in FIGURE 2. It can be seen that the extensive horizontal-vertical canal convergence in the VC population can help to generate similar convergence required at the motoneuronal level. On the other hand, we once again fail to find neurons that carry signals appropriate to activate muscles whose MAD vectors fall near the pitch plane. As in the case of the VOR, we assume that these are generated by convergent action of vestibulospinal neurons upon motoneurons of muscles like biventer cervicis and rectus capitis major. There are ample substrates for such convergence given the known ipsi- and contralateral projections of excitatory and inhibitory vestibulocollic neurons.[8–11]

The last point brings us back to the VCR model[27] described in the second section. The model suggests that neurons representing intermediate stages of the spatial transformation in the VCR would have MAD vectors aligned with neck muscle pulling and activation directions. The lack of vestibulospinal neurons whose MAD vectors align with the activation or pulling directions of biventer cervicis and rectus capitis major muscles indicates that the important three-neuron arc component of the VCR does not fully match the model predictions. Thus a different set of internal transformations must be sought to explain how the VCR system produces a motor output that has the optimization properties of the Moore-Penrose generalized inverse. Our morphological data do not provide the quantitative information on connections between vestibulospinal neurons and motoneurons that is required to address this problem. Its resolution must await more definitive experiments that identify the strengths of specific neuron-muscle connections.

REFERENCES

1. ROBINSON, D. A. 1982. The use of matrices in analyzing the three-dimensional behavior of the vestibulo-ocular reflex. Biol. Cybern. **46:** 53–66.
2. PELLIONISZ, A. & W. GRAF. 1987. Tensor network model of the "three-neuron vestibulo-ocular reflex arc" in cat. J. Theor. Neurobiol. **5:** 127–151.
3. GRAF, W., R. A. McCREA & R. BAKER. 1983. Morphology of posterior canal related secondary vestibular neurons in rabbit and cat. Exp. Brain Res. **52:** 125–138.

4. GRAF, W. & K. EZURE. 1986. Morphology of vertical canal related second order vestibular neurons in the cat. Exp. Brain Res. **63:** 35–48.
5. McCREA, R. A., K. YOSHIDA, A. BERTHOZ & R. BAKER. 1980. Eye movement related activity and morphology of second order vestibular neurons terminating in the cat abducens nucleus. Exp. Brain Res. **40:** 468–473.
6. McCREA, R. A., A. STRASSMAN, E. MAY & S. M. HIGHSTEIN. 1987. Anatomical and physiological characteristics of vestibular neurons mediating the horizontal vestibulo-ocular reflex of the squirrel monkey. J. Comp. Neurol. **264:** 547–570.
7. McCREA, R. A., A. STRASSMAN & S. M. HIGHSTEIN. 1987. Anatomical and physiological characteristics of vestibular neurons mediating the vertical vestibulo-ocular reflex of the squirrel monkey. J. Comp. Neurol. **264:** 571–594.
8. HIRAI, N. & Y. UCHINO. 1984. Superior vestibular nucleus neurones related to the excitatory vestibulo-ocular reflex of anterior canal origin and their ascending course in the cat. Neurosci. Res. **1:** 73–79.
9. UCHINO, Y., S. SUZUKI, T. MIYAZAWA & S. WATANABE. 1979. Horizontal semicircular canal inputs to cat extraocular motoneurons. Brain Res. **177:** 231–240.
10. UCHINO, Y., S. SUZUKI & S. WATANABE. 1980. Vertical semicircular canal inputs to cat extraocular motoneurons. Exp. Brain Res. **41:** 45–53.
11. UCHINO, Y., N. HIRAI & S. SUZUKI. 1982. Branching pattern and properties of vertical- and horizontal related excitatory vestibuloocular neurons in the cat. J. Neurophysiol. **48:** 891–903.
12. WILSON, V. J. & M. MAEDA. 1974. Connections between semicircular canals and neck motoneurons in the cat. J. Neurophysiol. **2:** 346–357.
13. WILSON, V. J. & M. YOSHIDA. 1969. Monosynaptic inhibition of neck motoneurons by the medial vestibular nucleus. Exp. Brain Res. **9:** 365–380.
14. GRAF, W., J. I. SIMPSON & C. S. LEONARD. 1988. Spatial organization of visual messages of the rabbit's cerebellar flocculus. II. Complex and simple spike responses of Purkinje cells. J. Neurophys. **60:** 2091–2121.
15. FUKUSHIMA, K., S. I. PERLMUTTER, J. F. BAKER & B. W. PETERSON. 1990. Spatial properties of second-order vestibulo-ocular relay neurons in the alert cat. Exp. Brain Res. **81:** 462–478.
16. PERLMUTTER, S. I., Y. IWAMOTO, J. F. BAKER & B. W. PETERSON. 1990. Spatial response properties and projection patterns of secondary medial vestibulospinal neurons. Soc. Neurosci. Abstr. **16:** 968.
17. PETERSON, B. W., W. GRAF & J. F. BAKER. 1987. Spatial properties of signals carried by second-order vestibuloocular relay neurons in the cat. Soc. Neurosci. Abstr. **13:** 1093.
18. BERTHOZ, A., G. MELVILL JONES & A. E. BEGUE. 1981. Differential visual adaptation of vertical canal-dependant vestibulo-ocular reflexes. Exp. Brain Res. **44:** 19–26.
19. KASPER, J., R. H. SCHOR & V. J. WILSON. 1988. Response of vestibular neurons to head rotations in vertical planes. I. Response to vestibular stimulation. J. Neurophysiol. **60:** 1753–1764.
20. CURTHOYS, I., R. BLANKS & C. MARKHAM. 1977. Semicircular canal functional anatomy in cat, guinea pig and man. Acta Otolaryngol. **83:** 258–265.
21. EZURE, K. & W. GRAF. 1984. A quantitative analysis of the spatial organization of the vestibulo-ocular reflexes in lateral- and frontal-eyed animals. I. Orientation of semicircular canals and extraocular muscles. Neuroscience **12:** 85–93.
22. WICKLAND, C. R., J. F. BAKER & B. W. PETERSON. 1991. Torque vectors of neck muscles in the cat. Exp. Brain Res. **84:** 649–659.
23. BAKER, J. F. & B. W. PETERSON. 1991. Excitation of the extraocular muscles in decerebrate cats during the vestibulo-ocular reflex in three-dimensional space. Exp. Brain Res. **84:** 266–278.
24. KASAHARA, M. & Y. UCHINO. 1974. Bilateral semicircular canal inputs to neurons in cat vestibular nuclei. Exp. Brain Res. **20:** 285–296.
25. BAKER, J. F., J. GOLDBERG & B. W. PETERSON. 1985. Spatial and temporal response properties of the vestibulocollic reflex in decerebrate cats. J. Neurophysiol. **54:** 735–756.
26. BANOVETZ J. M., S. A. RUDE, S. I. PERLMUTTER, B. W. PETERSON & J. F. BAKER. 1987. A

comparison of neck reflexes in the alert and decerebrate cat. Soc. Neurosci. Abstr. **13:** 1312.

27. PELLIONISZ, A. J. & B. W. PETERSON. 1988. A tensorial model of neck motor activation. *In* Control of Head Movement. B. W. Peterson & F. J. Richmond, Eds. 178–186. Oxford University Press. New York, N.Y.

28. PELLIONISZ, A. J. 1984. Coordination: a vector-matrix description of transformations of overcomplete CNS coordinates and a tensorial solution using the Moore-Penrose generalized inverse. J. Theor. Biol. **101:** 353–375.

29. BAKER, R., N. MANO & H. SHIMAZU. 1969. Postsynaptic potentials in abducens motoneurons induced by vestibular stimulation. Brain Res. **15:** 577–580.

30. BAKER, R., W. PRECHT & A. BERTHOZ. 1973. Synaptic connections to trochlear motoneurons determined by individual nerve branch stimulation in the cat. Brain Res. **64:** 402–406.

31. IWAMOTO, Y., T. KITAMA & K. YOSHIDA. 1990. Vertical eye movement-related secondary vestibular neurons ascending in the medial longitudinal fasciculus in the cat. I. Firing properties and projection pathways. J. Neurophysiol. **63:** 921–936.

32. PERLMUTTER, S. I., K. FUKUSHIMA, B. W. PETERSON & J. F. BAKER. 1988. Spatial properties of second order vestibuloocular relay neurons in the alert cat. Soc. Neurosci. Abtr. **14:** 331.

Spatial Transformation in the Vertical Vestibulocollic Reflex[a]

V. J. WILSON, P. S. BOLTON, T. GOTO, R. H. SCHOR,[b]
Y. YAMAGATA,[c] AND B. J. YATES

The Rockefeller University
1230 York Avenue
New York, New York 10021
[b] The Eye and Ear Institute
University of Pittsburgh
Pittsburgh, Pennsylvania 15213

INTRODUCTION

Movement of the body in space results in vestibulospinal reflexes acting on the head, limbs, and trunk. These reflexes, which have been studied extensively in neck and forelimb muscles of the decerebrate cat, are produced by converging otolith and canal inputs. The former predominate at low, the latter at high, frequencies of sinusoidal stimulation. Reflexly activated muscles have an associated response vector orientation that characterizes the stimulus plane evoking the maximal response. In forelimb muscles of decerebrate cats, vector orientation remains stable as the frequency of sinusoidal tilt increases: the reflex is apparently produced by convergent inputs from otolith and canal receptors that are spatially aligned.[1]

The situation is more complex for the vestibulocollic reflex (VCR) evoked by stimuli in vertical planes.[2] In contrast to the frequency-stable response vector orientation in limb muscles, the orientation of neck muscle vectors can shift considerably as frequency changes. Indeed, at frequencies around 0.1–0.2 Hz, it may be impossible to define a "best" stimulus direction as all directions may have comparable efficacy. This behavior of the VCR was called spatiotemporal convergence (STC behavior), as it was hypothesized to arise from canal and otolith influences (with inherently different temporal properties) having different spatial characteristics.[2] Because of STC behavior, response vector orientations of neck muscles have been measured only at higher stimulus frequencies, such as 0.5 Hz, at which they are mainly produced by canal inputs. The orientations of these vectors, some being near pitch, others near roll, others at planes in between,[2] suggest that many of them are produced by convergent inputs from more than one vertical canal. The question arises, where does the convergence that produces neck muscle vectors and STC behavior take place? More specifically, to what extent does it take place in the brain stem, i.e., in the vestibular nuclei and pontomedullary reticular formation? Our recent experiments have addressed this question.

[a]Supported by National Institutes of Health grants NS02619, NS24930, and DC00693.
[c]Present affiliation: Department of Ophthalmology, Hyogo College of Medicine, Nishinomiya-City, Hyogo 663, Japan.

CONVERGENCE IN THE VESTIBULAR NUCLEI

In a series of experiments on decerebrate cats,[3,4] we have studied the responses, to vertical vestibular stimulation, of neurons in regions of the vestibular nuclei that project to the spinal cord: Deiters' nucleus, the descending nucleus, and the rostral medial nucleus. Among other things, we looked for the convergence of afferent inputs needed to produce two specific aspects of the VCR: STC behavior, and convergence between bilateral vertical canals that would produce response vector orientations near pitch.

All neurons studied were spontaneously active. In some experiments they were identified by antidromic stimulation as projecting to the neck segments, but no further; these were called vestibulocollic, or VC, neurons.[4] Other neurons were not identified as to projection, or responded antidromically to stimulation with an electrode at C3–C4, but were not identified further.[3] The wobble stimulus,[5] a combination of roll and pitch sinusoids 90° out of phase, was used to determine each neuron's response vector orientation. Sinusoidal stimuli, usually at 0.05–1 Hz, oriented in a plane near the response vector, were then used to identify the dynamics, and therefore the receptor origin, of the neuron's labyrinthine input.[3] On

TABLE 1. Distribution of Vestibular Response Types among Vestibulospinal and Reticulospinal Neurons[a]

	Canal	Otolith + Canal	Otolith	STC	Unknown	Total
VS neurons	43 (38)	15 (13)	25 (22)	11 (10)	19 (17)	113
RS neurons	1 (1)	12 (18)	32 (48)	6 (9)	16 (24)	67

[a]VS population includes 86 vestibulocollic neurons (from Wilson *et al.*)[4] and 27 neurons responding antidromically to stimulation in C4, the only level tested.[3] RS data from Bolton *et al.*[12,13] Column labeled "unknown" contains neurons that responded to tilt but could not be classified. Numbers in parentheses are percentages.

the basis of these dynamics, most neurons could be classified as those receiving input from vertical canals only, from otoliths only, and those receiving convergent input. The latter could be divided into those having spatially aligned otolith + canal input, with vector orientations that did not change systematically with stimulus frequency, and those demonstrating STC behavior.

TABLE 1 shows the distribution of labyrinthine inputs for a population of vestibulospinal (VS) neurons; this includes neurons identified as vestibulocollic, as well as neurons antidromically identified from C3–C4 alone (there is no significant difference in the distribution of inputs between VC neurons and a larger vestibular population, including the C3–C4 neurons but mainly unidentified as to projection).[4] The largest group in this population receives vertical canal input only, and there is a substantial number of otolith neurons. The fraction of STC neurons is small (9%), and it is no bigger in the population consisting only of VC neurons (10%). This would seem to be insufficient to bring about the STC behavior observed at the reflex level.

As with the distribution of inputs from different receptor types, there was little difference between our various samples of vestibular nucleus neurons with regard to the distribution of response vector orientations. The upper part of FIGURE 1 shows these orientations for vestibulocollic neurons carrying vertical canal signals alone or in combination with otolith input. Most are ipsilateral, either near the planes of the anterior or posterior semicircular canals (iPC or iAC), or are displaced towards ear-down roll (iED). This shift in orientation towards roll, but not towards pitch (NU or ND), indicates the presence of convergence between vertical canals on one side, but not between bilateral anterior or posterior canals. These neck-projecting neurons could produce muscle response vectors in canal planes, or in planes shifted towards roll, by mono- or polysynaptic connections with neck motoneurons. By themselves, however, these neurons would not produce response vectors near pitch, which apparently requires convergence between bilateral vertical canals—such

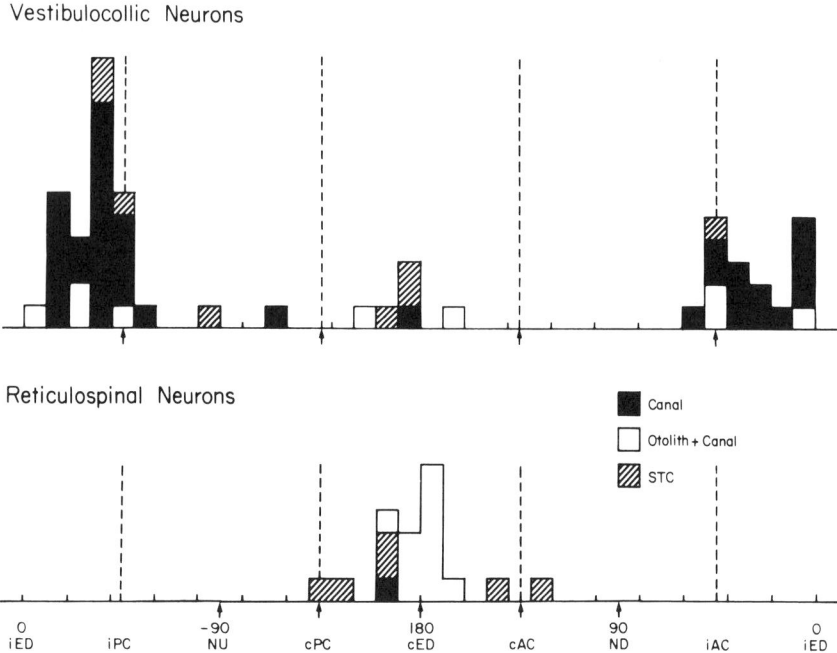

FIGURE 1. Response vector orientations of vestibulocollic and reticulospinal neurons receiving vertical canal input alone or in combination with otolith input. Orientations were typically measured with 0.5 Hz sinusoidal stimuli. Vestibulocollic data from Wilson *et al.*,[4] reticulospinal data from Bolton *et al.*[12,13] Abbreviations: iED and cED (0° and 180°), ipsi- and contralateral ear down; NU and ND (−90° and 90°), nose up and nose down; iPC and cPC, ipsi- and contralateral posterior canal; iAC and cAC, ipsi- and contralateral anterior canal. Dashed lines indicate the approximate planes of the vertical semicircular canals.

convergence of neck-projecting neurons must be taking place outside the vestibular nuclei.

CONVERGENCE IN THE PONTOMEDULLARY RETICULAR FORMATION

The reticular formation, which receives input from the labyrinth and transmits it to the spinal cord via reticulospinal fibers (reviewed in References 6 and 7), is a candidate locus for the interreceptor convergence absent from the vestibular nuclei. Reticulospinal neurons appear to play an important role in the horizontal VCR,[8-11] and we have recently explored their contribution to the VCR evoked by stimuli in vertical planes.[12,13] Study of responses to whole body tilt, i.e., determination of response vector orientation and classification of vestibular input type on the basis of dynamics, was as described above. Reticulospinal neurons, located within 2.5 mm of the midline in a region extending from the caudal medulla to caudal pons, were identified antidromically as projecting to the neck, cervical enlargement, thoracic cord, or lumbar cord. Relatively few reticulospinal neurons terminate in the neck,[14,15] and it appears that neck motoneurons receive most of their reticular input from branches of axons extending as far caudally as the thoracic cord.[16,17] Dynamics and response vector orientations of neurons projecting to neck, lower cervical, and thoracic segments did not appear to differ from the properties of those projecting to the lumbar cord, and for this paper we lump all of them together as reticulospinal (RS) neurons.

Inspection of TABLE 1 reveals a remarkable difference between RS and VS neurons. In contrast to VS neurons, almost no RS neurons get vertical input just from the semicircular canals (although about one-third of neurons not projecting to the spinal cord do), while the most prominent response class indicates otolith without canal input; neurons showing convergent canal and otolith input (including STC) occur with approximately the same frequency in the two populations. It would seem logical that if RS neurons make a significant contribution to the canal-driven component of the vertical VCR, and help determine the spatial and temporal properties of the reflex, then the spatial characteristics will largely reflect the distribution of vector orientations of RS neurons having convergent canal and otolith input; this distribution is illustrated in the lower part of FIGURE 1. The major difference in the orientation distribution between RS and VS neurons is that the latter are oriented towards the contralateral side—this tendency had previously been noted in studies using roll tilt.[18,19] The distribution of RS neurons shares a feature of the VS population in that the orientations lie between the canal planes and roll, with no tendency towards pitch. Thus as in the VS population, RS neurons, by themselves, do not have the correct spatial properties to directly produce the canal-dominated VCR response vectors shifted towards pitch seen in the neck. This result is consistent with recent studies of the properties of reticulospinal neurons:[16,17] one might not expect to find neck muscle vectors represented in a population of neurons that innervates the neck mainly by branches of longer fibers.

CONVERGENCE IN THE SPINAL CORD

We can now summarize the nature of the signals carried from the brain stem to the neck segments of the spinal cord by neurons whose axons end in these segments, or by neurons with longer axons that give off a collateral to the neck grey matter. There are many VC neurons receiving input from vertical canals only, or from otolith + canals, whose vectors are oriented in canal planes or closer to roll; a smaller number of STC neurons have similar vectors. RS neurons transmit little vertical canal information, whether alone or in combination with otolith input, and all response vectors, including those of STC neurons, are near canal planes or in the roll quadrants. Otolith input is transmitted by both VC and RS neurons.

It is obvious that canal-related activity is transmitted mainly by the vestibulospinal tracts. It is also obvious that for muscles whose response vectors lie near canal planes or in the roll quadrants, further spatial transformation in the spinal cord is not a necessity. This is not the case for muscles such as biventer cervicis and other vertical movers whose response vectors (in the VCR), or calculated torque vectors, are near pitch.[2,20]

The VCR is a complex reflex, involving a large number of muscles (cf. Reference 21). The complexity of motor control of the neck is increased by the fact that muscle torque vectors do not usually exactly oppose reflex response vectors, and are altered

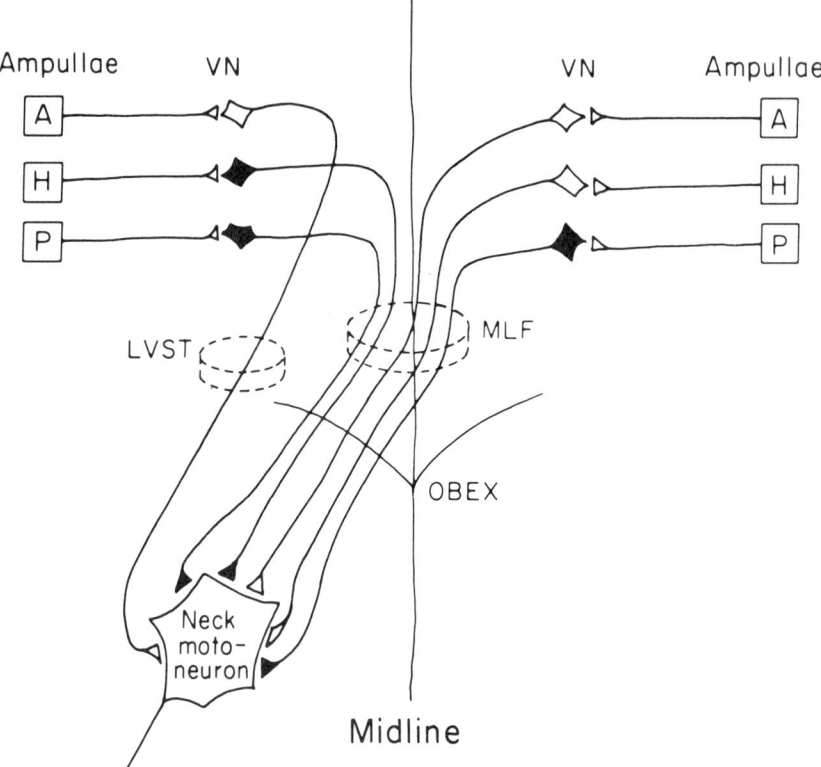

FIGURE 2. Schematic representation of disynaptic connections between semicircular canals and dorsal neck motoneurons. Connections were determined by electric stimulation of semicircular canal nerves and intracellular recording from biventer cervicis, complexus, and splenius motoneurons. Abbreviations: A, H, P, anterior, horizontal, posterior canal nerves; LVST, lateral vestibulospinal tract; MVST, medial vestibulospinal tract; VN, vestibular nuclei. Inhibitory neurons and terminals are black, excitatory are white. (From Wilson and Maeda,[23] with permission.)

by changing the angle of the head with respect to the neck.[20] It seems appropriate for much of the spatial transformation that produces appropriate reflex responses and STC behavior to take place at the segmental level, where signals transmitted by premotor neurons, each with its own response vector, can best interact with peripheral feedback.[22] The question arises, where are the premotor neurons? Vestibulospinal and reticulospinal fibers are premotor, as they make excitatory and inhibitory connections with neck motoneurons (for reviews see References 6 and 7). There are disynaptic pathways between canal nerves and these motoneurons via the medial and lateral vestibulospinal tracts[23] and, as FIGURE 2 shows, inputs from bilateral vertical canals converge on individual neck motoneurons. It remains to be determined to what extent spinal interneurons in the upper cervical segments are involved in transmitting vestibular signals to neck motoneurons. We have shown previously that the activity of neurons in C4 is modulated by sinusoidal tilt, with responses to the position and/or velocity of the stimulus.[24,25] The projections of these neurons were not known, however. Recent experiments have demonstrated that interneurons in the C2–C3 segments, projecting to the ventral grey matter of neighboring segments, may be excited or inhibited by electric stimulation of the labyrinth.[26] Further experiments are needed to reveal what role, if any, such interneurons play in vestibulocollic reflexes.

REFERENCES

1. WILSON, V. J., R. H. SCHOR, S. J. B. TIMERICK & B. R. PARK. 1986. Spatial organization of neck and vestibular reflexes acting on the forelimbs of the decerebrate cat. J. Neurophysiol. **55:** 514–526.
2. BAKER, J., J. GOLDBERG & B. W. PETERSON. 1985. Spatial and temporal properties of the vestibulocollic reflex in decerebrate cats. J. Neurophysiol. **54:** 735–756.
3. KASPER, J., R. H. SCHOR & V. J. WILSON. 1988. Response of vestibular neurons to head rotations in vertical planes. I. Response to vestibular stimulation. J. Neurophysiol. **60:** 1753–1764.
4. WILSON, V. J., Y. YAMAGATA, B. J. YATES, R. H. SCHOR & S. NONAKA. 1990. Response of vestibular neurons to head rotations in vertical planes. III. Response of vestibulocollic neurons to vestibular and neck stimulation. J. Neurophysiol. **64:** 1695–1703.
5. SCHOR, R. H., A. D. MILLER & D. L. TOMKO. 1984. Responses to head tilt in cat central vestibular neurons. I. Direction of maximum sensitivity. J. Neurophysiol. **51:** 136–146.
6. WILSON, V. J. & G. MELVILL JONES. 1979. Mammalian Vestibular Physiology. Plenum. New York, N.Y.
7. WILSON, V. J. & B. W. PETERSON. 1981. Vestibulospinal and reticulospinal systems. *In* Handbook of Physiology. The Nervous System. **2**(Section 1): 667–702. American Physiological Society. Bethesda, Md.
8. EZURE, K., S. SASAKI, Y. UCHINO & V. J. WILSON. 1978. Frequency-response analysis of the vestibular-induced neck reflex in cat. II. Functional significance of cervical afferents and polysynaptic descending pathways. J. Neurophysiol. **41:** 459–471.
9. WILSON, V. J., B. W. PETERSON, K. FUKUSHIMA, N. HIRAI & Y. UCHINO. 1979. Analysis of vestibulocollic reflexes by sinusoidal polarization of vestibular afferent fibers. J. Neurophysiol. **42:** 331–346.
10. PETERSON, B. W., K. FUKUSHIMA, N. HIRAI, R. H. SCHOR & V. J. WILSON. 1980. Responses of vestibulospinal and reticulospinal neurons to sinusoidal vestibular stimulation. J. Neurophysiol. **43:** 1236–1250.
11. BILOTTO, G., J. GOLDBERG, B. W. PETERSON & V. J. WILSON. 1982. Dynamic properties of vestibular reflexes in the decerebrate cat. Exp. Brain Res. **47:** 343–352.
12. BOLTON, P. S., T. GOTO, R. H. SCHOR, V. J. WILSON, Y. YAMAGATA & B. J. YATES. 1991. Reticulospinal neurons and vertical vestibulospinal reflexes. Soc. Neurosci. Abstr. **17:** 317.
13. BOLTON, P. S., T. GOTO, R. H. SCHOR, V. J. WILSON, Y. YAMAGATA & B. J. YATES. Response of pontomedullary reticulospinal neurons to vestibular stimuli in vertical planes, and their role in vertical vestibulospinal reflexes of the decerebrate cat. J. Neurophysiol. **67.** (In press.)
14. PETERSON, B. W., R. A. MAUNZ, N. G. PITTS & R. MACKEL. 1975. Patterns of projection and branching of reticulospinal neurons. Exp. Brain Res. **23:** 333–351.
15. IWAMOTO, Y. & S. SASAKI. 1990. Monosynaptic excitatory connexions of reticulospinal

        neurones in the nucleus reticularis pontis caudalis with dorsal neck motoneurons in the
        cat. Exp. Brain Res. **80:** 277–289.

16. WILSON, V. J. & B. W. PETERSON. 1988. Vestibular and reticular projections to the neck. *In* Control of Head Movement. B. W. Peterson & F. J. Richmond, Eds.: 129–140. Oxford University Press. New York, N.Y.

17. IWAMOTO, Y., S. SASAKI & I. SUZUKI. 1990. Input-output organization or reticulospinal neurones, with special reference to connexions with dorsal neck motoneurones in the cat. Exp. Brain Res. **80:** 260–276.

18. MANZONI, D., O. POMPEIANO, G. STAMPACCHIA & U. C. SRIVASTAVA. 1983. Responses of medullary reticulospinal neurons to sinusoidal stimulation of labyrinth receptors in decerebrate cat. J. Neurophysiol. **50:** 1059–1079.

19. POMPEIANO, O., D. MANZONI, U. C. SRIVASTAVA & G. STAMPACCHIA. 1984. Convergence and interaction of neck and macular vestibular inputs on reticulospinal neurons. Neuroscience **12:** 111–128.

20. WICKLAND, C., J. F. BAKER & B. W. PETERSON. 1991. Torque vectors of neck muscles in the cat. Exp. Brain Res. **84:** 649–659.

21. BAKER, J. F., S. I. PERLMUTTER & B. W. PETERSON. 1988. Comparison of spatial transformation in vestibulo-ocular and vestibulocollic reflexes. Ann. N.Y. Acad. Sci. **545:** 203–215.

22. WILSON, V. J. 1988. The tonic neck reflex: spinal circuitry. *In* Control of Head Movement. B. W. Peterson & F. J. Richmond, Eds.: 100–107. Oxford University Press. New York, N.Y.

23. WILSON, V. J. & M. MAEDA. 1974. Connections between semicircular canals and neck motoneurons. J. Neurophysiol. **37:** 346–357.

24. WILSON, V. J., K. EZURE & S. J. B. TIMERICK. 1984. Tonic neck reflex of the decerebrate cat: response of spinal interneurons to natural stimulation of neck and vestibular receptors. J. Neurophysiol. **51:** 567–577.

25. SCHOR, R. H., I. SUZUKI, S. J. B. TIMERICK & V. J. WILSON. 1986. Responses of interneurons in the cat cervical cord to vestibular tilt stimulation. J. Neurophysiol. **56:** 1147–1156.

26. BOLTON, P. S., T. GOTO & V. J. WILSON. 1991. Commissural neurons in the cat upper cervical spinal cord. Neuro Report **2:** 743–746.

Functional Synergies of Neck Muscles Innervated by Single Medial Vestibulospinal Axons[a]

YOSHIKAZU SHINODA,[b] TOHRU OHGAKI,[c] YURIKO SUGIUCHI,[c] TAKAHIRO FUTAMI,[b] AND SHINJI KAKEI[b]

[b]Department of Physiology
[c]Department of Otolaryngology
School of Medicine
Tokyo Medical and Dental University
1-5-45, Yushima, Bunkyo-ku
Tokyo 113, Japan.

INTRODUCTION

Head position control is an ideal paradigm for studying how the central nervous system (CNS) interacts to stabilize a multidimensional motor system. Head movement signals detected by the semicircular canals are mediated through vestibulocollic pathways that link each of the three semicircular canals to a set of neck muscles. For tasks necessitating compensatory head movements, the CNS will program muscles to respond in a specific spatial combination rather than to generate an infinite variety of muscle contraction patterns.

Stimulation of individual semicircular canals produces canal-specific head movement. The plane of the head movement produced by canal stimulation parallels that of the stimulated canal; they are almost coplanar.[1] Therefore, a signal from each semicircular canal must be distributed to a proper set of neck muscles to induce compensatory head movement in the same plane as the plane of the stimulated canal. Obviously a given motor control signal may be economically distributed by a single neuron with divergent branches to multiple target sites that participate in cocontraction of muscles to produce a purposeful movement. Only a little information is available concerning the divergent properties of vestibulospinal (VS) neurons, although the convergence of different inputs onto single neurons has been extensively analyzed in the vestibulocollic system.

This study was performed to characterize the divergent properties of single medial vestibulospinal tract (MVST) axons in the upper cervical cord, using the method of intraaxonal staining with horseradish peroxidase (HRP). The results show that single MVST neurons have multiple axon collaterals and innervate a functional set of multiple motor nuclei of the neck musculature. This divergent projection of the MVST neurons to multiple neck muscles strongly suggests that they implement a canal-dependent head movement synergy.

[a]This work was supported by a Human Frontier Science Program Organization Research Grant.

507

METHODS

Experiments were performed on 18 cats weighing 2.8–4.5 kg. The animals were anesthetized with pentobarbital sodium (Nembutal, Abott, Switzerland) (initial dose of 35 mg/kg, supplemented as required) and mounted in a stereotaxic apparatus. In four experiments, the cat was placed on a three-dimensional rotating table for identifying input from the semicircular canal to MVST neurons. The tympanic bulla was opened by a ventral approach, and fine Ag-AgCl wire electrodes (0.1 mm in diameter), insulated except for the spherical tips, were placed on the oval and the round windows on both sides, keeping the vestibular receptors intact to stimulate the vestibular nerves.[2] A laminectomy was performed between C1 and C4 to permit intraaxonal recording from MVST axons. Stimulating electrodes were inserted stereotaxically into the medial longitudinal fasciculus (MLF) on either side and fixed finally by monitoring characteristic fiber potentials evoked by stimulation of the vestibular nerves. Intraaxonal recording and HRP injection were made with a glass micropipette filled with 7% HRP (Toyobo Co., Osaka, Japan) in 0.05 M Tris:HCl buffer (pH = 8.6) with 0.2 M KCl. Electrode resistance varied from 30–80 MΩ. MVST axons were penetrated in the ventral funiculus on either side between C1 and C3, and HRP was injected iontophoretically through the recording electrode. Details of electrophysiological identification of MVST axons, HRP injection, and staining procedures have been reported before.[3]

RESULTS

Identification of Uncrossed and Crossed MVST Axons

The axons were identified electrophysiologically as second-order MVST axons by their monosynaptic responses to stimulation of the vestibular nerve and their direct responses to stimulation of the MLF.[4-6] The MVST axons were classified into two groups: uncrossed and crossed MVST axons. Uncrossed axons were activated by stimulation of the ipsilateral vestibular nerve and the ipsilateral MLF, whereas crossed MVST axons were activated by stimulation of the contralateral vestibular nerve and the ipsilateral MLF.

The Morphology of MVST Axons

Morphology of Collaterals and Their Arborizations

The stained distance of the stem axons was 2.5–14.4 mm [mean ± standard deviation (SD), 8.8 ± 3.3 mm] in 22 uncrossed MVST axons and 3.1–15.3 mm (9.3 ± 4.0 mm) in 19 crossed MVST axons at a level between C1 and C4. Over their courses in the upper cervical cord, the stem axons usually gave rise to multiple axon collaterals at different segments (FIGURES 1 and 2). All but two MVST axons (39/41) had at least one axon collateral. The maximum number of collaterals for a single uncrossed MVST axon was 9 (mean ± SD, 3.1 ± 2.0, $n = 22$), and that for a single crossed MVST axon was 8 (3.6 ± 2.1, $n = 19$) (see FIGURE 3). These collaterals arose at various intervals from the stem axons. The distances between branching points of adjacent primary collaterals from the stem axons ranged from 100 to 5100 μm (mean ± SD, 1620 ± 1380 μm, $n = 47$) for uncrossed MVST axons and from 100

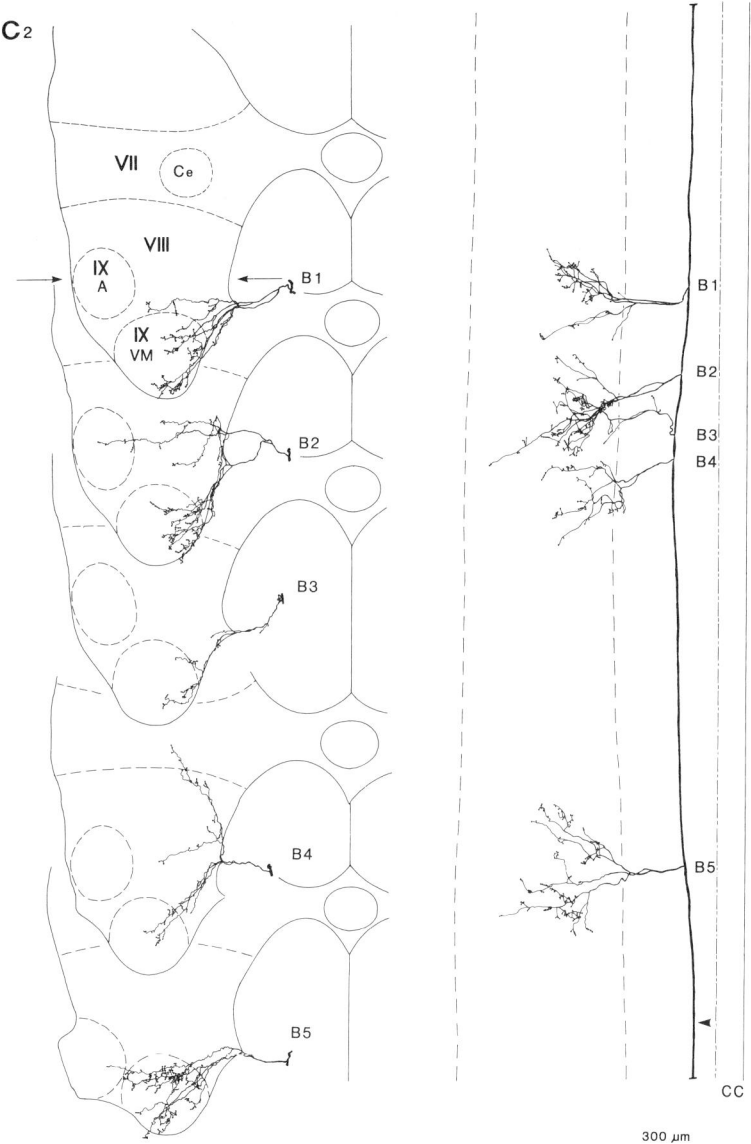

FIGURE 1. Reconstructions of an uncrossed MVST axon at C2. **Left:** Reconstructions in the transverse plane of individual axon collaterals indicated in B1–B5 on the right side. **Right:** Reconstruction in the horizontal plane made from 49 serial sections of 100 μm thickness. The solid line and the dotted line indicate the midline and the outer border of the central canal, respectively. Two broken lines indicate the inner and the outer borders of the cervical gray matter at a level shown by two arrows. The arrowhead shows the injection site. Short bars at the ends of the axon indicate the sites where the injected axons became invisible. All diagrams are drawn on the same scale. These symbols are applied to the other figures. VII, VIII, and IX: Rexed's laminae VII, VIII, and IX. A: The motor nucleus of n. spinalis accessorii. VM: The ventromedial nucleus. CC: Central canal. Ce: The nucleus cervicalis centralis. (From Shinoda *et al.*[6] with permission.)

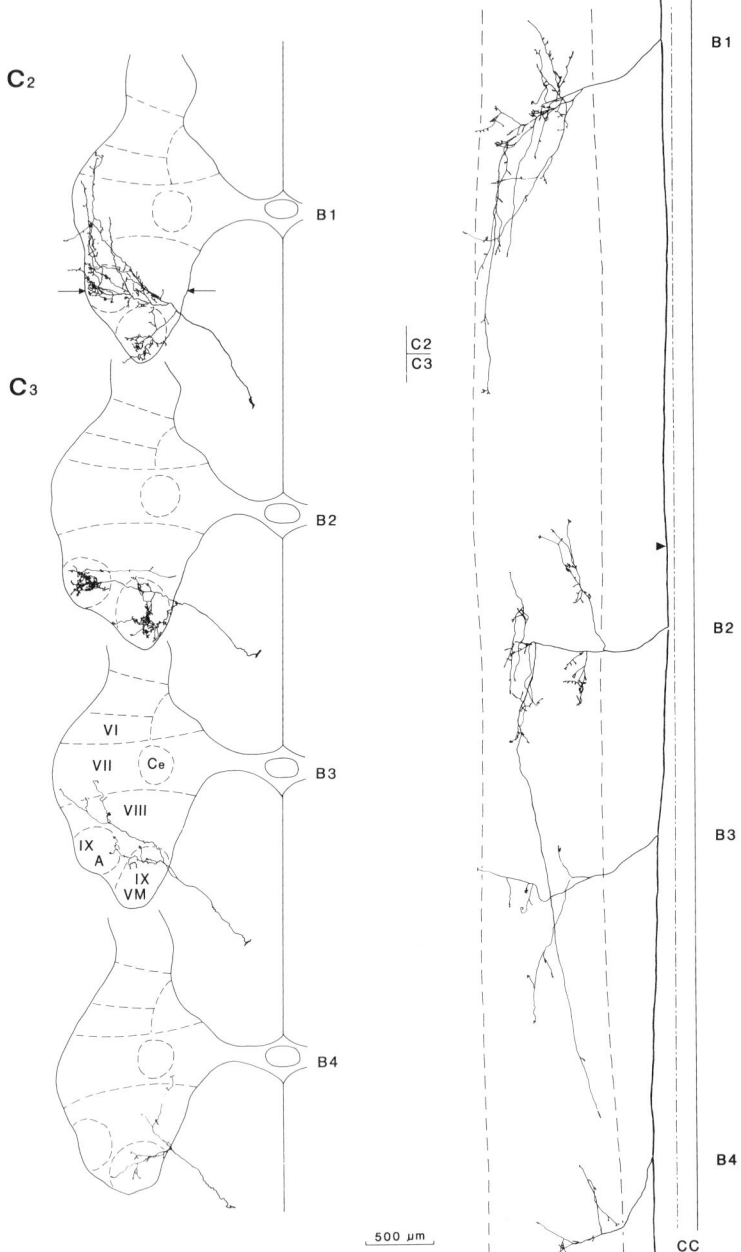

FIGURE 2. Reconstructions of a crossed MVST axon in the transverse plane (left) and in the horizontal plane (right). Axon terminals are abundantly distributed in both the VM and the SA nuclei. (From Shinoda *et al.*[6] with permission.)

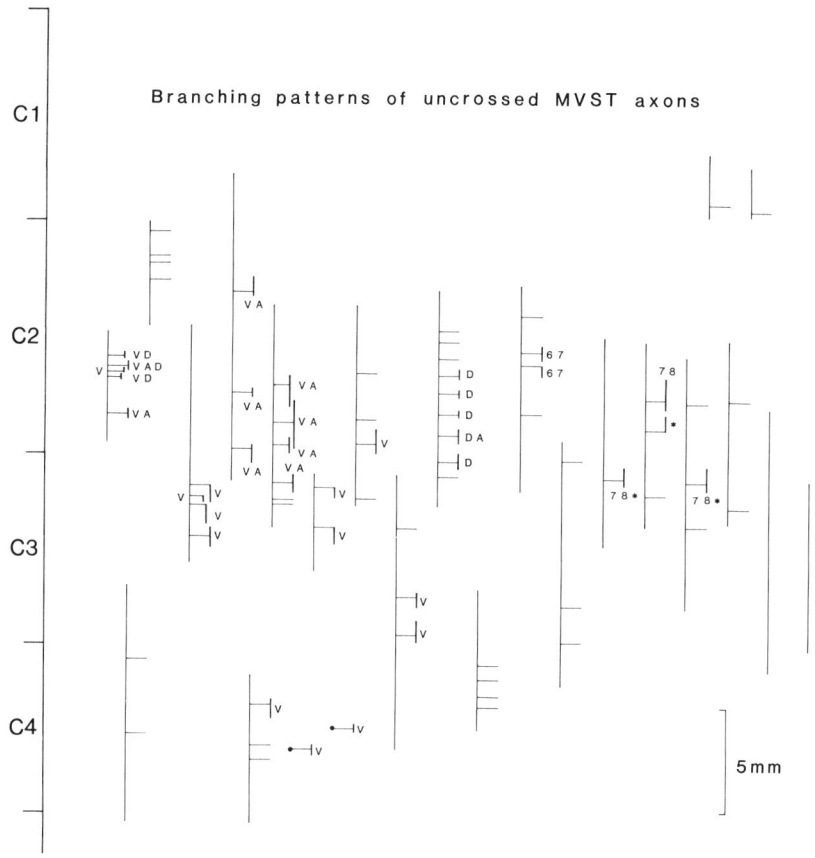

FIGURE 3. Summary diagram of the branching patterns of uncrossed MVST axons in the upper cervical cord. A vertical thin line represents the total length of each stem axon stained. Horizontal thin bars indicate the position of primary axon collaterals given off from the stem axons. Vertical thick bars indicate the extension of axon collaterals in the gray matter. Only when single axon collaterals could be traced to all terminal axons, were vertical thick bars drawn. The numbers and letters beside each collateral indicate the laminae of Rexed and the motor nuclei where the collateral terminates. The letters from left to right indicate the predominance of termination. V, the nucleus ventromedialis; A, the nucleus spinalis n. accessorii; D, the nucleus dorsomedialis. Asterisks indicate axon collaterals crossing the midline to the contralateral spinal gray matter. Closed circles indicate the axons in which collaterals were penetrated instead of stem axons. (Adapted from Shinoda *et al.*[6])

to 4300 μm (1420 ± 1040 μm, $n = 50$) for crossed MVST axons. The primary axon collaterals arose at more-or-less right angles from the stem axons (FIGURES 1 and 2) and ran some distance laterally or ventrolaterally in the white matter to the medial border of the ventral horn without branching, or often after bifurcating once. Then, the primary or secondary collaterals entered the ventral horn from the medial border at a level of a middle-third of the ventral horn, or from the ventromedial border of the ventral horn when the stem axons ran in the more ventral part of the ventral funiculus.

Collaterals of MVST axons have a distinctive morphology, and their branching patterns are very similar to those of lateral vestibulospinal tract (LVST) axons.[3] When viewed in the horizontal plane, primary collaterals arose at more or less right angles from stem axons and took almost straight courses to the ventral horn. Immediately after the entrance into the gray matter, the collaterals divided into several branches which formed a deltalike path in a rostrocaudal direction. The rostrocaudal extent of single-axon collaterals ranged from 300 to 2300 μm (mean ± SD, 750 ± 410 μm, $n = 33$) for uncrossed MVST axons and from 400 to 2500 μm (1010 ± 650 μm, $n = 12$) for crossed MVST axons. In contrast to the restricted spread of terminal arborizations in a rostrocaudal direction, the dorsoventral and mediolateral extension of individual collaterals was much wider. Immediately after the entrance into the gray matter, the primary or secondary collaterals divided into several branches and fanned out in a deltalike manner in a dorsoventral direction. After this division at the medial margin of the ventral horn, each branch further bifurcated or trifurcated one to three times before terminal arborizations.

Three groups of branches arising from single primary collaterals were identified at C2–C4 in terms of their courses and terminal sites in the gray matter. The first group of the branches ran ventrolaterally to the ventromedial (VM) nucleus (lamina IX) (Rexed),[7] terminating in different portions of this nucleus (FIGURE 1, B1, B3, and B5). Some of these branches turned dorsally and further extended to the nucleus spinalis n. accessorii (SA nucleus) (lamina IX) (Rexed).[7] Sometimes, the branches further extended dorsally into the lateral portions of laminae VII, VI, V, and even IV. The second group of branches ran almost laterally, giving off some short side branchlets in lamina VIII on their way, and reached the SA nucleus or its surrounding lamina VIII (FIGURE 1 B2). Some of these branches further extended dorsally into the lateral portions of laminae VIII, VII, and V. This group of branches often gave rise to extensive arborizations in lamina VIII dorsomedial to the VM nucleus just after the entrance into the ventral horn. In cases when the stem axons ran in the ventral funiculus below the level of the tip of the ventral horn, the above two groups of branches ran in a dorsolateral direction rather than in a ventrolateral or a lateral direction and the branches often entered the VM nucleus directly by penetrating its ventromedial border. The third group of branches ran dorsolaterally into lamina VIII and gave rise to terminal arborizations in the medial two-thirds of laminae VIII and VII (FIGURE 1 B4). This group of branches also sometimes gave off terminal branches in lamina VIII just dorsomedial to the VM nucleus on their way to the more dorsal gray matter. Some of the branches in this group further extended dorsally to lamina VI and sometimes even to lamina V. Among these three groups of branches, the axon branches to the VM nucleus were most prominent in both crossed and uncrossed MVST axons, whereas the dorsolateral branches to the medial portion of lamina VIII and its adjacent lamina VII were rarest.

Axon collaterals of single MVST axons usually tended to project to one or two common termination areas in the gray matter in the transverse plane. For example, in the uncrossed MVST axon shown in FIGURE 1, four axon collaterals (B1–B4) had a common termination area in the medial portion of the VM nucleus at different spinal levels, although the most caudal collateral (B5) had a different termination area which was more lateral and wider in the VM nucleus. As in this example, most collaterals from a single axon produced main terminal arborizations in the common area in the transverse plane that were in line with one another in the longitudinal axis of the cord.

Distribution of Terminal Boutons

The distribution of terminal swellings in transverse planes of the cervical gray matter is shown in FIGURE 4. Throughout the upper cervical cord, swellings were mainly distributed in lamina IX, laminae VIII and VII, and sometimes lamina VI. Only rarely, swellings were present in laminae IV and V. Although more terminals of the uncrossed MVST axons seemed to occupy laminae VII–VIII in the examples of FIGURE 4, other crossed MVST axons provided terminal arborizations to laminae VII–VIII. In the ventral horn, the swellings were most abundantly observed in the VM nucleus, the SA nucleus, and the medial part of lamina VIII just dorsomedial to the VM nucleus. According to Rexed,[7] lamina IX of the upper cervical cord consists

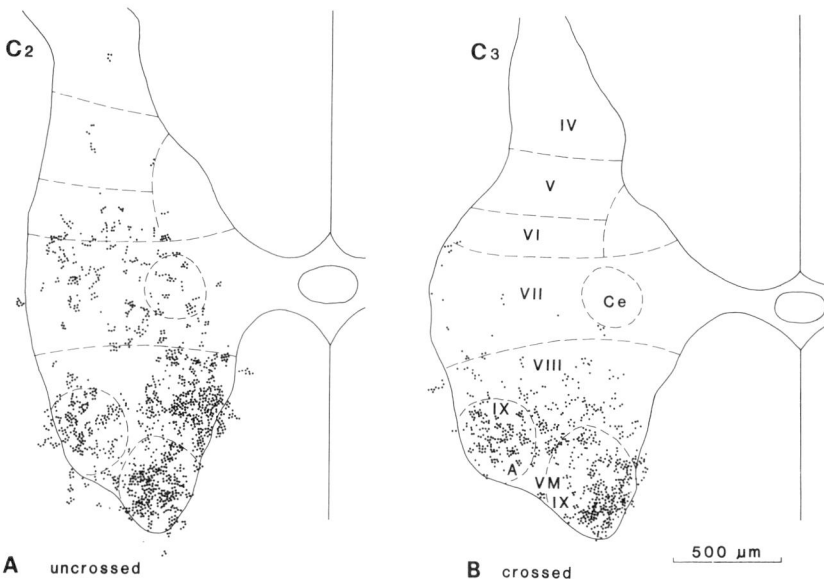

FIGURE 4. Distribution of synaptic boutons of uncrossed MVST axons at C2 (A) and crossed MVST axons at C3 (B). A is based on 14 axon collaterals from 5 MVST axons in one cat. B is based on 8 axon collaterals from 3 MVST axons in another cat. (Adapted from Shinoda *et al.*[6])

of two separate motor nuclei. One is the large VM nucleus, occupying the apex of the ventral horn, and the other is the SA nucleus, which lies laterally near the base of the ventral horn. However, in addition to these two nuclei, motoneurons are also located in the medial portion of lamina VIII which is adjacent to the medial border of the ventral horn and dorsomedial to the VM nucleus (the DM nucleus, see the Discussion for the nomenclature).[8] Abundant axon terminals were observed in this area (FIGURE 4A). The terminals in the DM nucleus were provided either by dorsolaterally running axon branches on their way to laminae VIII and VII or by laterally running axon branches on their way to the SA nucleus. The terminal arborizations in the VM nucleus were provided by rather thick ventrolateral branches. Individual MVST axons distributed their terminal arborization to a localized area in this nucleus, but when viewed as an ensemble, the terminal distribution of different MVST axons covered the whole area of the VM nucleus, indicating that all muscles innervated by motoneurons in the VM nucleus have effects from MVST axons (FIGURE 4). This was also the case for the SA nucleus. The terminal arborizations in the SA nucleus were provided by either laterally running axon branches or by ventrolaterally running axon branches. Although each MVST axon had a characteristic termination area in the SA nucleus, the axon terminals of different MVST axons as an ensemble covered the whole area of this nucleus. Among the MVST axons providing terminal arborizations to the motor nuclei, most frequently observed were the MVST axons projecting to more than one nucleus, and the rarest were the MVST axons projecting exclusively to one of the three nuclei. At least three types of MVST axons were found in relation to their terminations in the motor nuclei: (1) MVST axons distributing their terminations to both the VM and the SA nuclei; (2) those to both the DM and the SA nuclei; and (3) those to the three nuclei. In these axons, single collaterals usually provided terminal arborizations to two or more motor nuclei in the same transverse plane limited within about 700–1100 μm along the longitudinal axis of the cord.

Within the VM, SA, and DM nuclei, a large number of swellings appeared to make contact with the proximal dendrites or the cell bodies of counterstained cells (FIGURE 5), and apparent axosomatic or axodendritic contacts were observed on large, medium-sized, and even small neurons. A single or a few swellings of the terminal type or multiple en passant swellings of a single terminal branch were seen to be arranged in close association with the somata and proximal dendrites of lamina IX neurons. A single collateral gave usually about one to four swellings (up to 10 swellings) to a single neuron in lamina IX. Axon terminals of a single collateral rarely contacted more than one proximal dendrite of a neuron. Up to 8 neurons were identified as contacted on the proximal dendrites by a single primary collateral. The real number of target neurons innervated by a single collateral must be much larger, since only cells located near the surface were counterstained due to the thick slices. Moreover, any cell structures of the counterstained cells were not observed around most swellings, which suggested that these swellings probably make contact with distal dendrites of motoneurons or interneurons extending their distal dendrites in the motor nuclei.

MVST Axons Receiving Input from the Ipsilateral Posterior Canal

As exemplified in FIGURES 2 and 5, many MVST axons projected to more than one motor nucleus of the neck musculature. The important question arises as to whether individual MVST axons randomly innervate different motor nuclei or whether MVST axons may be classified into several groups based on the combinations of neck motor nuclei innervated by single MVST axons. To determine target motor nuclei, motoneurons of different neck muscles were retrogradely labeled with HRP and the map of motor nuclei for different neck muscles was made at C1 and C2. Then MVST axons were classified in terms of the input from the semicircular canal, by determining the maximal response plane of head rotation on the three-dimensional turntable. In this paper, we will deal with MVST axons receiving input from the ipsilateral posterior canal and projecting to the ipsilateral neck motor nuclei. After the above-mentioned physiological identification, MVST axons were injected with HRP in the upper cervical cord where two or three motor nuclei were retrogradely labeled with HRP in each experiment. Posterior-canal-related MVST axons commonly had a distinct innervation pattern on neck motor nuclei. Single axons innervated the motor nucleus for the sternomastoid-cleidomastoid, the motor nuclei for the semispinales, and the motor nucleus for the rectus capitis dorsalis. Single collaterals did not necessarily terminate on all of the above motor nuclei, but some other collaterals of the identical stem axons terminated on one or more of them. Therefore, single MVST axons innervated the above-described group of motor nuclei with their multiple collaterals. This spatial innervation pattern was found in almost all of the posterior-canal-related uncrossed MVST axons examined.

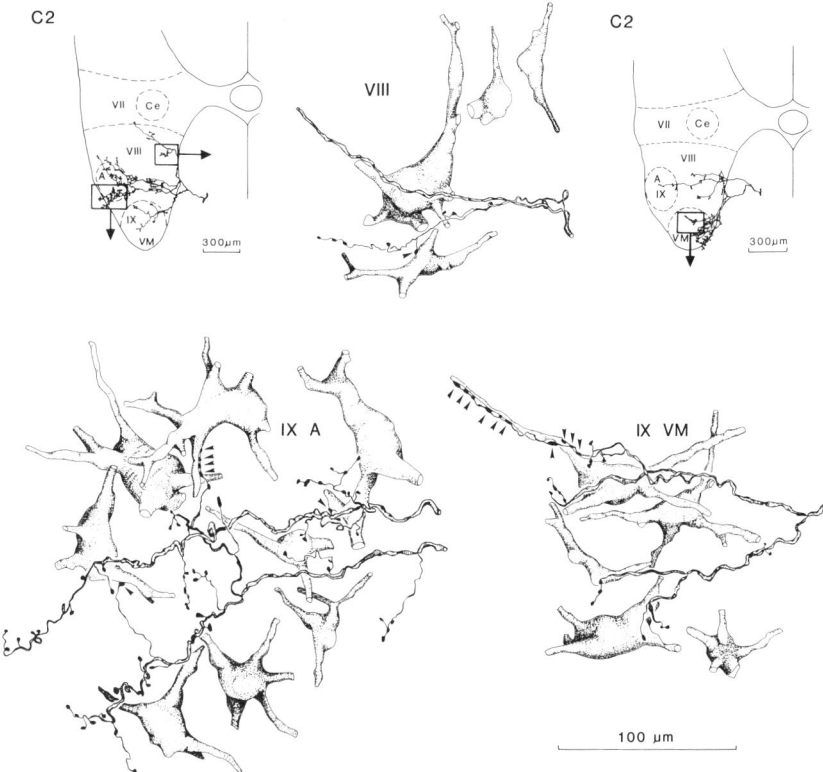

FIGURE 5. Camera lucida drawings of synaptic boutons from uncrossed MVST axons, from transverse sections at C2. **Left diagrams:** Terminal axons in lamina IX (Nucl. spinalis n. accessorii) (larger square) and (Nucl. dorsomedialis) (smaller square) of the axon collateral shown in the inset on the upper left. Arrowheads indicate apparent synaptic contacts of terminals with counterstained cells. **Right diagrams**: Terminal axons in lamina IX (Nucl. ventromedialis) of the axon collateral shown in the inset on the upper right. (From Shinoda *et al.*[6] with permission.)

DISCUSSION

The present study visualized branching patterns of single MVST axons in the upper cervical spinal cord with intraaxonal injection of HRP. It turned out that single MVST axons have multiple axon collaterals at a mean interval of 1510 μm and that individual axon collaterals have very narrow rostrocaudal extension with gaps free from axon terminals between adjacent collaterals. This result showed that both MVST and LVST axons have many common features as to the branching pattern and location of axon terminals in the cervical gray matter.[4] In the present study, many terminals of MVST axons were observed in lamina IX in the upper cervical cord. This gives morphological support for the electrophysiological finding of monosynaptic connections between VS neurons and neck motoneurons.[9–12] The axon terminals were distributed not only in the VM nucleus, but also in the SA nucleus. This finding fits well with the previous report that some motoneurons in the SA nucleus innervating sternocleidomastoid muscles receive disynaptic vestibular inputs from the semicircular canals via the MLF.[13] Within the VM and the SA nuclei, several motor nuclei for different neck muscles are involved.[8,14] Although individual VS axons have preferential terminal distributions within the VM and the SA nuclei, the ensemble distribution of axon terminals from several MVST axons covered the whole areas of the VM and the SA nuclei, which suggests that all neck muscles innervated by motoneurons in the VM and the SA nuclei are influenced directly by both crossed and uncrossed MVST axons (FIGURE 4). According to Rexed,[7] the commissural nucleus is located in lamina VIII dorsomedial to the VM nucleus. However, the gray matter along the medial border of the ventral horn just dorsomedial to the VM nucleus contains motoneurons innervating neck muscles.[8] Therefore, we name this area the dorsomedial (DM) nucleus. Motoneurons of neck ventroflexor muscles are involved in this nucleus. Many crossed and uncrossed MVST axons terminated in the DM nucleus. In lamina IX of the upper cervical cord, many axon terminals appeared to make contact with the proximal dendrites and cell bodies of various-sized counterstained neurons (FIGURE 5). Most of these neurons were probably motoneurons, since almost all neurons in this lamina were retrogradely labeled with HRP after injection of HRP into the accessory nerve and the spinal ventral root.[8] Our recent study confirmed that stained MVST axons really make contact with the proximal dendrites and cell bodies of different neck motoneurons labeled with HRP.[5,15]

Multiple axon collaterals of individual MVST axons tend to project to the common target areas in the gray matter, especially in lamina IX, at different spinal levels, although not all collaterals of the same MVST axons project to the common target areas. A series of terminal arborizations of multiple collaterals from a single axon are arranged along the longitudinal axis of the cord. Single axons distributed their terminals into longitudinally running columns of motoneurons in lamina IX at multisegmental levels. Many neck muscles are multisegmental and are innervated by motoneurons located over multiple segments.[8,14,16,17] Therefore, the above finding indicates that a single MVST axon innervates the motoneurons of a single multisegmentally innervated neck muscle by its multiple axon collaterals over several spinal segments. One of the most important findings in the present study is that single MVST axons projected to different neck motor nuclei and made contact with the proximal dendrites or cell bodies of motoneurons in those nuclei. Single axon collaterals innervated multiple motor nuclei in different combinations such as both the SA and the VM nuclei, both the SA and the DM nuclei, and all these three nuclei. Furthermore, single axon collaterals had wide terminal arborizations within each of those three motor nuclei. This finding suggests that axon collaterals terminate on motoneurons of more than one muscle, since each of these three nuclei contains motoneuron pools of a few neck muscles.[8,14] Taken together, the present data show that virtually all well-stained MVST axons innervate multiple motor nuclei of neck muscles in different combinations.

The plane of head movement produced by canal stimulation parallels that of the stimulated canal.[1] Therefore, a signal from each semicircular canal must be distributed to a proper set of neck muscles to induce compensatory head movement. Two possible models will explain the neural mechanisms underlying this spatial transformation in the vestibulocollic pathway. In one model, each MVST axon innervates one and only one neck motor nucleus. Primary afferents from a semicircular canal have divergent projection onto multiple MVST neurons, each of which innervates a different motor nucleus. In the other model, single MVST neurons have divergent projection to multiple motor nuclei and primary afferents originating from a certain semicircular canal innervate these MVST neurons. The present results exclude the former model and support the latter one, since virtually all well-stained MVST axons projected to more than one motor nucleus. More specifically, it was found that a single MVST axon receiving input from a semicircular canal diverged onto a functional group of multiple motor nuclei of neck muscles, which compose a functional synergy for a compensatory head movement.

To specify a functional synergy of neck muscles that are innervated by single MVST axons, the innervation pattern of posterior-canal-related MVST axons was analyzed and it turned out that individual axons had a common projection onto a set of neck motor nuclei that were specific to posterior-canal-related uncrossed MVST axons. Although these posterior-canal-related MVST axons were considered to be inhibitory,[18] excitatory MVST axons receiving input from other semicircular canals will also show that they implement a functional synergy for a compensatory head movement with their divergent branches. To further understand the neural mechanism of the spatial transformation in the vestibulocollic system, the muscle group of a functional synergy related to VS neurons receiving input from other semicircular canals will have to be determined.

REFERENCES

1. SUZUKI, J.-I. & B. COHEN. 1964. Head eye, body and limb movements from semicircular canal nerves. Exp. Neurol. **10:** 393–405.
2. SHINODA, Y. & K. YOSHIDA. 1974. Dynamic characteristics of responses to horizontal head angular acceleration in the vestibulo-ocular pathway in the cat. J. Neurophysiol. **37:** 653–673.
3. SHINODA, Y., T. OHGAKI & T. FUTAMI. 1986. The morphology of single lateral vestibulospinal tract axons in the lower cervical cord of the cat. J. Comp. Neurol. **249:** 226–241.
4. SHINODA, Y., T. OHGAKI, T. FUTAMI & Y. SUGIUCHI. 1990. Comparison of the branching patterns of lateral and medial vestibulospinal tract axons in the cervical spinal cord. Prog. Brain Res. **80:** 137–147.
5. SHINODA, Y., T. OHGAKI, Y. SUGIUCHI & T. FUTAMI. 1988. Structural basis for three-dimensional coding in the vestibulospinal reflex; morphology of single vestibulospinal axons in the cervical cord. Ann. N.Y. Acad. Sci. **545:** 216–227.
6. SHINODA, Y., T. OHGAKI, Y. SUGIUCHI & T. FUTAMI. The morphology of single medial vestibulospinal tract axons in the upper cervical spinal cord of the cat. J. Comp. Neurol. (In press.)
7. REXED, B. 1954. A cytoarchitectonic atlas of spinal cord in the cat. J. Comp. Neurol. **100:** 297–379.
8. SUGIUCHI, Y. & Y. SHINODA. Organization of the motor nuclei innervating epaxial muscles in the neck and the back. *In* The Head-Neck Sensory-Motor System. A. Berthoz, V. Graf & P. Vidal, Eds. The Oxford University Press. Oxford, England. (In press.)
9. WILSON, V. J. & M. YOSHIDA. 1969. Comparison of effects of stimulation of Deiters' nucleus and medial longitudinal fasciculus on neck, forelimb, and hindlimb motoneurons. J. Neurophysiol. **32:** 743–758.
10. WILSON, V. J. & M. YOSHIDA. 1969. Monosynaptic inhibition of neck motoneurons by the medial vestibular nucleus. Exp. Brain Res. **9:** 365–380.

11. WILSON, V. J. & M. YOSHIDA. 1969. Bilateral connections between labyrinths and neck motoneurons. Brain Res. **13:** 603–607.
12. AKAIKE, T., V. V. FANARDJIAN, M. ITO & T. OHNO. 1973. Electrophysiological analysis of the vestibulospinal reflex pathway of rabbit. II. Synaptic actions upon spinal neurones. Exp. Brain Res. **17:** 497–515.
13. FUKUSHIMA, K., B. W. PETERSON & V. J. WILSON. 1979. Vestibulospinal, reticulospinal and interstitiospinal pathways in the cat. Prog. Brain Res. **50:** 121–136.
14. RICHMOND, F. J. R., D. A. SCOTT & V. C. ABRAHAMS. 1978. Distribution of motoneurones to the neck muscles, biventer cervicis, splenius and complexus in the cat. J. Comp. Neurol. **181:** 451–464.
15. SHINODA, Y., T. OHGAKI, Y. SUGIUCHI & T. FUTAMI. Spatial innervation patterns of single vestibulospinal axons in neck motor nuclei. *In* The Head-Neck Sensory-Motor System. A. Berthoz, V. Graf & P. Vidal, Eds. The Oxford University Press. Oxford, England. (In press.)
16. REIGHARD, J. & H. S. JENNINGS. 1951. Anatomy of the Cat. Henry Holt and Company. New York, N.Y.
17. CROUCH, J. E. 1969. Text-Atlas of Cat Anatomy. Lea & Febiger. Philadelphia, Pa.
18. SUGIUCHI, Y., T. FUTAMI, N. ANDO, T. KAWASAKI, J. YAGI & Y. SHINODA. 1992. Patterns of connections between six semicircular canals and neck motoneurons. Ann. N.Y. Acad. Sci. (This volume.)

Noradrenergic and Cholinergic Modulations of Corticocerebellar Activity Modify the Gain of Vestibulospinal Reflexes[a]

O. POMPEIANO

Department of Physiology and Biochemistry
University of Pisa
Via S. Zeno 31
I-56127 Pisa, Italy

INTRODUCTION

In addition to the classical mossy fibers (MF) and climbing fibers (CF), which utilize excitatory amino acids as neurotransmitters, the cerebellar cortex receives other afferent systems of different neurochemical specificity. One of these systems is made by noradrenergic fibers, originating from the locus ceruleus (LC), which make synaptic contacts primarily on Purkinje (P) cell dendrites in the molecular layer and to a lesser extent on the P-cell body and superficial granule cell layers (cf. Reference 1). On the other hand another system is formed by cholinergic fibers originating from the caudal part of the medial and descending vestibular nuclei and the prepositus hypoglossi nucleus,[2] and probably also from the dorsolateral pontine tegmentum, where noradrenergic LC neurons are intermingled with cholinergic neurons in the cat (cf. Reference 3). In particular, a cholinoacetyltransferase immunoreactivity was found in a subpopulation of MF terminals in the granular layer,[4–6] in thin varicose fibers closely associated with the P cell and molecular layers[5,6] as well as in a subpopulation of Golgi cells.[6]

The neurochemical specificity of the noradrenergic afferents to the cerebellar cortex is supported by the results of *in vitro* autoradiographic studies showing in rats a high level of α_1 adrenoceptors in the molecular layer[7,8] (FIGURE 1A), a low level of α_2 adrenoceptors in the granular layer,[9,10] and β adrenoceptors in the molecular layer[11–13] as well as in "patches" surrounding small groups of P-cell somata[14] (FIGURE 1B). The possibility that β adrenoceptors, which are mainly of the β_2 subtype (cf. Reference 12), are not located postsynaptically on P cells, but rather presynaptically on the terminals of parallel fibers is documented by the fact that the neurons expressing β_2 adrenoceptors correspond to the granule cells, as shown by using *in situ* hybridization histochemistry[15] (FIGURE 1C).

As to the cholinergic fibers, different subtypes of muscarinic[16–21] and nicotinic receptors[20,22,23] have been identified in the cerebellar cortex. In particular, in rodents muscarinic receptors showed a higher density in the granular and P-cell layers than in the molecular layer[19] (FIGURE 1D). Moreover, parasagittal columns of very high

[a]This work was supported by National Institute of Neurological and Communicative Disorders and Stroke Research Grant NS 07685-23 and by grants from the Ministero dell'Universita' e della Ricerca Scientifica e Tecnologica and the Agenzia Spaziale Italiana, Rome, Italy.

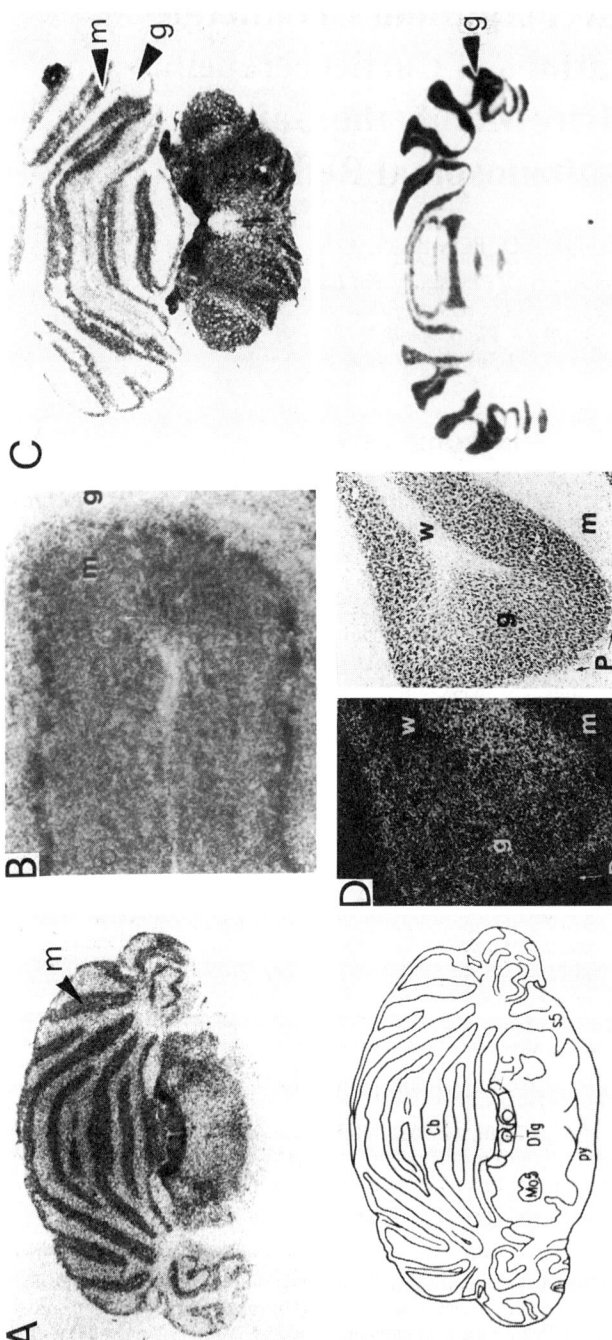

FIGURE 1. Distribution of noradrenergic and cholinergic receptors in rodent cerebellum. Abbreviations: g, granular layer; m, molecular layer; w, white matter; P, Purkinje (P) cell layer. **A:** Autoradiographic localization in rat cerebellum of α_1-adrenergic binding sites labeled with the α_1-adrenergic antagonist [^{125}I]-HEAT. (From Reference 7 with permission.) **B:** Autoradiographic localization in rat cerebellum of β-adrenergic receptors labeled with the nonselective β-adrenergic antagonist ligand [^{125}I]cyanopindolol [^{125}I]-CYP. (From Reference 14 with permission.) **C:** Upper part: autoradiographic localization in rat cerebellum of β_2-adrenergic receptors labeled with [^{125}I]-CYP in the presence of CGP20712A to block β_1 adrenoceptors. Lower part: localization of β_2-receptor mRNA visualized by *in situ* hybridization using cDNA-labeled oligonucleotide probe for β_2-receptor mRNA, NEN Du Pont. (By courtesy of Dr. G. Mengod.) There is a mismatch between the localization of the receptors in m and the RNA in g. **D:** Left side: autoradiographic localization in mouse cerebellum of muscarinic cholinergic receptors labeled by utilizing the muscarinic antagonist [^{3}H]quinuclidinyl benzilate (QNB). Right side: corresponding section of lobule VII stained with cresyl fast violet. (From Reference 19 with permission.)

density of receptors were found over the molecular layer of several vermal regions including lobules I–V, IX, and X. The nicotinic receptors, however, appeared to be located in the granular layer.[23,24]

The physiological actions of noradrenaline (NA) and acetylcholine (ACh) on individual neurons in the cerebellar cortex are rather controversial (see Discussion). After the demonstration that P cells located in the vermal cortex of the cerebellar anterior lobe and projecting to the lateral vestibular nucleus (LVN) respond to roll tilt of the animal leading to stimulation of labyrinth receptors,[25] evidence was presented indicating that functional inactivation of these P cells by intravermal microinjection of γ-aminobutyric acid (GABA) agonists decreased the amplitude of the electromiographic (EMG) responses of limb extensors to animal tilt.[26] This finding indicated that the P cells exert a positive influence on the gain of the vestibulospinal (VS) reflexes. The experiments summarized in the present report were aimed to investigate the effects of local administration of noradrenergic and cholinergic agonists and antagonists into the vermal cortex on the dynamics of these reflexes. In particular, we found that intravermal microinjection of noradrenergic and cholinergic agonists increased the gain of the VS reflexes, while injection of the corresponding antagonists decreased this gain.[27-30] It appears, therefore, that both the noradrenergic and cholinergic systems intervene in the gain regulation of the VS reflexes, probably by exerting a facilitatory influence on the cerebellar processing of labyrinth signals.

METHODS

The experiments were performed in 40 precollicular decerebrate cats, operated under ether anesthesia. The multiunit EMG activity of the medial head of the triceps brachii of both sides was recorded during roll tilt of the animal at 0.15 Hz, ±10° (cf. Reference 31). Sequential pulse density histograms (SPDHs) were obtained by averaging data of 6 sweeps, each containing the responses to two successive cycles (128 bins, 0.1-second bin width). These stimulation sequences were repeated at regular intervals of 4–10 minutes for several hours, before and after pressure microinjection of noradrenergic and cholinergic agonists or antagonists into the vermal cortex of the cerebellar anterior lobe of one side. The digital data of these averaged responses were processed on-line with a computer system (PET, 2001-8C or Commodore CBM 3032), which performed a fast Fourier transform. In particular, the gain (in impulses/second per deg) and the phase angle of the first harmonic component of the responses (in degrees with respect to the peak of the side-down displacement of the animal) were evaluated. The base frequency (in impulses/second), which corresponded to the DC value obtained from the harmonic analysis of the responses, was also evaluated; this value was comparable to the mean frequency of the multiunit discharge recorded at rest.

A vertically oriented stainless steel cannula with an outer diameter of 200–300 μm, connected to a Hamilton 1-μl syringe, was lowered into the cerebellar vermis passing through the third or the fourth folium rostrally to the fissura prima (culmen) at 1.4–1.8 mm laterally to the midline and at the depth of about 4 mm below the surface. Usually, 0.25 μl of the following agents solved in saline at the concentration indicated below were injected in separate experiments: the α_1-adrenergic agonist (metoxamine, Sigma Chemical Company, St. Louis, Mo: 4–8 μg/μl) or antagonist (prazosin, Pfeizer Italiana S.p.A., Rome: 8–16 μg/μl); the α_2-adrenergic agonist (clonidine, Sigma: 2–4 μg/μl) or antagonist (yohimbine, Sigma: 8–16 μg/μl); the nonselective β-adrenergic agonist ([±]-isoproterenol hydrochloride, Sigma: 8–16 μg/μl) or antagonist (dl-propranolol hydrochloride, Sigma: 8–16 μg/μl), which acts on both β_1 and β_2 receptors; the nonselective cholinergic agonist carbamylcholine chloride (carbachol, Sigma: 0.5 μg/μl), which acts on both muscarinic and nicotinic receptors; the muscarinic agonist carbamyl-β-methyl-choline chloride (bethanechol, Sigma: 0.1 μg/μl) or antagonist [(−)-scopolamine hydrobromide, Sigma: 4–8 μg/μl]; the nicotinic agonist [(−)-nicotine, Sigma: 0.1 μg/μl] or antagonists (hexamethonium bromide, Sigma: 4 μg/μl; D-tubocurarine chloride, Sigma: 7 μg/μl]; and finally, the acetylcholinesterase inhibitor (eserine sulfate, Sigma: 0.2–1 μg/μl). The injected solutions were also stained with the blue dye pontamine (5%) as a marker. The corticocerebellar area chosen for injection was identified by recording the inhibitory changes in the EMG activity of the ipsilateral triceps brachii induced by electrical stimuli (3 cathodal pulses at 300/second, 0.2 msecond, 0.1–10 V) applied monopolarly at the repetition rate of 0.5–2/second. A stainless steel wire of 300 μm in size, electrolytically sharpened and completely insulated except at the tip, was used for stimulation. At the end of each experiment, the localization of the tip of the cannula for each penetration, as well as the extent of the blue-stained tissue, was identified on cerebellar frontal sections counterstained with neutral red. The effects of the injected substances were tested for statistical significance (Student's t test). Because several of the changes observed were especially marked during the first two hours after the injection, only the measurements made during this period were used for statistical evaluation.

RESULTS

Unilateral microinjection of 0.25 μl of the α_1- and α_2-adrenergic agonists metoxamine and clonidine, as well as of the β-adrenergic agonist isoproterenol (at 4–16 μg/μl) into the vermal cortex of the cerebellar anterior lobe reduced the postural activity in the ipsilateral limbs, while that of the contralateral limbs either remained unmodified or slightly increased. Just the opposite changes in posture were obtained in other experiments after injection of the same doses of the α_1- and α_2-adrenergic antagonists prazosin and yohimbine, as well as of the β-adrenergic antagonist propranolol (at 8–16 μg/μl). Changes in posture similar to those elicited by adrenergic agonists and antagonists were also obtained after local administration of 0.25 μl of the cholinergic agonists carbachol, bethanechol, and nicotine (at 0.1–0.5 μg/μl) and the cholinergic antagonists scopolamine, hexamethonium, and D-tubocurarine (at 4–8 μg/μl), respectively.

Effects of Noradrenergic Agonists

Rotation around the longitudinal axis of the animal at 0.15 Hz, ±10° produced a sinusoidal modulation of the multiunit EMG activity of the triceps brachii, characterized by an increased activity during side-down tilt and a decreased activity during side-up tilt (cf. Reference 31). These responses were related to animal position, thus being attributed to stimulation of macular, utricular receptors. Occasionally, averaged traces did not show coherent responses to animal tilt in the control situation.

Unilateral microinjection of the α_1-*adrenergic agonist* metoxamine (0.25 μl at 4–8 μg/μl) as well as of the α_2-*adrenergic agonist* clonidine (0.25 μl at 2–4 μg/μl) into the vermal cortex of the cerebellar anterior lobe increased the EMG modulation and thus the response gain of the ipsilateral (FIGURE 2A and B) as well as of the contralateral forelimb extensor to labyrinth stimulation. Moreover, the proportion of averaged records that showed responses to tilt increased bilaterally. Only slight changes in the phase angle of the responses were observed in these conditions. In two representative experiments (experiments 7 and 8), the increase in gain of the

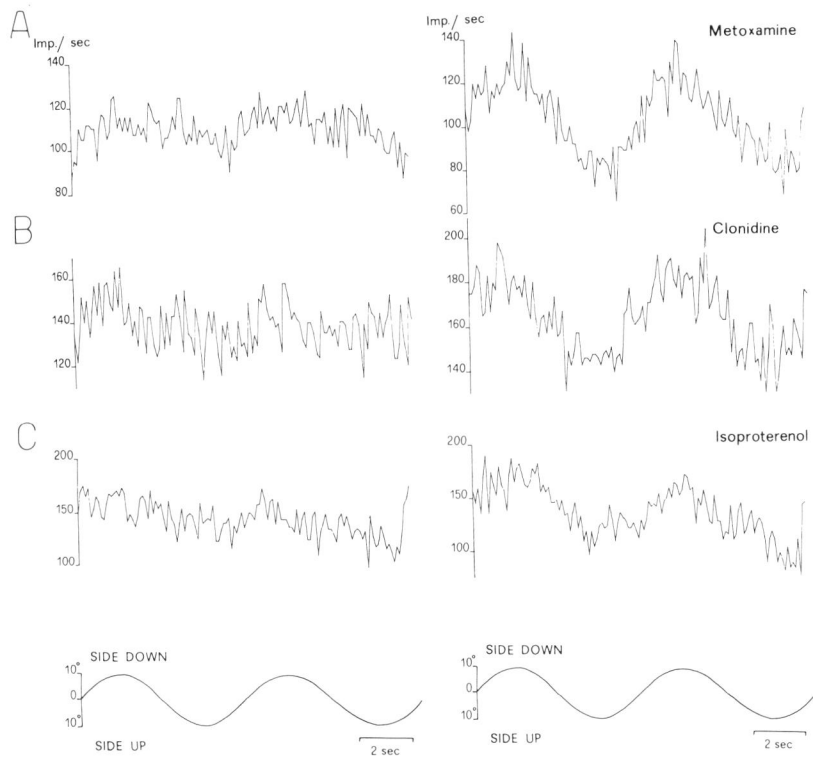

FIGURE 2. Increase of the response gain of the triceps brachii to animal tilt after microinjection of an α_1- (A), α_2- (B), and β- (C) adrenergic agonist into the ipsilateral vermal cortex of the cerebellar anterior lobe (culmen). Precollicular, decerebrate cats (A, experiment 8; B, experiment 2; C, experiment 16). SPDHs showing the averaged multiunit responses of the triceps brachii of one side to animal tilt at 0.15 Hz, $\pm 10°$. Each record is the average of 6 sweeps (128 bins, 0.1 second bin width). The lower traces indicate the animal displacement. The traces on the left side were taken before, while those on the right side after, individual or multiple injections of 0.25 μl of metoxamine (4 μg/μl), clonidine (4 μg/μl), and isoproterenol (16 μg/μl) solutions. The response gain increased on the average from 0.61 to 2.08 impulses/second per deg in A, from 0.60 to 1.94 impulses/second per deg in B, and from 1.24 to 2.72 impulses/second per deg. (From References 27 and 28 with permission.)

averaged responses elicited after metoxamine injection with respect to the control value was more prominent on the ipsilateral (180.3%; t test, $p < 0.001$) than on the contralateral side (137.4%; t test, $p > 0.05$, i.e., not significant). On the other hand in other experiments (experiments 1 and 2), the increase of the response gain induced by clonidine injection on the ipsilateral side (171.2%) was almost comparable to that of the contralateral side (190.4%) (*t* test, $p < 0.01$ and $p < 0.001$ for the ipsilateral and the contralateral responses, respectively). The gain changes of the EMG responses of the triceps brachii of both sides to animal tilt started 5–10 minutes after the injection and reached the highest values after 20–30 minutes; they were then followed for about 2 hours after the injections before disappearing.

Changes in gain similar to those described above, but smaller in amplitude, were obtained after unilateral microinjection of the β-*adrenergic agonist* isoproterenol (0.25 µl at 8–16 µg/µl) into the vermal cortex of the cerebellar anterior lobe (FIGURE 2C). In three representative experiments (Experiments 8, 14, and 16) individual or multiple injection of isoproterenol at the concentration indicated above increased the mean gain of the EMG responses of the ipsilateral and the contralateral triceps brachii to 156.7 and 148.3% of the control values, respectively (*t* test, $p < 0.001$ for both the ipsilateral and the contralateral responses). These gain changes, which were not associated with great changes in the phase angle of the responses, followed the same time course as that produced after injection of the α_1- or the α_2-adrenergic agonists (FIGURE 3A).

The results described above were obtained in the absence of any significant change in base frequency, which was kept constant due to appropriate changes in limb position. Moreover, neither changes in posture nor changes in the response gain of the triceps brachii to labyrinth stimulation were observed after local injection of an equal volume of saline prior to the effective injection. The results of the noradrenergic agonists were dose dependent and site specific, the effective area being located within the second and third or the third and fourth folium of the vermal part of the culmen, 1.4–1.8 mm lateral to the midline, and at the depth of 3–4 mm below the surface (see injection II in FIGURE 3A and FIGURE 4, left diagram). This area corresponded to the longitudinal zone B of the vermis which projects to the ipsilateral LVN.[36,37] Moreover, electrical stimulation of this corticocerebellar area prior to the injection inhibited the spontaneous EMG activity of the ipsilateral triceps brachii (FIGURE 5), a finding that can be attributed to suppression of the tonic facilitatory influence that the LVN exerts on posture (cf. References 32 and 33). Injections of noradrenergic agents into the same folia of the cerebellar vermis, but 0.2–0.5 mm more laterally or medially with respect to the effective zone, were almost ineffective (see injection I in FIGURE 3A and FIGURE 4, left diagram).

Effects of Noradrenergic Antagonists

Since in decerebrate cats the VS reflexes are of small amplitude, we decided to select experiments in which the response gain of the triceps brachii to animal tilt was in the control situation higher than usual and study the effects of local injection in the cerebellar vermis of the nonselective β-adrenergic antagonist. In four experiments (experiments 1, 6, 12, and 13) individual or multiple microinjections of the β-adrenergic antagonist propranolol (0.25–0.50 µl at 8–16 µg/µl) into the vermal cortex of the cerebellar anterior lobe of one side decreased the amplitude of modulation and thus the response gain of the ipsilateral forelimb extensor to animal tilt to 72.8% of the control value (*t* test, $p < 0.001$). There was also an increase in number of the averaged traces that did not show coherent responses to tilt with respect to the controls. On the contralateral side, however, the response gain decreased only to 84.5% of the control value (*t* test, NS). The effects described above followed the same time course as that obtained in previous experiments after local injection of isoproterenol.

The effects of unilateral microinjection of noradrenergic antagonists on the gain changes induced by a given agonist were also investigated. In particular, unilateral injection into the vermal cortex of the cerebellar anterior lobe of the α_1- or α_2-adrenergic antagonist prazosin or yohimbine (0.25 μl at 8–16 μg/μl) did not greatly modify the peak effects induced by previous injection into the same cortico-

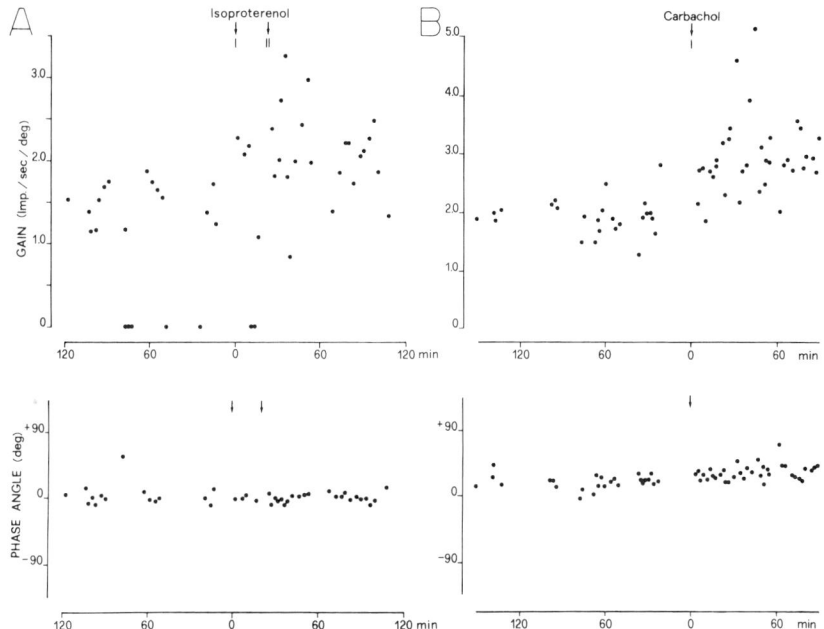

FIGURE 3. Effects of unilateral microinjection of nonselective β-adrenergic and cholinergic agonists into the vermal cortex of the cerebellar anterior lobe (culmen) on the response gain of the ipsilateral triceps brachii to animal tilt. Precollicular, decerebrate cats (A, experiment 16; B, experiment 46). The gain and the phase angle of averaged multiunit responses of the triceps brachii of one side to animal tilt at 0.15 Hz, ±10° were evaluated at different time intervals before and after local injections of isoproterenol (16 μg/μl in A) or carbachol (0.5 μg/μl in B) solutions. In **A** the first arrow (I) indicates the slight increase of the response gain of the triceps brachii to animal tilt after three injections of 0.25 μl of the isoproterenol solution into the ipsilateral hemivermis, 1.4 mm lateral to the midline (L1.4), and at the depths of 3, 4, and 5 mm below the surface. A more prominent effect occurred after similar doses had been injected a bit more laterally (L1.8), as indicated by the second arrow (II). In **B**, the arrow (I) indicates the prominent increase of the response gain of the triceps brachii to animal tilt after a 0.25-μl injection of the carbachol solution into the ipsilateral hemivermis, 1.6 mm lateral to the midline (L1.6), and at the depth of 5 mm below the surface. (From References 28 and 29 with permission.)

cerebellar area of the corresponding agonist metoxamine or clonidine (0.25 μl at 4–8 μg/μl), respectively. However, if the α_1- or the α_2-adrenergic antagonist was injected 2–3 hours after administration of the corresponding agonist, it prevented the occurrence of the gain increase induced by a successive injection of the corresponding agonist.

Isoproterenol   Carbachol

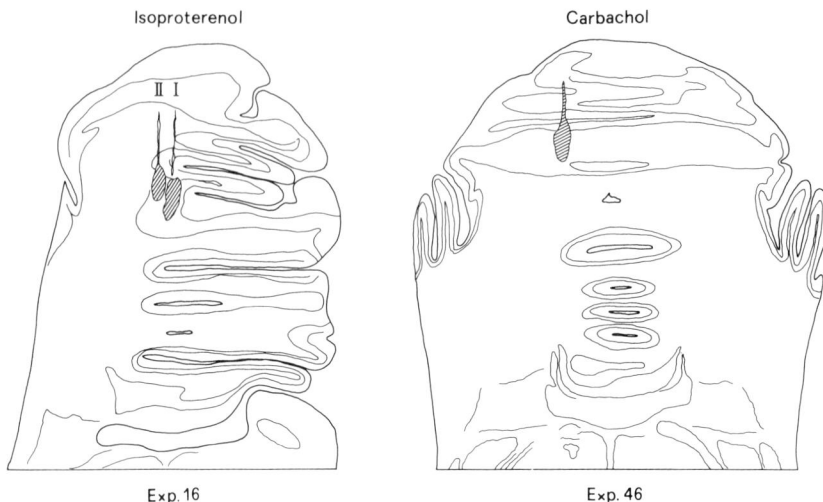

Exp. 16   Exp. 46

FIGURE 4. Frontal sections of the cerebellar cortex of the anterior lobe showing the localization of the sites that upon injection of nonselective β-adrenergic and cholinergic agonists increased the response gain of the ipsilateral triceps brachii to animal tilt. Same experiments as in FIGURE 3. Shaded areas indicate the localization of the injected solution stained with 5% pontamine into the right hemivermis (corresponding to the left side in each diagram). In Experiment 16 three successive doses of 0.25 μl of an isoproterenol solution (16 μg/μl), injected at the laterality of 1.4 mm (I), were slightly effective; a potentiation of the effect, however, was obtained when the same doses were injected 1.8 mm lateral to the midline (II). In experiment 46 a dose of 0.25 μl of carbachol solution (0.5 μg/μl), injected at the laterality of 1.6 mm, was quite effective. (From References 28 and 29 with permission.)

Cross interactions between α_1- and α_2-adrenergic agonists and antagonists were also observed. In particular, the α_2-adrenergic antagonist yohimbine suppressed both the ipsilateral and, if present, the contralateral effects induced by metoxamine, while the α_1-adrenergic antagonist prazosin suppressed only the ipsilateral, but not the contralateral, effects elicited by clonidine. Thus the α_1-adrenergic antagonist prazosin was more effective on the ipsilateral than on the contralateral responses, just as shown for the effects induced by the corresponding α_1 agonist, while the α_2-adrenergic antagonist yohimbine was almost equally effective on the limb extensors of both sides, in analogy with the results elicited by the α_2-adrenergic agonist. Additional findings indicated that microinjections of the β-adrenergic antagonist propranolol (0.25–0.50 μl at 8–16 μg/μl) reduced the response gain of the ipsilateral triceps brachii to animal tilt, which had been previously (1.5 hours before) enhanced by the isoproterenol injection. The same propranolol injection, however, did not prevent the increase in the response gain of the ipsilateral triceps brachii following successive administration into the same corticocerebellar area of either metoxamine (0.25 μl at 8 μg/μl) or clonidine (0.25 μl at 4–16 μg/μl). In this instance, however, the sudden increase in the response gain of the ipsilateral triceps brachii to animal tilt, following clonidine injection, was substituted by a slow increase in amplitude of modulation. Moreover, the clonidine-induced increase in the response gain of the contralateral triceps brachii to labyrinth stimulation was suppressed. This finding can be attributed to crossed blocking influences, exerted by the propranolol injection, which were not clearly detected in control preparations, but only after activation of the α_2-adrenergic system by clonidine.

Effects of Cholinergic Agonists

Unilateral microinjection of the *nonselective cholinergic agonist* carbachol (0.25 μl at 0.5 μg/μl) into the vermal cortex of the cerebellar anterior lobe (FIGURE 4, right diagram) increased the amplitude of the EMG modulation and thus the response gain of the ipsilateral forelimb extensor triceps brachii to labyrinth stimulation (FIGURE 6A). However, only slight changes in the phase angle of the responses were observed. Similar results were also obtained in a different experiment shortly after the injection of the *anticholinesterase* eserine sulfate (0.25 μl at 0.2–1 μg/μl).

Since carbachol acts on both muscarinic and nicotinic receptors, experiments were performed to study the effects of injection into the anterior vermis of selective muscarinic and nicotinic agonists. Unilateral injection either of the *muscarinic agonist* bethanechol or of *nicotine* (0.25 μl at 0.1 μg/μl) into the vermal cortex of the cerebellar anterior lobe increased the response gain of the ipsilateral (FIGURE 6B and C) as well as of the contralateral triceps brachii to labyrinth stimulation. Moreover, the number of averaged traces that did not show coherent responses to tilt decreased with respect to the controls, both ipsilaterally and contralaterally to the side of the injection. Only negligible changes in the phase angle of the responses were observed. The increased gain of the averaged responses elicited after bethanechol injection, with respect to the control values, as recorded in five experiments (experiments 35, 38, 39, 47, and 49), was more prominent on the ipsilateral (206.9%) than on the contralateral side (158.0%), while that elicited in four experiments (experiments 45 and 51–53) by nicotine injection was more prominent on the contralateral (205.7%) than on the ipsilateral side (146.0%) (*t* test, $p < 0.001$ for the ipsilateral and the contralateral responses in both groups of experiments). These changes in gain, which were associated only with slight changes in the phase angle of the responses, followed the same time course as that produced by injection of different adrenergic agonists (FIGURE 3B).

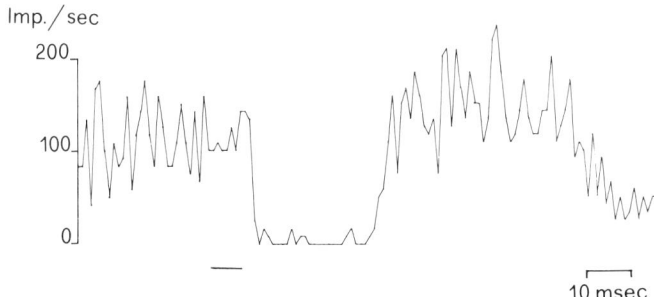

FIGURE 5. Inhibition of postural activity followed by rebound in the forelimb extensor triceps brachii after electrical stimulation of the ipsilateral vermal cortex of the cerebellar anterior lobe. Precollicular, decerebrate cat. SPDH showing the averaged multiunit response of the triceps brachii of one side to monopolar stimulation of the ipsilateral hemivermis (culmen) with 3 negative pulses at 300/second, 0.2 msecond duration, 10 V; the duration of the stimulus is indicated by the horizontal bar. The record (128 bins, 1 msecond bin width) is the average of 120 sweeps obtained at the repetition rate of 0.5/second. (From Reference 28 with permission.)

Effects of Cholinergic Antagonists

Observations performed in some experiments (experiments 47, 49, and 50) have shown that individual or multiple microinjections of the muscarinic antagonist scopolamine (0.25 μl at 4–8 μg/μl) into the vermal cortex of the cerebellar anterior

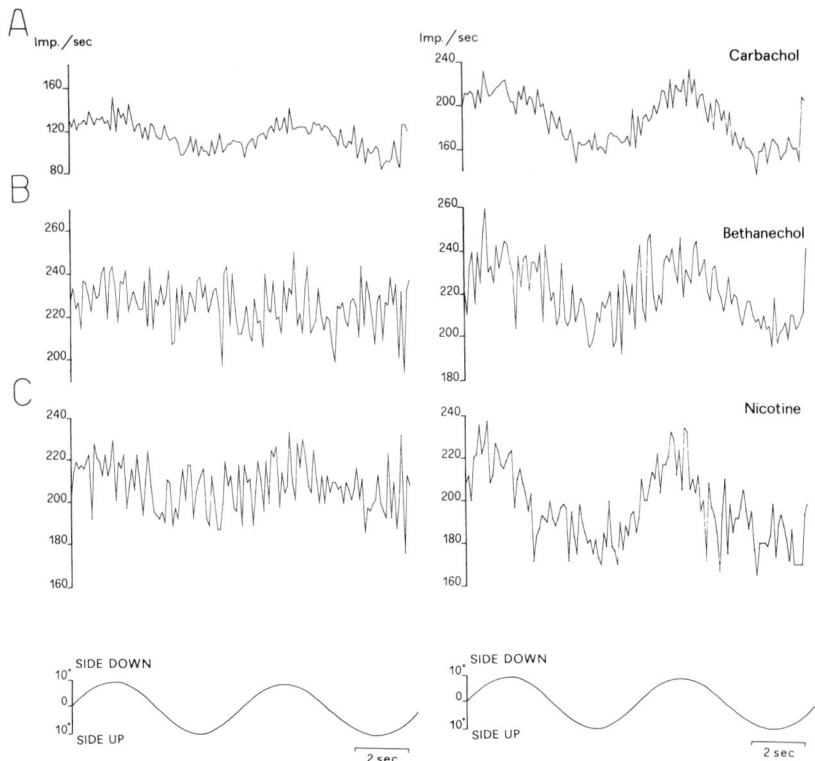

FIGURE 6. Increase of the response gain of the triceps brachii to animal tilt after microinjection of the cholinergic agonists carbachol (A), bethanechol (B), and nicotine (C) into the ipsilateral vermal cortex of the cerebellar anterior lobe (culmen). Precollicular, decerebrate cats (A, experiment 46; B, experiment 39; C, experiment 51). Responses as in FIGURE 2. The traces on the left side were taken before, while those of the right side after, individual injections of 0.25 μl of carbachol (0.5 μg/μl of saline), bethanechol (0.1 μg/μl), and nicotine (0.1 μg/μl) solutions. The response gain increased on the average from 1.46 to 2.87 impulses/second per deg in A, from 0 to 1.48 impulses/second per deg in B, and from 0.68 to 1.83 impulses/second per deg. (From References 29 and 30 with permission.)

lobe of one side decreased the amplitude of modulation and thus the response gain of both the ipsilateral and the contralateral triceps brachii to animal tilt to 65.7% (t test, $p < 0.001$) and 69.7% (t test, $p < 0.01$) of the control values, respectively. Smaller effects were obtained in other experiments (experiments 51–54) after injections into the cerebellar vermis of nicotinic antagonists of the ganglionic type hexamethonium and/or of the neuromuscular type D-tubocurarine (0.25 μl at 4 and 7 μg/μl, respectively), which decreased the response gain of the ipsilateral as well as of the contralateral triceps brachii to animal tilt to 79.5% (t test, NS) and 90.5% (t test, NS) of the control values, respectively. These effects appeared soon after the injection, reached a maximum in about 30 minutes, and recovered in about 60–90 minutes.

It is of interest that repetitive injections of scopolamine, performed at 100-minute intervals from each other, did not decrease but actually enhanced the effects of bethanechol injections performed 40 minutes after each scopolamine administration. This effect can be attributed to increased numbers of muscarinic receptors, which probably occurred after administration of muscarinic antagonists (muscarinic supersensitivity, cf. Reference 34) and led to an increased effect after injection of the same dose of the muscarinic agonist. On the other hand, microinjection into the cerebellar vermis of nicotinic antagonists slightly reduced the effects induced by an effective dose of nicotine.

DISCUSSION

The experiments summarized in the present report have shown that unilateral microinjection into the vermal cortex of the cerebellar anterior lobe of the selective α_1- (metoxamine) and the α_2-adrenergic agonist (clonidine) increased the gain of the VS reflexes, which affected not only the ipsilateral but also the contralateral triceps brachii. The latter effect, however, was more prominent after injection of α_2 than of the α_1 agonist. Moreover, bilateral effects similar in sign but smaller in amplitude than those induced by clonidine were elicited by the nonselective β-adrenergic agonist isoproterenol, which acts on both β_1 and β_2 adrenoceptors. There were only slight changes in the phase angle of the responses, which remained always related to the extreme animal position, as expected if the responses were mainly due to stimulation of macular receptors.

On the basis of the distribution of the different types of adrenoceptors in the different layers of the cerebellar cortex (see Introduction) we postulate that the ipsilateral effects are due to the adrenergic agents that act on the P-cell dendrites and somata either directly or through postsynaptic (local interneurons) and presynaptic (terminals of parallel fibers) mechanisms. On the other hand, the contralateral effects, particularly present after activation of the α_2 and to a lesser extent also of the β adrenoceptors, can be due to the activity of granule cells, whose parallel fibers interconnect the paramedial zones of the cerebellar vermis of both sides.

The specificity of the results is supported by the fact that the effects of metoxamine were greatly reduced or suppressed by previous injection into the cerebellar vermis of the α_1-adrenergic antagonist prazosin, while those induced by clonidine were prevented by local injection of the α_2-adrenergic antagonist yohimbine. The demonstration that the effects of metoxamine were also suppressed by previous injection into the cerebellar vermis of yohimbine, while injection of prazosin prevented the occurrence of the ipsilateral (but not of the contralateral) effects following injection of clonidine, can be attributed to colocalization of both α_1 and α_2 adrenoceptors on the same corticocerebellar neurons, as the P cells, so that administration of either one of the two α-noradrenergic blockers modified the responsiveness of the same neurons to the adrenergic agonist acting on the other subtype of α receptors. In addition to these findings, the nonselective β-adrenergic antagonist propranolol decreased the response gain of the ipsilateral and to a lesser extent of the contralateral triceps brachii to labyrinth stimulation and also depressed the effects of a previous injection of isoproterenol into the same corticocerebellar area. The possibility that at the concentration used in the present experiments the β-adrenergic agents acted also through α adrenoceptors (see Reference 28 for references) is excluded by the fact that a preinjection of propranolol did not prevent the increase in the response gain of the ipsilateral triceps brachii to labyrinth stimulation elicited by local injection of a selective α_1- or α_2-adrenergic agonist. However, the increase in the response gain of the contralateral muscle to labyrinth stimulation induced by α-adrenergic agents was impaired by the propranolol injection. These findings exclude that the β-adrenergic blocker propranolol displayed a local anesthetic action on the P cells (cf. Reference 35).

As to the cholinergic system, intravermal injection of the nonselective cholinergic agonist carbachol produced a significant increase in gain of the VS reflexes of both sides not associated with changes in the phase angle of the responses. Moreover, similar effects were elicited following injection of the anticholinesterase eserine sulfate, suggesting that the receptors activated by carbachol were indeed innervated by cholinergic elements and thus functionally significant.

The facilitatory influence of carbachol on the VS reflex was mediated through both muscarinic and nicotinic receptors. In fact, the gain of the VS reflex increased after local injection either of the muscarinic agonist bethanechol or of nicotine. This effect was bilateral. However, bethanechol produced a more prominent increase in gain of the ipsilateral triceps brachii to animal tilt, while nicotine increased the gain particularly on the contralateral side. These preferential effects can be attributed to the relative distribution of the muscarinic and nicotinic receptors on different neuronal populations in the cerebellar cortex (see Introduction). In addition to these findings, local administration either of the muscarinic antagonist scopolamine or of the nicotinic antagonists hexametonium and/or D-tubocurarine decreased the gain of the VS reflexes. Since hexametonium blocks the ganglionic-type nicotinic receptors, while D-tubocurarine blocks the neuromuscular-type nicotinic receptors, it appears that both types of receptors mediate the effects described above. The depression of the VS reflexes induced by the cholinergic, muscarinic, and nicotinic blockers is similar to that elicited by local injection into the cerebellar vermis of the β-adrenergic blocker propranolol, thus indicating that both the noradrenergic and cholinergic systems impinging on the cerebellar vermis are tonically active in the decerebrate cat and operate synergistically on the same neuronal systems.

The corticocerebellar area which, after injection of the noradrenergic and the cholinergic agonists, elicited the effects described above corresponded to a paramedial zone of the cerebellar anterior vermis at the level of the culmen. This area, which belongs to the forelimb region of the anterior vermis (cf. References 32 and 33), corresponds to zone B of the cerebellar cortex which projects to the ipsilateral LVN[36,37] and shows antidromic field potentials following single shock stimulation of the underlying LVN.[25] The same area actually exerts a direct inhibitory influence on the corresponding VS neurons (cf. References 32 and 33), which explains why repetitive electrical stimulation of this corticocerebellar area, performed prior to the microinjections, suppressed postural activity in the ipsilateral triceps brachii.

In order to understand the mechanisms by which the noradrenergic and the cholinergic afferent systems to the cerebellar cortex affect posture as well as the gain of the VS reflexes, we should consider the influence that both systems exert on the discharge of P cells. With respect to the *noradrenergic system, in situ* experiments have shown that microiontophoretic application of NA decreased the resting discharge of P cells.[38–40] This effect was associated with hyperpolarization of the P-cell membrane (cf. Reference 41), which was coupled with an increase in their membrane resistance, suggesting that these effects were probably mediated through cAMP.[42,43] These effects were originally attributed to β receptors.[38,40,41] However, investigations performed *in vitro*[44] as well as *in oculo*[45] have revealed that NA could induce not only inhibition but also excitation of P-cell spontaneous activity, the former effect being mediated by α receptors (cf. also Reference 35), while the latter by β receptors. The noradrenergic system may act postsynaptically not only on P cells but also on GABA-ergic inhibitory interneurons (cf. Reference 46). The same system may also exert a presynaptic role by increasing (through β_2 receptors) or decreasing (through α receptors) the release of glutamate at the level of parallel fibers and possibly also of CF terminals.[47]

As to the *cholinergic system,* the physiological action of ACh in the cerebellum is still unclear. In fact iontophoretic or bath application of this agent mildly excited P cells in the cat[48,49] (cf. Reference 50 in the rat), while pressure ejection of ACh inhibited P cells in the rat.[51] According to some authors, no effect of ACh was found for granule cells and interneurons,[48] while excitation was reported by others.[49,51,52] Also the reports on the receptor types involved in the postsynaptic ACh effects are inconsistent, the effects being attributed either to muscarinic[48,50] or to nicotinic receptors.[49] Moreover, De la Garza *et al.* showed that the inhibitory effects of ACh on P cells were mediated by mixed nicotinic (ganglionic type) and muscarinic receptors, while the excitatory action of ACh on interneurons was mediated by nicotinic (neuromuscular type) receptors.[51] The cholinergic system may also act on the cerebellar cortex not only postsynaptically but also presynaptically. In particular, it appears that the presynaptic receptors that regulate ACh release from cholinergic terminals are not of the muscarinic (M_2) type as originally thought (cf. Reference 16), but rather of the nicotinic type.[20]

In summary, the physiological actions of both NA and ACh on corticocerebellar units are to some extent controversial and it is not clear whether NE and ACh act as neurotransmitters.

The slight changes in postural activity that occur after unilateral injection of noradrenergic agonists, i.e., a reduced postural activity in the ipsilateral limbs and an increased activity of the contralateral limbs, can be attributed to local changes in the discharge of the P cells mediated by the injected agents either directly or via the related interneurons. As to the prominent changes in gain of the VS reflexes, it is known that the contraction of limb extensors that occurs during side-down tilt of the animal (cf. Reference 31) depends upon an increased discharge of LVN neurons, while just the opposite result occurs during side-up tilt (cf. Reference 53). This modulation of the discharge of VS neurons during tilt is submitted to a cerebellar control. In particular there is evidence that most of the P cells showed a modulation of their simple spike activity, which was opposite in phase with respect to that of LVN neurons, being characterized by a decreased discharge during side-down tilt and an increased discharge during side-up tilt.[25] This finding indicates that the increased activity of the VS neurons and thus the contraction of the corresponding limb extensors during side-down tilt depends not only upon an increased excitatory input originating from ipsilateral labyrinth receptors, but also on disinhibition of the same neurons resulting from a reduced discharge of the overlying P cells. This interaction would then lead to a positive influence of the cerebellar cortex on the VS reflex gain. Blocking experiments with GABA agonists clearly indicate that the vermal activity contributes to the gain of these VS reflexes.[26] Since the source of the simple spike modulation of the P cells during animal tilt is the MF input, particularly originating from the lateral reticular nucleus,[54] we postulate that the depth of this P-cell modulation is enhanced by the noradrenergic and the cholinergic systems.

Indeed, there is evidence that iontophoretic application of norepinephrine (NE), while depressing the spontaneous activity of P cells, enhanced the responses of these cells to both excitatory (MF and CF) and inhibitory (basket and stellate cells) inputs,[39,42,55,56] as well as to the corresponding excitatory (glutamate, aspartate) and inhibitory (GABA) transmitters.[40,41,55,57] It appears, therefore, that one of the main functions of the noradrenergic input in cerebellar operation is to augment target neuron responsiveness to conventional afferent systems, thus increasing the signal-to-noise ratio of the evoked versus spontaneous activity (cf. References 55 and 58). The same input could also act to gate the efficacy of subliminal synaptic inputs conveyed by the classical afferent systems.[56,58] As to the cholinergic system it was shown that electrically driven synaptic responses recorded in P cells were either unaffected or only slightly affected by local application of ACh.[50] We cannot exclude, however, that the cholinergic system might have a modulatory action on the corticocerebellar neurons, making the granular cells and possibly also the P cells more sensitive to specific MF and CF inputs elicited by naturally induced afferent signals.

In our experiments the increase in gain of the VS reflexes produced by local injection of noradrenergic and cholinergic agonists can be explained by assuming that both NA and ACh increased the amplitude of modulation of the P cells of the cerebellar vermis to given parameters of labyrinth stimulation, thus contributing to an increased gain of the VS reflex, as described above. The possibility that noradrenergic and cholinergic receptors are also located on granule cells, which may transmit the afferent signals from one side of the cerebellar vermis to the P cells of the opposite side, explains why local administration of noradrenergic and cholinergic agents may affect the response gain not only of the ipsilateral but also of the contralateral limb extensors to labyrinth stimulation.

In all the experiments reported above, we have hypothesized that the noradrenergic afferent system to the cerebellar cortex which originates from the LC (cf. Reference 1), and the cholinergic afferent system which originates in part at least from the vestibular and prepositus nuclei,[2] tonically modulate neuronal excitability, rather than transmit specific sensory signals. We cannot exclude, however, that these systems may also transmit specific sensory information from the labyrinth receptors, since a large proportion of presumably noradrenergic LC neurons,[59] as well as of vestibular nuclear neurons,[60] respond to animal tilt.

SUMMARY

In addition to mossy fibers and climbing fibers, the cerebellar cortex receives noradrenergic and cholinergic afferents. Since the Purkinje (P) cells of the cerebellar vermis (culmen) respond to roll tilt of the animal with a discharge pattern that is out of phase with respect to that of the related lateral vestibular neurons, thus exerting a facilitatory influence on the gain of the vestibulospinal (VS) reflex, we tested the effects of local microinjection into the anterior vermis of noradrenergic and cholinergic agents on these reflexes. In decerebrate cats, unilateral microinjection in the paramedial zone B of the culmen of 0.25 μl of small doses of α_1-, α_2-, and β-noradrenergic agonists (i.e., metoxamine, clonidine, and isoproterenol, respectively) increased the response gain (in impulses/second per deg) of the EMG response of the ipsilateral and to some extent also of the contralateral triceps brachii to animal tilt (at 0.15 Hz, ±10°). On the other hand local injection of the corresponding antagonists (i.e., prazosin, yohimbine, and propranolol) either decreased the gain of the ipsilateral triceps brachii to labyrinth stimulation or else prevented the occurrence of the effects induced by the corresponding agonists. An increase in gain of the VS reflexes was also elicited in other experiments by unilateral microinjection either of the nonselective cholinergic agonist carbachol or of the anticholinesterase eserine sulfate. Thus, the effects could be produced by increasing the naturally present amount of acetylcholine. Further experiments indicated that a bilateral increase in the response gain of the triceps brachii to labyrinth stimulation occurred after microinjection of a selective muscarinic (bethanechol) or nicotinic agonist (nicotine), while just the opposite result was obtained after microinjection of the corresponding muscarinic (scopolamine) and nicotinic (hexamethonium, D-tubo-curarine) blockers. The effects of the noradrenergic and cholinergic agonists, which persisted for about two hours after the injection, were site specific and dose dependent. It appears, therefore, that the noradrenergic and cholinergic afferents to the cerebellar vermis intervene in the gain regulation of the VS reflexes, possibly by increasing the amplitude of modulation of the P cells to labyrinth stimulation.

REFERENCES

1. FOOTE, S. L., F. E. BLOOM & G. ASTON-JONES. 1983. Nucleus locus coeruleus: new evidence of anatomical and physiological specificity. Physiol. Rev. **63**: 844–914.
2. BARMACK, N. H., R. W. BAUGHMAN, F. P. ECKENSTEIN & M. WESTCOTT-HODSON. 1986. Cholinergic projections to the cerebellum. Soc. Neurosci. Abstr. **12**: 577.
3. JONES, B. E. 1990. Immunohistochemical study of choline acetyltransferase immunoreactive processes and cells innervating the pontomedullary reticular formation in the rat. J. Comp. Neurol. **295**: 485–514.
4. KAN, K.-S. K., L.-P. CHAO & L. S. FORNO. 1980. Immunohistochemical localization of choline acetyltransferase in the human cerebellum. Brain Res. **193**: 165–171.
5. OJIMA, H., S. I. KAWAJIRI & T. YAMASAKI. 1989. Cholinergic innervation of the rat cerebellum: qualitative and quantitative analyses of elements immunoreactive to a monoclonal antibody against choline acetyltransferase. J. Comp. Neurol. **290**: 41–52.
6. ILLING, R. B. 1990. A subtype of cerebellar Golgi cells may be cholinergic. Brain Res. **522**: 267–274.
7. JONES, L. S., L. L. GAUGER & J. N. DAVIS. 1985. Anatomy of brain alpha-adrenergic receptors: in vitro autoradiography with [^{125}I]-HEAT. J. Comp. Neurol. **231**: 190–208.
8. PALACIOS, J. M., D. MOYER & R. CORTES. 1987. α_1-Adrenoceptors in the mammalian brain: similar pharmacology but different distribution in rodents and primates. Brain Res. **419**: 65–75.
9. UNNERSTALL, J. R., T. A. KOPAJTIC & M. J. KUHAR. 1984. Distribution of α_2-agonist binding sites in the rat and human central nervous system: analysis of some functional, anatomic correlates of the pharmacologic effects of clonidine and related adrenergic agents. Brain Res. Rev. **7**: 69–101.
10. BRUNING, G., P. KAULEN & H. G. BAUMGARTEN. 1987. Quantitative autoradiographic localization of α_2-antagonist binding sites in rat brain using [^{3}H] idazoxan. Neurosci. Lett. **83**: 333–337.
11. PALACIOS, J. M. & M. J. KUHAR. 1982. Beta adrenergic receptor localization in rat brain by light microscopic autoradiography. Neurochem. Int. **4**: 473–490.
12. RAINBOW, T. C., B. PARSONS & B. B. WOLFE. 1984. Quantitative autoradiography of β_1- and β_2-adrenergic receptors in rat brain. Proc. Natl. Acad. Sci. USA **81**: 1585–1589.
13. LORTON, D. & J. N. DAVIS. 1987. The distribution of beta-1 and beta-2-adrenergic receptors of normal and reeler mouse brain: an *in vitro* autoradiographic study. Neuroscience **23**: 199–210.
14. SUTIN, J. & K. P. MINNEMAN. 1985. Adrenergic beta receptors are not uniformly distributed in the cerebellar cortex. J. Comp. Neurol. **236**: 547–554.
15. PALACIOS, J. M. 1988. Mapping brain receptors by autoradiography. ISI Atlas Sci. Pharmacol. **2**: 71–77.
16. MASH, D. C. & L. T. POTT. 1986. Autoradiographic localization of M1 and M2 muscarine receptors in the rat brain. Neuroscience **19**: 551–564.

17. SPENCER, D. G., E. HORVÁTH & J. TRABER. 1986. Direct autoradiographic determination of M1 and M2 muscarinic acetylcholine receptor distribution in the rat brain: relation to cholinergic nuclei and projections. Brain Res. **380:** 59–68.
18. BUCKLEY, N. J., T. I. BONNER & M. R. BRANN. 1988. Localization of a family of muscarinic receptor mRNAs in rat brain. J. Neurosci. **8:** 4646–4652.
19. NEUSTADT, A., A. FROSTHOLM & A. ROTTER. 1988. Topographical distribution of muscarinic cholinergic receptors in the cerebellar cortex of the mouse, rat, guinea pig, and rabbit: a species comparison. J. Comp. Neurol. **272:** 317–330.
20. LAPCHAK, P. A., D. M. ARANJO, R. QUIRION & B. COLLIER. 1989. Presynaptic cholinergic mechanisms in the rat cerebellum: evidence for nicotinic, but not muscarinic autoreceptors. J. Neurochem. **53:** 1843–1851.
21. FUKAMAUCHI, F., C. HOUGH & D.-M. CHUANG. 1991. Expression and agonist-induced down-regulation of mRNAs of m2- and m3-muscarinic acetylcholine receptors in cultured cerebellar granule cells. J. Neurochem. **56:** 716–719.
22. HUNT, S. & J. SCHMIDT. 1978. Some observations on the binding patterns of α-bungarotoxin in the central nervous system of the rat. Brain Res. **157:** 213–232.
23. SWANSON, L. W., D. M. SIMMONS, P. J. WHITING & J. LINDSTROM. 1987. Immunohistochemical localization of neuronal nicotinic receptors in the rodent central nervous system. J. Neurosci. **7:** 3334–3342.
24. CLARKE, P. B. S., R. D. SCHWARTZ, S. M. PAUL, C. B. PERT & A. PERT. 1985. Nicotinic binding in rat brain: autoradiographic comparison of [^{3}H] acetylcholine, [^{3}H] nicotine, [^{125}I]-bungarotoxin. J. Neurosci. **5:** 1307–1315.
25. DENOTH, F., P. C. MAGHERINI, O. POMPEIANO & M. STANOJEVIĆ. 1979. Responses of Purkinje cells of the cerebellar vermis to neck and macular vestibular inputs. Pflügers Arch. **381:** 87–98.
26. ANDRE, P., P. D'ASCANIO, D. MANZONI & O. POMPEIANO. 1991. Depression of the vestibulospinal reflex by intravermal microinjection of GABA-A and GABA-B agonists in decerebrate cats. Proceedings of the XVIII Spring Meeting of the Italian Physiological Society, Florence, April 4–6.
27. ANDRE, P., P. D'ASCANIO, A. GENNARI, A. PIRODDA & O. POMPEIANO. 1991. Microinjections of α$_1$- and α$_2$-noradrenergic substances in the cerebellar vermis of decerebrate cats affect the gain of the vestibulospinal reflexes. Arch. Ital. Biol. **129:** 113–160.
28. ANDRE, P., P. D'ASCANIO, D. MANZONI & O. POMPEIANO. 1991. Microinjections of β-noradrenergic substances in the cerebellar vermis of decerebrate cats modify the gain of the vestibulospinal reflexes. Arch. Ital. Biol. **129:** 161–197.
29. ANDRE, P., P. D'ASCANIO, D. MANZONI & O. POMPEIANO. 1991. Microinjections of muscarinic cholinergic agents in the cerebellar vermis affect the gain of the vestibulospinal reflex in decerebrate cats. Proceedings of the XVIII Spring Meeting of the Italian Physiological Society, Florence, April 4–6.
30. ANDRE, P., P. D'ASCANIO, D. MANZONI & O. POMPEIANO. Nicotinic receptors in the cerebellar vermis modulate the gain of the vestibulospinal reflexes in decerebrate cats. Arch. Ital. Biol. (In press.)
31. MANZONI, D., O. POMPEIANO, U. C. SRIVASTAVA & G. STAMPACCHIA. 1983. Responses of forelimb extensors to sinusoidal stimulation of macular labyrinth and neck receptors. Arch. Ital. Biol. **121:** 205–214.
32. POMPEIANO, O. 1967. Functional organization of the cerebellar projections to the spinal cord. Prog. Brain Res. **25:** 282–321.
33. ITO, M. 1984. The Cerebellum and Neural Control. Raven Press. New York, N.Y.
34. NATHANSON, N. M. 1987. Molecular properties of the muscarinic acetylcholine receptor. Annu. Rev. Neurosci. **10:** 195–236.
35. PARFITT, K. D., R. FREEDMAN & P. C. BICKFORD-WILMER. 1988. Electrophysiological effects of locally applied noradrenergic agents at cerebellar Purkinje neurons: receptor specificity. Brain Res. **462:** 242–251.
36. CORVAJA, N. & O. POMPEIANO. 1979. Identification of cerebellar corticovestibular neurons retrogradely labeled with horseradish peroxidase. Neuroscience **4:** 507–515.
37. VOOGD, J. 1989. Parasagittal zones and compartments of the anterior vermis of the cat cerebellum. Exp. Brain Res. Ser. **17:** 3–19.

38. HOFFER, B. J., G. R. SIGGINS & F. E. BLOOM. 1971. Studies on norepinephrine-containing afferents to Purkinje cells of rat cerebellum. II. Sensitivity of Purkinje cells to norepinephrine and related substances administered by microiontophoresis. Brain Res. 25: 523–534.

39. FREEDMAN, R., B. J. HOFFER, D. J. WOODWARD & D. PURO. 1977. Interaction of norepinephrine with cerebellar activity evoked by mossy and climbing fibers. Exp. Neurol. 55: 269–288.

40. MOISES, H. C., D. J. WOODWARD, B. J. HOFFER & R. FREEDMAN. 1979. Interactions of norepinephrine with Purkinje cell responses to putative amino acid neurotransmitters applied by microiontophoresis. Exp. Neurol. 64: 493–515.

41. WATERHOUSE, B. D., H. C. MOISES, H. H. YEH & D. J. WOODWARD. 1982. Norepinephrine enhancement of inhibitory synaptic mechanisms in cerebellum and cerebral cortex: mediation by beta adrenergic receptors. J. Pharmacol. Exp. Ther. 221: 495–506.

42. SIGGINS, G. R., A. P. OLIVER, B. J. HOFFER & F. E. BLOOM. 1971. Cyclic adenosine monophosphate and norepinephrine: effects on transmembrane properties of cerebellar Purkinje cells. Science 171: 192–194.

43. HOFFER, B. J., G. R. SIGGINS, A. P. OLIVER & F. E. BLOOM. 1973. Activation of the pathway from locus coeruleus to rat cerebellar Purkinje neurons: pharmacological evidence of noradrenergic central inhibition. J. Pharmacol. Exp. Ther. 184: 553–569.

44. BASILE, A. S. & T. V. DUNWIDDIE. 1984. Norepinephrine elicits both excitatory and inhibitory responses from Purkinje cells in the in vitro rat cerebellar slice. Brain Res. 296: 15–25.

45. GRANHOLM, A.-C. E. & M. R. PALMER. 1988. Electrophysiological effects of norepinephrine on Purkinje neurons in intraocular grafts: α- versus β-specificity. Brain Res. 459: 256–264.

46. LANDIS, S. C. & F. E. BLOOM. 1975. Ultrastructure identification of noradrenergic boutons in mutant and normal mouse cerebellar cortex. Brain Res. 96: 299–305.

47. DOLPHIN, A. C. 1982. Noradrenergic modulation of glutamate release in the cerebellum. Brain Res. 252: 111–116.

48. CRAWFORD, J. M., D. R. CURTIS, P. E. VOORHOEVE & V. J. WILSON. 1966. Acetylcholine sensitivity of cerebellar neurones in the cat. J. Physiol. London 186: 139–165.

49. MCCANCE, I. & V. J. PHILLIS. 1968. Cholinergic mechanisms in the cerebellar cortex. Int. J. Neuropharmacol. 7: 447–462.

50. CREPEL, F. & S. S. DHANJAL. 1982. Cholinergic mechanisms and neurotransmission in the cerebellum of the rat. An in vitro study. Brain Res. 244: 59–68.

51. DE LA GARZA, R., P. C. BICKFORD-WIMER, B. J. HOFFER & R. FREEDMAN. 1987. Heterogeneity of nicotine actions in the rat cerebellum: an in vivo electrophysiologic study. J. Pharmacol. Exp. Ther. 240: 689–695.

52. CIARDO, A. & J. MELDOLESI. 1991. Regulation of intracellular calcium in cerebellar granule neurons: effects of depolarization and of glutamatergic and cholinergic stimulation. J. Neurochem. 56: 184–191.

53. MARCHAND, A., D. MANZONI, O. POMPEIANO & G. STAMPACCHIA. 1987. Effects of stimulation of vestibular and neck receptors on Deiters neurons projecting to the lumbosacral cord. Pflügers Arch. 409: 13–23.

54. KUBIN, L., P. C. MAGHERINI, D. MANZONI & O. POMPEIANO. 1980. Responses of lateral reticular neurons to sinusoidal stimulation of labyrinth receptors in decerebrate cat. J. Neurophysiol. 44: 922–936.

55. WOODWARD, D. J., H. C. MOISES, B. D. WATERHOUSE, B. J. HOFFER & R. FREEDMAN. 1979. Modulatory actions of norepinephrine in the central nervous system. Fed. Proc. 38: 2109–2116.

56. MOISES, H. C., R. A. BURNE & D. J. WOODWARD. 1990. Modification of the visual response properties of cerebellar neurons by norepinephrine. Brain Res. 514: 259–275.

57. YEH, H. H., H. C. MOISES, B. D. WATERHOUSE & D. J. WOODWARD. 1981. Modulatory interactions between norepinephrine and taurine, beta-alanine, gamma-aminobutyric and muscimol, applied iontophoretically to cerebellar Purkinje cells. Neuropharmacol. 20: 549–560.

58. WATERHOUSE, D. B., F. M. SESSLER, J. T. CHENG, J. D. WOODWARD, S. A. AZIZI & H. C.
     MOISES. 1988. New evidence for a gating action of norepinephrine in central neuronal
     circuits of mammalian brain. Brain Res. Bull. **21:** 425–432.
59. POMPEIANO, O., D. MANZONI, C. D. BARNES, G. STAMPACCHIA & P. D'ASCANIO. 1990.
     Responses of locus coeruleus and subcoeruleus neurons to sinusoidal stimulation of
     labyrinth receptors. Neuroscience **35:** 227–248.
60. BOYLE, R. & O. POMPEIANO. 1980. Reciprocal responses to sinusoidal tilt of neurons in
     Deiters' nucleus and their dynamic characteristics. Arch. Ital. Biol. **118:** 1–32.

Cyclorotation of the Eyes and Subjective Visual Vertical in Vestibular Brain Stem Lesions[a]

TH. BRANDT AND M. DIETERICH

Department of Neurology
University of Munich
8000 Munich, Germany

INTRODUCTION

In the "primary position" in the roll plane, the subjective vertical is aligned with the gravitational vertical and the axes of the eyes and head are horizontal and directed straight ahead (i.e., "primary position" of the vestibuloocular reflex in roll). Cyclorotation of the eyes (CR) and subjective visual vertical (SVV) were measured in a total of 111 patients, most of them suffering from acute vascular brain stem lesions, in order to determine the differential effects of central vestibular pathway lesions on ocular motor and perceptual control in the roll plane. Evidence will be presented in the first part of this chapter that abnormal CR and significant deviations of SVV reflect a pathological tilt in the roll plane and are among the most sensitive clinical signs of brain stem dysfunction. Pathological tilt in the roll plane is direction specific—ipsiversive with pontomedullary lesions; contraversive with pontomesencephalic lesions—and it may involve a complete ocular tilt reaction, the triad of lateral head tilt, skew deviation of the eyes, and cyclorotation.[1-3] In the second part of this paper special emphasis will be given to dorsolateral medullary infarctions because they constitute a common and clearly described clinical entity.

METHODS

Patients

A total of 111 patients took part in the study consisting of 43 females and 68 males with ages ranging from 17 to 84 (mean 54 years). Participation in the study was voluntary, and informed consent was obtained from all subjects prior to their inclusion. All subjects had clinical signs and symptoms of acute mostly vascular brain stem lesions (82 infarctions; 13 hemorrhages; 16 plaques of multiple sclerosis) from medulla oblongata to mesencephalon, some also involving cerebellar structures.

In 36 patients (12 female, 24 male; aged 25 to 79 years; mean age 54 years) the clinical diagnosis of *Wallenberg's syndrome* was set. The clinical syndromes varied according to the severity of the infarction, but all patients had at least 4 out of the following 6 signs and symptoms: ipsilateral limb dysmetria; Horner's syndrome; impairment of facial pain and temperature sensation; paralysis of the palate, pharynx, and larynx with dysphagia and dysphonia; contralateral impairment of pain

[a]The work was supported by Wilhelm Sander–Stiftung and Deutsche Forschungsgemeinschaft, SFB 220 D6 and Alfried Krupp–Stiftung.

537

and temperature sensation over the trunk and limbs; and a recognizable tendency to fall sideways (lateropulsion).

In the patients with Wallenberg's syndrome the severity of *body lateropulsion* and disturbance of upright stance was allocated to one of four grades independently by 2 investigators:

I. Moderate head and body tilt without considerable imbalance;
II. Head and body tilt with considerable imbalance, but without falls;
III. Head and body tilt and falls with eyes closed;
IV. Head and body tilt and falls with eyes open.[4]

Evaluation of fundus photographs and determinations of subjective visual vertical were performed at various times following the stroke. The data for the first investigation ranged from measurements on day 1 to week 9. Repeated control measurements were performed in 18 patients with Wallenberg's syndrome, the latest occurring after 240 days.

Subjective Visual Vertical

Each subject sat with his or her head fixed in the upright position by means of a bite board or a band fastened around the occiput, and looked into a hemispherical dome 60 cm in diameter which could be rotated about the line of sight. The surface of the dome extended to the limits of the observer's visual field, and was covered with a random pattern of colored dots and contained no cues to gravitational orientation. The center of the dome was fixed to the shaft of a DC torque motor; 30 cm in front of the observer was a circular target of 14 deg visual angle with a straight line through the center mounted on a coaxial shaft connected to a DC servo motor. The central test edge had to be adjusted to the vertical by the subject, using a potentiometer. The output of the potentiometer was recorded on a strip chart recorder.

Static visual vertical was determined by means of 10 adjustments of the target disc from a random offset position to the subjective vertical with the hemispherical dome stationary. Under these conditions the normal range [± 1 standard deviation (SD)] of the visual vertical is $\pm 2°$.[5]

Dynamic visual vertical was determined by means of 10 continuous tracking experiments of the visual vertical for a 20-second stimulation period of either clockwise or counterclockwise rotation of the dome at a constant angular velocity of 40 deg/second. A visual surround rotating around the observer's line of sight (roll motion) induced an apparent body rotation opposite in direction to pattern motion (rollvection) and caused a limited tilt of the subjective vertical of about 15° (range 6.5–40°).[5,6] The rationale for measurements of the *dynamic* visual vertical was that rollvection should make overt slight deviations of the visual vertical which would otherwise go undetected under purely static stimulus conditions.[2]

Static as well as dynamic visual vertical was measured *binocularly* and *monocularly* for both eyes (for comparison with cyclorotation of each eye).

Fundus Photographs

Cyclorotation (ocular torsion) in degrees was defined as the mean of 4–6 fundus photographs taken with the head upright. With this method the normal eye position in roll plane is an excyclotropia (counterclockwise rotation of the right eye, clockwise rotation of the left eye, from the viewpoint of the examiner). According to our own experience on 20 healthy subjects, the right eye shows an excyclotropia of $3.4° \pm 2.0°$ (mean $\pm$ SD) (range $1.0°$ to $7.0°$), the left eye an excyclotropia of $5.8° \pm 2.6°$ (range $2.0°$ to $9.1°$). The latter agrees with similar data obtained by others.[7,8]

RESULTS AND COMMENT

Subjective Visual Vertical and Cyclorotation Are Most Sensitive Clinical Brain Stem Signs

There are several reports in the literature of patients with brain stem lesions (particularely Wallenberg's patients) who experienced persistent tilting of the visual field.[9–14] Friedmann found a marked deviation of visual vertical and horizontal to the side of the lesion, in patients with acute unilateral brain stem lesions or following labyrinthectomy.[13,14] The deviation decreased during the course of recovery. In

TABLE 1. Differential Effects (in Percent) of Acute (Vascular) Brain Stem Lesions on Subjective Visual Vertical, Cyclorotation of One or Two Eyes, Skew Deviation, and Complete Ocular Tilt Reaction in 111 Patients

Lesion	Patients (n)	Subjective Visual Vertical	Cyclorotation	Skew Deviation	Ocular Tilt Reaction
Mesencephalic	16	94%	92%	37.5%	25%
Pontomesencephalic	12	92%	91%	25%	25%
Pontine	34	91%	86%	26.5%	12%
Pontomedullary	13	100%	90%	23%	7.7%
Medullary (Wallenberg syndrome)	36	94%	82%	44%	33%
Total	111				
Mean		94%	88%	31%	20%

contrast he found no deviation in patients with central cortical or cerebellar lesions, but did not specifically investigate Wallenberg's syndrome.

Our study is the first systematic investigation of SVV and CR in acute vascular lesions at different levels in the brain stem. The incidence of pathological findings was surprisingly high. Ninety-four percent of the patients had pathological deviations of the SVV from the true vertical as depicted in TABLE 1. Eighty-eight percent of the patients exhibited pathological cyclorotation of one or both eyes irrespective of the level of brain stem lesion. Vertical divergence of the eyes (skew deviation) was observed in 31% and was most frequent in Wallenberg's syndrome. Twenty percent of the patients presented with the eye-head synkinesis of a complete ocular tilt reaction (FIGURE 1).

It was striking that caudal brain stem lesions caused ipsiversive deviations of the SVV whereas upper brain stem lesions caused contraversive deviations. Correspondingly cyclorotation of the eyes (from the viewpoint of the examiner) was clockwise with a caudal left brain stem lesion or an upper right brain stem lesion. Cyclorotation of the eyes was counterclockwise with a caudal right brain stem lesion or an upper left brain stem lesion. This can be best demonstrated by the 31 out of 140 patients who presented with a complete ocular tilt reaction (FIGURE 1).

Evidence can be presented for a hypothetical explanation that concurrent tilting, of SVV and CR as well as the complete ocular tilt reaction is the perceptual, ocular motor, and postural consequence of a common lesion of central vestibular pathways that subserve the vestibuloocular reflex in the roll plane. Otolithic and vertical semicircular canal inputs to the interstitial nucleus of Cajal from the contralateral

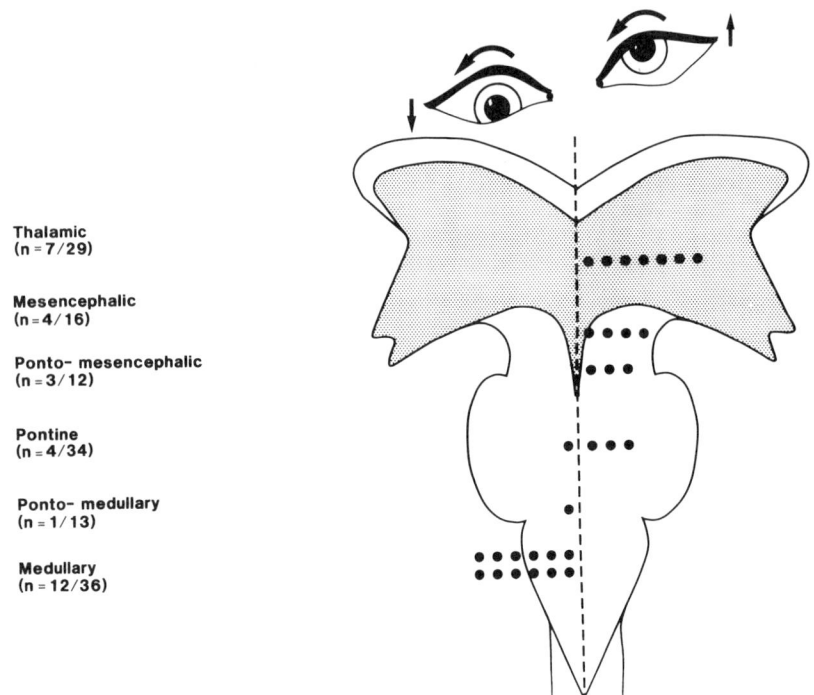

Thalamic
(n = 7/29)

Mesencephalic
(n = 4/16)

Ponto- mesencephalic
(n = 3/12)

Pontine
(n = 4/34)

Ponto- medullary
(n = 1/13)

Medullary
(n = 12/36)

FIGURE 1. Schematic representation of the frequency of ocular tilt reaction in acute vascular brain stem lesions from mesencephalon to medulla oblongata. Ocular tilt reaction is an eye-head synkinesis consisting of lateral head tilt in roll, vertical divergence of the eyes (skew deviation), and cyclorotation towards the undermost eye. Ocular tilt reaction occurs with caudal and upper brain stem lesions. The direction of head tilt and cyclorotation is ipsiversive with pontomedullary lesions and contraversive with upper pontine and mesencephalic lesions, which suggests a pontomedullary crossing of graviceptive pathways.

vestibular nucleus and motor outputs from the interstitial nucleus of Cajal to cervical and ocular motor neurons could be involved in the ocular tilt reaction.[3] Contraversive tilts of SVV and CR due to unilateral mesencephalic lesions point to the existence of a crossed graviceptive pathway between the vestibular nucleus and the contralateral interstitial nucleus of Cajal.

The restricted view that the vestibuloocular reflex (VOR) merely serves stabilization of gaze should be corrected. Its neuronal pathways also include ascending input to the thalamocortical projections for perception as well as descending input to vestibulospinal projections for adjustments of head and body posture (vestibulospinal reflexes). This means that physiologic stimulation or pathologic dysfunction of VOR pathways not only provokes eye movements but inevitably causes a direction-specific concurrent rotatory vertigo (or tilt) and postural imbalance.[15] That the VOR in roll is involved in pathological tilting of SVV and CR is best inferred from the data in the Wallenberg patients.[16]

Subjective Visual Vertical and Lateropulsion in Wallenberg's Syndrome

All but one patient showed deviations of the SVV as determined under static conditions. Thirty-four showed ipsilateral deviations of their adjustments of static SVV if a maximal normal deviation of ±2.5° is assumed. With optokinetically induced rollvection, all 36 patients showed significant ipsiversive deviations in comparison with the normal data.[2,5] Tilt angles at the first examination ranged from 2.7° to 53.3° for the static condition with different monocular values for each eye (TABLE 2, FIGURE 2).

All patients had lateropulsion to some degree. Three of the 36 patients showed severe lateropulsion of grade IV. Seventeen were classified as grade III, 11 as grade II, and 5 as grade I.

Net tilt values of SVV were greater in proportion to the severity of body lateropulsion (FIGURE 2).

Mean tilt angles ($n = 23$) were greater with monocular vision in the eye ipsilateral to the lesion: in 14 patients the deviation of SVV of the ipsilateral was greater than the deviation of the contralateral eye; in 5 patients the deviations were the same; and in 4 patients the deviation of the contralateral eye was greater (TABLE 2). Net tilt angles of the dynamic SVV all showed a strong asymmetry with clockwise and counterclockwise stimulation according to the deviation of the static SVV. The asymmetric deviations of the dynamic SVV range from a minimum of 3.7° in one direction and 17.6° in the other direction to a maximum of 9.7° to 96.6°, respectively.

Subjective Visual Vertical and Cyclorotation in Wallenberg's Syndrome

Eighteen out of the 22 patients in whom fundus photographs were taken showed pathological CR of one ($n = 6$) or both ($n = 12$) eyes (FIGURE 3, TABLE 2). With binocular involvement CR was always clockwise with left-sided lesions ($n = 5$) and counterclockwise with right-sided lesions ($n = 7$). Binocular CR was mostly dissociated. Net tilt angles were usually considerably greater for the eye ipsiversive to the lesion (predominant excyclotropia; 7 out of 12), but in rare cases (1 out of 12) incyclotropia of the contralateral eye could also exceed excyclotropia by some degrees. If only one eye was involved ($n = 6$), excyclotropia of the ipsilateral eye was more frequent ($n = 4$) than incyclotropia of the contralateral eye ($n = 2$). The net angles of CR ranged from 2° to 19°, if a normal range (mean ± SD) above excyclotropia is accepted for both eyes of 1.4°–6.0° for the right eye and 2.4°–9.1° for the left eye.

In all cases of binocular CR the directions of deviation of SVV were concordant with CR. However, net tilt angles of eye position and perception of vertical ($n = 22$) did not match (ipsilateral eye: 6 SVV = CR, 7 SVV > CR + 1, 9 SVV < CR − 1; contralateral eye: 4 SVV = CR, 12 SVV > CR + 1, 6 SVV < CR − 1). In a subgroup of 14 patients—with tilt angles of CR greater than 0—we found no correla-

TABLE 2. Results in 13 Patients with Wallenberg's Syndrome and the Time Course of Gradual Recovery[a]

Pat. No	Age/ Sex	Side of Lesion	Day after Onset	Latero-pulsion Grade	Skew Devia-tion[b]	Ocular Torsion		Subjective Visual Vertical		
						RE	LE	BIN	RE	LE
7	39 M	R	3	II	0	Ex 2°	Ex 13°	−0.2°	+2.2°	+2.7°
			10	I				+0.8°	+3.8°	+0.6°
			14	I		0°	Ex 21°	0°	+2.5°	+1.8°
			83	0				−0.3°	+1.8°	−3.3°
			12M	0				−0.9°	+0.9°	−1.4°
11	25 F	L	2	III	0	Ex 3.5°	Ex 15°	−6.8°	−5.6°	−7.2°
			5	II		Ex 6°	Ex 12°	−5.0°	−3.5°	−7.8°
12	56 F	L	24	III	+4-5°	In 9°	Ex 25°	−17°	−3.4°	−13°
			30	II	+3°			−4.0°	−3.1°	−7.4°
			34	II	+1-2°			−2.6°	−2.7°	−6.9°
			38	III	+3°	In 5°	Ex 20°	−8.3°	−8.8°	−13°
14	55 M	L	2	III	+7°	0°	Ex 20°	—	−9.3°	−12.5°
			15	II	+3°			—	−4.3°	−6.4°
			22	II	+3°	Ex 4°	Ex 15°	—	−1.6°	−7.8°
			36	I	+3°			—	+4.3°	−3.8°
			160	I	+2°	Ex 5°	Ex 10°	+1.2°	+1.7°	−0.3°
19	42 M	L	1	IV	+5°	0°-Ex 1°	Ex 12°	−53°		
			9	IV	+3°			−25°		
			19	III	+2°			−7.6°		
			68	II	0	Ex 5°	Ex 6°	−3.1°		
21	48 M	L	20	III	+6°	Ex 3°	Ex 12°	−14.4°		
			50	II	0	Ex 5°	Ex 25°	−7.2°		
			10M	I	0			−2.5°		
			4.5Y	0	0	Ex 6°	Ex 18°	−0.5°	−3.0°	−2.5°
23	75 M	R	12	II	−2-3°	Ex 13°	0°	+5.3°		
			60	I	0	Ex 9°	0°	+1.5°		
28	48 M	R	12M	III	0	Ex 14°	In 1°	−0.8°	+7.2°	+2.0°
			14M	II		Ex 12°	In 2°	−0.3°	+2.2°	+2.7°
			26M	II		Ex 7°	Ex 3°	+4.0°	−0.3°	+2.2°
29	61 M	R	3	II	−8°	Ex 15°	Ex 4°	+2.2°	+7.2°	+2.3°
			5	I	−3°	Ex 9°	Ex 5°	+2.9°	+1.9°	+2.7°
			11	0	0	Ex 10°	Ex 5°	+0.3°	+1.5°	−0.4°
30	42 F	L	10	III	+2°	In 8°	Ex 19°	−5.1°	−3.1°	−10.9°
			15	III	+2°	In 1°	Ex 10°	−3.8°	−8.5°	−6.3°
			21	II	0	Ex 2°	Ex 15°	−4.5°	−1.8°	−7.4°
34	30 M	R	8	III	−4°	Ex 25°	In 4°	+19.9°	+13.6°	+16.2°
			12	III	−2°	Ex 20°	Ex 2°	+13.7°	+13.1°	+11.2°
			20	II	0	Ex 12°	Ex 3°	+8.2°	+8.3°	+5.7°
			22	II	0	Ex 10°	Ex 6°	+5.5°	+4.8°	+3.3°
			7M	I	0	Ex 6°	Ex 9°	+0.9°	+2.5°	−0.2°
35	43 M	R	1	II	0	Ex 20°	In 6°	+4.6°	+5.9°	+1.4°
			3	II		Ex 16°	Ex 6°	−0.7°	+4.6°	+2.1°
			10	0-I		Ex 13°	Ex 13°	−2.8°	−1.3°	−1.7°
36	70 M	L	4	IV	+2°	In 7°	Ex 25°	−17.1°	−12.3°	−18.3°
			10	IV	0	In 3°	Ex 15°	−14.7°	−17.5°	−18.5°
			17	III	0	Ex 4°	Ex 15°	−7.5°	−5.2°	−7.2°
			24	III	0	Ex 6°	Ex 20°	−7.1°	−4.1°	−5.7°

[a]Spontaneous improvement of lateropulsion (grades I–IV), reduction of skew deviation, ocular torsion, and subjective visual vertical (in degrees) occur over periods ranging from weeks to months (M) with the most dramatic improvement within the first 4 weeks. (Skew deviation: + = right over left eye, − = left over right eye, 0 = no skew deviation; Ex = excyclotropia, In = incyclotropia; RE = right eye; LE = left eye; BIN = binocular).

[b]+ = R > L; − = L > R.

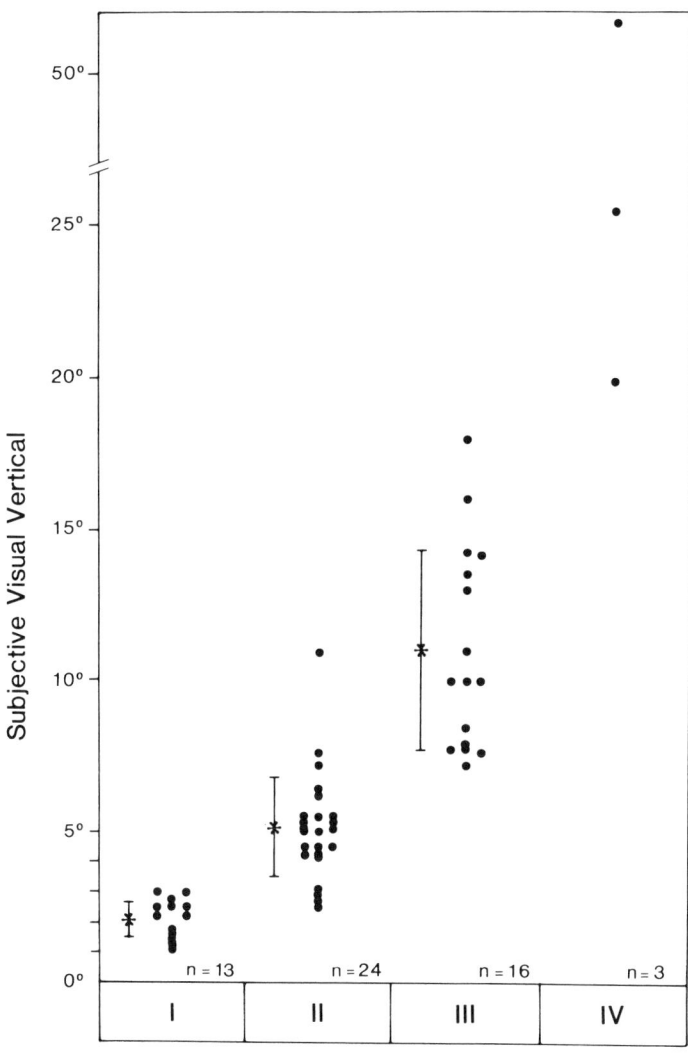

FIGURE 2. Deviations of the subjective visual vertical (tilt in degrees) in 36 patients with infarctions of the lateral medulla oblongata. Adjustments of the SVV (first measurements in the acute stage) are depicted in relation to the severity of body lateropulsion (abscissa). The dots represent single measurements (in some patients there were repeated measurements during the course of the disease); means and standard deviations are also depicted. The more pronounced the lateropulsion, the greater the deviations of SVV. SVV and lateropulsion are both ipsiversive to the side of the lesion.

tions between tilt angles of SVV and CR. There was no consistent relationship between CR and SVV in the 6 cases of monocular CR.

Vertical Semicircular Canal or Otolith Pathway Lesion in Wallenberg's Syndrome?

The combination of skew deviation with monocular or binocular cyclorotation—as is typical for Wallenberg's syndrome—can be best explained by a lesion of the vertical semicircular canal pathways.[2,16] However, a combined lesion of the

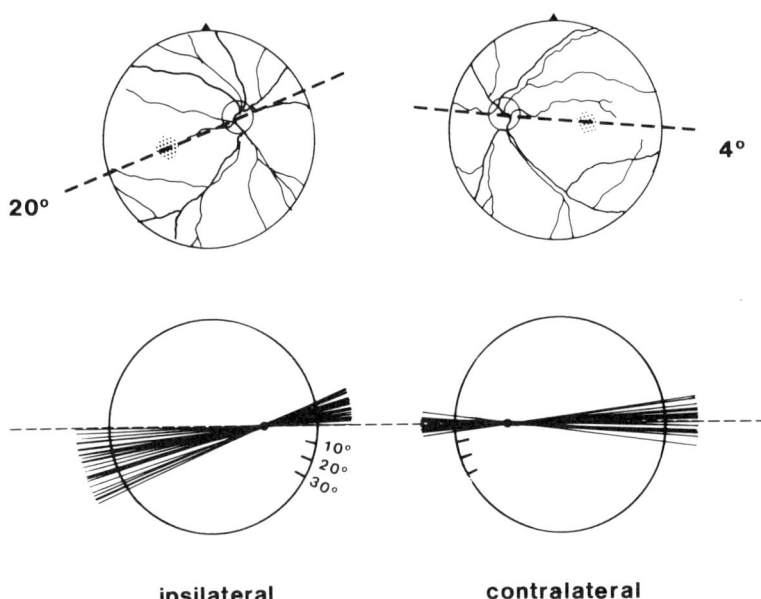

FIGURE 3. Diagram of typical fundus photographs with the head upright in a patient with an acute Wallenberg's syndrome of the right medulla oblongata (top). Note the pathological excyclotropia of 20° of the papilla/macula/meridian of the right eye, while the left eye remains in a normal position with an excyclotropia of 4°. Schematic representation of papilla/macula/meridians of both eyes in 22 Wallenberg's patients (bottom) with either monocular cyclorotation of the ipsilateral eye only or binocular ipsiversive cyclorotations. Ipsiversive cyclorotation of the ipsilateral eye is more pronounced than cyclorotation of the contralateral eye.

vertical semicircular canals and otolithic pathways is possible, since the fibers travel closely together (FIGURE 4). The excitatory ascending pathway from the vertical semicircular canals projects mainly via the *rostral medial vestibular nucleus* and the *contralateral medial longitudinal fasciculus (MLF)* to the motoneurons of the extraocular eye muscles.[17,18] The anatomical connections between the utricle and the extraocular eye muscles are not as well worked out as for the semicircular canals. Animal data suggest two pathways, the first via the *medial vestibular nucleus* (mainly the rostral part) along the *contralateral MLF* and the second via the lateral vestibular nucleus along the ipsilateral ascending tract of Deiter's.[19,20]

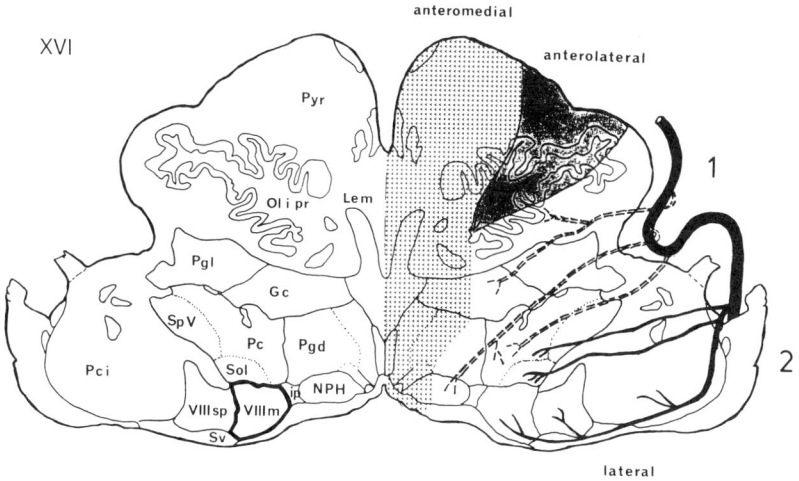

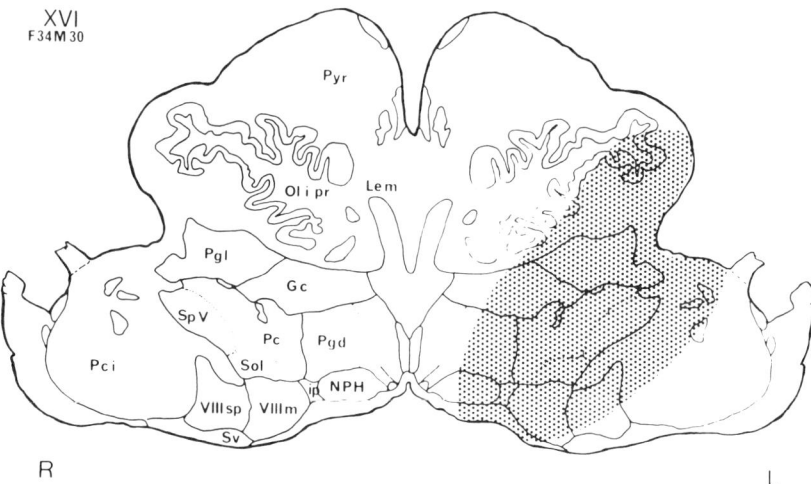

FIGURE 4. Transverse section of the medulla oblongata (top) with the internal courses of the supplying arteries (adapted from Duvernoy).[25] The region of the vestibular nuclei (VIII) is supplied either by vascular rami of the vertebral artery (1) or by posterior rami (2) of the anterior inferior cerebellar artery (AICA) or posterior inferior cerebellar artery (PICA). Typical distribution of the medullary infarction in one of our patients with left Wallenberg's syndrome (bottom) when the lesion of the magnetic resonance scan is projected onto a transverse section of the medulla oblongata (as adapted from Olszewski and Baxter).[26] Infarcted area involves the left medial vestibular nucleus (VIIIm). Abbreviations: Cn d, nucleus medullae oblongatae centralis, subnucleus dorsalis; Cor po, nucleus corporis pontobulbaris; Cu l, nucleus cuncatus lateralis; Le m, lemniscus medialis; NPH, nucleus praepositus hypoglossi; Ol i pr, nucleus olivaris inferior principalis; Pc, nucleus parvocellularis; Pci, pedunculus cerebelli inferior; Pg d, nucleus paragigantocellularis dorsalis; Pyr, pyramis; Sol, nucleus tractus solitarii; Sp V, nucleus tractus spinalis trigemini oralis; T, tractus solitarius; VIII m, nucleus vestibularis medialis.

There is reason to believe that different patterns of ocular tilt rotation (OTR) in Wallenberg's patients reflect different combinations of anterior and posterior semicircular canal and otolith dysfunction. The ocular motor pattern of skew deviation with monocular excyclotropia of the ipsilateral eye and hypertropia of the contralateral eye can best be explained by a circumscribed lesion of the ascending pathways of the ipsilateral posterior semicircular canal.[2] In the three-neuron vestibuloocular reflex arc between the posterior canal and extraocular eye muscles, an excitatory ascending pathway via the contralateral medial longitudinal fasciculus is linked to the ipsilateral superior oblique and the contralateral inferior rectus muscle (FIGURE 5). A lesion of these pathways near the vestibular nuclei complex causes both excyclotropia of the ipsilateral eye and hypertropia of the contralateral eye. Concurrent deviation of SVV could be mediated respectively by ascending perceptual pathways to the vestibular cortex. If there is an additional lesion of the fibers to the anterior semicircular canal,[17,18] a conjugate cyclorotation of both eyes and vertical divergence with the ipsilateral eye down and the contralateral eye up will occur (FIGURE 5). This combined pattern of binocular cyclorotation together with vertical divergence was seen in 2 out of 22 Wallenberg's patients. Different amounts of dysconjugation of cyclorotation (8 out of 22 patients) can be explained by unequal involvement of anterior or posterior canal pathways. Monocular incyclotropia of the contralateral eye—which we have seen in one patient—could reflect the possibility of involvement of the anterior semicircular canal pathways in Wallenberg's syndrome.

Why do we think that both otolith and semicircular canal inputs are involved? The fact that the torsional VOR is of low gain and does not show "velocity storage mechanism" in humans[21] suggests that the VOR in roll involves otolith input. The sustained skew deviation and cyclorotation in Wallenberg's patients are a further argument for otolithic influence. Convergence of semicircular canal and otolith input was demonstrated not only in abducens motoneurons[22] but especially in trochlear motoneurons in the cat.[23] In these experiments stimulations with different frequencies showed that in the high-frequency range the canal system serves to compensate for the lag present in the otolith system, whereas in the low-frequency range and during static head displacement the otolith input serves to compensate for the poor canal performance.[23] The short-latency utricular inputs to ipsilateral abducens[22] and contralateral trochlear motoneurons[24] corresponded to the latency of semicircular canal input, which suggests disynaptic linkage between both receptor types and ocular motoneurons. The otolith signal in the horizontal system was found to be much weaker than in the vertical/rotatory system.[23] The semicircular canal–ocular reflexes are disynaptic for both inhibitory and excitatory pathways; otolith-ocular reflexes are only disynaptic when excitatory. It is reasonable that both excitatory disynaptic pathways from the vertical semicircular canals[17,18] and the otoliths ascend closely together in the contralateral MLF. This nicely fits our findings of an ipsiversive OTR in pontomedullary (vestibular nuclei) lesions but a contraversive OTR in pontomesencephalic (MLF) lesions.

The general explanation that most of the vestibular symptoms in Wallenberg's disease are secondary to a lesion of the VOR in roll may be too simple since it is based on a two-dimensional analysis of perception, eye movements, and body tilt, namely, in the fore/aft and lateral direction. We have reason to believe that the distortion of spatial orientation as well as body sway involves the third dimension. At minimum, body sway is typically diagonal rather than purely lateral. It will be one of our further goals to measure disturbance of spatial orientation in three dimensions (roll, pitch, yaw). We are still in the early stages of appropriate understanding of complex vestibular function in all three spatial planes, and we are still lacking reliable techniques to measure the particular contribution of otolithic input.

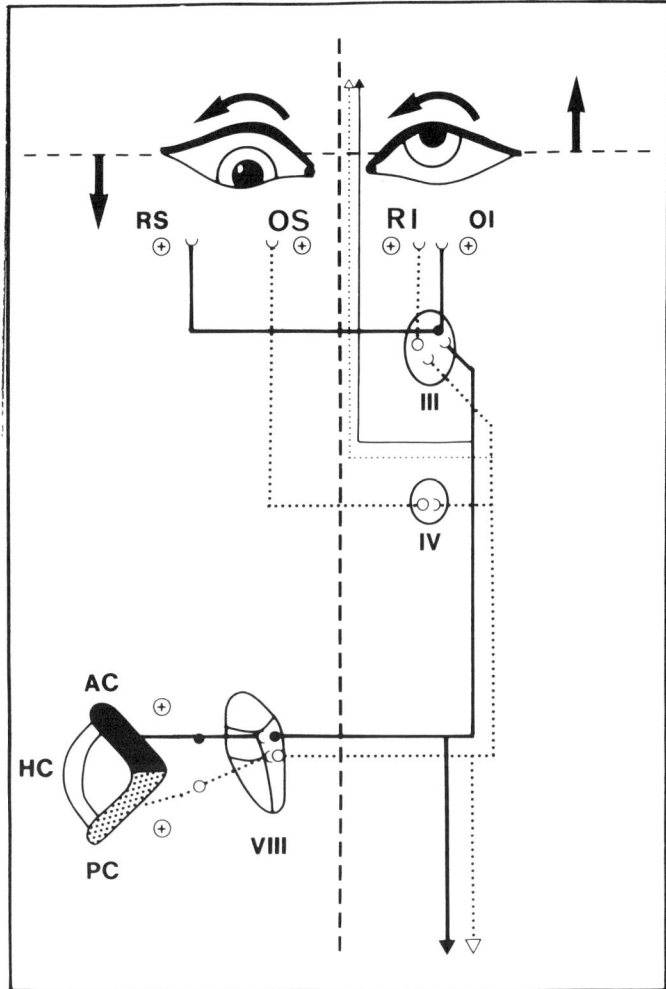

FIGURE 5. Hypothetical explanation of the combination of vertical divergence and cyclorotation due to lesions of the vertical semicircular canal pathways in Wallenberg's syndrome. Schematic drawing of the three-neuron vestibuloocular reflex arc between posterior (PC) and the anterior (AC) semicircular canal and the extraocular eye muscles. An excitatory ascending pathway is linked from the posterior semicircular canal to the ipsilateral superior oblique (OS) and the contralateral inferior rectus (RI). A lesion of this pathway causes excyclotropia of the ipsilateral and hypertropia of the contralateral eye. An excitatory ascending pathway projects from the anterior semicircular canal to the ipsilateral superior rectus (RS) and the contralateral inferior oblique muscle (OI). A lesion of this pathway will cause a hypotropia of the ipsilateral and incyclotropia of the contralateral eye. A combination of a lesion of both, anterior and posterior canal pathways, induces a combination of skew deviation with hypotropia of the ipsilateral eye and hypertropia of the contralateral eye and conjugate cyclorotation of both eyes.

REFERENCES

1. WESTHEIMER, G. & S. M. BLAIR. 1975. The ocular tilt reaction—a brainstem ocular motor routine. Invest. Ophthalmol. **14:** 833–839.
2. BRANDT, TH. & M. DIETERICH. 1987. Pathological eye-head coordination in roll: tonic ocular tilt reaction in mesencephalic and medullary lesions. Brain **110:** 649–666.
3. HALMAGYI, G. M., TH. BRANDT, M. DIETERICH, I. S. CURTHOYS, R. J. STARK & W. F. HOYT. 1990. Tonic contraversive ocular tilt reaction due to unilateral meso-diencephalic lesions. Neurology **40:** 1503–1509.
4. DIETERICH, M. & TH. BRANDT. 1990. Postural imbalance and subjective visual vertical in medullary infarctions. *In* Disorders of Posture and Gait, 1990. Th. Brandt, W. Paulus & W. Bles, Eds.: 419–422. Thieme. Stuttgart & New York.
5. DICHGANS, J., R. HELD, L. YOUNG & TH. BRANDT. 1972. Moving visual scenes influence the apparent direction of gravity. Science **178:** 1217–1219.
6. MAURITZ, K. H., J. DICHGANS & A. HUFSCHMIDT. 1977. The angle of visual roll motion determines displacement of subjective visual vertical. Percept. Psychophys. **22:** 557–562.
7. BIXENMAN, W. W. & G. K. VON NOORDEN. 1982. Apparent foveal displacement in normal subjects and in cyclotropia. Ophthalmology **89:** 58–62.
8. HERZAU, V. & E. JOOS. 1983. Untersuchungen von Bewegungen und Stellungsfehlern der Augen um die sagittale Achse. Z. prakt. Augenheilkd. **4:** 270–278.
9. SILFVERSKIÖLD, B. P. 1965. Skew deviation in Wallenberg's syndrome. Acta Neurol. Scand. **41:** 381–386.
10. BJERVER, K. & B. P. SILFVERSKIÖLD. 1968. Lateropulsion and imbalance in Wallenberg's syndrome. Acta Neurol. Scand. **44:** 91–100.
11. HAGSTRÖM, L., G. HÖRNSTEN & B. P. SILFVERSKIÖLD. 1969. Oculostatic and visual phenomena occurring in association with Wallenberg's syndrome. Acta Neurol. Scand. **45:** 568–582.
12. HÖRNSTEN, G. 1984. Vestibular dysfunction and visual illusions in Wallenberg's syndrome. *In* Functional Basis of Ocular Motility Disorders. G. Lennerstrand, D. S. Zee & E. L. Keller, Eds.: 347–353. Pergamon Press. New York, N.Y.
13. FRIEDMANN, G. 1970. The judgement of the visual vertical and horizontal with peripheral and central vestibular lesions. Brain **93:** 313–328.
14. FRIEDMANN, G. 1971. The influence of unilateral labyrinthectomy on orientation in space. Acta Otolaryngol. **71:** 289–298.
15. BRANDT, TH. 1991. Classification of vestibular brain stem disorders according to vestibulo-ocular reflex planes. Klin. Wochenschr. **69:** 121–123.
16. DIETERICH, M. & TH. BRANDT. 1992. Wallenberg's syndrome: lateropulsion, cyclorotation and subjective visual vertical in 36 patients. Ann. Neurol. **31:** 1–10.
17. GRAF, W., R. A. MCCREA & R. BAKER. 1983. Morphology of posterior canal–related secondary vestibular neurons in rabbit and cat. Exp. Brain Res. **52:** 125–138.
18. GRAF, W. & K. EZURE. 1986. Morphology of vertical canal related second order vestibular neurons in the cat. Exp. Brain Res. **63:** 35–48.
19. GACEK, R. R. 1971. Anatomical demonstration of the vestibulo-ocular projections in the cat. Laryngoscope **81:** 1559–1595.
20. GACEK, R. R. 1982. The anatomical-physiological basis for vestibular function. *In* Nystagmus and Vertigo: Clinical Approaches to the Patient with Dizziness. V. Honrubia & M. A. B. Brazier, Eds.: 3–23. Academic Press. New York & London.
21. SEIDMAN, S. H. & R. J. LEIGH. 1989. The human torsional vestibulo-ocular reflex during rotation about an earth vertical axis. Brain Res. **504:** 264–268.
22. SCHWINDT, P. C., A. RICHTER & W. PRECHT. 1973. Short latency utricular and canal input to ipsilateral abducens motoneurons. Brain Res. **60:** 259–262.
23. PRECHT, W., J. H. ANDERSON & R. H. I. BLANKS. 1979. Canal otolith convergence on cat ocular motoneurons. Prog. Brain Res. **50:** 459–468.

24. BAKER, R., W. PRECHT & A. BERTHOZ. 1973. Synaptic connections to trochlear motoneu-
    rons determined by individual vestibular nerve branch stimulation in the cat. Brain Res.
    64: 402–406.
25. DUVERNOY, H. M. 1978. Human Brainstem Vessels. Springer. Berlin, Heidelberg & New
    York.
26. OLSZEWSKI, J. & D. BAXTER, Eds. 1982. Cytoarchitecture of the Human Brain Stem. 2nd
    edit. Karger. Basel & New York.

An *In Vivo* and *In Vitro* Study of the Vestibular Nuclei Histaminergic Receptors in the Guinea Pig[a]

C. DE WAELE, M. SERAFIN,[b] A. KHATEB,[b] N. VIBERT,
T. YABE, J. M. ARRANG,[c] M. MULHETHALER,[b]
AND P. P. VIDAL

Laboratory of Neurosensory Physiology, CNRS
15 rue de l'Ecole de Médecine
75270, Paris Cedex 06, France

[b]Department of Physiology, CMU
1 Rue Michel Servet
1211, Geneva, Switzerland

[c]Neurobiology and Pharmacology Unit
INSERM Unit 109
Centre Paul Broca
Paris, France

INTRODUCTION

Since the first identification of histamine as a neuromodulator, numerous studies have been focused on the functional role of mammalian histaminergic neurons. In particular, it has been demonstrated that this amine plays an important role in a variety of physiological processes such as vigilance, thermoregulation, and control of cerebral circulation (see Reference 1 for a recent review).

In this study, we raise the question of the functional role and mode of action of histamine receptors in the vestibular nuclei. Indeed, several lines of evidence suggest that these receptors could modulate the function of the central vestibular neurons:

- First, a direct projection from the histaminergic tuberomammillary cells to the vestibular nuclei has been recently reported.[2–5]
- Second, both autoradiographic[6] and electrophysiological studies[7–10] have demonstrated the presence of histaminergic receptors in the vestibular nuclei.
- Third, histamine is well known to be implicated in the control of wakefulness. The vestibuloocular reflex (VOR) is highly dependent on the state of alertness; indeed, its gain becomes lower when wakefulness decreases.[11–21]
- Finally, the antagonists of the histaminergic receptors are frequently used in humans for the symptomatic treatment of vertigo and motion sickness.[22,23]

Therefore, we have combined *in vitro* and *in vivo* approaches in guinea pigs to

[a]This work was supported by grants from the Human Frontier Science program, the Swiss NSF (No. 3.288.0.85 and 3.560.0.86), the CNES, the Sandoz Foundation, INSERM, and the Ministère des Affaires Etrangères Français.

further investigate the histaminergic modulation of the vestibular function. As a first step, we have tested the effects of bath application of histamine and histaminergic related compounds (ligands of the three known types of histaminergic receptors) on medial vestibular nuclei neurons (MVNn) on slices. These cells were chosen because we[24,25] and others[26] have previously analyzed their membrane properties. Briefly, the MVNn could be segregated into two main cellular types, depending on their electrophysiological and pharmacological properties (FIGURE 1). Type A MVNn (FIGURE 1A) are characterized by a rather broad action potential, a large single after-hyperpolarization (AHP), and a transient A-like current. In contrast, type B MVNn have a thinner action potential followed by a double AHP, as well as long-lasting tetrodotoxin (TTX) sensitive sodium-dependent plateau potentials (FIGURES 1B and 2A). Furthermore, a minority of B MVNn exhibited low-threshold calcium-dependent spike bursts (LTS) and were therefore called B + LTS MVNn (FIGURE 1C). We confirm here that histamine depolarizes the MVNn via the H2

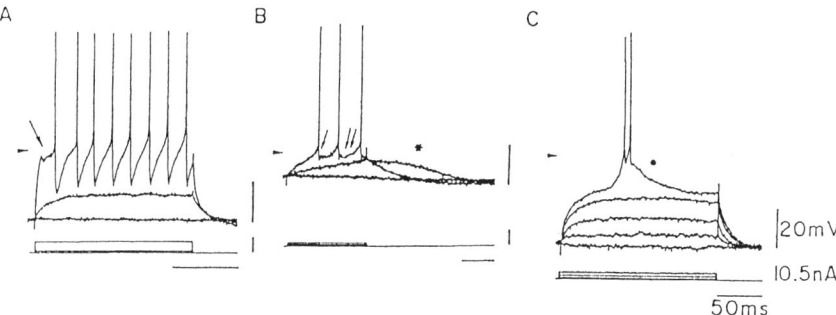

FIGURE 1. Characterization of MVNn. **A:** *Type A MVNn* are mainly characterized by the presence of a large AHP and a rectification due to the activation of an A-like conductance (arrow). **B:** *Type B MVNn's* main characteristic is the presence of an early fast (arrow) and a delayed slower AHP (double arrow) as well as a prolonged delayed depolarization (*) due presumably to a sodium-dependent persistent conductance (see also FIGURE 2A). **C:** Type B + LTS MVNn are similar to B MVNn, but under hyperpolarization they display in addition short duration low-threshold bursts (dot).

receptors.[27] In addition, we demonstrate that this depolarization is present in the three classes of MVNn and that, *in vitro,* the H3 ligands are without effect.

As a second step, we have tested the same histaminergic related compounds in alert guinea pigs. To this effect, guinea pigs were implanted with cannulae to infuse the vestibular nuclei with these drugs. We found that cimetidine, an H2 antagonist, and alphamethylhistamine (α-Me-His), an H3 agonist, induced a postural and oculomotor syndrome. These results very likely revealed the presence of vestibular presynaptic H3 receptors and help us to appreciate the functional weight of the H2 and H3 vestibular receptors. Finally, with potential clinical applications in mind, we have characterized the dynamic changes of the horizontal vestibuloocular reflex (HVOR) following intraperitoneal injection of thioperamide, a new selective H3 antagonist.

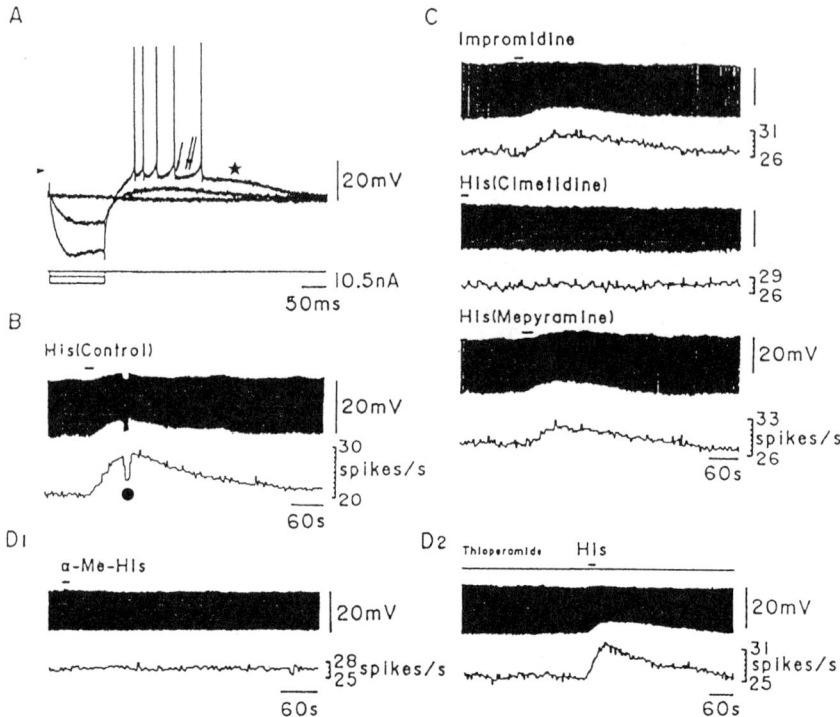

FIGURE 2. Effects of histamine and its related compounds on B MVNn. **A:** A characteristic B MVNn is shown to display a double AHP (single and double arrows) as well as a prolonged delayed depolarization (*) due presumably to a sodium-dependent persistent conductance. **B:** Histamine (10^{-4} M) depolarized a B MVNn and induced a reversible increase in firing rate. During the effect, the cell was manually clamped in order to monitor the change in membrane resistance (dot). **C:** The H2 agonist (and H3 antagonist) impromidine (2.10^{-7} M) mimicked the effects of histamine (two upper traces). In addition, whereas these effects were blocked by the H2 antagonist cimetidine at 10^{-4} M (two middle traces), they could not be suppressed by the H1 antagonist mepyramine at 10^{-4} M (two lower traces). **D:** Neither the specific H3 agonist (α-Me-His, 10^{-6} M) in D1 nor the specific H3 antagonist (thioperamide, 10^{-6} M) in D2 had any effect per se. Thioperamide, in addition, could not block (D2) the effect of histamine (10^{-4} M) applied for 15 seconds.

METHODS

In Vitro Study

Experiments were carried out on guinea pig brain stem slices using standard techniques.[28] In brief, following Nembutal anesthesia (30 mg/kg), the animal was decapitated, the skull opened, and a brain stem/cerebellum block was dipped into a cold (4°C) oxygenated saline. The cerebellum was then removed, the brain stem block fixed with cyanoacrylate to the stage of a vibrating microtome, and immersed in a cold saline. Several 500-μm-thick coronal slices containing the vestibular nuclei were then obtained and transferred to individually bubbled vials kept at room temperature. From there, slices were transferred to the recording chamber in which they were superfused (32°C) with an oxygenated (95% O_2/5% CO_2) modified saline solution containing 130 mM NaCl, 20 mM NaHCO$_3$, 1.25 mM KH$_2$PO$_4$, 1.3 mM MgSO$_4$, 5 mM KCl, 10 mM glucose, and 2.4 mM CaCl$_2$. Intracellular recordings were obtained with 3 M K-Acetate containing glass microelectrodes (resistance 70–100 MΩ). The recordings were obtained from MVNn, taking the border of the IVth ventricle as a landmark. Histamine (His) and mepyramine (an H1 antagonist) were obtained from Sigma, whereas cimetidine (an H2 antagonist) and impromidine (an H2 agonist and H3 antagonist) were obtained from Smith, Klein and French. The selective H3 agonist (R)-α-Me-His and the H3 antagonist, thioperamide were kindly provided. The duration of application of histamine was usually 15 seconds. Due to the dead space of the perfusion system, the drug reached the tissue within about 90 seconds. For synaptic uncoupling, low-Ca^{2+}, high-Mg^{2+} artificial cerebrospinal fluid (ACSF) was prepared by decreasing the concentration of CaCl$_2$ to 0.1 mM and increasing that of MgSO$_4$ to 6.3 mM.

In Vivo *Study*

Perfusion of Vestibular Nuclei

Implantation of Minipumps. The same method was used as described in the previous study on *N*-methyl-D-aspartate (NMDA) receptors in central vestibular neurons.[29] Briefly, guinea pigs were anesthetized with a nembutal (40 mg/kg)/droperidol (4 mg/kg) mixture delivered intraperitoneally. Two Teflon-coated silver electrodes exposed at their tips were placed over the round window of the middle ear cavity and in front of the horizontal and anterior semicircular ampullae, respectively. A craniotomy over the ipsilateral cerebellum was performed, and a T-shape screw was implanted along the midline. Vestibular extracellular field potentials, evoked by single shock applied through these labyrinthine electrodes, were then recorded with glass microelectrodes filled with 2 M NaCl. This procedure allowed us to precisely determine the location of the vestibular nuclei. Finally, a metal cannula of 0.5 mm diameter was lowered in the brain stem and fixed with dental cement to the T-shape screw at a depth corresponding to the superior border of the vestibular complex.

During the three following days, the posture and the eye position of the implanted guinea pigs were observed. Animals that presented any postural or eye position changes were discarded.

The vestibular nuclei were then perfused by using either chronically implanted osmotic minipumps (Alza 2002, delivering 1 μl/hour) or an external slow infusion pump (rate of perfusion 1 microliter/5 minutes). As, in both cases, the perfusion rate was constant, the total amount of drug injected in the vestibular nuclei could be approximately determined. Five solutions were tested:

- 0.9% saline.
- Impromidine (an H2-agonist and an H3-antagonist, SK&F) at 10^{-5} M.
- Cimetidine (a specific H2-antagonist, SK&F) at 10^{-4} M and 10^{-5} M.
- Alpha methylhistamine (α-Me-His, a specific H3-agonist) at 10^{-5} M.
- Thioperamide (a specific H3-antagonist) at 10^{-5} M.

Impromidine, cimetidine, and α-Me-His were dissolved in 0.9% saline. Thioperamide was dissolved in 0.9% saline and dimethilsulfoxide (DMSO). All the solutions were adjusted to a PH at 7.3.

Radiological Study. In order to quantify any skeletal geometry changes induced by the drug perfusion, guinea pigs were x-rayed using the same method as previously described in hemilabyrinthectomized guinea pigs.[30] X-ray exposures from above with the x-ray tube 60 cm away from the film, and from the side at a distance of 90 cm were taken before and at different times following the beginning of the perfusion. The thoracic and lumbar vertebrae rotation about the longitudinal axis of the animal was quantified by attributing, on top view exposures, a coefficient of rotation (CR) for each vertebra. CR values in arbitrary units were determined by using the position of the spiny process crest relative to the vertebral body (see Reference 30 for further explanations).

Histological Procedures. At the end of the experiments, the guinea pigs were sacrificed with a nembutal overdose. The brains were then removed and stored in 10% formaldehyde solution. Horizontal sections, cut at 75 μm thickness on a freezing microtome, were then processed and stained with cresyl violet acetate for the visualization of the vestibular nuclei. Sections were examined to determine the exact localization of the cannula tips and to detect any signs of lesion of vestibular nuclei neurons.

HVOR and Electroencephalograph Recording

Surgical Procedure. All surgical procedures were performed using general anesthesia with intraperitoneal administration of a nembutal (40 mg/kg)/droperidol (4 mg/kg) mixture.

A coil made of two turns (diameter of 9 mm) of Teflon-coated seven-stranded stainless steel wire (0.23 mm diameter, A-M systems, Inc.) was implanted subconjunctivally on the eye and connected to a connector cemented to the skull. To immobilize the head during the eye movement recordings, a head holder was cemented stereotaxically to three stainless steel T-shaped bolts cemented in the temporal and occipital bones. The stereotaxic plane was determined by a head rotation around the interaural axis until the calvarium was horizontal between anterior 6 and 14 mm from the stereotaxic 0. Finally, the electrocorticogram signal was recorded by using silver-wire electrodes attached to the T-shaped bolt.

Eye Movement Recordings. Horizontal and vertical eye movements were recorded using the magnetic search coil method.[31] In that respect, the animals were placed on the platform of a servo-controlled turntable for vestibular stimulation. The head-holder was attached to the turntable, such that the guinea pig's eyes were positioned in the center of a magnetic field generated by two pairs of field coils. At this place, the current induced in the coil was linearly related to its angular rotation. Finally, the head was placed at an angle of 35° (nose down) from the stereotaxic horizontal plane in order to align the plane of the horizontal semicircular canals with the horizontal plane.[32]

The horizontal and vertical components of the eye position were recorded by a SKALAR magnetic search coil system with a bandwidth of DC to 200 Hz. The calibration of the eye movements was obtained by rotating the field coils around the animal, in steps of 10 degrees in both the horizontal and vertical planes. The resolution of eye position was estimated to have an accuracy of 1 degree.

Vestibular Stimulation. The horizontal VOR was studied in darkness during sinusoidal rotations at four frequencies—0.05, 0.1, 0.5, and 2 Hz—and at a peak angular velocity of 40 degrees/second. This protocol was repeated before drug injection and ½ hour, 1 hour, 1½ hours, 24 hours, and 48 hours after the intraperitoneal injection. Each recording session lasted 1 hour, during which the guinea pigs were kept alert by the delivering of unexpected auditory stimuli.

Data Acquisition and Analysis. The table position and the EEG signals, together with the horizontal and vertical component of the eye position, were followed on-line on an oscilloscope and recorded on an FM magnetic tape (14-way Teac recorder, bandwidth DC to 1 kHz) for off-line analysis. They were also displayed on paper with an electrostatic chart recorder.

All calculations were made by hand from eye position traces. Fast movements were eliminated, and the cumulative slow phase was reconstructed by putting the successive slow phase segments end to end. The following parameters were calculated:

- Gain of the HVOR defined as the peak amplitude of the compensatory eye movement divided by the peak amplitude of the table movement.
- Phase defined as $0°$ when the eye movement was $180°$ out of phase with the table movement.

Drugs and Study Design

Thirty-five guinea pigs were used in this study. Out of these, 23 were administered the histamine-related compounds or saline by using minipumps or slow infusion pump and 12 by using an intraperitoneal injection. The skeletal geometry was analyzed with x-ray exposures in the former, whereas the HVOR was tested in the latter.

Skeletal Geometry Study. Impromidine was tested in four animals at 10^{-5} M, cimetidine in three animals at 10^{-4} M and in one animal at 10^{-5} M; α-Me-His and thioperamide were administered at 10^{-5} M in seven and five guinea pigs, respectively. Finally, three control animals were perfused with saline.

VOR Study. Out of the 12 animals used in the VOR study, 8 animals were subjected to intraperitoneal injection of thioperamide alone. The different concentrations used were 2 mg/kg ($n = 2$), 10 mg/kg ($n = 2$), or 20 mg/kg ($n = 4$). The other 4 guinea pigs were administered a combination of thioperamide and α-Me-His delivered at a concentration for thioperamide of either 5 mg/kg ($n = 2$) or 10 mg/kg ($n = 2$) and for α-Me-His of 20 mg/kg.

RESULTS

In Vitro *Studies*

Bath application of histamine at 10^{-4} M or 10^{-5} M induced an increase in the spontaneous firing rate in the large majority of the A, B (FIGURE 1B), and B+ LTS MVNn. Inhibitory effects were never observed. In all three MVN neuronal types, the excitatory effect was associated with membrane depolarization and a very small decrease in membrane resistance (dot in FIGURE 2B). In the presence of a low-Ca^{2+}/

high-Mg^{2+} medium used for synaptic uncoupling, the effect of histamine persisted in B and other neuronal cell types.

This depolarizing effect of histamine was mediated by H2 receptors. Indeed, a selective H2 agonist, impromidine, applied at 2.10^{-7} M mimicked the excitatory effect of histamine at 10^{-4} M in all three neuronal cell types (FIGURE 2C, two upper traces). In addition, the selective H2 antagonist, cimetidine, applied at 10^{-4} M, completely blocked the excitatory action of 10^{-4} M histamine (FIGURE 2C, two middle traces). This again was true for all three neuronal cell types. In contrast, the selective H1 antagonist, mepyramine, applied at 10^{-4} M, was unable to block the excitatory action of 10^{-4} M histamine in any of the three cell types (FIGURE 2C, two lower traces).

To test H3 receptors into the MVNn, α-Me-His, a selective agonist, and thioperamide, a selective antagonist, were both tested at a concentration of 10^{-6} M. No cells exhibited a response when α-Me-His (FIGURE 2D1) or thioperamide was applied in the bath. In addition, the H3 antagonist thioperamide (10^{-6} M) never blocked the excitatory action of histamine applied at 10^{-4} M (FIGURE 2D2).

In Vivo *Studies*

Perfusion Experiments

Control Experiments.   Control animals perfused with 15 μl saline did not present any postural or oculomotor symptoms. The resting posture was similar to the one observed in normal guinea pigs. The cervical column is oriented vertically parallel to the gravity vector and the anterior and inferior borders of the two tympanic bullae are superimposed, attesting the absence of any cephalic rotation (FIGURE 3).

Cimetidine and α-Me-His Perfusion.   Infusion in the vestibular nuclei of 10^{-4} M and 10^{-5} M cimetidine (an H2 antagonist) and of 10^{-5} M α-Me-His (an H3 receptor agonist) resulted in a postural and oculomotor syndrome similar, although milder, to the one observed after hemilabyrinthectomy. These syndromes occur in a few minutes with the slow infusion pump technique ($n = 7$) or in a few hours with osmotic minipumps ($n = 4$).

The postural deficits consisted of cervicocephalic and thoracolumbar vertebrae rotations associated with limb misplacements. The head was rotated ipsilaterally to the cannula side in both the frontal and the horizontal plane. In addition, a spiroid rotation of the thoracolumbar column was present. The spiny process crests of the first thoracic vertebrae are deviated ipsilaterally, whereas those of the lumbar and of the last thoracic vertebrae are rotated contralaterally (FIGURE 4). This corresponds to an ipsilateral rotation of the first thoracic vertebrae and a contralateral one of the lumbar vertebrae. The ipsilateral rotation of the first thoracic vertebrae is responsible for the lateral head tilt. Finally, the same limb misplacements as the ones that occur after hemilabyrinthectomy were observed: the ipsilateral forelimb and contralateral hindlimb were flexed, whereas the contralateral forelimb and ipsilateral hindlimb were extended.

In a minority of the tested guinea pigs, cimetidine and α-Me-His perfusion induced a more limited postural syndrome: the ipsilateral head rotation was confined to the horizontal plane. No ipsilateral rotation of the first thoracic vertebrae was present and consequently no lateral head-neck tilt. The oculomotor syndrome was characterized by a tonic downwards deviation of the eye ipsilateral to the perfused side and tonic upwards deviation of the contralateral eye. The tonic eye deviation

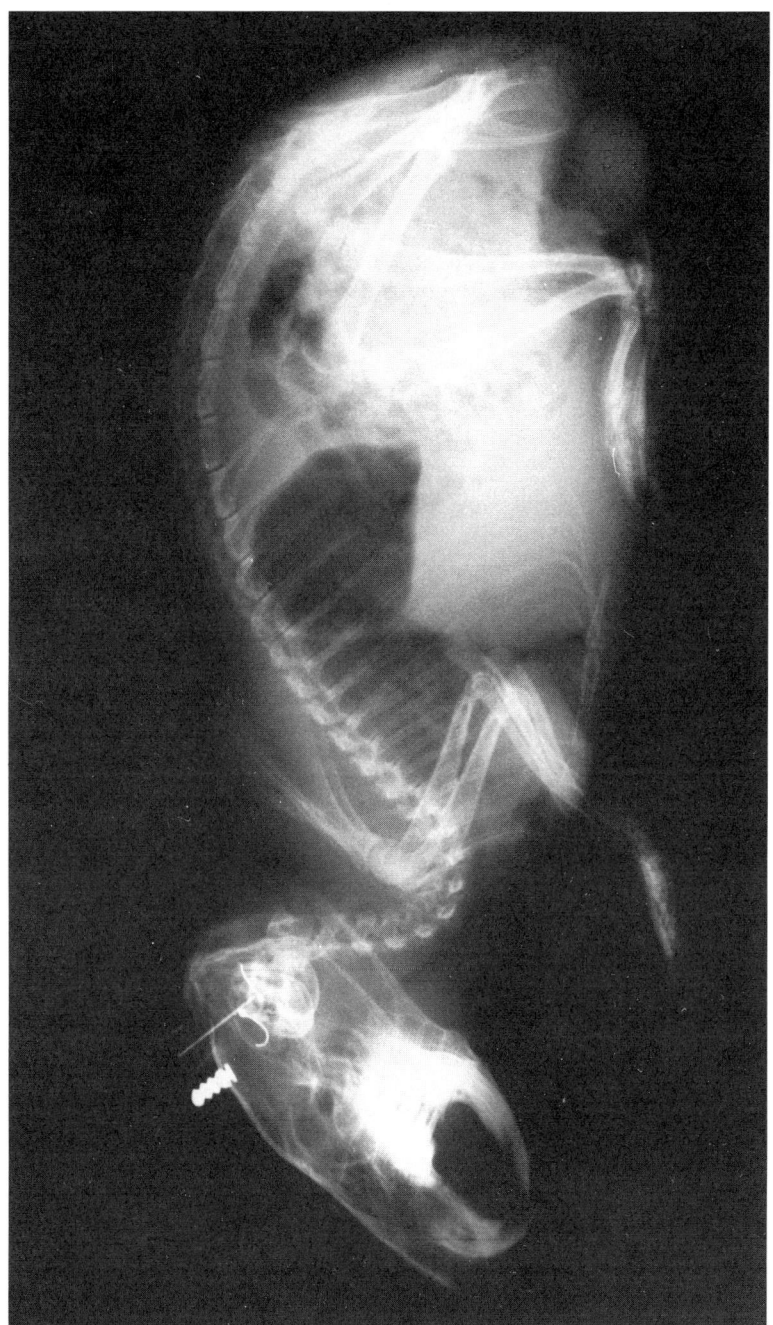

FIGURE 3. Lateral x-ray photograph of a normal guinea pig perfused with saline. Note the location of the metal cannula implanted at the superior border of the vestibular complex. The two tympanic bullae are superimposed attesting the absence of head rotation in the frontal and the horizontal plane.

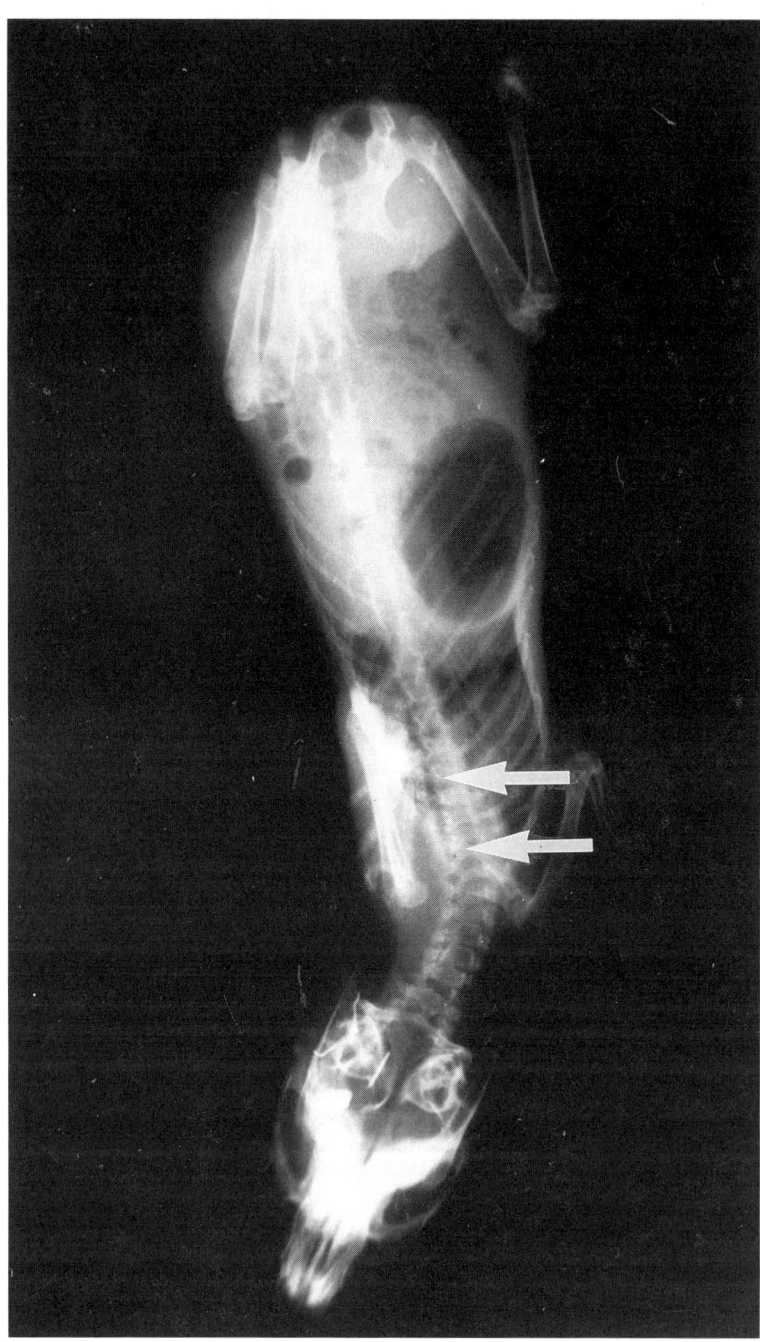

FIGURE 4. Top view x-ray exposure of a guinea-pig ten minutes after perfusion of the vestibular nuclei with α-Me-His. Note the progressive deviation of the spiny process crests of the first thoracic vertebrae (arrows) towards the cannula side which demonstrates the thoracic vertebrae rotation. This thoracic vertebrae rotation induced a head tilt in the frontal plane ipsilateral to the perfusion side.

was accompanied by a spontaneous nystagmus with the slow phase directed towards the perfusion side.

However, reversals of the postural and oculomotor syndromes were observed in half of the animals when α-Me-His-filled osmotic minipumps were used. Twenty-four hours after the beginning of perfusion, the head-neck ensemble became rotated towards the contralateral side of the cannula, in both the horizontal and frontal plane. In all the cases, all the syndromes disappeared in a few hours when the perfusion was discontinued.

Impromidine and Thioperamide Perfusion. In half of the tested guinea pigs, infusion of impromidine, an H2 agonist, induced the same postural and oculomotor syndrome as the one observed following cimetidine and α-Me-His infusion but milder and inversed. The head is rotated contralaterally to the cannula side in both the frontal and the horizontal plane. The first thoracic vertebrae are rotated contralaterally, whereas the last thoracic and the lumbar ones are rotated ipsilaterally. The same reversion was true for the discreet limb misplacement and the oculomotor syndrome.

Finally, no postural or oculomotor changes were ever observed in the thioperamide-perfused guinea pigs.

Histological Studies. The histological study showed that the cannula tips were always placed at the superior border of the vestibular complex. The absence of gliosis under the clearly visible cannula tract, and the lack of chromatolysis, strongly suggest that none of the applied drugs had a toxic effect on the vestibular nucleus neurons.

HVOR Recording Experiments

HVOR Gain and Phase. The horizontal vestibular nystagmus was investigated following intraperitoneal injection of thioperamide, a selective potent H3 antagonist. Compensatory eye movements interrupted by anticompensatory fast phases were still generated (FIGURE 5).

However, the HVOR gain was markedly lower than in normal guinea pigs at the four frequencies tested. For example, thioperamide injection at 20 mg/kg induced an approximate 40 to 50% HVOR gain decrease. As shown by the injection of 10 mg/kg and 5 mg/kg thioperamide, the decrease of the gain was dose dependent. In addition, the gain decrease was more important at low frequencies than at higher frequencies. The percentile values of the HVOR gain were plotted as a function of time for the concentrations of 5 mg/kg, 10 mg/kg, and 20 mg/kg of thioperamide. The effect of thioperamide on the HVOR gain was already noticeable half an hour after the injection, reached a maximum 1 or 1½ hours later, and then gradually returned to subnormal values in approximately 48 hours (FIGURE 6).

The phase characteristics remained unaltered: as in normal guinea pigs, HVOR phase lead decreased with increasing frequencies. No spontaneous nystagmus, VOR assymetry, or abnormal eye movements could be observed.

In order to determine if this HVOR gain reduction results from the specific blockade of H3 receptors, four guinea pigs were injected intraperitoneally with a thioperamide/α-Me-His mixture. The adjunction of the antagonist partially antagonized the effect of thiopcramide.

Alertness Level after the Drug Injection. The alertness level was continually recorded during the HVOR recordings sessions. No slow-wave periods were ever recorded in the 12 guinea pigs injected with thioperamide delivered intraperitoneally.

NORMAL

EEG

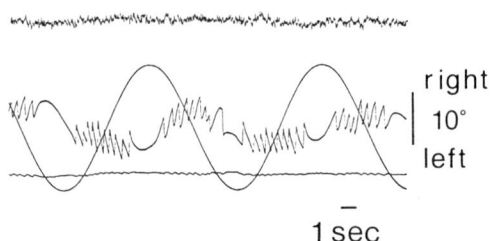

right
10°
left

–
1 sec

THIOPERAMIDE 20mg / kg

EEG

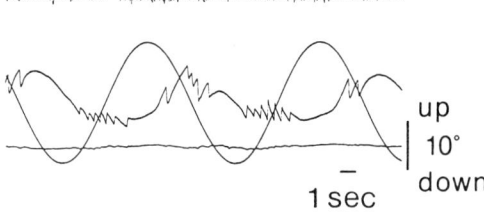

up
10°
–
down
1 sec

FIGURE 5. From top to bottom, EEG and horizontal and vertical eye movements during horizontal sinusoidal rotations records in the dark. Horizontal compensatory slow phases are interrupted by anticompensatory fast phases in both the normal and the thioperamide-injected guinea pigs. However, the maximal amplitude of the cumulative slow phase and the number of the fast phases of the HVOR are smaller in thioperamide-injected guinea pigs compared with normal animals. Note the absence of slow waves in the electrocorticogram signal following thioperamide injection.

DISCUSSION

In Vitro *Experiments*

The data clearly indicate that histamine activates the three cellular classes of MVNn tested. An increased resting discharge associated with a small decrease in resistance was always observed in A- and B-type MVNn (with and without LTS) following bath-applied histamine. This depolarizing effect results most probably from the activation of histaminergic receptors in the vestibular nuclei since it persists in synaptic uncoupling conditions, i.e., in low-Ca^{2+}/high-Mg^{2+} ACSF.

Three arguments led us to suppose that this excitatory effect was mediated by H2 receptors:

- First, the histamine excitation was mimicked by impromidine, a selective H2 agonist, and antagonized by cimetidine, a selective H2 antagonist.

- Second, bath application of mepyramine, an H1 receptor antagonist, could not block this histamine excitatory effect.
- Third, neither α-Me-His, a selective H3 agonist, nor thioperamide, a selective H3 antagonist, altered the resting discharge of the MVNn, excluding thus their direct involvement in the resting discharge increase.
- Finally, as mepyramine, thioperamide did not prevent the excitatory effect of histamine.

Therefore, we conclude that H2 receptors are present in the medial vestibular nuclei and that their activation induces an increase resting discharge of the MVNn. However, if these results agree with previous *in vitro* studies,[27] they differ from data collected during electrophysiological *in vivo* studies. Indeed, both an excitatory *and* inhibitory effect has been described following iontophoretic application of histamine in acute animals.[33,34] This discrepancy could be explained by the difference between these two types of preparations. For instance, contrary to what occurred *in vitro,* the presynaptic H3 receptors might have been activated *in vivo,* leading to a reduction of the MVNn resting discharge. Indeed, it has been well demonstrated that the activation of these autoreceptors induces a decrease of the synthesis and release of histamine from histaminergic terminals.[35-37] In addition, histamine could have, *in vivo,* a greater impact on the inhibitory vestibular neurons than on the excitatory ones.

The mode of action of histamine cannot be precisely determined in this study. However, the very weak increase in membrane conductance during histamine application suggests that the histaminergic receptors could be located quite remotely from the soma in the dendritic tree.

As mentioned above, neither mepyramine, a selective H1 antagonist, nor α-Me-His or thioperamide, respectively H3 agonist and antagonist, had a noticeable effect. This is surprising for H1 receptors, since autoradiographic studies have shown their presence at the level of the guinea pig vestibular nuclei.[38] However, our experimental

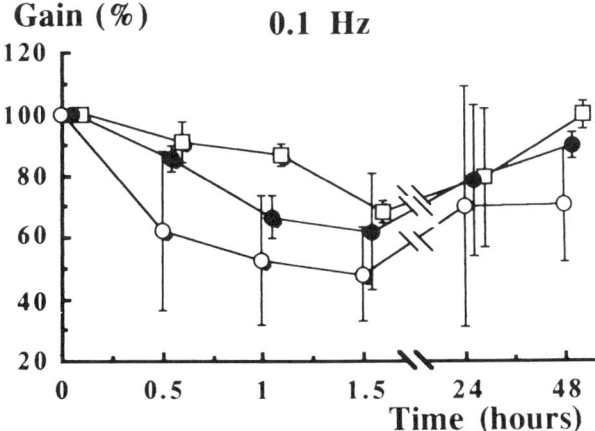

FIGURE 6. Average percentile values of the HVOR gain at 0.1 Hz as a function of time following the intraperitoneal administration of thioperamide. Open circles, filled circles, and open squares represent the average percentile values of the HVOR for 5 mg/kg, 10 mg/kg, and 20 mg/kg thioperamide concentrations, respectively.

protocol is probably not adequate to study them. This holds equally true for H3 receptors. These autoreceptors were located on the terminals of the vestibular histaminergic afferents, which were cut during the slicing procedure. Their role in the absence of a spontaneous release of neurotransmitter is therefore probably difficult to assess.

In Vivo *Experiments*

We have previously defined the criteria that could support the view that the diffusion of a drug was limited to the vestibular nuclei when a local injection through a cannula was performed:

- The signs observed must be similar to those observed following hemilabyrinth-ectomy or selective lesions of the vestibular apparatus.
- With the external slow infusion pump, the vestibular syndromes should appear in a matter of minutes. This is obviously not true with the osmotic minipump, which has a far lower perfusion rate.
- No other neurologic or neurovegetative signs should be present.
- Finally, the histological study should confirm the good placement of the cannula tip.

All of these criteria were met for the drugs perfused. We conclude thus that the signs observed most probably resulted from the drug infusion into the vestibular nuclei.

Cimetidine and α-Me-His Injection

The postural and oculomotor syndrome observed after cimetidine or α-methylhis-tamine infusion was similar but milder to those observed after hemilabyrinthectomy. First, this would tend to confirm that H2 receptors are present into the vestibular nuclei and would strongly suggest the presence of H3 vestibular receptors. Second, it allows us to speculate on the neuronal substrata of histamine effect on the vestibular nuclei. The deficits observed following hemilabyrinthectomy, which is mimicked by the H2 antagonist and the H3 agonist, is caused by a diminution of the resting discharge of the deafferented vestibular neurons. It is therefore very probable that the respective blockade of the H2 receptors and the activation of H3 receptors induced also a decrease in the spontaneous activity of the vestibular neurons. This view is supported by the *in vitro* results: the blockade of H2 receptors by cimetidine depressed the depolarizing effect of histamine. Moreover, the activation of H3 receptors by α-Me-His decreased the histamine release,[39] inducing a dysfacilitation of the vestibular neurons. Finally, bearing in mind that the postural and oculomotor syndrome observed was milder compared to the one following hemilabyrinthectomy or D-L-2-amino-5-phosphonovaleric acid (APV) perfusion, it confirms that histamine acts most probably as a neuromodulator of the MVNn and not as a neurotransmitter of the vestibular afferences, as was the case for glutamate.

However, it could be argue that our perfusion induced a firing rate decrease because of neurotoxic or mechanical damages. This can be most probably discarded because:

- the histological study did not reveal any signs of neuronal lesions;
- control animals perfused with saline did not exhibit any skeletal geometry changes.

In half of the guinea pigs perfused with an osmotic minipump, after a typical syndrome indicating a hypoactivity of the perfused vestibular neurons, α-Me-His induced a reverse postural syndrome. This result could be explained by the loss of selectivity of this agent at high concentrations. The initial H3-induced resting discharge decrease was probably overcome by a nonspecific activation of the H2 receptors, also known to occur *in vitro.*

Impromidine and Thioperamide Injection

Impromidine (H2 agonist) induced a very weak postural effect in only half of the four tested guinea pigs and thioperamide (H3 antagonist) perfusion had no observable effect. This means that a direct activation by impromidine is unable to recruit a significantly greater population of H2 receptors that are already activated in the alert unrestrained animal. This would tend to suggest that *in vivo,* contrary to what occurs *in vitro,* the level of histamine is sufficiently high to activate most of the vestibular H2 receptors. If this is the case, it is understandable that the blockade of H3 receptors, which should further increase the release of histamine, is without effect.

HVOR after Thioperamide Administration

This study demonstrates that intraperitoneal injection of thioperamide leads to significant alterations of VOR in the horizontal plane at both low and high frequencies of sinusoidal stimulation. The gain in HVOR was reduced, while the phase did not show any change.

The reduction in VOR gain may be attributed to specific blockade of the central H3 receptors since it is reversible, dose dependent, and partially antagonized by α-Me-His, a selective H3-agonist. Furthermore, this gain decrease could not be due to an alertness decrease as no slow-wave periods were recorded in the 12 guinea pigs injected. This result corroborates previous data,[40] which demonstrated that betahistine, a weak H1 agonist and a mildly specific H3 antagonist, used in the symptomatic treatment of vestibular disorders, depressed the HVOR gain.

It should not be consider as contradictory that thioperamide, which in theory should enhance the vestibular nucleus histamine level, depresses the VOR gain when injected intraperitoneally. Systemic injections obviously do not reveal the numerous neuronal targets of thioperamide; only the net result of their activation is observed.

Regardless, this new H3 antagonist behaves as a vestibuloplegic drug. This experimental result, if confirmed in humans, is interesting because it may pave the way to new therapeutic perspectives in peripheral vertigo, motion sickness, and central vestibular disorders. Indeed, thioperamide, compared to other commonly used histamine-related compounds, presents several advantages:

- This compound is a highly specific H3 antagonist. In particular, it is far more potent than betahistine since its affinity for the H3 receptor is in the nanomolar range.[41]
- It passes the blood-brain barrier.
- Finally, and more importantly, *it does not induce* somnolence or hypersomnia. In fact, in the cat *thioperamide enhances wakefulness in a dose-dependent manner.*[42]

ACKNOWLEDGMENTS

We thank M. Chat, S. Lemarchand, and D. Machard for excellent technical assistance.

REFERENCES

1. SCHWARTZ, J. C., J. M. ARRANG, M. GARBARG, H. POLLARD & M. RUAT. 1991. Histaminergic transmission in the mammalian brain. Physiol. Rev. **71**: 1–51.
2. TAKEDA, N., M. MORITA, T. KUBO, A. YAMATODANI, T. WATANABE, M. TOHYAMA, H. WADA & T. MATSUNAGA. 1987. Histaminergic projection from the posterior hypothalamus to the medial vestibular nucleus of rats and its relation to motion sickness. *In* Neurophysiologic and Clinical Research. M. D. Graham & J. L. Kemink, Eds.: 601–617. Raven Press. New York, N.Y.
3. AIRAKSINEN, M. S. & P. PANULA. 1988. The histaminergic system in the guinea pig central nervous system: an immunocytochemical mapping study using an antiserum against histamine. J. Comp. Neurol. **273**: 163–186.
4. PANULA, P., U. PIRVOLA, S. AUVINEN & M. S. AIRAKSINEN. 1989. Histamine immunoreactive nerve fibers in the rat brain. Neuroscience **28**(3): 585–610.
5. STEINBUSCH, H. W. M. 1991. Distribution of histaminergic neurons and fibers in rat brain. Comparison with noradrenergic and serotonergic innervation of the vestibular system. Acta Otolaryngol. Stockh. Suppl. **479**: 12–23.
6. BOUTHENET, M. L., M. RUAT, N. SALES, M. GARBARG & M. SCHWARTZ. 1988. A detailed mapping of histamine H1-receptors in guinea-pig central nervous system established by autoradiography with (I125) iodobolpyramine. Neuroscience **26**: 553–600.
7. KIRSTEN, E. B. & J. N. SHARMA. 1976. Microiontophoresis of acetylcholine, histamine, and their antagonists on neurons in the medial and lateral vestibular nuclei of the cat. Neuropharmacology **15**: 743–753.
8. SATAYAVIDAD, J. & E. B. KIRSTEN. 1977. Iontophoretic studies of histamine and histamine antagonists in the feline vestibular nuclei. Eur. J. Pharmacol. **41**: 17–26.
9. PHELAN, K. D., J. NAKAMURA & J. P. GALLAGHER. 1990. Histamine depolarizes rat medial vestibular nucleus neurons recorded intracellularly in vitro. Neurosci. Lett. **109**: 287–290.
10. SERAFIN, M., A. KATEB, N. VIBERT, C. DE WAELE, P. P. VIDAL & M. MUHLETHALER. 1990. In vitro and in vivo studies of the histaminergic receptors in the medial vestibular nuclei in the guinea-pig. Soc. Neurosci. Abstr. **16**: 401.8.
11. HOUGH, L. B. 1988. Cellular localization and possible functions for brain histamine: recent progress. Prog. Neurobiol. **30**: 469–505.
12. COLLINS, W. E., G. H. CRAMPTON & J. B. POSNER. 1961. Effects of mental activity on vestibular nystagmus and the encephalogram. Nature **190**: 194–195.
13. COLLINS, W. E. & F. E. GUEDRY. 1962. Arousal effects and nystagmus during prolonged constant angular acceleration. Acta Otolaryngol. Stockh. **54**: 349–363.
14. HODES, R. & J. I. SUZUKI. 1965. Comparative thresholds of cortex, vestibular and reticular formation in wakefulness, sleep and rapid eye movement periods. Electroencephalogr. Clin. Neurophysiol. **18**: 239–248.
15. BALDISSERA, F., G. BROGGI & M. MANCIA. 1967. Nystagmus induced by unilateral hemilabyrinthectomy affected by sleep-wakefulness cycle. Nature **215**: 62–63.
16. LENZI, G. L., O. POMPEIANO & T. SATOH. 1968. Input-output relation of the vestibular system during sleep and wakefulness. Pflugers Arch. Gen. Physiol. **299**: 326–333.
17. REDING, G. R. & C. FERNANDEZ. 1968. Effects of vestibular stimulation during sleep. Electroencephalogr. Clin. Neurophysiol. **24**: 75–79.
18. VON BERNUTH, H. & H. F. R. PRECHT. 1969. Vestibulo-ocular response and its state dependency in newborn infants. Neuropediatrie **1**: 11–24.
19. MELVILL JONES, G. & N. SUGIE. 1972. Vestibulo-ocular responses in man during sleep. Electroencephalogr. Clin. Neurophysiol. **32**: 43–53.
20. TAUBER, E. S., G. HANDELMAN, R. HANDELMAN & E. D. WEITZMAN. 1972. Vestibular stimulation during sleep in young adults. Arch. Neurol. Chicago **27**: 221–228.
21. FLANDRIN, J. M., J. H. COURJON, M. JEANNEROD & R. SCHMIDT. 1979. Vestibulo-ocular responses during the states of sleep in the cat. Electroencephalogr. Clin. Neurophysiol. **46**: 521–530.
22. GELLER, H. M., S. A. SPRINGFIELD & A. R. TIBERIO. 1984. Electrophysiological actions of histamine. Can. J. Physiol. Pharmacol. **62**: 715–719.

23. DOUGLAS, W. W. 1985. Histamine and 5-hydroxytryptamine (serotonine) and their antagonists. *In* Pharmacological Basis of Therapeutics. A. G. Gilman, L. S. Goodman, T. W. Rall & F. Murad, Eds.: 605–638. Macmillan. New York, N.Y.
24. SERAFIN, M., C. DE WAELE, A. KHATEB, P. P. VIDAL & M. MUHLETHALER. 1991. Medial vestibular nucleus in the guinea-pig. I. Intrinsic membrane properties in brainstem slices. Exp. Brain Res. **84:** 417–425.
25. SERAFIN, M., C. DE WAELE, A. KHATEB, P. P. VIDAL & M. MUHLETHALER. 1991. Medial vestibular nucleus in the guinea pig. II. Ionic basis of the intrinsic membrane properties in brainstem slices. Exp. Brain Res. **84:** 426–433.
26. LEWIS, M. R., J. P. GALLAGHER & P. SHINNICK-GALLAGHER. 1987. An in vitro brain slice preparation to study the pharmacology of central vestibular neurons. J. Pharmacol. Methods **18:** 267–273.
27. PHELAN, K. D., J. NAKAMURA & J. P. GALLAGHER. 1990. Histamine depolarizes rat medial vestibular nucleus neurons recorded intracellularly in vitro. Neurosci. Lett. **109:** 287–290.
28. LLINAS, R. & M. SUGIMORI. 1980. Electrophysiological properties of in vitro Purkinje cell somata in mammalian cerebellar slices. J. Physiol. London **305:** 171–195.
29. DE WAELE, C., N. VIBERT, M. BAUDRIMONT & P. P. VIDAL. 1990. NMDA receptors contribute to the resting discharge of vestibular neurons in the normal and hemilabyrinthectomized guinea pig. Exp. Brain Res. **81:** 125–133.
30. DE WAELE, C., W. GRAF, P. JOSSET & P. P. VIDAL. 1989. A radiological analysis of the postural syndroms following hemilabyrinthectomy and selective canal and otolith lesions in the guinea pig. Exp. Brain Res. **77:** 166–182.
31. ROBINSON, D. A. 1963. A method of measuring eye movement using a search coil in a magnetic field. IEEE Trans. Biomed. Eng. **10:** 137–145.
32. CURTHOYS, I. S., E. J. CURTHOYS, R. H. I. BLANKS & C. H. MARKHAM. 1975. The orientation of the semi-circular canals in the guinea pig. Acta Otolaryngol. Stockh. **80:** 197–205.
33. KIRSTEN, E. B. & J. N. SHARMA. 1976. Microiontophoresis of acetylcholine, histamine and their antagonists on neurons in the medial and lateral vestibular nuclei of the cat. Neuropharmacol. **15:** 743–753.
34. SATAYAVIDAD, J. & E. B. KIRSTEN. 1977. Iontophoretic studies of histamine and histamine antagonists in the feline vestibular nuclei. Eur. J. Pharmacol. **41:** 17–26.
35. ARRANG, J. M., M. GARBARG & I. C. SCHWARTZ. 1983. Auto-inhibition of brain histamine release mediated by a novel class (H3) of histamine receptor. Nature **302:** 832–837.
36. ARRANG, J. M., M. GARBARG & J. C. SCHWARTZ. 1985. Autoregulation of histamine release in brain by presynaptic H3-receptors. Neuroscience **15:** 553–562.
37. ARRANG, J. M., M. GARBARG & J. C. SCHWARTZ. 1987. Autoinhibition of histamine synthesis mediated by presynaptic H3-receptors. Neuroscience **23:** 149–157.
38. BOUTHENET, M. L., M. RUAT, N. SALES, M. GARBARG & M. SCHWARTZ. 1988. A detailed mapping of histamine H1-receptors in guinea-pig central nervous system established by autoradiography with (I125) iodobolpyramine. Neuroscience **26:** 553–600.
39. ARRANG, J. M., M. GARBARG, T. T. QUACH, M. D. T. TUONG, E. YERAMIAN & J. C. SCHWARTZ. 1985. Action of betahistine and histamine receptors in the brain. Eur. J. Pharmacol. **111:** 73–84.
40. OOSTERVELD, W. J. 1984. Betahistine dihydrochloride in the treatment of vertigo of peripheral vestibular origin. A double-blind placebo-controlled study. J. Laryngol. Otol. **98:** 37–41.
41. SMITS, R. P., H. W. M. STEINBUSCH & A. H. MULDER. 1990. The localization of histidine decarboxylase–immunoreactive cell bodies in the guinea pig brain. J. Chem. Neuroanat. **3:** 85–100.
42. LIN, J. S., K. SAKAI, G. VANNI-MERCIER, J. M. ARRANG, M. GARBARG, J. C. SCHWARTZ & M. JOUVET. 1990. Involvement of histaminergic neurons in arousal mechanisms demonstrated with H3-receptor ligands in the cat. Brain Res. **523:** 325–330.

Cholinergic Innervation of the Cerebellum of the Rat by Secondary Vestibular Afferents

N. H. BARMACK,[a] R. W. BAUGHMAN,[b]
AND F. P. ECKENSTEIN[c]

[a]Department of Ophthalmology
R. S. Dow Neurological Sciences Institute
Good Samaritan Hospital & Medical Center
1120 Northwest 20th Avenue
Portland, Oregon 97209

[b]Department of Neurobiology
Harvard University Medical School
Boston, Massachusetts 02115

[c]Department of Cell Biology and Anatomy
Oregon Health Sciences University
Portland, Oregon 97201

INTRODUCTION

The multiple anatomical origins of cerebellar mossy fibers suggest that they constitute a heterogeneous source of information. The axon terminals of these mossy fibers converge on granule cells in the cerebellar cortex forming glomerular synapses, "rosettes," that have a variety of morphologies.[1–3] These different afferent projections may use different synaptic transmitters which ultimately reflect what functions are required of the cerebellar circuitry activated by these transmitter-specific pathways.

Several putative cerebellar mossy fiber neurotransmitters have been identified primarily by using immunohistochemical and transmitter-uptake techniques. These putative neurotransmitters include glutamate,[4,5] aspartate,[6] serotonin,[7] enkephalin,[8] γ-aminobutyric acid (GABA),[9] and acetylcholine.[7,10] Classical acetylthiocholine stains have shown that the cerebellum is rich in both cholinesterase (AChE) and pseudocholinesterase (pAChE).[11–18] Both AChE and pAChE are distributed in a sagittal banding pattern in the cerebellar vermis.[12,18] These banding patterns may not necessarily be related to cholinergic function. AChE is also distributed differentially within separate divisions of the cerebellum. It is found in greatest density in the uvula-nodulus.[13–15,17] These same regions have high activity of the synthetic enzyme choline acetyltransferase (ChAT).[16] Furthermore, ChAT activity in the "archicerebellum" (uvula-nodulus) decreases following surgical deafferentation of "archicerebellum,"[16] suggesting that this activity can be attributed to afferent fibers of extrinsic origin.

We have used two polyclonal antibodies to ChAT to study putative cholinergic projections to the cerebellum of the rat. We have characterized regional differences in ChAT activity in the cerebellum using the radiochemical assay of Fonnum.[19] We have found that all lobules of the cerebellum receive a diffuse cholinergic afferent projection. Three regions receive a particularly dense cholinergic projection: (1) the uvula-nodulus (lobules 9 and 10), (2) the flocculus, and (3) lobules 1 and 2 of the anterior lobe vermis. Using double labeling for both retrogradely transported horseradish peroxidase (HRP) and immunostaining for ChAT we have traced the origin of the ChAT pathway to the uvula-nodulus from the medial vestibular nucleus and from the nucleus prepositus hypoglossi.

METHODS

Brain tissue from rats was prepared for either ChAT immunocytochemistry or for measurements of ChAT activity.

Choline Acetyltransferase Immunohistochemistry

The protocol for ChAT immunohistochemistry was similar to that reported previously.[20] Two polyclonal ChAT antisera were prepared. They were characterized previously and have been shown to specifically stain cholinergic neurons.[20,21]

Measurement of ChAT Activity

The radiochemical method of Fonnum, with slight modifications, was used for the measurement of ChAT activity in individual cerebellar lobules.[19]

Retrograde HRP

The cerebellum was exposed, and a 30% solution of HRP was pressure injected through a glass micropipette connected to a 1-μl syringe (Hamilton). The micropipette was inserted through the dorsal uvula into the ventral aspect of the uvula and into the nodulus at several medial-lateral locations, and a total of 0.5–2.5 μl was pressure injected. Following a 2–3 day postoperative survival, animals were deeply anesthetized with sodium pentobarbital (60 mg/kg). They were subsequently perfused transcardially with 0.9% saline, followed by 2.0–2.5% paraformaldehyde, 0.15% picric acid, and 0.05–0.20% glutaraldehyde in 0.1 M phosphate-buffered saline (PBS); pH = 7.2, lasting 20–30 minutes. The brains were removed, dehydrated, blocked, sectioned, and processed for both ChAT immunohistochemistry and HRP.[22,23]

RESULTS

ChAT-like Immunoreactivity in the Cerebellum

The entire cerebellar cortex of the rat contained a sparse network of thin, finely beaded ChAT-positive fibers that penetrated both the granule cell layer and the molecular layer (FIGURE 1C, D_3, and E). These thin, finely beaded fibers were similar morphologically, but less profuse than the cholinergic fibers of the rat cerebral cortex.[20] In addition to these finely beaded fibers, the cerebellar cortex was character-

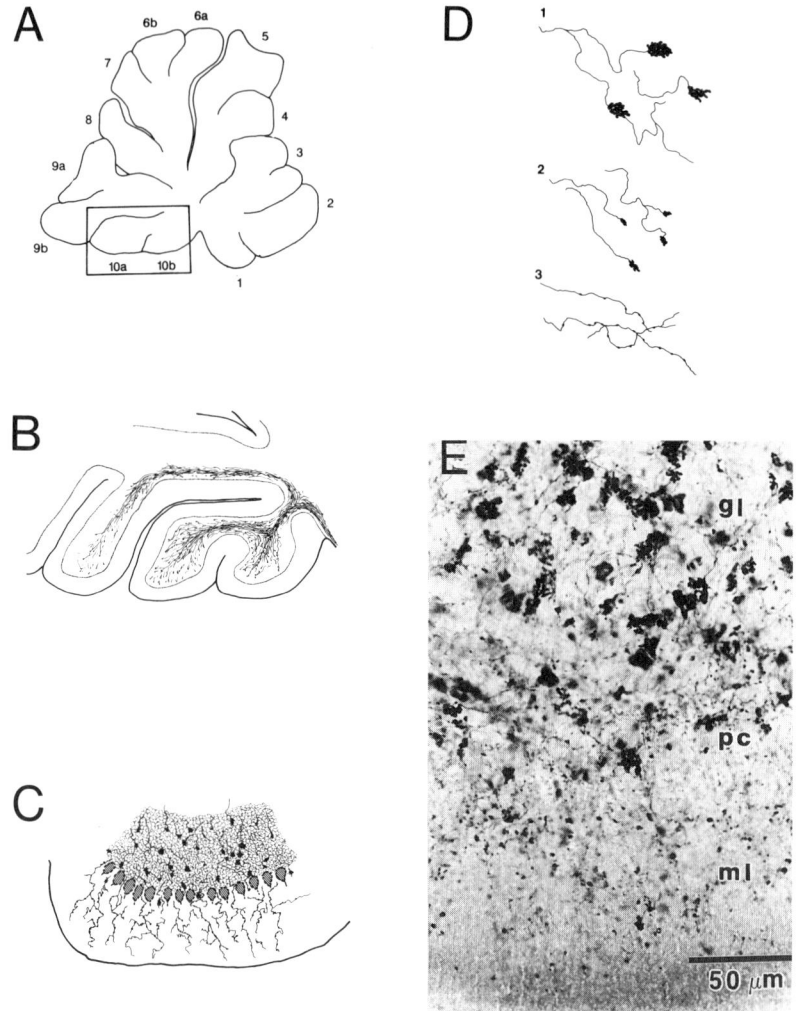

FIGURE 1. ChAT-like immunoreactivity in the cerebellar vermis. **A:** Schematic view of sagittal section through the midline of the cerebellar vermis. Lobules are numbered according to Larsell.[58] The area demarcated by the rectangle includes part of the uvula and all of the nodulus. The ChAT-positive mossy fiber afferents to this region are illustrated in B. The ChAT-positive innervation of the uvula-nodulus includes finely beaded fibers that terminate in the granule cell layer and molecular layer as well as the classical mossy fiber rosette terminals. These different types of terminals are illustrated in C and D. These different terminals are also seen in the photomicrograph in E. Abbreviations: gl, granule cell layer; pc, Purkinje cell layer; ml, molecular layer. Primary antibody: rat-αpig-ChAT.

ized by a nonuniform distribution of ChAT-positive mossy fibers with large, "grapelike" rosette terminals. These mossy fiber rosettes were concentrated in three distinct regions: (1) the ventral uvula and nodulus (lobules 9b and 10), (2) the anterior lobe vermis (lobules 1–2), and (3) the flocculus and ventral paraflocculus. Of these three regions, the ChAT-positive innervation of lobules 9b and 10 was the most prominent (FIGURE 1B).

Uvula-Nodulus

Lobules 9b and 10 contained large (5–12 μm) ChAT-positive mossy fiber rosettes that were distributed throughout the granule cell layer (FIGURE 1C and E). Although it was not possible to determine the frequency of branching of these ChAT-positive inputs, several examples of multiple mossy fiber rosettes stemming from a single axon were observed. Occasionally we could observe finely beaded axons that branched off from a large mossy fiber rosette in the granule cell layer. These fine fibers then penetrated past the Purkinje cell layer into the molecular layer. These observations suggest that the finely beaded axons and the large mossy fiber rosettes have a common anatomical origin. We observed no regions within the granule cell layers of lobules 9b and 10 that did not receive ChAT-positive afferent projections.

The more dorsal lobule of the uvula (lobule 9a) also received a ChAT-positive afferent innervation characterized by mossy fiber rosettes that were smaller (2–5 μm) than those observed in the ventral nodulus (lobule 10b).

Anterior Vermis

The granule cell layer of the anterior lobe vermis also received innervation from a cluster of small diameter ChAT-positive mossy fiber rosettes. These terminals were most densely concentrated at the junction between lobules 1 and 2.

Flocculus and Ventral Paraflocculus

The dual-lobed flocculus of the rat received a ChAT-positive afferent innervation which was less dense than that received by the nodulus, but comparable in density to that observed in the dorsal uvula (lobule 9a). This ChAT-positive afferent projection included the entire ventral lobule of the flocculus and the ventral aspect of the dorsal lobule. In addition, the flocculus was characterized by a particularly dense plexus of finely beaded fibers that extended into both the granule cell layer and the molecular layer. The ventral paraflocculus was similar in appearance to the other lobules of the vermis which did not have a particularly dense distribution of ChAT-like immunore-activity.

ChAT Activity in the Cerebellum

A radiochemical assay for ChAT activity was used to measure ChAT activity in individual cerebellar lobules of three rats. The highest ChAT activity in the cerebellum, 47.4 ± 7.6 Mmol of Ach/hour per g, was found in lobule 10b of the nodulus (FIGURE 2A). The second highest ChAT activity, 32.9 ± 12.0 Mmol of Ach/hour per g, was found in lobule 1 of the anterior lobe vermis (FIGURE 2A). The flocculus had a ChAT activity that was about 25% of that found in the nodulus (FIGURE 2B). The cerebellar hemispheres had a ChAT activity of 5.7 ± 2 Mmol of Ach/hour per g, about half the ChAT activity found in the flocculus. These quantitative measurements of ChAT activity were compared to estimates of the density of immunohistochemical staining for ChAT which are illustrated in Figure 2 and which were based

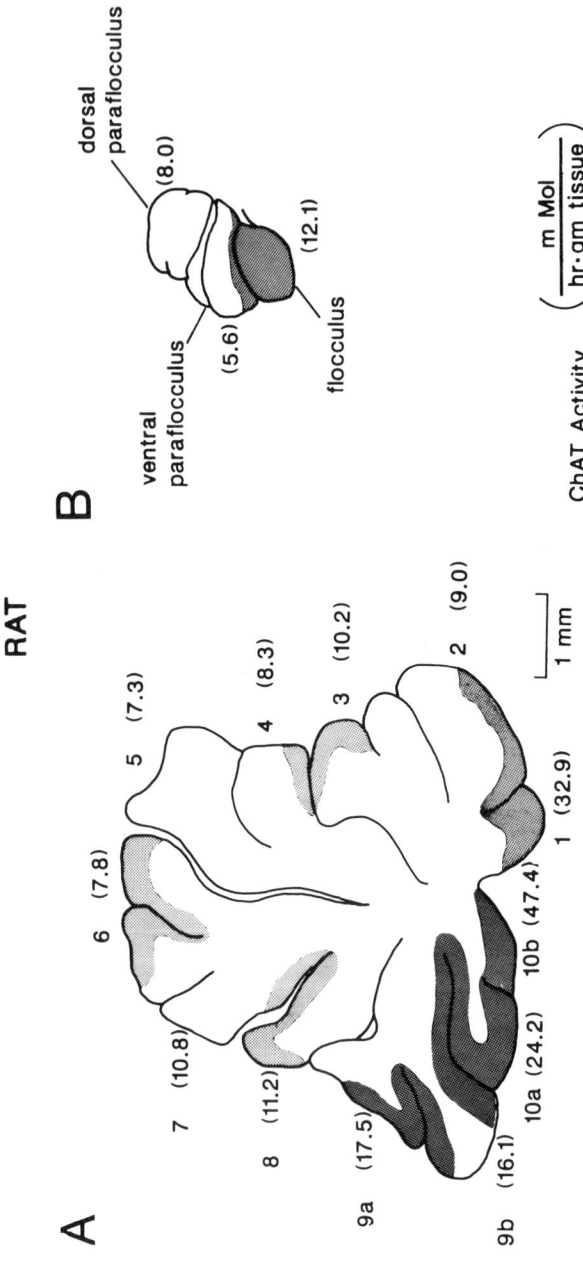

FIGURE 2. Distribution of ChAT-like immunostaining and ChAT activity in the cerebellum. ChAT-like immunostaining was graded, by eye, on a four-point scale. Using this scale, the areas of most dense ChAT-like immunoreactivity correspond to the darkest shading in the illustration. The numbers in parentheses are mean measurements of ChAT activity for each cerebellar lobule in three rats. ChAT activity was expressed as mmol of Ach synthesized/hours per g tissue at 37°C.

on a four-point scale; the ChAT staining of highest density corresponds to the darkest illustrated regions. These qualitative estimates of ChAT staining, based upon visual inspection, undoubtedly reflect more the presence of larger diameter mossy fiber rosettes than of the finely beaded fibers found in the granule and molecular layers. Nevertheless there is reasonable agreement between the ratings of histochemical staining density and the quantitative measurements of ChAT activity.

Absence of ChAT-like Immunoreactivity in Cerebellar-Related Nuclei

None of the intrinsic cerebellar cortical neurons were ChAT positive. Nor were any ChAT-positive neurons found in the locus ceruleus, the source of a diffuse noradrenergic cerebellar projection. No ChAT-positive neurons were observed in the inferior olive, the exclusive source of cerebellar climbing fibers. However, we have observed ChAT-positive synaptic terminals in the region of the dorsal cap of the inferior olive.[24]

ChAT-like Immunoreactivity in the Vestibular Nuclei

The brain stems of four rats were examined in both cross section and sagittal section for ChAT-like immunoreactivity. Secondary vestibular neurons were labeled in two main clusters within a rostral-caudal axis of the vestibular complex extending from a caudal level of the dorsal motor nucleus of the vagus (NX) to a rostral level of the abducens nucleus (FIGURE 3). Labeled neurons were found in the most caudal aspect of the medial vestibular nucleus (MVN) extending into the nucleus prepositus hypoglossi (NPH). Many smaller spindle-shaped neurons with cell body diameters of less than 8 μm were observed in both the caudal and rostral MVN near the ventricular surface (FIGURE 3). Larger ChAT-positive multipolar neurons with cell bodies exceeding 10 μm were located nearer the middle and ventral aspects of the MVN and the NPH.

ChAT- and HRP-Labeled Neurons in the Vestibular Complex following Pressure Injections of HRP into the Uvula-Nodulus

Pressure injections of HRP into the uvula-nodulus were made in four rats. The MVN contained the largest number of double-labeled neurons. These double-labeled neurons were concentrated in the caudal half of the MVN, but could be observed as far rostrally as the abducens nucleus. Double-labeled neurons were also observed in the NPH and to a lesser extent, the descending VN. The MVN, NPH, DVN, and superior VN also contained neurons that were clearly singly labeled for either HRP or ChAT (FIGURE 4). FIGURE 4A illustrates the largest extent of diffusion of HRP at the injection site. Only the brain stem ipsilateral to the intended injection is illustrated. Because of the difficulties in unambiguously identifying double-labeled neurons and because HRP injections into different regions of the cerebellum were always subtotal, it is likely that our counts of double-labeled neurons underestimate the true number cholinergic neurons with uvula-nodular projections.

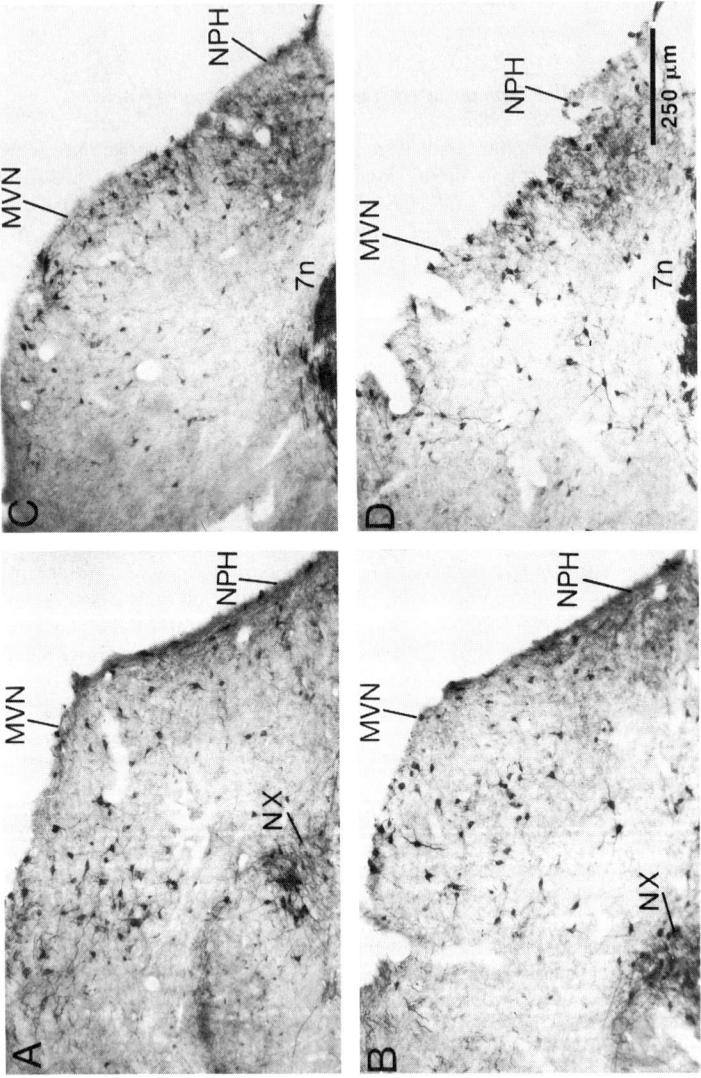

FIGURE 3. ChAT-like innervation of the medial vestibular nucleus and nucleus prepositus hypoglossi. Cross-sections of the rat brain stem extend from a level near the exit of the dorsal vagus nerve, A, to a more rostral level at the level of the genu of the facial nerve, D. The sections are spaced approximately 350 μm in the caudal-rostral axis. The ChAT-positive neurons are distributed throughout the MVN and NPH. Many ChAT-positive neurons were located near the ventricular surface of the MVN. Abbreviations: MVN, medial vestibular nucleus; NPH, nucleus prepositus hypoglossi; NX, dorsal motor nucleus of the vagus; 7n, genu of the facial nerve. Primary antibody: mouse-αrat-ChAT.

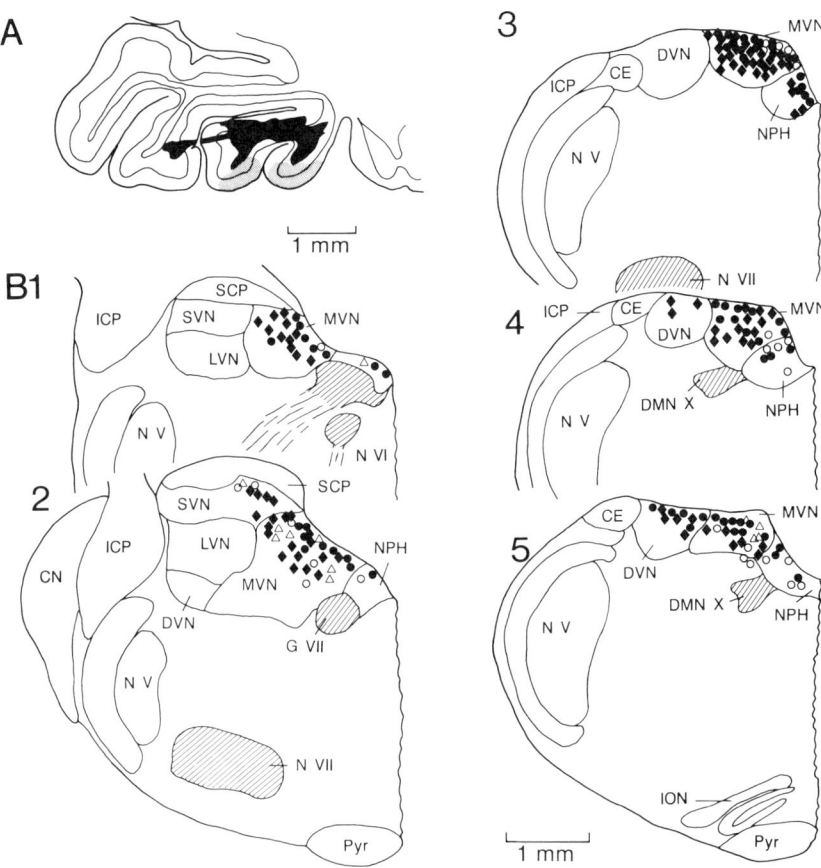

FIGURE 4. HRP- and ChAT-labeled neurons in the vestibular complex following an HRP injection into the uvula-nodulus. HRP, 1.2 μl, was pressure injected into the nodulus (rat 67) through the uvula. The injection was distributed over three separate injection tracks spaced mediolaterally about 500 μm apart in the left nodulus. **A:** Sagittal section through the largest injection site, located 300 μm from the midline. The black and stippled areas correspond to more and less dense HRP concentrations. **B1–5** are rostral-caudal brain stem sections through the left vestibular complex spaced approximately 500 μm apart. The filled circles correspond to HRP- and ChAT-labeled neurons. The open circles correspond to ChAT-labeled neurons only. The filled diamonds correspond to HRP-labeled neurons only. The open triangles correspond to neurons with such dense HRP labeling that it was not possible to determine if these neurons were also ChAT labeled. Abbreviations: MVN, SVN, LVN, DVN, medial, superior, lateral, and descending vestibular nucleus; NPH, nucleus prepositus hypoglossi; ICP, SCP, inferior and superior cerebellar peduncle; NV, trigeminal nucleus; ION, inferior olivary nucleus; DMN X, dorsal motor nucleus of the vagus; CE, external cuneate nucleus; CN, cochlear nucleus; G VII, genu of facial nerve; N VI, abducens nucleus; N VII, facial nucleus; Pyr, pyramidal tract.

ChAT- and HRP-Labeled Neurons in the Lateral Reticular Nucleus following Injections of HRP into the Uvula-Nodulus

The lateral reticular nucleus is a source of mossy fibers to the cerebellum.[25] In agreement with previous investigations, we observed ChAT-labeled neurons within the lateral reticular nucleus (LRN).[26,27] However, following HRP injections into the uvula-nodulus, few double-labeled neurons were observed in the LRN. This lack of double labeling could not be attributed to poor retrograde transport of HRP since appropriate groups of neurons in the vestibular complex and inferior olive were well labeled following uvula-nodular HRP injections.

Other Cholinergic Nuclei that Were Not Double Labeled following Pressure Injections of HRP into the Uvula and Nodulus

Several brain stem nuclei containing ChAT-positive neurons might be suspected as possible sources of cerebellar cholinergic afferents. These nuclei include: cranial nerve nuclei 3, 4, 5, 6, 7, 10, and 12; Edinger Westphal nucleus; and the laterodorsal tegmental nucleus. Each of these nuclei contained ChAT-positive neurons, but none of these neurons were labeled retrogradely by HRP.

DISCUSSION

Differences in ChAT-Positive Terminal Morphology

The rich variety of cholinergic endings within the cerebellum clouds the interpretation as to what constitutes a mossy fiber. It is possible that mossy fibers comprise a group of afferents that do not terminate exclusively upon granule cells. We cannot exclude the possibility that the cholinergic fibers revealed by the present immunohistochemical techniques also innervate Purkinje cell dendrites or the dendrites of cerebellar interneurons.

Large ChAT-positive mossy fiber rosettes comprise only a fraction of the cholinergic inputs to the cerebellum. The presence of the finely beaded fibers in both the granule cell and the molecular layers has also been observed by others using silver[2] as well as immunohistochemical stains.[28,29] These finely beaded fibers comprise a significant source of ChAT-positive fibers. Although it is possible that the finely beaded fibers ending within the molecular layer originate from a separate anatomical location or constitute a separate fiber class, this interpretation does not agree with the observation that a single stem fiber could give rise to both large and small mossy fiber rosettes as well as finely beaded terminals.

The functional significance of the different types of ChAT-positive terminals cannot be resolved by the present experiment. It is possible that these different fiber morphologies reflect different anatomical origins. Alternatively, the different morphologies of mossy fiber terminals might reflect different postsynaptic receptor specialization in the target neurons. Both muscarinic[30–33] and nicotinic[34] receptors have been identified in the cerebellum, and it is possible that the differences in terminal morphology may be associated with a particular receptor subtype.

It is possible that not all ChAT-positive fibers have an origin external to the cerebellum. A subpopulation of Golgi cells have been reported to be ChAT positive in the cat.[29] However, these ChAT-positive cells comprised less than 5% of the total Golgi cells. We have not observed ChAT-positive Golgi cells in the rat.

Functional Implications of ChAT-like Immunoreactivity in the Cerebellum

The cerebellum receives a ChAT-positive innervation that is concentrated in the nodulus (lobule 10) and extends into the lower lobule of the uvula (lobule 9b) (FIGURE 1). These findings agree with previous ChAT activity measurements[16] and immunohistochemical experiments that have examined the distribution of AChE.[14-17] The three areas of the cerebellum that have greatest density of ChAT-like immuno- reactive terminals and that have ChAT activities higher than the background activity of other cerebellar lobules also comprise recipient zones for primary and/or second- ary vestibular inputs. The uvula-nodulus receives both a primary[35-40] and a second- ary[41-43] vestibular projection. Although the existence of a primary vestibular projec- tion to the flocculus is controversial (see References 35–40 and 44), there is good agreement that the flocculus receives a secondary vestibular projection.[41-47] Anterior lobules 1 and 2 also receive a secondary vestibular projection.[43,48] The overlap of secondary vestibular projections with areas in the cerebellum that receive a cholin- ergic mossy fiber input and the evidence that at least a subset of the ChAT-positive innervation of the uvula-nodulus originates from the vestibular complex make it likely that regions of the cerebellum may also receive cholinergic secondary vestibu- lar fibers.

The Cholinergic Component of the Secondary Afferent Vestibular Projection to the Cerebellum

We have identified regions of the vestibular complex that contain cholinergic neurons as identified by ChAT immunohistochemistry. Furthermore we have shown that many of these neurons, particularly those in the caudal medial vestibular nucleus, are filled retrogradely by HRP following pressure injections of HRP into the uvula-nodulus.

The observation that the caudal medial vestibular nucleus contains a subpopula- tion of cholinergic neurons that project to the cerebellum contradicts some earlier surveys based upon ChAT immunohistochemistry which used different antibodies to ChAT and different protocols.[49,50] These investigations failed to find a significant population of cholinergic neurons in the vestibular complex of the rat. More recently, another immunohistochemical survey has found ChAT-positive neurons in the caudal MVN of the rat.[26] These positive findings are consistent with radiochemical measurements of ChAT activity in tissue samples obtained from the vestibular nuclei using the "micropunch" technique.[51] The ChAT activity of the MVN was higher by at least a factor of 10 than was the ChAT activity of the SVN, DVN, or LVN. These data were interpreted to indicate the presence of a cholinergic projection to the MVN rather than the existence of cholinergic neurons within the MVN.[51] There is preliminary electrophysiological evidence from MVN slice recordings that suggests that MVN neurons may have muscarinic, but not nicotinic receptors.[52,53] So it is possible that both cholinergic axon terminals and cholinergic cell bodies account for the relatively high levels of ChAT activity in the MVN. Some of the ChAT-containing axons in the MVN may be collaterals of ChAT-containing neurons that project to the cerebellum.

The uvula-nodulus has the appropriate anatomical connections to play an important role in the control of posture using both vestibular and visual information. The cholinergic mossy fiber pathway described in the present experiment may be particularly important in regulating the sensitivity of the uvula-nodulus to its extraordinarily large vestibular primary afferent input. The uvula-nodulus has been implicated in the genesis of motion sickness, since removal of the uvula-nodulus in dogs decreases the likelihood of vestibularly induced vomiting.[54] In this regard it is of interest to recall that the most effective anti-motion-sickness agents are H_1 receptor blockers that have antimuscarinic actions[55] and the belladonna alkaloids, such as scopolamine, which are also antimuscarinics.[56,57]

SUMMARY

The cholinergic innervation of the cerebellar cortex of the rat was studied by immunohistochemical localization of choline acetyltransferase, radiochemical measurement of ChAT activity, and double labeling of ChAT-positive neurons with HRP injected into the cerebellum. ChAT immunohistochemistry revealed large mossy fiber rosettes as well as finely beaded terminals with different morphological characterization, laminar distribution within the cerebellar cortex, and regional differences within the cerebellum.

Large "grapelike" ChAT-positive mossy fiber rosettes that were distributed primarily in the granule cell layer were concentrated, but not exclusively located, in three separate regions of the cerebellum: (1) the uvula-nodulus (lobules 9 and 10); (2) the flocculus, and (3) the anterior lobe vermis (lobules 1 and 2).

Regional differences in ChAT-positive afferent terminations in the cerebellar cortex demonstrated by immunohistochemistry were confirmed by regional biochemical measurements of ChAT activity.

Using ChAT immunohistochemistry in combination with HRP injections into the uvula-nodulus, we have studied the origin of the cholinergic projection. The caudal medial vestibular nucleus and to a lesser extent the nucleus prepositus hypoglossus contain ChAT-positive neurons that were double labeled following HRP injections into the uvula-nodulus.

We conclude that (1) there is a prominent cholinergic mossy fiber pathway to the vestibulocerebellum, (2) this pathway originates primarily in the caudal third of the medial vestibular nucleus, and (3) this cholinergic pathway likely mediates secondary vestibular information related to postural adjustment.

REFERENCES

1. Fox, C. A., D. E. Hillman, K. A. Siegesmund & C. R. Dutta. 1967. The primate cerebellar cortex: a Golgi and electron microscopic study. In Progress in Brain Research: the Cerebellum. C. A. Fox & R. S. Snider, Eds.: 174–225. Elsevier. New York, N.Y.

2. Brodal, A. & P. A. Drablos. 1963. Two types of mossy fiber terminals in the cerebellum and their regional distribution. J. Comp. Neurol. 121: 173–187.

3. Mugnaini, E., R. L. Atluri & J. C. Houk. 1974. Fine structure of granular layer in turtle cerebellum with emphasis on large glomeruli. J. Neurophysiol. 37: 1–29.

4. Somogyi, P., K. Halasy, J. Somogyi, J. Storm-Mathisen & O. P. Ottersen. 1986. Quantification of immunogold labelling reveals enrichment of glutamate in mossy and parallel fibre terminals in cat cerebellum. Neuroscience 19: 1045–1050.

5. Ottersen, O. P. & J. Storm-Mathisen. 1987. Localization of amino acid neurotransmitters by immunocytochemistry. TINS 10: 250–255.

6. Dorman, R. V., M. A. Schwartz & D. M. Terrian. 1986. Prostaglandin involvement in the evoked release of D-aspartate from cerebellar mossy fiber terminals. Brain Res. Bull. 17: 243–248.

7. Morales, E. & R. Tapia. 1987. Neurotransmitters of the cerebellar glomeruli: uptake and

release of labeled γ-aminobutyric acid, glycine, serotonin and choline in a purified glomerulus fraction and in granular layer slices. Brain Res. **420:** 11–21.

8. SCHULMAN, J. A., T. E. FINGER, N. C. BRECHA & H. J. KARTEN. 1981. Enkephalin immunoreactivity in Golgi cells and mossy fibres of mammalian, avian, amphibian and teleost cerebellum. Neuroscience **6:** 2407–2416.

9. HAMORI, J. & J. TAKACS. 1989. Two types of GABA-containing axon terminals in cerebellar glomeruli of cat: an immunogold-EM study. Exp. Brain Res. **74:** 471–479.

10. TERRIAN, D. M., E. L. NOISIN & W. E. THOMAS. 1986. Choline uptake by glomerular synapses isolated from bovine cerebellar vermis. Brain Res. **366:** 401–404.

11. BOEGMAN, R. J., A. PARENT & R. HAWKES. 1988. Zonation in the rat cerebellar cortex: patches of high acetylcholinesterase activity in the granular layer are congruent with Purkinje cell compartments. Brain Res. **448:** 237–251.

12. MARANI, E. & J. VOOGD. 1977. An acetylcholinesterase band-pattern in the molecular layer of the cat cerebellum. J. Anat. **124:** 335–345.

13. KAMEI, T., T. NAGAI, P. L. MCGEER & E. G. MCGEER. 1983. Evidence of an intracerebellar acetylcholinesterase-rich but probably non-cholinergic flocculo-nodular projection. Brain Res. **258:** 115–119.

14. BROWN, W. J. & S. L. PALAY. 1972. Acetylcholinesterase activity in certain glomeruli and Golgi cells of the granular layer of the rat cerebellar cortex. Z. Anat. Entwicklungs gesch. **137:** 317–334.

15. CSILLIK, B., F. JOO & P. KASA. 1963. Cholinesterase activity of archicerebellar mossy fibre apparatuses. J. Histochem. Cytochem. **11:** 113–114.

16. KASA, P. & A. SILVER. 1969. The correlation between choline acetyltransferase and acetylcholinesterase activity in different areas of the cerebellum of rat and guinea pig. J. Neurochem. **16:** 389–396.

17. SHUTE, C. C. D. & P. R. LEWIS. 1965. Cholinesterase-containing pathways of the hindbrain: afferent cerebellar and centrifugal cochlear fibres. Nature **205:** 242–246.

18. GORENSTEIN, C., M. C. BUNDMAN, J. L. BRUCE & A. ROTTER. 1987. Neuronal localization of pseudocholinesterase in the rat cerebellum: sagittal bands of Purkinje cells in the nodulus and uvula. Brain Res. **418:** 68–75.

19. FONNUM, F. 1975. A rapid radiochemical method for the determination of choline acetyltransferase. J. Neurochem. **24:** 407–409.

20. ECKENSTEIN, F. P., R. W. BAUGHMAN & J. QUINN. 1988. An anatomical study of cholinergic innervation in rat cerebral cortex. Neuroscience **25:** 457–474.

21. ECKENSTEIN, F., Y. A. BARDE & H. THOENEN. 1981. Production of specific antibodies to choline acetyltransferase purified from pig brain. Neuroscience **6:** 993–1000.

22. OLUCHA, F., F. MARTINEZ-GARCIA & C. LOPEZ-GARCIA. 1985. A new stabilizing agent for the tetramethyl benzidine (TMB) reaction product in the histochemical detection of horseradish peroxidase (HRP). J. Neurosci. Methods **13:** 131–138.

23. RYE, D. B., C. B. SAPER & B. H. WAINER. 1984. Stabilization of tetramethylbenzidine (TMB) reaction product: application of retrograde and anterograde tracing, and combination with immunohistochemistry. J. Histochem. Cytochem. **32:** 1145–1153.

24. BARMACK, N. H., A. BURLEIGH, P. ERRICO & M. FAGERSON. 1991. A cholinergic pathway to the dorsal cap of the inferior olive of the rat. Soc. Neurosci. Abstr. **17:** 919.

25. KUNZLE, H. 1975. Autoradiographic tracing of the cerebellar projections from the lateral reticular nucleus in the cat. Exp. Brain Res. **22:** 255–266.

26. TAGO, H., P. L. MCGEER, E. G. MCGEER, H. AKIYAMA & L. B. HERSH. 1989. Distribution of choline acetyltransferase immunopositive structures in the rat brainstem. Brain Res. **495:** 271–297.

27. AZIZI, S. A., A. J. PAINCHAUD & D. J. WOODWARD. 1990. Mapping of choline acetyl transferase (ChAT) cells in the rat brain: possible evidence for cholinergic input to the cerebellum. Soc. Neurosci. Abstr. **16:** 1057.

28. OJIMA, H., S. KAWAJIRI & T. YAMASAKI. 1989. Cholinergic innervation of the rat cerebellum: qualitative and quantitative analyses of elements immunoreactive to a monoclonal antibody against choline acetyltransferase. J. Comp. Neurol. **290:** 41–52.

29. ILLING, R.-B. 1990. A subtype of cerebellar Golgi cells may be cholinergic. Brain Res. **522:** 267–274.

30. KELLAR, K. J., A. M. MARTINO, D. P. HALL, R. D. SCHWARTZ & R. L. TAYLOR. 1985.
     High-affinity binding of [3H]acetylcholine to muscarinic cholinergic receptors. J. Neu-
     rosci. **6:** 1577–1582.

31. ROTTER, A., N. J. M. BIRDSALL, P. M. FIELD & G. RAISMAN. 1979. Muscarinic receptors in
     the central nervous system of the rat. II. Distribution of binding of [3H]propylbenzilyl-
     choline mustard in the midbrain and hindbrain. Brain Res. Rev. **1:** 167–183.

32. RAVIKUMAR, B. V. & P. S. SASTRY. 1985. Cholinergic muscarinic receptors in human fetal
     brain: ontogeny of [3H]quinuclidinyl benzilate binding sites in corpus striatum, brain-
     stem, and cerebellum. J. Neurochem. **45:** 1948–1950.

33. NEUSTADT, A., A. FROSTHOLM & A. ROTTER. 1988. Topographical distribution of musca-
     rinic cholinergic receptors in the cerebellar cortex of the mouse, rat, guinea pig, and
     rabbit: a species comparison. J. Comp. Neurol. **272:** 317–330.

34. HUNT, S. & J. SCHMIDT. 1978. Some observations on the binding patterns of alpha-bungaro-
     toxin in the central nervous system of the rat. Brain Res. **157:** 213–232.

35. KORTE, G. & E. MUGNAINI. 1979. The cerebellar projection of the vestibular nerve in the
     cat. J. Comp. Neurol. **184:** 265–278.

36. DOW, R. S. 1936. The fiber connections of the posterior parts of the cerebellum in the rat
     and cat. J. Comp. Neurol. **63:** 527–547.

37. CARPENTER, M. B., B. M. STEIN & P. PETER. 1972. Primary vestibulocerebellar fibers in
     the monkey: distribution of fibers arising from distinctive cell groups of the vestibular
     ganglia. Am. J. Anat. **135:** 221–241.

38. BRODAL, A. & B. HOIVIK. 1964. Site and mode of termination of primary vestibulocerebel-
     lar fibres in the cat. Arch. Ital. Biol. **102:** 1–21.

39. GERRITS, N. M., A. H. EPEMA, A. VAN LINGE & E. DALM. 1989. The primary vestibulocer-
     ebellar projection in the rabbit: absence of primary afferents in the flocculus. Neurosci.
     Lett. **105:** 27–33.

40. CARLETON, S. C. & M. B. CARPENTER. 1984. Distribution of primary vestibular fibers in
     the brainstem and cerebellum of the monkey. Brain Res. **294:** 281–298.

41. BRODAL, A. & P. BRODAL. 1985. Observations on the secondary vestibulocerebellar
     projections in the macaque monkey. Exp. Brain Res. **58:** 62–74.

42. KOTCHABHAKDI, N. & F. WALBERG. 1978. Cerebellar afferent projections from the
     vestibular nuclei in the cat: an experimental study with the method of retrograde axonal
     transport of horseradish peroxidase. Exp. Brain Res. **31:** 591–604.

43. MAGRAS, I. N. & J. VOOGD. 1985. Distribution of the secondary vestibular fibers in the
     cerebellar cortex. Acta Anat. **123:** 51–57.

44. LANGER, T., A. F. FUCHS, C. A. SCUDDER & M. C. CHUBB. 1985. Afferents to the flocculus
     of the cerebellum in the Rhesus macaque as revealed by retrograde transport of
     horseradish peroxidase. J. Comp. Neurol. **235:** 1–25.

45. BLANKS, R. H. I., W. PRECHT & Y. TORIGOE. 1983. Afferent projections to the cerebellar
     flocculus in the pigmented rat demonstrated by retrograde transport of horseradish
     peroxidase. Exp. Brain Res. **52:** 293–306.

46. YAMAMOTO, M. 1979. Topographical representation in rabbit cerebellar flocculus for
     various afferent inputs from the brainstem investigated by means of retrograde axonal
     transport of horseradish peroxidase. Neurosci. Lett. **12:** 29–34.

47. SATO, Y., T. KAWASAKI & K. IKARASHI. 1983. Afferent projections from the brainstem to
     the three floccular zones in cats. II. Mossy fiber projections. Brain Res. **272:** 37–48.

48. MATSUSHITA, M. & N. OKADO. 1981. Cells of origin of brainstem afferents to lobules I and
     II of the cerebellar anterior lobe in the cat. Neurosci. **6:** 2393–2405.

49. ARMSTRONG, D. M., C. B. SAPER, A. I. LEVEY, B. H. WAINER & R. D. TERRY. 1983.
     Distribution of cholinergic neurons in rat brain: demonstrated by the immunocytochem-
     ical localization of choline acetyltransferase. J. Comp. Neurol. **216:** 53–68.

50. CARPENTER, M. B., L. CHANG, A. B. PEREIRA & L. B. HERSH. 1987. Comparisons of the
     immunocytochemical localization of choline acetyltransferase in the vestibular nuclei of
     the monkey and rat. Brain Res. **418:** 403–408.

51. BURKE, R. E. & S. FAHN. 1985. Choline acetyltransferase activity of the principal
     vestibular nuclei of rat, studied by micropunch technique. Brain Res. **328:** 196–199.

52. UJIHARA, H., A. AKAIKE, M. SASA & S. TAKAORI. 1988. Electrophysiological evidence for

cholinoceptive neurons in the medial vestibular nucleus: studies on rat brain stem in vitro. Neurosci. Lett. **93:** 231–235.

53. UJIHARA, H., A. AKAIKE, M. SASA & S. TAKAORI. 1989. Muscarinic regulation of spontaneously active medial vestibular neurons in vitro. Neurosci. Lett. **106:** 205–210.

54. MONEY, K. E. 1970. Motion sickness. Physiol. Rev. **50:** 1–39.

55. DOUGLAS, W. W. 1985. Histamine and 5-hydroxytryptamine (serotonin) and their antagonists. *In* Goodman and Gilman's the Pharmacological Basis of Therapeutics. A. G. Gilman, A. L. S. Goodman, T. W. Rall & F. Murad, Eds. 7th edit.: 605–638. Macmillan. New York, N.Y.

56. WEINER, N. 1985. Atropine, scopolamine, and related antimuscarinic drugs. *In* Goodman and Gilman's the Pharmacological Basis of Therapeutics. A. G. Gilman, A. L. S. Goodman, T. W. Rall & F. Murad, Eds. 7th edit.: 130–144. Macmillan. New York, N.Y.

57. KOHL, R. L. & J. L. HOMICK. 1983. Motion sickness: a modulatory role for the central cholinergic nervous system. Neurosci. Biobehav. Rev. **7:** 73–85.

58. LARSELL, O. 1970. The Comparative Anatomy and Histology of the Cerebellum. J. Jansen, Ed. University of Minnesota Press. Minneapolis, Minn.

Peripheral Vestibular Transmission[a]

STEPHEN L. COCHRAN[b]

Department of Life Sciences
Indiana State University
Terre Haute, Indiana 47809

INTRODUCTION

Synaptic transmission within the vestibular labyrinth is an integral part of vestibular reflex function. Comprehension of the mechanisms of transmission and the transmitter substances leads to a better understanding of vestibular function and in addition may provide some insight into perturbations that can occur in pathological situations involving the inner ear.

The membranous labyrinth is fragile and is encased in bone, making it quite difficult to access in a viable state for the purposes of electrophysiological investigation. This difficulty is particularly evident in mammals and other higher vertebrates, as the fragility of the labyrinth is greater due to the high metabolic rate consequent from the homeothermic state of these vertebrates. However, the phylogenetic similarities of the vestibular labyrinth across the vertebrate phylum allow investigations of more accessible (and less fragile) labyrinths of lower vertebrates, such as the frog, to be extrapolated and generalized to those of higher vertebrates.

The frog possesses a more primitive inner ear than is found in higher vertebrates such as reptiles, birds, and mammals.[1-3] Within the frog's labyrinth is only one type of hair cell (the type II hair cell), which acts as the sensory transducer. There are supporting cells such as endolymph secreters, glia, and others. The hair cells establish synaptic contact with VIIIth nerve afferents (FIGURE 1) that conduct the information communicated from the transducers to the brain. Centrally arising efferents, which synapse exclusively upon the hair cells (FIGURE 2), provide some sort of central nervous system feedback. Basically the frog labyrinth possesses a reduced number of components over the labyrinths of higher vertebrates, making interpretations based upon electrophysiological investigations less equivocal. In addition, the labyrinth of the frog can be maintained *in vitro* for many hours allowing stable, prolonged intracellular recordings from afferents and other components.[4,5] The *in vitro* capability also provides the opportunity to experimentally manipulate the extracellular environment of the inner ear in order to investigate ionic and pharmacological aspects of transmission between its components.

This review will focus upon what is known about the transmission properties and transmitter substances at the two known synapses present in the frog labyrinth: the hair cell–afferent fiber synapse (FIGURE 1) and the efferent to hair cell synapse (FIGURE 2). It will also discuss the possibilities for afferent to hair cell and hair cell to hair cell interaction.

[a] This research was supported through the National Science Foundation (BNS88-96165) and the National Aeronautics and Space Administration (NAG2-498) grants to S.L.C.
[b] Present affiliation: Department of Otolaryngology, Rt. J63, University of Texas Medical Branch at Galveston, Galveston, Texas 77555–1063.

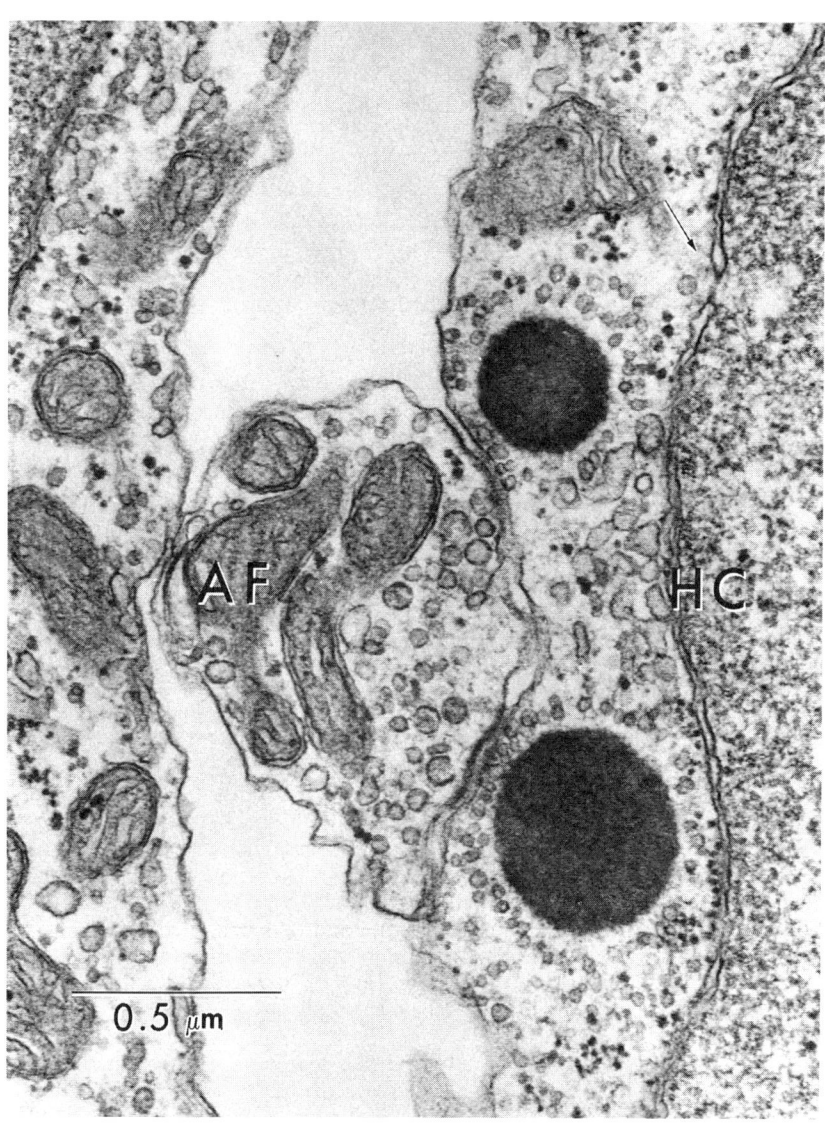

FIGURE 1. The hair cell–afferent fiber synapse of a bullfrog tadpole. A hair cell (HC) is shown establishing two contacts with a single afferent (AF). Two spherical dense bodies within the hair cell are rimmed with synaptic vesicles. There are clear pre- and postsynaptic thickenings at the active zones. Note also the presence of vesicles in the postsynaptic afferent.

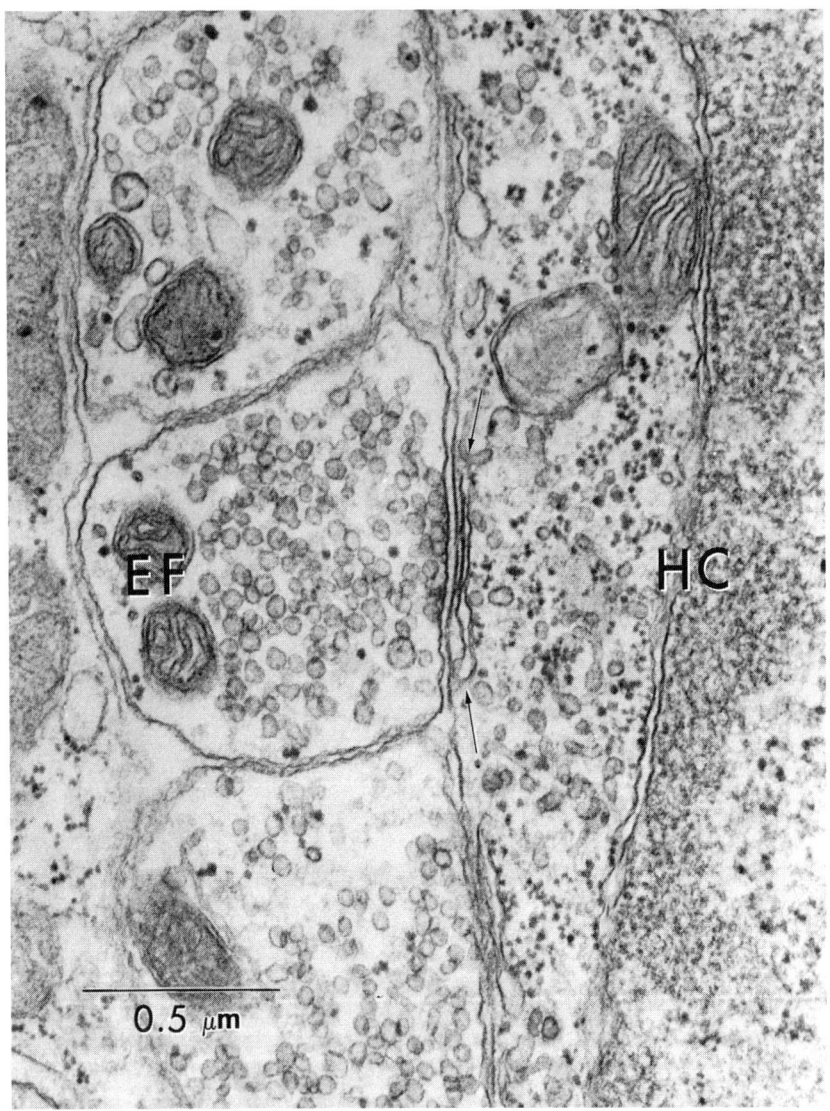

FIGURE 2. The efferent–hair cell synapse of a bullfrog tadpole. An efferent terminal (EF) is shown contacting a hair cell (HC). Note the presence of presynaptic vesicles in the efferent and the subsynaptic cistern (arrows) in the hair cell.

THE HAIR CELL–AFFERENT FIBER SYNAPSE

The pioneering work of Rossi and collaborators demonstrated that hair cells excite afferents,[4] in that small tetrodotoxin-insensitive potentials were found in intracellular recordings from canal afferents, and that the frequency of occurrence of these potentials was modulated by canal rotation. In the presence of high Mg^{+2}/low Ca^{+2} (to block transmitter release from hair cells), these potentials no longer were present, suggesting that they were in fact excitatory postsynaptic potentials (EPSPs) generated by release of transmitter from the hair cells. This conclusion was further strengthened by the finding that during the time course of the blockade of release by high Mg^{+2}/low Ca^{+2}, the amplitude of the potentials was constant, while their frequency of occurrence was unchanged,[5] indicating that the potentials were not the result of voltage (or ligand) dependent Ca^{+2} conductances, but were in fact chemically mediated EPSPs.

The Chemical Transmitter

Annoni and coworkers showed that the transmitter substance released from the hair cells resembled glutamate in its actions.[5] Both the endogenous transmitter and glutamate were shown to depolarize and excite afferents (FIGURES 3 and 4). Glutamate also was shown to depolarize the afferents when transmission between hair cell and afferent was blocked in high Mg^{+2}/low Ca^{+2}, indicating that glutamate acted independently from the Ca^{+2}-dependent depolarization-secretion coupling mechanisms of hair cell release of transmitter. This finding suggested that glutamate-gated channels were present in the postsynaptic membrane of the afferent. They also found that substances, reported in other systems to antagonize glutamate's actions and to antagonize excitatory synaptic transmission,[6] reduced glutamate-induced depolarizations (when hair cell release of transmitter was blocked in high Mg^{+2}/low Ca^{+2}) and reduced the amplitude of EPSPs in the afferent, without consistently affecting their frequency of occurrence. Since both the endogenous transmitter and glutamate depolarized the afferent, and since both actions were blocked by the same antagonists, it was suggested that glutamate could be the hair cell transmitter.

This criterion of identity of action of endogenous transmitter and exogenously applied transmitter candidate is a compelling piece of evidence for equating the endogenous substance with the candidate.[7] At the neuromuscular junction, for instance, that curare blocks end-plate potentials and exogenously applied acetylcholine is acceptable evidence supporting acetylcholine as the motoneuronal transmitter.[8] But this evidence is only so strong because there are no other known endogenous compounds that resemble acetylcholine in their actions. Such exclusivity is not applicable to glutamate. There are a plethora of compounds that have agonistic actions similar to glutamate and that are present in nerve cells including aspartate, homocystic acid, cystine sulfinic acid, and cystic acid, to name a few.[6] All of these compounds resemble the endogenous hair cell's transmitter in their ability to depolarize afferents, and all can be blocked by the same antagonists.[5,9,10] There are some who dispute that glutamate itself is the actual transmitter.[11-13] Therefore, the transmitter released by the hair cells may be glutamate, aspartate, or other such compounds, or compounds yet to be identified,[14] or there may be actually a number of such compounds released from the same hair cell. One can only refer to this synapse as of the glutamatergic genre in the sense referred by Dale,[15] until more definitive experimentation is performed.

The Subsynaptic Receptor

The availability of new acidic amino acid agonists and antagonists has led to the classification of three apparent types of "glutamate" receptor.[6] Because these

d:f07c2-.par nepsp= 11671

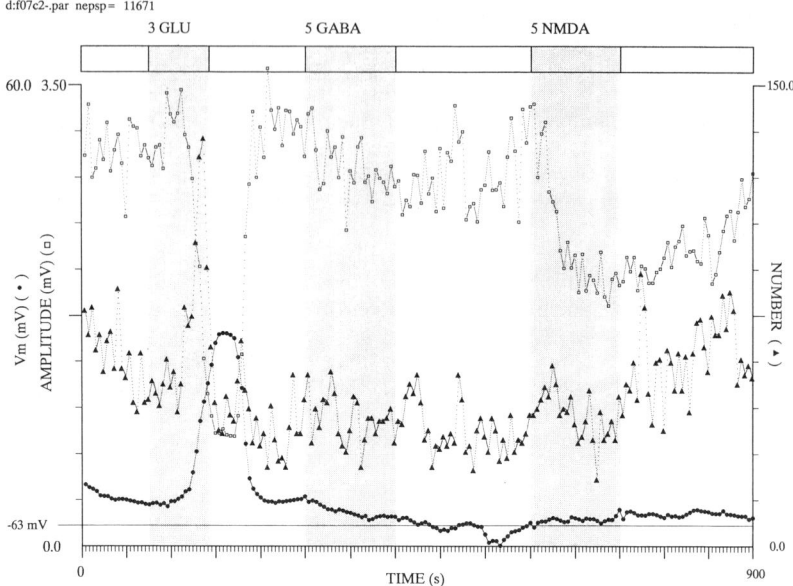

FIGURE 3. The effect of various pharmacological agents upon the activity of a vestibular afferent in the isolated frog labyrinth. Each data point represents a measurement from EPSPs detected from continuously sampled data (digitized at 50 kHz throughout the duration of the recording) over a five-second interval. Open squares depict the amplitude of EPSPs recorded in the afferent (action potentials have been excluded from these measurements). Filled triangles indicate the number of EPSPs occurring in the interval. Filled circles represent the mean transmembrane potential of the afferent relative to the indicated baseline (-63 mV). At 1 minute and 30 seconds into the recording 3 mM glutamate (3 GLU) was introduced into the solution bathing the labyrinth (see Reference 5) for the period of time indicated by the shading. The frequency of EPSPs increased and the cell depolarized as a result of the glutamate. The amplitude of the EPSPs also decreased due to the increased conductance of the postsynaptic membrane. These effects were reversed upon removal of the glutamate from the bath. In contrast to the effect of glutamate, neither 5 mM GABA (shaded region—5 GABA) nor 5 mM NMDA (shaded region—5 NMDA—the normal saline contains no added Mg^{+2}) influenced afferent activity when added to the bath. In this instance NMDA reversibly decreased EPSP amplitude.

pharmacological substances are relatively new and new substances are being synthesized, these classifications are still being established. At present there are three types of receptor. The N-methyl-D-aspartate (NMDA) receptor has a high affinity for NMDA [over quisqualate (QUIS) and kainate (KA)] and is selectively antagonized by low concentrations of D,2-amino-5-phosphonovaleric acid (D-APV). The QUIS receptor has a high affinity for QUIS over NMDA and KA, and is antagonized by low concentrations of 6-cyano-7-nitro quinoxaline-2,3-dione (CNQX).[16] The KA receptor has a high affinity for KA and is antagonized to a somewhat lesser degree than is the QUIS receptor by CNQX. At many synapses it is difficult to separate the actions of QUIS over those of KA. Kynurenic acid (KENYA) and a number of other antagonists have a concentration-dependent specificity of action, in that low concentrations are able to selectively block the NMDA receptor, while higher concentrations block all three receptors.[17] In many cases, Mg^{+2} ions in low millimolar concentrations block the channel coupled with the NMDA receptor in a voltage-dependent manner.[18] Again it should be stressed that such classifications as yet are still in the formative stages.

With respect to the characterization of the subsynaptic receptor at the hair cell–afferent fiber synapse (in the frog), it is clear that this receptor is not of the NMDA subtype.[9,10] NMDA depolarizes afferents inconsistently (FIGURE 3) and only at very high, millimolar concentrations, much higher than are necessary to activate this receptor in other systems. Both KA and QUIS are able to depolarize the afferents,[9] suggesting that the receptor could be of either subtype. However, both D-APV and CNQX reduce the amplitudes of EPSPs in the afferents. The concentrations required to block are, however, 10 to 50 times higher than are needed in other systems. Since D-APV at low micromolar concentrations has a selectivity for the NMDA receptor over the others, and since CNQX at low micromolar concentrations has a selectivity for the QUIS receptor, the fact that higher concentrations are needed to block at this synapse suggests that the hair cell–afferent fiber subsynaptic receptor may be of the KA subtype. Recently an antibody to the KA receptor has been generated.[19] This antibody has been shown to bind selectively to the subsynaptic receptor at the hair cell–afferent fiber synapse in the frog,[20] a finding that would confirm the electrophysiological conclusions that the subsynaptic receptor is of the KA type. Unfortunately, this antibody does not appear to bind to the KA-binding site, since it binds intracellularly and since it does not displace KA binding.[19] This antibody is consequently probably useless as a possible antagonist for electrophysiology.

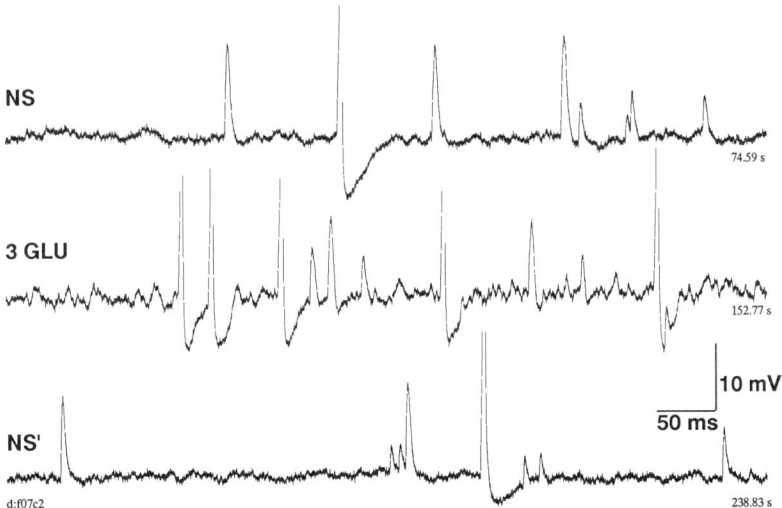

FIGURE 4. The effect of glutamate on afferent activity. Intracellular recordings from the same cell as shown in FIGURE 3. The top trace shows the afferent activity in normal saline (NS). Following the addition of 3 mM glutamate (3 GLU), EPSP and action potential frequency increased. After washout of glutamate and return to normal saline (NS'), afferent activity returned to control values. Each trace represents 32,765 points sampled at 50 KHz (i.e., 0.655 second). Numbers below each trace indicate the time after the beginning of the recording (0 in FIGURE 3). Action potentials have been truncated.

The Subsynaptic Iontophore

Since during the time course of blockade of release of transmitter in high Mg^{+2}/low Ca^{+2}, the amplitude of the EPSPs is unchanged,[5] Ca^{+2} is probably not the ion that enters the afferent to produce the depolarizing EPSP. More recent evidence in fact shows that lowering the concentration of Ca^{+2} bathing the *in vitro* labyrinth actually increases EPSP amplitude, while increasing the extracellular Ca^{+2} concentration results in a decrease in EPSP amplitude[21] (FIGURE 5). Rather, the subsynaptic iontophore most likely increases the afferent membrane's permeability to Na^+, since increasing the extracellular Na^+ concentration results in an increased amplitude of the EPSPs (probably consequent from the higher Na^+ equilibrium potential resultant from the higher extracellular Na^+ concentration, FIGURE 6). Ca^{+2} ions may to some extent block this channel, which would explain the modulation of EPSP amplitude by altered Ca^{+2} concentrations.

THE EFFERENT–HAIR CELL SYNAPSE

The role of the hair cell–afferent fiber synapse is clear. It is to pass on the sensory information to the central nervous system. The role of the centrally arising efferent–hair cell synapse remains a mystery. What the purpose is of the central command to the hair cell has yet to be determined. Investigations upon the mechanisms of

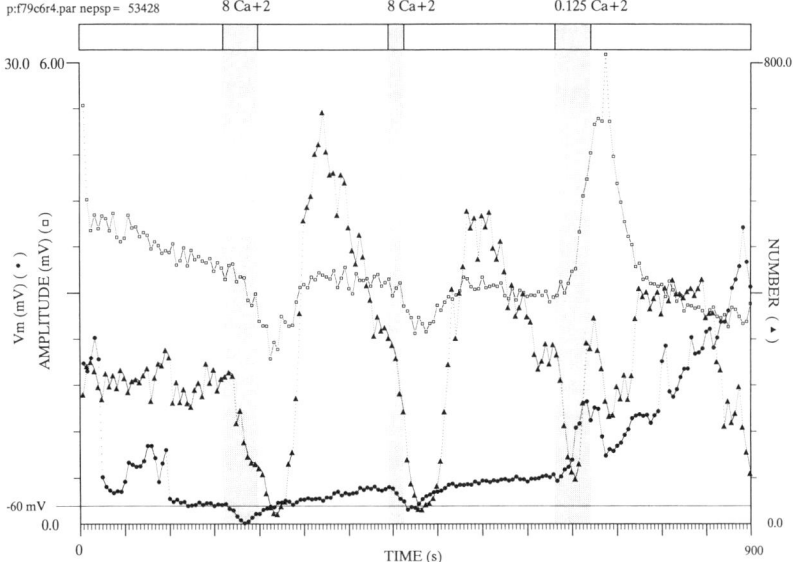

FIGURE 5. The contribution of Ca^{+2} to afferent activity. Data points are as indicated in FIGURE 3. The substitution of 8 mM Ca^{+2} (8 Ca^{+2}—shaded regions) over the normal 2 mM Ca^{+2} resulted in a reversible decrease in the amplitude and frequency of occurrence of the EPSPs. There was also a rebound increase in EPSP frequency in this cell. In contrast, reducing the Ca^{+2} concentration from 2 mM to 0.125 mM (0.125 Ca^{+2}) resulted in a reversible increase in the amplitudes of the EPSPs and a decrease in their frequency of occurrence.

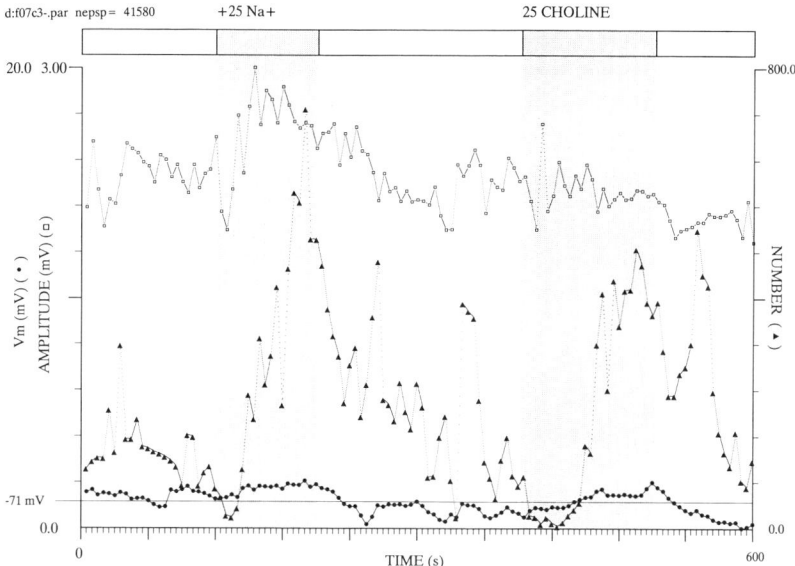

FIGURE 6. The contribution of Na^+ to afferent activity. Data points are as indicated in FIGURE 3. The addition of 25 mM Na^+ (+25 Na^+—shaded regions) to the normal 100 mM Na^+ resulted in a reversible increase in the amplitudes of the EPSPs, suggesting that the transmitter opens a subsynaptic Na^+ channel. In this cell there was also an increase in the frequency of occurrence of the EPSPs. As a control experiment, the addition of 25 mM Choline (25 CHOLINE) did not affect the amplitude of the EPSPs, although it did reversibly increase their frequency of occurrence.

transmission at this synapse have shed some light as to the possible purpose of this innervation.

In many systems, activation of the efferents has been shown to hyperpolarize hair cells and decrease the amplitude of the receptor potential (through increasing the overall conductance of the hair cell).[22] This reduction in the amplitude of the receptor potential ultimately results in a reduction in the gain of the vestibuloocular reflex.[23] What is somewhat puzzling however is that during intracellular recordings from vestibular afferents in the frog *in vitro*, electrical stimulation of the cut VIIIth nerve results in either an increase in EPSP frequency or a decrease in EPSP frequency[24,25] (that these effects were mediated by the efferents was determined by cutting the VIIIth nerve several days prior to isolation of the labyrinth, thereby allowing the efferents to degenerate, and demonstrating that neither effect then occurred with such electrical stimulation). Only a decrease in EPSP frequency would be expected if efferents universally hyperpolarized hair cells. How then can an increase in EPSP frequency be explained? Prior to answering this question, it is necessary to understand the nature of the mechanisms of transmission at this synapse.

It has been shown in all vertebrates investigated and in all efferent to hair cell contacts (e.g., lateral line, vestibular hair cells, cochlear hair cells) that the transmitter is acetylcholine (cf. References 26 and 27). In the frog both electrical stimulation and application of nicotinic agonists can result in either an increase or decrease in

the frequency of occurrence of EPSPs recorded intracellularly from afferents without changing EPSP amplitude.[25] Prolonged electrical stimulation or prolonged bath
application of nicotinic agonists results in a desensitization of the response, much as
is seen for the nicotinic receptor at the neuromuscular junction. Curare blocks both
the increase and decrease in EPSP frequency induced by either electrical stimulation
or by bath application of nicotinic agonists. Similarly, prior application of nicotinic
agonists blocks the effect of electrical stimulation, most likely by desensitizing the
receptor. Bath application of muscarinic agonists results in a slow increase in afferent
EPSP frequency that is antagonized by atropine. There is no effect of cholinergic
agonists on afferent activity when the hair cell–afferent synapse is blocked by high
Mg^{+2}/low Ca^{+2}, indicating that all cholinergic agents affect afferent activity indirectly
by modifying the membrane potential of the hair cell.

Cholinergic receptors on the hair cells are both nicotinic and muscarinic. The
nicotinic receptor most likely opens a K^+ channel, and consequently the membrane
potential of the hair cell approaches the K^+ equilibrium potential.[22] The muscarinic
receptor, by virtue of its apparent exclusive action to slowly increase afferent EPSP
frequency, probably acts to decrease the resting K^+ conductance of the hair cell in a
similar manner as has been shown in other systems.[28] The nicotinic action clearly
predominates as it is difficult to demonstrate a muscarinic effect with the electrical
stimulation of the efferents. Because it is slowly developing, the muscarinic component at this synapse might only be evident following prolonged efferent activation.

If the efferents act to increase the resting K^+ conductance of the hair cells
through a cholinergic, nicotinic synapse, the question remains, how can an increase
in the resting K^+ conductance of the hair cell increase the release of transmitter from
the hair cell? The extent of transmitter release from a nerve terminal (or from the
hair cell in this case) is dependent upon the concentration of intracellular Ca^{+2},
which in turn is dependent upon the number of open Ca^{+2} channels in the membrane, which in turn is dependent upon the level of polarization of the cell, such that
depolarization of the cell results in an increase in the number of open Ca^{+2} channels
and consequently an increase in transmitter release.[8] In other words, if the hair cell
increases release of transmitter, then it must have been depolarized. Therefore, in
the situations when efferent (or nicotinic agonist) activation increases EPSP frequency in the afferents, the resting potential of the hair cells innervating that afferent
must have been more negative than the equilibrium potential for the channel opened
by acetylcholine. Bernard and coworkers postulated that in the *in vitro* situation
(illustrated in FIGURE 7), the endolymphatic compartment became disrupted and
the normally high concentration of K^+ in the endolymph was replaced by the high
Na^+ present in the perfusate (or artificial perilymph).[25] Such an alteration in the
endolymph ionic composition should in fact result in a significant hyperpolarization
of the hair cells (cf. References 29 and 30), and consequently they could be
depolarized when the resting K^+ conductance was increased by efferent (nicotinic)
activation. Supportive of such a hypothesis was the finding that when the canal alone
was isolated over the entire, intact labyrinth, there was a greater incidence of
efferents and nicotinic agents producing an increase in afferent EPSP frequency,
while in the intact labyrinth, there was a greater incidence to produce a decrease in
afferent EPSP frequency.

From such investigations one can speculate as to the possible purposes of the
central command given by the efferents. If the stereocilia are deflected to such an
extent that either transmitter release is stopped (i.e., the cilia are deflected in the
hyperpolarizing direction so much so that the hair cell is maximally hyperpolarized—
resembling the state of the hair cell depicted in FIGURE 7B) or transmitter release is
maximal (i.e., the cilia are deflected in the depolarizing direction to a maximal

extent), increasing the resting K^+ conductance of the hair cell through the efferent nicotinic receptor would result in bringing these out-of-range signals back into a membrane-potential zone where additional cilial deflections would be able to continue to modulate transmitter release. Accompanying such a "recalibration" would be a decrease in the amplitude of the receptor potential resulting from a given

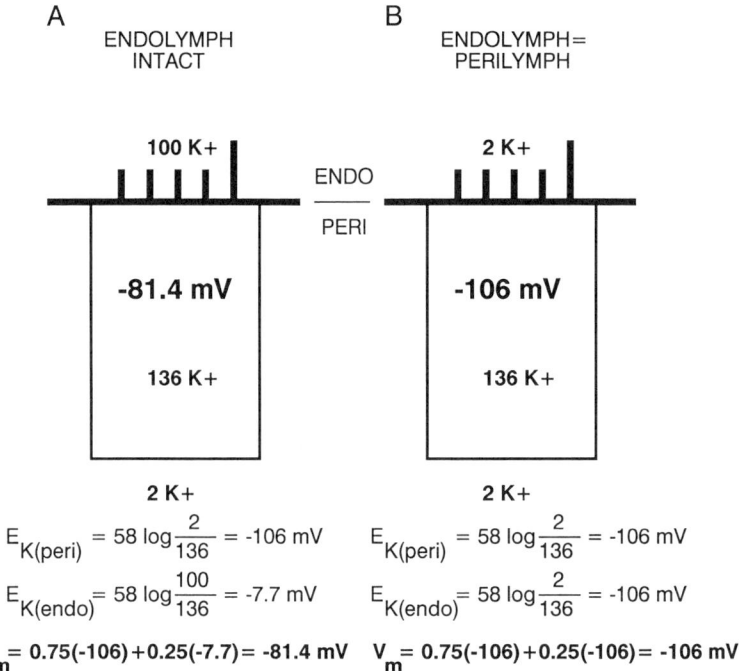

A
ENDOLYMPH
INTACT

B
ENDOLYMPH=
PERILYMPH

100 K+

ENDO

PERI

-81.4 mV

136 K+

2 K+

2 K+

-106 mV

136 K+

2 K+

$$E_{K(peri)} = 58 \log \frac{2}{136} = -106 \text{ mV}$$

$$E_{K(endo)} = 58 \log \frac{100}{136} = -7.7 \text{ mV}$$

$$V_m = 0.75(-106) + 0.25(-7.7) = -81.4 \text{ mV}$$

$$E_{K(peri)} = 58 \log \frac{2}{136} = -106 \text{ mV}$$

$$E_{K(endo)} = 58 \log \frac{2}{136} = -106 \text{ mV}$$

$$V_m = 0.75(-106) + 0.25(-106) = -106 \text{ mV}$$

FIGURE 7. Model of the influence of high endolymphatic K^+ on the resting membrane potential of the hair cell. In **A,** with the endolymphatic compartment intact, the resting membrane potential of the hair cell is -81.4 mV. In **B,** when the endolymph is washed out by perilymph, the hair cell hyperpolarizes to -106 mV. $E_{K(peri)}$ represents the potassium ion equilibrium potential across the perilymphatic compartment, while $E_{K(endo)}$ represents the potassium ion equilibrium potential across the endolymphatic compartment. The model assumes an insignificant contribution of sodium ion conductance to the resting state. Conductance ratios (0.25 across the endolymphatic compartment and 0.75 across the perilymphatic compartment) and intracellular concentrations are based upon references 29 and 30. Possible contributions to the resting membrane potential by electrogenic pumps are ignored. If the equilibrium potential for the channel opened by acetylcholine is between -81.4 mV and -106 mV, then in A, acetylcholine will hyperpolarize the hair cell and decrease transmitter release, while in B, acetylcholine will depolarize the hair cell and increase transmitter release.

stereocilial deflection due to the overall increase in the resting conductance of the hair cell. The reduction would then in turn be reflected in a decrease in the gain of the vestibuloocular reflex. Supportive of such a speculation are the findings of Goldberg and Fernandez that regardless of whether efferent activation results in an increase or decrease in afferent activity, there is a reduction of the gain of the vestibuloocular reflex in both circumstances.[23]

OTHER SYNAPSES(?)

Dunn has reported finding synaptic profiles from afferents to hair cells in the bullfrog[31] as have been also found in humans from afferents to outer cochlear hair cells.[32] These ultrastructural findings have suggested that afferent activity may effect hair cell activity. Electrophysiological support for such interaction is weak. Hair cells do increase the release of transmitter in response to bath application of agonists of the KA and QUIS receptors (but not NMDA)[5,9,10,12] (see also FIGURE 3). Since the afferent transmitter is glutamatergic (at its central termination),[17] if the afferent were to release transmitter upon the hair cell, one might expect that the released compound would be a glutamatelike substance. Thus afferents should depolarize hair cells and a positive feedback reciprocal synapse would be in action. However, after cutting the VIIIth nerve centrally and allowing the efferent fibers to degenerate, no alteration in afferent activity was seen following stimulation of the VIIIth nerve stump.[25] Furthermore, if such a positive feedback situation existed, one would expect to see an increase in the probability of an EPSP occurring after another EPSP or action potential occurred in the afferent. There is no such increase, however, as indicated by either statistical correlation or by just averaging action potentials[33] (FIGURE 8). It is unlikely then that there is a significant feedback interaction from the afferent to the hair cell, at least in terms of electrical function. If such a functional contact does exist, it may have more of a metabolic or maintenance nature. Alternatively, the contact may have some electrical function, but only have significance during prolonged excitation of the afferent.

Since the hair cells release an agonist of the KA or QUIS receptor, and since they have receptors for KA or QUIS,[9,10,33] it is likely that there is interaction between hair cells. The extent of interaction will be dependent upon how fast the reuptake of the transmitter is. The amphibian central nervous system has an extremely active uptake system,[34] but the actual details of such a system in the labyrinth are unclear. The extent of the possible hair cell to hair cell interaction remains to be determined.

In some vertebrates, efferents innervate afferents.[35] No such contacts have been reported in the frog, in either anatomical or electrophysiological investigations.

γ-Aminobutyric Acid

One of the most perplexing issues about transmission in the labyrinth is the role of γ-aminobutyric acid (GABA). Because GABA was found to increase the firing of cochlear afferents, it was proposed that GABA was the hair cell transmitter.[36] However, GABA does not always affect afferent activity (e.g., FIGURE 3) and does not depolarize afferents when depolarization secretion coupling is blocked in high Mg^{+2}/low Ca^{+2}, indicating that GABA acts presynaptically upon the hair cells to increase transmitter release.[5] It was suggested then that GABA could be a transmitter of a subpopulation of efferents. This hypothesis was strengthened by the finding that glutamic acid decarboxylase (GAD, the synthetic enzyme for GABA) was present in the labyrinth, albeit in a lower amount than is found for choline acetyltransferase (ChAT, the synthetic enzyme for acetylcholine).[37] In addition, GABA binds to a GABA-A receptor in the labyrinth.[38] However, when the VIIIth nerve was cut, thereby allowing the efferents to degenerate, there was a concomitant decrease in ChAT levels, but no change in GAD levels.[37] Consequently GABA is not the afferent transmitter[5,39,40] and not an efferent transmitter.[25,37] Recently it has been shown in mammals that both the afferent calyceal nerve endings[41,42] and the type I and type II hair cells[42] display GABA-like immunoreactivity. Both the afferent nerve

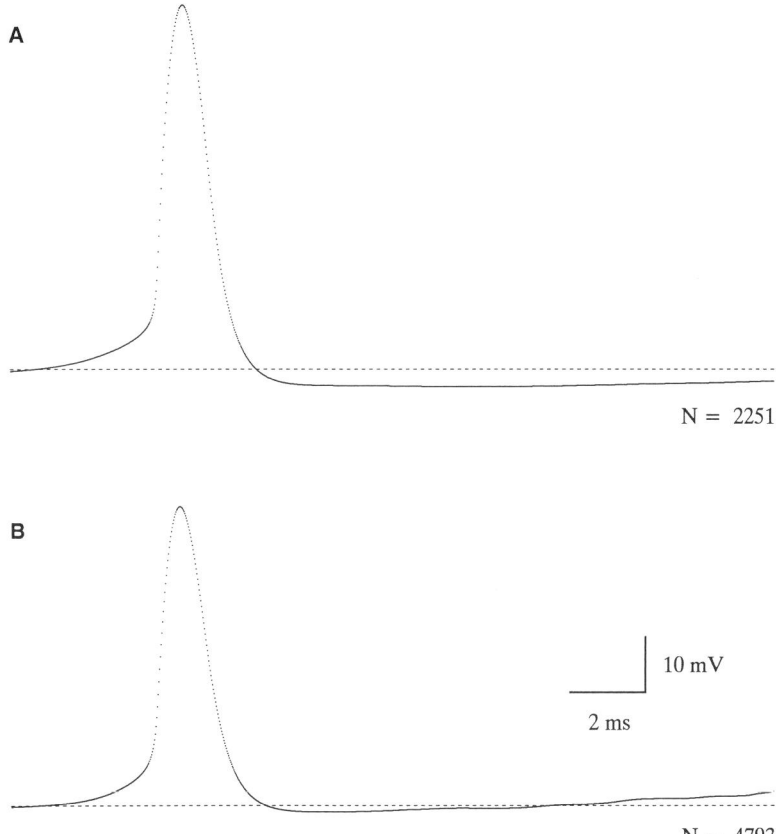

A

N = 2251

B

10 mV

2 ms

N = 4793

FIGURE 8. Averaged action potentials for one cell. In **A,** 2251 action potentials are shown averaged over a 2.5-minute period. In **B,** 3 mM glutamate was applied to the bath to increase the firing rate of the cell, and 4793 action potentials were averaged over the same period. Note the flat baseline in A following the action potential indicating no increase in excitability of the afferent subsequent to action potential generation. Although there is some increase in excitability in B, this increase is quite modest, and occurs at a latency much later than would be predicted from a disynaptic circuit.

endings and the hair cells also show glutamatelike immunoreactivity.[43] At present one can only speculate that GABA may be a feedback transmitter from the afferent to the hair cell. Because such feedback is difficult to demonstrate electrophysiologically, it is expected that this proposed interaction is weak.

ACKNOWLEDGMENTS

Electron microscopy was performed at the Department of Physiology at the University of Virginia Medical School. The author thanks Dr. John T. Hackett and Ms. Susan Purdy-Ramos for their support and advice.

REFERENCES

1. WERSALL, J., A. FLOCK & P. G. LUNDQUIST. 1965. Structural basis for directional sensitivity in cochlear and vestibular sensory receptors. Cold Spring Harbor Symp. Quant. Biol. **30:** 115–132.
2. GLEISNER, L., A. FLOCK & J. WERSALL. 1973. The ultrastructure of the afferent synapse on hair cells in the frog labyrinth. Acta Otolaryngol. **76:** 199–207.
3. HILLMAN, D. 1976. Morphology of peripheral and central vestibular systems. *In* Frog Neurobiology. R. Llinas & W. Precht, Eds.: 452–480. Springer-Verlag. New York, N.Y.
4. ROSSI, M. L., P. VALLI & C. CASELLA. 1977. Post-synaptic potentials recorded from afferent nerve fibers of the posterior semicircular canal in the frog. Brain Res. **135:** 67–75.
5. ANNONI, J.-M., S. L. COCHRAN & W. PRECHT. 1984. Pharmacology of the hair cell–afferent fiber synapse in the vestibular labyrinth of the frog. J. Neurosci. **4:** 2106–2116.
6. WATKINS, J. C. & R. H. EVANS. 1981. Excitatory amino acid transmitters. Annu. Rev. Pharmacol. Toxicol. **21:** 165–204.
7. ECCLES, J. C. 1964. The Physiology of Synapses. Springer-Verlag. New York, N.Y.
8. KATZ, B. 1966. Nerve, Muscle, and Synapse. McGraw-Hill. New York, N.Y.
9. COCHRAN, S. L. 1989. CNQX (6-cyano-7-nitro quinoxaline-2,3-dione) blocks excitatory synaptic transmission in the vestibular periphery and vestibular nucleus of the frog. Soc. Neurosci. Abstr. **16:** 502.
10. PRIGIONI, I., G. RUSSO, P. VALLI & S. MASETTO. 1990. Pre- and postsynaptic excitatory action of glutamate agonists on frog vestibular receptors. Hearing Res. **46:** 253–260.
11. NAGAI, T., S. OBARA & N. KAWAI. 1984. Differential blocking effects of a spider toxin on synaptic and glutamate responses in the afferent synapse of the acoustic-lateralis receptors of *Plotosus.* Brain Res. **300:** 183–187.
12. VALLI, P., G. ZUCCA, I. PRIGIONI, L. BOTTA, C. CASELLA, & P. S. GUTH. 1985. The effect of glutamate on the frog semicircular canal. Brain Res. **330:** 1–9.
13. GUTH, P. S., C. H. NORRIS & S. E. BARRON. 1988. Three tests of the hypothesis that glutamate is the sensory hair cell transmitter in the frog semicircular canal. Hearing Res. **33:** 223–228.
14. SEWELL, W. F. & E. A. MROZ. 1990. Purification of a low-molecular weight excitatory substance from the inner ear of goldfish. Hearing Res. **50:** 127–138.
15. DALE, H. H. 1933. Nomenclature of fibers in the autonomic nervous system and their effects. J. Physiol. **80:** 10–11.
16. HONORE, T., S. N. DAVIES, J. DREIER, E. J. FLETCHER, P. JACOBSEN, D. LODGE & F. E. NIELSEN. 1988. Quinoxalinediones: potent competitive non-NMDA glutamate receptor antagonists. Science **241:** 701–703.
17. COCHRAN, S. L., P. KASIK & W. PRECHT. 1987. Pharmacological aspects of excitatory synaptic transmission to second order vestibular neurons in the frog. Synapse **1:** 102–123.
18. ASCHER, P. & L. NOWAK. 1988. The role of divalent cations in the N-methyl-D-aspartate responses of mouse central neurones in culture. J. Physiol. **399:** 247–266.
19. HAMPSON, D. R., K. D. WHEATON, C. J. DECHESNE & R. J. WENTHOLD. 1989. Identification and characterization of the ligand binding subunit of a kainic acid receptor using monoclonal antibodies and peptide mapping. J. Biol. Chem. **264:** 13329–13335.
20. DECHESNE, C. J., K. WHEATON, G. GOPING, D. R. HAMPSON & R. J. WENTHOLD. 1988. Kainic acid receptor identified in the frog labyrinth using monoclonal and polyclonal antibodies. Soc. Neurosci. Abstr. **14:** 800.

21. COCHRAN, S. L. 1990. Sodium, calcium and potassium ion influences upon transmission at the hair cell–afferent fiber synapse in the frog. Soc. Neurosci. Abstr. **16**: 968.
22. ASHMORE, J. F. & I. J. RUSSELL. 1982. Effect of efferent nerve stimulation on hair cells of the frog sacculus. J. Physiol. **329**: 25–26P.
23. GOLDBERG, J. & C. FERNANDEZ. 1980. Efferent vestibular system in the squirrel monkey: anatomical location and influence on afferent activity. J. Neurophysiol. **43**: 986–1025.
24. ROSSI, M. L., I. PRIGIONI, P. VALLI & C. CASELLA. 1980. Activation of the efferent system in the isolated frog labyrinth: effects on the afferent EPSPs and spike discharge recorded from single fibers of the posterior nerve. Brain Res. **180**: 125–137.
25. BERNARD, C., S. L. COCHRAN & W. PRECHT. 1984. Pharmacology of the hair cell–afferent fiber synapse in the vestibular labyrinth of the frog. Brain Res. **338**: 225–236.
26. GUTH, P. S., C. H. NORRIS & R. P. BOBBIN. 1976. The pharmacology of transmission in the peripheral auditory system. Pharmacol. Rev. **28**: 95–125.
27. BLEDSOE, S. C., R. P. BOBBIN & J.-L. PUEL. 1988. Neurotransmission in the inner ear. *In* Physiology of the Ear. A. F. Jahn & J. Santos-Sacchi, Eds.: 385–405. Raven Press. New York, N.Y.
28. WEIGHT, F. F. & J. VOTAVA. 1970. Slow synaptic excitation in sympathetic ganglion cells: Evidence for synaptic inactivation of potassium conductance. Science **170**: 755–758.
29. BRACHO, H. & R. BUDELLI. 1978. The generation of resting membrane potentials in an inner ear hair cell system. J. Physiol. **281**: 445–465.
30. BRACHO, H., R. BUDELLI & F. GALEY. 1981. Ionic mechanisms in the vestibular apparatus: the resting state. *In* The Vestibular System: Function and Morphology. T. Gualtierotti, Ed.: 144–160. Springer-Verlag. New York, N.Y.
31. DUNN, R. F. 1980. Reciprocal synapses between hair cells and first order afferent dendrites in the crista ampullaris of the bullfrog. J. Comp. Neurol. **193**: 255–264.
32. NADOL, J. B., JR. 1990. Synaptic morphology of inner and outer hair cells of the human organ of Corti. J. Electron Microsc. Tech. **15**: 187–196.
33. COCHRAN, S. L. Do labyrinthine afferents excite hair cells. Soc. Neurosci. Abstr. **17**: 312.
34. SHANK, R. P., J. T. WHITEN & C. F. BAXTER. 1973. Glutamate uptake by the isolated toad brain. Science **181**: 860–862.
35. HIGHSTEIN, S. M. 1992. The vestibular efferent system. Ann. N.Y. Acad. Sci. (This volume.)
36. FELIX, D. & D. EHRENBERGER. 1982. The action of putative neurotransmitter substances in the cat labyrinth. Acta Otolaryngol. **93**: 101–105.
37. LOPEZ, I. & G. MEZA. 1988. Neurochemical evidence for afferent gabaergic and efferent cholinergic neurotransmission in the frog vestibule. Neuroscience **25**: 13–18.
38. MEZA, G., M. T. GONZALEZ-VIVEROS & M. RUIZ. 1985. Specific [3H]-gamma-aminobutyric acid binding to vestibular membranes of the chick inner ear. Brain Res. **337**: 179–183.
39. GUTH, S. L. & C. H. NORRIS. 1984. Pharmacology of the isolated semicircular canal: effect of GABA and picrotoxin. Exp. Brain Res. **56**: 72–78.
40. VEGA, R., E. SOTO, R. BUDELLI & M. T. GONZALEZ-ESTRADA. 1987. Is GABA an afferent transmitter in the vestibular system? Hearing Res. **29**: 163–167.
41. DIDIER, A., J. DUPONT & Y. CAZALS. 1990. GABA immunoreactivity of calyceal nerve endings in the vestibular system of the guinea pig. Cell Tiss. Res. **260**: 415–419.
42. LOPEZ, I, J. M. JUIZ, R. A. ALTSCHULER & G. MEZA. 1990. Distribution of GABA-like immunoreactivity guinea pig vestibular cristae ampullaris. Brain Res. **530**: 170–175.
43. DEMEMES, D., R. J. WENTHOLD, B. MONIOT & A. SANS. 1990. Glutamate-like immunoreactivity in the peripheral vestibular system of mammals. Hearing Res. **46**: 261–270.

Neurotransmitters in the Vestibular Commissural System of the Cat

NOBUHIKO FURUYA, TAKAO YABE,
AND TATUROU KOIZUMI[a]

Department of Otolaryngology
Teikyo University School of Medicine
Kaga, 2-11-2
Itabashiku
Tokyo 173, Japan

[a]*Nishiarai Hospital*
Nisiarai Hon-Mati 5-7-14
Adati-ku
Tokyo 123, Japan

INTRODUCTION

The role of mutual inhibition of bilateral vestibular type 1 neurons in turning up the gain of sensitivity to horizontal angular acceleration[1] is well known. This is referred to as vestibular commissural inhibition. Type 1 vestibular neurons have been shown to be inhibited by stimulation of the contralateral vestibular nerve in the cat. The mechanism producing this inhibition consists of postsynaptic inhibition via intercalated inhibitory neurons. It assumed that type 2 neurons are intercalated inhibitory neurons.[2] According to anatomical findings on commissural inhibition, the fibers are located in the superficial ventral portion of the brain stem.[3,4] Physioanatomical investigations of commissural inhibition have been intensively pursued,[2-7] but the corresponding neurotransmitters have yet to be identified. Several investigations using anticholinergic[8-10] and antihistaminic agents[11] have suggested a role for acetylcholine (Ach) (excitatory transmitter) and histamine (inhibitory transmitter) in vestibular neurotransmission. Recently, combined immunohistochemical and electron microscopic techniques have suggested that glutamate is one of the excitatory neurotransmitters on vestibular neurons.[12]

In addition to the above studies, there are a few reports in the literature claiming that systemic injections of bicuculline and strychnine are effective in blocking commissural inhibition, a finding that suggests γ-aminobutyric acid (GABA) and glycine as transmitter candidates.[13] Because these compounds were injected systemically, however, it was not possible to differentiate between a direct action on the vestibular neuron and an indirect action mediated via other central nervous system (CNS) sites. The precise action of GABA and glycine on vestibular commissural inhibition and their possible neurotransmitter function might best be determined by iontophoretic techniques.

The present study focused on the transmitters that inhibit the activity of vestibular type 1 neurons and, in particular, that regulate commissural inhibition.

METHODS

Experiments were performed on 32 adult cats. A tracheal cannula was introduced under ether anesthesia, and the cat was mounted in a stereotaxic apparatus on a turntable. Bipolar silver-ball electrodes for vestibular stimulation were implanted into bilateral middle ears by the ventral approach. The function of the labyrinth on the left side was preserved its for identification of the vestibular neurons, whereas on the right side of the labyrinth a part of the basal portion of the cochlea was removed and the stimulating electrodes were planted on the peripheral branches of the vestibular nerve. The position of the animal was changed to prone and its head inclined to 20 deg to make the position of the lateral canal approximately horizontal. A section of the occipital bone was removed and the medial portion of the cerebellum was aspirated in order to expose the floor of the fourth ventricle. The animal was decerebrated at the intercollicular level and immobilized by administration of gallamine triethiodide while being maintained on artificial respiration, and ether was discontinued. The initial recording was made at least 2 hours after discontinuation of ether anesthesia. FIGURE 1 represents a schematic drawing of the experimental condition described.

Animals were kept alert and the extracellular spikes of a single vestibular neuron were recorded during horizontal angular acceleration and deceleration. The seven barrels of the recording electrodes, with the exception of the center barrel, were filled with various transmitter candidates (GABA, glycine, and serotonin) and their specific inhibitors (bicuculline and strychnine),[14,15] while the center barrel was filled with 2 M NaCl for extracellular recording. Type 1 neurons were identified by their characteristic discharge pattern during rotatory stimulation as described elsewhere.[16,17] When the extracellularly recorded action potentials were positive and/or exhibited high thresholds to acceleration they were discarded, and only negative action potentials were used for analysis. Numbers of spike discharge were counted by a pulse-counter apparatus (Nihon Kohden Co.) every 10 seconds. After isolation of a type 1 neuron, a baseline recording was made and chemicals were iontophoretically applied to examine their effects on the activity of the neuron. Solutions of each of the chemicals were prepared less than 24 hours before being used, and iontophoretically ejected by means of anodal direct currents. Cathodal retaining currents (20–30 nA) were used to prevent leakage of chemicals from the micropipette tip. We used the following criteria to determine the effectiveness of the various chemicals: (1) The phenomenon had the proper latency after applying a current and would last several seconds after turning off the current. (2) When the same current was applied by means of an NaCl electrode, the phenomenon did not occur. (3) The phenomenon displayed qualitative changes compared with the baseline records. In the case of vestibular nerve stimulation, the stimulus current used was less than three times the strength of the threshold for the N_1 field potential. Within this range of stimulation intensity, field potentials in the contralateral vestibular nuclei were well developed, but no reticular evoked potentials were detected.[18]

RESULTS

One-hundred and thirteen vestibular neurons were recorded extracellulary under the conditions described. Most of the neurons recorded were located in the rostral part of the medial vestibular nucleus, and some of them were located in the superior vestibular nucleus. A few neurons were recorded at the border of the medial

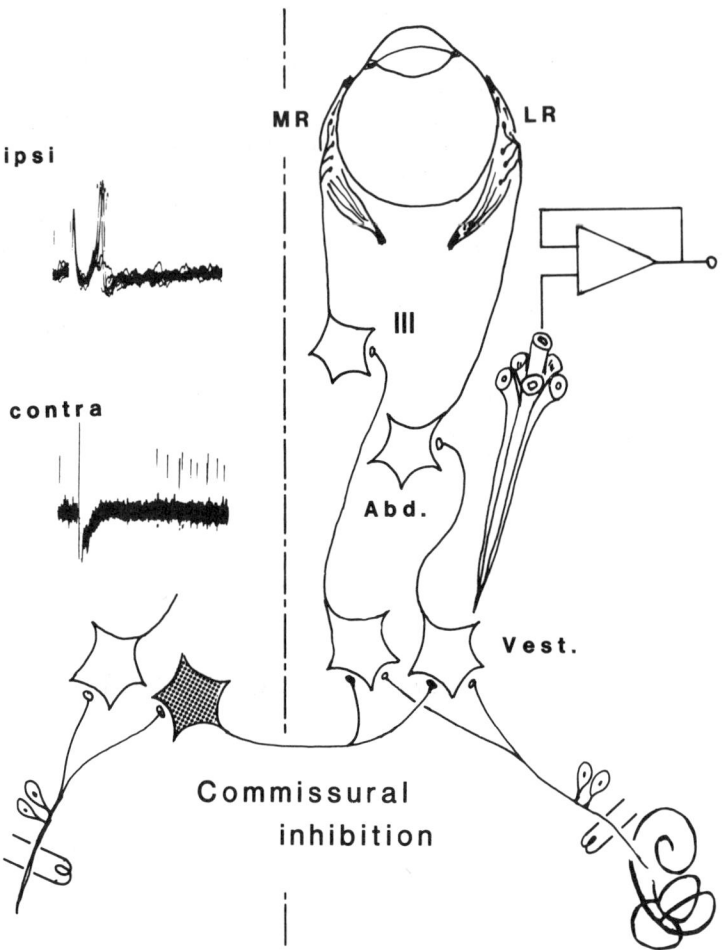

FIGURE 1. Schematic drawing of the experimental conditions. ipsi, identification of vestibular neuron following stimulation of the ipsilateral vestibular nerve; contra, commissural inhibition of the neuron when the contralateral vestibular nerve is stimulated; MR, medial rectus muscle; LR, lateral rectus muscle; III, oclomotor nucleus; Vest, vestibular nucleus; Abd, the abducens nucreus.

and lateral vestibular nuclei. One-third of them exhibited kinetic discharges, the rest displayed a tonic discharge pattern. FIGURE 2 shows the records for one of these neurons. The neuron exhibited tonic discharges and a sensitive response to acceleration, increasing activity during rotation of the turntable toward the recording side and decreasing activity toward the opposite side. Iontophoretic administration of GABA resulted in strong discharge inhibition. An injection current of 40 nA was the threshold, and a current of 80 nA produced marked changes in unit activity (FIGURE 2B). The neuron's discharges were also suppressed by administration of glycine

(FIGURE 2C), while administration of serotonin (60–200 nA current) did not alter the spike activity of the neuron. FIGURE 2D shows unit activity at an injection current of 200 nA through the serotonine electrode. GABA and glycine inhibited the spontaneous discharge of the neuron, but serotonin did not. This suppression was strong, and it was possible to analyze the effect qualitatively because of the marked suppression. The amount of effective current causing inhibition ranged from 40 to 200 nA depending on the neuron. Both tonic and kinetic neurons revealed similar behavior in response to these chemicals. GABA suppressed 65 out of 67 neurons, and glycine suppressed 42 out of 51 neurons. Forty-seven neurons were administered both GABA and glycine. Among them more than 76% of neurons had their spontaneous

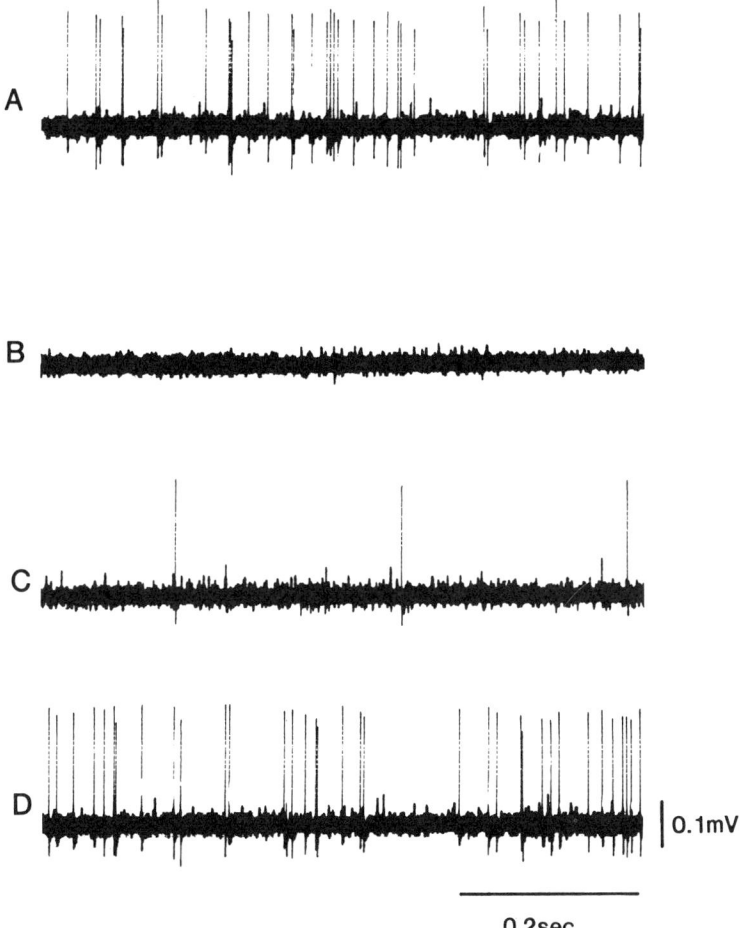

FIGURE 2. Recordings of a type 1 neuron (second-order vestibular neuron) and effects of iontophoretic application of GABA, glycine, and serotonin. **A:** Extracellular recording of type 1 activity before drug application. **B** and **C** indicate inhibitory effects of GABA and glycine with 80-nA and 100-nA currents, respectively. **D:** Effects of serotonin with 200 nA currents.

discharges suppressed by both GABA and glycine. Others were insensitive to just one of these chemicals. Some of the remaining neurons were administered one of the chemicals. There was no characteristic distribution of neurons within the vestibular nucleus.

Commissural inhibition was induced by stimulation of the contralateral vestibular nerve ($V_c$stim).[18] FIGURE 3 shows the recordings from one of them. In the control records taken before drug application (FIGURE 3A), a single stimulation of the contralateral vestibular nerve suppressed the spontaneous discharges with a latency of 4 mseconds and duration of about 25 mseconds (FIGURE 3C). When neurons were exposed to bicuculline or strychnine, their spontaneous discharges increased by

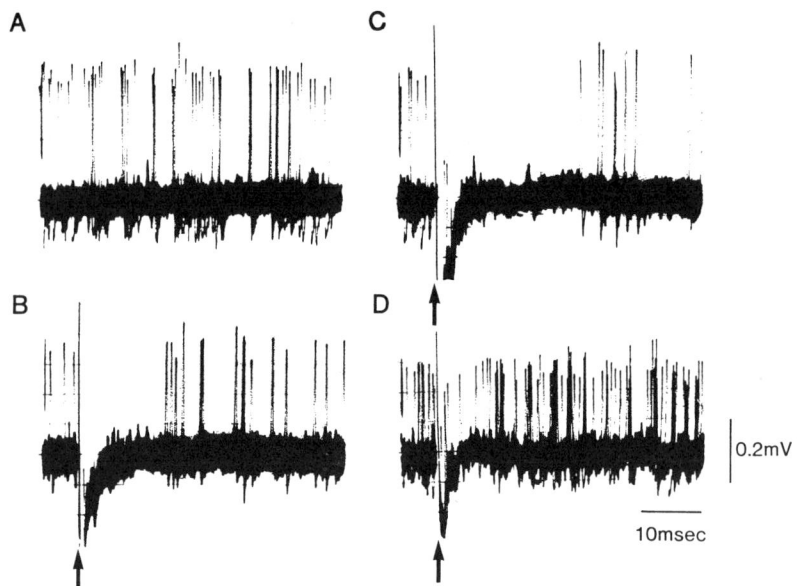

FIGURE 3. Effect of antagonists on commissural inhibition induced by electric stimulation of the contralateral vestibular nerve. **A:** Control spontaneous discharge. **B:** Commissural inhibition during iontophoretic application of strychnine (150 nA). **C:** Commissural inhibition before drug application. **D:** Commissural inhibition during iontophoretic application of bicuculline with 150-nA currents. Each tracing had 20 sweeps. Arrows show onset of contralateral vestibular nerve stimulation at 2.5 times the strength of the threshold for N_1 field potentials.

50–200%. This may represent disinhibition from GABA and glycine. This commissural inhibition is abolished following iontophoretic administration of bicuculline (FIGURE 3D). Subsequent administration of strychnine to the neuron did not suppress commissural inhibition (FIGURE 3B); however, the inhibitory effect of glycine on spontaneous discharges was abolished. Based on these results, it seems likely that strychnine is effective in blocking glycine receptors on type 1 neurons, but that glycine receptors do not participate in commissural inhibition itself. The effect of bicuculline was observed immediately after current injection, it peaked within 10–30 seconds, and lasted as long as 15 minutes after administration had been completed.

Commissural inhibition consists of disynaptic and polysynaptic routes from the contralateral vestibular nerve. Thirty-eight type 1 neurons, whose spontaneous discharges were suppressed by both GABA and glycine, were tested by commissural inhibition. The inhibition latency was determined by measuring from stimulation artifact to initial decrease of spike discharges using a computer technique and a pulse density variation program. Then a latency histogram was made consisting of two groups, i.e., a short-latency group (latency range: 2.0–3.2 mseconds) and a long-latency group (latency range: 3.2–4.7 mseconds). The neurons in both latency groups were exposed to bicuculline and/or strychnine administration. The commissural inhibition of all of these neurons was suppressed by administration of bicuculline, but not by strychnine.

DISCUSSION

The results of this study indicate that vestibular type 1 neurons have both glycine and GABA receptors. It is well known that secondary vestibular neurons play an important role in both the ocular system and commissural system. Moreover, previous correlative physiological findings suggest that ocular motoneurons are controlled by GABA in the vertical eye system[19] and glycine in the horizontal eye system.[20] It would seem convenient if the secondary vestibular neuron were to have both GABA and glycine receptors.

Double commissural inhibition latency peaks were first reported by Kasahara *et al.*[5] According to their data, tonic type 1 neurons have longer latencies than kinetic neurons. Our data indicate similar findings. In contrast to the wealth of neurophysiological data available on commissural inhibition, knowledge of the pharmacology of the system is poor. Our results in the present study show that commissural inhibition to type 1 neurons (secondary vestibular neuron) is abolished only by bicuculline and not by glycine. Thus these findings, taken as a whole, suggest that GABA is the major inhibitory neurotransmitter participating in commissural inhibition.

There is a report claiming that commissural inhibition of second-order vestibular neurons is blocked by the antagonist of one of the two presumed synaptic transmitters, GABA and glycine.[13] The discrepancy between previous findings and our own may be due to differences in experimental conditions. As discussed in the literature, the earlier data under other experimenters' conditions may have included other types of second-order vestibular neurons than type 1 neurons, which have intimate connection with the lateral semicircular canal. In addition, in other experimental protocols, these compounds were injected systemically and may have acted not only on commissural inhibition but other CNS sites as well.

Bicuculline markedly suppressed commissural inhibition. This finding indicates that the GABA receptor mediating commissural inhibition is the $GABA_A$ receptor, and that the $GABA_B$ receptor is unlikely. $GABA_A$ receptors are linked to Cl^- channels in the cell membrane, while $GABA_B$ receptors play a role in inhibitory synaptic transmission by way of presynaptic inhibition.[21] Intracellular recordings from vestibular neurons have shown that vestibular commissural inhibition is via inhibitory postsynaptic potentials (IPSPs) produced by stimulating the contralateral vestibular nerve.[7] Our present findings concerning the $GABA_A$ receptor support these previous pharmacological and physiological data.

Iontophoretically administered chemicals may act on adjacent neurons instead of the neurons being monitored. This possibility cannot be completely ruled out under the present experimental conditions. It may not be the case, however, because short-latency commissural inhibition (possibly disynaptic range)[5] was apparently suppressed by bicuculline.

Part of this work has been reported previously in *Acta Otolaryngol.* (*Stockholm*) *Supplement* 481.

SUMMARY

The present study focused on the transmitters that control the vestibular neural activity and, in particular, that regulate commissural inhibition. Extracellular spikes of a single vestibular neuron were recorded in decerebrate cats. The seven barrels of the electrode, with the exception of the center barrel, were filled with transmitter candidates and their specific antagonists, while the center barrel was filled with 2 M NaCl for extracellular recording. After isolation of a type 1 neuron, chemicals were iontophoretically applied to examine their effects on its activity. The results were as follows: (1) GABA and glycine markedly decreased spontaneous firing of the neurons, while serotonin did not affect their activity. (2) Bicuculline abolished the inhibitory effects of GABA on the neurons. (3) Strychine abolished the effects of glycine. (4) Commissural inhibition induced by electrical stimulation of the contralateral labyrinth was not abolished by strychine but was abolished by bicuculline. We conclud that (1) vestibular type 1 neurons are controlled by GABAergic and glycinergic but not serotoninergic neurons, and (2) commissural inhibition is activated by the $GABA_A$ receptor, but not by the $GABA_B$ receptor.

ACKNOWLEDGMENTS

The authors wish to thank Ms. Reiko Sekiguchi for skilled technical assistance.

REFERENCES

1. MARKHAM, C. H., T. YAGI & J. S. CURTHOYS. 1977. The contribution of the contralateral labyrinth to second order vestibular neuronal activity in the cat. Brain Res. **138:** 99–109.
2. SHIMAZU, H. & W. PRECHT. 1966. Inhibition of central vestibular neurons from the contralateral labyrinth and its mediating pathway. J. Neurophysiol. **28:** 467–492.
3. BRODAL, A., O. POMPEIANO & F. WALENBERG, EDS. 1962. Anatomy of the vestibular nuclei. *In* The Vestibular Nuclei and Their Connections, Anatomy and Functional Correlations: 27–53. The William Ramsay Henderson. London, England.
4. LADPLI, R. & A. BRODAL. 1968. Experimental studies of commissural and reticular formation projections from the vestibular nuclei in the cat. Brain Res. **8:** 65–96.
5. KASAHARA, M., M. MANO, T. OSHIMA, S. OZAWA & H. SHIMAZU. 1968. Contralateral short latency inhibition of central vestibular neurons in the horizontal canal system. Brain Res. **8:** 376–378.
6. KASAHARA, M. & Y. UCHINO. 1971. Selective mode of commissural inhibition induced by semicircular canal afferents on secondary vestibular neurons in the cat. Brain Res. **34:** 366–369.
7. MANO, M., T. OSHIMA & H. SHIMAZU. 1968. Inhibitory commissural fibers interconnecting the bilateral vestibular nuclei. Brain Res. **8:** 378–382.
8. JAJU, B. P., E. B. KIRSTEN & S. C. WANG. 1970. Effect of belladonna alkaloids on vestibular nucleus of the cat. Am. J. Physiol. **219:** 1248–1255.
9. KIRSTEN, E. B. & J. N. SCHOENER. 1973. Action of anticholinergic and related agents on single vestibular neurons. Neuropharmacology **12:** 1167–1177.

10. MATSUOKA, I. & E. F. DOMINO. 1975. Cholinergic mechanisms in the cat vestibular system. Neuropharmacology **14:** 201–210.
11. JAJU, B. P. & S. C. WANG. 1971. Effects of diphenhydramine and dimenhydriate on vestibular neuronal activity in the cat: a search for the locus of their antimotionsickness action. J. Pharmacol. Exp. Ther. **176:** 718–724.
12. RAYMOND, J., A. NIEOULLON, D. EMEMES & A. SANS. 1984. Evidence for glutamate as neurotransmitter in the cat vestibular nerve: radioautographic and biochemical studies. Exp. Brain Res. **56:** 528–531.
13. PRECHT, W., P. C. SCHWINDT & R. BAKER. 1973. Removal of vestibular commisseral inhibition by antagonists of GABA and glycine. Brain Res. **62:** 222–226.
14. CURTIS, D. R., A. W. DUGGAN, D. FELIX, G. R. A. JOHNSTON & H. MCLENNAN. 1971. Antagonism between bicuculline and GABA in supraspinal regions of the central nervous system of the cat. Brain Res. **33:** 57–73.
15. FELPEL, L. P. 1972. Effects of strychnine, bicuculline and picrotoxin on labyrinthine-evoked inhibition in neck motoneurons of the cat. Exp. Brain Res. **14:** 494–502.
16. PRECHT, W. & H. SHIMAZU. 1965. Functional connections of tonic and kinetic vestibular neurons with primary vestibular afferents. J. Neurophysiol. **28:** 1014–1028.
17. SHIMAZU, H. & W. PRECHT. 1965. Tonic and kinetic responses of cat's vestibular neurons to horizontal angular acceleration. J. Neurophysiol. **28:** 991–1013.
18. SHIMAZU, H. & W. PRECHT. 1966. Inhibition of central vestibular neurons from the contralateral labyrinth and its mediating pathway. J. Neurophysiol. **29:** 467–492.
19. OBATA, K. & S. M. HIGHSTEIN. 1970. Blocking by picrotoxin of both vestibular inhibition and GABA action on rabbit oculomotor neurones. Exp. Brain Res. **11:** 327–342.
20. SPENCER, R. F., R. J. WENTHOLD & R. BAKER. 1989. Evidence for glycine as an inhibitory neurotransmitter of vestibular, reticular, and prepositus hypoglossi neurons that project to the cat abducens nucleus. J. Neurosci. **9:** 2718–2736.
21. HILL, D. R. & N. G. BOWERY. 1981. 3H-baclofen and 3H-GABA bind to bicuculline-insensitive $GABA_B$ site in rat brain. Nature **290:** 149–152.

GABA and Glycine as Inhibitory Neurotransmitters in the Vestibuloocular Reflex[a]

ROBERT F. SPENCER[b] AND ROBERT BAKER[c]

[b]Department of Anatomy
Medical College of Virginia
1101 East Marshall Street
Richmond, Virginia 23298-0709

[c]Department of Physiology and Biophysics
New York University Medical Center
550 First Avenue
New York, New York 10016

Considerable knowledge has accumulated regarding the morphological, physiological, and functional organization of the vestibular system. The basic three-neurone chain, comprising primary and secondary vestibular neurones and motoneurones in the extraocular motor nuclei, is a necessary component of the vestibuloocular reflex (VOR). It is clear, however, that other neurones in the nucleus prepositus hypoglossi and the cerebellum, as well as vestibular commissural connections, have an important role in its normal operation. Inhibition mediated by second-order vestibular neurones is a fundamentally important aspect of the reciprocal excitatory and inhibitory synaptic inputs to extraocular motoneurones in the VOR. Although inhibition at the level of the motoneurones undoubtedly is important in providing a rapid, effective relaxation of antagonistic extraocular muscles, it appears not to be involved in determining, at least directly, the response properties (e.g., head velocity, eye position sensitivities) and discharge patterns (i.e., burst, burst/tonic) of these neurones. Rather, inhibitory mechanisms at all levels in the vestibuloocular circuitry may be important in shaping the excitatory networks that ultimately control the dynamic properties of the VOR. Even at the level of the extraocular motoneurones, however, inhibition is important to the extent that the reciprocal excitatory and inhibitory inputs overlap with respect to firing threshold, thus creating a larger dynamic response than otherwise would be present with either input alone. The types of neurotransmitters [i.e., γ-aminobutyric acid (GABA), glycine] utilized by inhibitory inputs and the postsynaptic receptors with which they are associated furthermore may translate into differences in the postsynaptic effects of interacting excitatory and inhibitory inputs. Consequently, the identification of the inhibitory neurotransmitters utilized in various circuits in the vestibuloocular system and the various types of receptors with which they are associated is of fundamental importance to further defining the role of such interactions in normal eye movements and the extent to which they are involved in eye movement deficits.

[a]This work was supported by U.S. Public Health Service MERIT Award EY02191 and Research Grant EY02007 from the National Eye Institute.

ROLE OF GABA AS AN INHIBITORY NEUROTRANSMITTER
IN THE VESTIBULOOCULAR SYSTEM

Electrophysiological, pharmacological, and biochemical studies have established that GABA is the inhibitory neurotransmitter utilized by second-order vestibular neurones that establish synaptic connections with motoneurones in the oculomotor and trochlear nuclei. Systemic administration of picrotoxin, an antagonist of GABA, abolishes the depression of the antidromic field potential recorded extracellularly in the oculomotor nucleus following IIIrd nerve stimulation and eliminates the extracellular positive field potentials that represent the inhibitory postsynaptic currents resulting from ipsilateral VIIIth nerve stimulation.[1] Iontophoresis of GABA in the vicinity of the oculomotor nucleus depresses the antidromic field potential elicited by IIIrd nerve stimulation and decreases or completely suppresses the spike generation of motoneurones in a manner similar to that produced by electrical stimulation of the ipsilateral VIIIth nerve.[2] The inhibitory responses elicited by VIIIth nerve stimulation and GABA iontophoresis furthermore are blocked by iontophoresis of picrotoxin in the vicinity of the motoneurones. Picrotoxin also blocks the slow muscle potential recorded from the extraocular muscles in a manner similar to removal of the second-order vestibular input to oculomotor motoneurones following lesions of the dorsolateral brain stem that effectively interrupt the inhibitory vestibular pathway.[3] These electrophysiological and pharmacological findings are supported further by anatomical studies that have demonstrated synaptic endings in the oculomotor nucleus labeled autoradiographically by high-affinity uptake of [3H]GABA[4,5] or immunohistochemically by the localization of GABA[5-7] or its synthesizing enzyme glutamate decarboxylase (GAD).[6] Within the oculomotor nucleus, GABA-immunoreactive synaptic endings are found predominantly within the inferior rectus, superior rectus, and inferior oblique subdivisions. Consistent with the absence of vestibular-evoked inhibitory postsynaptic potentials (IPSPs) in medial rectus motoneurones,[8] the medial rectus subdivision is virtually devoid of GABA-/GAD-immunoreactive synaptic endings. The density of GABA-immunoreactive synaptic endings in the various subdivisions of the oculomotor nucleus furthermore is higher in the Rhesus monkey than in the cat. Correlated at least in part with the high density of GABA-immunoreactive terminals within the oculomotor nucleus, GABA-immunoreactive axons also are observed predominantly in the medial portion of the medial longitudinal fasciculus (MLF) lateral to the nucleus (FIGURE 1A).

In the trochlear nucleus, systemic administration of picrotoxin significantly reduces or abolishes both the inhibitory synaptic current recorded extracellularly and the IPSPs recorded intracellularly from motoneurones following electrical stimulation of the ipsilateral VIIIth nerve.[9] A similar depressant action on vestibular-evoked inhibition is obtained by systemically administered bicuculline. Unilateral section of the MLF, which abolishes the vestibular-evoked inhibitory synaptic currents, reduces the concentration of GABA in the trochlear nucleus. Lesions of the superior vestibular nucleus also produce a marked decrease in GABA synthesis in the ipsilateral trochlear nucleus.[10] Of the three extraocular motor nuclei, the trochlear nucleus has the highest density of GABA-immunoreactive synaptic endings[6,7] (FIGURE 1C). The soma-dendritic distribution of many GABA-/GAD-immunoreactive synaptic endings, combined with the presence of multiple synaptic contact zones associated with individual synaptic endings, is a feature typical of the inhibitory second-order vestibular input to oculomotor and trochlear motoneurones identified previously by ultrastructural reconstructions of physiologically identified axons stained intracellularly with horseradish peroxidase (HRP).[11] Ventral to the trochlear nucleus, GABA-immunoreactive axons are found predominantly in the dorsal half of the MLF, suggesting that the axons are segregated in the MLF according to their origin (e.g., vestibular, abducens internuclear) and/or function (i.e., inhibitory vs. excitatory).

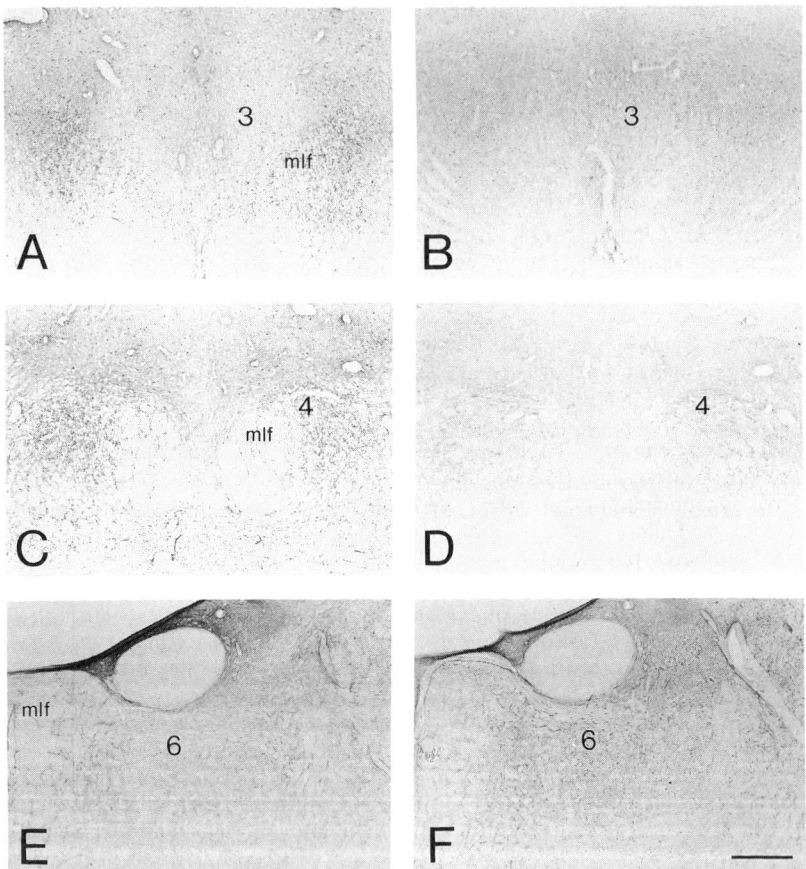

FIGURE 1. Immunohistochemical localization of GABA (A, C, and E) and glycine (B, D, and F) in the cat oculomotor (3; A and B), trochlear (4; C and D) and abducens (6; E and F) nuclei and the MLF (mlf). Calibration: 0.5 mm.

ROLE OF GLYCINE AS AN INHIBITORY NEUROTRANSMITTER IN THE VESTIBULOOCULAR SYSTEM

GABA also was thought initially to be the inhibitory neurotransmitter mediating vestibular-evoked inhibition in the abducens nucleus. Both the antidromic field potentials and orthodromic synaptic currents recorded extracellularly following VIth and VIIIth nerve stimulation, respectively, were reduced when picrotoxin was administered systemically in the rabbit.[12] A more recent comprehensive autoradiographic, immunohistochemical, and electrophysiological-pharmacological analysis in the cat, however, has revealed that, in contrast to the oculomotor and trochlear nuclei, inhibitory inputs to the abducens nucleus from second-order neurones in the vestibular nucleus, as well as from neurones in the dorsolateral medullary reticular formation and the prepositus hypoglossi nucleus, utilize glycine as a neurotransmitter.[7] The different populations of inhibitory premotor neurones are labeled selectively by retrograde transport of [³H]glycine, but not [³H]GABA, injected into the abducens nucleus. Correlated with these findings, the abducens nucleus exhibits a high density of glycine-immunoreactive synaptic endings (FIGURE 1F), but a paucity of GABA-immunoreactive terminals (FIGURE 1E). By contrast, the oculomotor (FIGURE 1B) and trochlear (FIGURE 1D) nuclei contain very few glycine-immunoreactive synaptic endings. In the oculomotor nucleus in the monkey, glycine-immunoreactive terminals are confined specifically to the superior rectus subdivision, whereas in the cat similar terminals appear to be distributed to the other (except medial rectus) motoneurone subdivisions as well. The source of this modest glycinergic input to the oculomotor nucleus presently is unknown. In further contrast to oculomotor and trochlear motoneurones, the vestibular inhibition of abducens motoneurones evoked by selective horizontal canal nerve electrical stimulation is abolished by strychnine, but is unaffected by picrotoxin or bicuculline administered systemically (FIGURE 2A–C). Most, if not all, of the glycine-immunoreactive synaptic endings in the abducens presumably are related to glycine-immunoreactive neurones that are located in the same areas as neurones labeled by retrograde transport of [³H]glycine from the abducens nucleus.

The complementary pattern of GABA and glycine localization in the extraocular motor nuclei is correlated with a distinctive pattern of immunoreactive staining in the MLF. GABA is associated predominantly with ascending axons that project to the oculomotor and trochlear nuclei (FIGURE 1A and C). By contrast, glycine is localized predominantly in descending axons that project to the abducens nucleus and the spinal cord (FIGURES 1F and 3F). Occasional GABA-immunoreactive axons that presumably are of vestibular commissural origin course transversely through the abducens nucleus, and only a few GABA-immunoreactive axons are observed in the MLF at this level of the brain stem. The paradoxical differences in inhibitory neurotransmitters utilized by vertical and horizontal canal-related vestibular neurones also may be correlated with the differential roles of GABA and glycine in vestibular commissural inhibition[13,14] and the differential association of GABA$_A$ and strychnine-sensitive glycine receptors with neurones in the vestibular nucleus.[15]

IMMUNOHISTOCHEMICAL LOCALIZATION OF GABA AND GLYCINE IN THE VESTIBULAR NUCLEI

Immunohistochemical studies of GABA localization only partially support the substantive physiological, biochemical, and pharmacological data cited above regarding the role of GABA as the inhibitory neurotransmitter in vestibuloocular reflex connections. For example, few or no GABA-immunoreactive neurones have been found in the superior vestibular nucleus,[16–19] despite evidence that both anterior and posterior vertical canal-related inhibitory second-order vestibular neurones are located in this region. By contrast, GABA-immunoreactive neurones have been observed predominantly in the medial and inferior vestibular nuclei. A population of

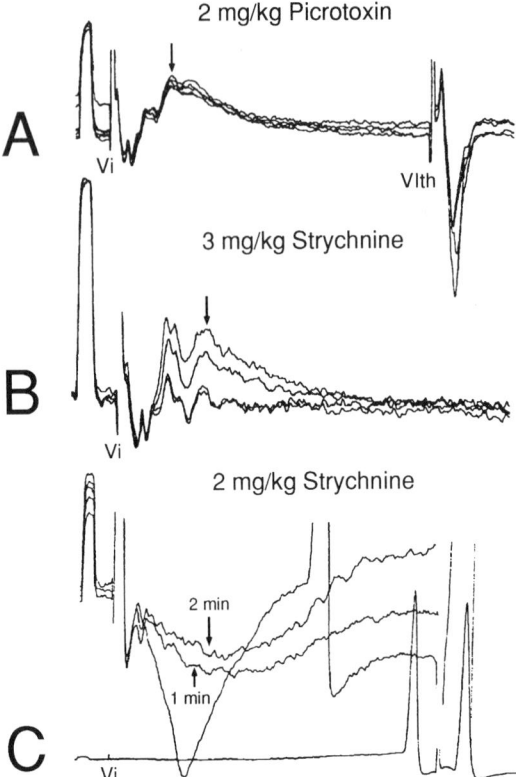

FIGURE 2. A and **B:** Extracellular recordings demonstrating the different effects of systemically administered picrotoxin (A) and strychnine (B) on inhibitory postsynaptic field potentials in the abducens nucleus following ipsilateral vestibular nerve stimulation. Arrows at 2 mseconds latency indicate the peak of the postsynaptic response. **C:** intracellular records from an abducens motoneurone following ipsilateral vestibular nerve stimulation demonstrating a rapid decrease in the rise time and peak amplitude of the IPSP following systemic administration of strychnine. Calibrations: 1 mV and 0.5 msecond. Negative polarity is downward.

intrinsic GABAergic neurones also has been described in the dorsal division of the lateral vestibular nucleus.[20]

Neurones in the medial and inferior vestibular nuclei that project to the spinal cord also are immunoreactive toward GAD.[21] Other evidence, however, suggests that glycine is involved in these vestibulospinal connections. For example, the vestibular-evoked disynaptic IPSPs in neck motoneurones are effectively blocked by strychnine, a glycine antagonist.[22] Presumed glycinergic inhibitory vestibular neurones that project to the ipsilateral abducens nucleus have axonal branches that descend in the ipsilateral MLF toward the spinal cord.[23,24] Possibly related, at least in part, to these descending inhibitory axons, spinal cord ventral horn motoneurones exhibit a high density of glycine receptors.[25,26]

At present, it is difficult to resolve these apparent disparate findings in regard to the locations of known populations of inhibitory vestibular neurones and the neurotransmitters with which they are associated. On the one hand, since vertical-canal-related inhibitory second-order vestibular neurones are projection neurones, it is possible that, like cerebellar Purkinje cells, the concentration of GABA within the somata of the neurones is significantly less than that at their synaptic endings in the oculomotor and trochlear nuclei and cannot be detected immunohistochemically. Consequently, the above studies may not have identified the total population of GABAergic neurones in the vestibular nuclei, particularly within the superior vestibular nucleus.

The specificity of various antibodies used for the immunohistochemical localization of GABA and glycine, as well as possible differences between species in the locations of neurotransmitter-specific populations of neurones, also may be a factor. In the cat, both GABA-immunoreactive and glycine-immunoreactive neurones are found in the superior, medial, and descending (inferior) vestibular nuclei. Within the superior vestibular nucleus, the two populations of neurones are located predomi-

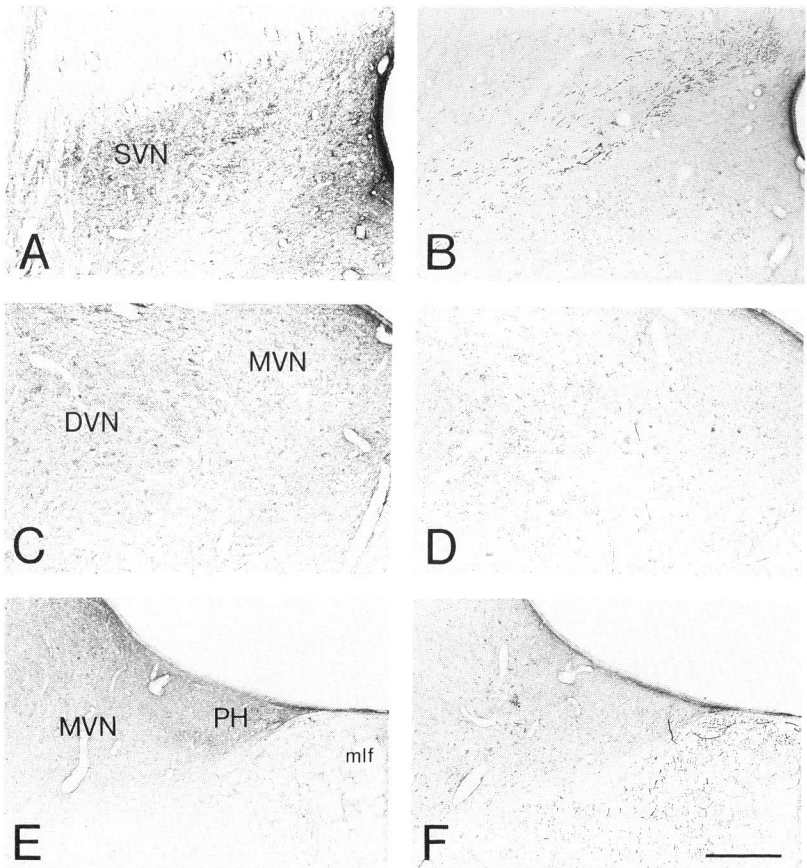

FIGURE 3. Immunohistochemical localization of GABA (A, C, and E) and glycine (B, D, and F) in the cat superior (SVN), medial (MVN), and descending (DVN) vestibular nuclei. Calibration: 1 mm.

nantly in the central region of the nucleus among immunoreactive axons that appear to be derived from the cerebellum and/or the vestibular nuclei (FIGURE 3A and B). GABA-immunoreactive neurones significantly outnumber glycine-immunoreactive neurones within this nucleus. Both GABA-immunoreactive (FIGURE 3C and E) and glycine-immunoreactive (FIGURE 3D and F) neurones also are distributed throughout the rostral-caudal extent of the medial and descending vestibular nuclei. Despite their coexistence within these regions, they appear to comprise separate populations of neurones on the basis of their size and morphology. Consistent with this premise, the descending limb of the MLF at caudal levels of the brain stem contains a large number of glycine-immunoreactive axons (FIGURE 3F), but few, if any, GABA-immunoreactive axons (FIGURE 3E). The GABA-immunoreactive neurones in the medial and descending vestibular nuclei thus appear to have different connections, possibly related to vestibular commissural interactions.

It is also possible that neurones may co-localize GABA and glycine[19] or that GABAergic synaptic endings are associated with glycine receptors.[27] Thus, neurones in all of the vestibular nuclei may exhibit GABA or GAD immunoreactivity irrespective of whether they utilize GABA as a neurotransmitter. In this regard, the co-localization of GABA and glycine, as well as their co-localization with putative excitatory amino acid neurotransmitters, may indicate a metabolic pool of one that is unrelated to the neurotransmitter function of another. Despite the co-localization of GABA and glycine in single vestibular neurones, in most instances only one or the other appears to have a synaptic effect, as indicated by the specificity of pharmacological antagonism.[15] This effect presumably is dictated by the type and presence of the postsynaptic receptor with which the input is associated. Indeed, the functional effects produced by a single neurotransmitter appear to be directly dependent on the type of postsynaptic receptor, as demonstrated by the differential roles of GABA acting on $GABA_A$ and $GABA_B$ receptors in neural integrator and velocity storage mechanisms, respectively, in the vestibular nuclei through vestibular commissural and cerebellar pathways.[28–30]

CONCLUSION

It is well established that GABA is the major inhibitory neurotransmitter utilized by premotor neurones involved in vertical vestibuloocular eye movements. By contrast, glycine is the inhibitory neurotransmitter of most premotor neurones that are related to horizontal eye movements. The significance of this dichotomy in inhibitory neurotransmitters utilized in the vertical and horizontal eye movement systems presently is unclear. On the one hand, it might reflect functional differences between different types of neurones, distinguishing, for example, between second-order vestibular neurones that participate only in eye movement (e.g., GABAergic inhibitory neurones in the superior vestibular nucleus that project only to the trochlear and/or oculomotor nuclei) versus those that are involved in gaze (e.g., glycinergic inhibitory neurones in the medial vestibular nucleus that project to both the abducens nucleus and the spinal cord). Although the postsynaptic effect of GABA or glycine acting on its respective receptor is the same, namely, inhibition of the motoneurones, the secondary effects of the two neurotransmitters, however, may be quite different. For example, the excitatory effects of glutamate activating N-methyl-D-aspartic acid (NMDA) receptors are augmented by glycine acting on strychnine-insensitive glycine receptors. These factors, however, probably do not translate into apparent differences in the way motoneurones produce vertical or horizontal eye movements.

On the other hand, these differences in neurotransmitters utilized in the vertical and horizontal eye movement systems may have an embryological basis, which, in the simplest case, might reflect that the medulla, in which most of the horizontal premotor neurones are located, is a rostral extension of the spinal cord where glycine is the major inhibitory neurotransmitter, while the midbrain, which is the location or site of termination of the vertical premotor neurones, is more closely associated with the forebrain where GABA is the major inhibitory neurotransmitter. The association of specific homeodomain proteins with specific neuromeric segments during embryo-genesis is correlated with the rhombomeric origin of ascending vestibuloocular and descending vestibulospinal neurones (reviewed in Reference 31). For example, Hox 2.9 is associated only with rhombomere 4 from which vestibulospinal neurones originate, while Wnt-1 is associated with rhombomeres 5 and 6 from which vestibu-loocular neurones originate. These DNA-binding proteins are likely to be causal factors that promote the expression by up or down regulation of families of genes that cause a cell to differentiate into a particular type of neurone. As part of this differentiation process, a neurone will express a particular neurotransmitter and will project its axon to a particular location. Thus, the organization of vestibuloocular and vestibulospinal projections into coherent groups is a type of regional specificity that further suggests an early embryonic specification of neurotransmitters. Given the findings from our studies in the adult, this hypothesis could be tested directly by comparing the developmental pattern of identified inhibitory neurones that utilize glycine or GABA as a neurotransmitter and that project to the abducens nucleus and spinal cord or the oculomotor and trochlear nuclei, respectively.

ACKNOWLEDGMENTS

We are grateful to Dr. Robert J. Wenthold for generously providing the antibod-ies to GABA and glycine. The excellent technical assistance of Lynn Davis also is greatly appreciated.

REFERENCES

1. ITO, M., S. M. HIGHSTEIN & T. TSUCHIYA. 1970. The postsynaptic inhibition of rabbit oculomotor neurones by secondary vestibular impulses and its blockage by picrotoxin. Brain Res. **17:** 520–523.
2. OBATA, K. & S. M. HIGHSTEIN. 1970. Blocking by picrotoxin of both vestibular inhibition and GABA action on rabbit oculomotor neurones. Brain Res. **18:** 538–541.
3. ITO, M., N. NISIMARU & M. YAMAMOTO. 1976. Postsynaptic inhibition of oculomotor neurons involved in vestibulo-ocular reflexes arising from semicircular canals of rabbits. Exp. Brain Res. **24:** 273–283.
4. LANOIR, J., J. J. SOGHOMONIAN & G. CADENEL. 1982. Radioautographic study of ³H-GABA uptake in the oculomotor nucleus of the cat. Exp. Brain Res. **48:** 137–143.
5. SOGHOMONIAN, J.-J., R. PINARD & J. LANOIR. 1989. GABA innervation in adult rat oculomotor nucleus: a radioautographic and immunocytochemical study. J. Neurocytol. **18:** 319–331.
6. SPENCER, R. F., S.-F. WANG & R. BAKER. 1992. The pathways and functions of GABA in the oculomotor system. Prog. Brain Res. **90:** 307–334.

7. SPENCER, R. F., R. J. WENTHOLD & R. BAKER. 1989. Evidence for glycine as an inhibitory neurotransmitter of vestibular, reticular, and prepositus hypoglossi neurons that project to the cat abducens nucleus. J. Neurosci. **9:** 2718–2736.

8. BAKER, R. & S. M. HIGHSTEIN. 1978. Vestibular projections to medial rectus subdivision of oculomotor nucleus. J. Neurophysiol. **41:** 1629–1646.

9. PRECHT, W., R. BAKER & Y. OKADA. 1973. Evidence for GABA as the synaptic transmitter of the inhibitory vestibulo-ocular pathway. Exp. Brain Res. **18:** 415–428.

10. ROFFLER-TARLOV, S. & E. TARLOV. 1975. Reduction of GABA synthesis following lesions of inhibitory vestibulo-trochlear pathway. Brain Res. **91:** 326–330.

11. SPENCER, R. F. & R. BAKER. 1983. Morphology and synaptic connections of physiologically-identified second-order vestibular axonal arborizations related to cat oculomotor and trochlear motoneurones. Soc. Neurosci. Abstr. **9:** 1088.

12. HIGHSTEIN, S. M. 1973. Synaptic linkage in the vestibulo-ocular and cerebello-vestibular pathways to the VIth nucleus in the rabbit. Exp. Brain Res. **17:** 301–314.

13. PRECHT, W., P. C. SCHWINDT & R. BAKER. 1973. Removal of vestibular commissural inhibition by antagonists of GABA and glycine. Brain Res. **62:** 222–226.

14. FURUYA, N., T. YABE & T. KOIZUMI. 1991. Neurotransmitters regulating vestibular commissural inhibition in the cat. Acta Otolaryngol. Stockh. Suppl. **481:** 205–208.

15. SMITH, P. F., C. L. DARLINGTON & J. I. HUBBARD. 1991. Evidence for inhibitory amino acid receptors on guinea pig medial vestibular nucleus neurons in vitro. Neurosci. Lett. **121:** 244–246.

16. NOMURA, I., E. SENBA, T. KUBO, T. SHIRAISHI, T. MATSUNAGA, M. TOHYAMA, Y. SHIOTANI & J.-Y. WU. 1984. Neuropeptides and γ-aminobutyric acid in the vestibular nuclei of the rat: an immunohistochemical analysis. I. Distribution. Brain Res. **311:** 109–118.

17. KUMOI, K., N. SAITO & C. TANAKA. 1987. Immunohistochemical localization of γ-aminobutyric acid- and aspartate-containing neurons in the guinea pig vestibular nuclei. Brain Res. **416:** 22–33.

18. CARPENTER, M. B., Y. HUANG, A. B. PEREIRA & L. B. HERSH. 1990. Immunocytochemical features of the vestibular nuclei in the monkey and cat. J. Hirnforsch. **31:** 585–599.

19. WALBERG, F., O. P. OTTERSEN & E. RINVIK. 1990. GABA, glycine, glutamate and taurine in the vestibular nuclei: an immunocytochemical investigation in the cat. Exp. Brain Res. **79:** 547–563.

20. HOUSER, C. R., R. P. BARBER & J. E. VAUGHN. 1984. Immunocytochemical localization of glutamic acid decarboxylase in the dorsal lateral vestibular nucleus: evidence for an intrinsic and extrinsic GABAergic innervation. Neurosci. Lett. **47:** 213–220.

21. BLESSING, W. W., S. C. HEDGER & W. H. OERTEL. 1987. Vestibulospinal pathway in rabbit includes GABA-synthesizing neurons. Neurosci. Lett. **80:** 158–162.

22. FELPEL, L. P. 1972. Effects of strychnine, bicuculline and picrotoxin on labyrinthine-evoked inhibition in neck motoneurons of the cat. Exp. Brain Res. **14:** 494–502.

23. MCCREA, R. A., K. YOSHIDA, A. BERTHOZ & R. BAKER. 1980. Eye movement related activity and morphology of second order vestibular neurons terminating in the cat abducens nucleus. Exp. Brain Res. **40:** 468–473.

24. ISU, N. & J. YOKOTA. 1983. Morphophysiological study on the divergent projection of axon collaterals of medial vestibular nucleus neurons in the cat. Exp. Brain Res. **53:** 151–162.

25. TRILLER, A., F. CLUZEAUD & H. KORN. 1987. Gamma-aminobutyric acid–containing terminals can be apposed to glycine receptors at central synapses. J. Cell Biol. **104:** 947–956.

26. GEYER, S. W., W. GUDDEN, H. BETZ, H. GNAHN & A. WEINDL. 1987. Co-localization of choline acetyltransferase and postsynaptic glycine receptors in motoneurons of rat spinal cord demonstrated by immunocytochemistry. Neurosci. Lett. **82:** 11–15.

27. TRILLER, A., F. CLUZEAUD, F. PFEIFFER, H. BETZ & H. KORN. 1985. Distribution of glycine receptors at central synapses: an immunoelectron microscopy study. J. Cell Biol. **101:** 683–688.

28. WAESPE, W., B. COHEN & T. RAPHAN. 1985. Dynamic modification of the vestibulo-ocular reflex by the nodulus and uvula. Science **228:** 199–202.

29. COHEN, B., D. HELWIG & T. RAPHAN. 1987. Baclofen and velocity storage: a model of the

effects of the drug on the vestibulo-ocular reflex in the Rhesus monkey. J. Physiol. London **393:** 703–726.

30. STRAUBE, A., R. KURZAN & U. BÜTTNER. 1991. Differential effects of bicuculline and muscimol microinjections into the vestibular nuclei on simian eye movements. Exp. Brain Res. **86:** 347–358.

31. BAKER, R. A contemporary view of the phylogenetic history of eye muscles and motoneurones. *In* Vestibular and Brain Stem Control of Eye, Head and Body Movements. H. Shimazu and Y. Shinoda, Eds. Japan Scientific Societies Press. Tokyo, Japan. (In press.)

Enhancement of Optokinetic and Vestibuloocular Responses in the Rabbit by Cholinergic Stimulation of the Flocculus

H. COLLEWIJN, H. S. TAN, AND J. VAN DER STEEN

Department of Physiology I
Faculty of Medicine
Erasmus University Rotterdam
Post Office Box 1738
3000 DR Rotterdam, the Netherlands

INTRODUCTION

The role of several neurotransmitter systems that may be, in principle, functional in cerebellar signal processing is only partly understood. There is abundant evidence for the role of excitatory amino acids (glutamate,[1] and possibly aspartate and homocysteic acid) and inhibitory amino acids [primarily γ-aminobutyric acid (GABA)] in the main signal flow from parallel and climbing fiber systems through the Purkinje cells and the inhibitory interneurons (see Ito).[2] Much less is known about the functionality of noradrenaline and acetylcholine, which are well established as neurotransmitters elsewhere in the nervous system, and for which the presence as such, as well as the appropriate enzyme systems and receptors, has been identified in the cerebellum. Our laboratory has recently begun to study the role of these substances by means of microinjections into the cerebellum, using the control of optokinetic and vestibuloocular responses by the cerebellar flocculus in the rabbit as a model.

There is good evidence for a crucial role of the floccular side loop in the rabbit in maintaining, partly through adaptive processes, optimal gain levels of the vestibuloocular reflex (VOR) and optokinetic reflex (OKR) in the rabbit. While the direct effects of floccular efferent fibers (Purkinje cell axons), which terminate mainly on vestibular nuclear cells, are inhibitory, the net effect of the floccular signal flow on the magnitude of the VOR and OKR is positive. This follows from the polarity of the Purkinje cell modulation during head or surround movements. Simple-spike discharge, the main source of output of Purkinje cells, is increased in the floccular "H zone" during head rotation to the side contralateral of the recording (modulation "out of phase"), and during rotation of an optokinetic drum towards the recording side (modulation "in phase").[3–5] Activation of this same floccular zone enhances ipsilateral smooth eye movements, as shown by electrical stimulation.[4,6–9] Adaptive changes in the magnitude (gain) of the VOR and OKR, induced by manipulation of the amount of retinal image slip,[10–12] have been correlated with shifts in the balance between these two classes of modulation.[3–4,13–15]

In agreement with the dominance of out-of-phase modulated Purkinje cells, found in single-unit recordings, lesions of the flocculus of the rabbit result in a decrease in the gain of the OKR and the VOR[2,7,15–16] or of the OKR alone.[17] Similarly, our laboratory obtained a marked, but reversible, reduction in the gain of both the

OKR and VOR in the rabbit by bilateral microinjections of the GABA agonist muscimol or baclofen into the flocculus.[18] These agonists probably caused a temporary functional ablation of the flocculus, due to massive inhibition of cerebellar circuits at several levels.

We subsequently investigated the role of noradrenergic agonists and antagonists in the floccular control of the OKR and VOR with similar injection techniques. Floccular microinjection of a β agonist (isoproterenol) facilitated the adaptation of the VOR, whereas similar injection of a β antagonist (sotalol) reduced the adaptability of the VOR. Neither of these substances acting on β receptors affected the basic gain of VOR or OKR in the absence of pressure for adaptation.[19] A complementary investigation with floccular injections of α_1 and α_2 (ant)agonists did not reveal any effect on the VOR and OKR or their adaptability,[20] thus supporting the hypothesis that noradrenergic innervation of the flocculus, mainly originating from the locus ceruleus, modulates adaptability of the VOR and OKR specifically through β receptors.

In the present paper, we address the possible role of acetylcholine (ACh) as a neurotransmitter or neuromodulator in the flocculus of the rabbit. A potential role of ACh as a cerebellar neurotransmitter has been suggested since the 1960s by the biochemical demonstration of substantial amounts of ACh, acetylcholinesterase (AChE), and choline-acetyltransferase (ChAT; for early references see Reference 21).

More recent histochemical and immunocytochemical techniques have revealed, notwithstanding considerable species differences, the archicerebellum (lobules IX and X) as a preferred site of high concentrations of AChE and ChAT.[22–23] Furthermore, ChAT, the more specific presynaptic marker, has been identified in a subpopulation of mossy fiber terminals in the granular layer of the cerebellar cortex.[23–26] Ojima *et al.* showed that ChAT-immunoreactive glomerular rosettes are most numerous in the vermal lobules IX and X and in the flocculus of the rat.[23] The mossy fibers in these areas of the cerebellum originate mainly from the vestibular complex (except from the lateral vestibular nucleus) and from the prepositus hypoglossi nucleus.[27] In an immunohistochemical double-labeling study, Barmack *et al.* demonstrated the cholinergic nature of a subset of secondary vestibulocerebellar neurons, located in the caudal half of the medial and descending vestibular nuclei and in the prepositus hypoglossi nucleus and projecting to the cerebellar flocculus and nodulus.[28] ChAT immunoreactivity in similar parts of the vestibular nuclei was also demonstrated by Carpenter *et al.*[29] Thus, the vestibular complex appears to be a source of cholinergic mossy fibers to the cerebellar flocculus.

In addition to mossy fibers, thin varicose fibers, closely associated with the Purkinje cell layer, similar fibers in the molecular layer in the rat[23] and cat,[26] and a subpopulation of Golgi cells in the cat[26] were found to be ChAT immunoreactive.

The functionality of the presynaptic cholinergic elements described above is further supported by the identification of muscarinic receptors of the M_2 type[30–32] and nicotinic receptors.[33–34] The nicotinic receptors appear to be located in the granular layer in the rat.[34] In the rabbit, muscarinic receptors, labeled by [^{3}H]quinuclidinyl benzilate[32] in lobules IX and X, including the flocculus, were most dense in the Purkinje cell layer; moderate and low densities were present over the granular and molecular layers, respectively. Furthermore, parasagittal columns of very high density were present over the molecular layer of several cerebellar cortical regions, including the vermis and hemispheres of lobules IX and X.

In comparison to these detailed histochemical studies, the physiological actions of acetylcholine in the cerebellum have remained unclear (for review see References 2 and 21) and no specific role of the cholinergic system in cerebellar signal processing has been demonstrated. The present study will show a marked potentiation of the OKR and a smaller enhancement of the VOR after floccular microinjections of a cholinergic agonist.

METHODS

We used 14 young adult, pigmented Dutch belted rabbits of either sex. They were permanently implanted with ocular sensor coils on each eye for eye movement recording with the magnetic induction method, based on phase detection.[35] The flocculi were localized electrophysiologically on the basis of visually induced direction-selective climbing fiber activity[5] and guide cannulas aimed towards the flocculi were implanted bilaterally (for further details see References 18 and 19). All surgical procedures were done under general anesthesia.

In a first experiment, optokinetic and vestibular stimulation was applied by sinusoidal oscillation of an optokinetic drum around the stationary rabbit, and oscillation of the platform in total darkness, respectively (0.15 Hz, 5 deg peak to peak). At the start of each session, three baseline measurements of OKR and VOR in darkness were made. The mean gain of these three measurements was taken as the reference gain at time $t = 0$. After the baseline measurements, a bilateral floccular injection was made of 1 µl of one of 5 solutions: (a) carbachol (1 g/l), a general (muscarinic and nicotinic) cholinergic agonist; (b) eserine (15 g/l), an acetylcholinesterase inhibitor; (c) mecamylamine (5 g/l), a nicotinic antagonist of the hexametonium type; (d) atropine sulfate (5 g/l), a muscarinic antagonist; and (e) the solvent (saline) only for the control experiments. Only one substance was tested in a single session, and successive sessions in the same animal were separated by intervals of one or several days, to allow complete recovery from the effects of earlier injections. The order in which the drugs were injected varied among rabbits. After the injection, samples of the OKR and the VOR in darkness were collected during a period of 2 hours.

In a second experiment, the platform (and rabbit) remained stationary while steady-state optokinetic nystagmus (OKN) was elicited by rotation of the drum at a constant speed about an earth-vertical axis. The rabbit was kept in darkness, and as soon as the drum velocity was stable, the light was turned on. The buildup of OKN was recorded. After 40 seconds of optokinetic stimulation, the lights were extinguished and optokinetic after-nystagmus (OKAN) was recorded during the following 30 seconds. Stimulation was applied in either direction (clockwise, CW = rightward and counterclockwise, CCW = leftward) at four drum-rotation speeds: 1, 5, 10, and 30°/second. After baseline measurements had been obtained for all of these stimuli, 1 µl of a 1-g/l solution of carbachol was injected on each side in the flocculus. Starting 20 minutes after these injections, all OKR measurements were repeated. Control sessions were run on the same animals with bilateral injection of 1 µl of saline.

The statistical significance of the effects of the injected substances in the experiments with sinusoidal stimulation were tested with a multiple analysis of variance (MANOVA), which allows the comparison of several variables in a single group of animals. Because some of the changes observed were especially prominent during the first hour, only the first 8 measurements following the injections were subjected to this statistical testing.

RESULTS

Sinusoidal Stimulation

Control Experiments

To obtain proper reference values, six rabbits subjected to sinusoidal stimuli were injected with saline alone into the flocculi. The results are plotted as reference curves (open circles) in FIGURE 1. The mean baseline gain values before the saline injections were 0.56 ± 0.23 for the OKR and 0.80 ± 0.19 for the VOR. After the injection, these values showed the tendency to rise slightly (by about 0.1) during the first hour, after which there was no further change. This increase probably represents a mild upward adaptation of the reflexes, due to the repeated optokinetic stimulation.[19] The values obtained in the control experiments served as the reference values in the statistical analysis of the effects of the subsequently injected drugs. The zero value at $t = 0$ in FIGURE 1 represents the normalized, average initial value, obtained in the three baseline measurements of the VOR in darkness and the OKR before an injection was made.

Effects of Carbachol

The mean baseline gains [± standard deviation (SD)] of the same six rabbits in the sessions in which carbachol was to be injected, were 0.63 ± 0.18 for the OKR and 0.73 ± 0.17 for the VOR. After injection of carbachol, all six rabbits showed a very pronounced increase in the gain of the OKR (FIGURE 1) and a smaller, but distinct, increase in the gain of the VOR in darkness. These increases were manifest in the first measurement after the injection, and after about 20 minutes, they reached a maximum, which persisted throughout the further duration of the session. By the next day, values had returned to baseline levels. At 25 minutes after the injection, the increase in the gain of the VOR in darkness, averaged over the six rabbits, was 0.14 ± 0.06 (SD) and the increase in the gain of the OKR was 0.46 ± 0.28 (SD). The highest increase in OKR gain, observed in a single rabbit, was 0.81. The increases in gain of both the VOR in darkness and the OKR were significant ($p = 0.033$ and $p = 0.013$, respectively) in the multiple analysis of variance.

In three other rabbits, the course of the VOR in the light was also assessed. The mean baseline values of the gains of OKR, VOR in darkness, and VOR in light were 0.52, 0.71, and 1.06, respectively. At 25 minutes after the injection of carbachol, the gain of the OKR had increased to 0.94; the gain of the VOR in darkness had increased to 0.92, and the gain of the VOR in the light had only slightly changed (gain 1.11). Thus, the strong increase in the gain of the OKR and moderate increase in the gain of the VOR in darkness were accompanied by an only slight increase in the gain of the VOR in the light. These changes were already prominent 6 minutes after the injection of carbachol, as shown in the representative eye movement recordings in FIGURE 2. (The gain values of the VOR in the light and the OKR in excess of unity were probably due to the inadvertent positioning of the rabbit's eyes slightly anterior to the center of rotation of the drum and platform, resulting in visual angles of rotation larger than the nominal rotation of the drum or platform).

Effects of Eserine

The mean baseline gains in these sessions were 0.73 ± 0.22 for the OKR and 0.66 ± 0.18 for the VOR. Shortly after the injection of the AChE inhibitor eserine, the

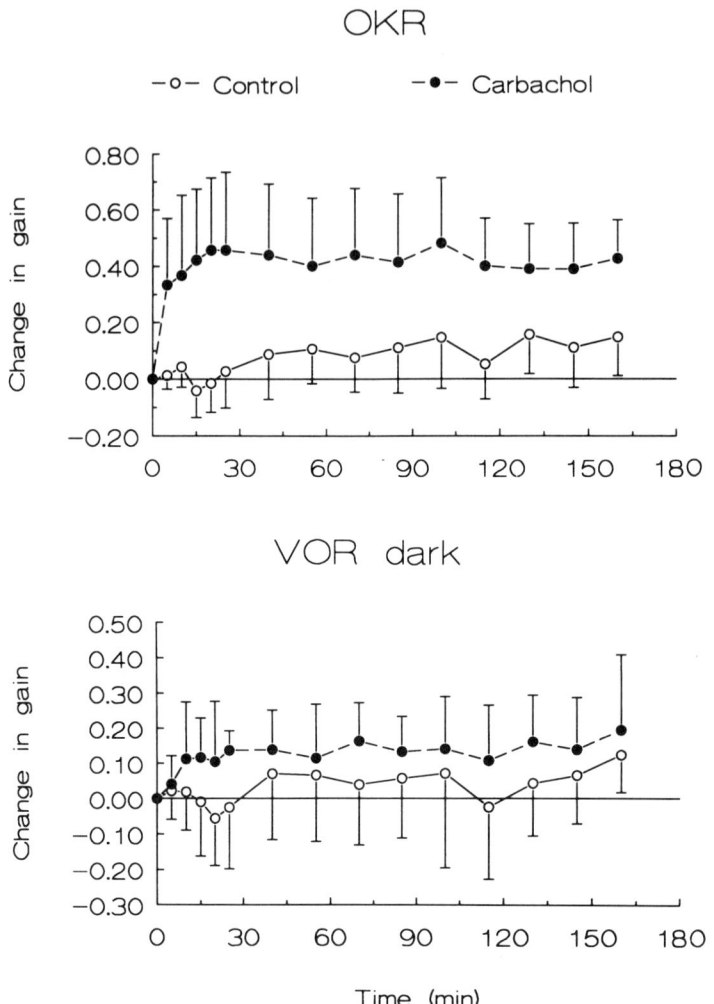

FIGURE 1. Time course of the changes in the gain of the OKR (upper panel) and the VOR in darkness (lower panel) in control experiments (floccular injection of saline; open circles) and after bilateral injection of carbachol in the flocculus at time $t = 0$ (filled circles). Mean and standard deviations (bars) of 6 rabbits. The zero level of the ordinate represents the mean baseline gains, measured just prior to the injection; see text for actual values. (From Tan and Collewijn, 1991. Reprinted by permission from *Experimental Brain Research*.)

gains of both the VOR in darkness and the OKR showed an increase that lasted for about 30 minutes. Similarly as after injection of carbachol, the increase in gain was larger for the OKR than for the VOR in darkness. The increase in the gain of the VOR in darkness at $t = 15$ minutes was 0.16 ± 0.20 (SD), while the increase in OKR gain was 0.22 ± 0.14 (SD). The increase in the OKR gain was significant ($p = 0.026$), but the increase in gain of the VOR in darkness was not significant ($p = 0.164$).

Effects of Atropine

The mean baseline gains in these sessions were 0.89 ± 0.25 for the OKR and 0.69 ± 0.20 for the VOR. Blockage of the muscarinic receptors in the flocculi by atropine resulted in a slow decrease in the OKR gain. After this initial decrease, which amounted to −0.18 ± 0.18 (SD) at t = 25 minutes, the OKR gain recovered partially within the first hour but remained lower than the gain in the control experiments during the next 1.5 hours of testing. In statistical testing, the OKR gain turned out to be significantly lower after atropine injection than after the injection of saline (p = 0.008). The gain of the VOR in darkness did not appear to be affected by the atropine injection, and did not differ significantly from the control experiment (p = 0.259).

Effects of Mecamylamine

The mean baseline gains in these sessions were 0.96 ± 0.26 for the OKR and 0.71 ± 0.19 for the VOR. Injection of the nicotinic, hexamethonium-type antagonist

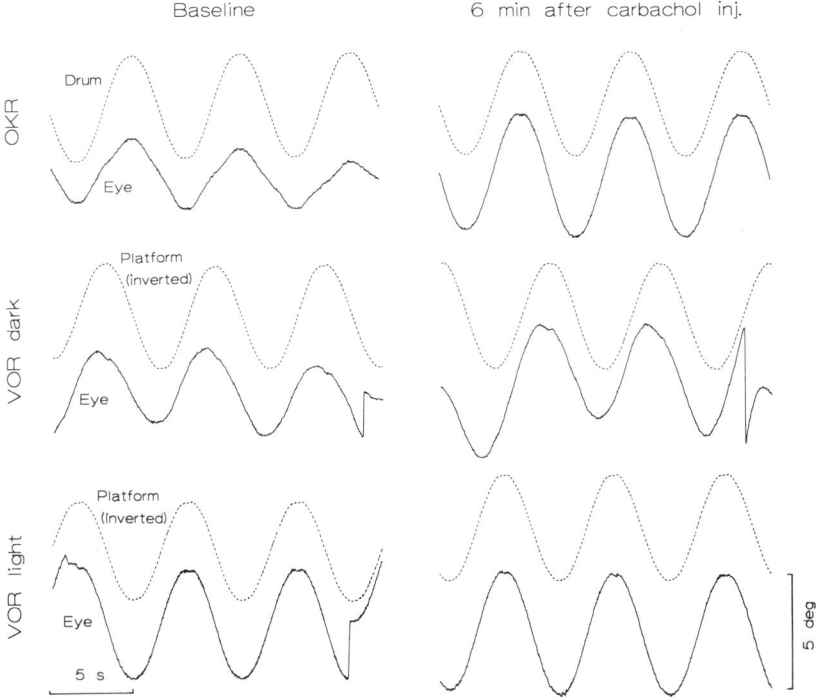

FIGURE 2. Representative recordings of horizontal eye movements of a rabbit in response to oscillation of the drum (OKR) or platform (VOR dark; VOR light) at 0.15 Hz, 5 deg (peak-to-peak) in baseline condition and 6 min after the bilateral floccular injection of carbachol. Note the marked enhancement of the OKR and VOR in darkness after the injection. (From Tan and Collewijn, 1991.[21] Reprinted by permission from *Experimental Brain Research.*)

mecamylamine resulted in an immediate decrease in the gains of both the VOR in darkness and the OKR. The mean change in gain at $t = 15$ minutes was -0.16 ± 0.17 (SD) for the VOR in darkness and -0.21 ± 0.26 (SD) for the OKR. This decrease lasted for only about 30 minutes. In statistical testing, however, these decreases in gain did not reach significance at the 5% level for any of the two reflexes ($p = 0.187$ for the VOR in darkness and 0.072 for the OKR).

Optokinetic Stimulation at Constant Velocities

Binocular Viewing

Typical recordings of OKN, elicited during binocular viewing by rotation of the drum at 10 deg/second in either direction, are shown for one rabbit in the upper panels of FIGURE 3. Although there was an immediate response after the lights were turned on (time marked by triangles), slow-phase velocity was built up only gradually, as is characteristic of the OKN in rabbits for velocities higher than a few deg/second.[36-38] In the rabbit shown in FIGURE 3, buildup was especially slow for drum rotation to the left side (counterclockwise = CCW), and did not reach a ceiling within the 20 seconds of stimulation illustrated in the figure. Bilateral floccular injection of 1 μl of carbachol solution (1 μg/μl) strongly accelerated the buildup (FIGURE 3, lower panels). In response to a similar stimulus of 10 deg/second, 20–30 minutes after the injection, OKN showed a rapid buildup to a maximum steady-state velocity, which was reached within about 5 seconds in either direction.

The changes in the OKN velocity buildup are illustrated systematically for the same typical rabbit in the velocity plots as a function of time, before and after carbachol, in FIGURES 4, 5, and 6 for stimulus velocities of 5, 10, and 30 deg/second, respectively.

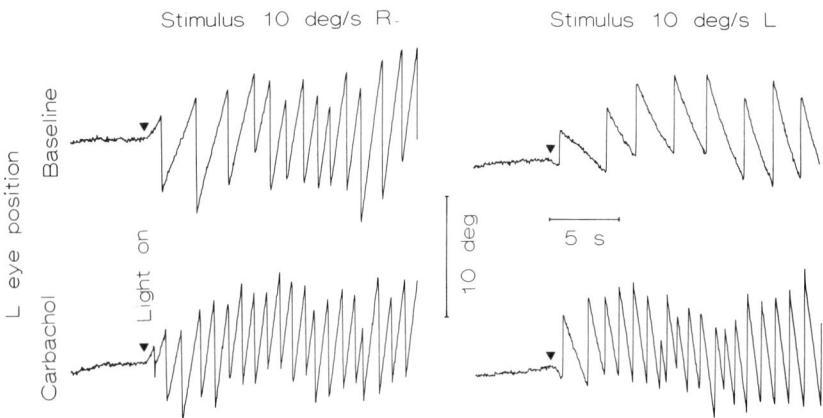

FIGURE 3. Representative recordings of binocularly elicited OKN (left eye position shown as a function of time) in one rabbit, elicited by drum rotation at 10 deg/second in either direction, in the baseline condition (upper panels) and about 20 minutes after bilateral intrafloccular injection of 1 μg carbachol (lower panels). Solid triangles mark the beginning of stimulation (light on).

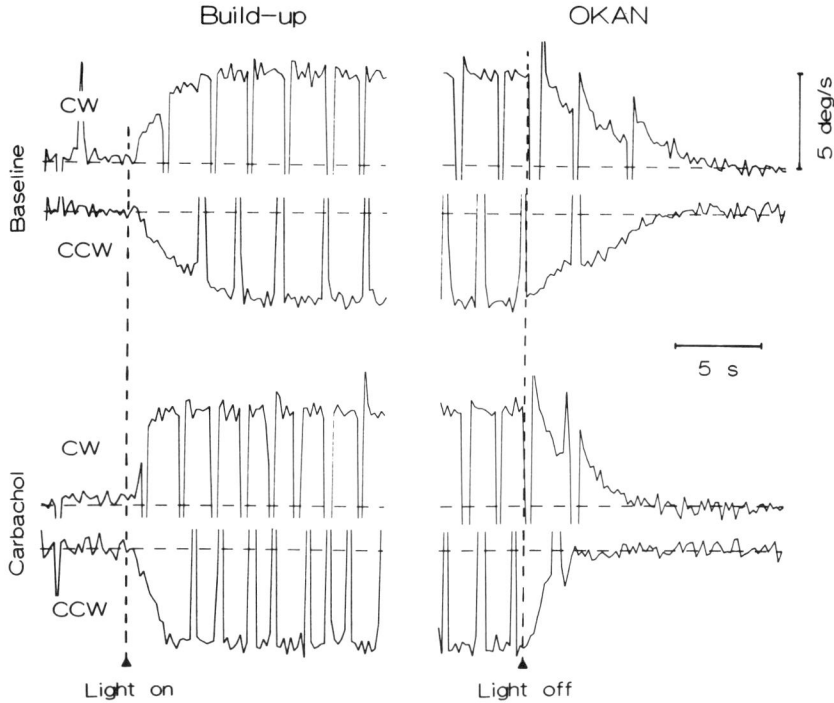

FIGURE 4. Slow-phase velocity of binocularly elicited OKN as a function of time, for a stimulus velocity of 5 deg/second in either direction (rightward = CW upward; leftward = CCW downward). Horizontal, dashed lines represent zero velocity; fast-phase velocities have been truncated. Vertical dashed lines represent onset and offset of stimulus. Notice faster buildup after carbachol (lower panels) compared to baseline conditions (upper panels).

Even for a stimulus of 5 deg/second, the acceleration of the buildup was distinct. Peak velocity was reached within about 6 seconds in the baseline condition, and within 2–3 seconds after injection of carbachol into the flocculus. Acceleration was about 0.72 deg/second^2 in the baseline condition, and about 2.15 deg/second^2 after carbachol. There was, however, no change in the level of the steady-state velocity.

The changes in the time course of slow-phase velocity for a stimulus of 10 deg/second are shown in FIGURE 5 (cf. FIGURE 3 for the position traces). In the baseline condition, buildup was slow, especially for CCW rotation, with an acceleration of about 0.65 deg/second^2. Steady-state velocities (reached after 20–25 seconds) were equal in either direction. After the carbachol injection, similar steady-state velocities were attained within 5–10 seconds, with accelerations of about 2.19 deg/second^2.

Similar effects were evident for a stimulus of 30 deg/second (FIGURE 6). In the baseline condition, buildup took considerably more than 30 seconds, especially in the CCW direction. The average acceleration was about 0.56 deg/second^2. The eventually reached velocity can be seen in the right panels of FIGURE 6. After carbachol, the same steady-state velocities were reached within about 25 seconds, with accelerations of about 0.89 deg/second^2. For all velocities, the steady-state velocities as well

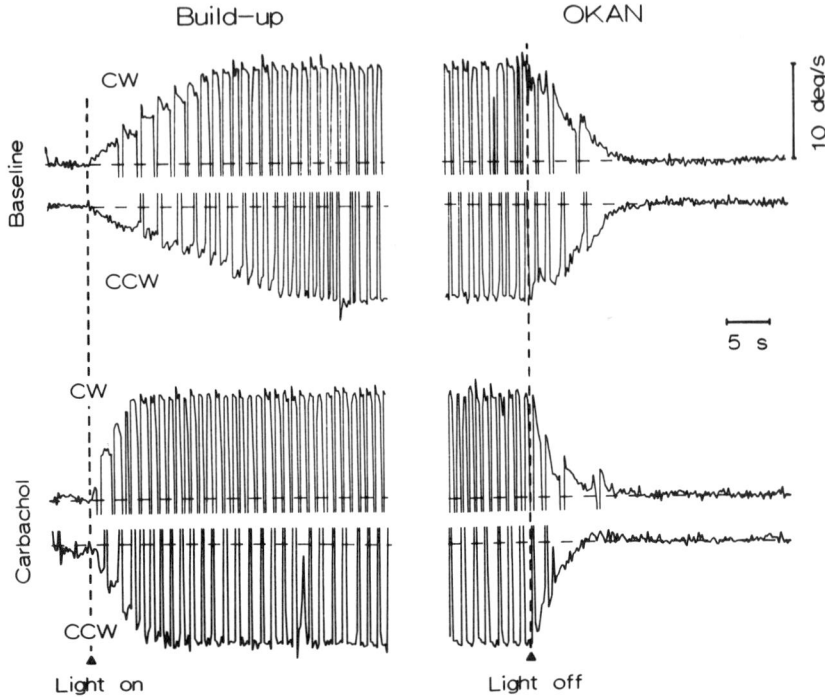

FIGURE 5. Slow-phase velocity of binocularly elicited OKN as a function of time, for a stimulus velocity of 10 deg/second in either direction (rightward = CW upward; leftward = CCW downward). Horizontal, dashed lines represent zero velocity; fast-phase velocities have been truncated. Vertical dashed lines represent onset and offset of stimulus. Notice faster buildup after carbachol (lower panels) compared to baseline conditions (upper panels).

as the general shape of the buildup remained unchanged. In particular, the linear increase of velocity as a function of time (i.e., constant acceleration), best seen for a stimulus of 30 deg/second (FIGURE 6), was maintained. Only the time course was compressed by carbachol.

The effects of carbachol on the mean buildup of slow-phase velocity in the five rabbits tested are summarized in FIGURE 7. These curves confirm that carbachol shortened the average buildup time to about half of the baseline values, without change in the steady-state velocities. The latter were on the order of 0.9 of the stimulus velocities and could, therefore, not be expected to increase substantially. (In this set of experiments, the rabbit's eyes were carefully positioned close to the axis of the drum, and gain would therefore be unlikely to rise above unity). For the lowest stimulus velocity (1 deg/second) no buildup is evident under normal conditions[37] and no effect of carbachol was seen.

The right panels of FIGURES 4–6 illustrate the time course of optokinetic after-nystagmus when the light was turned off (triangular mark) after a steady-state OKN velocity had been maintained for some time. There was a clear tendency towards shortening of the OKAN, paralleling to some extent the shortening of the buildup. Mean values of this shortening have been calculated, at present, for the five rabbits for stimulus velocities of 5 and 10 deg/second. After stimulation at 5 deg/second, the mean time constant of OKAN (time of reduction of velocity to $1/e$ of initial value) was shortened from 6.7 seconds (baseline) to 4.3 seconds (carbachol). After stimulation at 10 deg/second, these values were 7.1 seconds and 4.6 seconds.

Monocular Viewing

Monocularly elicited OKR in the rabbit and other lateral-eyed animals is well known to be asymmetric, with a strong preference for stimulus motion in the nasal (anterior) direction.[36–38] This normal asymmetry is illustrated for the baseline condition in the upper panels of FIGURE 8 (left eye viewing, right eye covered). Bilateral injection of carbachol into the flocculus markedly improved the monocularly elicited OKR in this rabbit. For the preferred direction (CW) the improvement consisted of a faster buildup (FIGURE 8, left panels), as described above for binocular viewing. For the nonpreferred direction (CCW) the effect was more dramatic, because carbachol induced a rapid buildup of slow-phase velocity to the left side, while such a buildup

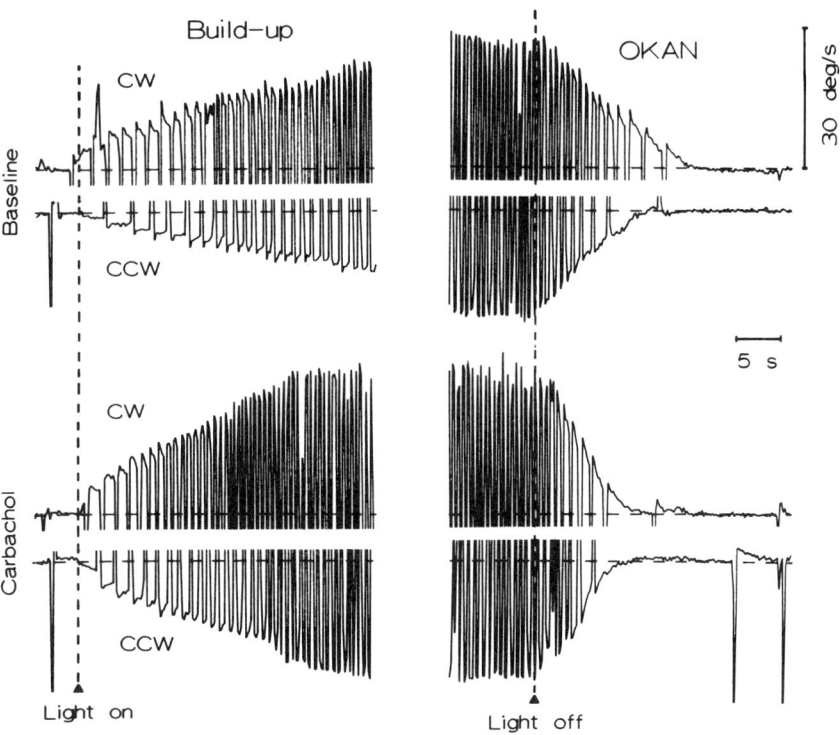

FIGURE 6. Slow-phase velocity of binocularly elicited OKN as a function of time, for a stimulus velocity of 30 deg/second in either direction (rightward = CW upward; leftward = CCW downward). Horizontal, dashed lines represent zero velocity; fast-phase velocities have been truncated. Vertical dashed lines represent onset and offset of stimulus. Notice faster buildup after carbachol (lower panels) compared to baseline conditions (upper panels).

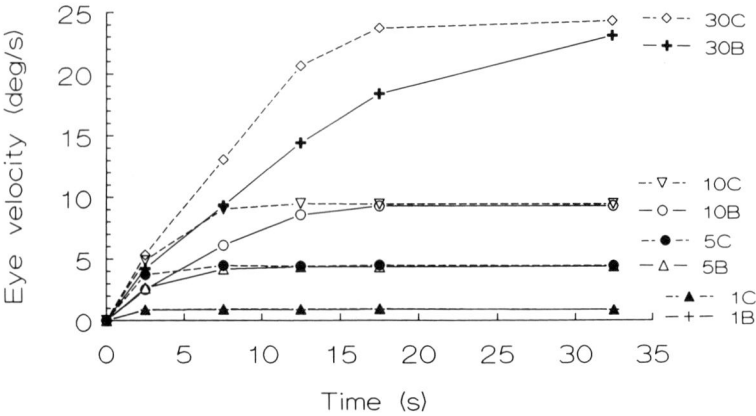

FIGURE 7. Average buildup of binocularly elicited OKN slow-phase velocity (means of five rabbits, two eyes, and two directions). Stimulus velocities are marked at the right side of the graphs (in deg/second). Solid lines represent baseline conditions (B); dashed lines represent buildup after carbachol (C).

was entirely absent in the baseline condition (FIGURE 8, right panels). The mean effects of carbachol on monocularly elicited OKR are shown in FIGURES 9 and 10.

FIGURE 9 shows the time course of slow-phase OKR velocity during stimulation in the preferred, nasal direction (means of five right eyes CCW and five left eyes CW). Comparison of FIGURE 9 with FIGURE 7 shows that buildup in the baseline condition was slower during monocular stimulation than during binocular stimulation, even in the preferred direction, although the eventually reached steady-state velocities were similar (end point not shown for 30 deg/second). The enhancement

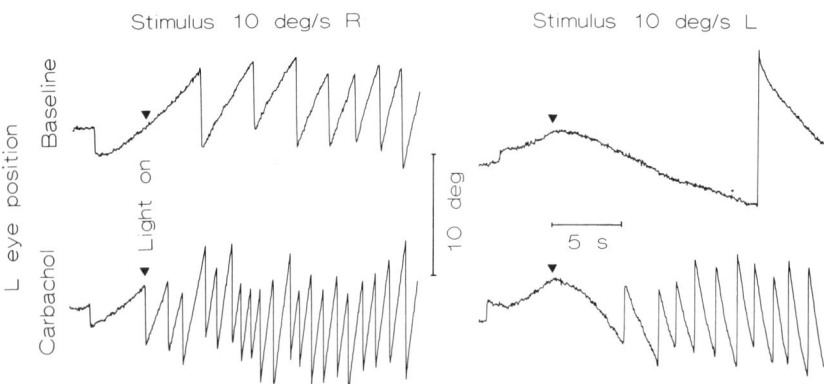

FIGURE 8. Representative recordings of monocularly elicited OKN (left eye position shown as a function of time) in one rabbit (same animal as shown in FIGURE 3), elicited by drum rotation at 10 deg/second in either direction, in the baseline condition (upper panels) and about 20 minutes after bilateral intrafloccular injection of 1 μg carbachol (lower panels). Solid triangles mark the beginning of stimulation (light on).

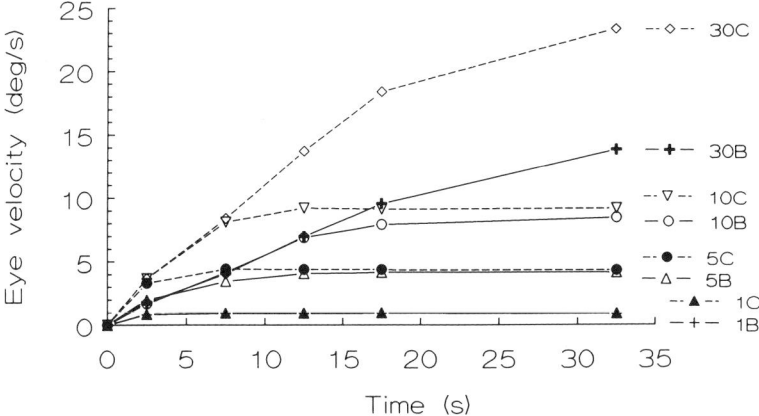

FIGURE 9. Average buildup of monocularly elicited OKN slow-phase velocity (means of five rabbits and two eyes). Stimulus velocities, presented in the preferred (nasal) direction, are marked at the right side of the graphs (in deg/second). Solid lines represent baseline conditions (B); dashed lines represent buildup after carbachol (C).

of the buildup induced by carbachol was relatively large in this condition, although the acceleration remained slightly lower than the acceleration reached after carbachol during binocular viewing.

Mean effects for monocular stimulation in the nonpreferred (temporal) direction are shown in FIGURE 10, at a magnified velocity scale. For stimulation at 1 deg/second, there was no difference with binocular responses[37] and there was no effect of carbachol. For the higher stimulus velocities, all mean baseline responses

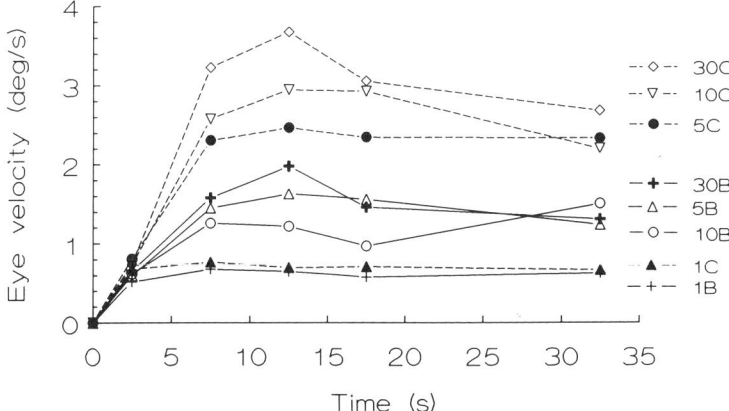

FIGURE 10. Average buildup of monocularly elicited OKN slow-phase velocity (means of five rabbits and two eyes). Stimulus velocities, presented in the nonpreferred (temporal) direction, are marked at the right side of the graphs (in deg/second). Solid lines represent baseline conditions (B); dashed lines represent buildup after carbachol (C).

saturated at about 1.5 deg/second slow-phase velocity, without systematic differentiation between stimulus velocities and without a systematic further buildup after about 10 seconds of stimulation. After carbachol, buildup was faster, but likewise not continued after 10–15 seconds. The steady-state velocities attained were about twice as high as before the carbachol injection, although differentiation for this group of stimulus velocities (5–30 deg/second) remained virtually absent. Thus, with few exceptions (FIGURE 8), the responses in the nonpreferred direction remained at a relatively low level, despite the improvement by a factor of about 2 after the carbachol injection.

It should be emphasized that, although OKN was monocularly elicited, floccular injections with carbachol were bilateral in all cases. Some pilot experiments with unilateral injection were done, but the effects were small and inconsistent.

Control Experiments

The specificity of the effects of carbachol on the time course of OKN velocity buildup was verified by bilateral floccular injections with the same volume (1 μl) of the solvent (saline) only. Such injections had no effect at all on OKN, either during the buildup or during the OKAN phase, as illustrated in FIGURE 11 for a stimulus velocity of 30 deg/second (same animal as in FIGURE 6). Thus, the increase in acceleration during buildup may be considered as specific for carbachol, and unrelated to the injection procedure as such, or the sequence in time of the recordings.

DISCUSSION

The results of the present study strongly suggest that activation of cholinergic receptors in the rabbit's flocculus markedly enhances the responsiveness of the optokinetic system. After intrafloccular injection of carbachol, the gain of OKR in response to sinusoidal stimuli (0.15 Hz) was markedly increased, while the buildup of OKN during stimulation with constant velocities was strongly accelerated, without a clear change in the eventually reached steady-state gain. The gain of the VOR in response to sinusoidal stimuli was moderately enhanced. While the range of stimuli analyzed in our present experiments is too small to allow a complete description of the changes in OKR and VOR dynamics due to cholinergic stimulation of the flocculus, our results are compatible with the hypothesis that a main effect consists in a shortening of the time constant of the buildup of OKN. Such a change would not only accelerate buildup to a steady state, but also facilitate the tracking of oscillatory motion.

This argument can be supported more quantitatively. Our sinusoidal optokinetic stimulus ($f = 0.15$ Hz, $A = 2.5$ deg) contained maximum velocities ($A\omega$) of 2.36 deg/second and maximum accelerations ($A\omega^2$) of 2.22 deg/second^2. In the baseline condition, the rabbit illustrated in FIGURES 4–6 showed buildup accelerations of 0.72, 0.65, and 0.56 deg/second^2 for constant stimulus velocities of 5, 10, and 30 deg/second, respectively. Such values are typical for the rabbit.[37] Taking the value for 5 deg/second as the most appropriate one, it is clear that it would be a limiting factor in the tracking of a sinusoidal motion with peak accelerations of 2.22 deg/second^2. Under the assumption of approximately linear behavior, the gain would be about $0.72/2.22 = 0.32$. After carbachol, peak acceleration for a 5-deg/second stimulus

(FIGURE 4) was increased to about 2.15 deg/second^2, just about enough to cover the acceleration of the sinusoidal stimulus, and to allow a nearly unity gain. These predicted values agree reasonably well with the actual responses. Thus, the increase in gain for sinusoidal optokinetic stimulation after carbachol is compatible with the faster buildup of OKN in response to constant-velocity stimuli. If the buildup is interpreted as a velocity-storage system behaving similarly to a leaky integrator,[39-41] then a more rapid charging should be accompanied by a more rapid discharge, i.e., a shorter OKAN.

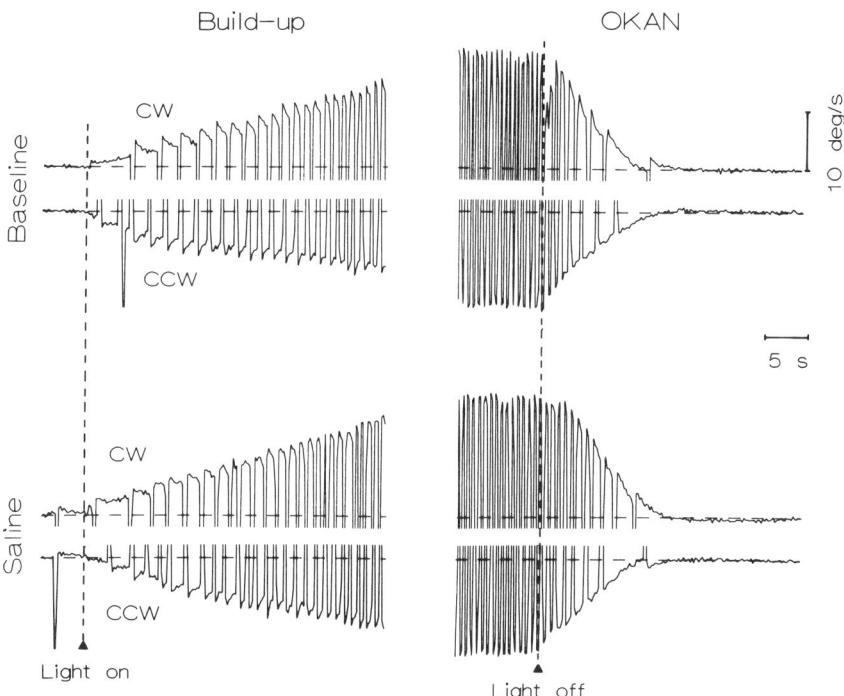

FIGURE 11. Slow-phase velocity of binocularly elicited OKN as a function of time, for a stimulus velocity of 30 deg/second in either direction (rightward = CW upward; leftward = CCW downward). Horizontal, dashed lines represent zero velocity; fast-phase velocities have been truncated. Vertical dashed lines represent onset and offset of stimulus. Control experiment, with intrafloccular injection of saline. Notice unchanged buildup after saline (lower panels) compared to baseline conditions (upper panels).

On the other hand, the increase in the gain of the VOR, elicited by sinusoidal stimuli, is difficult to explain in these terms. While a shortening in the time constant of the OKR, a low-pass system, would be expected to improve the response to a sinusoidal stimulus in a critical frequency range, a complementary effect, i.e., lowering of the gain, would be expected in the VOR, a high-pass system, if the VOR and OKR acted indeed as tightly coupled, complementary components of a synergic system.[42] Changes in time constants of velocity storage may, thus, offer only a partial explanation of our results. A result in the literature on primates may be related to our findings. Lisberger *et al.* changed the gain of the VOR (to sinusoidal stimulation) in monkeys in optical adaptation experiments, and studied concomitant changes in the OKN.[43] Interestingly, they found that in some conditions, a higher VOR gain was accompanied by a more rapid charging of OKAN.

The neurophysiological mechanism underlying our behavioral results has yet to be explored. The enhancing effect of carbachol on the OKR gain is probably mediated in part through muscarinic receptors, because the OKR gain to sinusoidal stimuli could be lowered significantly by specific blocking of these receptors with atropine. Although the effect of blocking of nicotinic receptors with mecamylamine did not reach statistical significance, an additional functional role of nicotinic receptors in the control of the dynamics of the OKR gain is suggested by our results. For better insight in the receptor types involved, specific agonists, and agonist-antagonist combinations, will have to be tested in future experiments.

The finding that the gain of the VOR in the light and of the steady-state OKN elicited by constant velocities remained virtually unaffected by carbachol has to be interpreted in the context of the high gain level measured for these responses under baseline conditions (close to unity gain), and the nature of the control of compensatory eye movements. A high gain of the VOR (a feed-forward system) in combination with a high gain of the OKR (a negative feedback system) will automatically result in a close-to-unity gain of the VOR in the light and of steady-state OKN.

The anatomical evidence on the cholinergic innervation of the cerebellum was briefly reviewed in the Introduction. In the rabbit's flocculus, muscarinic receptors are present at high density in the Purkinje cell layer, while moderate densities were found in the granular layer, and low densities in the molecular layer.[32] Nicotinic receptors were localized in the granular layer in the rat.[34] In combination with the strong evidence for cholinergic mossy fiber terminals,[23–26,28] these findings suggest the mossy fiber–granular cell transition as a likely site for the actions of the cholinergic drugs in our present experiments. The nature of this effect is not evident. It is well known that simple-spike activity in the rabbit's flocculus is profoundly modulated by optokinetic visual stimuli; this modulation is reciprocal with complex-spike activity.[5] As the depth of optokinetic modulation of simple-spike activity is not affected by the reversible blockade of climbing fibers,[44] the source of the simple-spike modulation has to be a mossy fiber input. We may speculate that this depth of modulation may be enhanced by the cholinergic system. The origin of the specific mossy fibers carrying an optokinetic signal is unknown. Neither is it known what signals are carried by the cholinergic subpopulation of mossy fibers originating in the vestibular and prepositus nuclei.[28–29] It seems possible that cholinergic inputs have a modulatory effect, making the granular cells, or the Purkinje cells, more sensitive to specific signals, carried by other fibers. An alternative possibility for a modulatory action of ACh in the cerebellum is via interference with the GABA system. Such a mechanism has been encountered in the hippocampus,[45] where ACh was found to cause disinhibition of hippocampal cells by blocking of the hippocampal GABA system. This effect of ACh was found to be mediated by nicotinic receptors. In the cerebellum, GABA is recognized as the neurotransmitter of several types of cerebellar interneurons.

SUMMARY

Bilateral microinjections into the cerebellar flocculus of the rabbit of carbachol, a general cholinergic agonist, profoundly affect vestibulooocular (VOR) and optokinetic (OKR) reflexes. For sinusoidal stimuli (0.15 Hz, 5 deg peak to peak), the gain of the OKR was strongly increased, while the gain of the VOR was moderately increased. These effects were partially mimicked by floccular injection of the acetylcholinesterase inhibitor eserine. Floccular injection of the muscarinic blocker atropine significantly lowered the gain of the OKR. The effects of the nicotinic blocker mecamylamine were not significant. Optokinetic nystagmus (OKN) in response to constant stimulus velocities (1–30 deg/second) showed a markedly accelerated buildup and a shortened optokinetic after-nystagmus (OKAN) after floccular injections of carbachol. The steady-state gain of OKN remained unaffected. None of the described effects occurred after floccular injection of the solvent, saline. It is postulated that cholinergic cerebellar afferents, one probable source of which are the vestibular nuclei, enhance the optokinetic and vestibular modulation of floccular Purkinje cells.

REFERENCES

1. OKAMOTO, K. & M. SEKIGUCHI. 1991. Synaptic receptors and intracellular signal transduction in the cerebellum. Neurosci. Res. **9:** 213–237.
2. ITO, M. 1984. The Cerebellum and Motor Control. Raven Press. New York, N.Y.
3. DUFOSSÉ, M., M. ITO, P. J. JASTREBOFF & Y. MIYASHITA. 1978. A neuronal correlate in rabbit's cerebellum to adaptive modification of the vestibulo-ocular reflex. Brain Res. **150:** 611–616.
4. NAGAO, S. 1988. Behavior of floccular Purkinje cells correlated with adaptation of horizontal optokinetic eye movement response in pigmented rabbits. Exp. Brain Res. **73:** 489–497.
5. GRAF, W., J. I. SIMPSON & C. S. LEONARD. 1988. Spatial organization of visual messages of the rabbit's cerebellar flocculus. II. Complex and simple spike responses of Purkinje cells. J. Neurophysiol. **60:** 2091–2121.
6. DUFOSSÉ, M., M. ITO & Y. MIYASHITA. 1977. Functional localization in the rabbit's cerebellar flocculus determined in relationship with eye movements. Neurosci. Lett. **5:** 273–277.
7. ITO, M., P. J. JASTREBOFF & Y. MIYASHITA. 1982. Specific effects of unilateral lesions in the flocculus upon eye movements in albino rabbits. Exp. Brain Res. **45:** 233–242.
8. NAGAO, S., M. ITO & L. KARACHOT. 1985. Eye field in the cerebellar flocculus of pigmented rabbits determined with local electrical stimulation. Neurosci. Res. **3:** 39–51.
9. SIMPSON, J. I., J. VAN DER STEEN, J. TAN, W. GRAF & C. S. LEONARD. 1989. Representations of ocular rotations in the cerebellar flocculus of the rabbit. Prog. Brain Res. **80:** 213–223.
10. ITO, M., T. SHIDA, N. YAGI & M. YAMAMOTO. 1974. The cerebellar modification of rabbit's horizontal vestibulo-ocular reflex induced by sustained head rotation combined with visual stimulation. Proc. Jpn Acad. **50:** 85–89.
11. ITO, M., P. J. JASTREBOFF & Y. MIYASHITA. 1979. Adaptive modification of the rabbit's horizontal vestibulo-ocular reflex during sustained vestibular and optokinetic stimulation. Exp. Brain Res. **37:** 17–30.
12. COLLEWIJN, H. & A. F. GROOTENDORST. 1979. Adaptation of optokinetic and vestibulo-ocular reflexes to modified visual input in the rabbit. Prog. Brain Res. **50:** 771–781.
13. ITO, M. 1976. Cerebellar learning control of vestibulo-ocular mechanisms. *In* Mechanisms in Transmission of Signals for Conscious Behavior. T. Desiraju, Ed.: 376–396. Elsevier. Amsterdam, the Netherlands.
14. NAGAO, S. 1989. Behavior of floccular Purkinje cells correlated with adaptation of vestibulo-ocular reflex in pigmented rabbits. Exp. Brain Res. **77:** 531–540.
15. NAGAO, S. 1989. Role of flocculus in adaptive interaction between optokinetic eye movement response and vestibulo-ocular reflex in pigmented rabbits. Exp. Brain Res. **77:** 541–551.
16. NAGAO, S. 1983. Effects of vestibulocerebellar lesions upon dynamic characteristics and adaptation of vestibulo-ocular and optokinetic responses in pigmented rabbits. Exp. Brain Res. **53:** 36–46.

17. BARMACK, N. H. & V. E. PETTOROSSI. 1985. Effects of unilateral lesions of the flocculus on optokinetic and vestibuloocular reflexes of the rabbit. J. Neurophysiol. **53:** 481–496.

18. VAN NEERVEN, J., O. POMPEIANO & H. COLLEWIJN. 1989. Depression of the vestibulo-ocular and optokinetic responses by intrafloccular microinjection of GABA-A and GABA-B agonists in the rabbit. Arch. Ital. Biol. **127:** 243–263.

19. VAN NEERVEN, J., O. POMPEIANO, H. COLLEWIJN & J. VAN DER STEEN. 1990. Injections of β-noradrenergic substances in the flocculus of rabbits affect adaptation of the VOR gain. Exp. Brain Res. **79:** 249–260.

20. TAN, H. S., J. VAN NEERVEN, H. COLLEWIJN & O. POMPEIANO. 1991. Effects of α-noradren-ergic substances on the optokinetic and vestibulo-ocular responses in the rabbit: a study with systemic and intrafloccular injections. Brain Res. **562:** 207–215.

21. TAN, H. S. & H. COLLEWIJN. 1991. Cholinergic modulation of optokinetic and vestibulo-ocular responses: a study with microinjections in the flocculus of the rabbit. Exp. Brain Res. **85:** 475–481.

22. KÁSA, P., K. BÁNSÁGHY, Z. RAKONCZAY & K. GULYA. 1982. Postnatal development of the acetylcholine system in different parts of the rat cerebellum. J. Neurochem. **39:** 1726–1732.

23. OJIMA, H., S. I. KAWAJIRI & T. YAMASAKI. 1989. Cholinergic innervation of the rat cerebellum: qualitative and quantitative analyses of elements immunoreactive to a monoclonal antibody against choline acetyltransferase. J. Comp. Neurol. **290:** 41–52.

24. KAREN KAN, K. S., L. P. CHAO & L. F. ENG. 1978. Immunohistochemical localization of choline acetyl-transferase in rabbit spinal cord and cerebellum. Brain Res. **146:** 221–229.

25. KAREN KAN, K. S., L. P. CHAO & L. S. FORNO. 1980. Immunohistochemical localization of choline acetyl-transferase in the human cerebellum. Brain Res. **193:** 165–171.

26. ILLING, R. B. 1990. A subtype of cerebellar Golgi cells may be cholinergic. Brain Res. **522:** 267–274.

27. EPEMA, A. H., N. M. GERRITS & J. VOOGD. 1990. Secondary vestibulocerebellar projec-tions to the flocculus and uvulo-nodular lobule of the rabbit—a study using HRP and double fluorescent tracer techniques. Exp. Brain Res. **80:** 72–82.

28. BARMACK, N. H., R. W. BAUGHMAN, F. P. ECKENSTEIN & H. WESTCOTT-HODSON. 1986. Cholinergic projections to the cerebellum. Soc. Neurosci. Abstr. **12:** 577.

29. CARPENTER, M. B., Y. HUANG, A. B. PEREIRA & L. B. HERSH. 1990. Immunocytochemical features of the vestibular nuclei in the monkey and cat. J. Hirnforsch. **31:** 585–599.

30. MASH, D. C. & L. T. POTTER. 1986. Autoradiographic localization of M1 and M2 muscarine receptors in the rat brain. Neuroscience **19:** 551–564.

31. SPENCER, D. G., E. HORVÁTH & J. TRABER. 1986. Direct autoradiographic determination of M1 and M2 muscarinic acetylcholine receptor distribution in the rat brain: relation to cholinergic nuclei and projections. Brain Res. **380:** 59–68.

32. NEUSTADT, A., A. FROSTHOLM & A. ROTTER. 1988. Topographical distribution of muscar-inic cholinergic receptors in the cerebellar cortex of the mouse, rat, guinea pig, and rabbit: a species comparison. J. Comp. Neurol. **272:** 317–330.

33. HUNT, S. & J. SCHMIDT. 1978. Some observations on the binding patterns of α-bunga-rotoxin in the central nervous system of the rat. Brain Res. **157:** 213–232.

34. SWANSON, L. W., D. M. SIMMONS, P. J. WHITING & J. LINDSTROM. 1987. Immunohistochem-ical localization of neuronal nicotinic receptors in the rodent central nervous system. J. Neurosci. **7:** 3334–3342.

35. COLLEWIJN, H. 1977. Eye and head movements in freely moving rabbits. J. Physiol. London **266:** 471–498.

36. TER BRAAK, J. W. G. 1936. Untersuchungen ueber optokinetischen Nystagmus. Arch. Néerl. Physiol. **21:** 309–376.

37. COLLEWIJN, H. 1969. Optokinetic eye movements in the rabbit: input-output relations. Vision Res. **9:** 117–132.

38. COLLEWIJN, H. 1981. The Oculomotor System of the Rabbit and Its Plasticity. Springer. Berlin, Germany.

39. COLLEWIJN, H. 1972. An analog model of the rabbit's optokinetic system. Brain Res. **36:** 71–88.

40. COHEN, B., V. MATSUO & T. RAPHAN. 1977. Quantitative analysis of the velocity characteristics of optokinetic nystagmus and optokinetic afternystagmus. J. Physiol. London **270:** 321–344.
41. RAPHAN, T., V. MATSUO & B. COHEN. 1979. Velocity storage in the vestibulo-ocular reflex arc. Exp. Brain Res. **35:** 229–248.
42. ROBINSON, D. A. 1977. Vestibular and optokinetic symbiosis: an example of explaining by modelling. *In* Control of Gaze by Brain Stem Neurons. R. Baker & A. Berthoz, Eds.: 49–58. Elsevier. Amsterdam, the Netherlands.
43. LISBERGER, S. G., F. A. MILES, L. M. OPTICAN & B. B. EIGHMY. 1981. Optokinetic response in monkey: underlying mechanisms and their sensitivity to long-term adaptive changes in vestibuloocular reflex. J. Neurophysiol. **45:** 869–890.
44. LEONARD, C. S. & J. I. SIMPSON. 1986. Simple spike modulation of floccular Purkinje cells during the reversible blockade of their climbing fiber afferents. *In* Adaptive Processes in Visual and Oculomotor Systems. E. L. Keller & D. S. Zee, Eds.: 429–434. Pergamon. Oxford, England.
45. FREUND, R. K., D. A. JUNGSCHAFFER, A. C. COLLINS & J. M. WEHNER. 1988. Evidence for modulation of GABAergic neurotransmission by nicotine. Brain Res. **453:** 215–220.

Modulation of Excitatory Transmission at the Rat Medial Vestibular Nucleus Synapse[a]

JOEL P. GALLAGHER, KEVIN D. PHELAN,[b] AND
PATRICIA SHINNICK-GALLAGHER

Department of Pharmacology and Toxicology
University of Texas Medical Branch
Room 1.101 Pharmacology Building
Galveston, Texas 77550

Information transfer from the peripheral vestibular apparatus to the central vestibular nuclear complex (VNC) and the intrinsic processing of this information within the VNC must first be understood in order to define normal vestibular function and the etiology of a variety of clinical vestibular disorders. It is only when the specific types of receptors responsible for the cellular chemical transduction at synapses within this complex are identified and characterized that an optimal rational therapeutic approach towards treating vestibular disorders will be realized. As a first step towards achieving this goal we have developed an *in vitro* brain slice preparation from rats[1,2] that contains components of the central VNC, specifically, the primary vestibular nerve (N. VIII) afferent inputs and second-order neurons in the medial and lateral vestibular nuclei (FIGURE 1A). We have coupled electrophysiological, primarily intracellular, recording techniques with this *in vitro* brain slice preparation to examine the cellular pharmacology at the primary afferent to second-order synapse in the medial vestibular nucleus (MVN).

The development of the *in vitro* brain slice preparation has allowed us to minimize many of the difficulties associated commonly with attempts to record intracellularly from the VNC in the intact animal. On the other hand, by reducing the preparation to the central components of the vestibular system, we have unfortunately removed all of the receptor-driven inputs that the brain normally receives from the peripheral vestibular apparatus as well as the extrinsic circuit inputs arising from other descending and ascending influences that are active in an intact animal. Nonetheless, we are still able to evoke "typical" excitatory synaptic transmission with our isolated *in vitro* preparation by stimulating the primary vestibular afferent fibers in the N.VIII rootlet and along its pathway to the MVN (FIGURE 1B). Furthermore, it is apparent that other nerve terminals (e.g., GABAergic) remain functionally intact in the slice and are fully capable of releasing their transmitters, since application of selective antagonists for various substances modifies spontaneous electrical and synaptic activity.

A significant finding associated with the development of the *in vitro* preparation was the observation that MVN neurons exhibited spontaneous action potentials,[1-6] i.e., they were not quiescent, despite the fact that they were no longer under the

[a]This research was supported by National Aeronautics and Space Administration Grant NAG2-260.

[b]Current affiliation: Department of Neurology, University of Pennsylvania School of Medicine, Philadelphia, Pennsylvania 19104.

influence of coordinated activity from their peripheral vestibular apparatus. In fact, we found that the majority of MVN neurons generated spontaneous action potentials with a frequency in the range of 15 to 40 Hz[1] (FIGURE 2). The rather surprising finding that vestibular neurons *in vitro* exhibit spontaneous high-frequency activity has been confirmed in other laboratories primarily using extracellular recording techniques from brain stem slice preparations containing the VNC from rat,[7-11]

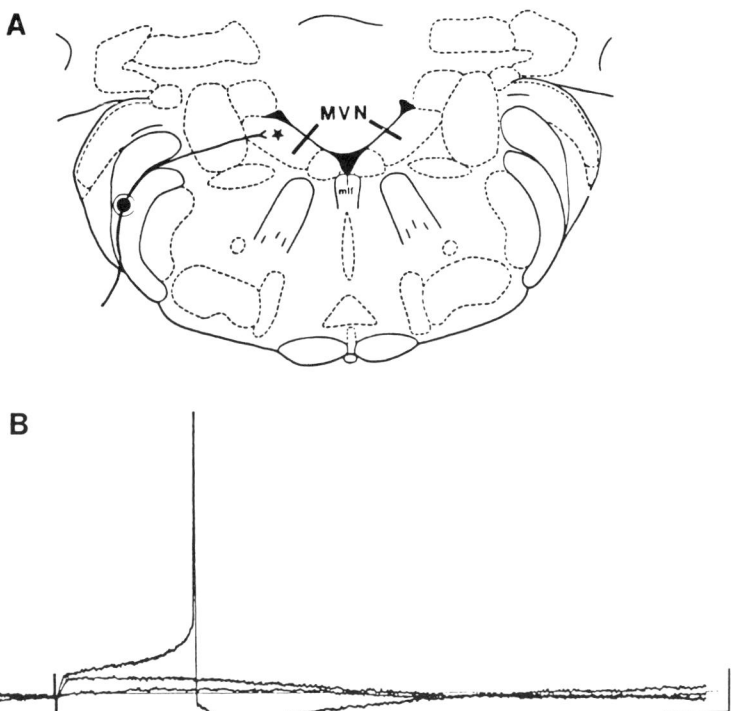

FIGURE 1. Typical anatomy of our slice preparation and synaptic response. **A:** Schematic diagram (adapted from Paxinos and Watson),[47] illustrating the position of the medial vestibular nucleus (MVN) in the rat brain stem with the position of the stimulating (filled circle) and recording (star) electrodes used for afferent stimulation and intracellular recording, respectively. MLF = medial longitudinal fasciculus. The solid line represents the path of the primary afferent fibers, N. VIII. **B:** Stimulus artifact followed by typical orthodromically induced EPSP/action potential of MVN neuron maintained at −65 mV. Calibration = 10 mV × 10 mseconds.

guinea pig,[12-17] or frog.[18] Moreover, Carpenter and Hori employing both extracellular and intracellular recording techniques from a rat MVN slice preparation have reported in this volume that MVN neurons exhibited an endogenous "pacemaker" discharge which was sensitive to membrane potential but insensitive to blockade of synaptic transmission.[11] Spontaneous action potential activity is a common feature of many other central nervous system (CNS) neurons[19] recorded *in vitro*.

The fact that MVN neurons are spontaneously active in a slice preparation suggests that these neurons possess intrinsic electrical activity that is dependent upon the distribution of specific ion channels along their membrane. These ion channels may be independent of or associated with specific receptors. The activation or blockade of these receptors could result in a voltage change that would activate or inactivate voltage-dependent, receptor-independent ion channels to alter the firing characteristics of the vestibular neurons. The existence of such receptors would have important implications for the primary afferent information transfer and processing that occurs within the MVN. We have found that the spontaneous firing frequency of rat MVN neurons is exquisitely sensitive to small changes in the voltage level of their membrane potential (FIGURE 2B). This feature endows MVN neurons with the special property that allows neurally active substances to exert a powerful effect on

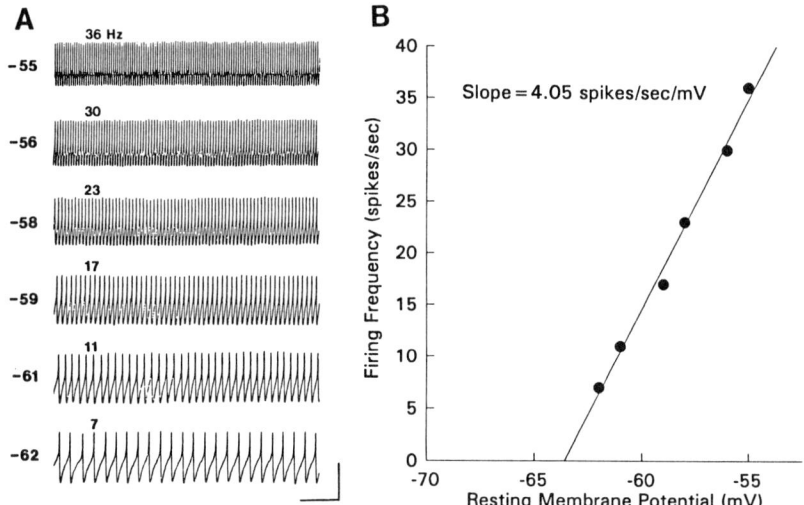

FIGURE 2. MVN neurons exhibit voltage-dependent spontaneous activity. **A:** Examples of spontaneous action potential frequency at various membrane potentials. Calibration = 20 mV × 500 mseconds. **B:** Plot of the cell in A showing the linear relationship between membrane potential and firing rate when examined over the range −62 to −55 mV.

the firing rate and ultimately the output of MVN neurons, simply by producing small depolarizations or hyperpolarizations in the MVN neuron.

We will summarize the effects of various pharmacological agents that we have applied to the *in vitro* preparation of the MVN in order to characterize the endogenous transmitter responsible for ipsilateral primary afferent transmission at the rat MVN synapse and to determine the cellular mechanisms that govern the normal modulation of this input during its translation into the output of the second-order MVN neuron.

We have employed standard intracellular electrophysiological methods to record from MVN neurons in the rat brain stem preparation.[1,2] Briefly, male Sprague-Dawley rats (100–200 grams) were decapitated and their brains removed rapidly and immersed in ice-cold artificial cerebrospinal fluid (ACSF) prebubbled with 95% O_2

and 5% CO_2 of the following composition (in mM): NaCl, 122; KCl, 5; $CaCl_2$, 2.5; $MgCl_2$, 1.2; $NaHCO_3$, 26; and D-glucose, 10; pH = 7.4. Serial, 500-μm thick sections from the brain stem containing the VNC were obtained and maintained at room temperature in the ACSF. A single slice was placed in the recording chamber where it was kept submerged and continuously superfused with oxygenated ACSF. The slice was allowed to stabilize for at least one hour at 34°C prior to the onset of recording. Intracellular recordings were obtained with glass microelectrodes filled with 4 M KAcetate and having a final electrode resistance of 60–150 MΩ. All drugs were applied in known concentrations either by bath superfusion, pressure ejection from a pipette, or microdrop application from a syringe directly into the recording chamber.[1-6]

EXCITATORY AMINO ACID–MEDIATED PRIMARY AFFERENT TRANSMISSION

We have reported that an excitatory amino acid (EAA), by activating a non-*N*-methyl-D-aspartate (NMDA) receptor, mediates ipsilateral synaptic transmission at the rat MVN synapse.[6] Early studies (for review, see Reference 20) concluded that acetylcholine (ACh) was the primary vestibular afferent transmitter. However, substantial support that an excitatory amino acid is the primary vestibular afferent transmitter has been obtained in a variety of species using multiple experimental approaches (for review, see Reference 21), e.g., electrophysiological,[1-3,6,10,13,22-24] biochemical,[25] anatomical,[26] and immunocytochemical.[27-29] It is interesting that the presence of multiple types of EAA receptors on rat MVN neurons[2] provides this synapse with an opportunity to exhibit plasticity by ipsilateral activation of non-NMDA receptors[6] and potential contralateral activation of NMDA receptors.[12,13,18,22-24]

MVN NEURONS, SPONTANEOUS ACTIVITY, AND ITS VOLTAGE DEPENDENCE

Our initial observations using the slice preparation indicated that neurons in the rat MVN are not homogeneous and could be separated into at least two major classes based on the differences in their action potential profile (see Figure 4 in Reference 1 and FIGURE 3 below). This classification scheme has been adopted in a recent analysis of MVN neurons in the guinea pig[16,17] where as many as three different types of MVN neurons have been suggested. In addition to different action potential profiles, rat MVN neurons could be distinguished by their voltage-dependent rectification properties (FIGURE 3B and C). One group of neurons displayed a voltage- and time-dependent rectification (FIGURE 3B and C; left), which may be due to a "Q"-type current.[19] On the other hand, a second type of MVN neuron exhibited only a voltage-dependent rectification (FIGURE 3B and C; right), which is most likely due to the typical potassium rectifier.[19] There is no statistical correlation between MVN neurons exhibiting different action potential profiles and those having different rectifying properties.

Despite the fact that rat MVN neurons displayed different action potential profiles and different rectifying properties, their mean firing frequency of 20 Hz $(n = 96)^2$ recorded intracellularly did not differ. Unfortunately, the average value of 20 Hz was obtained from neurons over a holding potential range of −52 to −69 mV, which, as can be seen from the tracings in FIGURE 2, is the range where the frequency

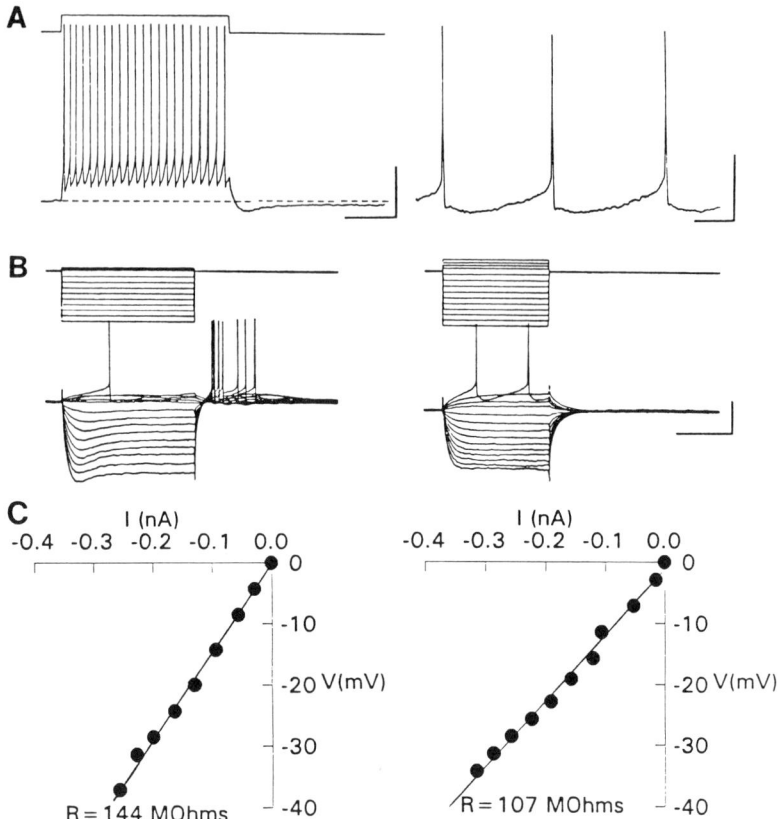

FIGURE 3. The MVN nucleus consists of a heterogeneous population of neurons. **A:** MVN neurons can be distinguished by exhibiting either a single or double (right) after-hyperpolarizing potential following each spike, while neither type of MVN neuron exhibits accommodation during a cathodal stimulus. Calibration = 20 mV × 150 mseconds, left; 20 mV × 50 mseconds, right) **B:** MVN neurons exhibit at least two different types of voltage-dependent rectification; one type (left) is also time dependent. Calibration = 20 mV × 100 mseconds. **C:** Plot of steady-state current-voltage relationship collected from the two different neurons in B.

is highly voltage dependent. In fact, based on a sampling of MVN neurons from which we determined spontaneous action potential frequency at various membrane potentials we have calculated that with each mV change in membrane potential, spontaneous action potential firing rate increases at 4.3 ± 0.4 spikes/second per mV [mean $\pm$ standard error (SE), $n = 19$; e.g., FIGURE 2B]. This spontaneous activity can be due to two interacting influences. One, MVN neurons possess intrinsic membrane properties that allow them to act as oscillators; and, two, synaptic activity can drive and/or inhibit second-order neuron output. Each of these variables can be altered by drugs. Our laboratory's long-term goal has been to identify the endogenous neuronal and hormonal mediators that are active chemical transducers within the central VNC.

CHOLINERGIC MODULATION OF MVN ACTIVITY

Acetylcholine has long been considered an important endogenous element of central vestibular function. Moreover, the role of ACh as an excitatory modulator of MVN activity has been highlighted following *in vitro* intracellular[4,5] and extracellular[7-9] electrophysiological studies using rat brain stem preparations. The excitatory effect of ACh on MVN neurons has been suggested to be mediated by activation of a subtype of muscarinic receptor other than the M_1 receptor.[9] There are currently five known subtypes of muscarinic receptors.[30] Our intracellular data have confirmed the excitation mediated by muscarinic receptors and have also revealed three additional actions of ACh not previously recognized in the MVN.

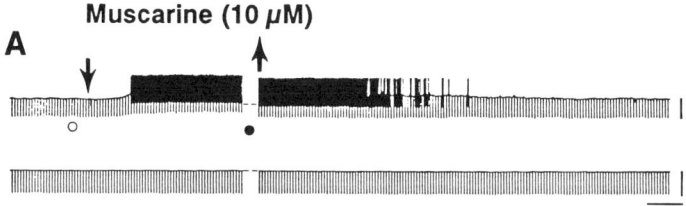

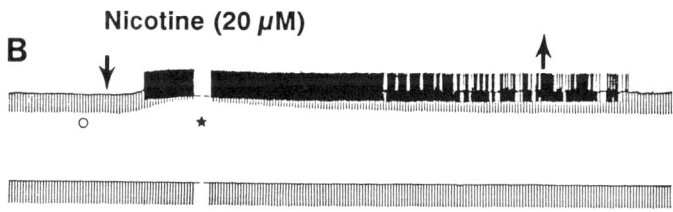

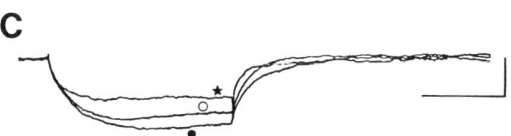

FIGURE 4. Actions of cholinergic agonists on a single rat MVN neuron. **A:** Muscarine depolarized and induced repetitive firing in an MVN neuron maintained at −65 mV. **B:** Nicotine applied to the same neuron also depolarized and induced repetitive firing. **C:** Comparison of the input resistance (1/conductance) in control (open circle) with changes during muscarine (filled circle) and nicotine (star). Down and up arrows in A and B indicate beginning and end of superfusion with respective agonists. Calibration = 10 mV, 0.15 nA × 30 seconds for A and B; ×50 mseconds for C.

Although we have demonstrated that ACh, and more specifically, muscarine, can depolarize and thus increase the firing rate of MVN neurons (FIGURE 4A),[4,5] we have also demonstrated that nicotine (FIGURE 4B), and the selective nicotinic agonist DMPP (1,1-dimethyl-4-phenylpiperazinium iodide), can produce similar effects, albeit through a different mechanism (FIGURE 4C). Muscarine activates postsynaptic muscarinic receptors on MVN neurons to cause a closure of a population of potassium channels (i.e., an increase in resistance) resulting in a subsequent depolarization of the neuron. Nicotinic agonists, on the other hand, also excite (depolarize) the same MVN neurons, but this depolarization is associated with an opening of cationic channels (i.e., a decrease in resistance). Thus, the endogenous release of ACh may depolarize MVN neurons by two different and perhaps opposing mechanisms involving both nicotinic and muscarinic receptors.

Second, we have demonstrated that muscarine indirectly hyperpolarizes an MVN neuron and decreases its rate of firing by inducing the release of γ-aminobutyric acid (GABA) from intrinsic neurons and/or intact terminals within the slice.[4,5] This inhibitory effect of muscarine is completely blocked by the GABA$_A$ antagonist bicuculline. An indirect inhibitory action for cholinergic agonists has been confirmed[7-9] and attributed to an excitatory action on muscarinic receptors to release inhibitory transmitters.[7]

Finally, we have discovered an additional inhibitory action of ACh at a presynaptic site through activation of muscarinic receptors located on the nerve terminals of primary afferent neurons releasing an EAA (FIGURE 5). Application of muscarine (10 μM) to the slice causes a marked decrease in the amplitude of an EAA-mediated excitatory postsynaptic potential (EPSP) recorded from a rat second-order MVN neuron. This inhibitory action of muscarine occurs when the membrane potential of the neuron is maintained at control levels to minimize changes associated with the conductance decrease caused by muscarine. ACh has been demonstrated to inhibit the release of excitatory amino acids from various CNS nuclei[31,32] by activating presynaptic muscarinic receptors. The precise subtype of muscarinic receptor underlying the presynaptic effect of muscarine at the MVN has not yet been determined, but could prove to be clinically relevant. For instance, atropine, which is a nonselective muscarinic receptor antagonist, would block both at the pre- and postsynaptic muscarinic receptors, regardless of the specific muscarinic receptor subtypes. On the other hand, if the muscarinic receptor located on the primary afferent EAA-releasing nerve terminals is different than the postsynaptic muscarinic receptor, then it is possible that the pharmacological blockade of the presynaptic receptor by a specific muscarinic receptor subtype antagonist would result in an enhancement of primary afferent transmission in the MVN.

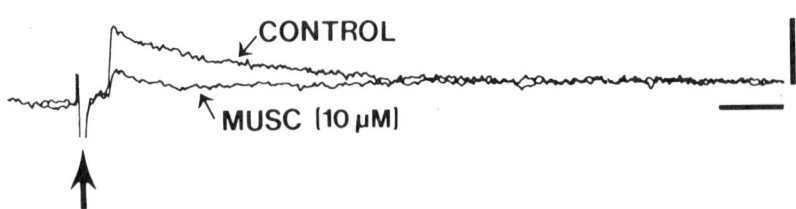

FIGURE 5. Muscarine depresses synaptically evoked, excitatory amino acid mediated, EPSPs. Initial vertical arrow points to stimulus artifact followed by a typical EPSP. The amplitude of the EPSP is depressed significantly while membrane potential is maintained at −65 mV. Calibration = 5 mV × 10 mseconds.

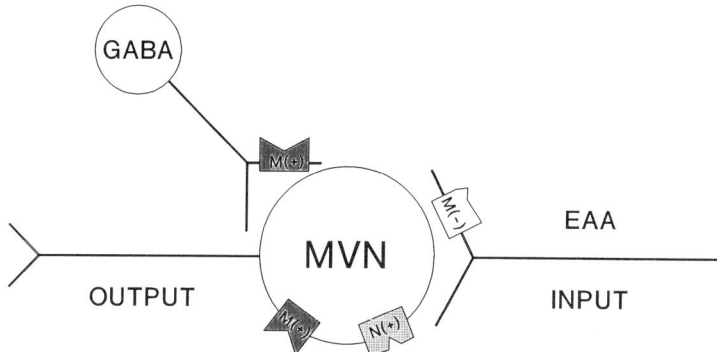

FIGURE 6. Summary of cholinergic receptors we have demonstrated within the MVN. Presynaptic inhibitory muscarinic receptor (M−) on primary afferent excitatory amino acid releasing terminal. Postsynaptic excitatory cholinergic receptors are both muscarinic (M+) and nicotinic (N+). An excitatory muscarinic receptor is located either on the cell body or nerve terminal of a GABAergic interneuron or projection neuron within the slice.

We have summarized the results of our work with cholinergic receptors at the MVN synapse in FIGURE 6. The EAA terminals representing the N.VIII input are depicted on the right with an inhibitory muscarinic receptor. Another pre- or postsynaptic muscarinic receptor is localized on a GABAergic interneuron, which may represent (1) the contralateral type I or type II inhibitory input; (2) a descending input from cerebellar projection neurons; or (3) an interneuron within the ipsilateral VNC. Finally, two excitatory cholinergic receptors, one nicotinic and the other muscarinic, are located postsynaptically on the MVN neuron; activation of either or both of these postsynaptic receptors may ultimately regulate the output from the MVN. A possible *local* source of endogenous ACh within the MVN could be from an axon collateral of cholinergic neurons described in this volume[33] to reside within in the MVN and terminate in the cerebellum.

GABAERGIC MODULATION OF MVN ACTIVITY

The role of GABA as an inhibitory neurotransmitter within the VNC has been known since the early work of Curtis *et al.*[34] Two major subtypes of GABA receptors are now recognized, $GABA_A$ and $GABA_B$ receptors. We have demonstrated that GABA can hyperpolarize MVN neurons (FIGURE 7A) and that this action can be antagonized by bicuculline, a selective $GABA_A$ antagonist.[35] This result has recently been confirmed in the guinea pig MVN.[36] We have further demonstrated that baclofen, a selective $GABA_B$ agonist, also hyperpolarizes rat MVN neurons (FIGURE 7B). Thus, our data demonstrate that GABA can inhibit MVN neuronal activity by activating either $GABA_A$ or $GABA_B$ receptors located postsynaptically. The role of similar GABA receptors at presynaptic sites, e.g., on GABA and/or EAA terminals, is yet to be determined. Also presented in this volume[37] is ultrastructural anatomical support with immunocytochemical probes for $GABA_B$ receptors, localizing discrete antibaclofen immunostains within both pre- and postsynaptic sites in the MVN of monkeys. The fact that muscarinic receptor activation induces only $GABA_A$-mediated inhibition[5] of MVN neurons raises the possibility that there is more than

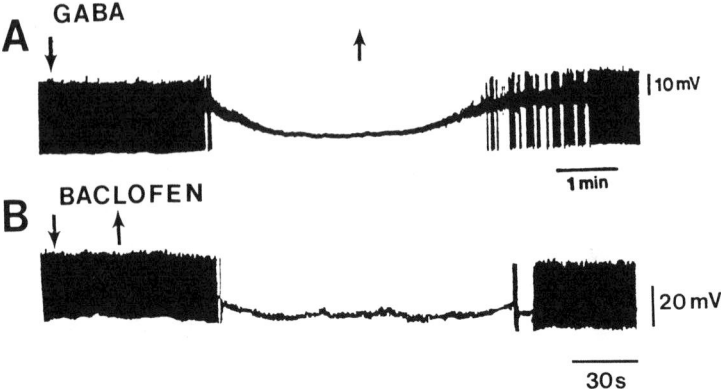

FIGURE 7. GABA inhibits the firing of MVN neurons. **A:** GABA (100 μM) hyperpolarizes and inhibits action potential firing of an MVN neuron. This inhibition was blocked by bicuculline (10^{-5} M). **B:** Application of baclofen (10 μM) to a different MVN also hyperpolarizes and inhibits MVN firing via GABA$_B$ receptor activation. Membrane potential is -60 mV in both neurons.

one GABAergic neuron with different originations and both may terminate on a single MVN neuron.

HISTAMINERGIC MODULATION OF MVN ACTIVITY

Antihistaminics are effective in preventing motion sickness,[38] and histamine receptors (H$_1$ subtype) have been identified in the rat MVN by the use of autoradiographic techniques.[39] In addition, a histaminergic projection to the rat MVN from the hypothalamus has been demonstrated using retrograde transport and immunocytochemical techniques.[40] Our data demonstrated that histamine receptors of the H$_2$ subtype are present on rat MVN neurons, and that activation of these receptors results in depolarization and increased firing of the neurons.[41] These data suggest that a possible therapeutic regimen to treat motion sickness, which may be more effective and cause less sedative side effects than current therapy, might be the use of an H$_2$ rather than an H$_1$ antagonist. The role of H$_1$ receptors in the MVN is not yet known, but may potentially be a site for presynaptic action of histamine. Vidal *et al.* have also demonstrated similar actions for histamine at H$_2$ receptors at the guinea pig MVN.[42] In addition, they have demonstrated H$_3$ receptor activity and suggest that manipulation of this latter presynaptic receptor may be useful in treating or controlling motion sickness.[42]

MULTIPLE AMINERGIC RECEPTORS AND MVN ACTIVITY

In addition to EAA receptors and receptors for ACh, GABA, and histamine, three endogenous catecholamines, namely, serotonin (5-HT), norepinephrine (NE), and dopamine (DA), all consistently depolarized MVN neurons (FIGURE 8) and usually induced action potential firing or an increase in the rate of firing when applied to the slice preparation. The receptor selectivity of these responses has not yet been determined. Nonetheless, these data raise the possibility that pharmacological manipulation of receptor subtypes responsible for the excitatory actions of the various catecholamines may prove to be clinically effective in treating vestibular related disorders. In fact, norepinephrine and dopamine have long been implicated to play a role in the amelioration or induction, respectively, of motion sickness.[43] In addition, Lucot has reported that emesis can be induced by several 5-HT receptor subtype agonists while being prevented by the selective 5-HT$_{1A}$ agonist 8-OH-DPAT.[44] His results and other related studies are summarized in a monograph dealing with motion and space sickness.[45] A critical question is the source of these endogenous amines. Since there is not a strong NE/DA input into the VNC, these aminergic receptors may be activated by circulating agents in the cerebrospinal fluid (CSF). Thus, the roles of specific catecholamine receptors in the VNC may prove to be an area worthy of further investigation.

SUMMARY OF RECEPTORS PRESENT AT THE RAT MVN SYNAPSE

The MVN represents one of several subnuclei of the VNC which receives and processes inputs from the peripheral vestibular sensory apparatus. Although the physiological processes responsible for vestibular function have been studied for more than 50 years (for review, see Reference 46), the cellular pharmacology of the central vestibular system is a relatively new area of investigation. We have attempted to review the state of our understanding of the cellular pharmacology in the central vestibular system as derived principally from data collected with *in vitro* preparations of the VNC. FIGURE 9 summarizes our findings at the rat second-order synapse in the MVN. It should be emphasized that specific receptor subtypes may differ depending upon species, so that pharmacological manipulations with the rat may not

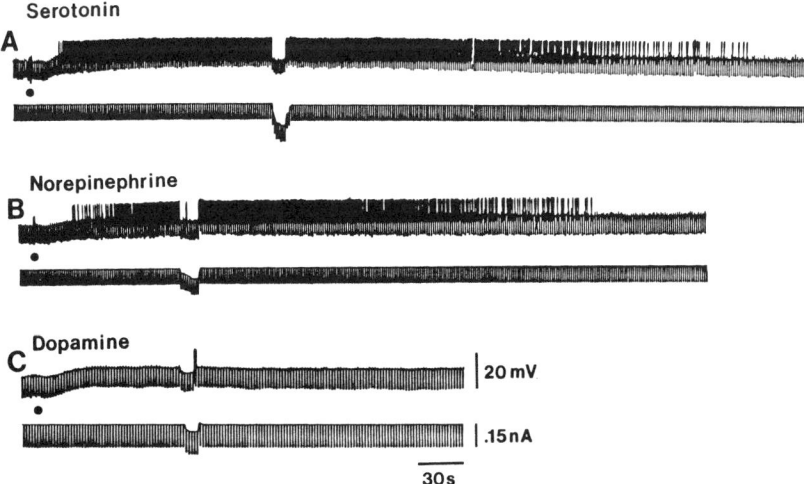

FIGURE 8. Catecholamines depolarize MVN neurons. **A:** Serotonin depolarizes and induces action potential firing. **B:** Norepinephrine also depolarizes and causes repetitive action potentials. **C:** Dopamine depolarizes, but does not generate, repetitive action potentials. All amines were applied as a drop (approximately 10 μM final bath concentration) to the same MVN neuron at a membrane potential of −65 mV.

necessarily apply to other animals or humans. Yet, without these investigations with the rat, the most appropriate choice for therapy or prophylaxis of vestibular disorders in the human is difficult. In addition, a number of receptors for other transmitters or modulators [e.g., peptidergic, opioid (but see Reference 11), etc.] are conspicuously absent from this diagram since their cellular mechanisms have not been examined with intracellular techniques. This diagram should therefore be considered only a simplification of the multiple modulations that could potentially occur between input and output at the primary afferent synapse. Much work remains to be done to expand our understanding of the physiology and pharmacology of the normal and pathological functioning of the central vestibular system.

THE MVN SYNAPSE: A MODEL FOR CENTRAL AFFERENT SYNAPTIC HOMEOSTASIS

We have been using the MVN synapse to study the cellular pharmacology of primary afferent transmission in the central vestibular system. Our working hypothe-

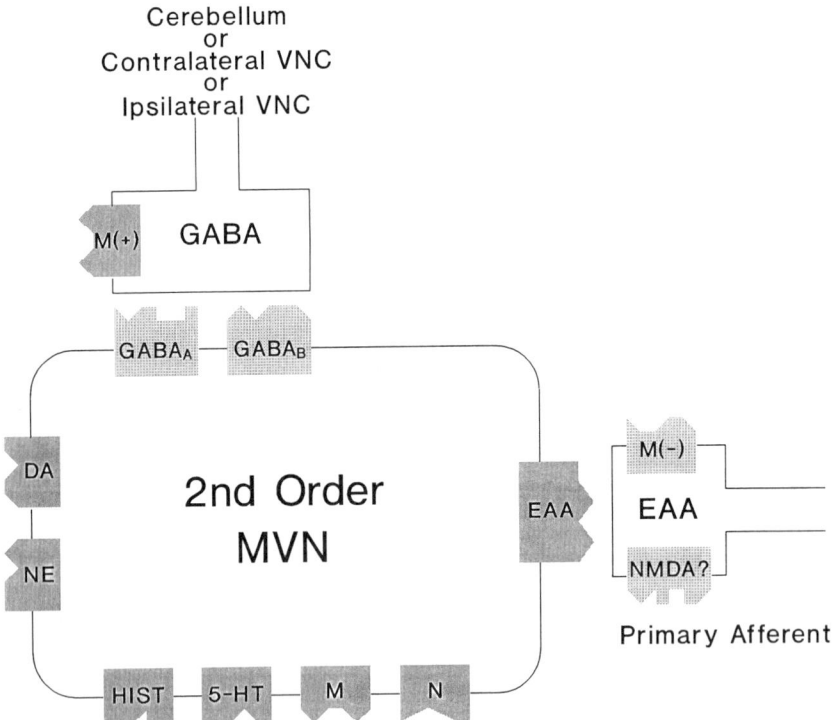

FIGURE 9. Summary of receptor sites within the MVN. Excitatory sites for muscarinic cholinergic, nicotinic cholinergic, excitatory amino acids, namely, quisqualate/kainate and NMDA; amines, namely, serotonin (5-HT), norepinephrine (NE), dopamine (DA), and histamine (Hist), have all been identified on second-order MVN neurones. An excitatory muscarinic site is also suggested on GABAergic neurons innervating the MVN. Inhibitory sites include postsynaptic GABA$_A$ and GABA$_B$ receptors, and a presynaptic muscarinic receptor on the N.VIII terminal.

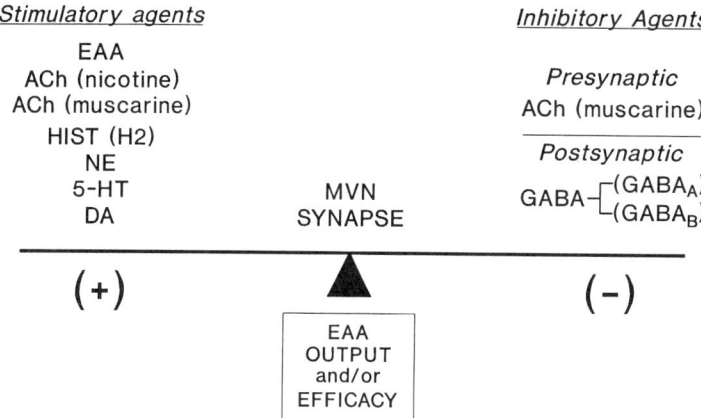

Stimulatory agents

EAA
ACh (nicotine)
ACh (muscarine)
HIST (H2)
NE
5-HT
DA

MVN SYNAPSE

Inhibitory Agents

Presynaptic
ACh (muscarine)

Postsynaptic

GABA $\left\lceil \begin{matrix} \text{(GABA}_A\text{)} \\ \text{(GABA}_B\text{)} \end{matrix} \right.$

(+)   ▲   (−)

EAA
OUTPUT
and/or
EFFICACY

FIGURE 10. Diagram depicting the stimulatory and inhibitory agents that modulate the output or efficacy at the rat second-order MVN synapse.

sis has been that under normal conditions EAA release onto second-order vestibular neurons is maintained by opposing inhibitory and stimulatory modulating influences (FIGURE 10). We postulate that a balance is normally maintained by an inhibitory tone due to activation of presynaptic muscarinic cholinergic and postsynaptic GABAergic receptor activation at the primary afferent vestibular synapse. The inhibitory tone may be offset by an excitatory drive established by activation of cholinergic (muscarinic or nicotinic receptors) and/or aminergic (NE, 5-HT, or DA) receptors localized at the same synapse. When this system is put into an unusual environment, is stressed, or is in the presence of pathological conditions, then an imbalance occurs, e.g., excessive inhibitory influences may dominate and a net deficiency of excitatory transmission occurs. By choosing drugs that can cause the unbalanced system to "adapt" or to block an apparent "mismatch,"[48,49] the system can respond in a quasi-normal manner. FIGURE 10 is our attempt to depict how the various receptors and their activation may maintain homeostasis at the vestibular primary afferent MVN synapse.

ACKNOWLEDGMENTS

We are grateful for comments and discussions about earlier versions of this manuscript with Drs. W. L. Lee, E. Sorenson, M. Tsurusaki, and F. Zheng.

REFERENCES

1. GALLAGHER, J. P., M. R. LEWIS & P. SHINNICK-GALLAGHER. 1985. An electrophysiological investigation of the rat medial vestibular nucleus *in vitro*. *In* Progress in Clinical and Biological Research. M. J. Correia & A. A. Perachio, Eds. **176:** 293–304. Alan R. Liss, Inc. New York, N.Y.
2. LEWIS, M. R., J. P. GALLAGHER & P. SHINNICK-GALLAGHER. 1987. An *in vitro* brain slice preparation to study the pharmacology of central vestibular neurons. J. Pharmacol. Methods **18:** 267–273.

3. GALLAGHER, J. P., M. R. LEWIS, K. D. PHELAN & P. SHINNICK-GALLAGHER. 1987. An investigation into the cellular mechanisms underlying space motion sickness using intracellular electrophysiological recordings from rat medial vestibular nucleus, *in vitro*. *In* Space Life Sciences Symposium: Three Decades of Life Sciences Research in Space: 206–208. NASA. Washington, D.C.

4. PHELAN, K. D. & J. P. GALLAGHER. 1988. The effects of cholinergic agonists on the passive membrane properties of rat medial vestibular neurons *in vitro*. Soc. Neurosci. Abstr. **14**: 331.

5. PHELAN, K. D. & J. P. GALLAGHER. Direct muscarinic and nicotinic receptor–mediated excitation of rat medial vestibular nucleus neurons *in vitro*. Synapse. (In press.)

6. LEWIS, M. R., K. D. PHELAN, P. SHINNICK-GALLAGHER & J. P. GALLAGHER. 1989. Primary afferent excitatory transmission recorded intracellularly *in vitro* from rat medial vestibular neurons. Synapse **3**: 149–153.

7. UJIHARA, H., A. AKAIKE, M. SASA & S. TAKAORI. 1988. Electrophysiological evidence for cholinoceptive neurons in the medial vestibular nucleus: studies on rat brain stem *in vitro*. Neurosci. Lett. **93**: 231–235.

8. UJIHARA, H., A. AKAIKE, M. SASA & S. TAKAORI. 1989. Muscarinic regulation of spontaneously active medial vestibular neurons *in vitro*. Neurosci. Lett. **106**: 205–210.

9. DUTIA, M. B., P. NEARY & D. S. McQUEEN. 1990. Effects of cholinergic agents on spontaneously active rat medial vestibular nucleus neurones *in vitro*. J. Physiol. London **425**: 90P.

10. DOI, K., T. TSUMOTO & T. MATSUNAGA. 1990. Actions of excitatory amino acid antagonists on synaptic inputs to the rat medial vestibular nucleus: an electrophysiological study *in vitro*. Exp. Brain Res. **82**: 254–262.

11. CARPENTER, D. O. & N. HORI. 1992. Neurotransmitter and peptide receptors on vestibular nucleus neurons. Ann. N.Y. Acad. Sci. (This volume.)

12. DARLINGTON, C. L., P. F. SMITH & J. I. HUBBARD. 1989. Neuronal activity in the guinea pig medial vestibular nucleus *in vitro* following chronic unilateral labyrinthectomy. Neurosci. Lett. **105**: 143–148.

13. SMITH, P. F., C. L. DARLINGTON & J. I. HUBBARD. 1990. Evidence that NMDA receptors contribute to synaptic function in the guinea pig medial vestibular nucleus. Brain Res. **513**: 149–151.

14. DARLINGTON, C. L., P. F. SMITH & J. I. HUBBARD. 1990. Guinea pig medial vestibular nucleus neurons *in vitro* respond to $ACTH_{4-10}$ at picomolar concentrations. Exp. Brain Res. **82**: 637–640.

15. SERAFIN, M., A. KHATEB, C. DE WAELE, P. P. VIDAL & M. MÜHLETHALER. 1990. Low threshold calcium spikes in medial vestibular nuclei neurones *in vitro*: a role in the generation of the vestibular nystagmus quick phase *in vivo*? Exp. Brain Res. **82**: 187–190.

16. SERAFIN, M., C. DE WAELE, A. KHATEB, P. P. VIDAL & M. MÜHLETHALER. 1991. Medial vestibular nucleus in the guinea-pig I. Intrinsic membrane properties in brainstem slices. Exp. Brain Res. **84**: 417–425.

17. SERAFIN, M., C. DE WAELE, A. KHATEB, P. P. VIDAL & M. MÜHLETHALER. 1991. Medial vestibular nucleus in guinea-pig II. Ionic basis of the intrinsic membrane properties in brainstem slices. Exp. Brain Res. **84**: 426–433.

18. COCHRAN, S. L., P. KASIK & W. PRECHT. 1987. Pharmacological aspects of excitatory synaptic transmission to second-order vestibular neurons in the frog. Synapse **1**: 102–123.

19. LLINAS, R. 1988. The intrinsic electrophysiological properties of mammalian neurons: insights into central nervous system function. Science **242**: 1654–1664.

20. MATSUOKA, I., J. ITO, H. TAKAHASHI, M. SASA & S. TAKAORI. 1985. Experimental vestibular pharmacology: a minireview with special reference to neuroactive substances and anti-vertigo drugs. Acta Otolaryngol. Stockh. Suppl. **419**: 62–70.

21. RAYMOND, J., D. DEMEMES & A. NIEOULLON. 1988. Neurotransmitters in vestibular pathways. Prog. Brain Res. **76**: 29–43.

22. KNOPFEL, T. 1987. Evidence for *N*-methyl-D-aspartic acid–receptor mediated modulation

of the commissural input to central vestibular neurons in the frog. Brain Res. **426:** 212–224.

23. KNOPFEL, T. & N. DIERINGER. 1988. The role of NMDA and non-NMDA receptors in the central vestibular synaptic transmission. Adv. Oto-Rhino-Laryngol. **42:** 229–233.

24. KNOPFEL, T. & N. DIERINGER. 1988. Lesion induced vestibular plasticity in the frog: are *N*-methyl-D-aspartate receptors involved? Exp. Brain Res. **72:** 129–134.

25. RAYMOND, J., A. NIEOULLON, D. DEMEMES & A. SANS. 1984. Evidence for glutamate as a neurotransmitter in the cat vestibular nerve. Exp. Brain Res. **56:** 523–531.

26. DEMEMES, D., J. RAYMOND & A. SANS. 1984. Selective retrograde labeling of neurons in the cat vestibular nuclei with [³H]-D-aspartate. Brain Res. **304:** 188–191.

27. RECASENS, M. & J. P. DELAUNOY. 1981. Immunological properties and immunohistochemical localization of cysteine sulfinate or aspartate aminotransferase isoenzymes in rat CNS. Brain Res. **205:** 351–361.

28. WALBERG, F., O. P. OTTERSEN & E. RINVIK. 1990. GABA, glycine, aspartate, glutamate and taurine in the vestibular nuclei: an immunocytochemical investigation in the cat. Exp. Brain Res. **79:** 547–563.

29. ANDERSON, K. J., D. T. MONAGHAN, C. B. CANGRO, M. A. A. NAMBOODIRI, J. H. NEALE & C. W. COTMAN. 1986. Localization of *N*-acetylaspartylglutamate-like immunoreactivity in selected areas of the rat brain. Neurosci. Lett. **72:** 14–20.

30. HULME, E. C., N. J. M. BIRDSALL & N. J. BUCKLEY. 1990. Muscarinic receptor subtypes. Annu. Rev. Pharmacol. Toxicol. **30:** 633–673.

31. HASUO, H. & J. P. GALLAGHER. 1990. Actions of muscarine on dorsolateral septal neurons. *In* Regulation of Potassium Transport across Biological Membranes. L. Reuss, J. M. Russell & G. Szabo, Eds.: 459–476. University of Texas Press. Austin, Texas.

32. SUGITA, S., N. UCHIMURA, Z.-G. JIANG & R. A. NORTH. 1991. Distinct muscarinic receptors inhibit release of gamma-aminobutyric acid and excitatory amino acids in mammalian brain. Proc. Natl. Acad. Sci. USA **88:** 2608–2611.

33. BARMACK, N. H. 1992. Cholinergic transmission in the cerebellum. Ann. N.Y. Acad. Sci. (This volume.)

34. CURTIS, D. R., A. W. DUGGAN & D. FELIX. 1970. GABA and inhibition of Deiters' neurones. Brain Res. **23:** 117–120.

35. GALLAGHER, J. P., M. R. LEWIS, J. NAKAMURA & P. SHINNICK-GALLAGHER. 1986. Synaptic transmission and its modulation at the rat medial vestibular nucleus studied intracellularly *in vitro.* Proc. Int. Union Physiol. Sci. **16:** 327.

36. SMITH, P. F., C. L. DARLINGTON & J. I. HUBBARD. 1991. Evidence for inhibitory amino receptors on guinea pig medial vestibular nucleus neurons *in vitro.* Neurosci. Lett. **121:** 244–246.

37. HOLSTEIN, G. R., G. P. MARTINELLI & B. COHEN. 1992. Visualization of baclofen-sensitive receptor sites in the medial vestibular nucleus. Ann. N.Y. Acad. Sci. (This volume.)

38. WOOD, C. D. & A. GRAYBIEL. 1970. A theory of motion sickness based on pharmacological reactions. Clin. Pharmacol. Ther. **11:** 621–629.

39. PALACIOS, J. M., J. K. WAMSLEY & M. J. KUHAR. 1982. The distribution of histaminergic H-1 receptors in the rat brain: an autoradiographic study. Neuroscience **6:** 15–37.

40. TOHYAMA, M., T. WATANABA, N. TAKEDA & H. WADA. 1987. Light-microscopical localization of histaminergic neurons using histidine decarboxylase as a marker. *In* Monoaminergic Neurons: Light Microscopy and Ultrastructure. H. W. M. Steinbusch, Ed.: 111–124. Wiley. New York, N.Y.

41. PHELAN, K. D., J. NAKAMURA & J. P. GALLAGHER. 1990. Histamine depolarizes rat medial vestibular neurons recorded intracellularly *in vitro.* Neurosci. Lett. **109:** 287–292.

42. DE WAELE, M. SERAFIN, A. KHATEB, N. VIBERT, T. YABE, J. M. ARRANG, M. MULHETHALER & P. P. VIDAL. 1992. An *in vivo* and *in vitro* study of the vestibular nuclei histaminergic receptors in the guinea pig. Ann. N.Y. Acad. Sci. (This volume.)

43. GRAYBIEL, A., C. D. WOOD, J. KNEPTON, J. P. HOCHE & G. F. PERKINS. 1975. Human assay of anti-motion sickness drugs. Aviat. Space Environ. Med. **46:** 1107–1118.

44. LUCOT, J. B. 1990. RU 24969–induced emesis in the cat: 5-HT$_1$ sites other than 5-HT$_{1A}$, 5-HT$_{1B}$ or 5-HT$_{1C}$ implicated. Eur. J. Pharmacol. **180:** 193–199.

45. CRAMPTON, G. H., Ed. 1990. Motion and Space Sickness. CRC Press. Boca Raton, Fla.
46. PRECHT, W. 1978. Neuronal Operations in the Vestibular System. Springer-Verlag. Berlin, Germany.
47. PAXINOS, G. & C. WATSON. 1986. The Rat Brain in Stereotaxic Coordinates. 2nd edit. Academic Press. New York, N.Y. (Figure 63.)
48. REASON, J. T. 1978. Motion sickness adaptation: a neural mismatch model. J. R. Soc. Med. **71:** 819–829.
49. KOHL, R. C. 1983. Sensory conflict theory of space motion sickness: an anatomical location for the neuroconflict. Aviat. Space Environ. Med. **54:** 464–465.

Oculomotor Effects of γ-Aminobutyric Acid Agonists and Antagonists in the Vestibular Nuclei of the Alert Monkey[a]

U. BÜTTNER, A. STRAUBE, AND R. KURZAN

Department of Neurology
Ludwig-Maximilian University
Grosshadern Clinic
Marchioninistrasse 15
D-8000 Munich 70, Germany

INTRODUCTION

Anatomical, physiological, and functional aspects of the vestibular nuclei have been studied in great detail. In contrast, comparatively little is known about the transmitters involved.[1] This applies particularly to the question of which specific functions are governed by the various transmitters. The inhibitory transmitter γ-aminobutyric acid (GABA) certainly plays an important role. The vestibular nuclei receive a large number of Purkinje cell (PC) afferents from various parts of the cerebellum,[2] all of which use GABA as the transmitter.

The commissural system, connecting the vestibular nuclei and the nucl. prepositus hypoglossi of both sides, is probably functionally GABAergic utilizing the activation of GABA interneurons.[3] Among other functions, the commissural system is thought to play a role in the horizontal common neural integrator of the oculomotor system[4] and the "velocity storage" mechanism.[5,6]

A neural integrator is required since all eye movements are initially encoded as an eye velocity signal in premotor structures, whereas motoneurons are modulated basically in proportion to eye position. The integrator has been located in the region of the medial vestibular nuclei and the nucl. prepositus hypoglossi.[7,8] Cerebellar lesions also affect neural integration.[9] However, the deficit is only partial. It is generally assumed that the final neural integrator in the vestibular nuclei–prepositus complex is leaky, and that the cerebellum improves its performance.

Velocity storage prolongs the dominant time constant of the canal afferents, which is 5–6 seconds in primates, to about 20 seconds found for the vestibuloocular reflex (VOR) and vestibular nuclei neurons.[10,11]

We investigated the influence of the GABA agonist muscimol and the GABA antagonist bicuculline on eye movements after unilateral microinjections into oculomotor related parts of the vestibular nuclei (i.e., the medial nucleus and medial parts of the superior and lateral nucleus) of the alert, behaving monkey. Bicuculline injections basically induced a vestibular imbalance with a temporary spontaneous nystagmus. Muscimol injections in contrast led to a reversible defect of the common neural integrator for horizontal eye movements with a time constant of the postsaccadic drift (T_i) as low as 250 mseconds. Both muscimol and bicuculline injections reduced the dominant time constant of the VOR (T_v) to 5–10 seconds.

[a] Supported by Deutsche Forschungsgemeinschaft SFB 220, D7.

METHODS

All results were obtained from Java monkeys (*Macaca fascicularis*) chronically prepared for single-unit recordings (for details see Reference 12). Eye position was recorded with implanted DC silver–silver chloride electrodes (DC-EOG) or (in different monkeys) with the scleral search-coil method. All eye movements shown in the figures are from search-coil recordings unless stated otherwise. During the experiments the monkeys sat upright in a primate chair with the head fixed to the chair. The chair was surrounded by an optokinetic cylinder. All vestibular and visual stimuli were applied in the horizontal plane. Eye position was recorded under the following conditions:

- Spontaneous eye movements in light and dark.
- Vestibular sinusoidal (0.5–0.2 Hz, ±20–100 deg/second) and ramp (40 deg/second^2 acceleration, 120 deg/second constant velocity for 60–120 seconds) stimulation in light and dark.
- Visual-vestibular conflict stimulation (0.2 Hz; ±40 deg/second). For this, the optokinetic cylinder and the vestibular turntable are mechanically coupled and move together. This stimulus reduces the vestibuloocular reflex and activates smooth pursuit mechanisms.[13]
- Sinusoidal (0.1–0.5 Hz, ±20–60 deg/second) optokinetic stimulation.

Prior to the microinjections, single-unit activity was recorded in the vestibular nuclei to aid the precise localization. After recording, a glass micropipette (diameter at the tip 10–20 μm) was advanced through the same guide tube used for the single-unit recordings. With a Hamilton pressure syringe, 1 μl NaCl (0.9%) alone (control), or containing 1 μg bicuculline, or muscimol was injected over a period of 40–60 seconds. Eye movements were recorded before and up to 180 minutes after the injection. The syringe remained in place throughout the experiment.

The following parameters were determined:

1. Time constant of the postsaccadic drift (T_i) with an exponential decay after horizontal and vertical saccades. It was taken as the time by which the eye drifted back 63% to the null position. The null position was taken as that eye position at which slow eye drifts, after saccades, reverse direction (FIGURE 1). It was seldom identical with the midposition of the eye in the orbit. Eye velocity increases with the eccentricity of the eyes from the null position[7] (FIGURE 1). To avoid the positional dependency of the exponential decay, only eye movements that drifted more than 70% back to the null position were measured. For vertical eye movements, no null position could be determined, since eyes only drifted up or down. In this case time constants (T_i) were estimated from the postsaccadic drift after a saccade with a long intersaccadic interval.
2. Slow-phase velocity was determined for slow eye movements with a linear decay.
3. The gain during horizontal sinusoidal vestibular and optokinetic stimulation was calculated by superposition of eye position traces after elimination of saccades. In instances with a short time constant T_i (i.e., after muscimol injections), such an analysis is no longer possible.[7]
4. The time constant of decay of vestibular nystagmus (T_v) during and after vestibular ramp stimulation, was determined as the time elapsed between maximal velocity and the point at which it had decreased to $1/e$ (37%) above

baseline level,[10] i.e., the velocity of preceding spontaneous nystagmus was taken into account.

At the end of all experiments small electrolytic lesions were made to facilitate the reconstruction of the injection sites.

RESULTS

General Observations

Injections were well tolerated. In no instance did sedative drugs have to be given. After some bicuculline injections, chewing movements lasting up to 20 minutes were

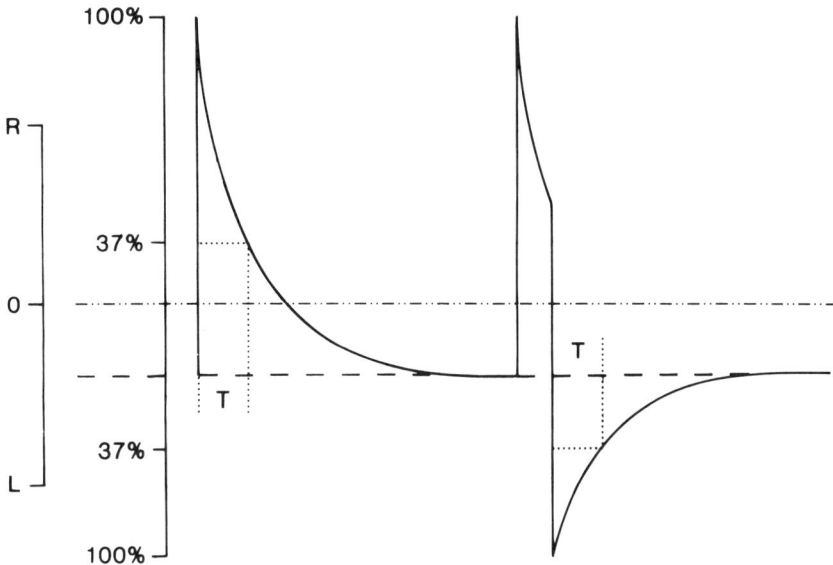

FIGURE 1. Schematic drawing of horizontal saccades and the postsaccadic drift after unilateral muscimol injection. The eyes drift after a saccade to an eye position (null-position, dashed line), which is different from the midposition of the eye in the orbit (dotted-dashed line). T = time constant of the postsaccadic drift. Note that the velocity of the postsaccadic drift increases with the eccentricity of the eye in relation to the null position.

observed. Back in their cages all monkeys with an eye movement deficit induced by the injection had a rotation and tilt of the head along with a falling tendency toward the side of the head rotation. Four to six hours after injection the monkeys showed normal behavior again.

Injection Sites

All injections were placed unilaterally on the left or right side at least 2 mm off the midline. Bicuculline and muscimol injections causing an eye movement deficit were located in the rostral third of the vestibular nuclei extending over an area 2–3.5 mm lateral to the midline and 0.5–2.5 mm caudal to the abducens nucleus. This area, hereafter to be called the "effective area," covered the medial nucleus and medial parts of the superior and lateral nucleus. No injection was within the nucl. prepositus hypoglossi or the inferior vestibular nucleus. NaCl injections within the effective area did not cause eye movement changes and served as controls. Bicuculline control injections lateral to (3.5–5 mm to the midline) and 3–4 mm above the effective area, i.e., in the cerebellar peduncles, also did not cause eye movement changes and also served as controls.

Spontaneous Eye Movements

Bicuculline Injections

All bicuculline injections into the "effective area" induced spontaneous nystagmus in the dark. The slow-phase velocity started to build up 30–150 seconds after the completion of the injection, reached a maximum within 5–10 minutes, and lasted between 15 and 90 minutes (FIGURE 2). The spontaneous nystagmus always had a horizontal component beating (fast phase) in 60% of the cases to the contralateral and in 40% of the cases to the ipsilateral side (FIGURE 3). There was a tendency that more laterally placed injections induced nystagmus to the contralateral side. Maximal slow-phase velocity could be as high as 95 deg/second. On average there was no statistical difference between contraversive and ipsiversive nystagmus.

In 80% of the cases there was an additional vertical nystagmus component, the time course of which was closely linked to the horizontal nystagmus component. The vertical nystagmus component in most instances beats downward, with slow-phase velocities of 5–40 deg/second.

The slow-phase velocity of the horizontal and vertical component had a linear slope, and the nystagmus followed Alexander's law.[14,15] Nystagmus velocity increased when the monkey looked in the direction of the fast phase and decreased when looking in the opposite direction. There was no nystagmus reversal on lateral or upward and downward gaze, although this could not be systematically tested in the dark for extreme eye positions. In no instance did bicuculline induce an exponential decay for slow phases.

In the light, nystagmus was generally completely suppressed. In a few cases, there was virtually no suppression, i.e., nystagmus velocity in the dark of 60 deg/second only decreased to 54 deg/second in the light.

Muscimol Injections

In contrast to the bicuculline injections, muscimol injections into the "effective area" led to a gaze-holding deficit with an exponential decay of the postsaccadic drift. For horizontal saccades the time constant of the postsaccadic drift (T_i) was as short as 250 mseconds (up to 800 mseconds) (FIGURE 4). After completion of injection, it took 2–3 minutes before the time constant T_i started to decrease. Minimal values were reached after 10–30 minutes, and normal eye movements occurred again after 3–4 hours. Besides this decrease of the time constant T_i, a shift of the null position up to 35 deg occurred in most cases (FIGURE 1). The shift of the null position was usually contraversive and less often ipsiversive. There was no correlation between the site of injection and the direction of the shift. This also

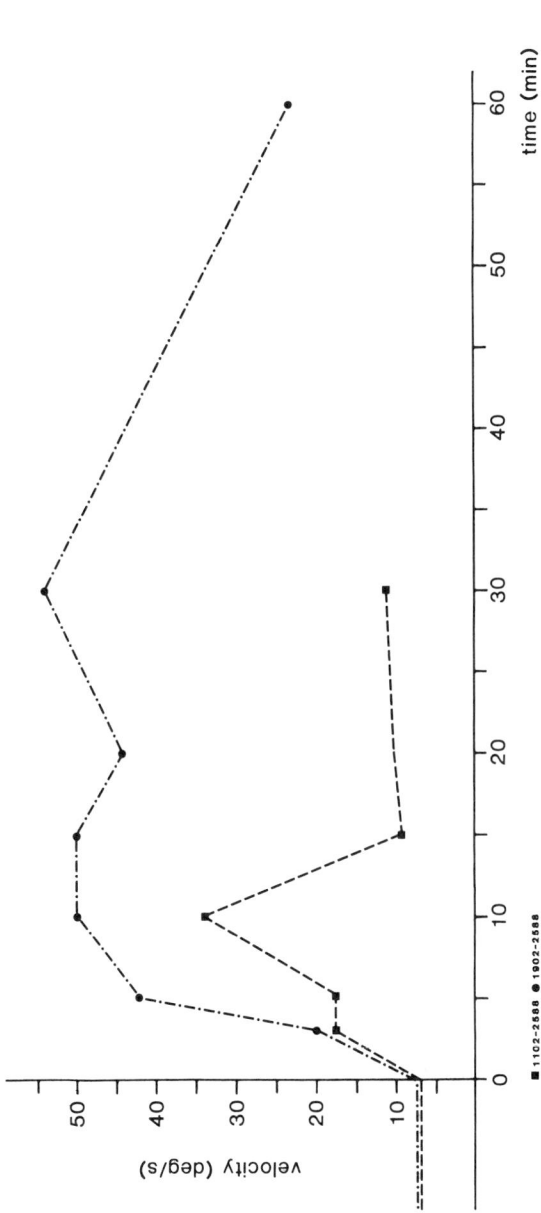

FIGURE 2. Time course of the horizontal slow-phase nystagmus velocity after unilateral bicuculline injections. Two experiments, one with a short, the other with a long, time course are shown.

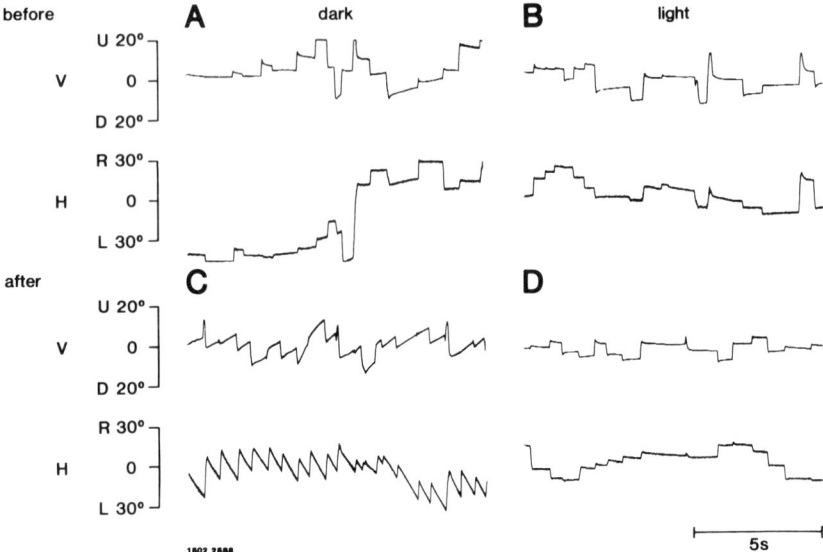

FIGURE 3. Spontaneous eye movements in the dark and light before and after a unilateral bicuculline injection into the right vestibular nuclei. H = horizontal and V = vertical eye position. EOG recording. Bicuculline induces spontaneous nystagmus with constant velocity slow phases, which is completely suppressed in the light.

applied to the amount of the null shift (deg), and the time constant T_i, i.e., shorter time constants were not associated with larger shifts.

For vertical eye movements, there was only a weak effect on the time constant of the postsaccadic drift. In nearly half of the cases, vertical eye movements were not affected at all. In the remaining cases, time constants T_i were > 2 seconds and vertical eye movements appeared as downbeat or, less frequently, as upbeat nystagmus (FIGURE 4). In the vertical plane, no null position could be determined, i.e., eyes always drifted in one direction.

Light had little effect on T_i (FIGURE 4). Only the width of the null position increased, i.e., the range of eye position around the null position in which the eyes did not drift.

Sinusoidal Vestibular and Optokinetic Stimulation

After *bicuculline* injection the responses became asymmetrical parallel to the development of the induced nystagmus (FIGURE 5). The gain (peak to peak) for the VOR in the dark at 0.2 Hz (± 40 deg/second) decreased only slightly on average. In 30% of the cases, actually an increase up to 1.5 was noted after injection.

The response after muscimol injections was dominated by the postsaccadic drift after each saccade, which prevents a gain analysis based on the superposition of slow phases induced by the stimulus (FIGURE 6). However, there was a head velocity dependent shift of the null position with a change of saccade direction (FIGURES 6 and 7). From the total amplitude of the null position shift, a gain of < 0.2 and a phase advance of >70 deg relative to stimulus position can be calculated for vestibular stimulation. For optokinetic stimulation the gain was less than 0.1. Vestibular stimulation in the light and dark gave similar results (FIGURE 6).

Visual-Vestibular Conflict Stimulation

Normally conflict stimulation reduces the gain to 30–60% of the response during VOR in the dark.[13] Bicuculline injections had no significant effect on this suppression by conflict stimulation. This also applied for muscimol injections, i.e., the eye movement pattern during VOR in the dark and visual-vestibular conflict stimulation was similar.

Vestibular Ramp Stimulation

In the dark, ramp stimulation normally induces nystagmus which is symmetrical after acceleration and deceleration, and the velocity of slow phases returns to baseline with a time constant T_v of 20 seconds.[10] After *bicuculline* injections, eye movements during and after vestibular ramp stimulation are altered mainly in two aspects. First, the nystagmus response becomes asymmetrical corresponding to the direction and the velocity of the induced spontaneous nystagmus. When related to

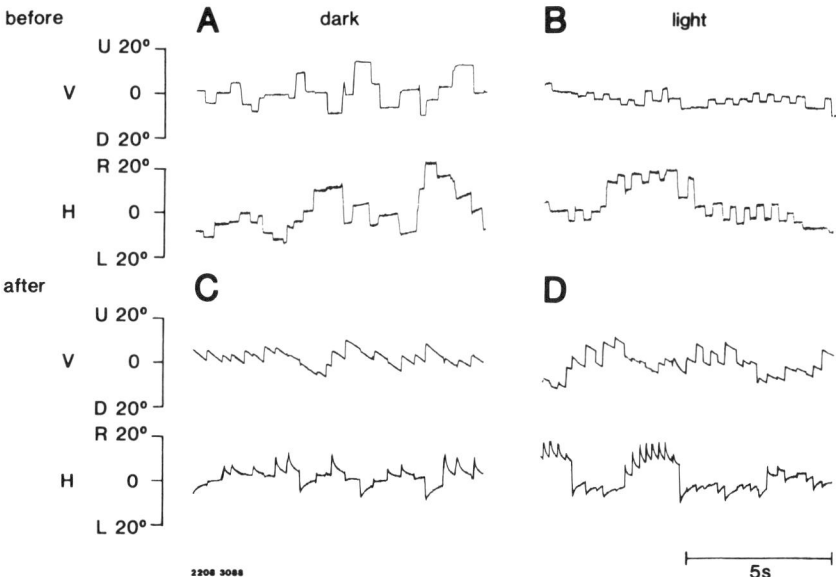

FIGURE 4. Spontaneous eye movements in the dark and light before and after a unilateral muscimol injection into the right vestibular nuclei. Muscimol leads in the dark and light to a severe horizontal gaze-holding deficit with a time constant of 300 mseconds. The null position is about 5 deg to the right. For vertical eye movements, an upbeat nystagmus with constant-velocity slow phases is induced.

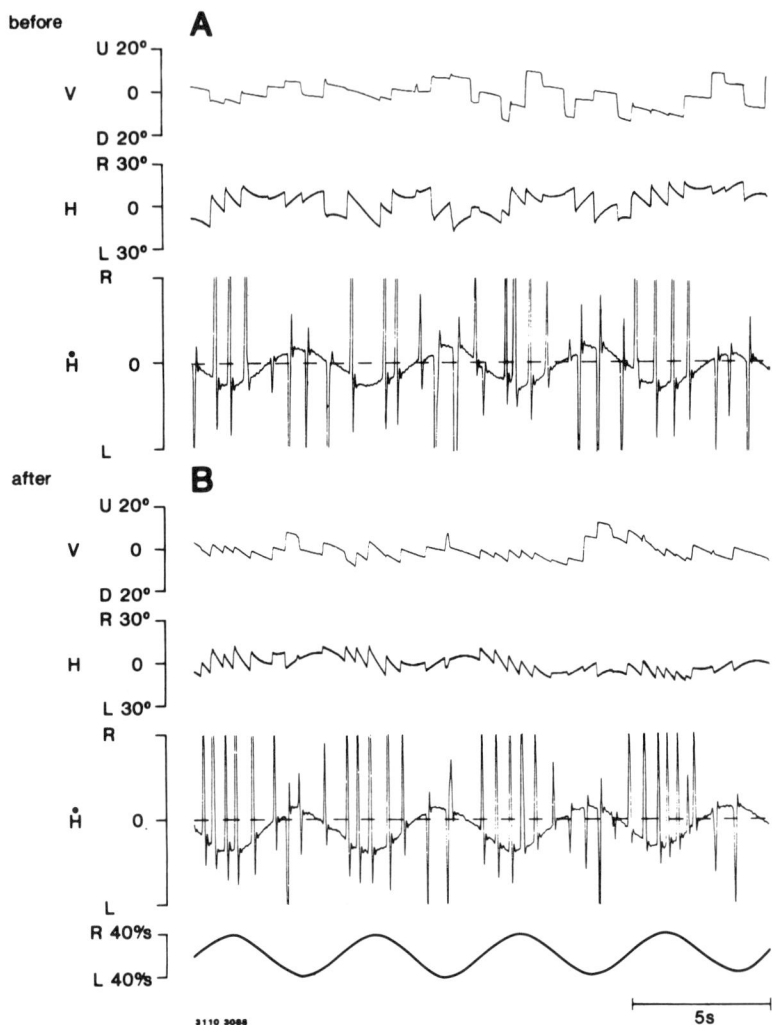

FIGURE 5. Vestibular stimulation (0.2 Hz, ±40 deg/second) in the dark before (A) and after (B) a bicuculline injection into the left vestibular nuclei, V = vertical, H = horizontal eye position, Ḣ = horizontal eye velocity. Bicuculline induces a vestibular imbalance with right nystagmus dominating without a gain deficit.

the velocity of the spontaneous nystagmus, responses are symmetrical without major changes in the response amplitude. After acceleration and decelaration, nystagmus velocity returns gradually to the prestimulus velocity with a time constant T_v of 6–10 seconds. The time constant T_v is also shortened (5–8 seconds) after *muscimol* injections (FIGURE 8) and often is as short as those of vestibular nerve afferents.

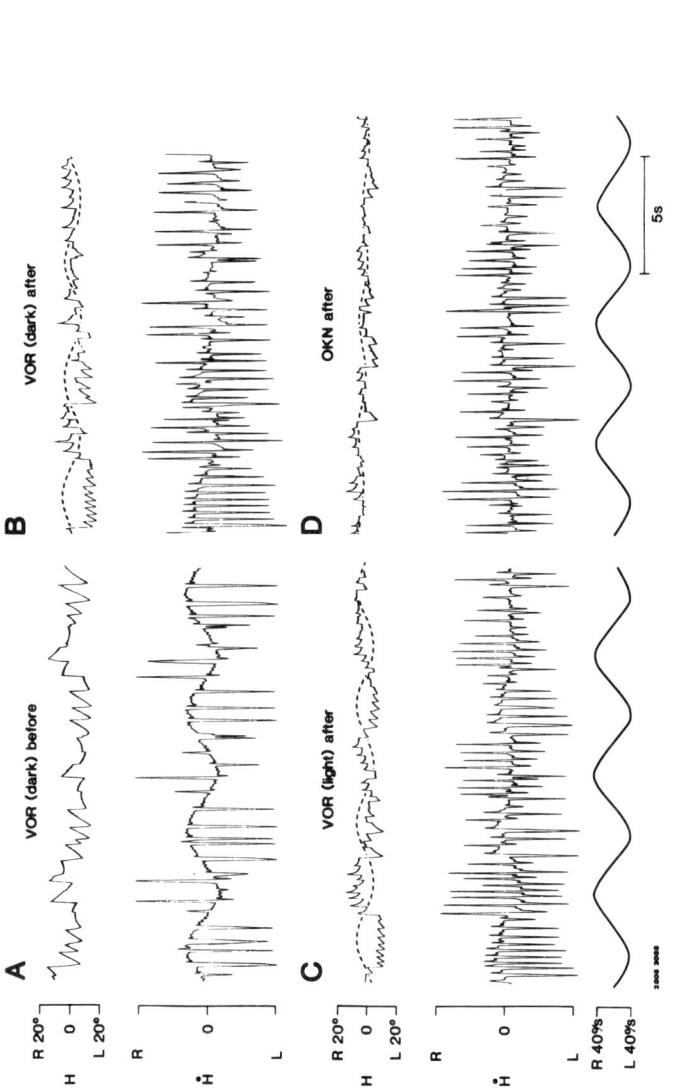

FIGURE 6. Horizontal eye position (H) and velocity (Ḣ) during vestibular (A, B, and C) and optokinetic (D) stimulation before (A) and after (B, C, and D) muscimol injection into the right vestibular nuclei. Stimulus velocity in all instances is ±40 deg/second at 0.2 Hz. There are no stimulus-dependent slow phases after muscimol injection. Saccades are always followed by a postsaccadic drift with an exponential decay. They alter their direction in relation to the stimulus. Dotted lines on the eye position trace indicate the stimulus-dependent shift of the null position.

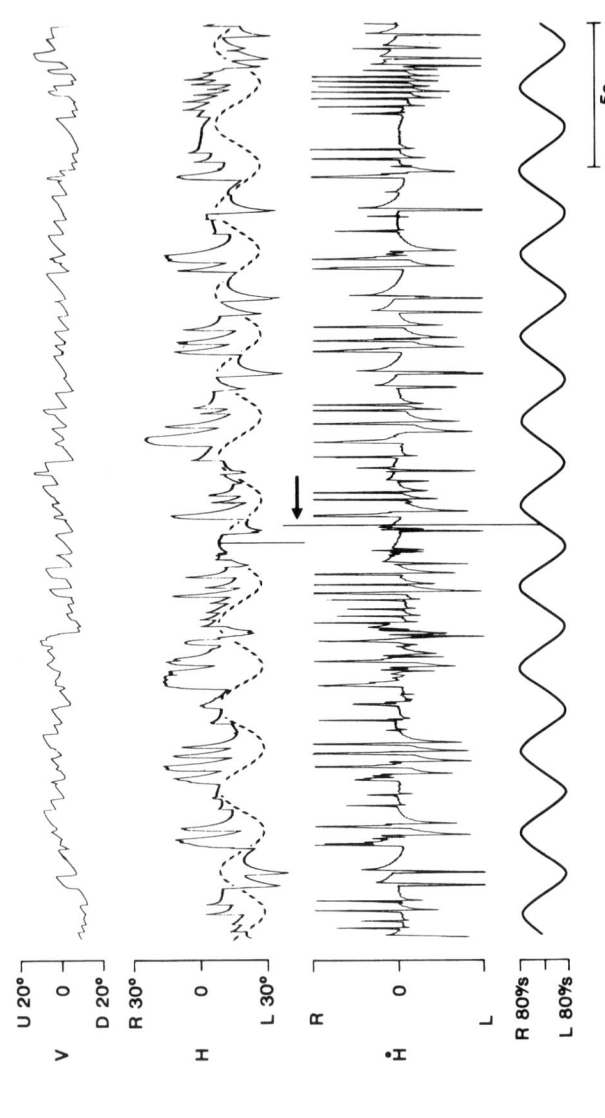

FIGURE 7. Eye movements during high-velocity vestibular stimulation (0.33 Hz, ±80 deg/second) in the dark after unilateral muscimol injection. Traces as in previous figures. Broken line indicates shift of the null position, which leads stimulus position by about 75 deg (arrow).

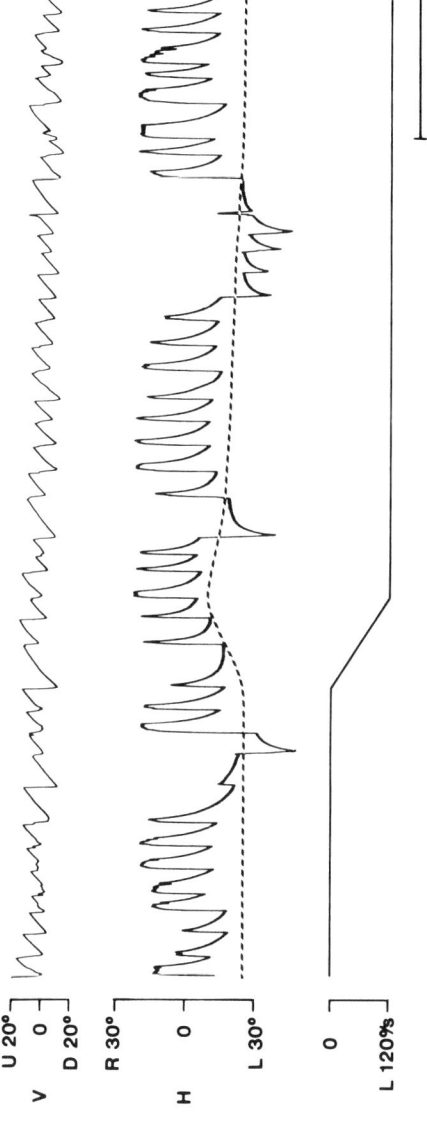

FIGURE 8. Vestibular ramp stimulation (acceleration 40 deg/second² for 3 seconds) in the dark after muscimol injection in the right vestibular nuclei. For horizontal eye movements (H) the null position is nearly 30 deg to the left (L). During acceleration, the null position moves to the right and returns with a time constant of decay (T_v) of 5–6 seconds to the prestimulus position. Broken line is inserted for better inspection. Horizontal saccades are followed by a postsaccadic drift with a time constant $T_i = 250$ mseconds. In contrast, after vertical saccades, slow phases with an exponential decay can only be seen for long saccadic intervals.

DISCUSSION

Small unilateral injections of the weak GABA antagonist bicuculline[16] and the strong GABA agonist muscimol[17] into the "effective area" of vestibular nuclei consistently induce fully reversible oculomotor deficits. The effects of these substances are quite different. Basically bicuculline induces a vestibular imbalance with spontaneous nystagmus, similar to the deficits seen after a peripheral vestibular lesion.[18] In contrast muscimol injections lead to a failure of the common neural integrator for horizontal eye movements with a severe gaze-holding deficit. Time constants of the postsaccadic drift can be as short as 250 mseconds, which are the shortest times possible allowed by the orbital tissue mechanics.[7]

Bicuculline and muscimol injections leave no visible histological changes, thus the extent of their pharmacological action can only be estimated as not exceeding a diameter of 1–2 mm. Similar values were reported after GABA injections into the substantia nigra of the rat.[19] This is also in agreement with the fact that our bicuculline injections 2–3 mm outside the "effective area" did not cause nystagmus.

Bicuculline

The spontaneous nystagmus induced by bicuculline injections follows Alexander's law. This implies a gaze-holding and hence a neural-integrator deficit, which is generally seen with higher nystagmus velocities.[15] The induced spontaneous nystagmus can beat to the contralateral or to the ipsilateral side, similar to the effect of electrolytic lesions.[20] There was a tendency that more laterally placed bicuculline injections caused ipsiversive nystagmus, although the proposed diameter of pharmacological action does not allow any detailed anatomical conclusions. However, it was quite obvious that injections only 1–2 mm apart could cause nystagmus of different directions.

Beside the spontaneous nystagmus, other oculomotor functions that were investigated were not much altered (for velocity storage see below). The spontaneous nystagmus was well suppressed in the light, the VOR gain in light and darkness was only slightly smaller on average, and there was little effect on VOR suppression during visual-vestibular conflict stimulation.

Muscimol

Unilateral muscimol injections into the effective area led to a failure of the common neural integrator for horizontal eye movements. All eye movements are affected. There is not only a severe gaze-holding deficit, but also the VOR and the optokinetic response are greatly deteriorated. The only residual effect of vestibular and optokinetic stimulation is a shift of the null position, i.e., vestibular stimulation to the left moves the null position to the right and vice versa. The VOR gain calculated on these measurements is less than 0.2. Sinusoidal optokinetic stimulation, known to activate smooth pursuit mechanisms,[13] only leads to a small null-position shift. Similar results showing an effect on all horizontal eye movements have been obtained in the monkey[7] and cat[8] using neurotoxic substances.

It is also quite obvious that injections into the vestibular nuclei have little effect on the integrator for vertical eye movements. The time constant T_i was never less than 2 seconds. This supports the assumption that the common neural integrator for vertical eye movements is outside the vestibular nuclei, probably in or near the interstitial nucleus of Cajal.[21]

Unilateral muscimol injections also lead to a shift of the null position in the horizontal plane. This reflects a vestibular imbalance induced by the injections. After bicuculline injections with the integrator intact, vestibular imbalance expresses itself as spontaneous nystagmus. With the failure of the integrator after muscimol injections, only a shift of the null position remains. Similarly to the ipsi- or contraversive nystagmus after bicuculline injections, the shift of the null position can be either to the ipsilateral or the contralateral side. Since the direction of the null-position shift is independent of the integrator deficit, it is not surprising that there is no correlation between the amount of the null-position shift (in deg) and the failure of the integrator (time constant T_i).

Velocity Storage

After both muscimol and bicuculline microinjections, velocity storage deteriorates. The time constant of decay (T_v) for vestibular nystagmus in both directions becomes as short as those found for vestibular nerve afferents. It is known that the velocity storage mechanism is under inhibitory control of PC afferents from the nodulus and uvula.[22] Lesions here prolong the time constant T_v.[23] Thus, potentially bicuculline, as a GABA antagonist, could prolong the time constant T_v. This, however, was not found. It is noteworthy that a vestibular imbalance due to a unilateral labyrinthectomy also impairs the velocity storage mechanism.[18] It appears that spontaneous nystagmus of higher velocity is always associated with a decrease of the VOR time constant.[15]

Commissural Pathways

Recent models suggest that both the common neural integrator and the velocity storage mechanism depend on commissural pathways.[4,5] However, midline sections caudal to the abducens nucleus interrupting commissural pathways affect the velocity storage mechanism but not the common neural integrator.[6] These differences are not fully understood.[b] It is feasible that midline sections affect all fibers essential for the velocity storage mechanism, but not all those for the common neural integrator. Some commissural pathways actually cross quite ventrally in the brain stem, others via the cerebellum,[24] and are certainly not affected by lesion studies so far. It is also feasible and probably more likely, that the common neural integrator depends on unilateral local circuits on each side, commissural pathways and cerebellar connections.[4,9] In this case a midline cut would leave the local circuits, the cerebellar connections, and the integrator basically intact. Unilateral injections in contrast would lead to dysfunction of all circuits on one side and the commissural pathways, which disables the integrator.

Bicuculline and Muscimol

Bicuculline and muscimol have quite different effects on eye movements. This could be related to the fact that bicuculline is a *weak* antagonist and muscimol a

---

[b]**Note added in proof:** But see Anastasio, T. J. & D. A. Robinson. 1991. Neurosci. Lett. **127:** 82–86.

strong agonist. According to this, bicuculline would only lead to modest changes of spontaneous activity for vestibular nuclei neurons. The difference of spontaneous activity on both sides would lead to a vestibular imbalance and spontaneous nystagmus, but neural integration and vestibular and optokinetic modulation would still be possible. In contrast the *strong* agonist muscimol would completely suppress spontaneous activity and prevent neural integration, which is in accordance with recent models.[25] Vestibular and optokinetic stimulation only lead to stimulus-dependent changes of the null position. According to this, one might expect that muscimol initially causes a vestibular imbalance with spontaneous nystagmus before the activity suppression is complete. This, however, was not found. Conversely one might also expect a gaze-holding deficit with higher bicuculline concentrations, which has to be tested in further experiments.

ACKNOWLEDGMENTS

The authors thank S. Langer for technical assistance and B. Pfreundner and I. Wendl for secretarial work.

REFERENCES

1. WALBERG, F., O. P. OTTERSEN & E. RINVIK. 1990. GABA, glycine, aspartate, glutamate and taurine in the vestibular nuclei: an immunocytochemical investigation in the cat. Exp Brain Res 79: 547–563.
2. BLANKS, R. H. I. 1988. Cerebellum. *In* Neuroanatomy of the Oculomotor System (Reviews of Oculomotor Research 2). A. Büttner-Ennever, Ed.: 225–272. Elsevier. Amsterdam, the Netherlands.
3. PRECHT, W. 1983. Physiopathologische Grundlagen des vestibulären Nystagmus. Akt. Neurol. 10: 123–127.
4. GALIANA, H. L. & J. S. OUTERBRIDGE. 1984. A bilateral model for central neural pathways in vestibuloocular reflex. J. Neurophysiol. 51: 210–241.
5. ANASTASIO, T. J. 1991. Neural network models of velocity storage in the horizontal vestibulo-ocular reflex. Biol. Cybern. 64: 187–196.
6. KATZ, E., I. M. B. V. DEJONG, J. A. BÜTTNER-ENNEVER & B. COHEN. 1991. Effects of midline medullary lesions on velocity storage and the vestibulo-ocular reflex. Exp. Brain Res. 87: 505–520.
7. CANNON, S. C. & D. A. ROBINSON. 1987. Loss of the neural integrator of the oculomotor system from brain stem lesions in monkey. J. Neurophysiol. 57: 1383–1409.
8. CHERON, G. & E. GODAUX. 1987. Disabling of the oculomotor neural integrator by kainic acid injections in the prepositus–vestibular complex of the cat. J. Physiol. 394: 267–290.
9. ROBINSON, D. A. 1974. The effect of cerebellectomy on the cat's vestibuloocular integrator. Brain Res. 71: 195–207.
10. BÜTTNER, U. & W. WAESPE. 1981. Vestibular nerve activity in the alert monkey during vestibular and optokinetic nystagmus. Exp. Brain Res. 41: 310–315.
11. WAESPE, W. & V. HENN. 1987. Gaze stabilization in the primate. The interaction of the vestibulo-ocular reflex, optokinetic nystagmus, and smooth pursuit. Rev. Physiol. Biochem. Pharmacol. 106: 37–125.
12. STRAUBE, A., R. KURZAN & U. BÜTTNER. 1991. Differential effects of bicuculline and muscimol microinjections into the vestibular nuclei on simian eye movements. Exp. Brain Res. 86: 347–358.
13. MARKERT, G., U. BÜTTNER, A. STRAUBE & R. BOYLE. 1988. Neuronal activity in the flocculus of the alert monkey during sinusoidal optokinetic stimulation. Exp. Brain Res. 70: 134–144.

14. ALEXANDER, G. 1912. Die Ohrenkrankheiten im Kindesalter. *In* Handbuch der Kinderheilkunde. M. Pfändler & A. Schlossmann, Eds.: 84–96. Vogel. Leipzig, Germany.

15. ROBINSON, D. A., D. S. ZEE, T. C. HAIN, A. HOLMES & L. F. ROSENBERG. 1984. Alexander's law: its behavior and origin in the human vestibulo-ocular reflex. Ann. Neurol. **16:** 714–722.

16. SMART, T. G. & A. CONSTANTI. 1986. Studies on the mechanism of action of picrotoxin and other convulsants at the crustacean muscle GABA receptor. Proc. R. Soc. London **227:** 191–216.

17. KROGSGAARD-LARSEN, P., T. HANORE & K. THYSSEN. 1979. GABA receptor agonists: design and structure activity studies. *In* GABA Neurotransmitters. P. Krogsgaard-Larsen, J. Scheel-Kruger & H. Kofod, Eds.: 201–216. Munksgaard. Copenhagen, Denmark.

18. FETTER, M. & D. S. ZEE. 1988. Recovery from unilateral labyrinthectomy in rhesus monkey. J. Neurophysiol. **59:** 370–393.

19. CHEVALIER, G., S. VACHER, J. M. DENIAU & M. DESBAN. 1985. Disinhibition as a basic process in the expression of striatal functions. I. The striato-nigral influence on tecto-spinal/tecto-diencephalic neurons. Brain Res. **334:** 215–226.

20. UEMURA, T. & B. COHEN. 1973. Effects of vestibular nuclei lesions on vestibulo-ocular reflexes and posture in monkeys. Acta Oto-Laryngol. Stockh. Suppl. **315:** 1–71.

21. FUKUSHIMA, K. 1991. The interstitial nucleus of Cajal in the midbrain reticular formation and vertical eye movement. Neurosci. Res. **10:** 159–187.

22. CARPENTER, M. B. & R. J. COWIE. 1985. Connections and oculomotor projections of the superior vestibular nucleus and cell group "y." Brain Res. **336:** 265–287.

23. WAESPE, W., B. COHEN & TH. RAPHAN. 1985. Dynamic modification of the vestibulo-ocular reflex by the nodulus and uvula. Science **228:** 199–202.

24. EPEMA, A. H., N. M. GERRITS & J. VOOGD. 1988. Commissural and intrinsic connections of the vestibular nuclei in the rabbit: a retrograde labeling study. Exp. Brain Res. **71:** 129–146.

25. ARNOLD, D. B. & D. A. ROBINSON. 1991. A learning network model of the neural integrator of the oculomotor system. Biol. Cybern. **64:** 447–454.

What is Motion Sickness?[a]

DOUGLAS G. D. WATT, LAURENT J. G. BOUYER, IGAL
T. NEVO, ALANNA V. SMITH, AND YANG TIANDE[b]

Aerospace Medical Research Unit
McGill University
3655 Drummond Street
Montreal, Quebec, Canada H3G 1Y6

[b]*Institute of Space Medico-Engineering*
Post Office Box 5104-(12)
Beijing, China 100094

INTRODUCTION

The symptoms of motion sickness have been experienced at least since the invention of ocean-going vessels. It seems extraordinary that several millenia later, the fundamental nature of the disorder is still not understood. In particular, it is not clear why it exists at all.

Through the years, many theories have been advanced as to the cause of motion sickness, and at least some of these have implied a purpose. For example, if a particular activity involving movement were being overdone, this would lead to motion sickness, which in turn should cause the sufferer to reduce that activity. In a broad sense, motion sickness was thought of as a form of pain. Many possible mechanisms were suggested to link motion exposure to motion sickness, including shifting of the abdominal contents, disturbances of the circulatory system, and overstimulation of the vestibular system. All of these ideas have since been disproven, to be replaced by Reason's "sensory conflict" or "sensory rearrangement" theory.[1] Unfortunately, as its author carefully points out, this theory describes where and when motion sickness occurs, but not why. In fact, Reason basically dismisses motion sickness as the coincidental result of technology outrunning evolution, making the question itself meaningless.

Others have not been quite so willing to accept that the signs and symptoms of motion sickness have no purpose, but they also have not been particularly successful at defining what that purpose is. One theory that has been considered proposes that the vomiting associated with motion sickness actually reflects a mechanism for getting rid of ingested poisons.[2-4] According to this theory, the fundamental mechanism is that poison reaches the inner ear (which is assumed to be particularly sensitive to that substance), resulting in vomiting and expulsion of whatever toxic material remains in the stomach contents. Motion, on the other hand, merely happens to activate this mechanism.

There are a number of problems with this theory, one being that there is no direct evidence to support it. For example, it has not been demonstrated that the localized application of an emetic poison to the inner ear, in suitable concentration, causes vomiting. Nor has it been shown that systemic administration of a substance that induces vomiting in any way affects the behavior of vestibular primary afferent fibers. The proposed mechanism itself would also seem to be indirect, complicated, and

[a]This work was supported by the Medical Research Council of Canada.

excessively slow. By the time poisons reach the inner ear, it is probably a bit late to be trying to get rid of them by vomiting. Vomiting is also the last, and a frequently absent, stage of motion sickness in humans. How will drowsiness, apathy, headache, yawning, and general malaise cause a poison to be expelled? Actually, some species (e.g., rats) do not vomit as part of motion sickness, but they still become ill. In fact, they cannot vomit, so why should poisons (and incidently, motion) cause other symptoms?

Is there a more plausible rationale for motion sickness? Over the past few years, the present authors have been developing a rather different theory as to the origin and purpose of at least some forms of motion sickness, beginning with several independent observations.[5] First, Guitton (personal communication) noted that substituting for a neck collar by voluntarily fixing the head to the torso during *ad lib.* movement could lead to motion-sickness-like symptoms. This technique has been called "torso rotation" because the head is turned by rotating the torso.[6] Second, a motor strategy that is similar to torso rotation is often adopted during weightlessness caused by orbital or parabolic flight. Of course, these are also situations that often lead to motion sickness. Finally, a very short exposure to torso rotation on the ground results in oscillopsia, an illusion that the world is moving back and forth when the head is oscillated. This suggests a temporary suppression of vestibular reflexes, although in practice the situation is not quite that simple.

The question addressed in these experiments was, can apparently normal, self-generated patterns of eye, head, and body movement cause motion sickness by forcing a rapid, and inappropriate, change in vestibular function?

METHODS

To date, nearly 30 series of closely related experiments have been performed. In general, these have begun by obtaining control measures of vestibular function, or the level of motion sickness signs and symptoms. This has been followed by 30 minutes of continuous, rhythmical movement. Finally, the original tests have been repeated immediately, 10 minutes, and 20 minutes after the period of voluntary movement. Between these latter tests, the subject has been encouraged to move about in a normal, nonrhythmical fashion.

The kind of rhythmical movement used in most experiments is illustrated in FIGURE 1. The subject was asked to sweep his gaze back and forth between two targets located 135 degrees to either side of straight ahead. Of necessity, the head and upper torso followed, rotating approximately ±90 degrees in the former case, and ±45 degrees in the latter. This activity was performed with reference to a sound cue adjusted to set the body oscillation frequency at 0.7 Hz, which made the task relatively easy for most subjects.

To obtain an objective measure of vestibular function, the vestibuloocular reflex (VOR) was tested using a modification of a technique developed by Gauthier and Robinson.[7] The subject was seated in a manually operated rotating chair to which the head and shoulders were firmly fixed (FIGURE 2). Eye position was recorded using conventional DC electrooculographic (EOG) techniques, and head and upper torso rotation were measured independently by means of goniometers. The subject was instructed to (1) look at a target located 11.5 ft straight ahead, (2) continue to "look" at the mental image (not afterimage) of the target when the lights go out, and (3) adjust gaze back to the target, if necessary, when the lights come back on. The operator (1) turned the lights out, (2) rotated the chair and subject 10 to 30 degrees,

FIGURE 1. Standard form of rhythmical movement used in these experiments. Subjects swept their gaze back and forth between two targets located 135 degrees to either side of straight ahead. This required head and body movements as illustrated here, performed at a rate of 0.7 oscillation cycles per second.

left or right, and (3) turned the lights back on. This was repeated 12 times, in random directions.

In other experiments, motion sickness signs were noted by the observer, and symptoms were described by the subject, using a detailed and systematically administered motion sickness questionnaire.

RESULTS

The subjective effects of 30 minutes of rhythmical movement were dramatic and quite consistent. Nearly all subjects became posturally unstable, experienced oscillopsia on head rotation (especially in yaw), and noted illusory sensations of self-movement if the head was displaced. In general, these effects lasted up to 20 minutes postmovement. Deliberate suppression of the VOR during the period of rhythmical movement produced stronger aftereffects, as did larger amplitude (and hence higher velocity) movement. The longer the period of movement, the greater was the effect, although significant subjective changes could be detected after a few tens of seconds in some cases. The presence or absence of vision during repetitive movement had little if any effect on the results, as did the particular orientation of the gravity vector relative to the subject.

Many subjects also became motion sick, in some cases quite severely as illustrated in FIGURE 3. This subject, who was evaluated at 10 minutes into the recovery period, was on the verge of vomiting. The motion sickness also occurred in a particularly interesting pattern. Symptoms usually developed after the subject had returned to a normal motor strategy. Nausea and other unpleasant sensations could appear during the rhythmical movement, especially if the subject was distracted by his thoughts or a conversation. However, refocusing attention on the movement task would then usually reduce or eliminate these symptoms.

FIGURE 4 shows records of head (re space) and eye (re head) rotation during VOR testing before and immediately after 30 minutes of rhythmical movement. VOR gain was calculated as the ratio of measurement A to measurement B, or in essence, how far the eye actually rotated versus how far it should have rotated. Note the significant undershoot after the period of rhythmical movement.

The effect of 30 minutes of rhythmical motor activity is summarized in FIGURE 5, which plots VOR gain as a function of time before and after the continuous movement. Each point is an average across 12 subjects, ± one standard error of the mean (SEM). Paired comparisons have been carried out between the points as indicated, using the *t*-test, and it should not be surprising that the A-B and B-C comparisons are very significantly different.

DISCUSSION AND CONCLUSIONS

FIGURE 5 confirms the subjective impression that VOR gain was decreased after 30 minutes of rhythmical motor activity, with a gradual return to normal during the

FIGURE 2. Subject seated on rotating platform used during measurement of VOR gain. The shoulders and chin were fixed to a frame as shown. EOG electrodes were used to measure lateral eye movements, and two goniometers recorded head and upper torso rotation.

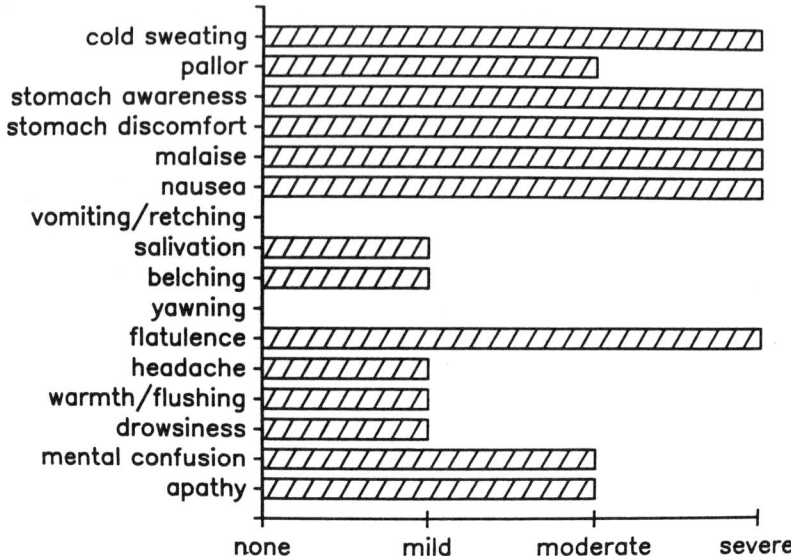

FIGURE 3. Motion sickness questionnaire results for one subject who was particularly sensitive to the effects of 30 minutes of rhythmical motion. This evaluation of signs and symptoms was carried out 10 minutes after resuming normal motor activity.

subsequent 20 minutes. The experiments also confirmed that this change could be accompanied by motion sickness. So, the answer to the question posed earlier is yes, normal-appearing, self-generated movement can cause motion sickness, apparently by forcing rapid and obviously inappropriate changes of vestibular function.

One point must be emphasized, however. Despite the use of VOR gain as a representative and objective measure of vestibular performance, and the fact that changes in VOR gain and motion sickness usually occurred in parallel, it is probably not correct to conclude that the rhythmical movement led to decreased VOR gain which in turn caused retinal image slip which then resulted in motion sickness. The entire experiment could be carried out with the eyes closed, producing if anything more severe motion sickness. Of course, we know that the blind can become motion sick, so this is not too surprising. It was also possible to produce a subject exhibiting oscillopsia only, by moving about with the eyes closed for 10 to 15 minutes after the rhythmical movement was completed. By the time the eyes were opened, all postural and perceptual problems had disappeared. The oscillopsia that remained was as strong as always, but was not provocative.

While the development of motion sickness in these experiments was probably not the result of retinal image slip, it may have been related to disrupted vestibular/neck interactions. At least two situations have been identified that did not result in symptoms. The first was rotating the *neck* only during the active, normal movement phase that followed the 30 minutes of rhythmic activity. Voluntary stabilization of the head in inertial space was a very common and very effective strategy adopted by many subjects to avoid motion sickness, when they were allowed to. The second was to rotate the *head* only in the period following the 30 minutes of rhythmic activity. In this case, a custom-molded and rigid neck cast was worn, which passively fixed the head to the body. Motion sickness has not occurred in those subjects tested so far. In contrast to these two situations, if the active, normal movement phase that followed rhythmical activity forced rotation of both *head and neck,* motion sickness did occur. It could be made even worse if visual and proprioceptive cues were eliminated or made unreliable, presumably forcing the use of a vestibular reference.

But why do we get motion sick? In these experiments, a wide variety of rhythmical motor strategies that suppressed vestibular function, and that utilized both head and

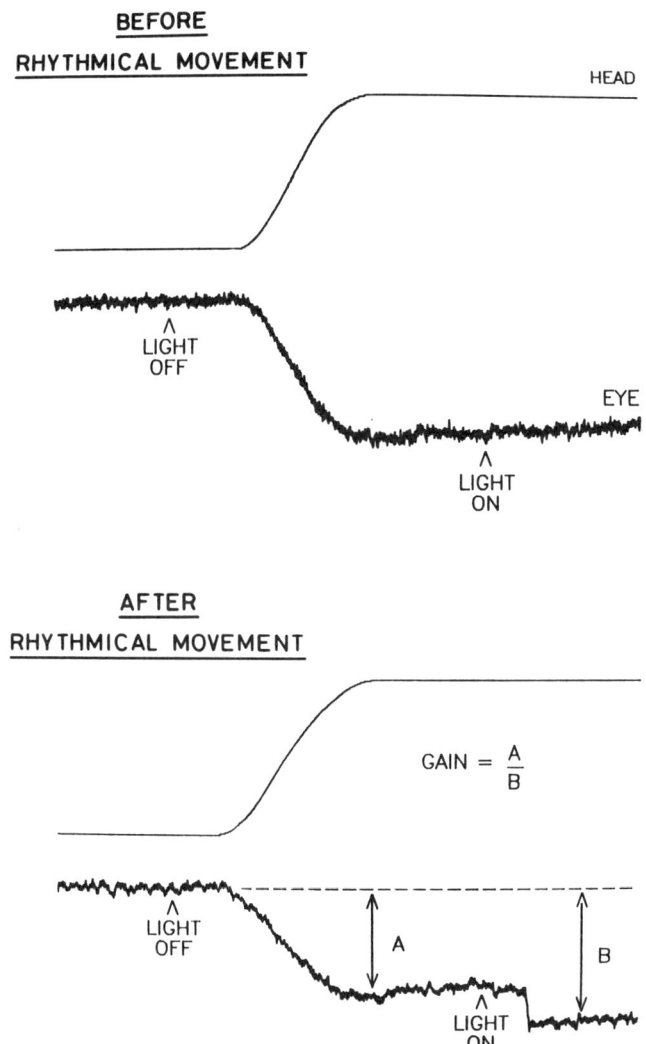

FIGURE 4. Head and eye movement during step rotations in the dark, before and after 30 minutes of rhythmical movement. VOR gain was calculated as the ratio of measurement A to measurement B.

neck rotation, resulted in subjective changes in postural, locomotor, and gaze control, objective changes in vestibular function, and motion sickness. Presumably, many other forms of "egocentric" motor activity (i.e., concentrating on a body frame of reference rather than the external world) would lead to similar effects, even if carried out inadvertently. Disordered postural, locomotor, and gaze control resulting from inappropriate motor strategies would have to be considered hazards to survival. However, the accompanying "motion sickness" symptoms of headache, nausea, drowsiness, etc., would tend to put an end to that activity. Thus, we may be dealing here with a phenomenon having considerable evolutionary significance, i.e., serving as a warning against inappropriate motor strategies that are causing undesired

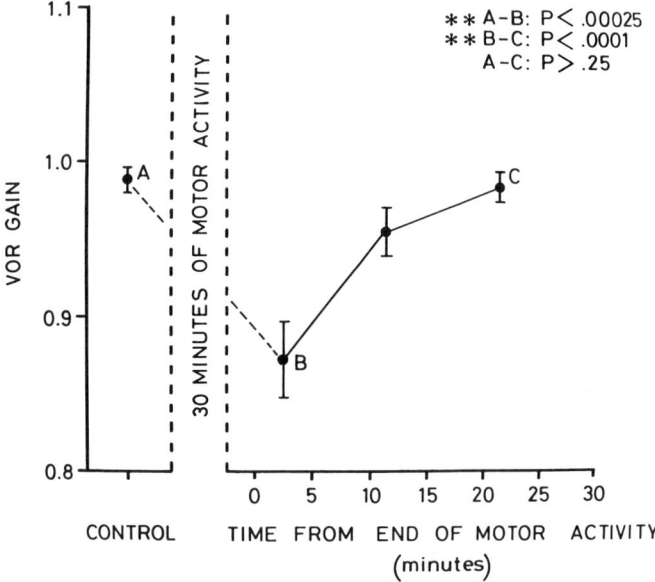

FIGURE 5. Average VOR gain of 12 subjects plotted as a function of time before and after 30 minutes of rhythmical motor activity, ± one SEM. The results of paired comparisons of combinations of points A, B, and C are shown, and the asterisks indicate statistically significant differences.

changes in vestibular function, and subsequent disruption of normal sensorimotor integration. Some of the modern "motion-sickness-provoking" environments may only be recently encountered, special cases that force a great deal of egocentric motor activity, leading to altered vestibular function. This would almost certainly be true in a situation such as reading in a moving car. Alternatively, there are many other situations that lead directly to an acute change in vestibular responses, and these could cause the same reactions. This might include cases of vestibular pathology, or moving the head in weightlessness. Thus, "motion sickness" may have a purpose after all, a reason quite unrelated to ships and cars and space shuttles, but one that extends far back into our development as a mobile and adaptive species.

ACKNOWLEDGMENTS

The authors would like to thank Luc Lefebvre and Walter Kucharski for their valuable assistance in these experiments. This study was made possible by our many volunteer subjects who chose to carry on despite considerable personal discomfort.

REFERENCES

1. REASON, J. T. & J. J. BRAND. 1975. Motion sickness. Academic Press. London, England.
2. CLAREMONT, C. A. 1931. The psychology of motion sickness. Psyche **11**: 86–90.
3. TREISMAN, M. 1977. Motion sickness: an evolutionary hypothesis. Science **197**: 493–495.
4. MONEY, K. E. & B. S. CHEUNG. 1983. Another function of the inner ear: facilitation of the emetic response to poisons. Aviat. Space Environ. Med. **54**: 208–211.
5. WATT, D. G. D., I. NEVO, T. YANG & A. SMITH. 1989. Motion sickness and motor strategy. *In* Proceedings 19th Annual Meeting. Society for Neuroscience. Phoenix, Ariz.
6. WATT, D. G. D. 1987. The vestibulo-ocular reflex and its possible roles in space motion sickness. Aviat. Space Environ. Med. **58**(9, Sect. II): A170–A174.
7. GAUTHIER, G. M. & D. A. ROBINSON. 1975. Adaptation of humans' vestibulo-ocular reflex to magnifying lenses. Brain Res. **92**: 331–335.

Neurotransmitter and Peptide Receptors on Medial Vestibular Nucleus Neurons[a]

DAVID O. CARPENTER AND NOBUAKI HORI

Wadsworth Center for Laboratories and Research
New York State Department of Health
and
School of Public Health
Albany, New York 12201-0509

INTRODUCTION

The vestibular nuclei are the site of termination of most of the primary afferent inputs from the peripheral vestibular organs and play a central role in the processing of vestibular information. While a considerable amount is known about the input-output relations in these nuclei, less is known about the neurotransmitters and neuropeptides involved in integration in these structures. The goal of the present studies is to contribute to this body of information by study of the medial vestibular nucleus of the rat, using isolated brain slices in which one can record from the neurons using either extra- or intracellular techniques, iontophoretic application of substances suspected of having biologic activity, and bath perfusion of antagonist drugs.

Several observations from this and previous studies are important in understanding the functioning of this nucleus. The neurons in the medial vestibular nucleus are all endogenous pacemakers, having an intrinsic rhythm which is modulated by excitatory or inhibitory synaptic inputs. It is relatively unusual for central nervous system (CNS) neurons to have such a regular pacemaker rhythm, but this property makes the neurons exquisitely sensitive to transmitter actions. The endogenous synaptic input from the eighth nerve is blocked by antagonists of the kainate/quisqualate type of excitatory amino acid receptor, consistent with the conclusion that the transmitter is glutamate or a related excitatory amino acid. Finally, there are potent excitatory receptors for acetylcholine (muscarinic), opiate peptides (mu and delta), and the excitatory amino acids. There is an interaction between the opiate and muscarinic receptors on these neurons that has not been seen elsewhere in the nervous system, but the functional significance of this interaction remains to be elucidated.

METHODS

Brain stem slices were prepared from male rats weighing 150–180 grams as shown diagrammatically in FIGURE 1. Animals were euthanized by cervical dislocation, and the brain rapidly removed and immediately placed in cold (4°C), oxygenated Krebs-Ringer solution containing (in mM) NaCl, 126; KCl, 5; $MgSO_4$, 1.3; $CaCl_2$, 2.4;

[a] Supported in part by a grant from the Aaron Diamond Foundation to the Capital District Center for the Study of Drug Abuse and Treatment.

NaHCO$_3$, 26; and glucose, 10. The brain stem was then blocked on wet filter paper, positioned on a vibratome, and 450-μm sections cut containing the vestibular nuclei in the planes indicated in FIGURE 1A and B. The details of the preparation of brain slices from any area as developed in our laboratory have been previously published.[1]

After the sections were cut they were incubated in oxygenated Krebs-Ringer at 37°C for at least two hours, after which a slice was positioned in the recording

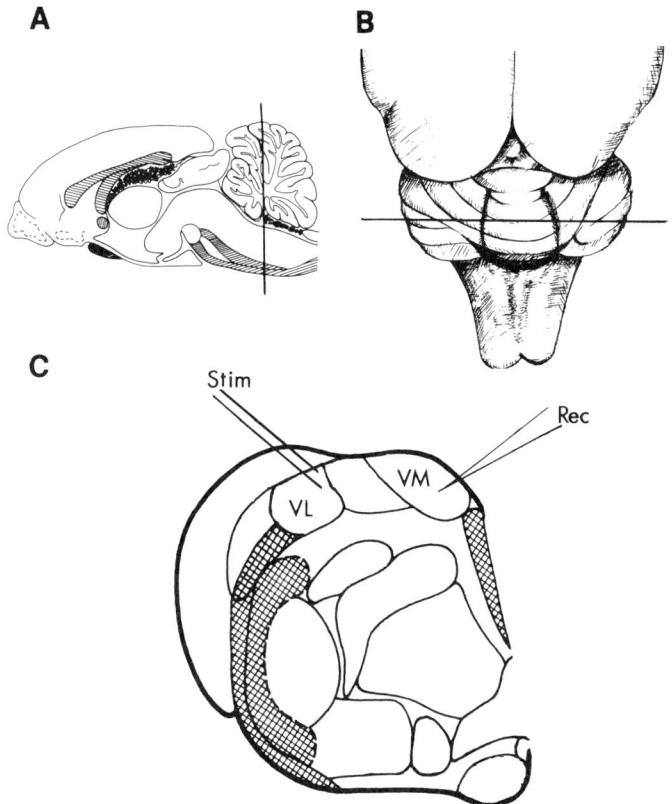

FIGURE 1. Schematic representation of the rat brain with indication of the preparation of the rat medial vestibular nucleus brain slice. **A** and **B** show the preparation of the slice in the vertical and horizonal planes. Part **C** shows the details of the slice, with the placement of the stimulating (Stim) and recording (Rec) electrodes. VL = lateral vestibular nucleus. VM = medial vestibular nucleus.

chamber, covered with a nylon mesh, and submerged under oxygenated Krebs-Ringer, which flowed through the chamber at a rate of approximately 3 mL/minute. Recording electrodes for extra- or intracellular studies were positioned in the medial vestibular nucleus as indicated in FIGURE 1C. Extracellular electrodes were 3–5 megohm glass pipettes filled with Krebs-Ringer solution, while intracellular electrodes were pipettes of 100–150 megohm filled with 3 M potassium acetate. In some experiments synaptic activation of the medial vestibular neurons was achieved by activation of the eighth nerve from a point at the medial edge of the lateral vestibular nucleus, using a bipolar electrode, as indicated in FIGURE 1C. Super maximal pulses (about 10 V) of 50-µs duration were used.

In other studies in which the responses of these neurons to iontophoretically applied transmitters and peptides were tested, a three-barreled iontophoretic electrode was prepared and positioned independently of the recording electrode in the dendritic tree (FIGURE 2A). Drugs were applied using a Neurophore unit and a regular sequence where each substance was applied once per 90 seconds with 30-second intervals between applications of the various substances. The schematic of the neuron was drawn by Ms. Naomi Hori from sections prepared from the medial vestibular nucleus in which single neurons were injected with horseradish peroxidase, as previously described.[2] All of the neurons ($n = 11$) injected with horseradish peroxidase that were examined had the general configuration shown in this drawing, with three or four major dendritic branches. The preparation and filling of all types of electrodes have been previously reported in detail.[1] Part B of FIGURE 2 gives the concentrations and pHs used for the various substances applied by iontophoresis in these experiments, and the polarity of application. Note that the excitatory amino acids and enkephalins were applied by electroosmosis, as previously reported.[3,4]

Recording procedures have also been described.[1] Signals were stored on tape, and hard records obtained by playing the tape at reduced speed onto a Gould pen recorder. Frequency histograms were obtained using a Princeton Applied Research Model 4203 signal averager in the time interval distribution mode. The bin width in all such recordings was 300 mseconds. In experiments in which active agents were applied by iontophoresis, extracellular recordings were obtained as illustrated in FIGURE 2C, and rate-meter records of the data obtained as previously described.[3]

RESULTS

The Origin of the Spontaneous Discharge of Medial Vestibular Neurons

All neurons recorded in the medial vestibular nucleus exhibited a regular, spontaneous discharge at a frequency varying from about 0.5 to 80 Hz, as was illustrated in FIGURE 2C in extracellular recordings. The majority of neurons in extracellular recordings, where there is less chance of injury from the electrode, showed discharge frequencies in the range of 10–30 seconds. Because of the very great regularity, this discharge might be a result of an endogenous pacemaker, rather than due to synaptic drive from external sources. In order to establish that this is the case, several experiments were done.

FIGURE 3 shows intracellular recordings from a medial vestibular neuron, and the effects of application of hyperpolarizing current through the intracellular electrode using a bridge circuit for current application. At rest this neuron showed a spontaneous discharge of approximately 60/second. The regularity and the absence of indication of subthreshold synaptic potentials is suggestive of this being an endogenous rhythm. With hyperpolarizing current of increasing amplitude there is a progressive slowing of the discharge. However, the rhythm remains regular, and no synaptic potentials are seen. This is very strong evidence that the discharge is due to an endogenous pacemaker.[5,6]

FIGURE 4 shows that the discharge of a vestibular neuron was not blocked by perfusion of a variety of drugs that antagonize various transmitter receptors. This figure shows extracellular action potentials. Perfusion of atropine, a muscarinic

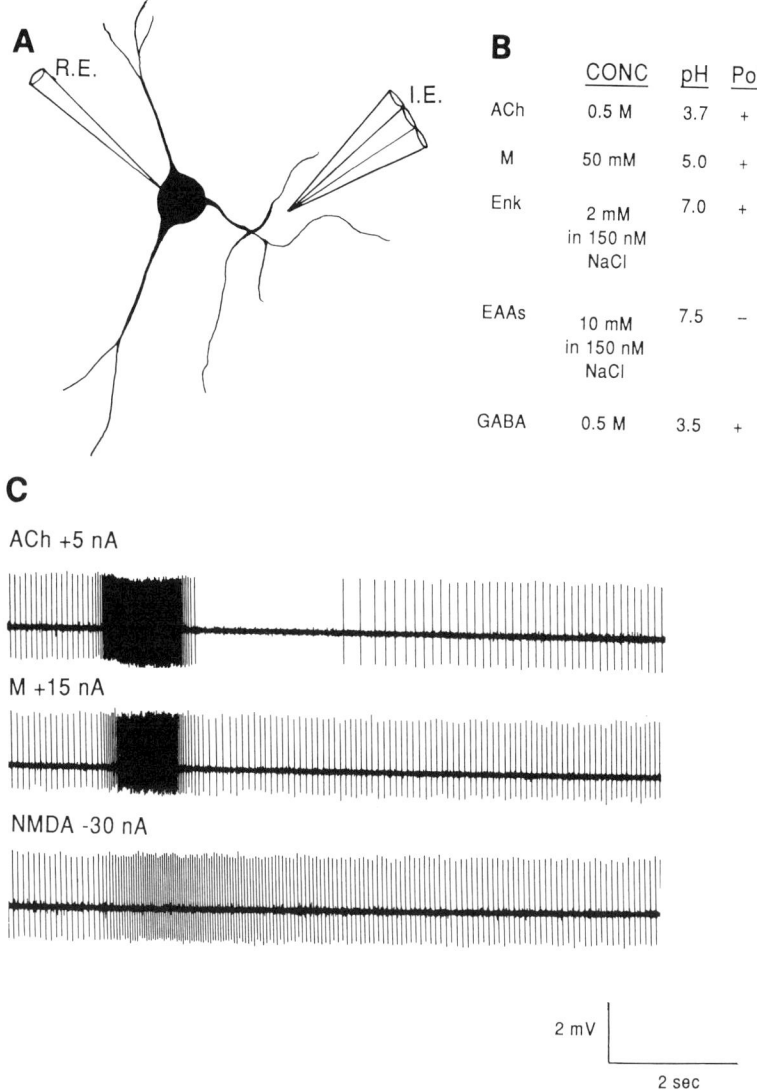

A

R.E.

I.E.

B

	CONC	pH	Pol
ACh	0.5 M	3.7	+
M	50 mM	5.0	+
Enk	2 mM in 150 nM NaCl	7.0	+
EAAs	10 mM in 150 nM NaCl	7.5	−
GABA	0.5 M	3.5	+

C

ACh +5 nA

M +15 nA

NMDA -30 nA

2 mV

2 sec

FIGURE 2. Recording techniques used in medial vestibular slices. Part **A** shows a camera lucida drawing of a medial vestibular neuron that was injected intracellularly with horseradish peroxidase. Recording electrodes (RE) were positioned near or in the cell soma for extra- or intracellular recordings of electrical activity, respectively. In those experiments in which neuroactive substances were applied, a three-barreled iontophoretic electrode was used and positioned in the dendritic tree, where responses to effectively all agents tested were found to be largest. Part **B** shows the substances used, the concentrations prepared, pH, and polarity of application. ACh = acetylcholine; M = morphine; Enk = methionine enkephalin; EAAs = excitatory amino acids (N-methyl-D-aspartate, quisqualate, or glutamate); and GABA = γ-aminobutyric acid. **C** shows extracellular recordings of a medial vestibular neuron and the response to acetylcholine, morphine, and N-methyl-D-aspartate (NMDA).

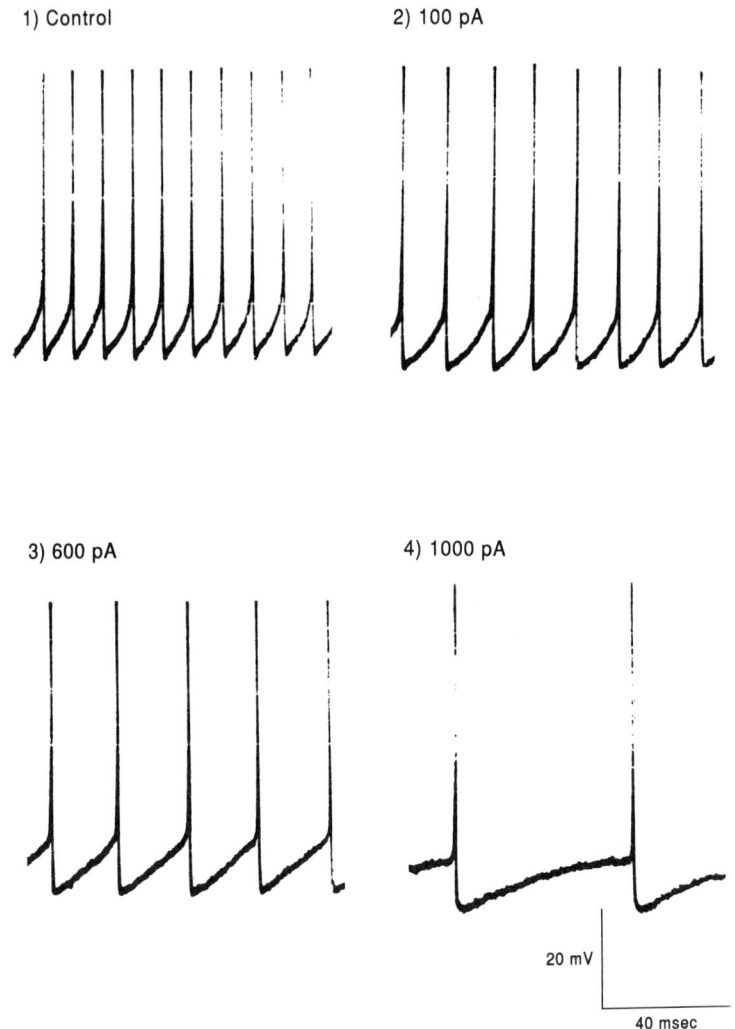

FIGURE 3. Pacemaker activity in medial vestibular neurons. All neurons recorded showed regular spontaneous activity the frequency of which was reduced by application of hyperpolarizing current. The records show the control (1), application of 100 pA of hyperpolarizing current (2), 600 pA (3), and 1000 pA (4). The fact that the discharge is smoothly slowed is strong indication that the discharge is endogenous and not due to synaptic input.

acetylcholine antagonist, or curare, a nicotinic acetylcholine antagonist, did not significantly affect discharge. There was also no effect of amino phosphonobutyric acid (APB) or amino phosphonovaleric acid (APV), antagonists of two types of excitatory amino acid receptors. In studies not shown but done on other cells, there was also no effect of 6-cyano-7-nitroquinoxaline-2,3-dione (CNQX), which is a selective antagonist of the quisqualate/kainate type of excitatory amino acid receptor.[7] Bicuculline, an antagonist of γ-aminobutyric acid (GABA) responses, had no effect, nor was there a significant alteration of discharge frequency by naloxone, an antagonist of opiate receptors. These observations imply not only that none of the above agonists is an excitatory transmitter driving the spontaneous discharge, but also that none of these agonists has a role in modulating the spontaneous discharge under the circumstances in which the experiment was done. This is particularly significant with regard to GABA, which is a common inhibitory transmitter, and implies that there is not a resting inhibitory background in the brain slices.

If this rhythm is truly endogenous it should not be blocked by ionic manipulations of the perfusing solution that block synaptic transmission. FIGURE 5 shows a frequency histogram of the discharge of a different neuron in the control Krebs-Ringers, and in low Ca^{2+}–high Mg^{2+} Krebs-Ringer, in which essentially all synaptic transmission is blocked through blockade of transmitter release at the presynaptic terminal. The figure shows a plot of the interspike intervals, recorded over a period of 30 seconds. Note the little variability in the control solution. In the low Ca^{2+}–high Mg^{2+} solution there is a slight acceleration of the discharge, rather than the slowing

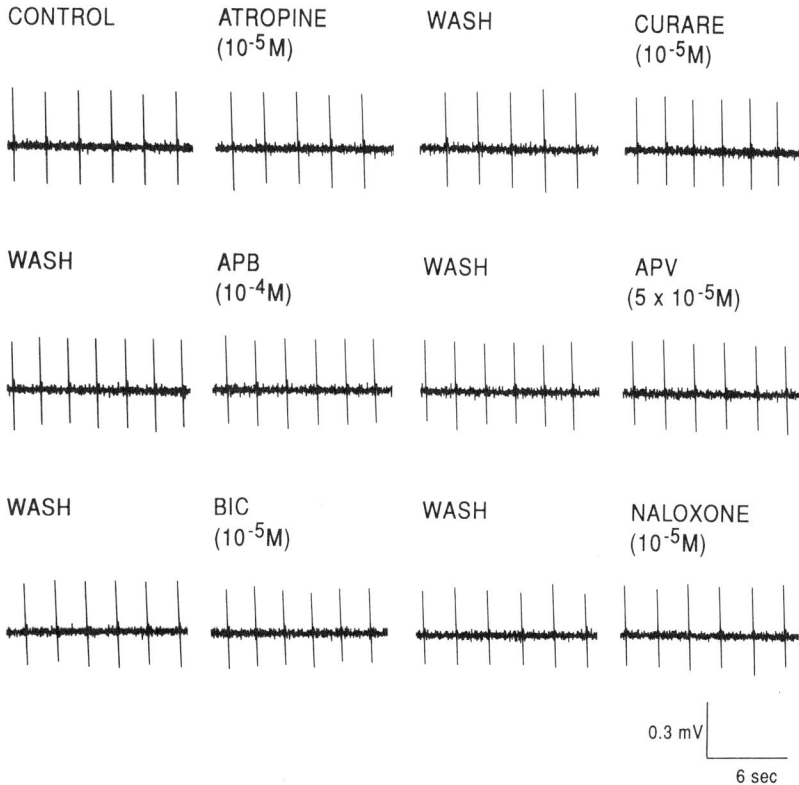

FIGURE 4. Lack of effect of various transmitter receptor antagonists on spontaneous activity of medial vestibular neurons recorded extracellularly. Each drug was perfused for 10 minutes at the concentration indicated, followed by a 15-minute wash with control Krebs-Ringer. APB = amino phosphonobutyric acid; APV = amino phosphonovaleric acid; BIC = bicuculline.

and cessation that would be expected if synaptic drive were the source of the discharge. The increase in frequency is almost certainly a reflection of the membrane instability and hyperexcitability induced by removal of calcium.[8] With wash, the discharge frequency returns to control values. These observations constitute proof that the spontaneous activity represents an endogenous pacemaker discharge of medial vestibular neurons.

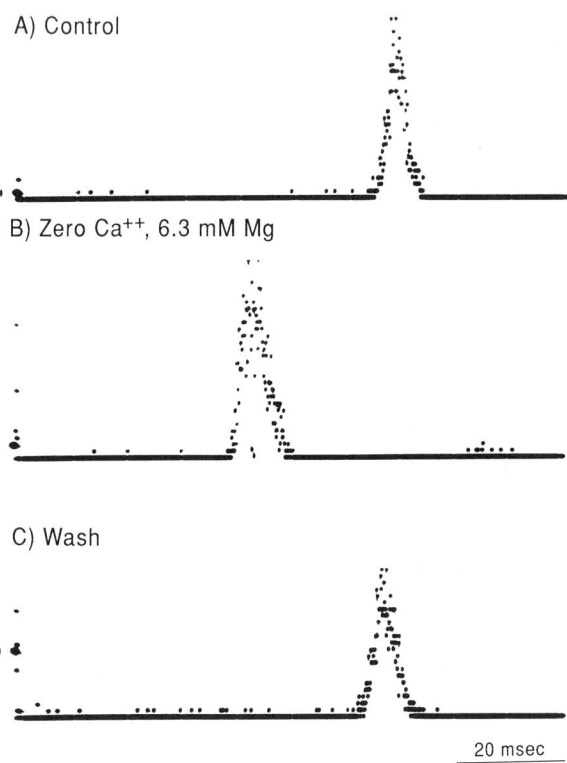

FIGURE 5. Effects of perfusion of a high-Mg^{2+}, no-added-Ca^{2+} Krebs-Ringer solution on the spontaneous activity of a medial vestibular neuron, recorded extracellularly. Records show a histogram of spike interval following the last spike at time zero. The zero Ca^{2+}, 6.3 mM Mg^{2+} solution, which should block all synaptic activity, resulted in a slight increase in frequency of spontaneous discharge as expected as a result of the effect of removing calcium on neuronal excitability. Each record represents accumulated spike intervals over a 30-second period.

The Endogenous Transmitter Released from the Fibers of the Eighth Nerve

Since there is considerable disagreement in the literature concerning what is the transmitter released from those afferent fibers projecting to the medial vestibular nucleus from the eighth nerve, we have performed a series of experiments in which synaptic excitation was evoked by electrical stimulation within the slice. The major problem in such studies is to find a site where excitatory drive is clearly not a result of direct current flow or the result of a relatively nonspecific activation of local interneurons. We have found that we could routinely drive these neurons by stimulation applied at the medial border of the lateral vestibular nucleus (shown diagrammatically in FIGURE 1C). FIGURE 6 shows results from one such study with extracellular recording. Following the stimulus artifact the cell is driven with a latency of about 3.5 mseconds. This is sufficiently long to indicate that the excitation is not due to current spread, and yet is sufficiently brief to be consistent with monosynaptic excitation.[9] The site of stimulation is consistent with what would be expected for the pathway of the eighth nerve. Therefore, we conclude that this stimulation is exciting eighth nerve fibers that are excitatory onto medial vestibular neurons.

The various antagonists that were studied in FIGURE 4 were tested for efficacy in blocking the synaptic response. FIGURE 6 illustrates one neuron in which CNQX, APV, and atropine were studied. The synaptic drive was reversibly blocked by CNQX, but not by APV, an N-methyl-D-aspartate (NMDA) receptor antagonist, or by atropine, a muscarinic acetylcholine receptor antagonist. There also was no effect of curare or of naloxone. These results are consistent with the conclusion that the excitatory transmitter released by eighth nerve fibers onto medial vestibular neurons is an excitatory amino acid, presumably glutamate, which acts at a kainate or quisqualate receptor.

Transmitter and Peptide Receptors on Medial Vestibular Neurons

FIGURE 7 shows the responses of one of these neurons to iontophoretic application of morphine (M), quisqualate (Q), and acetylcholine (ACh). All three substances excited this neuron. In this neuron the response to quisqualate quickly recovered to baseline discharge, while for both acetylcholine and morphine there was a rapid initial response followed by a prolonged excitation, which may reflect a second component. This figure shows the effect of increasing the iontophoretic current on the response to morphine. The threshold was to currents of less than 5 nA. There was little increase in the early peak discharge with currents beyond 15 nA, but the duration of discharge was prolonged when higher currents were used.

Of over 57 neurons recorded in the medial vestibular nucleus, all of those tested were excited by quisqualate and acetylcholine. The great majority of neurons were excited by morphine. Morphine is known to be primarily a mu receptor agonist, but does also act at delta receptors. We tested the effect of methionine enkephalin (which acts primarily at delta receptors) iontophoresis in five neurons, and found it to be excitatory in all cases. In two neurons excited by morphine, we tested the effects of the specific mu agonist DAGO and the specific delta agonist DPDPE and found that both were excitatory. These observations, although preliminary, suggest that the medial vestibular neurons have excitatory receptors for both mu and delta agonists. Occasionally (3 of 45 cells), pure inhibitory responses were seen to morphine. Biphasic (excitatory-inhibitory) responses were common for acetylcholine, and these are illustrated in the raw data records shown in FIGURE 2C. Previous studies by Gallagher et al.[10] suggest that the biphasic response to acetylcholine may be due to activation of inhibitory interneurons, releasing GABA, but we have not tested this possibility directly. The response of one of the neurons that showed pure inhibition by morphine is shown in FIGURE 8. Both the morphine inhibition and acetylcholine excitation were depressed by naloxone. We have not yet characterized the inhibitory responses in regard to receptor type.

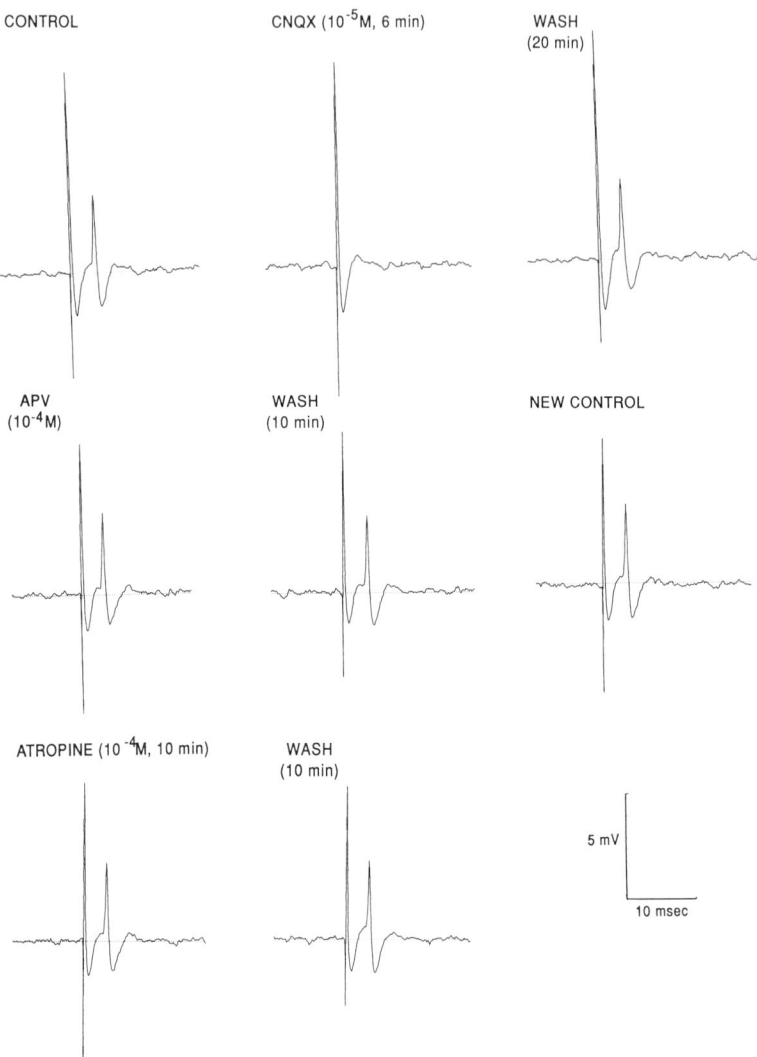

FIGURE 6. Blockade of synaptic activation of a medial vestibular neuron by stimulation of the eighth nerve by CNQX, an antagonist of excitatory amino acid receptors of the quisqualate/ kainate type. Records are extracellular recordings of responses following stimulation as illustrated in FIGURE 2. The stimulation was through a bipolar electrode, using an interval of 4 seconds and pulses of 50 μseconds duration and approximately 10 V intensity. The left peak in each recording is the stimulus artifact, while the biphasic response on the right is the neuronal spike. Note the response was blocked by CNQX but not by APV or atropine.

FIGURE 9 shows another neuron and the effects of naloxone on responses to acetylcholine, morphine, and quisqualate. All responses in this cell are pure excitation. Naloxone caused a depression of both the morphine and acetylcholine responses, although there was no effect on the response to quisqualate.

Naloxone has long been used as a relatively specific antagonist of opiate receptors, and therefore its effect on acetylcholine responses is surprising. However, this action was a consistent one on all neurons tested ($n = 14$). FIGURE 10 shows the effects of naloxone on NMDA, acetylcholine, and morphine responses. The NMDA response was unaffected, while the acetylcholine and morphine responses were depressed to effectively equal degrees. When a higher concentration of naloxone was used to give effectively total blockade of the morphine response, there was also effectively total blockade of the acetylcholine response (FIGURE 11).

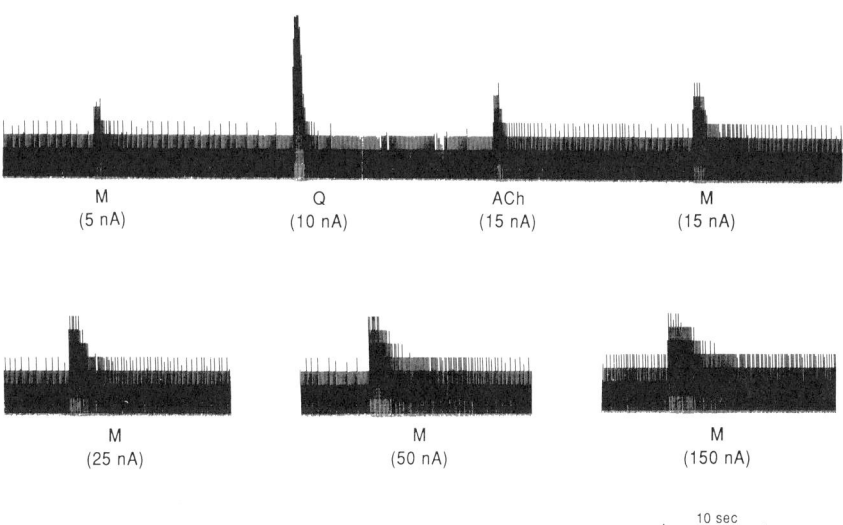

FIGURE 7. Rate-meter recordings of responses of a medial vestibular neuron to morphine (M), quisqualate (Q), and acetylcholine (ACh). The effects of dose on the response to morphine are also shown. The rate meter reset every 300 mseconds.

FIGURE 12 shows results obtained from one neuron in which it was possible to test the interactions of several agonists and antagonists on the acetylcholine and morphine responses. In this cell we iontophoresed acetylcholine, morphine, and quisqualate. The quisqualate response was selectively blocked by CNQX. Naloxone depressed the acetylcholine and morphine responses to an equal degree without effect on the response to quisqualate. When morphine was perfused in the bath there was clear and reversible potentiation of the acetylcholine response. Finally, when atropine, an antagonist of muscarinic acetylcholine responses, was applied there was an almost total blockade of both the acetylcholine and morphine responses without any effect on the response to quisqualate.

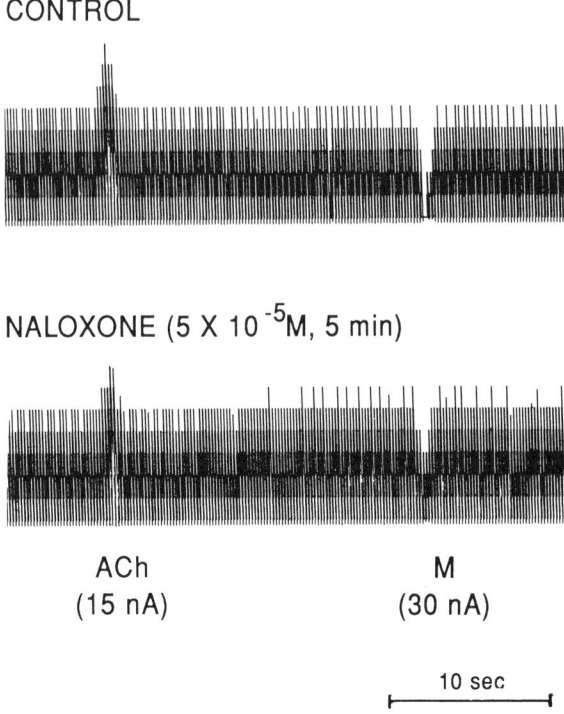

CONTROL

NALOXONE (5 X 10⁻⁵M, 5 min)

ACh
(15 nA)

M
(30 nA)

10 sec

FIGURE 8. An unusual hyperpolarizing response to morphine in a medial vestibular neuron. Both the depolarizing acetylcholine and the hyperpolarizing morphine responses were partially but not completely blocked by naloxone. The rate meter reset every 300 mseconds.

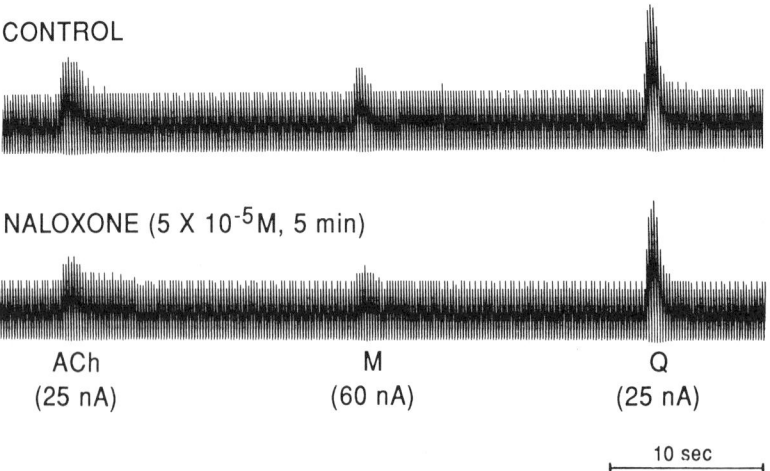

CONTROL

NALOXONE (5 X 10⁻⁵M, 5 min)

ACh
(25 nA)

M
(60 nA)

Q
(25 nA)

10 sec

FIGURE 9. Naloxone reduces both morphine and acetylcholine responses without effect on responses to quisqualate. The rate meter reset every 300 mseconds.

DISCUSSION

Medial Vestibular Neurons Exhibit Endogenous Pacemaker Activity

Endogenous pacemaker discharge in neurons was first definitively documented by Alving, who physically isolated neurons of *Aplysia* and demonstrated that sponta-

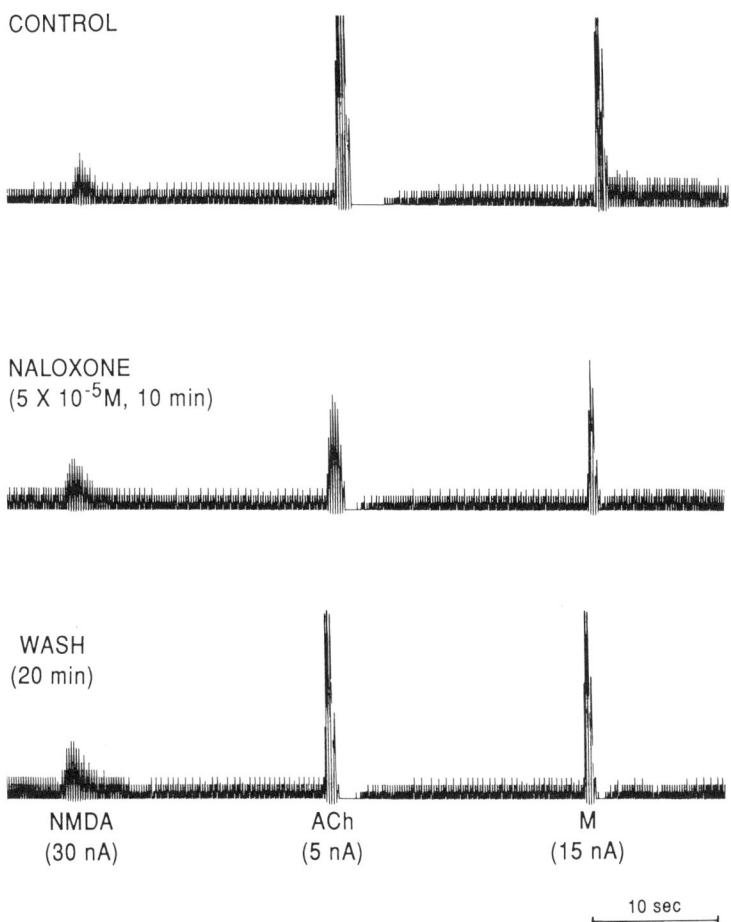

FIGURE 10. Naloxone reduces both morphine and acetylcholine responses without effect on responses to NMDA.

neous, patterned activity did not depend upon synaptic input.[11] Such pacemaker activity is commonly observed in invertebrate preparations[6] but definitive identification of pacemaker activity in the mammalian CNS has not previously been possible. It is not possible to prove that spontaneous activity is a result of endogenous activity in *in vivo* preparations where a total blockade of synaptic activity is incompatible with life, but the advent of use of isolated brain slices has allowed investigators new freedom to alter the environment without being dependent upon heart beat and respiration. Using brain slices, several authors have reported spontaneous activity, which they have ascribed to an endogenous pacemaker activity.[12-14] Previous studies using the medial vestibular nucleus have commented on the regular spontaneous activity and have presented the possibility that the discharge was endogenous.[9,15-17]

The present studies demonstrate that the regular spontaneous activity of medial vestibular neurons depends upon membrane potential, is independent of individual

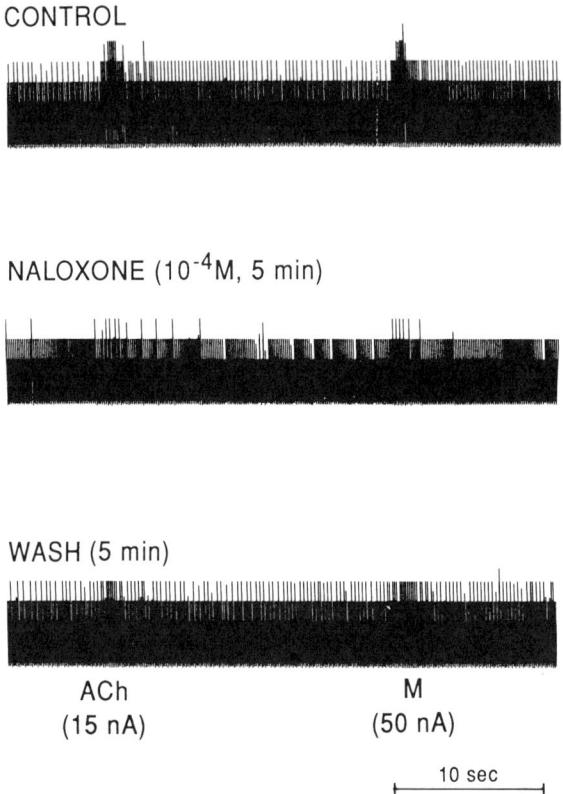

FIGURE 11. Concentrations of naloxone that give effectively total blockade of the morphine response also give almost total blockade of the acetylcholine response.

transmitter systems, and is present when all synaptic activity is blocked by low Ca^{2+}–high Mg^{2+} medium. Therefore, there can be no other conclusion than that these are true pacemaker neurons, having an endogenous rhythm resembling that of heart muscle. Thus, these observations represent the most convincing evidence to date that central neurons may exhibit endogenous pacemaker discharges similar to those characteristic of invertebrate neurons.

The presence of pacemaker activity, especially activity at a relatively high

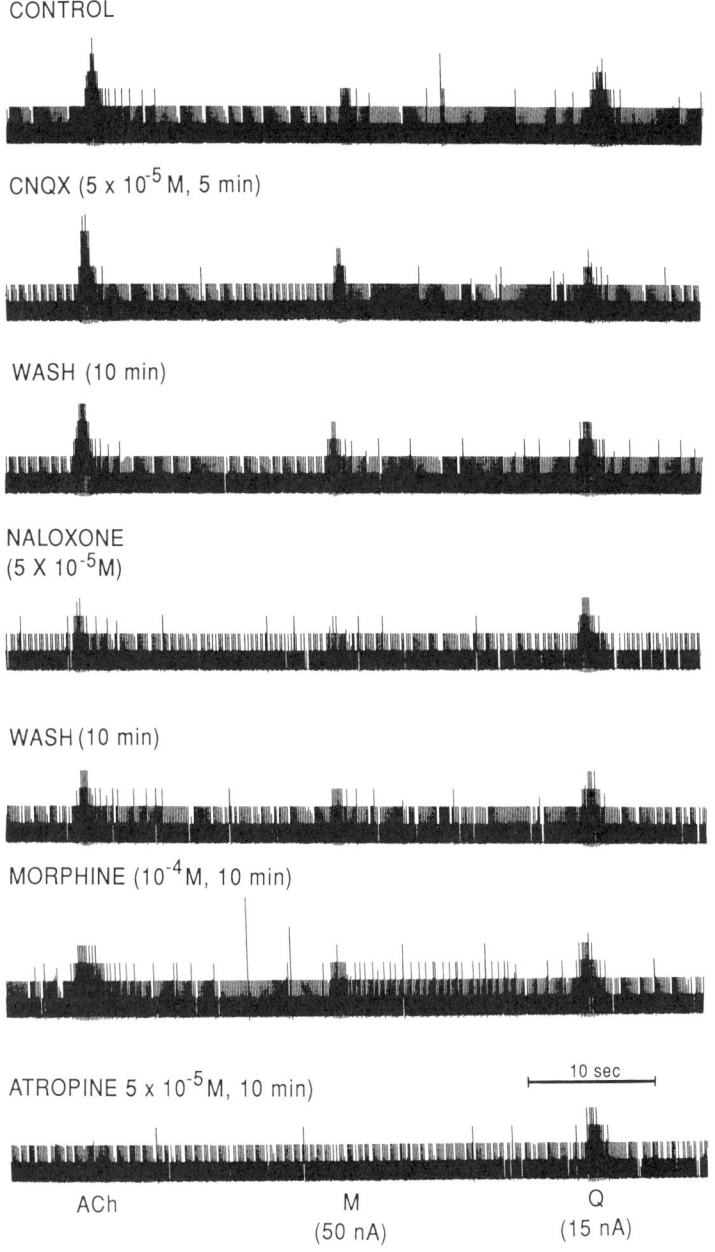

FIGURE 12. Effects of CNQX, naloxone, and atropine on responses to acetylcholine (ACh), morphine (M), and quisqualate (Q). While the quisqualate responses were selectively blocked by CNQX, both naloxone and atropine blocked both the acetylcholine and morphine responses. The rate meter resets every 300 mseconds.

discharge rate observed in these neurons (0.5 to 80 Hz), has considerable implications for understanding information processing in this nucleus. Clearly the simplistic expectation that all is quiet until a stimulus comes in, cannot apply here. Those neurons postsynaptic to this nucleus are constantly receiving input at a high rate of activity. Thus the information transfer must be in the patterning, not just in the presence or absence, of activity.

A second feature of pacemaker neurons is that they are exquisitely sensitive to small inputs. One characteristic of pacemaker neurons in invertebrates is that they tend to have high input resistances secondary to a relatively low resting potassium conductance.[5] The high input resistance and the fact that the discharge can be accelerated or decelerated by depolarizing or hyperpolarizing currents, respectively, make the discharge very sensitive to small synaptic currents, which would usually be totally subthreshold in a nonpacemaker neuron. There is, however, little or no understanding how the nervous system can decipher information encoded in these relatively subtle changes in discharge frequency. This problem poses a very interesting challenge to physiologists interested in the vestibular system.

The Transmitter in Vestibular Afferent Fibers

Until very recently it was believed that the transmitter of vestibular afferents was acetylcholine.[18–23] These studies, using *in vivo* preparations, resulted in conclusions that are not supported by more recent experiments using isolated brain slices, however. There has been suggestive evidence for some time that excitatory amino acids played a role in vestibular afferent fibers, based on biochemical[24,25] and immunocytochemical[26] evidence. In the frog there has been direct evidence that the excitatory postsynaptic potential (EPSP) evoked by stimulation of vestibular afferents is blocked by non-NMDA receptor antagonists.[27] Two recent studies have utilized brain slice preparations similar to ours, and both have concluded that the transmitter receptor was a non-NMDA excitatory amino acid.[9,28] The present results confirm this conclusion, even though all three studies used somewhat different methods for stimulation of vestibular afferents. Lewis *et al.* utilized local stimulation,[28] and obtained very short latency responses which were blocked by kynurenic acid, a relatively nonspecific excitatory amino acid antagonist, and amino phosphonobutyric acid, a substance closely related to other excitatory amino acid antagonists, which probably acts presynaptically.[29] Doi *et al.* attempted to include the root of the eighth nerve in their slices of the medial vestibular nucleus, but they had to use very long stimulation pulses to excite the neurons.[9] They were, however, able to distinguish mono- and polysynaptic excitation and to determine a synaptic delay to be of the order of 2 mseconds. Our method differs in that we could consistently obtain drive of medial neurons by placing the stimulating electrode at the medial margin of the lateral vestibular nucleus. While we have no direct proof that this is the pathway of the eighth nerve, the potent drive and the consistent brief latency are consistent with the conclusion that we are monosynaptically exciting the neurons by stimulation of the afferent fibers of the vestibular system. The consistency of conclusion among the three studies is perhaps the best evidence that the endogenous transmitter is glutamate or a related excitatory amino acid.

The lack of any effect of perfusion of the various transmitter antagonists on the endogenous discharge is of interest in light of reports that APV, microinjected into intact animals, blocks a major component of the resting discharge of vestibular neurons.[30] This observation is understandable in light of the fact that in the slice, the great majority of synaptic inputs are cut and therefore silenced.

Opiate Receptors in the Vestibular Nucleus, and the Interaction of Opiate and Muscarinic Receptors

Opiate receptors are widely distributed in the central nervous system, and both enkephalin-containing cell bodies and terminals are found in the vestibular nuclei.[31-33] Beitz *et al.* report that the medial vestibular nucleus contains the greatest numbers of enkephalin-labeled neurons among all brain stem structures.[33] The opiates have a number of important functions, but it is relatively unusual for there to be clear excitatory responses on the great majority of neurons in any specific population. In the first place, most responses to opiate peptides are inhibitory.[34,35] In addition, many if not most opiate receptors are on the presynaptic terminals and not the soma on neurons.[36] The fact that most neurons of the medial vestibular nucleus have receptors for opiate peptides was reported by Yasnetsov and Pravdivtsev,[37] who recorded in anesthetized cats. Our observations confirm their report that a very high percentage of vestibular neurons have both mu and delta opiate receptors, although the frequency of observing inhibitory responses was considerably less in our studies than theirs. These authors also reported that naloxone, a specific antagonist of opiate receptors, blocked both excitatory and inhibitory effects of morphine and enkephalin, an observation that our experiments confirm.

The interaction between morphine and acetylcholine responses that we have observed was unexpected. Early *in vivo* studies with iontophoresis of morphine and enkephalins showed that morphine could block some responses to excitatory amino acids and acetylcholine[38,39] by unknown mechanisms. In contrast, Bradley and Dray reported that morphine blocked excitation of unidentified brain stem neurons in the rat by acetylcholine, norepinephrine, and serotonin, but did not affect excitation by glutamate or homocysteic acid or inhibition by any of the above substances.[40] They did note that occasionally they saw a potentiation of the excitatory responses by morphine, but provided no further details in a preliminary report. While this reference could be to a process similar to the one that we observe, the interaction we see is both opposite in direction and more specific to one transmitter system than the common effect reported in all of these studies. Moreover, the interaction is demonstrated by the cross effectiveness of both naloxone, the specific opiate antagonist, and atropine, the specific muscarinic antagonist, and by the total lack of interaction with either the NMDA or the quisqualate receptors.

A very similar interaction between acetylcholine and opiate responses has been reported on Renshaw cells of the cat spinal cord.[41,42] These neurons are excited by acetylcholine, morphine, and leu-enkephalin as well as by the excitatory amino acids. The response to the opiate peptides and acetylcholine but not the excitatory amino acids is blocked by naloxone. Furthermore, the response to acetylcholine is enhanced by prolonged application of morphine. This action of morphine on acetylcholine responses of Renshaw cells had been previously reported by Lodge *et al.*[43] Atropine was tested by Duggan *et al.* on four neurons and found to reduce both morphine and acetylcholine excitation.[41] Therefore, in all regards this appears to be the same interaction as we observe. Nabatame *et al.* report that lateral vestibular neurons are excited by phencyclidine, and that this excitation is blocked by atropine.[23] This may reflect a similar effect to that observed here.

The molecular mechanisms responsible for this interaction are unknown and will be the subject of future investigations. Clearly this interaction is important for understanding vestibular function, particularly so since cholinergic mechanisms are so important in such vestibular-dependent events as motion sickness.[44] Further investigations will be required to elucidate the details of these interactions and their physiological implications.

The principal conclusions from this study are as follows:

1. Medial vestibular neurons, studied in rat brain slices, are endogenous pacemakers, and therefore an understanding of their physiological function necessitates study of the patterning of their output, not just discharge activity.
2. The neurotransmitter utilized by eighth nerve terminals is an excitatory amino acid, probably glutamate, acting at CNQX-sensitive receptors of the kainate/quisqualate type.
3. Opiate peptides are potent excitants of medial vestibular neurons, acting at both mu and delta receptors. There is an interaction between opiate and cholinergic receptors such that they both are blocked by the opiate antagonist naloxone and the muscarinic antagonist atropine.

REFERENCES

1. HORI, N., N. AKAIKE & D. O. CARPENTER. 1988. Piriform cortex brain slices: techniques for isolation of synaptic inputs. J. Neurosci. Methods **25:** 197–208.
2. ffRENCH-MULLEN, J. M. H., N. HORI, H. NAKANISHI, N. T. SLATER & D. O. CARPENTER. 1983. Asymmetric distribution of acetylcholine receptors and M channels on prepyriform neurons. Cell. Mol. Neurobiol. **2:** 153–182.
3. CARPENTER, D. O., D. B. BRIGGS, A. P. KNOX & N. STROMINGER. 1988. Excitation of area postrema neurons by transmitters, peptides and cyclic nucleotides. J. Neurophysiol. **59:** 358–369.
4. ffRENCH-MULLEN, J. M. H., N. HORI & D. O. CARPENTER. 1986. Receptors for the excitatory amino acids on neurons in rat pyriform cortex. J. Neurophysiol. **55:** 1283–1294.
5. CARPENTER, D. O. 1973. Ionic mechanisms and models of endogenous discharge of *Aplysia* neurons. *In* Proceedings of the Symposium on Neurobiology of Invertebrates: Mechanism of Rhythm Regulation: 35–58. Hungarian Academy of Sciences. Tihany, Hungary.
6. CARPENTER, D. O., Ed. 1982. Cellular Pacemakers, **1** and **2.** John Wiley and Sons. New York, N.Y.
7. HONORE, T., S. N. DAVIES, J. DREJER, E. J. FLETCHER, P. JACOBSEN, D. LODGE & F. E. NIELSEN. 1988. Quinoxalinediones: potent competitive non-NMDA glutamate receptor antagonists. Science **241:** 701–703.
8. CARPENTER, D. O. & R. GUNN. 1970. The dependence of pacemaker discharge of *Aplysia* neurons upon Na^+ and Ca^{2+}. J. Cell. Physiol. **75:** 121–127.
9. DOI, K., T. TSUMOTO & T. MATSUNAGA. 1990. Actions of excitatory amino acid antagonists on synaptic inputs to the rat medial vestibular nucleus: and electrophysiological study *in vitro.* Exp. Brain Res. **82:** 254–262.
10. GALLAGHER, J. P., K. D. PHELAN & P. SHINNICK-GALLAGHER. 1992. Modulation of excitatory transmission at the rat medial vestibular nucleus synapse *in vitro.* Ann. N.Y. Acad. Sci. (This volume.)
11. ALVING, B. O. 1968. Spontaneous activity in isolated somata of *Aplysia* pacemaker neurons. J. Gen. Physiol. **51:** 29–45.
12. ANDREW, D. 1987. Endogenous bursting by rat supraoptic neuroendocrine cells is calcium dependent. J. Physiol. London **384:** 451–465.
13. HATTON, G. I. 1982. Phasic bursting activity of rat paraventricular neurones in the absence of synaptic transmission. J. Physiol. London **327:** 273–284.
14. INOUYE, S. T. & H. HAWAMURA. 1979. Persistence of circadian rhythmicity in a mammalian hypothalamic "island" containing the suprachiasmatic nucleus. Proc. Natl. Acad. Sci. USA **76:** 5962–5966.
15. LEWIS, M. R., J. P. GALLAGHER & P. SHINNICK-GALLAGHER. 1987. An *in vitro* brain slice preparation to study the pharmacology of central vestibular neurons. J. Pharmacol. Methods **18:** 267–273.

16. SERAFIN, M., C. DE WAELE, A. KHATEB, P. P. VIDAL & M. MUHLETHALER. 1991. Medial vestibular nucleus in the guinea-pig. I. Intrinsic membrane properties in brainstem slices. Exp. Brain Res. **84:** 417–425.

17. SERAFIN, M., C. DE WAELE, A. KHATAB, P. P. VIDAL & M. MUHLETHALER. 1991. Medial vestibular nucleus in the guinea-pig. II. Ionic basis of the intrinsic membrane properties in brainstem slices. Exp. Brain Res. **84:** 426–433.

18. KIRSTEN, E. B. & J. SHARMA. 1976. Characteristics and response differences to iontophoretically applied norepinephrine, D-amphetamine, and acetylcholine on neurons in the medial and lateral vestibular nuclei of the cat. Brain Res. **112:** 77–90.

19. MATSUOKA, I. & E. F. DOMINO. 1975. Cholinergic mechanisms in the cat vestibular system. Neuropharmacology **14:** 201–210.

20. MATSUOKA, I., E. F. DOMINO & M. MORIMOTO. 1973. Adrenergic and cholinergic mechanisms of single vestibular neurons in the cat. Adv. Otorhinolaryngol. **19:** 163–178.

21. MATSUOKA, I., J. ITO, H. TAKAHASHI, M. SASA & S. TAKAORI. 1985. Experimental vestibular pharmacology: a minireview with special reference to neuroactive substances and antivertigo drugs. Acta Otolaryngol. Suppl. Stockh. **419:** 62–70.

22. SASA, M., S. FUJIMOTO, S. TAKAORI, J. ITO & I. MATSUOKA. 1981. Acetylcholine as a transmitter candidate in the lateral vestibular nucleus. Neurosci. Lett **6:** 570–575.

23. NABATAME, H., M. SASA, Y. OHNO, S. TAKAORI & M. KAMEYAMA. 1986. Activation of lateral vestibular nucleus neurons by iontophoretically applied phencyclidine. Jpn. J. Pharmacol. **42:** 117–122.

24. RAYMOND, J., A. NICOULLON, D. DEMEMES & A. SANS. 1984. Evidence for glutamate as a neurotransmitter in the cat vestibular nerve: radioautographic and biochemical studies. Exp. Brain Res. **56:** 523–531.

25. MONAGHAN, D. T. & C. W. COTMAN. 1985. Distribution of N-methyl-D-aspartate-sensitive-L-[^{3}H]-glutamate binding sites in rat brain. J. Neurosci. **5:** 2909–2919.

26. KANEKO, T., K. ITOH, R. SHIGEMOTO & N. MIZUNO. 1989. Glutaminase-like immunoreactivity in the lower brainstem and cerebellum of the adult rat. Neuroscience **32:** 79–98.

27. COCHRAN, S. L., P. KASIK & W. PRECHT. 1987. Pharmacological aspects of excitatory synaptic transmission to second-order vestibular neurons in the frog. Synapse **1:** 102–123.

28. LEWIS, M. R., K. D. PHELAN, P. SHINNICK-GALLAGHER & J. P. GALLAGHER. 1989. Primary afferent excitatory transmission recorded intracellularly *in vitro* from rat medial vestibular neurons. Synapse **3:** 149–153.

29. WHITTEMORE, E. R. & J. F. KOERNER. 1989. An explanation for the purported excitation of piriform cortical neurons by N-acetyl-L-aspartyl-L-glutamic acid (NAAG). Proc. Natl. Acad. Sci. USA **86:** 9602–9605.

30. DEWAELE, C., N. VIBERT, M. BAUDRIMONT & P. P. VIDAL. 1990. NMDA receptors contribute to the resting discharge of the vestibular neurons in the normal and hemilabyrinthectomized guinea pig. Exp. Brain Res. **81:** 125–133.

31. HOKFELT, T., R. ELDE, O. JOHANSSON, L. TERENIUS & L. STEIN. 1977. The distribution of enkephalin-immunoreactive cell bodies in the rat central nervous system. Neurosci. Lett. **5:** 25–31.

32. KHACHATURIAN, H., M. E. LEWIS & S. J. WATSON. 1983. Enkephalin systems in diencephalon and brainstem of the rat. J. Comp. Neurol. **220:** 310–320.

33. BEINZ, C. A. J., J. R. CLEMENTS, L. J. ECKLUND & M. M. MULLETT. 1987. The nuclei of origin of brainstem enkephalin and cholecystokinin projections to the spinal trigeminal nucleus of the rat. Neuroscience **20:** 409–425.

34. MADISON, D. V. & R. A. NICOLL. 1988. Enkephalin hyperpolarizes interneurones in the rat hippocampus. J. Physiol. London **398:** 123–130.

35. LOOSE, M. D. & M. J. KELLY. 1990. Opioids act at mu-receptors to hyperpolarize arcuate neurons via an inwardly rectifying potassium conductance. Brain Res. **513:** 15–23.

36. JIA, M. & P. G. NESON. 1987. A presynaptic locus of the action of met-enkephalin demonstrated in mouse spinal cord cultures. Peptides **8:** 565–568.

37. YASNETSOV, V. V. & V. A. PRAVDIVTSEV. 1986. Chemical sensitivity of medial vestibular nuclear neurons to enkephalins, acetylcholine, GABA and L-glutamate. Kosm. Biol. Aviakosm. Med. **20:** 53–57.

38. DOSTROSKY, J. & B. POMERANZ. 1973. Morphine blockade of amino acid putative transmitters on cat spinal cord sensory interneurons. Nature New Biol. **246:** 222–224.
39. SEGAL, M. 1977. Morphine and enkephalin interactions with putative neurotransmitters in rat hippocampus. Neuropharmacology **16:** 587–592.
40. BRADLEY, P. B. & A. DRAY. 1973. Actions and interactions of microiontophoretically applied morphine with transmitters substances on brain stem neurones. Br. J. Pharmacol. **4:** 642P.
41. DUGGAN, A. W., J. DAVIES & J. G. HALL. 1976. Effects of opiate agonists and antagonists on central neurons of the cat. J. Pharmacol. Exp. Ther. **196:** 107–120.
42. DAVIES, J. & A. DRAY. 1976. Effects of enkephalin and morphine on Renshaw cells in feline spinal cord. Nature **262:** 603–604.
43. LODGE, D., P. M. HEADLY, A. W. DUGGAN & T. J. BISCOE. 1974. The effects of morphine, etorphine and sinomenine on the chemical sensitivity and synaptic responses of Renshaw cells and other spinal neurons of the rat. Eur. J. Pharmacol. **26:** 277–284.
44. KOHL, R. L. & J. L. HOMICK. 1983. Motion sickness: a modulatory role for the central cholinergic nervous system. Neurosci. Biobehav. Res. **7:** 73–85.

The Role of Prediction in Head-Free Pursuit and Vestibuloocular Reflex Suppression

G. R. BARNES AND M. A. GREALY

MRC Human Movement & Balance Unit
National Hospital
Queen Square
London WC1N 3BG, England

In a recent experiment, the predictive mechanisms of the head-fixed ocular pursuit reflex were revealed by using the technique of repeated transient stimulation to show the changes in the temporal characteristics of the response.[1] Subjects were instructed to follow the motion of a constant-velocity target during repeated brief periods of stimulation that were separated by periods of darkness. Smooth eye velocity was observed to build up to an asymptotic level over the first three or four presentations of the target, while simultaneously becoming more phase advanced with respect to the onset of target illumination. When the target suddenly and unexpectedly changed velocity or frequency, an inappropriate predictive eye velocity trajectory was initiated that was highly correlated in peak velocity and timing with the characteristics of the preceding part of the stimulus. We have now used the same technique to examine the response during head-free pursuit of a moving target in order to establish whether the same predictive mechanisms govern the coordination of head and eye movements.

When the head and eyes are moved simultaneously, the motion of the head provides a very potent stimulus to the semicircular canals. The resultant vestibuloocular reflex (VOR) would normally drive the eyes in the opposite direction to head movement, but such a response would be counterproductive during head-free pursuit and the VOR must therefore be suppressed in order that gaze velocity (i.e., the sum of head and eye velocity) be allowed to match target velocity as closely as possible. In a number of previous experiments,[2–4] it has been shown that there is a remarkable similarity between the characteristics of the pursuit reflex and VOR suppression, suggesting that the processes are carried out by a common neurological mechanism. Prediction is one of the most important of these features of pursuit, and as we will now show, it appears to be present both in the control of head movement and the suppression of the VOR. In order to confirm the role of prediction in VOR suppression, a further experiment will be described in which the intermittent stimulation technique has been used to examine the response during whole-body passive stimulation on the turntable.

METHODS

Subjects were seated in a totally darkened room at the centre of a screen of two meter radius on to which a moving visual target was projected by a mirror galvanometer. The illumination of the moving target, which was a circle of radius 25 minutes of arc with fine cross hairs superimposed on it, was controlled by an electromechanical shutter that could produce pulse durations down to 8 mseconds. Eye movements were transduced by an infrared limbus tracking technique with a resolution of 5–10 minutes of arc (SKALAR IRIS). The eye movement recorders were rigidly coupled to the head by a helmet assembly, and rotation of the head was transduced by a potentiometer coupled to the crown of the helmet.

In order to examine the predictive component of the smooth eye movement response in experiment I, a novel target waveform was used in which sudden and unexpected changes were introduced into an otherwise regular periodic motion. The stimulus was split into five consecutive sequences of randomized and unequal duration that contained between 5 and 8 cycles of a sinusoidal waveform that varied in frequency from 0.26–0.78 Hz. At the transition between sequences, the target was suddenly blanked out for one complete cycle of the new sequence and eye movements were recorded in the absence of any visual stimulus. The target was illuminated for varying pulse durations of 120–320 mseconds at the time when the target passed through the midline position. The subject was required to pursue the small fixation target when it appeared using both the head and eyes together. In order to encourage head movements, the target moved at high velocity (96°/second). In experiment II, the subject sat with head fixed on a turntable that executed a sinusoidal motion in yaw. Two stimulus frequencies were used, 0.1 and 0.2 Hz with peak velocities of 48°/second and 64°/second respectively. A head-fixed target was presented for pulse durations of 10–160 mseconds at the time that the turntable executed peak velocity, and the subject was instructed to attempt to hold fixation during its exposure. After a random number of cycles of stimulation, the target was switched off and the subject continued to oscillate for two cycles in darkness. The response of the VOR in darkness was also recorded while the subject carried out mental arithmetic.

The experiments were performed on a group of eight normal subjects, one of whom required refractive correction in order to see the target clearly. All subjects participated with informed consent, and the experimental procedure was approved by the local ethical committee.

The head and eye movements were analyzed using an interactive computer graphics technique. Head and eye displacement signals were summated to obtain gaze displacement from which gaze velocity was derived. After removal of the fast-phase components from the gaze velocity trace, the slow-phase component was compared with target velocity. Cycle-by-cycle averages of slow-phase gaze and head velocity were obtained by overlaying and averaging successive cycles of the response (excluding the first two cycles) during each sequence of constant-frequency periodic stimulation. Time constants of transient decay were obtained by the fitting of an exponential function using an iterative, least-squares error curve-fitting procedure.

RESULTS

Experiment I: Head-Free Pursuit

When subjects attempted to follow the target motion during its brief period of exposure, a very characteristic pattern of head and eye movement was evoked, irrespective of the duration of exposure (FIGURE 1). Head displacement gradually built up over the first 3–4 presentations, inducing an ocular nystagmus that was frequently in the opposite direction to head movement, indicating incomplete VOR suppression. Summation of head and eye velocity revealed a similar buildup in slow-phase gaze velocity (FIGURE 2B), which was identical in form to that observed for eye velocity during head-fixed pursuit.[1] Detailed examination of the head and gaze velocity trajectories during the first four presentations (FIGURE 2A, upper traces) indicated the manner in which the timing of the responses changed with repeated stimulation. For example, when pulse duration was 240 mseconds and frequency was 0.39 Hz, gaze velocity was initiated on average 160 mseconds after the onset of the first target presentation and reached a peak 370 mseconds after target onset (FIGURE 3A). After four target presentations, gaze velocity passed through zero 280 mseconds before target onset and reached a peak only 240 mseconds after target exposure. This general trend was true for all frequencies of stimulation and

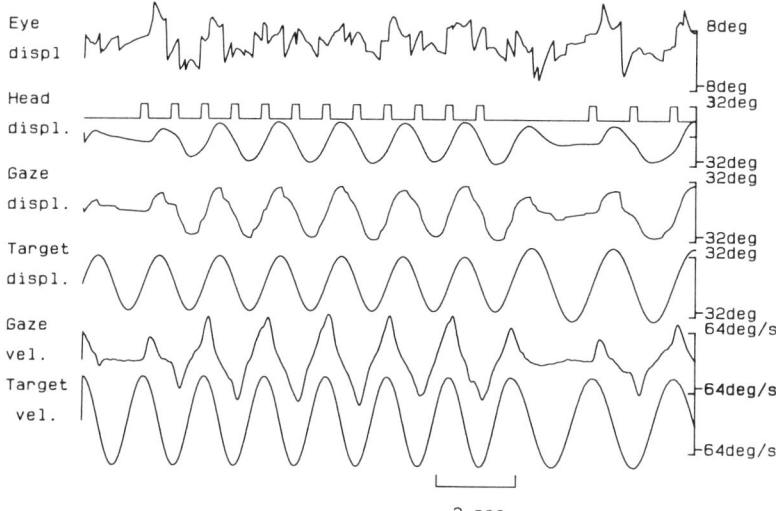

FIGURE 1. Examples of head and eye movements in response to the tachistoscopic presentation of a target moving at a near-constant velocity as it passed through the center of the visual field. Gaze displacement is obtained from the summation of head and eye displacement and fast-phase components have been removed from the derived gaze velocity trace. Pulses (trace 2) indicate the timing of target appearance which was on for 240 mseconds. The stimulus waveform was a sinusoid of peak velocity ±96°/second and a frequency that changed from 0.78 Hz to 0.5 Hz at the transition.

was also observed in the head velocity profiles (FIGURE 3B). Cycle-by-cycle averaging of the response trajectories revealed steady-state profiles that were remarkably similar for head velocity and gaze velocity in the majority of subjects, although some made a much lower proportion of head movement for the 120- and 160-msecond pulse durations. On average, gaze velocity had reached 45% of peak velocity at the time of target onset and 65% of the peak 100 mseconds after target onset when the visual feedback might be expected to take effect.[5]

When the moving target unexpectedly disappeared, both head and eye movements continued to be made, reaching a peak velocity at a variable time dependent on the periodicity of the previous sequence, that is, at a time when the next target would have appeared. It was actually very difficult for the subject to suppress this predictive head and eye movement response once it had been initiated. Peak gaze velocity for the target exposure prior to the blank period increased significantly ($p < 0.001$ by ANOVA) with pulse duration reaching a maximum value of 80°/second in response to the target velocity of 96°/second (FIGURE 4 left). Peak gaze velocity in the transition period (FIGURE 4 right) also increased with pulse duration, reaching a maximum of 56°/second. Peak head velocity exhibited a similar, but less-well-defined, change with pulse duration reaching a maximum of 53°/second in

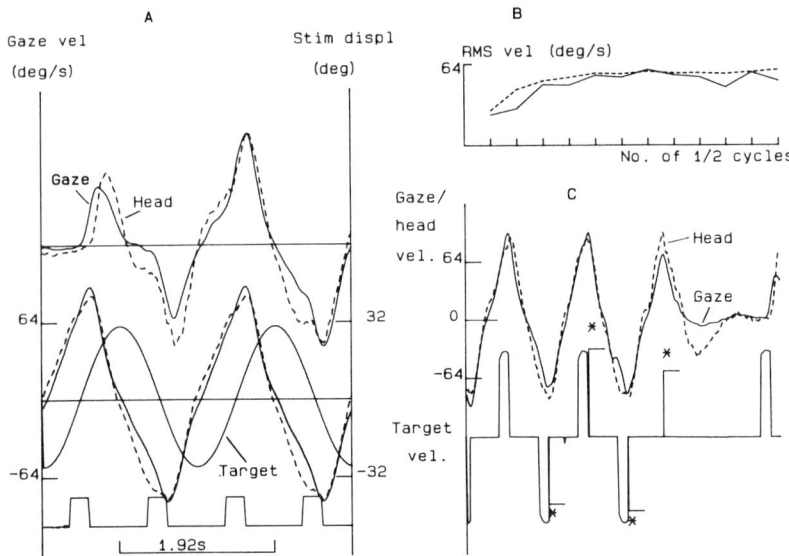

FIGURE 2. Center Traces in A: Repeated cycle-by-cycle averages of gaze velocity, head velocity (broken line), and target displacement (sinusoid) derived from the responses shown in FIGURE 1. Stimulus frequency 0.78 Hz. Upper Traces in A: The first two cycles of gaze velocity (solid line) and head velocity plotted to the same time scale as the average responses so as to indicate the manner in which response velocity changes during the first four target presentations. Lowest Trace in A: Timing of target exposure. B: Changes in root-mean-square gaze velocity (solid line) and head velocity (broken line) with number of half-cycles of stimulation. C: Changes in gaze and head velocity at the transition when the target suddenly disappeared. Bar lines in lower trace indicate the magnitude and timing of peak gaze velocity in relation to target velocity stimulus.

the transition. Head velocity tended to undergo a further reversal of direction that probably reflected its underdamped dynamic characteristics.

Experiment II: VOR Suppression

Examination of the cycle-by-cycle average of eye velocity during tachistoscopic illumination of the head-fixed target revealed a transient dip in eye velocity which was associated with the timing of target exposure (FIGURE 5 left). However, the departure of this response from that recorded in darkness could clearly be seen to

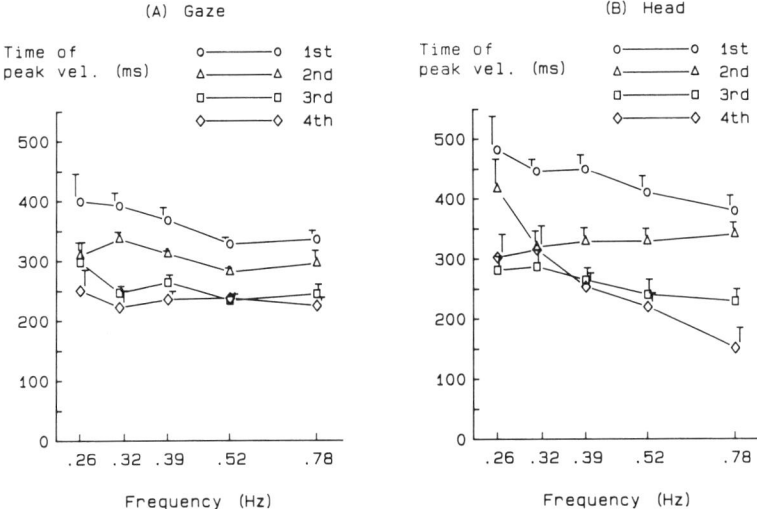

FIGURE 3. Changes in the timing of peak gaze velocity (A) and peak head velocity (B) in relation to the onset of target illumination during the first four presentations of the moving target as a function of the frequency of target motion. Pulse duration was 240 mseconds. Mean of 8 S's + 1 standard error of the mean (SEM).

have been initiated well before target onset, even during the briefest exposure (10 mseconds). This effect could be demonstrated by subtraction of the averaged suppressed response (VORS) from that recorded in darkness (VOR) as shown in the upper trace of FIGURE 5 (left) to give a measure of the suppression signal (SUPP).

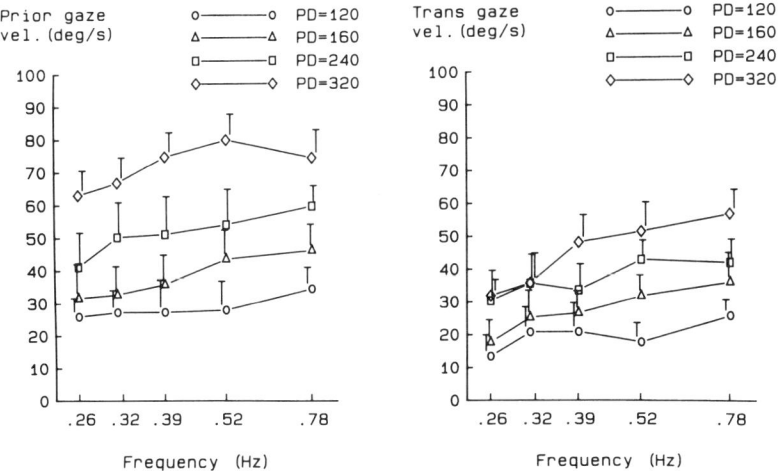

FIGURE 4. Peak gaze velocity just prior to (left) and at the time of (right) the transition stage at which the target was suddenly and unexpectedly blanked out, as a function of target frequency and exposure duration (PD). Mean of 8 S's + 1 SEM.

Comparison of the profile of this response with the average gaze velocity profile shown in FIGURE 2A revealed important similarities. Suppression clearly started before target onset and, on average, reached its peak 260 mseconds after target onset. Suppression had attained 67% of its peak level 100 mseconds after target onset and peak suppression velocity increased from 17 to 31°/second as pulse duration increased from 10–160 mseconds. Finally, when the target was unexpectedly switched off at the end of the sequence while the turntable was still in motion, the suppression response continued to be made in a predictive manner as shown in the upper trace of FIGURE 5 (right). In the majority of responses the normal VOR gain was reestablished after 1–2 half-cycles of oscillation in the dark.

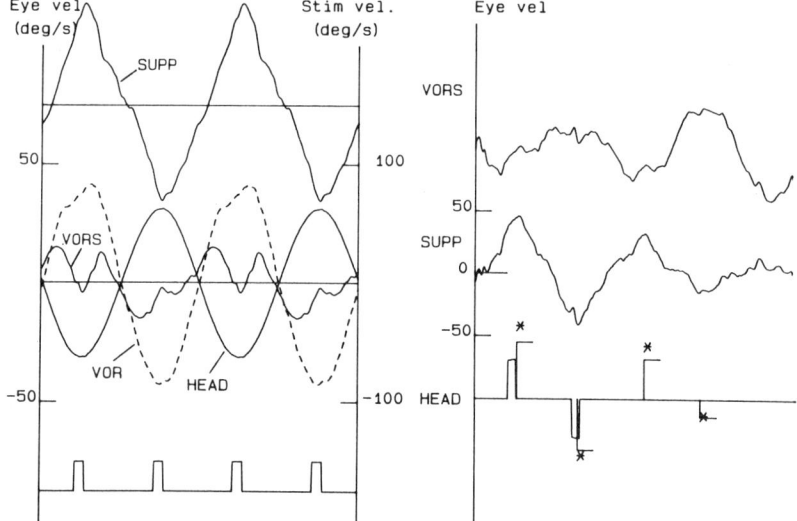

FIGURE 5. **Center Traces at Left:** Repeated cycle-by-cycle averages of eye velocity and head velocity during sinusoidal whole-body oscillation at a frequency of 0.2 Hz. VOR—response in darkness; VORS—response during attempted fixation of head-fixed target exposed for 120 mseconds as indicated by pulses in lowest trace. **Upper Trace at Left:** The degree of suppression (SUPP) of the dark response obtained as the difference between the average VOR and VORS signals. **Right:** Changes in gaze velocity (VORS) and the velocity of the suppression signal (SUPP) at the transition when the target suddenly disappeared but turntable motion continued. Bar lines in lower trace indicate the magnitude and timing of peak suppression velocity in relation to the head velocity stimulus.

DISCUSSION

There is a remarkable similarity in the features of the slow-phase gaze velocity recorded during head-free pursuit and those observed previously for head-fixed pursuit.[1] The buildup in the response with repeated stimulation combined with the changes in timing is not compatible with the simple linear summation of successive transient responses to each target presentation. This was most evident at the lowest frequencies of stimulation when, in the steady state, the individual transient responses could be seen to be quite separate, gaze velocity being clearly initiated well before target onset. Previously it was shown that the predictive response to each target presentation in the head-fixed response had a very stereotyped temporal characteristic, reaching a peak velocity that was dependent on pulse duration and target velocity, and subsequently decaying with a time constant of 0.5–1 second. When the repetition rate of target exposure was increased, a simple temporal overlay of successive predictive pulses could be demonstrated.[1] On the basis of these findings it was suggested that the predictive velocity estimate was derived by sampling of the efference copy of eye velocity and that this sample was stored internally and subsequently used to boost the drive to the oculomotor system. An important aspect of this mechanism is that the release of predictive velocity estimates must be controlled by a periodicity estimator, an entrainment system that becomes locked to the repetition frequency and thus enables the release to be timed so as to overcome the delays in the visual feedback. Although the stimulation technique outlined here is rather different to the normal pursuit process, we have been able to show by further experiments[6] and modeling techniques[1] that the observed response could represent a realistic component of the head-fixed pursuit response.

An analysis of the individual response to each target presentation induced during head-free pursuit has revealed the same type of stereotyped temporal characteristics in both gaze velocity and head velocity that were previously observed for eye velocity during head-fixed pursuit. The activity of the predictive mechanism is evident in the response that occurs when the target unexpectedly disappears. The timing and magnitude of the peak response for both head and gaze velocity did not simply represent the continuing decay pattern of a standard transient response but were dependent on the periodicity and amplitude of the preceding sequence. An important implication of the similarity in the peak velocity of head and gaze trajectories in this transition phase is that predictive suppression of the vestibuloocular response must be taking place in darkness at this time. If this were not so, the VOR would compensate for head velocity and gaze velocity would be close to zero if VOR gain were close to unity. The results of experiment II confirm that the same predictive process takes place during passive whole-body rotation on a turntable. Such predictive mechanisms might possibly be present during transient VOR suppression tasks[7] if the detection of vestibular mismatch were able to trigger the release of stored predictive estimates, thus modifying the apparent timing of the responses.

Robinson has suggested that VOR suppression could be carried out either by a central mechanism using an efference copy of the signal used to drive volitional head movement or by predictive mechanisms during passive stimulation.[8] In fact, the simplest interpretation of our findings is that the predictive pulse is derived not from an efference copy of eye velocity or from the central reconstruction of gaze velocity, but from an internal copy of the visual feedback signal that is used to drive the eye and/or suppress the VOR. This signal is representative of gaze velocity during head-free pursuit and eye velocity during head-fixed pursuit. It is proportional to head velocity during VOR suppression but would be absent during VOR stimulation in the dark. It is thus compatible with the features of "gaze velocity" cells recorded in the cerebellum.[9,10] The fact that head movements also exhibit the same type of predictive characteristics as eye movement indicates that the head is probably driven through the same visual feedback mechanisms as the eye. This conclusion is reinforced by the previous finding that head movements during head-free pursuit of pseudorandom target motion exhibit nonlinear frequency-dependent characteristics similar to those for eye movement control.[11]

An interesting observation in experiment II was that the response during fixation of the head-fixed target was remarkably consistent, even when there was little suppression, whereas the VOR recorded in darkness was highly variable as other observers have often noted. On occasions, peak velocity in the dark was actually less than that during the condition of least visual suppression. The mere presence of a minimal visual target may therefore have a general enhancing effect. It is possible that the apparent suppression that has been noted in darkness by some authors could be partly attributable to the absence of this enhancing effect, making comparisons of pursuit and VOR suppression difficult. It has been shown previously that nonvisual VOR suppression does not exhibit the predictive features associated with visual suppression,[12] suggesting that this specific type of prediction may be restricted to the visual feedback system.

REFERENCES

1. BARNES, G. R. & P. T. ASSELMAN. 1991. The mechanism of prediction in human smooth pursuit eye movements. J. Physiol. London **439:** 439–457.
2. BARNES, G. R., A. J. BENSON & A. R. J. PRIOR. 1978. Visual-vestibular interaction in the control of eye movement. Aviat. Space. Environ. Med. **49:** 557–564.
3. PAIGE, G. D. 1983. Vestibulo-ocular reflex and its interactions with visual following mechanisms in the squirrel monkey. I. Response characteristics in normal animals. J. Neurophysiol. **49:** 134–151.
4. KOENIG, E., J. DICHGANS & W. DENGLER. 1986. Fixation suppression of the vestibulo-ocular reflex (VOR) during sinusoidal stimulation in humans as related to the performance of the pursuit system. Acta Otolaryngol. Stockh. **102:** 423–431.
5. CARL, J. R. & R. S. GELLMAN. 1987. Human smooth pursuit: stimulus-dependent responses. J. Neurophysiol. **57:** 1446–1463.
6. ASSELMAN, P. T. & G. R. BARNES. 1989. The effects of intermittent exposure of a moving target on smooth pursuit in humans. (Abstr.) J. Physiol. London **423:** P69.
7. LISBERGER, S. G. 1990. Visual tracking in monkeys: evidence for short-latency suppression of the vestibulo-ocular reflex. J. Neurophysiol. **63:** 676–688.
8. ROBINSON, D. A. 1982. A model of cancellation of the vestibulo-ocular reflex. *In* Functional Basis of Ocular Motility Disorders. G. Lennerstrand, D. S. Zee & E. L. Keller, Eds.: 5–13. Pergamon Press. Oxford, England.
9. LISBERGER, S. G. & A. F. FUCHS. 1978. Role of primate flocculus during rapid behavioural modification of vestibuloocular reflex. I. Purkinje cell activity during visually guided horizontal smooth-pursuit eye movements and passive head rotation. J. Neurophysiol. **41:** 733–763.
10. SUZUKI, D. A. & E. L. KELLER. 1988. The role of the posterior vermis of monkey cerebellum in smooth-pursuit eye movement control. I. Eye and head movement–related activity. J. Neurophysiol. **59:** 1–18.
11. BARNES, G. R. & J. F. LAWSON. 1989. Head-free pursuit in the human of a visual target moving in a pseudo-random manner. J. Physiol. London **410:** 137–155.
12. BARNES, G. R. & R. D. EASON. 1988. Effects of visual and non-visual mechanisms on the vestibulo-ocular reflex during pseudo-random head movements in man. J. Physiol. London **395:** 383–400.

Self-Controlled Reorienting Movements in Response to Rotational Displacements in Normal Subjects and Patients with Labyrinthine Disease

TERESA METCALFE AND MICHAEL GRESTY

MRC Human Movement and Balance Unit
National Hospital for Neurology and Neurosurgery
Queen Square
London WC1N 3BG, England

METHODS

Subjects seated in a Barany chair with head restrained were passively displaced in darkness away from a resting position. The displacements had raised cosine velocity profiles with peak velocities of 30°/second and 60°/second and amplitudes varying from 30° up to 180° in 30° intervals. The stimuli were delivered in the rightwards and leftwards directions in a Latin square design. Following each displacement and still in darkness, the subjects were required to rotate themselves accurately back to the starting position using a directionally congruent steering wheel which gave a velocity demand to the chair motor. Following each perturbation, when the subjects said that they were satisfied with their response, the lights were turned on so that they could make corrective adjustments back to center. The subjects were given an initial practice session.

Normal subjects consisted of 6 male and female subjects with an age range of 20 to 62. *Subjects with unilateral loss of labyrinthine function* included patients with vestibular nerve section for intractable vertigo and patients with acoustic neurinectomies. The patients were studied one week to one year postoperatively. Some were tested longitudinally.

RESULTS

Normal subjects gave counterrotational responses which approximated the imposed displacement with standard deviations (SDs) ranging from 5° to 15° for the smallest to largest displacements. Normal subjects' and patients' responses tended to take the form of velocity trapezoids. The mean of normal responses is shown in the graphs as a continuous line with ±1 SD indicated by vertical bars.

Alabyrinthine subjects failed to respond because they could not sense the stimuli or could not sense the magnitude of their responses and failed to stop themselves rotating.

Most patients with *unilateral absence of vestibular function* who were tested within a month or so of operation gave similar responses which consisted of consistently low-amplitude counterrotations in response to displacements towards the side of the lesion and highly inaccurate responses in the counterdirection with amplitudes above

695

and below the normal range. Within several months this pattern tended to change to one in which responses were directionally symmetrical but mildly inaccurate with values distributed about and just beyond the outer upper and lower limits of the normal range (FIGURES 1 and 2).

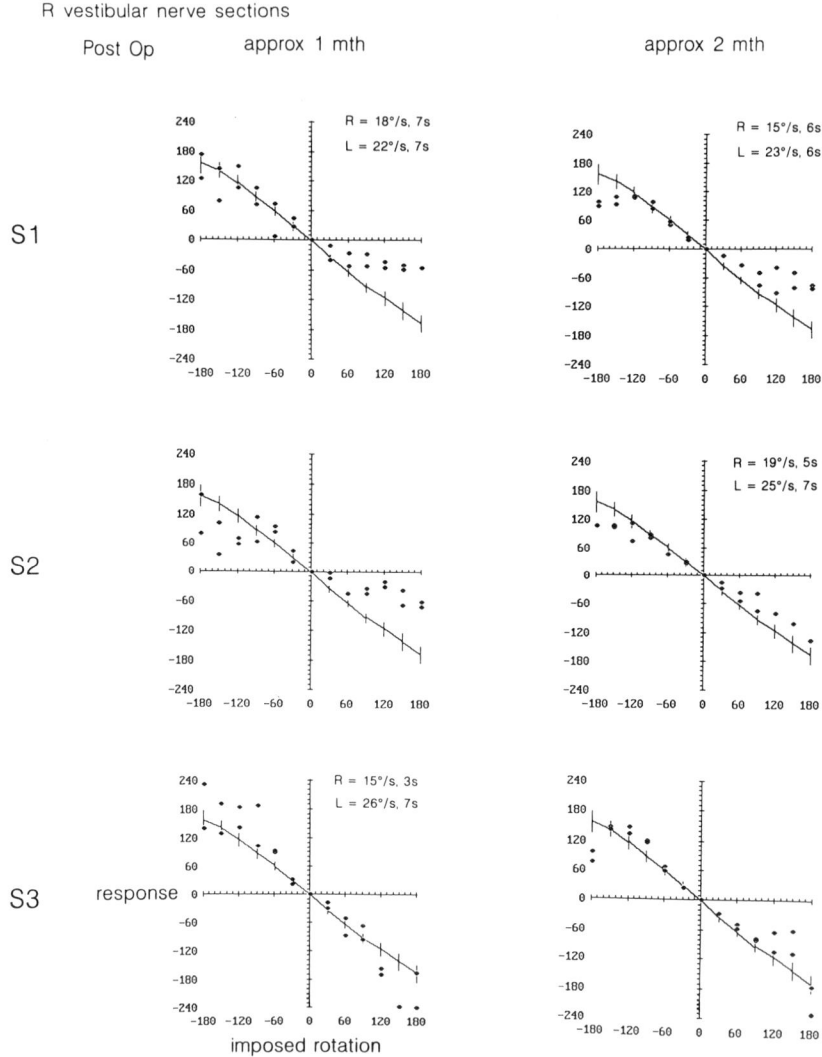

FIGURE 1. Counterrotational responses of patients with unilateral loss of labyrinthine function superimposed on means and ±1 SD of normal responses (shown as continuous diagonal line and vertical bars). Positive numbers are rightwards rotations, negative numbers are leftwards rotations. Initial velocities in °/second and time constants in seconds of rightwards and leftwards responses to 60°/second steps in rotational velocity are shown inset in the upper right quadrant of each graph.

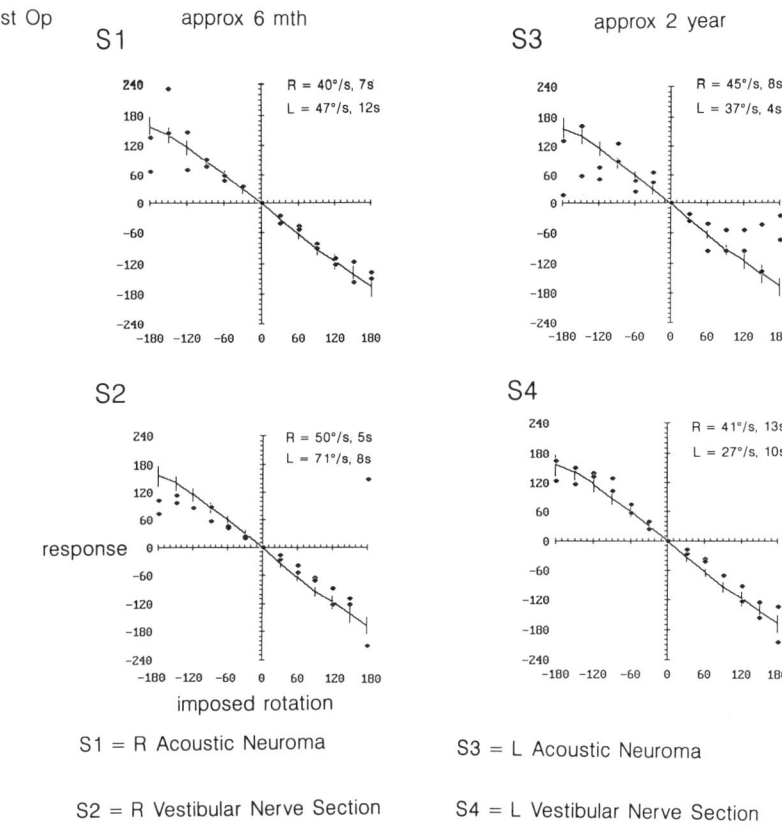

FIGURE 2. Legend as for FIGURE 1.

Occasionally patients whom we suspected to have good and rapid compensation had symmetrical responses with mild inaccuracy shortly after operation (FIGURE 1, S3). Some patients had persistently asymmetrical responses (FIGURE 2, S3) in which case we suspected poor compensation, possibly due to mild central nervous system disease (e.g., associated with acoustic neuroma).

The initial velocities and time constants of postrotatory nystagmus responses to 60°/second velocity steps (shown as insets in figures) did not correlate with self-controlled rotational responses and seemed less consistent within and between patients.

DISCUSSION

In this situation the self-controlled response to null an imposed rotational perturbation depends primarily upon memory and judgment of displacement ampli-tudes derived primarily from vestibular signals. Normal subjects are accurate to within ±10% (range) at this task with the stimulus parameters used. The asymmetrical responses seen in the early stages of unilateral loss of vestibular function reflect the uncompensated state. The subject is insensitive to rotation towards the lesioned side and accordingly makes a small counterrotational response. With time, the asymmetry becomes redressed which is probably a sign of good compensation but, of interest, the responses are still relatively inaccurate with respect to the normal range. It is possible that the results of this type of procedure may better reflect a patient's disability than nystagmus responses.

The Effects of Target Distance on Eye and Head Movement during Locomotion

J. J. BLOOMBERG,[a] M. F. RESCHKE,[a] W. P. HUEBNER,[b]
AND B. T. PETERS[b]

[a]National Aeronautics and Space Administration
Johnson Space Center
Space Biomedical Research Institute
Mail Code SD 5
Houston, Texas 77058

[b]KRUG Life Sciences
1290 Hercules, Suite 120
Houston, Texas 77058

INTRODUCTION

The vestibuloocular reflex (VOR) serves to maintain a stable retinal image by generating eye movements that compensate for head perturbations. It has been demonstrated that during natural behaviors such as walking and running, the VOR stabilizes gaze against *angular* head movement.[1] Indeed, individuals with loss of vestibular function experience impaired visual acuity and oscillopsia during locomotion, thus underscoring the importance of the VOR in maintaining gaze stability during locomotion.[2] However, during locomotion the head also experiences *linear* displacement[3–5] and must therefore generate a linear VOR (LVOR) if visual acuity is to be maintained. The response characteristics of the LVOR are dependent on the amount of ocular translation relative to the visual target position produced during linear head motion.[6–8] A target viewed at a relatively large distance requires only small compensatory eye movements. However, as the target of interest is brought closer to the eyes, larger compensatory eye movements are required to maintain ocular fixation. The LVOR is thought to be driven predominantly by otolith input and modulated by the state of ocular vergence; this provides larger compensatory eye movements during fixation of near targets.[7–9] Interestingly, angular head movements are also compensatory for linear head displacement experienced during locomotion.[10] However, it is not known how such head movements respond to changes in target distance from the eyes.

The aim of the present experiment was to characterize how the eyes and head compensate for vertical linear head displacement produced during locomotion and to determine how changes in target distance affect these compensatory responses. These studies represent ground-based investigations that were conducted prior to similar pre- and postflight testing of NASA astronaut subjects.

METHODS

Five normal subjects (aged 21–33) participated in this study. Subjects walked at a fast pace (6.4 km/hour) on a motorized treadmill while visually fixating an earth-fixed target positioned in the center of view either near (30 cm) or far (2 m) from the eyes. Each task was performed for a period of 25 seconds while eye and head movements were recorded. In addition, one subject repeated the experimental protocol while wearing up-down reversing lenses for one hour.

Head movements were recorded and analyzed using a video-based motion analyzing system (Motion Analysis Corp., Santa Rosa, Calif.). Passive spherical retroreflective markers were affixed to a lightweight headband and were located on the vertex and occipital areas of the head. Head movement data were collected using four video cameras sampling video images at 60 Hz. The video signal from each of the cameras was fed to a video processor where the outlines of the markers were extracted and passed to the system host computer (Sun SPARC Computer Workstation) for analysis. The centroids of the markers were tracked and their trajectories calculated, producing a three-dimensional reconstruction of head movement during locomotion.

DC electrooculography (EOG) was used to measure vertical eye position relative to the head. Eye movement data were digitally sampled at 300 Hz.

RESULTS

FIGURE 1 shows typical z-axis head position (vertical linear head movement) and the corresponding pitch head position (sagittal angular head movement) for one subject during gaze fixation of the far (FIGURE 1A) and near (FIGURE 1B) targets during locomotion. Note that pitch head movements were compensatory for z-axis head movement and increased in amplitude during gaze fixation of the near target (FIGURE 1B).

TABLE 1 presents mean peak-to-peak z-axis and pitch head movements during gaze fixation of the far and near targets for each of our subjects. Four out the five subjects demonstrated a significant ($p < 0.001$) increase in peak-to-peak pitch head movement during gaze fixation of the near target position. Interestingly, two subjects (subjects 4 and 5) also showed a significant ($p < 0.001$) reduction in peak to peak z-axis head movement during gaze fixation of the near target.

Typical Fourier spectra of z-axis and pitch head displacement during the far and near target conditions for one subject are depicted in FIGURE 2. The predominant frequency of both z-axis and pitch head displacement during walking for this subject was 2.2 Hz and was not affected by a change in target position. For all subjects combined, the mean predominant frequency of z-axis head movement was 2.15 ± 0.06 [mean ± standard error (SE)] and 2.17 ± 0.05 for the far and near target conditions, respectively. For pitch head movement, the mean predominant frequency for all subjects combined was 2.15 ± 0.06 and 2.17 ± 0.05 for the far and near targets, respectively. Thus target distance had no significant ($p > 0.10$) effect on the fundamental frequency characteristics of either z-axis or pitch head movements during locomotion.

FIGURE 3 shows typical vertical eye, pitch, and z-axis head position during locomotion and gaze fixation of the near target in one subject. As in FIGURE 1, pitch head movements were compensatory for z-axis head motion. Note that vertical eye movements were also compensatory for z-axis head movement. Thus it appears that vertical eye and pitch head movements act in a synergistic fashion during locomotion with near target fixation to stabilize gaze against z-axis head perturbations. Conversely, during the far target condition, vertical eye movements could not be reliably detected, presumably because they were within the resolving limit of EOG (e.g., 0.5–1 deg).

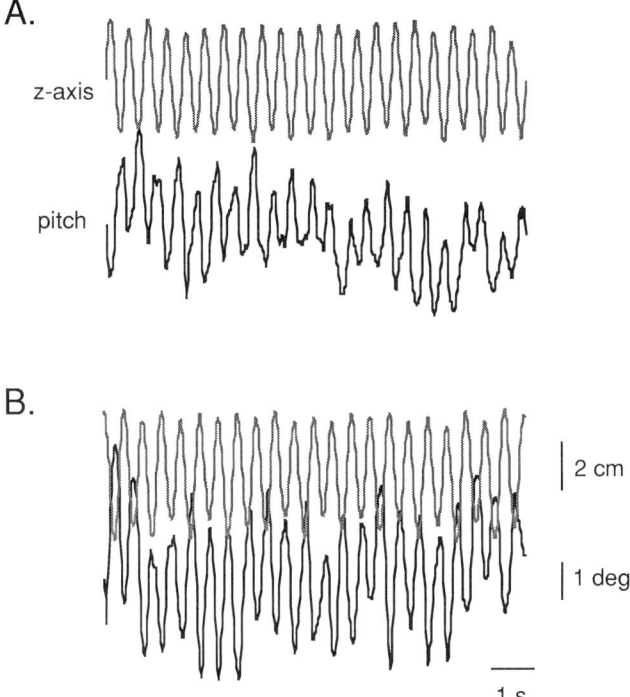

FIGURE 1. Example of z-axis head position and corresponding pitch head position traces during locomotion in one subject while visually fixating gaze on the far (A) and near (B) targets.

Effects of Up-Down Reversing Lenses on Compensatory Head Movement during Locomotion

Up-down reversing prism goggles vertically invert the visual field. FIGURE 4 shows the effects of up-down reversing prisms on z-axis and compensatory pitch head movements for one subject while walking at 4.0 km/hour and visually fixating the near target. Before exposure to this stimulus (FIGURE 4A), pitch head movements were compensatory for z-axis head displacement. Plotting pitch head position against z-axis head position (FIGURE 5A) produced regular, diagonal movement trajectories

TABLE 1. Individual Mean Peak-to-Peak z-Axis and Pitch Head Displacements (± 1 SE) for Far and Near Target Conditions

Subject	z Far (cm)	z Near (cm)	Pitch Far (deg)	Pitch Near (deg)
1	4.56 ± 0.04	4.74 ± 0.04	2.11 ± 0.09	3.55 ± 0.11
2	5.31 ± 0.10	5.13 ± 0.07	2.33 ± 0.13	3.51 ± 0.11
3	3.49 ± 0.05	3.41 ± 0.04	2.28 ± 0.11	2.86 ± 0.14
4	4.19 ± 0.04	3.72 ± 0.05	1.52 ± 0.08	2.40 ± 0.08
5	3.91 ± 0.05	3.59 ± 0.04	1.93 ± 0.12	2.25 ± 0.11

that confirm the compensatory nature of this response prior to donning the up-down reversing lenses. After 1 hour exposure to this stimulus and while still wearing the up-down prisms, note in FIGURE 4B that the relationship between z-axis and pitch head movements has changed markedly such that pitch and z-axis head motion appear unrelated. This change is further depicted in FIGURE 5B which shows an irregular, almost random, relationship between the pitch and z-axis head movement trajectories. Thus it appears that inverting the relationship between head perturbation and the corresponding visual input abolished the compensatory relationship

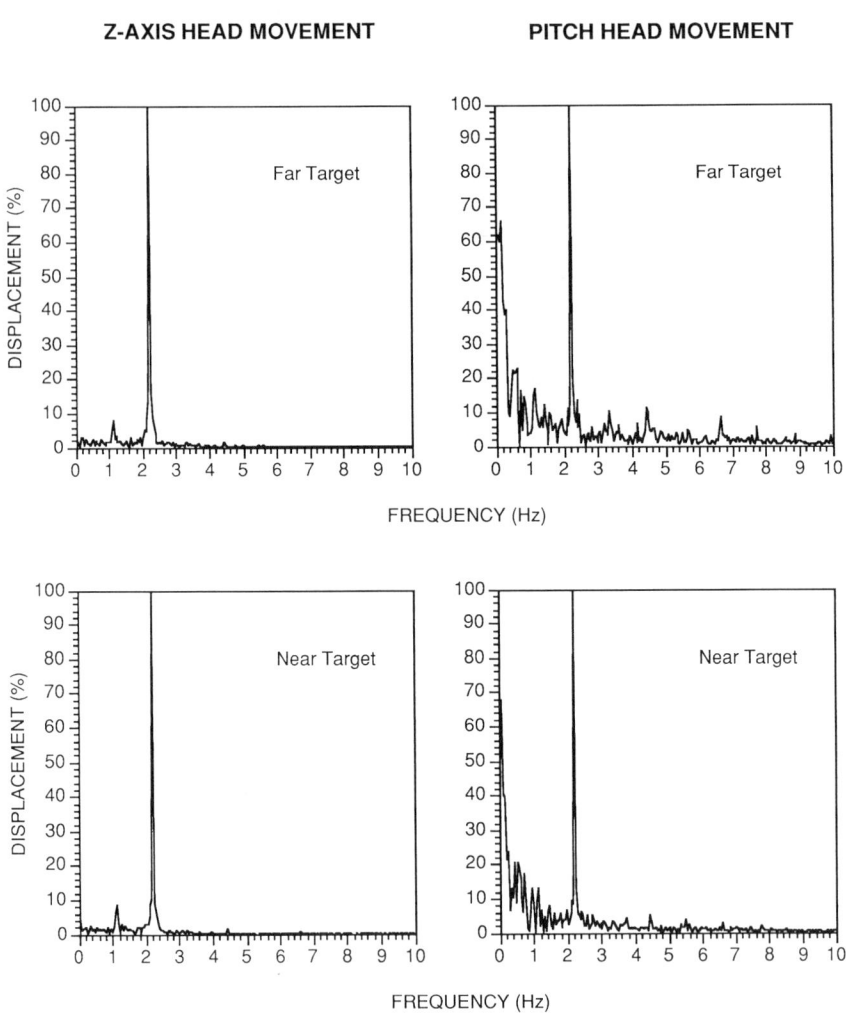

FIGURE 2. Fourier spectra of z-axis and pitch head displacements during locomotion while visually fixating gaze on the far and near targets. All frequency components are represented as percentages of the predominant frequency component which is assigned a value of 100%.

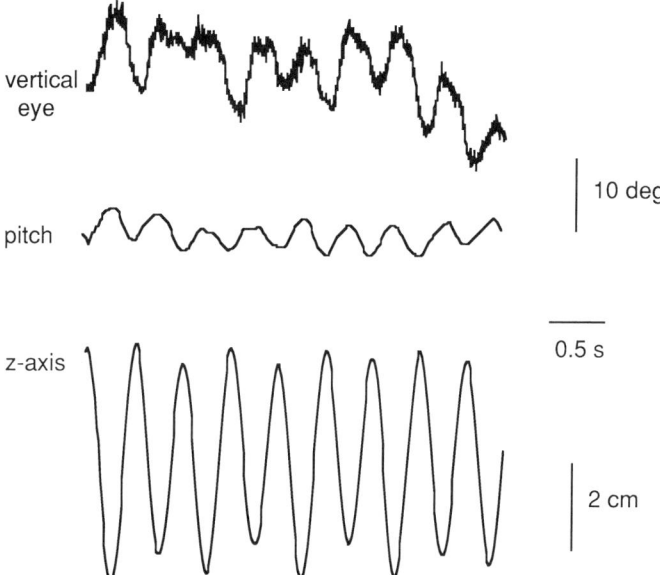

FIGURE 3. Examples of corresponding vertical eye, pitch, and z-axis head position traces during locomotion and visual fixation of the near target in one subject. Note that both vertical eye and pitch head movements are compensatory for z-axis head movement.

between z-axis and pitch head movements observed prior to donning the up-down reversing lenses.

DISCUSSION

The present results confirm the observation by Pozzo *et al.* that pitch head movements that occur during locomotion are compensatory for linear z-axis head motion.[10]

We have shown that these compensatory pitch head movements respond to changes in visual fixation target distance by increasing in amplitude when the target is brought closer to the eyes. This result supports the hypothesis that pitch head movements that occur during locomotion are under goal-directed neural mediation and are not only a result of the inertial and viscoelastic properties of the head-neck system. This interpretation is further confirmed by observing the effects on this response by exposure to up-down reversing lenses. Following 1 hour's exposure to this visual rearrangement, pitch head movements were suppressed during locomotion confirming the goal-directed nature of this response. Perhaps longer exposure periods (i.e., days rather than hours), equivalent to those used to adaptively reverse the VOR response with left-right reversing lenses,[11,12] would produce a reversal in compensatory pitch head movements produced during locomotion.

The results indicate that a combination of vertical eye and pitch head movements operate in a synergistic fashion to stabilize gaze against z-axis head motion during locomotion and visual target fixation. Given that the head is predominantly moving through the z-axis, parallel to gravity, the ocular compensatory response and perhaps the head pitch response may be driven in part by otolith stimulation, particularly by saccular input.[7] In addition, ocular vergence state may be the signal that initiates this compensatory eye-head response during near target fixation.

Two subjects showed a reduction in z-axis head motion during near target fixation. This strategy serves to reduce the amount of required compensatory vertical eye and pitch head movement required for gaze stabilization during locomotion.

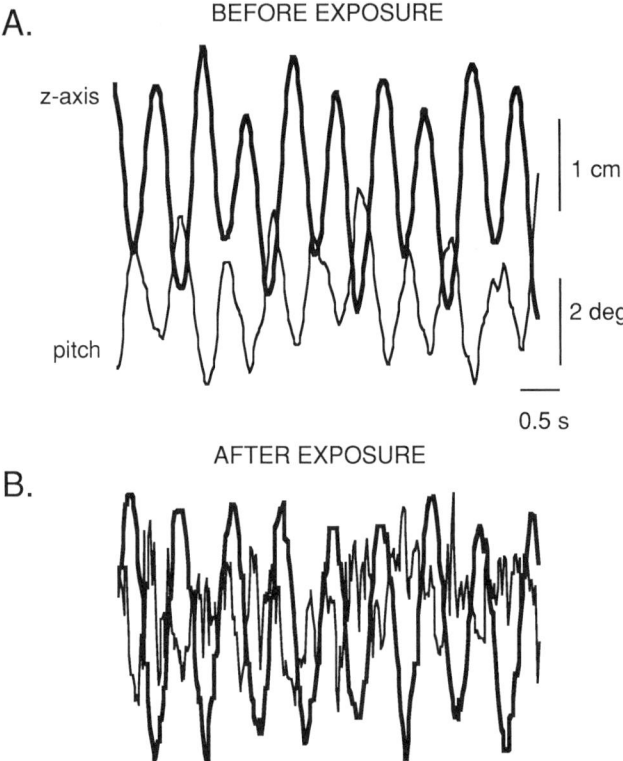

FIGURE 4. Effects of up-down reversing lenses on compensatory pitch head movements produced during locomotion. Examples of z-axis and pitch head position traces before (A) and after (B) exposure to up-down reversing lenses.

Reduction in z-axis head motion could be achieved by reducing torso z-axis motion by restricting hip or knee angular movement during locomotion. Regardless of the source, it is interesting to observe that gaze-stabilization strategies may include goal-directed compensatory body articulations that respond to changes in target distance.

It has been recently hypothesized that postural and gait motor control mechanisms may utilize a "top-down" control scheme to ensure head stability during body movement, thus aiding in the maintenance of gaze stability during locomotion.[10,13,14]

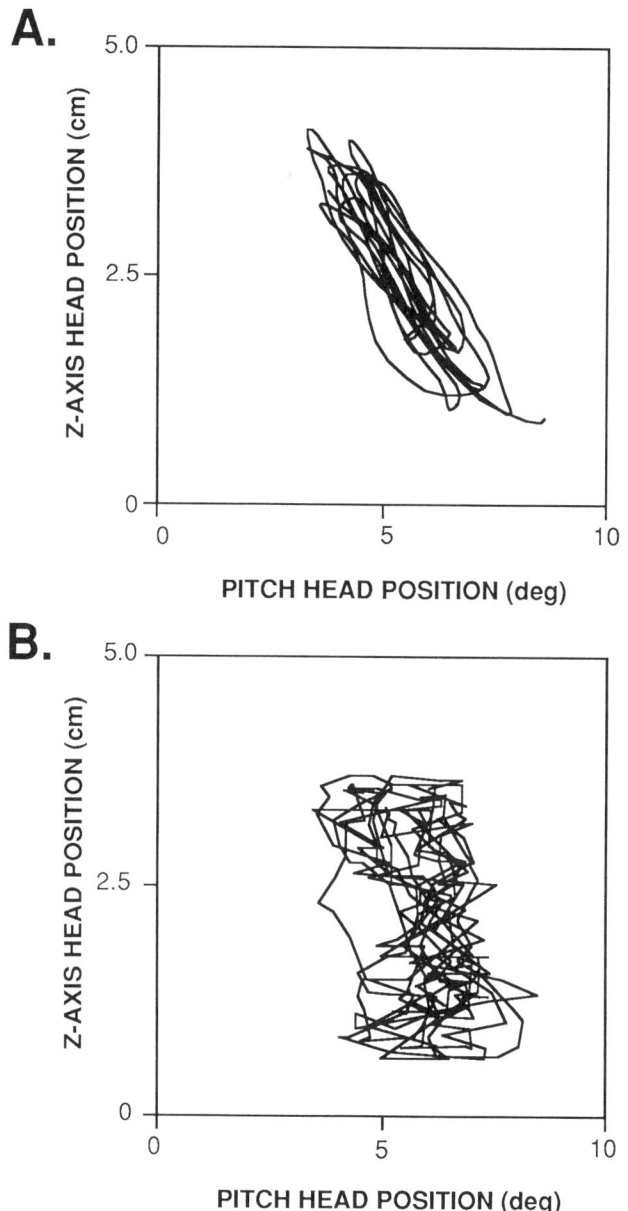

FIGURE 5. Plots of pitch versus *z*-axis head position before (A) and after (B) exposure to up-down reversing lenses.

The present experiments confirm this view by demonstrating that head movement during locomotion can be modified appropriately by varying ocular vergence angle.

Space Adaptation and Eye-Head Coordination during Locomotion

Astronauts and cosmonauts returning from spaceflights of even short duration (5–10 days) experience postural and gait instabilities on return to earth[15-18] along with disorienting illusions of self- and surround motion during head movement.[19]

During spaceflight, sensory-motor adaptation occurs enabling appropriate body movement in a microgravity environment. However, this new plastic sensory-motor state, tuned to the prevailing 0-G conditions, is inappropriate for a terrestrial 1-G environment producing motor and perceptual disturbances during the postflight readaptation period on return to earth. Gait and postural instabilities experienced by astronauts upon return to earth may be caused by in-flight adaptive acquisition of new "top-down" motor strategies designed to maintain head and gaze stability during body movement in microgravity. These strategies may be maladaptive for locomotion in a terrestrial environment leading to ataxic movement during initial reexposure to 1 G on return to earth.

REFERENCES

1. GROSSMAN, G. E., R. J. LEIGH, E. N. BRUCE, W. P. HUEBNER & D. J. LANSKA. 1989. Performance of the human vestibuloocular reflex during locomotion. J. Neurophysiol. **62:** 264–272.
2. GROSSMAN, G. E. & R. J. LEIGH. 1990. Instability of gaze during locomotion in patients with deficient vestibular function. Ann. Neurol. **27:** 528–532.
3. WATERS, R. L., J. MORRIS & J. PERRY. 1973. Translational motion of the head and trunk during normal walking. J. Biomech. **6:** 167–172.
4. CAPPOZZO, A. 1981. Analysis of the linear displacement of the head and trunk during walking at different speeds. J. Biomech. **6:** 411–425.
5. BLOOMBERG, J. J., M. F. RESCHKE, D. L. HARM, L. J. MICHAUD, W. B. KULECZ & B. T. PETERS. 1991. Head and gaze stability during locomotion. I. Analysis of linear head movement. Aviat. Space Environ. Med. **62:** 477.
6. SCHWARZ, U., C. BUSETTINI & F. A. MILES.1989. Ocular responses to linear motion are inversely proportional to viewing distance. Science **245:** 1394–1396.
7. PAIGE, G. D. 1989. The influence of target distance on eye movement responses during vertical linear motion. Exp. Brain Res. **77:** 585–593.
8. PAIGE, G. D. & D. L. TOMKO. 1991. Eye movement responses to linear head motion in the squirrel monkey. II. Visual-vestibular interactions and kinematic considerations. J. Neurophysiol. **65:** 1183–1196.
9. PAIGE, G. D. & D. L. TOMKO. 1991. Eye movement responses to linear head motion in the squirrel monkey. I. Basic characteristics. J. Neurophysiol. **65:** 1170–1182.
10. POZZO, T., A. BERTHOZ & L. LEFORT. 1990. Head stabilization during various locomotor tasks in humans. I. Normal Subjects. Exp. Brain Res. **82:** 97–106.
11. GONSHOR, A. & G. MELVILL JONES. 1976. Short-term adaptive changes in the human vestibulo-ocular reflex arc. J. Physiol. London **256:** 361–379.
12. GONSHOR, A. & G. MELVILL JONES. 1976. Extreme vestibulo-ocular adaptation induced by prolonged optical reversal of vision. J. Physiol. London **256:** 381–414.
13. BERTHOZ, A. & T. POZZO. 1988. Intermittent head stabilization during postural and locomotory tasks in humans. *In* Posture and Gait Development, Adaptation and Modulation. A. A. Berthoz & F. Clarac, Eds.: 189–198. Elsevier. Amsterdam, the Netherlands.

14. POZZO, T., A. BERTHOZ & L. LEFORT. 1989. Head kinematics during various motor tasks in humans. Prog. Brain Res. **80:** 377–383.
15. HOMICK, J. L. & M. F. RESCHKE. 1977. Postural equilibrium following exposure to weightless space flight. Acta Otolaryngol. **83:** 455–464.
16. KENYON, R. V. & L. R. YOUNG. 1986. MIT/Canadian vestibular experiments on Spacelab-1: 4. Postural responses following exposure to weightlessness. Exp. Brain Res. **64:** 335–346.
17. ANDERSON, D. J., M. F. RESCHKE, J. L. HOMICK & S. A. S. WERNESS. 1986. Dynamic posture analysis of Spacelab-1 crew members. Exp. Brain Res. **64:** 380–391.
18. PALOSKI, W. H., M. F. RESCHKE, D. L. HARM, F. O. BLACK & D. D. DOXEY. 1991. Increased reliance on visual feedback for postural equilibrium control following spaceflight. Aviat. Space Environ. Med. **62:** 478.
19. RESCHKE, M. F. & D. E. PARKER. 1987. The effects of prolonged weightlessness on self-motion perception and eye movements evoked by roll and pitch. Aviat. Space Environ. Med. **58:** A153–8.

Plastic Changes in the Human Cervicoocular Reflex

ADOLFO M. BRONSTEIN

MRC Human Movement and Balance Unit
Institute of Neurology
National Hospital
Queen Square
London WC1N 3BG, United Kingdom

The fact that neck torsion induces eye movements in neonates and in certain pathological conditions has been known for nearly a century although their role in normal man is not yet clear (see Reference 1 for review). The finding of Dichgans *et al.* that following labyrinthectomy in the monkey the cervicoocular reflex (COR) is enhanced and contributes significantly to gaze stability during movements of the head[2] brought a renewed number of questions. Clinicians were interested to know whether a similar enhancement of the COR would occur in patients with loss of vestibular function and whether the variable degree of clinical recovery from such loss could be related to higher or lower degrees of potentiation of the COR. Additional questions were concerned with more general problems of adaptive mechanisms in the oculomotor system, for instance, whether both aspects of vestibuloocular function, compensatory and "anticompensatory" eye movements, would be represented in the COR and what structures or mechanisms might mediate the plastic enhancement of the COR following loss of vestibular function. These three questions will be addressed successively in this paper.

THE COR IN PATIENTS WITH ABSENT VESTIBULAR FUNCTION

Twelve adult patients with acquired loss of vestibular function, in most cases due to ototoxic antibiotics and/or meningitis, underwent DC electrooculographic (EOG) recordings during trunk "yaw" motion in the dark while the head was kept earth fixed by means of a bite plate.[3] The results shown in FIGURE 1 are from a group of eight of those patients who, in addition to "trunk on head" experiments, had additional "head on trunk" stimuli and a questionnaire assessing the disability imparted by the oscillopsia (illusion of movement of the visual environment) due to loss of the vestibuloocular reflex (VOR). The results are presented as slow phase COR gain (peak eye velocity/peak trunk velocity) and the numbers next to each patient's curve represent a rank order of oscillopsia with "1" being asymptomatic (one case only) and "8" being the patient most disabled by this symptom. From the data plotted in FIGURE 1 it can be concluded that the slow phase COR is markedly enhanced in patients with total vestibular loss but that the degree of recovery from oscillopsia is not directly related to a more or less powerful COR; it is also apparent that COR gain decays markedly with increasing frequency of trunk oscillation. Complementary experiments showed that during the more natural "head on trunk" oscillation, however, COR gain was higher than during "trunk on head" oscillation and did not decay as a function of frequency.[4] Although this would make the slow phase COR more efficient in providing retinal image stability during natural head movements, again, no correlation between degree of oscillopsia and "head on trunk" gain was present. The main conclusion from these experiments was that although neck and visually induced eye movements generate robust compensatory eye movements during head movements, retinal stability remains permanently impaired. It would

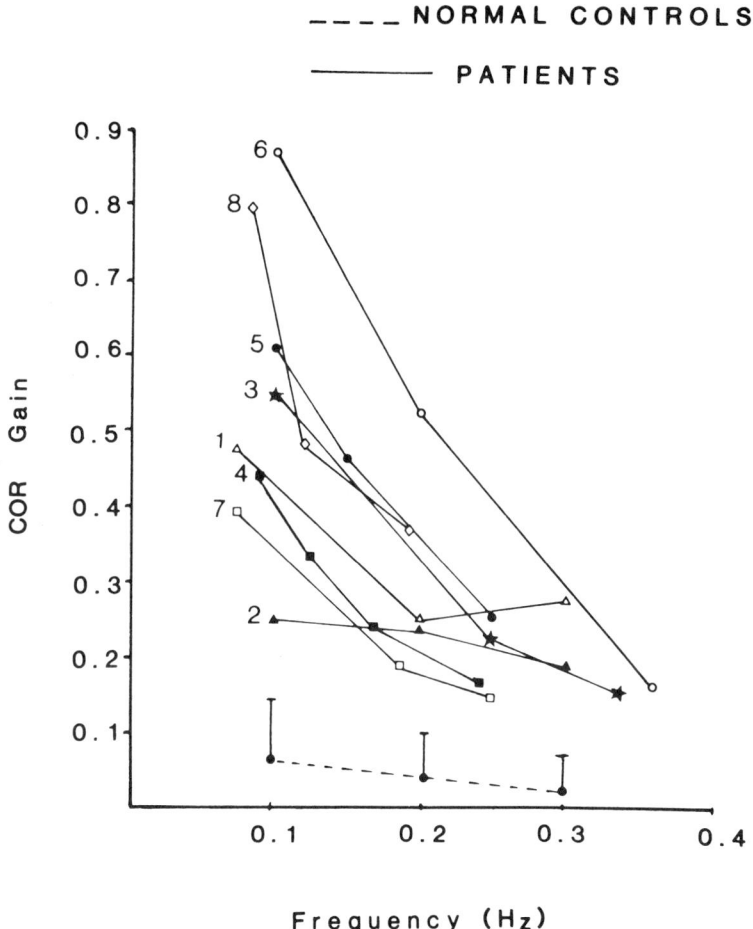

FIGURE 1. Slow phase COR gain at different frequencies in eight patients with absent vestibular function (solid lines) and in a group of 13 normal subjects [dashed lines, mean ± standard deviation (SD)]. The patients are ranked from 1 to 8 according to the subjective degree of oscillopsia. (From Reference 4 with permission.)

therefore seem that the process of recovery from oscillopsia due to loss of the VOR largely depends on some form of perceptual suppression of the moving retinal image, perhaps akin to that thought to occur in some forms of nystagmus or oculomotor palsies.[5] With this hypothesis in mind we have recently started to assess visual motion detection in these patients. Preliminary studies in a patient with absent vestibular function suggest that luminance thresholds for the detection of a moving target are normal during head-stationary measurements, but they are raised during head movements, particularly when the target moves against a background consisting of a grating of middle spatial frequencies (Brent, Bronstein, Ruddock, and Wooding, unpublished observations).

COMPENSATORY AND "ANTICOMPENSATORY" COR

Many natural head movements are fast ramp displacements occurring as part of an eye-head coordinated strategy aimed at refoveating newly appearing visual targets.[2] Although the initial eye saccade is usually visually triggered, animal[6] and human studies[7,8] have suggested that under certain circumstances the saccade is vestibular in origin, i.e., a quick component of nystagmus whose function is to shift the eyes in the direction of an ongoing head movement. Thus, by contrast to the compensatory, slow-phase-mediated function of the VOR, the vestibularly mediated fast eye shift in the same direction of the head movement has come to be known as the "anticompensatory" function of the VOR.

Partly, the notion of a vestibularly triggered pattern of coordinated eye-head refoveating movement came from the type of eye movements observed during ramp head movements carried out in the dark. In this condition, either active or passive head motion elicits an anticompensatory saccadic shift during the accelerative phase of head movement and a slow compensatory return of the eyes towards, but rarely reaching, primary gaze during the decelerative phase of head motion. Strikingly, the same pattern of eye movements is seen in patients with total loss of vestibular function (FIGURE 2), even if the head is moved passively and unpredictably in the

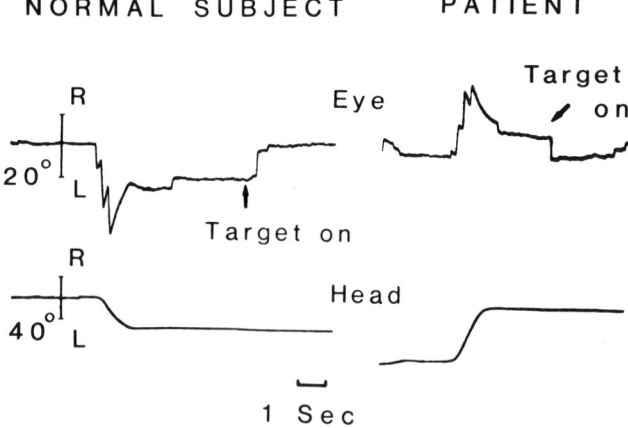

FIGURE 2. Voluntary ramp head displacements in the dark in a normal subject and in a patient with absent vestibular function. The arrow signals the illumination of a head-fixed target used to bring the eyes back to primary position. The recentering saccade following target presentation illustrates that the eyes were deviated in an "anticompensatory" direction as a result of the preceding head motion. Note the similarity of the eye movement in the two subjects. (From Reference 3 with permission.)

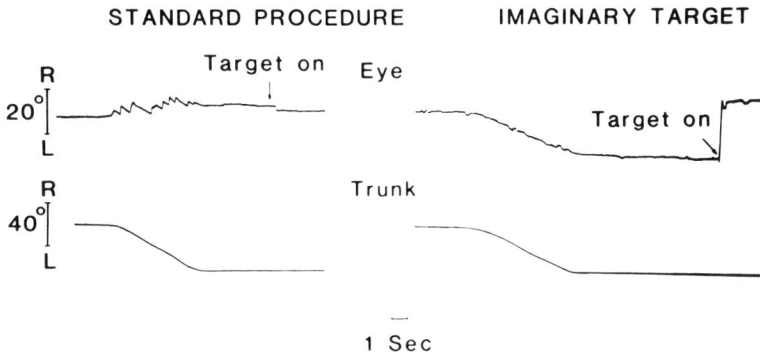

FIGURE 3. Ramp trunk displacement with the earth-fixed head in the dark in a patient with absent vestibular function. The arrow shows a recentering saccade in response to illumination of an earth-fixed target in line with primary gaze. During the standard procedure (left) there were no specific instructions, whereas during imaginary target (right) the subject had to try to keep her eyes on the (imaginary) earth-fixed target. (From Reference 3 with permission.)

dark, strongly suggesting that it is reflexively mediated by the COR.[3] Similarly, ramp trunk displacements around an earth-fixed head elicit, in patients with absent vestibular function, an active nystagmic pattern producing an overall eye shift in the direction of the (relative) head movement. If, however, patients are instructed to maintain fixation on an *imaginary* earth-fixed target, the quick phases of nystagmus are totally suppressed and the eyes deviate in a slow phase motion in the direction opposite to the (relative) head movement (FIGURE 3). These experiments indicate that the COR can generate, in the absence of vestibular function, either predominantly compensatory or anticompensatory patterns of eye movements and that the selection of one strategy or the other is under central control, influenced by mental set.

MECHANISMS MEDIATING COR PLASTICITY

We have recently encountered two patients with sporadic forms of adult onset idiopathic cerebellar degeneration in whom vestibular-oculomotor examination revealed total absence of response to caloric and rotational stimuli as well as severely impaired smooth pursuit and optokinetic eye movements. In one of the patients clinical and imaging evidence of central nervous system (CNS) lesion was restricted to the cerebellum; in the other patient neurological involvement was more widespread. We speculated that if the COR was not enhanced in these cases with absent vestibular function, it would be possible to postulate that the plastic enhancement of the COR is mediated by cerebellar structures or pursuit-optokinetic mechanisms.[9]

The COR was studied in the dark both by means of ramp and sinusoidal displacements of the trunk in yaw, around the earth-fixed head ("trunk on head"), and by sinusoidal motion of the head around the earth-fixed trunk ("head on trunk"). FIGURE 4 shows examples of eye movements during ramp trunk displacement in one of these patients. In the left traces, obtained under no specific instruction other than "stay alert with the neck relaxed," there is an initial "anticompensatory" gaze shift in the direction of the *relative* head motion, but

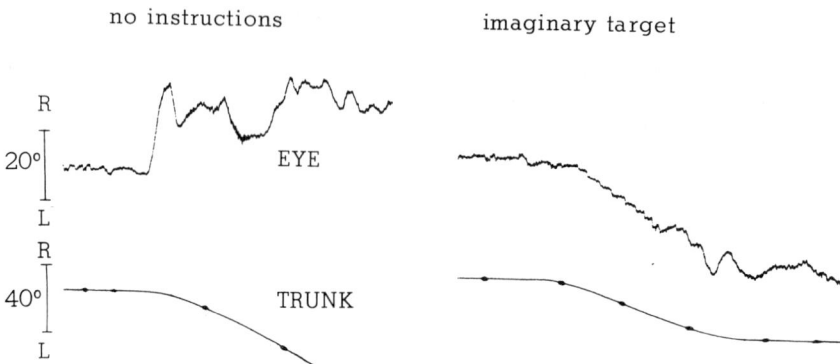

FIGURE 4. Ramp trunk displacement with the earth-fixed head in the dark in a patient with absent vestibular function and cerebellar atrophy. Compare the eye movements with those of FIGURE 3 and note, during "no instructions," the absence of a nystagmic pattern or a consistent slow phase eye movement and, during imaginary fixation of an earth-fixed target, the presence of small-amplitude saccades in the "compensatory" direction (i.e., direction opposite to *relative* head motion). (Modified from Reference 9.)

neither a nystagmic pattern nor slow phase eye movements in the opposite direction to the *relative* head motion were seen, in contrast to what occurs in ordinary patients with absent vestibular function (compare FIGURES 3 and 4). During instruction to maintain fixation on an imaginary earth-fixed target, the subject redirected gaze in a "compensatory" manner, i.e., in a direction opposite to *relative* head motion, which reproduced the stimulus waveform. However, such "compensatory" eye movements consisted of a succession of small amplitude saccades without any significant slow phase eye movement, again in contrast to what occurs in patients with pure vestibular loss. Slow phase eye velocity during ramp and sinusoidal movements of the trunk or the head was negligible. Oscillation of the head on the trunk, a more natural situation known to enhance the COR even further in patients with pure vestibular loss, did not generate any significant slow phase eye movement in the patient tested.[9]

In order to determine the type and origin of eye movements compensatory to head motion in these patients with absent vestibular-optokinetic function, additional experiments were carried out in the light. In condition i, to evaluate the contribution of visually guided eye movements (pursuit-optokinetic), the subjects had to follow the motion of a target attached to the inside of a full field optokinetic drum which surrounded them. In ii, the subjects maintained visual fixation on an earth-fixed target in a normally lit room during whole-body (i.e., head and trunk "en bloc") rotation, which is equivalent to the preceding condition plus a possible contribution from any remaining vestibuloocular function. And in iii, oscillation of the head on the earth-fixed trunk during fixation was delivered, which includes the two preceding conditions plus any possible contribution from the neck receptors. Sinusoidal oscillation was delivered at 0.3 Hz, 40 deg/second peak in all three conditions.

FIGURE 5 illustrates the eye movements obtained which were remarkably similar in all conditions and highly reminiscent of typical "broken pursuit" eye movements. The velocity of the slow phase eye movement was insufficient to match relative target velocity, so fixation was maintained by a rapid succession of small-amplitude saccades interspersed with the little slow phase eye movement present. The gain of the slow phase eye velocity component was approximately 0.40 in the two patients in the three conditions, with no statistically significant differences between conditions.

These experiments show that, in contrast to what is seen in patients with vestibular failure alone, the slow phase COR is not enhanced when there is additional cerebellar (presumably vestibulocerebellar) damage. The fact that head

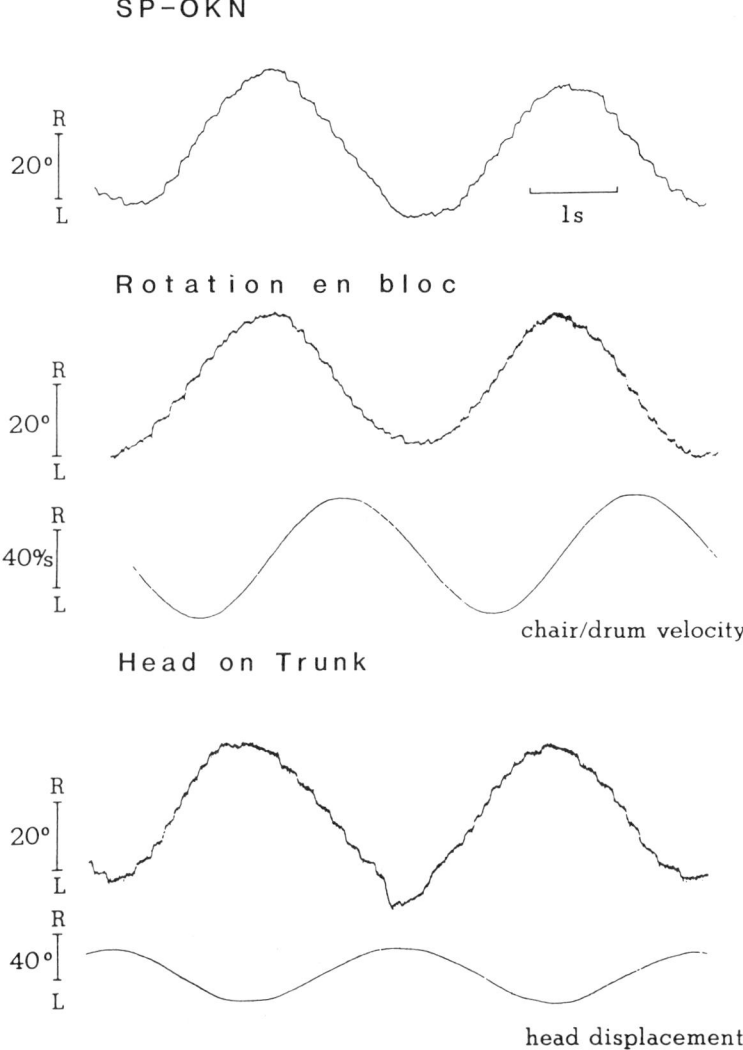

FIGURE 5. Eye movements elicited in a patient with absent vestibular function and cerebellar atrophy by oscillation of a full field optokinetic drum (SP-OKN), of the head and trunk en bloc, and of the head alone (on trunk) during optic fixation in a well-lit room. There are no appreciable differences in the eye movement elicited in these three conditions. (From Reference 9 with permission.)

movements in the light did not improve on the basic performance of visually guided eye movements is also indicative that there is no cervical contribution to gaze stability. Therefore the results suggest that the vestibullocerebellum is needed for the plastic enhancement of the COR in the absence of vestibular function and/or that such enhancement is mediated by pursuit mechanisms. In favor of the first of these possibilities is the fact that the vestibulocerebellum is also known to participate in other forms of neural plasticity involving slow phase eye movements, such as modification of the VOR by wearing telescopic lenses or prisms (see References 10 and 11 for review). For testing the second alternative the combination of absent vestibular function and a noncerebellar lesion affecting pursuit (i.e., parietal) is required.

Unfortunately, the type of lesions present in our patients do not rule out the possibility of a direct involvement of cervicoocular pathways by the degenerative process as the cause of lack of COR enhancement. However, this possibility seems less likely as both trunk on head rotation and head on trunk rotation were able to elicit good saccadic eye movements, especially during imaginary fixation when they adopted a typically compensatory appearance, which indicates that neck input clearly had access to the oculomotor system. Whatever the cause of the absence of potentiation of the slow phase COR in our patients, the findings also stress the independence of the cervically mediated fast phases and the fact that these could still be reoriented in a compensatory or "anticompensatory" manner according to a preselected strategy. It is likely that this "higher order" mechanism is under cortical control, as previously suggested by Jurgens et al.[12]

In the presence of intact vestibular function, lesions in more rostral areas of the cerebellar vermis also known to receive neck afferents[13] can occasionally give rise to a "disinhibited" COR, although this process might be mediated by enhanced saccadic components of this reflex[14] (and Bronstein and Hood, unpublished observations). Investigation of the COR in cases with positional nystagmus due to CNS disease did not show convincing abnormalities of the reflex that could explain the positional nystagmus (Bronstein and Hood, unpublished observations), as was suggested by Dix.[15]

Thus, the evidence gathered from the work in patients with pure absence of vestibular function indicates that the enhanced COR contributes to the generation both of slow phase compensatory eye movements during head movements and of fast "anticompensatory" gaze shifts with a possible role in target acquisition during fast ramp head movements. It would seem that the process of gain enhancement of the slow phase element is mediated by the vestibulocerebellum or by pursuit-optokinetic pathways.

IN MEMORIAM

This paper is dedicated to the memory of Margaret Ruth Dix, MD FRCS, who largely inspired some of the work herein described.

REFERENCES

1. BIEMOND, A. & J. M. B. V. DE JONG. 1969. On cervical nystagmus and related disorders. Brain **92:** 437–458.
2. DICHGANS, J., E. BIZZI, P. MORASSO & V. TAGLIASCO. 1973. Mechanisms underlying

recovery of eye-head coordination following bilateral labryinthectomy in monkeys. Exp. Brain Res. **18:** 548–562.

3. BRONSTEIN, A. M. & J. D. HOOD. 1986. The cervico-ocular reflex in normal subjects and patients with absent vestibular function. Brain Res. **373:** 399–408.

4. BRONSTEIN, A. M. & J. D. HOOD. 1987. Oscillopsia of peripheral vestibular origin: central and cervical compensatory mechanisms. Acta Oto-laryngol. **104:** 307–314.

5. DIETERICH, M. & TH. BRANDT. 1987. Impaired motion perception in congenital nystagmus and acquired ocular motor palsy. Clin. Vision Sci. **1**(4): 337–345.

6. ROUCOUX, A., M. CROMMELINCK, J. M. GUERIT & M. MEULDERS. 1981. Two modes of eye-head coordination and the role of the vestibulo-ocular reflex in these two strategies. *In* Progress in Oculomotor Research. A. F. Fuchs & W. Becker, Eds.: 309–315. Elsevier/North-Holland. New York, N.Y.

7. MELVILLE JONES, G. 1964. Predominance of anti-compensatory oculomotor response during rapid head rotation. Aerosp. Med. **35:** 965–968.

8. BARNES, G. R. 1979. Vestibulo-ocular function during coordinated head and eye movements to acquire visual targets. J. Physiol. **287:** 127–147.

9. BRONSTEIN, A. M., S. MOSSMAN & L. LUXON. 1991. The neck-eye reflex in patients with reduced vestibular and optokientic function. Brain **114:** 1–11.

10. ROBINSON, D. A. 1976. Adaptive gain control of vestibuloocular reflex by the cerebellum. J. Neurophysiol. **39:** 954–969.

11. MILES, F. A. & S. G. LISBERGER. 1981. Plasticity in the vestibulo-ocular reflex: a new hypothesis. Annu. Rev. Neurosci. **4:** 273–299.

12. JURGENS, R., T. MERGENER & W. SCHMID-BURGK. 1982. Modification of VOR slow and quick components by neck stimulation and turning sensation. *In* Physiological and Pathological Aspects of Eye Movements. A. Roucoux & M. Crommelinck, Eds.: 365–370. W. Junk. The Hague, the Netherlands.

13. BERTHOZ, A. & R. LLINAS. 1974. Afferent neck projection to the cat cerebellar cortex. Exp. Brain Res. **20:** 285–401.

14. BRONSTEIN, A. M. & J. D. HOOD. 1985. Cervical nystagmus due to loss of cerebellar inhibition on the cervico-ocular reflex: a case report. J. Neurol. Neurosurg. Psychiatry **48:** 128–131.

15. DIX, M. R. 1983. Positional nystagmus of central type and its neural mechanism. Acta Otolaryngol. **95:** 585–588.

Linear Acceleration Modulates the Nystagmus Induced by Angular Acceleration Stimulation of the Horizontal Canal[a]

IAN S. CURTHOYS,[b] SUSAN L. WEARNE,[b] MINGJIA DAI,[c]
G. MICHAEL HALMAGYI,[c] AND JOHN R. HOLDEN[b]

[b]Department of Psychology
University of Sydney
Sydney, New South Wales, Australia 2006

[c]Eye and Ear Research Unit
Neurootology Department
Royal Prince Alfred Hospital
Camperdown, New South Wales, Australia

INTRODUCTION

In 1965, Lansberg et al. reported that for human subjects concomitant linear acceleration directed along the interaural axis modulated the slow phase velocity (SPV) and the axis of eye rotation of the nystagmus produced by angular acceleration stimulation in yaw.[1] Both these results have since been confirmed in squirrel monkeys using torsional search coil recordings by Merfeld et al. (1990).[2,3] One particularly important feature of these results is that there is a marked decrease in the SPV when the person was oriented back to the motion as opposed to when the person was facing the motion, although the magnitudes of the angular and linear acceleration stimuli are the same in both cases.

Lansberg et al. used horizontal and vertical electrooculography (EOG) to measure eye movements, hence no records of torsion were obtained, although the subject received a large linear acceleration stimulus and the eye would be expected to tort in response to that linear acceleration stimulus alone.[4] We consider that torsion position records are necessary to understand the mechanism of the changes in SPV and eye rotation axis shift in these combined linear-angular acceleration situations, because the reported axis shift from yaw towards pitch could be achieved in either of two ways which imply totally different physiological bases. The eyeball could rotate around an oblique axis because of coactivation of horizontal and vertical recti without any ocular torsion, or the eye could rotate around an oblique axis because of activation of only the *horizontal* recti following an ocular torsion (see FIGURE 1). Neither EOG nor standard scleral search coils can provide the ocular torsion position information that is essential to discriminate between these alternatives. In the Lansberg et al. study, there was an axis shift but we do not know whether it was due to ocular torsion or not.

In the present study we used torsion search coils to determine the axis of eye rotation during combined linear-angular acceleration stimulation in order to identify

[a]This work was supported by NH&MRC of Australia.

716

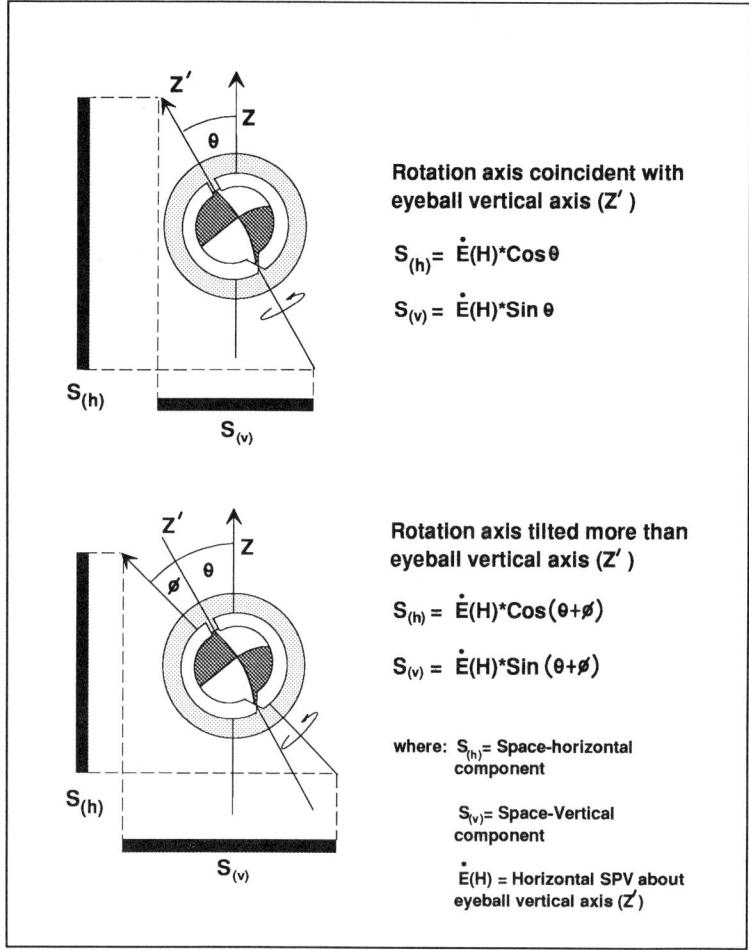

FIGURE 1. To show how torsion position signals can resolve between different causes of eyeball rotation axis shift. In the upper diagram the eye has torted and the solid black bars show the projection onto the space horizontal and vertical axes defined by the magnetic fields. The lower diagram shows that if there is an axis shift *additional* to this torsion, then the horizontal and vertical projections change accordingly, whereas the eye torsion position signal veridically signals the eye torsion position.

whether the axis shift and slow phase velocity modulation can be explained entirely on the basis of ocular torsion produced by the concomitant linear acceleration. Our preliminary data demonstrated a small but consistent vertical SPV component, more obvious when the subject was oriented back to motion, as was found by Lansberg *et al.*[1] The direction of the vertical component tended to align the axis of eye rotation with the gravitoinertial resultant force (GIF), as has been reported in the squirrel monkey. The two questions we sought to answer were:

1. Is there a reduction in SPV in the back to motion condition as opposed to facing the motion?
2. Is there a difference in ocular torsion when facing as opposed to back to motion that could account for *both* the reduced horizontal SPV and the larger vertical component in the back to motion condition?

METHODS

Eye movements were recorded by the torsional scleral search coil technique.[5] The subject was seated with his head and shoulders inside a chair-mounted 0.6-m^3 magnetic field and positioned one meter from the axis of rotation of a Servo-Med fixed chair human centrifuge. The subject could be oriented so that his interaural axis coincided with the radial direction of the centrifuge and was positioned either facing toward the motion or facing back to the motion. The transmitter coils were mounted on the chair, and they and the associated transmitting and detecting electronics all rotated with the subject. Calibration checks showed that this procedure did not materially affect the measured eye rotation. The eye was illuminated by an infrared light source and was monitored and videotaped by a closed-circuit TV system to allow qualitative verification of the orientation of the eye rotation axis. Horizontal, vertical, and torsional eye position signals were acquired to a PDP 11/73 with 12-bit A-D accuracy, sampled at 500 Hz, and digitally differentiated to yield horizontal, vertical, and torsional eye velocities. The digital records were desaccaded in UNIX using a fast desaccader written in C (by John Holden), and the appropriate calibrations and transformations enabled us to show the eye movement either in terms of space coordinates, determined by the fixed orientation of the magnetic field coils, or eyeball coordinates (using a right-hand rule in both cases). The absolute *in vitro* calibrations used a laser and a sighting device to ensure that the position of the torsion coil in the calibration jig matched the position of the coil on the eye.

A trial consisted of calibration and offset checks, followed by an angular acceleration (in darkness) of $10°/\text{second}^2$ from $0°/\text{second}$ to $200°/\text{second}$. During each trial there was a constant angular acceleration which entails that the linear acceleration stimulus changed in a parabolic fashion.

RESULTS

There is clear confirmation that during acceleration, SPV is smaller in the back to motion than the facing motion orientation (see FIGURE 2). The average SPV values (in °/second) across subjects were facing = 61.8 ± 11.2 (standard deviation), centered = 45.4 ± 9.7, back = 28.4 ± 7.0. Paired *t*-tests show there is a significant increase from centered to facing and a significant decrease from centered to back to the motion. During deceleration there were no significant differences and the mean SPV values were facing = 43.8 ± 10.3, centered = 43.4 ± 15.5, back = 43.4 ± 12.2. For each individual the torsion shift is similar in both facing the motion and back to the motion. The vertical component is very small, being about 10°/second maximum, whereas Merfeld *et al.* found about an average vertical velocity of 100°/second in monkeys with exactly the same angular and linear acceleration values.[2,3]

FIGURE 3 uses the desaccaded traces to show the combined effect of vertical and torsional movement on eye axis location, for accelerating stimuli. The upper row superimposes vertical and horizontal traces, and the very small vertical SPV compo-

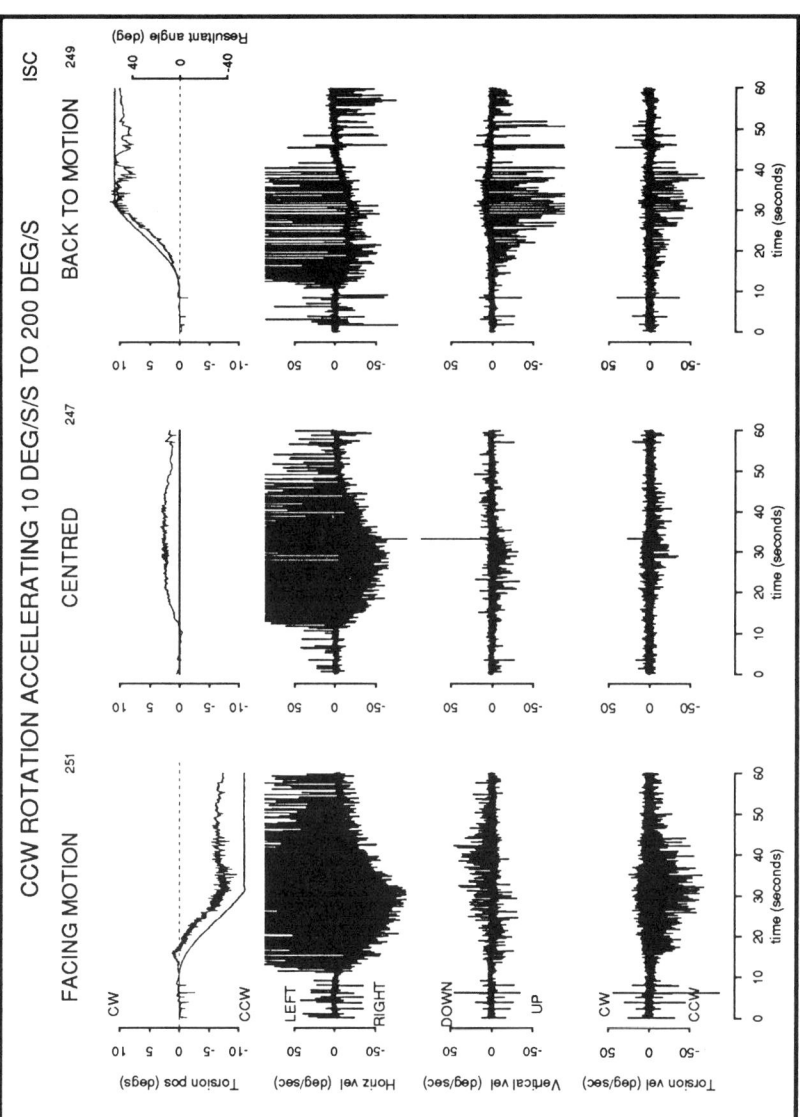

FIGURE 2. An example of raw velocity data to show how the direction of concomitant linear acceleration modulates the SPV. When the linear acceleration is in the same direction as the angular acceleration and increasing in magnitude, the horizontal nystagmus is reduced relative to its magnitude at center. The records shown are of raw slow phase velocity for horizontal, vertical, and torsional movements, together with torsional position. When the same conditions hold except the direction of the linear acceleration is reversed so that it is directed medially with respect to the canal receiving ampullopetal endolymph flow, the modulation is reversed and there is even some modest enhancement of the slow phase velocity.

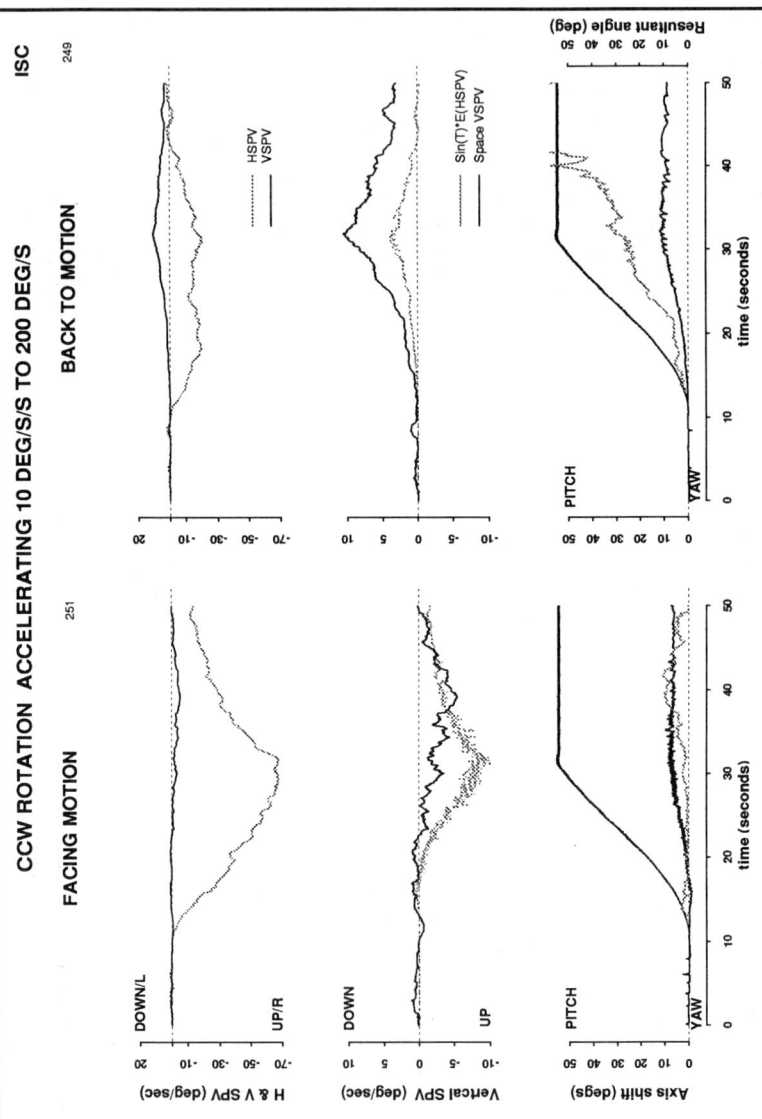

FIGURE 3. To show the orientation of the axis of eyeball rotation as derived from the vertical and horizontal measures compared with torsional position measures. The upper rows show desaccaded horizontal and vertical traces on the same scale to show the relative magnitudes of the vertical component. The middle rows show the actual and predicted vertical components (see text). The lower rows show the superimposed calculated values of eyeball axis of rotation together with the torsion position traces. If the torsion position had been sufficient to account for the axis shift, then the two traces would not systematically diverge, as they do in the back to motion condition shown here. Note that in the facing motion, the two traces almost overlap. These data show that torsional position information is not adequate to account for the axis shift during back to motion stimulation whereas it is adequate in facing the motion situation.

nent is apparent. The middle row plots the directly measured vertical SPV (solid line—note the different scales) together with the vertical component expected on the basis of pure eyeball yaw rotation and ocular torsion alone, i.e., the projection of the eyeball yaw axis onto the space pitch axis measured by the search coils (see FIGURE 1). The third row shows torsion position (solid line) and eyeball axis shift (dotted line) (in degrees of shift from yaw to pitch) plotted together with the resultant GIF angle. In the facing motion condition, the torsion position and axis shift curves are very similar indicating that for this subject eye torsion almost completely accounts for eye axis shift. In the back to motion condition, the two curves diverge showing that while eye torsion does take place (to the extent of about 10° in this observer), it is not sufficient to account for the eye axis shift that is observed because the axis continues to shift up to 40° after the torsion has reached a maximum of about 10°.

FIGURE 4 shows the same eye movement parameters plotted for the decelerating condition. Here the direction of the nystagmus is reversed and in both conditions, facing the motion and back to the motion, the eye axis shift can be accounted for by torsion alone; in this and in all 5 subjects there was no divergence as there was for the accelerating condition.

DISCUSSION

We confirm that the back to motion orientation produces a reduced horizontal slow phase velocity: during acceleration all our subjects showed a decrease in SPV in the back to motion as opposed to facing the motion orientation, and it is statistically significant. We also report a decrease in the time constant in the back to motion condition as opposed to the centered condition. The cause of these decreases is not immediately obvious because the direction of linear acceleration is the same in both conditions. It cannot be due to torsion alone since the eyeball torts in the same direction in both cases but the SPV is reduced in one case and increased in the other.

We also confirm the eye rotation axis shift reported by Lansberg *et al.*,[1] and while torsion does account for part of this axis shift, there is clearly an additional mechanism at work. In 1 of 5 subjects, the axis shift was the same amplitude and its time course closely matched the time course of ocular torsion. The remaining 4 subjects showed axis shifts that departed in both magnitude and time course from ocular torsion. FIGURE 3 demonstrates this departure. When the subject was oriented back to the motion, the vertical component (FIGURE 3, second row, solid line) was clearly larger than can be explained by ocular torsion alone (dotted line). This results in a shift of the axis of eyeball rotation from pure yaw towards pitch of about 40°, while ocular torsion reaches only 10°.

In accordance with Lansberg *et al.*'s result,[1] the vertical component was smaller when the subject faced the motion and followed a different time course to both horizontal SPV and ocular torsion, peaking at least 10 seconds after the end of the acceleration. In this case the axis shift from yaw to pitch *lags* ocular torsion until the vertical component peaks, and never exceeds the magnitude of ocular torsion. All of these effects were specific to accelerating trials only: neither the modulation of horizontal SPV nor the axis shift was present on deceleration.

These orientation-contingent differences are not directly attributable to ocular torsion, which followed a similar time course in both facing and back to motion conditions (e.g., FIGURE 2, top row). Nor are the marked differences between accelerating and decelerating trials due to ocular torsion, which was symmetric across all conditions in this subject.

In Merfeld *et al.*'s data in monkeys using the present testing configuration,[2,3] the

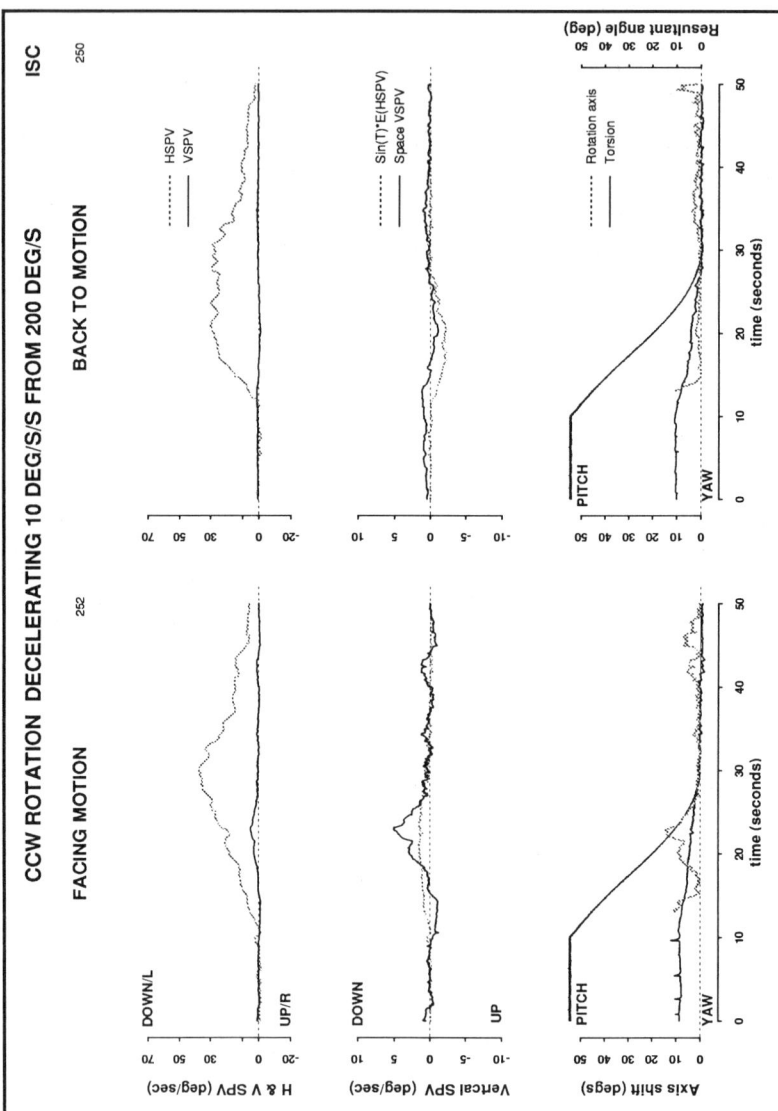

FIGURE 4. Conventions as for FIGURE 3 but now the records have been obtained during the deceleration phase. Notice that in both conditions for this subject the torsion position traces do overlie the axis shift traces.

new axis was achieved by a combination of horizontal and vertical rotations, with a large vertical component. Our data in man suggest that the vertical component is distinctly smaller than that found in the squirrel monkey, where peak vertical eye velocities of the order of 100°/second were found whereas the peak vertical eye velocities in man were of the order of only 10°/second. In some humans, ocular torsion seems to be the mechanism for achieving the change in eye axis. In other humans, a secondary mechanism appears to supplement the torsion in achieving a larger axis shift.

Scleral Coil Slippage

We used calibration checks at the beginning and end of each pair of trials to ensure that the offset in horizontal, vertical, and torsional position was almost identical; in other words that the eye signals returned to the same values they had started from. In some subjects the eye signals did not return to their initial values, and we consider that in such cases scleral coil slippage had occurred and the data from these trials are not presented or used.

CONCLUSION

In answer to the two questions posed in the Introduction: firstly, we confirm the SPV reduction in the back to motion condition in all subjects. Secondly, there is indeed a shift of the axis of eyeball rotation. However, the small differences found in ocular torsion between facing the motion and back to motion conditions were not sufficient in all subjects to explain either the reduced horizontal SPV or the more pronounced axis shift in the back to motion condition, relative to the facing motion condition. Finally, individual subjects varied in axis shift asymmetry between facing and back to motion conditions. In 1 of 5 subjects the mechanism of ocular torsion was entirely sufficient to explain the axis shift, whereas in the remaining 4 an additional mechanism must be inferred. The individual-specific response patterns we have observed present a challenge for recent models of the spatiotemporal organization of the VOR,[6] especially since the results with human observers show marked differences from monkey data.

ACKNOWLEDGMENTS

We are grateful to Dan Merfeld for many discussions via email and for giving us his analysis programs.

REFERENCES

1. LANSBERG, M. P., F. E. GUEDRY & A. GRAYBIEL. 1965. Effect of changing resultant linear acceleration relative to the subject on nystagmus generated by angular acceleration. Aerospace Med. **36:** 456–460.
2. MERFELD, D. M., L. R. YOUNG, D. L. TOMKO & G. D. PAIGE. 1991. Spatial orientation of the VOR to combined vestibular stimuli in squirrel monkeys. Acta Otolaryngol. Stockh. **481:** 287–292.
3. MERFELD, D. M. 1990. Spatial orientation in the squirrel monkey: an experimental and

theoretical investigation. PHD thesis. Massachusetts Institute of Technology. Cambridge, Mass.

4. MILLER, E. F. & A. GRAYBIEL. 1971. Effect of gravitoinertial force on ocular counterrolling. J. Appl. Physiol. **31:** 697–700.
5. COLLEWIJN, H., J. VAN DER STEEN, L. FERMAN & T. C. JANSEN. 1985. Human ocular counterroll: assessment of static and dynamic properties from electromagnetic scleral coil recordings. Exp. Brain. Res. **59:** 185–196.
6. RAPHAN, T. & D. STURM. 1991. Modelling the spatio-temporal organization of velocity storage in the vestibuloocular reflex by optokinetic studies. J. Neurophysiol. **66:** 1410–1421.

Validating the Hypothesis of Otolith Asymmetry as a Cause of Space Motion Sickness

SHIRLEY G. DIAMOND AND CHARLES H. MARKHAM

Department of Neurology
University of California at Los Angeles
School of Medicine
Los Angeles, California 90024-1769

Prior experiments in hypo- and hypergravity aboard NASA's KC-135 aircraft demonstrated in nine former astronauts a clear relation between ocular torsional instability in novel gravitational states and a history of space motion sickness (SMS).[1] When tested in the upright position during 0 G and 1.8 G in parabolic flight, five of these astronauts were found to have high scores of torsional asymmetry, a measure of disconjugate spontaneous torsional eye movements. These five persons had suffered SMS during their previous space missions, and their scores of torsional asymmetry on the KC-135 were directly related to the severity of their SMS symptoms in space. The four former astronauts who had low scores of torsional asymmetry on the KC-135 did not have SMS in space. Mean torsional asymmetry scores of the two astronaut groups differed significantly at $p = 0.01$ (two-tailed Wilcoxon ranked sums test).

In ground-based testing of torsional asymmetry in these same subjects, there were no differences among the nine astronauts.

These results supported the asymmetry hypothesis,[2,3] which asserts that a slight anatomical or physiological imbalance of the otolith organs may be well compensated in the usual 1-G environment on earth, but when such an asymmetric system is exposed to novel G states, the prior compensatory equilibration may be disturbed. This failure of compensation is believed to result in unaccustomed vestibular responses, giving rise to the sensory conflict proposed to underlie SMS.[4]

The present investigation focused on three questions suggested by the results of the previous study:

1. Would the earlier results be supported if more subjects were tested?
2. If subjects with high scores of torsional disconjugacy were tilted in the novel state of hypergravity, could the hypothesized asymmetry be equalized, i.e., could a position be found in which disconjugate torsion was nulled or minimized?
3. If ocular torsion on the KC-135 is to be used as a predictive test of SMS to learn which future astronauts would benefit from anti-motion-sickness medication or preflight adaptation training, would a fewer number of parabolas be as effective a test as the 10 to 20 parabolas used in the earlier study?

METHODS

Subjects were eight former astronauts, four of whom had not been tested previously. The new subjects' histories of SMS on previous space missions were

725

sealed and became known only after data analysis was completed. The other four subjects had flown in the previous study[1] and were selected for the present experiment because of their earlier high scores of torsional disconjugacy. All subjects gave informed consent to participate.

Apparatus consisted of a tilting chair mounted on the floor of the aircraft over the wings. Facing aft, the subject was strapped in securely. His head was fixed in position with a bite bar. Attached to the frame of the chair was an Olympus 35-mm single lens reflex camera, and 2 or 3 photographs were taken of both eyes at each episode of 1.8 G and 0 G during 20 parabolas. Measurements of eye torsion were made in our laboratory at UCLA with a two-projector system accurate to 0.1°. The apparatus is described more fully elsewhere.[5]

Protocol involved testing in five positions for four parabolas each, in the following order: with the chair upright, at 5° and 10° right ear down, and 5° and 10° left ear down. The same tests were performed on the ground for 1-G references.

Data analysis included a summary score of disconjugate eye torsion ("mean slope") for each subject. The calculation of this measure is detailed elsewhere.[1] Also examined was the effect of increasing numbers of parabolas on torsional disconjugacy.

RESULTS

The four new subjects had scores of torsional disconjugacy of 0.27, 0.34, 0.75, and 0.96. These are depicted in FIGURE 1, ranked in increasing order. When the SMS questionnaires were opened, the astronauts with the lowest scores, A-10 and A-11 [mean 0.30, standard deviation (SD) 0.05], were found to have not had SMS on their space missions. A-12 and A-13 had higher scores on the KC-135 test (mean 0.85, SD 0.15), and had suffered SMS in space. A-12 reported nausea and vomiting for the first two days in space but no symptoms on return to earth. A-13 had nausea and vomiting on the first flight day and severe vertigo lasting 12 hours after landing. Statistical tests on these two groups, each having two members, were performed. A one-tailed Wilcoxon ranked sum test of the significance of the difference in mean scores in these very small groups resulted in a $p = 0.06$, the lowest p value that this test can produce with only four subjects. A one-tailed t-test resulted in $p = 0.015$. The findings in these new subjects were consistent with the earlier study in which nine astronaut subjects showed a relationship of high torsional disconjugacy scores on the KC-135 to SMS in space.[1] In that larger study, the difference between the groups was significant at the 0.01 level in a two-tailed Wilcoxon ranked sum test. Because that study established the low score—low SMS susceptibility and high score—high SMS susceptibility predictions, one-tailed tests were deemed appropriate by our statistical consultants for the present study.

Tilting all eight subjects 5° and 10° to either side failed to discover a position in which torsional disconjugacy was nulled or minimized in 1.8 G. The results showed that torsional disconjugacy tended to increase when the subject was tilted. This was true in all three G states tested—1 G, 0 G, and 1.8 G—and was found in those who had SMS as well as those who did not. FIGURE 2 shows typical disconjugate eye torsion in the five tested positions in the three G states. Like virtually all the subjects, A-11 showed least disconjugacy in the upright position. The order of presentation of position was upright, 5° and 10° right ear down, and 5° and 10° left ear down. Amount of disconjugacy was greatest in 0 G, intermediate in 1.8 G, and least in 1 G.

Examining the consistency of torsional responses over the course of 10 or 20 parabolas revealed that in 0 G, disconjugacy increased with increasing number of

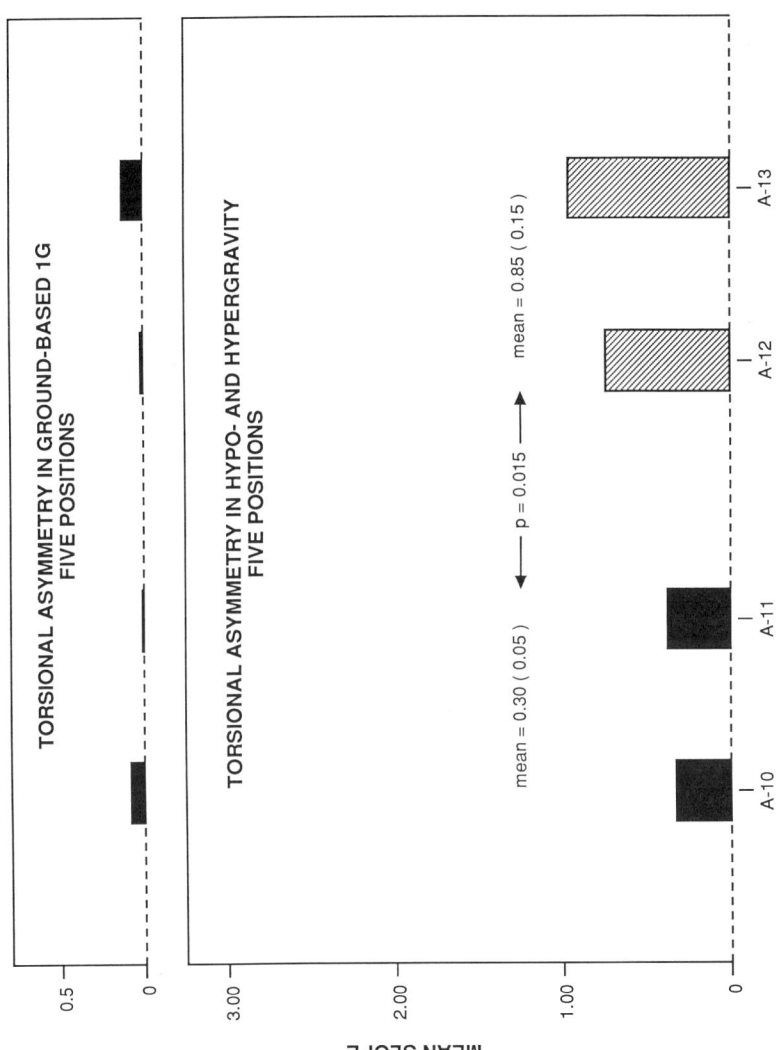

FIGURE 1. Four astronaut subjects are ranked in order of their scores of torsional disconjugacy in hypo- and hypergravity. The two with the lowest scores, shown with black bars, did not get sick in space. The two with the highest scores, shown in grey, did get sick. In 1-G testing, shown at the top of the figure, there were no differences between the subjects.

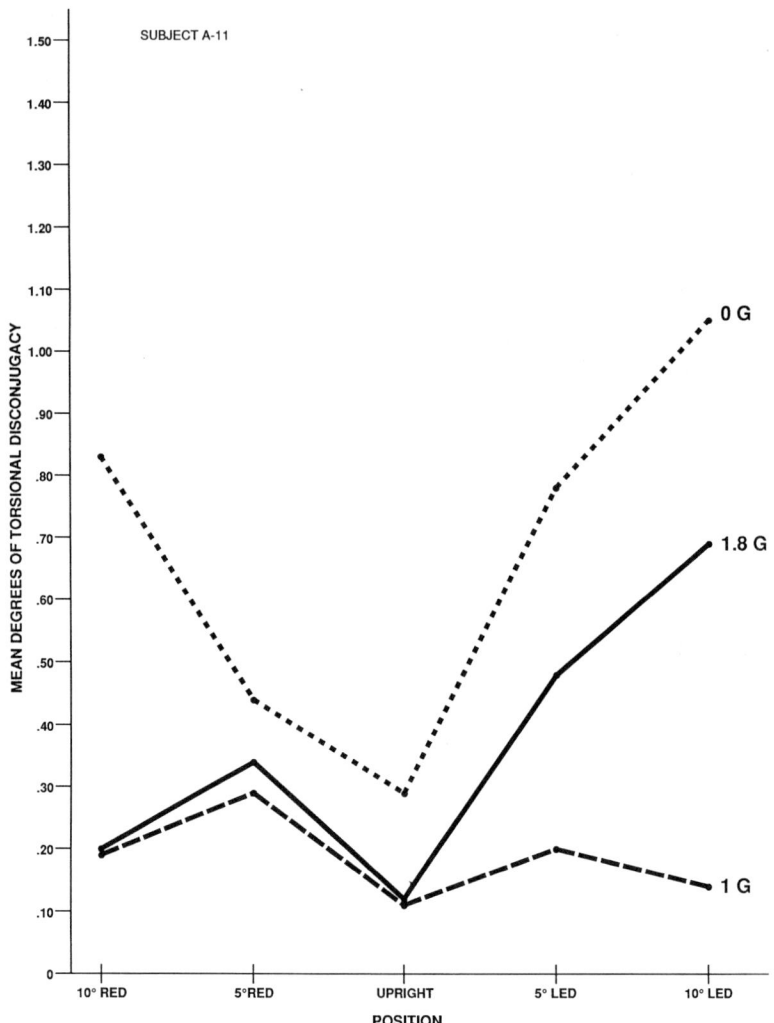

FIGURE 2. Disconjugate ocular torsion in various G states. Astronaut A-11 shows typical response to tilt in three gravitational states. The upright position showed least disconjugacy in all G states. Disconjugacy increased in the tilted positions and also increased as the flight progressed. Most disconjugacy was seen in 0 G, least in 1 G, consistent with the predictions of the asymmetry hypothesis.

parabolas. FIGURE 3 shows the combined results of the nine astronauts in the previous study[1] and the four astronauts in the present study. Results are grouped in sets of four parabolas and show that torsional disconjugacy in 0 G increased significantly as the flight progressed, in both those who had had SMS ($p = 0.001$) and in those who had not ($p = 0.02$, repeated measures ANOVA). The amount of torsional disconjugacy was far greater in the SMS group, and the difference between the groups became greater as the number of parabolas increased. Similarly, at 1.8 G, those who had been sick in space had more torsional disconjugacy than did those who had not been sick, but the differences between the groups were much less apparent. Disconjugacy at 1.8 G in both groups appeared to decrease after reaching a peak about midflight.

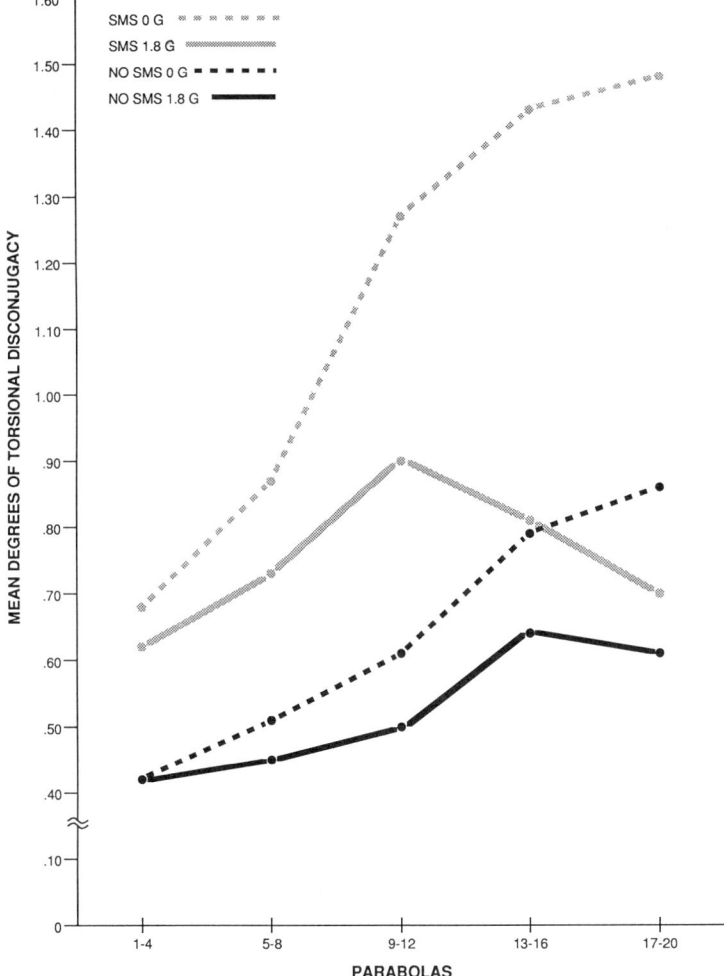

FIGURE 3. Torsional disconjugacy in 0 G and 1.8 G. Thirteen astronaut subjects in two combined studies were found to have significant increases in 0 G torsional disconjugacy with increasing numbers of parabolas. This effect was greater in astronauts who had been sick in space. In 1.8 G, the difference between the groups was less apparent and both groups seemed to improve in the latter half of the flight.

CONCLUSION

The results of this study offer additional support for the asymmetry hypothesis of otolith function. The astronauts who had been sick in space showed greater torsional disconjugacy when exposed to novel gravitational states on the KC-135 than did their cohorts who had been free of SMS on their shuttle missions. This finding is compatible with the proposed decompensation of an asymmetric system which has been calibrated to 1 G performance. In this and the previous study,[1] a total of 13 astronauts were tested with the experimenters blind to the SMS histories. While this is not a large number of subjects, the complete separation in torsional disconjugacy scores of the SMS group and the non-SMS group is reason to believe that ocular torsion on the KC-135 is a practical predictive test of SMS susceptibility.

This study failed to minimize disconjugate torsion by tilting subjects to one side or the other. The increasing disconjugacy in static tilted positions in 0 G and 1.8 G is similar to responses earlier reported in 1 G.[6] This finding in the novel G states tends to indicate that the hypothesized otolith asymmetry is not a simple planar difference about the x-axis. The possibility remains that the asymmetry may be in the z-axis[7] or a more complex difference such as unequal masses of otoconia[8,9] or number of hair cells in the two sides. Such conditions may result in unequal resting discharges, compensated in the usual 1 G but not yet recalibrated to a novel G state. SMS typically lasts a day or two or three before recovery. This pattern, as well as the common recurrence of symptoms on return to earth, supports the compensation-decompensation-recompensation process proposed to occur when an asymmetric system of gravity receptors meets changing G conditions.

The observation that disconjugacy in 0 G increases over time is consistent with the fact that SMS does not usually occur immediately on achieving orbit, but typically after an hour or more has passed. Ground-based motion sickness also has a variable latent period during which symptoms can be suppressed. After some threshold has been passed, an apparent accumulation of stimuli results in frank sickness.

In summary, the hypothesis of otolith asymmetry of SMS appears valid. A test of torsional disconjugacy while subjects are seated upright in hypo- and hypergravity of parabolic flight promises to be a reliable and relatively inexpensive predictive test of SMS susceptibility. Because torsional disconjugacy increases with increasing number of parabolas, 10 to 20 parabolas are a more certain discriminator than are a fewer number.

REFERENCES

1. DIAMOND, S. G. & C. H. MARKHAM. 1991. Prediction of space motion sickness susceptibility by disconjugate eye torsion in parabolic flight. Aviat. Space Environ. Med. **62:** 201–205.
2. VON BECHTEREW, W. 1909. Die Funktion der Nervenzentren 2. Gustav Fischer. Jena, Germany.
3. VON BAUMGARTEN, R. J. & R. A. THÜMLER. 1978. A model for vestibular function in altered gravitational states. Life Sci. Space Res. **17:** 161–170.
4. IGARASHI, M. 1988. New facets of space medicine. Acta Otolaryngol. Suppl. **458:** 103–107.
5. DIAMOND, S. G. & C. H. MARKHAM. 1983. Ocular counterrolling as an indicator of vestibular otolith function. Neurology **33:** 1460–1469.
6. DIAMOND, S. G., C. H. MARKHAM & N. FURUYA. 1982. Binocular counterrolling during sustained body tilt in normal humans and in a patient with unilateral vestibular nerve section. Ann. Otol. Rhino-Laryngol. **91:** 225–229.
7. MITTELSTAEDT, H. & S. GLASAUER. 1990. Determinants of spatial orientation in weightless-

ness. Proceedings of the Fourth European Symposium on Life Sciences Research in Space, Trieste, Italy, May 28–June 1. (ESA SP-307, Nov. 1990.)

8. ROSS, M. D. & K. M. DONOVAN. 1986. Otoconia as test masses in biological accelerometers: what can we learn about their formation from evolutionary studies and from work in microgravity? Scanning Electron Microsc. **4:** 1695–1704.

9. ROSS, M. D. 1987. Implications of otoconial changes in microgravity. Physiologist Suppl. **30**(1): S90-3.

The Human Vertical Vestibuloocular Reflex in Response to High-Acceleration Stimulation after Unilateral Vestibular Neurectomy[a]

G. M. HALMAGYI, S. T. AW, P. D. CREMER,
M. J. TODD, AND I. S. CURTHOYS[b]

Eye and Ear Research Unit
Royal Prince Alfred Hospital
Camperdown
Sydney, New South Wales, Australia 2050
[b]*Psychology Department*
University of Sydney
Sydney, New South Wales, Australia 2006

INTRODUCTION

The vertical vestibuloocular reflex (VVOR) normally originates from the integrated activity of all four vertical semicircular canals (VSCCs) in response to head acceleration in the pitch plane. During head accelerations in the upright position, the otoliths could also be contributing to the VVOR response. Nose-down head accelerations increase neural activity in anterior semicircular canal primary afferents, and at the same time decrease neural activity in the posterior semicircular canal primary afferents. Nose-up head accelerations have the opposite effect.

The question we tried to answer is, What difference does total unilateral vestibular deafferentation (UVD) make to the VVOR? Or in other words, How effective is the VVOR when it is generated by only 1 as opposed to the normal 2 labyrinths? We have previously shown that UVD produces a severe and permanent deficit in the horizontal vestibuloocular reflex (HVOR) response to high-acceleration horizontal head impulses directed away from the sole remaining HSCC.[8] Here we report our findings of the VVOR response to high-acceleration vertical head impulses 1 year or more after UVD.

MATERIALS AND METHODS

Details of the recording system and test protocols have been previously published;[8] only an outline is given here. Horizontal and vertical displacements of the head and the left eye were recorded using search coils (CNC Engineering, Seattle; Skalar, Delft). The system could accurately resolve head and eye rotations of 0.1 deg. Each patient sat fixating a solitary light-emitting diode in an otherwise totally darkened room. Without warning the light would be extinguished for about 3

[a]This study was supported by the Australian National Health and Medical Research Council and by the RPA Hospital Neurology Department Trustees.

732

seconds. This was the cue for one of the investigators, standing behind the patient, holding the patient's head, to deliver a rapid, passive, unpredictable, step displacement of head angular position (a head "impulse") by quickly rotating the patient's head around its interaural (pitch) or longitudinal (yaw) axis, through an angle of 10–20 degrees either leftward or rightward or upward or downward. The patient's instructions were to keep fixating the remembered position of the light and to refixate it as soon as it was reilluminated. Ten to 20 head impulses were generated in each direction. The peak velocity of head movements varied from 100 to 250 deg/second, and the peak acceleration from 1500 to 5000 deg/second per second. Head and gaze velocities were derived by analog differentiation of position signals. Eye position was derived by subtracting head position from gaze position, and eye velocity by subtracting head velocity from gaze velocity. The analog differentiators used to derive head and eye velocity were matched and had a bandwidth of 100 Hz. Signals were captured on-line at 500 Hz with 12-bit resolution using either a MINC 11/03 or an IBM-PC. The data were analyzed off-line using either an IBM-PC or a PDP 11/73.

To minimize possible contributions to compensatory eye movements from extravestibular sources such as the cervicoocular reflex, eye movement responses were only analyzed up to the peak of head velocity, since this always occurred within 100 mseconds of the onset of head rotation. Saccades and blinks usually did not occur until 100–150 mseconds after the onset of head movement.

We studied 10 patients who had only one functioning labyrinth. Each patient had undergone unilateral vestibular neurectomy at least 1 year before testing either as treatment for intractable vertigo or in the course of removal of an acoustic neuroma. All the acoustic neuromas were small, and none of the patients had symptoms or signs of brain stem dysfunction. Before operation all patients had reduced nystagmic response to bithermal caloric stimulation of the affected ear, but normal caloric responses from the other ear. Thirty normal volunteers served as normal controls. All patients and subjects gave informed consent and all protocols were approved by the RPA Hospital Human Ethics Committee.

More recently we studied 6 more UVD patients using a new data-acquisition system on a PDP 11/73. Position data were sampled at 1 kHz with 16-bit resolution, and both velocity and acceleration were derived by digital differentiation using a 2-point central-difference filter. The results described here are essentially the same on the two different systems except in that the signal resolution is superior with the 16-bit system.

RESULTS

In all normal subjects vertical head impulses produced an up-down symmetrical VVOR response, which had a gain close to unity at all head velocities tested (FIGURE 1A). For example at a head velocity of 100 deg/second, the upward VVOR gain was 0.97 and the downward VVOR gain was 1.02. The difference between these two values was not significant ($p > 0.05$). In all UVD patients, vertical head impulses produced an abnormal VVOR response (FIGURE 1B). At a head velocity of 100 deg/second, the mean upward VVOR gain was 0.52 whereas the downward VVOR gain was 0.72. The difference between these two values was significant ($p < 0.05$).

FIGURE 2 top shows a typical example of the HVOR and VVOR in responses of a UVD patient to horizontal and vertical head impulses 1 year after right vestibular neurectomy for acoustic neuroma. As previously reported,[8] there is a profound HVOR deficit in response to a head impulse toward the lesioned side, whereas the

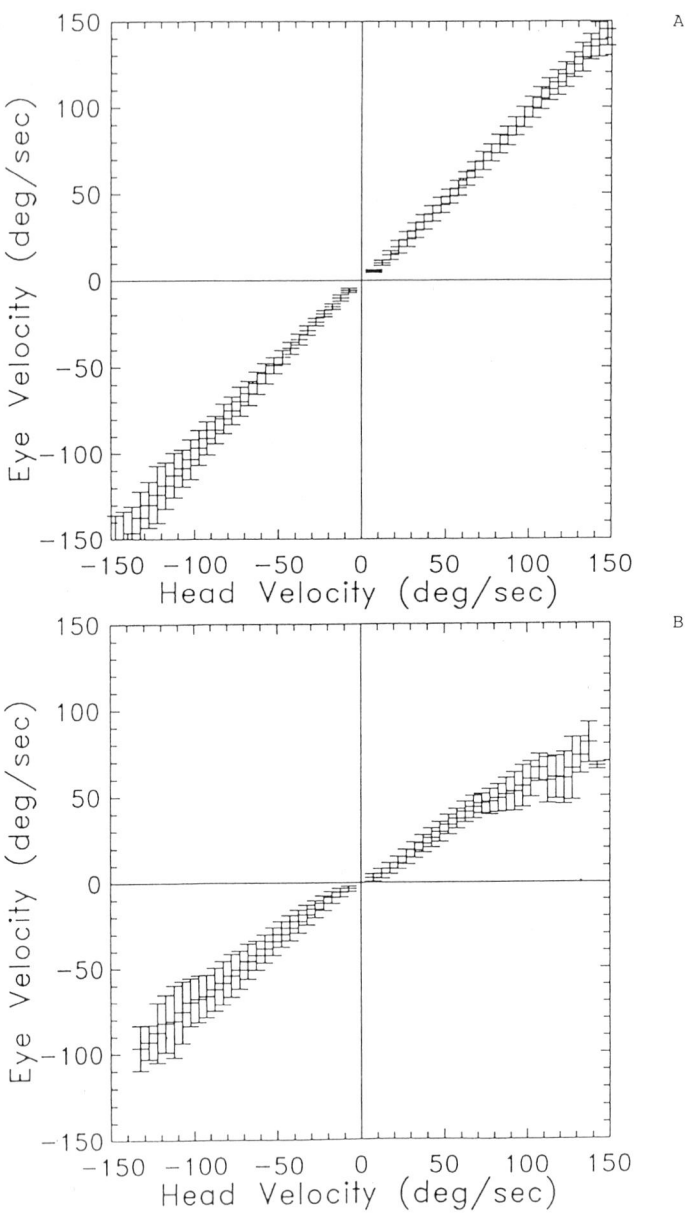

FIGURE 1. Averaged vertical eye velocity data as a function of vertical head velocity in response to vertical head impulses from 30 normal subjects (A) and from 10 patients (B) 1 year or more after unilateral vestibular neurectomy. Positive head velocity values indicate upward head movement. Following unilateral vestibular neurectomy, there is a bidirectional asymmetrical vertical vestibuloocular reflex (VVOR) deficit affecting the upward VVOR more than the downward VVOR. The data are in 5 deg/second velocity bins and shown with ±1 standard deviation.

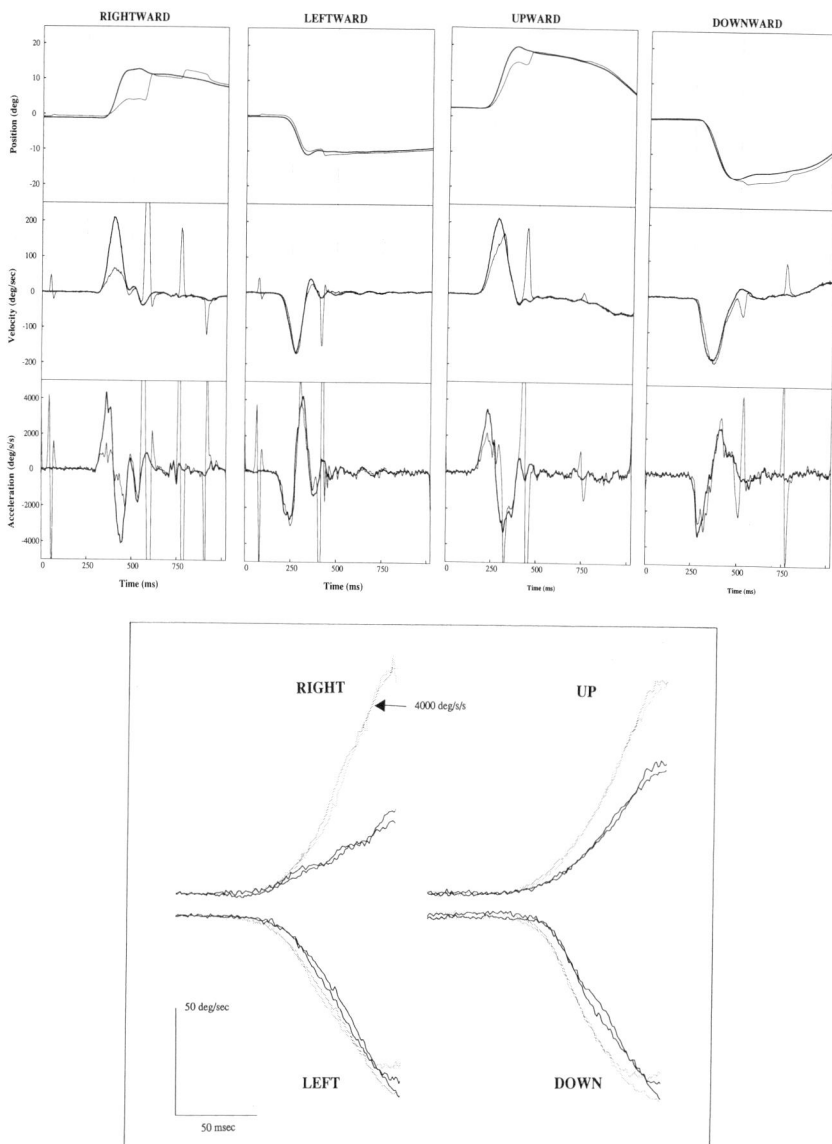

FIGURE 2. Typical examples of the eye responses to horizontal and vertical head impulses in a patient who had a right vestibular neurectomy. Both the combined position-velocity-acceleration records (top) and the expanded velocity records (bottom) show that whereas the vestibuloocular reflex in response to leftward and downward head impulses was close to normal, there was a marked deficit in the vestibuloocular reflex response to rightward and upward head impulses. In the top figure, head motion is shown in bold lines, eye motion in solid continuous lines. The polarity of eye motion has been reversed for ease of comparison with head motion. In each of the panels in the bottom figure, the eye velocity responses (shown in solid lines) to two identical head impulses (shown in faint lines) have been superimposed and shown to the peak of head velocity.

HVOR in response to a head impulse to the intact side is normal. There is also a clear VVOR deficit: whereas the VVOR in response to excitation of the single remaining anterior semicircular canal by a downward head impulse is normal, the VVOR in response to excitation of the posterior semicircular canal by an upward head impulse is significantly defective. FIGURE 2 bottom shows the onset of the eye velocity response in greater detail. Whereas the leftward HVOR and downward VVOR are normal, the rightward HVOR and the upward HVOR are defective. FIGURE 3 shows a plot of this patient's eye velocity as a function of head velocity in both the horizontal and vertical planes. At 100 deg/second the rightward HVOR gain at 0.25 was significantly less than normal and less than the leftward HVOR gain of 0.95. Also at 100 deg/second head velocity, the upward VVOR gain at 0.55 was significantly less than normal and significantly less than the downward VVOR with a gain of 0.75.

DISCUSSION

In this study we used high-acceleration passive unidirectional impulses to produce a VVOR response in upright patients with only one functioning labyrinth. Since this method has not been used before, our data are difficult to compare meaningfully with VVOR data from other studies.

Low-acceleration onside pitch stimulation in normal cat,[12] monkey,[10] and human[2,3] produces an asymmetric VVOR with the upward VVOR having a lower gain and time constant than the downward VVOR. More recently Baloh and Demer found in normal subjects a symmetric near-unity gain VVOR in response to 0.4–1.6 Hz active sinusoidal head oscillation both in the upright and in the onside positions.[4] From this they concluded that in humans, simultaneous otolithic stimulation normally makes little or no contribution to the VVOR produced by active pitch head movements in the upright posture. Upright pitch stimulation in humans produces a lower gain VVOR when produced passively than when produced actively.[3,6,9] Using passive low-acceleration onside pitch stimulation in UVD patients, Allum et al. found no long-term VVOR deficit.[1] The reason for this negative finding could have been that the acceleration used was too low to demonstrate the inherent on-off direction nonlinearity of the SCCs.

Using high-acceleration passive head impulses we have not found in normal subjects up-down VVOR asymmetries of the type that have been shown in cat and monkey.[7,10,11] What we have found is a consistent bidirectional deficit of the VVOR following UVD, indicating that a single labyrinth is not adequate to produce a normal VVOR.

Moreover we have also shown that the VVOR deficit is asymmetrical, more marked for upward head impulses that excite the sole functioning posterior SCC and disfacilitate the sole functioning anterior SCC, than for downward impulses that excite the sole functioning anterior SCC and disinhibit the sole functioning posterior SCC. There could be many reasons for this asymmetry. Perhaps the simplest explanation is in terms of the planar orientation of the VSCCs with respect to the saggital plane.[5] In humans the plane of the anterior canal makes an angle of 41.10 ± 4.82 (1 standard deviation) degrees with the saggital plane, whereas the plane of the posterior canal makes an angle of 55.84 ± 3.95 degrees. The projection of these mean VSCC planes to the saggital plane yields a posterior to anterior canal ratio of (cos 55.84/cos 41.10) = 0.74, a value that is close to the upward/downward VVOR gain asymmetry obtained in this study (0.52/0.72) = 0.72. The reason that the VVOR

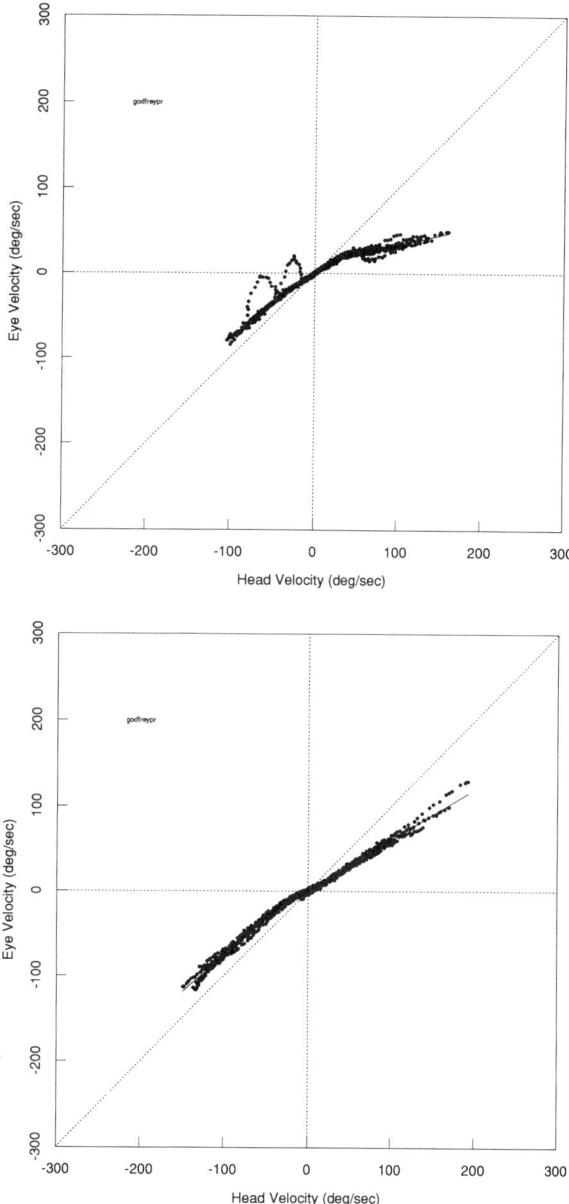

FIGURE 3. Horizontal (top) and vertical (bottom) eye velocity plotted as a function of horizontal and vertical head velocity from the patient whose raw records are shown in FIGURE 2. The deficit in rightward and upward vestibuloocular reflex gain is clearly seen. Only data to the peak of head velocity are shown.

asymmetry after UVD is not as marked as the HVOR asymmetry[8] could be that there are still two SCCs driving the VVOR as opposed to the one SCC driving the HVOR.

REFERENCES

1. ALLUM, J. H. J., M. YAMANE & C. R. PFALTZ. 1988. Long-term modifications of vertical and horizontal vestibulo-ocular reflex dynamics in man. Acta Otolaryngol. **105:** 328–337.
2. BALOH, R. W., V. HONRUBIA, R. D. YEE & K. JACOBSON. 1986. Vertical visual vestibular interaction in normal human subjects. Exp. Brain Res. **64:** 400–406.
3. BALOH, R. W., L. RICKMAN, R. D. YEE & V. HONRUBIA. 1983. The dynamics of vertical eye movements in normal human subjects. Aviat. Space Environ. Med. **54:** 32–38.
4. BALOH, R. W. & J. DEMER. 1991. Gravity and the vertical vestibulo-ocular reflex. Exp. Brain Res. **83:** 427–433.
5. BLANKS, R. H. I., I. S. CURTHOYS & C. H. MARKHAM. 1975. Planar relationships of the semicircular canals in man. Acta Otolaryngol. **80:** 185–196.
6. COLLEWIJN, H., A. J. MARTINS & R. M. STEINMAN. 1983. Compensatory eye movements during active and passive head movements: fast adaptation to changes in visual magnification. J. Physiol. **340:** 259–286.
7. CORREIA, M. J., A. A. PERACHIO & E. R. EDEN. 1985. The monkey vertical vestibulo-ocular response. J. Neurophysiol. **54:** 532–535.
8. HALMAGYI, G. M., I. S. CURTHOYS, P. D. CREMER, C. J. HENDERSON, M. J. TODD, M. J. STAPLES & D. M. D'CRUZ. 1990. The human horizontal vestibulo-ocular reflex in response to high-acceleration stimulation before and after unilateral vestibular neurectomy. Exp. Brain Res. **81:** 479–490.
9. JELL, R. M., C. W. STOCKWELL, G. T. TURNIPSEED & F. E. GUEDRY. 1988. The influence of active versus passive head oscillation and mental set on the human vestibulo-ocular reflex. Aviat. Space Environ. Med. **59:** 1061–1065.
10. MATSUO, V. & B. COHEN. 1984. Vertical optokinetic nystagmus and vestibular nystagmus in the monkey: up-down asymmetry and effects of gravity. Exp. Brain Res. **53:** 197–216.
11. SNYDER, L. H. & W. M. KING. 1988. Vertical vestibulo-ocular reflex in cat: asymmetry and adaptation. J. Neurophysiol. **59:** 279–298.
12. TOMKO, D. L., C. WALL, F. R. ROBINSON & J. P. STAAB. 1988. Influence of gravity on cat vertical vestibulo-ocular reflex. Exp. Brain Res. **69:** 307–314.

Optokinetic and Vestibular Interactions with Smooth Pursuit[a]

Psychophysical Responses

VICENTE HONRUBIA,[b] R. KHALILI,[b] AND R. W. BALOH[c]

[b]Goodhill Ear Center
Division of Head and Neck Surgery
[c]Department of Neurology
University of California at Los Angeles
School of Medicine
Los Angeles, California 90024-1624

This communication describes the effect of the interaction of a constant-velocity optokinetic stimulus and constant angular accelerations on the perception of movement of a small visual target. Both had a predictable effect on the perceived velocity of the visual signal.

The vestibular and visual systems provide the main information about the perception of self-motion, of the relative motion between objects, and of an object's motion relative to self. This information, key to human orientation, is generated in the brain through psychophysical and neurophysiological processes that are poorly understood. Often, in everyday life, visual and vestibular stimuli interact in unusual ways leading to illusions of motion that can adversely affect orientation and the performance of complex tasks, such as the operation of vehicles and aircraft.[1-4]

The vestibular system's semicircular canals operate effectively as velocity sensors in the frequency range of natural head motions.[5] Their stimulation produces compensatory eye movements with a velocity equal to but opposite from that of the head as well as a physiological sensation of motion.[6,7] This vestibularly induced sensation of motion has proven difficult to quantify, but qualitatively, it can be described in terms of head velocity (i.e., moving to the right or to the left) and in general follows, as the eye movement responses, the predictions of the pendulum model of vestibular function.[7-11]

The visual system has two subsystems to stabilize gaze and detect motion of objects—the smooth pursuit (SP) and the optokinetic (OK) systems.[12-14] The SP system predominates in foveate animals, while the phylogenetically older OK system predominates in afoveate animals. In afoveate animals, the OK system produces involuntary (reflexive) eye movements whose velocity matches that of the stimulus (a rotating surrounding drum) for low velocities only.[15] Signals from direction-sensitive neurons with large receptive fields travel through short subcortical neural pathways to the contralateral accessory optic system[15] which provides input to the "velocity storage" centers of the vestibular nuclei.[13,16,17] A similar system appears to exist in primates.[18-21] By contrast, the multiple cortical centers of the SP system can match high velocities of small targets.

The oculomotor responses of the vestibuloocular reflex (VOR), the SP, and OK

[a]Research supported by Grant DC00097 from the National Institute on Deafness and Other Communication Disorders, and a grant from National Medical Enterprises to the Victor Goodhill Ear Center at UCLA.

739

systems are known, but the psychophysical responses, particularly during interactive conditions, are not well characterized. There are a number of interesting and interrelated observations, but no common underlying process has been identified.[22-29] The present experiments point out the existence of a process, equivalent to the operation of an internal reference, controlling the perception of self-motion and of movement of objects in relation to self.

METHODS

Healthy volunteers between 21 and 26 years of age, who were not informed of the psychophysical aspect of the experiments, were instructed to follow a visual target (VT) generated by a helium-neon laser (Spectraphysics) which consisted of an intense (1.0 mW) red dot 1° in diameter. It was reflected from a motorized mirror-galvanometer (General Scanning). The VT was projected on the surface of a surrounding OK drum, with a sinusoidal trajectory of ±18° amplitude at frequencies of 0.2 and 0.4 Hz. Subjects were asked at the same time to indicate the direction of VT motion by pressing a switch with the right hand when the VT was perceived to be moving to the right and a left-hand switch for the perception of movement to the left. For each test, the ratios of the time the subject thought the VT was traveling to one direction to that in the opposite direction were used to compute the value of the VT velocity when the sensation was obtained that the light changed direction, either to the right or left. This VT "subjective velocity" was computed according to the trigonometric relation: VT velocity = $V_{max} \sin 2\pi \{[R - 1]/[4 (R + 1)]\}$. Where V_{max} is the peak velocity (22.6°/second for the 0.2 Hz and 45.2°/second for the 0.4 Hz trajectories), and R is the value of the ratio of the longer to the shorter interval of perceived VT motion.

Each subject was prepared for electrooculographic recordings of eye movements which were induced and analyzed using equipment and laboratory computer programs developed for these purposes.[30] In the first experiment ($n = 5$: 3 females, 2 males), the VT was projected on the inner surface of the surrounding OK drum, 130 cm in diameter and 145 cm high. The interior surface was black with 2.5-cm wide, vertical white stripes placed every 17.8°. The drum, driven by a DC motor with feedback servocontrol under computer command, was rotated at constant velocities of 10, 20, 30, 40, 50, 60, 80, and 100°/second in both clockwise (CW) and counterclockwise (CCW) directions. While the subject fixated on the stationary VT, the optokinetic drum began moving at each of the above velocities presented in random order. After 10 seconds, the VT moved for four and one-half cycles (data from the first half-cycle were eliminated). In addition, each subject was tested with 30°/second of OK drum velocity to ascertain their normal optokinetic response (in CW and CCW directions) but without the VT present. The mean value of the response gain, defined as the mean slow-component eye velocity (SCV) divided by the stimulus velocity, was 0.90 ± 0.11.

In the second experiment ($n = 6$: 3 females, 3 males), the influence of vestibular stimulation on the perceived velocity of the VT was investigated. For this the laser source mounted on the subject's chair was projected on the black surface of the same surrounding drum (without white stripes), appearing always in front of the subject, who was rotated in the dark with constant angular accelerations of 0.4, 0.8, 1.6, 2.0, 3.2, 4.0, 6.4, and 8.0°/second² in both CW and CCW directions under servo-loop control. Acceleration was maintained for 40 seconds before the VT was activated to allow the vestibular organ's viscoelastic damping to be overcome and its stimulation to reach a steady-state condition.[5] Nystagmus responses due to the constant-acceleration stimuli in the dark were recorded and later analyzed. Each acceleration continued thereafter for the time necessary to present four and one-half cycles of VT stimulation.

RESULTS

When the VT was projected against a stationary OK drum, the average ratios of the perceived time the target was moving to the left vs. right were: 1.02 ± 0.03 for the 0.2 Hz frequency and 1.05 ± 0.03 for 0.4 Hz, values that are not statistically different from the expected values of 1. The corresponding velocities of the VT at the time of reversal of direction were $0.34 \pm 0.44°$/second for the 0.2 Hz frequency and $0.83 \pm 1.16°$/second for 0.4 Hz both in the leftward direction.

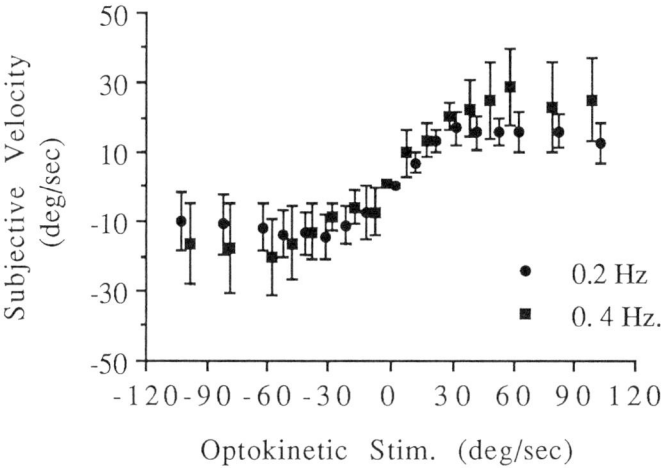

FIGURE 1. Graph shows in the ordinate the velocity of the visual target [mean ± standard deviation (SD)] at the time the subjects perceived the change in direction of motion and in the abscissa the different velocities of the optokinetic stimulus. Data are shown separately for the two different frequencies of the visual target trajectory as indicated (0.2 Hz, and 0.4 Hz).

During SP-OK interaction, subjects were under the illusion that the duration of each half-cycle of the VT became asymmetrical, appearing to be shorter when the VT moved in the direction of the OK drum. That is, the VT appeared to move in the direction of the OK drum only after reaching a significant velocity in the same direction. The estimated velocity at the time of direction reversal departed from zero, and its magnitude changed with the magnitude of the OK stimulus, reaching values that were a large fraction of the VT velocity (FIGURE 1). Note that the error in the velocity of the VT at the apparent time of reversal of direction is in direction opposite to that of the illusory self-motion associated with optokinetic stimulation, i.e., circularvection.[2,31] Similar effects were obtained during SP-VOR interaction (FIGURE 2). The velocity at the time the subjects thought the VT changed direction depended on the magnitude of the acceleration. The direction of the changes was similar to the previous experiment in that the increase in velocity error (the "subjective velocity") occurred when the VT motion was opposite to the velocity of the subjective sensation of self-motion.

The perception of the target trajectory during OK or vestibular sensory interactions can be grossly erroneous. For example, during 30°/second CW OK stimulation, the VT did not appear to begin moving to the right until its velocity reached 15.0 ± 5.8°/second. Approximately the same subjective sensation error was made during vestibular stimulation with CCW angular acceleration of 3.2°/second², when the subject perceived the VT to be traveling to the right only after its velocity reached 16.9 ± 6.2°/second. These two stimuli induced a subjective sensation of motion to the left, and the resulting velocity error is to the right.

An interesting observation resulted from a comparison of the subjective velocity error with the magnitude of the slow-component velocity of the nystagmus at the end of the VOR stimulation in the dark before the VT presentation. As shown in FIGURE 3, the nystagmus SCV increased with the magnitude of the acceleration. The SCV

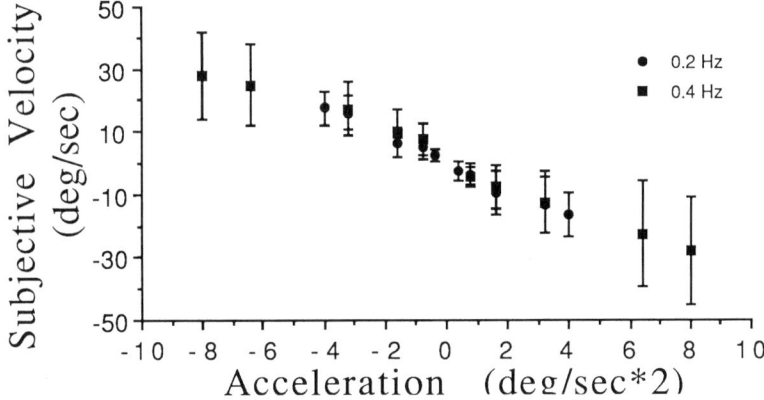

FIGURE 2. Graph shows in the ordinate the velocity of the visual target (mean ± SD) at the time the subjects perceived the change in direction of motion and in the abscissa the different angular accelerations. Data are shown separately for the two different frequencies of the visual target trajectory as indicated (0.2 Hz, and 0.4 Hz).

values were similar to those of the subjective sensation, as demonstrated by the scatter plot of FIGURE 4, in which a comparison is made between the two data sets. The relationship between these two variables was linear. A regression line drawn through the data had a slope of 0.99 and a correlation coefficient of 0.98. By contrast, the relationship between the subjective sensation error resulting from optokinetic stimulation and the magnitude of the stimuli is different. Only at low velocities (<20°/second) is the sensation a significant fraction of the stimulus. For higher velocities the effect decreased. This behavior is different from that of the OK nystagmus which can match the drum velocity at much higher stimuli.[32]

DISCUSSION

The brain judges the motion of objects in relation to the self in a relativistic manner, using internal references that are influenced by sensory stimuli. Two

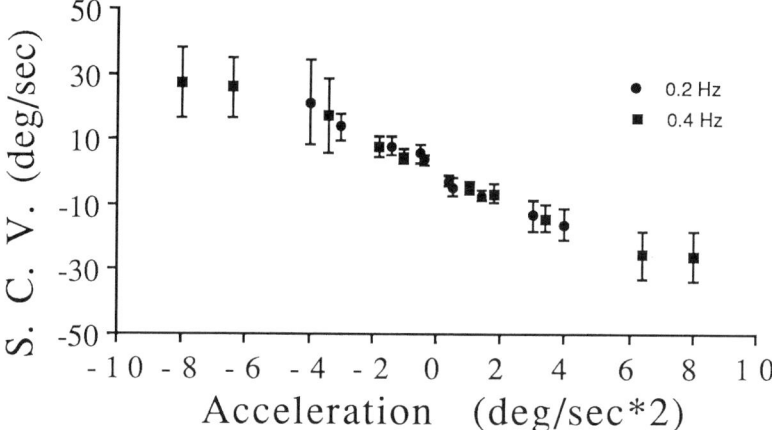

FIGURE 3. Graph shows the relationship between the slow-component velocity (SCV) of the VOR responses (mean ± SD) and the magnitude of the constant angular acceleration.

variables identified in these experiments are vestibular and optokinetic stimuli. The close relationship between the characteristics of the VOR responses and the errors in the perceived velocity of the VT is noteworthy. Normally, the vestibular nerve output from the semicircular canals is processed by the vestibular centers to produce eye movements that compensate for the brain's estimate of the head movement. The VOR characteristics reflect not only the vestibular nerve input but the integrating capacity of the velocity storage centers which increase the low-frequency capabilities of the vestibular organs.[5,6] Our data suggest that the information extracted by the cortical centers for estimating head motion is obtained from the velocity storage centers not from the vestibular nerve signal itself, because of its similarity with the eye movement responses.

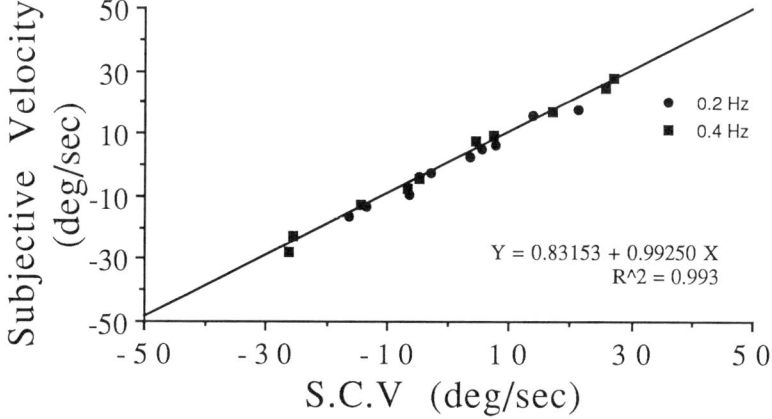

FIGURE 4. Graph shows the relationship between the mean vestibular subjective velocity (ordinate, from FIGURE 2) and the mean VOR response (abscissa, from FIGURE 3).

The present experiments point out that the perception of movement of visual targets rotating around a subject in the horizontal plane depends on the interaction of at least three variables, the real target velocity, the vestibular estimate of head velocity, and the optokinetic estimate of background velocity. These signals are evaluated in relation to an internal reference (IR) according to the relationship: $V_{imaginary} = V_{real} - IR$, where $V_{imaginary}$ is the perceived target velocity, V_{real} the actual (earth reference) target velocity, and IR is the velocity of the world as perceived by the brain during rotation. In normal subjects at rest, the value of IR is zero, but IR changes under experimental or possibly pathological conditions. With physiological vestibular stimulation, the value of IR is equal to the negative of the velocity of the head ($-V_h$) or its equivalent, the eye slow component velocity of the VOR response. As shown in FIGURE 4, the error in the perception of V_{real} (i.e., $V_{real} - V_{imaginary}$) is directly related to the VOR slow-component velocity. The value of IR can also be changed by optokinetic stimulation. In this case it takes the value of the velocity of the world (VOK) as estimated from the operation of the OK system (IR = VOK) and consequently $V_{imaginary} = V_{real} - VOK$.

These simple relationships predict the present experimental results as well as a number of other observations. For example, in the oculogyral illusion that is experienced by subjects fixating on a stationary light ($V_{real} = 0$) during rotation in the dark, $V_{imaginary}$ is in the direction of the sensation of motion (the perceived direction of the head velocity).[3] During the Filehne illusion in which stationary subjects look at a fixed target while the background moves (e.g., the stars or the moon in a cloudy sky), the apparent motion of the fixating object is opposite to that of the clouds. The existence of an internal command center to provide purpose to the various visual and vestibular oculomotor reflexes has been advocated on theoretical grounds to interpret psychophysical data[32] and cybernetic models of the SP system.[12] The data from these experiments confirm the existence of such a center and suggests a method to quantify its operation with physiological techniques.

A physiological consequence of the interaction of vestibular and visual stimuli during rotation of the head in the light is that VOR and VOK operate synergistically. This synergistic interaction would be particularly useful for low frequencies and low velocities of rotation for which the vestibular system is less sensitive. Finally, it can be anticipated that the sensation of vestibular motion is likely to be subjected to the same developmental and adaptive mechanisms that influence the VOR. This is an interesting aspect that will require further research.

ACKNOWLEDGMENTS

Ms. K. Jacobsen, M. Remba, M. Weinberger, and D. Aframian participated in the experiments. Karl Beykirch and C. Lau contributed to many theoretical discussions; Chris Coleman edited the manuscript.

REFERENCES

1. PROBST, T., S. KRAFCZYK, T. BRANDT & E. R. WIST. 1984. Interaction between perceived self-motion and object-motion impairs vehicle guidance. Science 225: 536–538.
2. BRANDT, TH., J. DICHGANS & E. KOENNING. 1973. Differential effects of central and peripheral vision for egocentric and exocentric motion perception. Exp. Brain Res. 16: 476–491.

3. WHITESIDE, T. C. D., A. GRAYBIEL & J. I. NIVEN. 1965. Visual illusions of movement. Brain **88**: 193–210.
4. FILEHNE, W. 1922. Uber das optische Wahrnehmen von Bewegungen. Zeitschrif fur Sinnesphysiologie **53**: 134.
5. GOLDBERG, J. M. & C. FERNANDEZ. 1975. Vestibular mechanisms. Annu. Rev. Physiol. **37**: 129–162.
6. ROBINSON, D. A. 1981. Control of eye movements. Handbook of Physiology, the Nervous Component. V. B. Brooks, ed. **2** (Part 2): 1275–1320. Williams and Wilkins. Baltimore, Md.
7. GUEDRY, F. E., JR. 1974. *In* Vestibular System: Psychophysics, Applied Aspects and General Interpretations. Handbook of Sensory Physiology. H. Autrum *et al.,* Eds. **VI/2** (Part 2). Springer-Verlag. Berlin, Germany.
8. YOUNG, L. R. 1969. The current status of vestibular system models. Automatica **5**: 369–383.
9. HUANG, J. & L. R. YOUNG. 1981. Sensation of rotation about a vertical axis with a fixed visual field in different illuminations and in the dark. Exp. Brain Res. **41**: 172–183.
10. ZACHARIAS, G. L. & L. R. YOUNG. 1981. Influence of combined visual and vestibular cues on human perception and control of horizontal rotation. Exp. Brain Res **41**: 159–171.
11. HONRUBIA, V., H. A. JENKINS, R. W. BALOH, H. R. KONRAD, R. D. YEE & P. H. WARD. 1982. Comparison of vestibular subjective sensation and nystagmus responses during computerized harmonic acceleration tests. Ann. Otol. Rhinol. Laryngol. **91**: 493–500.
12. LISBERGER, S. J., E. J. MORRIS & L. TYCHSEN. 1987. Visual motion processing and sensory-motor integration for smooth pursuit eye movements. Annu. Rev. Neurosci. **10**: 97–129.
13. COHEN, B., V. MATSUO & T. RAPHAN. 1977. Quantitative analysis of the velocity characteristics of optokinetic nystagmus and optokinetic after-nystagmus. J. Physiol. **270**: 321–344.
14. ROBINSON, D. A. 1981. The use of control systems analysis in the neurophysiology of eye movements. Annu. Rev. Neurosci. **4**: 463–503.
15. COLLEWIJN, H. 1981. The oculomotor system of the rabbit and its plasticity. Springer-Verlag. Berlin, Germany.
16. WAESPE, W. & V. HENN. 1987. Gaze stabilization in the primate. The interaction of the vestibulo-ocular reflex, optokinetic nystagmus, and smooth pursuit. Rev. Physiol. Biochem. Pharmacol. **106**: 37–123.
17. SIMPSON, J. I. 1984. The accessory optic system. Annu. Rev. Neurosci. **7**: 31–41.
18. KATO, I., K. HARADA, T. HASEGAWA & T. IGARASHI. 1988. Role of the nucleus of the optic tract of monkeys in relation to optokinetic nystagmus and optokinetic after-nystagmus. Brain Res. **474**: 116–126.
19. SCHIFF, D., B. COHEN, J. BUTTNER-ENNEVER & V. MATSUO. 1990. Effects of lesions of the nucleus of the optic tract on optokinetic nystagmus and after-nystagmus in the monkey. Exp. Brain Res. **79**: 225–239.
20. MUSTARI, M. & A. F. FUCHS. 1990. Discharge patterns of neurons in the pretectal nucleus of the optic tract (NOT) in the behaving primate. J. Neurophysiol. **64**: 77–90.
21. ZEE, D. S., R. J. TUSA, S. J. HERDMAN, P. H. BUTLER & G. GÜCER. 1985. The acute and chronic effects of bilateral occipital lobectomy upon eye movements in monkey. Adv. Bioscie. **57**: 267–274.
22. GRÜSSER, O.-J. & U. GRÜSSER-CORNEHLS. 1972. Interaction of vestibular and visual inputs in the visual system. Prog. Brain Res. **37**: 573–583.
23. YOUNG, L. R., J. DICHGANS, R. MURPHY & TH. BRANDT. 1973. Interaction of optokinetic and vestibular stimuli in motion perception. Acta Otolaryngol. **76**: 24–31.
24. BERTHOZ, A., B. PAVARD & L. R. YOUNG. 1975. Perception of linear horizontal self-motion induced by peripheral vision (linear vection): basic characteristics and visual-vestibular interactions. Exp. Brain Res. **23**: 471–489.
25. BUIZZA, A., A. LÉGER, J. DROULEZ, A. BERTHOZ & R. SCHMID. 1980. Influence of otolithic stimulation by horizontal linear acceleration on optokinetic nystagmus and visual motion perception. Exp. Brain Res. **39**: 165–176.

26. BLES, W. & J. M. B. V. DE JONG. 1982. Cervico-vestubular and visuo-vestibular interaction. Acta Otolaryngol. **94:** 61–72.
27. TOKUNAGA, O. 1977. The influence of linear acceleration on optokinetic nystagmus in human subjects. Acta Otolaryngol. **84:** 338–343.
28. RESCHKE, M. F. & D. E. PARKER. 1987. Effects of prolonged weightlessness on self-motion perception and eye movements evoked by roll and pitch. Aviat. Space Environ. Med. **58**(Suppl): A153–A158.
29. POST, R. B. 1986. Induced motion considered as a visually induced oculogyral illusion. Perception. **15:** 131–138.
30. BALOH, R. W., L. LANGHOFER, V. HONRUBIA & R. D. YEE. 1980. On-line analysis of eye movements using a digital computer. Aviat. Space Environ. Med. **51:** 563–567.
31. TEUBER, H-L. 1960. Perception. *In* Handbook of Physiology. J. Field, H. W. Magoud & V. E. Hall, Eds. **3**(Section 1): 1595–1668. American Physiology Society. Washington, D.C.
32. YEE, R. D., R. W. BALOH, V. HONRUBIA & H. A. JENKINS. 1982. Pathophysiology of optokinetic nystagmus. *In* Nystagmus and Vertigo: Clinical Approaches to the Patient with Dizziness. V. Honrubia & M. A. B. Brazier, Eds.: 251–275 Academic Press. New York, N.Y.

Recovery of Postural Equilibrium Control following Spaceflight[a]

WILLIAM H. PALOSKI, [b] MILLARD F. RESCHKE,[c]
F. OWEN BLACK,[d] D. DENINE DOXEY,[b]
AND DEBORAH L. HARM[c]

[b]Neurosciences Laboratory
KRUG Life Sciences, Inc.
1290 Hercules, Suite 120
Houston, Texas 77058

[c]Space Biomedical Research Institute
National Aeronautics and Space Administration
Johnson Space Center
Houston, Texas 77058

[d]R. S. Dow Neurological Sciences Institute
Good Samaritan Hospital
1040 Northwest 22nd Avenue
Portland, Oregon 97210

INTRODUCTION

Posture and gait instabilities have been reported in astronauts returning from spaceflight at least since the Apollo era.[1] Many studies have been performed to determine the causes of these instabilities. Data from these studies provide evidence that sustained exposure to microgravity (1) reduces the necessity for postural reflexes in major leg muscles, resulting in central nervous system (CNS) adaptation eliminating those reflexes;[2] (2) changes the "bias" on the otolith organs, resulting in CNS reinterpretation of signals from those acceleration sensors to provide an environmentally appropriate response;[3] (3) reduces static and dynamic postural inputs from the proprioceptive system;[4] and (4) reduces levels of tonic activity in both the soleus and anterior tibialis muscles.[5]

Postflight postural instabilities are thought to result from in-flight adaptive changes in CNS processing of sensory inputs from the vestibular, proprioceptive, and visual systems. These adaptive changes are a natural neurophysiological response to (1) the sudden loss of spatial orientation information provided by gravitational stimulation to the otolith organs[6] and (2) the modification of locomotion/propulsion strategies necessary to control movements from one location to another within the spacecraft.[7] These changes result in a CNS that is optimized for orientation and movement in the zero-g environment.

Upon return to the terrestrial gravitational environment, the adapted astronaut's CNS is no longer "tuned" properly for the task of maintaining his body center of gravity over his base of support. Before preflight stability levels can be reachieved,

[a]This work was supported by the NASA Extended Duration Orbiter Medical Project: PWC No. 106-70-0211. Dr. Paloski and Ms. Doxey were supported in part by NASA Contract No. NAS9-18492. Dr. Black was supported in part by his Javits Award: NIH NINCD Grant No. DC-00205.

747

the in-flight adaptive changes must be reversed via a postflight readaptation period. Anderson *et al.,* for instance, reported that Spacelab-1 crew members demonstrated an alteration in postural equilibrium control strategy from an ankle torque control strategy preflight to a less stable hip sway control strategy immediately postflight.[8] The hip sway strategy gradually diminished over the first few postflight days, as the crew members returned to their preflight ankle torque control strategy.

Because postflight postural instabilities may present operational hazards, particularly if emergency egress is required soon after landing, it is important to characterize the dynamics of the postflight readaptation period. To that end, the present study was designed to systematically evaluate postural stability in a group of astronauts before and after spaceflight. The data presented in this paper provide some quantitative insight into the magnitude of postflight postural instability and the time course of recovery of preflight postural equilibrium control levels.

METHODS

Thirteen crew members from six separate missions ranging from 4 to 10 days in duration participated in this study (TABLE 1). Each subject was studied at approximately 60, 30, and 10 days before launch (L-60, L-30, and L-10) in order to establish preflight postural stability levels. Postflight measurements began as soon as practical after return from space (R). Ten of the 13 subjects were studied on the day of landing (R + 0) within 3.75 hours of orbiter wheels stop (mean = 2.62 hours). Two of the 13 were flying aboard a vehicle that was directed to an alternate landing site and could not be studied on R + 0. The 13th subject (as well as 1 of the 2 subjects landing at the alternate site) suffered from earth readaptation sickness with symptoms debilitating enough to preclude R + 0 measurements. Of the 10 subjects studied on R + 0, 4 were studied twice in order to help establish the rapid time constant for readaptation. The second landing day measurement (R + 0b) was performed approximately 1.5 hours after the first. Subsequent to the landing day, each subject was studied at either 24 or 48 hours after wheels stop (R + 1 or R + 2), 96 hours after wheels stop (R + 4), and 192 hours after wheels stop (R + 8). Some deviations to this measurement schedule

TABLE 1. Subject Demographic Data and Postflight Measurement Times

Subject			Measurement Times (hours after wheels stop)				
Number	Age	Sex	R+0a	R+0b	R+1,2	R+4	R+8
1	44	M	3:00	—	47:15	—	215:45
2	34	M	3:00	—	25:00	101:30	192:00
3	44	M	1:50	—	24:30	—	192:45
4	41	M	2:00	3:15	48:00	96:45	191:00
5	40	M	2:30	3:40	47:00	95:30	192:00
6	42	M	2:45	—	17:45	97:00	192:30
7	43	M	3:10	—	17:20	97:30	193:00
8	44	M	3:45	—	18:10	96:30	193:30
9	34	M	—	—	24:30	91:15	187:30
10	46	M	—	—	24:00	91:00	189:30
11	31	F	—	—	49:30	94:30	168:30
12	39	M	1:40	3:15	48:30	95:00	167:30
13	44	M	2:35	4:15	47:30	95:30	166:30

TABLE 2. Sensory Organization Test Conditions

Test Number	Sensory Feedback Status		
	Visual	Proprioceptive	Vestibular
1	Normal	Normal	Normal
2	Absent	Normal	Normal
3	Sway referenced	Normal	Normal
4	Normal	Sway referenced	Normal
5	Absent	Sway referenced	Normal
6	Sway referenced	Sway referenced	Normal

were required in order to accommodate higher priority postflight crew member activities; the actual measurement times are listed in TABLE 1.

All pre- and postflight measurements were made using a modified commercial dynamic posturography system (EquiTest™; Neurocom International, Clackamas, Oregon). Modifications to the commercial system included addition of (1) an electromyographic (EMG) system for monitoring activity in the four major antigravity muscle groups of the left leg, (2) sway bars to monitor the anterior-posterior (A-P) excursions of the hips and shoulders, and (3) a head set containing sensors for monitoring angular velocity of the head in the pitch and roll planes, and headphones for providing a pink noise designed to block external auditory cues to the subject.

At each test session, the subject was first exposed to a set of motor tests in which his postural responses to sudden support surface movements were recorded. Following this, he was subjected to a series of six sensory organization tests (TABLE 2) in which his ability to maintain stable upright posture was assessed under normal and reduced/altered sensory feedback conditions. Only data from the sensory organization tests are presented here.

Each sensory organization test was presented to the subject for a 20-second period. A single trial of test 1 was always presented first, and a single trial of test 2 was always presented second. Following this, three trials of each of the other four tests were presented. The order of the four tests was randomized within each trial.

During each 20-second test period, the instantaneous A-P center of mass location was computed using data provided by the posture platform force transducers. During sensory tests 3–6, visual and/or ankle proprioceptive feedback was reduced or eliminated by "sway referencing" the visual surround and/or footplate to the subject's A-P center of mass displacements.[9] The peak-to-peak excursions of the center of mass sway (P-P sway) over each 20-second trial were extracted and used to compute a measure of postural stability known as the equilibrium score:

$$\text{Equilibrium Score} = \left[1 - \frac{P - P\,\text{sway}}{12.5}\right] \times 100$$

where 12.5 is the maximum stable sway amplitude expected in a normal population. Thus, the equilibrium score is directly related to postural stability. To provide an overall assessment of the subject's postural stability at each test session, a composite equilibrium score was computed by summing the subject's individual equilibrium scores from each of the 14 sensory organization test trials. Furthermore, to facilitate comparison of postflight readaptation time constants from subjects having widely varied preflight composite equilibrium scores, all scores for each subject were normalized by dividing each session score by the average of the L-60, L-30, and L-10 scores.

RESULTS

The mean preflight composite equilibrium score of 821 ± 12 (standard error of the mean) was similar to the mean of 798 reported by the platform manufacturer for a normal population; however, the individual preflight means spanned a relatively wide range (from 636 ± 24 to 899 ± 29). Thus, in order to facilitate comparison of temporal changes from subject to subject, all data for each subject were normalized to his/her preflight mean.

All pre- and postflight composite equilibrium data for a typical subject are shown in FIGURE 1. The three preflight scores demonstrate the stability of the measurement system; the subject's mean preflight score (879 ± 8) was above average for a normal population. The first postflight measurement was made 2.5 hours after wheels stop and indicated that the subject's postural stability would be considered clinically abnormal, falling below the ratio of the 95th percentile score to the mean score in a normative population. At this measurement time, the subject reported feeling far more stable than he had at wheels stop, suggesting that readaptation had already begun and was proceeding rapidly. This subjective report of rapid initial readaptation was corroborated quantitatively during the second postflight test session when the composite equilibrium score was found to have increased during the 1.16 hour period following the first postflight measurement. Despite the rapid initial readaptation, the subject's postural stability did not return to preflight levels until the measurement session at 96 hours after wheels stop. This suggests that the initial rapid phase gives way to a slower secondary phase before readaptation is complete.

Normalized composite equilibrium data from all 10 subjects having landing day

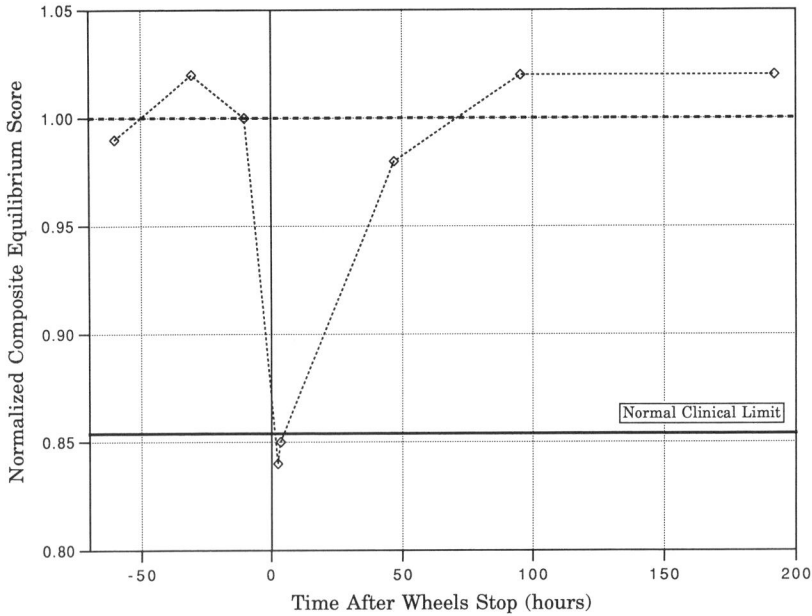

FIGURE 1. Pre- and postflight normalized composite equilibrium data for a typical subject. Note that time before flight is measured in days, not hours.

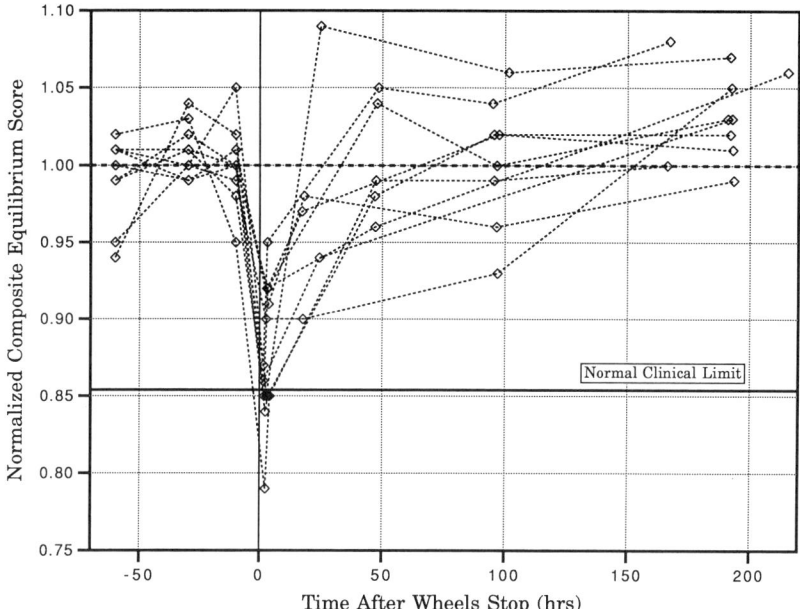

FIGURE 2. Normalized composite equilibrium data from all 10 subjects having landing day measurement sessions. Compared to their preflight measurements, every subject in this group exhibited a substantial decrease in postural stability on landing day; 4 of the 10 exhibited clinically abnormal scores. Note that time before flight is measured in days, not hours.

measurement sessions are presented in FIGURE 2. Compared to their preflight measurements, every subject in this group exhibited a substantial decrease in postural stability on landing day; 4 of the 10 had clinically abnormal scores. All subjects reported similar subjective feelings of rapidly increasing stability; many reported feeling much more stable ("almost back to normal") by the first postflight test session. As in the illustrative subject above, these reports of rapid initial readaptation were corroborated quantitatively in each of the 4 subjects studied twice on landing day. Although there was some variability in the time required to reachieve preflight stability levels, they were reachieved in all subjects by 8 days after wheels stop.

Based on these results, postflight readaptation was modeled as a double exponential process:

$$\text{Score} = K_0 - K_1 e^{-K_2 t} - K_3 e^{-K_4 t}$$

where K_0 was expected to be close to 1.0 (the normalized preflight composite equilibrium score), K_1 and K_3 represent the proportion of readaptation accounted for by the slow and fast phase processes, and K_2 and K_4 are the slow and fast rate constants. Postflight data from all 13 test subjects were fit to this model using the Levenberg-Marquardt nonlinear least-squares technique.[10] FIGURE 3 presents the results of this exercise, which suggest that (1) at wheels stop, the average returning crew member is far below the limit of clinical normality, (2) the initial rapid phase of readaptation has a time constant on the order of 2.7 hours, and accounts for about

50% of the postural instability, and (3) the slower secondary phase of readaptation has a time constant on the order of 100 hours and also accounts for about 50% of the postural instability.

DISCUSSION

Relative to preflight performance levels, postural stability was found to be disrupted early after landing in every subject studied. In 4 of the 10 subjects studied

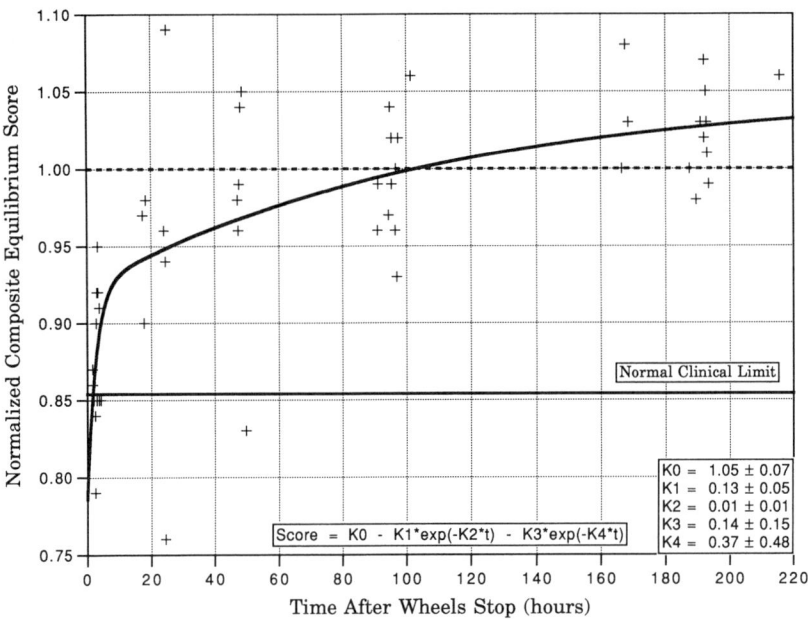

FIGURE 3. Results of fitting postflight readaptation data from all 13 test subjects to a double exponential model. The curve fit suggests that (1) at wheels stop, the average returning crew member's postural stability is far below the limit of clinical normality, (2) the rapid initial phase of readaptation has a time constant on the order of 2.7 hours and accounts for about 50% of the postural instability, and (3) the slower secondary phase of readaptation has a time constant on the order of 100 hours and also accounts for about 50% of the postural instability.

on landing day, postural stability measurements were clinically abnormal. Most subjects reported feeling even more unstable at wheels stop, approximately 2.5 hours before the first measurement session, suggesting that postural instabilities may indeed present postflight operational hazards.

Control of postural equilibrium is normally maintained through central nervous system integration of sensory inputs from the visual, vestibular, and proprioceptive systems.[9] During space flight, the microgravitational environment causes a distortion of the world observed by these neurosensory systems, and a modification in the requirements for maintaining spatial orientation and for locomotion/propulsion. Fortunately, the CNS is plastic, that is, it is able to adapt to new environments over time. The result of neurosensory adaptive changes during space flight is a CNS that is optimized for processing sensory inputs and controlling spatial orientation and locomotion/propulsion in the zero-*g* environment. Unfortunately, when sufficient time is spent on orbit for neurosensory adaptation to occur, then, upon return to the terrestrial environment, the world will again be distorted to the neurosensory systems of the astronaut. A period of neurosensory readaptation to one *g* must therefore occur. During the readaptation period, the ability of an astronaut to perform even simple tasks requiring stable postural equilibrium may be diminished. Complex tasks, such as rapidly egressing from the orbiter in an emergency, may be very difficult to achieve, particularly if visual system inputs are eliminated or distorted by darkness or smoke.

The degree to which in-flight alterations in postural equilibrium control systems present postflight operational hazards can only be assessed after sufficient quantitative data regarding the magnitudes, time constants, dependence on mission length, and dependence on previous space flight experience of these effects have been gathered and analyzed. Previously available data are scant; however, there is some indication that a considerable amount of readaptation occurs within the first 12 hours after return,[4,2,8] but that the readaptation process may not be complete even by the seventh postflight day.[11–13] The purpose of the present study was to obtain the data required to quantitatively assess the magnitude of postural instability immediately postflight and the time course of the terrestrial neurosensory readaptation process.

The results suggest that postflight readaptation of postural equilibrium control follows a two-phase time course. Initially, there is a very rapid recovery phase, having a time constant on the order of 3 hours and accounting for approximately 50% of the postural instability. This is followed by a slower recovery phase, having a time constant on the order of 100 hours. These quantitative results are consistent with the subjective reports of the crew members.

The choice of a double exponential recovery model to describe the observed postflight stability data was driven primarily by the statistical characteristics of the data and the subjective reports of the crew members. Although this model may accurately reflect readaptation time constants for underlying neurosensory posture control loops, correlation between model parameters and specific neurophysiological structures and/or processes was not attempted. While the most appropriate independent parameter in the model should be "environmental interaction" time, it was necessary to use elapsed time since wheels stop in the curve fits presented here. Some of the variability in the data may be explained by the difference between these two measures.

In summary, postural equilibrium control was found to be seriously disrupted immediately following spaceflight missions of 4 to 10 days duration, owing primarily to in-flight neurosensory adaptation to the zero-*g* environment. Readaptation to the terrestrial environment was found to begin immediately upon landing, and to proceed rapidly for the first 10–12 hours and then more slowly for the subsequent 2–4 days until preflight stability levels are reachieved. We conclude that, despite the likelihood of varying recovery time constants of the individual sensory systems involved in postural control, the overall postflight recovery of postural stability follows a predictable time course.

SUMMARY

Decreased postural stability is observed in most astronauts immediately following spaceflight. Because ataxia may present postflight operational hazards, it is important to determine the incidence of postural instability immediately following landing and the dynamics of recovery of normal postural equilibrium control. It is postulated that postflight postural instability results from in-flight adaptive changes in central nervous system (CNS) processing of sensory information from the visual, vestibular, and proprioceptive systems. The purpose of the present investigation was to determine the magnitude and time course of postflight recovery of postural equilibrium control and, hence, readaptation of CNS processing of sensory information. Thirteen crew members from six spaceflight missions were studied pre- and postflight using a modified commercial posturography system. Postural equilibrium control was found to be seriously disrupted immediately following spaceflight in all subjects. Readaptation to the terrestrial environment began immediately upon landing, proceeded rapidly for the first 10–12 hours, and then proceeded much more slowly for the subsequent 2–4 days until preflight stability levels were reachieved. It is concluded that the overall postflight recovery of postural stability follows a predictable time course.

REFERENCES

1. HOMICK, J. L. & E. F. MILLER. 1975. Apollo flight crew vestibular assessment. *In* Biomedical Results of Apollo. NASA SP-368.
2. RESCHKE, M. F., D. J. ANDERSON & J. L. HOMICK. 1986. Vestibulo-spinal response modification as determined with the H-reflex during the Spacelab-1 flight. Exp. Brain Res. **64**: 367–379.
3. RESCHKE, M. F., D. J. ANDERSON & J. L. HOMICK. 1984. Vestibulospinal reflexes as a function of microgravity. Science **225**: 212–213.
4. YOUNG, L. R., C. M. OMAN, D. G. D. WATT, K. E. MONEY & B. K. LICHTENBERG. 1984. Spatial orientation in weightlessness and readaptation to earth's gravity. Science **225**: 205–208.
5. CLEMENT, G., V. S. GURFINKEL & F. LESTIENNE. 1985. Mechanisms of posture control in weightlessness. *In* Vestibular and Visual Control on Posture and Locomotor Equilibrium. M. Igarashi & F. O. Black, Eds.: 158–163. Karger. Basel, Switzerland.
6. PARKER, D. E., M. F. RESCHKE, A. P. ARROTT, J. L. HOMICK & B. K. LICHTENBERG. 1985. Otolith tilt-translation reinterpretation following prolonged weightlessness: implications for preflight training. Aviat. Space Environ. Med. **56**: 601.
7. ROLL, J. P., K. POPOV, V. GURFINKEL, M. LIPSHITS, C. ANDRE-DESHAYS, J. C. GILHODES & C. QUONIAM. 1991. Body proprioceptive references in weightlessness as studied by muscle tendon vibration. *In* Life Sciences Research in Space. ESA SP-307.
8. ANDERSON, D. J., M. F. RESCHKE, J. L. HOMICK & S. A. S. WERNESS. 1986. Dynamic posture analysis of Spacelab-1 crew members. Exp. Brain Res. **64**: 380–391.
9. NASHNER, L. M., F. O. BLACK & C. WALL. 1982. Adaptation to altered support and visual conditions during stance: patients with vestibular deficits. J. Neurosci. **2**(5): 536–544.
10. MARQUARDT, D. W. 1963. J. Soc. Ind. Appl. Math. **11**: 431–441.
11. KOZLOVSKAYA, I. B., Y. V. KREIDICH & O. V. RAKHAM. 1981. Mechanisms of the effects of weightlessness on the motor system of man. Physiologist Suppl. **24**: 59–64.
12. KOZLOVSKAYA, I. B., I. F. ASLANOVA, L. S. GRIGORIEVA & Y. V. KREIDICH. 1982. Experimental analysis of motor effects of weightlessness. Physiologist Suppl. **25**: 49–52.
13. KOZLOVSKAYA, I. B., I. F. ASLANOVA, V. A. BARMIN, L. S. GRIGORIEVA, G. I. GEVLICH, A. V. KIRENSKAYA & M. G. SIROTA. 1983. The nature and characteristics of gravitational ataxia. Physiologist Suppl. **26**: 108–109.

Response of Otolith-Related Neurons in Bilateral Vestibular Nucleus of Acute Hemilabyrinthectomized Cats to Off-Vertical Axis Rotations[a]

Y. S. CHAN AND Y. M. CHEUNG

Department of Physiology
Faculty of Medicine
University of Hong Kong
5 Sassoon Road
Hong Kong

INTRODUCTION

The vestibular nuclei receive inputs from the ipsilateral and contralateral labyrinths.[1] The functional interaction of inputs from the bilateral labyrinths to the vestibular nucleus has been extensively studied in the semicircular canal system.[2–7] In comparison to the canal system, the interaction of bilateral otolith inputs at the level of the vestibular nucleus has received less attention. In cats, signals arising from the otolith organs can reach the contralateral vestibular nuclei via commissural connections[8] or a long-loop pathway involving the spinoreticulocerebellar connection.[9] Also, electrical stimulation of the labyrinth resulted in excitation of otolith-related neurons in the contralateral vestibular nucleus,[10–12] thus suggesting that cross-excitation operates between the bilateral vestibular nuclei of cats.[12] However, the relative contribution of inputs from otoliths on each side to the vestibular nucleus remains unclear. One approach is to eliminate the peripheral input from one side by hemilabyrinthectomy (HL) which serves to offset the balance between the bilateral labyrinthine inputs to the central nervous system[13,14] and creates severe motor asymmetries which usually disappear with time.[15] Under such a state, it becomes apparent that there is an increase in reliance on signals from the remaining intact labyrinth and other sensory inputs. It is also of considerable interest to elucidate how information processing is readjusted in the central nervous system, in particular in the vestibular nucleus, during compensation.

In acute HL cats, bilateral vestibular nuclear neurons have been shown to be responsive to roll tilts,[16–18] pitch tilts,[19] and off-vertical axis rotations (OVAR).[20,21] Compared with those of labyrinth-intact cats, these units showed a decrease in mean resting discharge which was significantly lower on the lesioned side than on the labyrinth-intact side.[17–19] This is in line with the decrease in glucose utilization (as indicated by ^{14}C-deoxyglucose activity) in the vestibular nuclei on the lesioned side as compared with that on the labyrinth-intact side of HL rats.[22] Also, central vestibular neurons on the lesioned[17] and labyrinth-intact[18] sides of HL cats showed an imbalance in response sensitivity to sinusoidal roll tilt. These asymmetric properties of

[a]This study was supported by research grants from the Lee Wing Tat Medical Research Fund, Sun Yat Sen Foundation Fund for Medical Research, Research Grants Committee of the University of Hong Kong, and University and Polytechnic Grants Committee (UPGC).

755

otolith-related neurons in the bilateral vestibular nuclei have been suggested to be the basis of asymmetric motor deficits induced by HL.

In cats with bilateral labyrinths intact, a map of sensory space near the horizontal plane, presumably representing inputs from the utricular receptors, was found at the level of the vestibular nucleus.[23-25] The distribution of these best-response orientations, deduced from OVAR[24,26] and wobble rotations,[23,27] covers 360° on the plane of rotation. This spatial coding property of otolith-related central vestibular neurons provides the basis for understanding the contribution of otolith inputs to the contralateral vestibular nucleus after HL. The dynamic behavior of otolith-related central vestibular neurons in response to low velocity OVAR[28] and sinusoidal vertical rotations of a higher frequency range[27] has been investigated in cats with bilateral labyrinths intact. The behavior of these neurons during OVAR after acute HL remains to be explored, although the behavior of a group of Deiters' neurons responding to sinusoidal roll tilt has been investigated in HL cats.[17,18] The aim of this study is to investigate whether central vestibular neurons on both the lesioned and labyrinth-intact sides of acute HL cats can encode spatial orientation of the head near the horizontal plane as perceived by the remaining intact otolith receptors. We also attempt to quantify the response properties of these neurons during slow OVAR and then compare the results with those obtained in control cats.

Some preliminary results have been presented in abstract form.[21]

METHODS

Experiments were performed on adult cats, the right labyrinth of which was extracranially destroyed on the day of electrophysiological experiment. Results obtained were compared with those obtained from cats with bilateral labyrinths intact (the control group). The cats were precollicularly decerebrated under light ketamine anesthesia. The head of the animal, mounted on a stereotaxic frame fixed on a tilt table system, was maintained on the earth's horizontal plane, while the cervical vertebral column of the animal was oriented perpendicular to gravity.[29] The C2 vertebra was clamped to the stereotaxic frame while the extended limbs were secured rigidly to the table. Relative movements of the neck, trunk and limbs were minimized with a plaster cast. The table was rotated with constant velocities (1.7–15°/second) in the clockwise (CW) and counterclockwise (CCW) directions at an axis (3 cm caudal to the midpoint of the interaural axis) tilted at a fixed angle (5–25°) from the earth's vertical.

Extracellular activities of neurons in the vestibular nucleus were recorded with tungsten microelectrodes inserted through the intact cerebellum. In HL cats, recordings were performed between 4 and 30 hours after HL on the lesioned side (on the right side) for one group of animals, and on the labyrinth-intact side (on the left side) for another group. Units were tested for their latencies to shocks delivered to the oval window on the labyrinth-intact side of HL animals or on the side ipsilateral to unitary recording in control animals. The response of each unit was studied during at least two successive 360° OVAR in both the CW and CCW directions. The positions of the CW and CCW discharge maxima and the response gain (half peak-to-peak discharge modulation per degree of head tilt) were evaluated off-line with an IBM-AT computer. The orientation of the best response of each unit on the plane of rotation was represented by the averaged position in the small sector delineated by the CW and CCW discharge maxima.[25,26] The spatial difference between the orientation of the best response and that of either discharge maximum was taken as the response phase of individual units.[28] The locations of neurons were reconstructed from cresyl violet–stained sagittal serial sections with respect to coordinates of the tracks and small electrolytic lesions made at the end of the experiment. The extent of the unilateral labyrinth destruction was confirmed at the end of each experiment by gross dissection.

RESULTS

In animals with intact bilateral labyrinths (the control), all the tilt-sensitive central vestibular units tested ($n = 83$) responded to OVAR ($1.7°/$second, $10°$ head tilt) in both the CW and CCW directions with position-dependent modulation in discharge rates.[25] In HL animals, however, only 42 out of 82 central vestibular units on the lesioned side[28] and 54 out of 77 units on the labyrinth-intact side responded to OVAR ($1.7°/$second, $10°$ head tilt) in both directions. The remaining units showed discharge modulation only in response to OVAR of one direction but not to both (FIGURE 1).

Bidirectionally Sensitive Units

In all preparations, about two-thirds of the bidirectionally sensitive units showed response lead at $1.7°/$second ($10°$ head tilt) and the remaining units showed response lag. A few units that showed response lead between $45°$ and $65°$ at such a stimulus parameter were not included in this study. This criterion was adopted to avoid possible inclusion of canal neurons which exhibited phase lead position by at least $90°$.[27] The mean $\text{gain}_{CCW}/\text{gain}_{CW}$ ratio of the bidirectionally sensitive units was 0.91 on the lesioned side of HL cats, 1.08 on the labyrinth-intact side of HL cats, and 1.14 for the control. In the control cats, 73 units had a $\text{gain}_{CCW}/\text{gain}_{CW}$ ratio of $<1.5:1$ while 10 units had a gain ratio of 1.5–$2:1$. Only one unit had a gain ratio of $2.1:1$. For HL cats, the mean gain value (of CW and CCW rotations) was 0.74 impulses/second per deg on the lesioned side and 0.79 impulses/second per deg on the labyrinth-intact side, while the corresponding value was 1.49 impulses/second per deg for the control.

In all preparations, the best-response orientations of the bidirectionally sensitive units were found to be distributed all over the plane of rotation (FIGURE 2). For units with their best-response orientations along the anteroposterior plane, a greater proportion of units showed type 1 (excited by head-up tilt) than type 2 (excited by head-down tilt) response in control cats, while a reverse pattern was observed on the labyrinth-intact side of HL cats. On the lesioned side of HL cats, an equal proportion of type 1 and type 2 units was found. For units with their best-response orientations along the transverse plane, there was a preponderance of units showing type α response (excited by ipsilateral ear-down tilt, i.e., left ear down in our preparation) over those showing type β response (excited by contralateral ear-down tilt) on both the labyrinth-intact side of HL cats and in control cats. A similar proportion (i.e., more units were excited with right-ear-down tilt) was also observed on the lesioned side (right side) of HL cats.

Within the velocity range tested (from $1.7°/$second to $15°/$second), these units were bidirectionally sensitive with a gain ratio $<2:1$. The best-response orientation of each of these units remained fairly stable. With increase in the velocity of OVAR,

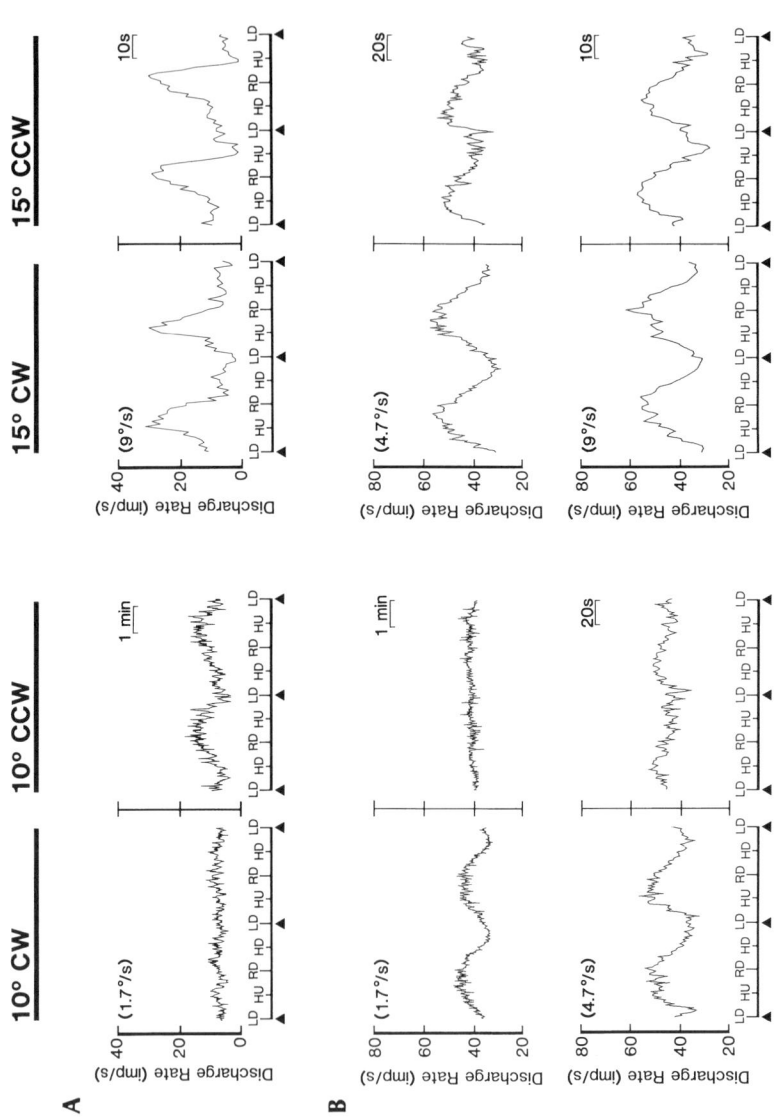

FIGURE 1. Responses of unidirectionally sensitive central vestibular units of HL cats to OVAR. At 1.7°/second and 10° tilt, the unit shown in A responded only to CCW rotation while that in B responded only to CW rotation. Both units became bidirectionally sensitive with increase in the stimulus parameter.

the gain of these units showed a flat response. In addition, most otolith-related units in control cats and on the labyrinth-intact side of HL cats showed progressive response lag while a smaller portion of units showed stable response lead. On the lesioned side of HL cats, however, a roughly equal proportion of units showed these two types of response pattern. In addition, the scatter of the response phase of units

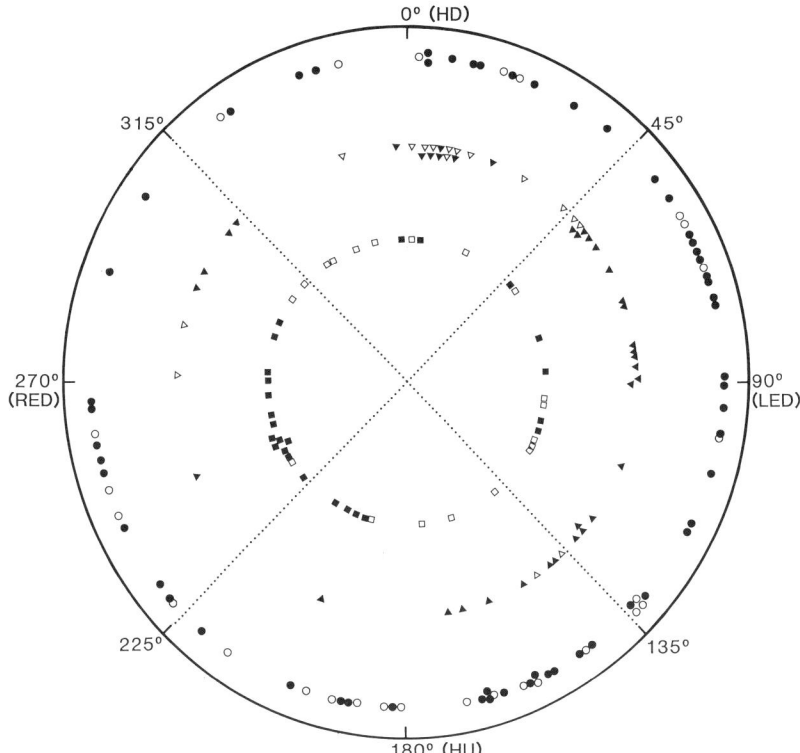

FIGURE 2. Polar diagram showing on the plane of rotation the distribution of the best-response orientations of bidirectionally sensitive central vestibular units in labyrinth-intact control cats (outer ring; circles), on the labyrinth-intact side of HL cats (middle ring; triangles), and on the lesioned side of HL cats (inner ring; squares). The response orientations were obtained during 1.7°/second OVAR (at 10° tilt). Dotted lines arbitrarily divide the plane of rotation into 4 quadrants, head down (HD), left-ear down (LED), head up (HU), and right-ear down (RED). Filled symbols indicate units with response lead while open ones indicate units with response lag. Most of the data on control animals is adapted from Reference 26.

on the lesioned side of HL cats was larger than those in control cats and on the labyrinth-intact side of HL cats. The response pattern of stable lead units of HL cats was similar to that of the controls. However, as compared with the control, progressive response lag units on the labyrinth-intact side of HL cats showed a parallel shift of 5° towards larger lag over the velocity range tested and those on the lesioned side showed a shift of 20° towards larger lag. (See FIGURE 3.)

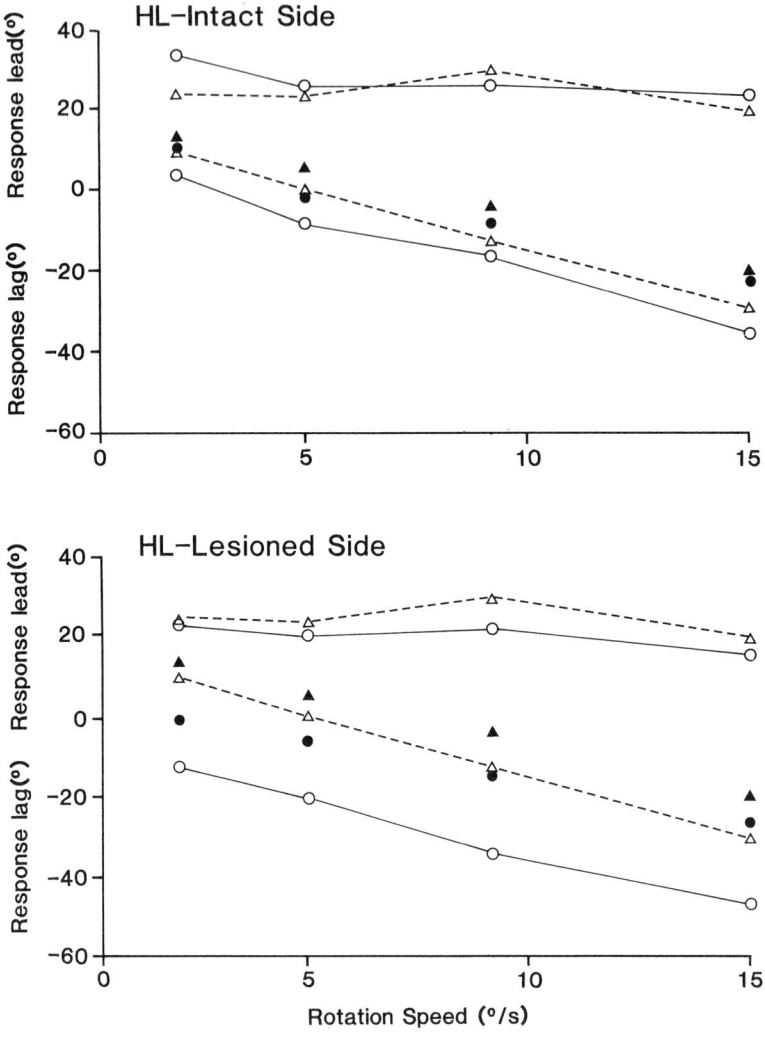

FIGURE 3. Averaged response phases of bidirectionally sensitive central vestibular units in HL cats (circles; solid lines) during OVAR (at 10° tilt). Data obtained from labyrinth-intact control cats (triangles; dotted lines) are included in each graph for comparison. 0° indicates the response is in phase with the best-response orientation of the unit. In each of the three preparations, two populations of units were found. One showed stable response lead (the upper set of solid and dotted lines in each graph) and the other showed progressive response lag (the lower set of solid and dotted lines in each graph). Units showing the former response pattern were 4 for control cats, 6 on labyrinth-intact side of HL cats, and 6 on lesioned side of HL cats; units showing the latter response pattern were correspondingly 16, 21, and 10. All units included were responsive throughout the velocity range tested. The filled symbols represent the mean response of all units in the respective preparation. The data on control animals are adapted from Reference 28.

Unidirectionally Sensitive Units

Some units on the lesioned side ($n = 27$) and labyrinth-intact side ($n = 16$) of HL cats responded only to 1.7°/second OVAR (10° head tilt) of one direction (either the CW or CCW direction) but not to both (FIGURE 1). Such a response pattern was not observed in the units tested in control cats. About 50% of the unidirectionally sensitive units of HL cats were also tested with higher velocity OVAR. With increase in the velocity of OVAR and/or amplitude of head tilt, some of these units remained unidirectionally sensitive throughout the velocity range studied while others became bidirectionally sensitive. This latter group of units exhibited (1) gain$_{CCW}$/gain$_{CW}$ ratio of <1.5:1, (2) stable best-response orientation, and (3) either stable response lead of <45° or progressive phase lag.

In HL cats, some units did not respond to 1.7°/second OVAR (10° head tilt) in either the CW or CCW direction, but became responsive to OVAR of one direction or both directions with increase in either the velocity of rotation or amplitude of head tilt. Such response pattern was not observed in control cats.

DISCUSSION

Central vestibular units of HL cats were found to respond to bidirectional OVAR which introduces a sequential change in the gravitational vector acting on the otolith organs. The response phase of most of these units was similar to those of otolith neurons in the vestibular nucleus.[23,27] As central vestibular units receiving canal inputs exhibited phase-led position ∼90° or more (cf. References 27 and 30), we excluded the few units with response lead >45° at low velocity from our data analysis so that only units receiving otolith inputs were analyzed in the present study. Our results thus suggest that inputs from the otoliths are operative at the level of the vestibular nuclei bilaterally. In HL cats, a fair number of central vestibular units responded to low-velocity OVAR of one direction but not to both. Such a pattern was not observed in control cats in which all central vestibular units tested were sensitive to OVAR in both the CW and CCW directions. With wobble rotations as the mode of stimulation, however, most central vestibular neurons of cats with bilateral labyrinths intact responded to CW and CCW rotations with amplitude disparity ratios of <2:1, 7/110 neurons responded to bidirectional rotations with amplitude disparity >2:1, and 2/110 neurons responded to wobble rotations of one direction but not to both.[27] Such asymmetric behavior of the neurons observed usually at 0.1–0.2 Hz[30,31] was suggested to result from convergent semicircular and otolith inputs carrying different spatial and temporal information (STC cells);[30] the unidirectionally sensitive behavior of our neurons was observed at a much lower velocity. Within the low-velocity range employed in our mode of stimulation, the relative contribution of canal inputs to the units studied should be minimal although canal afferents have been shown to exhibit position responses.[32,33] In addition, some of the unidirectionally sensitive units in our study became bidirectionally sensitive with increase in velocity and exhibited response properties similar to those of otolith neurons (cf. Reference 27). It is therefore reasonable to assume that at least some of the observed unidirectionally sensitive units receive otolith inputs within the low-velocity range tested. It is probable that the asymmetric response of these units resulted from convergent otolith inputs with very misaligned spatial and temporal information arising from the remaining intact labyrinth. One plausible deduction is that the response of these neurons in the labyrinth-intact state normally depends on bilaterally convergent inputs of the otoliths. Only with the elimination of the inputs arising from one of the labyrinths would such an asymmetric pattern be manifested. The complex information processing which might occur at the level of the otolith receptors[34] has also to be considered. In addition, the fidelity of some central neural mechanism serving as a comparator for symmetric inputs from the bilateral otolith organs could have been disturbed with HL (see Reference 35). Our results suggest that convergent inputs from the bilateral otoliths play an important role in the coding of head position in space.

On both the lesioned and labyrinth-intact sides of HL cats, the best-response orientations of central vestibular neurons pointed in all directions on the plane of rotation. Their spatial distribution is in agreement with that of the polarization vectors observed in otolith afferents[36,37] and best-response vectors of vestibular nuclear neurons[24,27] (see however Reference 23) of animals with bilateral labyrinths intact. Thus our data establish that the otoliths on one side can provide within the bilateral vestibular nuclei a coordinate frame of head positions with respect to gravity. Nevertheless, there was a significant reduction in response gain of these units with HL indicating the relative importance of the bilateral otoliths. With sinusoidal roll tilt, a predominance of Deiters' units showing type α response over those showing type β response was found in control cats[38] and on the lesioned side of HL cats.[17] On the labyrinth-intact side of HL cats, similar response pattern was observed in the rostroventral part of the Deiters' nucleus while a reverse pattern was found in the dorsocaudal part of the Deiters' nucleus which presumably is under the inhibitory control of the cerebellar vermis.[18] In our population of units, which were sampled throughout the Deiters' and inferior vestibular nuclei, a predominance of type α over type β response was observed in control cats as well as in both sides of HL cats. This finding suggests the existence of a predominant crossed contribution from units with type β response. The crossed otolith inputs may provide complementary and supplementary information about head positional changes along the transverse axis. For units with their best-response orientations along the anteroposterior axis, a greater proportion showed type 1 than type 2 response in control cats. The relative distribution of these unit types changed with HL. In alert cats, it has been shown that there was an asymmetry in the number of vertical nystagmus beats especially during low-frequency otolithic stimulation[39] (see also References 40 and 41). It was hypothesized that such eye movement pattern serves to compensate for the difference in the asymmetric inertia load of the head during upward and downward head movements.[39] Although speculative, this idea suggests an explanation for the occurrence of asymmetric distribution of central vestibular units along the anteroposterior axis. Whether or not such a neural pattern is being employed in the final adjustment of head motor control remains to be proven.

Using wobble rotations (0.01–2 Hz), it has been shown in canal-plugged cats that otolith-related vestibular nuclear neurons showed two types of dynamic behavior.[23] Some neurons were characterized by stable phase lead (otolith afferentlike pattern) while others showed progressive phase lag (otolith-forelimb reflexlike pattern). In control cats, the responses we observed during OVAR of a lower frequency spectrum are in general agreement with the otolith-related central vestibular neuronal responses to wobble rotations.[23,27] Such patterns were also found in our populations of HL units. In control cats as well as on the labyrinth-intact side of HL cats, most units showed progressive response lag and only a small proportion of units showed stable response lead. On the lesioned side of HL cats, however, there was a significant increase in the proportion of units exhibiting stable lead. This implies that in the intact state, both otolith afferentlike information and reflexlike information are conveyed to the contralateral vestibular nuclear complex during low-velocity OVAR. Also, the observation that the response of those HL units with progressive lag displayed a parallel shift towards greater lag with increase in the velocity of rotation suggests that bilateral otolith inputs are required to shape the dynamic properties of central vestibular neurons. It should be noted that in response to sinusoidal roll tilt (0.008–0.32 Hz), Deiters' units of control cats showed a progressive phase lag (re maximum animal displacement)[38] while those on both the lesioned side[17] and labyrinth-intact side[18] of HL cats displayed a shift in response phase towards position. These changes in vestibular nuclear neuronal activities after acute HL also suggest the importance of the functional interaction of convergent inputs from the bilateral otoliths.

ACKNOWLEDGMENTS

The authors wish to express their gratitude to Prof. J. C. Hwang for his encouragement. We also thank Mr. S. M. Chan and Mr. N. Y. S. Tan for their helpful technical assistance.

REFERENCES

1. GACEK, R. R. 1969. The course and central termination of first order neurons supplying vestibular endorgans in the cat. Acta Oto-Laryngol. Suppl. **254:** 1–66.
2. SHIMAZU, H. & W. PRECHT. 1966. Inhibition of central vestibular neurons from the contralateral labyrinth and its mediating pathway. J. Neurophysiol. **29:** 467–492.
3. MARKHAM, C. H., T. YAGI & I. S. CURTHOYS. 1977. The contribution of the contralateral labyrinth to second order vestibular neuronal activity in the cat. Brain Res. **138:** 99–109.
4. GALIANA, H. L., H. FLOHR & G. MELVILL JONES. 1984. A revaluation of intervestibular nuclear coupling: its role in vestibular compensation. J. Neurophysiol. **51:** 242–259.
5. REID, S., C. MAIOLI & W. PRECHT. 1984. Vestibular nuclear neuron activity in chronically hemilabyrinthectomized cats. Acta Oto-Laryngol. **98:** 1–13.
6. SMITH, P. F. & I. S. CURTHOYS. 1988. Neuronal activity in the contralateral medial vestibular nucleus of the guinea pig following unilateral labyrinthectomy. Brain Res. **444:** 95–307.
7. SMITH, P. F. & I. S. CURTHOYS. 1988. Neuronal activity in the ipsilateral medial vestibular nucleus of the guinea pig following unilateral labyrinthectomy. Brain Res. **444:** 308–319.
8. GACEK, R. R. 1978. Location of commissural neurons in the vestibular nuclei of the cat. Exp. Neurol. **59:** 479–491.
9. POMPEIANO, O. 1979. Neck and macular labyrinthine influences on the cervical spinoreticulocerebellar pathway. Prog. Brain Res. **50:** 501–514.
10. SHIMAZU, H. & C. M. SMITH. 1971. Cerebellar and labyrinthine influences on single vestibular neurones identified by natural stimuli. J. Neurophysiol. **34:** 493–508.
11. CHAN, Y. S., J. C. HWANG & Y. M. CHEUNG. 1977. Crossed sacculo-ocular pathway via the Deiters' nucleus in cats. Brain Res. Bull. **2:** 1–6.
12. WILSON, V. J., R. R. GACEK, Y. UCHINO & A. J. SUSSWEIN. 1978. Properties of central vestibular neurons fired by stimulation of the saccular nerve. Brain Res. **143:** 251–261.
13. PRECHT, W. 1986. Recovery of some vestibuloocular and vestibulospinal functions following unilateral labyrinthectomy. Prog. Brain Res. **64:** 381–389.
14. SMITH, P. F. & I. S. CURTHOYS. 1989. Mechanisms of recovery following unilateral labyrinthectomy: a review. Brain Res. Rev. **14:** 155–180.
15. SCHAEFFER, K.-P. & D. L. MEYER. 1974. Compensation of vestibular lesions. *In* Handbook of Sensory Physiology. Vestibular System. Psychophysics, Applied Aspects and General Interpretations. H. H. Kornhuber, Ed. **6**(Part 2): 463–490. Springer-Verlag. Berlin, Germany.

16. HOSHINO, K. & O. POMPEIANO. 1977. Crossed responses of lateral vestibular neurons to macular labyrinthine stimulation. Brain Res. **131:** 152–157.

17. XERRI, C., S. GIANNI, D. MANZONI & O. POMPEIANO. 1983. Central compensation of vestibular deficits. I. Response characteristics of lateral vestibular neurons to roll tilt after ipsilateral labyrinth deafferentation. J. Neurophysiol. **50:** 428–448.

18. LACOUR, M., D. MANZONI, O. POMPEIANO & C. XERRI. 1985. Central compensation of vestibular deficits. III. Response characteristics of lateral vestibular neurons to roll tilt after contralateral labyrinth deafferentation. J. Neurophysiol. **54:** 988–1005.

19. CHAN, Y. S., Y. M. CHEUNG & J. C. HWANG. 1983. The influence of unilateral otolith organs on central vestibular neuronal activities in the cat. Neurosci. Lett. Suppl. **12:** S31.

20. CHAN, Y. S., Y. M. CHEUNG & J. C. HWANG. 1988. Responses of vestibular nuclear neurons to bidirectional off-vertical axis rotations in normal and hemilabyrinthecto-mized cats. *In* Basic and Applied Aspects of Vestibular Function. J. C. Hwang, N. G. Daunton & V. J. Wilson, Eds.: 63–72. Hong Kong University Press. Hong Kong.

21. CHAN, Y. S., Y. M. CHEUNG & C. W. CHEN. 1990. Responses of bilateral vestibular nuclear neurons of acute hemilabyrinthectomized cats to off-vertical axis rotations. Soc. Neurosci. Abstr. **16:** 734.

22. LLINAS, R. & K. WALTON. 1978. Vestibular compensation: a distributed property of the central nervous system. *In* Integration in the Nervous System. H. Asanuma & V. J. Wilson, Eds.: 145–166. Igaku-shoin. Tokyo, Japan.

23. SCHOR, R. H., A. D. MILLER & D. L. TOMKO. 1984. Responses to head tilt in cat central vestibular neurons. I. Direction of maximum sensitivity. J. Neurophysiol. **51:** 136–146.

24. CHAN, Y. S., Y. M. CHEUNG & J. C. HWANG. 1985. Effect of tilt on the response of neuronal activity within the cat vestibular nuclei during slow and constant velocity rotation. Brain Res. **345:** 271–278.

25. CHAN, Y. S., Y. M. CHEUNG & J. C. HWANG. 1987. Response characteristics of neurons in the cat vestibular nuclei during slow and constant velocity off-vertical axes rotations in the clockwise and counterclockwise directions. Brain Res. **406:** 294–301.

26. CHAN, Y. S., Y. M. CHEUNG & J. C. HWANG. 1988. Unit responses to bidirectional off-vertical axes rotations in central vestibular and cerebellar fastigial nuclei. Prog. Brain Res. **76:** 67–75.

27. KASPER, J., R. H. SCHOR & V. J. WILSON. 1988. Responses of vestibular neurons to head rotations in vertical planes. I. Response to vestibular stimulation. J. Neurophysiol. **60:** 1753–1764.

28. CHAN, Y. S. & J. C. HWANG. 1991. Response characteristics of central vestibular neurons and compensatory mechanisms following hemilabyrinthectomy. *In* The Head-Neck Sensory-Motor System. A. Berthoz, W. Graf & P. P. Vidal, Eds. Oxford University Press. New York, N.Y.

29. VIDAL, P. P., W. GRAF & A. BERTHOZ. 1986. The orientation of the cervical vertebral column in unrestrained awake animals. I. Resting position. Exp. Brain Res. **61:** 549–559.

30. BAKER, J., J. GOLDBERG, G. HERMANN & B. PETERSON. 1984. Spatial and temporal response properties of secondary neurons that receive convergent input in vestibular nuclei of alert cats. Brain Res. **294:** 138–143.

31. WILSON, V. J., Y. YAMAGATA, B. J. YATES, R. H. SCHOR & S. NONAKA. 1990. Response of vestibular neurons to head rotations in vertical planes. III. Response of vestibulocollic neurons to vestibular and neck stimulation. J. Neurophysiol. **64:** 1695–1703.

32. ESTES, M. S., R. H. I. BLANKS & C. H. MARKHAM. 1975. Physiologic characteristics of vestibular first order canal neurons in the cat. I. Response plane determination and resting discharge characteristics. J. Neurophysiol. **38:** 1232–1249.

33. PERACHIO, A. A. & M. J. CORREIA. 1983. Response of semicircular and otolith afferents to small angle static head tilts in the gerbil. Brain Res. **280:** 287–298.

34. ROSS, M. D. 1985. Anatomic evidence for peripheral neural processing in mammalian graviceptors. Aviat. Space Environ. Med. **56:** 338–343.

35. RAPHAN, T. & C. SCHNABOLK. 1988. Modeling slow phase velocity generation during off-vertical axis rotation. Ann. N.Y. Acad. Sci. **545:** 29–50.

36. FERNANDEZ, C. & J. M. GOLDBERG. 1976. Physiology of peripheral neurons innervating otolith organs of the squirrel monkey. I. Response to static tilts and to long-duration centrifugal force. J. Neurophysiol. **39:** 970–984.
37. TOMKO, D. L., R. J. PETERKA & R. H. SCHOR. 1981. Responses to head tilt in cat eighth nerve afferents. Exp. Brain Res. **41:** 216–221.
38. BOYLE, R. & O. POMPEIANO. 1980. Reciprocal responses to sinusoidal tilt of neurons in Deiters' nucleus and their dynamic characteristics. Arch. Ital. Biol. **118:** 1–32.
39. TOMKO, D. L., C. WALL III, F. R. ROBINSON & J. P. STAAB. 1988. Influence of gravity on cat vertical vestibulo-ocular reflex. Exp. Brain Res. **69:** 307–314.
40. DARLOT, C., J. LOPEZ-BORNEO & D. TRACEY. 1981. Asymmetry of vertical vestibular nystagmus in the cat. Exp. Brain Res. **41:** 420–426.
41. MATSUO, V. & B. COHEN. 1984. Vertical optokinetic nystagmus and vestibular nystagmus in the monkey: up-down asymmetry and effects of gravity. Exp. Brain Res. **53:** 197–216.

Responses of Direction-Selective Neurons in Monkey Cortex to Self-Induced Visual Motion

ROGER G. ERICKSON[a] AND PETER THIER

Neurology Clinic
University of Tubingen
D-7400 Tubingen, Germany

In most vertebrates, visuospatial stability is preserved primarily via synergistic vestibuloocular and optokinetic reflexes (VOR and OKR) which attempt to stabilize the retina in space. Together, these reflexes provide visuospatial stability by minimizing low-velocity visual motion signals caused by voluntary movements. Although these measures alone are adequate for afoveate animals (e.g., rabbits) which lack voluntary eye movements, the evolution of voluntary tracking eye movements presented the visual system of foveate species with an additional challenge. Slow, pursuit eye movements override the OKR and VOR and force the visual system to receive and somehow discriminate the low-velocity visual motion of background images that is incident to every eye movement not made in complete darkness.[8] The facts that the OKR does not counteract smooth pursuit[10] and that spatial orientation is preserved during pursuit (otherwise the background would appear to move during slow eye movements) both indicate that the brain is able to discriminate self-induced visual motion signals. There are at least several methods by which this might be accomplished, including the possibility that one or more mechanisms act directly upon visual neurons to make them responsive only to externally induced retinal image motion. To test this hypothesis we compared the visual responses of directional and nondirectional cortical neurons from visual areas V4, MT, and MST within the superior temporal sulcus of awake, macaque monkeys while they performed two behavioral tasks. In one instance, referred to as the "passive" condition, an appropriate visual stimulus was moved through the receptive field of a visual neuron while the monkey maintained stable fixation on a stationary target. In the second task, referred to as the "active" condition, the receptive field was swept at the same speed across a stationary stimulus by requiring the animal to pursue a target moving at speeds of 20 deg/second or less. By this method we created in the laboratory conditions similar to those in which humans are usually unaware of the self-induced visual motion of images within the plane of fixation.

In each case, the animal sat in a standard primate chair facing a tangent screen where stimuli were displayed with projectors and moved using galvanometer-mounted mirrors. (Details of all procedures have been described previously, see Reference 5.) Briefly, during experiments the head was immobilized, metal microelectrodes for extracellular recordings were lowered through a chronically implanted cylinder, and eye movements were monitored with a magnetic search coil implanted under the sclera.[13]

Examples of six neurons tested by these procedures are shown in FIGURE 1. In

---

[a] Present affiliation: Laboratory of Neurophysiology, National Institute of Mental Health, Post Office Box 289, Poolesville, Maryland 20837.

each case, from A–F, the responses to stimulus motion back and forth across the visual field are shown, with the passive (eyes-stationary) responses on the left and the active (eyes-moving) responses on the right. The four neurons in parts A–D had similar responses to both passive and active testing, and the remaining two, E and F, responded only to passive visual stimulation. The first example is a nondirectional V4 cell, the next two are direction-selective MT cells with prominant excitation for motion in a preferred direction, and the last three are similar directional cells from MST, the last of which, however, was modulated primarily by inhibitory rather than excitatory influences (complete data not shown, but compare the passive and active conditions for this cell). These examples are typical of our overall result. All nondirectional cells, as well as many direction selective cells, were found to respond equally well to passively and actively generated retinal slip. A portion of directional cells, however, were found to be "passive only," i.e., selectively responsive only to externally induced retinal image motion (see Reference 5). These results therefore confirm our initial hypothesis that some visual neurons appear to be unresponsive to self-induced retinal slip, and also indicate that this property is observed only among directional cells (see also Reference 6).

One of the most interesting aspects of our finding concerns the distribution of this property among those cortical areas that contain large proportions of direction-selective neurons. To describe the incidence of passive-only visual cells in MT and MST we first sought to devise a single index summarizing the comparison between active and passive testing. The most simple approach, shown in FIGURE 2A, is to take the ratio of the active and passive visual responses in the preferred visual direction, i.e., Ar/Pr, where Pr indicates the passive response in the preferred visual direction (e.g., to the right) and Ar indicates the response to actively induced visual motion in the same direction (i.e., rightward visual motion induced by a leftward eye movement). In the course of the experiments, however, we became aware of several factors indicating that this approach might be too simplistic. For example, many visual cells express directionality as a combination of inhibitory (null direction = opposite to preferred) as well as excitatory (preferred direction) effects. In addition, some cells within the lateral portion of MST (see below) exhibited nonvisual activity related to the pursuit eye movements used to induce active visual responses. Nevertheless, the distribution of values shown in the top panel of FIGURE 2A is centered near 1.0, indicating that the response to passively and actively induced retinal image motion in the preferred direction is usually similar in MT cells. In contrast, the distributions for both the lateral and especially the dorsal portions of MST include more index values below 0.5, indicating that the active response in the preferred visual direction is often much smaller than the passive response. This nonuniform distribution of passive-only neurons immediately suggests the possibility that a sense of visuospatial stability could be derived from the activity of neurons from a restricted subset of cortical visual areas.

As noted above, the formula Ar/Pr ignores interesting and relevant information (see FIGURES 3–5, below) which, we found, could be revealed by more comprehensive indices (see also Reference 5). Since there are several mechanisms that could contribute to reducing responsiveness to actively induced visual motion, an index was needed that would be independent of the mechanism used. We therefore devised an alternative index, (Pm − Am)/Pm, which was sensitive to directionality as well as to changes in response rates. The variables Pm and Am indicate the differences between response rates in the preferred and null visual directions during passive and active testing respectively (e.g., Pm = passive preferred–passive null; for active testing the preferred visual direction was that determined in the passive trials, irrespective of whether direction preference appeared to have changed due to

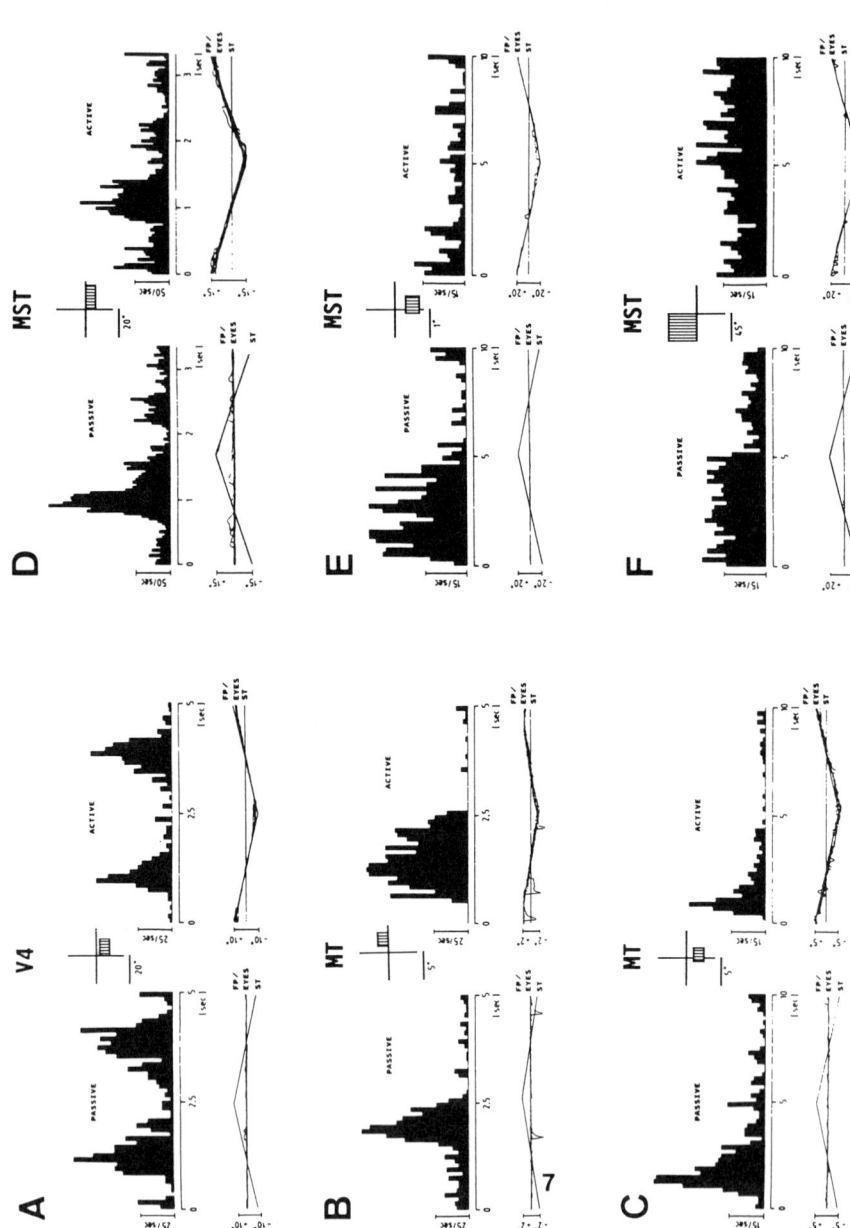

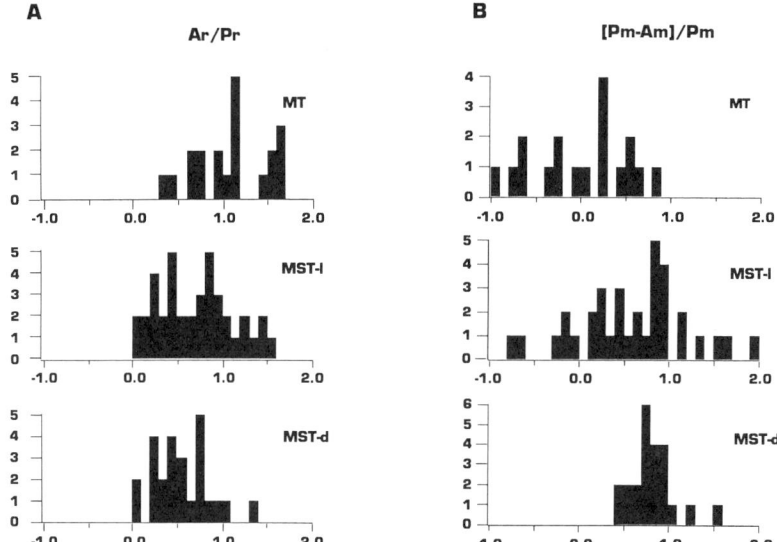

FIGURE 2. Distribution of p ~ a index values for areas MT, MST-1, and MST-d. **A:** Index calculated from Ar/Pr where Ar and Pr indicate average response rate in the preferred visual direction for the active and passive conditions respectively. **B:** Index calculated from (Pm − Am)/ Pm where Pm and Am indicate preferred response–null response for the passive and active conditions respectively. Horizontal axis = index value, vertical axis = number of cells.

factors such as extraretinal pursuit signals). For this index, a value near zero indicates similar modulation during active and passive testing, and values near 1.0 indicate negligible active modulation. The index values obtained (FIGURE 2B) are distributed between −1.0 and 0.8, for MT, between −0.8 and 2.0 for MST-I, and more tightly clustered about roughly 0.8 for MST-d. Comparison of these distributions confirms that cells preferentially responsive to passively induced retinal image motion are much more common in MST than MT.

The fact that index values above 1.0 in FIGURE 2B are found only in MST, and especially in MST-I, is of interest because these extreme positive index values identify cells with tonic directional responses during pursuit and call attention to one

---

FIGURE 1. Response histograms of visual neurons from the superior temporal sulcus to stimulus motion back and forth through the receptive field. Each section includes the cell response (top, histogram bin width = 100 mseconds in this and following figures) and records of stimulus (ST), target (FP), and eye position below. Scales at left indicate spikes per second (top) and position in degrees of visual angle (below). In each case, passive visual responses are shown on the left, active visual responses on the right, and the receptive field size is indicated at top center. **A:** Responses of a nondirectional V4 cell that is unable to discriminate passively from actively generated retinal image motion (speed 8.0 deg/second). **B,C,** and **D:** Similar responses to passive and active motion in directional cells from MT and MST (stimulus speed 1.6, 2.0, and 17 deg/second). **E** and **F:** MST cells directionally modulated only by passive visual motion (speed 8.0, deg/second). Eye-to-screen distance is 80 cm, and the screen subtends 90 × 90 deg.

possible explanation for passive-only behavior. When observed in the same cell, pursuit-related responses may interfere with visual responses to moving background images. Most MST tracking cells also have directional visual responses, and the visual and tracking responses nearly always have the same preferred direction.[4,15] In addition, pursuit in the nonpreferred or null direction can result in inhibition below resting response rates.[4,12] Therefore, antagonistic interactions may occur when eye movement in the null tracking direction is used to produce concurrent motion of background visual images in the preferred visual direction. FIGURE 3 shows an example of a passive-only neuron that appears capable of using this mechanism. First, a downward preferred pursuit direction during circular tracking in an otherwise totally dark room is demonstrated in part A. In parts B and C, however, downward motion of a long bar is shown to evoke a strong passive visual response (B) that is more vigorous than the response to downward linear pursuit (C). These two interactions may simply cancel during upward pursuit over a stationary background (part D, upward, null direction for pursuit evokes downward, excitatory direction of visual motion), resulting in the absence of active visual responses. Because the active response in the null visual direction (left side of FIGURE 3D) is slightly larger than the active response in the preferred visual direction (right side of FIGURE 3D), the term Am (pref-null) is negative, and (Pm − Am)/Pm becomes greater than 1.

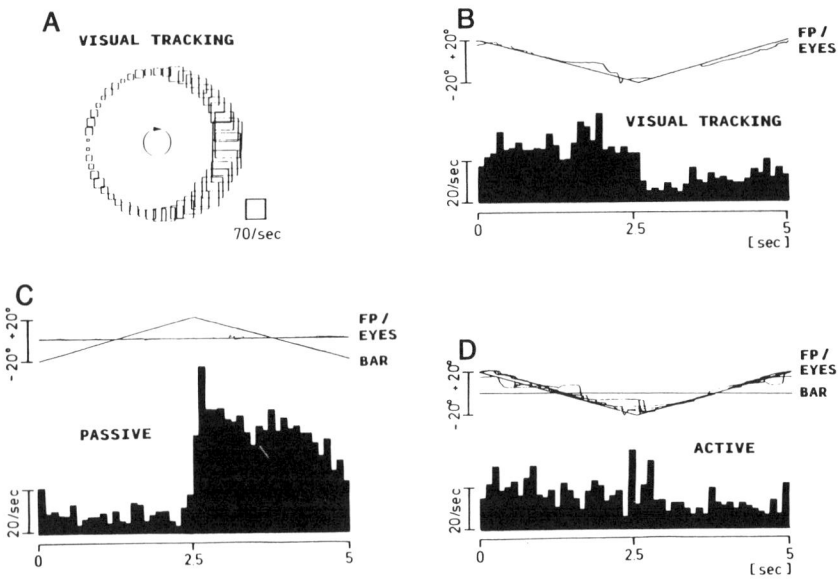

FIGURE 3. Responses of passive-only MST cell exhibiting pursuit responses in complete darkness. A: Plot of response rate as function of eye position during circular tracking (path radius 10 deg, period = 0.2 second, speed = 12.5 deg/second; preferred tracking direction is downward). B: Directional tracking response to downward phase of linear tracking (16 deg/second) in complete darkness except for dim target (projected red laser). C: Directional passive visual response to downward stimulus motion (16 deg/second) through receptive field (same preferred direction for visual and tracking responses). D: Absence, with combined visual stimulation and tracking, of expected active visual response during upward pursuit (16 deg/second) over a stationary pattern, and lack of expected pursuit response during downward eye movement over a stationary pattern.

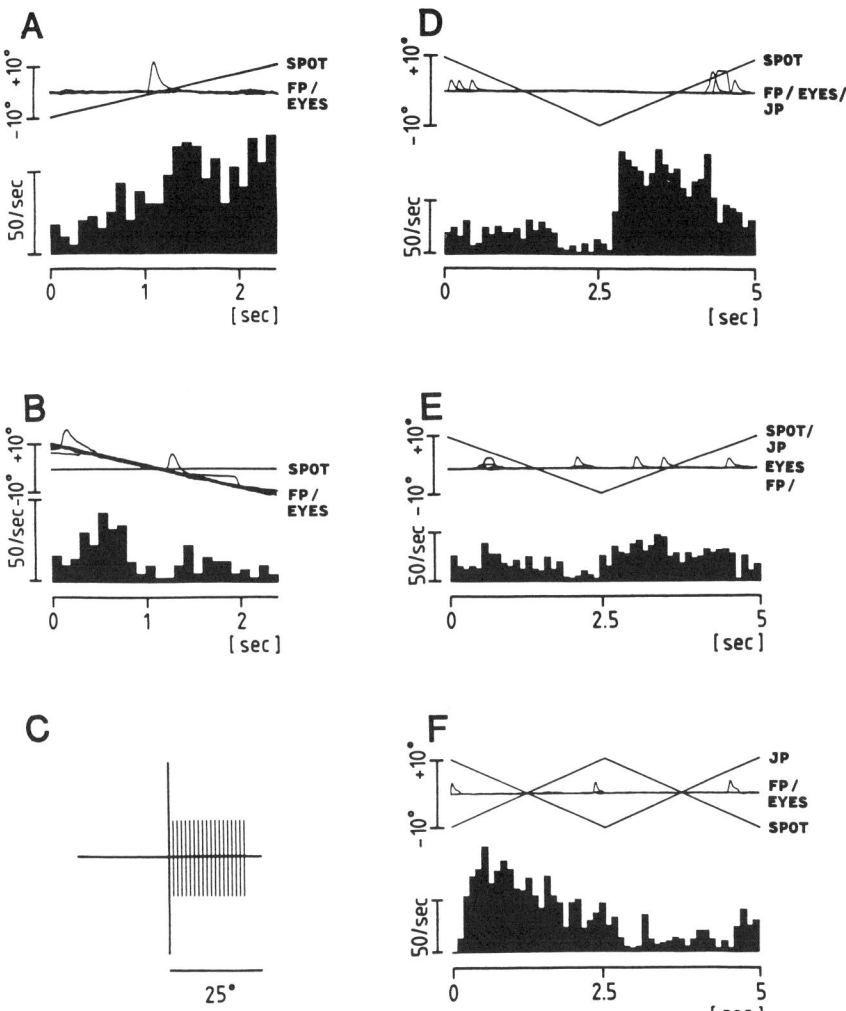

FIGURE 4. Responses of a passive-only MST cell exhibiting antagonistic directional center-surround interactions. **A:** Passive visual response in preferred direction (eyes stationary, spot moves at 8 deg/second). **B:** Lack of corresponding active visual response (spot stationary, eye moves at 8 deg/second). **C:** Receptive field plot. **D:** Passive visual response when a small spot crosses the field while the surround is covered by a stationary dot pattern (JP). **E:** Suppression of passive visual response when spot and pattern move together. **F:** Return of passive response to spot movement when spot and pattern move in opposite directions (eyes stationary in D–F).

Correspondingly larger pursuit-related responses would therefore result in index values increasingly larger than 1.

Tracking responses do not, however, represent the only mechanism capable of influencing responses to self-induced visual motion. Another possibility is the antagonistic direction-selective center-surround interaction found in some MST cells by Tanaka *et al.*[14] FIGURE 4 shows an example of a passive-only cell possessing this property. Part A shows the passive visual response to motion of a spot of light moving in the preferred direction (rightward) across the receptive field (shown in part C), and part B shows the much weaker response to active visual testing as the eyes move

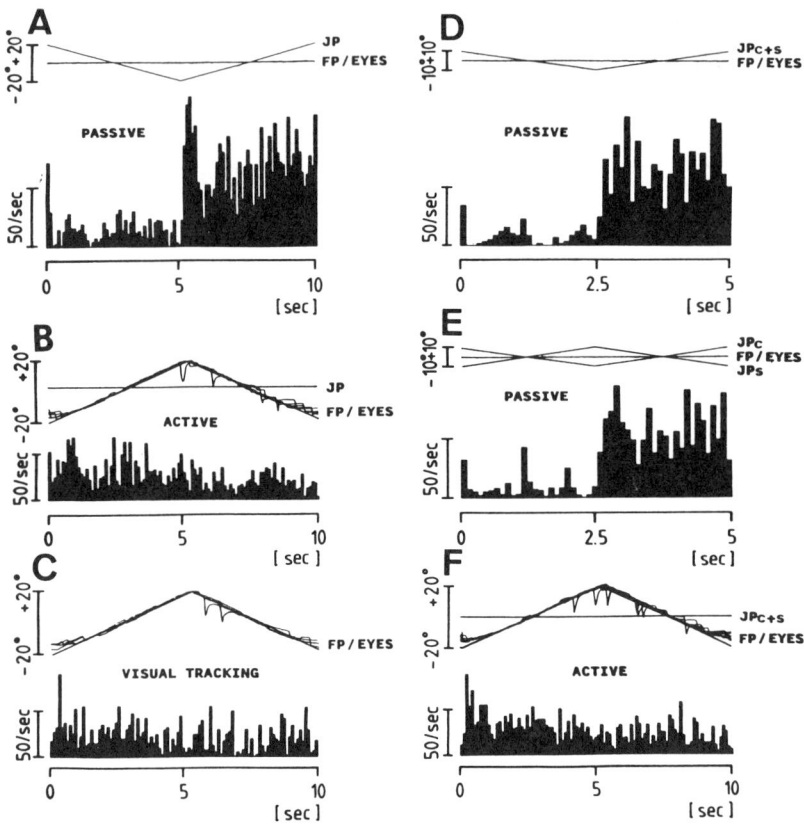

FIGURE 5. Responses of passive-only MST cell possessing neither center-surround visual antagonism nor responses to target motion or eye movement during pursuit. **A:** Passive visual response to upward motion (8 deg/second) of a small dot pattern. **B** and **C:** Absence of active visual response and pursuit-related responses. **D** and **E:** Similar responses to passive visual motion regardless of whether the central and surrounding dot patterns move (8 deg/second) in phase or counterphase. **F:** Absence of active response with both central and surrounding patterns stationary.

to the left and the receptive field moves leftward across the same stimulus. In part D, passive testing is repeated with the addition of a stationary random dot pattern which surrounds, but does not impinge upon, the receptive field. When the central spot and surrounding pattern both move in unison, however, the passive response is markedly attenuated (FIGURE 4, part E) even though the eyes remain stationary. Finally, by moving the spot and pattern simultaneously in opposite directions (part F), the passive visual response reappears. This, then, is an example of a directional cell whose passive-only response profile could arise entirely from visual interactions that prevent the cell from responding to full-field visual motion, thereby allowing discrimination of the visual conditions that usually prevail when the eyes pursue a small target moving past a stationary background.

Finally, some passive-only cells do not seem to use either of the mechanisms just mentioned. For the neuron described in FIGURE 5, strong passive visual responses (part A) and absence of active visual responses (part B) are not associated with overt tracking responses caused by either residual target motion within the receptive field or nonretinal input (part C). In addition, replacing the small dot pattern used for these tests with a receptive-field-sized pattern does not alter the passive response, regardless of whether the center and a surround move in phase (part D) or out of phase (part E). The active visual response also remains extremely weak when repeated with both center and surround patterns stationary (part F).

Together, these examples indicate that purely visual interactions alone can prevent directional neurons from responding to the retinal motion of large background images during pursuit. To maintain spatial orientation during pursuit against a sparsely textured background, however, it may be necessary to sometimes rely on interactions between visual motion of the background and pursuit-related signals. While it may be asked why visual interactions would be needed when a more reliable indicator (nonvisual pursuit signals)[12] is available, there are at least two plausible explanations. First, the presence of purely visual interactions may be necessary for the development and continuous calibration of visual-extraretinal interactions so that the two inputs remain mapped in register for both direction and speed. Second, it is not at all certain that the output of all neurons that display passive-only behavior is useful as a visual indicator of spatial stability. Some may have entirely unrelated uses. For example, index values in FIGURE 2B that are much larger than 1.0 identify cells in which pursuit-related influences outweigh the effect of visual input. Such cells are not useful for monitoring visuospatial stability during smooth pursuit since their responses do not discriminate between pursuit eye movement and movement of the surround. That function may be restricted to those neurons whose index in FIGURE 2B falls near to but below 1. These cells respond when portions of the surround actually move but not during any other circumstance (e.g., FIGURE 5).

Any directional cell exhibiting passive-only visual responses could, however, potentially contribute to selective activation of the OKR so that this reflex would not necessarily act to prevent pursuit. While laterally eyed, afoveate animals such as the rabbit rely primarily upon directionally selective retinal ganglion cells to provide the input to brain stem optokinetic circuits,[3] foveate species such as cats and primates make greater use of the direction selectivity recreated in the cortex in order to alter the characteristics of the OKR.[7,11,16] Since cortical areas containing directional neurons, including those areas studied in this experiment, project to subcortical OKR circuits,[1,2,9] it is possible that our passive-only cells may also contribute to the process of alleviating the potential conflict between those visual inputs driving OKR and pursuit.

REFERENCES

1. BERSON, D. M. & A. GRAYBIEL. 1980. Some cortical and subcortical fiber projections to the accessory optic nuclei in the cat. Neuroscience 5: 2203–2217.
2. CAMPOS-ORTEGA, J. A., W. R. HAYHOW & P. F. DE V. CLUVER. 1970. The descending

projections from the cortical visual fields of *Macaca mulatta* with particular reference to the question of a cortico-lateral geniculate-pathway. Brain Behav. Evol. **3**: 368–414.

3. OYSTER, C., J. I. SIMPSON, E. S. TAKAHASHI & R. E. SOODAK. 1980. Retinal ganglion cells projecting to the rabbit accessory optic system. J. Comp. Neurol. **190**: 49–61.

4. ERICKSON, R. & B. M. DOW. 1989. Foveal tracking cells in the superior temporal sulcus of the macaque monkey. Exp. Brain Res. **78**: 113–131.

5. ERICKSON, R. & P. THIER. 1991. A neuronal correlate of spatial stability during periods of self-induced visual motion. Exp. Brain Res. **86**: 608–616.

6. ERICKSON, R. G. & P. THIER. 1991. Responses of macaque V4 neurons to passively and self-induced retinal image motion. Soc. Neurosci. Abstr. **17**(180.4): 441.

7. GRASSE, K., M. CYNADER & R. M. DOUGLAS. 1984. Alterations of response properties in the lateral and dorsal terminal nuclei of the cat accessory optic system following visual cortex lesions. Exp. Brain Res. **55**: 69–80.

8. HELMHOLTZ, H. 1866. Handbuch der Physiologischen Optik, III. (English translation by J. P. C. Southall. 1962. Helmholtz's Treatise on Physiological Optics. **3**. The Perceptions of Vision. 3rd edit. Dover. New York, N.Y.)

9. HOFFMANN, K. P., C. DISTLER & R. G. ERICKSON. Functional projections from striate cortex and superior temporal sulcus to the nucleus of the optic tract (NOT) and dorsal terminal nucleus of the accessory optic tract (DTN) of Macaque monkey. J. Comp. Neurol. (In press.)

10. KELLER, E. L. & N. S. KHAN. 1986. Smooth-pursuit initiation in the presence of a textured background in monkey. Vision Res. **26**: 943–955.

11. LYNCH, J. C. & J. W. MCLAREN. 1983. Optokinetic nystagmus deficits following parieto-occipital cortex lesions in monkeys. Exp. Brain Res. **49**: 125–130.

12. NEWSOME, W. T., R. H. WURTZ & K. HIDEHIKO. 1988. Relation of cortical areas MT and MST to pursuit eye movements. II. Differentation of retinal from extraretinal inputs. J. Neurophysiol. **60**: 604–620.

13. ROBINSON, D. A. 1963. A method of measuring eye movement using a scleral search coil in a magnetic field. IEEE Trans. Biomed. Eng. **10**: 137–145.

14. TANAKA, K., K. HIKOSAKA, H.-A. SAITO, M. YUKIE, Y. FUKUDA & E. IWAI. 1986. Analysis of wide-field movements in the superior temporal visual areas of the macaque monkey. J. Neurosci. **6**: 134–144.

15. THEIR, P. & R. ERICKSON. 1992. Vestibular input to visual-tracking neurons in area MST of awake monkeys. Ann. N.Y. Acad. Sci. (This volume.)

16. WOOD, C. C., P. D. SPEAR & J. J. BRAUN. 1973. Direction-specific deficits in horizontal optokinetic nystagmus following removal of visual cortex in the cat. Brain Res. **60**: 231–237.

Cerebellar Uvula Involvement in Visual Motion Processing and Smooth Pursuit Control in Monkey

S. J. HEINEN AND E. L. KELLER

Smith-Kettlewell Eye Research Institute
2232 Webster Street
San Francisco, California 94115
and
Department of Electrical Engineering and Comparative Sciences
University of California
Berkeley, California 94720

The cerebellum has been shown to be involved in the generation of most types of eye movements. In particular, the flocculus and lobules VI and VII of the vermis, sometimes called just the posterior vermis, have been most extensively studied in this regard. Neurons have been found in both areas that code aspects of visual and motor signals important for the generation of smooth pursuit, and lesions to these areas produce pursuit deficits.[1,2] The posterior vermis seems to play an important role in saccade generation as well,[3–5] and the flocculus has been shown to be involved in the neural processing for all types of smooth eye movements including pursuit and vestibular.[6,7] Both the vermis and flocculus receive projections from the pontine nuclei, as well as from the nucleus reticularis tegmenti pontis (NRTP), brain stem regions that contain cells responding to visual stimuli and during oculomotor behavior.[8–12]

The dorsal portion of lobule IX of the vermal cerebellum (the dorsal uvula) receives at least as strong an input from visual areas of the pontine nuclei as do the posterior vermis and the flocculus.[13,14] Therefore the uvula might be expected to process information for the generation of visually guided eye movements as well. The uvula in rabbits has been studied for neural correlates of visual surround motion,[15] but no single-unit studies have previously been done to assess uvular involvement in smooth tracking or saccades. However, it has been shown that optokinetic after-nystagmus (OKAN) could no longer be suppressed or "dumped" by visual or tilt stimulation after combined nodulus/uvula lesions.[16]

We therefore set out to determine if the uvula plays a role in the generation of visually guided eye movements using behavioral paradigms, single-unit recording, and lesion techniques previously employed in clarifying the oculomotor relationships of the flocculus and vermis. Surprisingly we found very few cells in the uvula that responded either during smooth pursuit or for saccades. Instead, a large majority of uvular cells responded with relatively long latencies during prolonged optokinetic stimulation. In addition, our lesions to this area produced effects on pursuit different from those seen with lesions in the other two cerebellar oculomotor areas. This suggests that the uvula may not be in the direct pursuit pathway, a role hypothesized for the posterior vermis and the flocculus, but may instead have an indirect effect on the gain of other pathways actually producing smooth eye movements. The uvula then could be a structure that plays a role in the adaptive control of pursuit performance.

METHODS

Three *Macaca fascicularis* monkeys were used for all experiments. Each was surgically implanted with a stainless steel chamber located stereotaxically over the cerebellum for single-unit recording, and a coil of wire around the globe of the eye for monitoring eye position with the search coil technique. For single-unit recording, tungsten microelectrodes were lowered into the uvula through a guard tube which pierced both the dura and the tentorium.

For reversible lesions, 10–15 microliters of lidocaine were perfused through a cannula into the uvula. If an effect on pursuit or optokinetic eye movements was found, lidocaine administration was followed by an injection of 2–4 microliters of 15 micrograms/microliter ibotenic acid.

Standard behavioral paradigms were used to assess single-unit responses and lesion effects. Animals were required to smoothly track a small visual spot which moved at constant velocity over a dim homogeneous background in four directions at speeds of 10, 20, and 40 deg/second at random, or to make saccades to the spot as it was stepped to 5-, 10-, and 20-deg locations. Water was given when animals successfully performed these tasks. Visual responses of neurons were tested by moving small spots or sweeping a large textured background across the visual field while the animal fixated. A drum with random width vertical stripes rotated around the animal at speeds of 40 or 100 deg/second provided constant-velocity optokinetic stimulation.

Eye position signals were differentiated on-line by analog hardware to produce eye velocity traces (bandwidth = 150 Hz), then desaccaded either manually or with software based on an acceleration criterion. Eye position and velocity and target position signals were sampled at 500 Hz for short visual or pursuit trials and at 100 Hz for longer optokinetic trials.

Isolated single units were classified as Purkinje cells (P-cells) if they exhibited an irregular high-frequency simple-spike discharge accompanied by the concomitant low-frequency discharge of a complex spike. It was usually possible to determine when the electrode tip was located outside Purkinje cell layers by the lack of the characteristic, irregular, high-frequency discharge of these large cells, by their wider spike durations, and the absence of complex spikes. Units recorded outside Purkinje cell layers were characterized by more regular spontaneous discharge or by the lack of discharge and by more narrow spike durations. We classified these units as mossy fiber input units although this population may have also included other cerebellar cortical cell types. For the analysis of the unit data, the raw spike trains were convolved with a Gaussian time function resulting in a spike density profile.[17] Modulations of this spike density function were checked for correlation with eye velocity or retinal slip during pursuit and optokinetic trials (or target motion during fixation trials). For the analysis of the optokinetic data, the latency and the time constant of any neural response modulation as quantified by the spike density profiles were measured graphically. A sample of these measures was verified by fitting the spike density function with an exponential waveform using nonlinear optimization techniques.

Following the completion of the unit recordings and the lesions, several electrolytic marker lesions were placed in the cerebellum. Unit recording sites and the sites of chemical lesion placements were reconstructed from the location of these marking lesions.

RESULTS

We recorded from a total 78 P-cells and 14 mossy fibers in the uvula. We were surprised to find that very few (25%) of the units recorded carried signals relating to either pursuit or visual motion of a small spot. Most of the cells that could be modulated by these behaviors exhibited a long latency response to target or pursuit onset, although a few cells had more crisp, short-latency response typical of floccular or vermal units. Most of these (33%), however, were classified as mossy fiber inputs.

By far the most robust stimulus for activating uvular P-cells was prolonged constant-velocity drum rotation. This type of stimulation could either excite or inhibit spontaneous P-cell firing rate for one direction of drum rotation, and usually produced the opposite or no change for the other direction, and this modulation usually began some time after drum onset. These cells also usually did not begin to return to their spontaneous rate until well after the drum lights were turned off following prolonged (approx. 30-second) optokinetic stimulation. FIGURE 1 shows two examples of P-cell discharge during drum rotation, one (FIGURE 1a) shows an decrease in activity following the onset of optokinetic (drum) stimulation, and the other (FIGURE 1b) an increase.

We attempted to quantify the temporal relationship of such P-cell modulation to periods of optokinetic stimulation by determining the latency and time constant of modulation with respect to both drum onset and offset. An example of the results of the curve fit applied to the cell in FIGURE 1b (enclosed box) is shown in expanded form in FIGURE 1c. A summary of the results of similar fits from all cells can be seen in FIGURE 2. The latency of the neural response to drum onset averaged 3.5 seconds, and the mean time constant of the change in neural rate was 2.0 seconds for these cells. No speed tuning was evident for this population, at least for the 40- and 100-deg/second drum rotations that we used. There also was no apparent bias for a given direction of drum rotation in terms of response for the cells recorded on one side of the uvula.

Note that there is a somewhat bimodal distribution to both the latencies and time constants shown in FIGURE 2. Some of these neurons had extremely long temporal courses in their activity, and occasionally a cell would even continue to modulate away from its spontaneous rate throughout the entire period of OKAN in contrast to the cells shown in FIGURE 1.

After the collection of all unit data, we created reversible or more permanent lesions by the infusion of lidocaine and ibotenic acid into the uvula. We were surprised to find that the biggest effect on eye movements as a result of these lesions was on the pursuit tracking ability of the monkeys. Even more surprising was the nature of this effect which was very different from that previously demonstrated on pursuit following lesions to the posterior vermis or flocculus. Our animals showed an *increase* in velocity gain in tracking contralateral to the lesion, which was most dramatic in the open-loop period of pursuit. FIGURE 3 shows averaged tracking trials at 20 deg/second in a monkey before and after a left uvular lesion was created. This animal's open-loop eye acceleration (first 100 mseconds) during rightward pursuit was 461 deg/second^2, compared to its open-loop eye acceleration of 150 deg/second^2 before the lesion. Once steady state was reached the animals usually had a normal velocity gain, but brief periods of retinal slip continued to produce greatly exaggerated periods of eye acceleration in response to this slip. We saw evidence of this heightened open-loop gain in all three of our animals. However, one of our lesions only destroyed the tip of the most dorsal folia of the uvula, and a small portion of

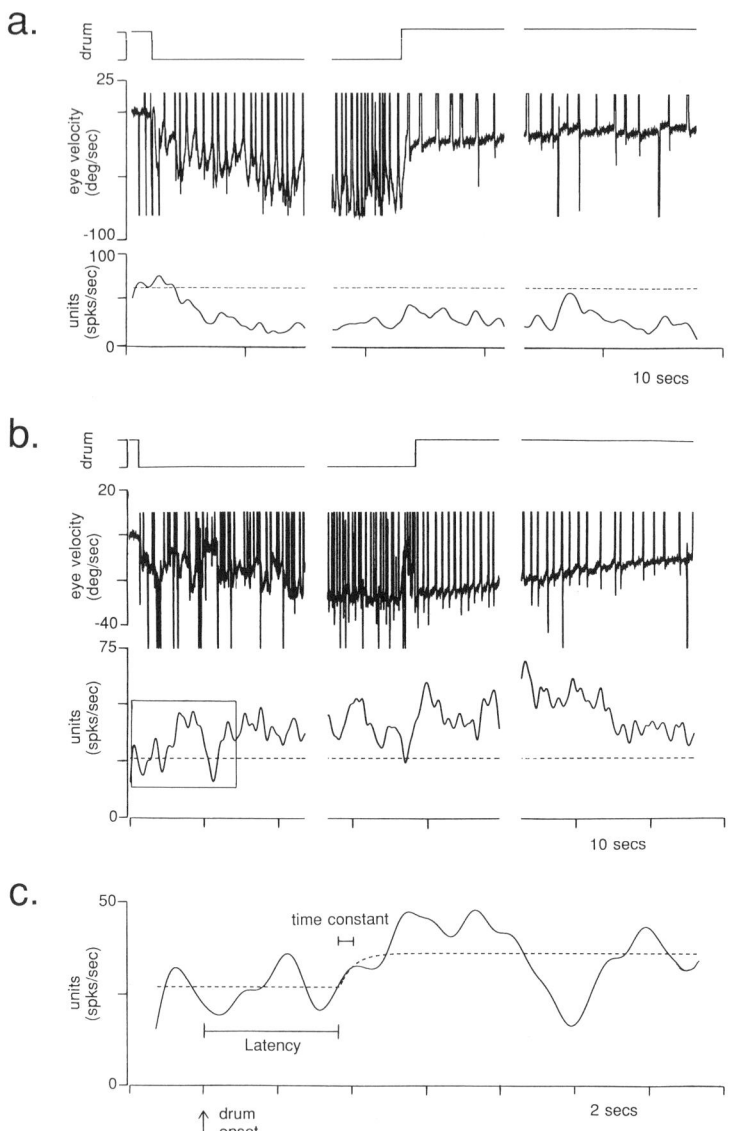

FIGURE 1. Purkinje cell discharge during constant-velocity drum rotation. In **a**, cell is inhibited by leftward drum rotation at 90 deg/second. In **b**, another cell is excited by leftward drum rotation at 40 deg/second. Top trace, drum state (downward steps indicate leftward drum motion, p); middle trace, eye velocity (positive signifies right); bottom trace, spike density. Dashed line is spontaneous discharge. In **c**, boxed-in segment of b is shown expanded to illustrate nonlinear fit of response latency and time constant.

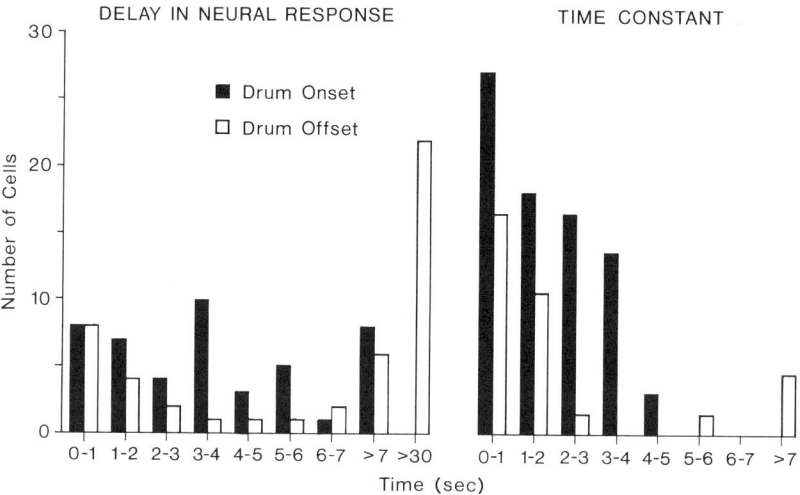

FIGURE 2. Summary of latency (left) and time constants (right) for population of Purkinje cells sampled.

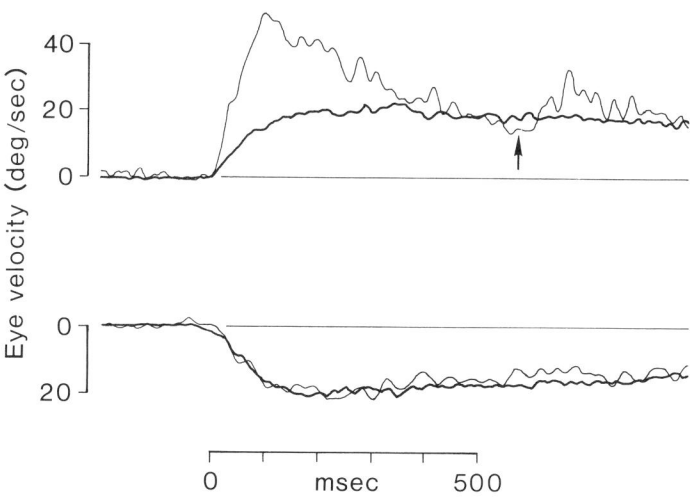

FIGURE 3. Pursuit eye velocity following left uvula lidocaine injection. At top, the animal is pursuing a target spot with a velocity 20 deg/second right. At bottom, the spot velocity is 20 deg/second left. Target motion begins about 100 mseconds before the zero time indication. Traces are average eye velocity responses before (thick line) and after (thin line) injection. Saccades have been removed. The arrow shows a brief period of rightward retinal slip that produced another large episode of rightward eye acceleration.

lobule VIII. In this animal the effects were minimal. In the third animal we saw a substantial increase in rightward open-loop gain with a left uvular lidocaine injection as in the first animal. Interestingly, another lidocaine and subsequent ibotenic acid lesion which affected the midline of the uvula resulted in increased open-loop pursuit response in *both* directions.

Another result of these lesions was an inability to suppress following eye movements with fixation of a small stationary spot during sweep of the large textured background. If background motion was to the right, the left uvular lesion animals were easily pulled off of the fixation point, whereas before the lesion they could almost suppress this motion totally. With background motion to the left, suppression was even better than normal. In FIGURE 4, averaged eye velocity from trials of this type for the same animal shown in FIGURE 2 is displayed, both before and after the lesion. Eye velocity for background motion to the right following the lesion was much greater than normal, as it was higher than 5 deg/second for most of the trial. To the left, eye velocity was less than normal, i.e., this animal suppressed optokinetic

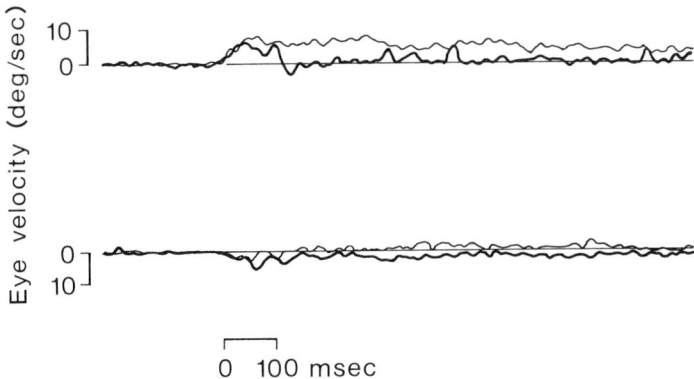

FIGURE 4. Eye velocity following left uvula lidocaine injection in suppression task. Background velocity is 20 deg/second right (top) and left (bottom). Background motion begins about 100 mseconds before the zero time indication. Other conventions as in FIGURE 3.

following movements better than before the lesion. Eye velocity actually switched directions and was opposite in direction to that of the background in this animal late in the trial.

Optokinetic nystagmus (OKN) and OKAN were less affected by the lesions than these other types of eye movements. For drum rotation directed contralateral to the lesion, eye velocity was almost normal during both OKN and OKAN in these animals. There was decreased OKN for drum rotation directed ipsilateral to the lesion; however we do not believe that this was due to an inability of slow eye velocity to build up during the prolonged stimulation. Our drum rotation apparatus had several peripheral fixed objects inside the drum which the animal occasionally attempted to fixate which decreased OKN even before the lesion. During these intermittent periods of attempted fixation, these objects would have become more effective following the lesion due to the monkeys' heightened suppression for this direction of optokinetic stimulation.

DISCUSSION

In summary, some uvular units showed pursuit and/or visual motion responses resembling those found on floccular or vermis units. These were rare, however, and most responses like this that we were able to record were classified as mossy fiber input units, probably reflecting dorsolateral pontine and other visual pontine nuclei projections to the uvula. Since P-cells almost never showed this type of immediate response, the uvula must utilize visual inputs for a different purpose than does the flocculus or vermis. Uvular P-cells seem to be associated with some other process, presumably some functional concomitant of long-duration large-field visual motion. What candidates for this function can be considered?

Velocity storage is one possibility, but the time course of change of most uvular P-cells during the induction of OKN was much different than that postulated for the velocity storage mechanism.[18] In addition, the time course of the unit modulation seldom paralleled the decay of OKAN, probably the single best measure of the storage dynamics. Also, the eye velocity built up during OKN is stored as evidenced by the presence of rather normal OKAN in our lesioned animals.

Another possibility is that this structure is involved in the visual dumping mechanism of velocity storage as hypothesized by Waespe et al.[16] However, dumping requires a rapid visual response which we rarely find on these cells. Furthermore, our lesions restricted to the uvula have very little effect on this function. We think that visual dumping of velocity storage is probably performed more by the nodulus and perhaps the ventral folia of the uvula which were not consistently affected by our lesions.

Our suggestion as to what function the uvula may perform concerns its anatomical projections to the vestibular nucleus, both directly and via the fastigial nucleus.[19] The vestibular nuclei have been shown to be a structure through which pursuit signals flow, and are also the posited site of the neural integrator.[20] Individual vestibular neurons project directly to eye muscle motor neurons, but the state of the integrator that represents expected eye position must be coded by the bilateral activity in the vestibular nuclei on each side of the brain stem. It seems likely that this push-pull arrangement would require an active balancing mechanism to provide stable drift-free operation of this final output stage of the oculomotor system. We hypothesize that the uvula is part of this balancing mechanism. Occasionally we have noticed abnormal temporary eye drift as a result of our lesions, and cerebellar patients sometimes show a gradual decline of their gaze-paretic nystagmus during prolonged periods of eccentric fixation. However, when this is followed with attempted central fixation, they show a period of rebound nystagmus in the opposite direction.[21] These phenomena and the results of our lesions could be explained by the removal of inhibitory input from the uvula on the neural integrator. The uvula then appears to be a structure that aids in adaptation to abnormal periods of unbalanced oculomotor output, such as that which results during prolonged optokinetic stimulation.

REFERENCES

1. ZEE, D. S. 1982. Ocular motor control: the cerebellum. *In* Neuro-ophthalmology. S. Lessell & J. T. W. van Dalen, Eds.: 136–147. Excerpta Medica. Amsterdam, the Netherlands.
2. KELLER, E. L. 1988. Cerebellar involvement in smooth pursuit eye movement generation:

flocculus and vermis. *In* Physiological Aspects of Clinical Neuro-opthalmology: 341–355. Chapman & Hall. London, England.

3. LLINAS, R., & J. W. WOLFE. 1977. Functional linkage between the electrical activity in the vermal cerebellar cortex and saccadic eye movements. Exp. Brain Res. **29:** 1–14.

4. KASE, M., D. MILLER & H. NODA. 1980. Discharges of Purkinje cells and mossy fibers in the cerebellar vermis of the monkey during saccadic eye movements and fixation. J. Physiol. London **300:** 539–555.

5. MCELLIGOTT, J. G. & E. L. KELLER. 1982. Neuronal discharge in the posterior cerebellum: its relationship to saccadic eye movement generation. *In* Functional Basis of Ocular Motility Disorders, G. Lennerstrand, D. S. Zee & E. L. Keller, Eds.: 453–461. Pergamon Press. Oxford, England.

6. MILES, F. A. & J. H. FULLER. 1975. Visual tracking and the primate flocculus. Science **189:** 1000–1002.

7. LISBERGER, S. F. & A. F. FUCHS. 1978. Role of primate flocculus during rapid behavioral modification of vestibulo-ocular reflex I. Purkinje cell activity during visually guided horizontal smooth pursuit eye movement and passive head rotation. J. Neurophysiol. **41:** 733–763.

8. KELLER, E. L. & W. F. CRANDALL. 1981. Neural activity in the nucleus reticularis tegmenti pontis in the monkey related to eye movements and visual stimulation. Ann. N.Y. Acad. Sci. **374:** 249–261.

9. CRANDALL, W. F. & E. L. KELLER. 1985. Visual and oculomotor signals in nucleus reticularis tegmenti pontis in alert monkey. J. Neurophysiol. **54**(5): 1326–1345.

10. MUSTARI, M. J., A. F. FUCHS & J. WALLMAN. 1988. Response properties of dorsolateral pontine units during smooth pursuit in the rhesus macaque. J. Neurophysiol. **60:** 664–686.

11. THIER, P., W. KOEHLER & U. W. BUETTNER. 1988. Neuronal activity in the dorsolateral pontine nucleus of the alert monkey modulated by visual stimuli and eye movements. Exp. Brain Res. **70:** 496–512.

12. SUZUKI, D. A., J. G. MAY, E. L. KELLER & R. D. YEE. 1990. Visual motion response properties of neurons in the dorsolateral pontine nucleus of the alert monkey. J. Neurophysiol. **63:** 37–59.

13. BRODAL, P. 1982. Further observations on the cerebellar projections from the pontine nuclei and the nucleus reticularis tegmenti pontis in the rhesus monkey. J. Comp. Neurol. **204:** 44–55.

14. ROBINSON, F., J. COHEN, J. MAY, A. SESTOKAS & M. GLICKSTEIN. 1984. Cerebellar targets of visual pontine cells in the cat. J. Comp. Neurol. **223:** 471–482.

15. PRECHT, W., J. I. SIMPSON & R. LLINAS. 1976. Responses of Purkinje cells in rabbit nodulus and uvula to natural vestibular and visual stimuli. Pflügers Arch. **367:** 1–6.

16. WAESPE, W., B. COHEN & T. RAPHAN. 1985. Dynamic modification of the vestibulo-ocular reflex by the nodulus and uvula. Science **228:** 199–202.

17. RICHMOND, B. J., L. M. OPTICAN, M. PODELL & H. SPITZER. 1987. Temporal encoding of two-dimensional patterns by single units in primate inferior temporal cortex. I. Response characteristics. J. Neurophysiol. **57:** 132–146.

18. COHEN, B., V. MATSUO & T. RAPHAN. 1977. Quantitative analysis of the velocity characteristics of optokinetic nystagmus and optokinetic after nystagmus. J. Physiol. London **270:** 321–344.

19. ANGAUT, P. & A. BRODAL. 1967. The projection of the "vestibulocerebellum" onto the vestibular nuclei in the cat. Arch. Ital. Biol. **105:** 441–479.

20. CANNON, S. & D. ROBINSON. 1987. Loss of the neural integrator of the oculomotor system from brain stem lesions in monkey. J. Neurophysiol. **57:** 1383–1409.

21. ZEE, D., R. YEE, D. COGAN, D. ROBINSON & K. ENGEL. 1976. Ocular motor abnormalities in hereditary cerebellar ataxia. Brain **99:** 207–234.

Three-Dimensional Eye Velocity Measurement following Postrotational Tilt in the Monkey [a]

DANIEL M. MERFELD AND LAURENCE R. YOUNG

Man-Vehicle Laboratory
Massachusetts Institute of Technology
77 Massachusetts Avenue
Cambridge, Massachusetts 02139-4307

Angular head accelerations are known to elicit reflexive movements of the eye called the angular vestibuloocular reflex (VOR). If an upright subject is rotated about an earth-vertical axis at a constant angular velocity and then decelerated after some time, horizontal postrotatory nystagmus follows. Recently a number of studies (monkey;[1] human;[2] cat[3]) have demonstrated the existence of an exponentially decaying vertical nystagmus as well as horizontal nystagmus following rotation about a nonvertical axis. Each of these studies indicated that the vertical response, combined with the horizontal nystagmus, tended to align the compensatory motion of the pupil with earth horizontal. Two earlier human studies reported the presence of a vertical nystagmus during centrifugation.[4,5] A later set of studies repeated this paradigm with monkeys and measured the extent of an axis transformation that moved the axis of eye rotation toward alignment with gravitoinertial force.[6,7]

Monkey studies have also demonstrated a shift in the axis of the optokinetic after-nystagmus (OKAN) response when the rotational axis of the stimuli did not align with gravity.[8] While the authors acknowledged the presence of cross-talk between the channels and could not quantitatively show the extent to which the response deviated from yaw, these data showed the presence of position dependent vertical or torsional responses which shifted the axis of the eye movement responses toward a spatial-vertical axis.

We have measured all three dimensions of eye rotation during and following vestibular stimulation. We also have developed and implemented an algorithm that allows us to minimize the cross-talk between the eye measurement channels and permits us to evaluate the change in the axis of eye rotation quantitatively.

METHODS

An adult male squirrel monkey (*Saimiri sciureus*) was the subject for these experiments. A stainless steel bolt was used to stabilize the head of the monkey while allowing repeatable head positioning. The bolt was anchored to the skull using dental acrylic and miniature stainless steel screws which were inverted and secured in keyhole slots drilled in the skull. Right eye position was measured using a search

[a] This work was supported by National Aeronautics and Space Administration contracts NASW-3651, NAG2-445, NAS9-16523, the NASA Graduate Student Researcher Program, and the GE Forgivable Loan Fund.

coil technique.[9] Prefabricated coils made of insulated stainless steel wire were implanted on the right eye. An 11-mm diameter 3-turn coil was sutured to the sclera concentric to the limbus in the eye's frontal plane, and an 8-mm diameter 4-turn coil was sutured to the globe laterally and posteriorly to the frontal coil in the eye's sagittal plane. The leads leave the orbit such that they do not interfere with normal movements of the eye. For a complete description of all surgical procedures, see Paige and Tomko.[10]

The upright monkey was rotated in yaw at a constant velocity of 160 degrees per second until the horizontal VOR was nearly extinguished (> 60 seconds) and then quickly decelerated (100 degrees/second^2) to a stop. While the postrotatory angular response was still strong, the monkey was tilted to the left or to the right 15 or 32 degrees. To maintain alertness, d-amphetamine (0.3 mg/kg) was administered intramuscularly 15 minutes before beginning a test session. The lights were turned on between trials and extinguished just prior to the start of a trial.

The eye measurement system was calibrated using orthogonal reference search coils, similar to those implanted in the monkey, which were mounted on a model eye secured to a nonmetallic three-axis calibration jig. This calibration procedure has been verified *in situ* by rotating subjects by hand in the presence of an earth-fixed visual field.[10,11]

Using two search coils we can accurately estimate the three-dimensional position of the eye. For eye movements that do not deviate far from the primary eye position, the detector circuits accurately indicate yaw, pitch, and roll of the eye. However, as the eye position deviates from the primary gaze position, a number of nonlinear errors become important.[9] These errors are enhanced by any misalignment of the search coils on the globe.[12] Haddad *et al.* evaluated the effect of coil placement errors and determined that with eye displacements of less than 10 degrees, the eye position error was less than 10%.[11]

During this study, eye displacement often exceeded 10 degrees, leading to errors between 20% and 30%. To minimize these errors we developed an algorithm that corrects for coil placement errors to yield better estimates of eye orientation and eye velocity.[6] This method of analysis is an exact solution of the kinematic and kinetic problems. The resolution is limited only by the accuracy of the calibration procedures and by the inherent measurement limitations (e.g., measurement noise).

The alignment of the right-handed coordinate scheme is shown in FIGURE 1A. The two reference frames used to define eye orientation are shown in FIGURE 1B. These reference frames share a common origin which is fixed at the center of rotation of the eye. A rotation matrix (C), also known as a direction cosine matrix, was defined to represent the rotation between the head fixed reference frame $[x\,y\,z]$ and the rotated reference frame $[x'\,y'\,z']$. To evaluate whether the response deviated from yaw, we calculated the deviation angles shown in FIGURE 1C and D.

Each measurement in a search coil system is proportional to the cosine of the angle between the direction of the magnetic field and the normal to the plane of the search coil.[9] Two spatially orthogonal magnetic fields in phase quadrature allow us to measure two direction cosines from each of the two search coils. In this configuration, where both the magnetic fields and the search coils are orthogonal or nearly orthogonal, three of the measurements yield the maximum possible sensitivity. Geometrical considerations show that the fourth measurement must be insensitive to changes in orientation. In the presence of noise and other measurement errors, this leads to difficulties since small signal changes indicate large changes in orientation. Therefore, we limited our calculations to the three maximally sensitive signals.

Notation

We used c_{ij} to represent the cosine of the angle between the ith direction in the rotated $[x'\,y'\,z']$ coordinate system and the jth direction in the $[x\,y\,z]$ coordinate system, and c'_{ij} to represent the cosine of the angle between the ith coil and the jth

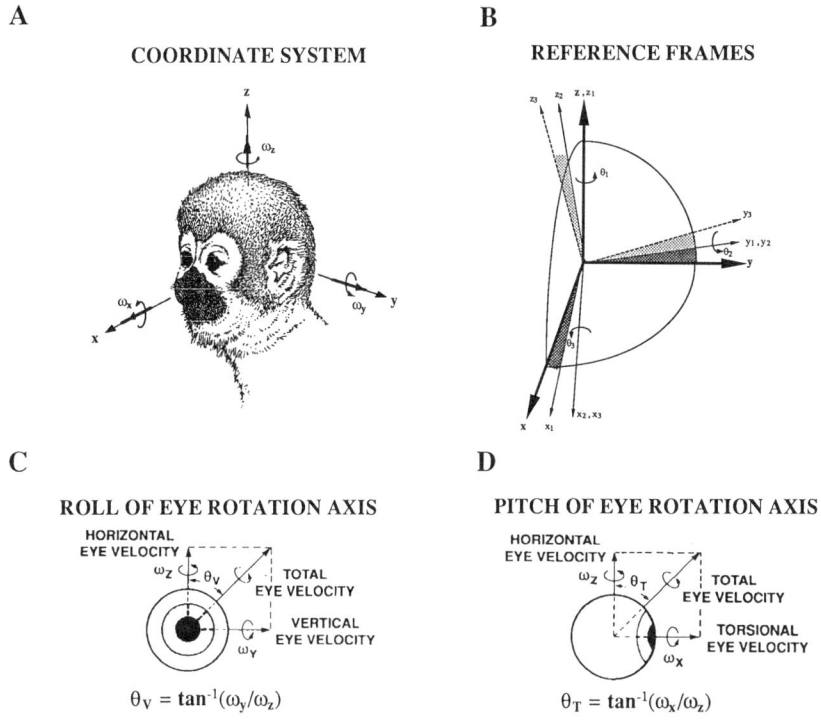

A   COORDINATE SYSTEM

B   REFERENCE FRAMES

C   ROLL OF EYE ROTATION AXIS

$$\theta_V = \tan^{-1}(\omega_y/\omega_z)$$

D   PITCH OF EYE ROTATION AXIS

$$\theta_T = \tan^{-1}(\omega_x/\omega_z)$$

FIGURE 1. **A** shows the head-fixed coordinate system used for this study. It also shows the vector representation of angular velocity. Horizontal (ω_z), vertical (ω_y), and torsional (ω_x) eye velocity are represented by vectors aligned with the specimen's z-, y-, and x-axes, respectively. Since this standard coordinate system is right handed, the right-hand rule is used to define the signs of all rotations. Eye movements to the left, down, and clockwise are defined to be positive. **B** shows the sequence of rotations that defines the relative orientation of the eye-fixed and head-fixed reference frames. $[x\,y\,z]$ is the head-fixed reference frame and is aligned with the coordinate system shown in A with the origin at the center of rotation of the eye. $[x_1\,y_1\,z_1]$ is the reference frame following a rotation of θ_1 about z, $[x_2\,y_2\,z_2]$ follows a rotation of θ_2 about y_1, and $[x_3\,y_3\,z_3]$ follows a rotation of θ_3 about x_2. $[x_3\,y_3\,z_3]$ is equivalent to $[x'y'z']$. **C** and **D** demonstrate how the shift in the axis of the eye response is calculated.

direction in the $[x\,y\,z]$ coordinate system. Coil 1 is defined to be the coil that encircles the iris of the right eye, and coil 2 is the other coil which is lateral to the right eye. Directions 1, 2, and 3 were defined by the x-axis, y-axis, and z-axis, respectively. With magnetic fields parallel to the y-axis and the z-axis, the three maximally sensitive measurements were c'_{12}, c'_{13}, and c'_{23}.

Calculations

From solid geometry we know that:

$$(c_{11}')^2 + (c_{12}')^2 + (c_{13}')^2 = 1.$$

Because the eye cannot rotate beyond 90 degrees, c_{11}' must be greater than 0. Therefore:

$$c_{11}' = +\sqrt{1 - (c_{12}')^2 - (c_{13}')^2}.$$

A unit vector pointing from the center 'of rotation through the center of the first search coil (c_1') is now completely defined. If we define the x'-axis in the rotated coordinate system to be aligned with the normal to the first search coil, we have:

$$c_1' = [c_{11}' \, c_{12}' \, c_{13}'] \equiv x' = [c_{11} \, c_{12} \, c_{13}].$$

Similarly, for the second coil we know that:

$$(c_{21}')^2 + (c_{22}')^2 + (c_{23}')^2 = 1.$$

We also know that since both of the coils are fixed to the eye, the angle between the coils (γ) is constant and that the dot product between the coils must be constant (k):

$$c_1' \cdot c_2' = c_{11}'c_{21}' + c_{12}'c_{22}' + c_{13}'c_{23}' = \cos(\gamma) = k.$$

If the angle between the coils (γ) is known, the two equations can be simultaneously solved for the two unknowns (c_{12}' and c_{22}'). There are two possible solutions, but since the eye movements are limited to rotations of less than 90 degrees, the solution is:

$$c_{22}' = \frac{c_{12}'k - c_{12}'c_{13}'c_{23}' + c_{11}' \sqrt{-k^2 + 2c_{13}'c_{23}'k + (c_{12}')^2 + (c_{11}')^2 - (c_{23}')^2}}{(c_{12}')^2 + (c_{11}')^2}$$

$$c_{21}' = \frac{k - c_{12}'c_{22}' - c_{13}'c_{23}'}{c_{11}'}$$

A unit vector pointing from the center of rotation through the second coil is now defined:

$$c_2' = [c_{21}' \, c_{22}' \, c_{23}'].$$

Since the search coils were not exactly orthogonal, we chose to develop an orthogonal coordinate system that is fixed with respect to the search coils. c_1 and c_2 define a plane. The unit vector that is perpendicular to this plane is defined to be the z'-axis:

$$z' = [c_{31} \, c_{32} \, c_{33}] = \frac{c_1' \times c_2'}{|c_1' \times c_2'|}.$$

Since we defined the coordinate system to be right handed the third axis is determined as:

$$y' = [c_{21} \, c_{22} \, c_{23}] = z' \times x'.$$

Therefore we have developed an orthogonal coordinate system $[x' \, y' \, z']$, which is fixed in the eye and is defined by the orientation of the coils. The orientation of

$[x'\ y'\ z']$ with respect to $[x, y, z]$ can be represented by a rotation matrix (C) where:

$$C = \begin{bmatrix} x' \\ y' \\ z' \end{bmatrix} = \begin{bmatrix} c_{11} & c_{12} & c_{13} \\ c_{21} & c_{22} & c_{23} \\ c_{31} & c_{32} & c_{33} \end{bmatrix}.$$

(The purpose of all of the previous calculations was to correct for nonorthogonality of the coils. If the search coils are orthogonal, only the following calculations are required to calculate eye velocity.) The Euler angle representation of coil orientation can be calculated from the following equations:

$$\theta_2 = \sin^{-1}(-c_{13}) \qquad \theta_1 = \sin^{-1}(c_{12}/\cos(\theta_2)) \qquad \theta_3 = \sin^{-1}(c_{23}/\cos(\theta_2)).$$

which represent a rotation of the eye frame $[x'y'z']$ with respect to the coil-orthogonal frame $[x\,y\,z]$ first by rotation (θ_1) about the z-axis, then by rotation (θ_2) about the intermediate y_1-axis, and finally by rotation (θ_3) about the intermediate x_2-axis (FIGURE 1B).

A number of approaches are available to determine the rate of angular rotation once angular orientation is determined. With additional manipulation, differentiation of the 9-element rotation matrix, the 4-element quaternion, or the 3-element Euler angles can yield angular velocity. We have chosen to differentiate the Euler angle representation because it is computationally efficient and because it offers a physically intuitive approach. By digitally differentiating each of the three previously defined Euler angles we obtained:

$$\omega_1 = \frac{d\theta_1}{dt} \qquad \omega_2 = \frac{d\theta_2}{dt} \qquad \omega_3 = \frac{d\theta_3}{dt}.$$

Each of these Euler rates is a vector quantity. ω_1 has a direction aligned with the head-fixed z-axis, ω_2 is aligned with the y_1-axis, and ω_3 is aligned with the x_2-axis (see FIGURE 1B). These nonorthogonal angular velocities can be easily converted to the head fixed orthogonal coordinate system $[x\,y\,z]$ by the transformation:

$$\omega = \begin{bmatrix} \omega_x \\ \omega_y \\ \omega_z \end{bmatrix} = \begin{bmatrix} -\omega_2 \sin(\theta_1) + \omega_3 \cos(\theta_1)\cos(\theta_2) \\ \omega_2 \cos(\theta_1) + \omega_3 \sin(\theta_1)\cos(\theta_2) \\ \omega_1 - \omega_3 \sin(\theta_2) \end{bmatrix}$$

To determine the angle between the coils we applied controlled rotational stimuli and measured the response. Since the response normally aligns with the stimulus (e.g., yaw rotations yield primarily horizontal responses while pitch yields primarily a vertical response), we determined the angle between the coils by calculating the eye response and minimizing the cross-talk between the orthogonal channels. Because the second search coil was located lateral to the eye, the cross-talk was primarily evident on the roll channel during pitch of the eye. We, therefore, minimized the indicated roll response during pitch by calculating the angle between the coils off-line. Once this cross-talk was minimized, we verified that the responses to yaw and roll were primarily aligned with the z (ω_z) and x (ω_x) axes, respectively.

This method was validated by placing two coils on the calibration jig such that the angle between the coils was nominally $80°$ with less than $2°$ of inaccuracy. FIGURE 2A and B show the measurements of pitch and roll during pitch of the calibration jig before any corrections were made. Notice the large amount of indicated roll despite

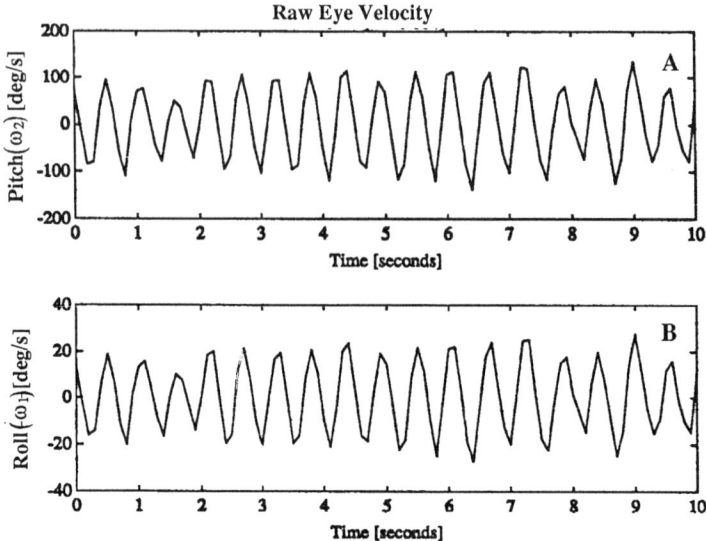

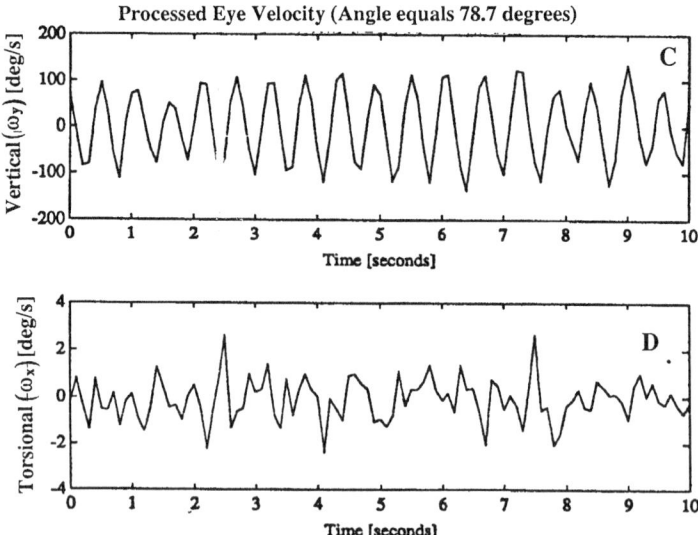

FIGURE 2. Test with calibration jig. **A** and **B** show the measurements of pitch rate (ω_2) and roll rate (ω_3), respectively, when the calibration jig was rotated in pitch with the angle between the coils nominally set at 80°. **C** and **D** show the calculated pitch rate (ω_y) and roll rate (ω_x), respectively, with the angle between the coils estimated at 78.7°. This value minimized the indicated roll rate. Note the change in roll rate scale between B and D.

the absence of roll from the actual motion. By changing the estimated angle between the coils and minimizing the root-mean-square (RMS) value of the roll channel, we estimated the angle between the coils as 78.7°. FIGURE 2D shows the estimated roll response. This process was repeated with the angle between the coils nominally set at values of 90° and 100°. The angle between the coils was estimated to be 89.3° and 98.6°, respectively. Since the estimates were within machining tolerances, the accuracy proved sufficient.

A similar process was used to estimate the angle between the coils *in vivo.* FIGURE 3 shows the uncorrected (A and B) response and the corrected (C and D) response with the angle between the coils estimated as 82 degrees. This value is physically realistic and was stable over time.

Fast-Phase Removal

Since we were primarily interested in the slow-phase eye velocity rather than eye position, we used a computer algorithm to identify and remove the fast phases of eye motion. This algorithm used digital filters to calculate three-dimensional eye velocity and eye acceleration. The movement was marked as a fast phase if the total acceleration of the eye exceeded a threshold set by the user. This simple algorithm removed more than 95% of the fast phases.[6] A limited number of fast phases were manually removed. All gaps were filled with the eye velocity just prior to the saccade. The three dimensions of eye velocity were then digitally filtered with a low-pass filter having a cutoff frequency of 2.5 Hz. The resulting slow-phase velocity was used for all further analysis.

RESULTS

A representative trial showing the postrotatory response during a right ear down tilt is shown in FIGURE 4. Shortly after a 32 degree roll tilt was applied, a strong vertical ocular response appeared. Its slow-phase velocity reached a peak of approximately 50 degrees per second. A relatively weak torsional response was also observed, but was not present on all trials. The sign of the vertical response always shifted the axis of eye rotation toward alignment with gravity.

Similar results were obtained for left ear down tilts during postrotatory responses, except that the direction of the vertical slow-phase eye velocity was always the reverse of that obtained for right ear down tilts. Again this shifted the axis of eye rotation toward alignment with gravity. The steady-state axis of eye rotation was calculated for all trials. FIGURE 5 shows a plot of the steady-state pitch (θ_V) of the axis of eye rotation versus the tilt angle for tilts to the left and to the right following constant-velocity yaw axis rotations. The slope of the least-squares fit, shown as the thin line, was 0.955. Statistics (t-test $p < 0.05$) indicated that this slope could not be distinguished from a slope of one. FIGURE 5 also shows a plot of the steady-state roll (θ_T) of the axis of eye rotation. The slope of the least-squares fit, shown as the thick line, was -0.026 which was statistically (t-test $p < 0.05$) indistinguishable from zero. Since slopes of one (θ_V) and zero (θ_T) indicate alignment of the axis of eye rotation with gravity, the axis of eye rotation appears to closely align with gravity following postrotational tilts. Similar results have been qualitatively observed in four other monkeys.

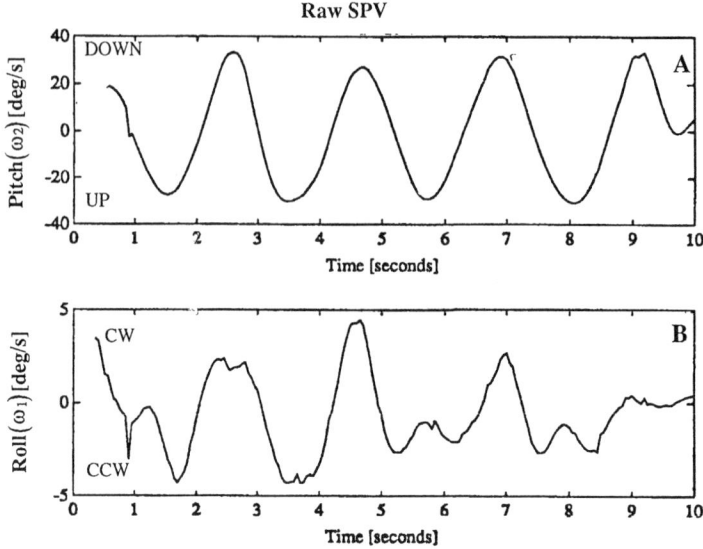

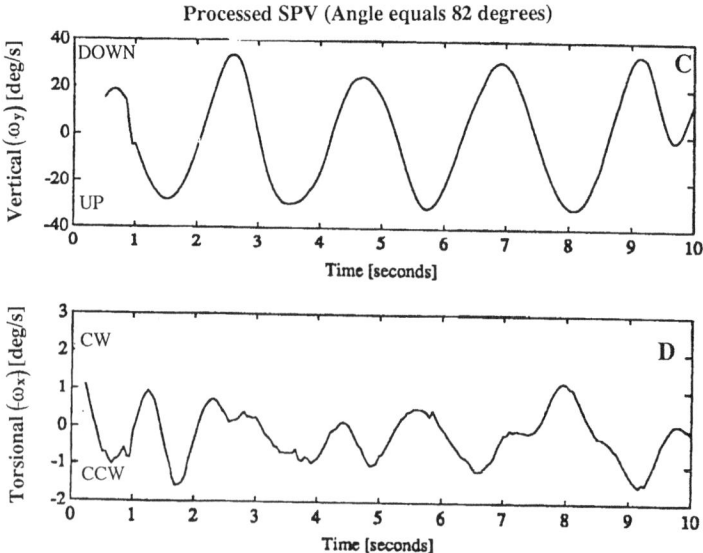

FIGURE 3. Test with slow-phase velocity. Similar to FIGURE 2, **A** and **B** show the measurements of slow-phase eye velocity for pitch (ω_2) and roll (ω_3), respectively. **C** and **D** show the calculated pitch (ω_y) and roll (ω_x) slow-phase eye velocity, respectively, with the angle between the coils estimated at 82°. This value minimized the roll slow-phase velocity (ω_x) during pitch rotations. Note the change in roll rate scale between B and D.

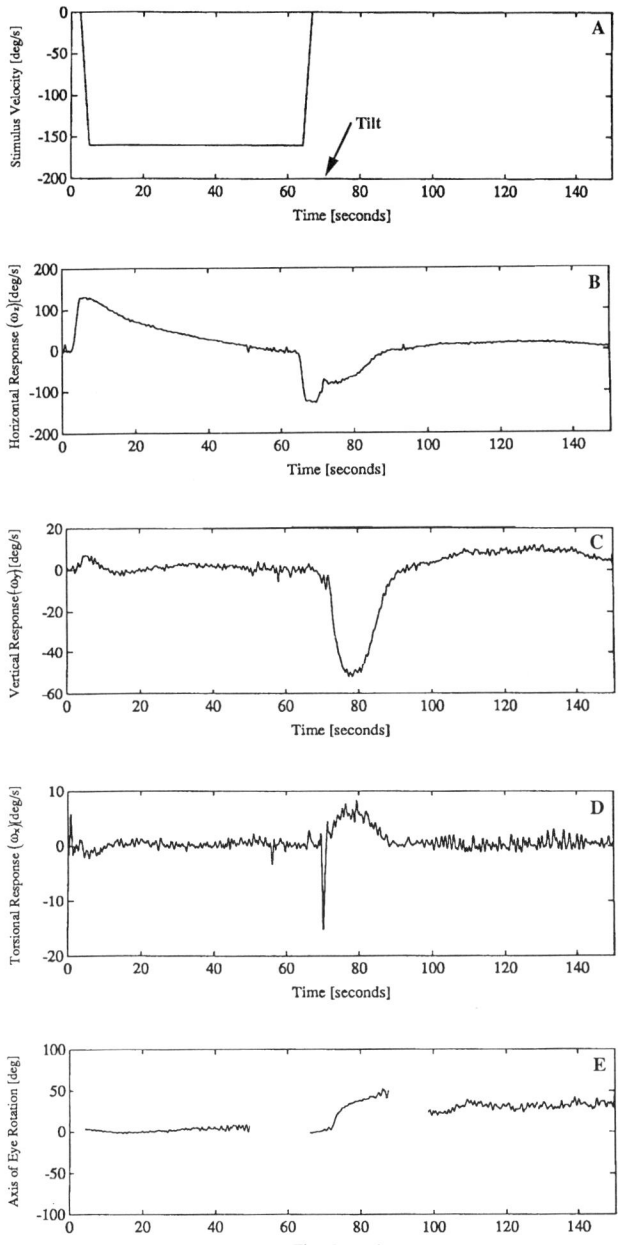

FIGURE 4. Right ear down dumping. Representative data from a single trial. **A** shows the rotational stimulation as well as when a 32-degree roll tilt to the right was initiated. **B** shows the horizontal ocular response. **C** shows a large vertical ocular response that is initiated immediately following the tilt. **D** shows a sharp torsional response to the roll and a small torsional response following the roll. **E** shows a shift (θ_V) in the axis of eye rotation. The shift in the axis of rotation is approximately equal to the tilt and both the horizontal and vertical responses reverse such that the axis of eye rotation stays near alignment with gravity. Discontinuities result from difficulty calculating the axis of eye rotation when the horizontal response is near zero.

DISCUSSION

A sensory conflict is induced by the postrotational tilt paradigm used in this study. When the subject is tilted with respect to gravity, the signals from the otoliths and other graviceptors contradict the inappropriate signal from the semicircular canals. Previous studies have shown that the postrotatory response was weaker when the subjects were rotated about a horizontal axis than when the same subjects were rotated about an earth-vertical axis.[13,14] Guedry suggested that this might be a direct effect of this sensory conflict since the conflict only exists while rotation is indicated.[15,16] The rapid decay might be a successful attempt to get the sensory information from the semicircular canals and graviceptors to concur rapidly.

An alternative method by which sensory concurrence is achieved is shown in FIGURE 6. In this case the sense of rotation is aligned with gravity. Since rotations about an earth-vertical axis will not result in graviceptor-sensed changes in gravity, a transformation of the sensed rotation from pure yaw as sensed by the canals to alignment with gravity might also minimize the conflict. The central nervous system might use this estimate of self-motion to generate the compensatory response, the VOR. This could lead to alignment of the axis of eye rotation with gravity. In the absence of other physiological or psychophysical measurements, the squirrel monkey's axis of eye rotation may provide a physiological measure of the monkey's central estimate of "down" (gravity).

It seems likely that a process of sensory conflict resolution is responsible for the axis transformation that has been observed in humans,[2,4,5] monkeys,[1,6–8] and cats.[3] This process may work in conjunction with the process described by Guedry[15,16] to minimize sensory conflict and might explain the observation[17,18] that velocity storage is altered by gravity.

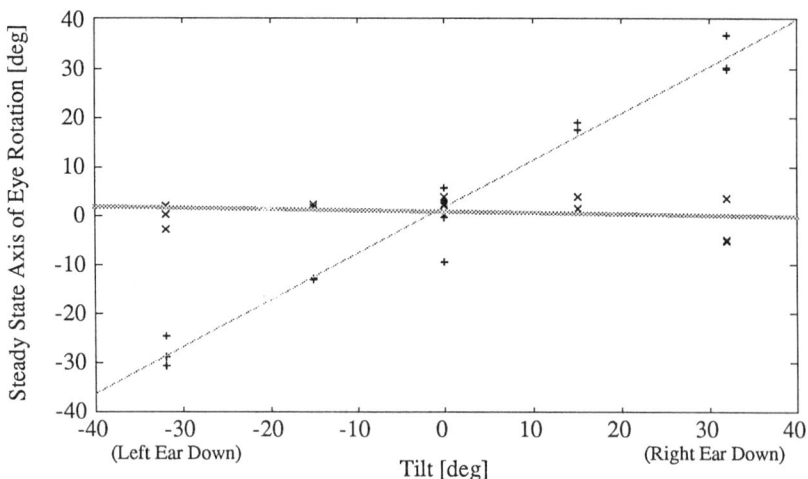

FIGURE 5. Steady-state axis shift (left/right tilts). The steady-state pitch of the axis of eye rotation (θ_V) for each trial (+) is plotted versus the magnitude of the tilt. The slope of the least-squares fit, shown as the thin line, was 0.955. The steady-state roll of the axis of eye rotation (θ_T) for each trial (x) is also plotted. The slope of the least-squares fit, shown as the thick line, was −0.026. Slopes of 1 and 0, respectively, indicate alignment of the axis of eye rotation with gravity.

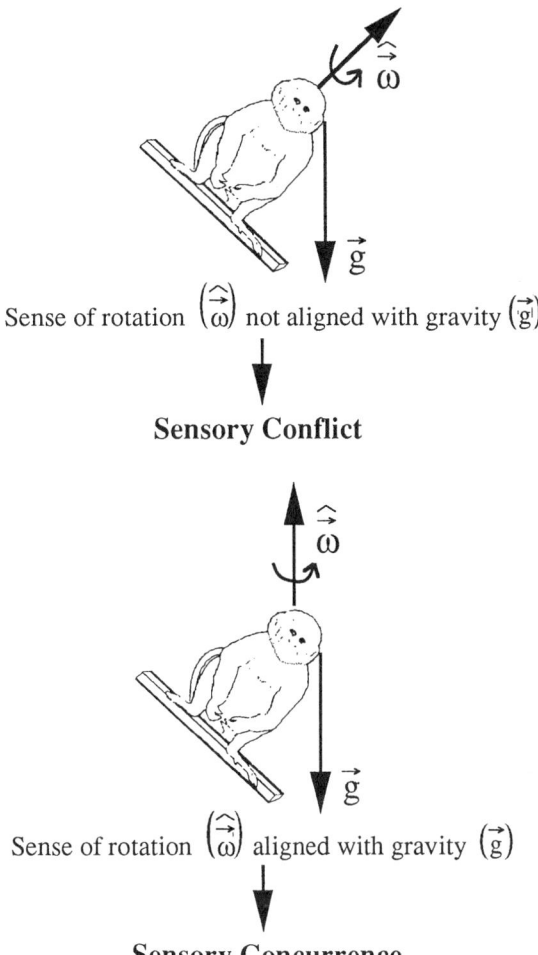

FIGURE 6. Conflict resolution during postrotational tilt. A sensory conflict results when the sense of rotation does not align with gravity. A resolution of this conflict is alignment of the sense of rotation with gravity. g represents gravitational force, while ω represents the central estimate of angular velocity.

ACKNOWLEDGMENTS

We would like to thank Dr. David Tomko, Dr. Gary Paige, Sherry Modestino, and the staff of the Vestibular Research Facility for their advice and assistance.

REFERENCES

1. RAPHAN, T., B. COHEN & V. HENN. 1981. Ann. N.Y. Acad. Sci. **374:** 44–55.
2. HARRIS, L. R. & G. R. BARNES. 1987. Orientation of vestibular nystagmus is modified by

head tilt. *In* The Vestibular System: Neurophysiologic and Clinical Research. Graham & Kemink, Eds. Raven Press. New York, N.Y.

3.   HARRIS, L. R. 1987. Exp. Brain Res. **66:** 522–532.
4.   LANSBERG, M. P., F. E. GUEDRY & A. GRAYBIEL. 1965. Aerosp. Med. **36**(5): 456–460.
5.   CRAMPTON, G. H. 1966. Does linear acceleration modify cupular deflection? 2nd Symposium of the Role of the Vestibular System in Space Exploration. NASA SP-115: 169–184.
6.   MERFELD, D. M. 1990. Spatial orientation in the squirrel monkey: an experimental and theoretical investigation. Ph.D. Thesis. Massachusetts Institute of Technology. Cambridge, Mass.
7.   MERFELD, D. M., L. R. YOUNG, G. D. PAIGE & D. L. TOMKO. 1991. Spatial orientation of VOR to combined vestibular stimuli in squirrel monkeys. Acta Otolarynol. Suppl. **481:** 287–292.
8.   RAPHAN, T. & B. COHEN. 1988. Ann. N.Y. Acad. Sci. **545:** 74–92.
9.   ROBINSON, D. A. 1963. IEEE Trans. Biomed. Electron., BME-10: 137–145.
10. PAIGE, G. D. & D. L. TOMKO. 1991. Eye movement responses to linear head motion in the squirrel monkey: I. Basic characteristics. J. Neurophysiol. **65**(5): 1170–1182.
11. HADDAD, F. B., G. D. PAIGE, M. J. DOSLAK & D. L. TOMKO. 1988. Practical method, and errors, in 3-D coil eye movement measurements. Invest. Ophthalmol. Vis. Sci. Suppl. **29:** 342.
12. FERMAN, L., H. COLLEWIJN, T. C. JANSEN & A. V. VAN DEN BERG. 1987. Vision Res. **27:** 811–827.
13. CORREIA, M. J. & F. E. GUEDRY. 1964. Influence of labyrinth orientation relative to gravity on responses elicited by stimulation of the horizontal semicircular canals. Report No. 905. Naval School of Aviation Medicine. Pensacola, Fla.
14. BENSON, A. J. & M. A. BODIN. 1966. Aerosp. Med. **37:** 144–154.
15. GUEDRY, F. E. 1965. Acta Otolaryngol. **60:** 30–49.
16. GUEDRY, F. E. 1965. Psychophysiological studies of vestibular function. *In* Contributions to Sensory Physiology. Academic Press. New York, N.Y.
17. HAIN, T. C. 1986. Biol. Cybern. **54:** 337–350.
18. RAPHAN, T. & B. COHEN. 1986. Multidimensional organization of the vestibulo-ocular reflex (VOR). *In* Adaptive Processes in Visual and Oculomotor Systems. Keller & Zee, Eds. Pergamon Press. Oxford, England.

A Model of Responses of Horizontal-Canal-Related Vestibular Nuclei Neurons that Respond to Linear Head Acceleration

A. A. PERACHIO,[a,b,c] G. A. BUSH,[a] AND D. E. ANGELAKI[d]

[a]Department of Otolaryngology
[b]Department of Physiology and Biophysics
[c]Department of Anatomy and Neurosciences
University of Texas Medical Branch
Galveston, Texas 77550

[d]Department of Physiology
The University of Minnesota
Minneapolis, Minnesota 55455

INTRODUCTION

Convergence of inputs to the vestibular nuclear complex (VNC) from otolith and semicircular canal related afferents has been demonstrated in a variety of studies using both natural vestibular stimuli and electrical stimulation of separate sensory end organs.[1-3] Most studies to date have addressed the question of how convergent inputs respond to linear and angular head motion in a series of investigations comparing the responses of central vestibular neurons in labyrinth-intact preparations to those in which canal inputs have been blocked by plugging of the ducts.[4,5] There is, however, a paucity of information concerning the functional characteristics of VNC cells with convergent inputs as defined by their responses to deterministic stimuli consisting of purely linear vs. purely angular acceleration. We have conducted a series of studies employing a device that provides linear head acceleration by a translational motion parallel to the earth-horizontal plane and angular acceleration about an earth-vertical axis to study neurons with convergent inputs. Data were obtained from neurons that exhibit responses to both types of stimuli. The functional characteristics of those neurons in terms of response dynamics, linear vectors of sensitivity, and classification according to their canal input have been described elsewhere.[6] Based on those characteristics, which are briefly described here, we present a set of models that predict certain relationships among horizontal-canal-related (HC), otolith-sensitive cells in and near the medial vestibular nucleus (MVN). We further propose that these types of neurons are candidates for the types of cells that mediate the otolith modulation of canal responses under conditions of coacting linear and angular acceleration.

METHODS

Experiments were conducted on decerebrated Long Evans strain rats prepared under a short-acting barbiturate anesthetic (Brevital, 65 mg/kg). The procedures for decerebration are described elsewhere.[6] A stimulation electrode was placed adjacent to the oval window. The animals were secured in a modified stereotaxic instrument that was mounted to a circular frame constructed as a set of concentric rings. This frame could be manually rotated about two axes. One axis was oriented parallel to the earth-horizontal plane so that rotation provided angular motion in a vertical head plane. The inner ring to which the stereotaxic instrument was attached could be rotated within the outer ring to adjust the plane of vertical rotation. The horizontal rotational axis was coupled to a U-shaped brace that was itself attached to the shaft of a 23 lb. ft. DC torque motor. The axis of the motor was oriented parallel to an earth-vertical axis and aligned so as to intersect the horizontal rotational axis. The animal was positioned so that the intersection of the head nasooccipital and interaural axes coincided with the intersection of the rotational axes. The motor was controlled by a velocity servo system. The rat's head was inclined so as to position the major plane of the horizontal semicircular canal near the plane of rotation about the vertical (yaw) axis. All of the above hardware was attached to a cart which was mounted on a set of circular bearings riding on parallel rails. The cart was driven along the rails by a 40 lb. ft. DC torque motor to which it was coupled by a cable/capstan arrangement. The linear track motor was controlled by a position servo system. Oscillation of the capstan attached to the motor shaft resulted in a translational horizontal motion of the cart along the rails which were oriented parallel to the earth-horizontal plane. By static repositioning of the animal's head about the vertical or horizontal rotational axes, the linear force vector generated by translational motion could be oriented in different directions through the head. The linear motion was sinusoidal with a defined period and peak acceleration (ranging from ± 0.1 to $\pm 0.145 \times g$) at a single frequency selected over a range from 0.2–1.4 Hz.

Labyrinth electrical stimulation was applied at current levels that were determined at the outset of the experiment to be twice the threshold for producing a vestibulocollic response. During recordings, the rats were paralyzed and artificially respirated. Average response latencies to electrical stimulation were measured at suprathreshold stimulus levels from an average of approximately 10 consecutive stimulations. The neurons were functionally classified using a sequence of manually applied head rotations and tilts. Beginning at a standard head position (as described above), cells with a response to rotation about the yaw axis were identified as type I HC if they responded to ipsilateral angular motion. Type II HC cells had opposite phase characteristics. For angular acceleration about the yaw axis, HC responses diminished as the head was statically tilted nose up or down or rolled to either side. These cells did not respond to angular acceleration in vertical head planes. Vertical-canal-related cells usually exhibited type II responses to horizontal head rotation and their response increased as the head was statically tilted nose upward or downward. Those neurons also were responsive to rotation in vertical head planes especially when the plane of rotation was aligned with one of the bilateral pairs of vertical canals. Neurons that did not exhibit responses to canal stimulation but did respond to static tilt or dynamic linear acceleration were classified as otolith cells. All otolith-sensitive neurons were tested with different vectors of linear acceleration applied in the horizontal head plane by repositioning the head statically at fixed angles (increments ranging from 15° to 45°) relative to the applied linear acceleration vector. Data were analyzed off-line using special purpose software designed to acquire and digitize the neural data and the stimulus. Average cycle histograms of the action potential activity and the stimulus were subjected separately to a least-squares fit to the sum of two sinusoids representing the first and second harmonics of the fundamental stimulus frequency. Gain was defined as the ratio of the response (impulses/second) to the stimulus (measured in *g*) following subtraction of the average (DC level) discharge activity. The response phase was defined as the difference, in degrees, between the response maximum and the stimulus maximum. For each frequency, the spatial and temporal properties of each cell were assessed by fitting the gain and phase simultaneously, from data obtained at each head orientation angle, using the response ellipse algorithm as described by Angelaki.[7] This allowed calculation of the vectors of maximum (S_1) and minimum (S_2) sensitivities and their respective gains and phases. The phase of S_2 is always 90° (either phase leading or lagging) from that of S_1. We are especially concerned here with data obtained from HC neurons with convergent otolith input that exhibit spatial/ temporal response characteristics to dynamic linear head acceleration.

RESULTS

Spatial Harmonic Characteristics

Data were obtained from a sample of 658 neurons of which 215 were identified as type I HC neurons and 106 were type II HC cells. Ipsilateral electrical stimulation applied to the labyrinth produced a complex field potential composed in part of two negative potentials with respective peaks at average latencies of approximately 1.0 (N1) and 2.0 (N2) mseconds. Average [± standard deviation (SD)] latency for type I HC cells was 2.2 ± 0.9 mseconds and for type II HC cells 2.4 ± 1.2 mseconds. Among HC cells, all type I neurons responded to ipsilateral electrical labyrinth stimulation. Among type II HC cells 64% responded. Based on earlier reports[8–10] it is assumed that the type II cells were responding to an input representing afferents that innervate the otolith organs. The majority of HC cells were also found to exhibit responses to time-varying linear acceleration (73% of type I, 72/98 and 79% of type II, 42/53). Not all cells could be tested with a sufficient number of linear vectors to assess the response profiles. From the neurons that were tested at more than two directions, only a small portion of the sampled cells exhibited *narrowly tuned* responses that were comparable to those previously reported for otolith afferents[11] and of VNC cells tested with dynamic head tilts.[4,5] The response gain of the vast majority of HC cells was found to vary with stimulus vector angle relative to that producing the maximum response (S_1), but the gain was not proportional to the force applied at S_1. Unlike most otolith afferents, *broadly tuned* HC cells also exhibited a significant response sensitivity (S_2) to linear force vectors orthogonal to S_1. Moreover, the phase of the response varied systemically as a function of vector angle. We have designated the former type of response as *narrowly tuned* because it represents a sensitivity that is analogous to a one-dimensional linear accelerometer in spatial/ temporal characteristics. The latter response profile was defined as *broadly tuned* because it encodes two-dimensional linear forces in the horizontal head plane.

By examining the relative spatial/temporal properties of the S_1 and S_2 vectors, we found that HC cells could be grouped into two major categories. The S_2 vector which was defined as that which temporally led the S_1 vector [i.e., $\Theta (S_2) = \Theta (S_1) + 90°$] could be spatially oriented at an angle that was 90° rotated clockwise to the right (CW) or to the left counterclockwise (CCW) of S_1. When *broadly tuned* HC cells were classified according to this scheme, type I HC cells were designated as 59.3% CW and

40.7% CCW; the distribution of *broadly tuned* type II cells was 40.9% CW and 59.1% CCW. Among the CCW neurons, there were no statistically significant differences between their respective S_1 and S_2 vectors in terms of response magnitude, phase, and direction. In contrast, the average gain of CW type II HC neurons for both S_1 and S_2 vectors was significantly greater than that of CW type I cells. Comparing CCW and CW response categories, CCW type I neurons had greater S_1 response gains than

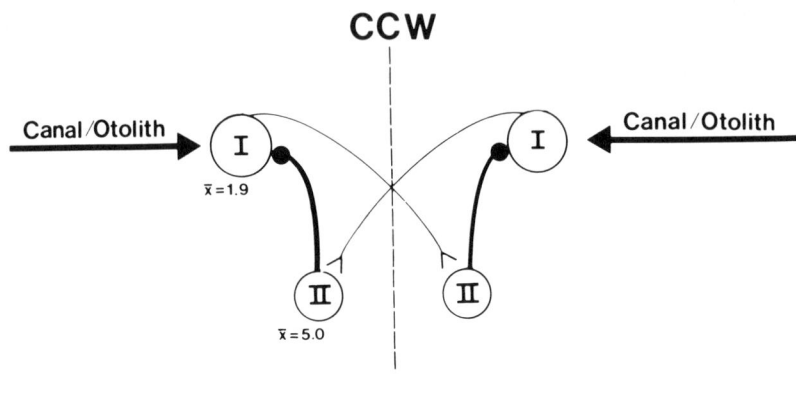

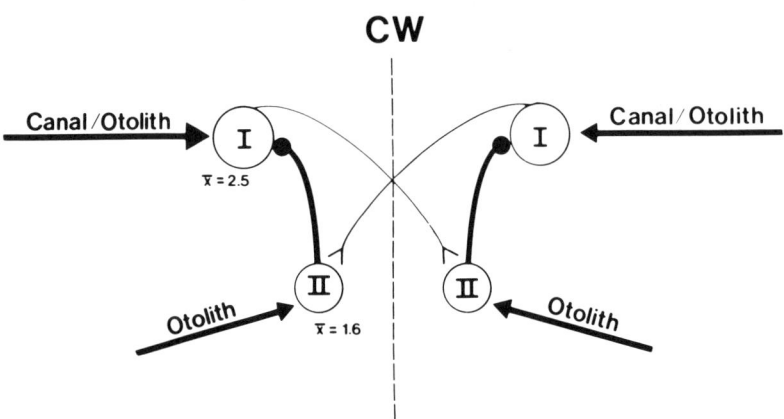

FIGURE 1. Two configurations of closed-loop systems for HC otolith-convergent neurons. **Top panel:** CCW neurons are connected so that all vestibular inputs to type II cells are derived from the contralateral side. **Bottom panel:** CW neurons receive converging otolith inputs from both labyrinths.

did CW type I cells; S_2 gains were equivalent. For type II cells, the opposite was observed. The two response categories also differed in terms of how each type of neuron responded to galvanic stimulation of the ipsilateral labyrinth. The shortest average response latencies were observed in CW type II cells (mean = 1.6 mseconds) and CCW type I neurons (mean = 1.9 mseconds). Markedly longer average response latencies were found for CCW type II neurons (5.0 mseconds) and CW type I cells (2.5 mseconds).

Commissural Models Distinguishing CCW and CW Neurons

Experimental evidence has led to the long-standing proposal that HC cells are connected bilaterally via commissural fibers.[8,9] The most commonly drawn circuit designates the type II cell as an inhibitory interneuron that receives a crossed excitatory input from a type I HC cell and in turn provides inhibitory input to an ipsilateral type I cell. The canal input to type II cells is presumed to derive from the contralateral side, since type II responses are largely eliminated by transection of fibers crossing the midline.[8] However, since short latency responses in type II cells can be elicited by electrical stimulation of the ipsilateral labyrinth, it has also been suggested that such input derives from an ipsilateral otolith source.[10] The connections among type I and type II cells are not identified anatomically; however, two general arrangements of connections have been proposed that are based on an assumption of functional bilateral reciprocity. These are commonly diagramed as open-loop or closed-loop connections. The closed-loop model has been proposed and evaluated by different groups of investigators especially as it is applied to questions related to vestibular compensation[10,12-15] and velocity storage.[16] Our findings on the response characteristics of HC neurons to linear head acceleration suggest an organization of HC neurons that is more parsimoniously described as a closed-loop commissural system.

Two schema of closed-loop circuits are depicted in FIGURE 1. These circuits are based on several assumptions and considerations based on our experimental findings. First, we maintain that CW and CCW neurons are segregated assuming a closed-loop circuit. Second, considering the CCW *broadly tuned* HC cells, we propose that ipsilateral canal and otolith inputs to the circuit derive from afferent connections (that may be monosynaptic) onto type I HC cells. The type I average response latency extends beyond the peak of the N1 potential, however, this may be due to the reduced conduction velocity at distal branches of the afferent neurons as they arborize within the VNC.[17] The type II response latencies are long enough to represent recurrent inputs via the contralateral neurons in the loop. In contrast, both type I and type II CW neurons may receive ipsilateral otolith-related input. The average response latency for type II neurons is near the N1 peak. Type I average latencies are about 50% greater and likely to be polysynaptic.

The operation of this model system can be described in a set of simplified equations. In the general case, the output F of each type (1 and 2 represent type I and type II cells) of neurons in the model system can be estimated as follows:

$$F_2 = G_1F_1 + O_2 \tag{1}$$

$$F_1 = -G_2F_2 + C + O_1 \tag{2}$$

where G are the gains and O and C are the otolith and canal vectorial inputs, respectively. The equations can be rewritten as:

$$F_1 = \frac{-G_2O_2 + C + O_1}{1 + G_1G_2} \tag{3}$$

$$F_2 = \frac{G_1C + G_1O_1 + O_2}{1 + G_1G_2} \tag{4}$$

For CCW neurons in which O_2 does not exist:

$$F_1 = \frac{C + O_1}{1 + G_1G_2} \tag{5}$$

$$F_2 = \frac{G_1(C + O_1)}{1 + G_1G_2} = G_1F_1 \tag{6}$$

For CW neurons, to have similar responses to linear acceleration under conditions where the canal input is not activated such as during translational motion:

$$-G_2O_2 + O_1 = G_1O_1 + O_2$$

$$(G_2 + 1)O_2 = (1 - G_1)O_1$$

$$O_2 = \left(\frac{1 - G_1}{1 + G_2}\right)O_1 \tag{7}$$

Therefore, for CCW neurons we would expect that, based on Equation 6 with a gain (G_1) of 1, the response vectors (F) of both type I and type II cells would be similar in orientation, magnitude, and phase. In contrast, Equations 3, 4, and 7 suggest that CW type I and type II neurons' responses to linear acceleration would differ unless their otolith inputs were the same. These estimations are consistent with the observed results (see above).

A further consequence of the two circuits, combined with the observed relationship between the type I and type II gains for CW and CCW neurons, is the prediction of how the otolith-related response vectors should be oriented among cells wired in a closed-loop system. For CCW neurons, the response vectors for type I cells should be opposite in direction and parallel in orientation relative to those of the ipsilateral type II cells. Similarly, since we are considering a bilaterally symmetric system, the same rule should apply between paired type I and paired type II cells bilaterally. Thus, the response vectors of CCW type I cells and the contralateral type II cells to which they project should be similar. This prediction is also consistent with the statistical similarity of S_1 and S_2 vectors among the entire sample of type I and type II CCW HC cells.

We have observed, for CW neurons, that the type II response vectors are greater in magnitude compared to those of type I cells. Therefore, since type II neurons are believed to provide a direct inhibitory input to type I cells, the difference in response gains suggests that the orientation of the vectors should be nearly parallel. Slight misalignments could exist, since the model depicts separate otolith-related inputs to each cell type. In addition to differences in the relative orientation of S_1 and S_2 that distinguish CW from CCW HC neurons, the average ratio of the gains of the two vectors (S_2/S_1) differs. The high ratio for CW response vectors is representative of *broader tuning.* The closed-loop model for CW neurons (lower panel in FIGURE 1) is constructed with four potential otolith inputs as opposed to two for CCW neurons (upper panel in FIGURE 1). The broader spatial sensitivity of CW neurons may be a consequence of increased vector summation that implied by the model. An additional possibility is that CW and CCW neurons are interconnected via open-loop pathways. Thus, broader spatial tuning of CW neurons may derive from additional inputs from CCW neurons.

The commissural model of connections among HC neurons with convergent otolith input may be considered in terms of its relevance to vestibular functions that have been associated with the vestibular commissures. A positive feedback closed-loop model was proposed by Galiana *et al.*[12] based on the observations that commissurotomy reversed postural compensation in hemilabyrinthectomized (HL) frogs[18] and, in the same species, that synaptic efficacy is increased in the excitatory commissural pathway after compensation.[19] Although some of the predictions of the model have not been supported by behavioral studies on HL mammalian species,[14,15] an elaboration of the commissural loop model, based on data recorded from MVN neurons in compensating animals, has strengthened the argument for a role of the commissural system in compensation.[10] The present model suggests that HL would acutely drastically reduce the number of CCW neurons that are type I HC ipsilateral and type II HC contralateral to the damaged labyrinth, in terms of their responses to *both* linear and angular acceleration. For CW ipsilateral type I and contralateral type II HC neurons canal-related responses would also be acutely impaired. Responses to otolith input, however, would persist in contralateral CW type II neurons. HL results in a loss of velocity storage as evidenced by a shortening of the time constant of the VOR and optokinetic after-nystagmus that only partially recovers with compensation.[13] Commissurotomy results in a more profound impairment of velocity storage.[20,21] Transection of the commissural pathways would affect both CW and CCW neurons as connected in a closed-loop configuration. Though differing in numbers of neurons and connections, an open-loop commissural system would also be affected by midline lesions. At present, the relevance of convergent HC neurons to the velocity storage mechanism can not be demonstrated. It is of interest to note however that in the rat, type II HC neurons have been implicated in pathways mediating the component of optokinetic nystagmus associated with velocity storage.[22] Future studies may determine if the responses characteristic of convergent HC neurons include low-frequency rotational response dynamics and/or optokinetic sensitivity that is more consistent with velocity-storage-related properties.

REFERENCES

1. CURTHOYS, I. S. & C. H. MARKHAM. 1971. Brain Res. **35:** 469–490.
2. MARKHAM, C. H. & I. S. CURTHOYS. 1972. Brain Res. **43:** 383–396.
3. SEARLES, E. J. & C. D. BARNES. 1977. Brain Res. **125:** 23–36.
4. SCHOR, R. H., A. D. MILLER & D. L. TOMKO. 1984. J. Neurophysiol. **51:** 136–146.
5. SCHOR, R. H., A. D. MILLER & D. L. TOMKO. 1985. J. Neurophysiol. **53:** 1444–1452.
6. ANGELAKI, D. E., G. A. BUSH & A. A. PERACHIO. 1992. Biol. Cybern. **66:** 231–240.
7. ANGELAKI, D. E. 1991. IEEE Trans. Biomed. Eng. **38**(11): 1053–1060.
8. SHIMAZU, H. & W. PRECHT. 1966. J. Neurophysiol. **29:** 467–492.
9. PRECHT, W. & R. LLINAS. 1972. Prog. Brain Res. **37:** 89–109.
10. NEWLANDS, S. D. & A. A. PERACHIO. 1990. Exp. Brain Res. **82:** 373–383.
11. FERNANDEZ, C. & J. M. GOLDBERG. 1976. J. Neurophysiol. **39:** 985–995.
12. GALIANA, H. L., H. FLOHR & G. MELVILL JONES. 1984. J. Neurophysiol. **51:** 242–259.
13. FETTER, M. & D. S. ZEE. 1988. J. Neurophysiol. **59:** 370–393.
14. SMITH, P. F., C. L. DARLINGTON & I. S. CURTHOYS. 1986. Neurosci. Lett. **65:** 209–213.
15. NEWLANDS, S. D. & A. A. PERACHIO. 1990. Neurosci. Abstr. **12:** 254.
16. ANASTASIO, T. J. 1991. Biol. Cybern. **64:** 187–196.
17. PERACHIO, A. A., J. D. DICKMAN & M. J. CORREIA. 1988. *In* Basic and Applied Aspects of Vestibular Function. J. C. Hwang, N. G. Daunton & V. J. Wilson, Eds.: 13–25. Hong Kong University Press. Hong Kong.
18. BIENHOLD, H. & H. FLOHR. 1978. J. Physiol. London **284:** 178.
19. DIERINGER, N. & W. PRECHT. 1979. Prog. Brain Res. **50:** 607–615.
20. BLAIR, S. M. & M. GAVIN. 1981. Acta Otolaryngol. **91:** 1–8.
21. KATZ, E., B. COHEN, H. COHEN & J. BUETTNER. 1989. Neurosci. Abstr. **211.3:** 513.
22. CAZIN, L., J. LANNON & W. PRECHT. 1984. Exp. Brain Res. **54:** 337–348.

Cytotoxic Effects of Somatostatin in the Cerebellum

CAREY D. BALABAN AND WALTER B. SEVERS[a]

Department of Otolaryngology
University of Pittsburgh and
The Eye & Ear Institute of Pittsburgh
203 Lothrop Street
Pittsburgh, Pennsylvania 15213

[a]Department of Pharmacology
Milton S. Hershey Medical Center
Post Office Box 850
Hershey, Pennsylvania 17033

The actions of neurotransmitters in the central nervous system are generally considered from the perspective of effects on postsynaptic potentials and spike generation. The actions of the transmitters, then, are viewed as the substrate for producing appropriate physiologic and behavioral responses of organisms. However, the concept that neurotransmitters can have specific neurotoxic actions has emerged from the findings that N-methyl-D-aspartic acid (NMDA) analogues are neurotoxic in a variety of sites within the central nervous system. These cytotoxic effects are believed to reflect (1) damage mediated by high intracellular Ca^{2+} levels that result from activation of NMDA receptors (possibly through Ca^{2+}-calmodulin-mediated activation of nitric oxide synthetase) and (2) osmotic damage that is initiated by Na^+ influx through kainate/quisqualate-type receptors and successive influxes of Cl^- and water.[1-5] Thus, neurotransmitter-induced cytotoxicity is regarded as a potential etiologic factor in neuropathologic conditions.[1]

This communication reviews the evidence that somatostatin (SRIF-14) is a regionally selective neurotoxin in central vestibular pathways. Somatostatinlike immunoreactivity is prominent in some cerebellar Purkinje cells, Golgi cells, and climbing fiber afferents[6] and in the vestibular nuclei; moderate levels of SRIF-14 binding have been reported in the vestibular nuclei. There is also behavioral evidence that SRIF-14 can affect vestibular function. Cohn and Cohn first reported that rats display a characteristic pattern of acute vestibular dysfunction, which they termed *barrel rotation,* after intracerebroventricular (icv) injections of this endogenous neuropeptide.[7] This finding has been confirmed repeatedly (reviewed in Reference 8), and Burke and Fahn later reported that low-dose microinjections of SRIF-14 into the vestibular nuclei produced barrel rotation.[9] During the course of studies of the temporal dynamics of barrel rotation elicited by icv injections of SRIF-14 and arginine-vasopressin (AVP), we noted that rats remained ataxic for 1–2 days after a single injection of SRIF-14. The initial histologic analyses of brains from these rats revealed that SRIF-14 selectively damages sagittal groups of Purkinje cells in the cerebellar anterior lobe (lobules I–III) and lobules IX–X that send axons to the fastigial and vestibular nuclei.[10] This finding suggested, then, that SRIF-14 is a potentially toxic neurotransmitter/neuromodulator in central vestibular pathways.

The regional distribution and dose dependence of SRIF-14 toxicity was investigated subsequently in a larger population of rats.[11] Different groups of alert rats were given a 5-μl icv bolus injection containing a 20- or 40-μg dose of SRIF-14 alone or a combined dose of 20 or 40 µg SRIF-14 and a 1-µg dose of either AVP or a potent antivasopressor vasopressin antagonist, (1-(β-mercapto-β,β-cyclopentamethylene propionic acid) 2-(O-methyl)-tyrosine)-arginine[8]-vasopressin (mcAVP), through an implanted cannula. After a survival time of 4–5 days, the rats were euthanized by a pentobarbital overdose [100 mg/kg, intraperitoneally (ip)] and central toxic effects of the peptide doses were assessed with the cupric-silver degeneration method of Carlson and DeOlmos.[12] This selective silver degeneration technique produces an opaque black reaction product in degenerating neuronal somata, dendrites, and axons and has proven to be an extremely sensitive and selective method for detecting actions of neurotoxins on both axonal and somatodendritic regions of neurons.[13] In addition, some rats were euthanized after longer survival times (2 weeks to one

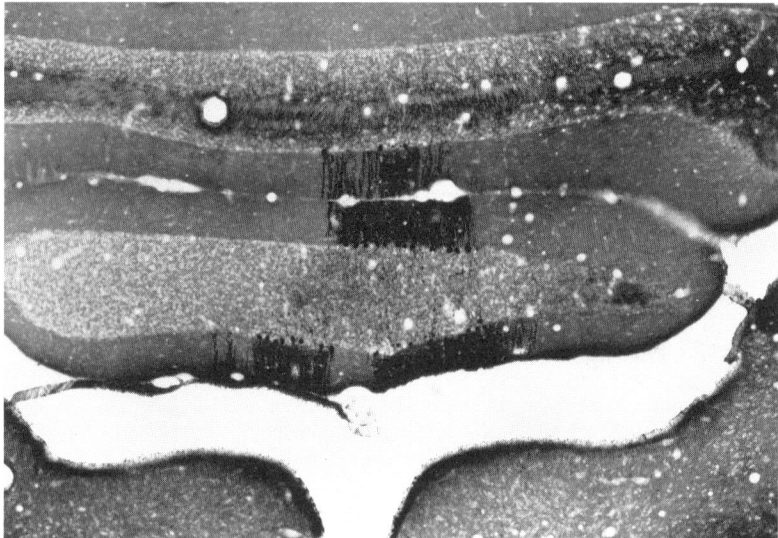

FIGURE 1. Photomicrograph of SRIF-induced neurotoxicity in the cerebellum. This micrograph shows typical degenerating Purkinje cells in the medial aspect of lobules II and III of the anterior lobe. Note the intense argyrophilia of Purkinje cell dendrites, somata, and axons in parasagittal bands. The rat was euthanized four days after receiving a 20-µg icv bolus of SRIF; the brain was sectioned in the transverse plane.

month), so that toxicity could be assessed in paraffin-embedded sections stained with cresyl violet.

After exclusion of degeneration produced by implantation of the chronic lateral ventricular cannula, the primary evidence of consistent SRIF-14 toxicity was the appearance of degenerating neuronal Purkinje cell somata in the cerebellar cortex (FIGURE 1). Degenerating cells were encountered less frequently (i.e., in fewer animals) in the vestibular nuclei, fastigial nucleus, and dorsal cochlear nuclear nucleus. In the cerebellar cortex, the degeneration showed both cellular and regional selectivity. First, only Purkinje cells showed degenerative changes. Cupric-silver preparations (FIGURE 1A) revealed the Golgi-like argyrophilia of sagittal bands of Purkinje cell somata, dendrites, and axons after SRIF-14 administration that is characteristic of the effects of central neurotoxins such as trimethyltin and 3-acetyl-pyridine.[13–15] These bands of degenerating cells were flanked by normal, undamaged Purkinje cells. The axons of these degenerating Purkinje cells could also be traced to the fastigial and vestibular nuclei, confirming the regional selectivity of the lesions and their relationship to vestibular function. The examination of Nissl-stained sections underscored the striking selectivity of this effect to sagittal subsets of Purkinje cells. For example, the medial $\frac{2}{3}$ of lobule I were virtually devoid of Purkinje cells after SRIF-14 exposure (FIGURE 2A and B); this region was flanked by apparently normal Purkinje cells (FIGURE 2A and C). The regional differences in susceptibility of Purkinje cells to SRIF-14 intoxication were also consistent and striking: degeneration was dose dependent and confined to lobules I–III, VIII–X, and restricted longitudinal bands within the cerebellar hemispheres. The maximum extent of regions susceptible to SRIF-14 damage in these experiments is summarized in FIGURE 3. It is important to note that degeneration was highly restricted in the cerebellum. For example, the vermis of lobules V–VII and large portions of paravermal cortex, the flocculus, and the cerebellar hemispheres were resistant to SRIF-14 intoxication at the administered doses. Although the summarized pattern of SRIF-14-induced degeneration is symmetric in FIGURE 3, it was almost invariably asymmetric in individual rats and rarely occupied the entire susceptible region in individual animals. Furthermore, the regional distribution of Purkinje cell degeneration was dependent solely on the dose of SRIF-14; the incidence of degeneration in each region is summarized in FIGURE 4. Lobules I–III in the anterior lobe were most susceptible to SRIF intoxication. By contrast, degeneration of Purkinje cells in the cerebellar hemisphere (copula pyramis, paramedian lobule and paraflocculus) showed a clear dose dependence which indicated a lower sensitivity to SRIF intoxication. Degeneration in the cerebellar hemisphere was present in 84% (26/31) of rats given 40 µg SRIF either alone or with mcAVP or AVP; hemispheric degeneration was present in only 12% (4/33) of rats receiving the lower dose of SRIF (χ^2 test, $p < 0.05$). Finally, Purkinje cells in lobules IX–X displayed an intermediate susceptibility to SRIF intoxication. Although the incidence of degeneration in lobules IX–X did not vary significantly as a function of the SRIF dose (χ^2 test, $p > 0.05$), with an estimated incidence of 81% across treatment groups, the severity of degeneration varied in a dose-dependent manner. These data suggest that sagittal groups of Purkinje cells differ in their susceptibility to SRIF-induced cell death. Cells in the vermal region of lobules I–III are most sensitive, cells in lobules IX–X have intermediate susceptibility, and neurons in the copula pyramis, paramedian lobule, and paraflocculus are least sensitive to SRIF toxicity.

In addition to the highly consistent Purkinje cell damage, there were less consistent, relative slight cytotoxic effects of icv SRIF-14 in the fastigial, vestibular, and dorsal cochlear nuclei. These effects appeared to depend strictly on the dose of SRIF-14 and were almost exclusively found at the 40-µg dose. Since the few degenerating cells observed in the vestibular and fastigial nuclei were often surrounded by degenerating Purkinje cell axons, the relative contributions of direct cytotoxicity and anterograde transneuronal degeneration[13] to this effect cannot be discriminated at present. These experiments also revealed that a small proportion of pyramidal neurons along the medial aspect of the dorsal cochlear nucleus are susceptible to SRIF intoxication; however, in contrast to anterior lobe cerebellar degeneration in virtually all rats, it was present in only 35% of rats in the groups receiving 40 µg SRIF. This finding is of interest because Mugnaini et al. have noted that the organization of the rat dorsal cochlear nucleus is analogous to the cerebellar cortex, with pyramidal cells receiving parallel fiber inputs from cochlear nuclear granule cells.[16] The existence of selective susceptibility to SRIF-14 intoxication, then,

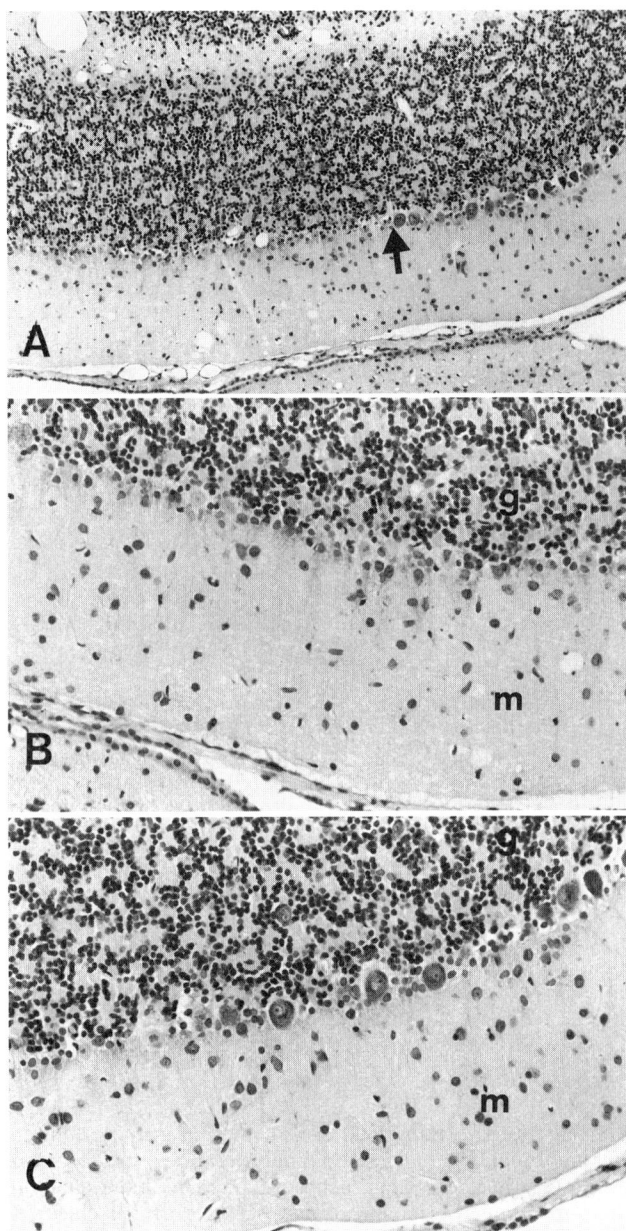

FIGURE 2. Photomicrographs of SRIF-induced cytotoxicity in the cerebellar cortex. These photomicrographs illustrate the appearance of SRIF-induced Purkinje cell damage in cresyl violet-stained, transvere sections of the anterior lobe. Panel A is a low-magnification micrograph showing the abrupt transition between cortex lacking Purkinje cells and cortex with a normal appearance. Higher magnification views of these regions are shown in panels B and C, respectively, with granular (g) and molecular (m) layers indicated.

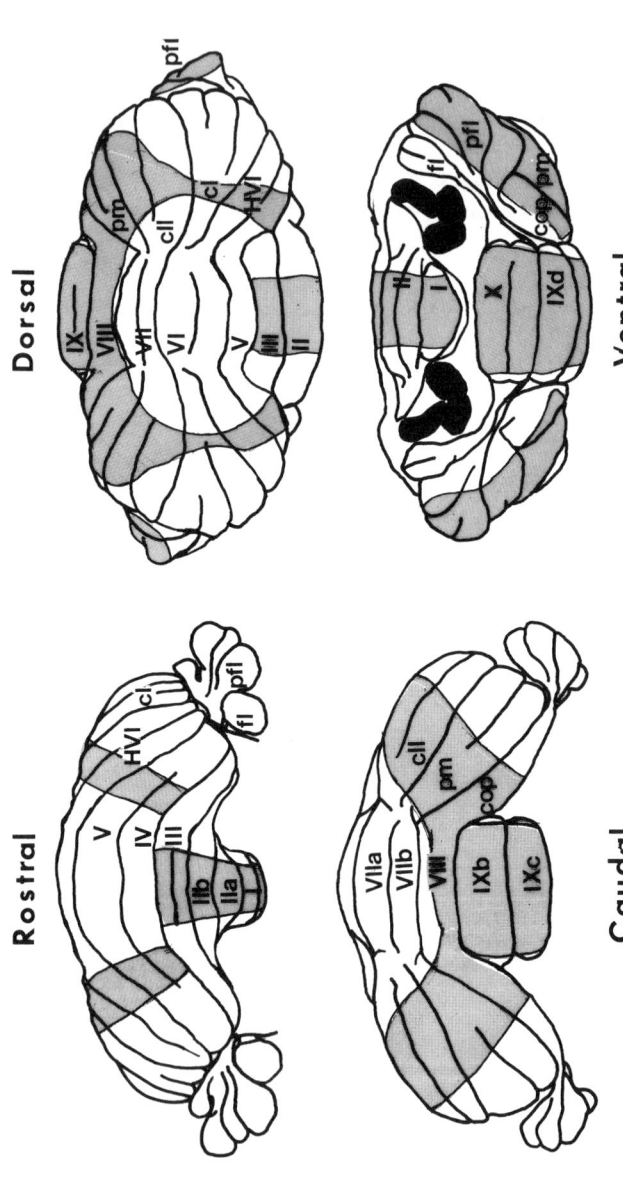

FIGURE 3. Diagram of the regional distribution of SRIF-induced Purkinje cell loss in rat cerebellar cortex. The maximum extent of Purkinje cell damage is shaded on rostral, caudal dorsal and ventral views of the cerebellar cortex. Roman numerals designate cerebellar folia according to Larsell's terminology; the copula pyramis (cop), paraflocculus (pfl), crus I (cI), crus II (cII) and the paramedian lobule (pm) are also labeled.

may be indicative of common neurochemical features of cerebellar Purkinje cells and dorsal cochlear nucleus pyramidal neurons.

It is important to note that peptide injections into the lateral ventricle produce damage to cerebellar Purkinje cells that contact cerebrospinal fluid (CSF) in the cerebellomedullary cistern and confluent subarachnoid space. Since the peptide dose must traverse the cerebral aqueduct, fourth ventricle, and enter the subarachnoid space to reach these sites of damage, it is improbable that asymmetries in Purkinje cell degeneration can be attributed to asymmetric peptide delivery to the cerebellar cortex. Thus, the data suggest that the sensitivity to SRIF intoxication is independently regulated at the level of sagittal zones in cerebellar cortex.

Previous studies indicate that the central neurotoxicity of SRIF-14 and its analogues extends to other mammals, including primates. For example, LeBlanc *et al.* reported truncal ataxia and dysmetria during intracisternal infusions of a long

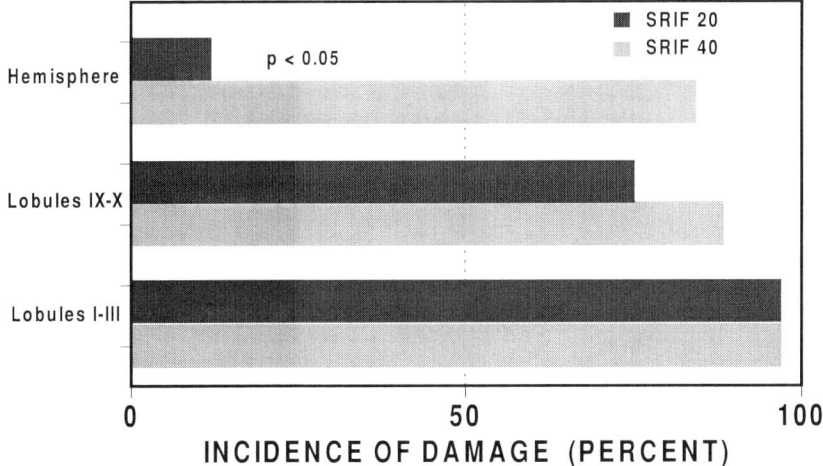

FIGURE 4. Dose dependence of the spatial distribution of SRIF toxicity in cerebellar cortex. This histogram illustrates the percentage of animals displaying Purkinje cell damage (incidence of damage) in different cerebellar regions as a function of the icv dose of SRIF-14. The difference in the hemispheric regions is significant by a χ^2 test.

acting octapeptide SRIF-14 analogue (Sandostatin®) in primates (*Circopithecus aethiops*),[17] which is suggestive of cerebellar toxicity. In addition, direct intrathecal infusions of SRIF-14 produce local spinal cord neuron loss in rats,[18] mice, and cats,[19] which may limit its proposed therapeutic value as an epidural or intrathecal analgesic. These documented toxic effects of this peptide are of interest because the cytological features and regional distribution of SRIF-14-induced Purkinje cell degeneration are remarkably similar to the pathological changes that have been reported in cases of *late onset* (or *parenchymatous*) *cortical cerebellar atrophy*.[20–24] This disorder of unknown etiology has an adult onset and is characterized by gross atrophy of the medial aspect of cerebellar cortex, particularly the anterior lobe. The results of histological analyses of cerebella revealed a picture that is virtually identical to the appearance of SRIF-14-intoxicated rats: there is a regional loss of cerebellar Purkinje cells with sparing of other neuronal elements (see Figure 2 of

Thomas;[20] Figures 3–5 of Archambault;[21] Figures 3 and 10 of Marie *et al.*;[23] and Figures 1–6 in Parker and Kernohan[24]). The description by Parker and Kernohan (1933) of the cortex is particularly compelling:

> In the outer portions of several foliæ there were normal Purkinje cells, and in the deeper recesses of the foliæ they had completely disappeared. The transition between these two portions was abrupt.[24] (p. 202)

This is also a precise description of the appearance of SRIF-14-intoxicated cortex in rats (FIGURE 2). Although the regional distribution of Purkinje cell loss in late cortical cerebellar atrophy has not been described in great detail, it is reminiscent of the regional distribution of SRIF-14 intoxication; the most severe effects were observed in the anterior lobe (superior vermis and quadralateral lobule), lesser effects appeared in the posterior vermis, and the least marked effects were reported in the cerebellar hemisphere. It is also noteworthy that the disorder presents first with truncal ataxia, followed by upper limb ataxia and dysarthria,[20–25] which suggests that the initial degenerative effects occur within the anterior lobe. Thus, it is possible that cellular mechanisms of SRIF-14 neurotoxicity can elucidate etiologic factors for this disorder.

The cellular mechanisms that are responsible for the differential susceptibility of sagittal groups of Purkinje cells to SRIF intoxication are unknown. However, a possible mechanism can be proposed by analogy with the toxic actions of excitatory amino acids in the brain stem and hippocampus. Amino acid cytotoxicity is believed to result from both a massive influx of Ca^{2+} through NMDA-receptor-linked channels and an osmotic load imposed by Na^+ influx through quisqualate/kainate channels.[1–5] The hypothesis of a role of Ca^{2+} fluxes in SRIF-14 toxicity is suggested by several lines of evidence. First, SRIF-14 and its analogues can affect Ca^{2+} flux across neurons. SRIF-14 has been reported variously to produce Ca^{2+}-dependent neuronal depolarization in central neurons,[26] increased intracellular Ca^{2+} in hippocampal neurons,[27] or block L-type channels in cultured cell lines[28,29] or isolated pituitary somatotrophs.[30] Second, Purkinje cells express voltage-gated Ca^{2+} channels that mediate both slow depolarization and fast dendritic spikes during spontaneous and evoked activity.[31–33] The data from dihydopyridine (L-type channel) and ω-conotoxin (N-type channel) binding studies[34,35] are consistent with these results. Third, Purkinje cells express a rich repertoire of Ca^{2+}- and Ca^{2+}/calmodulin-binding proteins (e.g., References 36–39), indicating an important role in Ca^{2+} as a second messenger. Fourth, we have previously reported that Purkinje cell expression of one calmodulin-binding protein, Ca^{2+}/calmodulin-dependent cyclic nucleotide phosphodiesterase, appears to be regulated transsynaptically by climbing fibers.[40] Since climbing fibers define sagittal connections in cerebellar cortex,[41] this latter result suggests that intracellular mechanisms related to Ca^{2+} fluxes can be regulated at the level of sagittal bands in the cerebellar cortex. These factors, then, make Ca^{2+} fluxes a logical candidate for a mediator of a regionally selective pattern of SRIF-14 toxicity in the cerebellum. The demonstration that SRIF-14 cerebellar toxicity can be attenuated by pretreatment with diazepam[42] is not inconsistent with this hypothesis; in addition to the benzodiazepine binding site on $GABA_A$ (γ-aminobutyric acid) receptors, diazepam has been reported to be both a selective L- and T-Ca^{2+} channel blocker[24] and a selective noncompetitive NMDA receptor antagonist (i.e., a blocker of chemically gated Ca^{2+} channels).[43] Clearly, it is important to determine whether the neuroprotective effects of diazepam against SRIF-14 toxicity can be attributed to the latter two effects. In particular, the elucidation of mechanisms of endogenous neurotransmitter toxicity in central vestibular circuits offers the possibility of (1) understanding etiologic factors in vestibular dysfunction (particularly progressive dysfunction during aging) and (2) designing neuroprotective therapies to arrest degenerative processes.

REFERENCES

1. COTMAN, C. W., R. J. BRIDGES, J. S. TAUBE, A. S. CLARK, J. W. GEDDES & D. T. MONAGHAN. 1989. The role of the NMDA receptor in central nervous system plasticity and pathology. J. NIH Res. **1:** 65–74.
2. MCCASLIN, P. P. & T. G. SMITH. 1990. Low calcium-induced release of glutamate results in autotoxicity of cerebellar granule cells. Brain Res. **513:** 280–285.
3. WEISS, J. H., D. M. HARTLEY, J. KOH & D. W. CHOI. 1990. The calcium channel blocker nifedipine attenuates slow excitatory amino acid neurotoxicity. Science **247:** 1474–1477.
4. ELLRÉN, K. & A. LEHMANN. 1989. Calcium dependency of N-methyl-D-aspartate toxicity in slices from the immature rat hippocampus. Neuroscience **32:** 371–379.
5. LYSKO, P. G., J. A. COX, M. A. VIGANO & R. C. HENNEBERRY. 1989. Excitatory amino acid neurotoxicity at the N-methyl-D-aspartate receptor in cultured neurons: pharmacological characterization. Brain Res. **499:** 258–266.
6. VINCENT, S. R., C. H. S. MCINTOSH, A. M. J. BUCHAN & J. C. BROWN. 1985. Central somatostatin systems revealed with monoclonal antibodies. J. Comp. Neurol. **238:** 169–186.
7. COHN, M. L. & M. COHN. 1975. Barrel rotation induced by somatostatin in the non-lesioned rat. Brain Res. **96:** 138–141.
8. BALABAN, C. D., V. P. STARCEVIC & W. B. SEVERS. 1989. Neuropeptide modulation of central vestibular circuits. Pharmacol. Rev. **41:** 53–90.
9. BURKE, R. E. & S. FAHN. 1983. Studies of somatostatin-induced barrel rotation in rats. Regul. Peptides **7:** 207–220.
10. BALABAN, C. D., D. A. FREDERICKS, J. N. D. WURPEL & W. B. SEVERS. 1988. Motor disturbances and neurotoxicity induced by central administered somatostatin and vasopressin in conscious rats: interactive effects of two neuropeptides. Brain Res. **445:** 117–129.
11. BALABAN, C. D. & W. B. SEVERS. 1991. Toxic effects of somatostatin in the cerebellum and vestibular nuclei: multiple sites of action. Neurosci. Res. **12:** 140–150.
12. CARLSEN, J. & J. S. DEOLMOS. 1981. A modified cupric-silver technique for the impregnation of degenerating neurons and their processes. Brain Res. **208:** 426–431.
13. BALABAN, C. D. The use of selective silver degeneration stains in neurotoxicology: lessons from studies of selective neurotoxicants. In The Vulnerable Brain: Nutrition and Toxins. K. F. Jensen & R. L. Isaacson, Eds. Plenum Press. New York, N.Y. (In press.)
14. BALABAN, C. D. 1985. Central neurotoxic effects of intraperitoneally administered 3-acetylpyridine, harmaline and nicotinamide in Sprague-Dawley and Long-Evans rats: a critical review of central 3-acetylpyridine neurotoxicity. Brain Res. Rev. **9:** 21–42.
15. BALABAN, C. D., J. P. O'CALLAGHAN & M. L. BILLINGSLEY. 1988. Trimethyltin-induced neuronal damage in the rat brain: comparative studies using silver degeneration stains, immunohistochemistry and immunoassay for neurotypic and gliotypic proteins. Neuroscience **26:** 337–361.
16. MUGNAINI, E., W. B. WARR & K. K. OSEN. 1980. Distribution and light microscopic features of granule cells in the cochlear nuclei of cat, rat and mouse. J. Comp. Neurol. **191:** 581–606.
17. LEBLANC, R., S. GAUTHIER, M. GAUVIN, R. QUIRION, R. PALMOUR & H. MASSON. 1988. Neurobehavioral effects of intrathecal somatostatinergic treatment in primates. Neurology **38:** 1887–1890.
18. LONG, J. B. 1988. Spinal subarachnoid injection of somatostatin cause neurological deficits and neuronal injury in rats. Eur. J. Pharmacol., **149:** 287–296.
19. GAUMANN, D. M., T. L. YAKSH, C. POST, G. L. WILCOX & M. RODRIGUEZ. 1989. Intrathecal somatostatin in cat and mouse: studies on pain, motor behavior and histopathology. Anesth. Analg. **68:** 623–632.
20. THOMAS, A. 1905. Atrophie lamellaire des cellules de Purkinje. Rev. Neurol. **18:** 917–924.

21. ARCHAMBAULT, L. 1918. Parenchymatous atrophy of the cerebellum. J. Nerv. Ment. Dis. **48:** 273–312.
22. LHERMITTE, M. J. 1922. L'astasie-abasie cérébelleuse par atrophie vermienne chez le vieillard. Rev. Neurol. **38:** 313–316.
23. MARIE, P., C. FOIX & T. ALAJOUANINE. 1922. De l'atrophie cérébelleuse tardive à pré-dominance corticale (atrophie parenchymeuse primitive des lamelles du cervelet, atrophie paléocérébelleuse primitive). Rev. Neurol. **38:** 1082–1111.
24. PARKER, H. L. & J. W. KERNOHAN. 1933. Parenchymatous cortical cerebellar atrophy (chronic atrophy of Purkinje's cells). Brain **56:** 191–212.
25. BALOH, R. W., R. D. YEE & V. HONRUBIA. 1986. Late cortical cerebellar atrophy: clinical and oculographic features. Brain **109:** 159–180.
26. TWERY, M. J. & J. P. GALLAGHER. 1989. Somatostatin hyperpolarizes neurons and inhibits spontaneous activity in the rat dorsolateral septal nucleus. Brain Res. **497:** 315–324.
27. MIYOSHI, R., S. KITO, S. KATAYAMA & S. U. KIM. 1989. Somatostatin increases intracellular Ca^{2+} in cultured rat hippocampal neurons. Brain Res. **489:** 361–364.
28. JANIS, R. A., P. J. SILVER & D. J. TRIGGLE. 1987. Drug action and cellular calcium regulation. Adv. Drug Res. **16:** 309–591.
29. TSUNOO, A., M. YOSHII & T. NARAHASHI. 1986. Block of calcium channels by enkephalin and somatostatin in neuroblastoma-glioma hybrid NG108-15 cells. Proc. Natl. Acad. Sci. USA **83:** 9832–9836.
30. NUSSINOVITCH, I. 1989. Somatostatin inhibits two types of voltage-activated calcium currents in rat growth hormone secreting cells. Brain Res. **504:** 136–138.
31. TANK, D. W., M. SUGIMORI, J. D. CONNOR & R. LLINAS. 1988. Spatially resolved calcium dynamics of mammalian Purkinje cells in cerebellar slices. Science **242:** 773–777.
32. LLINAS, R. & M. SUGIMORI. 1980. Electrophysiological properties of *in vitro* Purkinje cell somata in mammalian cerebellar slices. J. Physiol. London **305:** 171–195.
33. LLINAS, R. & M. SUGIMORI. 1980. Electrophysiological properties of *in vitro* Purkinje cell dendrites in mammalian cerebellar slices. J. Physiol. London **305:** 197–213.
34. DOOLEY, D. J., M. LICKERT, A. LUPP & H. OSSWALD. 1988. Distribution of [^{125}I]ω-conotoxin GVIA and [^{3}H]isradipine binding sites in the central nervous system of rats of different ages. Neurosci. Lett. **93:** 318–323.
35. KERR, L. M., F. FILLOUX, B. M. OLIVERA, H. JACKSON & J. K. WALMSLEY. 1988. Autoradiographic localization of calcium channels with [^{125}I]ω-conotoxin in rat brain. Eur. J. Pharmacol. **146:** 181–183.
36. MCGUINNESS, T. Y., Y. LAI & P. GREENGARD. 1985. Ca^{2+}/calmodulin-dependent protein kinase II. Isozymic forms from rat forebrain and cerebellum. J. Biol. Chem. **260:** 1696–1704.
37. BAIMBRIDGE, K. G. & J. J. MILLER. 1982. Immunohistochemical localization of calcium-binding protein in the cerebellum, hippocampal formation and olfactory bulb of the rat. Brain Res. **245:** 223–229.
38. ERONDU, E. & M. B. KENNEDY. 1985. Regional distribution of type II Ca^{2+}/calmodulin dependent protein kinase in rat brain. J. Neurosci. **5:** 3270–3277.
39. KINCAID, R. L., C. D. BALABAN & M. L. BILLINGSLEY. 1987. Differential localization of calmodulin-dependent enzymes in the rat brain: evidence for selective expression of cyclic nucleotide phosphodiesterase. Proc. Natl. Acad. Sci. USA **84:** 1118–1122.
40. BALABAN, C. D., M. L. BILLINGSLEY & R. L. KINCAID. 1989. Evidence for transsynaptic regulation of calmodulin-dependent cyclic nucleotide phosphodiesterase in cerebellar Purkinje cells. J. Neurosci. **9:** 2374–2381.
41. ITO, M. 1984. The Cerebellum and Neural Control. Plenum. New York, N.Y.
42. BALABAN, C. D., A.-J. ROSKAMS & W. B. SEVERS. 1988. Diazepam attenuation of somatostatin-induced motor disturbances and neurotoxicity. Brain Res. **458:** 91–96.
43. COLLINGRIDGE, G. L. & R. A. J. LESTER. 1989. Excitatory amino acid receptors in the vertebrate central nervous system. Pharmacol. Rev. **40:** 143–210.

Preservation of the Electrical-Evoked Vestibuloocular Reflex and Otolith-Ocular Reflex in Two Patients with Markedly Impaired Canal-Ocular Reflexes[a]

R. W. BALOH,[b] J. OAS,[b] V. HONRUBIA,[c]
AND D. M. MOORE[d]

[b]Department of Neurology
[c]Department of Surgery/Head and Neck
University of California at Los Angeles
School of Medicine
Los Angeles, California 90024

[d]Department of Otolaryngology
Northwestern University Medical School
303 East Chicago Avenue
Searle Building, 12-561
Chicago, Illinois 60611

INTRODUCTION

Unlike caloric and rotational tests, the electrical-evoked vestibuloocular reflex (EVOR) offers the potential of site-of-lesion testing. Since the current activates peripheral afferents independent of peripheral receptors, the vestibular nerve and central pathways can be evaluated separately from the vestibular receptors. Furthermore, adaptive changes within the central VOR can be assessed even in patients with damaged peripheral receptors. We identified two patients with absent response to caloric stimulation and markedly decreased response to horizontal angular rotation who had preserved EVOR and otolith-ocular reflex (OOR) responses.

METHODS

Caloric and rotational stimulation of the horizontal semicircular canal was conducted with standard techniques.[1] For the EVOR, constant-current sinusoidal stimuli (± 4 mA) were applied through Ag-AgCl electrodes placed on each mastoid as the subject sat in a stationary chair in a darkened room.[2] For OOR testing, subjects sat with eyes open in darkness on a parallel swing facing the side so that the linear acceleration occurred along the interaural axis (0.3 Hz, peak acceleration $0.48 \times g$).[3] Data were digitized at a rate of 200 S/second, fast components were removed, and a fast fourier transform was performed giving the magnitude, phase, and DC bias of the fundamental and first 5 harmonics.[1]

[a]This work is supported by National Institutes of Health Grant DC00097.

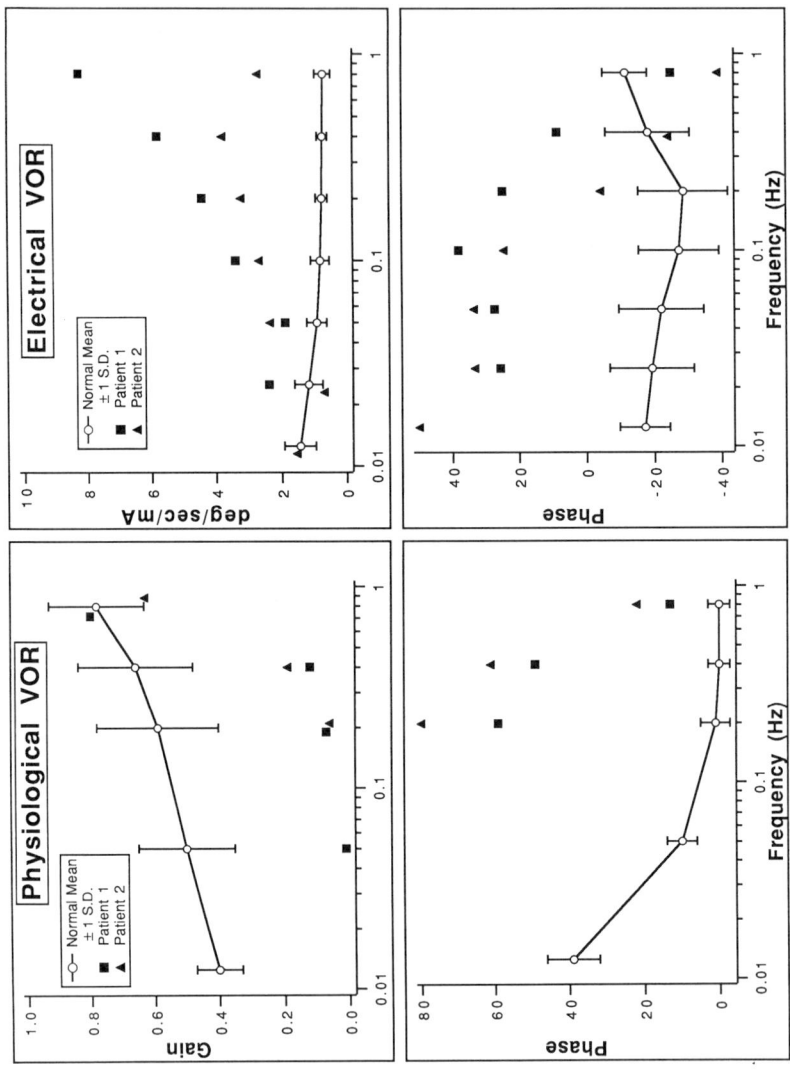

FIGURE 1. Plots of gain and phase versus frequency for the physiologic horizontal canal–ocular reflex and the EVOR in 2 patients with absent caloric responses. Normal mean (±1 standard deviation) is given for comparison.

TABLE 1. Horizontal Linear Vestibuloocular Reflex in Two Patients with Idiopathic Bilateral Vestibulopathy (0.3 Hz, peak amplitude 0.48 × *g*)

	Gain		Phase Lead
	(deg/meter)	(deg/sec/g)	(deg)
Patient No. 1	8.2	47.0	+85
Patient No. 2	2.2	12.5	+60
Normal (*n* = 10)	4.7 ± 1.4	26.9 ± 8.0	+20 ± 9

RESULTS

Both patients exhibited absent response to horizontal angular rotation at low frequencies but normal responses at higher frequencies (FIGURE 1, left side). By contrast, the EVOR gain was normal at low frequencies and progressively increased at higher frequencies, reaching values several times normal (FIGURE 1, right side). The OOR gain was increased in patient 1 and low normal in patient 2 (TABLE 1). All 3 responses in both patients had an increased phase lead at low frequencies.

CONCLUSION

The fact that these two patients with absent response to caloric stimulation and markedly decreased response to horizontal angular acceleration at low frequencies had hyperactive EVORs indicates that some afferent fibers were functioning normally. The gain and phase changes seen in the EVOR and OOR are probably the result of central compensation.

REFERENCES

1. BALOH, R. W., L. LANGHOFER, V. HONRUBIA & R. D. YEE. 1980. On-line analysis of eye movements using a digital computer. Aviat. Space Environ. Med. **56:** 563–567.
2. MOORE, D. M., L. F. HOFFMAN, K. BEYKIRCH, V. HONRUBIA & R. W. BALOH. 1991. The electrically evoked vestibulo-ocular reflex. I. Normal subjects. Otolaryngol. Head Neck Surg. **83:** 427–433.
3. BALOH, R. W., K. BEYKIRCH, V. HONRUBIA & R. D. YEE. 1988. Eye movements induced by linear acceleration on a parallel swing. J. Neurophysiol. **60:** 2000–2013.

Eye Movements in Response to Canal and Otolith Signals in Opposing Directions

ADOLFO M. BRONSTEIN AND MICHAEL A. GRESTY

MRC Human Movement and Balance Unit
Institute of Neurology
National Hospital
Queen Square
London WC1N 3BG, England

Otolith-ocular reflexes (OOR) can generate powerful slow-phase compensatory eye movements when, in addition to linear head acceleration, there is concurrent co–directional angular head acceleration.[1–3] This is achieved with a rotating chair by placing the subject's head in front of the axis of rotation. By contrast, the present

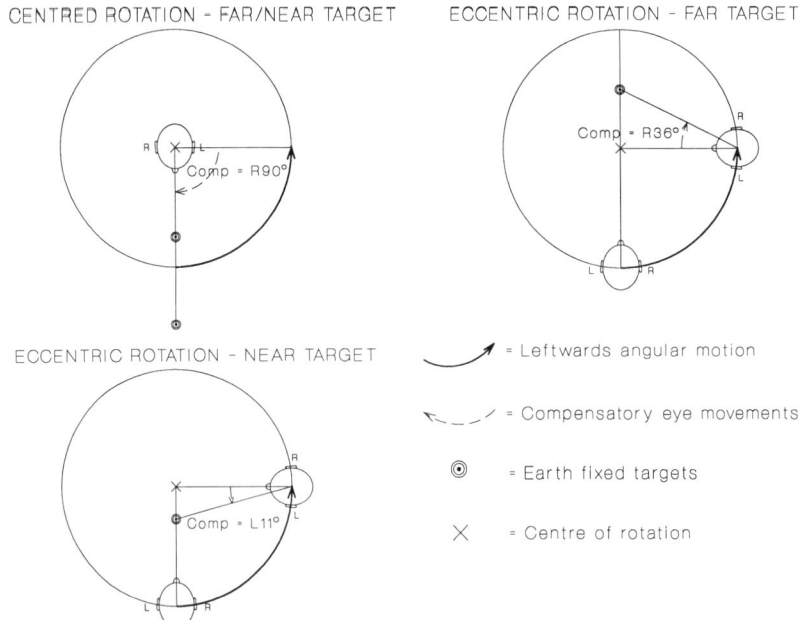

FIGURE 1. Diagram showing the geometrical relations during rotation with head centred and eccentric with the face pointing inwards towards the axis of rotation. For clarity a hypothetical 90° leftwards stimulus is shown for all conditions. With the head centred the required compensatory eye movement is 90° rightwards. With head eccentric while fixating a target beyond the rotational axis ("far target"), the required eye movement is less than 90° rightwards. Fixating a target proximal to the rotational axis ("near target") requires a leftwards compensatory eye movement, i.e., in the same direction of head rotation and in the opposite direction to head linear translation.

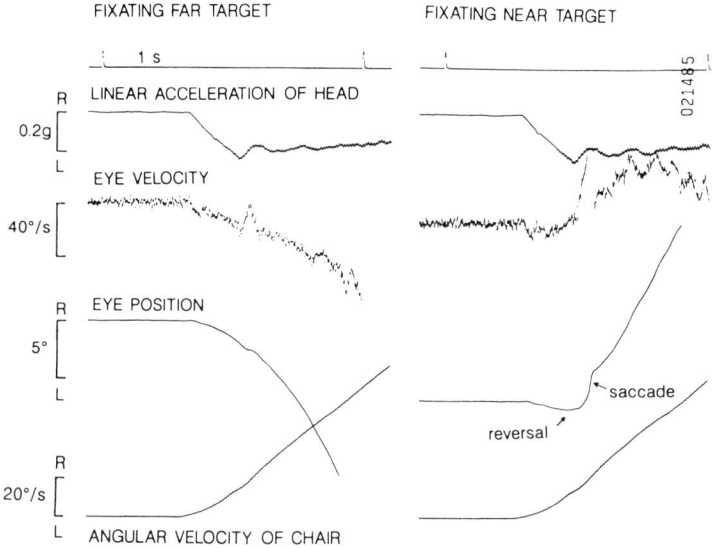

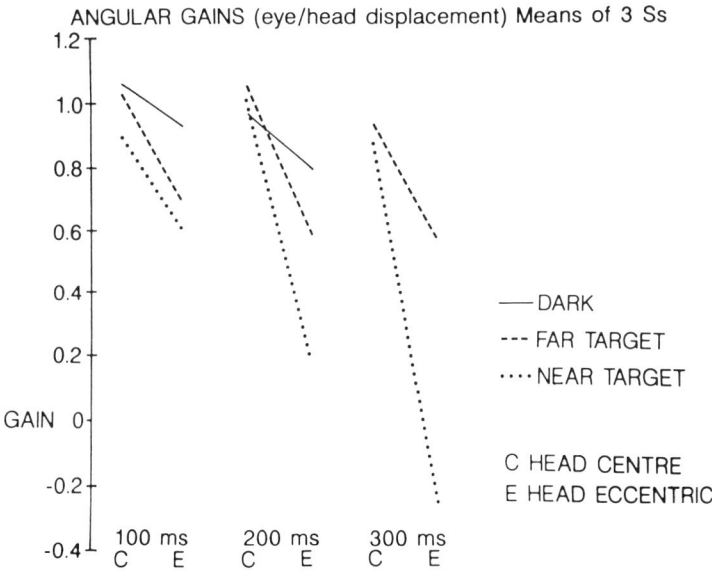

FIGURE 2. TOP: Eye movements elicited during head-eccentric rotation showing short latency (ca. 40 mseconds) slow-phase responses in the opposite direction of head rotation during fixation of both the far and near targets. After 150 ms, during near target fixation, the eye movement reverses direction and shows a "catch-up" saccade. **BOTTOM:** Angular gains (eye displacement/angular head displacement) during rotation in the dark and fixating the near and far targets (means of three subjects) calculated at 100-msecond intervals from the onset of the stimulus. Gains with head eccentric are always below unity but a true reversal of the eye movement (negative gain), as required for foveating the near target, does not occur until the 300-msecond interval. Values in the dark at 300 mseconds are not given because nystagmic beats had reset eye position. (Modified from Reference 4.)

experiments describe the eye movements elicited by placing the subject's head behind the axis of rotation of the turntable with the face pointing inwards so that the canal and otolith signals are in opposite directions in the rightwards-leftwards sense. Clinical assessment of vestibuloocular function in infants is often performed with this type of motion, by holding the baby by the shoulders and swinging him from side to side. During this maneuver, in which as one rotates the subject, say, to the right the head is linearly displaced to the left, the relative position of the fixation point can be expected to markedly influence the eye movement response, as explained in FIGURE 1. If the subject is fixating an object beyond the axis of rotation ("far target") the eye movement required is less than that required if there was purely rotational head movement. If the subject fixates an earth-fixed object placed between the head and the axis of rotation, the compensatory eye movement required would have to be in the same direction of head rotation and in the opposite direction to head translation. Thus, the experimental paradigm assesses, as in a "tug of war," the relative strengths of otolith and canal ocular reflexes.

Stimuli were delivered with a 120 Nm turntable with a chair positioned eccentrically so that the subject faces inwards, with his eyes 70 cm away from the axis of rotation. The three normal subjects tested fixated self-illuminated earth-fixed targets either at 35 cm or at 185 cm from the eyes in a dimly lit room. The stimulus demand consisted of a single cycle of sinusoidal motion (0.3 Hz, 80°/second peak) which due to torque limitations produced an initial 230-msecond acceleration ramp from 0 to 0.2g, followed by another 200 mseconds of 0.16–0.2g constant acceleration. Eye movements were recorded with infrared oculography with emphasis on the first 200 mseconds of the ocular response. The head was held with a dental bite.

FIGURE 2 (top) shows typical responses during head eccentric rotation. During fixation of the near target the response begins with a slow-phase movement in the opposite direction to chair rotation and, therefore, in the wrong direction for compensation. After 200 mseconds a reversal in direction occurs, with saccades frequently contributing to target foveation, giving the typical appearance of a visually guided (pursuit) response. During fixation of the far target the amplitude of the eye movement was reduced with respect to that required during purely angular stimulation which resulted in the compensatory response being scaled appropriately at short latency (FIGURE 2 top and bottom).

Therefore, during "conflicting" angular-linear head motion when fixating a target beyond the axis of rotation the OOR subtracts from the canal reflex to produce a compensatory movement of appropriate gain. However, for a target positioned between the subject and the axis of rotation, which requires a compensatory eye movement in the same direction as head rotation, the OOR is not able to operate at a sufficiently high gain to counteract the canal-ocular reflex.

REFERENCES

1. GRESTY, M. A., A. M. BRONSTEIN & H. BARRATT. 1987. Eye movement responses to combined linear and angular head movement. Exp. Brain Res. **65:** 377–384.
2. TAKEDA N., M. IGARASHI, I. KOIZUKA, S. CHAE & T. MATSUNAGA. 1990. Recovery of the otolith-ocular reflex after unilateral deafferentation of the otolith organs in squirrel monkeys. Acta Otolaryngol. **110:** 25–30.
3. VIIRRE, E., D. TWEED, K. MILNER & T. VILLIS. 1986. A re-examination of the gain of the VOR. J. Neurophysiol. **56:** 439–450.
4. BRONSTEIN, A. M. & M. A. GRESTY. 1991. Compensatory eye movements in the presence of conflicting canal and otolith signals. Exp. Brain Res. **85:** 697–700.

The Effects of Unidirectional Visual Surround Translation on Detection of Physical Linear Motion Direction[a]

A Psychophysical Scale for Vection

T. R. CARPENTER-SMITH AND D. E. PARKER

Department of Psychology
Miami University
Oxford, Ohio 45056

Most previous attempts to assess perception of self-motion while the body is stationary (vection) have employed the magnitude estimation procedure.[1] This approach encounters serious difficulties for several reasons including lack of stimulus control and motion reference variability. The goal of our research is development and validation of a linear vection scale, based on acceleration level (dB AL re 10^{-3} cm/second^2), that would be analogous to the SONE and PHON scales commonly used in acoustics.[3]

METHODS

Two subjects were trained until they achieved 95% correct detection in darkness of one-half cycle sinusoidal linear translations at 85 dB AL using the apparatus illustrated in FIGURE 1. Five types of trials were presented to each subject during data collection. These were (1) D (darkness) cart motion only, no illumination, eyes closed; (2) CF (congruent forward) belt motion simulating forward motion of the subject paired with forward motion of the cart; (3) CB (congruent backward) belt motion simulating backward motion of the subject paired with backward motion of the cart; (4) IF (incongruent forward) belt motion simulating backward motion of the subject paired with forward motion of the cart; (5) IB (incongruent backward) belt motion simulating forward motion of the subject paired with backward motion of the cart. The cart motion half-cycle period was 2.5 seconds, and belt velocity was constant at 46 deg/second. Between trials the cart was repositioned to the middle of the track at a subthreshold velocity.

Subjects reported direction of cart (self) motion for each trial using a two-category ("forward" or "backward") forced-choice procedure. An experimental session consisted of three congruent trials and seven incongruent trials in randomized order for each direction at three different ALs.

[a] Supported by National Aeronautics and Space Administration Grant NAG 9-446/Basic and National Aeronautics and Space Administration Training Grant NGT-50427.

817

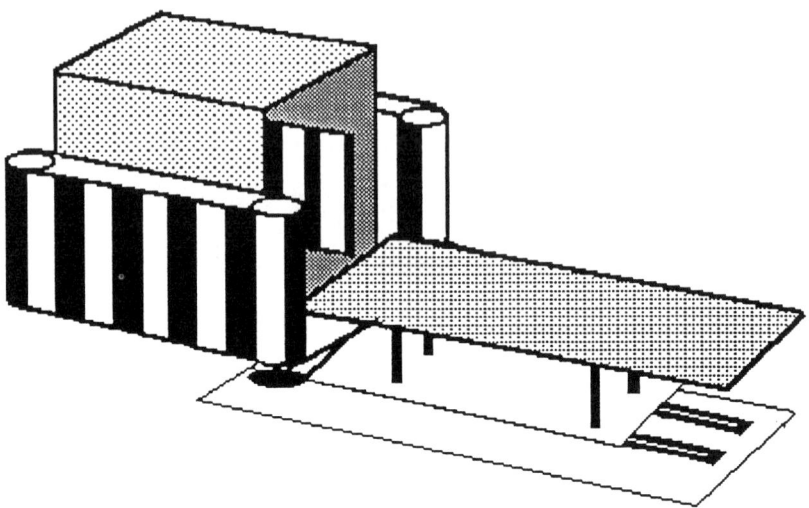

FIGURE 1. The stimulus apparatus consisted of a gurney cart and a visual surround made of a pair of movable belts mounted on the cart. The cart provided translational motion of both the subject and the belts. The subject lay prone with the head dorsal flexed. The visual surround viewed through lateral openings in a masking box was 60 deg vertical and 65 deg horizontal. The belts were covered with a pattern of black and white vertical stripes (0.037 cycles/deg), and illumination was provided by incandescent lamps mounted between the masking box and the belts.

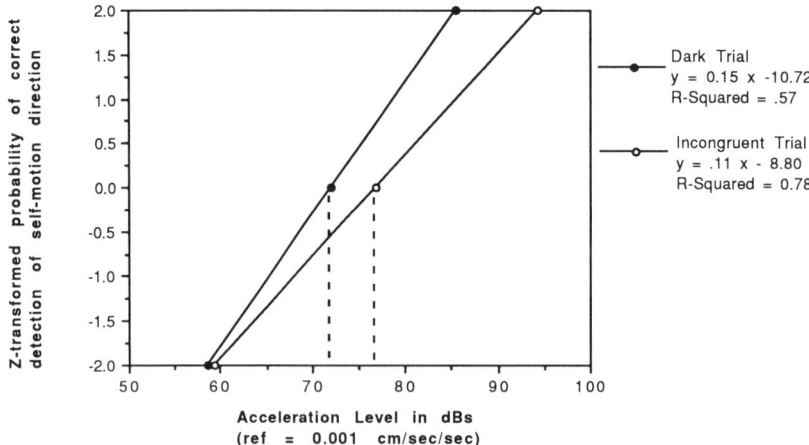

FIGURE 2. Z-transformed probabilities of correct detection of cart motion direction are plotted against AL. The plot summarizes the results obtained in darkness and during exposure to incongruent cart and visual surround motion for one subject. For the dark data, a z-score of zero corresponds to a probability of 0.75; for the incongruent data, the zero z-score value corresponds to a probability of 0.5. For this subject, the vection experienced was 4.8 dB AL.

RESULTS

The average magnitude of linear vection across the two subjects under the conditions employed in this experiment was 5.13 dB AL. The data for one of the subjects are illustrated in FIGURE 2.

DISCUSSION

The results of this experiment suggest that a VEC scale of linear vection can be developed. VEC scale data could be used to inform decisions regarding development and implementation of any motion simulator, including the Preflight Adaptation Trainers being developed at Johnson Space Center.[2]

REFERENCES

1. HOWARD, I. P. 1986. The perception of posture, self-motion and the visual vertical. *In* Handbook of Perception and Human Performance. K. Boff, L. Kaufman & J. Thomas, Eds. **1:** 18-1 to 18-62. John Wiley and Sons. New York, N.Y.
2. PARKER, D. E. & K. L. PARKER. 1990. Adaptation to the stimulus rearrangement of weightlessness. *In* Motion and Space Sickness. G. Crampton, Ed.: 247–262. CRC Press. Boca Ratan, Fla.
3. PARKER, D. E., T. R. CARPENTER-SMITH, J. SHERMAN & S. DOUGLAS. 1990. Acceleration level as a measure of linear vection: an alternative to magnitude estimation. Aviat. Space Environ. Med. **61**(12): A14.

Variation of Gravitoinertial Force and Its Influence on Ocular Torsion and Caloric Nystagmus

A. H. CLARKE, W. TEIWES, AND H. SCHERER

Department of Otorhinolaryngology
Steglitz Medical Center
Free University of Berlin
Berlin, Germany

Experimentation in the hypo- and hypergravic G continuum made available by centrifuge and parabolic flight facilities has provided much relevant information on the performance of the vestibulooculomotor function. This has been complemented by the findings of those experiments conducted in the extended microgravity of spaceflight.

In the light of more recent considerations on the three-dimensionality of spatial orientation, adequate recording and measurement of the vestibulooculomotor response in all three orthogonal planes has become a precondition for such studies. As an alternative to the semiinvasive search coil technique, videooculography (VOG) has proved to be particularly suitable for applications in these more difficult environments. The findings reported here are intended to contribute to our understanding of the synergistic interaction between the otolithic and canalicular systems. Ocular torsion is of particular interest here, since it is elicited both by stimulation to the utricles and to the vertical canals; thus, by selection of translational and/or angular acceleratory stimuli, the contributions of one or other of the receptor systems can be examined.

OCULAR TORSION

Within this context, two experiments on ocular torsion were carried out. Firstly, the variation of gravitoinertial force during the parabolic flight maneuver was exploited to provide dynamic modulation of the translatory acceleration along the mean interutricular plane. Thus, each test subject ($n = 9$) was positioned for optimal lateral utricular stimulation by the G_Z force (i.e., left-ear down and facing forward). Eye movements were recorded throughout repeated parabolae by means of videooculography. G_Z acceleration was recorded synchronously on an auxiliary channel.

Thus the linearity between ocular counterrolling (OCR) and translatory G force could be verified and, using the (quasi-) step functions at the 2 G–0 G and 0 G–2 G transitions as input, the transfer function of the otolith-ocular response could be estimated. The estimated frequency responses from these experiments facilitate to some extent the separation of the otolithic and canal contributions. The differential response of the irregular utricular cells is reflected by a steady increase in gain at lower frequencies (< 2 Hz).

The second study involved simultaneous sinusoidal stimulation to the otolithic and vertical canal receptors by means of passive roll at a number of frequencies (0.05, 0.075, 0.10, 0.15 Hz, with a peak amplitude of 45°, axis of rotation 30 cm caudal to

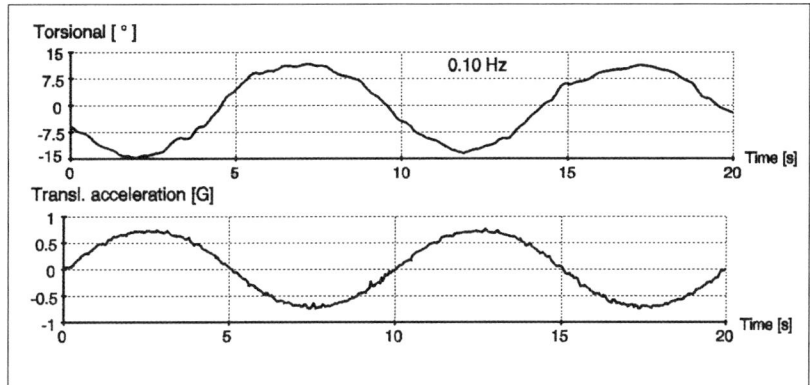

FIGURE 1. Dynamic ocular counterrolling during passive roll (0.10 Hz, ±45° about horizontal). Upper trace, torsional component of eye movement. Center trace, reconstructed slow-phase component. Lower trace, imposed G force.

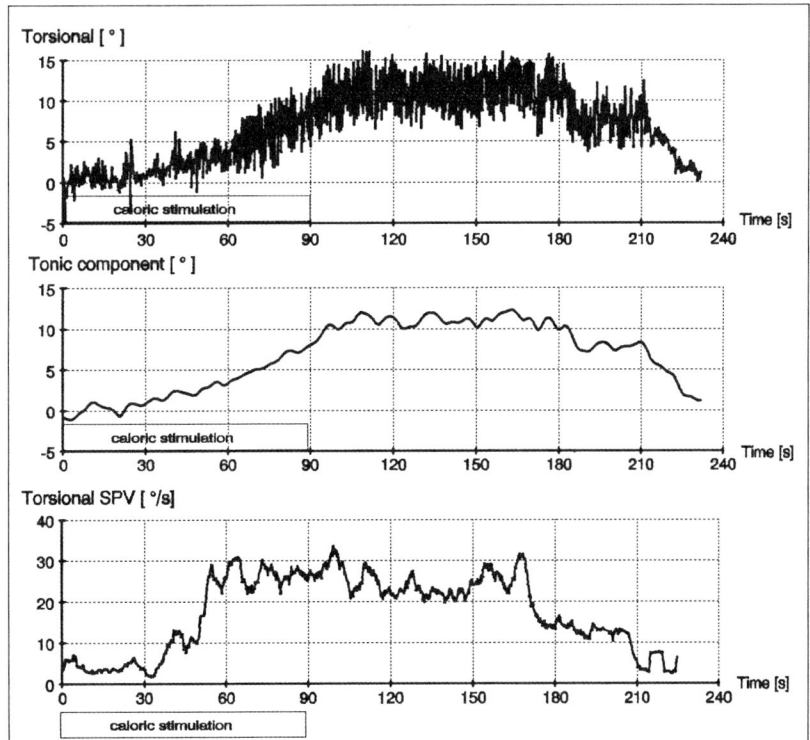

FIGURE 2. Caloric response—the torsional component: analysis of the composite torsional eye movement (top) is low-pass filtered (1.25 Hz) to extract the systematic "tonic" component (center); after simple differentiation of the composite eye movement, a first estimate of torsional SPV is obtained (bottom).

interaural axis). Here again the frequency response could be estimated. The differential response of the irregular cells of the utricle and/or the velocity-related canal output is reflected by the phase lead in the "roll" response.

THREE-DIMENSIONAL CALORIC RESPONSE

The dispute on the peripheral mechanisms involved in the caloric response remains unresolved. The present study was primarily intended to reexamine earlier findings on the influence of hypergravic gravitoinertial force on the intensity of the caloric nystagmus response. The tests were carried out in a human centrifuge. G levels of 1.5, 2.0, 2.5, and 3.0 G, with the subject pitched back to 70° (i.e., presumed optimum position) or forward 20° (pessimum position). Caloric stimulation was presented by unilateral air insufflation (right ear, 90 seconds @ 18°C). Eye movements were recorded and evaluated by the VOG technique.

The findings verify the presence of a nystagmus response with horizontal, vertical, and torsional components at all G levels tested. However, the relationship between nystagmus intensity (SPV) and G force was not found to be linear.

Of considerable interest was the systematic observation—in addition to the torsional nystagmus—of a tonic ocular torsion. This is interpreted as evidence of a utricular response to caloric stimulation, and would therefore represent a means of examining unilateral utricular function.

Age-Related Changes in Human Smooth Pursuit Responses to Horizontal Step-Ramp Target Trajectories

IAN S. CURTHOYS,[a] SUSAN L. WEARNE,[a]
MELISSA S. STAPLES,[a] SWEE T. AW,[b]
MICHAEL J. TODD,[b] AND G. MICHAEL HALMAGYI[b]

[a]Department of Psychology
University of Sydney
Sydney, New South Wales 2006, Australia

[b]Eye and Ear Research Unit
Neurootology Department
Royal Prince Alfred Hospital
Camperdown, New South Wales, Australia

The evidence from monkey studies is that time domain analysis of smooth pursuit eye movements, especially the onset of smooth pursuit to step-ramp target trajectories, can index motion processing in extrastriate cortex (MT and MST areas) (see Reference 1 for a review). Thurston *et al.* using step-ramp target trajectories have reported that human patients with lesions in cortical areas encompassing the human homologue of MT and MST show deficits in smooth pursuit[2] that correspond to those found in monkeys, and we sought to investigate such deficits further. The present study aimed (1) to develop norms for different age groups for horizontal smooth pursuit in the time domain using step-ramp target trajectories documenting eye position, eye velocity, eye position error and eye velocity error and (2) to compare the performance of human patients having lesions in areas encompassing the human homologue of MT and MST with the performance of age-matched normal healthy controls.

The head was held firmly by a sturdy wooden head holder. The target was a back-projected laser dot on a screen 1.0 meters from the subject's eye moving in one of 10 target waveforms: 3° step; ramp at 0°/second, 5°/second, or 10°/second towards or away from the center of the screen; the waveforms were given in random order and each waveform was repeated 4 times. Standard search coils yielded eye position; eye velocity was obtained by analog differentiation. Other parameters: 16-bit A-D conversion precision, 1 kHz sampling, filtering (16 Hz low-pass FIR filter), and desaccading. The desaccaded data were analyzed using a graphical-statistical analysis package in UNIX: S and NewS on a Microvax III. The present results have been obtained from 59 healthy people ranging from 18 to 86 years: 18–34 ($n = 20$), 35–49 ($n = 17$), 50–64 ($n = 14$), 65+ ($n = 8$). In addition we tested a number of stroke patients some of whom had well-documented lesions of human cortical areas homologous to MT and MST in the monkey.

The plots show the group data as the range ± 1.96 standard deviations (SD) from the mean, within which shaded area we expect 95% of normal healthy subjects' responses should lie, and the patients' results are superimposed on this range (see FIGURE 1).

The eye position and eye position error plots show that older subjects, as

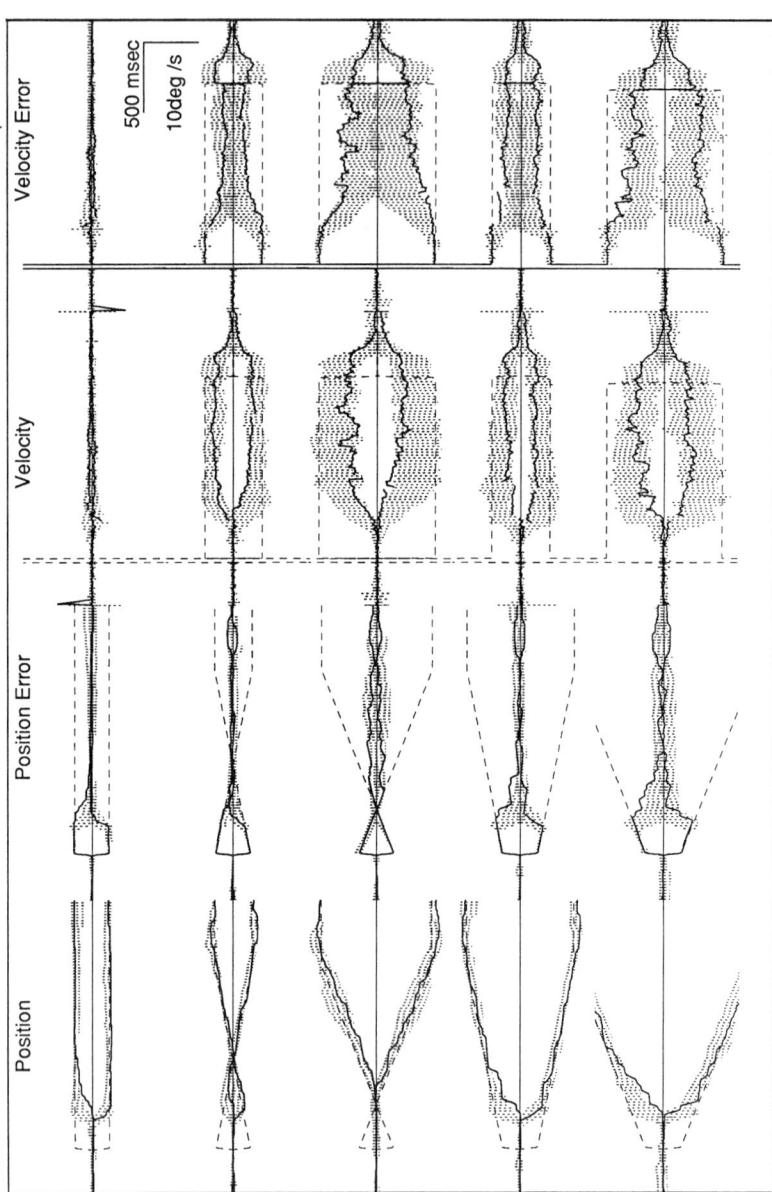

FIGURE 1. Patient JR, age 52, re age-matched healthy controls 50–65. The 4 panels of graphs are different time domain representations of subjects' performances to the different stimulus waveforms. Each row shows the group mean ± 1.96 SDs (hatched areas), and the average responses of the patient are shown in relation to the age-matched controls for the one target trajectory (dashed line) directed either to the subject's right (down), or left (up).

expected, tend to use more saccades in their pursuit and that result is complemented by the eye velocity and eye velocity error traces which show that older subjects generate significantly smaller peak eye velocities to 10 deg/second ramps than young subjects. Some patients with known cortical lesions that incorporate the probable human homologue of MT and MST show deficits of smooth pursuit. But other patients with large lesions show performance on this test that is not detectably different from their age-matched controls.

Wurtz *et al.* note that the distinctive smooth pursuit deficits found in the monkey consequent on ibotenic acid lesion of MT and MST are short-lived.[1] It would seem that in some human stroke patients the pursuit deficit may also be short-lived and such rapid recovery, together with the rather wide range of normal performance on this test by humans as opposed to monkeys, will limit the clinical usefulness of smooth pursuit as an index of human visual motion processing.

REFERENCES

1. WURTZ, R. H., H. KOMATSU, D. S. G. YAMASAKI & M. R. DÜRSTELER. 1990. Cortical visual motion processing for oculomotor control. *In* Vision and the Brain. B. Cohen & I. Bodis-Wollner, Eds.: 211–231. Raven Press. New York, N.Y.
2. THURSTON, S. E., R. J. LEIGH, T. CRAWFORD, A. THOMPSON & C. KENNARD. 1988. Two distinct deficits of visual tracking caused by unilateral lesions of cerebral cortex in humans. Ann. Neurol. **23:** 266–273.

VTM—a New Method of Measuring Ocular Torsion Using Image-Processing Techniques

IAN S. CURTHOYS,[a] STEVEN T. MOORE,[b]
STEVEN G. McCOY,[b] G. MICHAEL HALMAGYI,[b]
CHARLES H. MARKHAM,[c] SHIRLEY G. DIAMOND,[c]
STEVEN W. WADE,[a] AND STUART T. SMITH[a]

[a]Department of Psychology
University of Sydney
Sydney, New South Wales 2006, Australia

[b]Eye and Ear Research Unit
Neurootology Department
Royal Prince Alfred Hospital
Camperdown, New South Wales, Australia

[c]Department of Neurology
University of California at Los Angeles
School of Medicine
Los Angeles, California 90024

We have developed and validated a new, relatively inexpensive method of measuring ocular torsion.[1] The luminance distribution of a sector of the iris of the observer at rest is cross-correlated with the same sector during test, and the shift in the peak of the cross-correlation function is a measure of torsion. Our realization of this method[4] is called VTM, and this paper reports the method and validation.

THE METHOD

An IBM/AT-compatible microcomputer with an image processing card (Matrox MVP-AT) is used for capturing and storing single video frames. The video source is an IR sensitive monochrome CCD video camera fitted with a zoom lens (18–108 mm) to enable a close-up image of the iral pattern of the eye to be obtained. The iris is illuminated with an IR source. The subject's pupil is constricted by 2% pilocarpine hydrochloride. Testing is done in darkness with the exception of a small red fixation point. During the testing phase the VTM system acquires an image of the iris, thresholds it in order to identify the pupil; calculates the pupil area and locates the center of the pupil; records the gray level distribution along an arc 256 pixels long at an operator-selected radius from the pupil center; carries out a fast Fourier transform (FFT) on this (interpolated) gray level distribution; stores the parameters of this reference FFT. The above operations take about 4 seconds. During the test the gray-level distribution of the selected sector of the iris is stored for up to 50 images for off-line cross-correlation calculations, 70 of which take about 2 minutes. The magnitude of this cross-correlation is the measure of torsion. VTM has a

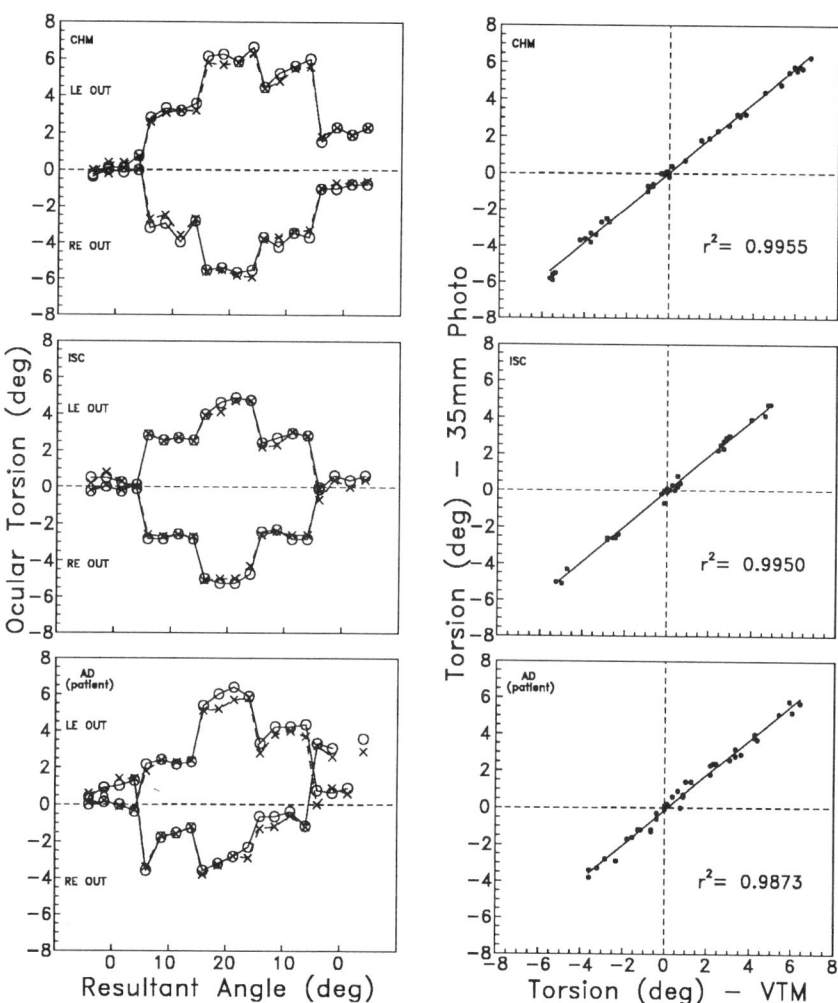

FIGURE 1. Comparison of the torsion values measured by VTM and those obtained from simultaneous measures by the standard 35-mm photographic method of the same iris of subjects undergoing linear acceleration stimulation on the human centrifuge. The stimulus values correspond to the angle of the resultant force re gravitational vertical, and the graphs show torsion for stimuli directed to the left (positive torsion) and right (negative torsion). There is very close correspondence between the torsion measured by VTM (open circles) and the torsion measured by the 35-mm photography (crosses).

resolution of the order of 0.1 deg depending on the magnification of the video camera and the arc radius selected.

VTM was validated by using it to measure known torsional rotations of an irislike pattern held in a mechanical rotator which had a torsional rotational accuracy of better than 0.01°. The results show that VTM measures torsion to the desired 0.1° accuracy: the standard deviation of the differences between real torsion and VTM-measured torsion was 0.09°.

We also validated VTM *in vivo* by directly comparing the torsion measures it provided of a human eye with the measures obtained by the standard 35-mm photographic technique[2] by having two cameras mounted on a human centrifuge and triggering the 35-mm photo by the same keypress that acquired the VTM image. The aim of this experiment was to obtain images by the two systems as close in time as possible and to compare the magnitude of the torsion measured by these 2 systems on each measurement occasion. The stimulus to induce torsion was a modest linear acceleration stimulus directed along the interaural axis which would, on the basis of previous studies, be expected to generate up to about 7° of torsion.[3] We tested both normal healthy subjects and patients in this comparison.

The subject or patient was seated 1 meter from the axis of rotation on a Servo-Med human centrifuge to which was attached a fixation light which revolved with the observer, 600 mm from their eyes. The head was firmly held, and the subjects were accelerated (at a very low angular acceleration of $1°/\text{second}^2$) to constant velocities that provided resultant angles of 10° and 20° and 10° with respect to earth vertical. After attaining these constant velocities, there was a 60-second rest period following which 5 pairs of measures (VTM and 35 mm) were taken. On successive trials the subject was positioned so that the resultant vector rotated around their nasooccipital axis toward their left or right ear.

The 35-mm photos were measured without knowledge of the VTM results. The results of this very direct comparison were almost identical (see FIGURE 1), confirming that VTM does measure human ocular torsion. We are presently using this method to measure ocular torsion produced by linear acceleration on a centrifuge in normal healthy subjects and in patients before and after unilateral vestibular neurectomy.

REFERENCES

1. MOORE, S. T., I. S. CURTHOYS & S. G. McCOY. 1991. VTM—an image-processing system for measuring ocular torsion. Comp. Methods Prog. Biomed. **35:** 219–230.
2. DIAMOND, S. G., C. H. MARKHAM, N. E. SIMPSON & I. S. CURTHOYS. 1979. Binocular counterrolling in humans during dynamic rotation. Acta Otolaryngol. **87:** 490–498.
3. MILLER, E. F. & A. GRAYBIEL. 1971. Effect of gravitoinertial force on ocular counterrolling. J. Appl. Physiol. **31:** 697–700.
4. PARKER, J. A., R. V. KENYON & L. R. YOUNG. 1985. Measurement of torsion from multitemporal images of the eye using digital signal processing techniques. IEEE Trans. Biomed. Eng. **BME-32:** 28–36.

Characterization of Yaw to Roll Cross-Coupling in the Three-Dimensional Structure of the Velocity Storage Integrator

MINGJAI DAI,[c] THEODORE RAPHAN,[a,b,c]
AND BERNARD COHEN[c,d]

[b]Department of Computer and Information Sciences
Brooklyn College of CUNY
Bedford Avenue and Avenue H
Brooklyn, New York 11210

[c]Department of Neurology
[d]Department of Physiology and Biophysics
Mount Sinai School of Medicine
One Gustave Levy Place
New York, New York 10029

INTRODUCTION

Work on the spatial organization of velocity storage has shown that the yaw axis eigenvector of its system matrix lies close to the spatial vertical and remains approximately invariant with regard to roll tilts.[1] This was determined by inducing animal yaw optokinetic nystagmus (OKN) in various roll tilts and comparing the ensuing optokinetic after-nystagmus (OKAN) with the output of a three-dimensional model of velocity storage.[2]

In this study we analyzed OKAN induced when animals were in prone and supine positions. The purpose was to determine whether principles of organization determined during roll tilt apply when there is yaw to roll axis cross-coupling during pitch tilt. The methods of data acquisition, modeling, and analysis are as described in Raphan and Sturm[2] and Dai et al.[1]

RESULTS

When animals were upright and OKN was induced about the yaw axis, the compensatory eye velocity was about the yaw axis with no compensatory pitch or roll components (FIGURE 1A, top). In darkness, the ensuing OKAN also had only a yaw component that decayed along the yaw axis towards zero in the yaw-roll state space (FIGURE 1A, bottom). When animals were placed in the prone position and yaw axis OKN was induced, there was pronounced cross-coupling to the roll axis both during

[a]Author to whom correspondence should be addressed.

OKN and OKAN (FIGURE 1B, top). In contrast to the upright position, slow-phase eye velocity approached the origin of the yaw-roll state space in a curved trajectory (FIGURE 1B, bottom). The computed eigenvector for this trajectory was tilted forward 20° relative to the spatial vertical.

When a combined yaw-roll stimulus was given about an axis along the computed eigenvector angle, i.e., 20°, yaw and roll eye velocity decayed during OKAN with approximately the same time course (FIGURE 2A, top). The trajectory was close to a straight line (FIGURE 2A, bottom left) as indicated by the narrow autocorrelation function of the residual sequence about the line and its associated broad power spectrum (FIGURE 2A, bottom right, bottom panel). This indicated that the residual sequence about the straight line was noise.

Yaw-roll OKN at 40° relative to the spatial vertical was not followed by OKAN that declined with the same time course (FIGURE 2B, top). The yaw-roll state space plot had a residual curve with a broad autocorrelation function (FIGURE 2B, bottom right, middle panel) and a narrow power spectrum (FIGURE 2B, bottom right, bottom panel), indicating a significant signal component in the residual sequence. Thus, it did not correspond to an eigenvector. This supports the calculated results that the eigenvector was at 20° relative to the spatial vertical.

Similar eigenvector angles were obtained for supine positions of the animal. However, there was an asymmetry in that there was less roll eye velocity during OKN when animals were supine than when they were prone.

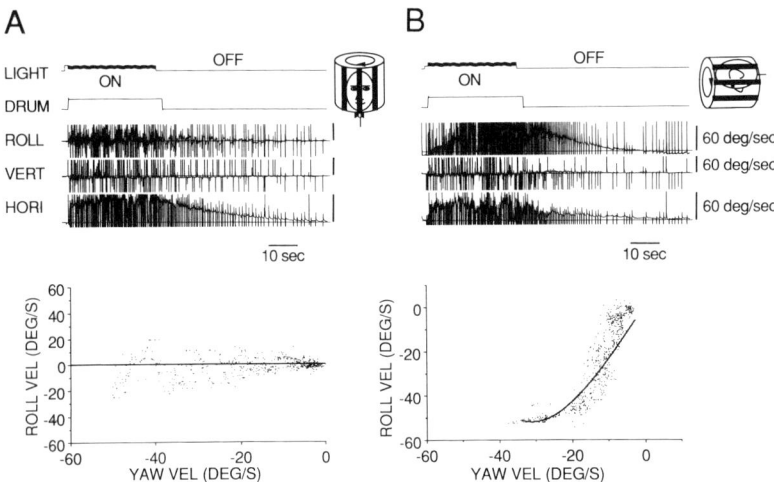

FIGURE 1. A. Top: OKN and OKAN with monkey in the upright position and receiving yaw axis stimulation. **Bottom:** Slow-phase eye velocity trajectory during OKAN. It is close to the yaw axis with little variation along the roll axis. **B. Top:** OKN and OKAN with monkey in prone position and receiving yaw axis stimulation. There was a considerable roll component during OKN, indicating an inability to suppress the cross-coupled component from yaw to roll. During OKAN the roll eye velocity decayed along with the yaw component. **Bottom:** Eye velocity trajectory during OKAN. It follows a curved trajectory which decays toward the origin of the yaw-roll state space. The computed eigenvector for this trajectory was 20° relative to the spatial vertical.

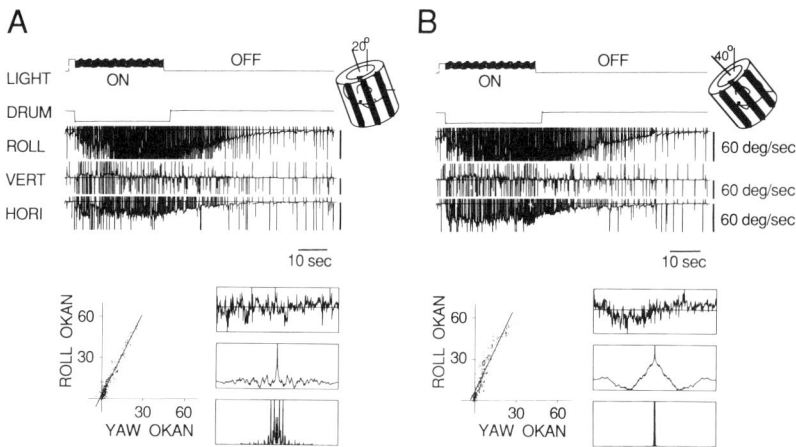

FIGURE 2. OKN and OKAN for yaw-roll stimulation at 20° (A) and 40° (B) relative to the spatial vertical with the animal in a prone position. **A. Top:** During OKN, eye velocity followed the direction of the stimulus with appropriate roll and yaw components, and during OKAN, these eye velocities decayed to zero. **Bottom left:** The trajectory of OKAN, shown in A (top) in yaw-roll state space, was close to a straight line. The residual sequence about the line (bottom right, top panel) had a narrow autocorrelation (bottom right, middle panel) and a broad power spectrum (bottom right, bottom panel). **B:** In response to 40° yaw-roll stimulation, yaw and roll eye velocities decayed at different rates. **Bottom left:** The trajectory of OKAN in yaw-roll state space, shown in B (top), was curved. The residuals of the trajectory about the straight line fit to the data (bottom right, top panel) had a broad autocorrelation function (bottom right, middle panel) and a narrow power spectrum (bottom right, bottom panel). This indicated that there was a significant signal component to the residuals and the linear fit of data was not adequate.

CONCLUSION

This study indicates that the yaw axis eigenvector lies close to the spatial vertical and remains approximately invariant as a function of head position with regard to gravity during prone-supine tilts as during roll tilts.

REFERENCES

1. DAI, M. J., T. RAPHAN & B. COHEN. 1991. J. Neurophysiol. **66:** 1422–1439.
2. RAPHAN, T. & D. STURM. 1991. J. Neurophysiol. **66:** 1410–1421.

Visual-Vestibular Interaction during High-Frequency, Active Head Movements in Pitch and Yaw[a]

JOSEPH L. DEMER,[b] JOHN G. OAS,[c]
AND ROBERT W. BALOH[c]

[b]Jules Stein Eye Institute
University of California at Los Angeles
School of Medicine
100 Stein Plaza
Los Angeles, California 90024-7002

[c]Department of Neurology
University of California at Los Angeles
School of Medicine
10833 Le Conte Avenue
Los Angeles, California 90024

Linear interaction of the vestibuloocular reflex (VOR) with pursuit is commonly assumed to mediate visual-vestibular interaction (VVI). If correct, this implies decreasing VVI at the frequencies of natural head movement (> 1.0 Hz). Without effective VVI, wearing magnifying spectacles would be expected to produce oscillopsia and reduced visual acuity during high-frequency head movements.

Evidence for VVI at high frequencies was sought by measuring VOR gain in darkness, and the visually-enhanced VOR (VVOR) during active pitch and yaw head movements in 9 normal adults (age 27 ± 5 years, mean $\pm$ standard deviation). Extreme VVI was tested for head movements during wearing of $1.9\times$ binocular telescopic spectacles (field of 23°). Subjects were trained to make sinusoidal head movements in synchrony with an audible tone modulated in pitch at discrete or continuously increasing frequencies from 0.5–6.0 Hz. Gaze (eye position in space) and head position were digitally sampled using magnetic search coils. Position data were differentiated and quick phases were replaced with interpolated values based on linear regression of slow-phase eye velocity against head velocity. Data were also analyzed by Fourier techniques.

Subjects readily learned to make nearly sinusoidal head movements (amplitude 50–100°/second, harmonic distortion $\sim 15\%$) to maximum frequencies of 4–6 Hz. For all subjects, VOR and VVOR responses under all conditions for both axes had phase $\sim 0°$. Typical data for frequency sweeps in a single subject are illustrated in FIGURE 1. Data obtained at single frequencies were consistent with swept frequency data and are summarized in FIGURE 2. At several frequencies, VOR gain was significantly lower in yaw than in pitch ($p < 0.05$); gain in both axes was < 1.0 at low frequencies, rising to a maximum at 1–3 Hz. Yaw VOR gain was directionally symmetrical; pitch VOR gain was slightly but not significantly greater for down than up slow phases at low frequencies. At all frequencies, gain during viewing without spectacles approximated the ideal value of 1.0. With $1.9\times$ telescopic spectacles, gain

[a]Supported by U.S. Public Health Service grants EY-08656 and NS-10940.

832

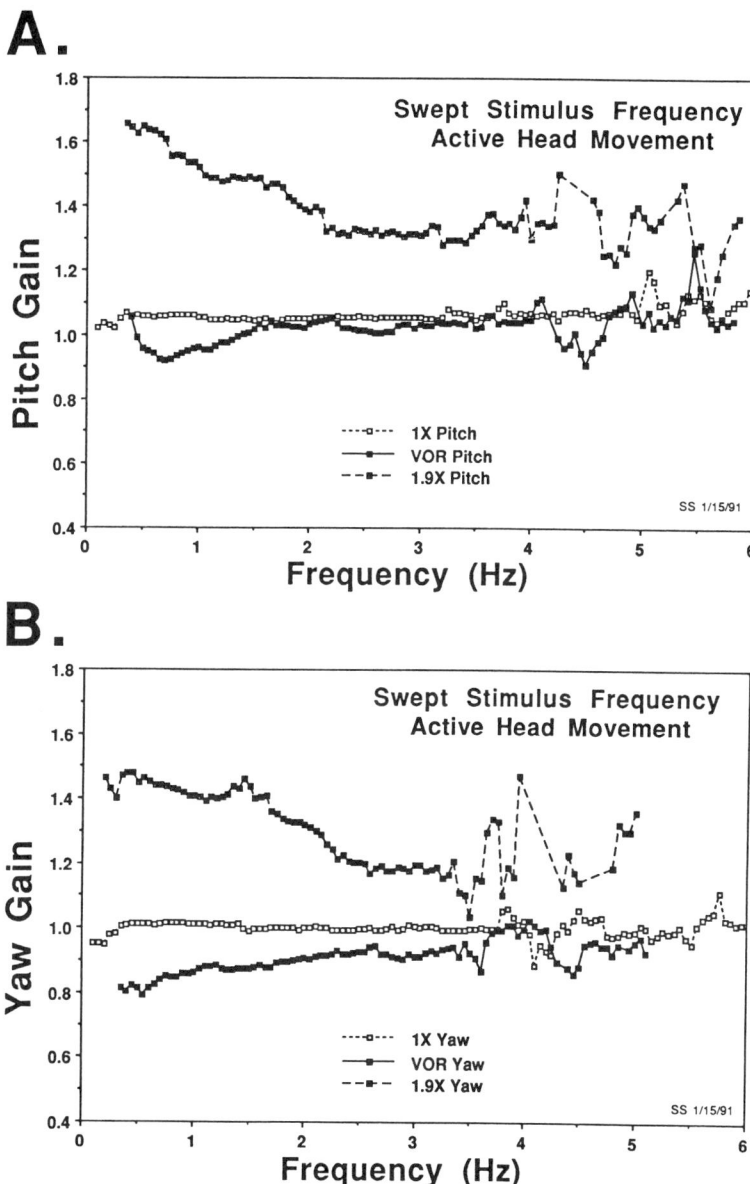

FIGURE 1. VVI data from a representative subject during active, sinusoidal head rotation in synchrony with an audible tone modulated at frequencies continuously increasing from 0.4–6.0 Hz over 20 seconds. Gains were computed by Fourier transform. VOR gain was measured in darkness. VVOR gain was measured in light with a distant fixation target during normal vision (1×) and wearing of 1.9× telescopic spectacles. Graphs plot Fourier spectrum of gain (eye velocity/head velocity) for each condition at frequencies where coherence exceeded 0.95. Despite high coherence, gain values became erratic above 4 Hz. Gain is modified by viewing condition to even the highest frequencies of head movement. **A:** Pitch. **B:** Yaw.

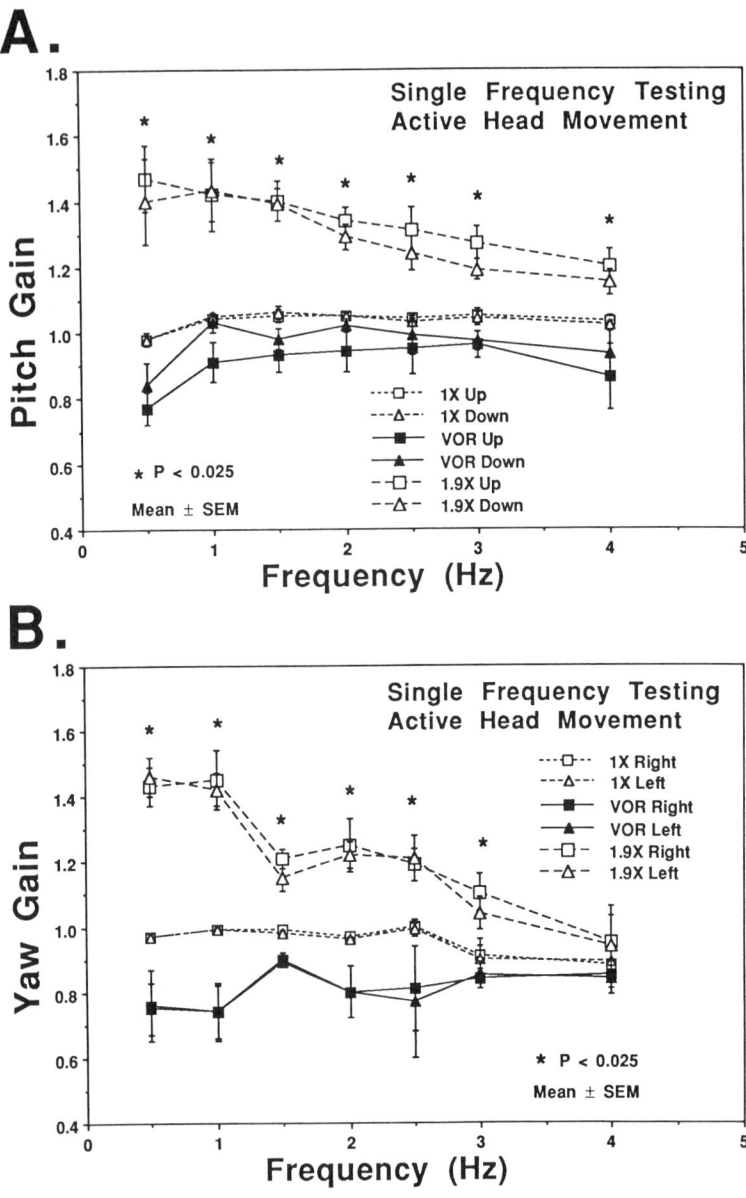

FIGURE 2. Pooled VVI data from 9 subjects during active, sinusoidal head rotation in synchrony with an audible tone modulated at discrete frequencies. Data were obtained by regression of the slow-phase eye velocity response against head velocity over 3–16 cycles at each frequency, with separate computation for each direction. Visual conditions were as in FIGURE 1. Not every subject contributed data at each frequency, due to individual variations in frequency of head shaking. SEM, standard error of the mean. Asterisks indicate significant differences, for both directions in each case, between VOR and 1.9× VVOR gains (Student's *t*-test). **A:** Pitch. **B:** Yaw.

at 0.5 Hz was a maximum of ~1.4 for pitch and yaw, and decreased but remained significantly greater than VOR gain up to 3.0 Hz for yaw and to ≥ 6.0 Hz for pitch.

Human horizontal[1] and vertical[2] pursuit has very low gain at frequencies > 1.0 Hz. The present finding of significant VVI during active head movements at frequencies up to 6.0 Hz suggests a role for other mechanisms besides interaction of VOR and pursuit. A contribution is possible from prediction, known to act during pursuit of repetitively moving targets.[3] Motor efference copy from neck musculature assists in gaze stabilization in labyrinthine-defective patients,[4] and probably contributes to VVI observed here. The effect of these nonvisual inputs is apparently modified by visual condition. Contrary to assumptions of vestibular autorotation clinical testing,[5] even at high frequencies, VVI prevents accurate VOR testing during visual fixation. While significant, VVI is insufficient to reach optimum gain for 1.9× telescopic spectacles, so oscillopsia and reduced visual acuity are likely during head movements with these devices.

REFERENCES

1. LISBERGER, S. G., C. EVINGER, G. W. JOHANSON & A. W. FUCHS. 1981. Relationship between eye acceleration and retinal image velocity during foveal smooth pursuit in man and monkey. J. Neurophysiol. **46:** 229–249.
2. BALOH, R. W. & J. DEMER. 1991. Gravity and the vertical vestibulo-ocular reflex. Exp. Brain Res. **83:** 427–433.
3. BARNES, G. R., S. F. DONNELLY & R. D. EASON. 1987. Predictive velocity estimation in the pursuit reflex response to pseudo-random and step displacement stimuli in man. J. Physiol. **389:** 111–136.
4. KASAI, T. & D. S. ZEE. 1978. Eye-head coordination in labyrinthine-defective human beings. Brain Res. **144:** 123–141.
5. O'LEARY, D. P. & L. L. DAVIS. 1990. High-frequency auorotational testing of the vestibulo-ocular reflex. Neurol. Clin. **8:** 297–312.

Segregation of Visual Features Is Not the Basis of Visual Orientation Relative to Gravity

THOMAS EGGERT

Max-Planck Institut für Verhaltensphysiologie
Abteilung Mittelstaedt
8130 Seewiesen, Germany

A strong deviation of the subjective vertical (SV) of humans lying in a horizontal body position can be caused by an inducing pattern (IP) presented at different orientations. The dependence of the effect on time and space relations between the test pattern (TP), which is used to measure the SV, and the IP[1,2] as well as the insensitivity of the effect to changes of the geometric form of the TP[3] furnishes evidence that besides pure visual interactions between TP and IP, the connection of visual and otolith afferents is the essential mechanism. The visual input extracted from the IP for this mechanism is considered.

A first experiment shows that a homogeneous texture consisting of randomly positioned small lines (length, 0.24°; mean distance, 0.34°) with two orthogonal orientations is an effective IP. In this pattern statistic dependence is limited to a local neighborhood of 0.5°. Thus the hypothesis arises that the dependence of second order (i.e., the power spectrum of the pattern) contains the information to affect the SV. By addition of a low-frequency orientation (low-pass filtered randomly positioned large lines, frequency cutoff at 3 cpd) to the IP of the first experiment, it was found that in the event of conflict between low and high frequency the orientation of structures of low frequency is dominant.

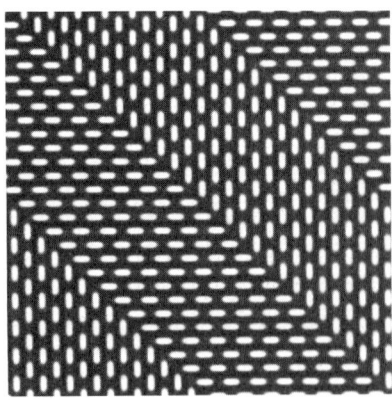

FIGURE 1. Patterns with texture borders used to affect the SV. Elements are positioned on a grid (shortest distance between lines: 0.34°) with the orientation of the texture border such that the orientation of frequency components below 3 cpd, which have been removed in the high-pass filtered pattern at the left side, is on the right side equal to that of the texture border and the high-frequency components have different orientations (45° shift). The pattern covered figure shows details from a circular disk with a diameter of 26° of visual angle.

S1 [deg] (texture borders, high and low frequency range)

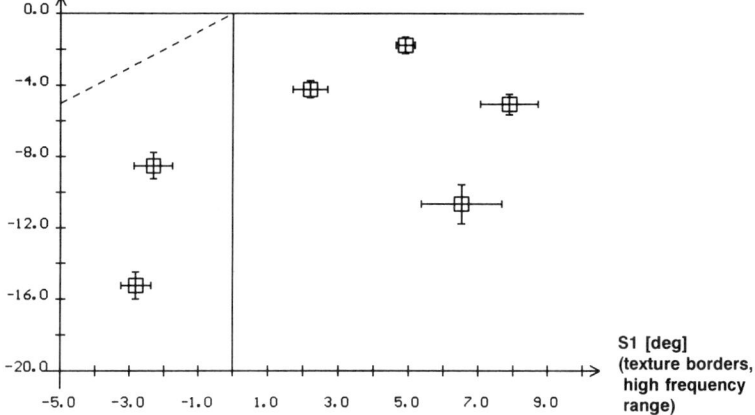

FIGURE 2. Both axes show the sine component S1 (period 90°) of the SV deviation curve (i.e., the deviation of the SV from its mean versus orientation of line elements related to the mean SV) of 6 subjects. A positive value of S1 indicates a strong attraction of the SV by the lines, while a negative S1 means an attraction by an orientation shifted 45°. High-pass filtering of the pattern (axis of abscissas) with frequency cutoff at 3 cpd (see FIGURE 1) causes a distinct increase of S1.

By a second group of experiments it was tested whether the inhomogeneity of the power spectrum contributes to an explanation of the modulation of the SV by patterns that include statistic dependence of higher order. The superposition of low-frequency orientation was repeated by high-pass filtering of a pattern with texture borders (see FIGURE 1). The complete energy of the frequency range below and above 3 cpd as well as the orientation of the two frequency bands was the same as in the first two experiments. If the texture border alone would attract the SV, all points in FIGURE 2 should lie along the diagonal $y = x$ (cf. dashed line). If the isolated high-frequency components alone were sufficient to attract it, all points should lie along the counterdiagonal $y = -x$. In fact most lie at intermediate positions, yet closer to those expected from the first experiment, indicating a compromise in favor of the spectral properties.

Obviously it can be concluded that local inhomogeneity of *intensity* is an important factor for the global orientation of visual structures with respect to gravity. This inhomogeneity of intensity is also exploited if inhomogeneities of higher order like texture borders are present.

REFERENCES

1. WENDEROTH, P. & S. JOHNSTONE. 1988. The different mechanisms of the direct and indirect tilt illusions. Vision Res. **28**(2): 301–312.
2. WENDEROTH, P. & R. V. D. ZWAN. 1989. The effect of exposure duration and surrounding frames on direct and indirect tilt aftereffects and illusions. Percept. Psychophy. **46**(4): 338–344.
3. MITTELSTAEDT, H. 1986. The subjective vertical as a function of visual and extraretinal cues. Acta Psychol. **63**: 63–85.

The Influence of Head Reorientation on the Axis of Eye Rotation and the Vestibular Time Constant during Postrotatory Nystagmus

M. FETTER, D. TWEED, H. MISSLISCH,
W. HERMANN, AND E. KOENIG

Department of Neurology
Eberhard-Karls University
Hoppe-Seyler Strasse 3
D-7400 Tübingen, Germany

While it has been shown that reorienting the head with respect to gravity during postrotatory nystagmus (PRN) affects the decay of PRN, usually reducing the vestibular time constant (TC),[1-3] little is known about the influence of a change in static otolith input on the axis of eye rotation during PRN. We therefore investigated 10 normal human subjects by means of three-dimensional magnetic field–search coil recordings using a variety of head reorientation paradigms in a randomized order during PRN after stop from a 90°/second rotation about earth vertical (1) starting 30° nose down and changing rapidly 2 seconds after the stop to either 120° nose down or 60° nose up; (2) starting 90° nose down and changing to the upright position, (3) starting 90° left-ear down and changing to the upright position or to 90° right ear down. PRN was also recorded in all initial and final positions without head reorientation. Average eye velocities were calculated over 2 time intervals: from second 1 to 2 and 7 to 8 after the stop. TC was estimated as one-third of the duration of PRN.

For most conditions, a reorientation of the head with respect to gravity 2 seconds after the rotation had stopped did not alter significantly the eye velocity vector of PRN (FIGURE 1). Apart from sporadic runaways, only changes from 90° nose down to upright produced a pronounced change in the eye velocity vector which, however, never exceeded the amount of change of the eye velocity vector occasionally found when the head was not moved. The results regarding TC are summarized in FIGURE 2.

In conclusion, the finding that there were under most circumstances no significant changes of the direction of the eye movements if compared before and about 4 seconds after reorientation of the head with respect to gravity in comparison to the conditions when no head reorientation took place indicates that PRN is mainly stabilized in head coordinates and not in space coordinates even if the otolith input changes. This finding falsifies the notion that the shortening of PRN due to reorientation of the head could be due to a change of the eye velocity vector towards a direction (torsion) that is not detectable with the eye recording methods used in earlier studies (ENG). The results regarding TC basically confirm earlier findings[1,2] showing a strong dependence on static head position with TC being lowest if mainly the vertical canals are stimulated (60° nose up and 90° left-ear down) and, secondly, a drastically shortened TC for tilts in the pitch plane when starting from 30° nose down. There was also a slight but not significant reduction of TC for taking up the

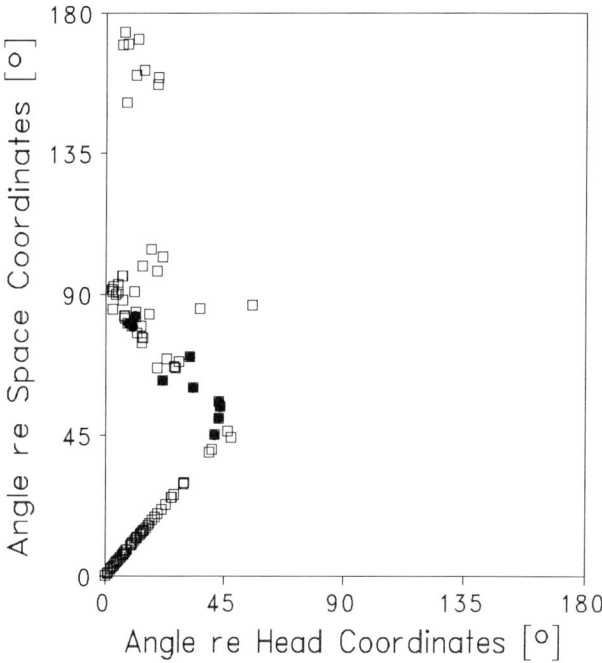

FIGURE 1. This figure shows the results of the calculation of the angle of the change in the orientation of the two eye velocity vectors before and after reorientation of the head in space versus in head coordinates for all trials. Every data point indicates a single experiment. All open squares lying on the diagonal (slope of 1) are experiments without a change in head orientation showing that even under these circumstances the orientation of the eye velocity vector can change up to 45° in some individuals (most pronounced with 30° nose up). If after a head reorientation of 90° or 180° the eye velocity vector would have been stabilized mainly in space coordinates, all data points would have clustered around the x-axis. But in fact most data points are close to the y-axis, indicating that the eye movement vectors change only little their orientation in the head compared to their orientation in space and by that being mainly stabilized in head coordinates (filled squares indicate the condition: change from 90° nose down to upright).

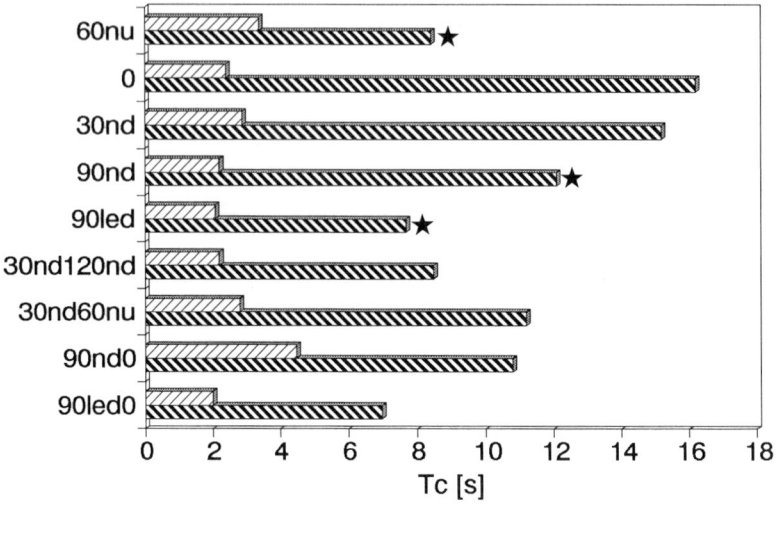

NNNN Tc avg ▨▨▨ Tc std

FIGURE 2. This figure shows the average TC of PRN for all subjects (nu = nose up, nd = nose down, led = left-ear down). The upper five rows show the TC under static conditions, the time constant being lowest if mainly the vertical canals are stimulated (60° nose down and 90° left-ear down) [significant difference compared to the upright position ($p < 0.005$) is indicated by a star]. The lower four rows show the TC when the head has been reorientated (i.e., 30nd120nd = transition from 30° nose down to 120° nose down). While both tilting forward (30nd120nd) or backward (30nd60nu) from 30° nose down reduced the TC significantly, tilting backward was significantly less effective than tilting forward. A change from 90° nose down to upright (90nd0) or from 90° left ear down to upright (90led0) reduced the TC compared to the initial position even further with, however, no significant difference.

upright position, confirming the earlier notion that a combination of "dumping" and "static" mechanisms contribute to the overall behavior of the vestibular velocity storage mechanism.[4]

REFERENCES

1. BENSON, A. J. & M. A. BODIN. 1966. Interaction of linear and angular accelerations on vestibular receptors in man. Aerosp. Med. **37:** 144–157.
2. BENSON, A. J. & M. A. BODIN. 1966. Comparison of the effect of the direction of gravitational acceleration on postrotational responses in yaw, pitch and roll. Aerosp. Med. **37:** 889–897.
3. RAPHAN, T., B. COHEN & V. HENN. 1981. Effects of gravity on rotatory nystagmus in monkeys. Ann. N.Y. Acad. Sci. **374:** 44–55.
4. HAIN, T. C. & U. W. BUETTNER. 1990. Static roll and the vestibulo-ocular reflex (VOR). Exp. Brain Res. **82:** 463–471.

Multidimensional Descriptions of the Optokinetic and Vestibuloocular Reflexes

M. FETTER, D. TWEED, H. MISSLISCH,
D. FISCHER, AND E. KOENIG

Department of Neurology
Eberhard-Karls University
Hoppe-Seyler Strasse 3
D-7400 Tübingen, Germany

The vestibuloocular and optokinetic reflexes (VOR and OKR) help to stabilize the retinal image by rotating the eyes to compensate for movements of the head or visual field. In the past both reflexes have been studied one-dimensionally and characterized by their gain, which is output (eye velocity) divided by input (head or visual field velocity). The 3-D generalization of gain is a 3×3 matrix whose 9 components describe the dependence of all 3 components of the output vector on all 3 components of the input. But this description ignores the dependence of eye velocity on eye position, which should theoretically be negligible for OKR and VOR; e.g., for the VOR, eye velocity should be equal and opposite to head velocity, regardless of eye position. In this study we used linear fitting to describe the 3×3 matrix of VOR and OKR and examined both reflexes for any eye position dependence.

For OKR, 3-D eye motion was recorded using the magnetic field–search coil technique in 3 normal human subjects as they tracked a field of light spots rotating about various axes (horizontal, vertical, torsional) with a constant velocity of 60°/second. For VOR, 3 normal subjects were rotated about body-vertical, nasooccipital, and interaural axes in complete darkness with a constant velocity of 150°/second by means of a computer-controlled rotating chair.

TABLE 1 shows the gain matrix of the OKR averaged over the 3 subjects. To see whether slow phases follow Listing's law, we used an analysis program that assigned scores to periods of horizontal tracking: 0 when eye velocity was independent of eye position (as required for optimal tracking of a full-field stimulus), 0.5 for a Listing's law strategy (pursuit and saccades).[1,2] For full-field tracking (stare), scores averaged 0.35 (range 0.25–0.40) showing a smaller but still strong dependence on eye position. TABLE 2 shows the gain matrices of the VOR averaged over the 3 subjects separate

TABLE 1[a]

Eye Velocity	Field Velocity		
	Torsional	Vertical	Horizontal
Torsional	0.07	0.02	0.02
Vertical	0.02	0.36	−0.02
Horizontal	0.07	−0.04	0.56

[a]This table shows the gain matrix of the OKR averaged over the 3 subjects: T-T gain (dependence of torsional eye velocity on torsional field velocity), 0.07 (range 0.05 to 0.12); V-V gain, 0.36 (0.30 to 0.44); and H-H gain, 0.56 (0.47 to 0.69). Nonzero off-diagonal elements indicate cross-coupling, e.g., a purely torsional field rotation elicits vertical and horizontal eye velocities—the latter approximately as large as the torsional response (first column).

Eye Velocity	Head Velocity		
	Torsional	Vertical	Horizontal
Otolith Input Constant			
Torsional	−0.22	0.04	0.06
Vertical	0.00	−0.24	0.02
Horizontal	0.02	0.00	−0.37
Otolith Input Variable			
Torsional	−0.19	0.04	0.00
Vertical	0.01	−0.26	0.00
Horizontal	0.02	−0.02	−0.37

[a]This table shows the gain matrices of the VOR averaged over the 3 subjects: one matrix for an earth-vertical rotation (constant otolith input) and one for earth horizontal (variable otolith input). With constant otolith input, responses were lowest in the torsional direction with an average T-T gain of −0.22 (range −0.14 to −0.29); V-V gain, −0.24 (−0.17 to −0.29); and H-H gain, −0.37 (−0.19 to −0.53), with small amounts of cross-coupling (small off-diagonal elements). With variable otolith input, T-T gain averaged −0.19 (−0.12 to −0.29); V-V gain, −0.26 (−0.16 to −0.33); and H-H gain, 0.37 (−0.17 to −0.50).

for two cases: one with the rotation axis earth vertical (constant otolith input) and one with it earth horizontal (changing otolith input). Subjects were instructed to adopt 5 different eye positions before and during rotation about each axis (straight ahead, 20° up, down, right, and left). For the calculations the average eye velocity from second 1–2 after the start was used. The data analyzed so far show that the VOR responses in the dark are small and highly variable from subject to subject. The average score for eye position dependence was 0.26 (range 0.19–0.35).

In conclusion, the gain matrix of the OKR shows a very small torsional gain and larger vertical and horizontal sensitivities. VOR matrixes, for both earth-vertical and earth-horizontal axes, show more nearly equal torsional, vertical, and horizontal gains. Surprisingly, eye velocity outputs of horizontal OKR and VOR show a pattern of eye position dependence lying about halfway between the optimal (no dependence) and that characteristic for a Listing's law type strategy. This finding for OKR may be in part attributable to coactivation of the pursuit system. For VOR, it may indicate a strategy to restrict ocular torsion like the saccadic and pursuit systems.

REFERENCES

1. TWEED, D. & T. VILIS. 1990. Geometric relations of eye position and velocity vectors during saccades. Vision Res. **30:** 111–127.
2. TWEED, D., M. FETTER, S. ANDREADAKI, E. KOENIG & J. DICHGANS. Three-dimensional properties of human pursuit eye movements. Vision Res. (In press.)

The Representation of the Spatial Vertical in Human Optokinetic Nystagmus

MARTIN GIZZI,[a] STEVEN RUDOLPH,[a,b] BERNARD
COHEN,[a,c] AND THEODORE RAPHAN[d]

[a]Department of Neurology
[b]Department of Ophthalmology
[c]Department of Biophysics and Physiology
Mount Sinai School of Medicine
One Gustave Levy Place
New York, New York 10029

[d]Department of Computer and Information Sciences
Brooklyn College
Bedford Avenue and Avenue H
Brooklyn, New York 11210

Modeling of optokinetic nystagmus (OKN) and after-nystagmus (OKAN) has included an integrator as an essential element to explain the slow changes in slow-phase eye velocity.[1–3] The observation has recently been made that visual stimulation that is horizontal with respect to an animal's head (yaw axis) can lead to OKN and OKAN that has vertical (pitch axis) or roll components if the animal is tilted during stimulation.[4,5] In particular, the axis of the nystagmus shifts toward the spatial vertical. A three-dimensional model for velocity storage has been developed which includes this phenomenon, known as cross-coupling.[6,7] We examined OKN in humans under conditions of roll axis head tilt to study gravitational influences on cross-coupling.

METHODS

Visual stimulation was provided by a binocular goggle system (developed by G. Clement and A. Berthoz, University of Paris; Centre National de la Recherche Scientifique and the National Aeronautics and Space Administration) which presented a square wave grating of an approximately 5° period subtending roughly 125° vertically and horizontally. Subjects were studied with head or body tilt 45° to either side. We tested four stimulus orientations: vertical, horizontal, and the two 45° obliques. Horizontal and vertical eye movements were measured with electrooculography bandpass filtered between 0.05 and 30 Hz. Each trial consisted of stimulation at 20, 50, and 35°/second for 45 seconds each. With these stimulation techniques, none of our subjects had any appreciable OKAN. Four normal subjects participated in this experiment; two were investigators (MG and SR) and two were volunteers naive to the nature of the study.

Fast phases of nystagmus were eliminated from the eye position traces using a maximum likelihood criterion[8] and the remainder differentiated. Using the horizontal and vertical slow-phase velocities, the vector defining the axis of the OKN response was calculated. The axis of rotation of the eyes (as well as the stimulus) was defined using a right-hand rule. In this manner, horizontal leftward slow phases are described by a vector pointing directly upward and downward slow phases by a leftward vector.

RESULTS

Visual stimulation that produced horizontal (yaw-axis) OKN with subjects upright gave rise to oblique nystagmus when the subject's head was tilted (FIGURE 1). The shift in axis of response was characteristically towards the spatial vertical. In the upright condition, the average deviation of the response axis from the stimulus axis was less than 1.5° with a standard error of 2.5°. In the head tilted condition, the average shift of axis due to upward slow phases was 4° (standard error = 5.5°); the average shift due to downward slow phases was 11° (standard error = 6.0°). This up/down asymmetry of cross-coupling is demonstrated for one subject in FIGURE 2. The tendency to shift the response axis was more pronounced in the head-tilt condition than in the body-tilt conditions for the two subjects tested in both conditions.

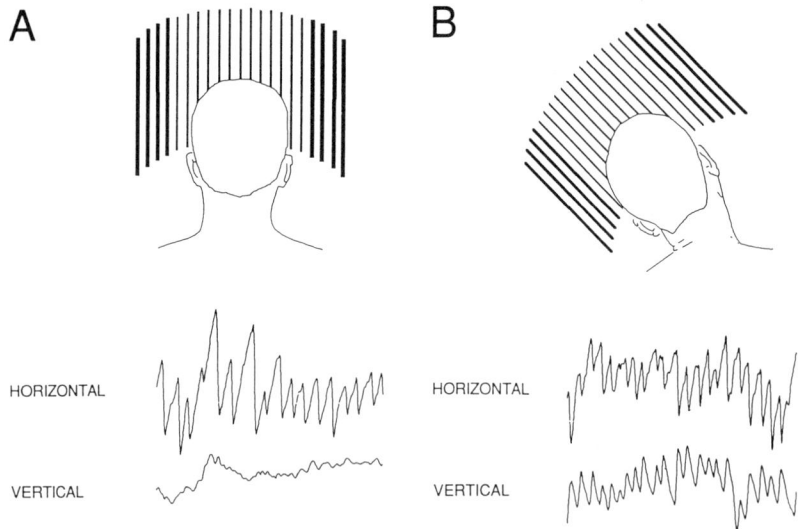

FIGURE 1. The upper portion of each panel represents the stimulus conditions. The lower portions contain tracings of horizontal and vertical eye position. For the horizontal tracings, upward deflection of the pen indicates rightward motion of the eye; for vertical tracings upward deflection of the pen indicates upward motion of the eye. **A:** Horizontal visual stimulation with the subject upright gave rise to rightward slow phases with no significant vertical component. **B:** Horizontal visual stimulation with the head tilted to the left. The eye position traces demonstrate both a rightward and a downward component to the slow phases.

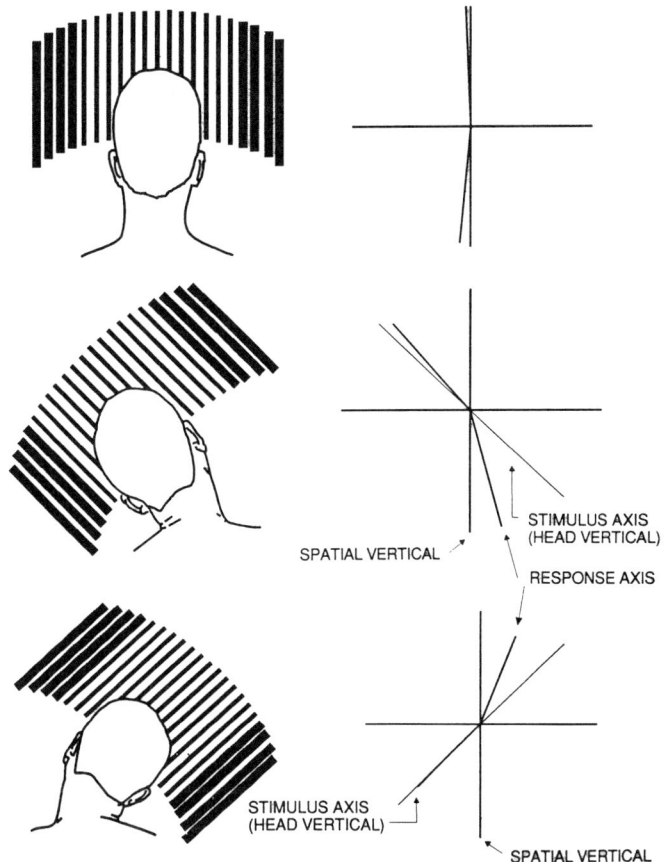

FIGURE 2. The left-hand panel indicates the stimulus conditions and the right-hand panel shows the axes of responses defined by a right-hand rule. For each condition both directions of stimulus rotation are represented, therefore each plot illustrates two responses. Since the OKN goggles were attached to the subject's head, the axis of rotation of the stimulus also moved with the subject's head. The axis of response, however, tended to remain closer to the spatial vertical. This was more evident when downward slow phases were introduced than when upward slow-phases were required.

SUMMARY

With our stimulus conditions we were unable to record more than 2–3 beats of OKAN; therefore direct comparison to the data recorded from monkeys[4,6] is not possible. We did, however, see cross-coupling in OKN. In monkeys, cross-coupling predominates in OKAN, indicating that velocity storage underlies this phenomenon. We consistently saw the axis of response shift towards the spatial vertical. This implies that although OKAN was weak, velocity storage contributed a representation of the spatial vertical to OKN that is dependent on the axis of the head or body with respect to gravity.

REFERENCES

1. COHEN, B., V. MATSUO & T. RAPHAN. 1977. J. Physiol. London **270:** 321–344.
2. RAPHAN, T., V. MATSUO & B. COHEN. 1979. Exp. Brain Res. **35:** 229–248.
3. RAPHAN, T. & B. COHEN. 1985. *In* Adaptive Processes in Visual and Oculomotor Systems. E. L. Keller & D. S. Zee, Eds.: 285–292. Pergamon Press. New York, N.Y.
4. RAPHAN, T. & B. COHEN. 1988. Ann. N.Y. Acad. Sci. **545:** 74–92.
5. HARRIS, L. 1987. Exp. Brain Res. **66:** 522–532.
6. DAI, M. J., T. RAPHAN & B. COHEN. 1991. J. Neurophysiol. **66:** 1422–1439.
7. RAPHAN, T. & D. STURM. 1991. J. Neurophysiol. **66:** 1410–1421.
8. SINGH, A., F. E. THAU, T. RAPHAN & B. COHEN. 1981. 34th Annual Conference on Engineering in Medicine and Biology, Sept 21, Houston, Texas. (p. 136.)

Interaction of Semicircular Canals and Otoliths in the Processing Structure of the Subjective Zenith

STEFAN GLASAUER

Max-Planck-Institut für Verhaltensphysiologie
Abt. Mittelstaedt
D-8130 Seewiesen, Germany

In order to discriminate between translatory and gravitational linear acceleration as well as to locate the direction of "up" (the subjective zenith, SZ) during and after movements, an interaction of otoliths (linear acceleration sensors) and canals (angular velocity sensors) for human spatial orientation is necessary. Three basic assumptions underlying the present model of interaction are confirmed by experiments in which subjects were asked to adjust dynamically a display to their subjective vertical (SV) during and after motion stimuli.

1. Since translatory and gravitational acceleration cannot be measured separately, the processing structure proposed here is based on the assumption, first made by Mayne,[1] that in a low-frequency range linear acceleration is interpreted as a change of gravitational acceleration. The slow increase of the subjective vertical in the "oculogravic illusion" experiment (see FIGURE 1 or, e.g., Reference 2) validates the assumption that estimation of gravitational acceleration uses the low-pass filtered otolith afferents.
2. The function of the semicircular canals is to rotate the SZ contrary to the rotation of the head and accomplish a constant orientation. Thus high-frequency changes of the direction of gravity are distinguished from translational acceleration. As found experimentally, after a brisk tilt around the x-axis of the head the SV reaches its static value very quickly.
3. Problems that arise from the dynamics of the semicircular canals (which indicate angular velocity fairly accurate only in a frequency range of approx. 0.01 to 10.0 Hz) are solved in the proposed model by testing the canal output for plausibility and by blocking inappropriate information. This hypothesis is corroborated by (a) the time course of the "oculogravic illusion" (see FIGURE 1; without blocking an oscillation of the SV would be expected) and (b) our own centrifuge experiments with "free" cabin (stimulation of the canals equivalent to a tilt of 60° without tilting the resulting acceleration vector has almost no effect on the SV). It is supposed that canal information is only used to rotate the actual estimate of gravity if this rotation would decrease the angle between the estimated gravity vector and the measured linear acceleration vector (otolith afferents).

The model (see FIGURE 2 for block diagram), which also incorporates the static model proposed by Mittelstaedt,[4] is able to reproduce all of the mentioned experimental results and to explain some other phenomena like "orbiting" during barbecue spit rotation (see Reference 5).

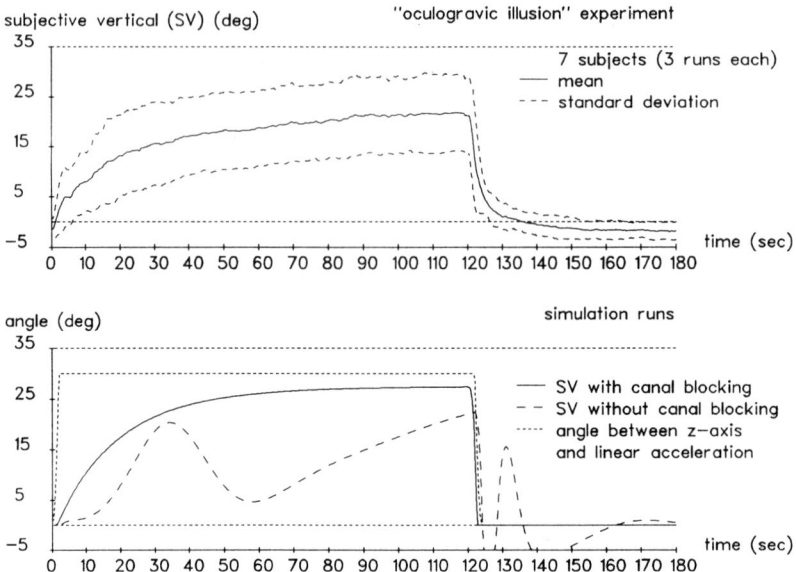

FIGURE 1. "Oculogravic illusion" experiment (upper part) and model simulations (lower part) with and without blocking of inappropriate canal information. The simulation with blocking fits the results, whereas an oscillating time course like that of the simulation without blocking has not been found experimentally (centrifuge data: radius 10 m; angle between acting acceleration and z-axis of the head during run 30°, time from start to constant velocity approx. 2 seconds).

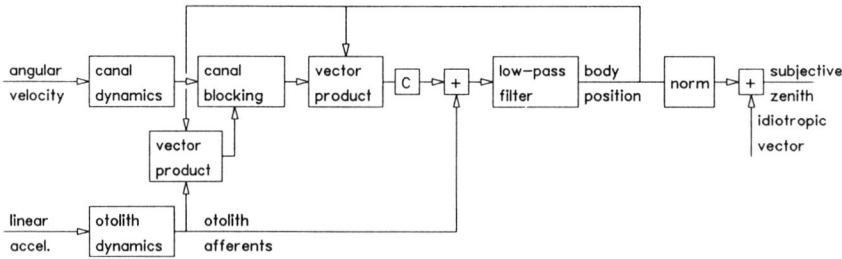

FIGURE 2. Block diagram of the three-dimensional processing structure of the subjective zenith. All model states are vectors. Sensor dynamics are modeled after Ormsby.[3] Short description of the diagram: "C" denotes a constant factor; "norm" means normalization of the input vector (for details on the last two parts of the model see Mittelstaedt);[4] "canal blocking" symbolizes a switch depending on both inputs, the output is either zero or equal to the output of the "canal dynamics." In the latter case the structure performs the rotation of a vector called "body position" (since head and body are aligned here); in the former case "body position" becomes equal to the low-pass filtered otolith afferents.

REFERENCES

1. MAYNE, R. 1974. A systems concept of the vestibular organs. *In* Handbook of Sensory Physiology. H. H. Kornhuber, Ed. **VI/2:** 493–580. Springer-Verlag: Berlin, Heidelberg & New York.
2. GRAYBIEL, A. & B. CLARK. 1965. Validity of the oculogravic illusion as a specific indicator of otolith function. Aerosp. Med. **36:** 1173–1181.
3. ORMSBY, C. C. 1974. Model of human dynamic orientation. Ph.D. Thesis. Department of Aeronautics and Astronautics. Massachusetts Institute of Technology. Cambridge, Mass.
4. MITTELSTAEDT, H. 1983. A new solution to the problem of the subjective vertical. Naturwissenschaften **70:** 272–281.
5. MITTELSTAEDT, H., S. GLASAUER, G. GRALLA & M.-L. MITTELSTAEDT. 1989. How to explain a constant subjective vertical at constant high speed rotation about an earth horizontal axis. Acta Otolaryngol. Suppl. **468:** 295–299.

How Does the Otolith System Detect Three-Dimensional Head Angular Velocity?[a]

B. J. M. HESS

Department of Neurology
University Hospital
8091 Zurich, Switzerland.

The otolith system appears to be able to detect not only head orientation relative to gravity but also head angular velocity in space.[1] The capacity of this system as angular position and velocity detector is most strikingly documented in generating three-dimensional compensatory ocular nystagmus due to constant-velocity rotation about an off-vertical axis (OVAR) while the subject is in complete darkness (i.e., in the absence of visual and semicircular canal input). An example of such three-dimensional nystagmus is illustrated in FIGURE 2 showing the steady-state ocular responses due to constant-velocity rotation (50 deg/second) about an axis that is tilted in a diagonal head plane by 45 deg from the vertical. It is apparent from these data that otolith ocular movements not only are compensatory in direction with respect to the instantaneous head orientation relative to gravity but also with respect to the imposed head angular velocity in space. To produce such eye movements, what kind of otolithic sensory information processing is required? Otolith afferents that encode information about the spatial pattern of shear forces and their rate of change over time (= jerk) provide information about head tilt and head tilt velocity.[2] The geometrical relationship of these directional specific sensory signals during off-vertical axis rotation can best be studied from the standpoint of an outside observer (i.e., relative to space-fixed coordinates, FIGURE 1B). First, note that rate of change of force of gravity (jerk) is the vector product of angular velocity and force of gravity. As a consequence, head angular velocity components along the direction of gravity do not contribute to jerk. Second, the plane of head rotation (i.e., the plane perpendicular to the head angular velocity vector) is determined by the jerk (J) and its time derivative (K). From the standpoint of the subject, these vectorial signals change in space and time relative to its egocentric (head-fixed) reference. Tonic otolith signals that encode the instantaneous spatial pattern of shear forces in the saccular and utricular maculae (i.e., head tilt information) are transmitted to the brain stem mainly by regularly firing otolith units. These units are characterized by flat response dynamics.[2] The expected spatiotemporal evolution of these signals in the utricular and saccular plane during one revolution of the above described OVAR paradigm is illustrated in FIGURE 1C and D (assuming force transduction with zero phase lag and unity gain). The beating field modulations that are observed in the horizontal, vertical, and torsional movement plane (see FIGURE 2) presumably reflect a rather direct coupling of tonic otolith signals to the oculomotor plant. Head tilt velocity information, on the other hand, is obtained by processing signals carrying information about the time rate of change of shear forces (= jerk). Such phasic

[a]Supported by the Swiss National Science Foundation (Grant numbers 3.503-086 and 31-25239.88)

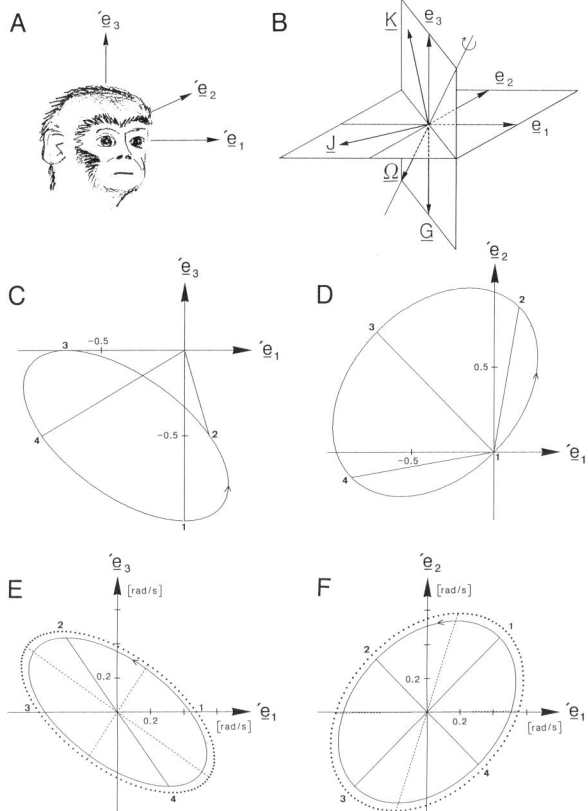

FIGURE 1. Spatiotemporal evolution of gravity and jerk signals in steady state due to rotation of the subject about an axis that is tilted in a diagonal head plane by 45 deg relative to vertical. **A:** Head-fixed reference defined by a right-handed orthogonal coordinate system with axis $'e_1$ in the midsagittal plane (directed 15 deg up relative to the stereotaxic plane), axis $'e_2$ parallel to the interaural axis, and axis $'e_3$ upward. **B:** Geometrical relationship between force of gravity (G), rate of change of force of gravity (jerk: $J = \Omega \wedge G$), rate of change of jerk ($K = \Omega \wedge J$), and associated angular velocity (Ω) in space-fixed coordinates. Note that the rotation axis (see Ω) is oriented in a diagonal plane from right anterior up to left posterior down. At movement onset, the monkey was upright such that head- and space-fixed reference axes were aligned. **C** and **D:** Time course of gravity vector component 1 versus 3 (saccular plane) and 1 versus 2 (utricular plane) relative to head-fixed coordinates during one cycle of revolution. **E** and **F:** Time course of jerk vector component 1 versus 3 (saccular plane) and 1 versus 2 (utricular plane) relative to head-fixed coordinates during one cycle of revolution. The transduced jerk signal is indicated by the dotted ellipses (sensitivity of velocity transduction $v = 0.4$). The phases of rotation are labeled from 1 (upright position) to 4 in 90-deg steps. For more details see text.

signals are transmitted mainly by irregularly firing otolith units and characterized by some gain increase and phase leads of up to 40 deg over the frequency range tested.[2] Thus, tilt velocity transduction results in rather large phase lags (relative to head tilt velocity) and some gain compression or enhancement, depending on the particular velocity sensitivity of the afferent channel (gain enhancement occurs for angular velocities $k < 1$ rad/second since tilt velocity gain is proportional to k^{v-1}). Apart from this change in magnitude and phase shift, the transduced jerk signals exhibit the same spatiotemporal characteristics as jerk, i.e., rate of change of gravity (dotted

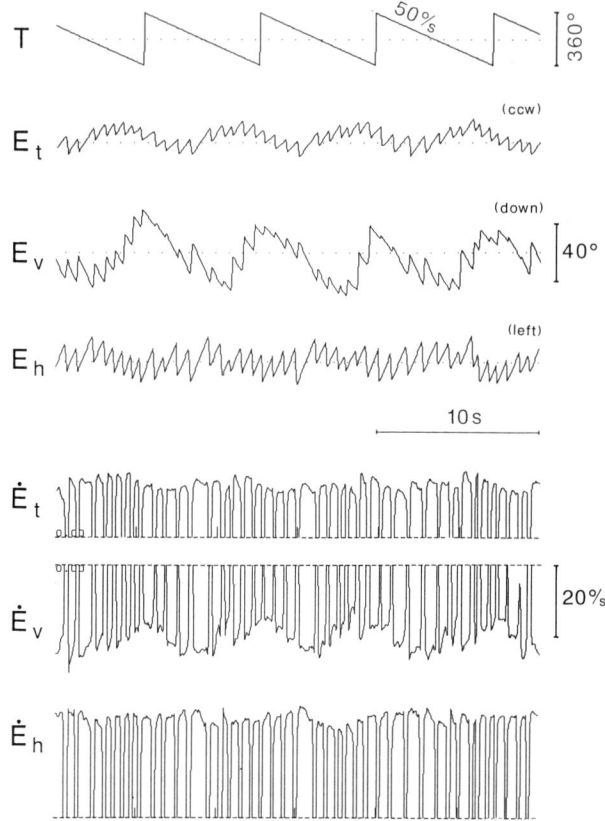

FIGURE 2. Ocular responses during constant-velocity rotation of 50 deg/second about an axis that is tilted by 45 deg with respect to vertical and oriented in a diagonal head plane [i.e., angular velocity vector $\Omega = k$ (−0.50, 0.50, −0.71) with $k = 50$ deg/second; see FIGURE 1B]. The magnitude of slow-phase angular velocity is 38.5 deg/second (gain = 0.77) along a direction closely aligned with the head rotation axis [i.e., $w = k$ (0.40, −0.52, 0.78) with $k = 38.5$ deg/second]. Note bias velocities and cyclic modulations of mean eye position (beating field). E_t, E_v, E_h for torsional (rotation axis along $'e_1$), vertical (along $'e_2$), and horizontal (along $'e_3$) position of the recorded right eye. $\dot{E}_t$, $\dot{E}_v$, $\dot{E}_h$: torsional, vertical, and horizontal angular velocity, T: table position. Data record obtained 2 minutes after onset of rotation.

ellipses in FIGURE 1E and F). The gain change and phase shift may be centrally compensated by a complementary filtering process that is matched to the particular velocity sensitivity of the respective afferent channel. Since head tilt velocity signals do not include information about head angular velocity components parallel to earth vertical, a further stage of central processing is required to reconstruct the complete head angular velocity in space. In fact, the head angular velocity signals (denoted by w) may be extracted from the jerk signals and their time derivatives by a process that can formally be described by the vector product $w = j \wedge k_n$ (the normalized k vector is a weighted time derivative of jerk, i.e., $k_n := d/dt(j)/\|j\|^2$). This relation holds true independent of the particular tilt angle. The normalized k vector signals do not depend on frequency and can, therefore, only be distinguished from the shear force signals by their different spatial characteristics. Thus, in contrast to the rather straightforward computational relationship between tonic otolith afferent information and head tilt relative to gravity, reconstruction of head angular velocity in space does require significant processing of phasic otolith signals that encode head tilt velocity.

REFERENCES

1. GUEDRY, F. E. 1965. Orientation of the rotation axis relative to gravity: its influence on nystagmus and the sense of rotation. Acta Otolaryngol. **60:** 30–48.
2. FERNANDEZ, C. & J. M. GOLDBERG. 1976. Physiology of peripheral neurons innervating otolith organs of the squirrel monkey. III. Response dynamics. J. Neurophysiol. **39:** 970–984.

Vestibular-Somatosensory Interaction in Rapid Responses to Head Perturbations

F. HORAK, C. SHUPERT, V. DIETZ, G. HORSTMANN,
AND F. O. BLACK

R. S. Dow Neurological Sciences Institute of
Good Samaritan Hospital
1120 Northwest 20th Avenue
Portland, Oregon 97209

This study examines automatic electromyographic (EMG) responses to perturbations of the head in order to determine whether (1) vestibular inputs normally trigger responses to head displacements, (2) cervicospinal inputs could substitute for absent vestibular inputs, and (3) the sensitivity to head sensory inputs is increased when somatosensory information from the foot support surface is disrupted in standing subjects.

To investigate the role of vestibular and cervicospinal inputs in short latency responses to head perturbations, EMG responses to translations of the head were compared in 13 healthy adults and 4 patients with profound bilateral vestibular loss. Anterior and posterior head perturbations were imposed randomly with a backpack-mounted torque motor.[1] Two patients, aged 57 and 65 years, lost vestibular function 6–8 years previously as adults and two patients, aged 17 and 19 years, lost vestibular function within their first year of life. All patients were profoundly deaf and had bilaterally absent caloric responses, and vestibuloocular reflex (VOR) gains were less than 0.01 for rotation stimulus frequencies between 0.01 and 0.5 Hz. To determine

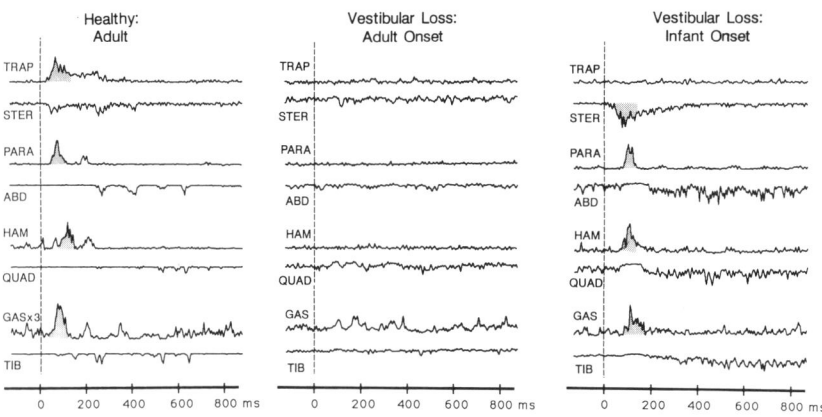

FIGURE 1. Typical EMG responses to forward head perturbations in a healthy adult (left); absent responses from a patient with adult-onset bilateral vestibular loss (center); and normal responses from patient with infant-onset bilateral vestibular loss (right). Responses are averaged from 10 trials with onset of head perturbation at dashed line and dorsal compensatory EMGs plotted upgoing and ventral EMGs plotted downgoing.

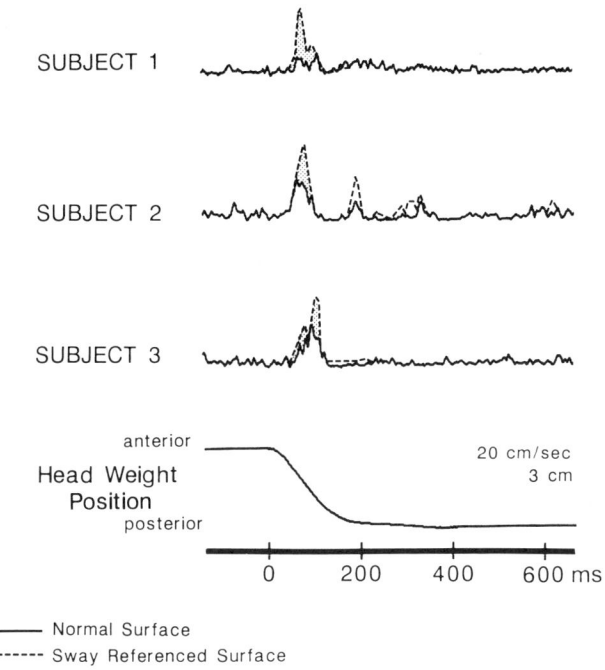

FIGURE 2. Examples of enhanced gastrocnemius EMG response to the same head perturbation when three subjects stood on a sway-referenced surface (dashed line) versus on a normal surface (solid line).

whether responses to head perturbations would be magnified when somatosensory information from the surface was disrupted, subjects were tested with eyes closed on both a normal surface and on a sway-referenced surface.[2]

Mean EMG responses to $0.1 \times g$ forward head translations in healthy subjects were 45 mseconds for trapezius (TRAP), 51 mseconds for sternocleidomastoid (STER), 54 mseconds for lumbar paraspinals (PARA), 91 mseconds for hamstrings (HAM), and 64 mseconds for gastrocnemius (GAS) (FIGURE 1, left). The two patients who lost vestibular function as adults showed absent or greatly reduced responses to head perturbations, consistent with a vestibular trigger for the normal response (FIGURE 1, center). However, the two patients who lost vestibular function as infants showed responses similar to normal, suggesting that cervicospinal or other somatosensory inputs from upper body perturbations can substitute for an absent vestibular trigger (FIGURE 1, right). The magnitude of responses to head perturbations was increased when subjects stood on a sway-referenced surface. Thus, sensitivity to head perturbations can be increased by disrupting somatosensory information from the foot support surface (FIGURE 2).

In conclusion, unlike short-latency responses to surface perturbations, which are usually triggered by lower body somatosensory signals, responses to head perturbations appear to be normally triggered by vestibular signals. Since young adults who suffered vestibular loss as infants showed robust responses to head perturbations, cervicospinal and/or other upper body somatosensory signals must be able to substitute for the missing vestibular trigger, possibly by increasing somatosensory loop gains.[3,4] Further studies are required to determine whether developmental plasticity, duration of vestibular loss, or other factors are responsible for the greater somatosensory compensation for absent vestibular function in patients who lost vestibular function as infants. Just as the gain of somatosensory pathways may be enhanced when vestibular inputs are lost, this study also demonstrates that the gain of vestibulospinal pathways may be enhanced when somatosensory inputs from the surface are disrupted by standing on a sway-referenced surface.[5] These results are compatible with the concept that vestibular and somatosensory inputs interact to allow substitution or reweighing among sensory pathways for automatic head and trunk postural stabilization when one sense is absent or disrupted.

REFERENCES

1. HORSTMANN, G. A. & V. DIETZ. 1988. Neurosci. Lett. **95:** 179–184.
2. NASHNER, L. M., F. O. BLACK & C. WALL III. 1982. J. Neurosci. **2:** 536–544.
3. HORAK, F. 1990. *In* Disorders of Posture and Gait. T. Brandt, I. O. Paulus & W. Bles, Eds.: 370–373. George Thieme Verlag. Stuttgart, Germany.
4. BLES, W., J. M. B. VIANNEY DEJONG, G. DEWIT. 1984. Acta Otolaryngol. **97:** 213–222.
5. NASHNER, L. M. & P. WOLFSON. 1974. Brain Res. **67:** 255–268.

Dynamic Modulation of Vestibuloocular Reflex Gain during Passive Head Rotation[a]

WILLIAM P. HUEBNER[b] AND R. JOHN LEIGH[c]

[b]JSC Neurosciences Laboratory
KRUG Life Sciences, Inc.
1290 Hercules Drive, Suite 120
Houston, Texas 77058

[c]Case Western Reserve University
University Hospitals of Cleveland and
Department of Veterans Affairs Medical Center
Cleveland, Ohio 44106

To measure the performance of the visually enhanced vestibuloocular reflex (VE-VOR), we exposed four normal subjects to the onset and subsequent offset of ±15 deg/second velocity steps of passive horizontal head rotation while they viewed a stationary light spot. Gaze and head movement data were collected using the search coil technique, and, after saccade removal and digital filtering, the measured position waveforms were differentiated to obtain gaze and head velocity.

We used a modeling approach to analyze the data. Based on currently accepted schemes,[1,2] a simple model of the VOR was created that incorporated elements characterizing semicircular canal dynamics and VOR latency, as well as an element providing a constant gain value. We coupled this with a model describing how visual inputs may augment VOR signals to maintain target fixation during rotation (based on Robinson et al.).[3] Optimal parameter estimation techniques were employed to determine values for model parameters that caused model simulations to accurately reflect measured data.[4]

If the VOR acts perfectly with a gain of near 1.0, one would expect that gaze would not be significantly perturbed, despite perturbations of the head, because the generated eye movements would be almost completely compensatory. However, as depicted in FIGURE 1, the VE-VOR data showed substantial gaze perturbations when the head begins to move from rest (at 0.0 seconds), but much lower level perturbations when the head was subsequently stopped (at 2.0 seconds). This asymmetric degree of gaze perturbation was typical for *all* of our subjects and could not be predicted using the aforementioned model. The improved performance when the head was stopped could not have arisen due to augmentation of the VOR with signals derived from visual inputs; the latency of visual processing (requiring at least 50 mseconds for visual processing)[3] would delay such visual contributions until long after the observed compensatory eye motion had already occurred. Also, if vision did play a role when the head stopped, one might expect it to play a similar role at the *onset* of head motion, which it clearly does not.

We reasoned that if the internal VOR gain were initially at some value less than 1.0 when a head rotation is initiated, the magnitude of the induced eye rotations will

[a]Supported in part by National Institutes of Health Grant EY06717 and the Department of Veterans Affairs.

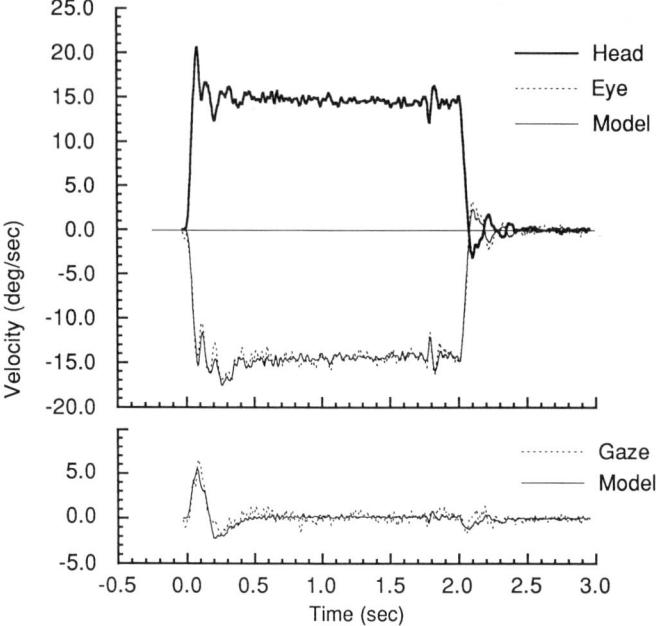

FIGURE 1. Performance of the visually enhanced VOR (VE-VOR) in response to the onset and offset of a 15-deg/second velocity step of head motion. The rightward head motion (Head) induces a leftward eye rotation (Eye); the degree to which the eye rotation deviates from being completely compensatory is indicated by deflections in the Gaze waveform, where Head + Eye = Gaze. Notice that gaze perturbation is much greater at the onset of head motion than at the offset. Waveforms generated using the model depicted in FIGURE 2 (Model) are plotted for comparison with the Eye and Gaze waveforms that they simulate.

be less than the magnitude of the corresponding head rotation. In this case, gaze will be perturbed in the direction of head motion because the compensatory eye rotation cannot keep up with the head movement. Then, when the brain senses head motion, if the VOR gain is increased to near 1.0, the resulting eye rotations become equal and opposite to the head rotation, yielding a near-zero change in gaze. Next, because the VOR gain has increased to near 1.0, when the head is stopped the VOR causes the eyes to stop at virtually the same rate. This will result in only a small perturbation of gaze. Finally, because the head has returned to rest, the brain can subsequently reduce the VOR gain back to its resting level. Thus, we postulate that VOR gain changes *dynamically* in response to the onset and offset of transient head rotation. This hypothesis of dynamically changing VOR gain relative to the onset and offset of head motion is consistent with the variation in degrees of gaze perturbation shown in FIGURE 1.

After modifying the VE-VOR model to provide dynamic modulation of VOR gain (FIGURE 2), we could satisfactorily predict the observed gaze perturbations in the measured responses. FIGURE 1 demonstrates the close similarity of eye and gaze velocity data with their corresponding simulations derived using the revised VE-VOR model. Results of parameter estimation provided a group mean for resting VOR gain of 0.76 and a mean gain value for active VOR of 0.96. Also, we confirmed that the rise in VOR gain started before visual mechanisms could contribute: the average latency for the initiation of VOR gain change (τ_{Avor} in FIGURE 2) was about 80 mseconds, which is 29 mseconds shorter than the average optimal latency of the visual contribution to the VE-VOR (sum of τ_R and τ_f).

From this research, we postulate that when the VOR is not being used (i.e., when the head is stationary), its gain approaches some quiescent level, perhaps near 0.7. However, when the head begins to move and gaze stabilization is required, the brain

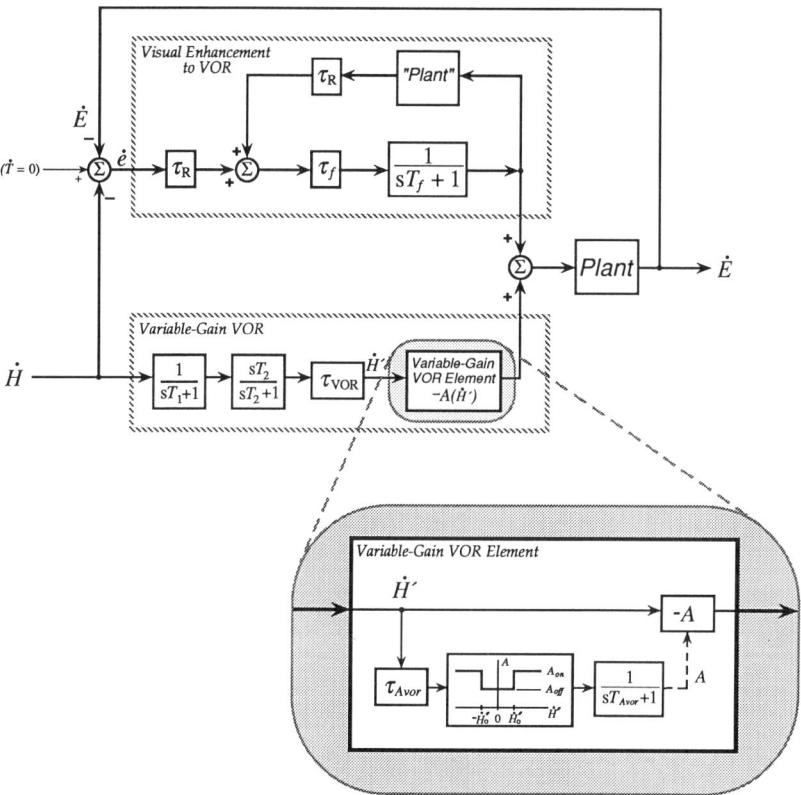

FIGURE 2. Model of VE-VOR that provides for dynamic modulation of VOR gain. The signal from the section representing the VOR is augmented with a signal derived from visual inputs based on retinal slip velocity, $\dot{e}$. Model input: head velocity, $\dot{H}$; model output: eye velocity, $\dot{E}$ (input target velocity, $\dot{T}$, equals zero). The various τ's represent pure time delays: τ_R, retinal processing delay; τ_f, central delay of visual enhancement; τ_{VOR}, VOR latency. The T's are time constants of linear elements: T_1, "short" time constant of the cupula; T_2, "long" time constant of the VOR; T_f, time constant governing the dynamics of visual enhancement. The element responsible for modulating VOR gain proposes that the brain monitors the internal head velocity signal, $\dot{H}'$. If the magnitude of $\dot{H}'$ is less than some threshold value $\dot{H}'_0$, then the VOR gain settles to its *resting* gain value A_{off}. If the magnitude of $\dot{H}'$ is greater than $\dot{H}'_0$, then the brain increases VOR gain up to the *active* value A_{on}. A delay element τ_{Avor} was added to allow the brain time to make decisions about which level to set the VOR gain. Also, we included a phase-lag element, governed by the time constant T_{Avor}, to make the VOR gain change gradual.

realizes the need to derive maximum utility from the VOR and, thus, increases the VOR gain to near 1.0.

REFERENCES

1. WILSON, V. & G. MELVILL JONES. 1979. Biophysics of the peripheral end organs. *In* Mammalian Vestibular Physiology. Plenum Press. New York, N.Y.
2. ROBINSON, D. A. 1981. Control of eye movements. *In* Handbook of Physiology—the Nervous System II. American Physiological Society. Bethesda, Md.
3. ROBINSON, D. A., J. L. GORDON & S. E. GORDON. 1986. A model of the smooth pursuit eye movement system. Biol. Cybern. **55:** 43–57.
4. HUEBNER, W. P., G. M. SAIDEL & R. J. LEIGH. 1990. Nonlinear parameter estimation applied to a model of smooth pursuit eye movements. Biol. Cybern. **62**(4): 265–273.

A Neural Network for Computing Surface Curvature from Optic Flow[a]

G. LEONE, J. DROULEZ, AND V. CORNILLEAU-PÉRÈS

Laboratory of Neurosensory Physiology, CNRS
15 rue de l'Ecole de Médecine
75006 Paris, France

The ability of an observer to identify an object in any position seems to require a coding of intrinsic characteristics of the object's structure. The *surface curvature* is of particular interest since it is a geometrical invariant of 3-D objects. The visual perception of 3-D structure is based on many cues, among which motion parallax plays a major role.[1] As an object moves, its 3-D velocity field is projected in a 2-D velocity field (also known as the optic flow) on the retina. Many physiological experiments have suggested that the perception of 3-D structure from motion parallax was mediated by the coding of the optical flow in the visual pathway, and theoretical schemes have been proposed to account for such a process. It has been shown[2] that the curvature of the object's surface is directly related to one of the second spatial derivatives of the optic flow: the *spin variation* (SV). Our goal was to build an artificial neural network able to extract the mean surface curvature from the optic flow, and to test the predictions of the SV model with our network.

THE SPIN VARIATION MODEL: THEORETICAL BACKGROUND

The spin variation is a second spatial derivative of the image velocity field and quantifies the bending of an image line on the retina during motion. It varies linearly with the curvature of the surface S and the norm of the frontal translation T, and depends on the relative orientation of S and T. It can be computed along every direction in a given image point. We call SV the average value of such a computation. For a cylinder of frontal translation T, FIGURE 1 illustrates the fact that SV is higher when T and the cylinder axis are parallel (cylinder P), rather than orthogonal (cylinder O). Therefore, if the visual perception of surface curvature is mediated by the coding of spin variation, the cylinder curvature should be detected more easily for cylinder P than for cylinder O. The psychophysical experiments confirmed that the sensitivity to curvature was always larger for cylinder P than for cylinder O.[3]

THE NETWORK

As stated above, the task is to extract the mean curvature of a surface (that is the mean of the 2 principal curvatures) from the optic flow. Each unit is connected to every unit in the subsequent layer, but not to units in the same layer. The input layer consists in two sets of 9×9 units coding the velocity components along 2 orthogonal image axes. In the subsequent layers, the activity of units is determined by usual

[a]This research was supported by CEE Esprit BRA no. 3149 and Ecole Polytechnique.

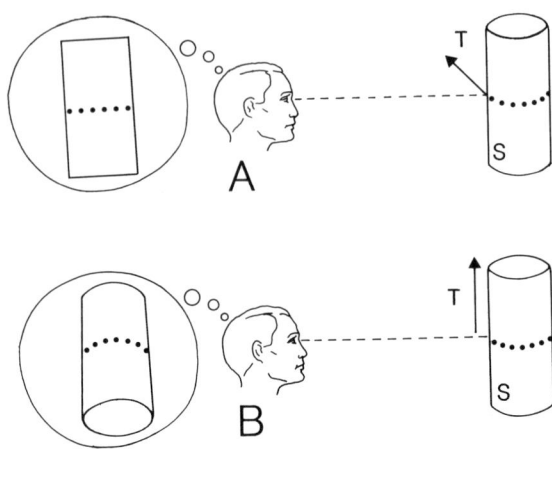

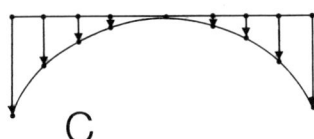

FIGURE 1. S is a patch of cylinder moving in front of the subject. Initially, the line of S, shown as the dotted line on S, projects onto the retina as a straight line. **Panel A:** When S is submitted to a frontal translation, T, orthogonal to the axis of S, the image line remains a straight line on the subject's retina throughout the motion (cylinder O). **Panel B:** On the opposite, when T is parallel to the axis of S, the dotted line turns into an ellipse arc during the motion (cylinder P). **Panel C:** The bending of the image line during the motion is quantified by the spin variation which is a second derivative of the image velocity field.

feedforward network algorithm. The activities of the units can take any value between -1 and $+1$. We used the *modified back-propagation learning algorithm*.[5] As compared to the classical back-propagation,[4] it presents 3 differences that have been shown to improve the convergence rate: (1) the synaptic weights are modified only after the propagation of the whole learning set; (2) the value of the learning rate is adjusted automatically rather than fixed; (3) the momentum factor is rejected each time the error (difference between desired and actual output) increases. We use no masks or sharing weights in our network. The learning set is composed of 1400 noisy optic flows similar to those used in the psychophysical experiments, and represents quadratic surfaces in motion (sphere, plane, cylinders O and P of radius 10 or 100 cm for a viewing distance of 72 cm and an image area of 10 cm × 10 cm). The proportional noise is fixed to 5%, and the background noise represents 5% of the mean norm of the optic flow.

RESULTS

For a 5-layer network (162 units in input layer, then 20, 15, 10 units in the subsequent hidden layers and one output coding the mean curvature), we obtain a high correlation (0.93) between actual and desired outputs, indicating that *the network is successfully extracting the desired parameter* (FIGURE 2). However, we never obtain a good correlation with a 3- or 4-layer network. According to Palm,[6] our problem can be solved with 3 layers, but probably at the price of a larger number of units in the hidden layer. Our 5-layer network was also successfully (0.91) tested on several sets, distinct from the learning set, of 1400 noisy optic flows, for each surface and each radius (FIGURE 2). The estimated curvature is larger for cylinder O than for cylinder P, this being significant ($p < 0.02$). The estimation of surface curvature is more accurate for cylinder P than for cylinder O. However, the results show that a

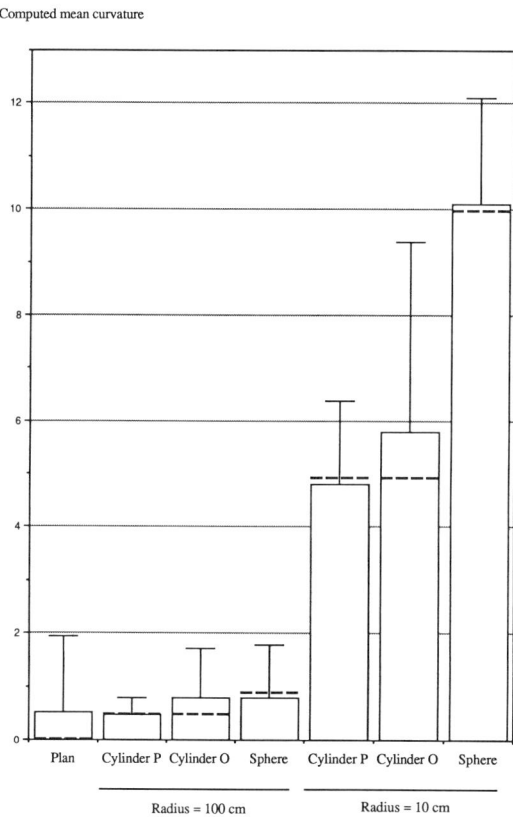

FIGURE 2. Computed mean curvature during the generalization task. A generalization set is composed of 1400 noisy optic flows, distinct from the learning set, representing one quadratic surface in motion (among sphere, plane, and cylinders O and P) of fixed radius. The standard errors are shown by the vertical bars, and the actual mean curvature is represented by a dotted line for each surface.

task of discrimination between cylinder and plane would be easier for our network in case O than in case P. This is not in agreement with the theoretical and psychophysical findings. Nevertheless, the curvature of 1 m^{-1}, which seems to be close to the discrimination threshold of our network, is of same magnitude as the threshold found in psychophysical experiments (between 0.69 m^{-1} and 1.97 m^{-1}).

Our study showed that *an artificial neural network can extract the mean curvature of quadratic surfaces from optic flow.* The results of this network were not always consistent with the psychophysical results, but we must bear in mind that the specific task of our network was to extract mean surface curvature, while the psychophysical task used in Reference 2 was the discrimination between cylinders and planes.

REFERENCES

1. GIBSON, J. J. 1957. Psychol. Rev. **64:** 288–295.
2. CORNILLEAU-PÉRÈS, V. & J. DROULEZ. 1989. Percept. Psychophys. **46:** 351–364.
3. DROULEZ, J. & V. CORNILLEAU-PÉRÈS. 1990. Biol. Cybern. **62:** 211–224.
4. RUMELHART, D. E. 1986. Nature **323:** 533–536.
5. VOGL, T. P., J. K. MANGIS, A. K. RIGLER, W. T. ZINK & D. C. ALKON. 1987. Biol. Cybern. **59:** 257–263.
6. PALM, G. 1988. Biol. Cybern. **34:** 49–52.

Visual Direction Is Corrected by a Hybrid Extraretinal Eye Position Signal[a]

WENXUN LI AND LEONARD MATIN

Department of Psychology
Columbia University
New York, New York 10027

Although neural signals regarding changes in eye position are required in order to explain the fidelity of spatial orientation and spatial localization in the presence of voluntary saccadic eye movements, whether these signals arise from proprioceptors in the orbit of the eye (inflow),[1] from the command to turn the eye (outflow),[2] or from both (hybrid)[3] has remained unclear. The results of experiments involving voluntary saccades described below require a hybrid source with 80% inflow and 20% outflow.

The experiments were conducted in a darkened room in which the only source of illumination came from the stimulus display which was presented on a 23-inch CRT with a short persistence phosphor, P15. The observer, whose head position was stabilized by a forehead rest and a bite bar, viewed the display with the right eye from a distance of 93 cm while the left eye was occluded by an eye patch. The horizontal position of the observer's viewing eye was continuously monitored by an infrared reflection eye movement monitor (Gulf and Western model 200). Voluntary saccades were initiated from a fixation target (A, FIGURE 1) and were directed toward a saccadic target 10° distant (B, FIGURE 1). When the saccade reached a point 2.5° from the fixation target, the initial 2-target display was extinguished and, following a 70-msecond delay, a single 10-msecond test target (B': 4' high and 1' wide) was presented whose distance and direction from the saccadic target (*d*) was randomly varied across trials. The observer reported the direction of the displacement of the test target from the saccadic target. Separate blocks of trials were run with rightward-going and leftward-going saccades.

For the three observers across a total of more than 10,000 trials, actual saccade length ranged from 7.5° to 12°. Psychometric functions were fitted to the data in half-degree bins separately for leftward and rightward saccades for each of the observers. PSEs (point of subjective equality) for the saccadic target, the mean of the best-fitting cumulative normal distribution, were calculated for each of the psychometric functions and are displayed in FIGURE 2.

For all six sets of results, the distance of the saccadic target PSE from the fixation target increased with saccade length. However, this variation was small, ranging in slope between +0.13 and +0.29. Since the inflow model predicts that perceived visual direction should be corrected for the inaccuracy of the oculomotor behavior (slope of 0) and the outflow model predicts that the perceptual error equals the oculomotor error (slope of 1.0),[4] the results of the present experiments in FIGURE 2 show that a hybrid model, in which inflow plays a major role and outflow a minor role, is required in order to account for visual localization during voluntary saccades.

[a]The research was supported by grants AFOSR 91-0141 from the Air Force Office of Scientific Research, BNS 8617059 from the National Science Foundation, and EY 05929 from the National Eye Institute, National Institutes of Health.

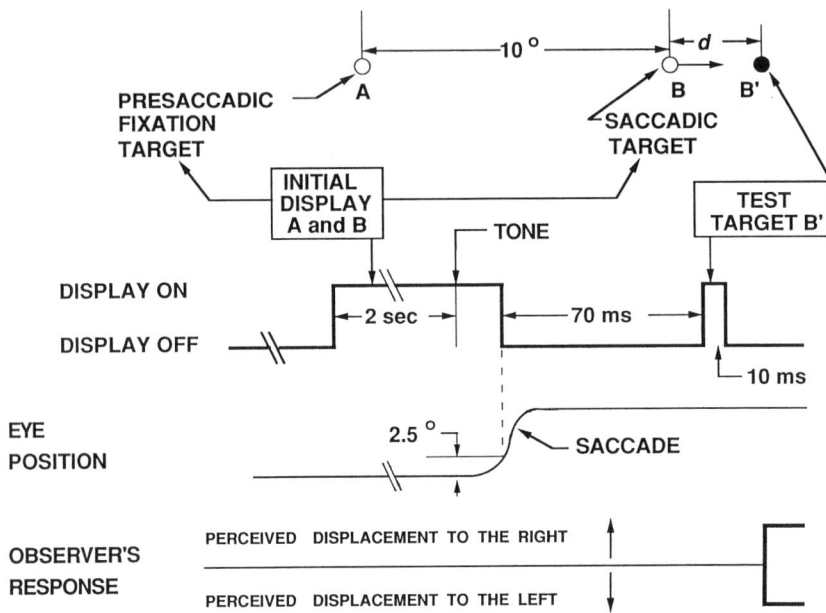

FIGURE 1. The spatial and temporal outline of a single experimental trial. Each trial began with the simultaneous onset of the initial visual display which consisted of two identical targets separated horizontally by 10°: fixation target (A) and saccadic target (B). The observer fixated target A; 2.0 seconds following the onset of the initial display a tone was presented which signaled to the observer that he was free to execute a saccade from A to B, rightward going for one set of conditions, leftward going for a second set of conditions. When the eye crossed the trigger point, a distance of 2.5° from A, the initial display was extinguished. Following a dark interval of 70 mseconds, a test target (B') was presented for a duration of 10 mseconds. B' was identical to target B, but was displaced from it by a distance d. On each trial, d was set at one of 17 possible values ranging from −4° (displacement to the left) to +4° (displacement to the right), spaced at 0.5° intervals (this included trials in which the test target was presented at the same location as the saccadic target: $d = 0$). Each trial ended with the observer's psychophysical report of whether the test target appeared to the right or left of the previously viewed saccadic target.

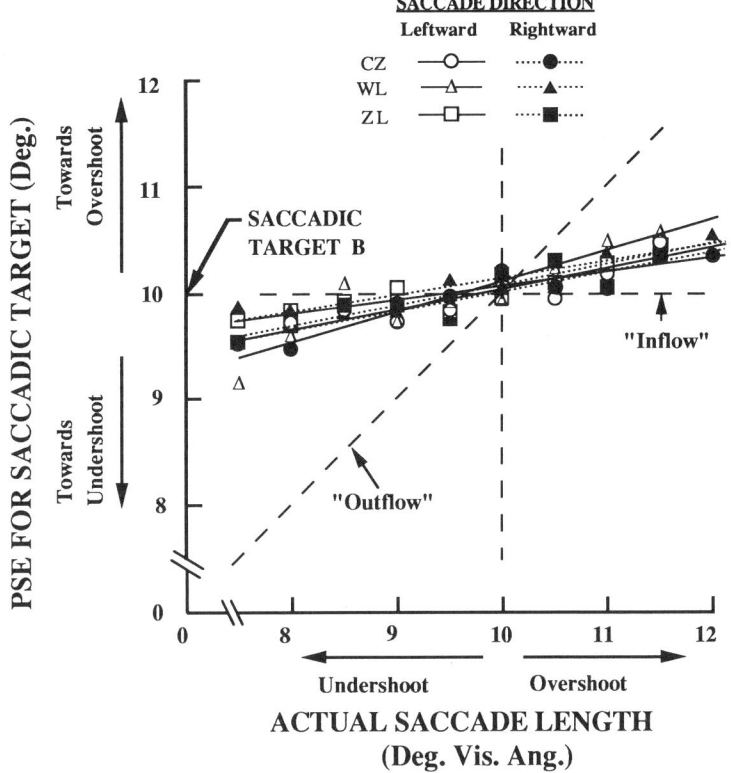

FIGURE 2. The changes in the PSE of the saccadic target for each of the three observers for each of the two saccade directions are plotted directly against the corresponding changes in actual saccade length. Each point plots the horizontal location of the PSE (on the ordinate) against saccade length (on the abscissa) for a single combination of actual saccade length and saccade direction for each observer. The best-fitting straight line is drawn through the data for each saccadic direction for each observer. A horizontal line in this figure represents perceptual accuracy at all saccade lengths, a result predicted by an inflow model. A straight line with a slope of 1.0 represents the result predicted by an outflow model (that the perceived visual direction of the saccadic target is predicted to appear in the foveated direction regardless of actual saccade length). The slopes of the regression lines for the six sets of experimental data fell between +0.13 and +0.29. The results require a hybrid model, in which inflow plays a major role and outflow a minor role, in order to account for visual localization during voluntary saccades.

REFERENCES

1. SHERRINGTON, C. S. 1918. Observations on the sensual role of the proprioceptive nerve supply of the extrinsic ocular muscles. Brain **41:** 332–343.
2. HELMHOLTZ, H. 1866. A Treatise on Physiological Optics **3.** Voss. Leipzig, Germany.
3. MATIN, L. 1976. A possible hybrid mechanism for modification of visual direction associated with eye movements—the paralyzed eye experiment reconsidered. Perception **5:** 233–239.
4. LI, W. 1989. A quantitative study of the accuracy and the precision of the visual perception of direction immediately after voluntary saccades. Ph.D. Dissertation. Columbia University. New York, N.Y.

The Interaction of Angular and Linear Optokinetic Stimuli with an Angular Vestibular Stimulus

S. S. MOSSMAN, A. M. BRONSTEIN, J. D. HOOD,
AND P. SACARES

MRC Human Movement and Balance Unit
and Institute of Neurology
National Hospital
Queen Square
London WC1N 3BG, United Kingdom

In this paper we investigate the selectivity of angular and linear optokinetic (OK) stimuli in modulating angular vestibular nystagmus in man. This was prompted by the existence of different vection illusions occurring with linear and angular visual stimuli and by the observation that slow buildup of OK nystagmus with some cerebellar and parietal lesions occurs with angular but not with linear OK stimuli.[1]

Optokinetic-vestibular interaction was assessed by means of *sequential* stimuli; a prolonged OK stimulus was followed immediately by a vestibular stimulus in the dark. The linear or angular OK stimuli moved rightwards or leftwards for 30 seconds at constant velocity (20°/second). At this point the lights were extinguished and, simultaneously, the chair started rightwards rotation (a 20°/second velocity step immediately followed by constant acceleration at 1.5°/second per second for 17 seconds, FIGURE 1). The rotational stimulus was designed to counteract the normal exponential decay of vestibular nystagmus and maintain a relatively constant eye velocity. Ten normal adult subjects took part in the experiment and viewed passively OK stripes at intervals of 9° at a distance of 0.8 meters provided by a motorized full-field drum or a linear screen (subtending 74° of vertical; 70° of vertical and 86° of horizontal visual angle respectively). Three vestibular stimuli alone in each subject provided a baseline with which to compare the interaction of the vestibular-OK responses. Eye movements were recorded by DC electrooculography (EOG) and displayed upon an ink jet recorder. Sequential responses of slow phase eye velocity (SPV) were measured by a semiautomated procedure from the onset of the vestibular stimulus. Differences in mean SPV between paradigms were assessed by one-way analysis of variance using Dunnett's multiple comparison procedure.

As shown in FIGURE 2, when the preceding OK stimuli were in the opposite direction to the rotational stimulus ("summating SPV"), the initial SPV of the vestibular response was increased; when the OK and rotational stimuli were in the same direction ("opposing SPV"), SPV was reduced. This effect was larger and more prolonged with angular than with linear OK stimuli; statistical differences ($p < 0.05$) between additive and opposing angular OK stimuli lasted up to 15 seconds but for only 1.25 seconds with the linear OK stimuli. The mean time to return to "baseline" vestibular values with opposing vestibular-OK stimuli was 12 seconds for an angular OK stimulus, which was significantly longer ($p < 0.004$) than the 4 seconds taken by the linear OK stimulus. As shown in previous studies,[2] opposing responses had a stronger effect than summating responses (FIGURE 2).

In conclusion, the modulation of rotational vestibular nystagmus is larger and

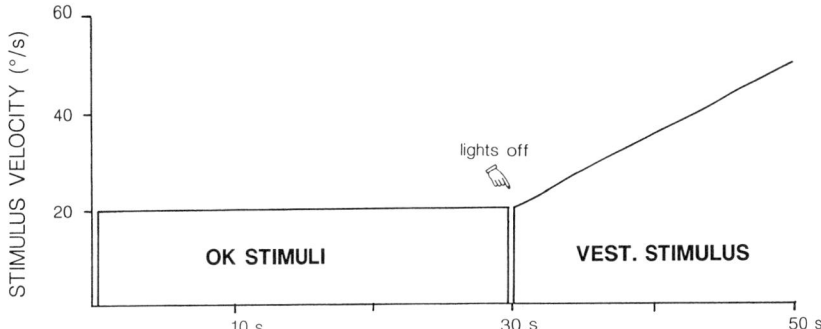

FIGURE 1. Schematic diagram of the experimental protocol, showing sequential stimulation with an initial optokinetic (OK) velocity step for 30 seconds, immediately followed by step-ramp whole-body rotation in the dark. The optokinetic stimuli were angular or linear, rightwards ("opposing") or leftwards ("summating") at 20 deg/second. Vestibular stimulus was 20 deg/second + 1.5 deg/second per second rightwards.

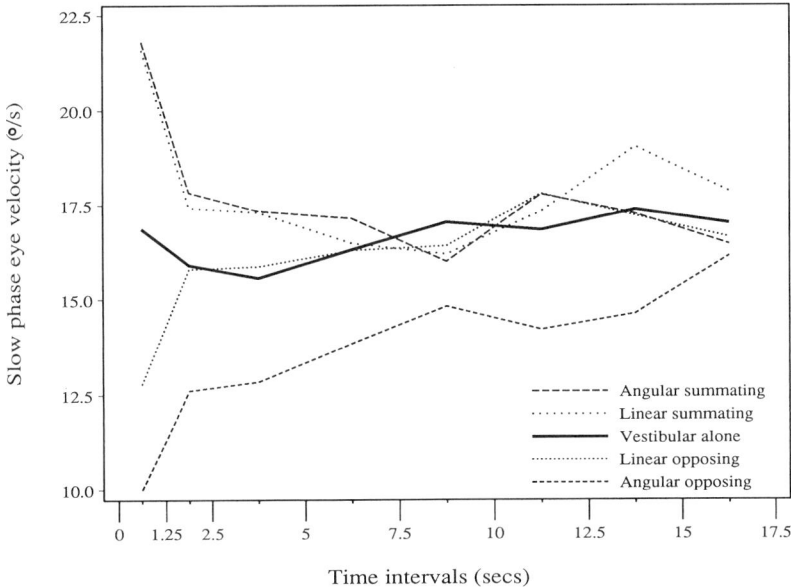

FIGURE 2. Mean results of slow phase eye velocity in 10 subjects. The full line is the result of pure vestibular stimulation in the dark *not preceded* by optokinetic (OK) stimulation and shows reasonably constant eye velocity. Interrupted lines show the result of vestibular stimulation *preceded* by different optokinetic stimuli. Note the marked and slowly decaying suppressive effect of the opposing angular OK stimulus.

longer lasting with an angular than with a linear preceding OK stimulus. This may be due to preferential activation of "velocity storage" mechanisms by angular stimuli and suggests that pathways with visuovestibular convergence show some degree of canal-angular OK, and presumably otolith-linear OK, selectivity. Opposing stimuli are more effective than additive stimuli,[2] which may have a function in reducing postrotational nystagmus and sensation.[3]

REFERENCES

1. MOSSMAN, S. S., A. M. BRONSTEIN & J. D. HOOD. 1991. Linear and angular optokinetic nystagmus in labyrinthine and CNS lesions. Acta Otolaryngol. Stockh. Suppl. **481:** 352–356.
2. BARRATT, H. J. & J. D. HOOD. 1988. Transfer of optokinetic activity to vestibular nystagmus. 1988. Acta Otolaryngol. Stockh. **105:** 318–327.
3. COHEN, B., V. HENN, T. RAPHAN & D. DENNETT. 1981. Velocity storage, nystagmus and visual-vestibular interactions in humans. Ann. N.Y. Acad. Sci. **374:** 421–433.

Optical and Somesthetic Biases of Postural Orientation in the Pitch Dimension

KENNETH NEMIRE AND MALCOLM M. COHEN

National Aeronautics and Space Administration
Ames Research Center
Moffett Field, California 94035

Some problems in aviation and in space flight can be attributed to underspecification of the optical environment and to conflicts between sensory cues describing the internal and external environments. We examined the contributions of multiple sensory inputs to body orientation under a number of stimulus conditions.

Each experiment employed a within-subject factorial design, and a task requiring each of 12 subjects (S) to set his/her body erect or 45° back from erect ("diagonal") while restrained in a movable bed. Before each trial, S was exposed to a starting position of 15° in front of or behind the vertical or diagonal goal.

In experiment I, a box, which was level or pitched ±20°, surrounded the bed. The box provided an illuminated grid, two luminous lines, or a dark environment. Both the illuminated grid and luminous lines biased settings of body position when the box was pitched; the pitched grid was three times more effective than the pitched lines (FIGURE 1). The difference in optic bias observed for the pitched luminous lines and the the pitched grid in this experiment may appear to conflict with results obtained with similar pitched stimuli in Matin's laboratory.[1,2] We suggest, instead, that the results from the two studies indicate that the mechanisms underlying perceived body orientation and perceived eye level may use different aspects of the available optic input, or may weigh the optical inputs differently. For example, although the pitch of the box in our experiment biased orientation to both goals, the effect was greater for the diagonal goal than for the erect goal. The increased optical bias of postural orientation at the diagonal goal may be due to decreased precision of the somesthetic cues contributing to orientation to the diagonal goal and consequent increased weighting of the optical inputs. This analysis is in agreement with a "modality precision" model of intersensory interaction[3] and models postulating combination of multiple ambiguous inputs.[4]

In experiment II, we examined the effects of starting position on postural orientation in the dark. Ss spent 10, 30, or 50 seconds at the starting position. Increased time spent at a starting position caused greater errors in orienting to the diagonal goal that to the vertical goal (FIGURE 2).

Because, for blindfolded, water-immersed Ss, the precision of orienting to the erect or supine positions does not differ,[5] we postulate that the proprioceptive and cutaneous pressure cues provide more precise information about body orientation when the body is near the upright position, and that these cues are responsible for the diminished optical bias of body position when orienting to the vertical goal in experiment I.

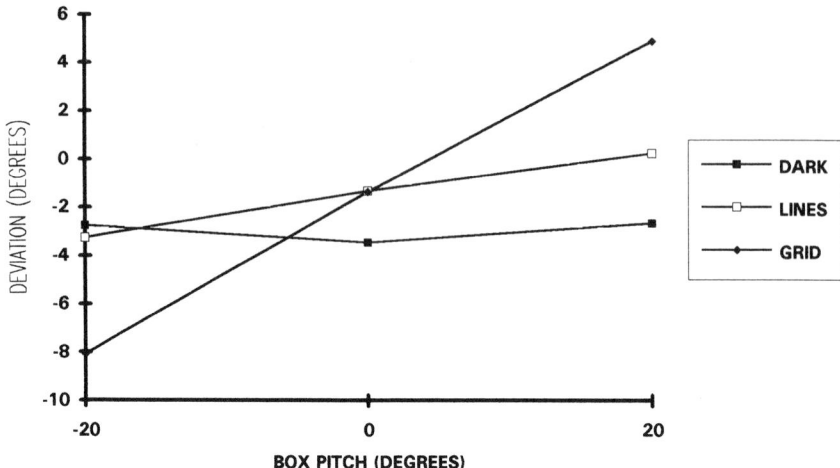

FIGURE 1. The influence of the pitched optical array on perceived body orientation was greater when the optical structure consisted of an illuminated grid than when it consisted of two luminous lines.

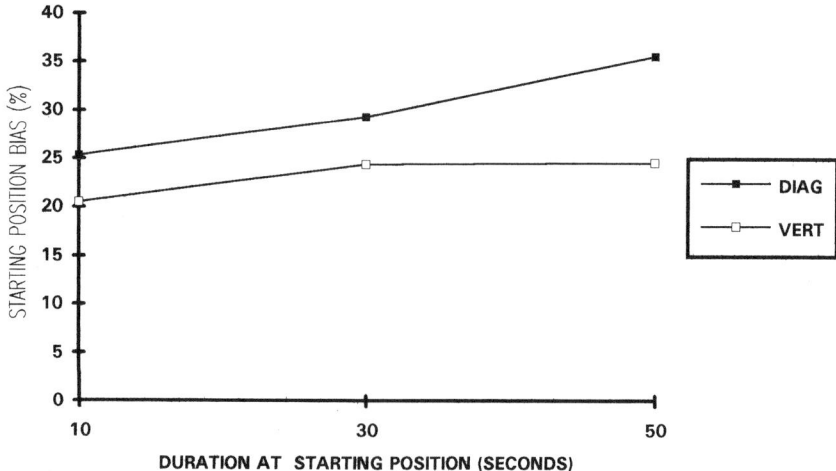

FIGURE 2. In the dark, the influence of starting position on perceived body orientation was greater when Ss oriented to the diagonal goal than to the vertical goal. Increasing the delay at the initial position from 10 to 30 seconds resulted in increased error when orienting to both the diagonal and the vertical goals; further increase to the 50-second delay adversely influenced orientation to the diagonal, but not to the vertical, goal.

REFERENCES

1. MATIN, L. & C. R. FOX. 1989. Vision Res. **29:** 315–324.
2. MATIN, L., C. R. FOX & Y. DOKTORSKY. 1987. Invest. Ophthalmol. Visual Sci. **28** (Suppl.): 300.
3. WELCH, R. B. & D. H. WARREN. 1986. *In* Handbook of Perception and Human Performance. Sensory Processes and Perception. K. R. Boff, L. Kaufman, & J. P. Thomas, Eds. **1.** John Wiley and Sons. New York, N.Y.
4. MASSARO, D. W. 1987. Speech Perception by Eye and Ear: a Paradigm for Psychological Inquiry. Lawrence Erlbaum Associates, Publishers. Hillsdale, N.J.
5. NELSON, J. G. 1968. Aerosp. Med. **39:** 806–811.

The Effect of Target Distance and Stimulus Frequency on Horizontal Eye Movements Induced by Linear Acceleration on a Parallel Swing

JOHN G. OAS,[a] ROBERT W. BALOH, JOSEPH L. DEMER,
AND VICENTE L. HONRUBIA

University of California at Los Angeles Medical Center and
Jules Stein Eye Institute
Los Angeles, California

INTRODUCTION

With the goal of developing a clinical test of otolith-ocular function, we have added springs to a parallel swing and assessed the dynamics of the horizontal linear vestibuloocular reflex (LVOR) in seven normal human subjects across three frequencies (0.41, 0.65, and 0.77 Hz). Eye movements were measured with a scleral search coil, and data analyzed with newly developed digital signal processing techniques. We compared the sensitivity to translation (ST, defined by slow phase eye velocity/ swing velocity) and phase measurements as subjects performed mental arithmetic, fixated on near and far earth-fixed targets, and imagined earth-fixed targets at these distances.

METHODS

The parallel swing was modified from the design used in our earlier study[1] and could be oscillated at frequencies above the natural frequency of the swing by varying the number of extension springs attached at the base. Lateral swing displacement was held constant while small added forces were applied manually at each peak displacement to accommodate the damping due to airflow drag and bearing friction. Orthogonal field coils were attached rigidly to the head holder and swing chair. A bite bar, adjustable head-fixation device, and special high-impact resilient foam held the subject at the center of the coil field. Dark circular targets subtending 5 minutes of arc were placed on a neutral background at 2.1 and 1.05 meters from each subject. Seven normal subjects were instructed to perform mental arithmetic while they were oscillated in the dark without any prior visual cues. They were then oscillated in the light while attempting to maintain fixation on the earth-fixed target, then immediately thereafter placed in the dark and instructed to imagine this same target. Horizontal eye displacement and horizontal swing acceleration signals were recorded and later digitized for computer analysis at 409.5 Hz. Fast eye components were identified by characteristic velocity profiles at threshold displacements, and the

[a]Address correspondence to UCLA Medical Center, Department of Neurology, 710 Westwood Plaza, Los Angeles, California 90024-1769.

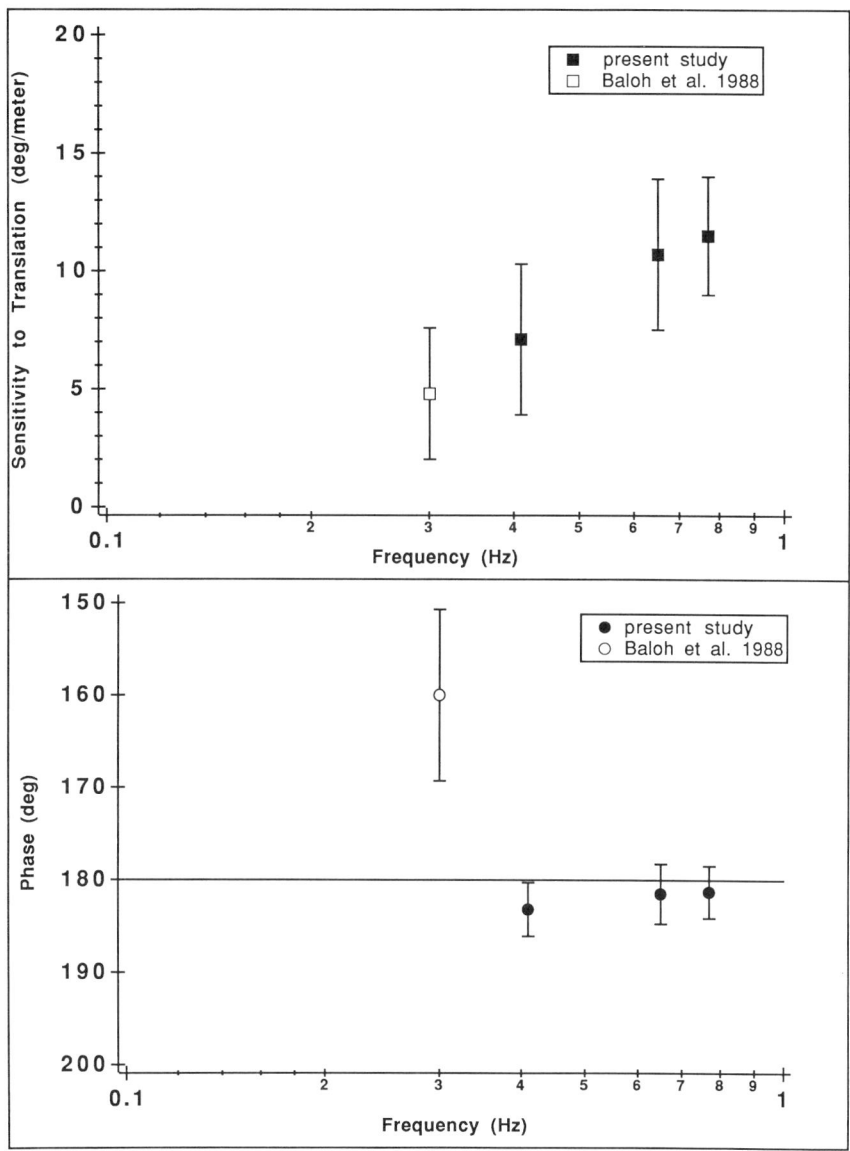

FIGURE 1. Frequency response of the horizontal LVOR. Each value represents the mean response induced by linear acceleration along the interaural (y) axis in seven normal subjects while performing mental arithmetic. ST (squares, upper plot) and phase (circles, lower plot) were computed by cross spectral density methodology after digital removal of fast component eye movements. Error bars represent ± one standard deviation from the mean values. The open square and circle represent ST and phase values obtained at the natural frequency of the parallel swing as reported in an earlier study.[1]

TABLE 1. The Effect of Target Distance on Vis-LVOR and LVOR Gains[a]

Frequency (Hz)	Mental Arithmetic	Imagined Far Target	Imagined Near Target	Visualized Far Target	Visualized Near Target
0.77	11.4 ± 3	14.3 ± 3	17.5 ± 5	27.7 ± 2	54.1 ± 6
0.65	10.7 ± 3	13.0 ± 3	16.9 ± 7	27.3 ± 2	54.0 ± 5
0.41	7.1 ± 3	10.6 ± 8	15.8 ± 12	27.1 ± 2	55.1 ± 8

[a]Gain as measured by ST for each of the five experimental conditions at the given mean stimulus frequencies. Each value represents the mean ± one standard deviation and was obtained from the responses in seven normal subjects.

gaps in the slow eye velocity record were filled with points based on the least-squares-fit of eye velocity versus stimulus velocity. Cross spectral analysis of horizontal eye and head translational velocities produced values for phase and ST in degrees/meter.

RESULTS

Horizontal LVOR gain increased with frequency when normal subjects performed mental arithmetic without specific instructions (FIGURE 1). The eye movements were 180 degrees out of phase with respect to head translation for all frequencies and target distances in all normal subjects. The gain of the LVOR in the light while viewing targets (Vis-LVOR) remained near 1.0 for all swing frequencies (0.41 to 0.77 Hz), peak swing velocities (0.2 to 1.6 meters/second), and target distances (1.05 to 2.1 meters). This was consistent even at near target conditions where eye velocities were greater than 80 degrees/second. Horizontal LVOR gain while imagining earth-fixed targets was more variable and depended on individual strategies. In was inversely related to target distance and increased only slightly with frequency (TABLE 1).

CONCLUSIONS

As in prior studies,[1-3] we found that the horizontal LVOR is highly dependent on the distance of both real and imagined visual targets. If measured in the dark without specific instructions, ST gradually increases with frequency. Clinical tests of otolith function using the horizontal LVOR must control for each of these variables.

REFERENCES

1. BALOH, R. W., K. BEYKIRCH, V. HONRUBIA & R. D. YEE. 1988. Eye movements induced by linear acceleration on a parallel swing. J. Neurophysiol. **60**(6): 2000–2013.
2. SKIPPER, J. J. & G. R. BARNES. 1989. Eye movements induced by linear acceleration are modified by visualisation of imaginary targets. Acta Oto-Laryngolo. **468:** (Suppl) 289–293.
3. PAIGE, G. D. 1989. The influence of target distance on eye movement responses during vertical linear motion. Exp. Brain. Res. **77**(3): 585–593.

Response Characteristics of the Human Torsional Vestibuloocular Reflex[a]

ROBERT J. PETERKA[b]

R. S. Dow Neurological Sciences Institute and
Clinical Vestibular Laboratory
Good Samaritan Hospital & Medical Center
Portland, Oregon

The torsional vestibuloocular reflex (TVOR) is less well studied than the horizontal or vertical VOR. Due to limitations in recording techniques, early studies concentrated on the static properties of the TVOR. Static TVOR gains of about 0.1 were identified,[1] leading to the widespread assumption that the TVOR is unimportant in gaze stabilization. More recent work identified moderate TVOR gains during active head movements[2,3] and identified other properties of voluntary control and visual-vestibular interactions associated with torsional eye movements.[4] In comparison to

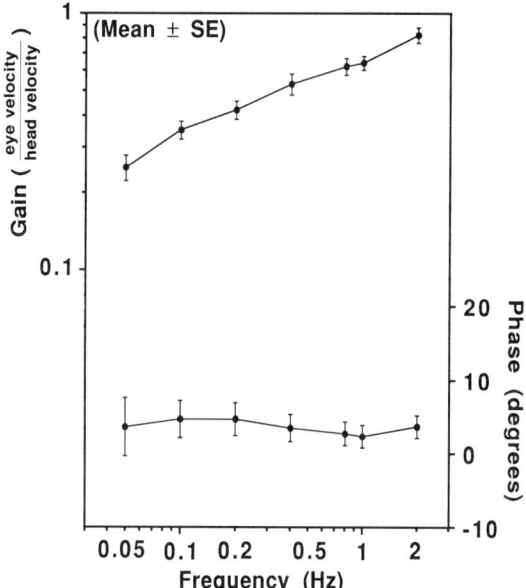

FIGURE 1. Average torsional VOR gain and phase measures of eight normal subjects as a function of stimulus frequency.

[a]Work supported by National Aeronautics and Space Administration grant NAG 9-117.
[b]Send correspondence to Clinical Vestibular Lab, N010, Good Samaritan Hospital & Medical Center, 1040 NW 22nd Avenue, Portland, Oregon 97210.

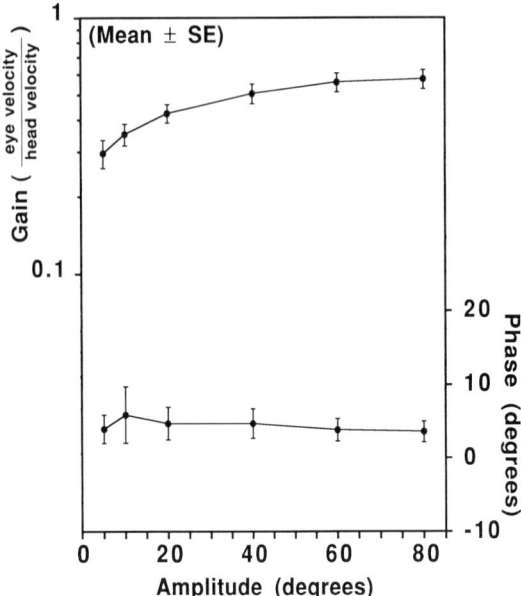

FIGURE 2. Average torsional VOR gain and phase of eight normal subjects as a function of stimulus amplitude. Stimulus frequency was 0.2 Hz for all tests.

vertical and horizontal VOR, TVOR gains are influenced much less by visual cues and by voluntary attempts to enhance or suppress torsional eye movements. The present study characterizes the response dynamics of the TVOR during controlled rotations about an earth-horizontal axis, extends the frequency range of test results to 2 Hz, and identifies an amplitude response nonlinearity.

TVOR was measured in eight subjects. The stimulus consisted of passive, whole-body sinusoidal rotations about an earth-horizontal axis aligned with the subject's nasooccipital axis at the level of the intraaural axis (stimulating vertical semicircular canals and otolith organs). The subject fixated on a single dim light-emitting diode located on the rotation axis that provided no visual orientation reference. Two data sets were collected. First, frequency was varied from 0.05 to 2.0 Hz at an amplitude of ±20° (exceptions: ±10° at 1.0 Hz and ±5° at 2.0 Hz). Second, amplitude was varied from ±5° to ±80° at a stimulus frequency of 0.2 Hz. A small bite-plate-mounted video camera recorded eye movements under infrared illumination. Off-line digital image processing was used to calculate ocular torsion (60 samples/second). Slow phase torsional eye velocity was compared to the stimulus velocity in order to calculate TVOR gain and phase. Unity VOR gain and 0° phase represent perfect compensatory response dynamics.

TVOR gain increased monotonically with increasing frequency (FIGURE 1). Gains at 0.05 Hz ranged from 0.15–0.38 and at 2.0 Hz from 0.60–1.0. Four of the eight subjects had gains greater than 0.8 and 2.0 Hz. On average, phase showed about a 5° lead across the bandwidth of test frequencies.

At a fixed frequency of 0.2 Hz, VOR gain increased with increasing stimulus amplitude while VOR phase was relatively unchanged (FIGURE 2). On average, the gain at ±80° was 1.9 times greater than the gain at ±5°. This gain nonlinearity and the high gains observed at high frequencies suggest that the torsional VOR can play a significant roll in stabilizing retinal image motion during rapid head movements.

REFERENCES

1. MILLER, E. F. 1962. Acta Otolaryngol. Stockh. **54:** 480–501.
2. FERMIN, L., H. COLLEWIJN, T. C. JANSEN & A. V. VAN DEN BERG. 1987. Vision Res. **27:** 811–828.
3. VIÉVILLE, T. & D. MASSE. 1987. Acta Otolaryngol. Stockh. **103:** 280–290.
4. LEIGH, R. J., E. F. MASS, G. E. GROSSMAN & D. A. ROBINSON. 1989. Exp. Brain Res. **75:** 221–226.

The Effect of Scopolamine on the Vestibuloocular Reflex, Gain Adaptation, and the Optokinetic Response

LEX W. SCHULTHEIS[a] AND DAVID A. ROBINSON[b]

[a] Department of Anesthesiology and Critical Care Medicine
[b] Department of Ophthalmology
The Johns Hopkins Medical School
Baltimore, Maryland 21205

INTRODUCTION

The etiology of motion sickness is widely believed to result from conflicting efforts of visual and vestibular systems to stabilize retinal images. Space adaptation syndrome is a related condition that may be caused by unresolved differences in interpretation of head motion by the vestibular canals and otolith organs in the absence of gravity.[1] Scopolamine, a nonspecific muscarinic antagonist that crosses the blood-brain barrier, is a particularly effective drug in preventing symptoms of nausea at sea and in space. Discontinuation of scopolamine therapy in previously asymptomatic sailors and astronauts has resulted in a new onset illness.[2,3] Histological studies have revealed choline acetyltransferase in important vestibuloocular control sites including the vestibular nuclei, flocculus, and inferior olive,[4] but the mechanism by which scopolamine prevents symptoms of motion sickness is unresolved. Our investigation seeks to measure quantitative effects of intravenously administered scopolamine on gain adaptation of the vestibuloocular reflex (VOR) and on the optokinetic response.

METHODS

Five adult female cats were fitted with a skull plate to restrain the hed and a subconjunctival search coil to detect eye movements. The gain of the VOR was measured by passive rotation in the dark $\pm15°$ at 0.25 Hz as the ratio of eye velocity to head velocity. Adaptation of the VOR was accomplished with the animal inside a translucent drum upon which alternating black and white stripes were back projected. Stripe motion was driven by a mirror galvanometer to precisely equal had velocity (fixed field) or to move in the opposite direction at twice the speed of the head ($\times2$ viewing). An intravenous infusion (1.0 μg/cc) was begun 1 hour prior to experimentation. The rate was controlled by a mechanical pump to within 0.1 cc/hour. In experiments to test the effect of scopolamine, the infusion was administered at 1–8 μg/kg per hour. Several animals were given supplementary amphetamines to limit sedation. Animals in control experiments received Ringer's lactate at the same rate. All procedures were approved by the Animal Care and Use Committee of our institution. Data are graphed as mean $\pm$ standard error of the mean (SEM), and significance was reported when $p < 0.05$.

880

RESULTS

1. The gain of the VOR was measured in five cats during oscillation in yaw and pitch hourly at 0.25 Hz while they were infused with Ringer's lactate (control) or scopolamine solution. There was no difference in VOR gain in cats given scopolamine (1–4 µg/kg per hour) compared to controls for up to 6 hours of drug administration. Two cats were allowed to sleep with their heads restrained in darkness after 2 hours of adaptation with scopolamine infusion. After 3 hours of sleep, the animals were vigorously awakened and the VOR gain remeasured. In both cats, the gain was the same as immediately prior to their sleep interval indicating that the gain reduction was retained. Two cats were also studied during high-dose scopolamine infusions (6–8 µg/kg per hour). At these high doses, the animals appeared sedated and displayed

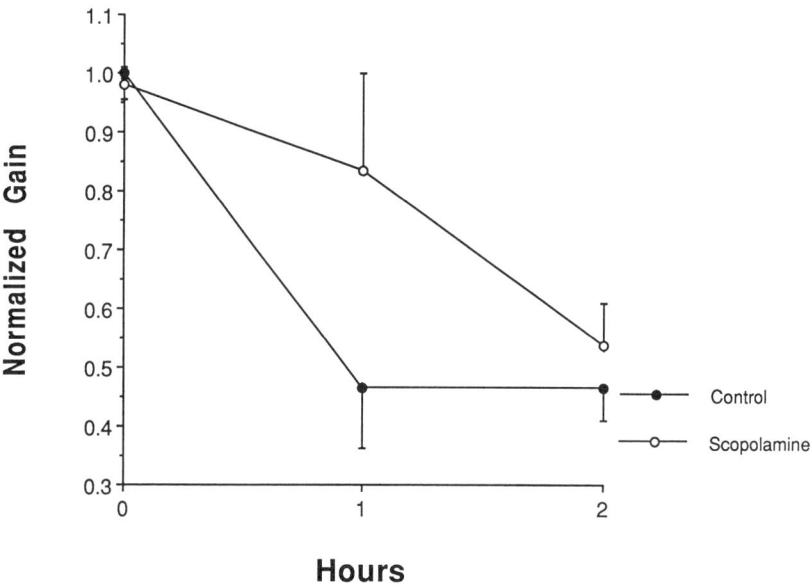

FIGURE 1. Scopolamine does not prevent VOR adaptation to a fixed visual field.

infrequent low-velocity quick phases. VOR gain appeared depressed prior to attempts at adaptation.

2. Five cats were adapted for 2 hours to a fixed visual field during passive oscillation in yaw. An infusion of Ringer's lactate (control) or scopolamine (4 µg/kg per hour) was begun 1 hour prior to oscillation and maintained throughout the experiment. There was no difference in VOR gain reduction after 2 hours in animals given scopolamine compared to control. The VOR was significantly lower ($p < 0.5$, paired Student's t-test) after 1 hour of adaptation with Ringer's infusion than with scopolamine. (See FIGURE 1.) Addition of k-amphetamine (0.25 mg/kg per hour) to the scopolamine infusion did not change the magnitude of gain reduction after 1 or 2 hours.

3. Three cats were adapted to ×2 viewing during scopolamine infusion (3 μg/kg per hour) or Ringer's lactate. As with the fixed field experiments, infusions were begun 1 hour prior to oscillation in yaw and continued throughout 5 hours of adaptation and hourly testing. The VOR gain appeared to increase less when animals were given scopolamine than Ringer's solution, but the difference was not significantly different (ANOVA). (See FIGURE 2.) Addition of *d*-amphetamine (0.25 mg/kg per hour) to the scopolamine infusion did not change the magnitude of VOR gain increase.
4. The optokinetic response to constant-velocity rotation of a patterned drum was compared in 3 cats infused with scopolamine (3μg/kg per hour) or Ringer's solution. Infusions were begun 1 hour prior to testing. At low velocities, the ratio of eye velocity to drum velocity was unchanged by

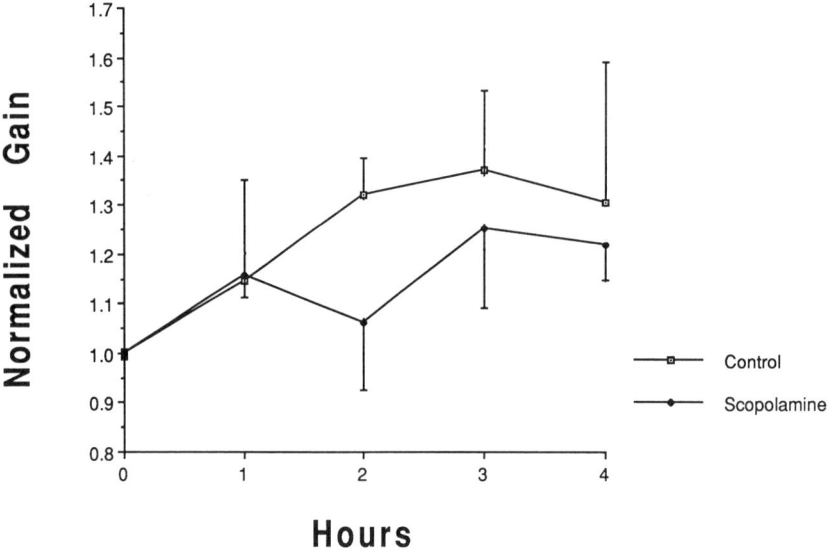

FIGURE 2. Scopolamine does not prevent an adaptive increase in VOR gain.

scopolamine. At drum velocities 30–45° second, peak eye velocity appeared to be less than in control animals, but the difference wsa not significant (ANOVA).

DISCUSSION

Low-dose scopolamine does not reduce VOR gain. Higher doses may reduce VOR indirectly by sedation. The VOR gain may be adaptively increased or decreased in animals pretreated with scopolamine. The rate of VOR adaptation may be reduced in the presence of scopolamine although the reduction is usually not significant. Adaptive changes in VOR gain in animals treated with scopolamine are retained, i.e., plastic. The maximal eye velocity response to optokinetic stimulation is unchanged by scopolamine. Although muscarinic antagonists may directly interfere with specific vestibuloocular processing centers when applied locally, systemic administration does not appear to prevent VOR gain adaptation.[5] Scopolamine may prevent motion sickness by limiting the effect of adaptive sensory conflict on the emesis centers in the brain.[6]

REFERENCES

1. VON BAUMGARTEN, R., A. BENSON, A. BERTHOZ, T. BRANDT, U. BRAND, W. BRUZEK, J. DICHGANS, J. KASS, T. PROBST, H. SCHERER, T. VIEVILLE, H. VOGEL & J. WETZIG. 1984. Effects of rectilinear acceleration and optokinetic and caloric stimulations in space. Science 225: 208–212.
2. THORNTON, W. E., J. J. URI, T. MOORE & S. POOL. 1989. Studies of the horizontal vestibulo-ocular reflex in spaceflight. Arch. Otolaryngol. Head Neck Surg. 115: 943–949.
3. VAN MARION, W. F., M. C. BONGAERTS, J. C. CHRISTIAANSE, H. G. HOFKAMP & W. VAN OUWERKERK. 1985. Influence of transdermal scopolamine on motion sickness during 7 days' exposure to heavy seas. Clin. Pharmacol. Ther. 38: 301–305.
4. KIMURA, H., P. L. MCGEER, J. H. PENG & E. G. MCGEER. 1981. The central cholinergic system studied by choline acetyltransferase immunohistochemistry in the cat. J. Comp. Neurol. 200: 151–201.
5. TAN, H. S. & H. COLLEWIJN. 1991. Cholinergic modulation of optokinetic and vestibulo-ocular responses: a study with microinjections in the flocculus. Exp. Brain Res. 85: 475–481.
6. PEDIGO, N. W. & K. R. BRIZZEE. 1985. Muscarinic cholinergic receptors in area postrema and brainstem areas regulating emesis. Brain Res. Bull. 14: 169–177.

The Error Signal for Modification of Vestibuloocular Reflex Gain

CHARLES A. SCUDDER[a] AND ALBERT F. FUCHS[b]

[a]Department of Otolaryngology
University of Pittsburgh
Pittsburgh, Pennsylvania 15213

[b]Department of Physiology and Biophysics and
Regional Primate Center
Seattle, Washington

In the early 1970s, Ito proposed an attractive neural mechanism to explain how the gain of the vestibuloocular reflex (VOR) became adaptively modified when human and animal subjects viewed the world through various optical devices.[1] He proposed that climbing fibers to the cerebellum conveyed an "error signal," correlated with whole-field motion of the visual scene across the retina (retinal slip), which altered the efficacy of the VOR pathways through the cerebellum. Using rabbits, many of the physiological tenets of this theory have been confirmed. But unlike rabbits, humans and monkeys have a fovea and can accurately track small moving targets (smooth pursuit). This introduces two other possible error signals: (1) slip of a small earth-fixed target across the fovea during VOR in the light, and (2) a nonzero gaze-velocity signal conveyed by floccular Purkinje cells during head rotations when there should be no actual gaze velocity.[2] More specifically, the normally equal and opposite head velocity and eye velocity inputs to floccular Purkinje cells[3] become imbalanced when subjects move their heads while viewing the world through the appropriate optical devices.

A factorial experiment was designed to compare the efficacy of these potential error signals. VOR gain was measured in the dark in three monkeys before and after three hours of forced sinusoidal rotation at 0.5 Hz, ±10 deg. During this time, monkeys simultaneously viewed an optokinetic background and tracked a small light-emitting diode (LED) target. The LED could be (1) fixed with respect to the monkey (×0 condition), (2) fixed in space (×1 condition), or (3) moved in space in a direction opposite to head motion (×2 condition). Motion of the optokinetic background was chosen from the same three conditions but independently of the LED. Adaptation was also obtained with the monkeys tracking the LED in the dark.

The results from one monkey are shown in TABLE 1. Data from the other two were similar except for larger variances. Each cell in the table corresponds to a unique combination of LED and background motion during adaptation, and data are expressed as the percent difference in VOR gain (final − initial) relative to initial gain. When monkeys were exposed to synergistic conditions (e.g., LED and background both fixed with respect to the head), gain changed rapidly. When exposed to conflicting conditions, gain changed slowly if at all. Combinations of adaptive (×0 or ×2) and neutral (×1) conditions caused intermediate gain changes. The results show that whole-field slip and pursuit-related error signals were both effective in causing VOR gain changes and were quantitatively about equal.

To differentiate a foveal-slip error signal from a gaze-velocity error signal, monkeys were rotated in the dark while viewing a chair-fixed LED which flashed at 8/second. This was designed to eliminate motion of the target LED across the fovea.

TABLE 1[a]

Target LED	Background			
	Dark	×0	×1	×2
×0	−14.3%	−27%	−7.7%	−2.7%
×1	+3.0%	−14%	+8.3%	X
×2	+8.1%	−1.7%	X	+12.6%

[a]Each cell contains the data from one combination of target LED and background motion used during the VOR adaptation period. Numbers represent percent change in VOR gain [100 × (final-initial)/initial]. X's denote conditions that could not be run. Initial VOR gain was 0.85 ± 0.021 (standard deviation) for this monkey so gain increased toward 1.0 in some ×1 conditions. Gain changes can be compared to a control condition in which monkeys tracked a sinusoidally moving LED without head rotation: gain changed −2.1% ± 2.6% (SD).

The gain drop over 3 hours was the same as when the LED did not flash (14.2%), indicating that the gaze-velocity signal is probably the pursuit-related error signal that causes VOR gain to change.

REFERENCES

1. ITO, M. 1982. Cerebellar control of the vestibulo-ocular reflex: around the flocculus hypothesis. Annu. Rev. Neurosci. **5:** 275–296.
2. MILES, F. A. & S. G. LISBERGER. 1981. Plasticity in the vestibulo-ocular reflex: a new hypothesis. Annu. Rev. Neurosci. **4:** 273–299.
3. LISBERGER, S. G. & A. F. FUCHS. 1978. Role of primate flocculus during rapid behavioral modification of vestibulo-ocular reflex. I. Purkinje cell activity during visually guided horizontal smooth pursuit eye movements and passive head rotation. J. Neurophysiol. **41:** 733–763.

The Torsional Vestibuloocular Reflex Can Be Canceled but Not Enhanced by Visual Stimuli[a]

S. H. SEIDMAN AND R. J. LEIGH[b]

Departments of Biomedical Engineering, Neurology, Otolaryngology,
and Neuroscience
Department of Veterans Affairs and
Case Western Reserve University
Cleveland, Ohio

The torsional vestibuloocular reflex (VOR) in humans shows different properties from the VOR in the horizontal or vertical plane; the gain is lower, and "velocity storage" appears to be absent.[1] Despite the low gain of the torsional optokinetic system, it is possible to reduce the gain of the torsional VOR by viewing a head-fixed visual display during head rotations in roll. In the horizontal plane, the ocular motor system can nullify more than 90% of the VOR during combined eye-head tracking, and a significant portion of this reduction is believed to be due to cancellation by smooth pursuit signals.[2] It has also been shown that these visual inputs can enhance the horizontal VOR if a subject attempts to track a target moving opposite to the direction of head rotation.[2] In the torsional plane it has been shown that the VOR can be reduced by about 30% while viewing a head-fixed visual display.[3] However, attempts to enhance or reduce the gain of the torsional VOR with visual stimuli moving at a variety of speeds with respect to the head have not been reported.

Torsional eye and head movements were measured in four subjects using the magnetic search coil technique, and digitized for later analysis. Analog head position signals were differentiated in real time to yield velocity information, which was then fed to the command input of a torsional optokinetic stimulator. This stimulator employed a servo motor to rotate a circular (70 cm diameter) randomly spotted disk subtending > 70 deg. Using this servo system we could rotate the disk at positive or negative multiples of head velocity. Subjects were instructed to fixate a small red spot in the center of the visual display, and to roll their head ear-to-shoulder continuously in both directions at approximately 0.5 Hz. During this head rotation, visual feedback ranging between −1 and +3 in gain was provided to the subject. Following data collection the cross- and autopower spectral densities of the input and output were then estimated using Welch periodogram techniques. This allowed us to estimate gain and coherence as a function of frequency. Coherence was always > 0.86.

Results from one subject are shown in FIGURE 1, and cumulative results are shown in TABLE 1. In all subjects, the visually enhanced VOR (i.e., ×0 visual feedback) was of greater gain than the VOR in the dark. Subjects were not able to further increase VOR gain with visual stimuli. In fact, for two subjects, VOR gain actually decreased by more than 25% when the display rotated in a direction

[a]This work was supported by National Institutes of Health EY06717, the Department of Veterans Affairs, and Evenor Armington Fund.

[b]Author to whom correspondence should be addressed: Department of Neurology, University Hospitals, Cleveland, Ohio 44106.

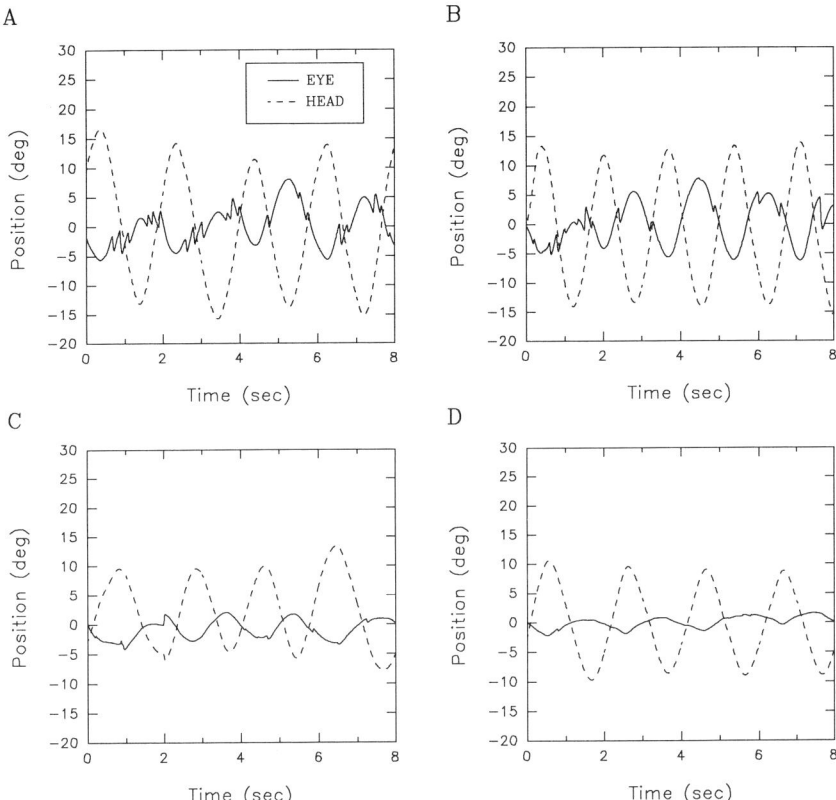

FIGURE 1. Representative results of one subject with visual feedback gain of (A) ×−1, (B) ×0, (C) ×1, and (D) ×2. Positive gain indicates display rotation in the same direction as head rotation. Note decreased VOR gain for higher positive values of visual feedback.

TABLE 1. Torsional Vestibuloocular Reflex Gains for Various Visual Feedback Gains

Subject	Visual Display Feedback Gain					
	×−1	×0	×1	×2	×3	Dark
1	0.67	0.62	0.27	0.18	0.12	0.42
2	0.57	0.78	0.35	0.42	0.28	0.42
3	0.36	0.51	0.39	0.42	0.22	0.40
4	0.63	0.65	0.56	0.53	0.54	0.62

opposite to that of the head. When the display rotated in the same direction as the head, gain was decreased in all cases, with the most reduction for three out of four subjects occurring when the display moved 3 times as fast as the head (the largest visual feedback). The greatest reduction in gain observed was 80% from the visually enhanced VOR and 70% from the VOR in the dark.

We conclude that the gain of the torsional VOR can be reduced but not increased by viewing a display that moves relative to the head during active head roll. Since smooth pursuit is nonexistent for torsional eye movements, and the optokinetic response is feeble, the reduction in gain of the VOR cannot be attributed to visual mechanisms. These properties differ considerably from those of eye movements in the horizontal plane. While the mechanisms by which the gain of the torsional VOR is reduced remain unknown, suppression, nonvisual cancellation, proprioceptive, and predictive mechanisms are possible explanations for the observed responses.

REFERENCES

1. SEIDMAN, S. H. & R. J. LEIGH. 1989. The human torsional vestibulo-ocular reflex during rotation about an earth-vertical axis. Brain Res. **504:** 264–268.
2. LISBERGER, S. G. 1990. Visual tracking in monkeys: evidence for short-latency suppression of the vestibulo-ocular reflex. J. Neurophysiol. **63:** 676–688.
3. LEIGH, R. J., E. F. MAAS, G. E. GROSSMAN & D. A. ROBINSON. 1989. Visual cancellation of the torsional vestibulo-ocular reflex in humans. Exp. Brain Res. **75:** 221–226.

Context-Specific Gain Switching in the Human Vestibuloocular Reflex[a]

MARK SHELHAMER, DAVID A. ROBINSON,
AND HENDRA S. TAN[b]

Department of Ophthalmology
The Johns Hopkins University School of Medicine
Room 355 Woods Research Building
The Wilmer Institute
601 North Broadway
Baltimore, Maryland 21205

[b]Department of Physiology I
Erasmas University Rotterdam
Post Office Box 1788
3000 DR Rotterdam, the Netherlands

It is well established that altered visual feedback (e.g., magnifying spectacles) can effect a short-term change in the gain of the vestibuloocular reflex (VOR) in humans.[1,2] It is also presumably true, though previously unproven, that subjects can store two different VOR gains simultaneously, using each one under the appropriate circumstances. Such gain switching would occur, for example, when eyeglasses are donned and doffed, or when bifocals are worn. In each of these cases, a nonvisual cue accompanies the required gain adjustment (frames on/off for eyeglasses, looking up/down for bifocals). We set out to show that a subject can establish two VOR gains simultaneously, and also to determine if an associated nonvisual cue alone (vertical gaze deviation, in our experiment) is sufficient to subsequently determine which gain to employ.

Each of three subjects sat in a rotating chair inside an optokinetic nystagmus (OKN) drum during 2 hours of sinusoidal rotation at 0.2 Hz, 30 deg/second peak velocity. For 10 minutes the chair and drum counterrotated, driving VOR gain toward 1.7, while subjects looked up 20 deg. The chair and drum were then coupled for 10 minutes, driving VOR gain toward 0, during which subjects looked down 20 deg. This sequence was repeated for 2 hours. Immediately before and after adaptation, VOR gains were measured using 20- to 40-deg step rotations, as follows. In the dark, a fixation target appeared, was extinguished, the chair performed a step of rotation, and the target reappeared. The subject made a corrective saccade to the target after the rotation, providing a measure of the proportion of head rotation compensated for by eye rotation (VOR) during the step. This was repeated while subjects looked alternately up and down 20 deg.

For steps that took the eye through primary gaze position, there was consistent reduced VOR gain looking downward (average 6%) and increased gain looking upward (average 8%). (TABLE 1). This shows that subjects had stored two VOR gains and could use vertical eye position to implement one or the other. "Eccentric" steps, which either began or ended with the eyes centered (and not passing through

[a]Supported by National Research Service Award grant T32 EY07047 and National Institutes of Health grant EY00598.

TABLE 1. Vestibuloocular Reflex Changes after Two Hours of Alternating
1.7× and 0× Viewing[a]

Subject	Up Gaze (1.7×)			Down Gaze (0×)		
	Before	After	p	Before	After	p
1	1.04	1.12	0.0001	1.03	0.96	<0.0001
	±0.038	±0.050		±0.030	±0.016	
2	1.05	1.08	0.1	1.01	0.97	0.0024
	±0.049	±0.037		±0.021	±0.026	
3	1.03	1.16	<0.0001	1.04	0.99	0.0024
	±0.037	±0.053		±0.039	±0.032	
Grand	1.04	1.12		1.03	0.96	
Mean	±0.039	±0.046		±0.030	±0.024	
Changes		+8%			−6%	

[a]Means and standard deviations of gains of the three subjects looking up and down before
and after adaptation. Fourth row shows means and standard deviations for all subjects grouped
together. Last row shows percentage changes. p values are for one-tailed t-tests.

the midline), showed a different pattern (FIGURE 1): gain increases were more
prominent when the eyes began eccentrically and moved to the center. We suggest
that this may be due to a decrease in the time constant of the neural velocity-to-
position integrator,[3] which then predisposes the eyes to move toward the center
position upon eccentric gaze. We have demonstrated in one subject that the time
constant of gaze holding (which is mediated by this integrator) decreases after one
hour of 1.7× adaptation, from 23 to 12 seconds.

We conclude that humans can adjust their VOR gain dependent on a situational
context, and speculate that this context can take many forms.

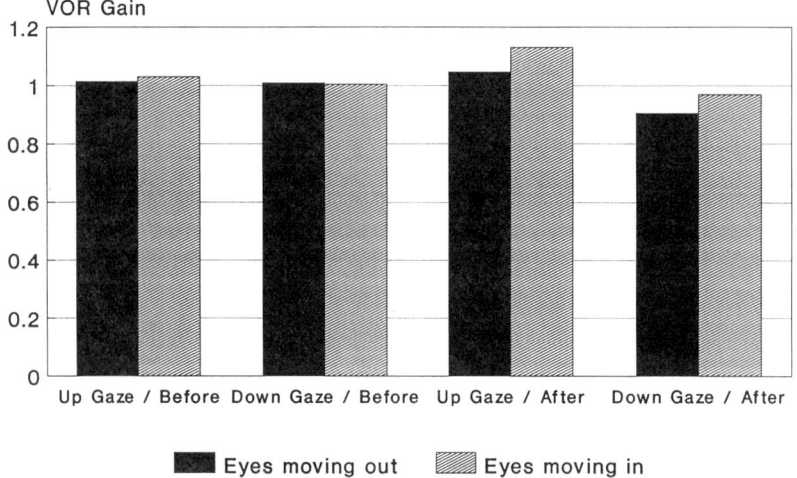

FIGURE 1. Difference between steps that begin with the eyes centered and those that begin
with the eyes eccentric, showing larger gains when the eyes move towards the center in the 1.7×
situation.

REFERENCES

1. ITO, M. 1982. Cerebellar control of the vestibulo-ocular reflex—around the flocculus hypothesis. Annu. Rev. Neurosci. **5:** 275–296.
2. COLLEWIJN, H., A. J. MARTINS & R. M. STEINMAN. 1983. Compensatory eye movements during active and passive head movements: fast adaptation to changes in visual magnification. J. Physiol. London **340:** 259–286.
3. ROBINSON, D. A. 1981. Control of eye movements. *In* Handbook of Physiology. V. B. Brooks, Ed. **2:** 1275–1320. Williams & Wilkins. Baltimore, Md.

Negative Relationships between Responses from Head-Shake and Caloric Tests

JOHN SPINDLER

Otolaryngology/Head and Neck Surgery
Division of General Surgery
University of California Medical Group
La Jolla, California 92092-06271

The negative relationship between caloric and head-shake responses throws sharper focus on the head-shake mechanism, which cannot be explained exclusively in terms of unilateral canal paresis, Ewald's second law and velocity storage.[1] The following patterns of head shake responses call for a revision of aforementioned concepts.

1. The reversed type of biphasic head-shake response (BHSR) is found not only in patients with Meniere's disease, but also in patients with other vestibulopathies.[2] In this type of response, nystagmus in the first phase beats toward the affected hypofunctional ear and contradicts Ewald's second law. By definition, this is an irritative nystagmus and it mimics the events of depolarization occurring during the patient's vertiginous attack. The reversed type of BHSR may be modified in its intensity by medication (dilantin), but the response does not change its direction. A patient who would exhibit the reversed type of BHSR first and later a BHSR of the "Kamei" type has not been found.

2. Immediately after a sudden, unilateral total loss of function, the head shake cannot be elicited in some patients. The response does not increase the preexisting spontaneous nystagmus, or the spontaneous nystagmus is suppressed or even absent. The amplitude of eye deviation during the head-shake administration is reduced. The subsequent partial compensation leads to restoration of the second phase first with nystagmus beating toward the affected ear. One can speculate that the velocity storage is either absent or the response is modified by a velocity storage "clamp."

3. Intense head-shake responses can be evoked in patients with perfectly symmetrical caloric responses who recovered from peripheral injury incurred two or three years before. Without knowledge of the patient's history, such response could be wrongly interpreted as a false positive. The head shake is the last provocation to disappear, and in that lies its medicolegal value.

4. Prolonged head-shake responses show a linear time course and are obtained from the patients with bilaterally diminished caloric responses. These responses extend over a period of many minutes and do not show the second phase.

False-negative head-shake responses may be obtained from rotation of the head by the subject (active), who preprograms the efferent copy, thereby canceling the response. The passive head shake done by the examiner results in unopposed exafference, and a response is evoked that confirms the canal paresis obtained with the caloric test. Therefore, a passive test yields stronger specific (rotatory) and unspecific (alerting) stimuli than those of the active test.

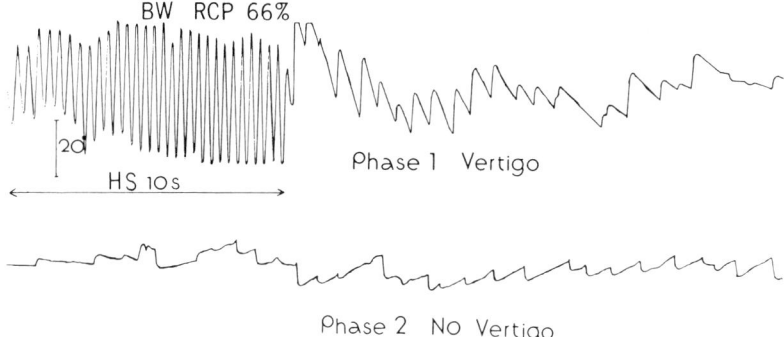

FIGURE 1. Reversed type of biphasic head-shake response from a patient with Meniere's disease with 66% right canal paresis. Right-beating nystagmus in the first phase beats toward the hypofunctional ear and contradicts Ewald's second law. By definition this is an irritative nystagmus and it mimics the events of depolarization evoked by warm irrigation. Right-beating irritative nystagmus was also observed during the patient's vertiginous attacks.

Besides a plausible explanation that a caloric test represents the low frequency and the head shake the high frequency of the spectrum of stimuli, additional concepts are needed to explain the variations in responses. Presumably the internal feedbak loops modify the assymetries of the peripheral organ, the central inputs, and velocity storage. These may all contribute the the variations in intensity and time course of responses. Clinically, the head-shake test is not a screening test, but it

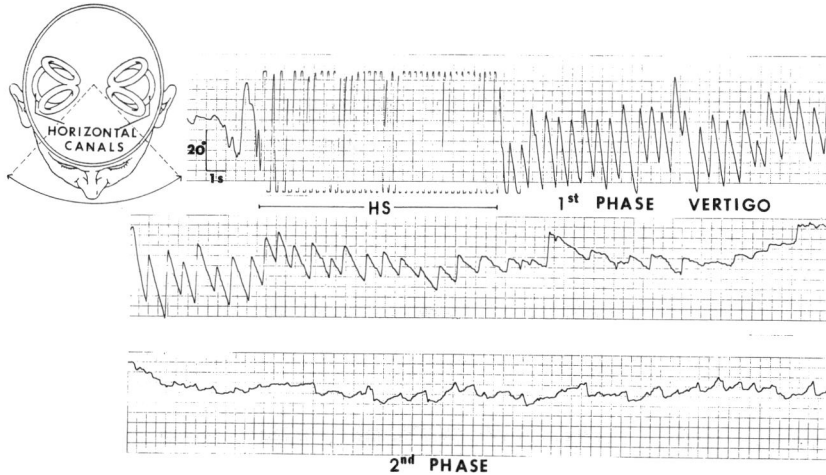

FIGURE 2. Despite strong and symmetrical caloric responses, this patient exhibited an intense vertigo elicited by a head shake. Head-shake provocation decompensated the imbalance in latent vestibular tonus caused by a peripheral injury to the left ear that had occurred three years before. The injury ended the career of the professional soccer player, but the head shake resolved the medicolegal case.

JM 43F

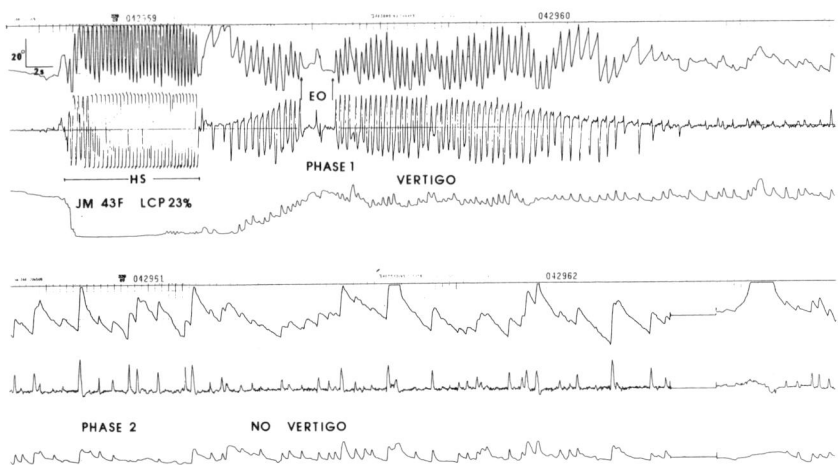

FIGURE 3. The reversed type of biphasic response is not observed only in Meniere's disease. This patient was a diabetic with Cushing's disease. The response was unusual: (1) the time course showed a delayed response with a crescendo-decrescendo pattern of intensity; (2) the reversed type of nystagmus beats toward the hypofunctional ear in the first phase; (3) the velocity of the maximal slow phase exceeds that of the fast phase.

complements other findings. A positive head shake ascertains underlying vestibulopathy. But, a negative one does not confirm the absence of vestibulopathy.

REFERENCES

1. HAIN, T. C. & D. S. ZEE. 1987. J. Otolaryngol. **8:** 36–47.
2. SPINDLER, J. & M. SCHIFF. 1988. Adv. Oto-Rhino-Larynogol. **42:** 95–103.

Three-Dimensional Optokinetic Responses in Man and Rabbit

A Comparison between a Frontal and a Lateral Eyed Species

J. VAN DER STEEN, H. S. TAN, AND L. J. VAN RIJN

Department of Physiology I
Faculty of Medicine
Erasmus University
Post Office Box 1738
3000 DR Rotterdam, the Netherlands

In order to maintain a perceptually stable visual image, vestibular and visual control mechanisms cooperate to stabilize the eyes during involuntary head movements.

In humans, the fovea plays a prominent role in the uptake of visual information, and the vestibuloocular and optokinetic reflexes (VOR and OKR) operate at almost unity gain in horizontal and vertical directions in order to maintain the orientation of this structure with respect to a point of interest and to minimize retinal image motion across this central region of the retina.[1] This concentric organization of the retina allows for a relatively large tolerance for retinal image motion about the visual axis, which is reflected in a considerably lower gain for torsional vestibular compensatory eye movements (up to 0.6).[2] Recently, we found that in humans also the gain of optokinetic responses for image motion about the visual axis (torsion) is low (between 0.2 and 0.3), even under optimal conditions (i.e., presenting the binocularly viewing subjects with a large-field, highly textured pattern). Apart from this distinct quantitative difference in the optokinetic responses between horizontal, vertical, and torsional directions, in human OKR, in contrast to the rabbit, only small qualitative differences exist in directional asymmetries or differences between monocular and binocular conditions.

Rabbits have laterally placed eyes and an almost panoramic visual field. Therefore they have no need to control gaze in terms of maintaining a preferred position of the eyes, Yet, retinal image speed has to be kept low enough to prevent blurring. Simpson and colleagues have shown that the rabbit processes optical flow information through the accessory optic system (AOS) to detect self-motion.[3,4] The different cell groups involved in this processing are direction selective and organized in a reference frame similar to that of the semicircular canals, and the pulling axes of the eye muscles, i.e., a vertical axis and two axes lying in the horizontal plane at azimuthal angles of approximately 45° and 135° (with the nose taken as the 0° azimuthal angle). The presence of this reference frame and the direction selectivity agree with the well-established fact that during monocular viewing in the rabbit, the optokinetic responses for stimulation about the VA axis (horizontal eye movements) have a preference for temporal-nasal stimulus motion.[5] So far, little is known about optokinetic responses in other planes.

In this study we investigated to what extent the reference frame for the processing of optical flow is reflected in the gain and response axis of the eye movements elicited by optokinetic stimulation about axes in the horizontal plane (HA axis stimulation). Rabbits were implanted with dual search coils on both eyes,[6] and eye movements were recorded in a VA, 45° and 135° reference frame. The stimulus consisted of an optokinetic drum driven by constant-velocity steps of ±2°/second

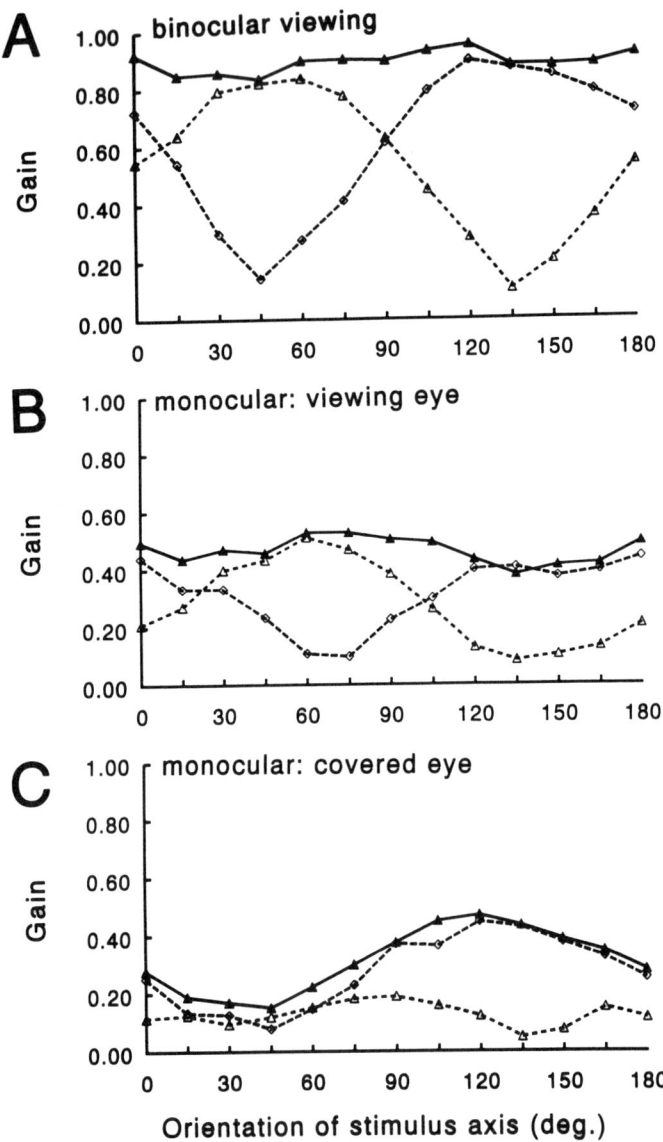

FIGURE 1. Gain of the optokinetic responses during triangular stimulus motion as a function of the orientation of the stimulus axis in the horizontal plane. The figure shows the gain of the eye movement responses about the 45° axis (dashed line with triangles), the 135° axis (dashed lines with diamonds), and the total response gain taken as the vectorial sum of the 45° and 135° axis (solid line with filled triangles). **A:** Binocular viewing conditions. **B:** Monocular viewing conditions, showing the response of the viewing eye. **C:** Monocular viewing, showing the response of the covered eye.

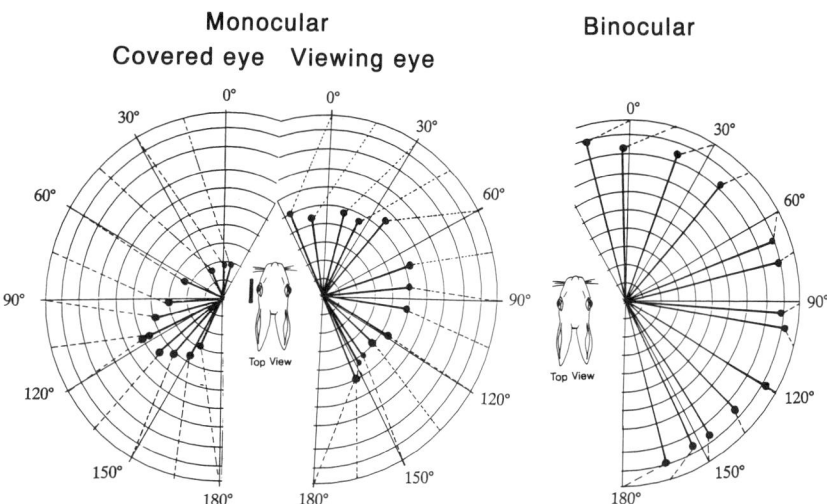

FIGURE 2. Polar diagram showing the orientation and magnitude of the optokinetic eye movement response relative to the orientation of the stimulus axis. The response vectors (solid lines) are interconnected with the corresponding orientation of the stimulus axis by the dashed lines. The position of the stimulus axis was incremented in steps of 15 degrees going in clockwise direction from the 0° position. The gain and the deviation of the responses under binocular conditions are indicated on the right panel; those of both the viewing and covered eye under monocular conditions are indicated on the left panel.

($A = 5°$). The angular orientation of the stimulus axis in the horizontal plane with respect to animal was incremented in steps of 15° azimuthal angle.

During binocular viewing the gain of the HA optokinetic responses was about 0.9, independent of stimulus axis orientation (FIGURE 1A). Under monocular conditions the OKR gain for the *viewing* eye was about 0.5 (FIGURE 1B). The responses of the *covered* eye were anisotropic, with an optimum when the drum axis was at 45°, and a minimum with the drum axis positioned at 135° re the viewing eye (FIGURE 1C). In addition, the rotational axis of the eye movements deviated from the stimulus axis (up to 30° under monocular and up to 15° under binocular conditions). The deviation had its maximum at 90° and its minimum when the stimulus axis was at about 45° or 135° azimuthal angle (FIGURE 2).

The anistropies in the monocular responses of HA-axis OKR provide further support to the idea that in the rabbit, optokinetic responses are represented in an intrinsic reference frame as put forward by Simpson *et al.*[3] and in which each eye is controlled by the class of cells that provides its dominant visual input.

REFERENCES

1. VAN DEN BERG, A. V. & H. COLLEWIJN. 1988. Exp. Brain Res. **70:** 597–604.
2. COLLEWIJN, H., J. VAN DER STEEN, L. FERMAN & T. C. JANSEN. 1985. Exp. Brain Res. **59:** 185–196.
3. SIMPSON, J. I. 1984. Annu. Rev. Neurosci. **7:** 13–41.
4. SIMPSON, J. I., C. S. LEONARD & R. E. SOODAK. 1988. J. Neurophysiol. **60:** 2055–2072.
5. COLLEWIJN, H. 1969. Vision Res. **9:** 117–132.
6. VAN DER STEEN, J. & H. COLLEWIJN. 1984. Exp. Brain Res. **56:** 263–274.

Interactions between Otoliths and Vision Revealed by the Response to Z-Axis Linear Movements[a]

CONRAD WALL III,[b] LAURENCE R. HARRIS,[c]
AND CORINNA E. LATHAN[d]

[b]Vestibular Testing Laboratory
Massachusetts Eye and Ear Infirmary,
Boston, Massachusetts 02114

[c]Department of Psychology
York University
North York, Ontario M3J 1P3, Canada

[d]Department of Psychology
Massachusetts Institute of Technology
Cambridge, Massachusetts 02139

Compensatory eye movements evoked by linear movement in the dark are rather ineffective compared to the response to angular movements.[1] However, the response to horizontal, linear, visual movement (L nystagmus, x-axis) is enhanced if the subject is simultaneously moving.[2,3] This suggests interactive effects between the visual and the otolith systems. To maintain clear vision during linear movement, compensatory eye movements have to take into account not only the physical movement itself but also the distance of regard.[4] So it is perhaps not too surprising that the otoliths' contribution is greatest when other systems (e.g., vision) can provide the necessary distance cues. We have investigated the effect of otolith-visual interactions in the z-axis (through the top of the head) with subjects in a supine position. Linear motion along this axis is transduced primarily by the otoliths of the saccule.[1]

Subjects lay on a sled and viewed an optokinetic stimulus during sinusoidal z-axis movement. The stimulus arrangement is shown in FIGURE 1. There were four stimulus conditions: vision only (in which the sled did not move), otolith only (sled movement in the dark), vision and otolith signals complementary (e.g., visual movement upwards accompanying sled motion towards the feet—as in natural vertical movements), and finally, vision and otoliths opposed (i.e., visual and physical motion in the same direction). A range of otolith/visual velocity ratios were used (FIGURE 1).

The vertical eye movement response to the otolith-only condition was, as expected, small. The visual velocity was chosen so that for the vision-only condition the closed-loop gain of vertical eye movements [peak slow-phase eye velocity (deg/second)/peak stimulus velocity (deg/second at zenith)] was about 0.5. When the otoliths were stimulated to complement the visual information, the gain was enhanced impressively to above 0.8 (FIGURE 2). Surprisingly, when the signals were opposed, the gain was not reduced below that found in the vision-only condition. There is a nice symmetry around the otolith/visual velocity ratio of unity. A ratio of unity represents the normal correspondence of otolith and visual signals.

[a]Supported by National Aeronautics and Space Administration grant NCC 2-602.

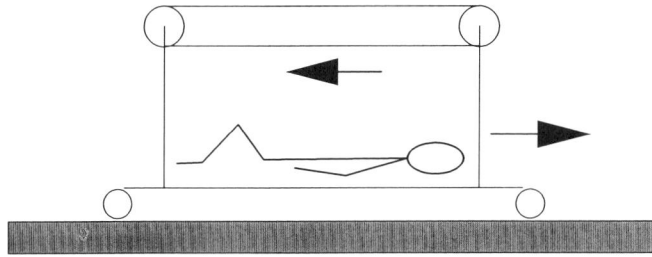

Otolith stimulus: 0.6 g sines @ 0.3, 0.5 & 1.0 Hz
Visual stimulus: 61 d/s sines viewed at 56 cm

Otolith/Visual velocity ratio:
Otolith peak velocity [m/s]
Visual peak velocity [m/s]

FIGURE 1. Methods. Subjects lay supine on a sled and looked up at a high-contrast striped pattern (84 × 84 cms, spatial frequency about 0.1 c/deg at zenith) on an endless belt suspended 56 cm above their heads. The stripe display was attached to the sled and moved with it physically. Horizontal and vertical eye movements were measured using scleral search coils. Stimuli were always sinusoidal with frequencies of 0.3, 0.5 and 1 Hz. The frequency of the visual movement was always the same as that of the sled. All visual stimuli had a peak velocity of 0.6 m/second whereas the sled had a peak acceleration of 0.6 × g. This resulted in peak sled velocities of from 0.9 to 3.1 m/second and a range of otolith/visual velocity ratios [otolith peak velocity (m/second)/visual peak velocity (m/second)].

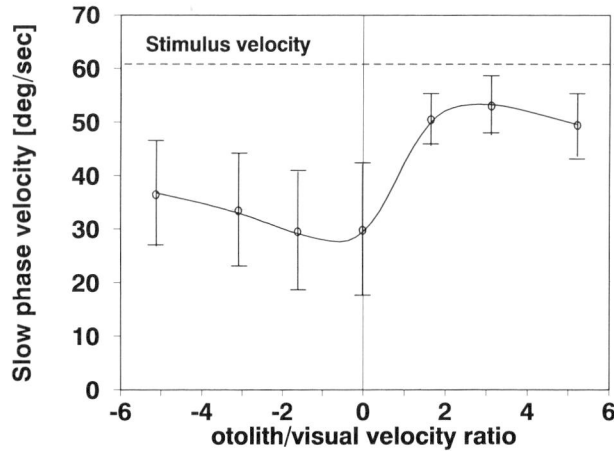

FIGURE 2. Interactive effects between the otolith and visual signals in the production of vertical eye movements. The vertical axis is the peak slow-phase eye velocity (deg/second). The horizontal axis represents the degree of agreement between the otolith and visual signals as a simple ratio of the stimulus velocities in m/second. When the signals are in agreement, this produces a positive ratio; when they are in opposition it produces a negative ratio.[6] The line is a spline interpolation through the mean data points of the five subjects. Standard deviations are also shown. Note that the response at an otolith/vision ratio of zero (the vision-alone condition) represents examples of all the frequencies that were used to generate all the other ratios: clearly the effect cannot be explained as an influence of frequency or speed.

These data add support to the emerging view that the response to physical motion cannot be regarded as a set of responses evoked from independent systems.[5] The response to movement is an indicator of the integrative nature of the nervous system.

REFERENCES

1. WILSON, V. J. & G. MELVILL JONES. 1979. Mammalian Vestibular Physiology. Plenum Press. New York, N.Y.
2. TOKUNAGA, O. 1977. Acta Otolaryngol. **84:** 338–343.
3. BUIZZA, A., A. LEGER, J. DROULEZ, A. BERTHOZ & R. SCHMID. 1980. Exp. Brain Res. **39:** 165–176.
4. PAIGE, G. D. 1989. Exp. Brain Res. **77:** 585–593.
5. COLLEWIJN, H. 1989. Prog. Brain Res. **80:** 197–209.
6. HARRIS, L. R. 1988. Exp. Brain Res. **71:** 147–152.

Gaze Control during Head-Free Pursuit in Patients with Loss of Vestibular Function

J. A. WATERSTON, G. R. BARNES, M. A. GREALY,
AND L. M. LUXON[a]

MRC Human Movement and Balance Unit
[a]Neurootology Unit
National Hospital
Queen Square
London WC1N 3BG, United Kingdom

Studies of gaze control in labyrinthine-deficient (LD) patients have revealed a variety of adaptive mechanisms that may be used to compensate for loss of the vestibuloocular reflex (VOR), including central preprogramming of eye movements and the potentiation of the pursuit, optokinetic, and cervicoocular reflexes. To investigate the mechanisms that compensate for loss of the VOR during head-free pursuit we examined the responses to pursuit of pseudorandom target motion in LD patients under head-free and head-fixed conditions, using a stimulus that covers the frequency range of normal head-free pursuit movements. Previous studies in normal subjects over limited ranges of motion have shown little difference in tracking performance between head-free and head-fixed pursuit.[1,2] Five well-compensated patients with isolated bilateral loss of vestibular function and a similar number of age-matched controls were tested. Eye movements were recorded using an infrared limbus reflection technique. The recorders were attached to a light helmet which was fixed firmly to the subject's head, and a single turn potentiometer, attached to the top of the helmet via a flexible assembly, was used to record head movements. Slow-phase eye velocity was calculated using a computer graphics procedure to remove saccadic components and then summed with head velocity to give gaze velocity gain. A pseudorandom stimulus composed of the sum of 4 sinusoids (0.11, 0.24, 0.37, and 1.56 Hz) was used for the pursuit experiments. The peak velocity of the 3 lower frequency sinusoids remained constant at 8 deg/second, while the velocity of the 1.56-Hz component was varied as a ratio of the lower frequency velocity between 0 and 2. Previous studies have demonstrated a breakdown in gain at the lowest frequencies when a sinusoid whose frequency is greater than a critical level of 0.4 Hz is added to the stimulus, and further breakdown occurs when its velocity is increased with respect to the other components (FIGURE 1).[3] The breakdown in the response occurs essentially as a result of making the stimulus less predictable, and provides a reproducible method for comparison of responses at normal frequencies of eye and head movement. Subjects were asked either to track the target using eye movements only or to use the eyes and head together in a natural fashion. In addition, the gain of the compensatory eye movement (CEM) response was assessed by recording the eye movements produced by active head movements made in the dark in response to a stationary auditory cue, at each of the 3 lowest frequencies used in the pseudorandom stimulus (0.11, 0.24, and 0.37 Hz) using a peak velocity for each stimulus of 16 deg/second to approximate the root-mean-square velocity of the corresponding components in the pseudorandom waveform. CEM gains were also measured in an additional two patients with degenerative neurological diseases characterized by relatively pure loss of vestibular and cerebellar function. Vestibular function was assessed during whole-body sinusoidal oscillation at the same frequencies.

There were no significant differences in gaze velocity gain for head-fixed and head-free pursuit at any of the velocity ratios (FIGURE 1). Head displacement gains were not significantly different for patients and controls, ranging between 0.61 and 0.86 for the 3 lowest frequencies and falling to a mean of 0.38 at 1.56 Hz in the patient group. The VOR gains obtained in the dark using active head movements (mean 0.68) approximated those obtained during whole-body turntable motion (mean 0.66) for the control subjects. Despite the absence of any significant turntable VOR response in the LD subjects (mean 0.11), CEM of near normal gain (mean 0.56) could be demonstrated during active head movements (FIGURE 2). These results demonstrate that patients with bilateral loss of vestibular function are able to produce CEM of sufficient gain to influence gaze control during head-free pursuit. These responses were absent in the 2 patients with additional cerebellar dysfunction, suggesting that the cerebellar pathways are necessary for the genesis of the compensatory response.[4] The pattern of eye movement response seen in the LD subjects during active head movements in the dark is very similar to that reported by others[2,5,6] and has been attributed in part to enhancement of the normal cervicoocular reflex. A previous study demonstrated that smooth head-free tracking gains were greater in LD subjects under head-free conditions, but only at the highest frequency tested (1.0 Hz).[2] However these differences were not marked and the gains never

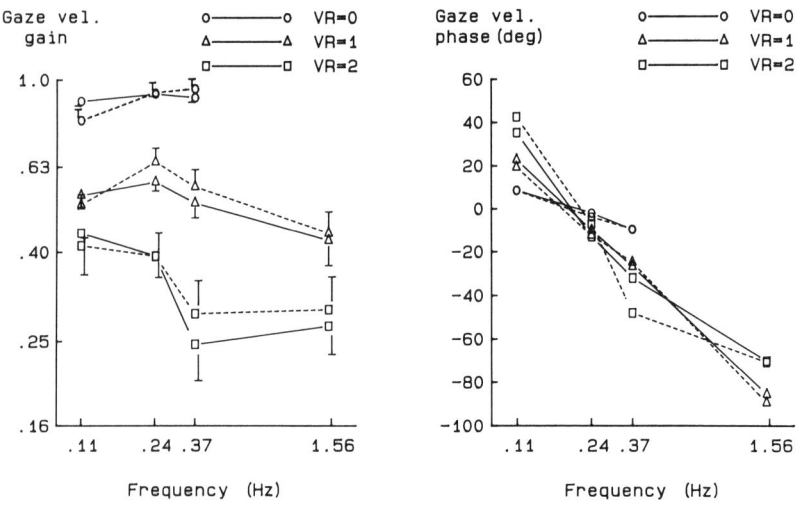

LD subjects: HEAD-FIXED vs. HEAD FREE

FIGURE 1. Gaze velocity gain and phase for labyrinthine-deficient (LD) patients during head-fixed and head-free pursuit of a pseudorandom stimulus containing 0.11, 0.24, 0.37 and 1.56 Hz sinusoids. Peak velocity of the 3 lowest frequency components was fixed at 8 deg/second, while the velocity of the 1.56-Hz component was varied as a ratio (velocity ratio = VR) of the lower frequency velocity between 0 and 2. Note that each plot represents a separate stimulus. Mean + 1 standard error (SE) ($n = 5$).

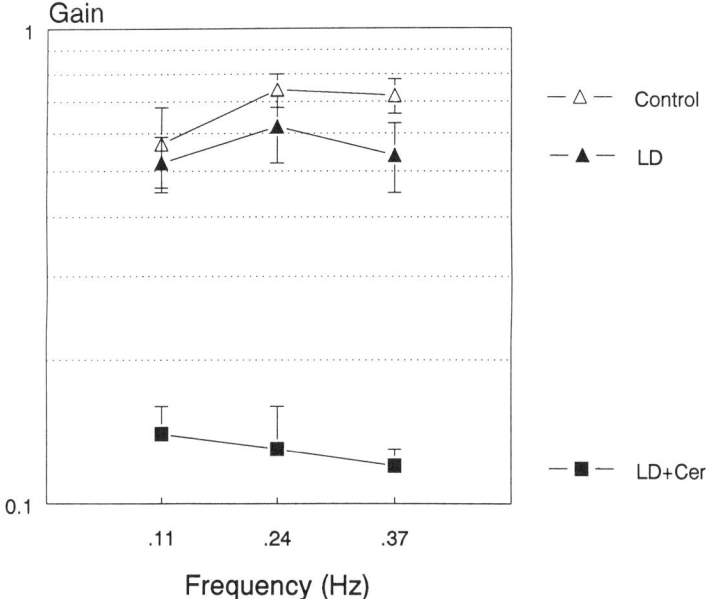

FIGURE 2. VOR gain during active head movements in the dark at each frequency, for control subjects, labyrinthine-deficient (LD) patients, and patients with combined labyrinthine deficiency and cerebellar degeneration (LD + Cer). Mean ± 1 SE ($n = 7$).

exceeded unity because of the presence of CEM, simulating the normal VOR. The results of our study, using pseudorandom target motion containing lower frequencies, showed no significant difference under head-free conditions, even at peak target velocities of up to 40 deg/second. In compensating for the loss of the VOR, LD patients apparently modify their gaze control in such a way as to reduce any difference in the performance characteristics under head-free and head-fixed conditions, a finding that has been noted previously in normal subjects.[1,2]

REFERENCES

1. COLLEWIJN, H., P. CONIJN & E. P. TAMMINGA. 1982. *In* Functional Basis of Ocular Motility Disorders. G. Lennerstrand, D. S. Zee & E. L. Keller, Eds.: 369–378. Pergamon Press. Oxford, England.
2. LEIGH, R. J., J. A. SHARPE, P. J. RANALLI, S. E. THURSTON & M. A. HAMID. 1987. Exp. Brain. Res. **66:** 458–464.
3. BARNES, G. R. & C. J. S. RUDDOCK. 1989. J. Physiol. London **408:** 137–165.
4. BRONSTEIN, A. M., S. MOSSMAN & L. M. LUXON. 1991. Brain **114:** 1–11.
5. KASAI, T. & D. S. ZEE. 1978. Brain. Res. **144:** 123–141.
6. TAKAHASHI, M., T. UEMURA & T. FUJISHIRO. 1981. Ann. Otol. Rhinol. Laryngol. **90:** 241–245.

Velocity Characteristics of Smooth Eye Movements in Patients with Cerebellar Ataxia

MARK J. MORROW[a] AND ROBERT W. BALOH

Department of Neurology
University of California, Los Angeles
Los Angeles, California 90024

INTRODUCTION

Patients with cerebellar damage have impairment of smooth pursuit, optokinetic nystagmus (OKN) slow phases, and fixation suppression of the vestibuloocular reflex (VOR).[1] Pursuit, optokinetic, and vestibular eye movements can be described as linear systems with gain elements relating eye velocity outputs to stimulus velocity inputs. At high stimulus velocities, however, smooth pursuit and OKN become nonlinear as velocity saturation occurs.[2] The neural substrates of gain and saturation properties of these smooth eye movements have not been defined.

METHODS

We used DC electrooculography to investigate horizontal eye movements in 19 patients with cerebellar degenerative diseases, aged 21–53 years (mean 38). Four normal subjects, aged 27–46 years (mean 34), served as controls. Smooth pursuit, OKN slow phases, and VOR cancellation were measured as previously described.[3] Subjects were tested with sinusoidal stimuli at 3 different frequencies and peak velocities (TABLE 1). A digital algorithm yielded plots of smooth eye velocity vs. stimulus velocity, from which we analyzed smooth eye velocity gain (slope) around the zero velocity crossing, and peak smooth eye velocity.

RESULTS

Normals had velocity gains near the ideal values of unity for pursuit and OKN and zero for VOR cancellation, but patients had reduced gain of smooth pursuit and OKN and increased gain of smooth eye velocity during VOR cancellation, at all stimulus frequencies tested (TABLE 1, FIGURE 1). In addition to abnormal gains, patients showed abnormal velocity saturation at the highest stimulus velocity, 60°/second (TABLE 1, FIGURE 1). There was no dissociation between abnormal smooth eye tracking and VOR cancellation in patients with normal VOR. OKN and smooth pursuit peak eye velocities and gains correlated inversely with the values for VOR cancellation.

[a] Address correspondence to the Department of Neurology, 2B-182, Olive View Medical Center, Sylmar, California 91342.

TABLE 1. Peak Eye Velocity and Eye Velocity/Stimulus Velocity Slope for Patients and Normals[a]

	0.2 Hz, 22.6°/Second Peak		0.4 Hz, 45.2°/Second Peak		0.05 Hz, 60°/Second Peak	
	Peak Eye Velocity (°/second)	Velocity Slope (gain)	Peak Eye Velocity (°/second)	Velocity Slope (gain)	Peak Eye Velocity (°/second)	Velocity Slope (gain)
VOR cancellation:						
Patient mean	4.1 ± 3.3[b]	0.15 ± 0.15[b]	26.8 ± 20.7[b]	0.56 ± 0.47[b]	18.8 ± 17.4[c]	0.19 ± 0.25[b]
Normal mean	0.3 ± 0.3	0.02 ± 0.02	3.1 ± 1.2	0.06 ± 0.03	1.5 ± 0.7	0.01 ± 0.01
OKN slow phases:						
Patient mean	17.3 ± 2.7[c]	0.79 ± 0.10[c]	21.3 ± 10.2[c]	0.50 ± 0.25[b]	23.4 ± 8.5[c]	0.63 ± 0.18[c]
Normal mean	23.2 ± 1.1	1.03 ± 0.05	43.4 ± 1.8	0.96 ± 0.04	54.6 ± 3.2	0.98 ± 0.08
Smooth pursuit:						
Patient mean	16.6 ± 2.9[c]	0.75 ± 0.13[c]	20.7 ± 7.8[c]	0.47 ± 0.20[c]		
Normal mean	22.6 ± 0.4	1.00 ± 0.02	40.9 ± 2.2	0.91 ± 0.05		

[a]Data are shown ±1 standard deviation. At target velocities up to 45°/second, eye velocity saturation was not consistently observed, and gain changes determined poor patient performance. At 60°/second stimulus velocities, however, observed peak eye velocity deviated from eye velocity as predicted by (velocity slope × target velocity). For VOR cancellation, observed peak velocity exceeded predicted velocity by 7.4°/second; observed OKN velocity fell 14.4°/second short of predicted. These data reflect velocity saturation.
[b]$p < 0.05$.
[c]$p < 0.001$.

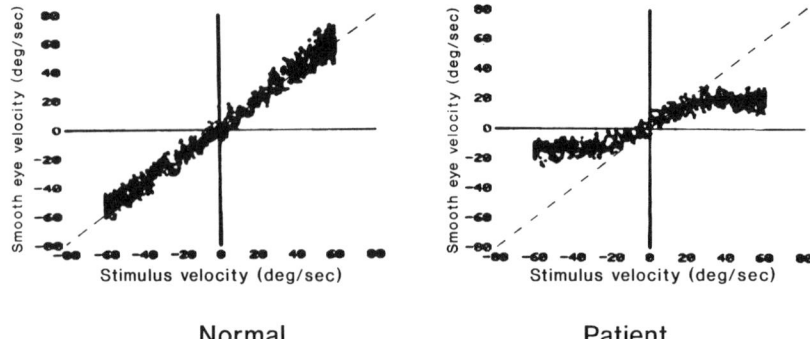

Normal   Patient

FIGURE 1. Velocity characteristics of OKN slow phases in a patient with cerebellar disease and a control, responding to 0.05 Hz, 60°/second peak velocity rotation of a full-field drum. Plots of smooth eye velocity vs. drum (stimulus) velocity; saccades and phase shifts between eye and target motion have been removed. Dashed line represents ideal slope of 1.0. Normal subject (left plot) shows mean gain of 0.95 and mean peak velocity of 57°/second, while patient shows abnormal gain of 0.76, calculated at zero velocity crossing, and abnormal velocity saturation, with peak velocity of 18°/second.

CONCLUSIONS

Abnormal velocity gain accounts for impairment of smooth pursuit, OKN slow phases, and VOR cancellation in patients with cerebellar damage. At higher stimulus velocities, abnormal velocity saturation also contributes to degraded smooth eye movements in these patients. This indicates that cerebellar circuits contribute both to gain and saturation limits. Parallel dysfunction of VOR cancellation and smooth tracking eye movements implies that common inputs from the cerebellum drive both systems.

REFERENCES

1. ZEE, D. S., R. D. YEE, D. G. COGAN, D. A. ROBINSON & W. K. ENGEL. 1976. Ocular motor abnormalities in hereditary cerebellar ataxia. Brain **99:** 207–234.
2. MEYER, C. H., A. G. LASKER & D. A. ROBINSON. 1985. The upper limit of human smooth pursuit velocity. Vision Res. **25:** 561–563.
3. BALOH, R. W., L. LANGHOFER, V. HONRUBIA & R. D. YEE. 1980. On-line analysis of eye movements using a digital computer. Aviat. Space Environ. Med. **51:** 563–567.

Implications of Vestibular Nucleus Neuron Rectification for Signal Processing in the Horizontal Vestibuloocular Reflex

THOMAS J. ANASTASIO

Department of Physiology and Biophysics
Beckman Institute
University of Illinois
405 North Mathews Avenue
Urbana, Illinois 61801

For the horizontal vestibuloocular reflex (VOR), velocity storage (VS) transforms the canal afferent time constant into the longer VOR time constant.[1] VS is nonlinear in that VOR time constant decreases as head rotation magnitude increases.[2] Also, VOR[2] and vestibular nucleus neurons (VNNs)[3] exhibit skew in response to sinusoidal stimuli. In response to step stimuli, VNNs have short rise but longer fall time constants.[4]

To study this nonlinear behavior, the VOR is modeled as an adaptive neural network composed of nonlinear units (FIGURE 1). It has forward as well as feedback connections. The network is trained to transform afferent into motoneuron impulse responses with equal and opposite amplitudes but longer time constants.[5] (Only the velocity component of motoneuron discharge[6] is modeled.) The network develops realistic connections and produces VS using inhibitory feedback connections that correspond to closed-loop vestibular commissures.[7] Model[5] and real[3] VNNs are similar in that they have lower spontaneous rates, higher gains, and longer time constants than the afferents. Their combined activity produces a seemingly linear transformation of the training inputs. However, network responses to untrained inputs can be nonlinear.

If the system were linear, the response of a motoneuron to sinusoidal input should have a phase lag commensurate with VS (FIGURE 2A, dotted). Instead, the actual response at higher magnitude (solid) alternates between the expected response and a response that has no phase lag relative to the afferents (dashed). This nonlinear behavior results from VNN rectification.

The responses of model VNNs rectify (FIGURE 2C). This breaks the closed commissural loops and disrupts VS, thereby reducing time constant and, consequently, phase lag. Skew results as VS is disrupted at peak and reactivated again midrange throughout each cycle of the response. Due to VNN rectification, VS never fully charges and at no point is phase fully lagged. VNN rectification increases as stimulus magnitude increases. This further decreases phase lag and increases skew.[5] Thus VS magnitude dependence and skew in the VOR[2] and VNNs[3] may both be due to VNN rectification.

The motoneuron step response (FIGURE 2B, solid) has a rise time constant near that of the afferents (dashed), but a fall time constant closer to that expected of the VOR (dotted). This again is due to VNN rectification (FIGURE 2D), which breaks the VS loop at step onset and reduces rise time constant. At step offset, VNNs and VS are reactivated. This prolongs the fall responses of VNNs and motoneurons.

907

Thus VNN rectification can also explain why VNN step responses have short rise but longer fall time constants.[4]

VNN rectification, which results from low VNN spontaneous rate and high gain, may produce functionally useful nonlinearities in the VOR. Most natural head movements probably consist of a short step acceleration followed by a period of constant velocity. VNN rectification at step onset would disrupt VS and allow the

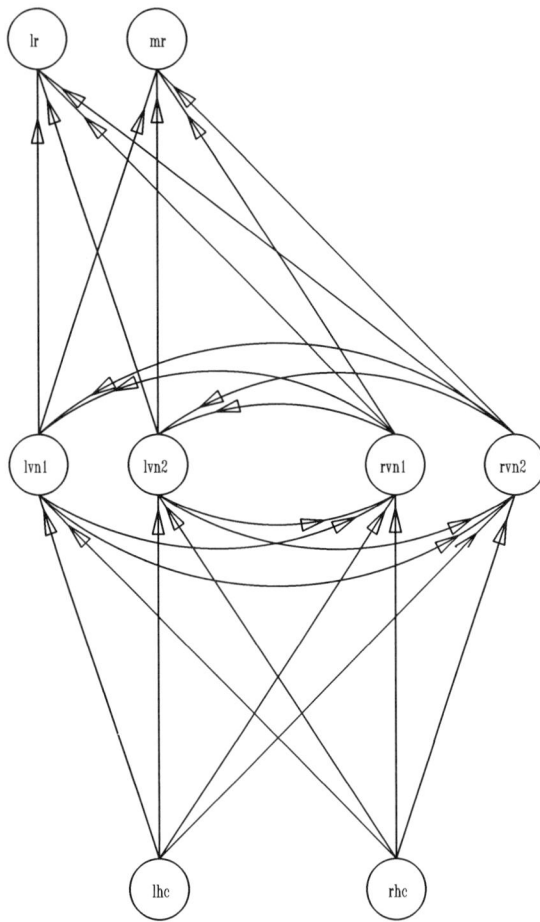

FIGURE 1. Recurrent neural network model of the horizontal vestibuloocular reflex (VOR) (redrawn from Reference 5). Input units represent afferents from the left and right horizontal semicircular canals (*lhc, rhc*). Output units correspond to motoneurons of the lateral and medial rectus muscles of the left eye (*lr, mr*). Middle layer units represent vestibular nucleus neurons (VNNs) on the left and right (*lvn1, lvn2, rvn1, rvn2*). All units are nonlinear, having firing rates that are sigmoidally bounded from zero to one. Forward connections reflect the primary three-neuron arc of the VOR. Recurrent connections are made between VNNs on opposite sides, and correspond to closed-loop commissures. The recurrent (feedback) connections allow the network to learn VOR dynamics.

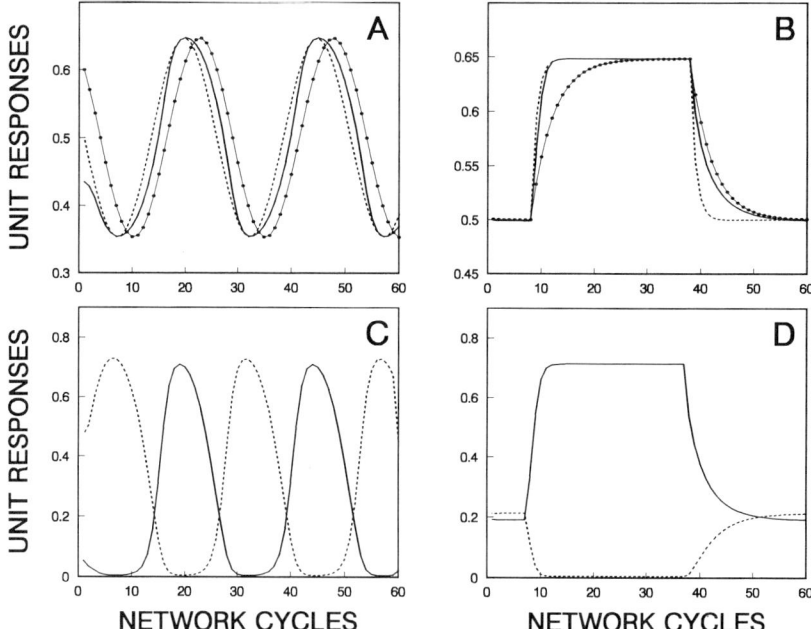

FIGURE 2. Nonlinear responses of model motoneurons and VNNs. Responses of *lr* are shown in A and B; those of *mr* are the same but inverted. Responses of *lvn1* (dashed) and *rvn1* (solid) are shown in C and D; those of *lvn2* and *rvn2* show less rectification. **A:** The actual sinusoidal response of *lr* (solid) alternates between a response at canal afferent phase (dashed) and another at the expected VOR phase (dotted). **C:** Alternation results because velocity storage (VS) is disrupted at the peaks due to VNN rectification, which breaks the closed commissural loops. **B:** The actual step response of *lr* (solid) rises near a response at the afferent time constant (dashed) and falls closer to a response at the expected VOR time constant (dotted). **D:** This is again due to VNN rectification, which inactivates VS at step onset, causing short rise time constants. At step offset, VNNs come out of rectification, reactivating VS. This prolongs fall time constants as the network relaxes from the push-pull initial condition in which it is left after the step.

VOR to respond to its canal input with minimum delay. At step offset, VNNs would come out of rectification and reactivate VS. This would then prolong the VOR response during the period of constant velocity, the period during which the canals become inactive.

REFERENCES

1. RAPHAN, TH., V. MATSUO & B. COHEN. 1979. Exp. Brain Res. **35:** 229–248.
2. PAIGE, G. C. 1983. J. Neurophysiol. **49:** 134–151.
3. FUCHS, A. F. & J. KIMM. 1975. J. Neurophysiol. **38:** 1140–1161.
4. WAESPE, W. & V. HENN. 1979. Exp. Brain Res. **37:** 337–347.
5. ANASTASIO, T. J. 1991. Biol. Cybern. **64:** 187–196.
6. SKAVENSKI, A. A. & D. A. ROBINSON. 1973. J. Neurophysiol. **36:** 724–738.
7. GALIANA, H. L. & J. S. OUTERBRIDGE. 1984. J. Neurophysiol. **51:** 210–241.

Vestibular Neurons Encoding Two-Dimensional Linear Acceleration Assist in the Estimation of Rotational Velocity during Off-Vertical Axis Rotation

DORA E. ANGELAKI[a]

Department of Physiology
University of Minnesota
Minneapolis, Minnesota 55455

It has been demonstrated that some otolith-sensitive vestibular nuclei neurons have spatial properties that can be described by *two* response vectors (S_1 and S_2) which are in temporal and spatial quadrature[1-4] [broadly tuned (BT) or two-dimensional neurons]. These neurons exhibit phase dependance on stimulus direction and gains that do not follow a simple cosine function of stimulus orientation during stimulation with pure linear acceleration. Depending on the relative orientation of the two vectors, BT neurons can be separated into counterclockwise (CCW) and clockwise (CW) units. CW BT neurons have vectors whose phase lead increases in the CW direction. The opposite is true for CCW BT neurons. The temporal quadrature between the two vectors in BT neurons and the observation that S_2 is proportional to the product of S_1 and the angular frequency of stimulation[3] ($S_2 = k\omega S_1$) suggests that S_2 is proportional to the derivative of S_1. Therefore, BT neurons encode both the linear acceleration vector and its time derivative (jerk vector).

The precise computation of the linear acceleration and jerk vectors by BT neurons could be accomplished by spatiotemporal convergence (STC)[5] of regular (tonic) and irregular (phasic-tonic) otolith afferents. Different temporal dynamics[6] accompanied by differences in the orientation of the polarization vectors of the converging afferents result in different dynamic characteristics for the two response vectors of BT target neurons. An example is shown in FIGURE 1: S_1 is almost flat and S_2 exhibits approximately a tenfold increase per decade of frequency. This gain behavior and the phase difference of 90° indicate a derivative relationship between S_1 and S_2. Similar transfer functions for S_1 and S_2 have been observed experimentally in both type I and type II convergent otolith–horizontal canal neurons.[3,4] Note that STC convergence of otolith afferents not only accomplishes this computation but also creates dynamic characteristics in central neurons which are very different from the afferent population (without the need for additional cascade or parallel filtering). Therefore, the diversity in the spatial and temporal properties of otolith afferents might provide significant computational power for efficient information processing in the central vestibular system.

Off-vertical axis rotation (OVAR) stimulates otolith neurons because it introduces a rotating gravity vector around the head. A narrowly tuned (NT) neuron (characterized by a single vector) exhibits a sinusoidal response during OVAR with equal magnitudes for both directions of rotation (FIGURE 2A, left). In contrast, as a

---

[a]Address for correspondence: ENT Research, Medical Research Bldg. Rm 2.104C, Route J63, University of Texas Medical Branch, Galveston, Texas 77555.

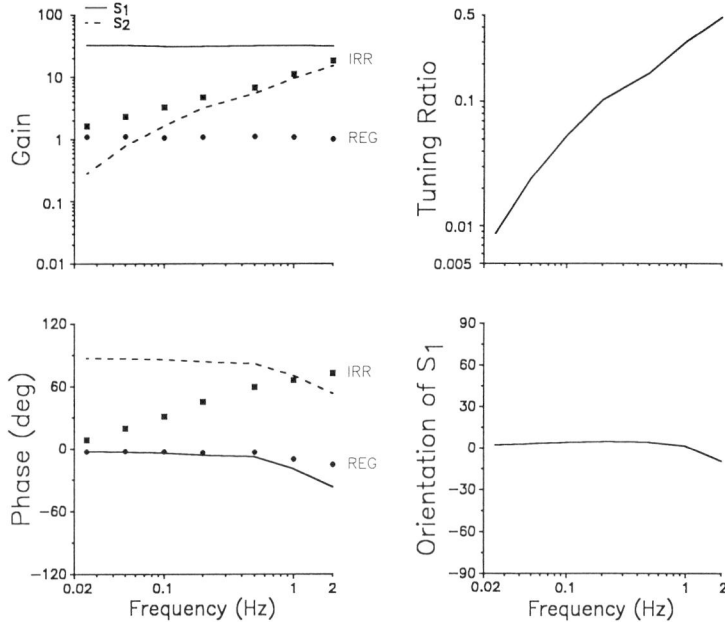

FIGURE 1. Simulations of STC from two otolith afferent inputs with an orientation difference of 120°. The gain and phase values of the irregular (squares) and regular (circles) otolith afferents[6] used in the simulations are included. The gain (top left traces) and phase (bottom left traces) of S_1 (solid lines) and S_2 (dashed lines), the tuning ratio (S_2/S_1; top right trace) and the orientation of S_1 (bottom right trace) of the target neuron are plotted as a function of frequency. Note that the orientation of S_1 remained relatively constant as a function of frequency. The predicted phase of S_1 exhibited moderate (as in the figure) or more pronounced phase lags.

result of the existence of two vectors in spatial and temporal quadrature, the response of a BT neuron during OVAR has unequal magnitudes to CW and CCW rotations. For example, the response of a CW BT neuron (having S_1 with orientation ωt_0 and phase θ; $A_{CCW} = \omega t_0 + \theta$ and $A_{CW} = \omega t_0 - \theta$) would be:

$$\text{CW OVAR: } S_1 \cos(\omega t + A_{CW}) + S_2 \sin(\omega t + A_{CW} + 90)$$
$$= (S_1 + S_2) \cos(\omega t + A_{CW})$$
$$= S_1(1 + k\omega) \cos(\omega t + A_{CW}),$$
$$\text{CCW OVAR: } S_1 \cos(\omega t - A_{CCW}) + S_2 \sin(\omega t - A_{CCW} - 90)$$
$$= (S_1 - S_2) \cos(\omega t - A_{CCW})$$
$$= S_1(1 - k\omega) \cos(\omega t - A_{CCW}). \tag{1}$$

Thus, CW BT neurons exhibit the largest response magnitude during CW rotation (FIGURE 2A, right), whereas CCW BT neurons exhibit the largest response magnitude during CCW rotation. Equations 1 further show that the response magnitudes

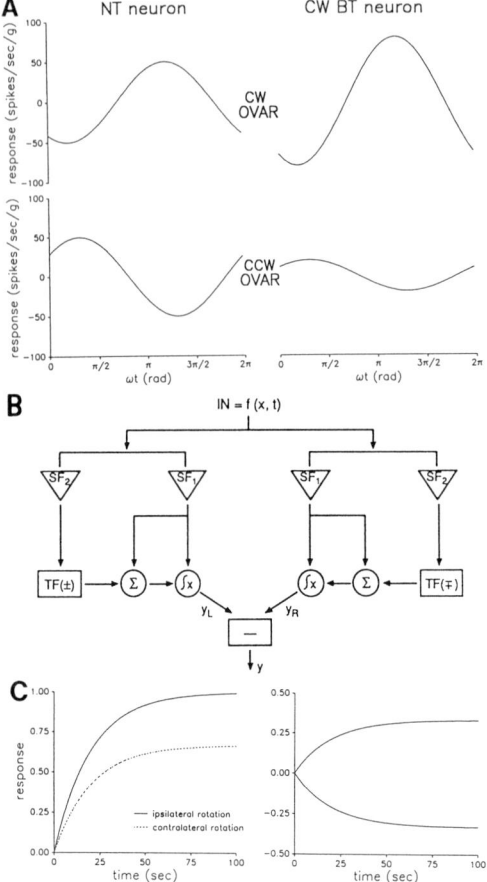

FIGURE 2. A: Simulated responses of vestibular otolith-sensitive neurons to CW and CCW OVAR. The orientation of S_1 is $\omega t_0 = 100°$ CCW to the direction of the gravity vector at the start of rotation and the neurons' dynamic characteristics along S_1 exhibit a phase lead of $\theta = 45°$. Only a single rotation cycle is plotted. *Left:* Predicted response of an NT neuron with $S_1 = 50$ spikes/second per *g*. *Right:* Predicted response of a CW BT neuron with $S_1 = 50$ spikes/second per *g* and $S_2 = 30$ spikes/second per *g*, i.e., a tuning ratio of $S_2/S_1 = 0.6$. **B:** An algorithmic model proposed to explain the motion detection of the gravity vector in the vestibular system. The output of one of the spatial filters (SF_1) corresponds to the first term of Equations 1, i.e., that due to the maximum sensitivity vector of the unit. The second filter, SF_2, introduces a spatial phase shift of 90° and gives the direction of minimum sensitivity. The output of SF_2 goes through a temporal filter $TF(\pm)$ which represents either a phase advance or lag of 90°, depending on the direction of rotation (note that the two subunits have opposite temporal filters). Summation of the outputs of SF_1 and the temporal filter produces signals that are equivalent to the responses of broadly tuned neurons to CW and CCW off-vertical axis rotation (Equations 1). The outputs of the NT and BT neurons are cross-correlated (symbol $\int X$; however, the integration is leaky) to produce the output y_L/y_R of each subunit. Algebraic subtraction of y_L and y_R gives an antisymmetric response that is proportional to the rotational velocity. **C:** Simulated responses (normalized relative to the maximum steady-state value) of y_L and y_R (left) and y (antisymmetric response; right). For these simulations, a sudden instantaneous tilt (from the earth vertical) of the rotational axis at time zero occurred while the constant-velocity rotation was maintained (the velocity storage time constant was 20 seconds). Solid lines are used for ipsilateral rotation (CCW rotation for type I HC cells in the left brain stem or CW rotation for type I HC cells in the right brain stem) and dashed lines are used for contralateral rotation. Zero response amplitude represents spontaneous firing rate (left) and zero eye velocity (right).

of BT neurons during OVAR are not only directionally selective, but also proportional to the rotational velocity ω.

It is proposed that BT neurons perform a first step in the computations necessary for the generation of steady-state eye velocity during constant-velocity OVAR. Similarly to models of motion detection in the visual system,[7] an additional step involving a nonlinear operation is necessary to produce sustained eye velocity during OVAR. It is suggested that the sinusoidal output of BT neurons during OVAR is transformed into the appropriate command signal for oculomotor neurons through operations that involve mathematical multiplication followed by leaky integration with the velocity storage mechanism. An algorithmic model incorporating responses from CW and CCW BT neurons and capable of producing the bias eye velocity during OVAR is shown in FIGURE 2B. Simulations of the subunit outputs (y_L and y_R) are shown in FIGURE 2C (left). Simulations of the detector output (y) are included in FIGURE 2C (right).

The three-dimensional angular velocity vector can be centrally reconstructed and the generated three-dimensional steady-state eye velocity is compensatory under the assumption that there is plane-specific convergence of canal and otolith signals in the vestibular nuclei: the plane defined by the S_1 and S_2 vectors is in close alignment to the respective canal plane. In other words, the gravity and jerk signals are encoded by the central vestibular system in canal coordinates. Thus, the computations for the estimation of the steady-state horizontal eye velocity during OVAR would take place within the horizontal canal system. Similarly, steady-state vertical and torsional eye velocity during OVAR would be computed within the vertical canal systems. The proposed model for the generation of the maintained eye velocity during OVAR is in agreement with the known anatomical and physiological properties of vestibular nuclei neurons and capable of predicting the experimental characteristics of steady-state eye velocity.

REFERENCES

1. ANGELAKI, D. E. 1991. IEEE Trans. Biomed Eng. **38**(11): 1053–1060.
2. ANGELAKI, D. E., G. A. BUSH & A. A. PERACHIO. 1992. Biol. Cybern. **66**: 231–240.
3. BUSH, G. A., A. A. PERACHIO & D. E. ANGELAKI. Ann. N.Y. Acad. Sci. (This volume.)
4. BUSH, G. A., A. A. PERACHIO & D. E. ANGELAKI. 1991. Soc. Neurosci. Abstr. **17**: 315.
5. ANGELAKI, D. E. Biol. Cybern. (In press.)
6. GOLDBERG, J. M., G. DESMADRYL, R. BAIRD & C. FERNANDEZ. 1990. J. Neurophysiol. **63**: 781–790.
7. VAN SANTEN, J. P. H. & G. SPERLING. 1985. J. Opt. Soc. Am. A **2**: 300–321.

Vestibular Afferents Innervating the Posterior Ampullae in a Turtle, *Pseudemys scripta*

A. M. BRICHTA AND E. H. PETERSON[a]

Neurobiology Program
Ohio University
Athens, Ohio 45701

Physiological studies in many vertebrates indicate that vestibular primary afferents are not a homogeneous population. Such data raise the question of what mechanisms underlie observed physiological differences in vestibular primaries and what functional role is played by afferents of each type. We have approached these questions by characterizing the functional architecture of vestibular primaries and their afferent hair cells, using the aquatic turtle *Pseudemys scripta* as an experimental model. In the present study we assessed the peripheral processes of 110 primaries innervating the posterior canal. These neurons were visualized by extracellular injections of horseradish peroxidase, their terminal arborizations fully reconstructed from serial sections, and their processes traced to somata in the vestibular ganglion. We characterized each primary afferent in terms of morphological parameters that are likely to have functional significance, with special emphasis on their spatial organization because physiological evidence indicates that the response properties of these neurons exhibit significant spatial heterogeneity.

The sensory surface of the vertical canals in *Pseudemys* comprises two triangular patches of neuroepithelium (*hemicristae*) with their apices opposed at the center of the canal (FIGURE 1). This surface bears type I and type II hair cells, which are presynaptic to vestibular primary afferents. Our data indicate that these vestibular primaries can be divided into two broad types based on their morphology and spatial organization. (**1**) *Bouton afferents* accounted for 66% of our sample. They are found over the entire surface of the hemicrista (FIGURE 1). All bear a terminal spray studded with varicosities (presumed synaptic boutons), but they differ significantly in the shape and size of their collecting areas, number and density of boutons per terminal, soma size, and axon diameter. This morphological variation is systematically related to the position on the afferent on the hemicrista: all measures increase significantly with distance from the walls to the center of the canal. Multivariate statistical analyses suggest that bouton afferents may comprise two types: α *afferents* (FIGURE 1A and B) are relatively delicate and are found throughout the hemicrista; β *afferents* (FIGURE 1B) are more robust and are located preferentially near the canal center. Whether these two types represent distinct subgroups or are the ends of a continuum remains to be clarified. (**2**) *Calyx-bearing afferents* comprise two morphological types: dimorphs (13% of our sample) bear both calyceal (cuplike) and bouton endings, and calyceal afferents (21%) bear calyceal endings only. Both types occur exclusively in an elliptical region near the center of the hemicrista (FIGURE 1). Within this region, both types vary structurally with radial distance from the center of

[a]Author to whom correspondence should be addressed, at Department of Biological Sciences, Irvine Hall, Ohio University, Athens, Ohio 45701.

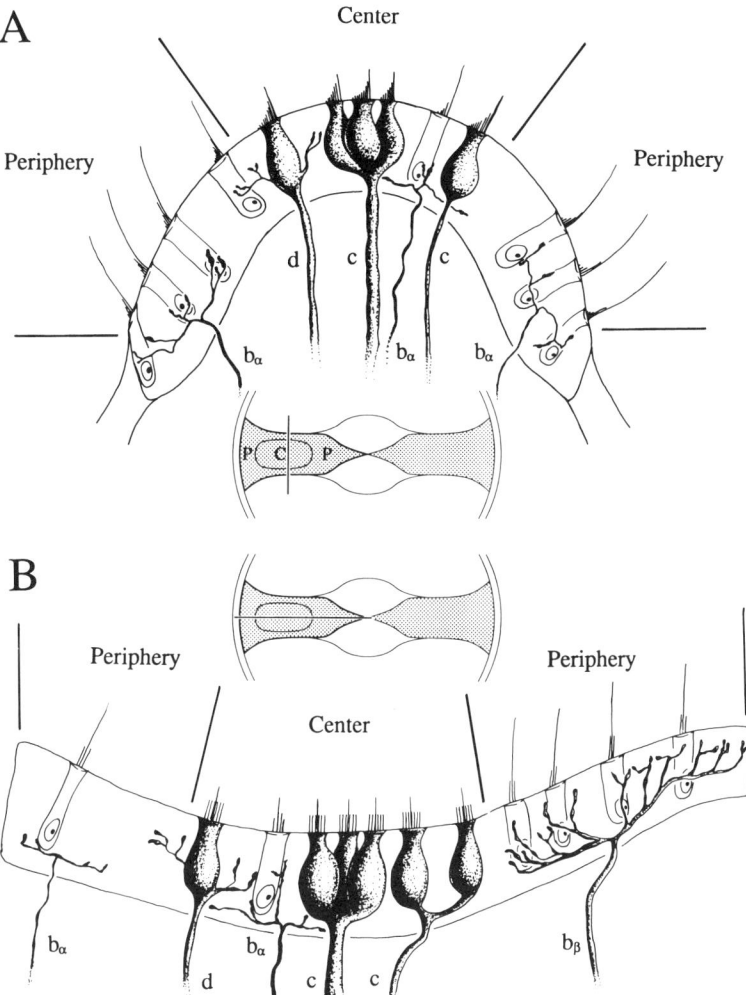

FIGURE 1. Sensory epithelium of the posterior canal in *Pseudemys scripta*. In each figure, a small dorsal view of the crista shows the shape of the neuroepithelium (shaded). The central region (C) is distinguished from the periphery (P) by the presence of type I hair cells and calyx-bearing primary afferents. A line drawn through each epithelium indicates the plane of the section shown at higher magnification. The diameter of the canal is approximately 1 mm. **A:** Transverse section through the central region of the hemicrista. Bouton afferents (b) contact type II hair cells in both peripheral and central regions. Only α-type bouton afferents are illustrated. In the central region, one dimorphic afferent (d) and two calyceal afferents (c) contact type I hair cells. Complex calyces are more numerous at the center of this region. **B:** Longitudinal section through the hemicrista. Two α afferents are illustrated, one near the wall of the canal and one within the central region. A β afferent (b_β) is shown near the center of the canal. Beta primaries have thick processes and large collecting areas.

the hemicrista. For example, the frequency of complex calyces decreases significantly toward the periphery of the central zone.

Our data provide evidence that, in *Pseudemys,* the spatial sampling characteristics of vestibular primaries are highly structured and are distinctive for each type. Bouton afferents sample the entire hemicrista, and their structure is organized around a canal-centered coordinate frame. In contrast, calyx-bearing afferents sample a restricted region of the neuroepithelium, and they are organized around a polar coordinate frame that is centered on the hemicrista. These spatial patterns may shed light on regional differences in physiological profiles of vestibular afferents, and they raise questions about the role of this spatial heterogeneity in signaling head movement.

Quantification of Different Classes of Canal-Related Vestibular Nuclei Neuron Responses to Linear Acceleration

GEOFFREY A. BUSH,[a] ADRIAN A. PERACHIO,[a,b,d]
AND DORA E. ANGELAKI[c]

[a]*Department of Otolaryngology*
[b]*Departments of Physiology & Biophysics
and Anatomy & Neurosciences
University of Texas Medical Branch
Galveston, Texas 77550*

[c]*Department of Physiology
University of Minnesota
Minneapolis, Minnesota 55455*

A change in otolith activity modifies the dynamic responses of both the horizontal[1-3] and the vertical[4] vestibuloocular reflexes (VOR). In response to rotations in vertical planes, dynamic otolith activity is necessary for compensatory eye movements in the rabbit[5] and the cat.[6] Therefore, significant convergence of otolith and canal information in the VOR pathway must occur.

The activity of single vestibular nuclei neurons in the decerebrate rat were recorded extracellularly during sinusoidal linear translation in the horizontal head plane. Details of the experimental procedure are presented elsewhere.[7] Neurons from the four groups—(1) type I and (2) type II horizontal canal related, (3) vertical canal related, and (4) purely otolith—were systematically tested for their responses to translation at various horizontal head orientations. These responses were then used to describe a *response ellipse*[8,10] in which the semimajor axis (S_1) defined the cell's direction of maximum sensitivity and its associated gain and phase and the semiminor axis (S_2) defined the minimum sensitivity of the cell in the horizontal head plane. When the magnitude of the S_2 vector was zero, the response was referred to as narrowly tuned and was characterized by gain values that were proportional to the cosine of the angle between S_1 and the stimulus direction and phase values that were constant with respect to stimulus direction. Whereas, a response with a nonzero magnitude of the S_2 vector was referred to as broadly tuned and was characterized by a response phase that varied as a function of stimulus angle. The accuracy with which the response ellipse quantitatively described the data was assessed by comparing the direction, gain, and phase values of the maximum response determined empirically with those calculated from the fitted curves (compare the data points with the curve in FIGURE 1). The calculated and experimentally measured responses had high linear regression coefficients ($r = 0.9301–0.9976$) and slopes close to unity.

Broadly tuned neurons were observed in each of the four groups of neurons studied. The ratio of S_2 and S_1 response magnitudes (tuning ratio) was calculated for all neurons. The distribution of ratios was similar for all neuron groups. In addition,

[d]Author to whom correspondence should be addressed at ENT Research, Medical Research Bldg Rm 2.104C, Route J63, University of Texas Medical Branch, Galveston, Texas 77555.

917

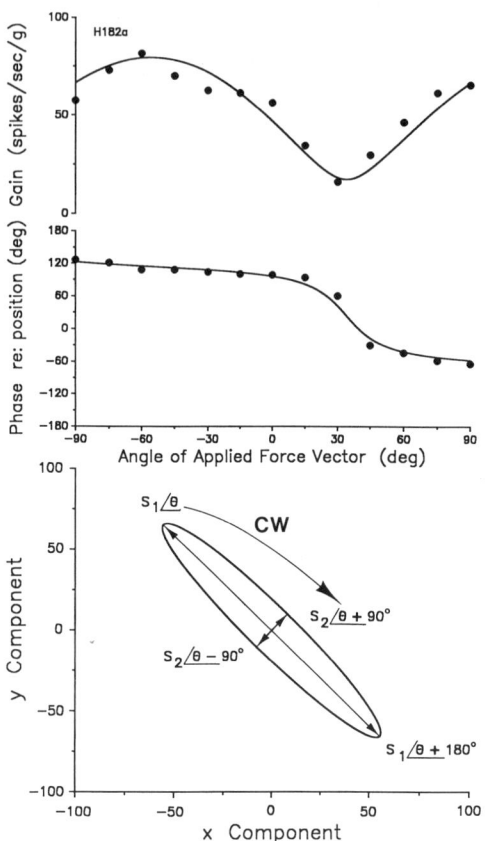

FIGURE 1. Response gain (top panel) and phase (middle panel) are shown as a function of the applied force vector angle for a neuron exhibiting a broadly tuned response. The gain and phase data points were measured experimentally and then simultaneously fit with continuous functions based on Angelaki's response ellipse formulation.[8,10] S_1 = 71.2 spikes/second per g with a response phase of $\theta = -62°$, while S_2 = 22.1 spikes/second per g. The angle of the maximum sensitivity vector calculated for this cell was 126°. The response ellipse for this cell is shown in the lower panel. The semimajor axis (S_1) of the ellipse defines the direction and magnitude of the maximum sensitivity vector. The semiminor axis (S_2) of the ellipse defines the vector of minimum sensitivity. Neurons exhibiting broadly tuned responses were classified according to the relative orientations of their S_1 and S_2 vectors in the horizontal head plane. At each stimulus frequency, the maximum sensitivity vector was defined to have a magnitude of S_1 and a phase of θ ($-90° < \theta < 90°$). Similarly, the minimum sensitivity vector had a magnitude of S_2, a phase of $\theta + 90°$ and was oriented perpendicular to the maximum sensitivity vector in the horizontal plane. Based on this definition, each broadly tuned neuron could have the S_2 vector located either clockwise (CW) or counterclockwise (CCW) in the horizontal head plane relative to S_1. The broadly tuned neuron illustrated here was classified as a CW neuron.

the tuning ratio was frequency dependent.[9] The type I and type II horizontal canal neurons were classified according to the relative orientation of their S_1 and S_2 vectors (FIGURE 1). The clockwise (CW) and counterclockwise (CCW) characterization was not frequency dependent and therefore provided a means of classifying the broadly tuned horizontal canal neurons. Further, there was a highly significant, positive linear correlation between S_1 and S_2 for the CW and CCW type I and CCW type II horizontal canal neurons. The linear regression lines of S_1 plotted versus S_2 also showed an inverse frequency dependence that was present for both the CW and CCW horizontal canal neurons. These results along with a phase difference of 90° between the two vectors suggest a derivative relationship between the S_1 and S_2 response vectors. This was further supported by the frequency response of both type I and type II horizontal canal neurons.[9] When Bode plots of the S_1 and S_2 responses were compared, for those neurons in which the S_1 response remained flat across frequencies, the S_2 response showed large gain enhancements.[9]

REFERENCES

1. MINOR, L. B. & J. M. GOLDBERG. 1990. Exp. Brain Res. **82:** 1–13.
2. ANGELAKI, D. E. & J. H. ANDERSON. 1991. Exp. Brain Res. **86:** 40–46.
3. HAIN, T. C. & U. W. BUETTNER. 1990. Exp. Brain Res. **82:** 463–471.
4. ANGELAKI, D. E., J. H. ANDERSON & B. W. BLAKLEY. 1991. Exp. Brain Res. **86:** 27–39.
5. BARMACK, N. H. & V. E. PETTOROSSI. 1988. J. Neurosci. **8:** 2827–2835.
6. TOMKO, D. L., C. WALL III, R. F. ROBINSON & J. P. STAAB. 1988. Exp. Brain Res. **69:** 307–314.
7. PERACHIO, A. A., G. A. BUSH & D. E. ANGELAKI. 1992. Ann. N.Y. Acad. Sci. (This volume.)
8. ANGELAKI, D. E. 1991. IEEE Trans. Biomed. Eng. **38**(11): 1053–1060.
9. BUSH, G. A., A. A. PERACHIO & D. E. ANGELAKI. 1991. Soc. Neurosci. Abstr. **17:** 315.
10. ANGELAKI D. E., G. A. BUSH & A. A. PERACHIO. 1992. Biol. Cybern. **66:** 231–240.

Participation of Secondary Vestibular Neurons in Nonvisual Mechanisms of Vestibuloocular Reflex Cancellation

K. E. CULLEN, C. CHEN-HUANG, AND R. A. McCREA

Committee on Neurobiology
University of Chicago
947 East 58th Street
Chicago, Illinois 60637

Our recent studies have suggested that squirrel monkeys can utilize a fast, nonvisual mechanism for canceling their vestibuloocular reflex (VOR) when they are fixating a visual target and their head is moving.[1] We have investigated the neural basis of this nonvisual suppression of the VOR by recording from neurons in the vestibular nuclei in alert, trained squirrel monkeys during smooth pursuit and cancellation of the VOR (VORC).

Secondary vestibular neurons were identified by their short latency (0.9–1.3 mseconds) response to electrical stimulation of the vestibular nerve. They were further classified on the basis of their spiking behavior during horizontal sinusoidal smooth pursuit eye movements and VOR suppression. Two main types of ipsilateral head movement sensitive secondary neurons were identified: (1) neurons sensitive only to horizontal head movements (*pure vestibular*) and (2) neurons whose firing increased during contralateral smooth pursuit and paused during saccades [position-vestibular-pause (*PVP*) neurons]. The head velocity sensitivity of the neurons was investigated during two types of vestibular stimulation, sinusoidal head rotations and unpredictable steps in head acceleration.

The head movement sensitivity of most *pure vestibular* neurons ($n = 20$) did not change when squirrel monkeys canceled their VOR during sinusoidal turntable rotation [mean response gain (0.8 sp/second)/(deg/second)]. The data from one of these units, averaged over 10 cycles of head rotation, is shown in FIGURE 1A. In contrast, the head movement sensitivity of most *PVP* neurons ($n = 19$) was significantly attenuated during cancellation of the VOR (mean attenuation in response gain 30%). The response of a typical *PVP* neuron during sinusoidal VORC and VOR (corrected for eye position and desaccaded) was averaged over 10 cycles of head rotation and is displayed in FIGURE 1B. The modulation of this neuron decreased by more than 50% during VORC. The amount of attenuation in the head velocity sensitivity of *PVP* neurons was poorly correlated with the eye velocity signal of the neuron during smooth pursuit, suggesting that the smooth pursuit eye velocity sensitivity of the neuron is not responsible for this attenuation head velocity sensitivity of the *PVP* neurons.

The amplitude of the response of most *PVP* neurons to unexpected passive perturbations of the head ($400°$/second^2 head acceleration steps) was attenuated in trials during which the monkeys were canceling their VOR, while the response of the *pure vestibular* neurons was the same regardless of the monkey's behavior. The short latency suppression of the VOR, which is generated by an unpredictable head acceleration step when the squirrel monkeys were initially canceling their VOR, is demonstrated in FIGURE 2A. The averaged population responses (corrected for each

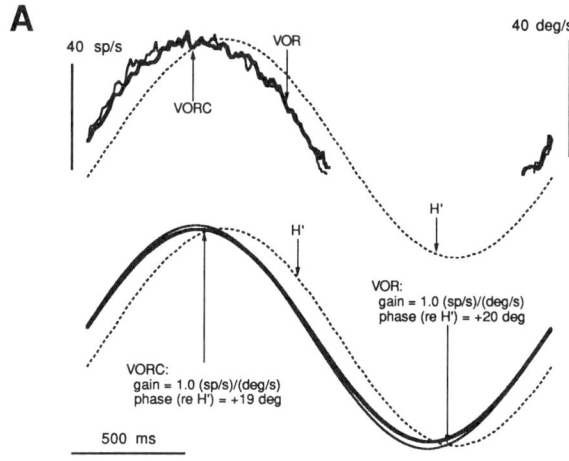

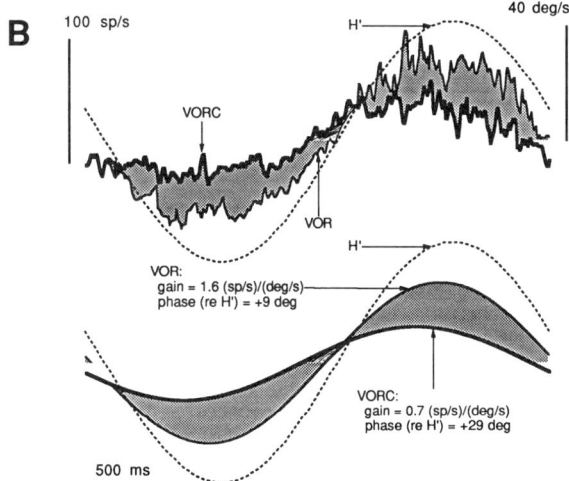

FIGURE 1. A: The response of a horizontal type I *pure vestibular* neuron during horizontal sinusoidal VOR in the light (light line) and when the monkey attempted to cancel the reflex by viewing a head-fixed target (heavy line). Each response was averaged over 10 cycles of table rotation. In the top trace, two averaged unit responses are superimposed on the vestibular stimulus (H', an upward deflection is ipsilateral). The sine wave fits for the actual averaged data are illustrated in the bottom trace. The gains and phases of the unit response were calculated with respect to head velocity. **B:** The averaged response of a *PVP* neuron during VOR and VOR cancellation. The averaged data and sine wave fits are displayed in the top and bottom traces respectively as in A.

neuron's eye position sensitivity) of all of the *pure vestibular* neurons and *PVP* neurons that were recorded in our experiments are shown in FIGURE 2B and C respectively. The attenuation in the *PVP* neurons' firing rates was apparent at short latencies after the perturbation (<30 mseconds), which suggests that the attenuation was mediated by a nonvisual input. The mean change in the firing rate of each of the neurons studied with this paradigm was calculated over the 30–80 mseconds that

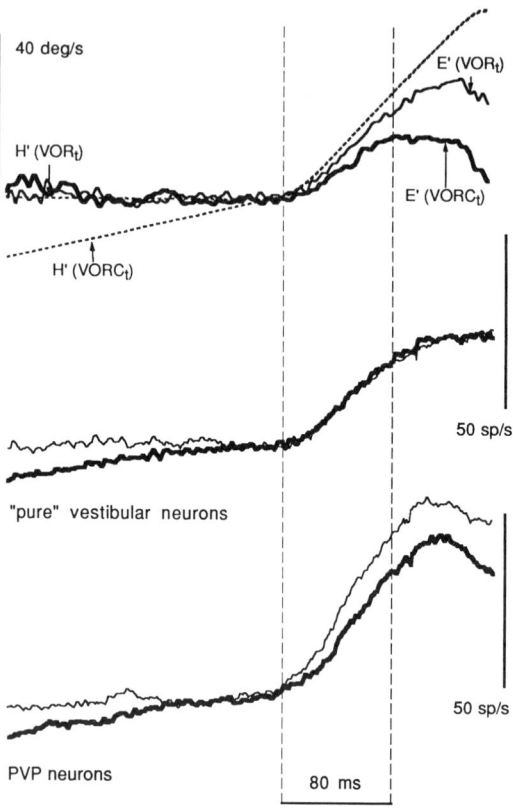

FIGURE 2. The summed responses of the *pure vestibular* and *PVP* neurons recorded during head acceleration steps. The top trace illustrates the eye velocity and corresponding head velocity generated while the monkey was fixating a target and was stationary (VOR$_t$), or was initially canceling its VOR (VORC$_t$). The bottom two traces represent the corresponding averaged response of the *pure vestibular* and *PVP* neurons respectively during VOR$_t$ (light line) and VORC$_t$ (heavy line).

followed the initiation of the head acceleration step. Although there was no significant change in the firing rate of the *pure vestibular* neurons by this measure, the firing rate of the *PVP* neurons was significantly attenuated by 30% during short latency suppression of the VOR.

The attenuation of the firing rates of the population of *PVP* neurons mirrored the decrease in the eye velocity generated by the head acceleration step during VORC.

Since several previous studies have shown that secondary *PVP* neurons project to the extraocular motor nuclei, it is likely that the nonvisual mechanism for VOR suppression is primarily mediated by modifying the head movement sensitivity of secondary vestibuloocular pathways.

REFERENCES

1. CULLEN, K. E., T. BELTON & R. A. MCCREA. 1991 A non-visual mechanism for voluntary cancellation of the vestibulo-ocular reflex. Exp. Brain Res. **83:** 237–252.

Antagonistic Otolith-Visual Units
in Cat Vestibular Nuclei

NANCY G. DAUNTON[a] AND CAROL A. CHRISTENSEN[b]

[a]Life Science Division
National Aeronautics and Space Administration
Mail Stop 261-3
Moffett Field, California 94035

[b]Department of Psychology
Vassar College
Poughkeepsie, New York

The vestibular nuclei serve as one center in which multisensory interactions necessary for the control of posture, locomotion, and gaze and for the perception of self-motion and orientation take place. We have studied various aspects of neural coding of visual and vestibular information about translational motion in the region of these nuclei, using extracellular single-unit recordings in alert adult cats, maintained under Flaxedil.[c] Responses were recorded and averaged over 60 cycles of stimulation (0.59 Hz, 0.15 G) in the vertical (up-down) and horizontal (fore-aft and side-to-side) planes. Conditions of stimulation included: vestibular (Vst—movement of the animal in the dark); visual (Vis—movement of the lighted visual surround); combined visual and vestibular (Vis + Vst—movement of the animal within the lighted stationary visual surround). Data reported are for responses to stimulation along the axis showing maximum sensitivity.

As we[1] and others[2] have reported previously, most linear-acceleration-sensitive units in and around the vestibular nuclei also respond to linear motion of the visual field. Most of these visually sensitive units respond to Vis and Vst stimulation in a synergistic fashion. That is, units with peak excitatory responses to Vst stimulation in one direction show maximum excitation to Vis stimulation in the opposite direction within a given axis. (This type of stimulation occurs when an individual moves through a stationary visual environment.) However, we have identified a small number of units that showed an antagonistic relationship between their Vis and Vst responses, since they were maximally excited by Vis and by Vst stimulation in the same direction. This antagonistic response was seen in units having a difference between the phase of peak excitation to Vis and to Vst stimulation of less than 36° [mean = 17°; standard error (SE) = 4.1]. In these units, Vis + Vst stimulation resulted in a decrease in gain, rather than an enhancement of gain, as found in synergistic units. Synergistic responses were seen in units with excitatory peaks to Vis and to Vst stimulation that differed by 136–178° (mean = 156°, SE = 3.2). Synergistic responses were also seen in most units in which the difference in peak excitation ranged from 50–134° (mean = 97°, SE = 5.2).

The antagonistic and synergistic units did not differ in terms of any of their basic characteristics (see TABLE 1). However, unlike synergistic responses, which were equally likely to be found in units maximally sensitive to stimulation in the vertical

---

[c]Procedures were in accordance with all guidelines for animal care and use in force at the time these data were collected.

TABLE 1. General Characteristics of Antagonistic and Synergistic Units

	Antagonistic		Synergistic
Visual units	9		44
($n = 53$)	17%		83%
Mean firing rate (AP/second)	33		36
(SE)	(6.3)		(3.5)
Mean coefficient of variation	0.98		1.07
(SE)	(.29)		(.26)
Response to: Vertical plane	0%	a	52%
Horizontal plane	100%		48%

[a]$p < 0.01$

and the horizontal plane, the preponderance of antagonistic units were responsive only to stimulation in the horizontal plane.

As reflected in FIGURE 1, the mean gain increase found in synergistic units was 57 action potentials (AP)/second per G (SE = 15.95), while the mean gain decrease in antagonistic units was 50 AP/second per G (SE = 21.37). Thus, the two types of units were affected equally, but oppositely by the combined Vis + Vst stimulation ($F = 8.44$, $df = 1$, 73, $p = 0.005$). The phase relationship of unit responses to maximum stimulus velocity also differed for the two types of units. With Vst stimulation more antagonistic than synergistic units lagged maximum velocity (78% vs. 34%, $p = 0.02$). With combined Vis + Vst stimulation a smaller proportion of antagonistic than synergistic units showed a shift in phase towards maximum stimulus velocity (33% vs. 86%, $p = 0.001$).

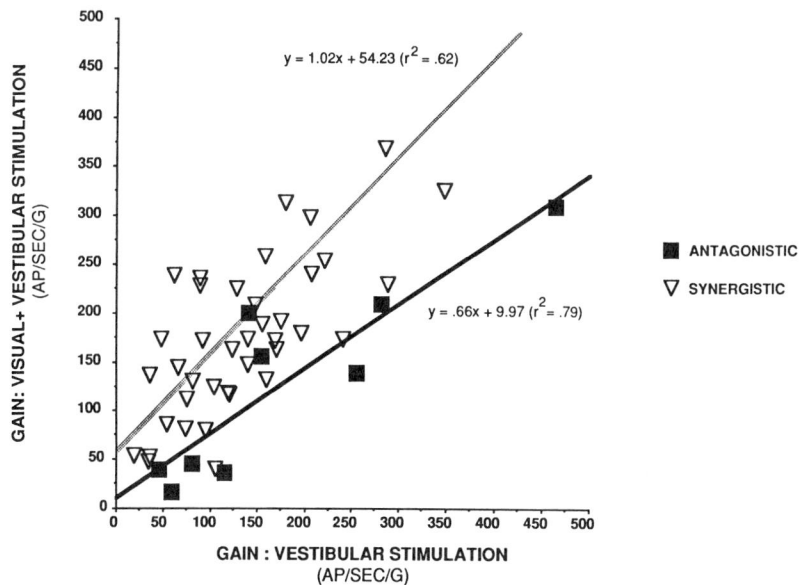

FIGURE 1. Vestibular gain vs. visual + vestibular gain.

These data suggest that antagonistic units may belong to an infrequently encountered, but functionally distinct, class of neurons. Their responses do not appear to be aberrant responses, found only in one or two abnormal subjects, since a few antagonistic units were found in most of the animals tested. In addition, the general characteristics of these units are not abnormal. Most importantly, the antagonistic units show predictable responses to Vis + Vst stimulation, differing from those of synergistic units only in the sign of the gain change. The fact that a small number of antagonistic-type units have been found by investigators studying visual-vestibular interactions during rotational motion[3] lends support to the possibility that such neurons have a distinct role to play.

Some clues to this functional role are provided by our results. First, the function supported by these units should be reliant for optimal performance on enhanced gain during self- and visual motion in the same direction (e.g., when an overtaking or passing visual stimulus is encountered). In addition, this function should be restricted primarily to the horizontal plane, and should not require a better response to stimulus velocity than that provided by vestibular stimulation alone, at least at the frequency of stimulation used in this experiment. Further experimentation is required, however, before the exact role of antagonistic units can be identified.

REFERENCES

1. DAUNTON, N. & D. THOMSEN. 1979. Visual modulation of otolith-dependent units in cat vestibular nuclei. Exp. Brain Res. **37:** 173–176.
2. BARTHELEMY, J., C. XERRI, L. BOREL & M. LACOUR. 1988. Neuronal coding of linear motion in the vestibular nuclei of the alert cat. II. Response characteristics to vertical optokinetic stimulation. Exp. Brain Res. **70:** 287–298.
3. WAESPE, W. & V. HENN. 1977. Neuronal activity in the vestibular nuclei of the alert monkey during vestibular and optokinetic stimulation. Exp. Brain Res. **27:** 523–538.

Vestibular Efferent System in Pigeons

Anatomical Organization and Effect upon Semicircular Canal Afferent Responsiveness

J. D. DICKMAN[a,c] AND M. J. CORREIA[b]

[a]Departments of Surgery (Otolaryngology)
and Anatomy
University of Mississippi Medical Center
Jackson, Mississippi

[b]Departments of Otolaryngology and
Physiology & Biophysics
University of Texas Medical Branch
Galveston, Texas

The existence of efferent neurons innervating the vestibular labyrinth in a number of species, including pigeons, has been widely reported. However, important questions remain regarding patterns of efferent innervation and the effect of efferent activation upon primary afferent responsiveness. In previous investigations, it was difficult to ascertain the specific effect of efferent activation upon ipsilateral semicircular canal afferent responsiveness produced by discrete contralateral semicircular canal stimulation. By utilizing the unique advantages offered by a newly developed mechanical stimulation method, the present study addressed this issue.

NEURAL TRACING STUDIES

Horseradish peroxidase crystals (HRP) were placed on the exposed neuroepithelium of the left horizontal semicircular canal ampulla for 2 hours, followed by a 48-hour survival in two pigeons. The animals were sacrificed, perfused with aldehyde fixatives, and serial coronal sections were processed for HRP reaction using the standard tetramethylbenzidine procedure. Retrograde transport of HRP by efferent fibers was obtained in both animals, with most of the efferent cell bodies being located bilaterally in the reticular pontine nucleus just ventral to the abducens nucleus. Approximately 50–60 efferent cells were observed in each animal.

ELECTROPHYSIOLOGICAL STUDIES

The effect of vestibular efferent stimulation upon semicircular canal afferent responsiveness was investigated by extracellular recordings of horizontal semicircular canal afferent (HCA) fiber responses in awake, decerebrate animals that were paralyzed (pancuronium bromide) and ventilated (250 ml/minute, O_2/CO_2). Step and sinusoidal (0.1 Hz) mechanical displacements[1] of the contralateral horizontal

[c]Address correspondence to Department of Surgery, Division of Otolaryngology, University of Mississippi Medical Center, 2500 North State Street, Jackson, Mississippi 39216.

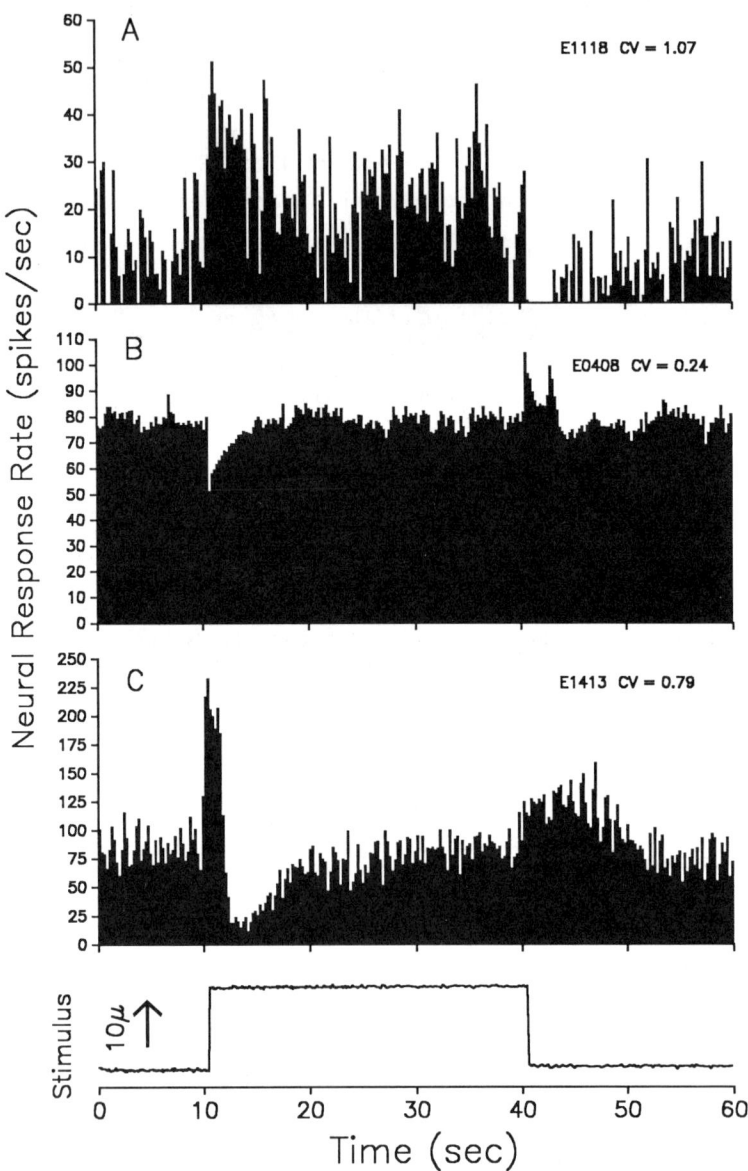

FIGURE 1. Neural response of horizontal semicircular canal afferent fibers to mechanical stimulation of the contralateral horizontal canal. **A–C:** Poststimulus time histograms represent excitatory, inhibitory, and mixed-burst type responses of three different left horizontal canal single afferent fibers, respectively. Bottom trace represents a 10-μm step mechanical displacement of the exposed right horizontal membranous duct.

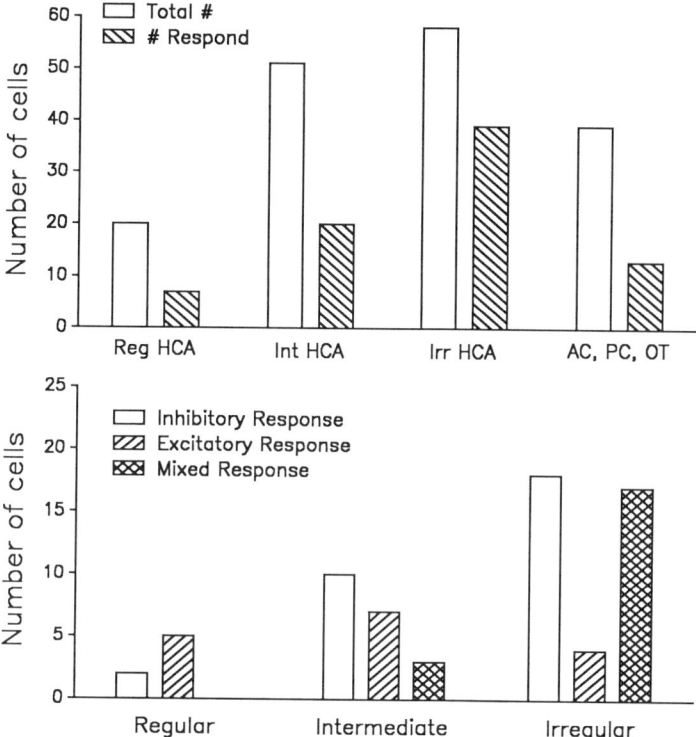

FIGURE 2. Number and response type of all HCA fibers recorded ($n = 129$). **Top:** Open bars indicate the total number of regular (Reg), intermediate (Int), and irregular (Irr) HCA fibers recorded as well as anterior canal (AC), posterior canal (PC), and otolith fibers (OT). The diagonal bars indicate the number of each type of fiber that responded to the 10-μm step mechanical displacement delivered to the contralateral horizontal membranous duct. **Bottom:** For the HCA fibers, the open, diagonal, and cross-hatched bars represent the number of neurons that responded with inhibitory, excitatory, and mixed types of responses (see FIGURE 1), respectively.

membranous duct (2.5 μm–30 μm) were delivered in order to determine if discrete vestibular inputs from the semicircular canals stimulate efferent neurons to produce modulation of HCA fibers innervating the *contralateral* labyrinth. Responses to step stimulations were obtained from 129 HCA fibers, with 51% (66/129) showing significant modulations in the firing rate above or below the resting discharge level. Both excitatory (16/66) and inhibitory (30/66) responses have been observed, but mostly only to large step stimulations (i.e., 10 μm–20 μm) as shown in FIGURE 1A and B. A separate type of response, termed a mixed-burst response, was also observed in many HCA fibers (20/66) that consisted of an initial rapid increase in firing rate followed by a longer lasting inhibitory response, as shown in FIGURE 1C. The response amplitudes to a 10-μm step displacement for these HCA fibers ranged between 4 and 130 spikes/second above or below the resting discharge level. Time constant values for the inhibitory and excitatory responses ranged between 0.04 and 32 seconds. In addition to the HCA fibers, 33% of the 39 anterior canal, posterior canal, and otolith afferent fibers that were recorded had significant responses to contralateral *horizontal* canal mechanical step stimulation. When HCA fibers were grouped according to their coefficient of variation (CV) values, it was observed that the units with lower firing rates and irregular discharge patterns were much more sensitive to the contralateral canal stimulation than were regular discharging cells, as shown in FIGURE 2. As also shown in FIGURE 2, the types of responses observed differed for the three CV classes of HCA fibers, with regular firing cells exhibiting excitatory responses and irregularly firing cells exhibiting inhibitory or mixed-burst types of responses. Sinusoidal mechanical stimulation elicited responses in the HCA units; however, most of the discharge patterns contained significant harmonic distortion (e.g., 10–50%). Stimulus-response intensity functions were generated by sinusoidal stimulation with various displacement magnitudes (± 2.5 μm–± 30 μm) for 13 HCA units. At lower stimulus amplitudes, many cells were nonresponsive, and at higher stimulus amplitudes, most fibers exhibited response saturation.

The present results are intriguing and do provide evidence that effective communication between complementary canal pairs via the vestibular efferent system exists. Still, the specific efferent mechanisms that produce the observed afferent responses and their functional role remain unclear.

REFERENCES

1. DICKMAN, J. D. & M. J. CORREIA. 1989. Responses of pigeon horizontal semicircular canal afferent fibers. I. Step, trapezoid, and low-frequency sinusoid mechanical and rotational stimulation. J. Neurophysiol. **62:** 1090–1101.

Does Counterrolling Violate Listing's Law?[a]

Th. HASLWANTER, D. STRAUMANN,
B. J. M. HESS, AND V. HENN

Neurology Department
University Hospital
CH-8091 Zürich, Switzerland

For the head erect and not moving, eye movements in man and monkey are implemented in such a way that only two of the three possible degrees of freedom are used. The eye torsion is specified by "Listing's law:" describing every eye position by a single rotation from a reference position to the current position, one obtains a set of rotation vectors that are approximately confined to a plane, called "Listing's plane."[1]

We have investigated the effects of different static roll and pitch positions on this plane in three alert, chronically prepared rhesus monkeys. Roll positions were reached by rotation about the nasooccipital axis, pitch positions by rotation about the interaural axis. Experimental details are described in Henn *et al.* (this volume).

For different roll positions, Listing's plane is shifted along the axis representing

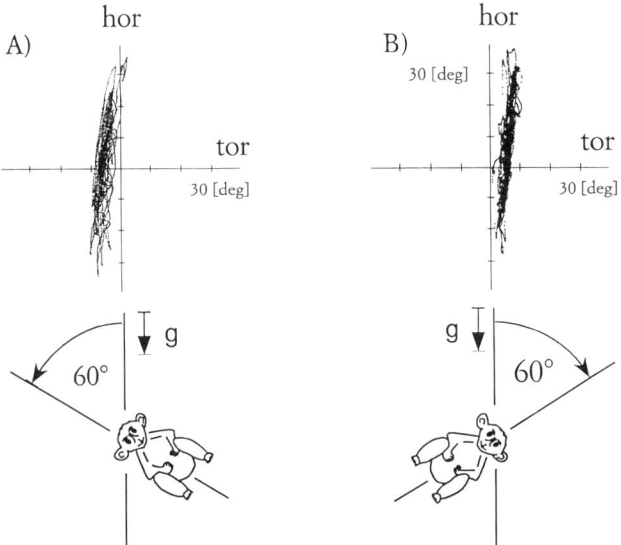

FIGURE 1. Horizontal vs. torsional eye positions, with the monkey (A) in a 60-deg right-ear-down roll position, and (B) in a 60-deg left-ear-down roll position. Eye positions were recorded for 90 seconds during spontaneous eye movements in the light.

[a] Supported by SNF 3199-25239 (ESPRIT, MUCOM 3149).

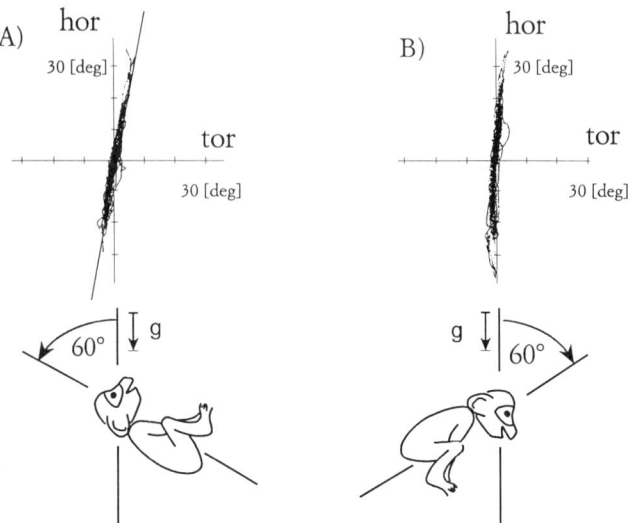

FIGURE 2. Horizontal vs. torsional eye positions, with the monkey (A) pitched 60 deg backwards, and (B) pitched 60 deg forwards. Eye positions were recorded for 90 seconds during spontaneous eye movements in the light.

eye torsion by an amount proportional to the sine of the roll angle. (See FIGURE 1.) In the light, the maximum eye torsion was 6.4 ± 1.2 deg (14 trials). This ocular torsion is the same for all eye positions, during saccades as well as during fixations. Our values are in close agreement with data reported for gaze straight ahead, i.e., counterrolling about the line of sight.[2]

For different pitch positions, the orientation of Listing's plane changes in a direction opposite to the rotation of the body: when the monkey is pitched backwards, the plane tilts forwards and vice versa. (See FIGURE 2.) During a typical pitch experiment in the light, i.e., stepwise forward or backward rotation through a total angle of 360 deg, the maximum change of the orientation of Listing's plane was 11 ± 2.5 deg (9 experiments).

The "thickness" of Listing's plane, i.e., the standard deviation of data points from a perfect plane, remained small (between 0.6 and 1.4 degrees).

The experiments have shown that Listing's law, as defined above, is valid for all static body positions. Thus, eye rotation vectors stay confined to a plane. Static otolith inputs shift or tilt this plane, but do not affect the general organizational principle of Listing's law.

REFERENCES

1. HAUSTEIN, W. 1989. Considerations on Listing's law and the primary position by means of a matrix description of eye position control. Biol. Cybern. **60:** 411–420.
2. DIAMOND, S. G. & C. H. MARKHAM. 1983. Ocular counterrolling as an indicator of vestibular otolith function. Neurology **33:** 1460–1469.

Immunocytochemical Visualization of L-Baclofen-Sensitive GABA$_B$ Binding Sites in the Medial Vestibular Nucleus[a]

GAY R. HOLSTEIN,[b,c,f] GIORGIO P. MARTINELLI,[d] AND
BERNARD COHEN[b,e]

*Departments of Neurology,[b] Cell Biology/Anatomy,[c]
Surgery,[d] and Physiology/Biophysics[e]
Mount Sinai School of Medicine
New York, New York 10029*

Velocity storage is responsible for producing slow changes in eye velocity that enhance the low-frequency characteristics of the vestibuloocular reflex (VOR).[1] Neural activity related to velocity storage is manifest in cells of the medial vestibular nucleus (MVN).[2] In addition, MVN neurons are sensitive to gamma-aminobutyric acid (GABA), and velocity storage is inhibited by the GABA$_B$ receptor agonist L-baclofen.[3] This suggests that inhibitory control of velocity storage is exerted by GABAergic, GABA$_B$ receptor-mediated synaptology. The objective of the present work was to visualize immunocytochemically the L-baclofen-sensitive GABA$_B$ binding sites in MVN, in an effort to characterize these synaptic interactions.

Mouse monoclonal antibodies were raised against the L-enantiomer of gamma-amino-beta-(p-chlorophenyl)butyric acid (L-baclofen) conjugated by glutaraldehyde to keyhole limpet hemocyanin. Preembedding peroxidase-antiperoxidase immunocytochemical studies of the ultrastructural localization of L-baclofen were conducted on tissue sections from baclofen-preinjected rats and monkeys (see Reference 4 for methodological details). Saline-injected control animals exposed to the primary antibody showed no immunoreactivity. In addition, no immunostaining was observed in sections from both saline-injected and L-baclofen-injected animals that were incubated in the absence of primary antiserum.

Immunoreactivity was present in discrete neuronal elements of MVN from the drug-injected animals. There were many immunostained dendrites present in the neuropil (FIGURE 1A), often participating in symmetric synapses with nonimmunoreactive axonal profiles (FIGURE 1B). The presynaptic elements in these synapses usually contained loosely packed spherical or ellipsoid vesicles. The presence of L-baclofen-sensitive GABA$_B$ binding sites on dendritic elements suggests that the presynaptic boutons are GABAergic. This synaptology represents a morphologic basis for postsynaptic inhibition involving GABA$_B$ receptors.

Axon terminals with small mitochondria and densely packed spherical or pleomorphic synaptic vesicles were occasionally immunostained (FIGURE 2A). These boutons formed synapses with unlabeled dendrites, which also received contacts from unlabeled profiles. In addition, synapses between two vesicle-containing profiles were also observed, in which only one of the two elements was immunostained (FIGURE 2B). This synaptology represents two possible morphologic substrates for presynap-

[a]Aided by National Institutes of Health Grants No. NS-24656 and NS-00294.

[f]Address correspondence to Box 1140, Mount Sinai School of Medicine, One Gustave Levy Place, New York, New York 10029.

933

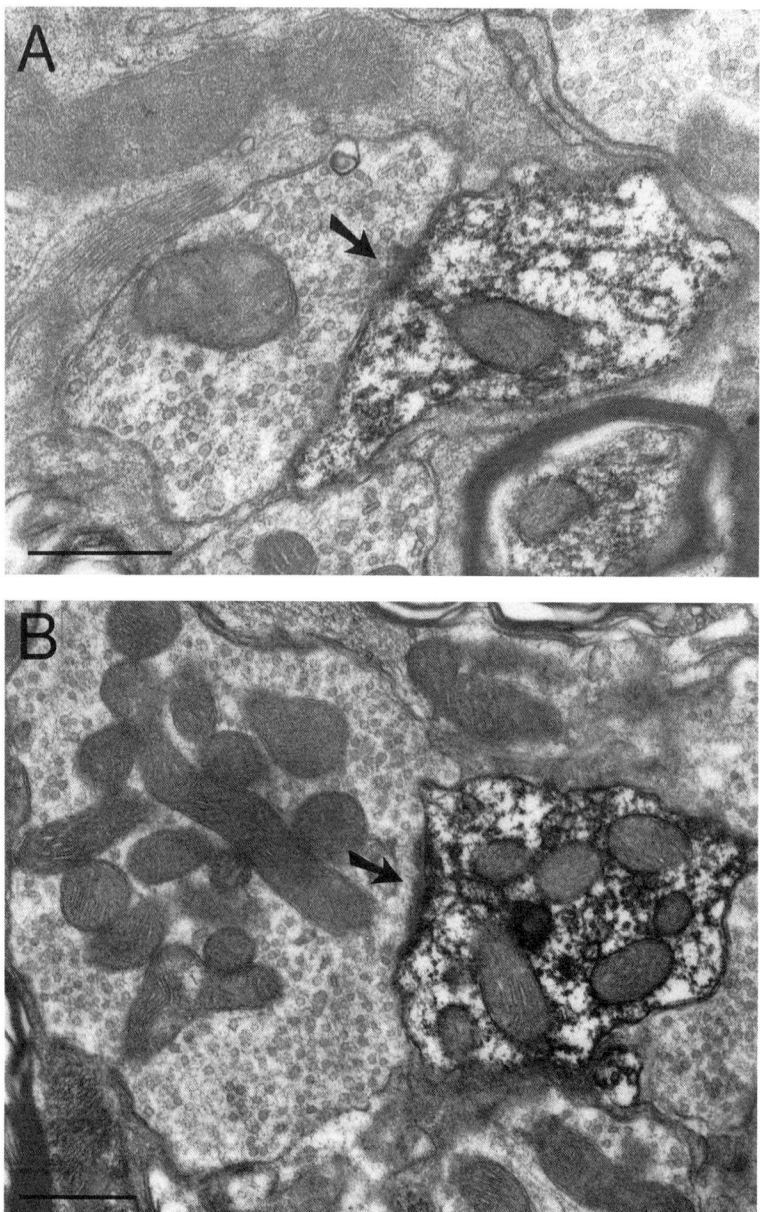

FIGURE 1. Two examples of immunoreactive dendrites in contact with unlabeled boutons. In **A,** the bouton contains loosely packed clear vesicles, which are mostly spherical in shape. A possible synaptic contact is apparent at arrow. In **B,** the bouton contains primarily pleomorphic vesicles, and forms a symmetric contact at arrow. Scale bars: 0.5 μm.

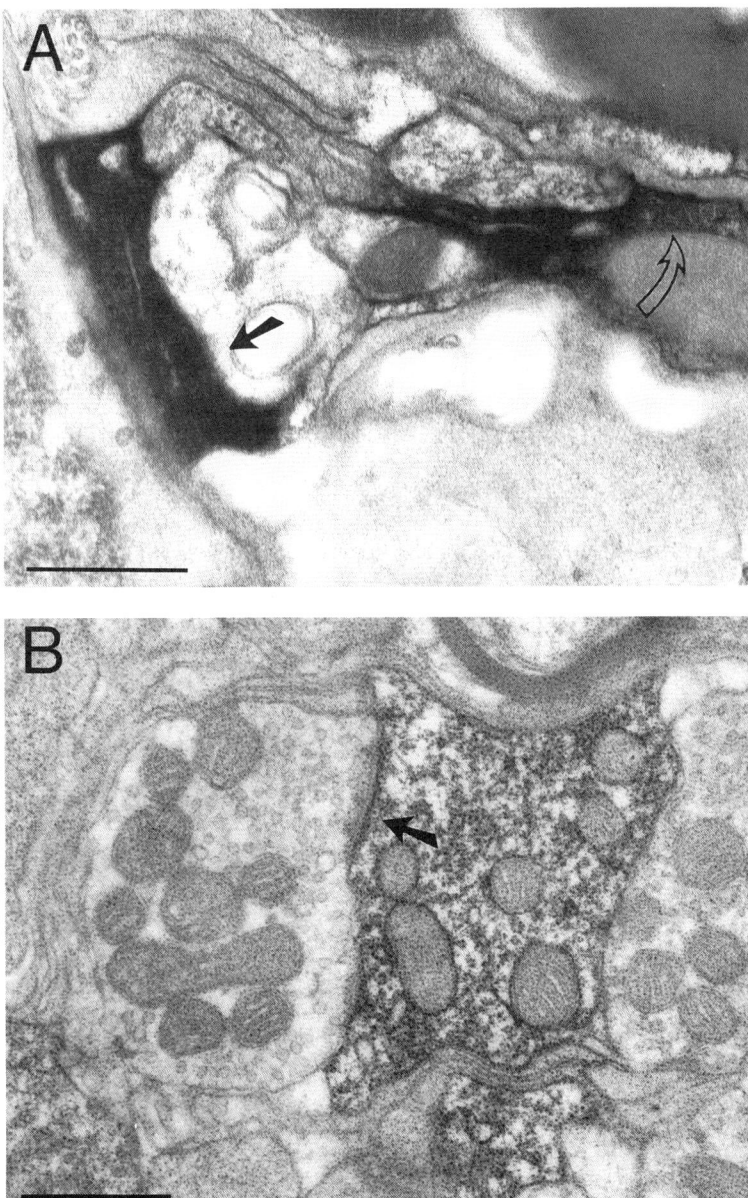

FIGURE 2. Two examples of immunoreactive vesicle-containing profiles. In **A,** the print was intentionally underexposed to aid visualization of the densely packed round synaptic vesicles. There is a suggestion of a synapse at arrow. The preterminal portion of the axon is apparent to the right of the figure (open arrow). In **B,** a vesicle-containing immunostained element is postsynaptic to a bouton containing clear round vesicles at arrow. Scale bars: 0.5 μm.

tic inhibition, one involving axoaxonic, and the other axodendritic connectivity. In both cases, the presence of $GABA_B$ binding sites in the axon terminal suggests GABAergic inhibition of those terminals. No synapses were observed between two immunostained elements.

The results present preliminary evidence for both pre- and postsynaptic GABAergic inhibition in MVN, mediated by $GABA_B$ receptors. Since L-baclofen administration results in a specific, dose-dependent inhibition of velocity storage in the VOR, some of the immunoreactive profiles may provide the morphologic basis for the mediation of this inhibition.

ACKNOWLEDGMENTS

L-Baclofen was provided by Ciba-Geigy, Ltd.

REFERENCES

1. RAPHAN, T., V. MATSUO & B. COHEN. 1979. Velocity storage in the vestibulo-ocular reflex (VOR). Exp. Brain Res. **35:** 229–248.
2. KATZ, E., J. M. B. V. DE JONG, J. BUETTNER-ENNEVER & B. COHEN. 1991. Effects of midline medullary lesions on velocity storage and the vestibulo-ocular reflex. Exp. Brain Res. **87:** 505–520.
3. COHEN, B., D. HELWIG & T. RAPHAN. 1987. Baclofen and velocity storage: a model of the effects of the drug on the vestibulo-ocular reflex. J. Physiol. London **393:** 703–725.
4. MARTINELLI, G. P., G. R. HOLSTEIN, P. PASIK & B. COHEN. 1992. Monoclonal antibodies for ultrastructural visualization of L-baclofen-sensitive $GABA_B$ receptor sites. Neuroscience **46:** 23–33.

Mechanisms of Head Stabilization during Random Rotations in the Pitch Plane[a]

E. A. KESHNER,[b] R. L. CROMWELL,[c] G. ROVAI,[c]
AND B. W. PETERSON[c]

[b]Department of Physical Therapy (M/C 898)
University of Illinois at Chicago
1919 West Taylor Street
Chicago, Illinois 60612

[c]Department of Physiology
Northwestern University Medical School
Chicago, Illinois 60611

The hypothesis that mechanisms controlling posture are frequency dependent has emerged from studies of dynamic head stability. Measures of head and trunk angular velocity during sagittal plane postural reactions have revealed a normal 2–3 Hz peak in the power spectrum that is absent in patients with bilateral labyrinthine deficit.[1] Normal human subjects rotated about the vertical axis of motion with a random white noise[2] (0–1 Hz) or sum-of-sines[3] (SSN) stimulus (0.185 to 4.115 Hz) exhibited excellent voluntary head stabilization in space at frequencies below 1 Hz, reaching a resonant peak at about 3 Hz. When subjects performed a mental arithmetic task so that attention was diverted from the task of stabilizing the head, stability was poor at frequencies below 1 Hz, but approached voluntary stabilization conditions above 1 Hz. Coincident increases in neck muscle electromyographic (EMG) activity suggested a contribution of vestibulocollic (VCR) and cervicocollic (CCR) reflexes at frequencies between 1–2 Hz. Thus, voluntary, reflex, and mechanical mechanisms contributed to final head position, each dominant in a different frequency range.

Pitching motions of the head need to be examined for they are both more biomechanically demanding and more crucial for maintaining posture and gaze than are movements in the horizontal plane. In this study, seated subjects underwent pitch rotations in the dark about a horizontal, interaural axis. A triaxial angular rate sensor and a laser pointer, affixed to a helmet worn by the subjects, were positioned at the interaural axis of the head. A random SSN stimulus (0.35 to 3.05 Hz), high enough to elicit VCR and CCR responses, was presented. Bilateral surface EMGs from the semispinalis (SEMI) and sternocleidomastoid (SCM) muscles, and chair and head velocities were recorded. Gain and phase of head velocity and EMG responses were calculated using a best-fit sinusoid and analyzed with respect to chair (i.e., trunk) velocity to derive changes at the neck. Four conditions were presented: (1) VS = active head stabilization on a stationary spot of light with visual feedback as the body was rotated; (2) NV = active head stabilization during body rotations without visual feedback; (3) MA = performing a mental arithmetic task during body rotations in the dark; (4) VT = body stationary, head tracks a projected light.

One subject's responses of the neck with respect to the trunk can be seen in FIGURE 1. At *0.35–1 Hz* the head was well stabilized in VS and NV with gains close to one, and a compensatory phase of 180°. Gains in MA were reduced by half, but with

[a]Supported by grants NS22490 and DC01125.

937

phases still compensatory to movement of the trunk. SEMI responded in phase with trunk position, and SCM with trunk velocity (see FIGURE 2). Tracking in VT occurred predictively (180° phase) with gains close to one. At *1–2 Hz,* good head stability was seen in both VS and NV. In MA, gains ascending towards one and phase holding around 180° imply improved head stability. Descending gains and phases in VT suggest a poorer match between the head and the target light. Muscle gains began a gradual ascension in this frequency range. Unlike voluntary muscle responses which plateau, phase of the reflex EMG (in MA) advances with increasing frequency. Above *2 Hz,* descending gains and 90° phase lags characterize head motion in VS, NV, and MA. In VT, response gains diminished, and responses occurred 180° out of phase with the target light. Larger EMG gains and advancing phases suggest continued reflex and voluntary attempts to compensate for trunk and head movements.

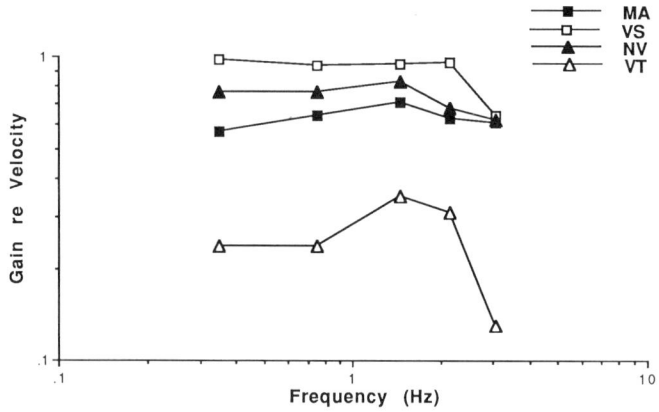

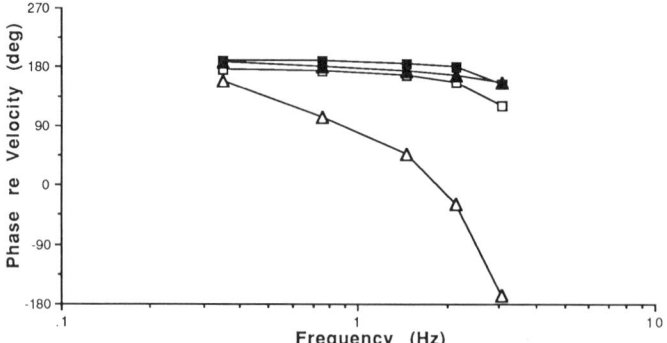

FIGURE 1. Bode plots of gain and phase of neck velocity with respect to trunk velocity or head velocity with respect to the target light velocity (in VT) during a random SSN (0.35, 0.75, 1.45, 2.15, 3.05 Hz) in the four conditions of rotation (see text for abbreviations). Perfect neck stability has a gain of 1 and a phase of 180°.

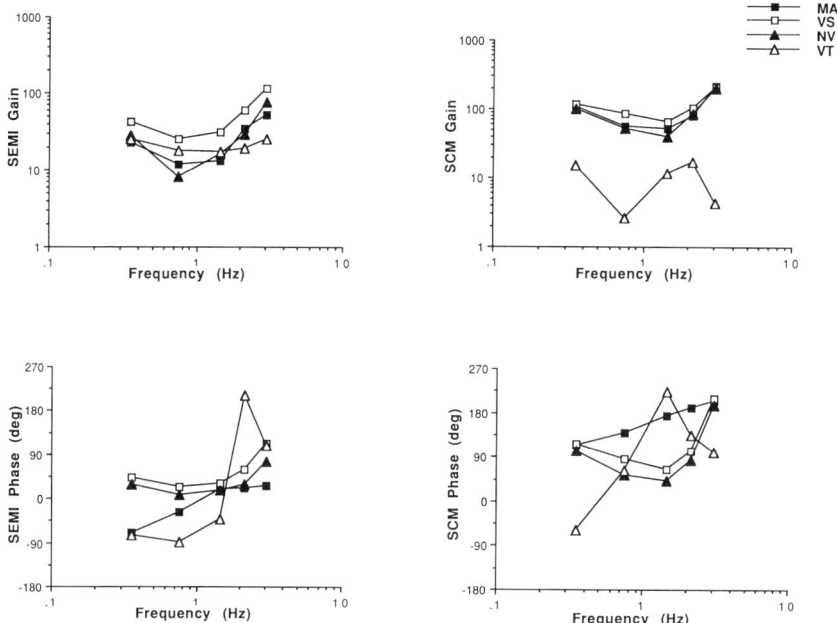

FIGURE 2. Bode plots of gain and phase of right semispinalis (extensor) and right sternoclei-domastoid (flexor) muscle EMG responses with respect to trunk and target light velocity.

In the vertical plane, both reflex and voluntary attempts at stabilizing the head extended over a wider frequency range than in the horizontal plane of motion. There was no evidence of the increased gains with descending phases that characterized mechanical resonant activity of the head with horizontal rotations.[3] Voluntary head movement (VT) and reflex head stabilization (MA) demonstrated reversed dynamics: head movement was performed more accurately at low frequencies, and reflex stabilization improved as frequency increased. Muscle frequency characteristics during head stabilization were similar to those seen during neck reflex activity in alert cats.[4] Thus it appears that reflexes are stronger in the vertical plane, and in order to test the mechanics of the system we must record higher frequencies of rotation than in the horizontal plane.

REFERENCES

1. KESHNER, E. A., J. H. J. ALLUM & C. R. PFALTZ. 1987. Exp. Brain Res. **69:** 66–72.
2. GUITTON, D., R. E. KEARNEY, N. WERELEY & B. W. PETERSON. 1986. Exp. Brain Res. **64:** 59–69.
3. KESHNER, E. A. & B. W. PETERSON. 1988. Soc. Neurosci. Abstr. **14:** 1235.
4. BILOTTO, G., J. GOLDBERG, B. W. PETERSON & V. J. WILSON. 1982. Exp. Brain Res. **47:** 343–352.

Some Excitatory Transmitters in the Central Vestibular Pathways in the Gerbil

GOLDA ANNE KEVETTER

Departments of Otolaryngology and
Anatomy and Neurosciences
University of Texas Medical Branch at Galveston
Galveston, Texas 77550

The vestibular system relies on its main ascending and descending pathways to the oculomotor nuclei and the spinal cord in order to facilitate the reflexlike responses to changes in head accelerations. Although several anatomical characteristics of these pathways have been investigated, the neurotransmitters used in these systems have not been well explored. Correlation of the chemical anatomy of pathways and their morphological organization should lead to a greater appreciation of functional and behavioral aspects of the vestibular system.

The goal of the present research was to determine which excitatory neurotransmitter is utilized by vestibulospinal (VST) and vestibuloocular (VOR) neurons. Excitatory amino acids, considered to be the major class of excitatory transmitters used in the nervous system, were studied. The possibility of acetylcholine being used was also investigated.

VST neurons were identified with a retrograde tracer injected into the spinal cord of anesthetized gerbils. VOR neurons transported tracers that were injected into the oculomotor complex. After identification of the retrograde marker, sections were incubated in polyclonal antisera to either aspartate or glutamate, followed by an avidin-biotin procedure to visualize the immunocytochemical reaction. Sections from some gerbils were also reacted with antisera to γ-aminobutyric acid (GABA), substance P, choline acetyltransferase (ChAT), serotonin, or vasoactive intestinal peptide (VIP).

Many cells that were retrogradely labeled from the oculomotor nuclei also stained with either glutamate-like immunoreactivity (GLU-lir) or aspartatelike reactivity (ASP-lir). Cells retrogradely labeled from the cervical spinal cord (VST neurons) also stained with ASP-lir and GLU-lir.

Some retrogradely identified VOR neurons also stained with ChAT-like immunoreactivity (ChAT-lir). Fewer VOR neurons were labeled with both the retrograde marker and ChAT than with ASP-lir or GLU-lir. This was not surprising because ChAT cells were restricted to only a portion of the medial and dorsal descending vestibular nuclei.

When an antibody to GABA was used, many retrogradely labeled VST neurons were surrounded by boutonlike profiles that were stained with immunocytochemistry. Although many substance P fibers and terminals were identified in the vestibular nuclei, especially in the medial vestibular nucleus and dorsal descending vestibular nucleus; few, if any substance P positive cell bodies were identified. The relationship of substance P fibers with projection neurons is being determined. Neither neurons

TABLE 1. Percent of Labeled Neurons that Also Contain ASP-lir and GLU-lir Reactivity

V. Medial	Medial	Lateral	Descending	Superior	y
		Vestibulospinal Neurons ASP-lir			
64.3% (70)	39.6% (48)	59.7% (196)	27.8% (54)		
		Vestibuloocular Neurons ASP-lir			
57.1% (77)	54.3% (394)	42.3% (26)	54.5% (11)	40.0% (270)	51.9% (79)
		GLU-lir			
25.0% (8)	56.6% (76)	—	54.2% (24)	33.7% (199)	66.7% (15)

nor fibers and terminals stained when antibodies to VIP, serotonin, or m-enkephalin were used.

CONCLUSIONS

1. Many VOR neurons contain GLU-lir and ASP-lir. The percentage of VOR neurons that stain GLU-lir and ASP-lir is about the same.

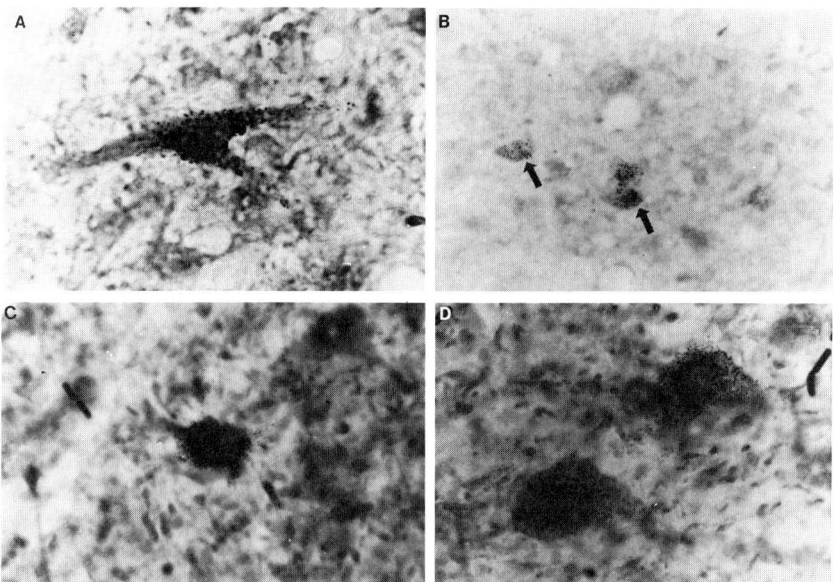

FIGURE 1. **A:** VOR neuron retrogradely labeled with horseradish peroxidase (HRP, granular reaction product) which also labels with GLU-lir (uniform reaction product). **B:** Four retrogradely labeled VOR neurons. The two cells pointed out with arrows also contain ASP-lir (uniform reaction product). **C:** VST neuron retrogradely labeled with HRP that also contains GLU-lir reaction product. **D:** Two VST neurons containing ASP-lir.

2. Of VOR neurons in the medial vestibular nucleus, more than 50% also stain with immunocytochemistry for excitatory amino acids.

3. Of VOR neurons in the superior vestibular nucleus, less than 50% also stain with immunocytochemistry for excitatory amino acids.

4. Some VOR neurons also stain with ChAT-lir, but fewer than the number that stain with antibodies for excitatory amino acids.

5. Many VST neurons contain GLU-lir and ASP-lir.

6. Of VST neurons in the lateral vestibular nucleus, more than 50% also stain with immunocytochemistry for ASP-lir.

7. Some VST neurons are surrounded by boutons that stain with GABA-lir.

GABA Is an Afferent Vestibular Neurotransmitter in the Guinea Pig[a]

Immunocytochemical Evidence in the Utricular Maculae

G. MEZA, J-Y. WU,[b] AND I. LÓPEZ

Department of Neuroscience
IFIC, UNAM
Apartado Postal 70-600
04510 México, D.F. México

[b]*Department of Physiology and Cell Biology*
University of Kansas
Lawrence, Kansas 66045-2106

INTRODUCTION

The control of posture and balance in man is multimodal and depends on the optimum and concerted functioning of receptors responding to a variety of stimuli: retinal photoreceptors, "self" or propiosensors in muscle, skin, and joints, and gravity detectors of these latter and of otolithic organs (utricle and saccule) of the inner ear. Changes in linear accelerations, mainly those caused by gravity, are "sensed" by hair cells located in utricular and saccular specialized epithelial areas (maculae). This information is there processed and translated to a chemical language, then sent to upper centers via neurotransmitters acting upon the synaptic contact of the afferent fiber originated in the Scarpa ganglion. In the absence of gravity, these sensors lack its messages whereas the rest remain functioning; thus conflicting information reaches superior centers, causing unwanted symptoms.

As the crucial step in all these processes is the conveying of the correct information to the central nervous system (CNS) via neurotransmitters released by the vestibular-otolithic organ's hair cell, the identification of these neurotransmitters is of vital importance. In mammals, very few studies have tried to shed light on this matter and they have led to contradictory results: some postulate γ-aminobutyric acid (GABA)[1] whereas others propose glutamate as neurotransmitters.[2,3]

With the purpose of adding more direct evidence in favor of the GABA postulation, immunocytochemical techniques were used to localize simultaneously, within their respective cell compartment, some GABA-system-related parameters, i.e., GABA itself and its synthesizing and catabolizing enzymes: glutamate descarboxylase, GAD, and GABA transaminase, GABA-T, respectively.

The methodology followed and the results obtained are hereby described.

---

[a]This project was financed in part by Grants D113-903570 (CONACyT) and IN201791 (DGAPA, UNAM), Mexico to G.M.

943

METHODOLOGY

Six pigmented guinea pigs (*Cavia cobaya*) (both sexes) (250–300 g weight) were used in this study. They were deeply anesthetized with chloral hydrate (1.5 g/kg weight) and perfused transcardially with isotonic saline solution, followed by 4% paraformaldehyde–0.1% glutaraldehyde in 0.1 M sodium cacodylate buffer (SCB) pH 7.3. The auditory bullae were opened, and utriculae (both sides) were isolated and otoliths removed. Organs were postfixed in the same fixatives for 90 minutes, after which they were dehydrated to 100% ethanol and embedded in paraffin or epoxy resin. Organs were cut in 1–3 μm sections and mounted on glass slides and let stand at 37°C overnight. Some sections were deparaffinized to water, then briefly immersed in 1 mM β-mercaptoethanol in 0.1 M phosphate buffer (PBS), pH 7.3. Others were subjected to a previously reported etching process to remove resin.[4] In both cases, after rinsing, sections were sequentially incubated in (1) 10% normal goat serum in PBS–Triton X-100 0.01% for 30 minutes, (2) GAD, GABA-T, or GABA antisera diluted in PBS 1:500, 1:1500 or 1:2000 respectively, for 48 hours at 4°C, and developed by the peroxidase technique.[4]

RESULTS AND DISCUSSION

Immunocytochemically visualized GAD (IV-GAD) was exclusively observed in hair cell cytoplasm throughout the utricular macula sensory epithelium. Both type I and II hair cells were intensively stained and in good contrast to unreactive supporting cells or nerve fibers including chaliceal afferent terminals (FIGURE 1A). In contrast, GABA-like immunoreactivity (GABA-LIR) was found in cytoplasm of both type I and II hair cells as well as in some nerve fibers running through the adjacent stroma; supporting cells were definitely unstained (FIGURE 1B).

GABA-T-like immunoreactivity (GABA-T-LIR) distribution closely resembled GABA-LIR's for it was found within the afferent chalices surrounding type I hair cells and in some nerve fibers running through the stroma and it was occasionally observed in some hair cell cytoplasm. Supporting cells were, again, unreactive (FIGURE 1C). Our findings, the first report in the literature using immunocytochemistry to simultaneously visualize three parameters of the GABA system in a mammalian utricular macula, are in agreement with indirect evidence in higher vertebrates suggesting the localization of GAD in the hair cells of vestibular organs[5,6,7] and conform with the more direct evidence of IV-GAD and GABA-LIR hair-cell localization and GABA-LIR and GABA-T presence both in chaliceal endings and afferent fibers,[4,8] although controverted with GABA-LIR efferent bouton localization.[9] Thus, according to these distributions, our results mean that GABA can be synthesized in the hair cell cytoplasm and, after its release and action upon its receptor, it can be transported to the postsynaptic element in the afferent fiber where it is degraded. This strongly supports its afferent neurotransmitter role in vestibular organs of higher vertebrates. Clinical implications of these finding include the selective use of GABA-mimetic drugs to substitute for the unreleased natural neurotransmitter in the absence of gravity stimuli in manned spaceflights.

ACKNOWLEDGMENTS

Thanks are due to Mrs. Teresa Cortés and Mr. Rodolfo Paredes for valuable technical help and Miss Lorena López-Griego for secretarial assistance.

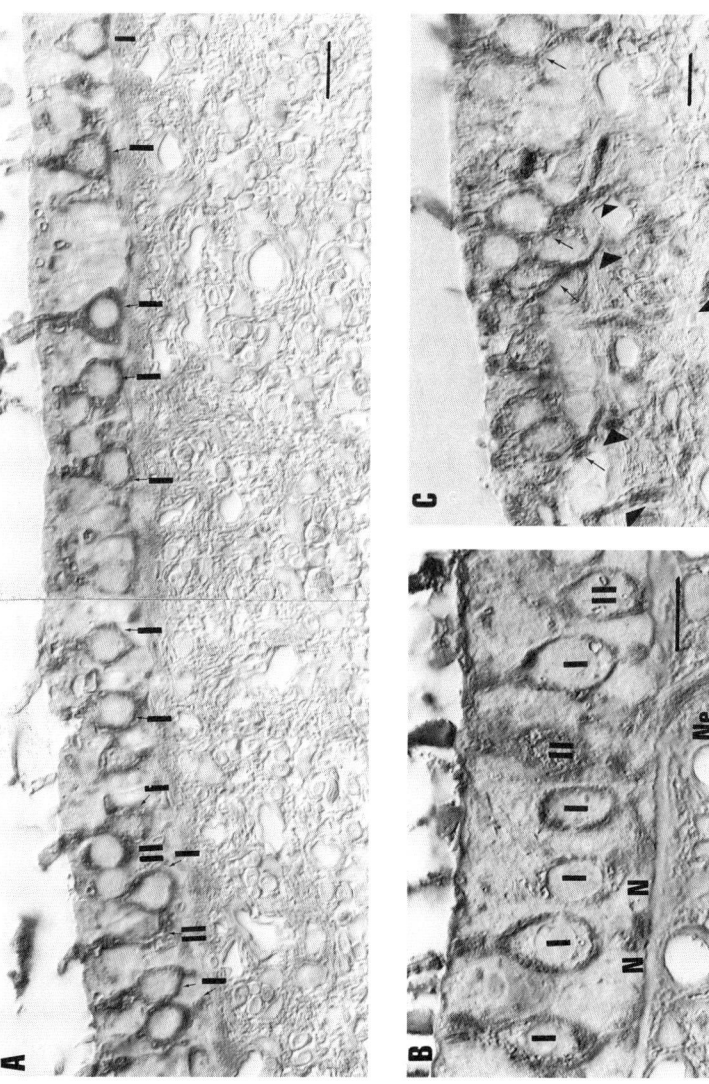

FIGURE 1. Immunocytochemically visualized GAD (IV-GAD) and GABA- and GABA-T-like immunoreactivity (GABA-LIR and GABA-T-LIR) in the guinea pig utricular macula. **A:** IV-GAD is exclusively present in cytoplasm (arrows) of both type I (I) and type II (II) hair cells. Bar = 10 μm. **B:** GABA-LIR is observed in the cytoplasm of hair cell types I (I) and II (II), in nerve fibers of the sensory epithelium (N), and of the adjacent stroma (Ne). Bar = 10 μm. **C:** GABA-T-LIR is mainly found around type I hair cells (arrows) and in the afferent chaliceal terminal and fiber (arrowheads). Bar = 10 μm. Note that in all cases, immunoreactivity is absent from supporting cells.

REFERENCES

1. FELIX, D. & K. ENHRENBERGER. 1982. Acta Otolaryngol. **93:** 101–105.
2. DECHESNE, C., J. RAYMOND & A. SANS. 1984. Ann. Otol. Rhinol. Laryngol. **93:** 163–173.
3. DEMEMES, D., R. J. WENTHOLD, B. MONIOT & A. SANS. 1990. Hear. Res. **46:** 261–270.
4. LÓPEZ, I., J. M. JUIZ, R. A. ALTSCHULER & G. MEZA. 1990. Brain Res. **530:** 170–175.
5. LÓPEZ, I. & G. MEZA. 1990. Comp. Biochem. Physiol. **95B**(2): 375–379.
6. MEZA, G. & R. HINOJOSA. 1987. Hear. Res. **28:** 73–85.
7. MEZA, G., I. LÓPEZ, M. A. PAREDES, Y. PENALOZA & A. POBLANO. 1989. Acta Otolaryngol. Stockh. **107:** 406–411.
8. USAMI, S., J. HOZAWA, M. TAZAWA, M. IGARASHI, G. C. THOMPSON, J.-Y. WU & R. J. WENTHOLD. 1989. Brain Res. **503:** 214–218.
9. USAMI, S., M. IGARASHI & G. C. THOMPSON. 1987. Brain Res. **417:** 367–370.

Acceleration Detection by Vestibular Hair Cells

Hair Bundles as Spatially Distributed Phased-Array Antennas

DALE H. MUGLER

Department of Mathematical Sciences
The University of Akron
Akron, Ohio 44325-4002

The stereociliary bundle at the apex of a vestibular hair cell is a mechanoelectrical transducer of the signal of acceleratory motion. That signal propagates to the hair bundle when motion of the otoconia is communicated through the endolymph to the array of stereocilia. Array antennas are the preferred kind of receiver for many types of signals, transducing sound waves for ultrasound or sonar applications or receiving radar signals in modern radar equipment. Array antennas are more efficient when sensors are hexagonally distributed, as are stereociliary arrays, and they exhibit directional sensitivity in that they respond only to signals from certain directions, as do stereociliary arrays, making a natural analogy in both form and function to the stereociliary hair bundle.

In previous work, the author and M. D. Ross have investigated the directional sensitivity of stereociliary bundles using a simplified *planar* phased-array model.[1-3] A goal to extend that analysis to a three-dimensional positioning of the sensors is initialized in this note. Such a model follows evidence, e.g., Hudspeth,[4] that points to the initial transduction site as the tip of each stereocilium, although other evidence, Ohmori,[5] indicates that the base is also important in gating.

The phased-array model is based on the assumption that the different lengths of stereocilia in the array cause proportional phase shifts in the response of each stereocilium. Modulation of the ionic flow through the stereocilium is assumed to be accomplished by stereociliary movement, so that even responses to static displacements are included. The lengths of the stereocilia are assumed to increase linearly towards the kinocilium, and be constant across each rank. This is based on data such as in TABLE 1, which lists the results of measuring lengths of stereocilia from electron micrographs of a representative hair bundle from rat utricle, made available to the author by M. D. Ross from her laboratory at NASA-Ames. The general trend of these lengths is linear, although the measurements were not precise and may contain a large amount of error.

TABLE 1. Measured Lengths of Stereociliary Heights from a Representative Hair Bundle

	Row of Hair Bundle							
	1	2	3	4	5	6	7	8
Height (in μm)	4.8	5.9	7.1	8.0	10.0	12.6	13.7	15.1

For an analysis of the array directional sensitivity, the standard engineering tool is the array pattern function. This describes the relative field strength of the array in terms of both azimuth (ϕ) and elevation (θ) angles. The direction from which the signal is coming is identified by these angles, measured relative to the base of the array. This model uses the position of the nth sensor at $\mathbf{r}_n = (x_n, y_n, z_n)$, where (x_n, y_n) describe the hexagonal distribution of the base of the hair bundle, and z_n the height of the corresponding stereocilium. A uniform excitation of each element is also assumed. If $\mathbf{u} = (\sin(\theta)\cos(\phi),\sin(\theta)\sin(\phi),\cos(\theta))$ specifies the direction of the incoming signal, then the normalized array pattern function is $F(\theta, \phi) = 1/N \, \Sigma_n \exp[-i \, (\mathbf{r}_n \cdot \mathbf{u})]$. Its magnitude, $|F(\theta,\phi)|$, describes the relative field strength of the array to a signal from that direction. Since magnitude, and not sign, is important here, this model cannot explain the hyperpolarizing or depolarizing effect, but only the magnitude of that response.

An example for a hair bundle array containing forty-three stereocilia is given in FIGURE 1. A micrograph of a cross-section through the bundle was presented earlier[3]

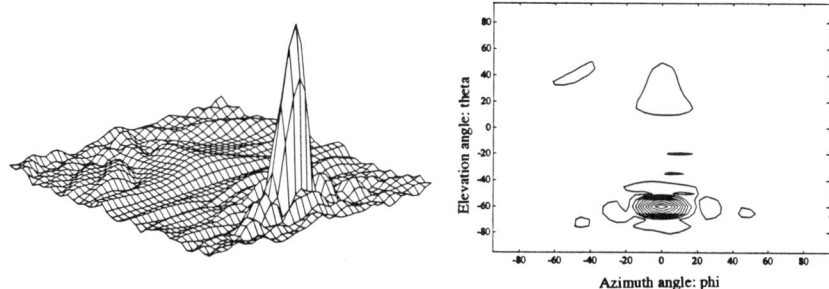

FIGURE 1. Array pattern surface sketch and contour plot. Array with 43 stereocilia.

along with an array pattern based on planar sensors. The graphs in FIGURE 1 are based on the analysis presented above, and indicate that the array would be sensitive to waves pushing the array towards the kinocilium (see the large pattern function values over $-40° < \theta < -80°$) with a reduced but clear response to waves pushing the array in the opposite direction. This pattern would indicate no response for signals coming at right angles to the array.

These results match the general features of the measured experimental response of stereociliary arrays. Experimental methods of measuring the directional sensitivity have been different than those presented here, to make precise comparisons difficult. The field strength of the phased array indicates a generally focused array, more focused than indicated by a direct cosine relation between direction and response. Lack of space in this note prevents more detailed analysis, but it seems likely that experimental methods may soon be developed to provide more exact descriptions of the directional sensitivity of vestibular hair cells for comparison.

REFERENCES

1. MUGLER, D. H. & M. D. ROSS. 1990. Vestibular receptor cells and signal detection: bioaccelerometers and the hexagonal sampling of 2-D signals. Math. Comput. Modelling **13:** 85–92.

2. MUGLER, D. H. & M. D. ROSS. 1990. Phased array characteristics and the directional sensitivity of vestibular hair cells. Proc. Int. Conf. IEEE/EMBS **12:** 1893–1894.
3. MUGLER, D. H. & M. D. ROSS. The directional sensitivity of vestibular hair cells based on planar phased array design. (Unpublished.)
4. HUDSPETH, A. J. 1985. Models for mechanoelectrical transduction by hair cells. Contemp. Sens. Neurobiol. **176:** 193–205.
5. OHMORI, H. 1985. Mechano-electrical transduction currents in isolated vestibular hair cells of the chick. J. Physiol. **359:** 189–217.

Two-Dimensional Eye Movement and Vestibular Responses in Primate Brain Stem[a]

J. O. PHILLIPS,[b] F. R. ROBINSON, AND A. F. FUCHS

Regional Primate Research Center and
Departments of Physiology and Biophysics
and Psychology
University of Washington
Seattle, Washington 98195

We investigated the responses of neurons in monkey vestibular nuclei during whole-body rotation and tilt as well as while the monkeys tracked a small moving visual target. We oscillated two rhesus monkeys sinusoidally (0.1 to 1.4 Hz, ±10 deg) about either the vertical axis (yaw) or the interaural axis (pitch). Vestibular sensitivity was determined with the monkeys in complete darkness and/or during active suppression of compensatory eye movements. We also pitched these monkeys statically (±25 deg). In addition, the monkeys tracked a spot that oscillated sinusoidally (±10 deg) or that stepped to new positions either vertically or horizontally. In some cases, the target spot was briefly extinguished during pursuit tracking to determine if the response was due to visual or eye movement sensitivity.

We recorded our neurons in a region that contained labeled cells after horseradish peroxidase conjugated wheat germ agglutinin (WGA-HRP) injections in the C2 segment of the cervical spinal cord. Labeled cells were present in the ventral-lateral aspect of the medial vestibular nucleus and in the lateral vestibular nucleus. This region also contains cells that project to the abducens nucleus.

Most neurons in this region responded to both yaw and pitch oscillation, but none responded to static pitch. Most responses were in phase with chair velocity, and the magnitude of the response increased monotonically with stimulus frequency, and, therefore, peak velocity. For many units, the relative magnitude of the pitch and yaw responses depended strongly on stimulus frequency. Units often preferred yaw or pitch at low frequencies, and then switched their preference at high frequencies. These changes appeared to be random from unit to unit, so there was little change in the overall direction preference of the population of units. Cells that responded to ipsilateral yaw rotation (i.e., type I) often had little, if any, pitch sensitivity, whereas type II neurons (which responded to contralateral rotation) almost invariably also responded during pitch. Indeed, in type II neurons, the pitch sensitivity at the standard 0.5-Hz test frequency often exceeded the yaw sensitivity.

Many neurons in the same regions of the vestibular nuclei also responded during eye movements to smoothly moving targets. These were tonic vestibular pause (TVP) and tonic neurons. All cells that were tested with blanking of the target spot during pursuit eye movements showed no decrement in response during the brief interval

[a]This work was supported by National Institutes of Health grants EY00745, RR00166, EY07991, and a grant from the Virginia Merrill Bloedel Hearing Research Center.

[b]Address correspondence to Department of Physiology and Biophysics, SJ-40, School of Medicine, University of Washington, Seattle, Washington 98195.

when the target spot was off, suggesting that these responses were to eye movement per se. Similarly, cells that responded to rotational stimuli while the animal was visually suppressing eye movements, also responded to the same passive rotations in the dark. Visual slip does not, therefore, appear to be an important input to these eye movement related cells.

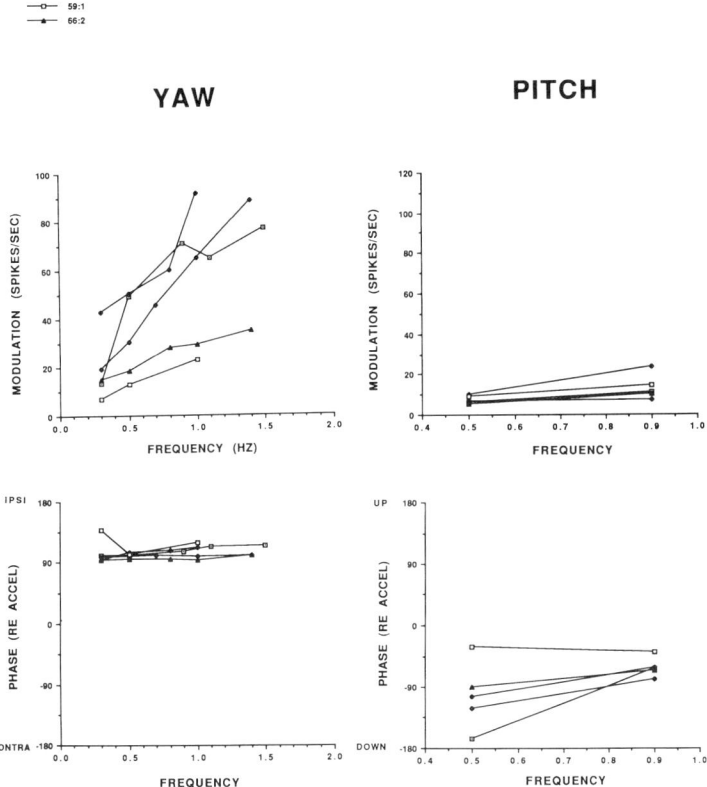

FIGURE 1. Representative type 1 vestibular responses to yaw and pitch over a series of stimulus frequencies (in Hz) at a fixed amplitude of ±10 deg. Modulation and phase values were obtained from fast Fourier transform of 10 cycles of binned and desaccaded unit data. Modulation is defined as peak frequency of the fit minus mean frequency. Phase is defined relative to head acceleration.

All of the TVP neurons that were tested had pursuit sensitivity in only one direction (e.g., vertical), but the majority had both pitch and yaw vestibular responses. Tonic neurons had no vestibular sensitivity, but were modulated during smooth pursuit such that they increased their discharge during movement in a preferred (on) direction, and decreased their discharge for movements in the opposite (off) direction. They also paused for saccades in their off direction.

The above results suggest the following conclusions:

1. Type I and type II cells have fundamentally different sensitivities to pitch and yaw rotation, and therefore are best suited for different tasks. Type I cells are well suited for driving pure horizontal movers (e.g., abducens motoneurons) and horizontal movement synergies (e.g., combinations of oblique neck muscle motoneurons). Type II cells are well suited for driving individual oblique

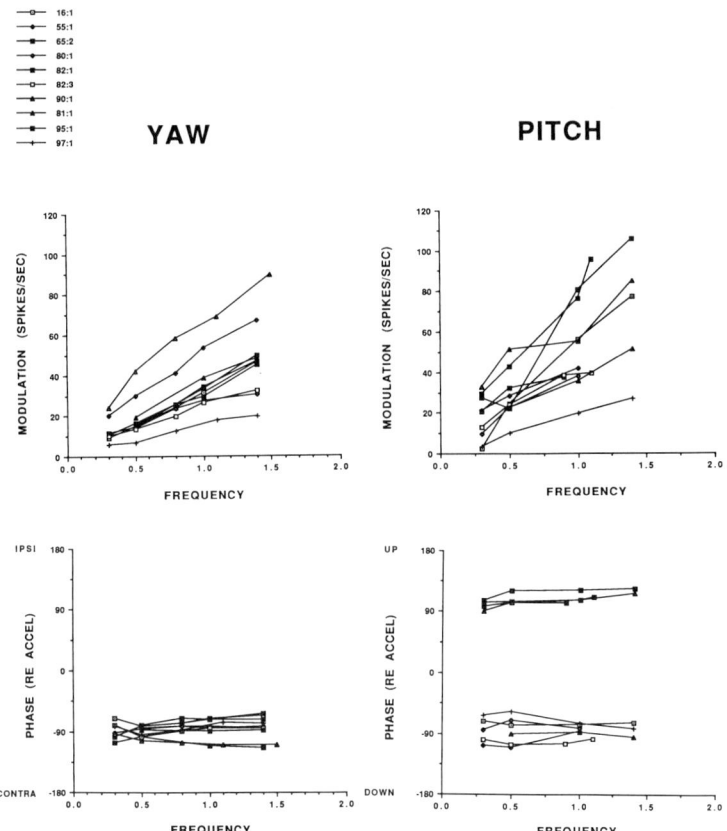

FIGURE 2. Representative type II vestibular responses to yaw and pitch over a series of stimulus frequencies (±10 deg.). Conventions are the same as those used in FIGURE 1.

movers (e.g., cervical neck motoneurons) or oblique synergies of horizontal and vertical movers (e.g., combined extraocular motoneuron pools).

2. The preferred direction tuning of the vestibular responses is not static. These systematic changes in preferred direction may parallel biomechanical changes that would require the activation of different muscle synergies to produce accurate compensatory responses at different frequencies and/or velocities (especially for head movement responses).

3. There is no evidence of otolith or visual slip input to these vestibular neurons.
4. Eye movement and vestibular responses of many vestibular nucleus neurons are not in register, suggesting that the smooth pursuit and vestibular systems use these neurons differently to solve the same problems. The pursuit system may operate strictly in horizontal and vertical coordinates, simplifying its coordination with other volitional movement systems. At the level of these neurons, the vestibuloocular/colic systems may operate either in plant coordinates or in an intermediate between motor and sensory coordinates. For example, both systems could drive purely leftward horizontal movements, but the pursuit system command would be a purely horizontal left eye movement signal to the appropriate neurons, while the summed peripheral vestibular end organ responses would produce the appropriate synergies by activating sets of horizontally sensitive neurons with offsetting vertical actions.

Unit Activity in the Vestibular Nuclei of Monkeys during Off-Vertical Axis Rotation

HARVEY REISINE[a] AND THEODORE RAPHAN

Department of Computer and Information Science
Brooklyn College
Bedford Avenue & Avenue H
Brooklyn, New York, 11210

Department of Neurology
Mt. Sinai School of Medicine
One Gustave Levy Place
New York, New York 10029

INTRODUCTION

Yaw rotation about an axis tilted from the spatial vertical induces horizontal eye velocity that is comprised of two components: a steady-state or bias component and an oscillating component dependent on head position with regard to gravity.[1-4] There is considerable evidence that both components arise from activity generated by the otoliths.[5-7] Models have been developed that can estimate head velocity from the otolith signals and produce eye velocity through activation of the velocity storage integrator.[8-10] However, the neural basis for many of the theoretical predictions has not been explored. The purpose of this study was to record single units in the

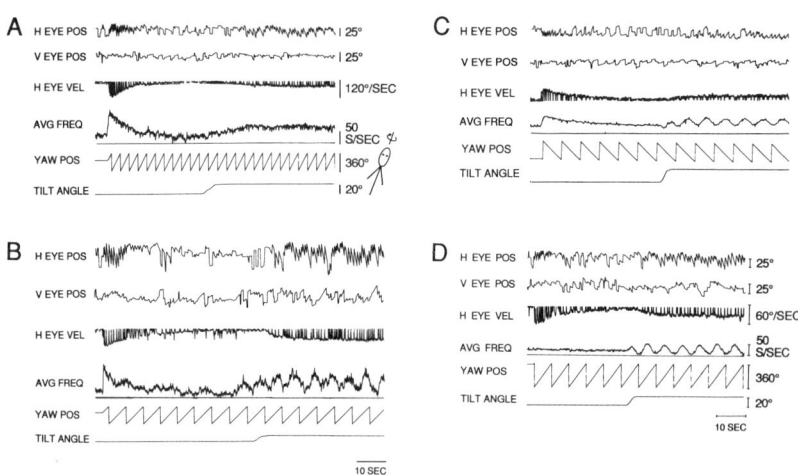

[a] Address correspondence to Dr. Reisine at Mt. Sinai, Box 1135.

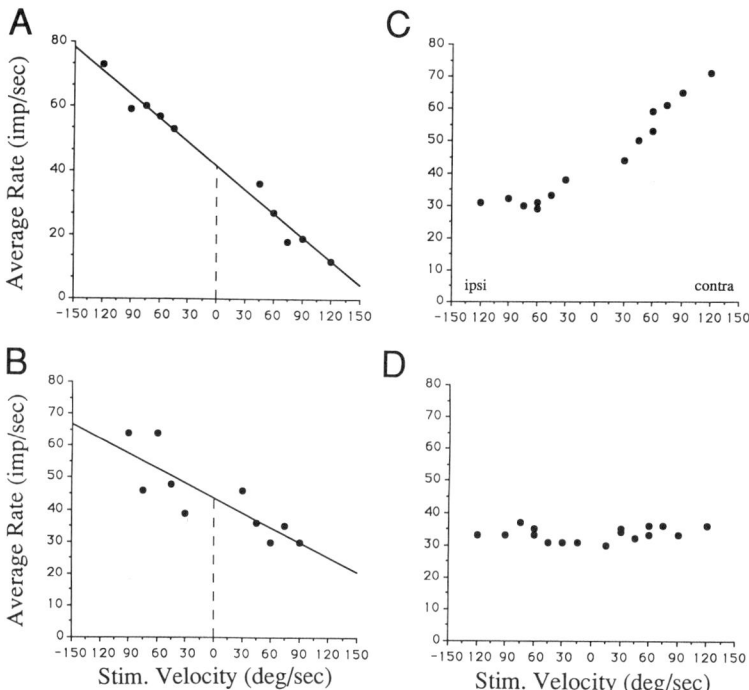

FIGURE 2. Steady-state firing rates as a function of stimulus velocity for the various unit classes. Lateral-canal-related units that were not modulated during OVAR (A) as well as those that were modulated (B) had average firing rates linearly related to stimulus velocity over a range of ±120 deg/second. Vertical-canal-related units (C) had firing rates that were linearly related to stimulus velocity during OVAR that increased their firing rates. However, for stimulation to the ipsilateral side, the unit saturated at 33 impulses/second. Otolith-related units (D) had a spontaneous firing rate of 33 impulses/second independent of stimulus velocity. Each point on the graphs represents an average of eight cycles of rotation during steady-state OVAR.

FIGURE 1. **A:** Lateral-canal-related central unit without activity modulation during OVAR (VO unit). In response to a velocity step of rotation about the animal yaw axis in the upright position, unit activity had a rapid jump followed by an approximately exponential decline to the spontaneous level. When the axis of rotation was tilted, OVAR was induced and frequency of unit firing rose to a level proportional to eye velocity. In all instances the dynamics were closely related to what would be expected from the behavior of velocity storage. **B:** Response to ipsilateral rotation during steps of velocity and OVAR for a lateral-canal-related central unit with activity modulation during OVAR (VPS unit). The unit had a bidirectional bias component. There were also oscillations in unit activity related to eye position. **C:** Response to contralateral rotation of a vertical-canal-related central unit. The unit increased its firing rate for a step of velocity about a vertical axis. During OVAR the bias level increased in conjunction with the increase in eye velocity. The unit firing rate oscillated during OVAR in relation to head position. **D:** Response to rotation of an adapting otolith afferentlike central unit. During steps of velocity about a vertical axis, the unit did not change its firing rate. During OVAR there were oscillations in the firing rate about the spontaneous rate, and there was no relation of the bias level to stimulus velocity.

vestibular nuclei and determine how individual neuron classes are related to eye velocity during off-vertical axis rotation (OVAR).

RESULTS

Activity of neurons related to OVAR could be classified as "vestibular only" (VO) and "vestibular plus saccade" (VPS) units. The optimal plane of activation of recorded units ($n = 79$) was related to the mean plane of reciprocal semicircular canals, and all displayed a type I response.

In response to steps of angular velocity about a vertical axis, lateral-canal-related VO and VPS units had rapid changes in firing rate and increased their firing rate during ipsilateral OVAR (FIGURE 1A and B). Vertical-canal-related VO and VPS units behaved similarly but were excited by contralateral rotation (FIGURE 1C). These neuron types were bidirectionally activated and the average firing rate was approximately linearly related to stimulus velocity (FIGURE 2 A, B, and C). In addition to the bias level, vertical canal and lateral VPS units also modulated their activity as a function of head position (FIGURE 1B and C). The long rising and falling time constants of their responses to rotation and the fact that these units were also activated during OKN and OKAN indicated that they were closely linked to velocity storage.

The otolith-related units on the other hand did not respond to a step of velocity about a vertical axis (FIGURE 1D), and their average frequency was not related to stimulus velocity (FIGURE 2D). The firing frequency only oscillated as a function of head position (FIGURE 1D).

The close association of firing frequency of VO and VPS lateral and vertical-canal-related units in the vestibular nuclei suggests that they are probably responsible for the velocity storage component of eye velocity during OVAR. The otolith units are not related to velocity storage and probably contribute to the oscillatory component of eye velocity. Thus, the vestibular nuclei contain the requisite neural classes for generating both the bias and oscillating components of eye velocity during OVAR.

REFERENCES

1. GUEDRY, F. E., JR. 1965. Acta Otolaryngol Stockh. **60**: 30–48.
2. BENSON, A. J. & M. A. BODIN. 1966. Aerosp. Med. **37**: 144–154.
3. YOUNG, L. R. & V. HENN. 1975. Fortschr. Zool. **23**(1): 235–246.
4. RAPHAN, T., B. COHEN & V. HENN. 1981. Ann. N.Y. Acad. Sci. **374**: 44–55.
5. CORREIA, M. J. & K. E. MONEY. 1970. Acta Otolaryngol Stockh. **69**: 7–16.
6. JANECKE, J. B., L. B. JONGKEES & W. J. OOSTERVELD. 1970. Acta Otolaryngol. Stockh. **69**: 1–6.
7. COHEN, B., J. I. SUZUKI & T. RAPHAN. 1983. Brain Res. **276**: 159–164.
8. HAIN, T. C. 1986. Biol. Cybern. **54**: 337–350.
9. RAPHAN, T. & C. SCHNABOLK. 1988. Ann. N.Y. Acad. Sci. **545**: 29–50.
10. FANELLI, R., T. RAPHAN & C. SCHNABOLK. 1990. Neural Networks **3**: 265–276.

Patterns of Connections between Six Semicircular Canals and Neck Motoneurons[a]

Y. SUGIUCHI, T. FUTAMI, N. ANDO, T. KAWASAKI,
J. YAGI, AND Y. SHINODA

Department of Physiology
School of Medicine
Tokyo Medical and Dental University
1-5-45, Yushima
Bunkyoku
Tokyo 113, Japan

The pattern of connections between different semicircular canals and dorsal neck motoneurons was determined by Wilson and Maeda.[1] Dorsal neck motoneurons (MNs) receive excitation from bilateral anterior canals and the contralateral horizontal canal and inhibition from bilateral posterior canals and the ipsilateral horizontal canal. It is assumed that this organized pattern of short-latency postsynaptic potentials is seen for all dorsal neck MNs. However, Suzuki and Cohen showed that stimulation of individual semicircular canal nerves produces canal-specific neck movements.[2] This strongly suggests that there is more than one pattern of connections between six semicircular canals and MNs of different neck muscles.

We investigated patterns of input from six semicircular canals to neck MNs in chloralose-pentobarbital sodium anesthetized cats, by recording intracellular potentials evoked by separate electrical stimulation of individual ampullary nerves. In order to stimulate ampullary nerves, bipolar stimulating electrodes were implanted near the ampullary nerves through small holes made in bony canals. The stimulating electrodes were implanted on six ampullary nerves in bilateral labyrinths in each experiment. Electrode positions were determined by monitoring characteristic eye movements elicited by stimulation of individual ampullary nerves. Muscle nerves to three neck muscles, the complexus (COMP), the obliquus capitis caudalis (OCA), and the rectus capitis dorsalis (RD) muscles at C1, and the accessory nerve were dissected for stimulation.

A typical pattern of inputs from six semicircular canals to a COMP MN is shown in FIGURE 1. Stimulation of the anterior ampullary nerve on either side and the horizontal ampullary nerve on the contralateral side produced disynaptic excitatory postsynaptic potentials (EPSPs), and stimulation of the posterior ampullary nerve on either side and the horizontal ampullary nerve on the ipsilateral side produced disynaptic inhibitory postsynaptic potentials (IPSPs). A typical pattern of inputs from six ampullary nerves in an RD MN is shown in FIGURE 2. Stimulation of either the ipsi- or the contralateral anterior ampullary nerves produced EPSPs. Stimulation of the ipsi- and contralateral horizontal ampullary nerves produced IPSPs and EPSPs, respectively. Stimulation of the posterior ampullary nerves on either side produced IPSPs. These input patterns exemplified in FIGURES 1 and 2 were observed

[a]This research was supported by a grant from the Human Frontier Science Program Organization.

in most COMP and RD MNs examined. In OCA MNs, stimulation of either the ipsilateral anterior or posterior ampullary nerves produced EPSPs, whereas stimulation of either the contralateral anterior or posterior ampullary nerves produced IPSPs. Stimulation of the ipsilateral and the contralateral horizontal ampullary nerves produced disynaptic IPSPs and EPSPs, respectively. In MNs of the accessory nerve, stimulation of any ipsilateral ampullary nerve produced disynaptic IPSPs, whereas stimulation of any contralateral ampullary nerve produced disynaptic EPSPs. These results showed that MNs of each neck muscle have their own characteristic pattern of input from six semicircular canals.

To determine the central pathways linking six semicircular canals and neck MNs, a lesion was made in the medial longitudinal fasciculus (MLF) at the medulla and the effect of ampullary nerve stimulation was compared before and after the lesion. In COMP MNs, after MLF sectioning on the ipsilateral side to the recording site, the IPSPs evoked from the ipsilateral horizontal and posterior ampullary nerves completely disappeared, but the EPSPs evoked from the ipsilateral anterior ampullary nerves remained unaffected, whereas the disynaptic PSPs evoked from three contralateral ampullary nerves disappeared. Similar lesion experiments were performed to determine the pathways from six semicircular canals to RD and OCA MNs. In summary, disynaptic EPSPs and IPSPs evoked from the contralateral canals and disynaptic IPSPs evoked from the ipsilateral canals were conveyed to neck MNs via the ipsilateral MLF.

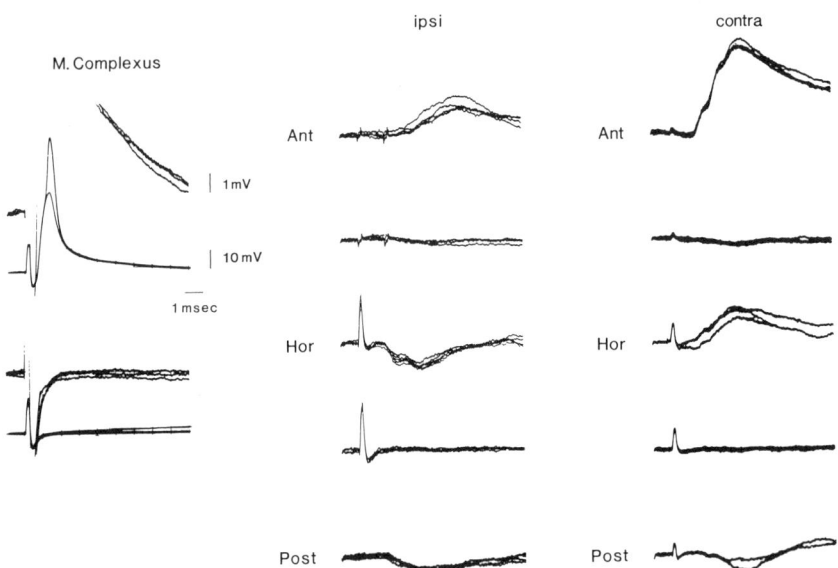

FIGURE 1. Typical pattern of inputs from six semicircular canals in a COMP motoneuron. **Left:** Antidromic potentials evoked by stimulation of the COMP muscle nerve. **Center:** Postsynaptic potentials evoked by stimulation of the ipsilateral anterior (Ant), horizontal (Hor), and posterior (Post) ampullary nerves. **Right:** Postsynaptic potentials evoked by stimulation of the contralateral Ant, Hor, and Post ampullary nerves. In each panel, the upper trace indicates the intracellular potential and the lower trace the juxtacellular field potential.

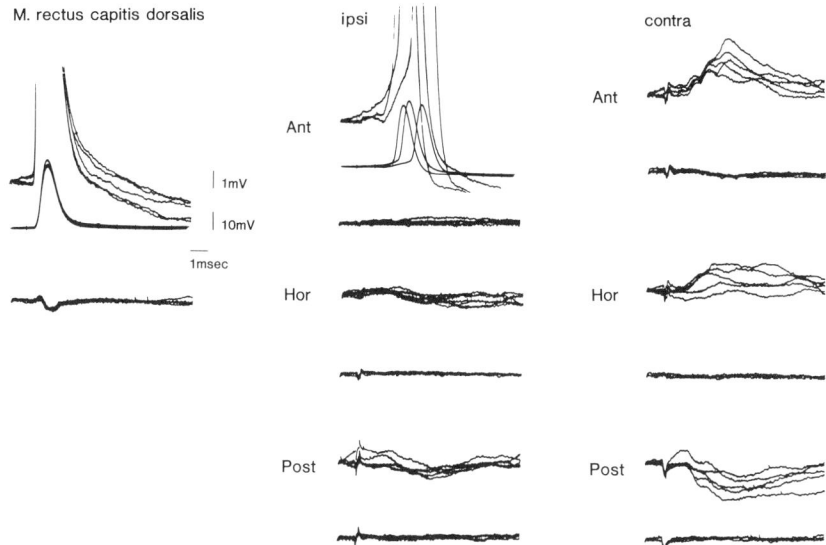

FIGURE 2. Typical pattern of inputs elicited by stimulation of six semicircular canal nerves in an RD motoneuron. The arrangement of the records and the abbreviations are the same as in FIGURE 1.

REFERENCES

1. WILSON, V. J. & M. MAEDA. 1974. Connections between semicircular canals and neck motoneurons in the cat. J. Neurophysiol. **37:** 346–357.
2. SUZUKI, J-I. & B. COHEN. 1964. Head, eye, body and limb movements from semicircular canal nerves. Exp. Neurol. **10:** 393–405.

Vestibular Input to Visual-Tracking Neurons in Area MST of Awake Rhesus Monkeys

P. THIER[a] AND R. G. ERICKSON

Department of Neurolology
University of Tübingen
D-7400 Tübingen, Germany

Area MST in the superior temporal sulcus (STS) of the monkey cortex has been implicated in the organization of smooth-pursuit eye movements. This view is based both on the results of single-unit recordings and the effects of experimental lesions. Previous work has demonstrated that the directional responses of many MST neurons (VT neurons) during smooth pursuit reflect integration of retinal image motion with a nonvisual pursuit-related signal,[1,2] and it has been proposed that the latter might be an eye movement efference copy useful for the maintenance of ongoing smooth pursuit.[2] We have now found that the pursuit-related discharge of such neurons is the same during head tracking [suppression of the vestibuloocular reflex (VOR)] and eye tracking (with the head stationary). This observation indicates that nonvisual input to VT neurons includes both eye- and head-movement-related components. Eighteen out of 29 VT neurons recorded from the lateral part of area MST (MST-1) on the lower anterior bank and the floor of the STS in 2 adult rhesus monkeys were studied during alternating sets of both eye and head tracking. Fifteen of these had the same preferred directions for eye tracking and retinal image slip, the latter as determined by probing with behaviorally irrelevant visual stimuli during stationary fixation. Of the remaining 3, 1 was visually unresponsive and the other 2 so weakly responsive that directionality could not be adequately determined. Eye tracking was elicited using step ramps and circular motion of a target rear-projected onto a tangent screen. Head tracking was achieved by suppression of the horizontal (yaw oscillation) or vertical (pitch oscillation) VOR by fixation of a light-emitting diode (LED) target rotated in synchrony with the animal in an otherwise dark room. The nonvisual nature of the tracking response was verified by the lack of significant responsiveness to brief periods of target blinks during eye tracking or head tracking and in addition by the lack of responsiveness to brief periods of retinal target image stabilization during eye tracking. The discharge of all VT neurons tested was modulated similarly by both eye and head tracking (see FIGURE 1 for an example). Complete data sets for both suppression of the horizontal and the vertical VOR were obtained for 14 neurons. This allowed us to estimate the preferred direction for head tracking by vectorial addition of the responses to suppression of the horizontal VOR and vertical VOR and then to compare to this the preferred direction for eye tracking. Without exception, the preferred directions for eye and head tracking proved to be well aligned (see FIGURE 2 for examples), with a range of differences not statistically different from 0. Moreover, preliminary analysis suggests that speed preferences for eye and head tracking were also the same.

---

[a] Address correspondence to Neurologische Universitätsklinik, Hoppe-Seyler-Strasse 3, 7400 Tübingen, Germany.

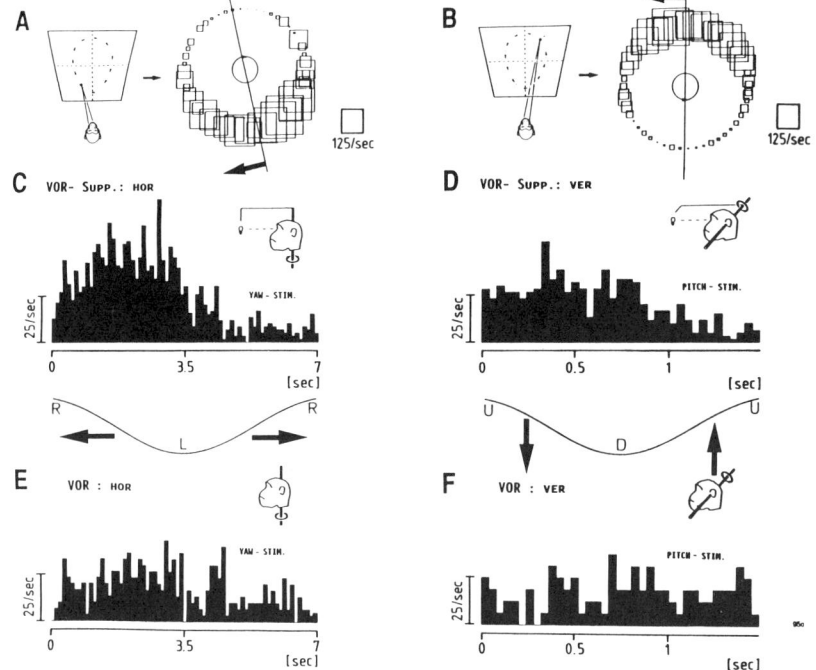

FIGURE 1. VT neuron recorded from the lateral part of area MST (MST-1) in the superior temporal sulcus (STS) of the rhesus monkey. Comparison of responses to eye tracking (A and B), head tracking (suppression of the vestibuloocular reflex (C and D), and vestibular stimulation in darkness (E and F). **A** and **B:** Eye tracking: two-dimensional histograms plotting mean discharge rate per bin (bin width 100 mseconds) for smooth-pursuit eye movements (head fixed) elicited by target movement along a circular trajectory (radius 15 deg, angular velocity 72 deg/second) in the frontoparallel plane either clockwise (CW) (A) or counterclockwise (CCW) (B). The squares representing mean discharge rate per bin are centered at the respective target position on the circular trajectory. The linear extension of the squares is a measure of the average number of spikes per bin. The lines describing the discharge gradient hit the high-response regions of the two-dimensional histograms at positions on the circle that are roughly 180 deg out of phase for CW and CCW tracking. A phase shift of 180 deg indicates that the unit is sensitive to the direction of movement rather than to target (or eye) position, a salient feature of VT neurons in MST-1. The preferred directions (arrows in A and B) correspond to the directions of momentary target (or eye) velocity in the center of the high-response segment as determined by the intersection with the lines describing the discharge gradients for CW and CCW tracking. The estimates of the preferred eye-tracking direction of this cell were 192 deg and 179 deg for CW and CCW tracking respectively, i.e., this unit preferred mainly leftward eye tracking with an at best small downward component visible for CW tracking. **C** and **D:** An overall preference for leftward tracking was also observed when the monkey was oscillated sinusoidally around either the yaw axis or the pitch axis while fixating a small LED target which was stationary relative to the monkey's head (suppression of the VOR, head-tracking). **C:** Suppression of the horizontal VOR (amplitude and period of oscillation 30 deg and 7 seconds respectively), **D:** Suppression of the vertical VOR (amplitude and period of oscillation 12 deg and 1.5 seconds respectively). Note the strong discharge enhancement elicited by leftward movement of target, head and eyes (C) and the much weaker response to downward movement (D). **E** and **F:** Oscillating the animal in complete darkness without the visual target available for fixation caused a much smaller response to leftward movement of the head while no response to pitch stimulation was detectable. Eye movements in all tests were monitored using the search-coil technique. The room was completely dark except for the tiny and dim target used for fixation in the tracking tests. Accurate tracking performance was enforced by applying standard reward contingencies.

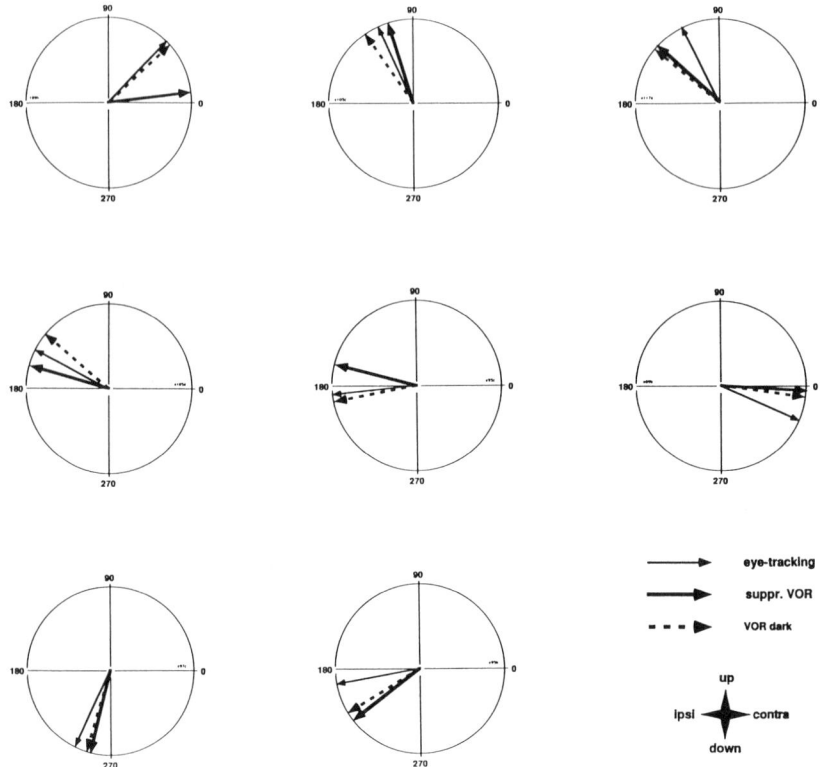

FIGURE 2. Vector plots comparing the preferred directions for eye tracking (thin arrows), head tracking (thick arrows), and head movement in darkness (VOR dark, dashed arrows) for 9 VT neurons recorded from MST-1. For eye-tracking, the preferred directions were determined by analyzing unit responses to circular smooth pursuit as exemplified in FIGURE 1 (see also Reference 4). The preferred directions for head tracking and head movement in darkness were estimated by adding vectorially responses to yaw and pitch stimulation. To aid comparison of small (VOR dark) and large (eye and head tracking) responses, the arrow lengths were scaled to unit length 1.0 independent of the absolute magnitude. The majority of VT neurons were also visually responsive. Receptive fields were usually of medium size (in the order of 20 × 20 deg), localized in the more central parts of the contralateral visual field and usually touching or crossing the vertical meridian. As determined by probing with small, behaviorally irrelevant patterns during stationary fixation, each cell was directionally selective and the preferred visual direction corresponded closely to the preferred tracking directions. Visual responses were in all cases directional. The preferred directions for retinal image slip as determined by probing with small, behaviorally irrelevant patterns during stationary fixation corresponded in all cases to the preferred tracking directions.

The directionality of the responses to head movements can be demonstrated even in the absence of a visual target. For 8 cells it was possible to determine not only the preferred directions for eye and head tracking but also the preferred axis of head movement by adding vectorially the responses to yaw and pitch oscillation in complete darkness. As shown in FIGURE 2, the 3 vectors representing the preferred directions for eye tracking, head tracking, and head movement in darkness were well aligned in all cases. Our view that the head-movement-related ("vestibular") input impinging on VT neurons in MST-1 is specifically related to the task of tracking a target is supported by the consistently weaker responses (on the average, by a factor of 2) to head movement in total darkness as compared to the response during VOR suppression. The responses were reduced even further during VOR enhancement by fixation of an earth-fixed target (3 cells).

In summary these results demonstrate that, during tracking of a visual target, many VT neurons in a well-defined part of the monkey cortex integrate information about retinal image motion with nonvisual directional signals related to the movement of the eyes and the head. These various input signals are aligned in such a way as to allow these cells to reconstruct a representation of the target movement in (at least) body-centered coordinates as has been proposed by models on eye-head coordination during smooth-pursuit.[3] While comparable nonretinal representations of target movement have previously been observed in both the dorsolateral pontine nucleus[4] and the cerebellar vermis,[5] it appears that such signals are first elaborated in cortical areas such as MST-1 before being relayed to these subcortical centers by way of the well-established corticopontocerebellar pathway.

REFERENCES

1. ERICKSON, R. G. 1985. Representation of the fovea and identifaction of foveal tracking cells in the superior temporal sulcus of the macque monkey. PhD thesis. State University of New York. Buffalo, N.Y.
2. NEWSOME, W. T., R. H. WURTZ & H. KOMATSU. 1988. Relation of cortical areas MT and MST to pursuit eye movements. II. Differentiation of retinal from extraretinal inputs. J. Neurophysiol. **60**(2): 604–620.
3. BIZZI, E. 1981. Eye-head coordination. *In* Handbook of Physiology. The Nervous System. J. M. Brookhart & V. B. Mountcastle, Eds. (Part II): 1321–1336. American Physiological Society. Bethesda, Md.
4. THIER, P., W. KOEHLER, U. W. BUETTNER & J. DICHGANS. 1990. Does the system for smooth-pursuit eye movements rely on a representation of target motion in space? *In* From Neuron to Action. An Appraisal of Fundamental and Clinical Research. L. Deeke, J. C. Eccles & V. B. Mountcastle, Eds. Springer-Verlag. Berlin & Heidelberg.
5. KASE, M., H. NODA, D. A. SUZUKI & D. C. MILLER. 1979. Target velocity signals of visual tracking in vermal Purkinje cells of the monkey. Science **205**(4407): 717–720.

The Role of GABA in Inhibitory Synaptic Inputs on Inhibitory Burst Neurons in the Cat

TAKAO YABE AND NOBUHIKO FURUYA

Department of Otolaryngology
Teikyo University School of Medicine
Kaga 2-11-1, Itabashi-ku
Tokyo, Japan

In the pontine neural networks that govern vestibular nystagmus in the cat, inhibitory burst neurons (IBNs) are known to fire in bursts during the quick phase, and to suppress firing of the abducens motoneurons on the contralateral side.[1] IBNs have direct inhibitory synaptic connections with contralateral abducens motoneurons which terminate their slow-phase activity and play an important role in the generation of fast eye movements. The inhibitory input of IBNs consists of bilateral monosynaptic axonal connections with pause neurons (PNs), and their excitatory input consists of ipsilateral projections from excitatory burst neurons. As mentioned above, the circuitry and functional role of premotor burst neurons have been investigated in detail, but there have been few systematic studes of potential neurotransmitters of these nystagmus-related neurons[2,3] and the neurotransmitter controlling their burst activity has not yet been identified. The present experiments were designed to identify the neurotransmitter that controls the burst firing activity of IBNs.

The experiments were performed on 30 adult cats. Inhibitory burst neuron activity was recorded extracellularly, and various chemicals were applied to the IBNs iontophoretically through seven-barrel micropipettes. Iontophoretic application of γ-aminobutyric acid (GABA) strongly suppressed IBN activity and eliminated the burst pattern. In order to examine GABA receptor subtypes, $GABA_A$-receptor agonist and antagonist were administered. Iontophoretic application of muscimol (a $GABA_A$-receptor agonist) showed the same suppression of IBN activity as GABA application. Bicuculline (a $GABA_A$-receptor antagonist) increased the firing of IBNs and suppressed the inhibitory effect of GABA when applied simultaneously. Neither glycine nor serotonin, other inhibitory transmitter candidates in the central nervous

TABLE 1. Results of Iontophoretic Administration of GABA, the GABA Agonist, the GABA Antagonist, Glycine, and Serotonin[a]

Drug	Inhibition +	Inhibition −	Total
GABA	39	2	41
Muscimol	11	3	14
Bicuculline	0	18	18
Glycine	1	8	9
Serotonin	3	13	16

Inhibition +: inhibition of IBN activity. Inhibition −: no effect on IBN activity or suppression of GABA-induced inhibition of IBN. Total: number of neurons iontophoretically treated with the respective chemicals.

system, nor their respective antagonists, strychnine and methysergide, had any effect. The results of iontophoretic administrations are summarized in TABLE 1. Systemically administered picrotoxin (a $GABA_A$-receptor antagonist) prevented the GABA-induced suppression of IBNs.

These results suggested that IBNs possess $GABA_A$ receptor and are controlled by GABAergic afferent neurons and Cl^- channels. Based on recent physiological studies of fiber connections centering on the IBN, it appears that GABAergic afferent neurons may be PNs and that the inhibition of IBNs is caused by PNs.

REFERENCES

1. HIKOSAKA, O. & T. KAWAKAMI. 1977. Inhibitory reticular neurons related to the quick phase of vestibular nystagmus—their location and projection. Exp. Brain Res. **27:** 377–396.
2. FURUYA, N., H. ASHIKAWA, T. YABE & J-I. SUZUKI. 1988. The effect of serotonin on the pontine pause neorons in the cat. Adv. Otol. Rhinol. Laryngol. **42:** 224–228.
3. SPENCER, R. F., R. J. WENTHOLD & R. BAKER. 1989. Evidence for glycine as an inhibitory neurotransmitter of vestibular, reticular and prepositus hypoglossi neurons that project to the cat abducens nucleus. J. Neurosci. **9:** 2718–2736.

Stimulation and Single-Unit Studies of Velocity Storage in the Vestibular and Prepositus Hypoglossi Nuclei of the Monkey[a]

JUN-ICHI YOKOTA, HARVEY REISINE,
AND BERNARD COHEN

Departments of Neurology and Physiology
Mount Sinai School of Medicine
One Gustave Levy Place
New York, New York 10029

By modeling the vestibuloocular reflex (VOR) and optokinetic nystagmus (OKN), it has been suggested that several mathematical integrations are necessary in the oculomotor system to maintain stable images on the retina during head movements. However, the neural circuits underlying these mechanism are still unknown. A

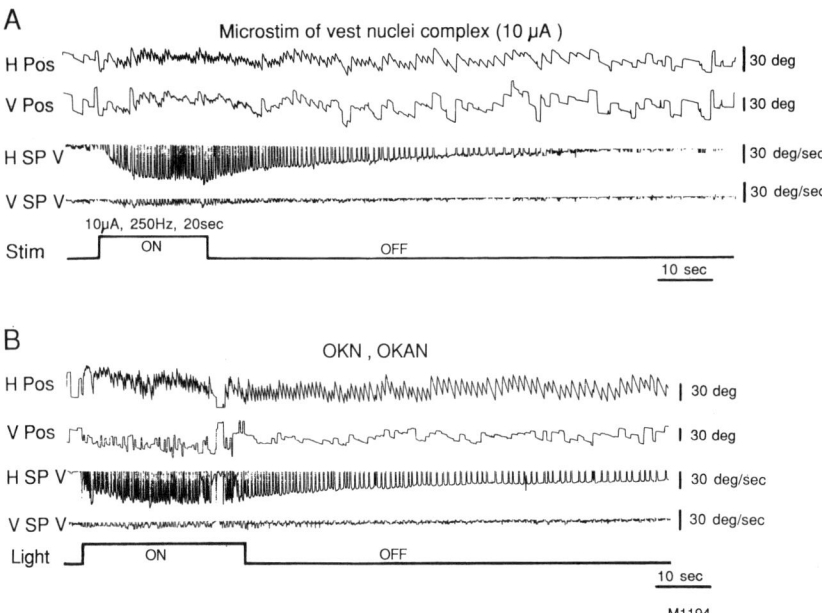

FIGURE 1. A: Nystagmus induced by stimulation of the vestibular nuclei in darkness. The slow-phase velocity of the horizontal nystagmus was contralateral, followed by after-nystagmus in the same direction. **B:** The characteristics of the induced nystagmus shown in A were similar to OKN and OKAN in the same animal.

[a] Supported by NS00294, EY01867.

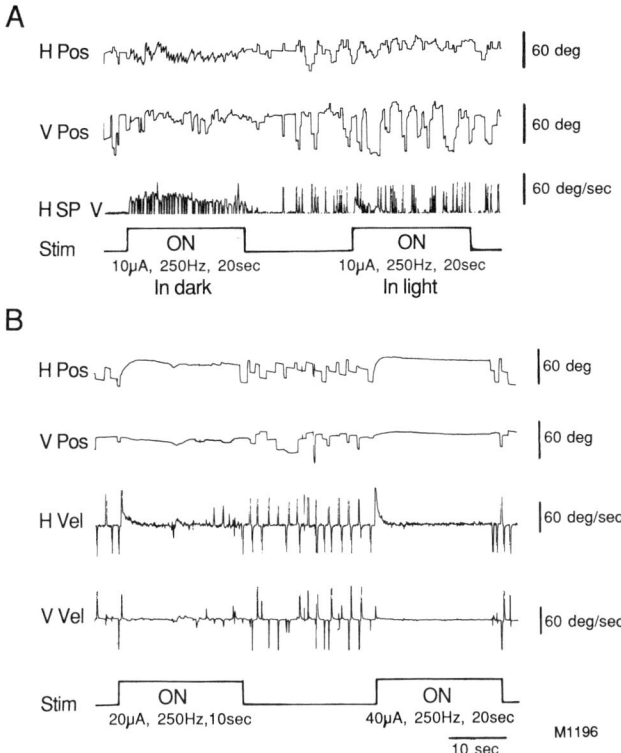

FIGURE 2. A, left: Nystagmus induced by stimulation of PPH. The slow-phase velocity was ipsilateral, and the stimulus-induced nystagmus was not followed by after-nystagmus in dark. **A, right:** The stimulus-induced nystagmus was completely suppressed in light. **B:** At another location in PPH, stimulation induced a shift in eye position to the ipsilateral side both at 20 and 40 μA of current.

chemical lesion study suggested that the final common integrator is located in the region of the medial vestibular nucleus (MVN) and the nucleus prepositus hypoglossi (PPH).[1] Functions attributable to velocity storage, i.e., optokinetic after-nystagmus (OKAN), are lost after midline section of vestibular commissural connections caudal to the abducens nucleus, suggesting that the velocity storage integrator is represented in the vestibular nuclei.[2] In the present study, we recorded single units in the vestibular nuclei (VN) and PPH and electrically stimulated at these sites to determine the location of neural structures associated with the various integrators.

Eye movements of cynomolgus monkeys were monitored with search coils. Currents of 40 μA or less evoked horizontal or rotatory nystagmus in dark with contralateral slow phases followed by after-nystagmus in the same direction.[3] Positive stimulus sites for the nystagmus were mainly located at the boundary between MVN, vLVN, SVN, and DVN, just caudal to the abducens nucleus and 3–4 mm lateral from the midline (FIGURE 1A).[3] The rising and falling time courses of the evoked nystagmus and after-nystagmus were similar to OKN and OKAN (FIGURE 1B), and the steady-state slow-phase velocities [mean 52.0 ± standard deviation

(SD) 11.3°/second] were as same as those of OKAN (mean 48.7 ± SD 6.0°/second). Characteristic nystagmus was commonly elicited by microstimulation at the sites where vestibular-only type I or type II neurons were recorded.

In contrast, burst-tonic or tonic neurons related to eye position were recorded in PPH. The slow-phase velocity of nystagmus induced by PPH stimulation was ipsilateral, and there was no after-nystagmus (FIGURE 2A). Eye position shifts to the ipsilateral side were also elicited (FIGURE 2B). The steady-state velocities (mean 40.5 ± SD 7.7°/second) were less than from VN stimulation.

These data suggest that the two different neural integrators associated with oculomotor function lie in the rostral medulla. One is the "velocity-to-position" integrator,[4] probably closely associated with PPH and possibly with the vestibular nuclei.[1,5] The other is the velocity storage integrator,[6] which appears to lie in the vestibular nuclei.

REFERENCES

1. CANNON, S. C. & D. A. ROBINSON. 1987. Loss of the neural integrator of the oculomotor system from brain stem lesions in monkey. J. Neurophysiol. **57:** 1383–1409.
2. KATZ, E., J. M. B. V. DE JONG, J. BÜTTNER-ENNEVER & B. COHEN. 1991. Effects of midline medullary lesions on velocity storage and the vestibulo-ocular reflex. Exp. Brain. Res. **87:** 505–520.
3. YOKOTA, J., H. REISINE & B. COHEN. 1990. Velocity storage and the ocular response to microstimulation of vestibular nuclei in alert monkey. Soc. Neurosci. Abstr. **16:** 733.
4. ROBINSON, D. A. 1975. Oculomotor control signals. *In* Basic Mechanisms of Ocular Motility and Their Clinical Implications. G. Lennerstrand & P. Bach-y-Rita, Eds.: 337–374. Pergamon Press. Oxford, England.
5. STRAUBE, A., R. KURZAN & U. BÜTTNER. 1991. Differential effects of bicuculline and muscimol microinjections into the vestibular nuclei on simian eye movements. Exp. Brain. Res. **86:** 347–358.
6. RAPHAN, T. & B. COHEN. 1985. Velocity storage and the ocular response to multidimensional vestibular stimuli. *In* Adaptive Mechanisms in Gaze Control. A. Berthoz & G. Melvill Jones, Eds. **1:** 123–143. Elsevier. Amsterdam, the Netherlands.

Index of Contributors

Subject Index

Visual signals, 266–275
Visual stimulation, 400–402
Visual surround linear vestibuloocular reflex (VSLVOR), 235–240
Visual system, 152, 178
  and velocity storage, 140, 142, 153–154, 290
Visual-tracking (VT) neurons, 960–963
Visual-vestibular conflict stimulation, 651
Visual-vestibular interaction, 140, 142
  NOT in, 287–288
  in rapid eye movement, 396–406
  and velocity storage, 153
Visual-vestibular interaction (VVI), 178, 240, 832–835
Visual-vestibular mechanisms, 220–221, 223–224
Visuospatial stability, 766
VN, *see* Vestibular nucleus
VNN, *see* Vestibular nucleus neuron; Vestibular nucleus neurons
VOC, *see* Vestibulooculocollic neurons; Vestibulooculospinal neurons
VOG, *see* Videoculography
VOR, *see* Vestibuloocular reflex

VORC, *see* Vestibuloocular reflex (VOR) cancellation
VPEL, *see* Visually perceived eye level
VPV, *see* Visually perceived vertical
VS, *see* Vestibulospinal
VSLVOR, *see* Visual surround linear vestibuloocular reflex
VTM system, 826–828
VVOR, *see* Vertical vestibuloocular reflex

Wallenberg's syndrome, 540–543
  lateropulsion in, 538, 541
  otolith pathway lesion in, 544–547
  OTR in, 539, 546
  SCC lesion and, 544–547
  skew deviation in, 539
  SVV in, 541
  symptoms of, 537
  vertical SCC lesion in, 544–547
Weightlessness
  experienced orientation in, 337–338
  reaction cues to, 342–343
  and visually induced self-motion, 340

Zona incerta, 271